Cotton

2nd edition

Cotton

2nd edition

David D. Fang and Richard G. Percy, Editors

American Society of Agronomy | Crop Science Society of America | Soil Science Society of America

American Society of Agronomy, Inc.
Crop Science Society of America, Inc.
Soil Science Society of America, Inc.
5585 Guilford Road, Madison, WI 53711-1086 USA

agronomy.org | soils.org | crops.org
dl.sciencesocieties.org
SocietyStore.org

ISBN: 978-0-89118-625-0 (print)
ISBN: 978-0-89118-626-7 (digital)
doi:10.2134/agronmonogr57
Library of Congress Control Number: 2015944544

Agronomy Monographs
ISSN 0065-4663 (print)
ISSN 2156-3276 (online)

Cover design: Patricia Scullion
Cover photo: James Frelichowski

Printed in the United States of America.

Contents

Foreword

Cotton has been cultivated and used to make fabrics for at least 7,000 years, and evidence indicates it may have been cultivated in the Middle East as early as 12,000 BCE and used to produce clothing and a number of other products. Cotton, however, is much more than clothing and ear swabs. The farm value of US cotton and cottonseed production is approximately $5 billion. Annual business revenue for cotton products in the US economy exceeds $120 billion, making cotton America's number one value-added crop. China and India are world leaders in annual cotton production.

Contrary to popular belief, US "paper" currency is not actually made of paper; it's a blend of 75% cotton and 25% linen. Cotton is a critical component of baseballs. The Wright Brothers used cotton to cover the wings of their aircraft for the first powered flight in 1905. This small assortment of trivia is representative of the breadth of cotton's use and economic importance of this perennial, yet annually propagated, agronomic crop.

Editors David Fang and Richard Percy have complied a prestigious list of contributing authors in a series of chapters reviewing the history, improvement and cultivation of this amazing and versatile crop. *Cotton*, 2nd edition, Agronomy Monograph 57, is sure to become the go-to publication for professors, students, and practitioners.

Roch Gaussoin
President, Crop Science Society of America

Jean Steiner
President, American Society of Agronomy

Carolyn Olson
President, Soil Science Society of America

Preface

Three decades have elapsed since the publication of the first edition of *Cotton*. During the past 30 years, many changes have occurred in almost every aspect of the cotton industry. New species were discovered. New germplasm repositories were established in some countries, older collections were consolidated, and established germplasm collections in the United States and other countries adopted new methodologies and techniques for better germplasm preservation, characterization, and utilization. Development of genetically modified (transgenic) cotton varieties revolutionized cotton production. Selections assisted by molecular markers and genomic tools are being implemented in both public and private breeding programs. Genome sequences of two diploid species have been released, and two *Gossypium hirsutum* Upland sequences are becoming available as this book goes to press. A *G. barbadense* Pima cotton genome sequence is on the horizon. New technologies such as remote sensing and robotics are aiding in the development of precision agricultural practices that will help growers to more effectively and efficiently manage fertilization, irrigation, and pest control operations and grow profitable, sustainable crops. Broad use of transgenic varieties has brought tremendous changes in growing practices and pest management. Newer high speed spinning instruments have been developed that demand longer, stronger and finer fibers. Consequently, genetic improvement in fiber quality has been stressed, and improved harvesting and ginning instruments have been developed to minimize damage to fibers. With improvement in fiber quality has come the realization that further fiber improvement will require a better understanding of fiber development and biology. As a consequence, developmental biology and genetics have become focal points in the cotton research community. All of these advances have been incorporated into this second edition. Eight chapters are new topics, and 15 chapters are completely rewritten. This book consists of four parts. The first part starts with the origin and history of cotton industry, followed by taxonomy and germplasm management of the genus *Gossypium*, botanical characteristics of cotton plants, fiber biology, and cytology. The next part discusses qualitative and quantitative traits with agronomic importance, breeding methods and strategies, and advances in marker-assisted selection, genome sequences, databases and bioinformatics tools. The third set of chapters describe growing practices and pest management. The fourth part deals with harvesting, postharvest handling of fibers and seeds, and commerce of world cotton industry.

We would like to thank the excellent work of the Societies' Book and Multimedia Publishing Committee Chair, Dr. April Ulery; Books Acquisition Editor, Dr. Nicole Sandler; and Managing Editor, Lisa Al-Amoodi. Their encouragement, patience, and assistance ensured timely publication of this book. Our appreciation also goes to the 50 authors who contributed to this book, and more than 60 experts who critically reviewed the manuscripts.

David D. Fang and Richard G. Percy, Editors

Contributors

Abidi, Noureddine	Dep. of Plant and Soil Sci., Fiber and Biopolymer Research Inst., Texas Tech Univ., 1001 East Loop 289, P.O. Box 45019, Lubbock, TX, 79403 (n.abidi@ttu.edu)
Adams, Gary M.	National Cotton Council, 7193 Goodlett Farms Parkway, Cordova, TN 38016 (gadams@cotton.org)
Auld, Dick	Plant and Soil Science Dep., Texas Tech Univ., MS 2122, Lubbock, TX 79409-2122 (dick.auld@ttu.edu)
Barnes, Edward M.	Cotton Incorporated, 6399 Weston Parkway, Cary, NC 27513 (EBarnes@cottoninc.com)
Bauer, Philip J.	USDA-ARS, Coastal Plains Soil, Water, and Plant Research Center, 2611 West Lucas St., Florence, SC, 29501 (Phil.bauer@ars.usda.gov)
Bechere, Efrem	USDA-ARS, Crop Genetics Research Unit, 141 Experiment Station Road, Stoneville, MS 38776 (Efrem.bechere@ars.usda.gov)
Boman, Randal K.	Southwest Research and Extension Center, Dep. of Plant and Soil Sciences, Oklahoma State Univ., 16721 U.S. Highway 283, Altus, OK 73521 (randy.boman@okstate.edu)
Bourland, Fred	Univ. of Arkansas Northeast Research and Extension Center, 1241 West County Road 780, Keiser, AR 72351 (bourland@uark.edu)
Brewer, Michael J.	Texas A&M AgriLife Research, 10345 Hwy. 44, Corpus Christi, TX 78406 (mjbrewer@ag.tamu.edu)
Brodrick, Rose	CSIRO, Australian Cotton Research Institute, Locked Bag 59, Wee Waa Road, Narrabri, NSW, Australia 2390 (Rose.Brodrick@csiro.au)
Campbell, B. Todd	USDA-ARS Coastal Plains Soil, Water, and Plant Research Center, 2611 W. Lucas St., Florence, SC 29501 (todd.campbell@ars.usda.gov)
Conaty, Warren	CSIRO, Australian Cotton Research Institute, Locked Bag 59, Wee Waa Road, Narrabri, NSW, Australia 2390 (warren.conaty@csiro.au)
Dowd, Michael K.	Southern Regional Research Center, USDA-ARS, 1100 Robert E. Lee Blvd., New Orleans, LA 70124 (michael.dowd@ars.usda.gov)
Endrizzi, John E.	Retired, Dep. of Plant Sciences, Univ. of Arizona, Tucson, AZ 85721
Ethridge, Dean	Fiber and Biopolymer Research Institute, Texas Tech Univ., 1001 East Loop 289, P.O. Box 45019, Lubbock, TX, 79403 (dean.ethridge@ttu.edu)
Fang, David	Cotton Fiber Bioscience Research Unit, USDA-ARS-SRRC, 1100 Robert E Lee Blvd., New Orleans, LA 70124 (David.Fang@ars.usda.gov)
Frelichowski, James	USDA-ARS, Crop Germplasm Research Unit, 2881 F&B Rd., College Station, TX 77845 (James.Frelichowski@ars.usda.gov)
Grover, Corrinne E.	Dep. of Ecology, Evolution, and Organismal Biology, Bessey Hall, Iowa State Univ., Ames, IA 50011 (corrinne@iastate.edu)
Hendon, Bralie	Plant and Soil Science Dep., Texas Tech Univ., MS 2122, Lubbock, TX 79409-2122 (bralie.hendon@ttu.edu)
Hequet, Eric F.	Fiber and Biopolymer Research Institute, Texas Tech Univ., 1001 East Loop 289, P.O. Box 45019, Lubbock, TX, 79403 (eric.hequet@ttu.edu)
Holt, Greg	USDA-ARS Cotton Production and Processing Research, 1204 E FM 1294, Lubbock, TX 79403 (greg.holt@ars.usda.gov)
Huang, Yanbo	USDA-ARS, Crop Production Systems Research Unit, 141 Experiment Station Rd., P.O. Box 350, Stoneville, MS 38776 (yanbo.huang@ars.usda.gov)
Hughs, Ed	USDA-ARS Southwestern Cotton Ginning Research Laboratory, 300 E. College Drive, Mesilla Park, NM 88047 (ed.hughs@ars.usda.gov)
Kelley, Mark S.	Texas A&M AgriLife Extension Service, 1102 E. FM 1294, Lubbock, TX 79403 (m-kelley@tamu.edu)
Kelly, Brendan	Fiber and Biopolymer Research Institute, Texas Tech Univ., 1001 East Loop 289, P.O. Box 45019, Lubbock, TX, 79403 (brendan.kelly@ttu.edu)

Kemerait, Robert C. Jr.	Dep. of Plant Pathology, 104 Research Way, Univ. of Georgia, Coastal Plain Experiment Station, Tifton, GA 31793 (kemerait@uga.edu)
Kim, Hee Jin	USDA-ARS, Southern Regional Research Center, 1100 Robert E. Lee Blvd, New Orleans, LA 70124 (Heejin.kim@ars.usda.gov)
Lee, Joshua A.	Retired, USDA-ARS and North Carolina State Univ., Raleigh, NC
Luttrell, Randall G.	USDA-ARS, Southern Insect Management Research Unit, 141 Experiment Station Rd., Stoneville, MS 38776 (randy.luttrell@ars.usda.gov)
Mahan, James	USDA-ARS Cropping Systems Research Laboratory, 3810 4th Street, Lubbock, TX (james.mahan@ars.usda.gov)
Main, Dorrie	Dep. of Horticulture, Washington State Univ., 45 Johnson Hall, Pullman, WA 99164-6414 (dorrie@wsu.edu)
Mauney, Jack R.	Retired, USDA-ARS, Mesa, AZ
Myers, Gerald O.	Louisiana State Univ., School of Plant, Environmental, and Soil Sciences, 104 M.B. Sturgis Hall, Baton Rouge, LA 70803 (gmyers@agcenter.lsu.edu)
Oosterhuis, Derrick M.	Univ. of Arkansas, Dep. of Crop, Soil, and Environmental Sciences, 1366 Altheimer Drive, Fayetteville, AR 72704 (oosterhu@uark.edu)
Payton, Paxton	USDA-ARS Cropping Systems Research Laboratory, 3810 4th Street, Lubbock, TX (paxton.payton@ars.usda.gov)
Percy, Richard	USDA-ARS, Crop Germplasm Research Unit, 2881 F&B Rd., College Station, TX 77845 (Richard.Percy@ars.usda.gov)
Rothrock, Craig S.	Dep. of Plant Pathology, Plant Science 217, Univ. of Arkansas, Fayetteville, AR 72701 (Rothrock@uark.edu)
Snider, John L.	Univ. of Georgia, Dep. of Crop and Soil Sciences, 115 Coastal Way, Tifton, GA 31794 (jlsnider@uga.edu)
Teague, Tina G.	Univ. of Arkansas, Division of Agriculture and Arkansas State Univ., 245 Agriculture Bldg., Jonesboro, AR 72467 (tteague@astate.edu)
Thomson, Steven J.	USDA-ARS, Crop Production Systems Research Unit, 141 Experiment Station Rd., P.O. Box 350, Stoneville, MS 38776 (steve.thomson@ars.usda.gov)
Wang, Maojun	National Key Laboratory of Crop Genetic Improvement, Huazhong Agricultural Univ., Wuhan, Hubei 430070, China (mjwang1234@gmail.com)
Wanjura, John D.	USDA-ARS Cotton Production and Processing Research Unit, 1604 E. FM 1294, Lubbock, TX 79403 (John.wanjura@ars.usda.gov)
Weaver, David B.	Dep. of Crop, Soil and Environmental Sciences, Auburn Univ., Auburn, AL 36849-5412 (weavedb@auburn.edu)
Wendel, Jonathan F.	Dep. of Ecology, Evolution, & Organismal Biology, Bessey Hall, Iowa State Univ., Ames, IA 50011 (jfw@iastate.edu)
Woodward, Jason E.	Texas A&M AgriLife Extension Service, 1102 East FM 1294, Lubbock, TX 79403 (jewoodward@ag.tamu.edu)
Yu, Jing	Dep. of Horticulture, Washington State Univ., 45 Johnson Hall, Pullman, WA 99164-6414 (jing.yu@wsu.edu)
Yuan, Daojun	National Key Laboratory of Crop Genetic Improvement, Huazhong Agricultural Univ., Wuhan, Hubei 430070, China (robert@mail.hzau.edu.cn)
Zhang, Jinfa	Dep. of Plant and Environmental Sciences, New Mexico State Univ., MSC 3Q, Skeen Hall, 945 College Drive, Las Cruces, NM 88003 (jinzhang@nmsu.edu)
Zhang, Tianzhen	Nanjing Agricultural Univ., Weigang No. 1, Nanjing, Jiangsu 210095, China (cotton@njau.edu.cn)
Zhang, Xianlong	National Key Laboratory of Crop Genetic Improvement, Huazhong Agricultural Univ., Wuhan, Hubei 430070, China (xlzhang@mail.hzau.edu.cn)

Cotton as a World Crop: Origin, History, and Current Status

Joshua A. Lee and David D. Fang*

Abstract

Archaeological evidence indicates that cotton has been used by humans for more than 4000 yr. The history of cotton cultivation is at least 3000 yr old. There are four cultivated cotton species, two diploid species and two tetraploid species. In this chapter, the origin and history of these four species as a crop were discussed based on theoretical consideration and archaeological evidence. Domestication of wild tetraploid cotton as an annual day-neutral crop was a big step toward cotton agriculture. Spread of cotton to the United States and its wide cultivation in this country facilitated cotton becoming a world crop. Technology advancements in cultivation, ginning, and spinning solidified the important role of cotton in the world economy. New technologies such as transgenic varieties are revolutionizing the cotton industry. This chapter provides a historical view of how cotton has become as a world crop.

Cottons (from the Arabic *quotn*) are useful in that some species produce spinnable fibers (lint) on the seed coats. Such fibers begin as elongated cells growing outward from the surface of the ovule. As the ovule enlarges, successive layers of cellulose are laid down in a helical pattern by the protoplast. As the fiber matures, the protoplast dies, and the cell wall, which is virtually pure cellulose, collapses inward to form a convoluted ribbon. The flattening and convolution of the dried cell wall promotes adhesion when the fibers are twisted together in yarn bundles during spinning.

There are four domesticated species of cotton. *Gossypium arboreum* L. and *G. herbaceum* L. (Fig. 1 and 2), both diploids, are native to the Old World. *Gossypium arboreum* remains an important crop in India, whereas *G. herbaceum*, important in earlier times, is today grown mostly for local use in the drier areas of Africa and Asia. *Gossypium barbadense* L. and *G. hirsutum* L. (Fig. 3 and 4) evolved in the New World. Both are allotetraploids. *Gossypium barbadense*, commonly known as extralong-staple, Egyptian, and Pima cotton, and other names, supplies about 3

Joshua A. Lee, USDA-ARS and North Carolina State University, Raleigh, NC.

David, D. Fang, Cotton Fiber Bioscience Research Unit, USDA-ARS-SRRC, 1100 Robert E Lee Blvd., New Orleans, LA 70124. *Corresponding author (david.fang@ars.usda.gov).

Portions of this chapter were previously published as Lee, J.A. 1984. Cotton as a World Crop. In: R.J. Kohel and C.F. Lewis, editors, Cotton. Agron. Monogr. 24. ASA, CSSA, and SSSA, Madison, WI. p. 1–25. doi:10.2134/agronmonogr24.c1.

doi:10.2134/agronmonogr57.2013.0019

Cotton. 2nd. ed. David D. Fang and Richard G. Percy, editors. Agronomy Monograph 57.

Fig. 1. ***Gossypium herbaceum*** **L. This figure and Fig. 2 to 4 from Parlatore (1866).**

Fig. 2. ***Gossypium arboreum*** **L.**

Fig. 3. ***Gossypium barbadense*** **L.**

Fig. 4. ***Gossypium hirsutum* L.**

to 5% of the current world production of fiber. The fiber is used mostly for the production of luxury fabrics and sewing thread. *Gossypium hirsutum*, known most widely as Upland cotton, contributes about 95% of the current world production of 118 million bales of fiber weighing about 225 kg/bale. Upland cotton fibers are used in the manufacture of a variety of textile products, cordage, and other non-woven products. Linters, the short fibers removed from seeds before crushing, are an important source of industrial cellulose.

Although cotton is grown mostly for fiber, the seeds are also important. Cottonseed oil is used for culinary purposes, and the oilcake residue is a protein-rich feed for ruminant livestock. Cotton is grown virtually around the world in tropical latitudes and as far north as 43°N latitude in Uzbekistan and 45°N in China.

The Origins of Cotton

Theoretical Considerations and Archaeological Evidence

Seemingly, the presence of spinnable seed fibers should have provided sufficient reason for using wild cottons and the eventual domestication of the four cultivated species. However, Chowdhury and Burth (1971) pointed out that a diploid species—possibly *G. arboreum*—might have been domesticated in Nubia for the seeds as a nutritious feed for domesticated animals. Hutchinson et al. (1947) argued that the presence of seed fibers was the chief impetus for the domestication of *G. herbaceum* in the Old World but hypothesized that spinnable fiber represented a mutant, presumably rare, phenotype in a wild, or perhaps early domesticated, form of the cotton. These workers concluded further that the New World tetraploid cottons evolved under domestication after the introduction of a cultivated Old World species, *G. arboreum* being the most likely candidate. The cultivated diploid then crossed with a New World diploid taxon, one similar to, or identical with, the extant species, *G. raimondii* Ulb., thereby spawning the prototype from which the tetraploid species of *Gossypium* diverged. Presumably the event occurred in South America. Modern views support the hypothesis of primitively wild linted cottons in both the Old and the New Worlds and that the presence of lint provided the chief impetus for domesticating the cottons, perhaps the only reason for domestication in the New World (Stephens, 1967; Fryxell, 1979).

Evidence for the beginnings of agriculture in the Near East date with certainty from about 7000 BCE (Harlan, 1975). No remains of cotton have been found in the Near Eastern Center. Indeed, there is no evidence that cotton ever grew wild in the region. The chief interest of the Near Eastern Center in relation to the domestication of cotton in the Old World is that the agriculture evolving in the area—a grain, pulse, livestock, and linen economy—supported the societies into which cotton was later introduced as a wild, or perhaps newly domesticated, plant. Thus, Hutchinson et al. (1947) reasoned that the growing and spinning of cotton evolved in economies that already had a technology for the spinning and weaving of flax (*Linum usitatissimum* L.) and wool. This conclusion was based largely on the fact that the only known wild form of a domesticated diploid cotton, *G. herbaceum* var. *africanum* (Watt) Hutchinson and Ghose, has fiber length in the range of 15 mm or less, well below the staple length of modern *G. herbaceum*. Presumably cotton was grown initially for uses other than textile yarn, for wound dressings, wadding, and the like. Longer staples had to be selected before cotton fiber could become competitive with either flax or wool.

The findings of cotton remains in Nubia (Chowdhury and Burth, 1971) and at Mohenjo-Daro (Fig. 5) dating from about 2700 BCE (Gulati and Turner, 1928) give credence to the idea that cottons were domesticated at sites that already knew a productive agriculture and the arts of spinning and weaving. The Nubia finds might represent cotton at the interface of domestication; the Mohenjo-Daro finds certainly do not. The cottons of Mohenjo-Daro, which seems to have been the hub city of an ancient Indus civilization, were advanced diploids, very likely *G. arboreum*. There is no clue as to whether the cottons had advanced to an annual habit of production, as is standard for modern cottons of all species, but the latitude where discovered—about 27°N lat.—suggests that the cottons might have evolved toward an annual habit.

Mohenjo-Daro was a flourishing metropolis of commerce and industry. Irrigation was practiced and the agriculture obviously sustaining. Spinning and weaving were practiced arts. Cotton lint was separated from seeds on the churka, a kind of hand-cranked roller gin, and the fibers processed into fabrics with drop spindles and primitive looms. The quality of the textiles produced appears to have been excellent.

Gossypium herbaceum var. *africanum* grows as a wild plant in the bushveld of southern Africa and possibly in the Sahel of north Africa (Hutchinson et al., 1947) and, according to these workers and Fryxell (1979), is the wild form from which the domesticated species most likely arose. The cotton is linted, but the seed fibers are sparse, short, and dingy gray in color. Nevertheless, the fiber of *africanum* is spinnable.

South Africa is distant from the Near East and an unlikely place for *G. herbaceum* to have been domesticated. Fryxell (1979) points out that the cotton could have been introduced to the Near East in ancient times where the plant was subsequently domesticated. Trade between Arabia and southern Africa is very old (Hutchinson et al., 1947) but just how ancient is not known. Alternatively, the

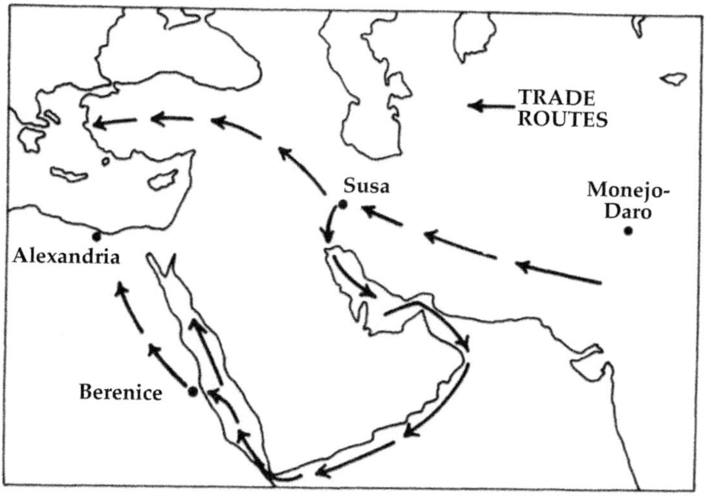

Fig. 5. The Near East in Alexandrian and post-Alexandrian times with points of archaeological interest, trade routes, and commercial centers.

domestication of *G. herbaceum* might have begun in north Africa, or wild *G. herbaceum* might once have been distributed to Arabia, Baluchistan, and perhaps the Sind.

No forms of *G. arboreum* are extant in the wild. Hutchinson et al. (1947) proposed that the species evolved from *G. herbaceum* after the latter was domesticated. Other than through ploidy manipulation, the derivation of one species from another, or others, under domestication is unprece dented. Hybrids between *G. arboreum* and *G. herbaceum* have apparently regular chromosomal pairing, except for the presence of one translocation (Gerstel, 1953), and an uncomplicated meiosis. There is evidence of developmental incompatibility ("genetic breakdown") in ensuing generations. Where the two species overlap in distribution as cultivars, as in India, the identity of each is maintained, presumably through the combined effects of genetic and temporal barriers (Hutchinson et al., 1947). Fryxell (1979) reasoned that the time allowable for such a degree of divergence between the two taxa seemed insufficient to indicate speciation in the domesticated state. Modern molecular marker technology does support Fryxell's view that *G. arboreum* did not derive from *G. herbaceum* (Hinze et al., 2015).

Today *G. barbadense* grows as a wild plant on the coasts of Peru and Ecuador and perhaps the Galapagos Islands. Commensal and marginally cultivated forms of the species occur through much of northern South America, the West Indies, and parts of Middle America (Fig. 6). The latter site, along with the warmer parts of the Old World, were populated with *G. barbadense* in post-Columbian times (Hutchinson et al., 1947). The center of diversity for the species is northwestern South America. *Gossypium darwinii* Watt grows wild in the Galapagos Islands. This taxon is a near relative of *G. barbadense* (Fryxell, 1979) and is regarded as being conspecific with the cultivated species by some authorities, Hutchinson et al. (1947) for example.

Stephens and Moseley (1974) examined fruits, foliar parts, and fiber of cottons taken from archaeological contexts, dating from about 2500 to 1750 BCE, in the Ancon-Chillon district of Peru. Based on fiber characteristics and colors, fruit shape and size, and the absence of trichomes from the floral nectaries, the authors concluded that the remains were from primitive cultivated forms of *G. barbadense*, some barely beyond the threshold of domestication. The only other cultivated species recovered was the hard gourd, *Lagenaria* spp. The impetus for the utilization of cotton in the area seems to have been the need for cordage for fishing lines and nets.

Ancon-Chillon thus appears to be a site where cotton was domesticated. Of special interest was the fact that the cotton was *G. barbadense* and not an Old World diploid species. This finding does much to negate the possibility that one of the ancestors of New World tetraploid cottons was introduced by agricultural people. Moreover, the primitive state of the Ancon-Chillon cottons suggest that a linted form of *G. barbadense* grew wild in the area. The Ancon-Chillon people, even if they showed no evidence of farming for food, were, nonetheless, a settled folk who practiced a kind of agriculture suited to their needs, namely, the gathering of seafoods. *Gossypium barbadense* thus seems to have evolved in South America as a wild, and later as a cultivated, species.

Gossypium hirsutum in its wild and commensal forms is distributed in the drier areas of Middle America, northern South America, the West Indies, the southern tip of Florida, Polynesia, and through introduction, north Africa and

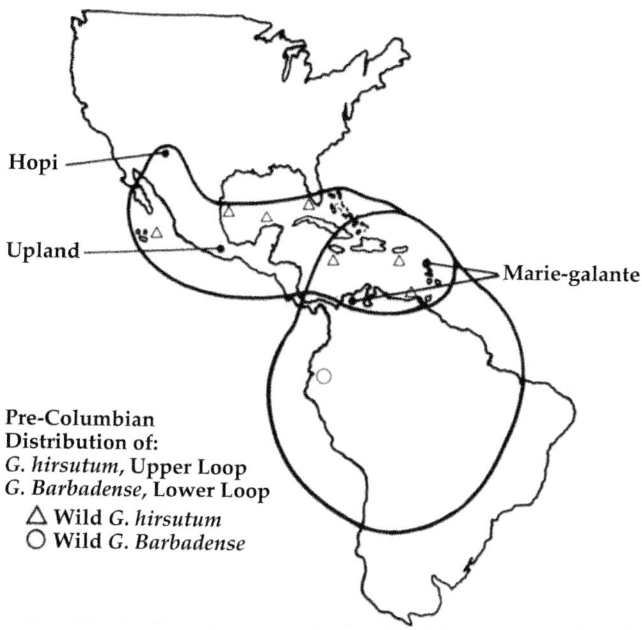

Fig. 6. Approximate distribution of *G. barbadense* **and** *G. hirsutum* **in the New World in pre-Colombian times with notations on the distribution of wild populations, and sites of domestication.**

southern Asia. The wild populations are rare and widely dispersed. All grow on beach strands, or else are confined to small islands. Archaeological remains of *G. hirsutum* have been found mostly in Mexico, the oldest from the Tehaucan Valley contexts (Smith and Stephens, 1971). These were dated from the Abejas phase, about 3500 to 2300 BCE, and appeared to have been from domesticated stocks. The Tehaucan cottons were believed to have been introductions, inasmuch as there is no evidence that *G. hirsutum* ever grew wild in the valley.

Evidently *G. hirsutum* was domesticated in at least two, perhaps three, separate areas. Modern Upland cottons stem from a center of diversity near the border of Mexico and Guatamala (Hutchinson et al., 1947). The domesticated forms of *G. hirsutum* var. *marie-galante* (Watt) Hutch. might have been derived from a center of diversity for the race in northern Colombia (Stephens, 1974a) or from introgression between West Indian wild forms of *G. hirsutum* and introduced *G. barbadense* (Stephens, 1967). Wild and commensal forms of *marie-galante* along the Pacific coast of Middle America overlap the distribution of other races of *G. hirsutum* intruding from the north only slightly. That, plus the fact that domesticated stocks of *marie-galante* are distributed in northern South America and the West Indies, argues for a separate domestication for the race and the Upland cottons.

The Hopi cottons of the American Southwest are annual forms placed in *G. hirsutum* var. *punctatum* (Schumacher) Hutch. by Hutchinson et al. (1947). The *punctatum* cottons are the most widely distributed of the wild and commensal forms of *G. hirsutum*. The distinctiveness of the Hopi cottons, when compared with Upland cultivars, suggests that the Hopi forms had an origin independent of modern Uplands and *marie-galante*. No form of *G. hirsutum* as a perennial plant has a history of growing north of the Mexican border in what is now the United States. Therefore, it is likely that the Hopi cottons were part of a northward migration of what was basically a corn, bean, and squash agriculture into the American Southwest, about 100 to 500 CE.

Domestication and Spread of Cotton

Most wild cottons inhabit regions where rainfall is strictly seasonal, occurring in most such places in summer. Characteristically such cottons are perennial shrubs or small trees. Mauney and Phillips (1963) showed that a series of wild cottons required 15 or more nodes along the main plant axis before reaching a transition from the vegetative to a fruitful phase of growth. Therefore, a seedling emerging during the rains would rarely have time to develop the plant stature required for flowering at the end of the wet season but could flower seasonally thereafter.

Typically a wild cotton, such as the Jackson Bay population of *G. hirsutum* of the Caribbean strand of Jamaica, lapses into dormancy during the winter dry season and becomes vegetative with the onset of rain in early summer. At the end of the rainy season, vegetative recruitment ceases, and the plants enter the reproductive phase. There is a short flush of flowering, followed by fruit maturation, and re-entry into the dormant phase.

At least three environmental factors interact to trigger an orderly flowering process in wild cottons. First, flowering does not proceed as long as there is sufficient soil moisture to sustain vegetative increase (Hutchinson et al., 1947; Mauney and Phillips, 1963). Second, a differential between day and night temperatures is needed. During the wet season in the tropics the temperature differential between day and night is minimized by the presence of high humidity and cloud cover. At the onset of the dry season radiation on clear nights increases the differential between daily maximum and nightly minimum temperatures. According to Mauney and Phillips (1963) a differential of at least 12°C is needed to maintain orderly bud set and subsequent flowering in many wild cottons. The third requirement is for a decreasing photoperiod. However, it is difficult to argue for a photoperiod requirement for a taxon such as *G. darwinii* which grows within 1° of the equator. Nevertheless, when the cotton was subjected to variable photoperiods by Mauney and Phillips (1963), *G. darwinii* was found to require short days for flowering and setting fruit.

Based on the present distribution of the wild forms in the New World, *G. hirsutum* must have diversified several degrees north of the equator, at latitudes where photoperiod varied seasonally. Thus photoperiodic response has evolved as an important criterion in timing the onset of reproduction in many of these cottons (Stephens, 1967).

Most modern cottons behave as annuals; that is to say, the crop is harvested within the same year planted. However, "annual" cottons are not true annuals in the sense that death of the plant is a natural consequence of seed ripening. Modern cultivars of cotton can be maintained indefinitely under greenhouse culture. As one might expect with taxa that have been transformed from long-lived perennials with a slow rate of plant development to cultivars that can be cropped annually from seeding, a broad spectrum of fruiting habits, relative to timing after seedling emergence, can be observed. Thus, Hutchinson et al. (1947) spoke of early annuals, late annuals, and perennials. Stephens (1974b) listed a group of cottons as annuals apparently because the cultivars shared the property of fruiting at low enough node for the crop to be harvested within the same year planted. Some of the cottons listed by Stephens are not "day neutral" as far as photoperiodic requirement for fruiting is concerned. For example, the *G. barbadense* cultivar Lengupa does not set fruit when grown in the field at College Station, Texas (Lewis and Richmond, 1960), but sets fruits at low nodes under the short-day conditions of northern South America. Therefore Stephens (1974b) seemed to be defining annual habit as the trait of setting fruit precociously under at least some environmental conditions. In a later paper Stephens (1976a) chose to reserve the term *annual habit* for some cultivars of *G. hirsutum* that cropped precociously in the tropics and then failed to survive the dry season. Under the revised definition cottons such as the Sea Island and Egyptian forms of *G. barbadense* would be regarded as perennials that are grown as annuals in subtropical latitudes.

From the standpoint of understanding how cottons fitted for extratropical production might have evolved, the definition implicit in the Stephens (1974b) reference seems most meaningful. Therein annual habit is interpreted as the consequence of a downward shift in the number of nodes required along the main plant axis for the onset of fruiting. Such a change must precede, or at least accompany, loss of short-day response if the cotton is to adapt to latitudes that have killing frosts.

The loss of the short-day requirement for flowering and fruiting might have accompanied the shift to annual habit in some cultivated cottons, but most certainly not in others. Lengupa is one such example, and some accessions of *G. hirsutum* from the Mexican center are annual in fruiting habit but retain the short-day requirement for flowering (Kohel and Richmond, 1962). The evidence shows that the shift to "day neutrality" in *G. hirsutum* involved multigenic changes (Kohel and Richmond, 1962), whereas the one case where contrasts were compared in *G. barbadense* showed evidence for a monogenic change (Lewis and Richmond, 1960). Obviously, both changes to annual habit and day neutrality were required before cottons could be grown in extra-tropical environments with cold winters.

The cottons of the Old World might have begun the route to domestication as wild plants introduced into what were sophisticated agricultural economies. In the New World the indigenous peoples doubtless took from the wild cotton plants the materials needed for trivial domestic uses, whether fiber or seeds. From such utilization the cultivation of cotton might have started as an adaptation to "dump heap" agriculture as described by Anderson (1954). The dropping of seeds on such middens, likely with no initial intent to cultivate, could have evolved into a handy way to grow a few plants away from the cotton's native ground. The wild forms of the domesticated New World cottons seem opportunistic toward disturbed land. Thus I once noted that Jackson Bay wild cotton, which typically grows at the interface between saline marsh and xeric scrub in Jamaica, had spread inland a short distance along roads cut by charcoal gatherers.

Cottons commensal with human settlement, arising perhaps from vicarious plantings of wild material, evolved into forms with somewhat more fibers on seeds, and frequently, but not always, larger fruits (Hutchinson et al., 1947; Fryxell, 1979). Human selection is indicated inasmuch as there seems to be no conceivable way that increased fibers could be advantageous to cottons naturally propagating, but would be of value to the incipient, or actual, cultivator. Such plants remained perennial in growth and reproductive habit, and they fitted in well with small-plot subsistence agriculture. A few plants supplied a family's needs, and perenniality meant a definite savings in management.

The citizens of Mohenjo-Daro were clothed in cotton raiment, as were those of southern Mexico in the days of Toltec, Aztec, and Maya. The Inca of the Andes also used cotton lint extensively. Annual habit in cotton was selected in conjunction with intensive cultivation of the crop by these peoples. Annual production allowed for the transition of cotton-growing from a marginal, home-based enterprise to production on a commercial scale (Hutchinson et al., 1947).

The loss of the short-day sensitivity requirement for fruiting in the various cotton species remains, as to time and place, a mystery, but apparently occurred in tropical, or near-tropical, latitudes. Such an inference seems sound because planting cottons with stringent photoperiodic requirements in latitudes such as those of the American Cotton Belt could not result in production. However, it is possible, inasmuch as the photoperiodic trait of *G. hirsutum* is multigenic and expressed in an additive manner (Kohel and Richmond, 1962), that some of the earlier introductions of Upland cottons to the American Cotton Belt were only partly converted to day neutrality, that is to say, were perhaps capable of producing a crop as daylength decreased in late summer. Such cottons could then have been selected for earlier performance in the new environment. The introduction of the Sea Island form of *G. barbadense* into the south Atlantic coastal region of the American colonies is documented (Stephens, 1976b). With these cottons, annual habit coupled with day neutrality was definitely a matter of preadaptation.

Stephens (1976a) proposed a model whereby the Sea Island cottons might have lost their short-day response and simultaneously acquired the annual habit. A cross between a West Indian commensal form of *G. barbadense* and wild accession of *G. hirsutum*, WH-24, produced some segregates in later generations that set fruits under the long-day summer environment of North Carolina. Neither parent was day neutral in flowering habit. Earlier Kohel et al. (1974) recovered a few summer-flowering plants from the F_2 of a cross between an accession of *G. hirsutum* var. *marie-galante* and the *G. barbadense* cultivar, Lengupa.

Some of the Stephens segregates resembled Sea Island in both plant form and fiber properties. Thus Stephens reasoned that Sea Island cotton might have acquired both fitness for production in South Carolina and Georgia and the cotton's distinctive fiber properties from introgression of germplasm from a short-day-flowering form of *G. hirsutum* into a coarse fibered form of *G. barbadense*. If such introgression occurred, the event must have taken place in the West Indies before Sea Island cotton was introduced into a region where it gained such renown as a crop.

There is no evidence that *G. hirsutum* in the Mexican center of diversity was ever exposed to interspecific introgression. The potential for interracial mixing is evident. Hutchinson et al. (1947) hypothesized that as the need for cotton grew in Middle America, the crop was gradually introduced into altitudinal zones not well suited for annual cottons adapted for short-day flowering and fruiting during the dry season. The result was selection for a tendency to set fruit during a part of the vegetative cycle of growth, a trait of modern Upland cotton cultivars. The short-day response diminished under the new regime of growth. Table 1 compares wild and modern cultivated *G. hirsutum*.

Table 1. Comparison of wild and modern cultivated forms of G. hirsutum.

Characteristic	Wild	Domesticated
Distribution	Mostly tropical; populations rare; insular and strand habitats.	Tropical to temperate; widespread as a crop; cultivated land.
Germination response	Impervious seeds and sporadic germination.	Seeds imbibe water and germinate rapidly.
Phenology	Sharply seasonal pattern of vegetative increase followed by fruit production and dormancy.	Short vegetative phase followed by overlap in vegetative and fruit increase and eventual "cut-out" fruit ripening; plants usually destroyed after producing.
Fruiting habit	Perennial fruiting; 15, or more, nodes to first fruiting branch.	Annual fruiting; 5 to 7 nodes to first fruiting branch.
Fruiting response	Requires short days, cool nights, and increasing drought stress for fruiting.	"Day-neutral"; prefers cool nights and ample soil misture for best production.
Fruit size	Small; 2 g of seed and lint.	Large; 6 to 8 g of seed and lint.
Lint quantity	Sparse; 15 to 20% gin out-turn.	Copious; 35 to 40% gin out-turn.
Fiber properties	Lint short and fine; dingy brown in color.	Lint length variable up to 28 mm; coarse or fine; most often white.

Gossypium arboreum spread to China, and *G. herbaceum* to China, Turkestan, and Persia before the 13th century (Fryxell, 1979), where some very precocious annual forms of both species were selected. *Gossypium hirsutum* became a crop of local importance in the southeastern United States by the mid-18th century. The early stocks, known as "Georgia Green Seed," were probably of Mexican origin (Hutchinson et al., 1947). Similar stocks were taken to southern Asia at about the same time. These cottons became known as Cambodias, and, unlike Georgia Green Seed, were short-day sensitive in flowering response. The Cambodia cottons are virtually obsolete today. Later introduction of Mexican stocks to the American Cotton Belt were "white-seeded" stocks with larger fruits and longer lint than the "green-seeded" cottons. Hybridization between the stocks produced the explosion of Upland cottons that are the foundation of the world's cotton industry today.

The first day neutral annual forms of *G. barbadense* were the Sea Island cottons of South Carolina and Georgia. In the early 19th century, Sea Island hybridized with Jumel's tree cotton, apparently a perennial form of Peruvian *G. barbadense,* to spawn the distinctive annuals known today as Egyptian cottons (Balls, 1912). Egyptian cottons later gave rise through further hybridization to the Pima cottons of the southwestern United States (Bryan, 1955) and to the limbless cultivars of *G. barbadense* grown today in the former Soviet Union countries in central Asia.

The Advance of Cotton in World Trade and Manufactories

From Alexander to the Industrial Revolution

Much in the following sections was extracted from J.A.B. Scherer's (1916) *Cotton as a World Power.* Scherer pointed out that Alexander of Macedonia introduced cotton and cotton goods to the then-known Western World. Evidence for the existence of a cotton industry in the Meroitic empire of the Sudan dates from about 400 BCE (Hutchinson et al., 1947). Nevertheless, cotton was apparently unknown in ancient Egypt. Nearchus, a general in Alexander's army during the march to the

Indus, described cotton apparel thus: "...a shirt, or a tunic, reaching the middle of the leg, a sheet folded over the shoulders, and a turban wound around the head."

Theophrastus (373–288 BCE) described cotton plants from material brought to Greece by Alexander. Cotton had spread to Persia in pre-Hellenic times, cotton textiles being mentioned in the Book of Esther. That cotton was grown and utilized by the Assyrians in the time of King Sennacherib (706–681 BCE) is attested to by a cylinder now in the British Museum. The cylinder reads: "The trees that bore wool they clipped, and they carded it into garments." Neither Persia nor Assyria seemed to have traded with the civilizations to the west. As mentioned before, the spinning of cotton was an ancient art in India by the time of Alexander. Some of the fabrics produced on the looms of the Indus people have been described as "webs of woven wind."

Alexander established trade routes between East and West. Susa, one of the capital cities of Persia, was selected to serve as the hub of a trading empire with a highway overland through Asia Minor and water routes through the Euphrates and Tigris to the Persian Gulf and thence to the eastern Mediterranean (Fig. 5).

After Alexander, the Ptolemaic pharoahs made Alexandria the chief city of the eastern Mediterranean and the leading center for dispersing goods from the East. Ptolemy Philadelphus envisioned a Suez canal, but built instead the port of Berenice on the Red Sea coast. Goods from India were landed at Berenice and transported overland by caravan to the Nile whence they were floated down river to Alexandria. Cotton goods thus became commonplace, although luxurious, items for the ruling classes of Greco-Roman times.

The growing of cotton in the lands about the eastern Mediterranean followed the introduction of cotton goods. The crop became established in lower Egypt and in the Levant. Cotton (most likely *G. herbaceum*) was introduced into Spain by the Moors in 712 CE where it remained an important crop until the final expulsion of Islam from Iberia in 1492.

Cotton fabrics developed to the point of promoting style in clothing during Renaissance times in Italy. The state of Venice forthwith became the leading commercial center of the Mediterranean, depending largely on cotton goods from India as a staple of commerce. From Venice cotton cloth found markets as far away as northern Europe.

In 1492 Christopher Columbus sailed under the flag of Isabella the Catholic westward, seeking India, but discovered the New World instead. Among the first products of the New World examined by the Admiral were cotton fabrics woven by the indigenous people, his "Indians." In 1497 Vasco DaGama sailed around Africa to India, thereby establishing the sovereignty of Portugal over the sea routes to the East. Lisbon replaced Venice as purveyor of cotton goods and spices to Europe, and the Alexandrian trade routes collapsed into disuse.

England eventually became predominant in cotton trade and manufactories, and one can justly state that cotton rode to prominence on the back of wool. As a result of royal guarantees concerning private property, English yeomen of post-medieval times had no fear that the crown, or the nobility, would confiscate such readily purloinable property as sheep, apparently a genuine hazard for the husbandmen of continental Europe at the time. Thus did England come to enjoy a virtual monopoly on wool production in Europe. However, England had no commercial spinning and weaving industry until Edward III (1327–1377) invited Flemish weavers, who worked virtually as serfs for the continental guilds, to settle in the Midlands. Shortly thereafter England became not only the chief producer of European wool but a leading manufacturer of woolens as well.

The incursion of cotton into the domain of wool was, at first, neither welcomed nor tolerated. However, the utility of cotton apparel for summer wear overwhelmed the opposition so that by the turn of the 18th century the cottage spinning and weaving industry of Lancashire was heavily involved in the processing of the "tree wool," and the West Indies had become a prominent source of supply.

Cotton and the Industrial Revolution

Up to the late 18th century the cottage textile manufactories in England comprised a labor-intensive industry employing hand cards, spinning wheels, and manually operated looms. Two kinds of spinning wheels have evolved through the ages (Anderson, 1976), the Old, or Jersey, wheel and the Saxony wheel. The Jersey wheel uses a system of spinning hand-drawn roving (slightly twisted strands) followed by winding of the spun yarn onto the base of an open spindle. This kind of spinning is an intermittent process and represents a kind of mechanization of the ancient drop-spindle technique of spinning yarn. The Saxony wheel spins and winds continuously, the roving being twisted by a U-shaped flyer while being wound onto a bobbin inside the flyer. Both the intermittent and continuous spinning principles live on in modern textile machinery.

In the latter half of the 18th century problems of supply in the cottage weaving industry spurred inventions which, in turn, begot further supply problems. The concatenation of events that followed revolutionized the production of textiles, and, indeed, spawned the factory system and the Industrial Revolution. Warp for the loom was most commonly produced on the Saxony wheel, whereas shuttle weft, requiring fewer twists and lower counts, was produced largely on the Jersey wheel. Because the Saxony wheel, with its continuous process of spinning was more productive than the Jersey wheel, there was frequently a shortage of weft. The shortage was made acute for both warp and weft after John Kay's invention of the fly shuttle in 1738, a system whereby the shuttle was routed rapidly through the warp by a system of springs (Anderson, 1976).

As if in answer to the shortage of yarn, James Hargreaves patented the spinning jenny in 1770 (Aspin and Chapman, 1964). The jenny (Fig. 7) employed the same intermittent spinning system as the Jersey wheel, the difference being that the jenny had a bank of 12 or more spindles, and spinning was accomplished by twisting against the draft; that is, a bank of rovings, each attached to a spindle, was attenuated by drawing the clamped rovings away from the spindles with a sliding clasp, or clove. Once enough twists had been applied, the spindles were reversed to "back off" the twists on the spindles knives, and a "faller" wire was dropped to depress the spun yarn onto the body of the spindles. The yarn was then wound up by pushing the clove toward the spindles as the spindles were cranked slowly by hand.

In 1738, Lewis Paul built a machine that multiplied yarn through a bank of flyers and bobbins, the principle of the Saxony wheel. The machine was apparently not successful. However, Paul was granted a patent for roller drafting, a system of attenuating roving by passing it through sets of rollers traveling at progressively higher speeds (Anderson, 1976). In 1769, Richard Arkwright led a group that built

Fig. 7. The Hargreaves spinning jenny. (Courtesy of Smithsonian Institution Archives.)

a spinning frame employing roller drafting and the flyer and bobbin principle. The machine was successful and came to be called the "water frame" after it was adapted to water powering. The coarse yarns spun by the frame were useful as warp, whereas the jenny spun fine-count weft at an accelerated rate.

The spinning mule became obsolete in recent times. This machine was a kind of hybrid between the water frame and the jenny; that is to say, rovings were drafted by rollers and spinning was of the open-knife, intermittent kind as in the jenny (Fig. 8). Samuel Crompton, who invented the mule, was operating the prototype by 1779 (Catling, 1970). The mule differed from the jenny in that the spindles traveled on a carriage that ran outward on the spinning and drafting cycle, and inward on the wind.

The jenny, used as late as 1926 in the woolen industry of Yorkshire, was always hand-cranked. The spinning frame and the mule were originally hand-cranked machines. Obviously, the amount of yarn one worker could spin on such machines was limited. The perfection of the steam engine by James Watt around 1790 changed that. By the beginning of the 19th century powered mules and spinning frames, as well as powered looms, perfected by Edmund Cartwright around 1792, were the tools of the burgeoning textile industry in England, and increasingly, abroad. But where were the mills to get the cotton needed to keep the machines operating?

Eli Whitney and the First "White Revolution"

The grievances that spawned the American Revolution were many, prominent among which were attempts by Britain to exclude manufactories from the colonies, to levy import taxes on certain goods bound for the colonies, and to restrict coinage. Britain envisioned the colonies as producers of raw materials

Fig. 8. Crompton's mule. (Courtesy of Harold Catling.)

and consumers of finished goods and little more. Be that as it may, the American Revolution was fought and won, and by 1793, the year Eli Whitney invented the saw gin, the newly united states were well on their way to self-sufficiency in both agriculture and manufactories.

Eli Whitney was an educated man, seemingly destined for a career as a schoolmaster. While visiting Mrs. Nathaniel Greene near Savannah, Georgia, he became intrigued with the problem of separating the lint of Upland cotton from the seed. The naked seed of Sea Island cotton could be separated with a churka, but that device did not work well with the fuzzy seed of Upland. Sea Island cotton, the basis of a modest industry along the coasts of Georgia and South Carolina at the time, was poorly adapted for growing on inland sites; by contrast, the potential for growing Upland cotton seemed unlimited.

Whitney's gin employed a system of rollers. One roller was equipped with teeth for plucking lint from seeds, the lint then being doffed by a counter-rotating roller equipped with brushes. That Whitney envisioned and developed the basis principles of the saw gin is a matter of history. Less well known is the fact that a mechanic in Augusta, Georgia, Hodgen Holmes, developed a roller equipped with saw teeth, and the Holmes saws proved to be more serviceable than the Whitney pluckers. Litigation over patent rights to the various particulars of the saw gin's development seem endless. Suffice it to say that both Whitney and Holmes were well rewarded for their efforts.

In 1793 the Southern states produced about 10,000 bales of cotton of some 227 kg each. By 1810, 177,824 bales were produced, and southern cotton had become important in world trade. The result of the invention of the saw gin was to reverse the trend toward emancipation of slavery in the South and to turn the region from a growing system of manufactories to a nearly total agricultural economy based on the production of cotton and slaves. The South changed from a region strongly supporting the concept of the Federal Union to one espousing the narrowest conception of States Rights. The North, and particularly the New England States, became centers of manufactories and began to demand protective tariffs. The South, as producer and supplier of cotton both to the North and Britain, found it good economics to load vessels on the return voyage from Britain with British goods. As the North gained in population and political strength, tariffs became a reality, and the profit margin from the cotton trade began to shrink. The doctrine of nullification of Federal statutes, if the law in question did not serve a particular state's interests, was advocated by John C. Calhoun of South Carolina. That, growing sectionalism, and the question of the status of slavery in western lands, exacerbated tensions between the North and the South and did much to precipitate the War Between the States.

That the Civil War was ruinous to the South can scarcely be exaggerated. The Confederacy had hoped that a shortage of cotton would compel the industrial nations of Europe to recognize the soveriegnty of the Confederate States and lend military assistance. That did not happen. Apparently the working classes of Britain came to despise slavery more than they craved bread. More subtle opinions held that Britain preferred a united and strong English speaking nation in the western hemisphere to prevent incursions by other European nations into what the British had come to accept as their sphere of influence. Be that as it may, cotton, if it had caused the rise and fall of the Confederacy, most certainly rescued the economy of the Southern states after reunion. After a short hiatus of cotton delivery and consumption during and after the war, demand rebounded for the fiber for which there was no substitute. Cotton production continued to climb, both in the South and abroad.

Cotton into Modern Times—The Second White Revolution

The Development of Modern Spinning Technology

William Kelly of Lanark first applied power to the spinning mule around 1790 (Catling, 1970). However, power was used only with the drawing and spinning cycle. Human labor was still required to "put up" the mule, that is, to back yarn twists off the spindles and then wind the spun yarn onto spindle cops, or packages. The invention of the quadrant arm, and the attendant nosing and strapping actions, by Richard Roberts between 1830 and 1836 provided for automatic cop winding, and the "self-acting" mule was born (Catling, 1970). Mules reached the zenith of their development in Britain near the end of the 19th century when units driving 1000 or more spindles were commonplace. The mule was used widely in Britain up to modern times.

With the advent of steam power, the speed possible on spinning frames increased dramatically, thereby creating the need for more durable equipment. The wooden flyer was replaced by one of metal, a device that warbled as it ran at high speeds. Mill workers likened the sound to a babel of song thrushes, the "throstle" of Shakespeare, and forthwith christened the new machines throstles.

The throstle was to provide the prototype from which sprang a series of important inventions in America. In 1828 Charles Danforth and John Thorp independently patented a cap-spinning throstle. The cap sat over the bobbin, replacing the flyer of the spinning frame. A guide at the base of the cap fed spun yarn to the bobbin which traveled up and down on a lifter. Air resistance caused the yarn to "balloon" free of the cap.

Although the cap throstle increased spinning speed, virtually doubling per-spindle output in comparison with the mule, the cap had to be removed for doffing the filled spools. Thus the major importance of the cap throstle principle was that it provided the basic impetus for the development of modern ring spinning (Anderson, 1976).

In 1828 John Thorp, one of the inventors of cap spinning, patented the basic elements of ring spinning. Thorp's mechanism featured a stationary ring fitted with a rotating wire band carrying a hook to guide the ballooning yarn onto the bobbin. The wire band rotated within the ring. The major advantage of the ring over the cap was that the bobbin was easily doffed. Indeed, automatic doffing became commonplace in later years.

The advantage of the ring frame over the mule was that the continuous spinning processes of the frames were about twice as productive as the mule. A disadvantage was that the yarn quality was generally inferior to that produced on the mule. More recent developments have erased much of the difference so that ring spinning is the method most widely used in America today. Moreover, ring spinning has largely replaced mule spinning in Britain, the place of the mule's birth (Anderson, 1976). Newer methods promise to replace the ring frame. As of today, ring spinning accounts for about 70% of global spinning capacity.

In 1807 a patent was granted to a British inventor, Samuel Williams, for a spinning system that foreshadowed modern open-end, or break, spinning (Anderson, 1976). Apparently Williams' proposal met with little reception, hence no success. By 1975 the concept of twisting without rotating the take-up package—bobbin or cop—was being pursued seriously in many textile laboratories. An open-end "rotor" technique has been the result.

In the open-end technique fibers are plucked from a continuous sliver by a wire-covered cylinder. The fibers are then lifted from the cylinder and transported inside a rotor by an air stream. Within the rotor the fibers are continuously reassembled in a groove in the rim of the rotor and twisted onto the end of a yarn skein exiting at the center of the rotor.

A much higher draft ratio (attenuation of roving) is possible with the open-end process than with the mule or with the ring frame. Hence finer yarns can be spun at a more rapid rate. Open-end rotor spun yarns comprise about 20 to 25% of the world's supply of yarns from all methods of spinning. These yarns are spun mostly in Europe and Japan.

In 1970s, a new spinning technology, air-jet spinning, the so-called Vortex technology from Japan-based Murata Machinery Ltd. was introduced into cotton textile industry. This technology enables faster spinning and results in higher productivity. However, less than 5% yarn was spun using air-jet spinning technology at the present time.

Development of Fiber Evaluation

Over the millennia cotton farmers have made impressive strides in fitting the various taxa of cotton for production under domesticated conditions. Indeed, selection at the hands of people innocent of any kind of formal training transformed cottons from wild plants closely adapted to seasonally arid conditions to productive crop plants fitted for enriched soils and abundant soil moisture. Moreover, some of the best cotton fibers known—those of the Sea Island growths—were selected largely by the senses of sight and touch. Nonetheless, processors of cotton over the centuries have suspected that measures of cotton quality needed to depend on more than "feel" and "pull."

W.L. Balls (1928) recognized the problems involved in attempting to spin variable lots of cotton and developed the micro-spinning test for the purpose of evaluating small lots of fiber. Aside from its usefulness in mills, the micro-spinning test inaugurated the breeding of cottons, particularly Egyptian cotton, fitted for specific end uses.

Research and development in cotton fiber evaluation began in the United States in the 1930s (Lewis and Richmond, 1968). Robert W. Webb organized a fiber and spinning laboratory in the USDA Bureau of Agricultural Economics. H.W. Barre, Head of the Cotton Division of the Bureau of Plant Industry at the time, saw the potential usefulness of fiber technology in cotton breeding and entered into close cooperation with Webb. To these workers must go the lion's share of the credit for the subsequent establishment of fiber technology laboratories by cotton merchandising firms, plant breeding establishments, and universities around the United States. Notable among the achievements in fiber technology in modern times are those of Hertel (summarized by Duckett, 1974). Among the instruments developed to modern design in Hertel's laboratory were the digital fibergraph, the Stelometer breaker, and the Micronaire (Arealometer) instrument. In Britain the Shirley Institute, organized primarily to serve the needs of the British textile industry, has made great contributions to cotton textile technology.

Modern cotton merchants routinely sample and extract a full round of information on the fiber properties of each bale of cotton that enters the trade. Such meticulous operations indicate the strides that have been made in relating the properties of raw cotton to spinnability and yarn appearance.

Nearly every bale of cotton produced in the United States is classed by the USDA, Agricultural Marketing Service by high volume instruments that rapidly measure fiber physical properties including fiber length, length uniformity, strength, micronaire (a measure of fiber fineness and maturity), and trash content (a measure of non-fiber contaminants). The quality measurement information along with the origin of the cotton bale is available for the merchants.

Using fiber evaluation (mainly by high volume instruments) as a guide, plant breeders can now set goals for fiber quality in their cultivars with a reasonable expectation of achieving them. That is because basic genetic studies have shown that fiber properties are highly heritable and are thus readily selectable (Miller and Rawlings, 1967). Unfortunately, desirable fiber properties may be associated with low fiber yield. However, some of these correlations have been broken using breeding methods, such as random mating (Jenkins et al., 2008). Modern fiber technology made possible the breeding for substantial increase in fiber yield in Pima cottons while maintaining acceptable fiber quality (Feaster and Turcotte, 1968; Feaster et al., 1979), and the combining of acceptable lint yield with increased fiber strength in Upland cotton (Culp et al., 1979). Thus can cotton breeders, cotton merchants, and processors and spinners of cotton appreciate differences in the quality of cotton from the breeding nursery through to the finished textile product.

Development of Transgenic Cotton

Development and commercialization of transgenic cottons has been one of the revolutionary events in the cotton industry during the past 30 yr. In 1996, the first transgenic cotton variety containing a Bt gene from bacterium *Bacillus thuringiensis* was introduced to the US market. Since then, the transgenic cotton has been grown in more than 15 countries. As of today, transgenic cotton accounts for more than 90% of US cotton acreage and about 70% worldwide, respectively. There are two major traits: Bt toxin (a protein from *Bacillus thuringiensis*) expressed in Bt cotton varieties to protect fruit from lepidopteran insects such as boll worms and herbicide tolerance that enables easy management of weeds using herbicides such as glyphosate. Bt cotton includes Bollgard (Monsanto) that expresses Cry1Ac toxin, Bollgard II (Monsanto) that expresses Cry1Ac and Cry2Ab, WideStrike (Dow AgroSciences) that expresses Cry1F and Cry1Ac, TwinLink (Bayer CropScience) that expresses Cry1Ab and Cry2Ae insecticidal proteins, and WideStrike 3 (Dow AgroSciences) that expresses Cry1F, Cry1Ac, and Vip3A(a). Bollgard III cotton expressing Cry1Ac, Cry2Ab, and Vip3A(a) insecticidal proteins has been registered for use in Australia. The first herbicide tolerant gene to be commercialized in cotton conferred tolerance to the herbicide bromoxynil (BXN by Stoneville Pedigree Company). The BXN varieties were soon replaced by those containing genes that confer tolerance to glyphosate under the name of RoundUp Ready (Monsanto). Later, RoundUp Ready Flex (Monsanto) and Liberty-link (Bayer CropScience) cottons were introduced. Many cotton varieties contain both Bt and herbicide tolerance genes.

Modern Cotton Production

Cotton production recovered at a remarkable pace after 1865, but production techniques changed little during the following 60 yr. Planting and cultivation continued to be performed with mule-drawn gear, and the hoe remained the nemesis of the weed. The use of fertilizers was sporadic, and the crop was picked by hand. Production in the United States hovered around 224 kg/ha, or slightly less, from around 1870 to about 1935 (Lewis and Richmond, 1968).

Beginning in the mid-1930s, yields began to rise. That was because of incentives offered by the newly enacted price support programs coupled with restrictions in acreage. Beyond that, the use of fertilizers increased. There were, in at least some regions, attempts at insect control, and production rose in the irrigated West where production practices could be controlled more precisely than in the rainfed areas to the east.

The period just before World War II saw the first tentative steps toward mechanization of the crop. The post-war era saw a move toward total mechanization, not only in the United States, but in many of the other cotton producing areas of the world. Multi-row planting and cultivating equipment became standard. Weeds are controlled today largely through the use of herbicides, and insect control through the use of various insecticides gives a degree of plant protection earlier cotton growers could not have dreamed of. By 1966 yields of lint in the United States had reached 561 kg/ha, a twofold increase in yield in 30 yr. Since that time the yield has plateaued until around 2000. The reasons for the leveling off in yield have been a vexing problem. Some reasons frequently cited are that insect control has become more difficult, especially with the *Heliothis* (bollworm) complex, which has adapted to resist pesticides, and that larger areas of traditionally low productivity per acre—the High Plains of Texas for example—have come increasingly into production. Since 2000, US cotton average yield has been growing from about 750 kg/ha in 2000 to about 900kg/ha in 2012. One major reason behind the yield increase is likely the wide use of transgenic varieties.

On a world basis, the future of cotton growing and utilization seems bright. The crop continues to increase in volume as the world's population grows even though the cotton market share is decreasing. Utilization of transgenic varieties and hybrids in India dramatically increased average yield and total production in this country.

References

Anderson, E. 1954. Plants, life, and man. Andrew Melrose, London. p. 120–133.

Anderson, G.C. 1976. The evolution of spinning. The Edward R. Schwartz Lecture. Conf. Amer. Soc. Mech. Eng., New York.

Aspin, C., and S.D. Chapman. 1964. James Hargreaves and the spinning jenny. The Guardian Press, Preston, Great Britain.

Balls, W.L. 1912. The cotton plant in Egypt. McMillan and Co., Ltd., London.

Balls, W.L. 1928. Studies of quality in cotton. McMillan and Co., Ltd., London.

Bryan, W.E. 1955. Prospects for Pima S-1 cotton. Prog. Agric. Ariz. 7:6.

Catling, H. 1970. The spinning mule. David L. Charles Ltd., Newton Abbott, Britain.

Chowdhury, K.A., and G. M. Burth. 1971. Cotton seeds from the Neolithic of Egyptian Nubia and the origin of Old World cotton. Linn. Soc. London. Biol. J. 3:303–312.

Culp, T.W., D.C. Harrell, and T. Kerr. 1979. Some genetic implications in the transfer of high fiber strength genes in Upland cotton. Crop Sci. 19:481–484. doi:10.2135/cropsci1979.0011183X001900040013x

Duckett, K. E. 1974. Cotton fiber instrumentation research in Tennessee. Univ. of Tennessee Agric. Exp. Stn. Bull. 536.

Feaster, C.V., and E.L. Turcotte. 1968. Association of color, yield, and quality of lint in Pima cotton. Crop Sci. 8:197–199. doi:10.2135/cropsci1968.0011183X000800020019x

Feaster, C. V., J. E. Ross, E. L. Turcotte, and C. K. Bragg. 1979. Effects of fiber length uniformity on processing Pima cotton. USDA Prod. Res. Rep. 176. U.S. Gov. Print. Office, Washington, DC.

Fryxell, P.A. 1979. The natural history of the cotton tribe. Texas A&M Univ. Press, College Station.

Gulati, A.M., and A. J. Turner. 1928. A note on the early history of cotton. Indian Central Cotton Committee, Tech. Lab. Bull. No. 17.

Gerstel, D.U. 1953. Chromosomal translocations in interspecific hybrids of the genus Gossypium. Evolution 7:234–244. doi:10.2307/2405734

Harlan, J.R. 1975. Crops and man. ASA and CSSA, Madison, WI.

Hinze, L.L., D.D. Fang, M.A. Gore, B. Scheffler, J.Z. Yu, J. Frelichowski, and R.G. Percy. 2015. Molecular characterization of the *Gossypium* diversity reference set of the U.S. National Cotton Germplasm Collection. Theor. Appl. Genet. doi:10.1007/s00122-014-2431-7.

Hutchinson, J.B., R.A. Silow, and S.G. Stephens. 1947. The evolution of *Gossypium* and the differentiation of the cultivated cottons. Oxford Univ. Press, London.

Jenkins, J.N., J.C. McCarty, O.A. Gutierrez, R.W. Hayes, D.T. Bowman, C.E. Watson, and D.C. Jones. 2008. Registration of RMUP-C5, a random mated population of upland cotton germplasm. J. Plant Regist. 2:239–242.

Kohel, R.J., and T.R. Richmond. 1962. The genetics of flowering response in cotton. IV. Quantitative analysis of photoperiodism of Texas 86, *Gossypium hirsutum* race *latifolium*, in a cross with an inbred line of cultivated American Upland cotton. Genetics 47:1535–1542.

Kohel, R.J., T.R. Richmond, and C.F. Lewis. 1974. Genetics of flowering response in cotton. VI. Flowering behavior of *Gossypium hirsutum* L. and *Gossypium barbadense* L. hybrids. Crop Sci. 14:696–699. doi:10.2135/cropsci1974.0011183X001400050026x

Lewis, C.F., and T.R. Richmond. 1960. The genetics of flowering response in cotton. II. Inheritance of flowering response in a *Gossypium barbadense* cross. Genetics 45:79–85.

Lewis, C.F., and T.R. Richmond. 1968. Cotton as a crop. In: F.C. Elliot, M. Hoover, and W.K. Porter, Jr., editors, Advances in the production and utilization of quality cotton. Iowa State University Press, Ames. p. 1–21.

Mauney, J.R., and L.L. Phillips. 1963. Influence of daylength and night temperature on flowering in Gossypium. Bot. Gaz. 124:278–283. doi:10.1086/336206

Miller, P.A., and J.O. Rawlings. 1967. Selection for increased lint yield and correlated responses in Upland cotton, *Gossypium hirsutum* L. Crop Sci. 7:637–640. doi:10.2135/cropsci1967.0011183X000700060024x

Parlatore, F. 1866. Le specie dei cotoni. Stamperia Reale, Firenze, Italy.

Scherer, J.A.B. 1916. Cotton as a world power. F.A. Stokes Co., New York.

Smith, C.E., Jr., and S.G. Stephens. 1971. Critical identification of Mexican archeological cotton remains. Econ. Bot. 25:160–168. doi:10.1007/BF02860076

Stephens, S.G. 1967. Evolution under domestication of the New World cottons (*Gossypium* spp.). Cienc. Cult. (Sao Paulo) 19:118–134.

Stephens, S.G. 1974a. Geographical distribution of cultivated cottons relative to probable centers of domestication in the New World. In: A.M. Srb, editor, Genes, enzymes, and population. Plenum Publishing Corp., New York. p. 239–254.

Stephens, S.G. 1974b. The use of two polymorphic systems, nectary fringe hairs and corky alleles, as indicators of phylogenetic relationships in New World cottons. Biotropica 6:194–201. doi:10.2307/2989652

Stephens, S.G. 1976a. Some observations on photoperiodism and the development of annual forms of domesticated cottons. Econ. Bot. 30:409–418. doi:10.1007/BF02904663

Stephens, S.G. 1976b. The origin of Sea Island cotton. Agric. Hist. 50:391–399.

Stephens, S.G., and M.E. Moseley. 1974. Early domesticated cottons from achaeological sites in central coastal Peru. Am. Antiq. 39:109–122. doi:10.2307/279225

Taxonomy and Evolution of the Cotton Genus, *Gossypium*

Jonathan F. Wendel* and Corrinne E. Grover

Abstract

We present an overview of the taxonomy of *Gossypium* L. (the cotton genus) and its evolutionary history. *Gossypium* contains more than 50 recognized species, including several recently described, distributed in arid to semiarid regions of the tropics and subtropics. Diversity in *Gossypium* has been promoted by two seemingly unlikely processes: transoceanic, long-distance dispersal and wide hybridization among lineages that presently are widely separated geographically. Included are four species that were independently domesticated for their seed fiber—two diploids from Africa–Asia and two allopolyploids from the Americas. This repeated domestication of different wild progenitors represents a remarkable case of human-driven parallel evolution. Morphological variation in *Gossypium* is extensive; growth forms in the genus range from sprawling herbaceous perennials to ~15-m-tall trees, representing a notable array of reproductive and vegetative characteristics. Equally impressive is the striking cytogenetic and genomic diversity that emerged as *Gossypium* diversified and spread worldwide, ultimately spawning eight groups of closely related diploid (n = 13) species (i.e., genome groups A through G, and K). DNA sequence data place the origin of *Gossypium* at about 5 to 10 million years ago (mya), which rapidly diversified into these major genome groups shortly thereafter. Allopolyploid cottons appeared within the last 1 to 2 million years, a consequence of the improbable transoceanic dispersal of an A genome taxon to the New World and subsequent hybridization with an indigenous D genome diploid. Diversification of the nascent allopolyploid gave rise to three modern lineages containing seven species, including the agronomically important *G. hirsutum* L. and *G. barbadense* L.

Because of its economic importance, the cotton genus (*Gossypium* L.) has been of interest to agricultural scientists, taxonomists, and many other kinds of biologists. Accordingly, a considerable amount is understood regarding the origin and diversification of the genus, its basic plant biology, and its properties as a crop plant (Paterson, 2009; Stewart et al., 2010; Wendel et al., 2009, 2010, 2012). Increasingly over the past two decades, classic taxonomic questions such as the origin of the polyploid species, the relationships among species and species

---

Abbreviations: mya, million years ago.

Jonathan F. Wendel and Corrinne E. Grover (corrinne@iastate.edu), Dep. of Ecology, Evolution, and Organismal Biology, Bessey Hall, Iowa State University, Ames, IA 50011. *Corresponding author (jfw@iastate.edu).

doi:10.2134/agronmonogr57.2013.0020

Cotton. 2nd. ed. David D. Fang and Richard G. Percy, editors. Agronomy Monograph 57.

groups, and the origins of the domesticated forms from their wild progenitors have been revisited using modern molecular technologies (Wendel et al., 2009, 2010, 2012; Wendel and Cronn, 2003). One of the salient features of the genus is that its history is so encompassing in scope, involving a global phylogenetic diversification and a repeated history of transoceanic dispersal. A remarkable consequence of this global spread and natural diversification has been that it provided the opportunity for different ancient human cultures on several continents to independently domesticate four different cotton species—two allopolyploids from the Americas, *G. hirsutum* and *G. barbadense*, and two diploids from Africa–Asia, *G. arboreum* L. and *G. herbaceum* L. The canvas for this human history is the natural evolutionary diversification of the genus, one that involves multiple, seemingly impossible hybridization events, often between lineages that presently are geographically disjunct or even on different continents, as well as multiple, remarkable cases of transoceanic, long-distance dispersals. These twin themes of cryptic intergenomic introgression or hybridization and long-distance dispersal have repeatedly come into play as the genus diversified and colonized much of the arid to semiarid regions of the tropics and subtropics.

A wealth of recent studies have provided new insights into *Gossypium* taxonomy and diversity, suggesting that it is an appropriate time to update our understanding of the evolution, taxonomy, and biogeographic history of the genus. Here we provide such an update, drawing on our earlier reviews (Wendel et al., 2009, 2010, 2012; Wendel and Cronn, 2003) and with new syntheses.

Emergence of the Genus *Gossypium*

The cotton genus belongs to the *Gossypieae*, a small taxonomic tribe that includes only nine genera, eight of which are classically recognized (Fryxell, 1968, 1979) and one that is a new segregate from *Cienfuegosia* (Phuphathanaphong, 2006). Five of these genera are small with restricted geographic distributions (Fryxell, 1968, 1979; Phuphathanaphong, 2006) including *Lebronnecia* (Marquesas Islands), *Cephalohibiscus* (New Guinea, Solomon Islands), *Gossypioides* (East Africa, Madagascar), *Thepparatia* (known only from the type locality in Northern Thailand) and *Kokia* (Hawaii). The tribe also includes four moderately sized genera with broader geographic ranges: *Hampea* (21 neotropical species), the relatively diverse *Cienfuegosia* (25 species from the neotropics and parts of Africa), *Thespesia* (17 tropical species), and *Gossypium*, whose 50+ species (Fryxell, 1992) make it the largest and most widely distributed genus in the tribe. Thus, in aggregate this small taxonomic tribe has achieved an expansive geographic range, though many of the included taxa are quite rare and/or narrowly distributed.

Molecular phylogenetic analyses have shed light on three important aspects of the evolutionary history of the tribe (Cronn et al., 2002; Seelanan et al., 1997). First and foremost is that despite their extensive distribution and extraordinary diversity, the group of species that belong to *Gossypium* do constitute a single natural lineage. A second important revelation has been the identity of the closest relatives of *Gossypium*, that is, the African–Madagascan genus *Gossypioides* and the Hawaiian endemic genus *Kokia*. This is unsuspected finding, given their biogeography and lower chromosome number ($n = 12$), yet it is quite important in that they may serve as outgroups in phylogenetic analyses, which provides an essential context both for studying evolutionary patterns and processes within

Gossypium and for understanding genome evolution. The third important insight gained from these molecular analyses was the provision of a temporal component to the major divergences. Using sequence divergence data as a proxy for time (i.e., a "molecular clock"), Seelanan et al. (1997) suggested that *Gossypium* diverged from *Kokia* and *Gossypioides* circa 12.5 mya, an estimate bolstered by a later, more extensive data set containing 10 different nuclear genes (Cronn et al., 2002). Collectively, all molecular data suggest that *Gossypium* diverged from its closest relatives approximately 10 to 15 mya during the Miocene and acquired its modern, worldwide distribution by multiple transoceanic dispersals combined with regional speciation.

Diversification of the Diploid Cotton Species

From the time of origin 5 to 10 mya, *Gossypium* experienced rapid speciation and diversification, while also achieving a nearly worldwide distribution containing several primary centers of diversity in the arid or seasonally arid tropics and subtropics (Table 1). Species-rich regions include Australia, especially the Kimberley region in NW Australia; the Horn of Africa and southern Arabian Peninsula; and the western part of central and southern Mexico. The taxonomy of the genus has been well studied (Cronn et al., 2002; Fryxell, 1979, 1992; Hutchinson et al., 1947; Saunders, 1961; Seelanan et al., 1997; Watt, 1907), with the taxonomic classification of Fryxell (1979, 1992) being the most modern and widely followed. In this treatment, species are grouped into four subgenera and eight sections (Table 1). The

Table 1. Diversity and geographic distribution of the major lineages of *Gossypium*. Genomic placements and taxonomic status of species enclosed by parentheses are yet to be validated by cytogenetic and other means.

Genome (no. of species)	Recognized species	Geographic distribution
A (2)	*G. arboreum*, *G. herbaceum*	Africa, Asia
B (3–4)	*G. anomalum*, *G. triphyllum*, *G. capitis-viridis*, (*G. trifurcatum*)	Africa, Cape Verde Islands
C (2)	*G. sturtianum*, *G. robinsonii*	Australia
D (13–14)	*G. thurberi*, *G. armourianum*, *G. harknessii*, *G. davidsonii*, *G. klotzschianum*, *G. aridum*, *G. raimondii*, *G. gossypioides*, *G. lobatum*, *G. trilobum*, *G. laxum*, *G. turneri*, *G. schwendimanii*, (*Gossypium* sp. nov.)	primarily Mexico, with range extensions into Peru, Galapagos Islands, Arizona
E (5–9)	*G. stocksii* Mast., *G. somalense* (Gurke) J.B. Hutch., *G. areysianum* Deflers, *G. incanum* (O. Schwartz) Hillc., *G. trifurcatum*, (*G. benadirense*), (*G. bricchettii*), (*G. vollesenii* Fryxell), (*G. trifurcatum*)	Arabian Peninsula, Northeast Africa, Southwest Asia
F (1)	*G. longicalyx*	East Africa
G (3)	*G. bickii*, *G. australe*, *G. nelsonii*	Australia
K (12)	*G. anapoides*, *G. costulatum*, *G. cunninghamii*, *G. enthyle* Fryxell et al., *G. exiguum* Fryxell et al., *G. londonderriense* Fryxell et al., *G. marchantii*, *G. nobile* Fryxell et al., *G. pilosum* Fyxell, *G. populifolium* (Benth.) F. Muell. ex Tod., *G. pulchellum* (C.A. Gardner) Fryxell, *G. rotundifolium* Fryxell et al.	Northwest Australia, Cobourg Peninsula, Northern Territory, Australia
AD (7)	*G. hirsutum*, *G. barbadense*, *G. tomentosum*, *G. mustelinum*, *G. darwinii*, *G. ekmanianum*, *Gossypium* sp. nov.†	New World tropics and subtropics, including Hawaii, the Wake Atoll, and the Galapagos Islands

† J.F. Wendel, C.E. Grover, J. Jareczek, and J.P. Gallagher, unpublished data, 2014.

classification represents the accumulated scientific understanding that emerged both from basic plant exploration and taxonomic and evolutionary study. While these species groupings are based primarily on morphological and geographical evidence, most infrageneric alignments are congruent with subsequent cytogenetic and molecular data sets (Endrizzi et al., 1985; Wendel and Albert, 1992; Wendel et al., 2009, 2010).

The current circumscription of *Gossypium* presently includes approximately 50 species (Fryxell, 1992), although new species continue to be discovered (Alvarez and Wendel, 2006; Grover et al., 2014; Krapovickas and Seijo, 2008; Stewart et al., 2015; Ulloa et al., 2006). Morphologically, the genus is exceptionally diverse; species morphologies range from fire-adapted, herbaceous perennials with massive underground rootstocks in NW Australia to trees in SW Mexico that escape the dry season by dropping their leaves. Corolla colors span a rainbow of blue to purple [*G. triphyllum* (Harv.) Hochr.], mauves and pinks (*G. sturtianum* J.H. Willis, or "Sturt's Desert Rose," the official floral emblem of the Northern Territory, Australia), whites and pale yellows (NW Australia, Mexico, Africa–Arabia), and even a deep sulfur-yellow (*G. tomentosum* Nutt. ex Seem., Hawaii). Seed coverings are equally diverse; these range from nearly glabrous to the naked eye (e.g., *G. klotzschianum* Andersson and *G. davidsonii* Kellogg), to short, stiff, dense, brown hairs that aid in wind dispersal (*G. australe* F. Muell and *G. nelsonii* Fryxell), to the long, fine, white fibers that characterize highly improved forms of the four cultivated species (Fig. 1). There are even seeds that produce fat bodies to facilitate ant-dispersal (Seelanan et al., 1999).

In addition to the remarkable morphological diversity, the genus also experienced extensive chromosomal evolution during diversification (Baudoin et al., 2009; Endrizzi et al., 1985). Despite sharing the same chromosome number (n = 13), genome sizes among the diploid species vary more than threefold (Hendrix and Stewart, 2005). Among closely related species, chromosome morphology is quite similar, as is demonstrated by the ability of closely related species to form hybrids that undergo normal meiotic pairing and that often experience high F_1 fertility. In contrast, crosses between more distantly related species are often difficult to generate, with the few successful hybrids typically experiencing meiotic abnormalities. Extensive observations concerning the pairing behavior, chromosome sizes, and relative fertility in interspecific hybrids ultimately led to the clustering of closely related species into "genome groups," designated by single-letter genome symbols (Beasley, 1941). Currently, there are eight recognized diploid genome groups (A through G, plus K) (Endrizzi et al., 1985; Stewart, 1995), whose species inclusions are largely congruent with taxonomic and phylogenetic placements.

A temporal framework for the origin of *Gossypium* and its diversification is provided by sequence divergence data serving as a proxy for time (Senchina et al., 2003). These analyses indicate that *Gossypium* diverged from its closest relatives during the Miocene approximately 10 to 15 mya to originate approximately 5 to 10 mya, and these estimates of divergence times have recently been supported by an enormous amount of new sequence data derived from a global assembly of expressed sequence tags (Flagel et al., 2012). A consideration of the near worldwide distribution of species combined with their phylogenetic history (Fig. 2), relative to plate tectonic history, indicates that the evolution of *Gossypium* has involved multiple episodes of transoceanic dispersal. Minimally, these include

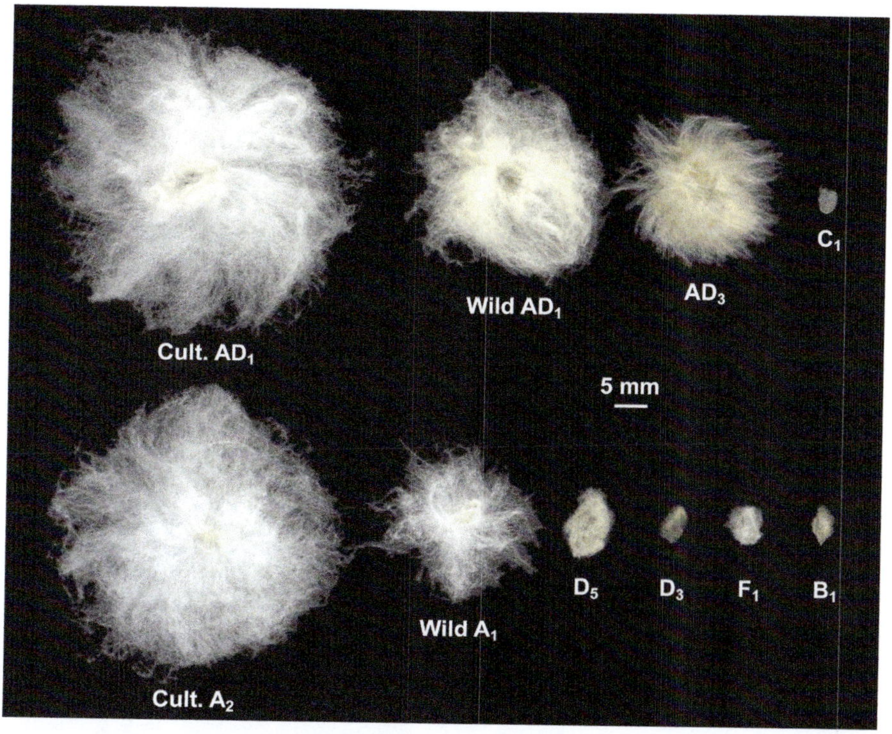

Fig. 1. Representative seed and trichome diversity in *Gossypium*. Seed and trichome size and morphology are exceedingly variable in the genus. Most wild species have relatively small seeds (<5 mm in any dimension) with equally short fibers. Long (spinnable) fiber evolved only once, in the ancestor of modern A genome cottons, which subsequently donated this capacity to modern allopolyploid species, including the commercially important *G. hirsutum* and *G. barbadense*, at the time of allopolyploid formation. Key to species: Cult. AD_1, *G. hirsutum* TM1; Wild AD_1, *G. hirsutum* Tx2094 from the Yucatan Peninsula; AD_3, *G. tomentosum* WT936 from Hawaii; C_1, *G. sturtianum* C_1-4 from Australia; Cult. A_2, *G. arboreum* AKA8401; Wild A_1, *G. herbaceum* subsp. *africanum* (Watt) Vollesen from Botswana; D_5, *G. raimondii* from Peru; D_3, *G. davidsonii* D_{3d}-32 from Baja California; F_1, *G. longicalyx* F_1-3 from Tanzania; B_1, *G. anomalum* B_1-1 from Africa.

at least one dispersal between Australia and Africa; another to the Americas, which led to the evolution of the D genome diploids; and a second, much later colonization of the New World by the A genome ancestor of the AD genome allopolyploids (see below).

Out of Africa, the Cytogenetic Center of Diversity in the Genus (A, B, E, and F Genomes)

An important implication of the phylogenetic data for *Gossypium* is that it supports the proposition that Africa is the ancestral home for the genus (Fig. 2). This is not strictly true from a purely cladistic standpoint, in that the basal phylogenetic split is between the New World and Old World diploids; however, the supposition that *Gossypium* originated in Africa is supported by the genomic

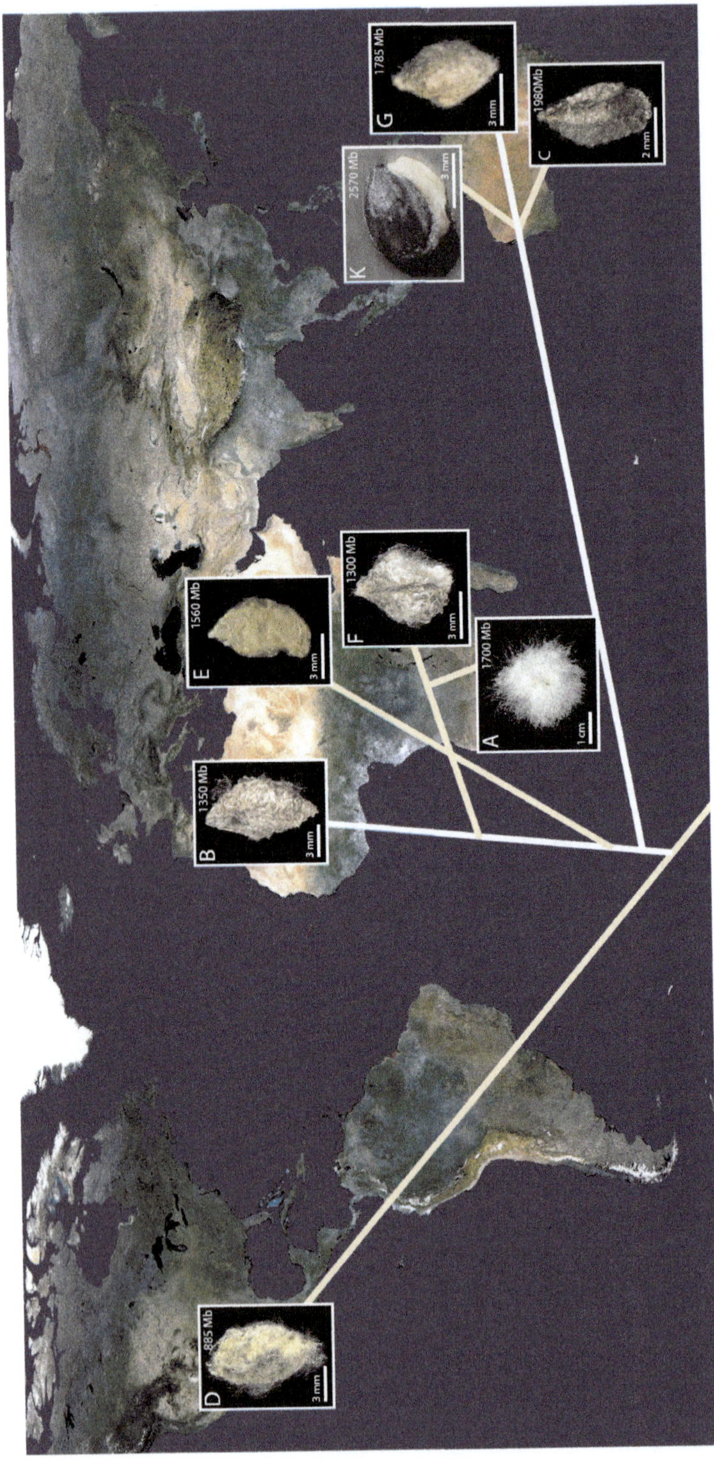

Fig. 2. Biogeographic history of *Gossypium*. Following its likely origin 5 to 10 mya, *Gossypium* split into three major diploid lineages: the New World clade (D genome); the African–Asian clade (A , B, E, and F genomes); and the Australian clade (C, G, and K genomes). The diversity of genome sizes and seed morphologies in the genus are shown with exemplars from each genome group. See text for additional details.

diversity found in the region (see below), versus the relative homogeneity found in the New World. Fourteen species from Africa and Arabia are recognized in the most recent taxonomic treatment of the genus (Fryxell, 1992) and together compose the African–Arabian subgenus *Gossypium*. Recognized within the subgenus *Gossypium* are two taxonomic sections: section *Gossypium*, which contains four subsections, and another African–Arabian section (*Serrata*), which contains the single species *G. trifurcatum* Vollesen (found in desert area of eastern Somalia). The presence of dentate leaves has raised the question of whether *G. trifurcatum* belongs in the genus *Gossypium* at all, as dentate leaves are more characteristic of *Cienfugosia*; however, molecular work has established that this poorly known specimen is indeed a cotton species, if unusual in form (Rapp and Wendel, 2005). This example underscores the provisional nature of much of the taxonomy of the African–Arabian species of *Gossypium*, which are sorely in need of basic plant exploration and systematic study. Other examples include subsection *Pseudopambak*, which contains several species whose recognition and definition are based solely on limited herbarium material (e.g., *G. benadirense* Mattei, *G. bricchettii* (Ulbr.) Vollesen, *G. vollesenii*). As no seeds have been collected and no living specimens are available, there exists no information on the cytogenetic characteristics and molecular phylogenetic placement for these species. It is nearly certain that new taxonomic diversity remains to be discovered in this region of the world.

Cytogenetically, the African–Arabian species are rather diverse and account for four of the eight genome groups (A, B, E, and F), which neatly correspond to their taxonomic subsections. The A genome group contains the two cultivated cottons of subsection *Gossypium*, *G. arboreum* and *G. herbaceum*, while the three African species in the B genome (*G. anomalum* Wawra, *G. capitis-viridis* Mauer, and *G. triphyllum*) compose subsection *Anomala*. It bears noting that *G. trifurcatum* (taxonomic section *Serrata*) may in fact belong to the B genome, but this has not been established. The sole F genome species, *G. longicalyx* J.B. Hutch. & B.J.S. Lee, represents subsection *Longiloba*, and, of the section *Gossypium*, it is considered cytogenetically distinct (Phillips, 1966), morphologically isolated (Fryxell, 1971, 1992), and according to Fryxell may be better adapted to more mesic conditions than any other diploid *Gossypium* species. Further adding to the interest in this species is the phylogenetic placement of *G. longicalyx* as sister to the A genome taxa (Fig. 2), which provides a window into the evolutionary origin of "long" fiber. The remaining African–Asian species, those of subsection *Pseudopambak*, are considered to possess E genomes, although this has yet to be verified.

To Australia, with Genomic Gigantism and New Ecological Specializations

Australia represents another center of diversity for the cotton genus. The Australian cottons (subgenus *Sturtia*) are represented by 3 taxonomic sections containing 17 total species, including one, *G. anapoides* J.M. Stewart et al., nom. inval., that is newly described (Stewart et al., 2015). Cytogenetically, these taxa are divided into the C, G, and K genome groups, having 2, 3, and 12 species, respectively. DNA sequence data (Liu et al., 2001; Seelanan et al., 1997, 1999) support these groups as natural lineages, consistent with their formal alignments into the taxonomic sections *Sturtia* (C genome), *Hibiscoidea* (G genome), and *Grandicalyx* (K genome), although the relationships among these three are yet unresolved. Some data support a west to east radiation of *Gossypium* in Australia by placing *G. robinsonii* F. Muell. (C genome) as the basal-most branch of the Australian clade (DeJoode and Wendel, 1992). This basal placement of the C genome has not been observed in the most recent analyses (Liu et al., 2001; Seelanan et al., 1997, 1999), which provide no resolution among the Australian lineages and thereby highlight the uncertainty in the phylogenetic history of the Australian cottons.

Taxonomically, the C and G genome groups are well understood, as these are amply represented in collections and have been thoroughly studied (DeJoode and Wendel, 1992; Fryxell, 1979, 1992; Liu et al., 2001; Seelanan et al., 1997, 1999; Tiwari et al., 2014; Wendel et al., 1991). Species in the G genome are interfertile and apparently hybridize commonly (Tiwari et al., 2014; Wendel et al., 1991). The taxonomy of the K genome (section *Grandicalyx*), however, is less certain. Collecting expeditions to the Kimberley region have doubled the number of recognized K genome species in the past two decades through the discovery of at least seven new species (Fryxell, 1992; Stewart et al., 2015) and have expanded our understanding of the diversity in this section. The species of this section are geographically, morphologically, and ecologically distinct from other cotton species, frequently exhibiting features characteristic of fire adaptation. Specifically, these are herbaceous perennials that exhibit a biseasonal growth pattern, where vegetative growth dies back during the dry season and the plants survive as underground rootstocks until fire or the return of the wet season initiates a new cycle of growth. Also unique to section *Grandicalyx* are the flower and fruit morphologies. K genome species possess upright flowers (like other cotton species) that subsequently become pendant following pollination. This change in floral positioning is thought to aid in these uniquely ant-dispersed cotton seeds. That is, at maturity the pendant capsules release sparsely haired seeds, which bear elaiosomes (fat bodies) that attract the ants who participate in their dispersal. Nearly all of K genome species are poorly represented in collections (Campbell et al., 2010), leading to much of this taxonomic uncertainty. These species have enormous genomes, relative to others in the genus (Fig. 2), reflecting a relatively recent proliferation of *gypsy*-like transposable elements (Hawkins et al., 2006). Phylogenetic analyses have yielded conflicting results regarding interspecific relationships in section *Grandicalyx* (Liu et al., 2001; Seelanan et al., 1999), although there is some support for a fundamental division of the section into a group of prostrate species and a clade of more upright plants.

One final comment about the Australian species concerns their apparent propensity to hybridize, a somewhat surprising observation given their present status as species whose populations are mostly small, scattered, and geographically highly disjunct from those of related species. These geographical and often phenological barriers notwithstanding, there is abundant evidence for multiple episodes of historical introgression in the group, involving *G. bickii* Prokh., *G. sturtianum*, and *G. cunninghamii* Tod. (Cronn and Wendel, 2004; Seelanan et al., 1999; Tiwari et al., 2014; Wendel et al., 1991).

To the New World, Part I: The D Genome Diploids

A watershed moment in the evolutionary history of *Gossypium* followed a seemingly impossible transoceanic (either Atlantic or Pacific) dispersal of an African ancestor to the shores of Central America, possibly on what is now the western coast of Mexico. This dispersal event, dating to about 5 to 10 mya, led to the evolution and diversification of the New World D genome clade. This assemblage of species have unusually small genomes for the genus (Fig. 2), due at least in part to deletional clearance of much of the transposable element fraction (Hawkins et al., 2009), and none bear spinnable fiber (Fig. 1). Taxonomically placed into subgenus *Houzingenia*, the species are aligned into two sections and six subsections. These species have been more thoroughly studied than most, and consequently their taxonomy is reasonably well understood. Nonetheless, there remains sufficient cryptic variability in the group that much remains to be learned about diversity and relationships (Alvarez and Wendel, 2006; Cronn et al., 2003; Feng et al., 2011; Ulloa et al., 2006, 2013; Wendel and Cronn, 2003; Wendel et al., 1995). This is particularly true for the "arborescent clade" comprising the group of small trees that includes the widespread *G. aridum* (Rose & Standl.) Skovst. and the more narrowly distributed *G. laxum* L. Ll. Phillips, *G. lobatum* Gentry, and *G. schwendimanii* Fryxell & S.D. Koch. Evidence indicates that cryptic variation exists within this clade, possibly including new species related to *G. aridum* (Alvarez et al., 2005; Feng et al., 2011) and *G. laxum* (Alvarez et al., 2005; Ulloa et al., 2006). There even are populations of *G. aridum* from Colima that have a plastid genome much more similar to those from *G. klotzschianum* (from the Galapagos Islands) and *G. davidsonii* (from Baja California) than to all other *G. aridum* populations, likely reflecting historical interspecific gene flow (Alvarez and Wendel, 2006; Wendel and Albert, 1992). *Gossypium gossypioides* (Ulbr.) Standl. also has a somewhat mysterious introgressive ancestry (Cronn et al., 2003; Wendel et al., 1995), pointing out once again the significance of ancient, natural interspecific hybridization in the generation of diversity in the cotton genus (Cronn and Wendel, 2004).

The D genome cottons also have received considerable phylogenetic attention (Alvarez et al., 2005; Cronn et al., 2002, 2003; Feng et al., 2011; Ulloa et al., 2006, 2013; Wendel and Albert, 1992), which provides support for the naturalness of most of the subsections. Evolutionary relationships among the apparently natural subsections are less certain, however (Alvarez et al., 2005; Ulloa et al., 2013), although available evidence suggests that *G. gossypioides* is basal-most within the subgenus (Cronn et al., 2003; Feng et al., 2011; Yu-xiang et al., 2013). All analyses support the monophyly of the arborescent clade, the grouping of the Baja California species (*G. harknessii* Brandegee–*G. armourianum* Kearney–*G. turneri* Fryxell) as a distinct clade, and the naturalness of the *G. thurberi* Tod.–*G. trilobum* (DC.) Skovst. species pair, but the relative phylogenetic placements of these three groups with *G. raimondii* Ulbr. remain uncertain.

Twelve of the 14+ D genome diploid species are endemic to western Mexico, indicating this area is the center of diversity of the D genome. It seems likely, therefore, that this lineage became established and initially diversified in this region. From this we can deduce later range extensions, probably relatively recently (during the Pleistocene), following long-distance dispersals leading to the evolution of endemics in Peru (*G. raimondii*) and the Galapagos Islands (*G. klotzschianum*). The latter species is closely related to *G. davidsonii* from Baja California, representing a clear and unusual case of this floristic connection (Wendel and Percival, 1990). As for *G. raimondii*, it is interesting that the sole South American representative of the New World diploid cottons is implicated as the closest ancestor of the D genome of allopolyploid cotton (more on this below); the implication, perhaps ironic, is that if allopolyploidy occurred in northern South America, as suggested (but not shown) by this observation, the possibility exists that they never would have evolved had colonization of the New World by an A genome propagule not followed closely on the heels of an earlier interhemispheric North American to South American dispersal of the ancestor of what we now recognize as *G. raimondii*. This aspect of the origin of the allopolyploids makes an improbable story all the more remarkable and yet again points to the recurring theme of the evolutionary significance of long-distance dispersal in cotton diversification.

To the New World, Part II: Origin and Diversification of the Polyploid Cottons

A rich body of cytogenetic and experimental evidence now has demonstrated convincingly that the tetraploid cotton species, which are entirely New World in their distribution, are allopolyploids containing two coresident genomes: one from an African or Asian A genome species and the other from a species similar to the American D genome diploids (Endrizzi et al., 1985; Wendel, 1989; Wendel and Cronn, 2003; Wendel et al., 2012). The hemisphere-scale allopatry of these two diploid genome groups led to many decades of mystery surrounding the timing and parentage of the New World allopolyploids. The question of "when" has been convincingly addressed through gene sequence data that indicate allopolyploid *Gossypium* originated approximately 1 to 2 mya in the mid-Pleistocene, before the evolution of modern humans but relatively recently in geological terms (Flagel et al., 2012; Senchina et al., 2003; Wendel, 1989). The question of parentage has also been clarified through a wealth of DNA sequence data that indicate: (i) both extant A genome species (*G. arboreum*, *G. herbaceum*) are equally distant from the allopolyploid A genome and (ii) *G. raimondii* is the closest extant relative of the actual D genome donor of the allopolyploid (Endrizzi et al., 1985; Li et al., 2014; Wendel, 1989; Wendel and Cronn, 2003; Wendel et al., 2012; Yu-xiang et al., 2013). Additionally, all allopolyploids contain an A genome cytoplasm, as evidenced from analysis of both mitochondrial and plastidial genomes (Galau and Wilkins, 1989; Li et al., 2014; Small and Wendel, 1999; Wendel, 1989; Xu et al., 2012). Finally, the studies mentioned above, and additional, multilocus DNA sequence data (Grover et al., 2012), support a single origin for allopolyploid cotton.

As noted above, one truly remarkable feature of the origin of allopolyploid cottons is that their genesis likely was dependent on the following series of highly improbable events, which when considered in their totality amount to a biological marvel: (i) an ancient colonization of the New World by the ancestor of the D genome diploids, following an initial transoceanic dispersal; (ii) a more recent interhemispheric North American to South American dispersal of the ancestor of modern *G. raimondii*; (iii) a third long-distance dispersal of an African–Asian A genome ancestor to the New World, likely in the mid-Pleistocene; and (iv) the chance biological encounter between the immigrant A genome taxon and the endemic D genome entity, giving rise to hybridization and genome doubling, and hence the genomic reunion among genomes that had been isolated on different continents for 5 to 10 million years.

Given this now-justified scenario of a Pleistocene origin, the morphological diversification and spread of the allopolyploid cotton species must have been relatively rapid following polyploidization. Classically, five allopolyploid species were widely recognized, but a sixth species (*G. ekmanianum* Wittm.) was recently resurrected (Krapovickas and Seijo, 2008) and subsequently validated (Grover et al., 2014) by molecular sequence data. In addition, ongoing studies provide justification for recognition of a seventh species that is relatively closely related to *G. hirsutum*, from two islands (Wake, Peale) in the Wake Atoll in the Pacific Ocean, *Gossypium* sp. nov. (J.F. Wendel, C.E. Grover, J. Jareczek, and J.P. Gallagher, unpublished data, 2014). Included among the classically recognized species are two island endemics, *G. darwinii* and *G. tomentosum*, that have differing population structures. *Gossypium darwinii*, a native of the Galapagos Islands, forms large and continuous populations in some areas (Wendel and Percy, 1990), whereas the Hawaiian endemic, *G. tomentosum*, tends to be more sparse, primarily occurring as scattered individuals and small populations on several islands. The latter species, *G. tomentosum*, has recently received some attention with respect to its reproductive biology and diversity (DeJoode and Wendel, 1992; Hawkins et al., 2005; Pleasants and Wendel, 2010). A third classical allopolyploid, *G. mustelinum* Miers ex G. Watt, is restricted to a relatively small region (Bahia state) of northeast Brazil, where it survives in apparently relictual and highly homozygous scattered populations (de Menezes et al., 2014; Wendel et al., 1994). In addition to these three truly wild species, there exist two species with cultivated forms (*G. barbadense* and *G. hirsutum*). Both species are characterized by large indigenous ranges that in the aggregate encompass a wealth of morphological forms spanning the wild-to-domesticate continuum (Brubaker et al., 1999; Brubaker and Wendel, 1993, 1994; Fryxell, 1979; Hutchinson, 1951, 1959; Hutchinson et al., 1947; Lubbers and Chee, 2009; Percy and Wendel, 1990; Stephens, 1963). *Gossypium hirsutum* is widely distributed in Central America and northern South America, the Caribbean, and even reaches distant islands in the Pacific (Solomon Islands, Marquesas). Natural diversity in this species is underscored by its complex nomenclatorial and taxonomic history (Fryxell, 1976, 1979). It is therefore unsurprising that a new species from the Dominican Republic was recently discovered both in natural settings and within the *Gossypium* germplasm collection (Grover et al., 2014; Krapovickas and Seijo, 2008) and that other undescribed species likely are encompassed by the diversity presently subsumed under the name *G. hirsutum*, as exemplified also by *Gossypium* sp. nov., as noted above. *Gossypium hirsutum* has a more northerly indigenous distribution than *G. barbadense*, with wild populations occurring as far north as Tampa Bay, Florida (27°38′ N) (J.M. Stewart, personal observation). *Gossypium barbadense* has an indigenous range centered in the northern third of South America but with a large region of range overlap with *G. hirsutum* in the Caribbean (Percy, 2009).

Consideration of the ecological and geographic distribution of the allopolyploid species has led to the suggestion that polyploidy created the opportunity to invade an ecological niche that is distinct from that occupied by most diploids. Fryxell (1965, 1979) noted that the truly wild forms of the allopolyploid species typically occur in coastal habitats, in contrast to most of their diploid relatives. Three species (*G. darwinii*, *Gossypium* sp. nov., and *G. tomentosum*) are island endemics that typically reside near coastlines, and two of the other species (*G. barbadense* and *G. hirsutum*) possess wild forms that typically occupy littoral habitats tracing the Gulf of Mexico, northwest South America, and reaching distant Pacific Islands. Fryxell speculated that the adaptations that allowed the newly evolved allopolyploid to survive, and even thrive, in littoral habitats ultimately facilitated the establishment of new polyploid species and provided a means for rapid dispersal of the saltwater tolerant seeds. These ideas gain additional support from recent ecological niche modeling studies (d'Eeckenbrugge and Lacape, 2014), which provide an elegant framework for understanding past and present ecogeographic distributions of wild polyploid cottons.

Phylogenetic Relationships in the Genus

The genealogy of *Gossypium* has been evaluated by multiple molecular phylogenetic investigations (reviewed in Wendel and Cronn, 2003; Wendel et al., 2012), each of which demonstrated that the major clades of species (i.e., genealogical lineages) are largely consistent with geographical distributions, cytogenetic relationships (Baudoin et al., 2009; Endrizzi et al., 1985), and genome designations. Thus, each taxonomically and cytogenetically established genome group corresponds to a single natural lineage, which also, in most cases, is geographically cohesive. This information is summarized in a depiction of our present understanding of these relationships (Fig. 2 and 3).

Gossypium phylogenetic history has several broad aspects worth highlighting. First, four major diploid lineages exist that correspond to three continents: Australia (C, G, and K genomes), the Americas (D genome), and Africa/Arabia (itself containing two lineages: one comprising the A, B, and F genomes, and a second solely containing the E genome species). Second, the earliest divergence event separated the New World D genome lineage from that containing the ancestor of all Old World taxa, approximately 5 to 10 mya, making the New World and Old World diploids phylogenetic sister groups. Subsequently, the Old World lineage itself diversified into three groups, namely, the Australian cottons (C, G, and K genome species), the African–Arabian E genome species, and the African A, B, and F genome cottons. Third, analyses identify the sole F genome species (*G. longicalyx*, Africa) as sister to the A genome clade (*G. arboreum* and *G. herbaceum*), where the first emergence of fiber suitable for domestication evolved. This information diagnoses the wild forms that best represent the ancestral lintless condition, and provides a foundation for ultimately understanding the genetic basis of the origin of useful, long lint (cf. Fig. 2). Fourth, the major lineages of *Gossypium* are characterized by rapid radiation and diversification that occurred shortly after the genus originated and diverged from the *Kokia–Gossypioides* clade. Fifth, and finally, the phylogenetic depiction summarizes our current understanding and uncertainties regarding relationships among the seven allopolyploid species.

From Dispersal Ecology to Human Opportunity

As emphasized in this review, a consideration of the phylogeny of Fig. 2 and 3 in a temporal context leads to the realization that the genus has had an inordinate fondness for transoceanic travel. As already noted, this includes at least one dispersal between Australia and Africa, another to the Americas (probably Mexico) leading to the evolution of the D genome diploids, and a second, much later, colonization of the New World by the A genome ancestor of the AD genome allopolyploids. Long-distance dispersal played a role not only in diversification of major evolutionary lines but also in speciation within *Gossypium* genome groups. Examples include dispersals from southern Mexico to Peru (*G. raimondii*), from northern Mexico to the Galapagos Islands (*G. klotzschianum*), from western South

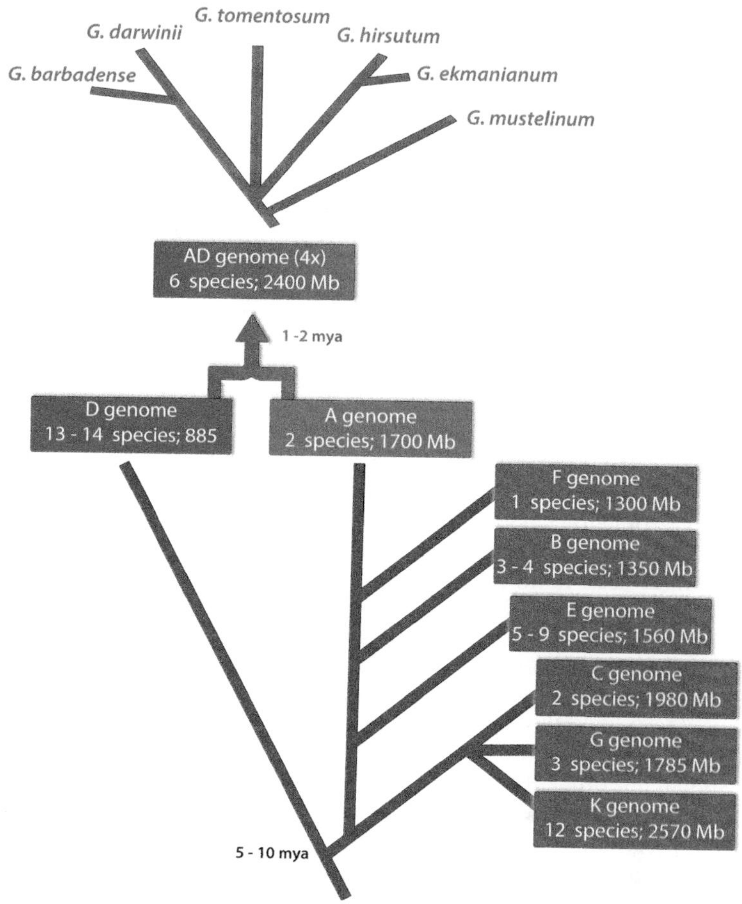

Fig. 3. Evolutionary history of *Gossypium*, as inferred from multiple molecular phylogenetic data sets. The closest relative of *Gossypium* is a lineage containing the African–Madagascan genus *Gossypioides* and the Hawaiian endemic genus *Kokia*. Following its likely origin 5 to 10 mya, *Gossypium* split into three major diploid lineages: the New World clade (D genome); the African–Asian clade (A, B, E, and F genomes); and the Australian clade (C, G, and K genomes). This global diversification involved multiple transoceanic dispersal events and was accompanied by morphological, ecological, and chromosomal differentiation (genome sizes shown). Interspecific hybridization has been exceptionally common during the evolution of the genus. Allopolyploid cottons formed following transoceanic dispersal of an A genome diploid to the Americas, where the new immigrant underwent hybridization, as female, with a native D genome diploid similar to modern *G. raimondii*. Polyploid cotton probably originated during the Pleistocene (1–2 mya), with the modern species representing the descendants of an early and rapid colonization of the New World tropics and subtropics.

America to the Galapagos Islands (*G. darwinii*), from Africa to the Cape Verde Islands (*G. capitis-viridis*), and from the neotropics to the Hawaiian Islands (*G. tomentosum*). What is so striking about this history is not just its biographically fantastical nature, but also the lack of obvious adaptations for saltwater dispersal

(Stephens, 1958, 1966). Not only that, but the life history bauplan for *Gossypium* is of arid-zone adapted plants that one might assume at first blush do not possess any capacity for overseas journeys. This aspect of the natural history of the cotton genus represents a fascinating conundrum, one that lends itself more to speculation than to experimentation. Yet it also highlights the potential evolutionary significance of the seemingly most improbable and rare events.

Given the absence of obvious adaptations for water dispersal, what can we say about the evolutionary significance of the single-celled epidermal trichomes for which cotton is justifiably so famous? The seeds and their coverings are extraordinarily diverse in *Gossypium*, as illustrated in Fig. 1, which provides a visual comparison of the various lineages of wild species and domesticated cotton species. Even though wild and cultivated cottons both produce fiber on the seed coat, there are striking morphological and structural differences between these fibers, the most obvious of which is their size. In contrast to the canonical (domesticated) cotton fiber, some D genome species (*G. thurberi*, *G. trilobum*, *G. davidsonii*, and *G. klotzschianum*) do not possess obvious seed hairs; however, the presence of these developmentally repressed structures is apparent under microscopy (Applequist et al., 2001). Similarly, while the three D genome species of subsection *Caducibracteolata* appear hairless to the unaided eye, these species do in fact have seed hairs that are simply tightly appressed to the seed. Cultivated lint fiber is a model for cell wall developmental biology, as it is single cell of almost pure cellulose that experiences remarkable elongation during development in cultivated species, achieving a final length up to 6 cm (Arpat et al., 2004; Haigler et al., 2012; Kim and Triplett, 2001). In contrast, the wild-type fiber cell is composed mostly of a combination of cellulose and suberin, which elongates to <1 cm (Applequist et al., 2001; Haigler et al., 2009; Ryser and Holloway, 1985). These and many other differences reflect both natural evolutionary processes as well as human-mediated selection during domestication. The duration of the elongation phase and the timing of onset of secondary wall synthesis appear to be key determinants of the final length of the fiber in both wild and cultivated plants (Applequis et al., 2001). Phylogenetic analysis of growth rates has shown that the evolutionary innovation of prolonged elongation arose in the F genome/A genome lineage, which may have facilitated the original domestication of the A genome cottons. This trait of prolonged elongation was passed on to the allopolyploids, which in turn was a key component of their eventual domestication. Relatively little is known about the developmental and genetic underpinnings of the diverse morphologies illustrated in Fig. 2, but an improved understanding of the changes that occurred during evolution and domestication may have implications for crop improvement (Hovav et al., 2007, 2008; Hu et al., 2013; Rapp et al., 2010; Yoo and Wendel, 2014).

In nature, seed dispersal in *Gossypium* often follows a "shaker" model, where erect, mature capsules dehisce along the sutures and the seeds are distributed near the parental plant as wind shakes the branches. This likely is the ancestral method of seed dispersal in the genus, as it occurs in all species of the B, C, E, F, and (in part) D genome groups and in one species (*G. bickii*) of the G genome group. The A, AD, (some) D, G, and K genome species have evolved other mechanisms of seed dispersal, with perhaps the most noteworthy innovation being the fat bodies to facilitate ant dispersal on the seeds of the K genome species (Seelanan et al., 1999). The development of spinnable fibers apparently has occurred only once in the history of *Gossypium*, in the ancestor of the two A genome species that became the progenitor of the A genome of the tetraploids.

The evolutionary "purpose" of epidermal seed hairs, or trichomes, is a matter of speculation. Fryxell (1979) suggested that elongated fibers aid dispersal by birds, a hypothesis that gains credibility from sporadic mention of cotton seeds in birds' nests as well as a collection of *G. darwinii* from a finch's nest in the Galapagos Islands (Wendel, unpublished data, 1990). One might also speculate that fibers serve to inhibit germination unless there is sufficient moisture to saturate the fibers; should germination occur following a light rain, there might not be sufficient water for subsequent survival of the seedling. In this respect the waxy coating of the fibers would repel water, to a point, and thus prevent premature germination. A related possibility is that seed hairs function as "biological incubators" to facilitate germination only when ecological conditions are appropriate, by recruiting particular microbial communities under appropriate moisture regimes. Finally, two G genome species (*G. australe, G. nelsonii*) have evolved stiff straight seed hairs that facilitate wind dispersal in that the stiffening hairs function to extrude seeds from the locules of the dehisced capsule.

It is against this natural history backdrop that ancient peoples discovered that the unique properties of cotton fibers made them useful for ropes, textiles, and other applications. This history has involved several cultures on different continents and a parallel plant improvement process starting from divergent and geographically isolated wild ancestors. Each of these crop species (*G. hirsutum, G. barbadense, G. arboreum, G. herbaceum*) has its own history of domestication, diversification, and utilization (Brubaker et al., 1999; Hutchinson, 1951, 1959; Hutchinson et al., 1947; Lubbers and Chee, 2009; Percival et al., 1999; Percy, 2009; Percy and Wendel, 1990; Tyagi et al., 2014; Ulloa et al., 2013; Wendel et al., 1992, 2009, 2010). Cotton domestication, thus, has been a replicated, and in many respects parallel experiments for the four domesticated species. Many papers describe various aspects of patterns of genetic diversity, the shape and severity of genetic bottlenecks that accompanied the development of landraces and cultivars, and the influence of recent human history on geographic patterns of cultivation (Brubaker et al., 1999; Brubaker and Wendel, 1994, 2001; Fang et al., 2013; Hutchinson, 1951, 1954, 1959; Hutchinson et al., 1947; Percy and Wendel, 1990; Tyagi et al., 2014; Wendel et al., 1989, 1992). It is worth reiterating that this history involved human shaping and molding of naturally occurring diversity that originated through the process of evolutionary diversification over a period of millions of years, a legacy we continue to exploit today through deliberate introgression of exotic germplasm from diverse, wild gene pools (Mehetre, 2010; Stelly, 2014; Wu et al., 2010; Chapala et al., 2012; He et al., 2011a, 2011b; McCarty and Percy, 2001; Nacoulima et al., 2012; Niles and Feaster, 1984; Zhang et al., 2011, 2014). Moreover, new species of cotton and previously unrecognized taxonomic diversity continue to be discovered today (Alvarez and Wendel, 2006; Feng et al., 2011; Grover et al., 2014; Krapovickas and Seijo, 2008; Stewart et al., 2015; Ulloa et al., 2006), a remarkable realization when one considers that *Gossypium* ranks among our most important and hence frequently studied plant genera. The wild species of cotton, consequently, represent an ample genetic repository for potential exploitation by cotton breeders. The status of much of this diversity in the world's germplasm collections was recently reviewed (Campbell et al., 2010). Given current threats to global ecosystems, it has never been more critical than it is today to recognize this legacy of speciation and diversification for humankind.

Acknowledgments

Research in the Wendel lab has largely been funded by the NSF Plant Genome Program, with additional support from other NSF programs, the USDA NRI, and Cotton Incorporated.

References

Alvarez, I., R. Cronn, and J.F. Wendel. 2005. Phylogeny of the New World diploid cottons (*Gossypium* L., Malvaceae) based on sequences of three low-copy nuclear genes. Plant Syst. Evol. 252:199–214. doi:10.1007/s00606-004-0294-0

Alvarez, I., and J.F. Wendel. 2006. Cryptic interspecific introgression and genetic differentiation within *Gossypium aridum* (Malvaceae) and its relatives. Evolution 60:505–517. doi:10.1111/j.0014-3820.2006.tb01132.x

Applequist, W.L., R.C. Cronn, and J.F. Wendel. 2001. Comparative development of fiber in wild and cultivated cotton. Evol. Dev. 3:3–17. doi:10.1046/j.1525-142x.2001.00079.x

Arpat, A., M. Waugh, J.P. Sullivan, M. Gonzales, D. Frisch, D. Main, et al. 2004. Functional genomics of cell elongation in developing cotton fibers. Plant Mol. Biol. 54:911–929. doi:10.1007/s11103-004-0392-y

Baudoin, J.-P., A. D'Hont, and G. Mergeai. 2009. Bridging classical and molecular cytogenetics of *Gossypium*. In: A.H. Paterson, editor, Genetics and genomics of cotton. Springer, New York. p. 257–281.

Beasley, J.O. 1941. Hybridization, cytology, and polyploidy of *Gossypium*. Chron. Bot. 6:394–395.

Brubaker, C.L., F.M. Bourland, and J.F. Wendel. 1999. The origin and domestication of cotton. In: C.W. Smith and J.T. Cothren, editors, Cotton; origin, history, technology and production. John Wiley & Sons, New York. p. 3–31.

Brubaker, C.L., and J.F. Wendel. 1993. On the specific status of *Gossypium lanceolatum* Todaro. Genet. Resour. Crop Evol. 40:165–170. doi:10.1007/BF00051121

Brubaker, C.L., and J.F. Wendel. 1994. Reevaluating the origin of domesticated cotton (*Gossypium hirsutum*; Malvaceae) using nuclear restriction fragment length polymorphisms (RFLPs). Am. J. Bot. 81:1309–1326. doi:10.2307/2445407

Brubaker, C.L., and J.F. Wendel. 2001. RFLP diversity in cotton. In: J.N. Jenkins and S. Saha, editors, Genetic improvement of cotton: Emerging technologies. Science Publishers, Inc., Enfield, NH. p. 81–102.

Campbell, B., S. Saha, R. Percy, J. Frelichowski, J. Jenkins, W. Park, et al. 2010. Status of the global cotton germplasm resources. Crop Sci. 50:1161–1179. doi:10.2135/cropsci2009.09.0551

Chapala, M.M., D.B. Weaver, B.T. Campbell, E. van Santan, and R.R. Sharpe. 2012. Exotic germplasm introgression effect on agronomic and fiber properties of Upland cotton. In: Proceedings of the National Cotton Council Beltwide Cotton Conference, NCC, Orlando, FL. 3–6 Jan. 2012. NCC, Memphis, TN. p. 803–807.

Cronn, R.C., R.L. Small, T. Haselkorn, and J.F. Wendel. 2002. Rapid diversification of the cotton genus (*Gossypium*: Malvaceae) revealed by analysis of sixteen nuclear and chloroplast genes. Am. J. Bot. 89:707–725. doi:10.3732/ajb.89.4.707

Cronn, R.C., R.L. Small, T. Haselkorn, and J.F. Wendel. 2003. Cryptic repeated genomic recombination during speciation in *Gossypium gossypioides*. Evolution 57:2475–2489. doi:10.1111/j.0014-3820.2003.tb01493.x

Cronn, R., and J.F. Wendel. 2004. Cryptic trysts, genomic mergers, and plant speciation. New Phytol. 161:133–142. doi:10.1111/j.1469-8137.2004.00947.x

d'Eeckenbrugge, G.C., and J.M. Lacape. 2014. Distribution and differentiation of wild, feral, and cultivated populations of perennial Upland cotton (*Gossypium hirsutum* L.) in mesoamerica and the Caribbean. PLoS ONE 9:e107458. doi:10.1371/journal.pone.0107458

DeJoode, D.R., and J.F. Wendel. 1992. Genetic diversity and origin of the Hawaiian Islands cotton, *Gossypium tomentosum*. Am. J. Bot. 79:1311–1319. doi:10.2307/2445059

de Menezes, I.P.P., F.A. Gaiotto, L.V. Hoffmann, A.Y. Ciampi, and P.A.V. Barroso. 2014. Genetic diversity and structure of natural populations of *Gossypium mustelinum*, a wild relative of cotton, in the basin of the De Contas River in Bahia, Brazil. Genetica (The Hague) 142:99–108.

Endrizzi, J.E., E.L. Turcotte, and R.J. Kohel. 1985. Genetics, cytology, and evolution of *Gossypium*. Adv. Genet. 23:271–375. doi:10.1016/S0065-2660(08)60515-5

Fang, D.D., L.L. Hinze, R.G. Percy, P. Li, D. Deng, and G. Thyssen. 2013. A microsatellite-based genome-wide analysis of genetic diversity and linkage disequilibrium in Upland cotton (*Gossypium hirsutum* L.) cultivars from major cotton-growing countries. Euphytica 191:391–401. doi:10.1007/s10681-013-0886-2

Feng, C., M. Ulloa, C. Perez-M, and J.M. Stewart. 2011. Distribution and molecular diversity of arborescent *Gossypium* species. Botany 89:615–624. doi:10.1139/b11-042

Flagel, L.E., J.F. Wendel, and J.A. Udall. 2012. Duplicate gene evolution, homoeologous recombination, and transcriptome characterization in allopolyploid cotton. BMC Genomics 13:302. doi:10.1186/1471-2164-13-302

Fryxell, P.A. 1965. Stages in the evolution of *Gossypium*. Adv. Frontiers Plant Sci. 10:31–56.

Fryxell, P.A. 1968. A redefinition of the tribe *Gossypieae*. Bot. Gaz. 129:296–308. doi:10.1086/336448

Fryxell, P.A. 1971. Phenetic analysis and the phylogeny of the diploid species of *Gossypium* L. (Malvaceae). Evolution 25:554–562. doi:10.2307/2407355

Fryxell, P.A. 1976. A nomenclator of *Gossypium*: The botanical names of cotton. USDA Agricultural Research Service, Washington DC.

Fryxell, P.A. 1979 The natural history of the cotton tribe Texas A&M Univ. Press, College Station, TX.

Fryxell, P.A. 1992. A revised taxonomic interpretation of *Gossypium* L. (Malvaceae). Rheedea 2:108–165.

Galau, G.A., and T.A. Wilkins. 1989. Alloplasmic male sterility in AD allotetraploid *Gossypium hirsutum* upon replacement of its resident A cytoplasm with that of D species *G. harknessii*. Theor. Appl. Genet. 78:23–30. doi:10.1007/BF00299748

Grover, C.E., K.K. Grupp, R.J. Wanzek, and J.F. Wendel. 2012. Assessing the monophyly of polyploid *Gossypium* species. Plant Syst. Evol. 298:1177–1183. doi:10.1007/s00606-012-0615-7

Grover, C.E., X. Zhu, K. Grupp, J. Jareczek, J.P. Gallagher, E. Szadkowski, et al. 2014. Molecular confirmation of species status for the allopolyploid cotton species, *Gossypium ekmanianum*. Genet. Resour. Crop Evol. doi:10.1007/s10722-014-0138-x (in press)

Haigler, C.H., L. Betancur, M.R. Stiff, and J.R. Tuttle. 2012. Cotton fiber: A powerful single-cell model for cell wall and cellulose research. Front. Plant Sci. 3:104.

Haigler, C.H., B. Singh, G. Wang, and D. Zhang. 2009. Genomics of cotton fiber secondary wall deposition and cellulose biogenesis. In: A.H. Paterson, editor, Genetics and genomics of cotton. Springer, New York. p. 385–417.

Hawkins, J.S., H. Kim, J.D. Nason, R.A. Wing, and J.F. Wendel. 2006. Differential lineage-specific amplification of transposable elements is responsible for genome size variation in *Gossypium*. Genome Res. 16:1252–1261. doi:10.1101/gr.5282906

Hawkins, J.S., J. Pleasants, and J.F. Wendel. 2005. Identification of AFLP markers that discriminate between cultivated cotton and the Hawaiian island endemic, *Gossypium tomentosum*. Genet. Resour. Crop Evol. 52:1069–1078. doi:10.1007/s10722-004-6115-z

Hawkins, J.S., S.R. Proulx, R.A. Rapp, and J.F. Wendel. 2009. Rapid DNA loss as a counterbalance to genome expansion through retrotransposon proliferation in plants. Proc. Natl. Acad. Sci. USA 106:17811–17816. doi:10.1073/pnas.0904339106

He, S., J. Sun, and X. Du. 2011. Genetic and epigenetic status of triple exotic consanguinity cotton introgression lines. Genet. Mol. Res. 10:4063–4072. doi:10.4238/2011.October.3.1

He, S.-P., J.-L. Sun, C. Zhang, and X.-M. Du. 2011. Identification of exotic genetic components and DNA methylation pattern analysis of three cotton introgression lines from *Gossypium bickii*. Mol. Biol. 45:204–210. doi:10.1134/S002689331102018X

Hendrix, B., and J.M. Stewart. 2005. Estimation of the nuclear DNA content of *Gossypium* species. Ann. Bot. (Lond.) 95:789–797. doi:10.1093/aob/mci078

Hovav, R., J.A. Udall, B. Chaudhary, E. Hovav, L. Flagel, G. Hu, et al. 2008. The evolution of spinnable cotton fiber entailed prolonged development and a novel metabolism. PLoS Genet. 4:E25. doi:10.1371/journal.pgen.0040025

Hovav, R., J. Udall, E. Hovav, R. Rapp, L. Flagel, and J. Wendel. 2007. A majority of cotton genes are expressed in single-celled fiber. Planta 227:319–329. doi:10.1007/s00425-007-0619-7

Hu, G., J. Koh, M.J. Yoo, K. Grupp, S. Chen, and J.F. Wendel. 2013. Proteomic profiling of developing cotton fibers from wild and domesticated *Gossypium barbadense*. New Phytol. 200:570–582. doi:10.1111/nph.12381

Hutchinson, J.B. 1951. Intra-specific differentiation in *Gossypium hirsutum*. Heredity 5:161–193. doi:10.1038/hdy.1951.19

Hutchinson, J.B. 1954. New evidence on the origin of the Old World cottons. Heredity 8:225–241. doi:10.1038/hdy.1954.20

Hutchinson, J.B. 1959. The application of genetics to cotton improvement. Cambridge Univ. Press, Cambridge, UK.

Hutchinson, J.B., R.A. Silow, and S.G. Stephens. 1947. The evolution of *Gossypium* and the differentiation of the cultivated cottons.Oxford Univ. Press, London.

Kim, H.J., and B.A. Triplett. 2001. Cotton fiber growth in planta and in vitro. Models for plant cell elongation and cell wall biogenesis. Plant Physiol. 127:1361–1366. doi:10.1104/pp.010724

Krapovickas, A., and G. Seijo. 2008. *Gossypium ekmanianum* (Malvaceae), algodon silvestre de la Republica Dominicana. Bonplandia 17:55–63.

Li, P., Z. Li, H. Liu, and J. Hua. 2014. Cytoplasmic diversity of the cotton genus as revealed by chloroplast microsatellite markers. Genet. Resour. Crop Evol. 61:107–119. doi:10.1007/s10722-013-0018-9

Liu, Q., C.L. Brubaker, A.G. Green, D.R. Marshall, P.J. Sharp, and S.P. Singh. 2001. Evolution of the *FAD2-1* fatty acid desaturase 5′ UTR intron and the molecular systematics of *Gossypium* (Malvaceae). Am. J. Bot. 88:92–102. doi:10.2307/2657130

Lubbers, E.L., and P.W. Chee. 2009. The worldwide gene pool of *G. hirsutum* and its improvement. In: A.H. Paterson, editor, Genetics and genomics of cotton. Springer, New York. p. 23–52.

McCarty, J.C., and R.G. Percy. 2001. Genes from exotic germplasm and their use in cultivar improvement in *Gossypium hirsutum* L. and *G. barbadense* L. In: J.N. Jenkins and S. Saha, editors, Genetic improvement of cotton: Emerging technologies. Science Publishers, Inc., Enfield, NH. p. 65–80.

Mehetre, S.S. 2010. Wild *Gossypium anomalum*: A unique source of fibre fineness and strength. Curr. Sci. 99:58–71.

Nacoulima, N.L., J.-P. Baudoin, and G. Mergeai. 2012. Introgression of improved fiber fineness trait in *G. hirsutum L.* from *G. longicalyx* Hutch. & Lee. *Commun. Agric. Appl. Biol. Sci.* 77(1):207–211

Niles, G.A., and C.V. Feaster. 1984. Breeding. In: R.J. Kohel and C.E. Lewis, editors, Cotton. ASA, Madison, WI. p. 201–231.

Paterson, A.H., editor. 2009. Genetics and genomics of cotton. Springer, New York.

Percival, A.E., J.M. Stewart, and J.F. Wendel. 1999. Taxonomy and germplasm resources. In: C.W. Smith and J.T. Cothren, editors, Cotton; origin, history, technology and production. John Wiley and Sons, New York. p. 33–63.

Percy, R.G. 2009. The worldwide gene pool of *Gossypium barbadense L.* and its improvement. In: A.H. Paterson, editor, Genetics and genomics of cotton. Springer, New York. p. 53–68.

Percy, R.G., and J.F. Wendel. 1990. Allozyme evidence for the origin and diversification of *Gossypium barbadense* L. Theor. Appl. Genet. 79:529–542. doi:10.1007/BF00226164

Phillips, L.L. 1966. The cytology and phylogenetics of the diploid species of *Gossypium*. Am. J. Bot. 53:328–335. doi:10.2307/2439872

Phuphathanaphong, L. 2006. *Thepparatia* (Malvaceae), a new genus from Thailand. Thai For. Bull. (Bot.) 34:195–200.

Pleasants, J.M., and J.F. Wendel. 2010. Reproductive and pollination biology of the endemic Hawaiian cotton, *Gossypium tomentosum* (Malvaceae). Pac. Sci. 64:45–55. doi:10.2984/64.1.045

Rapp, R., C. Haigler, L. Flagel, R. Hovav, J. Udall, and J. Wendel. 2010. Gene expression in developing fibres of Upland cotton (*Gossypium hirsutum* L.) was massively altered by domestication. BMC Biol. 8:139. doi:10.1186/1741-7007-8-139

Rapp, R.A., and J.F. Wendel. 2005. Epigenetics and plant evolution. New Phytol. 168:81–91. doi:10.1111/j.1469-8137.2005.01491.x

Ryser, U., and P. Holloway. 1985. Ultrastructure and chemistry of soluble and polymeric lipids in cell walls from seed coats and fibres of *Gossypium* species. Planta 163:151–163. doi:10.1007/BF00393501

Saunders, J.H. 1961. The wild species of *Gossypium* and their evolutionary history.Oxford Univ. Press, London.

Seelanan, T., C.L. Brubaker, J.M. Stewart, L.A. Craven, and J.F. Wendel. 1999. Molecular systematics of Australian *Gossypium* section *Grandicalyx* (Malvaceae). Syst. Bot. 24:183–208. doi:10.2307/2419548

Seelanan, T., A. Schnabel, and J.F. Wendel. 1997. Congruence and consensus in the cotton tribe (Malvaceae). Syst. Bot. 22:259–290. doi:10.2307/2419457

Senchina, D.S., I. Alvarez, R.C. Cronn, B. Liu, J.K. Rong, R.D. Noyes, et al. 2003. Rate variation among nuclear genes and the age of polyploidy in *Gossypium*. Mol. Biol. Evol. 20:633–643. doi:10.1093/molbev/msg065

Small, R.L., and J.F. Wendel. 1999. The mitochondrial genome of allotetraploid cotton (*Gossypium* L.). J. Hered. 90:251–253. doi:10.1093/jhered/90.1.251

Stelly, D. 2014. Facilitating the use of wild species germplasm for cotton improvement via CS lines. In: Proceedings of the National Cotton Council Beltwide Cotton Conference, NCC, New Orleans, LA. 6–8 Jan. 2014. NCC, Memphis, TN.

Stephens, S.G. 1958. Salt water tolerance of seeds of *Gossypium* species as a possible factor in seed dispersal. Am. Nat. 92:83–92. doi:10.1086/282014

Stephens, S.G. 1963. Polynesian cottons. Ann. Mo. Bot. Gard. 50:1–22. doi:10.2307/2394914

Stephens, S.G. 1966. The potential for long range oceanic dispersal of cotton seeds. Am. Nat. 100:199–210. doi:10.1086/282413

Stewart, J.M. 1995. Potential for crop improvement with exotic germplasm and genetic engineering. In: G.A. Constable and N.W. Forrester, editors, Challenging the future: Proceedings of the World Cotton Research Conference, Brisbane Australia. 14–17 Feb. 1994. CSIRO, Melbourne, Australia. p. 313–327.

Stewart, J.M., L.A. Craven, C.L. Brubaker, and J.F. Wendel. 2015. *Gossypium anapoides* (Malvaceae), a new species from Western Australia. Novon 23(4):447–451. doi:10.3417/2007140

Stewart, J.M., D. Oosterhuis, J.J. Heithholt, and J.R. Mauney. 2010. Physiology of cotton. Springer, Dordrecht, The Netherlands.

Tiwari, R.S., J. Zhang, and J.M. Stewart. 2014. Molecular diversity based on AFLP markers and possible natural hybridization among the Australian arid zone *Gossypium* species. Aust. J. Crop Sci. 8:674–679.

Tyagi, P., M.A. Gore, D.T. Bowman, B.T. Campbell, J.A. Udall, and V. Kuraparthy. 2014. Genetic diversity and population structure in the US Upland cotton (*Gossypium hirsutum* L.). Theor. Appl. Genet. 127:283–295. doi:10.1007/s00122-013-2217-3

Ulloa, M., I.Y. Abdurakhmonov, C. Perez-M, R. Percy, and J.M. Stewart. 2013. Genetic diversity and population structure of cotton (*Gossypium* spp.) of the New World assessed by SSR markers. Botany 91:251–259. doi:10.1139/cjb-2012-0192

Ulloa, M., J.M. Stewart, E.A. Garcia-C, S. Godoy-A, A. Gaytan-M, and N.S. Acosta. 2006. Cotton genetic resources in the western states of Mexico: In situ conservation status and germplasm collection for ex situ preservation. Genet. Resour. Crop Evol. 53:653–668. doi:10.1007/s10722-004-2988-0

Watt, G. 1907. The wild and cultivated cotton plants of the world. Longmans, Green and Co., London.

Wendel, J.F. 1989. New World tetraploid cottons contain Old World cytoplasm. Proc. Natl. Acad. Sci. USA 86:4132–4136. doi:10.1073/pnas.86.11.4132

Wendel, J.F., and V.A. Albert. 1992. Phylogenetics of the cotton genus (*Gossypium* L.): Character-state weighted parsimony analysis of chloroplast DNA restriction site data and its systematic and biogeographic implications. Syst. Bot. 17:115–143. doi:10.2307/2419069

Wendel, J.F., C.L. Brubaker, J.P. Alvarez, R.C. Cronn, and J.M. Stewart. 2009. Evolution and natural history of the cotton genus. In: A.H. Paterson, editor, Genetics and genomics of cotton. Springer, New York. p. 3–22.

Wendel, J.F., C.L. Brubaker, and A.E. Percival. 1992. Genetic diversity in *Gossypium hirsutum* and the origin of Upland cotton. Am. J. Bot. 79:1291–1310. doi:10.2307/2445058

Wendel, J.F., C.L. Brubaker, and T. Seelanan. 2010. The origin and evolution of *Gossypium*. In: J.M. Stewart, D. Oosterhuis, J.J. Heithholt, and J.R. Mauney, editors, Physiology of cotton. Springer, Dordrecht, The Netherlands. p. 1–18.

Wendel, J.F., and R.C. Cronn. 2003. Polyploidy and the evolutionary history of cotton. Adv. Agron. 78:139–186. doi:10.1016/S0065-2113(02)78004-8

Wendel, J.F., L.E. Flagel, and K.L. Adams. 2012. Jeans, genes, and genomes: Cotton as a model for studying polyploidy. In: P.S. Soltis and D.E. Soltis, editors, Polyploidy and genome evolution. Springer-Verlag, Heidelberg. p. 181–207.

Wendel, J.F., P.D. Olson, and J.M. Stewart. 1989. Genetic diversity, introgression and independent domestication of Old World cultivated cottons. Am. J. Bot. 76:1795–1806. doi:10.2307/2444478

Wendel, J.F., and A.E. Percival. 1990. Molecular divergence in the Galapagos Island–Baja California species pair, *Gossypium klotzschianum* and *G. davidsonii* (Malvaceae). Plant Syst. Evol. 171:99–115. doi:10.1007/BF00940598

Wendel, J.F., and R.G. Percy. 1990. Allozyme diversity and introgression in the Galapagos Islands endemic *Gossypium darwinii* and its relationship to continental *G. barbadense* . Biochem. Syst. Ecol. 18:517–528. doi:10.1016/0305-1978(90)90123-W

Wendel, J.F., R. Rowley, and J.M. Stewart. 1994. Genetic diversity in and phylogenetic relationships of the Brazilian endemic cotton, *Gossypium mustelinum* (Malvaceae). Plant Syst. Evol. 192:49–59. doi:10.1007/BF00985907

Wendel, J.F., A. Schnabel, and T. Seelanan. 1995. An unusual ribosomal DNA sequence from *Gossypium gossypioides* reveals ancient, cryptic, intergenomic introgression. Mol. Phylogenet. Evol.. 4:298–313. doi:10.1006/mpev.1995.1027

Wendel, J.F., J.M. Stewart, and J.H. Rettig. 1991. Molecular evidence for homoploid reticulate evolution in Australian species of *Gossypium*. Evolution 45:694–711. doi:10.2307/2409921

Wu, J., J. McCarty, J. Jenkins, and W. Meredith. 2010. Breeding potential of introgressions into Upland cotton: Genetic effects and heterosis. Plant Breed. 129:526–532.

Xu, Q., G. Xiong, P. Li, F. He, Y. Huang, K. Wang, et al. 2012. Analysis of complete nucleotide sequences of 12 *Gossypium* chloroplast genomes: Origin and evolution of allotetraploids. PLoS ONE 7:E37128. doi:10.1371/journal.pone.0037128

Yoo, M.-J., and J.F. Wendel. 2014. Comparative evolutionary and developmental dynamics of the cotton (*Gossypium hirsutum*) fiber transcriptome. PLoS Genet. 10:E1004073. doi:10.1371/journal.pgen.1004073

Yu-xiang, W., C. Jin-hong, H. Qiu-ling, and Z. Shui-jin. 2013. Parental origin and genomic evolution of tetraploid *Gossypium* species by molecular marker and GISH analyses. Caryologia 66:368–374. doi:10.1080/00087114.2013.857830

Zhang, J., R.G. Percy, and J.C. McCarty, Jr. 2014. Introgression genetics and breeding between Upland and Pima cotton: A review. Euphytica 198:1–12. doi:10.1007/s10681-014-1094-4

Zhang, Z., J. Rong, V.N. Waghmare, P.W. Chee, O.L. May, R.J. Wright, et al. 2011. QTL alleles for improved fiber quality from a wild Hawaiian cotton, *Gossypium tomentosum*. Theor. Appl. Genet. 123:1075–1088. doi:10.1007/s00122-011-1649-x

Germplasm Resources Collection and Management

James Frelichowski and Richard Percy*

Abstract

The US National Cotton Germplasm Collection (NCGC) is distinct in that it is the largest publicly available cotton (*Gossypium* L.) collection in the world, with a history of support and interaction from federal, state, private, university, and international institutions. In the last century, the rapid rise of mechanization and the resulting profitability of cotton production in the United States contributed directly to the early and rapid development of the NCGC, which in turn has been a foundation for continued success in cotton research and breeding, nationally and worldwide. Coordination and collaboration with national and international collections and institutions have contributed significantly to the growth, maintenance, distribution, and conservation of germplasm resources of the NCGC. Efforts to characterize the genetic and phenotypic diversity of the collection have contributed greatly to its utility as a resource in cotton improvement efforts. Extensive literature on germplasm discovery and germplasm collections exists elsewhere, and this chapter will only provide a rudimentary introduction to these topics. An in-depth review of the NCGC, its contents, operations, and improvements follows. The final sections deal with the current goals and challenges faced by the NCGC.

Discovery and Collections

The use of cotton in the old and new world has been dated to at least 4000 yr BP from archaeological evidence (Gulati and Turner, 1928; Bird and Mahler, 1952; Lee, 1984; Percival and Kohel, 1990). Although genetic studies have suggested pre-Columbian human dispersal of cotton in the New World (Percy and Wendel, 1990), historical records report the transport and trade of diploid cottons along Alexanderian trade routes of the Old World dating to the third century B.C. (Scherer, 1916), and post-Columbian transport and subsequent trade of tetraploid cottons almost simultaneous with their European discovery in the 16th century

---

Abbreviations: APHIS, Animal and Plant Health Inspection Service; CCGC, Cotton Crop Germplasm Committee; CGC, Crop Germplasm Committees; CWN, cotton winter nursery; GRIN, Germplasm Resources Information Network; NCC, National Cotton Council; NCGC, National Cotton Germplasm Collection; NCGRP, National Center for Genetic Resources and Preservation; PGOC, Plant Germplasm Operations Committees.

James Frelichowski (James.Frelichowski@ars.usda.gov) and Richard Percy, USDA-ARS, Crop Germplasm Research Unit, 2881 F&B Rd., College Station, TX 77845. *Corresponding author (Richard.Percy@ars.usda.gov).

doi:10.2134/agronmonogr57.2013.0041

Cotton. 2nd. ed. David D. Fang and Richard G. Percy, editors. Agronomy Monograph 57.

(Lewis and Richmond, 1968). This transport of the new world tetraploid cottons extended throughout the world, even to islands of the Pacific (Stephens, 1963). Cotton was one of many plants to be introduced, characterized, and conserved in new botanical gardens during the Age of Discovery and colonization. Among the gardens were the Kew Gardens in Great Britain and Royal Botanic Gardens in Calcutta, India, from exploits of the British East India Company (Pandey, 2012; Arora, 1991). While many of these botanical gardens had interests in discovering the potential uses and profitability of new plant species, there was also a growing interest in the scientific study and classification of new species. One of the earliest efforts at creating a purposeful and systematic collection of cotton occurred in 1784 when J.B.P. von Rohr established a cotton garden on St. Croix in the Danish West Indies (now the Virgin Islands) under a commission from the Danish King. His work resulted in herbarium specimens that enabled study by modern taxonomists (Fryxell, 1969) and a summary of agricultural methods used and cotton grown (von Rohr, 1791–1793). A second documented systematic collection of cotton was established by Agostino Todaro in the Royal Botanical Gardens in Palermo, Italy around 1860. Todaro was one of the leading taxonomists of his time, and his interest in cotton was sparked by its recent introduction into Italy as a crop plant. Todaro studied a wide range of *Gossypium*, represented by more than 200 accessions, and wrote the first major taxonomic treatment of *Gossypium*, the Relazione Monografia (Todaro, 1877). Other large-scale categorization works that followed included those of Hooker (1875–1897) and Watt (1885). Recognition of natural variation and selective responses in species by Darwin (1859) and Mendel (1866), as well as the rediscovery of Mendelian genetics further propelled the scientific investigation and utilization of plant germplasm. Eventually the documentation and taxonomic treatment of geographically diverse germplasm would be coupled with genetics, one example being Vavilov (1926), which is frequently used as basis for conservation of plant genetic resources. Rapid improvement in cotton would involve collection and preservation of germplasm from the centers of origin of *G. hirsutum* L. and *G. barbadense* L. in the new world.

The first major breeding effort to incorporate the development and maintenance of a cotton germplasm collection was implemented in Trinidad in 1926 by the Empire Cotton Growing Corporation. This collection was dispersed in 1947 to other collections (Hutchinson et al., 1947). During the same time period, key collections were established in the United States, USSR (Mauer, 1930), India, Sudan (Saunders, 1961), Argentina, and Czechoslovakia (Fryxell, 1984). Maintenance of cotton collections have generally occurred in regions and by nations associated with cotton production and commerce. The current status of eight major cotton collections in the United States, Russia, Uzbekistan, China, India, Brazil, Australia, and France have recently been summarized in several publications (Campbell et al., 2010; Abdullaev et al., 2013; Percy et al., 2014). Each collection has its own individual history, but all were basically established with a combination of collection trips to many areas of the tropics, solicitation from agricultural officials, botanists, or others, and exchanges among collections to fill in gaps. The collective coverage of germplasm by these collection is extensive, but gaps remain because of production areas and collections not represented (e.g., Pakistan), native ranges not well explored (e.g., many regions of Africa and the Middle East), and the unknown status of smaller collections (Bettencourt and Konopka, 1988). The NCGC has been fortunate to have had germplasm exchanges with several of the larger collections and the support from the cotton community in its acquisition efforts. Economic, national, and scientific support, involving federal, state, private, and even international institutions, collectively known as the cotton community, has helped to expand and successfully utilize cotton germplasm. A condensed synopsis of events contributing to the establishment of the NCGC follows.

Growth of US Cotton Genetic Resources and Infrastructure

Native Americans have long used *G. hirsutum* (Percival and Kohel, 1990) and *G. barbadense* (Percy, 2009) before European "discovery," and thus European colonists encountered a mix of wild, feral and domesticated cotton on their arrival. Columbus reportedly found cotton in the Bahamas. New world cotton was fairly rapidly distributed throughout the New and Old World by European trading interests and colonizing efforts that adapted it to field cropping systems. However, these systems required intense labor resources. Cotton establishment in the developing United States was slowed initially by protectionism by England of its wool industry. Initial cotton production in the United States was most successful in areas with long growing seasons, such as offshore islands and onshore coastlands of the east coast (Gray, 1933). Cottons grown at this time were primarily Sea Island *G. barbadense* types. With the growth westward, pioneers of new upland environments required a different type of cotton. Upland *G. hirsutum* cottons were imported and became the predominant cotton. The early 1800s saw the introduction of Burling's Mexican hybrid, from high plateaus of central Mexico, and this cotton provided improvement in maturity and fiber quality over contemporary Upland breeds such as Georgia Green Seed and Creole Black Seed (Moore, 1956; Hague et al., 2009). The federal government began to set national priorities for agricultural improvement and support activities for crop germplasm acquisition, distribution, and research. As early as 1827, President John Adams instructed foreign consuls to collect seeds and rare plants. The patent office was established in 1849, and patents were issued. Varietal development, summarized elsewhere (Smith et al., 1999; Lubbers and Chee, 2009; Moore, 1956; Ware, 1951), and the marketing of new cotton varieties was a profitable venture, with a colorful history. In 1862, President Abraham Lincoln started the United States Department of Agriculture (USDA), which would open the Plant Introduction Office in 1898 (or Bureau of Plant Industry) (Baranski, 2013; Williams, 2005). The Morrill Land-Grant College and Hatch Experiment Station Acts would establish a solid foundation of agricultural education and research. Collection, distribution, and evaluation of crop species germplasm became a national priority. Typically the US government would promote exploration and characterization of germplasm, create and maintain herbarium specimens, provide identifying designations, and distribute seed to experiment stations for research and use in varietal development. Cotton had long been a labor intensive crop, but this would be moderated by various agricultural breakthroughs and increased mechanization. Expansion of the Cotton Belt across the United States would encounter buildup of pests such as the boll weevil, which shifted cropping of cotton to earlier yielding varieties to outpace the seasonal buildup of the pests (Ware, 1937). Other advances such as area wide pest control practices and continued utilization of diverse, typically earlier maturing, germplasm would extend cotton further westward. Introductions to meet needs of earlier maturity are exemplified by germplasm collected in the Acala Valley of Chiapas, Mexico, brought back by G.N. Collins and C.B. Doyle (Ware, 1937).

New germplasm was acquired from many sources and collaborators for taxonomic and botanical investigations and introgression of specific traits related to resistance to the boll weevil and other pests. Long-term care of these accumulating germplasm materials was delegated to three subcollections, set up with the Regional Research Project S1 (USDA-ARS, 1974). Subcollections were established in three locations to serve as points to receive new germplasm and disseminate existing cotton germplasm and information. Obsolete cultivars were kept at the Delta Branch Experimental Station in Stoneville, MS, a fertile region for managing even long-season cotton. *Gossypium barbadense* accessions were kept at the USDA station in Sacaton (relocated to Maricopa), AZ for breeding heat-resistant long-staple cotton for the Southwest. Landraces and tropically collected *G. hirsutum*, and other species of *Gossypium*, were kept at Texas A&M University in College Station, TX, typically for investigations into the physiology and genetics of *Gossypium*. More details on their contents and consolidation into the NCGC are in a later section. Considerable collaboration and research occurred with these materials as evidenced in a wide range of published works, some of which were summarized in bulletins of the Southern Cooperative Series (1956, 1968, 1981, 1997). The genus was well defined taxonomically, geographically, and cytologically. A major difficulty encountered in the maintenance of these collections was the tropical, short-day photoperiodism of much of the materials, particularly in the *G. barbadense* materials maintained in Arizona and the *G. hirsutum* race stocks maintained in College Station. Simultaneously, breeding efforts in cotton were being slowed by the long-season nature of cotton and an inability to grow more than a single generation a year. The National Cotton Council (NCC) addressed both of these concerns with the creation of a cotton winter nursery (CWN) in 1950. The nursery was first established in Iguala, Mexico, then relocated to a research facility of the Instituto Nacional de Investigaciones Forestales y Agropecuarias (INIFAP) in Tecoman, Mexico (Percival et al., 1999). Cooperative agreements that allowed import and export of cottonseed back to the United States were established. The nursery became the backbone for seed increase and maintenance of the cotton collections of College Station, TX and Arizona, and ultimately the NCGC.

Plant introduction centers were established with the Agricultural Marketing Act of 1946 and a National Seed Storage Laboratory (now the National Center for Genetic Resources and Preservation [NCGRP]) was established in 1958 for long-term backup storage of germplasm. Improved seed storage protocols (Justice and Bass, 1978) and facilities created a permanent backup of all genetic resources and greatly extended the viability of seed. It was within this timeframe that much of the earliest collected cotton germplasm was incorporated into the NCGC, and the use of the CWN (essential in preserving germplasm) was initiated. However, much of the germplasm collected or developed before the 1940s is poorly represented in the current NCGC, and this gap is a reflection of the need for a national center. This gap in the NCGC is particularly true of improved cultivar materials and reflects the widespread abandonment of numerous local varieties in favor of a select few nationally marketed varieties and the lack of an adequate mechanism to preserve the obsolete materials before the establishment of the NCGRP. In 1953 the Agricultural Research Service (ARS) was created to be the leading research wing of the USDA. Much of the germplasm curation activities, research, and personnel were absorbed by the ARS. Germplasm collection and preservation, an activity practiced since the Age of Exploration, has gained impetus in the modern age as concerns have grown over the narrowing of the genetic base of crop germplasm and recognition of the loss native genetic resources has increased.

The Food and Agricultural Organization (FAO), formed in 1945 with the goal of achieving food security, grew more and more concerned with loss of plant genetic diversity both in the wild and domesticated relatives of food crops. In 1967 the FAO and the International Biological Program in England held the Technical Conference on the Exploration, Utilization and Conservation of Plant Genetic Resources in Rome, Italy. Concerns over genetic erosion led to the development of standards and plans for ex situ conservation of genetic resources. The United States would follow suit when the National Research Council reported the genetic vulnerability of major crops (National Research Council, 1972). In 1974, FAO and CGIAR founded the International Board for Plant Genetic Resources (IBPGR, now Bioversity International), which would help set standards (Toll, 1995) and coordinate and fund genetic resource projects, seed banks, and agricultural research. Start-up of these activities proved easier than long-term care, because crop relatives and wild species require recurring and often increasing funds, along with significant modifications of cultivation practices, which are often much different and incompatible with those used by plant breeders. Fortunately in the United States, the federal government continued to recruit personnel and develop the infrastructure to handle the expanding germplasm genetic resources and increase access to users.

The ARS underwent reorganization in 1972 and in the process developed the National Plant Germplasm System (NPGS) to improve coordination and organization of genetic resources activities while maintaining the existing cooperative network among federal agencies, state agricultural stations, and public and private representatives. The structures are well explained by the Board on Agriculture National Research Council (1991, 1993). Sharing of technical and financial resources and developing priorities were essential to maintain and utilize rapidly growing genetic resources. Under the new NPGS structure, the three cotton subcollections were consolidated at College Station, where a single curator would have full responsibility for maintenance, characterization, and distribution activities of the NCGC. Walk-in cold vaults were built to house the working collection and for handling all distribution activities. The ARS also established the Database Management Unit (DBMU) to create a centralized database for the curator and users, which migrated online (www.ars-grin.gov) as the Germplasm Resources Information Network (GRIN). The Plant Exchange Office (PEO) was created to facilitate germplasm exploration and exchange efforts through the development of exploration and exchange guidelines, provide funding for exploration/exchange efforts, communicate with foreign government authorities to negotiate terms of access to genetic resources, and assist in complying with Animal and Plant Health Inspection Service (APHIS) plant import regulations. Various other oversight, technical, and advisory committees were created within the NPGS structure, with membership including stakeholders and officials from public and private sectors to provide input to unite crop germplasm interests with national priorities. The Plant Germplasm Operations Committees (PGOC) and Crop Germplasm Committees (CGC) were formed. The Cotton Crop Germplasm Committee (CCGC) has been a long-standing committee that customarily has held its meetings and reported to the cotton community at the annual NCC Beltwide Meetings. Examples of outcomes from the CCGC include cotton crop vulnerability status reports (Wallace et al., 2009), planning and approval of recent cotton exploration trips, and a worldwide cotton germplasm status report (Campbell et al., 2010). Reports by the CCGC, along with reports of other crop CGCs, are relayed to Congress and weigh in on their policy and budget decisions affecting the NPGS and ARS (USDA, 1997).

Contents of the NCGC

The NCGC has nearly 10,000 accessions of *Gossypium* (Tables 1 and 2). These are accessible at www.ars-grin.gov or by contacting the curator or technician at gossypium.collection@ars.usda.gov. Description and management of the NCGC is complex because of the diversity inherent in the genus and the collection's composite nature due to its assemblage from multiple sources. The NCGC has multiple identifications for each accession, but has a single Plant Introduction number (PI) assigned to each accession at the time the accession enters the collection. Often materials have been received by the collection with identifying collector numbers assigned by the explorer or donor, and these have been maintained in collection records as well. The PI number system has been in effect since the late 1800s to track all plant germplasm, collected or developed. This is a unique number and is the key value for sourcing an accession in the GRIN databases. In addition to identification numbers, accessions carry various designations such as cultivar or local names that also are kept in the database. The curator maintains and assigns an additional name, called the inventory identification (ID), which primarily assists in retrieval of accessions in a cold storage vault with physical separation according to species and subcollection. In other words, simply knowing the inventory ID is sufficient to locate and subsequently return seed to the vault. When known, cultivar name, local name, collector ID, and developer or collection site data are entered in the database. When submitting requests in the GRIN database or to the NCGC, the best results are obtained by using the PI and/or inventory ID because of variability in cultivar or local names that is encountered and may complicate search strings (e.g., Deltapine, Delta Pine&Land, or DP are all synonyms of the same developer).

Organization of the NCGC maintains a subcollection structure that serves a useful purpose but is in part a vestige of the composite assemblage of the NCGC. Cultivars of *G. hirsutum* continue to be added to a preexisting obsolete cultivar subcollection formerly maintained in Stoneville, MS. Inventory IDs of the accessions are "SA" (short for Stoneville Accession) followed by a sequential number, and the most recent accession is SA-3887. Current additions to the SA subcollection are most frequently acquired from NCGRP and are elite germplasm developed by public breeders and registered with the Journal of Plant Registration (formerly Crop Science Registration), cultivars with expired or abandoned Plant Variety Protection certificates (with developer's permission), or donations or exchanges with other collections. Currently only materials developed through conventional plant breeding are accepted in the collection because of the cost and difficulties of segregating genetically modified (GM) and non-GM plants and in securing permits for planting transgenic cotton. A few mutants and specialty bred lines are present in the collection and are available to the public, contingent upon their ease of germination and self-pollination. Materials that fall into this category include monosomic chromosome stocks (Stelly et al., 2005), daylength

Table 1. *Gossypium* species represented in the NCGC, also indicating inventory ID, quantities, environment of seed increase, flowering season, and approximate seed yields per accession per year.†

Species	Genome ID	NCGC ID	NCGC number	Increase environment‡	Increase amount per year§	Flower time¶
Gossypium hirsutum L. obs. Cv.	$(AD)_1$	SA-	3520	CWN, CS	>1000	neutral
G. hirsutum obsolete cultivar (obs. Cv.) landraces	$(AD)_1$	TX-	2140	CWN, GH	variable	1–5, 10–12
G. barbadense L. obs. Cv.	$(AD)_2$	GB-	1608	CWN, CS	>1000	neutral
G. barbadense obs. Cv. commensal	$(AD)_2$	GB-	n/a	CWN, GH	variable	1–5, 10–12
G. tomentosum Nutt. ex Seem.	$(AD)_3$	$(AD)_3$-	16	GH	<100	1–6, 8, 10–12
G. mustelinum Miers ex G. Watt	$(AD)_4$	$(AD)_4$-	23	GH	<1000	1–5, 11,12
G. darwinii G. Watt	$(AD)_5$	$(AD)_5$-	138	GH	<1000	1– 5, 12
G. herbaceum L.	A_1	A_1-	194	CWN	variable	neutral
G. arboreum L.	A_2	A_2-	1729	CWN	>1000	neutral
G. anomalum Wawra	B_1	B_1-	7	GH	<1000	1–5, 9–12
G. triphyllum (Harv.) Hochr.	B_2	B_2-	2	GH	n/a	n/a
G. capitis-viridis Mauer	B_3	B_3-	1	GH	<1000	1–5, 10–12
G. stocksii Mast.	E_1	E_1-	4	GH	<100	1–6, 8–12
G. somalense (Gurke) J.B. Hutch.	E_2	E_2-	3	GH	<1000	1–5, 8–12
G. areysianum Deflers	E_3	E_3-	2	GH	<100	3–12
G. incanum (O. Schwartz) Hillc.	E_4	E_4-	4	GH	<100	3–8, 10–12
G. longicalyx J.B. Hutch. & B.J.S. Lee	F_1	F_1-	4	GH	<100	1–6, 9–12
G. thurberi Tod.	D_1	D_1-	41	GH, CS	variable	1–12
G. armourianum Kearney	D_{2-1}	D_{2-1}-	10	GH	<10	4, 5, 11
G. harknessii Brandegee	D_{2-2}	D_{2-2}-	19	GH	<10	2, 4–12
G. davidsonii Kellogg	D_{3-d}	D_{3-d}-	32	GH	<1000	1–6, 9–12
G. klotzschianum Andersson	D_{3-k}	D_{3-k}-	59	GH	<1000	1–5, 10–12
G. aridum (Rose & Standl.) Skovst.	D_4	D_4-	20	GH	<100	1–5, 12
G. raimondii Ulbr.	D_5	D_5-	56	GH, CWN	variable	1–5
G. gossypioides (Ulbr.) Standl.	D_6	D_6-	8	GH	<1000	1–5, 11, 12
G. lobatum Gentry	D_7	D_7-	4	GH	<100	2–5, 12
G. trilobum (DC.) Skovst.	D_8	D_8-	11	GH	<1000	1–5, 10–12
G. laxum L. Ll. Phillips	D_9	D_9-	5	GH	<10	1–4, 12

Table continued.

Table 1. Continued.

Species	Genome ID	NCGC ID	NCGC number	Increase environment‡	Increase amount per year§	Flower time¶
G. turneri Fryxell	D_{10}	D_{10}-	8	GH	<10	1, 3–11
G. schwendimanii Fryxell & S.D. Koch	D_{11}	D_{11}-	2	GH	<10	2–4, 12
G. sturtianum J.H. Willis	C_1	C_1-	8	GH	<1000	1–12
G. nandewarense (Derera) Fryx.	C_{1N}	C_{1N}-	6	GH	<1000	1–9, 11
G. robinsonii F. Muell.	C_2	C_2-	3	GH	n/a	n/a
G. bickii Prokh.	G_1	G_1-	5	GH	<100	1, 6–9, 12
G. australe F. Muell.	G_2 (C_3)	G_2-	11	GH	<1000	1–12
G. nelsonii Fryxell	G_3 (C_9)	G_3-	4	GH	<1000	1, 3–12
G. costulatum Tod.	K_1 (C_5)	K_1-	2	GH	<1000	2–12
G. populifolium (Benth.) F. Muell. ex Tod.	K_2 (C_6)	K_2-	4	GH	<10	1, 4–12
G. cunninghamii Tod.	K_3 (C_7)	K_3-	1	GH	n/a	n/a
G. pulchellum (C.A. Gardner) Fryxell	K_4 (C_8)	K_4-	1	GH	<100	1–12
G. pilosum Fryxell	K_5 (C_{10})	K_5-	1	GH	n/a	n/a
G. anapoides J.M. Stewart et al., nom. inval.	K_6	K_6-	n/a	n/a	n/a	n/a
G. enthyle Fryxell et al.	K_7	K_7-	1	GH	n/a	n/a
G. exiguum Fryxell et al.	K_8	K_8-	1	GH	<10	2–5, 7–11
G. londonderriense Fryxell et al.	K_9	K_9-	1	GH	n/a	n/a
G. marchantii Fryxell et al.	K_{10}	K_{10}-	2	GH	<100	1, 3–12
G. nobile Fryxell et al.	K_{11}	K_{11}-	2	GH	<100	1–12
G. rotundifolium Fryxell et al.	K_{12}	K_{12}-	1	GH	<10	2–7, 9–12

† n/a, not currently in the NCGC or not currently grown; NCGC, National Cotton Germplasm Collection.

‡ CWN, cotton winter nursery; CS, College Station TX field; GH, greenhouses in College Station, TX.

§ variable: ranges from <10 to >1000, depending on the accession and/or the environment.

¶ neutral: flowering is not seasonally dependent. Month of recorded flowering is represented by a number: January = 1, December = 12.

Table 2. Representation in the National Cotton Germplasm Collection of accessions according to continent or area, of origin or development, irrespective of the donor, separated by species, whether cultivated or wild relative, and number of nations covered in the region.

Continent or region	Number of accessions	*Gossypium hirsutum* L., cv. (relatives) †	*G. barbadense* L., cv. (unk)‡	A1/A2§	Other species (quantity)¶	Number of nations or territories
Atlantic Ocean Islands	3	0 (1)	1 (0)	0	1(1)	3
Indian Ocean Islands	17	9 (5)	0 (1)	2	0	6
Pacific Ocean Islands	57	0 (13)	0 (16)	0	1 (28)	9
Australia	106	25 (9)	0 (0)	0	12 (72)	1
Africa	284	147 (27)	64 (13)	20	3 (13)	29
Asia	2782	712 (44)	238 (38)	1747	3 (3)	18
North America	3591	2434 (669)	291 (22)	0	11 (175)	2
Caribbean	769	50 (606)	1 (110)	0	0	35
Central America	347	6 (286)	0 (55)	0	0	7
South America	1469	43 (490)	33 (620)	0	4 (283)	12
Mediterranean	62	56 (3)	0 (7)	1	0	6
Middle East	96	34 (7)	5 (0)	39	4 (11)	6

† cv., cultivated.

‡ unk: status of cultivation or wild collected needs to be verified.

§ A1/A2: *G. herbaceum* L./*G.arboreum* L. accessions.

¶ Other species: number of other *Gossypium* species collected or received.

conversion lines (McCarty and Jenkins, 2001), and interspecific cytoplasms (Meyer, 1973). Earliest accessions include a few *G. barbadense* cultivars and are left in this subcollection because of familiarity by users with the inventory ID.

The consolidation of a collection maintained at Texas A&M University in College Station into the NCGC required reassignment of accessions into subcollections because of the vast diversity and many species in the College Station collection. Previous to the consolidation, materials in the College Station collection were inventoried with TEX- (currently shortened to TX-) numbers. The vast majority of these materials were landraces of *G. hirsutum*, and they have kept their TX inventory IDs. The material in this subcollection is poorly characterized morphologically and agronomically. These accessions have diverse origins from collections, exchanges, or donations from tropical areas worldwide and range from wild to feral or escaped to dooryard cottons maintained by locals (USDA-ARS, 1974; Percival and Kohel, 1990; Percival et al., 1999). Typically they are photoperiodic in flowering and may require multiple seasons of growth for maturation. Challenges with many landrace accessions are slower growth and larger and lower yielding plants, often with hard seed coats that result in delayed germination. Seed scarification is often used to improve germination of hard seeded TX accessions. Therefore, the TX designation indicates special needs in cultivation, typically at the CWNs and/or greenhouses. As with the SA subcollection, a few *G. barbadense* accessions have been included and left there for references, because of user familiarity with our inventory IDs.

When *G. barbadense* accessions were consolidated into the NCGC from a collection in Arizona, the accessions were given a GB- prefix and a number, with the most recent addition numbered GB-1608. Previous identification codes, such as AZ, K, or B numbers assigned to these materials, represented prior collection numbers, before these materials were consolidated into the Arizona collection.

These identification numbers have been treated as collector numbers and have been kept in the records for reference by users familiar with the original Arizona subcollection. Currently, efforts are underway to verify and categorize the photoperiodic status of *G. barbadense* accessions of the NCGC. A few *G. barbadense* accessions remain in other subcollections as noted earlier, and in some cases, suspected duplicates have been left in place, until in depth molecular marker analyses can confirm them as being identical. Some accessions of *G. barbadense* are particularly challenging with tall growth, late maturity, and low yields, as well as low germination. Hot water treatment (typically 80°C for 1 minute) is used to improve germination of tropically collected *G. barbadense,* and this procedure softens the wax covering of the chalazal plug, thus allowing imbibition by the seeds. Mutant stocks of both the *G. hirsutum* and *G. barbadense* have been maintained separately from the general collection because of difficulty in their maintenance and their need for special handling.

The remainder of the NCGC is composed of Asiatic Diploid cotton, 3 tetraploid, and 38 diploid species of *Gossypium,* and their inventory IDs are shown in Table 1. The Asiatic cottons and the non-cultivated *Gossypium* species accessions have been assigned alphanumeric inventory IDs on the basis of their genome type (Fryxell, 1965, 1992; Percival and Kohel, 1990). The genome types are a shorthand way to separate the species, according to research on taxonomic, breeding, and cytogenetic studies, and are shown in Table 1. Species with the same base letter(s) possess the same genome types and usually some degree of inter-fertility. Success rates of crossing between genome types drop dramatically (Harland, 1932). The Asiatic A genome cottons have a long history of cultivation, particularly in India, and have been acquired by the NCGC through germplasm exchanges with India, the Russian VIR collection, and other collections. A surprising range of variation is seen in the Asiatic diploid cotton collection (Kebede et al., 2007), and most can be grown along with regular cotton for observation and seed increase.

Although the cultivated tetraploid species and the A genome *G. arboreum* L. are well represented in the collection, representation of the remaining species drops off dramatically. Representation of these species is a reflection of their absence from most cotton breeding programs, scarce occurrence in natural habitats, and not least of all, the difficulty of their acquisition and seed increases ex situ. Distribution of seed of these "wild" tetraploid and diploid species to users of the NCGC (Percy et al., 2014) fluctuates because of irregular flowering habits (Table 1) and challenges in optimizing cultivation practices for seed production in greenhouses. Because of gaps in the numerical string of the inventory IDs, possible in situ species misidentification, and even receipt of non-viable seed, the NCGC makes sure that new additions are successfully grown, identified, and seed increased before they are officially assigned a PI number and an inventory ID for the NCGC. This extra step insures that the accession becomes a viable addition to the NCGC and avoids the post facto deactivation of an inventory ID of the NCGC if it should fail to generate seed ex situ. It is not unusual for wild *Gossypium* to be nearly leafless when cotton or seed is present on plants, and this could account for the original misidentification of species (Fryxell, 1992).

Users depend on the database of the NCGC, which starts with passport and provenance information and when available pedigree, descriptor, and even evaluation data. Contact is frequently made with the curator or technician in accessing this information. The NCGC is continually challenged to provide as much of this information online for viewing and dissemination to the public and for management purposes. GRIN, managed by the USDA-ARS-DBMU, has online databases for the NPGS accessible to the public and the curators. Interaction between them will be streamlined with the newest release of GRINglobal (Postman et al., 2010), and their various spreadsheet building options of the platform promises many shortcuts to data entry, retrieval, comparisons, list building, etc.

Activities of the NCGC

Increase, distribution, acquisition, and characterization are the main activities of the NCGC. The intensity and emphasis given to these activities vary across species and even among accessions within a species or subcollection. Seed increase is influenced greatly by demand, and demand by users varies greatly according to their needs. All NCGC activities are recorded as much as possible in the GRIN database, and status updates are annually reported to CCGC and PGOC members. Feedback loops among the activities and input from the committees helps in prioritizing management activities and maximizing utility for the users.

Seed Increase

Few long-term studies have been conducted on the longevity of cottonseed in storage, but cottonseed is considered an orthodox seed and is known to retain viability long after 10 yr in proper storage. At least 10% of the NCGC is planned for seed increase each year to keep most accessions on a 10-yr cycle. Accessions with 10 or more years since their last increase, newly deposited accessions, or those indicated by users or by the NCGRP as low in seedling germination/vigor (<50% germination rate) have the highest priority for seed regeneration. Our goal in seed increase is to generate no more than 6000 seed, with 3000 being used to replenish the seed in the working collection, and another 3000 available for the NCGRP. Seed quantities of exotic landraces, lines with mutant characters, most wild species, and highly requested *Gossypium* accessions frequently fall far below the optimum and require that these accessions be grown repeatedly within the normal 10-yr timeframe to generate seed. Frequent contact is necessary with NCGRP to make sure viable seed of all accessions exists at both locations.

The major part of the NCGC is comprised of four cotton producing species (*G. hirsutum*, *G. barbadense*, *G. arboreum*, and *G. herbaceum* L.) that grow and reproduce readily in a tropical "winter nursery" at Tecoman, Mexico. This CWN has been used by the germplasm collection for seed increase for over 50 yr and has been structured to provide annual seed increases for improved annual accessions and multi-year increases for perennial materials. An onsite manager for all aspects of seed production has been provided by the NCC. Wes Malloy is the most recent and longest tenured manager of the CWN, having supervised the CWN for over 30 yr, maintaining links between Mexican and the US Embassy officials, and ensuring that phytosanitary regulations of import and export of cotton with the CWN were met. The CWN crews have consistently grown and maintained the nursery, self-pollinated, harvested the plants, and ginned and packaged cottonseed of every line at the CWN. Unfortunately, after over 60 yr of operation, the CWN at Tecoman, Mexico was discontinued in 2014. New arrangements for a tropical nursery will have to be made to maintain a large portion of the NCGC. Operations at the CWN specific to the NCGC will be replicated at a new site to ensure adequate and accurate seed production. Standard operations of the NCGC previously used in the CWN follow.

Seed of accessions to be grown at the CWN are acid delinted and fungicide treated for shipment to the nursery in September. Often a portion of these accessions require special treatment such as seed scarification or hot water treatment before shipment to ensure germination. Mexican import laws require seed to be fumigated before entry into Mexico. At the nursery, individual accessions are hand planted into seven hill plots at a rate of 3 seed per hill and are thinned to two plants per hill after germination. Typically, accessions of the SA, GB, and TX subcollections are alternated with the Asiatic diploid accessions in the CWN plot design to provide sharp visual breaks between the plots and simplify harvesting. Sometime between late January and early March, the curator visits the CWN to observe accession uniformity, record descriptors, and relay any special instructions to the nursery manager. Examples of special instructions would include rouging of off types in each plot, requesting separate harvest of rare sports or segregants of interest, and requesting that lintless accessions have their flowers or bolls enclosed in cotton or mesh bags to prevent loss of naked dehiscent seeds.

Accession purity of materials increased in the winter nursery is maintained by forced self-pollination of flowers. All plants of *Gossypium* have a perfect flower that can be mechanically prevented from opening at anthesis. In the CWN, on the day before opening, the tips of the corolla are kept closed by wrapping a 6-in malleable wire around the pedicel while the opposite end of the wire is wrapped on the tip of the flower bud. Alternatively, the unopened tip is glued shut and a labeled tag is attached to the pedicel. Selfed cotton is hand harvested in the winter nursery from March until May. Total boll numbers and seedcotton weights are recorded for each accession, and then the samples are ginned onsite. The fuzzy seed and a ~30-g (a compact handful) sample of the lint was separately bagged for each accession, bundled, and sent to a designated US port of entry staffed with APHIS officers and equipped for fumigation treatments. Treated seed is sent to College Station, TX, and once received all samples are barcode labeled with inventory and CWN plot IDs. Careful handling through all steps has ensured that clean and pure seed was generated and received yearly from the CWN for the NCGC.

The procedures above, when relevant to acquisition of adequate seed amounts and genetic purity, should be maintained at any successor to the CWN. Other procedures are subject to change under new circumstances. Recent communications have revealed that APHIS facilities are no longer providing fumigation services, and any new counter season winter nursery will have to make arrangements of their own for this required procedure.

A subset of accessions in the collection has critically low seed quantities or has proven to be unreliable seed producers in the tropical winter nursery. Reasons for these difficulties vary but include characteristics such as a protruding stigma that prevent self-pollination, delayed maturation that interferes with normal crop rotation, or other factors preventing culture in the CWN such as dwarf size, growth as vines, low square set, partial self-incompatibility, etc. Accessions that have flowers with stigmas that protrude well beyond the anthers must be hand-pollinated in insect-free greenhouses or the enclosed flowers must be briefly exposed for pollen transfer and closed again. These critical or challenging accessions as well as all of the *Gossypium* wild species must be grown in College Station, in greenhouses, until sufficient, viable seed is produced for distribution and long-term backup. Here, horticultural know how, "tricks of the trade," experiences passed down through curators, technicians, etc. are used to micromanage accessions. A few of the procedures, used successfully for the NCGC, are listed here to be of possible assistance to users of the NCGC, or other curators.

Successful early establishment and growth of plants of many species produces best results in quick seed production, because of negative effects of extended pot culture such as water stress, and micronutrient deficiencies. For most species of the collection, temperature controls set between 17°C (nighttime low) and 35°C (daytime high), with fan cooling and reflective shade cloth used in the summer months, produce the best results. When quarantine grade greenhouses are not available, then pest control is paramount in greenhouses. Damage from unchecked pests populations cause square drop and even plant death (Leonard et al., 1999). Spider mites, whiteflies, mealy bugs, aphids, and thrips are controlled with spraying, limited fertilization to avoid lush growth and, if necessary, whole plant replacement. Plants are often being grown continuously to bulk enough seed for distribution and storage; therefore, plant replacement is the last option, particularly when starting with very limited seed. When only one pest species is present, then selective insecticides are effective. Because several pests are often present and of the possible development of resistance to pesticides, and even health concerns, mineral oil solutions are often sprayed at least once a month as a means to smother and suppress these pests. When pesticide burn of foliage is a problem with species, systemic insecticides and natural predators are used as an alternative. Tap water is purified by reverse osmosis and connected to a programmable watering system, with metered siphoning of diluted fertilizer solution into a network of 3/4-in irrigation tubing and an emitter line connected to each pot. Low phosphorus fertilizer with micronutrients, typically fish emulsion, is used because Australian species are reported as being sensitive to high phosphorus levels. Tall, well-drained pots are best used because *Gossypium* does poorly in wet soils. The frequency and duration of watering is reduced in winter months and helps to mimic the drought patterns in the geographical origins of many photoperiodic species and landraces (Fryxell, 1992). Pruning is only done in summer on photoperiodic plants or after the last seed set during flushes of growth. Single potting of large plants and repotting is recommended before each flush or season of growth to reduce micronutrient deficiencies and sustain boll formation. Poor attention to these details can lead to empty or incomplete seeds on otherwise healthy plants.

Because of restricted seed quantities of a subset of accessions, particularly wild diploid species accessions, germination success is critical. Although standard protocols for germination are well developed internationally (ITSA, 2014), they are based on cultivated germplasm and create low germination rates when applied to several wild *Gossypium* species. Protocols for each species remain to be formalized, but a few procedures to boost germination are available and proven in particular circumstances. Although not completely understood in cotton, there is an apparent dormancy in new or green seed that inhibits germination and disappears with "aging" in a storage vault and acclimated to room temperature for several weeks before planting. Pretreatments such as scarification or hot water treatments further improve germination in some species, particularly those associated with extremely arid environments (Christiansen and Rowland, 1986). The use of smoke water (Light et al., 2005; Dixon et al., 1995) has improved germination of some fire adapted Australian species of *Gossypium*. Heating mats boosted germination of some D genome species, even when they failed to germinate for at least 6 mo in the same soil. Germination of seedlings in vitro or on some type of absorbent paper are best transplanted temporarily into vermiculite to reduce seedling diseases and pests such as fungus gnats, often found in organic soils. Australian species are difficult to transplant and benefit from individual sowing or in a medium-like sand or vermiculite that washes easily from intertwined seedling roots. The seasonal flowering and variable seed yields for most of the *Gossypium* species for the NCGC are listed in Table 1. Species that yield poorly need more research to trigger greater flowering in pot culture. In some instances several seasons may be involved in plant maturation to a reproductive state, and a dedicated tropical nursery is required to generate ample seed. A priority increase of select accessions of the diploid *G. raimondii* Ulbr. required multiple seasons of growth in a perennial nursery at the CWN with special instructions for manual self-pollinations of the protruding stigmas, with reclosing of the flower, and bagging of each boll to produce significant quantities of seed. Permanent access to a tropical nursery is ideal to best manage the full diversity of *Gossypium* germplasm. CIRAD (Cameroon), China (Hainan Island), and Pakistan use tropical nurseries for species seed regeneration.

Accurate records are essential to maintaining the germplasm collection. Seed inventories are first updated in our local excel databases and then uploaded into GRIN. Greenhouse produced seed is harvested daily, bulked once weekly, and promptly stored in the cold vault. Samples from the CWN accessions are first barcoded and use an electronic seed check-in and weighing system. The inventory and CWN plot ID barcodes are scanned, a seed index is calculated (typically weight per 200 seed), and the remaining seed weighed, using the index to estimate the quantity of seed. The weights and quantities are automatically entered into a computer from the connected scales and new barcode labels are printed, summarizing this same information, for the seed index sample and the bottle for holding the remaining seed. Seed from the previous seed increase is discarded, holding back a two tablespoon quantity or the seed index subsample, which is kept in a coin envelope as a backup. The new seed is kept in a 500-mL screw top, Nalgene bottle. The bottles and seed envelopes are stored in a walk-in cold vault kept at 20 to 23% relative humidity and 4°C, which is optimum for seed longevity in a medium-term storage facility (Vertucci and Roos, 1990), designed for an easy access working collection. Ideally, enough seed has been produced so that about 3000 seed, or half of the harvest, may be sent to NCGRP for long-term storage, at ~5% moisture content and –20°C with airtight, aluminum sealed bags. The NCGRP conducts routine germination tests and when results are low, new seed is requested from the NCGC. Because germination protocols for divergent germplasm may be suboptimal, viability test are also conducted, and if these tests fail the accession maybe inactivated if the NCGC cannot regenerate the accession. Only accessions with at least 1000 readily germinating seed are made available for distribution. Rare exceptions are made if a fresh harvest is imminent or a cooperator agrees to return pure selfed seed to the NCGC.

Distribution

Demand for seed of the NCGC is high and packaging seed for users is a regular activity, often taking up substantial time before planting activities in the spring. Typical distribution for a year is over 2000 accessions sent to meet over 100 orders, with most demand originating in the United States. Users vary widely from researchers to teachers and even to community gardeners. Typically, a few large orders come from regular users such as cotton scientists and breeders associated with federal and state institutions. The distribution policy is 25 seeds to bona fide users of cotton germplasm and they must arrange for their own seed increase and not request seeds simply for resale. Difficulties are encountered with foreign requests as the users typically need import permits or they must provide proof that their nation does not require one. Import permits require that APHIS supply a phytosanitary certificate and APHIS must receive two copies of this import permit, with at least one translated into English. When additional declarations (or stipulations related to disease or insect presence) are added to the permit, the requestor must arrange for a separate lab analysis to declare the seeds as free from specified pests, or have seed sent to their quarantine facility, or have their officials remove such declarations from the terms of the permit. Users are encouraged to submit feedback on the germination and uniformity of the received seed as assistance to the quality control of the NCGC. If time permits, users are encouraged to submit additional observations in the formats used by the NCGC (Table 3) and, once published, their positive and negative data resulting from their research. Both results can assist future users in avoiding costly and duplicate research, as well as limit the volume of requests from the NCGC. In the past few years, users of the collection wishing to do molecular genetic investigations have on occasion been given access to tissue from plants growing in NCGC greenhouses. This practice has preserved seed of rare, hard to reproduce species and saved users time and effort invested in germinating and growing plants.

Acquisition

Much of the recent growth of the NCGC has come from public germplasm registered with the Journal of Plant Registration (formerly Crop Science Registration) or from cultivars with expired Plant Variety Protection that have been received by NCGC from the NCGRP. Deposition of materials into NCGRP follows guidelines and protocols that include deposition of 5000 seed for cultivars, breeding lines, and elite germplasm, with lesser amounts required for RIL populations, genetic stocks, landrace/crop wild relatives, and species (http://www.ars.usda.gov/Main/docs.htm?docid=23333, accessed 6 Apr. 2015). Registered materials must be accompanied by a detailed description of the material being deposited. An alternate route for germplasm acquisition is for collection exchanges or donations to occur directly with the NCGC, in which case the curator then secures a duplicate seed increase for backup at the NCGRP and ensures that PI numbers are assigned by DBMU. Traditional means of germplasm acquisition through germplasm exchanges between collections and exploration trips have been curtailed or inhibited in recent years by proprietary issues regarding derivatives of the germplasm and by international treaties. However, these two means of germplasm acquisition have provided the bulk of diversity of the NCGC. Germplasm exchange and exploration efforts continue through bilateral agreements between countries in which both countries benefit.

Table 3. Standardized descriptors and rating scale used for the National Cotton Germplasm Collection.

Character	Rating scale
Boll color	green (1), dark green (2), light green (3), red (4), sun red (5), seg/off type (9)
Boll pitting	smooth (1), lightly pitted (2), pitted (3), very pitted (4), seg/off type (9)
Boll glanding	glandless (0), medium (1), light (2), heavy (3), seg/off type (9)
Boll point	moderately pointed (1), pointed (2), blunt (3), seg/off type (9)
Boll shape	oval (1), round (2), cone (3), cone-oval (4), seg/off type (9)
Boll size	medium (1), large (2), small (3), extra small (4), seg/off type (9)
Locule number	1 (1), 2 (2), 3 (3), 4 (4), 5 (5), >5 (6), seg/off type (9)
Bract color	green (1), red (2), sun red (3), seg/off type (9)
Bract teeth number	medium (1), many (2), few (3), seg/off type (9)
Bract teeth size	none (0), medium (1), small (2), large (3), seg/off type (9)
Bract type	normal (1), frego (2), flared (3), recurved (4), seg/off type (9)
Lint color	white (1), cream (2), brown (3), green (4), tan (5), rust (6), off white (7), seg/off type (9)
Boll nectaries	absent (0), present (1), reduced (2), inactive (3), seg/off type (9)
Petal color	cream (1), yellow (2), light yellow (3), red (4), white (5), light blue (6), golden (7), pink (8), seg/off type (9)
Petal spot	none (0), light (1), medium (2), heavy (3), seg/off type (9)
Pollen color	yellow (1), cream (2), dark yellow (3), orange (4), seg/off type (9)
Stigma	normal (1), protruding (2), extreme (3), short buried (4), seg/off type (9)
Leaf color	green (1), red (2), dark red (3), seg/off type (9)
Leaf shape	normal (1), okra (2), sub okra (3), super okra (4), laciniate (5), ovate (6), cordate (7), seg/off type (9)
Leaf size	medium (1), small (2), large (3), extra small (4), seg/off type (9)
Leaf glands	glandless (0), medium (1), light (2), heavy (3), seg/of type (4)
Leaf hair	none (1), few (2), moderate (3), hairy (4), very hairy (5), pilose (6), seg/off type (9)
Leaf nectaries	absent (0), present is one of the following: one, main vein (1-1), two (1-2), three (1-3), four (1-4), five (1-5), reduced (2), seg/off type (9)
Seed fuzz	none (0), medium (1), high (2), sparse (3), tufted (4), seg/off type (9)
Seed fuzz color	lintless (0), white (1), cream (2), brown (3), green (4), tan (5), rust (6), off white (7), seg/off type (9)
Seed type	free (1), semi kidney (2), kidney (3), seg/off type (9)
Stem color	green (1), sun red (2), red (3), seg/off type (9)
Stem glands	glandless (0), medium (1), light (2), heavy (3), seg/of type (9)
Stem hair	none (1), few (2), moderate (3), hairy (4), very hairy (5), pilose (6), seg/off type (9)
Fruiting type	normal (1), cluster (2), seg/off type (9)
Growth habit	normal (1), spreading (2), prostrate (3), pyramid (4), stovepipe (5), seg/off type (9)
Leaf canopy	typical (1), open (2), dense (3), compact (4), seg/off type (9)
Height in meters	actual measurement
Photoperiodic rating	no squares (0), squares (1), flowers (2), bolls (3), open bolls (4), seg/off type (9)
Maturity	not flowering (1), flowering (2), less than 1/2 open (3), more than 1/2 open (4), complete (5), seg/off type (9)
Productivity	none (0),<12/plant (1), 13–24/plant (2), 25–36/plant (3), 37–48/plant (4), >48/plant (5), seg/off type (9)

Exploration for Germplasm

The species' ranges of *Gossypium* covers tropical areas of five continents, of which significant portions remain to be explored for cotton. Successful collection efforts of the past have depended on expert knowledge of taxonomy, geography, ethnobotany, agricultural history, consultation with locals, etc. (Collins and Doyle, 1911). USDA collecting expeditions began in the 1800s, and records of collecting trips from the early 1900s to the 2012 have been documented in several publications

(USDA-ARS, 1974; Percival, 1987; Percival and Kohel, 1990; Percival et al., 1999; Percy et al., 2014). Many of the US collecting efforts have been conducted in cooperation with other national institutions and home country efforts. The French organization CIRAD has been a particularly productive partner in collecting efforts. Table 4 shows a few of the more notable collecting trips, collectors, and their contributions to the NCGC. It also reveals weaknesses in the NCGC collecting efforts. A cursory glance at Table 4 shows a significant gap in collecting the diploid species of East Africa and the Arabian Peninsula. Unfortunately, these areas have demonstrated political volatility in recent years that precludes collecting efforts. Trip reports, particularly of earlier collecting efforts, have rarely been formally published and are difficult to find. As a result, details of these explorations are rare (Schwendiman et al., 1986) and few collecting expeditions have modern GPS coordinates for collection sites (Ulloa et al., 2006). The involvement of FAO has been critical to collection of landraces and species and their conservation ex situ, because of the threat of habitat loss or landrace abandonment due to modern cotton cultivation.

Successful trips of exploration have become increasingly more problematic because of species habitat loss, replacement of local germplasm with modern cultivars, difficulty in identifying previous collecting sites from limited information of prior exploration trips, and complications and ambiguities created by recent international treaty agreements (Percy et al., 2014). Habitat loss due to conversion of wild lands to agricultural use is probably the greatest challenge and is most problematic in developing nations (Rubenstein et al., 2005; Holubec, 2004). A native diploid *G. trilobum* (DC.) Skovst. is considered endangered because of its co-habitation with areas exploited for Guava production in Mexico (Ulloa et al., 2006). In addition, *G. armourianum* Kearney, *G. klotzschianum* Andersson, *G. darwinii* G. Watt, *G. tomentosum* Nutt. ex Seem., and *G. capitis-viridis* Mauer are known only from islands, and continued development threatens their existence (Fryxell, 1992). Percival et al. (1999) and later Ulloa et al. (2006) reported the loss of Acala cotton germplasm from its native habitat in the state of Chiapas, Mexico. Historically, wild or feral cotton populations have been the targets of eradication programs in some locations, where they were perceived to be primary sources of insect and disease pests that attack adjoining cultivated cotton. An example of eradication efforts occurred in Arizona in the 1930s when a program was instituted against *G. thurberi* Tod. in a misguided effort to control the cotton boll weevil (Kearney and Peebles, 1942). A similar eradication process occurred in Florida in the 1930s to eliminate and wild *G. hirsutum* species as hosts for pink bollworm and boll weevil (Cook, 1935; Nelson, 1996). Unfortunately eradication programs still occur in some countries today. Losses of local landrace or dooryard cottons have occurred as subsistence or basic economies have disappeared and modern agricultural economies have taken their place with improved cultural practices and improved cultivars (Singh et al., 2003). The challenges of negotiating successful in situ explorations in foreign nations, and decreasing probabilities of success on the ground, have led to increasing emphasis on negotiated germplasm exchanges or solicitation of germplasm from other collections as a means to expand the NCGC (Percival and Kohel, 1990). Since 1985, 10 acquisition trips have been made by US scientists for in situ explorations, 2 of which were for collection-to-collection exchanges with India, China, Russia, and Uzbekistan (Wallace et al., 2009). Germplasm exchanges also provide the additional advantage of

Table 4. Key exploration trips and germplasm exchanges for the National Cotton Germplasm Collection, showing dates, when known, number of accessions received, number of species, and countries represented.†

Explorers	Dates	Number of accessions	Number of species	Number of countries
T.R. Richmond	1946	22	1	1
Richmond and Manning	1946	96	1	2
S.G. Stephens	1947–1972	523	4	48
Manning and Ware	1948	293	2	3
W. Koelz	1949	28	1	2
W.H. Cross	1965, 1966, 1982	27	2	2
M.J. Lukefahr	1970–1983	118	2	12
James Vreeland	1978	34	1	1
Simpson and Vreeland	1983	56	2	1
Jose Giles	unk	36	1	2
C.M. Rick	1971	14	1	2
Winters and Clark	1976, 1979	7	1	6
IRCT-UN-FAO (Ano, Schwendiman)	1982–1986	621	8	19
A.E. Percival	1983–1988	195	3	9
Percival and Stewart	1984–1991	186	7	2
Percival and Wilson	1985	59	3	1
P.A. Fryxell	1971–1999	73	3	5
J.M. Stewart	1979–1988	40	13	2
Ulloa and Stewart	2003	86	7	1
Stewart and Bertoni	2007	3	2	1
India (NBPGR, CICR)	various	1415	3	4
Russia (VIR)	various	867	7	21
Uzbekistan	unk	552	4	unk
People's Republic of China	1985, 1987	113	5	1
Brazil (Embrapa, Cenergen)	1982	20	2	1
Peru (Instituto Peruano Algodon [Peruvian Cotton Institute])	unk	6	1	1

† 141 more donors, trips, or institutions giving another 418. Various: multiple exchanges, donations that occurred over time, and total accessions received are reported; unk: status of cultivation or wild collected needs to be verified.

allowing inter-collection redundancy and therefore backup of valuable germplasm resources.

While the NCGC continues to pursue germplasm acquisition through collecting or exchange with other nations by means of mutually beneficial bilateral agreements, opportunities exist for collecting within US territories with beneficial results. Most recently, an exploration trip to the US territory of Puerto Rico was conducted in 2013. Available literature (Grover et al., 2012; Schwendiman et al., 1986), accessions in the NCGC, and accounts of collaborators indicated that wild cotton was plentiful in Puerto Rico and an exploration trip would succeed. Additional inducements were the territorial status of Puerto Rico and therefore the lack of impediments to collecting and the presence of ARS collaborations in Puerto Rico. Further, a new tetraploid species (*G. ekmanianum* Wittm.) has been proposed from the neighboring country of the Dominican Republic on the basis of herbarium records (Krapovickas and Seijo, 2008). It was thought that similar genetic diversity might reside in Puerto Rico and should be documented and collected. From 8 to 17 February in 2013, James Frelichowski and Louis Prom explored Puerto Rico for wild cotton. They tracked their routes with GPS (Montana 650T, Garmin) and tagged digital images of cotton with GPS coordinates (EOS 5D, Canon). Frelichowski and Prom traveled over 900 mi by vehicle, always near the coast or short distances inland. Assistance in exploration and location were provided by the Tropical Agricultural Research Station in Mayaguez, PR (TARS), notably David Jenkins, Veronica Rodriguez, and Brian Irish, and particularly with remote roads in the southwest portion of Puerto Rico. Only two routes diverged from the planned pattern into the interior (Highway 10 from Arecibo to Utuado and roads 31, 30, and 181 at the base of El Yunque) at much higher elevations. Cotton was not sighted on either deviation. Most of the wild cotton was associated with vegetation typical of seashore areas or a short distance inland. Commercial cotton plantings were not found in the surrounding areas, evidence that the cottons were naturally occurring and either wild or feral from dooryard plantings. In several communities, cotton trees were established as ornamentals and used as a natural, chemical-free source of fiber. In one instance, a red leaved *G. barbadense* was seen in a courtyard but was not accessible. This particular plant was seen nowhere else and assumed to be a clear case of an ornamental, and perhaps an imported, cotton cultivar.

In total, 29 cotton plants or populations were photographed and tagged with a fitted GPS unit (both Canon brand) to give an excellent survey of Puerto Rico. Images were taken of whole plant, leaf, unopened and open boll, and raw cottonseed, for each discovered plant or population. Seed was collected from 74 plants and was left with TARS collaborators (Timothy Porch and Richard Goenaga) for proper fumigation according to APHIS protocols before being shipped to the NCGC in College Station, TX. A quick survey of the island was possible by driving roads close to the coast, but it is expected that more sites could be found with a saturated survey of hotspots, off-road trails, and further questioning of locals on wild or dooryard cotton. Open, dry, disturbed areas away from cattle were better prospects for finding wild cotton and perhaps serve as a guide to narrow the search in other areas of Puerto Rico or other tropical regions. The preponderance of cotton, particularly with coarser fiber, in salt flat areas of the south and southwest tip of Puerto Rico was noteworthy. The southern portion of Puerto Rico was much more populated with cotton than the north, but morphological diversity (e.g., leaf and stem pubescence, seed fuzz, and general fiber quality) was seen throughout and justified the wide search for wild cotton. Future surveys should include the southeast coast of Puerto Rico, and should saturate the SW coast, because of the presence there of plants there with unimproved primitive characters such as short coarse fiber and a preference for harsh salt flat habitats.

Wild cotton in Puerto Rico exhibited the paradox of being dependent on, and at the same time at risk from, activities that disturb the environment, as observed by Francis (2004). Many individual and colony sites were in fence rows, vacant lots, or brush piles, areas subject to frequent burns. The tidal flats of the southwest corner of Puerto Rico are replete with wild cotton, but typically at the very margins of drier soil. Lack of competing and shading vegetation appears to be one requirement for wild cotton. Abundance of cotton appeared to be related to the drier regions of Puerto Rico, when comparing collection sites with Puerto Rico precipitation charts. These observations gave clues to finding wild and feral cotton in Puerto Rico and may extend to other regions for increasing the chances of success. Discovery of dooryard cotton was by chance, and greater success will depend on consultation with locals. Digital images and GPS tracking will help with in situ characterization of wild cotton in Puerto Rico. Further characterization with full range of morphological descriptors and other tools will be accomplished ex situ as part of routine operations of the NCGC and help put into perspective the contribution of the new germplasm.

Characterization

For the variability of the NCGC to be useful in scientific and improvement efforts, it must be adequately described and evaluated. Characterization also serves the essential purpose of providing a phenotypic baseline to monitor purity and uniformity of accessions being maintained in collection. Continuing characterization over time also allows the NCGC to track environmental influences on trait expression, particularly with agronomic characters. The CCGC (Wallace et al., 2009) summarized the need to expand the NCGC databases with morphological, agronomic, molecular characterizations, and priority trait evaluations to improve user utility of the NCGC.

Characterization of the collection with standardized descriptor sets has been practiced for over 45 yr. Currently the NCGC utilizes a standard set of descriptors that is strongly aligned with previous rating systems described by Percival (1987) and the IBPGR (1980) and primarily based on general taxonomic morphology of *G. hirsutum*. In response to requests from users of the collection, descriptors of a more agronomic nature have been added. The set of descriptor traits has been extended to include trait categories that are found in *G. barbadense* and the two Asiatic diploid cotton species (e.g., deep petal spot, cream lint color, kidney seed, etc.), and efforts are being made to include descriptor categories useful in describing accessions of other species of the genus. The descriptors and their rating categories and scales have been built into an Excel file that can receive data from handheld electronic tablets. The portability of the file is greatly enhanced with field scoring on the electronic tablets, followed by instant uploading into local databases. Scoring the collection with a full range of descriptors is a complicated process that is lengthened by germplasm lines with varying maturation rates, being grown in a tropical counter-season nursery. A number of strategies have been adopted to counter this difficulty, including longer visits to tropical nurseries, repeated plantings in College Station for a full season of observation, help from cooperators, and adding phenotype neutral (i.e., molecular) tools. Probably the most significant tool in these efforts, however, has been the creation of a permanent digital image library of accessions in the collection.

Digital imaging using standardized images and formats has proven to be a very useful and permanent tool in characterizing germplasm collections. Mackay and Alercia (2011) have described the use of digital imaging to document variation in germplasm, and this approach has been used in several collections (Puzone and Hazekamp, 1998; Mincione et al., 1994). Digital imaging can be tailored with software to detect minute differences among accessions and to determine duplicates (Smykalova et al., 2012). Digital photography has been used previously to document *Gossypium* diversity (Dever and Sheehan, 2011; Sheehan et al., 2010), but not utilizing systematic imaging or on a large scale. In the NCGC, digital images have been made of several descriptor traits simultaneously, using fully expanded leaves (top and bottom views), open flowers (side and cross-section views), non-cleft but mature bolls (side and basal views), and stem apexes (with leaves removed). Photographs were taken with several Canon 5D MII cameras, using either a 100-mm or 180-mm macro lens mounted on a tripod in field, or with camera stands set indoors in College Station. Aperture priority was used at f14 with both lenses. Consistent control of lighting (shading when outdoors), consistent size reference (with metered grid backgrounds and fixed focusing distances), and improved manual focusing to allow zooming in to characters like hairs, nectar, and gossypol glands were practiced. Images were annotated with descriptor ratings and organized and formatted in Adobe Photoshop. Imaging equipment has been taken to the CWN and images made to directly pair with the recorded descriptors taken in the nursery. When possible, the oldest lots of seed in the NCGC have been planted to retroactively establish standard descriptor and digital image baselines for the accessions, typically with field plantings in College Station, TX or greenhouses. To date, digital imaging has been acquired on 5320 accessions, and these images have been uploaded and formatted for viewing on the CottonGen database. The high resolution of the current digital cameras has enabled the creation of "virtual" voucher specimens and a virtual herbarium for preservation of the morphological variation in the genus.

High-resolution digital images allow for future collection of data and assessment with metrics of fine scale such as variation in trichome ("hairs") structure and density or gland density (leaf hairs/mm^2 area, glands/mm^2, etc.).

The consolidation of several collections to create the NCGC created difficulties in uniform characterization of the collection that have not been satisfactorily resolved even to this day. The subcollections each had their own methods and descriptor rating systems that have proven difficult to merge and rationalize (Percival, 1987). Further, instrumentation to measure certain traits, particularly fiber traits, have changed over the years. Often the measurements these instruments make are not directly relatable. Fiber strength as measured by stelometer and HVI is an example. The result is that for some traits of the collection there is a multiplicity of ways of measuring the trait and multiple descriptors employed to describe the same trait. These present a problem to the user of the collection for scientific studies or improvement efforts. Expert users of the NCGC often know the pedigree information (Bowman et al., 2006; Smith et al., 1999) and general phenotypic information of elite germplasm or obsolete cultivars, but are totally unfamiliar with foreign collected or exchanged material. This situation requires that standard descriptors be collected de novo on all materials so that direct comparisons can be made. Another difficulty is that in earlier efforts to describe the diversity of the collection, pseudo-taxonomic designations were assigned to portions of the collection, often on the basis of geographic origin and a few simply inherited traits. Such is the case of seven geographic races of *G. hirsutum* designated as latifolium, marie-galante, morrilli, palmeri, punctatum, richmondii, and yucatenense (Hutchinson, 1951). A similar division was described for *G. arboreum* (Watt, 1907; Kulkarni et al., 2009), and while these designations relate well with geographical diversity, they need to be aligned with the morphological and agronomic descriptors. Without consistent reference to morphological characters and clear photographic evidence, it becomes difficult to evenly apply landrace designations to portions of the NCGC. Trait characterizations of accessions can blur when accessions are hybrids among landraces and dooryard cotton, or products of introgressions across species. Capture of hard to describe variation with imaging and addition of phenotype neutral (e.g., molecular) tools can improve characterization.

Evaluations

Characterizations of the collection for simply inherited and expressed traits have yielded significant benefits in cotton improvement. Identification and breeding of materials with high pubescence by cotton breeders in Africa conferred a level of resistance to the jassid (Knight, 1952; Ridgeway, 1984). The okra leaf ($L^o{}_2$) trait has been demonstrated to enhance earliness, decrease boll rot disease (Andries et al., 1969), enhance pesticide spray coverage (James and Jones, 1985), and reduce harvest trash content (Novick et al., 1990). The glandless character has been investigated as a potentially useful trait in developing edible cottonseed for niche markets, or as universal livestock feed (McMichael, 1959; Kohel, 1979). However, numerous traits of agronomic interest in cotton are often quantitatively inherited, or exhibit complex expression, or are significantly affected by environmental interactions. Such traits go beyond simple characterization to more complicated evaluation of germplasm, often involving multiple locations and environments.

Numerous evaluations have been conducted on the NCGC for phenotypic characteristics of agronomic importance. Portions of the collection have been evaluated for seed protein, seed oil, and seed gossypol content (Kohel et al., 1985; Kohel, 1978; Hinze et al., 2015; Dilday and Shaver, 1976; Percy et al., 1996), boll weevil resistance (Jenkins, 1978), Cercospora leaf spot and Verticillium wilt (Jenkins and Parrott, 1978), root knot nematodes (Shepherd, 1983), etc.

Numerous traits presently available in commercial tetraploid cotton have their origins in the diverse species of the genus (Percy and Kohel, 1999). Resistance to bacterial blight has been found in *G. arboreum*, *G. anomalum* Wawra, *G. herbaceum*, and *G. barbadense*, as well as in *G. hirsutum* (Percy and Kohel, 1999), and resistance to Fusarium wilt has been observed in wild Australian species (McFadden et al., 2004). Resistance to various nematode species has been transferred to *G. hirsutum* from *G. barbadense* and *G. longicalyx* J.B. Hutch. & B.J.S. Lee (Robinson et al., 2004, 2005; Dighe et al., 2009). Drought and salt tolerance have been identified in non-cultivated species (Quisenberry et al., 1981; Ashraf, 2002). Considerable variation in the content and chemical form of gossypol has been identified in the seed of wild cotton species, and a delayed gossypol gland development trait has been introgressed from the diploid species *G. bickii* Prokh. to *G. hirsutum* (Stipanovic et al., 2005; Zhu et al., 2005). Resistance to the cotton leaf curl virus has been demonstrated in the diploid species *G. arboreum* and *G. herbaceum*, and efforts to transfer resistance to *G. hirsutum* are ongoing (Amjad et al., 2011). The above examples are cited to demonstrate the utility of a diverse germplasm collection, and the need to maintain and distribute its contents to improvement efforts.

More recently, a range of evaluations of germplasm for disease, and pest resistance, drought, heat, and salt tolerance, and numerous evaluations for general agronomic performance and fiber quality were reported in a status report of the NCGC by the CCGC (Wallace et al., 2009). A deeper look at this report indicates that the evaluations are ad hoc in nature, varying in accession number, often overlapping in the inclusion of germplasm, utilize non-standardized procedures, and often only report "positive" results. Coordinated priority evaluations of the NCGC are essential in saving breeders time and resources by identifying key genetic sources for cotton breeding and avoiding duplicative requests and research. Summarizing and displaying all results of evaluation efforts (negative as well as positive) for future use by researchers in pre-screening the NCGC is not just a challenge for the NCGC curator and the DBMU. The challenge extends to individual researchers to make their research results available to the collection so that the utility of the collection can be maximized. Future use of the NCGC will be streamlined with a database incorporating morphological, agronomic, and molecular marker characterization and references to evaluation results and when possible use of a diversity core set for multiple evaluations and characterizations (Hinze et al., 2014).

Recent advances in cotton molecular genetics have provided the molecular markers appropriate for phylogenetic investigations and for measuring diversity in various portions of the *Gossypium* genus (Lacape et al., 2007; Frelichowski et al., 2006; Park et al., 2005; Yu et al., 2012; Hinze et al., 2012; Fang et al., 2013). However, many of these investigations have used ad hoc marker sets and small or taxonomically limited accession sets. In recent efforts to define the diversity of the NCGC, a representative accession subset composed of 272 diploid and 1984 tetraploid accessions, spanning all available species, geographical regions, developmental status, and development time periods, was created. To characterize this collection subset, designated the *Gossypium* Diversity Reference Set (GDRS), a core marker set was created using the criteria of uniform coverage of the tetraploid genome with two markers per chromosome arm (total of 104), representation of major marker collections, and demonstrated polymorphism of the markers on a multi-species panel (Yu et al., 2012). Aside from revealing the structure of diversity within the genus and identifying sources of variability within the collection, the characterization effort will provide information integral to the management of the collection. Molecular characterization of the NCGC will assist the curator by revealing accession redundancy, species misclassifications, and introgression within accessions, and will be a powerful tool in monitoring the integrity of accessions. Instances of long suspected species misidentifications and introgressions have been confirmed by analyses of species specific alleles and PCA analyses (Hinze et al., 2014). These studies also confirm that the genetic diversity present in wild or landrace accessions of cultivated species or in the wild diploid species of *Gossypium* greatly exceeds that encompassed with the cultivated germplasm. This observation also mirrors the estimate that less than 1% of the genetic variability of the *G. hirsutum* species currently resides in improved US commercial cotton cultivars (Esbroeck and Bowman, 1998). This genomic approach gives a platform to investigate further within species variation (Hinze et al., 2012; Fang et al., 2013; Kuester and Nason, 2012; Kantartzi et al., 2009; Alvarez and Wendel, 2006). These robust molecular tools can give an in-depth diagnosis of genomic diversity and gaps in the NCGC and other collections. Users will benefit from seeing more detail on introgressions and uniqueness of accessions in marker profiles (Tanksley and McCouch, 1997; Kohel and Yu, 2001) and possible relationship with phenotypic characters (Kalivas et al., 2011; Abdurakhmonov et al., 2008; Hoffman et al., 2006).

The expanded, standardized descriptor list and companion images will greatly improve the visibility, understanding, and utility of the collection to the users. Images and genetic sequence based markers also may overcome characterization barriers between collections due to differences in methodology, thereby enhancing the ability to make cross-collection comparisons and improve exchange efforts (Abdullaev et al., 2013; Bhale and Narayanan, 1988). Data and images collected by the NCGC are being uploaded to www.cottongen.org, which has expanded server capabilities to handle high-resolution images. This site already supports extensive genetic and genomic data for cotton (Yu et al., 2014).

Challenges and Improvements

The NCGC and other collections in ARS and worldwide are facing rising costs of operations with plateaued or reduced budgets. The distribution of germplasm and worldwide demands for germplasm and associated information continues to rise. Technological investments and greater efficiency in operations have been essential to keep up with demand without incurring additional time and labor costs. A variety of collaborator support has been essential to help the NCGC meet some of its goals. Cotton Incorporated has provided timely funds for travel and camera equipment. Collaborators in a joint project with ARS in Stoneville, MS assisted in the seed increase, descriptor scoring at the nursery, and screening of a portion of the NCGC for Cotton Leaf Curl Virus resistance, currently recognized as the biggest threat to cotton in Asia (Briddon and Markham, 2000). Electronic field tablets and digital barcoding and seed weighing systems were also upgraded with funding, and expedited processing and inventorying of NCGC seed.

Uniform seed increases of the entire NCGC face serious obstacles. Unexpected was the recent cessation of operations at the Tecoman CWN, and a new location is being sought to reestablish a counter season nursery. In the meantime, limited numbers of non-photoperiodic accessions that do not require a tropical nursery to reproduce will be increased in the field at College Station, TX. Greenhouse space available for the increase of photoperiodic germplasm is limited, and seed production is inferior to that obtained from a nursery. Greenhouse space is best utilized for seed production in rare species that may require several years to reproduce. Currently, more seed needs to be generated of critical accessions and rare species so they do not remain underutilized. A variety of measures may collectively help to reduce this workload. Extension of seed storage times and maximizing seed viability and germination will help prioritize each planting and seed increase. Germination protocols are being tested across species to provide best recommendation and chances of success to the users. Collaboration with scientists at the NCGRP is underway to better predict the longevity of *Gossypium* seed in storage. Other curators will be contacted in reference to methods of cultivating difficulty species.

Large portions of the NCGC are underutilized because of the lack of adequate characterization or the scarcity of seed. Often the former circumstance arises from the latter circumstance. Diploid species and hard to reproduce tetraploid landrace materials have not been adequately characterized or evaluated by users of the collection because of scarce seed quantities and the NCGC's inability to distribute seed. The NCGC needs to regularly attempt germination, then characterize the exotic, albeit critical germplasm, before they are permanently lost and if not amenable to current germination strategies, be clearly marked as unavailable, and placed into long-term storage. This will conserve the seed for future use and better define the extent of collection gaps and identify unique germplasm in the NCGC. A second step involves active investigations on the germination, growth, and reproductive requirements of rare accessions when seed allows. If seed availability is such that cultural investigations of rare accessions cannot occur, then planning should proceed for reacquisition of these materials. Finally, the NCGC needs to encourage more long-term breeding programs, or what is now being called "pre-breeding," aimed at the introgression of traits from exotic germplasm (Culp and Harrell, 1974; Fryxell, 1976) because they are the essential link to unlocking genetic potential in the NCGC. Academic interaction with the NCGC is essential to introduce aspiring plant breeders to the wealth of genetic resources (Kothari et al., 2011).

Campbell et al. (2010) already showed that all major collections have gaps in a number of wild diploid *Gossypium* species, most notably the most recently described species from East Africa and the Middle East. The abundance of cotton in just the narrow coastal areas of Puerto Rico alone, suggest that, with extrapolation to tropical regions worldwide, a wide range of *Gossypium* resources still exist to be explored and documented. Within species characterization is severely undermined with low ex situ conservation and limited in situ documentation (Kuester, 2012; de Almeida et al., 2009; Rudgers et al., 2004). The NCGC, like all collections, cannot afford to become oversized and must constantly strive to be streamlined and cost effective while maximizing utility and relevance for the users. Further characterization and database development of the NCGC will help highlight uniqueness of each accession and more clearly show gaps and simplify exchanges and explorations when such opportunities arise. However, characterization of diversity within rare species remains challenging with such poor representation in collections. Diverse germplasm in each collection and access to production areas worldwide is essential to finding resistances to major diseases, pests, and abiotic stresses that limit cotton production. Recent worldwide recession and consequent governmental budgets restrictions have impacted the maintenance of numerous collections. The decline or loss of any collection is an unfavorable outcome to the cotton community and may result in permanent loss of valuable germplasm. The efforts of the collection curators participating in the International Cotton Germplasm Initiative (ICGI) (Campbell et al., 2010) have demonstrated that coordination and collaboration among the major cotton collections and associated scientists worldwide was possible to document the worldwide status of the cotton collections. Hopefully this cooperation can be extended to cooperation in exchange and preservation.

References

Abdullaev, A., A.A. Abdullaev, I. Salakhutdinov, S. Rizaeva, Z. Kuryazov, D. Ernazarova, and I.Y. Adburakhmonov. 2013. Cotton germplasm collection of Uzbekistan. Asian Aust. J. Plant Sci. Biotechnol. 7(Special Issue 2):1–15.

Abdurakhmonov, I.Y., R.J. Kohel, J.Z. Yu, A.E. Pepper, A. Abdullaev, F.N. Kushanov, I.B. Salakhutdinov, Z.T. Buriev, S. Saha, B. Scheffler, J.N. Jenkins, and A. Abdukarimov. 2008. Molecular diversity and association mapping of fiber quality traits in exotic *G. hirsutum* L. germplasm. Genomics 92:478–487. doi:10.1016/j.ygeno.2008.07.013

Alvarez, I., and J.F. Wendel. 2006. Cryptic interspecific introgression and genetic differentiation within *Gossypium aridum* (Malvaceae) and its relatives. Evolution 60:505–517. doi:10.1111/j.0014-3820.2006.tb01132.x

Amjad, F., F. Jehanzeb, M. Abid, S. Amir, R. Abdul, B. Asia, R. Muhammad, T. Muhammad, S. Hussain, and M. Saira. 2011. An overview of cotton leaf curl virus disease (CLCuD) a serious threat to cotton productivity. Aust. J. Crop Sci. (13):1823–1831.

Andries, J.A., J.E. Jones, L.W. Sloane, and J.G. Marshall. 1969. Effects of okra leaf shape on boll rot, yield, and other important characters of Upland cotton, *Gossypium hirsutum* L. Crop Sci. 9:705–710. doi:10.2135/cropsci1969.0011183X000900060009x

Arora, R.K. 1991. Plant exploration and germplasm collection. In: R.S. Paroda, editor, Plant genetic resources: Conservation and management. Concepts and approaches. IBPGR, New Delhi, India. p. 55–93.

Ashraf, M. 2002. Salt tolerance of cotton: Some new advances. Crit. Rev. Plant Sci. 21(1):1–30. doi:10.1016/S0735-2689(02)80036-3

Baranski, M. 2013. Seed collection and plant genetic diversity, 1900–1979. Embryo Project Encyclopedia, Arizona State University. http://embryo.asu.edu/handle/10776/6468 (accessed 4 April 2015).

Bettencourt, E., and J. Konopka. 1988. Directory of germplasm collections. Bind 5. II: Industrial crops: Beet, coffee, oil palm, cotton and rubber. IBPGR, Rome.

Bhale, N.L., and S.S. Narayanan. 1988. Need for augmenting genetic resources in cotton—Indian perspective. In: R.S. Paroda, R.K. Arora, and K.P.S. Chandel, editors, Plant genetic resources—Indian perspective. National Symposium on Plant Genetic Resources, New Delhi, India. 3–6 Mar. 1987. NBPGR, New Delhi. p. 220–231.

Bird, J.B., and J. Mahler. 1952. America's oldest cotton fabrics. Am. Fabr. 20:73–78.

Board on Agriculture National Research Council. 1991. Managing global genetic resources. The U.S. National Plant Germplasm System. Natl. Acad. Press, Washington, DC.

Board on Agriculture National Research Council. 1993. Managing global genetic resources. Agricultural crop issues and policies. Natl. Acad. Press, Washington, DC.

Bowman, D.T., O.A. Gutierrez, R.G. Percy, D.S. Calhoun, and O.L. May. 2006. Pedigrees of upland and Pima cotton cultivars released between 1970 and 2005. Bull. 1155. Mississippi Agric. and For. Exp. Stn., Mississippi State, MS.

Briddon, R.W., and P.G. Markham. 2000. Cotton leaf curl virus disease. Virus Res. 71:151–159. doi:10.1016/S0168-1702(00)00195-7

Campbell, B.T., S. Saha, R. Percy, J. Frelichowski, J.N. Jenkins, W. Park, C.D. Mayee, V. Gotmare, D. Dessauw, M. Giband, X. Du, Y. Jia, G. Constable, S. Dillon, I.Y. Abdurakhmonov, A. Abdukarimov, S.M. Rizaeva, A. Abdullaev, P.A.V. Barroso, J.G. Padua, L.V. Hoffmann, and L. Podolnaya. 2010. Status of the global cotton germplasm resources. Crop Sci. 50:1161–1179. doi:10.2135/cropsci2009.09.0551

Christiansen, M.N., and R.A. Rowland. 1986. Germination and stand establishment. In: J.R. Mauney and J.McD. Stewart, editors, Cotton physiology. Chap. 34. The Cotton Foundation, Memphis, TN. p. 535–541.

Collins, G.N., and C.B. Doyle. 1911. Notes on southern Mexico. Natl. Geogr. Mag. 22:301–320.

Cook, O.F. 1935. How shall we know plants? J. Hered. 26:35–31.

Culp, T.W., and D.C. Harrell. 1974. Breeding quality cotton at the Pee Dee Experiment Station. Florence, S.C. ARS-S-30. USDA-ARS, New Orleans, LA.

Darwin, C. 1859. The origin of species. John Murray, London.

de Almeida, V.C., L.V. Hoffmann, G.K.I. Yokomizo, J.N. Costa, M. Giband, and P.A.V. Barroso. 2009. In situ and genetic characterization of *Gossypium barbadense* populations from the states of Para and Amapa, Brazil. Pesqui. Agropecu. Bras. 44(7): 719–425.

Dever, J.K., and M.A. Sheehan. 2011. Update of cotton race-stock screening and phenotypic characterization. In: Proceedings of the Beltwide Cotton Conferences, Atlanta, GA. 4–7 Jan. 2011. Natl. Cotton Counc. of Am., Memphis, TN. p. 718–719.

Dighe, N.D., A.F. Robinson, A.A. Bell, M.A. Menz, R.G. Cantrell, and D.M. Stelly. 2009. Linkage mapping of *Gossypium longicalyx* resistance to reniform nematode during introgression into cotton *Gossypium hirsutum*. Crop Sci. 49:1151–1164. doi:10.2135/cropsci2008.03.0129

Dilday, R.H., and T.N. Shaver. 1976. Survey of the regional *Gossypium hirsutum* L., primitive race collection for flowerbud gossypol. ARS-S-146. USDA-ARS, New Orleans, LA.

Dixon, K.W., S. Roche, and J.S. Pate. 1995. The promotive effect of smoke derived from burnt native vegetation on seed germination of Western Australian plants. Oecologia 101:185–192. doi:10.1007/BF00317282

Esbroeck, G.V., and D.T. Bowman. 1998. Cotton germplasm diversity and its importance to cultivar development. J. Cotton Sci. 2:121–129.

Fang, D.D., L.L. Hinze, R.G. Percy, P. Li, D. Deng, and G. Thyssen. 2013. A microsatellite-based genome-wide analysis of genetic diversity and linkage disequilibrium in Upland cotton (*Gossypium hirsutum* L.) cultivars from major cotton-growing countries. Euphytica 191:391–401. doi:10.1007/s10681-013-0886-2

Francis, J.K. 2004. Wild cotton. In: J.K. Francis, editor, Wildland shrubs of the United States and its territories: Thamnic descriptions: Volume 1. Gen. Tech. Rep. IITF-GTR-26. USDA Forest Service, International Institute of Tropical Forestry, San Juan, PR. p. 355–358. http://www.fs.fed.us/global/iitf/pdf/shrubs/Gossypium%20hirsutum.pdf (accessed 4 April 2015).

Frelichowski, J.E., Jr., M.B. Palmer, D. Main, J.P. Tomkins, R.G. Cantrell, D.M. Stelly, J. Yu, R.J. Kohel, and M. Ulloa. 2006. Cotton genome mapping with new microsatellites from Acala 'Maxxa' BAC-ends. Mol. Genet. Genomics 275(5):479–491. doi:10.1007/s00438-006-0106-z

Fryxell, P.A. 1965. Stages in the evolution of *Gossypium* L. Adv. Front. Plant Sci. 10:31–56.

Fryxell, P.A. 1969. The West Indian species of *Gossypium* of von Rohr and Rafinesque. Taxon 18:400–414. doi:10.2307/1218471

Fryxell, P.A. 1976. Germpool utilization: *Gossypium*, a case history. ARS-S-137. USDA-ARS, New Orleans, LA.

Fryxell, P.A. 1984. Taxonomy and germplasm resources. In: R.G. Kohel and C.F. Lewis, editors, Cotton. Agron. Monogr. 24. ASA, CSSA, and SSSA, Madison, WI. p. 27–57.

Fryxell, P.A. 1992. A revised taxonomic interpretation of *Gossypium* L. (Malvaceae). Rheedea 2(2):108–165.

Gray, L.C. 1933. History of agriculture in the southern United States to 1860. Vol. 2. The Carnegie Inst. of Washington, Washington, DC.

Grover, C.E., K.K. Grupp, R.J. Wanzek, and J.F. Wendel. 2012. Assessing the monophyly of polyploidy *Gossypium* species. Plant Syst. Evol. 298:1177–1183. doi:10.1007/s00606-012-0615-7

Gulati, A.M., and A.J. Turner. 1928. A note on the early history of cotton. Tech. Lab. Bull. No. 17. Indian Cent. Cotton Comm., Bombay.

Hague, S., L. Hinze, and J. Frelichowski. 2009. Cotton. In: J. Vollman and I. Rajcan, editors, Oil crops. Handb. Plant Breed. 4. Springer-Verlag, New York. p. 257–285.

Harland, S.C. 1932. The genetics of *Gossypium*. Bibliogr. Genet. 9. Martinus Nijhoff, The Hague, the Netherlands.

Hinze, L.L., J.K. Dever, and R.G. Percy. 2012. Molecular variation among and within improved cultivars in the U.S. Cotton Germplasm Collection. Crop Sci. 52:222–230. doi:10.2135/cropsci2011.04.0202

Hinze, L.L., D.D. Fang, M.A. Gore, B.E. Scheffler, J.Z. Yu, J. Frelichowski, and R.G. Percy. 2014. Molecular characterization of the *Gossypium* diversity reference set of the US National Cotton Germplasm Collection. Theor. Appl. Genet. 128:313–327.

Hinze, L.L., P.J. Horn, N. Kothari, J.K. Dever, J. Frelichowski, K.D. Chapman, and R.G. Percy. 2015. Nondestructive measurements of cottonseed nutritional trait diversity in the U.S. National Cotton Germplasm Collection. Crop Sci. 55:770–782. doi:10.2135/cropsci2014.04.0318

Hoffman, S.M., R.J. Kohel, M. Ulloa, J. Cho, Z. Xu, W. Smith, A. Pepper, R. Cantrell, and J.Z. Yu. 2006. Association analysis of morphological characters with molecular markers in a TM-1 x 3-79 RIL population grown in three diverse environments. In: Proceedings of the Beltwide Cotton Conferences, San Antonio, TX. 3–6 Jan. 2006. Natl. Cotton Counc. of Am., Memphis, TN. p. 785.

Holubec, V. 2004. Genetic erosion and extinction threat to old world cottons, *Gossypium* L.—A case study. In: B. Ford-Lloyd, S.R. Dias, and E. Bettencourt, Genetic erosion and pollution assessment methodologies: Proceedings of PGR Forum Workshop 5, Terceira Island, Autonomous Region of the Azores, Portugal, 8–11 Sept. 2004. Bioversity International, Rome. p. 65–78.

Hooker, J.D. 1875–1897. The flora of British India (7 vols.). L. Reeve and Co., London.

Hutchinson, J.B. 1951. Intra-specific differentiation in *Gossypium hirsutum*. Heredity 5:161–193. doi:10.1038/hdy.1951.19

Hutchinson, J., B.R.A. Silow, and S.G. Stephens. 1947. The evolution of *Gossypium*. Oxford Univ. Press, London.

IBPGR. 1980. Descriptors for cotton. AGP:IBPGR/80/10. IBPGR, Rome.

ITSA. 2014. International rules for seed testing. ITSA, Bassersdorf, Switzerland.

James, D., and J.E. Jones. 1985. Effects of leaf and bract isolines on spray penetration and insecticidal efficacy. In: Proceedings of the Beltwide Cotton Production Research Conferences, New Orleans, LA. 6–11 Jan. 1985. Natl. Cotton Counc. of Am., Memphis, TN. p. 395–396.

Jenkins, J.N. 1978. Evaluation of primitive races of *Gossypium hirsutum* L. for resistance to boll weevil. Tech. Bull. 91. Mississippi Agric. and For. Exp. Stn., Mississippi State, MS.

Jenkins, J.N., and W.L. Parrott. 1978. Field evaluation of primitive races of *Gossypium hirsutum* L. for resistance to Cercospora leaf spot and Verticillium wilt. Tech. Bull. 92. Mississippi Agric. and For. Exp. Stn., Mississippi State, MS.

Justice, O.L., and L.N. Bass. 1978. Principles and practices of seed storage. Agric. Handb. 506, US Gov. Print. Office, Washington, DC.

Kalivas, A., F. Xanthopoulos, O. Kehagia, and A.S. Tsaftaris. 2011. Agronomic characterization, genetic diversity and association analysis of cotton cultivars using simple sequence repeat molecular markers. Genet. Mol. Res. 10:208–217. doi:10.4238/vol10-1gmr998

Kantartzi, S.K., M. Ulloa, E. Sacks, and J.M. Stewart. 2009. Assessing genetic diversity in *Gossypium arboreum* L. cultivars using genomic and EST-derived microsatellites. Genetica (The Hague) 136:141–147.

Kearney, T.H., and R.H. Peebles. 1942. Flowering plants and ferns of Arizona. US Gov. Print. Office, Washington, DC.

Kebede, H., G. Burow, R.G. Dani, and R.D. Allen. 2007. A-genome cotton as a source of genetic variability for Upland cotton (*Gossypium hirsutum*). Genet. Resour. Crop Evol. 54:885–895. doi:10.1007/s10722-006-9157-6

Knight, R.L. 1952. The genetics of jassid resistance in cotton. I. The genes H_1 and H_2. J. Genet. 51:47–66. doi:10.1007/BF02986704

Kohel, R.J., and J. Yu. 2001. Molecular characterization of *Gossypium* germplasm for cotton improvement. In: J.M.M. Engels, V.R. Rao, A.H.D. Brown, and M.T. Jackson, editors, Managing plant genetic diversity. CABI Publishing, Oxford, UK. p. 67–76.

Kohel, R.J. 1978. Survey of *Gossypium hirsutum* L. germplasm collections for seed-oil percentage and seed characteristics. ARS-S-187. USDA-ARS, New Orleans, LA.

Kohel, R.J. 1979. Cotton. In: G. Robbelen, R.K. Downey, and A. Ashri, editors, Oil crops of the world. Their breeding and utilization. McGraw-Hill, New York. p. 404–415.

Kohel, R.J., J. Glueck, and L.W. Rooney. 1985. Comparison of cotton germplasm collections for seed-protein content. Crop Sci. 25:961–963. doi:10.2135/cropsci1985.0011183X002500060015x

Kothari, N., S.S. Hague, J. Frelichowski, R.L. Nichols, and D.C. Jones. 2011. Utilization of cotton germplasm in the winter nursery at Tecoman, Mexico for plant breeding training and research. J. Cotton Sci. 15:271–273.

Krapovickas, A., and G. Seijo. 2008. *Gossypium ekmanianum* (Malvaceae), algodon Silvestre de la Republica Dominicana. Bonplandia. 17(1):55–63.

Kuester, F. 2012. Variation in plant defense traits and population genetics within a Sonoran Desert cotton endemic, *Gossypium davidsonii* and boll weevil, *Anthonomus grandis*. Graduate thesis and dissertation. Paper 12371. Iowa State Univ., Ames.

Kuester, A.P., and J.D. Nason. 2012. Microsatellite loci for *Gossypium davidsonii* (Malvaceae) and other D-genome, Sonoran desert endemic cotton species. Am. J. Bot. 99:e91–e93. doi:10.3732/ajb.1100421

Kulkarni, V.N., B.M. Khadi, M.S. Maralappanavar, L.A. Deshapande, and S.S. Nraryanan. 2009. The worldwide gene pools of *Gossypium arboreum* L. and *G. herbaceum* L. and their improvement. In: A.H. Paterson, editor, Genetics and genomics of cotton. Springer-Verlag, New York. p. 69–97.

Lacape, J.-M., D. Dessauw, M. Rajab, J.-L. Noyer, and B. Hau. 2007. Microsatellite diversity in tetraploid *Gossypium* germplasm: Assembling a highly informative genotyping set of cotton SSRs. Mol. Breed. 19:45–58. doi:10.1007/s11032-006-9042-1

Lee, J.A. 1984. Cotton as a world crop. In: R.J. Kohel and C.F. Lewis, editors, Cotton. ASA, CSSA, and SSSA, Madison, WI. p. 1–25.

Leonard, B.R., J.B. Graves, and P.C. Ellsworth. 1999. Insect and mite pests of cotton. In: C.W. Smith and J.T. Cothren, editors, Cotton: Origin, history, technology, and production. John Wiley & Sons, New York. p. 489–551.

Lewis, C.F., and T.R. Richmond. 1968. Cotton as a crop. In: Advances in production and utilization of quality Cotton: Principles and practices. Iowa State Univ. Press, Ames. p. 1–21.

Light, M.E., B.B. Burger, and J.V. Staden. 2005. Formation of a seed germination promoter from carbohydrates and amino acids. J. Agric. Food Chem. 53:5936–5942. doi:10.1021/jf050710u

Lubbers, E.L., and P.W. Chee. 2009. The worldwide gene pool of *G. hirsutum* and its improvement. In: A.H. Paterson, editor, Genetics and genomics of cotton. Springer-Verlag, New York. p. 23–52.

Mackay, M., and A. Alercia. 2011. Processing of germplasm, associated material and data. In: L. Guarino, R.V. Ramanatha, and E. Goldberg, editors, Collecting plant genetic diversity: Technical guidelines—2011 Update. Chap 28. Bioversity Int., Rome. p. 1–4.

Mauer, F.M. 1930. The cottons of Mexico, Guatemala, and Colombia. (Russian version, with illustrations, on p. 425–464.) Bull. Appl. Bot. Genet. Plant Breed. 47(Suppl.):543–553.

McCarty, J.C., and J.N. Jenkins. 2001. Primitive cotton germplasm: Yield and fiber traits for 16 day-neutral accessions. Miss. Agric. For. Exp. Stn. Res. Rep. 22(19):1–5.

McFadden, H., D. Beasley, and C.L. Brubaker. 2004. Assessment of *Gossyium sturtianum* and *G. australe* as potential sources of fusarium wilt resistance to cotton. Euphytica 138:61–72. doi:10.1023/B:EUPH.0000047076.38747.81

McMichael, S.C. 1959. Hopi cotton, a source of cottonseed free of gossypol pigments. Agron. J. 51:630. doi:10.2134/agronj1959.00021962005100100025x

Mendel, G. 1866. Versuche über pflanzen-hybriden. Verh. Naturforsch. Ver. Brünn 4:3–47 (In English in 1901). J. R. Hortic. Soc. 26:1–32.

Meyer, V. 1973. A study of reciprocal hybrids between upland cotton (*Gossypium hirsutum* L.) and experimental lines with cytoplasms from seven other species. Crop Sci. 13:439–444. doi:10.2135/cropsci1973.0011183X001300040015x

Mincione, A., E. Filippone, and L.M. Monti. 1994. Construction of an image-based germplasm data base for *Phaseolus* spp. Euphytica 75:149–159. doi:10.1007/BF00024543

Moore, J.H. 1956. Cotton breeding in the old South. Agric. Hist. 30:95–104.

National Research Council. 1972. Genetic vulnerability of major crops. Natl. Acad. of Sci., Washington, DC.

Nelson, G. 1996. The shrubs and woody vines of Florida. Pineapple Press, Sarasota, FL.

Novick, R.G., J.E. Jones, W.S. Anthony, W. Aguillard, and J.I. Dickson. 1990. Seedcotton cleanability and non-lint trash at the gin as affected by morphological traits. In: Proceedings of the Beltwide Cotton Production Research Conferences, Las Vegas, NV. 9–14 Jan. 1990. Natl. Cotton Counc. of Am., Memphis, TN. p. 80–81.

Pandey, S. 2012. 225 years of botanic history. Sci. Rep. 49:8–13.

Park, Y.-H., M.S. Alabandy, M. Ulloa, T.A. Wilkins, J.Z. Yu, D.M. Stelly, R.J. Kohel, O.M. Elshihy, and R.G. Cantrell. 2005. Genetic and QTL mapping of new fiber EST-derived microsatellites on an interspecific recombinant inbred line (RIL) cotton population. Mol. Genet. Genomics 274:428–441. doi:10.1007/s00438-005-0037-0

Percival, A.E. 1987. The National Collection of *Gossypium* Germplasm. S. Coop. Ser. Bull. 321. Texas A&M Univ., College Station, TX.

Percival, A.E., and R.J. Kohel. 1990. Distribution, collection, and evaluation of *Gossypium*. Adv. Agron. 44:225–256. doi:10.1016/S0065-2113(08)60823-8

Percival, A.E., J.E. Wendel, and J.M. Stewart. 1999. Taxonomy and germplasm resources. In: C.W. Smith and J.T. Cothren, editors, Cotton: Origin, history, technology, and production. John Wiley & Sons, New York. p. 33–63.

Percy, R.G. 2009. The worldwide gene pool of *Gossypium barbadense* L. and its improvement. In: A.H. Paterson, editor, Genetics and genomics of cotton. Springer-Verlag, New York. p. 53–68.

Percy, R.G., M.D. Arnold, B.T. Campbell, J.T. Dever, D.D. Fang, L.L. Hinze, D. Main, J. Scheffler, M.A. Sheehan, M. Ulloa, J. Yu, and J. Yu. 2014. The U.S. National Cotton Germplasm Collection—Its contents, preservation, characterization and evaluation. In: I.Y. Abdurakhmonov, editor, World cotton germplasm resources. InTech, Rijeka, Croatia. p. 167–201.

Percy, R.G., M.C. Calhoun, and H.L. Kim. 1996. Seed gossypol within *Gossypium barbadense* L. cotton. Crop Sci. 36:193–197. doi:10.2135/cropsci1996.0011183X003600010034x

Percy, R.G., and R.J. Kohel. 1999. Qualitative genetics. In: C.W. Smith and J.T. Cothren, editors, Cotton: Origin, history, technology, and production. John Wiley & Sons, New York. p. 319–360.

Percy, R.G., and J.F. Wendel. 1990. Allozyme evidence for the origin and diversification of *Gossypium barbadense* L. Theor. Appl. Genet. 79:529–542. doi:10.1007/BF00226164

Postman, J., K. Hummer, T. Ayala-Silva, P. Bretting, T. Franko, G. Kinard, M. Bohning, G. Emberland, Q. Sinnott, M. Mackay, P. Cyr, M. Millard, C. Gardner, L. Guarino, and B. Weaver. 2010. GRIN-GLOBAL: An international project to develop a global genebank information management system. Acta Hortic. 859:49–55.

Puzone, L., and T. Hazekamp. 1998. Characterization and documentation of genetic resources utilizing multimedia databases. Proceedings of a workshop. 19–20 December 1996. University of Naples Federico II, Naples, Italy. IPGRI, Rome.

Quisenberry, J.E., W.R. Jordan, B.A. Roark, and D.W. Fryrear. 1981. Exotic cottons as genetic source for drought resistance. Crop Sci. 21:889–895. doi:10.2135/cropsci1981.0011183X002100060022x

Ridgeway, R.L. 1984. Cotton protection practices in the USA and the world. Insects. In: R.J. Kohel and C.F. Lewis, editors, Cotton. Agron. Mongr. 24. ASA, CSSA, and SSSA, Madison, WI. p. 265–287.

Robinson, A.F., A.A. Bell, D.M. Stelly, N.D. Dighe, and M.A. Menz. 2005. Introgression of resistance to *Rotylenchulus reniformis* into *Gossypium hirsutum* from *G. longicalyx*. J. Nematol. 37:391.

Robinson, A.F., A.C. Bridges, and A.E. Percival. 2004. New sources of resistance to the reniform (*Rotylenchulus reniformis* Linford and Olivereira) and root-knot (*Meloidogyne incognita* (Kofoid and White) Chitwood) nematode in Upland (*Gossypium hirsutum* L.) and Sea Island (*G. barbadense* L.) cotton. J. Cotton Sci. 8:191–197.

Rubenstein, K.D., P. Heisey, R. Shoemaker, J. Sullivan, and G. Frisvold. 2005. Crop genetic resources. An economic appraisal. Econ. Info. Bull. No. EIB-2. USDA-ERS, Washington, DC. http://www.ers.usda.gov/publications/eib-economic-information-bulletin/eib2.aspx (accessed 4 April 2015).

Rudgers, J.A., S.Y. Strauss, and J.F. Wendel. 2004. Trade-offs among anti-herbivore resistance traits: Insights from Gossypieae (Malvacaeae). Am. J. Bot. 91(6):871–880. doi:10.3732/ajb.91.6.871

Saunders, J.H. 1961. The wild species of *Gossypium*. Oxford Univ. Press, London.

Scherer, J.A.B. 1916. Cotton as a world power. Frederick A. Stakes Co., New York.

Schwendiman, J., A.E. Percival, and J.L. Belot. 1986. Cotton collecting on Caribbean Islands and South Florida. FAO/IBPGR. Plant Genet. Resour. Newsl. 66:20–25.

Sheehan, M.A., J.K. Dever, and M.D. Arnold. 2010. Phenotypic documentation of *Gossypium* species from the US Cotton Germplasm Collection, USDA-ARS, Using Digital Photography. In: Proceedings of the Beltwide Cotton Conferences, New Orleans, LA. 4–7 Jan. 2010. Natl. Cotton Counc. of Am., Memphis, TN. p. 753–756.

Shepherd, R.L. 1983. New sources of resistance to root-knot nematodes among primitive cottons. Crop Sci. 23:999–1002. doi:10.2135/cropsci1983.0011183X002300050044x

Singh, V.V., P. Mohan, S.J. Baitule, B.R. Pathak, and V.N. Kulkarni. 2003. Explorations within India for collection of cotton species germplasm. Plant Genet. Resour. Newsl. 136:40–46.

Smith, C.W., R.G. Cantrell, H.S. Moser, and S.R. Oakley. 1999. History of cultivar development in the United States. In: C.W. Smith and J.T. Cothren, editors, Cotton: Origin, history, technology, and production. John Wiley & Sons, New York. p. 99–171.

Smykalova, I., D. Hampel, and M. Pavelek. 2012. Determination of duplicates of accessions in a germplasm collection of flax/linseed by means of digital image analysis. Ind. Crops Prod. 36:177–187. doi:10.1016/j.indcrop.2011.09.004

Southern Cooperative Series. 1956. Genetics and cytology of cotton. 1948–55. S. Coop. Ser. Bull. 47. Texas A&M Univ., College Station, TX.

Southern Cooperative Series. 1968. Genetics and cytology of cotton. 1956–67. S. Coop. Ser. Bull. 139. Texas Agric. Exp. Stn., College Station, TX.

Southern Cooperative Series. 1981. Preservation and utilization of germplasm in cotton. 1968–80. S. Coop. Ser. Bull. 256. Texas A&M Univ., College Station, TX.

Southern Cooperative Series. 1997. Preservation and utilization of germplasm in cotton. 1981–92. S. Coop. Ser. Bull. 386. Mississippi Agric. and For. Exp. Stn., Mississippi State, MS.

Stelly, D.M., S. Saha, D.A. Raska, J.N. Jenkins, J.C. McCarty, and A.O. Gutierrez. 2005. Registration of 14 Upland (*Gossypium hirsutum*) germplasm lines disomic for different *G. barbadense* chromosome or arm substitutions. Crop Sci. 45(6):2663–2665. doi:10.2135/cropsci2004.0642

Stephens, S.G. 1963. Polynesian cottons. Ann. Mo. Bot. Gard. 50:1–22. doi:10.2307/2394914

Stipanovic, R.D., L.S. Puckhaber, A.A. Bell, A.E. Percival, and J. Jacobs. 2005. Occurrence of (+)- and (-)- gossypol in wild species of cotton and in *Gossypium hirsutum* var. mariegalante (Watt). Hutchinson. J. Agric. Food Chem. 53:6266–6271. doi:10.1021/jf050702d

Tanksley, S.D., and S.R. McCouch. 1997. Seed bank and molecular maps: Unlocking genetic potential from the wild. Science 277:1063–1066. doi:10.1126/science.277.5329.1063

Todaro, A. 1877. Relazione sulla cultura dei Cotoni in Italia seguita da una Monografia del genere *Gossypium*. Molina, Rome.

Toll, J.A. 1995. Processing of germplasm, associated material and data. In: L. Guarino, R.V. Ramanatha, and R. Reid, editors, Collecting plant genetic diversity: Technical guidelines. CABI, Wallingford, UK. p. 577–595.

Ulloa, M., J.M. Stewart, C.E.A. Garcia-C., S. Godoy-S., A. Gaytan-M., and S. Acosta N. 2006. Cotton genetic resources in the western states of Mexico: In situ conservation status and germplasm collection for ex situ preservation. Genet. Resour. Crop Evol. 53:653–668. doi:10.1007/s10722-004-2988-0

USDA. 1997. Information on the condition of the National Plant Germplasm System. GAO/RCED-98-20. US Gov. Print. Office, Washington, DC.

USDA-ARS. 1974. The regional collection of *Gossypium* Germplasm. ARS-H-2. USDA-ARS, Hyattsville, MD.

Vavilov, N.I. 1926. Tzentry proiskhozhdeniya kulturnykh rastenii. (The centers of origin of cultivated plants.) Bull. Appl. Bot. Plant Breed. 16:1–248.

Vertucci, W.E., and E.E. Roos. 1990. Theoretical basis of protocols for seed storage. Plant Physiol. 94:1019–1023. doi:10.1104/pp.94.3.1019

von Rohr, J.B.P. 1791–1793. Anmerkungen iiber den Cattunbau, zum Nuzen der Danischen Westindischen Colonien. 2 Volumes. Hammerich, Altona, Germany.

Wallace, T.P., D. Bowman, B.T. Campbell, P. Chee, A. Gutierrez, R.J. Kohel, J. McCarty, G. Meyers, R. Percy, F. Robinson, W. Smith, D.M. Stelly, J.M. Stewart, P. Thaxton, M. Ulloa, and D.B. Weaver. 2009. Status of the USA cotton germplasm collection and crop vulnerability. Genet. Resour. Crop Evol. 56:507–532. doi:10.1007/s10722-008-9382-2

Ware, J.O. 1937. Plant breeding and the cotton industry. In: USDA Yearbook. U.S. Gov. Print. Office Washington, DC. p. 657–744.

Ware, J.O. 1951. Origin, rise, and development of American Upland cotton varieties and their status at present. Dep. of Agron. Mimeogr., Univ. of Arkansas, Fayetteville.

Watt, G. 1885. A dictionary of economic products of India (6 vols.). Gov. of India Cent. Print. Off., Calcutta, India.

Watt, G. 1907. The wild and cultivated cotton plants of the world. Longmans, Green, and Co., London.

Williams, K. 2005. An overview of the U.S. National Plant Germplasm System's exploration program. HortScience 40(2):297–301.

Yu, J.Z., D.D. Fang, R.J. Kohel, M. Ulloa, L.L. Hinze, R.G. Percy, J. Zhang, P. Chee, B.E. Scheffler, and D.C. Jones. 2012. Development of a core set of SSR markers for the characterization of *Gossypium* germplasm. Euphytica 187:203–213. doi:10.1007/s10681-012-0643-y

Yu, J., S. Jung, C.H. Cheng, S.P. Ficklin, T. Lee, P. Zheng, D. Jones, R.G. Percy, and D. Main. 2014. CottonGen: A genomics, genetics and breeding database for cotton research. Nucleic Acids Res. 42 (D1):D1229–D1236.

Zhu, S.J., N. Reddy, and Y.R. Jian. 2005. Introgression of a gene for delayed pigment gland morphogenesis from *Gossypium bickii* into Upland cotton. Plant Breed. 124:590–594. doi:10.1111/j.1439-0523.2005.01151.x

Anatomy and Morphology of Cultivated Cottons

Jack R. Mauney

Abstract

Cotton is a typical woody perennial shrub. As such, its flowering structures are sympodial axillary branches. The plant produces both vegetative and flowering structures throughout the productive season. The vegetative axis is composed of a tap root and vertical stem with alternate leaves, with a 3/8 turn, which continues to produce nodes as long as moisture and temperature allow. Flowering branches are sympodial and arise from the first (and sometimes second) axillary branch position at leaves above about the fifth node on the stem. Flowers develop with three floral bracts, a calyx whorl, five petals, and a 4- or 5-locule capsule in which six to nine ovules develop. Seeds require 35 to 60 d to mature. The epidermis of the seed develops two types of fibers: short (5 mm or less) linters and long (25–35 mm.) lint fibers.

To fully comprehend the limitations to the productive capacity of the cotton plant, it is necessary to understand its structures: the vegetative framework, the reproductive branches, flowers, fruit, and seed. These structures in cotton, a typical woody perennial shrub, are botanically relatively simple. The difficulty in analyzing cotton morphology arises from the fact that the plant simultaneously develops both vegetatively and reproductively. Because of its indeterminate growth habit cotton has the most complex morphology of any major field crop grown as an annual.

A complete discussion of the developmental patterns in cotton is given in J.M. Hector's *Introduction to the Botany of Field Crops* (Hector, 1936). This work by Hector and the Russian monograph, *The Structure and Development of the Cotton Plant* (Baranov and Maltzev, 1937), as well as published reports of Gore (1935), Mauney and Ball (1959), Quintanilha et al. (1962), will, with additional observations made in my laboratory, form the basis for this chapter.

Abbreviations: HFI, horizontal flowering interval; PA, post anthesis; VFI, vertical flowering interval.

Jack R. Mauney, Consultant, USDA-ARS, retired, Mesa, AZ.

The main text of this chapter was previously published as Mauney, J.R. 1984. Anatomy and morphology of cultivated cottons. In: R.J. Kohel and C.F. Lewis, editors, Cotton. Agron. Monogr. 24. ASA, CSSA, and SSSA, Madison, WI. p. 59–80. doi:10.2134/agronmonogr24.c3

Contribution by USDA-ARS.

doi:10.2134/agronmonogr57.2013.0020.5

Cotton. 2nd. ed. David D. Fang and Richard G. Percy, editors. Agronomy Monograph 57.

Development of Primary Axis

The primary axis results from elongation and development of the embryo. In the dormant seed, the primary axis consists of a radicle, a hypocotyl, and a poorly developed epicotyl. The epicotyl contains one true leaf initial and a dome of meristematic cells (Fig. 1). Thus, all differentiation of the vegetative framework above the cotyledons must take place subsequent to germination. Cotton remains as a seedling longer than grasses and legumes because it has no differentiated true leaves, only cotyledons, to expand immediately after germination.

The growing cotton plant actively proliferates new cells on many fronts. To describe the plant completely, each facet of growth should be examined apart from the whole. A picture of both the simplicity of the part and complexity of the whole emerges from this examination.

Four principles of morphological correlation are necessary for a full understanding of the architecture of a cotton plant:

1. Apical meristems make four organs: leaves, stems, roots, and flowers.
2. There are three types of leaves: cotyledons, prophylls, and true leaves. Cotyledons, the kidney shaped seed-leaves, ordinarily reach a breadth of approximately 50 mm. True leaves range from entire to deeply palmately lobed and are sometimes more than 200 mm in breadth. The prophyll is inconspicuous and not ordinarily observed. It is the first leaf on any branch, only 5 mm or less wide and may be mistaken for a stipule, which it resembles (Fig. 2).

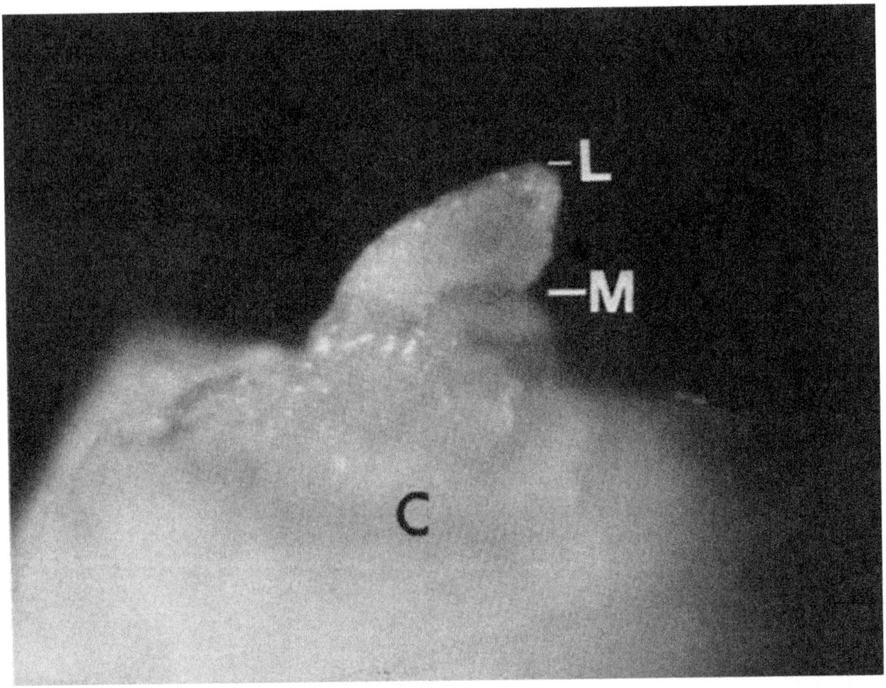

Fig. 1. Growing point 1 day after planting. C—attachment of one cotyledon (attachment of other cotyledon in background behind growing point), L—primordium of first true leaf, M—apical meristem.

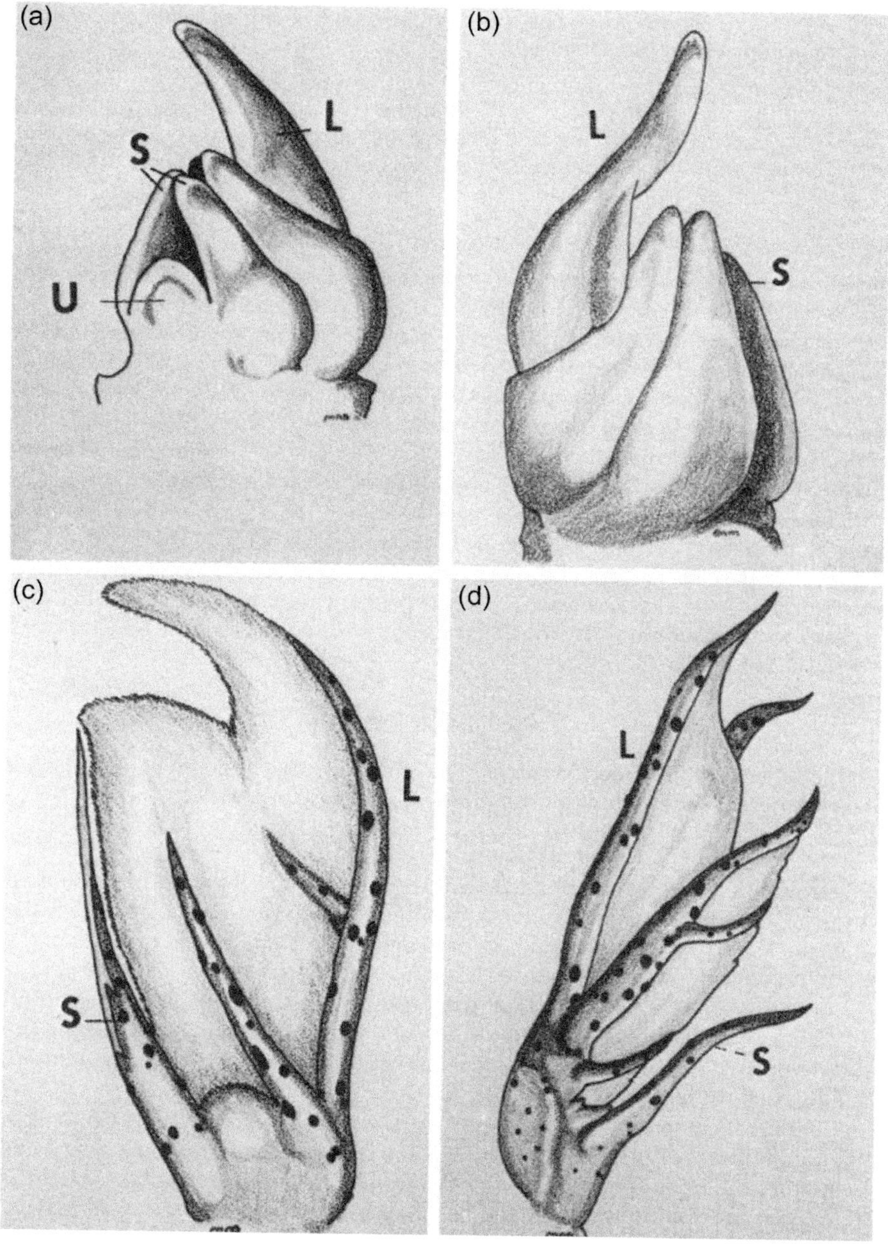

Fig. 2. Stages in leaf development. (a) U—unlobed primordium on edge of apical meristem and 0.1-mm L—second youngest leaf with stipules (S) differentiated. (b) 0.2-mm L—third leaf behind apex with 0.2-mm petiole. (c) 1.0-mm L—fifth or sixth leaf behind apex with 1.0-mm petiole. (d) 2.0-mm L—oldest unexpanded leaf. (a, b) 40× magnification. (c, d) 10× magnification. (Drawings by Edward Mulrean.)

3. In the axil of each cotyledon, prophyll, or true leaf one and only one branch apical meristem develops.
4. The only difference between a vegetative branch (monopodium) and a fruiting branch (sympodium) is that the fruiting branch apical meristem, after making a prophyll and a true leaf, terminates in a flower. A vegetative branch continues to make leaves until some stress causes it to cease growth.

Roots

Because the radicle is more massive in the seed than the epicotyl, the root develops more rapidly during germination. During the period before true leaves begin expansion, the primary root is penetrating deeply into the soil and branch roots are being formed. The normal root system in cotton is a typical tap root. Secondary root primordia develop endogenously at the juncture of the phloem and the cortex about 12 cm behind the primary growing point. Tertiary roots begin developing about 5 cm behind the secondary growing points. These branch roots are arranged in four rows along the tap root, one row in each quadrant.

Should the apex of the primary root be killed, there is an immediate flush of growth of secondary roots behind the necrotic area. One of these secondary roots may assume a dominant growth pattern and penetrate downward as would the tap root. Generally, however, when the primary root dies, the secondary and tertiary roots proliferate to form a lateral root system which never penetrates as deeply as tap roots.

Aerial Organs

The aerial portions are more complex. The remaining discussion will describe the developmental pattern and interrelationships of these organs.

Leaf Development

As germination begins, the first true leaf is nothing more than a primordium on the growing point between the two cotyledons (Fig. 1). At the end of this stage the first true leaf expands to the sun. By the time the first leaf begins unfolding, the plant has developed six or seven additional leaf initials.

The first evidence of a new leaf is a protuberance on the edges of the apical meristem. The leaf primordium elongates and becomes lobed in developing the stipule (Fig. 2a). During the early development the stipules may elongate more rapidly than the leaf itself (Fig. 2b, 2c, 2d).

The arrangement of the leaves on a stem is known as phyllotaxy. Cotton has a spiral phyllotaxy with each leaf being 3/8 turn above the last. A stem may have a counter-clockwise (dextrorse) or a clockwise (sinistrorse) phyllotaxy. Half the stems tend to spiral to the right and half to the left. Infrequently, a stem may be seen in which the phyllotaxy has reversed direction.

Ordinarily there are 7 to 11 leaf initials above the youngest unfolded leaf (Fig. 3). I have chosen to characterize the developmental stages of these leaves by the length of the petioles, which range from approximately 0.1 mm in the youngest leaf in which lobing can be observed to 2 to 4 mm in the oldest unexpanded leaf. Cotton leaves have several distinctive features. They have two pulvini, one at each end of the petiole, which allow a wide arc of movement of the leaf blade

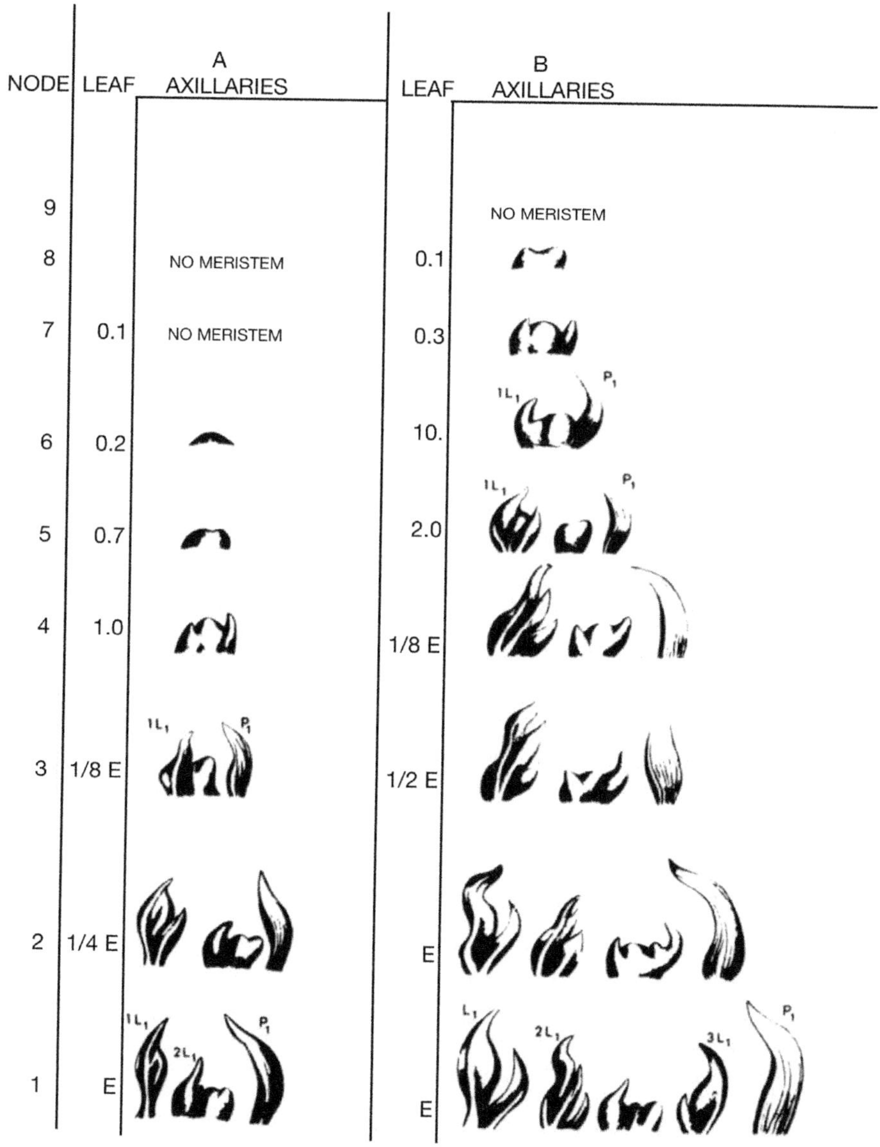

Fig. 3. Progressive development of axillary branch axes. Column A represents axillaries of a plant with first true leaf fully expanded. Column B represents axillaries of a plant with three leaves expanded. Leaves above unfolded leaves are designated by petiole length (mm). Unfolded leaves are designated by proportion of full expansion (1/8 E, 1/2 E, 3/4 E, E). The progress of the branch at any particular node can be seen by comparing that node on Plant A and B. A similar progression is observed in comparing progressively older branches on each plant. The prophyll (P1) of the axillary differentiates first, followed by the first true leaf (1L1). Additional leaves (2L1, 3L1) develop as the branch ages. (Drawings by David Moreno.)

to follow the movement of the sun. They have leaf nectaries on the dorsal veins, and pigment glands scattered randomly in all tissues. These glands, which are also present in roots, stems, and flowers, are the repository of gossypol and other toxic compounds distinctive of *Gossypium* and closely related genera. Stomates are located two to three times more frequently on the lower surface than on the upper surface. Dale (1961) observed 37 to 50 stomates per mm^2 on the upper surface and 107 to 121 stomates per mm^2 on the lower leaf surface.

The first true leaf of the primary axis is usually not lobed. Its stipules are small and often abscise early. As the plant matures the leaves formed at the higher nodes are more and more deeply lobed until the climax leaf shape is formed at the 6th to the 10th node.

Typical Branch Development

A branch meristem develops in the axil of each leaf (Fig. 3). The initial developmental stages of monopodia and sympodia are indistinguishable. The meristem which produces the branch is first visible in the axil of the third leaf below the apical meristem. Esau (1962) states that branch meristems which develop close to the apex are residues of the apical meristem itself. This probably happens in cotton.

The branch meristem begins development by producing a prophyll, an internode and a true leaf. The prophyll is a strap shaped leaf without stipules (Fig. 4). Its appearance is so like that of stipules that it can be mistaken for one. The prophyll almost always occupies the position on the branch axis away from the next higher node on the main stem. That is, if the plant has a counter-clockwise phyllotaxy, the prophyll will be found on the left side of the branch axis. The node represented in Fig. 4 was from a plant with a clockwise phyllotaxy.

After differentiating a prophyll, a branch meristem develops stem tissue of the first internode followed by the second leaf, a true leaf. This leaf is the first conspicuous leaf of the branch, and in the remaining discussion, it will be designated the "first true leaf."

Vegetative Branches

At the lower nodes of the primary axis the branch is a monopodium, a vegetative limb. Though the branch is technically a secondary axis the vegetative limb behaves essentially like the primary axis. It assumes an upright orientation and, as long as it is active, produces additional true leaves and internodes. At the time a main stem leaf begins unfolding, the vegetative branch at its base ordinarily has three or four true leaf initials. Frequently the monopodium becomes dormant at this stage.

The leaves on a vegetative branch arrange themselves with a 3/8 phyllotaxy. There is no correlation between the direction of turn of branch phyllotaxy and that of the main stem.

An axillary meristem develops in the axil of each leaf on the branch. This is true of a prophyll as well as the true leaves. Figure 5 represents schematically the vegetative branch and its axillary meristems. Because there is no internode below the prophyll, the axillary bud of the prophyll seems to lie in the axil of the main stem leaf. Thus, the bud appears to be an additional axillary of the primary leaf

The naming of the prophyll axillary varies. It has been called the "axillary bud," the "extra axillary bud," and the "accessory bud." Mauney and Ball (1959)

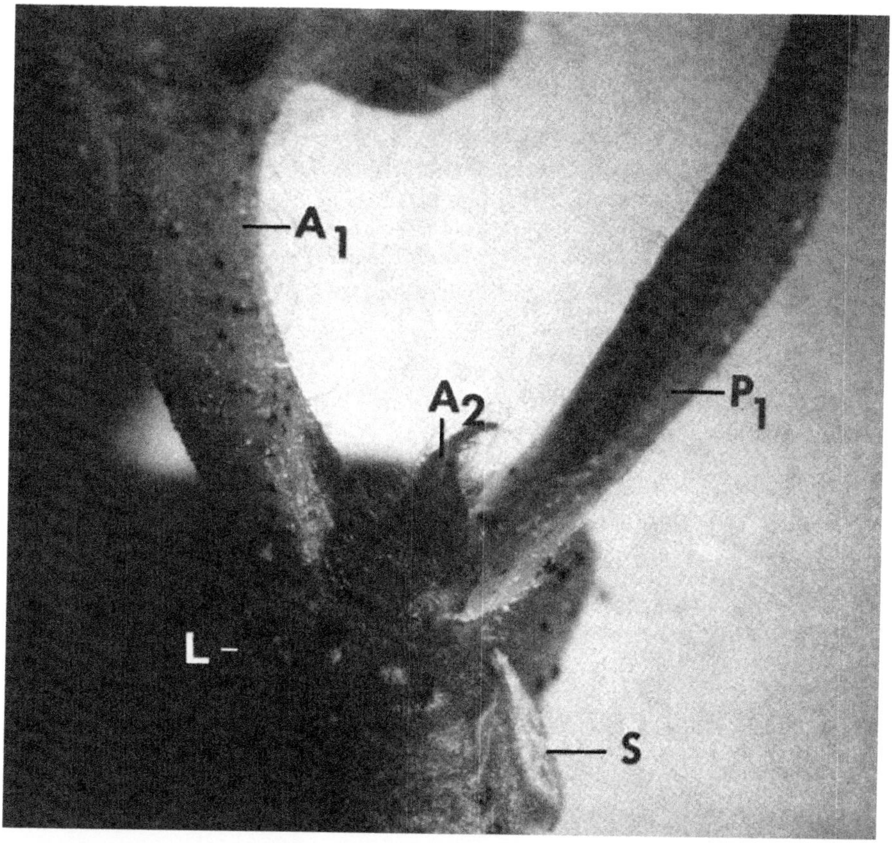

Fig. 4. Photograph of vegetative axillary branch bud showing prophyll (P_1), first axillary (A_1), and dormant second axillary (A_2). The main stem leaf has been removed to obtain this view and scars of its leaf (L) and stipules (S) are visible. This arrangement of parts is typical of plants with left-hand phyllotaxy. Other plants may have a reversed arrangement.

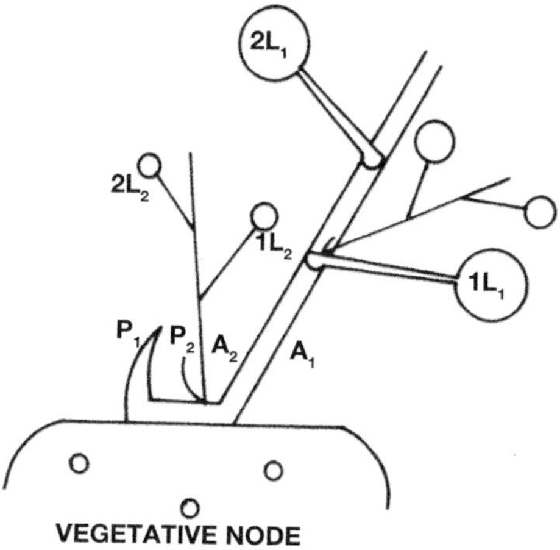

Fig. 5. Schematic representation of a vegetative axillary branch. Sawtooths at left-prophylls; open circles—true leaves; P_1—prophyll of first axillary (A1) axis; P_2—prophyll of second axillary (A_2) axis; $1L_1$—first true leaf of first axillary; $2L_1$—second true leaf of first axillary; $1L_2$—first true leaf of second axillary.

have called the major axillary the first axillary and the axillary of its prophyll the second axillary. The first axillary is responsible for most branching. In its developmental stages the second axillary is identical to the first axillary. It produces a prophyll, internodes, and true leaves. The second axillaries at the base of monopodia are invariably vegetative and usually dormant.

At nodes on which the second axillary elongates into a branch, a third axillary is observed (Fig. 6). This branch bud is in the axil of the prophyll of the second axillary. It is dormant unless forced to elongate by pruning of the stems and branches. Presumably, a plant which was forced by pruning to continue to make new branches could develop an indefinite number of axillaries in this manner.

Reproductive Branches

Early development of sympodial branches is identical to that described for monopodia (compare nodes 5, 4, and 3, Fig. 3B with nodes 11, 10, and 9, Fig. 7D). A prophyll, an internode, and a true leaf are differentiated. These are indistinguishable from those of a vegetative axis. All axillary primordia, then, begin development as vegetative meristems. Fruiting branches result from the transformation of these initially vegetative meristems into flower primordium.

The differentiation of the flower begins with formation of a whorl of three bracts after development of the first true leaf. Bract differentiation is first visible on the branch at the fifth node behind the growing point. The floral differentiation consumes the meristem of the axis, and thus the axis terminates with the flower. There is no other difference between a vegetative and reproductive axis.

With the extinction of terminal growth by the floral development, the first axillary primordium at the base of the true leaf is stimulated. Very rapidly the

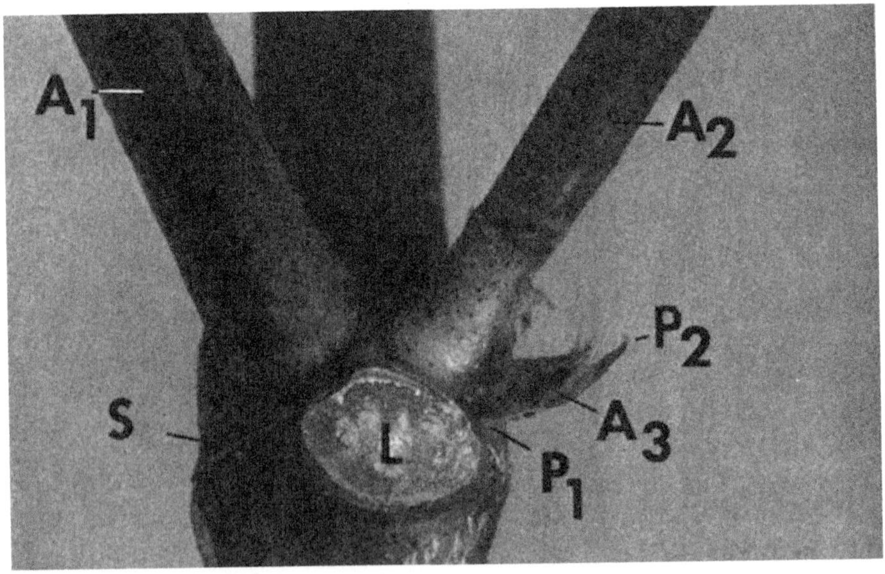

Fig. 6. Node showing development of third axillary (A3) after elongation of first (A1) and second (A2) axillary branches. P1—scare of prophyll of first axillary; P2—prophyll of second axillary; L—leaf stump; S—scare of stipule.

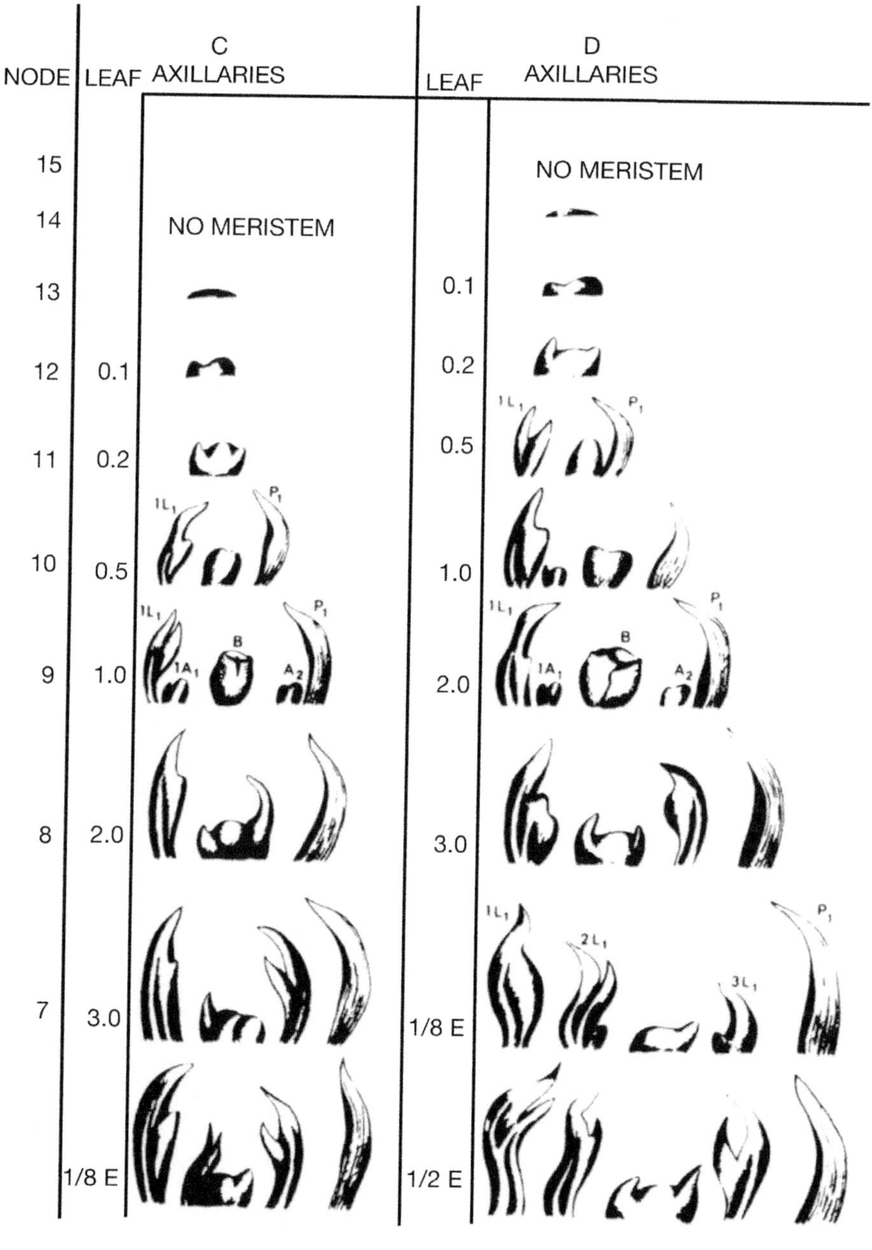

Fig. 7. Progressive development of axillary branch axes after floral initiation. Column C represents axillaries of plant with 6 unfolded leaves; column D pictures axillaries of plant with 7 unfolded leaves. Leaf designations and other symbols the same as in Fig. 3. Visible development is evident at the base of the second leaf below the apex. After the first axillary (A1) differentiates the prophyll (P1) and first true leaf (1L1), the three bracts (B) of the flower bud develop. The flower bud terminates the first axillary. The second axillary (A2) and first axillary of the first true leaf (1A1) develop more actively than on vegetative branches. (Drawings by David Moreno.)

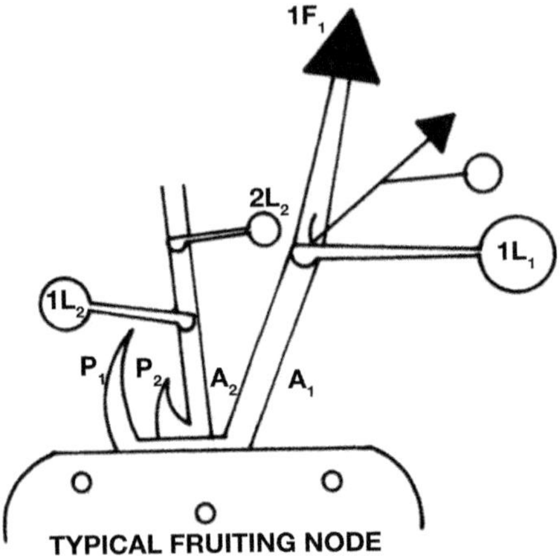

Fig. 8. Schematic representation of reproductive axillary branch, sympodium. Black triangles—flower buds; 1F1—first flower on first axillary. All other symbols the same as Fig. 5.

primordium develops a prophyll, an internode, a true leaf and a terminal flower. These are the structures associated with the next flower on the sympodium.

Development of the sympodium continues—termination of each new branch by a flower bud and development of the axillaries to produce the next branch (Fig. 8). The fact that each internode is a new branch axis gives the sympodium a typical zigzag appearance.

Second axillaries at the base of sympodia may be vegetative or reproductive, dormant or active. The second axillaries at the base of early fruiting branches on the main stem often develop into vigorous vegetative branches. Less frequently, the second axillaries accompanying sympodia develop into fruiting branches.

The second axillaries at the base of the true leaves on the sympodium are ordinarily dormant vegetative buds. Infrequently, they differentiate into floral branches and add additional flowers to the sympodial axis.

The elements of the system described are basically simple. The plant has a complex branching habit because: 1) both the first and second axillary may be vegetative or reproductive, 2) any axillary may be dormant or active, and 3) each segment of a sympodium is a new first axillary axis. Each of the various branches on the plant are the result of an individual apical meristem responding to physiological forces within the plant. These forces determine whether the meristem will continue to produce leaf initials or terminate in a flower and whether the branch axis will elongate or remain dormant.

Additional Observations

Some of the implications of the system and some of the frequently observed variations on the general theme are as follows:

1. Because both monopodia and sympodia begin by making a prophyll- internode-true leaf, it is not until the bracts of a flower bud are differentiated that we see differences that would allow diagnosis of the kind of branch, vegetative or reproductive, a given meristem will produce.

   In Fig. 9 the early development of the two branch types is compared. Figures 9a and 9b show that the two are almost identical in appearance until the second true leaf becomes lobed or until the second of the floral bracts differentiates.

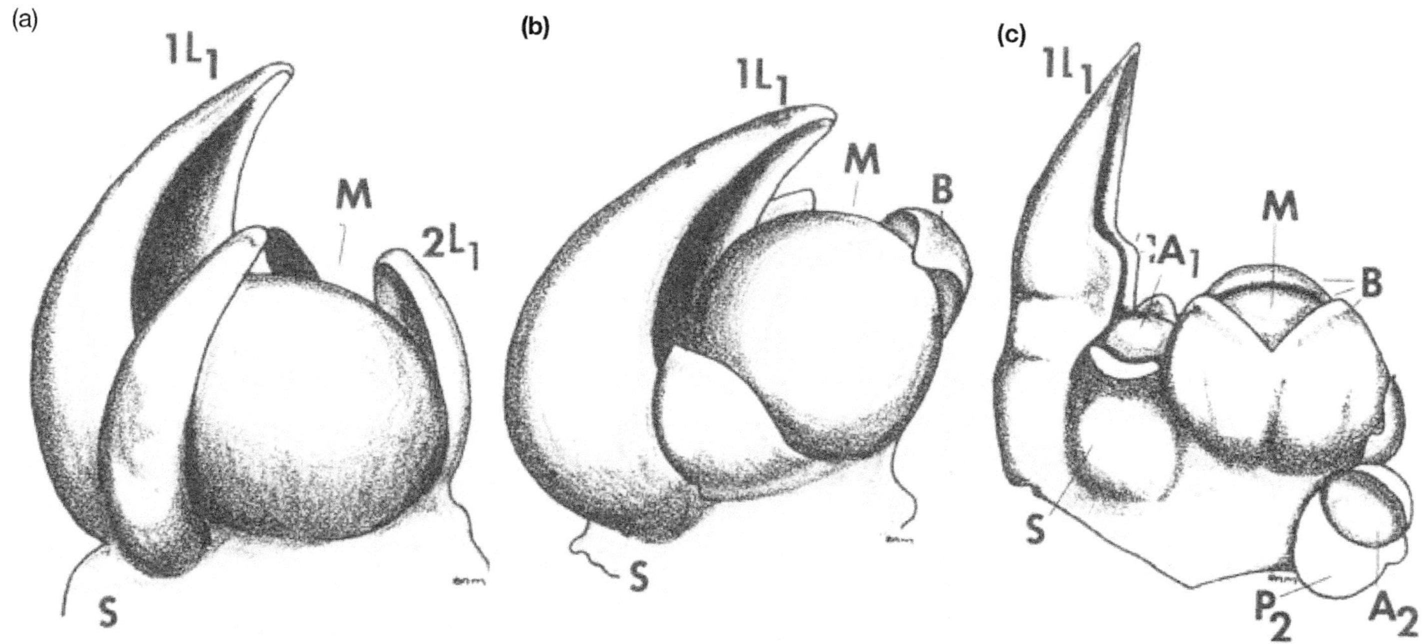

Figure 9. Comparison of early stages in differentiation of vegetative and reproductive axillary branches. One stipule (S) of the first true leaf (1L1) has been removed from each branch to show the apical meristem (M). 40X magnification. (a) Vegetative branch with differentiation of the second true leaf (2L1) beginning. Subsequently, this branch would continue production of leaves in sequence identical to that shown in Fig. 2. (b) Reproductive branch with differentiation of the floral primordium beginning. The first of the three bracts (B) has developed on the meristem (M). (c) Subsequent development of the floral branch resulting in the three bracts (B) of the square. Development of the second axillary (A2) and the first axillary (1A1) of the first true leaf is apparent at this stage. P1 is the base of the prophyll removed so that A2 could be seen. (Drawings by Edward Mulrean.)

2. Flowering branches grow more rapidly than monopodia. This difference is often a better characteristic to use in distinguishing the two types of primordia than their appearance. Apical dominance seems to diminish as the plant enters the flowering cycle. Flowering branch meristems approach the primary apex more closely than do the vegetative branch meristems (compare Fig. 3A with 7D). Likewise the second axillary is more active at the base of a fruiting branch, and the axillary at the base of the first true leaf is ordinarily more active on a fruiting branch than on a vegetative branch (Fig. 9c).
3. The forces causing the development of fruiting branches seem to act throughout the plant at about the same time. If we examine the fruiting branches that occur on vegetative limbs, we find them progressively on lower nodes as we approach the first fruiting branch of the main stem (Fig. 10). From the cotyledonary node we find approximately equal number of nodes (leaf units in Fig. 10) to the first fruiting branch on all the vegetative axes. At the time of floral induction, branches at low nodes have more nodes already formed than branches at the higher nodes.
4. The first two fruiting branches are less vigorous than subsequent ones. Frequently the flower at these nodes develops more slowly than is typical later and may blossom on the same day or the day after the flower on the third fruiting branch. This slight inhibition of the early reproductive branches is probably a residual trait based on the dormant fruiting spurs observed in some ancestral desert species such as *G. aridum* (Mauney, 1968).
5. The rate of development of all organs on the plant is a function of temperature. However, the time interval between flowers on successive nodes along an individual sympodium, the horizontal flowering interval (HFI), is several days longer than the interval between first flowers on successive flowering

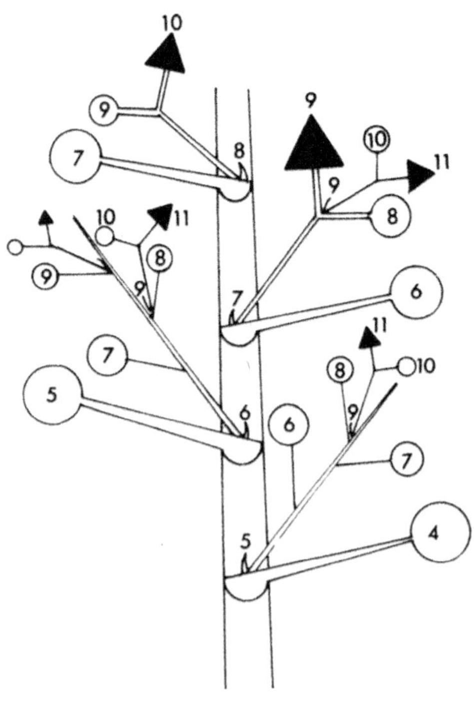

Fig. 10. Schematic representation of squares on sympodia on vegetative limbs and those on successive sympodia on the main stem. Numerals represent succession of leaf units (true leaves, prophylls, and bracts of squares). Other symbols same as Fig. 5 and 8. The plant represented had first flower at node 6. Though all leaf units are not precisely equal amounts of time. the events with equivalent leaf units should happen within a few days.

branches, the vertical flowering interval (VFI). Hesketh et al. (1972) calculated the ratio HFI:VFI to be 2.7. Though the ratio was not constant over a wide range of temperature, at 30°C VFI was 3 d and HFI about 8 d.

The reason for this difference can be understood by analyzing the number of leaf units between flowers in the vertical and horizontal dimensions (Fig. 10). Because the structures of two successive sympodia can develop simultaneously, vertically there is only one additional leaf, that on the main stem, between two flowers. On the other hand, the second node on each sympodium cannot develop until after the formation of the first sympodial leaf (1L1). Thus, horizontally there are two leaves, prophyll 2P1 and leaf 2L1 between successive flowers (Fig. 8). First flowers on successive sympodia occur as parallel events separated by one leaf, whereas successive flowers along each sympodium are series events separated by two leaves. Hence the longer time for HFI.

Atypical Branch Development

Certain branching patterns are rarely seen or occur only on certain genetic material or in particular environments. These patterns can be explained as extreme, but consistent, variations of the general scheme.

Sympodia may develop from the second axillary position (Fig. 11 and 12). If the forces causing flower differentiation are sufficiently strong to reach the second axillary, the plant can be induced to flower. The tendency to develop sympodia from this location seems to be dependent on both genetics and environment. Moderate day and cool night temperatures favor development of this trait. *Gossypium barbadense* cultivars show the trait more often than those of *G. hirsutum.*

Fig. 11. Node at which two normal sympodia have developed.

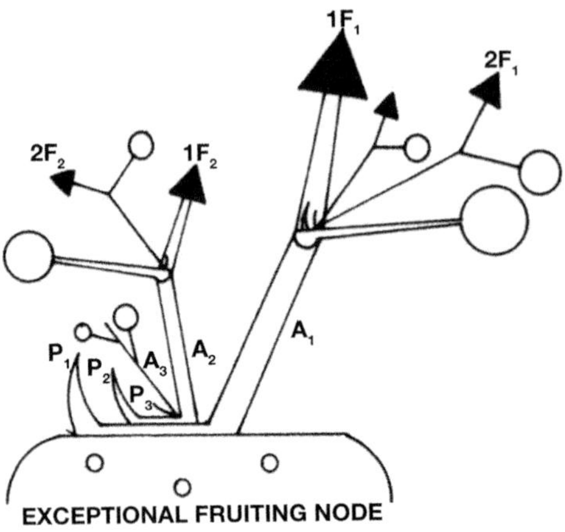

Fig. 12. Schematic representation of second axillary sympodium (A2) and the resulting development of the third axillary (A1). Symbols same as in Fig. 5 and 8.

Though I have not systematically studied branching among cultivars the early fruiting Upland cultivars seem to branch atypically more than others.

When the second axillary is a sympodium it is usually abnormal in that no true leaf is developed. Thus, the branch is single-jointed because of the absence of a leaf axil from which additional nodes can develop. Less frequently the second axillary produces a many-jointed flower branch, an important factor in developing additional fruiting positions in heavily fruited plants.

The fact that second axillary sympodia are uncommon would seem to indicate that floral induction can occur only in a limited area behind vegetative apices. This area includes the three youngest nodes below the meristem (Nodes 14, 13, 12, Fig. 7D). Here the first axillary begins differentiation. The second axillary ordinarily differentiates on nodes five and six below the meristem (Nodes 10, 9, Fig. 7D). Should the induction area expand to include the differentiation site of the second axillary or should the second axillary differentiate within the narrow limits of induction it could become floral.

Photoperiodic species and cultivars may revert to production of vegetative branches when the inductive condition ceases. This action can be explained from the fact that each flowering node, whether first or last on a sympodium, is the result of a new branch meristem. If induction ceases, the plant must make vegetative axillaries at each new leaf. The primary axis will revert to making vegetative limbs, and the sympodia may continue growth as monopodia or become dormant.

While this reversion is ordinarily observed only in selections which are extremely dependent on photoperiod or other environmental features, it is my belief that it also happens occasionally among cultivars.

Several genetically determined, atypical branching habits have been described. The cluster or short branch trait results in the shortening of internodes on the second joint of the sympodia and an accompanying development of the second axillaries into flowers. The foreshortening of the distances between the flower primordia frequently causes them to coalesce into a fasciated, malformed fruit. The whorl mutant of *G. herbaceum* is the only species in the genus in which I have observed a true terminal flower. Sympodia are initiated normally for 6 to 12 nodes, but then the vegetative apex differentiates into the terminal flower. This extinction of the apex permits vegetative branches lower on the stem to develop vigorously, but they, in turn, also terminate in flowers.

Gausmann et al. (1977) described a branching trait which they named "side-by-side." As the name implies, on these plants two adjacent branches occur opposed at a single node. Not all leaves and branches on these plants are arranged side-by-side. Most retain the alternate pattern with 3/8 phyllotaxy. Nodes with side-by-side branches usually occur early in the season. Because the two branches develop simultaneously, the first flowers on these branches typically blossom within 1 day of each other. The short flowering interval for these branches contributes significant earliness to genetic stocks that have a high frequency of the side-by-side expression (Namken et al., 1979).

Flower, Fruit, and Seed

Flower

The first true differentiation of the flower bud on the branch axis is production of three bracts rather than another true leaf. Subsequently, the other flower parts differentiate from the broad dome meristem that is characteristic of most plant flower buds. Quintanilha et al. (1962) estimated that a flower bud requires 36 d to develop (Fig. 13). However, the cross-section of a flower bud 36 d before anthesis

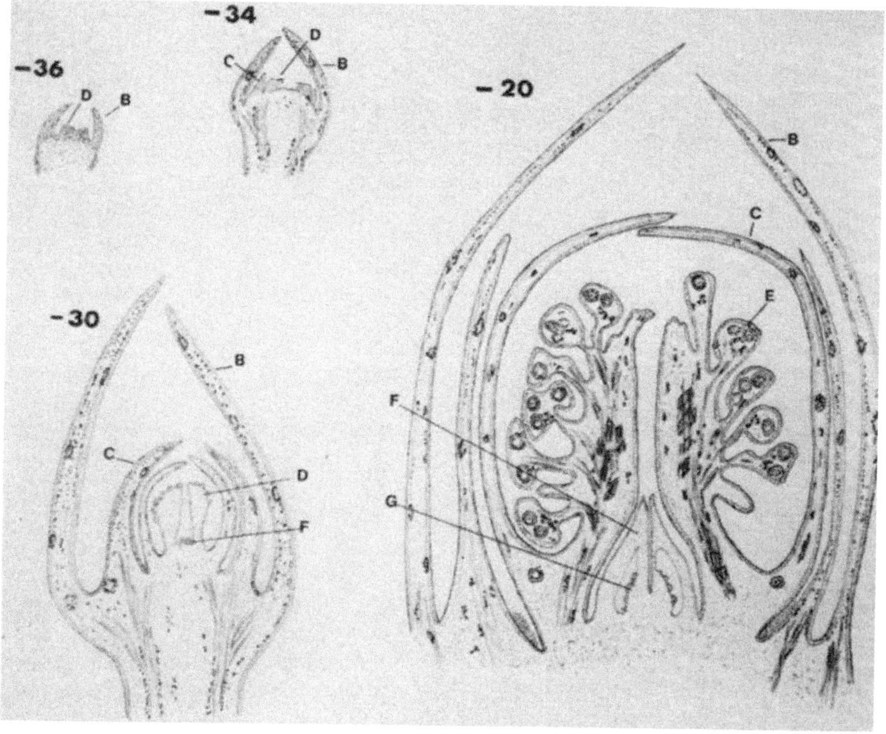

Fig. 13. Development of flower bud. Numbers refer to estimated days before anthesis (Quintanilha et al., 1962). B—bracts; C—corolla; D—androecium; E—anthers; F—ovary (locules); G—ovules. Drawings adapted from those of Baranov and Maltzev (1937).

(Fig. 13A) shows well developed bracts. Thus, the total time for flower development is more nearly 40 to 50 d from differentiation to anthesis.

Stewart (1986) has outlined the coordination of events during flower morphogenesis and the effects which environment may have on stages of flower development. Formation of the carpels occurs about 35 d before anthesis. This is about 10 to 14 d before the flower bud (square) becomes visible. The bud is well protected by the terminal plumule at this stage. Nonetheless, environmental influences apparently determine ovule number at this stage. Pollen mother cells are undergoing meiosis about 22 to 23 d before anthesis. At this stage subsequent pollen viability is particularly sensitive to environmental stresses (Stewart, 1986).

Floral nectaries begin development about 25 d before anthesis. They are located internally on a ring at the inner base of the fused calyx and externally at the base of each of the three bracts.

Fertilization

After anthesis the pollen is deposited on the stigma, germinates, and the pollen tube begins its growth down into the ovary. Growth of the pollen tube takes 12 to 30 h (Pollock and Jensen, 1964; Stewart, 1986) after which fertilization is accomplished. Cell division of the zygote begins 4 to 5 d after anthesis (Fig. 14).

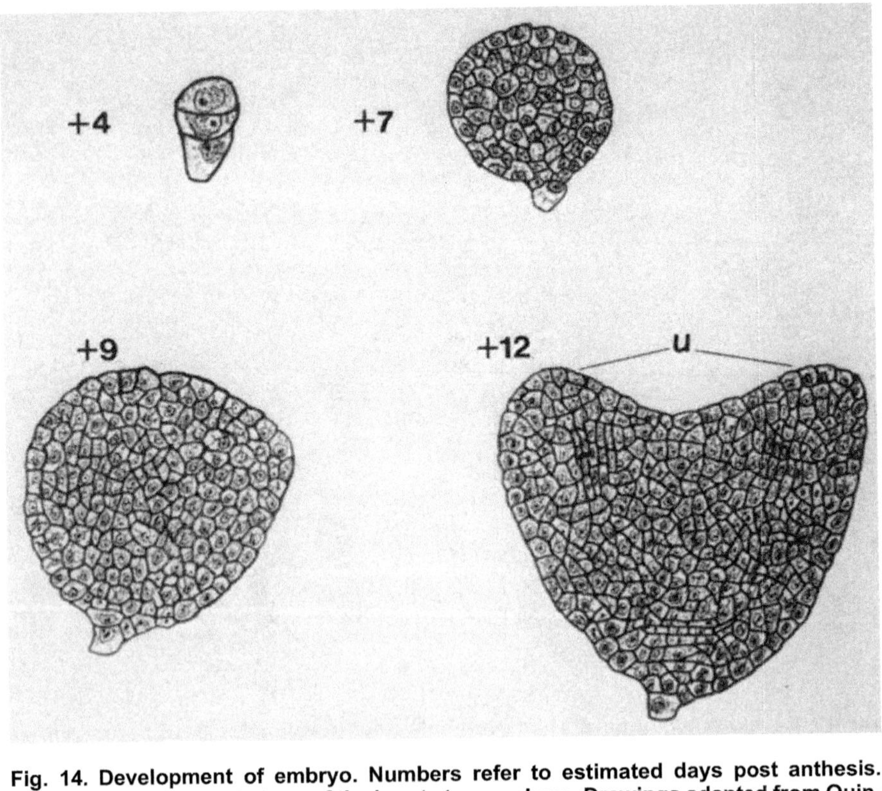

Fig. 14. Development of embryo. Numbers refer to estimated days post anthesis. u—differentiating cotyledons of the heart stage embryo. Drawings adapted from Quintanilha et al. (1962).

Enlargement of the young capsule (boll) begins with fertilization. The growth follows a sigmoid curve with the most rapid enlargement during the period 7 to 18 d post anthesis (PA). At 20 d PA the boll will have attained 90% of its final size. The mature size is reached about 25 d PA.

Fiber

Enlargement of the ovules and elongation of the ovule epidermal cells which result in the fiber, parallel closely the growth of the boll. The unfertilized ovule is about 1 mm in length. After fertilization it enlarges to about 8 to 9 mm in 15 to 18 d. The average size of mature seed is about 10 mm.

Initiation and development of the fiber of cultivated cottons have been described and supported by excellent electron photomicrographs by Stewart (1975), Beasley (1975), and Berlin (1986). These observers agree that the seed epidermal hairs, which result in the fiber, begin development within 24 h PA. The epidermal cells forming these fibers are located randomly over the surface of the ovule, though elongation of those near the chalazal and micopyle ends may lag by a few hours or days (Fig. 15). The fiber cell elongates at a linear rate for about 20 d and eventually achieves a length of about 25 to 35 mm depending on cultivar and environment (Quisenberry and Kohel, 1975). At about 16 to 19 d PA, secondary thickening of the cell wall begins and continues until fiber maturity at about 35 to 60 d PA. Schubert et al. (1973) showed that secondary wall deposition begins 2 to 5 d before cessation of primary wall elongation. Delmar et al. (1980) suggested that elongation during this period might be limited to the fiber tip or that the secondary wall deposit might be more flexible than the heavy deposits which occur subsequent to 21 d PA. The rate and duration of boll and fiber development are highly influenced by temperature (Gipson, 1986). At maturity the fiber dries, the central vacuole of the cell collapses and the fiber assumes the characteristic spiral ribbon that contributes to its spinability.

At 5 to 10 d PA a second group of epidermal hairs initiate development over the surface of the ovule. These hairs result in very short (about 5 mm or less) fibers that adhere to the seed during ginning and are generally referred to as fuzz or linters. Berlin (1986) observed that the secondary wall thickening of fibers is restricted at the base of the fiber in the epidermis. Fuzz fibers, on the other hand, have a well developed secondary wall at the juncture with the epidermis. This difference in thickening causes the difference in tendency to break away during ginning. Stomata are a feature of the ovule epidermis observed by both Beasley (1975) and Stewart (1975) (Fig. 15). The significance of these openings has not been elucidated.

Embryo

Growth of the embryo lags behind that of the ovule and fruit (Fig. 14). The size of the zygote decreases during the early cell division 4 to 5 d after anthesis (Pollock and Jensen, 1964). At 7 to 9 d the embryo is in the globular stage, approximately 0.1 mm in diam. At 10 to 12 d the differentiation of the cotyledon begins.

Pollock and Jensen (1964) found that the pattern of early cell divisions in the zygote is irregular. No definite pattern of cell lineage is established.

Turcotte and Feaster (1967) observed chimeras which can be traced to mixtures of genetically different cells in the young zygote. These cell mixtures

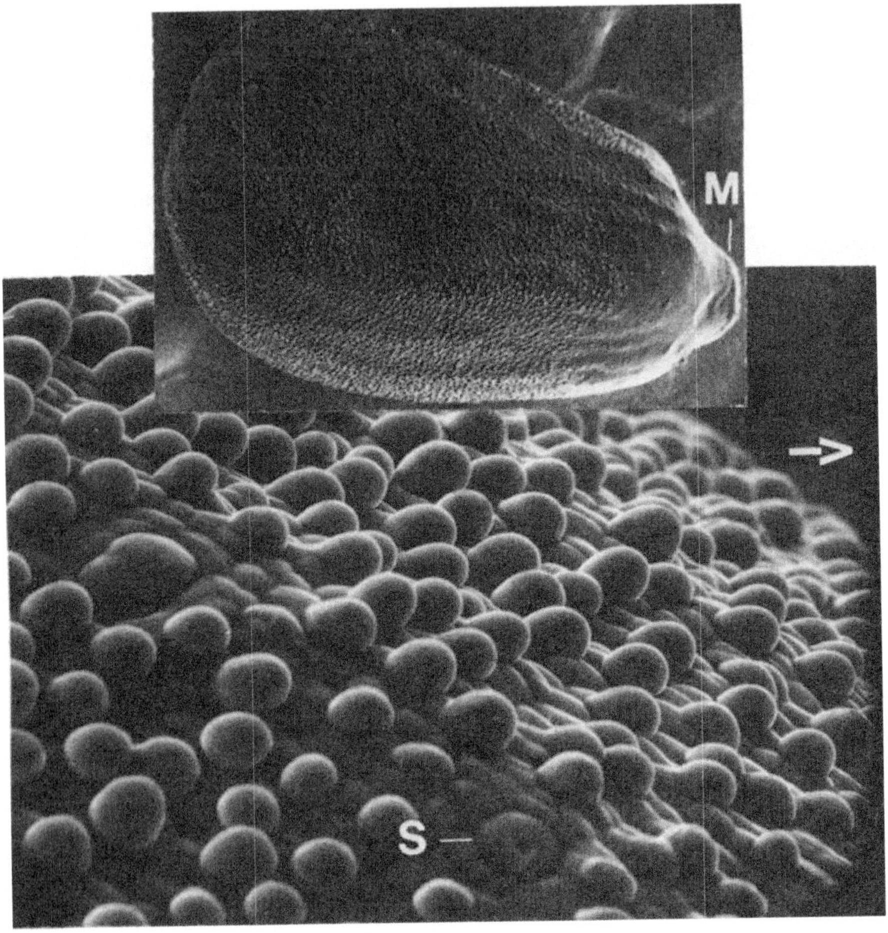

Fig. 15. Development epidermal fibers on ovules 1 day postanthesis showing tendency of fiber initials to grow toward micropyle (arrow) (X211), inset (X51) showing delay of development near micropyle (M). Note stomata (S). (Photographs by J. McD. Stewart.)

behave as a unit in the embryonic stage, and the mature seed develops normally. After germination, sectors develop and indicate the presence of chimera. The fact that cell division patterns are irregular and that genetic mixtures present at the globular stage proceed to differentiate normally suggests that the organization of the embryo develops in response to forces external to the zygote. The physiology of the ovular tissues and the orientation of the embryonic mass of cells probably generate the forces that initiate differentiation.

The development of the cotyledons is the most striking feature of embryo maturation. The cotyledons enlarge rapidly 15 to 25 d after anthesis, fill the embryo sac at about 30 d, then mature and become capable of dormancy. At maturity the boll dehisces and the carpels (burrs) flex open. Subsequent rapid drying of the fibers and seed causes the funiculus to be severed from the placenta and the chalazal cap of the seed to be cemented in place. Christiansen and Moore (1959) confirmed earlier observations that imbibition of water for germination occurs through the chalazal cap. Density and impermeability of the chalazal plug may prevent water entry in certain *Gossypium* species and genetic strains. We have now completely described the life cycle, which began with the germinating seed.

Summary

Even though the plant is limited to making leaves, stems, roots, and flowers, it can demonstrate a wide array of habits by controlling the arrangement and placement of these organs. Plant habit is determined largely by events at the developing apical meristems. The root system is typically a deeply penetrating tap root. However, if growth restrictions destroy the apical meristem of the primary root, a shallow, fibrous system may result from the proliferation of the secondary roots.

The aerial branching habit is determined by the differentiation and elongation of axillary meristems. The apex of the branch may continue to develop new true leaves or it may differentiate into a flower after the formation of the first true leaf. The branch may elongate rapidly or may be dormant. Flowering branches are less likely to remain dormant than vegetative branches.

Additional variations in branching habit arise from the development of the second axillary position. These positions are in the axils of the prophylls at the base of first axillary branches. The only restriction on the character of a second axillary is that it is more likely to be vegetative and dormant than the first axillary it accompanies.

References

Baranov, P.A., and A.M. Maltzev. 1937. The structure and development of the cotton plant: An atlas. Ogis-Isogis Moscow-Leningrad.

Beasley, C.A. 1975. Developmental morphology of cotton flowers and seed as seen with the scanning electron microscope. Am. J. Bot. 62:584–592 10.2307/2441936.

Berlin, J.D. 1986. The outer epidermis of the cottonseed. In: J.R. Mauney and J.McD. Stewart, editors, Cotton physiology. The Cotton Foundation, Memphis, TN. p. 375–413

Christiansen, M.N., and R.P. Moore. 1959. Seed coat structural differences that influence water uptake and seed quality in hard seed cotton. Agron. J. 51:582–584 doi:10.2134/agronj1959.00021962005100100003x.

Dale, J.E. 1961. Investigation into the stomatal physiology of Upland cotton. Ann. Bot. (Lond.) 25:38–50.

Delmar, D., N. Carpita, and D. Montezinos. 1980. Cellulose synthesis in fibers of *G. hirsutum* Acala SJ-1. Proc. Beltwide Cotto Res. Prod. Conf. St. Louis, Mo. p. 351–351.

Esau, K. 1962. Anatomy of seed plants. John Wiley & Sons, New York.

Gausmann, H.W., L.N. Namken, M.D. Heilman, and R.H. Dilday. 1977. Histological evidence for a cotton genotype with opposite leaves and fruiting branches at a main stem node. Crop Sci. 17:659–661. doi:10.2135/cropsci1977.0011183X001700040047x.

Gipson, J.R. 1986. Temperature effects on growth, development, and fiber properties. In: J.R. Mauney and J.McD. Stewart, editors, Cotton physiology. The Cotton Foundation, Memphis, TN. p. 47–56.

Gore, U.R. 1935. Morphogenetic studies on the inflorescence of cotton. Bot. Gaz. 97:118–138. doi:10.1086/334540.

Hector, J.M. 1936. Introduction to the botany of field crops. Vol. II. Non-cereals. Central News Agency Ltd., Johannesburg.

Hesketh, J.D., D.N. Baker, and W.G. Duncan. 1972. Simulation of growth and yield of cotton: II. Environmental control of morphogenesis. Crop Sci. 12:436–439 10.2135/cropsci1972 .0011183X001200040011x.

Mauney, J.R. 1968. Morphology. In: F.C. Elliot, M. Hoover, and W.K. Porter, editors, Advances in production and utilization of quality cotton: Principles and practices. Iowa State Univ. Press, Ames.

Mauney, J.R., and E. Ball. 1959. The axillary buds of *Gossypium*. Bull. Torrey Bot. Club 86:236–244. doi:10.2307/2482840.

Namken, L.N., M.D. Heilmann, and R.H. Dilday. 1979. Arrangement of sympodia and earliness potential of cotton. Crop Sci. 19:620–622. doi:010.2135/cropsci1979.0011183X0019 00050018x

Pollock, E.G., and W.A. Jensen. 1964. Cell development during early embryogenesis in *Capsella* and *Gossypium*. Am. J. Bot. 51:915–921. doi:10.2307/2440240.

Quintanilha, A.L., D'Eca Salazar, and A. Cabral. 1962. Desenvolvimento de botao floral do algodoeiro em funcao do tempo. Bol. Sco. Brot. 36:189–215.

Quisenberry, J.R., and R.J. Kohel. 1975. Growth and development of fiber and seed in Upland cotton. Crop Sci. 15:463–467. doi:10.2135/cropsci1975.0011183X001500040005x.

Schubert, A.M., C.R. Benedict, J.D. Berlin, and R.J. Kohel. 1973. Cotton fiber development—Kinetics of cell elongation and secondary wall thickening. Crop Sci. 13:704–709. 10.2135/cropsci1973.0011183X001300060035x.

Stewart, J.McD. 1975. Fiber initiation on the cotton ovule (*G. hirsutum*). Am. J. Bot. 62:723–730. doi:10.2307/2442061.

Stewart, J.McD. 1986. Integrated events and their response to environment in cotton bolls. In: J.R. Mauney and J.McD. Stewart, editors, Cotton physiology. The Cotton Foundation, Memphis, TN. p. 261–297.

Turcotte, E.L., and C.V. Feaster. 1967. Semigamy in Pima cotton, *Gossypium barbadense* L. J. Hered. 58:54–57.

Fiber Biology

Hee Jin Kim

Abstract

Cotton (*Gossypium hirsutum* L.) fiber cells arising from the seed epidermis are considered the most important agricultural textile commodity in the world. Approximately 2 mo of developmental process are required to produce fully mature fibers. The temporal occurrences of four distinct fiber developmental stages consisting of initiation, elongation, secondary cell wall (SCW) biosynthesis, and maturation are tightly regulated and determine the yield and physical properties of the cotton fibers. With the advent of cutting edge genomics along with conventional genetic and breeding tools, our understanding of the molecular mechanisms and regulatory networks controlling fiber development has been dramatically expanded in recent years. The acquired knowledge is being used for improving cotton fibers. To overview the recent advances and provide potential strategies for improving cotton fiber, the chapter covers (i) commonly used techniques for elucidating regulatory mechanisms on fiber development; (ii) morphological and physiological changes during cotton fiber development; and (iii) recent advances in cotton functional genomics.

Cotton fiber is the most important agricultural textile commodity in the world (Wakelyn et al., 2010). In biological terms, a cotton fiber is a seed trichome arising from the protodermal cells of the outer integument layer of the seed coat. Cotton fiber trichome cells have been used as a good biological model for studying elongation and cell wall biogenesis because of the formation of its elongated structure with exaggerated size, distinct developmental stages between primary cell wall (PCW) and SCW biosynthesis, and a unicellular growth without interruption of cell division (Haigler et al., 2012; Kim and Triplett, 2001). Cotton fibers grow inside of a cotton boll, which is the ovary of the flower that develops into a mature seed-containing fruit (Fig. 1). The boll of *G. hirsutum* grown at typical US fields grows rapidly from 0 to 16 d post anthesis (DPA) and reaches full length at approximately 20 to 24 DPA. The boll is mature and open around 40 to 60 DPA depending on genotype and environmental conditions. Each cotton boll contains three to five

Abbreviations: AFIS, Advanced Fiber Information System; DEG, differentially expressed gene; DPA, days post anthesis; GA, gibberellic acid; HVI, High-Volume Instrument; IAA, indole-3-acetic acid; IAM, image analysis microscopy; miRNA, microRNA; MR, maturity ratio; NIL, near isogenic line; PCW, primary cell wall; RNA-seq, RNA sequencing; ROS, reactive oxygen species; SCW, secondary cell wall; TIP, tonoplast intrinsic protein; VIGS, virus-induced gene silencing; VLCFA, very long-chain fatty acid.

H.J. Kim, USDA-ARS, Southern Regional Research Center, 1100 Robert E. Lee Blvd, New Orleans, LA 70124 (Heejin.kim@ars.usda.gov).

doi:10.2134/agronmonogr57.2013.0022

Cotton. 2nd. ed. David D. Fang and Richard G. Percy, editors. Agronomy Monograph 57.

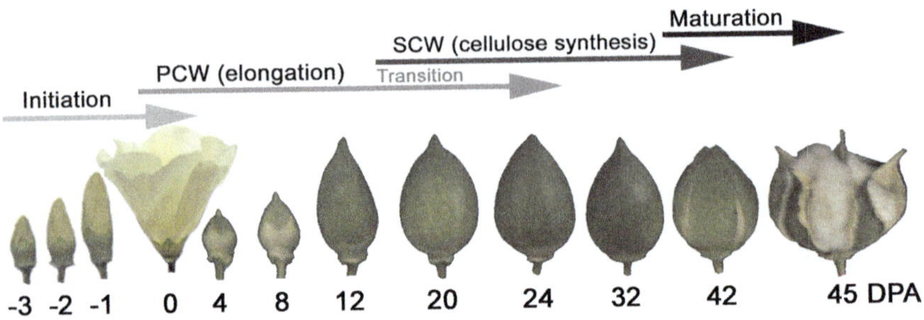

Fig. 1. Cotton boll development of *Gossypium hirsutum*. Major fiber developmental stages are classified according to the physical properties (length, cellulose content, and cell wall thickness) of cotton fibers grown at New Orleans, LA. DPA, days post anthesis; PCW, primary cell wall biosynthesis stage; SCW, secondary cell wall biosynthesis stage.

locules (compartments of the boll), and each locule has eight or nine ovules. Fiber development occurs fairly synchronously on 20 to 30 ovules in each boll. Cotton fiber development is divided into four overlapping stages: (i) initiation, (ii) PCW biosynthesis for fiber elongation, (iii) SCW biosynthesis for cellulose production and cell wall thickening, and (iv) maturation and cell death process before the boll opens (Fig. 1) (Benedict et al., 1999; Kim and Triplett, 2001; Naithani et al., 1982). The transition period that is overlapping between PCW and SCW stages was classified as a separate stage of fiber development due to its importance of fiber cell wall biosynthesis (Haigler et al., 2012; Stiff and Haigler, 2012).

At the fiber initiation stage (–1 or 0 DPA) of *G. hirsutum*, lint fibers growing up to 2.2 to 3.6 cm are differentiated from the ovular epidermal cells. Around 5 to 10 DPA, another class of fiber cells, fuzz fibers or linters, growing up to 1.2 to 1.5 cm starts growing (Lang, 1938). Fiber elongation continues from 0 DPA to approximately 21 to 26 DPA. A large central vacuole becomes prominent at early stage of cotton fiber development (Berlin, 1986; Ramsey and Berlin, 1976). During formation of the PCW, a thin PCW is deposited in elongating fibers. Approximately 14 to 16 DPA, fiber cells synchronously enter the SCW stage overlapping the final PCW stage. At the transition from PCW to SCW biosynthesis in cotton fiber development, synthesis of other cell wall polymers ceases and the rate of cellulose synthesis in cotton fibers dramatically increases (Meinert and Delmer, 1977). Mature fibers exhibit thickened SCW composed of nearly pure cellulose. Little or no lignin are detected from the mature fibers by cytochemical analyses (Macmillan et al., 2013; Weis et al., 1999). There was a report showing that very low levels (0.4–1% of dried weight) of lignin-like phenolics were detected from various cotton fibers by chemical analyses (Fan et al., 2009).

Cotton is also a good model for understanding the evolution of genomic duplications leading to chromosomal polyploidy. With the accumulated cotton genome sequences by emerging molecular biology techniques (Li et al., 2014; Paterson et al., 2012; Wendel et al., 2012), the diploid cotton species *G. raimondii* Ulbr. (D_5 genome), which only produces short and nonspinnable fibers, is the putative pollen parent for the allotetraploid species *G. hirsutum* (AD_1 genome) and *G. barbadense* L. (AD_2 genome) that are the agriculturally important cotton species producing spinnable fibers (Applequist et al., 2001; Wendel and Cronn, 2003). Two other diploid cotton species, *G. herbaceum* L. (A_1 genome) and *G. arboreum* L. (A_2 genome), are the closest relatives of the maternal parent for the allotetraploid cotton species. Among various cotton species, physical properties of cotton fibers are different from one another. The diploid cotton genomes from *G. raimondii* (D_5 genome, 885 Mb) and *G. arboreum* (A_2 genome, 1746 Mb) have been fully sequenced and assembled (Li et al., 2014; Paterson et al., 2012; K. Wang et al., 2012). The *G. raimondii* genome sequence (Paterson et al., 2012) is extensively used as a reference for the polyploidy cottons (2400 Mbp) due to its high quality, usefulness, and accessibility. Those genome sequences can be used within diverse experimental frameworks to answer biological questions on how fiber genes and networks affect the cotton fiber physical properties and yield, how cultivated cotton species evolved, how environmental condition affect fiber property, and how fiber properties and yield that are often negatively correlated with each other can be improved together.

Extensive lists of cotton transcriptome profiles during the last decade have dramatically expanded our understanding of the potential molecular mechanisms involved in cotton fiber development and can guide potential strategies for improving cotton fiber yield and properties. Thus, recent advances of cotton fiber development by genomics techniques have been reviewed by several cotton researchers (Aleman and Allen et al., 2010; Guan and Chen, 2013; Haigler et al., 2012; Kim and Triplett, 2001; Stiff and Haigler, 2012; Udall, 2009). This chapter describes (i) commonly used research tools for cotton fiber biology including genetic tools, genomics techniques, and phenotypic assays, (ii) morphological changes of *G. hirsutum* fibers during fiber development, and (iii) recent and new understanding of molecular mechanisms on fiber initiation, elongation, and cellulose biosynthesis whose biological processes directly affect the yield and physical properties like length, thickness (maturity), fineness, and strength of cotton fibers.

Common Tools for Studying Cotton Fiber Biology

To study cotton fiber biology, both conventional and innovative methods are required for evaluating genotypic and phenotypic traits in quantitative manners. Among them, genetic, genomic, and phenotypic approaches that are commonly used in cotton biology are summarized here.

Genetic Tools for Comparative Analyses

Several natural cotton mutants with their isogenic wild-type cotton are powerful genetic tools for comparative analyses identifying genes responsible for fiber initiation, elongation, and secondary wall biosynthesis (Fig. 2). A fuzzless and lintless (*fl*) mutant (also called a fiberless mutant), Xu-142 *fl*, which is also referred to as Xuzhou 142 lintless–fuzzless, XZ142WX, fl-xz-142, and XZ142w in different reports with its wild-type near isogenic line (NIL), has been extensively used to identify cotton genes involved in the fiber initiation and elongation process (Ji et al., 2003; Seki et al., 2002; Shi et al., 2006; Wang et al., 2010d; Wu et al., 2006, 2007; Zhang et al., 2007; Zhang and Pan, 1991). Other fiberless mutants such as MD17 and SL 1-7-1 (Turley and Kloth, 2002, 2008) and fuzzless mutants such as N_1 (Kearney and Harrison, 1927; Lee et al., 2006) and n_2 (Ware et al., 1947) have

Fig. 2. Phenotypes of *Gossypium hirsutum* wild-type (WT) and its near isogenic mutant lines defective in lint initiation (Xu-142 *fl*), linter initiation (N_1), fiber length (Li_1 and Li_2), and fiber cell wall thickness (*im*).

been frequently used for studying the process of fiber initiation. Two short fiber mutants, *Ligon-lintless 1* and *2* (Li_1 and Li_2), with fiber lengths less than 0.6 cm, have been used for identifying fiber genes involved in the fiber elongation process by comparison with their wild-type NILs that have a normal fiber length (average 3.0 cm) (Bolton et al., 2009; Ding et al., 2014; Gilbert et al., 2013, 2014; Hinchliffe et al., 2011; Kohel, 1973; Naoumkina et al., 2013; Zhao et al., 2010). An immature fiber (*im*) mutant was originally identified as nonfluffy cotton boll phenotype with a low micronaire value and lint percentage (Kohel and McMichael, 1990). The lower degree of fiber cell wall thickness of the *im* mutant over a wild-type NIL was subsequently confirmed (Kim et al., 2013a), and the *im* mutant has been used for identifying cotton genes involved in fiber thickness and secondary wall biosynthesis (Kim et al., 2013b; Wang et al., 2013a, 2014a).

In addition to natural cotton mutants, a pair of cotton germplasm lines having one or a few distinctive physical properties were derived by backcross breeding and used for comparative genomics analyses (Hinchliffe et al., 2010; Islam et al., 2014). For studies of the evolution of cotton fiber formation, various diploid and polyploid cotton species producing either spinnable or nonspinnable fibers were compared to one another (Applequist et al., 2001; Hovav et al., 2008a; Li et al., 2014; Yoo and Wendel, 2014).

Genomics Tools for Functional Analyses

The genomics revolution has greatly affected what is currently known about cotton fiber biology. Emerging cutting edge techniques of genomics with the reduced cost of such analyses have enabled more comprehensive genomics analyses than were previously possible with single gene studies. The sequence data obtained from cotton expressed sequence tags constructed from a variety of tissues and organs from multiple cotton species enabled to perform genome-wide transcriptome analyses on developing cotton fibers with cotton microarray chips or next generation sequencing techniques like RNA sequencing (RNA-seq) (Udall, 2009; Udall et al., 2006). The fully sequenced and assembled cotton genome of two diploid species, *G. raimondii* (D_5 genome) and *G. arboreum* (A_2 genome), have become reference genome sequences for study of the agriculturally important allotetraploid cotton species, *G. hirsutum* and *G. barbadense* (Li et al., 2014; Paterson et al., 2012; K. Wang et al., 2012). Despite the fact that the genome sequence of *G. hirsutum* remains to be fully sequenced, the genome physical map and subgenome specific UniGene sequences are available (Jiang et al., 2012; Jin et al., 2013).

To determine physiological roles and functions of each cotton gene, *Agrobacterium tumefaciens*–mediated cotton transformation is considered as the ultimate method although it requires lengthy tissue culture procedures (Trolinder, 2009; Zhang, 2013). In attempts to mitigate its laborious and time-consuming procedures, several alternative methods have been developed. First, the in vivo transformation methods, which are also called as pollen or floral-type transformation methods, were developed for functional analyses of cotton genes. The methods directly inoculating naked DNAs or *Agrobacterium* containing genes in cotton embryos through a pollen tube pathway have a great potential and have been successfully used in multiple laboratories (Burke et al., 1999; Chen et al., 2010; Zhou et al., 1983). However, the transformation efficiencies and ease of performing these procedures for the in vivo transformation methods remain to be further optimized (Duncan, 2010). Second, virus-induced gene silencing (VIGS) was developed as a tool for cotton functional genomics and gained acceptance as a rapid alternative knockdown method that allows silencing of individual genes or gene families. Two different cotton VIGS vectors from cotton leaf crumple virus and tobacco rattle virus were developed (Gao et al., 2011; Qu et al., 2012; Tuttle et al., 2008). Both cotton VIGS vectors were shown to silence fiber genes and consequently altered fiber phenotypes (Qu et al., 2012; Tuttle, 2011). Despite the potential of VIGS to have a higher throughput and more rapid evaluation of gene function than the *A. tumefaciens*–mediated stable transformation, some limitations of the VIGS were reported for other plants. In *Arabidopsis*, the silencing efficiency of the VIGS is affected by growth conditions, host plant age, and chosen gene fragment for silencing; thus, the silencing effect by the VIGS was not uniformly detected (Padmanabhan and Dinesh-Kumar, 2009). In tobacco and tomato, there are several reports describing an inability to silence gene during seed germination, a low silencing efficiency in seed coat, and a nonstable nature of silencing (Senthil-Kumar and Mysore, 2011). Third, *Agrobacterium rhizogenes*–induced cotton hairy root transformation was developed for determining promoter activities of a cellulose synthase (*GhCesA4*) responsible for SCW cellulose biosynthesis in developing fibers and root xylem (Kim et al., 2011). The *A. rhizogenes*–induced transformation can be used in a limited way for functional analysis of cotton genes that are expressed in the root (Kim, 2013).

Cotton ovule culture has been used as a powerful tool for numerous cotton researches (Triplett, 2000) since the culture methods for cotton ovules were optimized more than four decades ago (Beasley and Ting, 1973, 1974). The fibers cultured from unfertilized ovules with auxin and gibberellic acid (GA) showed a substantial similarity in fiber development and chemical composition to those from developing fibers grown in planta (Fig. 3a) (Meinert and Delmer, 1977). The development of both PCW and SCW in cultured fibers is also similar to that in field grown fibers (Fig. 3b). However, the fiber length (~1.7 cm) and cellulose content (70–80%) grown in vitro are shorter and lower than those grown in planta (Triplett, 2000). Cotton ovule culture has been used as a functional tool for evaluation of the effect of temperature on cellulose biosynthesis (Haigler et al., 1991; Martin and Haigler, 2004; Wuzi et al., 1993), the effect of ethylene on fiber elongation (Shi et al., 2006), the effect of brassinosteroids (Sun et al., 2005), and the effect of peptide hormones on fiber development (Han et al., 2014). It has also been used for studies determining subcellular localizations of cotton fiber proteins (Kim et al., 2008) and gene silencing studies (Bu et al., 2014).

Phenotypic Assays for Cotton Fibers

Accurate assessments of cotton fiber phenotypes driven by each gene are required for cotton biologists to determine functions of cotton fiber genes. Fiber phenotype is mainly determined according to fiber length, thickness (maturity), fineness, and strength (tenacity). The fiber length can be determined by a relatively easy manual method (Schubert et al., 1973) or other instruments, whereas other properties like thickness, fineness, and strength require several sophisticated instruments. Since cotton produces a highly variable fiber population with different fiber physical properties depending on genetic and environmental factors (Bradow and Davidonis, 2000; Paterson et al., 2003), cotton biologists often face challenges to identify minor phenotypic differences driven by an individual cotton gene. The

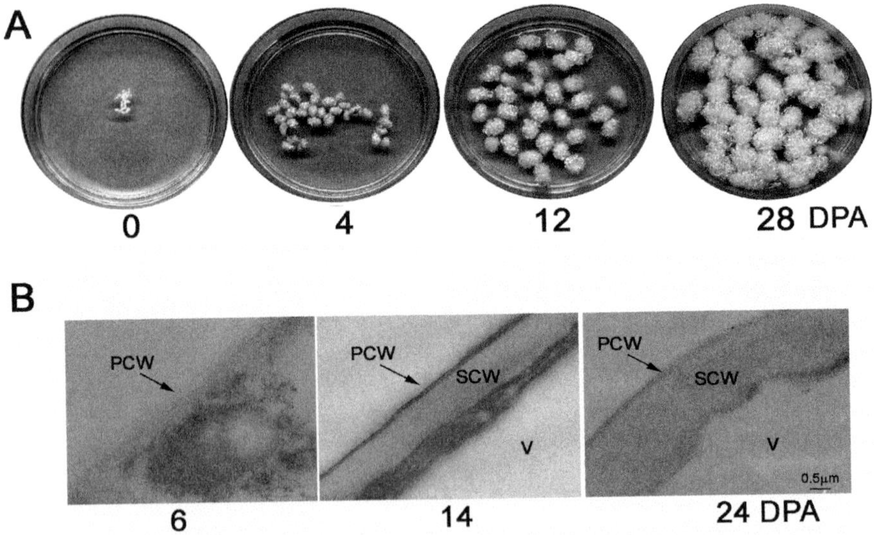

Fig. 3. Fiber development in cotton ovule culture. (a) Developing cotton ovules grown in culture. Unfertilized ovules (TM-1, 0 DPA) were cultured on basal medium with phytohormones (5 μM IAA and 0.5 μM GA_3) for 0, 4, 12, and 28 d. (b) Transmission electron microscopic images of fiber cell walls from developing fibers cultured at 6, 14, and 24 DPA (Kim et al., 2008). DPA, days post anthesis; IAA, indole-3-acetic acid; PCW, primary cell wall; SCW, secondary cell wall; V, vacuole.

High-Volume Instrument (HVI) was developed in 1969 to classify marketing values of fully developed commercial fibers, and the Advanced Fiber Information System (AFIS) was developed in the 1980s to predict spinning performance and yarn quality (Wakelyn et al., 2010). Both instruments have become important research tools for cotton breeding programs and functional genomics researches although they are not originally designed as science research instruments. The HVI is currently used to assess cotton fibers by the USDA's Agricultural Marketing Service and defined as a Standardized Instrument for Testing Cotton by the International Cotton Advisory Committee. High-Volume Instrument data provide major fiber physical properties such as length, bundle strength, and micronaire value representing a combination of fiber maturity (degree of cell wall thickness) and fineness of cotton fibers (Bradow and Davidonis, 2000; Wakelyn et al., 2010). To measure fiber maturity and fineness separately, the AFIS measurement is broadly accepted as a standard method among cotton biologists (Shofner et al., 1988). The AFIS measurement of fiber maturity produces a value called maturity ratio (MR), which represents the relative amount of cellulose in the fiber cross section in general. A higher MR indicates a more mature fiber. Fiber fineness is measured in millitex (mtex), which is milligrams per kilometer (mg km^{-1}) of fibers. The HVI and AFIS can assess the quality of fully developed and mature cotton fibers from commercial cotton varieties that have a narrow range of variations. However, there are some practical limits in their use for biological research because such studies often deal with developing fibers, immature fibers, and short and nonspinnable fibers composed of a wide range of variability (Kim et al., 2014).

With the advent of genomics research in which all transcriptome profiles are determined from transcripts extracted from developing fibers, there is a growing need to measure the physical properties of developing fibers that are difficult to be measured by automated instruments due to their short length, immaturity, and stickiness in developing fibers. For these analyses, the maturity can be determined by fiber cross-section image analysis microscopy (IAM), fineness by a gravimetric method, and strength by stelometry. However, these traditional methods are not frequently used due to their being slow and labor intensive (Thibodeaux and Evans, 1986; Wakelyn et al., 2010). Newly developed techniques such as Cottonscope, FTIR, and near infrared spectroscopy have been introduced to measure maturity and/or fineness (Abidi et al., 2014; Brims and Hwang, 2010; Liu, 2013; Paudel et al., 2013; Rodgers et al., 2010, 2013). Because of variations of the underlying principles and the practical limitations of each method, the quantitative values of fiber maturity measured from developing fibers by AFIS are not always matched to those measured by other methods such as IAM and Cottonscope (Fig. 4a). The MR measured by all AFIS, IAM, and Cottonscope is quantitative and represents the actual degrees of fiber wall thickness from mostly high maturity fibers at 40 DPA. In contrast, IAM and Cottonscope determine MR more accurately from low maturity fibers at 24 and 28 DPA than AFIS (Fig. 4a). Physiological sugars that are commonly found in developing fibers affect quantitative measurement of fiber fineness than fiber maturity since it causes stickiness among individual fibers. Fineness values measured by AFIS are affected by the stickiness caused by the high quantity of the physiological sugars in the early stage of developing fibers (Fig. 4b). Fineness values measured by Cottonscope are much more severely affected by the stickiness than those measured by AFIS (Kim et al., 2014).

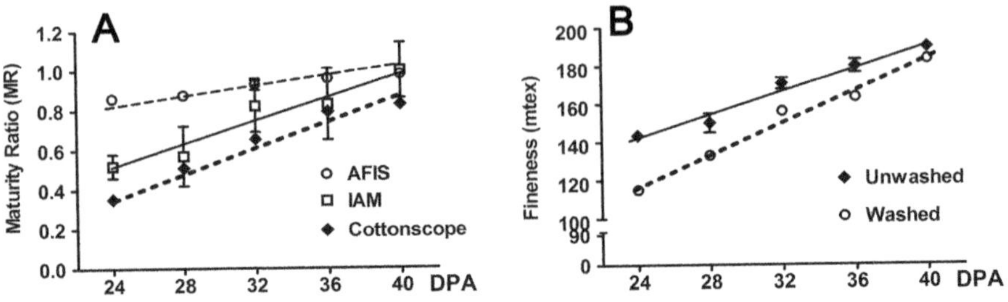

Fig. 4. Variations of fiber property measured by different methods and treatment. A. Comparisons of maturity ratio of developing fibers measured by three different methods. Maturity ratio of developing fibers (24–40 days post anthesis [DPA]) was quantitatively measured by the Advanced Fiber Information System (AFIS), image analysis microscopy (IAM), and Cottonscope (CS). (b) Effect of physiological sugars causing stickiness among individual fibers on fineness value of developing fibers. Fineness value of developing fibers was measured by AFIS from developing fibers in the presence (unwashed) or absence (washed) of physiological sugars.

In addition to the tools of measuring physical properties of fibers, a method of determining changes in fiber cell wall chemistry was developed using immunological techniques (Avci et al., 2013). Antibody-based glycome profiling of cotton fibers showed that *G. barbadense* fibers contained lower levels of xyloglucan than *G. hirsutum* fibers.

Recent Advances in Cotton Fiber Biology

This section mainly focuses on morphological changes during the four distinctive cotton fiber developmental stages and the recent progress made in understanding the molecular biology of cotton fiber biology using the research tools described in the previous section.

Differentiation of Fiber Initials from *G. hirsutum* Ovules

Morphological Changes during Fiber Initiation Stage

Fiber initials are differentiated at the chalazal end of the ovules around the day of anthesis (or 0 DPA), and new fiber initials are progressively differentiated toward the micropylar end until 4 DPA (Fig. 5a) (Lang, 1938; Ryser, 1999; Stewart, 1975). Approximately 15 to 30% of epidermal cells on cotton ovules become fiber initials (Fig. 5b and 5c) (Applequist et al., 2001; Lee et al., 2007; Stewart, 1975). The differentiated fiber initials rapidly elongate (Fig. 5d).

Biochemical analyses revealed that fiber cell fate determination occurs before the formation of morphologically visible fiber initials at 0 DPA (Graves and Stewart, 1988; Lee et al., 2007). Fiber initial density (the ratio of epidermal cells differentiating into fiber initials) or the number of fiber initials of each ovule was suggested to be an indicator of predicting cotton yield and considered as a potential target for improving yield by biotechnological manipulations (Lee et al., 2007). The fiber initial density had positive correlations with lint percentage affecting yield when five elite Upland cotton cultivars were compared to one another (Li et al., 2009). Transgenic cotton with higher fiber initial density caused by higher auxin levels in the ovule epidermis at the fiber initiation stage produced higher

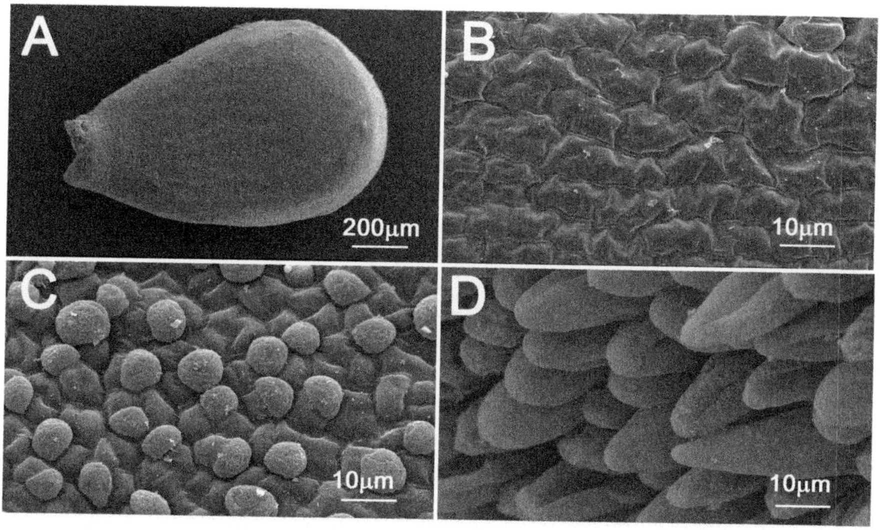

Fig. 5. Scanning microscopic images of differentiating fiber initials from epidermal tissues of cotton ovules. (a) *Gossypium hirsutum* TM-1 ovule (−1 days post anthesis [DPA]) with no detected fiber initials. (b) Magnified image of the epidermal surface of −1 DPA ovule. (c) Magnified image of fiber initials differentiating from 0 DPA ovule. (d) Magnified image of rapidly elongating fibers from +1 DPA ovule.

yield and finer fibers (Zhang et al., 2011). In contrast, no relationship between fiber initial densities and lint percentage was also reported by another research performed with 18 cotton cultivars (Romano et al., 2011).

Cotton Genes Involved in Fiber Initiation

In an effort to identify candidate genes responsible for differentiation of fiber initials, multiple transcriptome profiles were determined from fiber initials of wild-type ovules or compared between wild-type and fiberless mutants (Taliercio and Boykin, 2007; Wang et al., 2010d; Wu et al., 2006, 2007). From these studies it was found that multiple transcription factors such as *GhMYB25* and *GhMYB25-like* (Wu et al., 2006), *GaMYB2* (Guan et al., 2011; Wang et al., 2004), *GhMYB109* (Pu et al., 2008; Suo et al., 2003), *GhHD-1* (Wu et al., 2006), *GaHOX1* (Guan et al., 2011), *GbML1* (Zhang et al., 2010b), and a cotton protodermal factor 1 (*PDF1*) (Deng et al., 2012) were differentially expressed in fiber initials at the initiation stage.

Among the differentially expressed genes (DEGs), several genes such as a *GhMYB25-like* transcription factor (Walford et al., 2011), a *GhHD-1* transcription factor (Walford et al., 2012), and a vacuolar invertase, *GhVIN1* (Wang et al., 2014b) were confirmed to be involved in differentiation of fiber initials by functional analyses. RNAi-mediated suppression of *GhMYB25-like* transcription factor completely shut down differentiation of fiber initial trichomes, but not affect epidermal trichomes in other tissues (Walford et al., 2011). *GhMYB25-like* transcripts are predominantly detected in the epidermal layers of cotton ovules, and its temporal expression pattern is developmentally regulated during fiber initiation stage. *GhMYB25-like* is upsteam of *GhMYB25* and *GhMYB109*. Overexpression of *GhMYB25-like* did not increase the number of fiber initials, suggesting that there may be other factors interacting with GhMYB25-like for differentiating fiber initials from epidermal cells. The suppression of *GhHD-1* homeodomain-leucine zipper transcription factor that is structurally similar to the *Arabidopsis* GLABRA2 transcription factor responsible for leaf trichome differentiation delayed the timing of fiber initiation (Walford et al., 2012). Its overexpression increased the number of fiber initials on the seed but did not affect leaf trichomes. *GhHD-1* transcripts are expressed predominantly in epidermal tissues during early fiber development, and in other tissues bearing epidermal trichomes. Microarray analysis of silencing and overexpression lines of *GhHD-1* indicated that it potentially regulates the levels of ethylene and reactive oxidation species (ROS) through a WRKY transcription factor and calcium-signaling pathway. In contrast to *GhMYB25-like* and *GhHD-1* that play an indispensable roles in differentiation of fiber initials from ovules, many other identified DEGs have been suggested to involve in fiber elongation in addition to fiber initiation. A cotton vacuolar invertase GhVIN1 had known as a major regulator in fiber elongation since *GhVIN1* expression level in cotton fibers matched closely with VIN activity and fiber elongation rate, RNAi-mediated suppression of *GhVIN1* reduced fiber elongation, and its overexpression enhanced fiber elongation (Wang et al., 2010c). Interestingly, further functional analysis shows that GhVIN1 is also required for differentiation of fiber initials from the ovule epidermis (Wang et al., 2014b). Suppression of *GhVIN1* expressed highly in both elongating fibers and fiber initials resulted in a significant dosage-dependent reduction of VIN activity and consequently produced a fiberless seed phenotype (Wang et al., 2014b). As results, GhVIN1-derived hexose signaling was suggested to be involved in fiber cell differentiation from the ovule epidermis.

Similarly, *GhPDF1, GaHOX1, GhMYB109, GaRDL1,* and *GaMYb2* that are highly expressed at the late initiation (3–5 DPA) and early elongation stages (6–10 DPA) are likely to involve in both fiber elongation and initiation. RNAi-mediated suppression of *GhMYB25* leads to only a slight reduction in the number of fiber initials but reduces fiber length (Machado et al., 2009). Ectopic overexpression of *GhMYB25* increases in cotton fiber initiation and leaf trichome number. Similarly, antisense-mediated suppression of *GhMYB109* leads to a substantial reduction in fiber length (Pu et al., 2008). RNAi-mediated suppression of *PDF1* produces shorter length of fibers in addition to retardation of fiber initiation (Deng et al., 2012). The phenotypes of these transgenic cotton plants containing the suppressed levels of *MYB25, MYB109,* or *PDF1* indicate that they may play roles in the timing of fiber elongation although they were originally indentified as the genes involved in fiber initiation. Involvements of those genes in both fiber initiation and elongation are consistent with the previous report that suppression of sucrose synthase inhibited both fiber initiation and elongation (Ruan and Chourey, 1998; Ruan et al., 2003).

Regulators on Fiber Initiation

Phytohormonal regulation on differentiation of fiber initial from cotton ovules is known from cotton ovule culture. In cultured fibers, two phytohormones, indole-3-acetic acid (IAA) and GA are required for fiber differentiation from unfertilized cotton ovules and fiber elongation (Beasley and Ting, 1974). Comparisons of transcript abundance in cotton ovules at fiber initiation stage (–3 to 3 DPA) with those in other cotton tissues implicate that networks related to biosynthesis and signaling of auxin, GA, and brassinosteroids are involved in early fiber development (Yang et al., 2006). Increased IAA levels by the upregulation of the *iaaM*, an IAA biosynthetic gene in the epidermis of cotton ovules at the fiber initiation stage enhance fiber initial density, lint percentage, and fiber yield (Zhang et al., 2011). Increased GA levels by upregulation of $GhGA20_{ox}1$ in cotton cause more fiber initials and elongating fiber at 0 to 3 DPA ovules (Xiao et al., 2010).

Reactive oxygen species was also suggested to have a potential role on fiber initiation because exogenous applications of hydrogen peroxide at 0 DPA induced fiber initiation from naked seed (N_1) and fuzzless (XinFLM) cotton mutants (Zhang et al., 2010a). The expression of a peroxidase, *GhPOX* involved in regulating ROS levels was significantly suppressed in the transgenic line in which *GhHD-1* responsible for fiber initiation was silenced, but upregulated in the transgenic line in which *GhHD-1* was overexpressed (Walford et al., 2012). Transgenic plants in which fiber initiation and elongation processes are retarded by the suppression of *PDF1* also showed a higher accumulation of hydrogen peroxide compared to the wild-type cotton (Deng et al., 2012).

Genome-wide analysis of small RNAs from wild-type cotton during fiber initiation stage (–3, 0, and +3 DPA) suggested that microRNAs (miRNAs) may play important roles in the fiber initiation process (Pang et al., 2009). Significant differences in expression abundance of miRNAs between wild-type Xu-142 and its fiberless mutant NIL suggested that miRNAs targeted to genes involved in auxin signal transduction, GA signal transduction, actin bundles, and lignin biosynthesis potentially regulate the fiber initiation process (Kwak et al., 2009; Z.M. Wang et al., 2012).

Fiber Elongation in *G. hirsutum*

Morphological and Physiological Changes in Elongating Fibers

The differentiated fiber initials elongate rapidly during active PCW stage (0 to 14–16 DPA) at a rate of more than 2 mm d^{-1} (Fig. 6a and 6b). In elongating fibers, a thin PCW is deposited. As SCW synthesis begins at approximately 14 to 16 DPA, the fiber elongation rate slows down and stops at approximately 24 to 28 DPA (Fig. 6b and 6c). Cotton fibers elongate as tissue-like bundles, not as individual cells (Singh et al., 2009). At the end of the fiber elongation stage, cotton fibers become separated from one another and elicit high enzyme activities related to cell wall hydrolysis.

Previously, it was suggested that fiber elongation occurred solely via a diffuse growth mode in which fiber cell elongation occurs by extension of all over its surface, but not by tip growth mode in which elongation is localized at the apex of the cell (Seagull, 1990; Tiwari and Wilkins, 1995). Recent observations support a notion that fiber cells may expand via a linear-growth mode, which is the combination of the diffuse growth modes and tip-growth (Qin and Zhu, 2011). During fiber development, microtubules exhibit specific changes in orientation, organization, number, length, and proximity to the plasmalemma. In expanding cotton fibers, the patterns of microtubule deposition correlate precisely with the wall microfibril arrays (Seagull, 1986, 1993).

The rapid fiber elongation is driven by high turgor pressure that is regulated by developmentally reversible gating of plasmodesmata and a coordinated activity of solute transporters and wall-loosening expansins (An et al., 2007; Ruan et al., 2001, 2004; Shi et al., 2006). Aquaporins consisting of plasma membrane

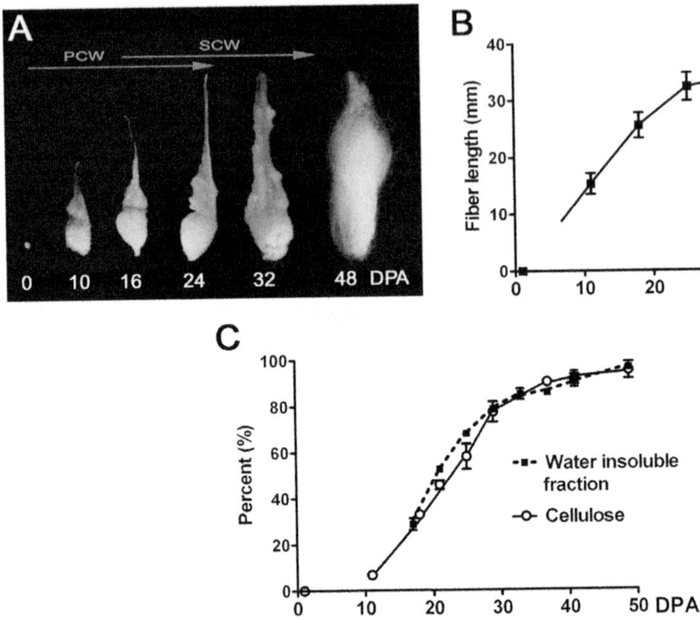

Fig. 6. Fiber elongation and cellulose biosynthesis during fiber development. (a) Single seed (*Gossypium hirsutum*, TM-1) associated with developing fibers at different time points. (b) Average fiber lengths at different time points. (c) Contents of cellulose and water-insoluble fraction in developing fibers. DPA, days post anthesis; PCW, primary cell wall biosynthesis stage; SCW, secondary cell wall biosynthesis stage.

intrinsic proteins (PIPs) and tonoplast intrinsic protein (TIPs) facilitate osmotic water influx in elongating cotton fibers (Liu et al., 2008; Park et al., 2010). A significant amount of sucrose enters into vacuoles by high activity of vacuole invertase (GhVIN1) during fiber elongation, which may play a key role in driving water influx into vacuoles mediated by TIPs (Ruan, 2014; Wang et al., 2010c).

Identification of Candidate Genes Involved in Fiber Elongation

Transcriptome profiles of elongating fibers compared with the developing fibers at different developmental stages identified DEGs that are related to GA, ethylene, and brassinosteroid mediated signaling, vesicle transport, cytoskeletion, expansin metabolism, cell wall biosynthesis, lipid metabolism, water channel activity, ubiquitin-dependent protein catabolism, and phenylpropanoid biosynthesis (Arpat et al., 2004; Gou et al., 2007; Hovav et al., 2008b). Those DEGs were also identified by comparisons of trascriptome profiles of developing fibers of the Li_1 and Li_2 short fiber mutants with their wild-type NILs (Bolton et al., 2009; Ding et al., 2014; Gilbert et al., 2013, 2014; Hinchliffe et al., 2011; Liu et al., 2012). Several genes involved in vesicle transport were commonly identified among the DEGs from both Li_1 and Li_2 short mutants regardless of growth conditions (Gilbert et al., 2014). Comparative transcriptome analysis of *G. longicalyx* J.B. Hutch. & B.J.S. Lee producing nonspinnable short fibers and *G. herbaceum* producing spinnable long fibers showed that the evolution of spinnable long fibers in cotton was accompanied by ROS regulating genes (Hovav et al., 2008a). Drought conditions changing the turgor pressure in cotton fibers can affect fiber elongation (Bradow and Davidonis, 2000; Ruan et al., 2001). Transcriptomic analysis of cotton under drought stress showed that drought has relatively little impact on fiber initiation but has profound effect on fiber elongation by downregulating important genes involved in cell wall loosening and expansion process (Padmalatha et al., 2012). When fiber elongation was inhibited by drought, genes involved in osmoprotectants, ion transporters, heat shock proteins, phytohormone (ABA, ethylene, and jasmonic acid) biosynthesis, and signal transduction were upregulated, whereas genes involved in phenylpropanoid and flavonoid biosynthesis, pentose and glucuronate interconversions, and starch and sucrose metabolism pathways were downregulated (Padmalatha et al., 2012; Park et al., 2012).

Functions of Genes Involved in Fiber Elongation

The genes described in the fiber initiation such as *GhVIN1* (Wang et al., 2010c), *GhMYB25* (Machado et al., 2009), *GhMYB109* (Pu et al., 2008), *PDF1* (Deng et al., 2012), $GhGA20_{ox}1$ (Xiao et al., 2010), and sucrose synthase (Ruan et al., 2003) are involved in fiber elongation since their suppressions causes shorter fiber lengths.

Genes involved in cytoskeleton formation are important for fiber elongation. A cotton actin, GhACT1, plays an important role in fiber elongation, but not fiber initiation (Li et al., 2005). RNAi-mediated suppression of *GhACT1* predominantly expressed in fiber cells disrupted the actin cytoskeleton network in fibers, and inhibited fiber elongation. A cotton annexin, AnxGb6, interacting with actin regulated fiber elongation (Huang et al., 2013b). Downregulation of another annexin gene, *GhAnn2*, also inhibited cotton fiber elongation and decreased Ca^{2+} influx at the cell apex (Tang et al., 2014a). Overexpression of a fiber-specific profilin (*GhPFN2*), an actin-binding protein, caused preterminated cell elongation, resulting in a marked decrease in the length of mature fibers (Wang et al., 2010b). Downregulation of the actin-depolymerizing factor gene (*GhADF1*) increased fiber length and fiber strength (Wang et al., 2009). A VIGS-mediated silencing of a cotton microtubule severing protein (KATANIN) generating internal breaks along microtubules produced shorter fibers and elevated weight ratio of seed oil to endosperm (Qu et al., 2012).

A cotton xyloglucan endotransglycosylase/hydrolase (GhXTH) mediating fiber cell wall extensibility by cleaving and reattaching xyloglucan polymers is involved in fiber elongation. *GhXTH1* is predominantly expressed in elongating cotton fibers. Overexpression of the *GhXTH1* produced mature cotton fibers that were between 15 and 20% longer than wild-type cotton plants under both greenhouse and field growth conditions (Lee et al., 2010). Both transgenic works and and segregation analyses confirm that GhXTH1 plays important roles in fiber elongation process (Lee et al., 2010).

Arabinogalactan proteins (AGPs) are also involved in fiber initiation and elongation. RNAi-mediated suppression of *GhAGP4*, a fasciclin-like arabinoglactan protein (FLA) inhibited fiber initiation and elongation (Li et al., 2010). Overexpression of a fasciclin-like arabinogalactan protein gene (*GhFLA1*) in cotton promoted fiber elongation, leading to an increase in fiber length. In contrast, suppression of *GhFLA1* expression in cotton slowed down fiber initiation and elongation, causing significantly shorter fiber length as compared with a wild-type cotton fibers (Huang et al., 2013a).

Pectin also plays an important role on fiber elongation. A cotton pectate lyase (GhPEL) mediating cell wall loosening by depolymerization of the de-esterified pectin is preferentially expressed in elongating fibers. Suppression of GhPEL reduced cell wall loosening activity and repressed fiber elongation (Wang et al., 2010a)

Fatty acids are involved in fiber elongation. *GhCER6* significantly upregulated in elongating fibers is determined as a functional 3-ketoacyl-CoA synthase catalyzing the initial condensation reaction during fatty acid elongation using malonyl-CoA and long-chain acyl-CoA as substrates (Qin et al., 2006). Exogenous applications of saturated very long-chain fatty acids (VLCFAs) in cotton ovule culture significantly promoted cotton fiber elongation, whereas the treatment of an inhibitor (acetochlor) of VLCFA biosynthesis abolished fiber growth. A long-chain fatty acid like lignoceric acid (C24:0) induced ethylene production by upregulation of an ethylene producing enzyme. Thus, the VLCFA was concluded to involve in maximizing the extensibility of elongating fibers, possibly by activating ethylene biosynthesis (Qin et al., 2007).

Plant-specific TEOSINTE-BRANCHED1/CYCLOIDEA/PCF (TCP) transcription factors also involve in fiber elongation. A *G. hirsutum* TCP transcription factor (*GhTCP14*) is expressed predominantly in fiber cells, especially in initiating and elongating fibers, and its expression increases in response to exogenous auxin (Wang et al., 2013b). The association of GhTCP14 with auxin responses was also evidenced by the enhanced expression of auxin response gene *IAA3*, a gene in the AUXIN/INDOLE-3-ACETIC ACID (Aux/IAA) family. A TCP transcription factor (*GbTCP*) is also preferentially detected in elongating fibers from *G. barbadense* (Hao et al., 2012). RNAi silencing of *GbTCP* produced shorter fiber and a reduced lint percentage than the wild-type plants. *GbTCP* positively regulates the level of jasmonic acid and, as a result, activates downstream genes involved in reactive oxygen species, calcium signaling, and ethylene biosynthesis and response.

Several genes in brassinosteroid signaling are also involved in fiber elongation. GhDET2, a steroid 5α-reductase catalyzing a major rate-limiting in brassinosteroid biosynthesis, plays an important role in fiber elongation (Luo et al., 2007). Antisense-mediated suppression of *GhDET2* inhibited both fiber initiation and fiber elongation. Furthermore, seed coat–specific expression of *GhDET2* increased fiber number and length. A cotton *pagoda1* (*pag1*) mutant exhibiting dwarfism and reduced fiber length was identified among activation-tagged T-DNA cotton lines (Yang et al., 2014). The PAG1 encodes a homolog of *Arabidopsis* CYP734A1 that inactivates brassinosteroids via C-26 hydroxylation. A comparison of transcript levels of elongating fibers between *pag1* and wild-type by RNa-seq analyses showed that the constitutive expression of PAG1 downregulated the expression of genes involved in VLCFA biosynthesis, ethylene-mediated signaling, response to cadmium, cell wall development, cytoskeleton organization, and cell growth.

Calcium-mediated signal transduction has been suggested to involve in cotton fiber elongation since calcium signaling genes including calmodulin (*CaM*), glutamate decarboxylase (*GAD*), calcineurin B-like (CBL) protein-interacting protein kinases (*CIPKs*), and calcium-dependent protein kinase (*GhCPK1*) are preferentially expressed in elongating fibers (Gao et al., 2007; Huang et al., 2008). A functional analysis of a calcium sensor (*GhCaM7*) shows that calcium signaling involves in fiber elongation by modulating ROS production (Tang et al., 2014b).

Overexpressing *GhCaM7* promotes early fiber elongation, whereas *GhCaM7* suppression by RNAi delays fiber initiation and inhibits fiber elongation. *GhCaM7* overexpression fiber cells show increased ROS concentrations compared with the wild-type, while *GhCaM7* RNAi fiber cells reduce concentrations.

LIM domain proteins involved in a variety of cellular processes in animal cells play roles in fiber elongation. In elongating fibers, two LIM domain proteins (GhWLIM5 and WLIM1a) have been characterized. *GhWLIM5* had a peak in expression during fiber elongation and GhWLIM5 facilitated fiber elongation by modulating the dynamic F-actin cytoskeleton (Li et al., 2013). Another LIM domain protein, WLIM1a, was preferentially expressed during the elongation and secondary wall synthesis stages in developing fibers (Han et al., 2013). Overexpression of WLIM1a in cotton led to significant changes in fiber length and secondary wall structure. Compared with the wild-type, fibers of WLIM1a overexpressing plants grew longer and formed a thinner and more compact SCW. WLIM1a may act as an actin bundler and also function as a transcription factor to activate expression of Phe ammonia-lyase box genes involved in phenylpropanoid biosynthesis to build up the SCW. WLIM1a localizes in the cytosol and nucleus and moves into the nucleus in response to hydrogen peroxide.

GhH2A12, a replication-dependent histone H2A gene, is involved in fiber elongation (Hao et al., 2014). No significant change was detected when *GhH2A12* was suppressed. However, overexpression of *GhH2A12* caused retardation of fiber initiation, shorter fibers, and lower lint percentages.

Fiber elongation was retarded by flavonoid naringenin (NAR), which is a substrate of flavanone 3-hydroxylase (F3H) (Tan et al., 2013). Silencing the *F3H* gene significantly increased the NAR content of fiber cells and reduced fiber length.

Apyrase enzyme removing the terminal phosphate from extracellular nucletoside tri- and diphosphates was suggested to promote fiber elongation by regulating ethylene signaling pathway (Clark et al., 2010). Two cotton ectoapyrases (*GhAPY1* and *GhAPY2*) were upregulated in elongating fibers. In a cotton ovule culture system, high extracellular ATP levels by blocking the ectoapyrase activity suppressed fiber growth, whereas the reduced concentrations stimulate growth. Treatment of cotton ovule cultures with the poorly hydrolyzable nucleotides ATPγS induces increased levels of ethylene.

Regulators on Fiber Elongation

Several phytohormones and their signaling cascades regulate fiber elongation. Exogenous application of GA is required for inducing elongation of fibers in cotton ovule culture, whereas GA biosynthesis inhibitor paclobutrazol reduces fiber elongation (Beasley and Ting, 1973; Liao et al., 2009). Upregulation of $GhGA20_{ox}1$ promoting GA levels in cotton resulted in more fiber initials and elongating fiber at 0 to 3 DPA ovules (Xiao et al., 2010). Several DELLA transcriptional regulators repressing GA-responsive growth are likely to involve in regulating fiber elongation. Overexpression of two cotton DELLA protein genes such as *GhSLR1b* (Aleman et al., 2008) and *GhRGL* (Liao et al., 2009) reduced growth in transgenic *Arabidopsis*. Exogenous applications of growth stimulating phytohormones like auxin and brassinosteroids on developing fibers also reduced transcript levels of *GhGAI1* DELLA gene, and consequently promote fiber elongation (Hu et al., 2011).

Brassinosteroids (BR) also stimulate fiber elongation in culture, whereas a BR inhibitor (BRZ) inhibits fiber elongation (Shi et al., 2006; Sun et al., 2005; Yang et al., 2014). Functional analyses of GhDET2 and PAG1 demonstrated that brassinosteroid is a critical regulator of fiber elongation, which may in turn be linked to effects on VLCFA biosynthesis, ethylene signaling, and cell wall– and cytoskeleton-related gene expression (Luo et al., 2007; Yang et al., 2014).

Ethylene gas known as a stress-related phytohomone in most plants stimulates fiber elongation, whereas an ethylene biosynthesis inhibitor, L-(2-aminoethoxyvinyl)-glycine (AVG), inhibits fiber elongation (Shi et al., 2006). Upregulation of each of the three ACOs (1-aminocyclopropane-1-carboxylic acid oxidases), the key enzymes in ethylene biosynthesis, was shown to occur in the fiber elongation stage in *G. hirsutum*. The VLCFA stimulated ethylene biosynthesis by upregulation of ACO (Qin et al., 2007). Activity of another key enzyme, 1-aminocyclopropane-1-carboxylic acid synthase (ACS) for ethylene biosynthesis was post-transcriptionally regulated by a Ca^{2+}–dependent protein kinase 1 (CPK1) during fiber elongation (Wang et al., 2011). Phosphorylated ACS2 significantly increased ACS activity, leading to elevated ethylene production. Comparative analyses of fiber elongation by *G. arboreum* (A_2 diploid genome) and *G. hirsutum* (AD_1 allotetraploid) suggested that the lower levels of ACO transcripts in elongating fibers in *G. arboreum* might be responsible for the shorter fiber length phenotype of *G. arboreum* (Li et al., 2014). Interestingly, the highest ACO transcript level was found in *G. raimondii* (D_5 diploid genome), a species producing the shortest fiber of the three cotton species. Thus, the extremely high level of ethylene produced by the substantial upregulation of ACOs in *G. raimondii* appears to act as a stress phytohormone inhibiting fiber elongation (Li et al., 2014).

Exogenous jasmonic acid inhibits cotton fiber elongation (Tan et al., 2012). Functional analysis of *GbTCP* transcription factor also suggested that jasmonic acid is involved in activating downstream genes involved in ROS, calcium signaling, and ethylene response (Hao et al., 2012). A peptide hormone, phytosulfokine-a (PSK-a), known to regulate plant cell proliferation and elongation in other plants, stimulates fiber elongation in *G. hirsutum* (Han et al., 2014).

Reactive oxygen species was suggested to involve in fiber development because ROS was significantly generated during cotton fiber initiation and elongation (Mei et al., 2009), exogenous application of hydrogen peroxide promoted early fiber elongation (Tang et al., 2014b), and exogenous application of physiological level of hydrogen peroxide into cotton ovule culture ended fiber elongation (Potikha et al., 1999). Comparative transcriptome analysis between *G. longicalyx* producing short and nonspinnable fibers and *G. herbaceum* producing long spinnable fibers showed that ROS regulating genes were expressed in the latter, but not the former, species (Hovav et al., 2008a). Cu/Zn superoxide dismutases (SOD) regulating levels of hydrogen peroxide in cytoplasm and leucoplast in developing fibers were upregulated in elongating fiber (Kim and Triplett, 2008; Kim et al., 2008). *GhPOX1* peroxidase is predominantly expressed in elongating fibers (Mei et al., 2009). Functional analysis of the calcium sensor *GhCaM7* showed ROS production is modulated in elongating fibers (Tang et al., 2014b).

In addition, eight miRNAs potentially targeting genes involved in calcium and auxin signal transduction, fatty acid metabolism, anthocyanin synthesis, and the xylem tissue differentiation are predicted to involve in regulating fiber elongation (Xue et al., 2013).

Secondary Cell Wall Cellulose Biosynthesis in *G. hirsutum* Fibers

At the transition (14–16 DPA) from PCW to SCW biosynthesis in cotton fiber, the rate of cellulose production in developing fibers increases substantially (Fig. 6c) (Meinert and Delmer, 1977) and the first layer of SCW that is referred to as the S1 or winding layer is deposited (Seagull, 1993). During the active SCW stage from 24 to 40 DPA, the vacuole is progressively filled as SCW cellulose is deposited. The molecular weight of cellulose in secondary walls is greater than that in primary walls (Marx-Figini, 1982). Differences in physical properties between primary and secondary walls are due, in part, to the numbers of glucose residues per cellulose molecule (Delmer and Amor, 1995; Timpa and Triplett, 1993). The SCW cellulose deposition in developing fibers increases the degree of fiber cell wall thickness that is measured as the MR by AFIS or Cottonscope (Fig. 7a). Thus, the degree of fiber cell wall thickness of developing fibers of *G. hirsutum*, TM-1, increases as the β-1,4-glucan chains form the cellulose microfibrils of the secondary wall during the SCW stage. High levels of variability on fiber thickness are detected among individual fibers (Fig. 7a). The HVI micronaire values, which indirectly measure a combination of fiber maturity and fineness, proportionally increase during the SCW thickening stage (24–40 DPA). Fiber cross-section image analysis microscopy (IAM) analysis shows that average perimeter (55 μm) of developing fibers is not changed during SCW stage, whereas the cell wall area proportionally increases from 41 to 148 μm^2 (Fig. 7b and 7c).

Changes of microtubule orientation are most apparent in the transition from rapid elongation and primary wall synthesis to the onset of SCW synthesis and

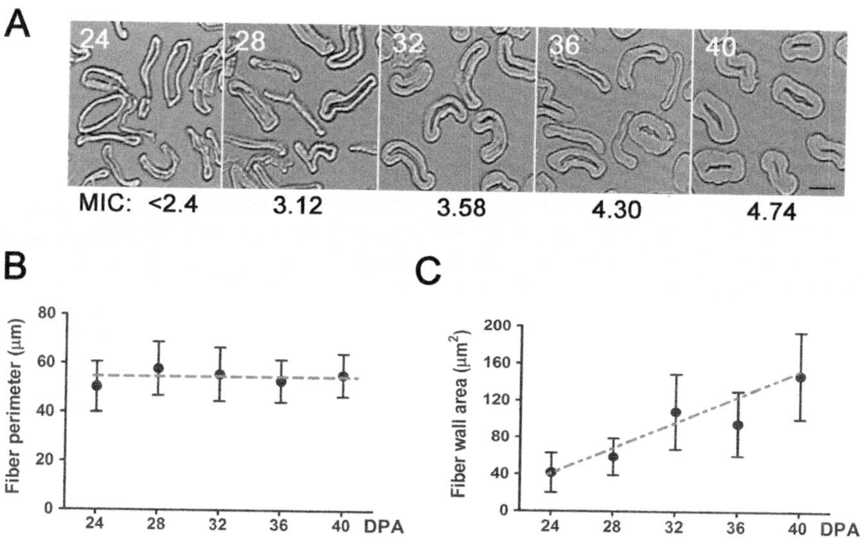

Fig. 7. Fiber perimeter and wall area of developing fiber during secondary cell wall biosynthesis. (a) Fiber cross-section image analysis microscopy of dried fibers (*Gossypium hirsutum* TM-1) at different developmental stages (24–40 days post anthesis [DPA]) with micronaire (MIC) values measured by the High-Volume Instrument. A scale bar represents 10 μm. (b) Average perimeters at different developmental stages measured from 300 fibers. (c) Average cell wall area at different developmental stages measured from 300 fibers.

slowing of elongation. Cellulose microfibrils are arranged helically around the growing fiber with periodic changes in the deposition angle (Seagull, 1986, 1993). The differently oriented cellulose microfibrils in the winding layer compared with the primary wall was suggested to play an important role in fiber strength (Haigler et al., 2012; Hinchliffe et al., 2010; Hsieh et al., 1995). Crystallinity of the cellulose microfibrils produces birefringence that is detected by polarized microscopes (Seagull et al., 2000). As the amount of ordered cellulose in the fiber wall increases, birefringence, a measure of cellulose microfibril organization, increases in the fibers. A developing fiber at 12 DPA having little crystalline cellulose shows little birefringence, whereas a developing fiber at 20 DPA having ordered crystalline cellulose produces birefringence (Fig. 8). As cotton fibers become mature, fibers are twisted due to the reversal regions where cellulose microfibril orientation changes.

During the active SCW biosynthesis stage, transcripts involved in SCW and cellulose biosynthesis are greatly upregulated (Haigler et al., 2005, 2009; Hovav et al., 2008b). Cellulose is synthesized by cellulose synthase complex (CSC), which is an active unit containing catalytic subunits (CesAs) and possibly CesA-associated proteins associated with the plasma membrane (Delmer and Amor, 1995; McFarlane et al., 2014). The first plant cDNAs encoding *CesA*, *GhCesA1* and *GhCesA2*, originally named *CelA1* and *CelA2*, were identified from developing cotton fibers at the SCW stage (Pear et al., 1996). A study on rapid turnover of the zinc-binding domain of GhCesA1 in vitro and in vivo suggested that active cellulose synthase complexes contain CesA proteins in dimerized form, and turnover and degradation of the complexes are mediated through reductive zinc insertion by metallothionein and subsequent proteolysis involving a cysteine protease

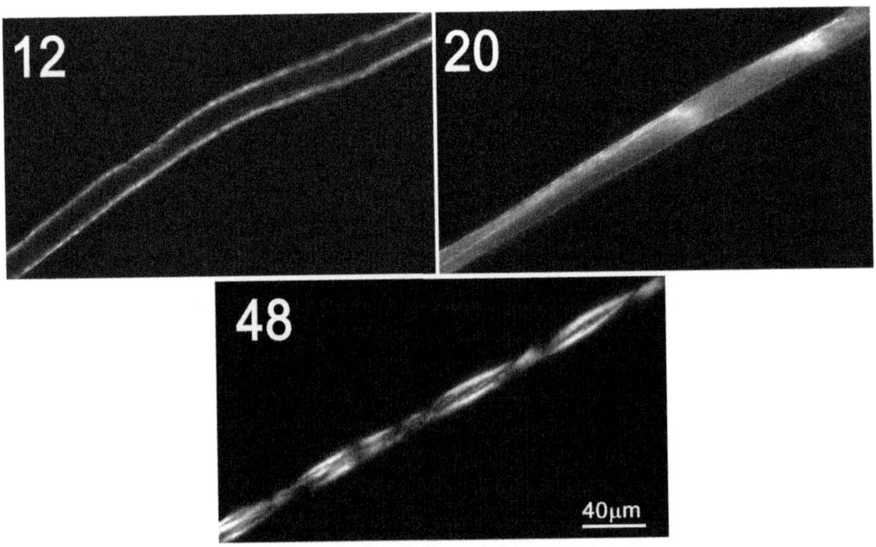

Fig. 8. Polarized microscopic images of cotton fibers. The intensity of white color represents birefringence produced from crystalline cellulose from developing (12 and 20 days post anthesis [DPA]) and fully developed *Gossypium hirsutum*, DP90 fiber (48 DPA) showing convolutions.

(Jacob-Wilk et al., 2006). Multiple proteins including KORRIGAN, sucrose synthase, COBRA, CSI1, CTL1, and unidentified proteins are known to interact with CesAs in *Arabidopsis* (McFarlane et al., 2014). To respond to a wide variety of signals for developmental and environmental transitions causing changes in plant cell wall biosynthesis, the CesAs are regulated at both transcriptional and posttranslational levels.

Hydrogen peroxide participates in dimerization of GhCesA1 cellulose synthases responsible for the SCW in cotton fibers. A sucrose synthase (Amor et al., 1995; Brill et al., 2011; Salnikov et al., 2003) and a chitinase-like gene (Zhang et al., 2004) were co-regulated with *GhCesA1* specifically at the SCW stage of development. Functional analyses showed that several genes such as *WLIM1a* and *GhADF1* involved in fiber elongation also appear to affect secondary wall biosynthesis. Overexpression of *WLIM1a* caused a thinner and more compact secondary wall (Han et al., 2013), whereas underexpression of *GhADF1* generated thicker secondary wall (Wang et al., 2009).

Phytohormones like cytokinins and ABA, which inhibit fiber elongation in cotton ovule culture, may be candidates for regulation of SCW biosynthesis in cotton fibers although direct evidence is lacking (Beasley and Ting, 1973, 1974). A synthetic auxin, NAA suppresses cellulose biosynthesis as compared with the natural auxin, IAA (Singh et al., 2009). Transcript levels of *GhCesA4*, the gene responsible for cellulose biosynthesis at the SCW stage, increased after exogenous applications of GA, kinetin, and brassinolide, whereas it decreased after NAA treatment (Kim et al., 2011).

In addition to the phytohormones, cellular H_2O_2 levels mediated by small GTPases were suggested to be involved in regulating SCW biosynthesis during fiber development. At the transition from elongation to the SCW stage, small *Rac9* and *13 GTPases* were specifically upregulated, whereas high levels of *GhRac1 GTPase* were detected during the elongation stage (Delmer et al., 1995; Gapper and Dolan, 2006; Kim and Triplett, 2004). This involvement was deduced from the observation that fiber birefringence increased after exogenous H_2O_2 treatment and decreased when treated with diphenyleneidonium (DPI), an inhibitor of NADPH oxidase producing ROS (Han et al., 2013; Potikha et al., 1999).

Secondary cell wall cellulose biosynthesis in developing fibers determines the degree of fiber wall thickness (fiber maturity). Cellulose is a polymer of glucose that is produced by photosynthesis. Thus, any environmental factors affecting photosynthesis are expected to modulate fiber maturity at the SCW biosynthesis stage. The micronaire values from cotton fibers grown with higher heat-unit accumulation and favorable photosynthesis were usually higher than those grown with cooler temperatures and unfavorable photosynthesis conditions (Bradow and Davidonis, 2000). Cool temperature is reported to inhibit the flux from glucose to sucrose that is required for SCW cellulose biosynthesis in developing fibers (Martin and Haigler, 2004). Comparative analysis between the *im* mutant fibers and the wild-type NIL fibers showed that the reduced fiber cell wall thickness of the *im* mutant fibers is associated with deregulation of the genes involved in stress responses and cellular respiration in addition to suppression of the genes involved in SCW synthesis like a sucrose synthase (Kim et al., 2013a; Wang et al., 2014a).

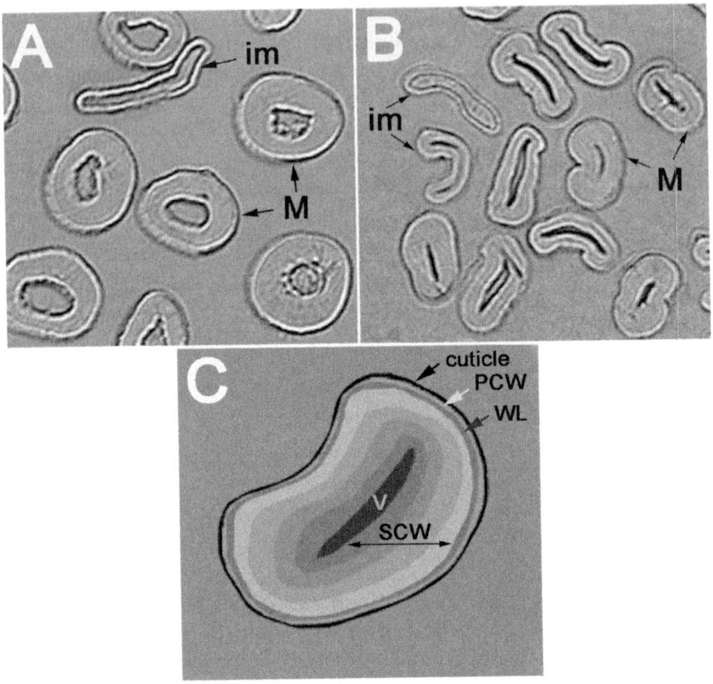

Fig. 9. Transition from cylindrical to bean shapes as fibers dehisces. (a) Cylindrical shapes of cross-sectioned never-dried developing fibers at 37 DPA. (b) Bean shapes of cross-sectioned dried mature fibers at 48 DPA (Kim et al., 2014). (c) Compartment of cross-sectioned dried and mature fibers. DPA, days post anthesis; im, immature fibers; M, mature fibers; PCW, primary cell wall; SCW, secondary cell wall with multiple layers; V, vacuole or lumen; WL, winding layer or transition layer.

Maturation Process in *G. hirsutum* Fiber

Approximately 40 to 45 DPA, *G. hirsutum* bolls dehisce at maturity and fiber are exposed to air and sunlight. As the fibers dry, the vacuole collapses and the fiber changes from a cylindrical shape to a flattened and twisted ribbon structure with a kidney bean–shaped cross section (Fig. 8, 9a, and 9b). The cross-sectioned fiber images also show that immature fibers containing thin SCW are found together with mature fibers containing thick SCW among the fully developed fiber bundles (Fig. 9b). Variations of fiber maturity (degree of wall thickness) among individual fibers cause problems for the textile industry such as yarn breakage and uneven dye uptake.

A dried fiber is covered by cuticle and primary cell wall components such as waxes, proteins, randomly organized microfibrils, and noncellulosic components (Fig. 9c). Between PCW and SCW, there is the winding layer that is deposited at the transition stage between the PCW and SCW stage (Fig. 9c). The SCW containing multiple layers determines the wall thickness and fiber maturity affecting the micronaire values and determines the quality and value of raw cotton fibers for textile processors.

The molecular mechanisms governing fiber maturation have been little studied because of difficulties of extracting high quality RNA and DNA from the mature fiber cells composed of almost pure cellulose. Programmed cell death is speculated to occur after SCW biosynthesis stage or during the maturation stage, but the mechanism for this process at this stage is unknown.

Conclusions

Using conventional breeding tools, innovative genomics techniques, and fully sequenced cotton genome sequences, our understanding on regulations of cotton fiber development has been greatly expanded. The acquired knowledge has been successfully adopted to improve cotton yield and physical properties of cotton fibers grown in fields (Abdurakhmonov et al., 2014; Jiang et al., 2012; Zhang et al., 2011). For a rapid development of cotton cultivars with improved yield, properties, adaptability for production, management and utilization, more research is needed to further determine physiological roles and functions of the candidate genes identified by comparative transcriptomic analyses and understand the molecular mechanisms and the gene networks that regulate fiber development.

Acknowledgments

This research was supported by the USDA–ARS CRIS project no. 6435-21000-016-00D and Cotton Incorporated–sponsored project no. 12-199. The author is grateful to Dr. Barbara Triplett, who introduced cotton fiber research as a mentor, and acknowledges contributions from former and current members of the Cotton Fiber Bioscience Research Unit and the Cotton Structure & Quality Research Unit, Southern Regional Research Center, USDA-ARS. The author thanks Dr. Kenneth Ehrlich of USDA-ARS-SRRC and unknown reviewers for critically reviewing the manuscript.

References

Abdurakhmonov, I.Y., Z.T. Buriev, S. Saha, J.N. Jenkins, A. Abdukarimov, and A.E. Pepper. 2014. Phytochrome RNAi enhances major fibre quality and agronomic traits of the cotton *Gossypium hirsutum* L. Nat. Commun. 5:3062. doi:10.1038/ncomms4062

Abidi, N., L. Cabrales, and C.H. Haigler. 2014. Changes in the cell wall and cellulose content of developing cotton fibers investigated by FTIR spectroscopy. Carbohydr. Polym. 100:9–16. doi:10.1016/j.carbpol.2013.01.074

Aleman, L., and R. Allen. 2010. Research in cotton fibre improvement. In: B.P. Singh, editor, Industrial crops and uses. CABI, Cambridge, MA. p. 277–307.

Aleman, L., J. Kitamura, H. Abdel-Mageed, J. Lee, Y. Sun, M. Nakajima, M. Ueguchi-Tanaka, M. Matsuoka, and R.D. Allen. 2008. Functional analysis of cotton orthologs of GA signal transduction factors GID_1 and SLR_1. Plant Mol. Biol. 68:1–16. doi:10.1007/s11103-008-9347-z

Amor, Y., C.H. Haigler, S. Johnson, M. Wainscott, and D.P. Delmer. 1995. A membrane-associated form of sucrose synthase and its potential role in synthesis of cellulose and callose in plants. Proc. Natl. Acad. Sci. USA 92:9353–9357. doi:10.1073/pnas.92.20.9353

An, C., S. Saha, J.N. Jenkins, B.E. Scheffler, T.A. Wilkins, and D.M. Stelly. 2007. Transcriptome profiling, sequence characterization, and SNP-based chromosomal assignment of the EXPANSIN genes in cotton. Mol. Genet. Genomics 278:539–553. doi:10.1007/s00438-007-0270-9

Applequist, W.L., R. Cronn, and J.F. Wendel. 2001. Comparative development of fiber in wild and cultivated cotton. Evol. Dev. 3:3–17. doi:10.1046/j.1525-142x.2001.00079.x

Arpat, A., M. Waugh, J.P. Sullivan, M. Gonzales, D. Frisch, D. Main, T. Wood, A. Leslie, R. Wing, and T. Wilkins. 2004. Functional genomics of cell elongation in developing cotton fibers. Plant Mol. Biol. 54:911–929. doi:10.1007/s11103-004-0392-y

Avci, U., S. Pattathil, B. Singh, V.L. Brown, M.G. Hahn, and C.H. Haigler. 2013. Cotton fiber cell walls of *Gossypium hirsutum* and *Gossypium barbadense* have differences related to loosely-bound xyloglucan. PLoS ONE 8:e56315. doi:10.1371/journal.pone.0056315

Beasley, C.A., and I.P. Ting. 1973. The effects of plant growth substances on *in vitro* fiber development from fertilized cotton ovules. Am. J. Bot. 60:130–139. doi:10.2307/2441099

Beasley, C.A., and I.P. Ting. 1974. Effects of plant growth substances on *in vitro* development from unfertilized cotton ovules. Am. J. Bot. 61:188–194. doi:10.2307/2441189

Benedict, C.R., R.J. Kohel, and H.L. Lewis. 1999. Cotton fiber quality. In: C.W. Smith and J.T. Cothren, editors, Cotton: Origin, history, technology, production. John Wiley & Sons, New York. p. 269–288.

Berlin, J.D. 1986. The outer epidermis of the cotton seed. In: J.R. Mauney and J.McD. Stuart, editors, Cotton physiology. Cotton Foundation, Memphis, TN. p. 375–414.

Bolton, J.J., K.M. Soliman, T.A. Wilkins, and J.N. Jenkins. 2009. Aberrant expression of critical genes during secondary cell wall biogenesis in a cotton mutant, Ligon lintless-1 (*Li-1*). Comp. Funct. Genomics. 28:659301. doi:10.1155/2009/659301

Bradow, J.M., and G.H. Davidonis. 2000. Quantitation of fiber quality and the cotton production-processing interface: A physiologist's perspective. J.Cotton Sci. 4:34–64.

Brill, E., M. van Thournout, R.G. White, D. Llewellyn, P.M. Campbell, S. Engelen, Y.L. Ruan, T. Arioli, and R.T. Furbank. 2011. A novel isoform of sucrose synthase is targeted to the cell wall during secondary cell wall synthesis in cotton fiber. Plant Physiol. 157:40–54. doi:10.1104/pp.111.178574

Brims, M., and H. Hwang. 2010. Introducing Cottonscope: A rapid and precise measurement of cotton fibre maturity based on Siromat. In: Proceedings of the National Cotton Council Beltwide Cotton Conference, NCC, New Orleans, LA. 4–7 Jan. 2010. NCC, Memphis, TN. p. 1417–1423.

Bu, Y.P., N.H. Chua, Y.F. Geng, J. Qu, and J. Ye. 2014. Virus induced gene silencing (vigs) for functional analysis of genes in cotton. US Patent 8853493. Date issued: 7 October.

Burke, J.J., M.J. Oliver, and J.P. Velten. 1999. Pollen-based transformation system using solid media. US Patent 5929300. Date issued: 27 July.

Chen, T., W. ShenJie, Z. Jun, G. WangZhen, and Z. TianZhen. 2010. Pistil drip following pollination: A simple in planta *Agrobacterium*-mediated transformation in cotton. Biotechnol. Lett. 32:547–555. doi:10.1007/s10529-009-0179-y

Clark, G., J. Torres, S. Finlayson, X. Guan, C. Handley, J. Lee, J.E. Kays, Z.J. Chen, and S.J. Roux. 2010. Apyrase (nucleoside triphosphate-diphosphohydrolase) and extracellular nucleotides regulate cotton fiber elongation in cultured ovules. Plant Physiol. 152:1073–1083. doi:10.1104/pp.109.147637

Delmer, D.P., and Y. Amor. 1995. Cellulose biosynthesis. Plant Cell 7:987–1000. doi:10.1105/tpc.7.7.987

Delmer, D.P., J.R. Pear, A. Andrawis, and D.M. Stalker. 1995. Genes encoding small GTP-binding proteins analogous to mammalian rac are preferentially expressed in developing cotton fibers. Molecular Gen. Genet. MGG 248:43–51. doi:10.1007/BF02456612

Deng, F., L. Tu, J. Tan, Y. Li, Y. Nie, and X. Zhang. 2012. $GbPDF_1$ is involved in cotton fiber initiation via the core cis-element $HDZIP_2ATATHB_2$. Plant Physiol. 158:890–904. doi:10.1104/pp.111.186742

Ding, M., Y. Jiang, Y. Cao, L. Lin, S. He, W. Zhou, and J. Rong. 2014. Gene expression profile analysis of Ligon lintless-1 (Li_1) mutant reveals important genes and pathways in cotton leaf and fiber development. Gene 535:273–285. doi:10.1016/j.gene.2013.11.017

Duncan, D. 2010. Cotton transformation. In: U.B. Zehr, editor, Cotton: Biotechnological advances. Springer, Berlin. p. 65–77.

Fan, L., W.J. Shi, W.R. Hu, X.Y. Hao, D.M. Wang, H. Yuan, and H.Y. Yan. 2009. Molecular and biochemical evidence for phenylpropanoid synthesis and presence of wall-linked phenolics in cotton fibers. J. Integr. Plant Biol. 51:626–637. doi:10.1111/j.1744-7909.2009.00840.x

Gao, P., P.-M. Zhao, J. Wang, H.-Y. Wang, X.-M. Wu, and G.-X. Xia. 2007. Identification of genes preferentially expressed in cotton fibers: A possible role of calcium signaling in cotton fiber elongation. Plant Sci. 173:61–69. doi:10.1016/j.plantsci.2007.04.008

Gao, X., T. Wheeler, Z. Li, C.M. Kenerley, P. He, and L. Shan. 2011. Silencing *GhNDR1* and *GhMKK2* compromises cotton resistance to *Verticillium wilt*. Plant J. 66:293–305. doi:10.1111/j.1365-313X.2011.04491.x

Gapper, C., and L. Dolan. 2006. Control of plant development by reactive oxygen species. Plant Physiol. 141:341–345. doi:10.1104/pp.106.079079

Gilbert, M.K., H.J. Kim, Y. Tang, M. Naoumkina, and D.D. Fang. 2014. Comparative transcriptome analysis of short fiber mutants Ligon-lintless 1 and 2 reveals common mechanisms pertinent to fiber elongation in cotton (*Gossypium hirsutum* L.). PLoS ONE 9:e95554. doi:10.1371/journal.pone.0095554

Gilbert, M.K., R.B. Turley, H.J. Kim, P. Li, G. Thyssen, Y. Tang, C.D. Delhom, M. Naoumkina, and D.D. Fang. 2013. Transcript profiling by microarray and marker analysis of the short cotton (*Gossypium hirsutum* L.) fiber mutant Ligon lintless-1 (Li_1). BMC Genomics 14:403. doi:10.1186/1471-2164-14-403

Gou, J.-Y., L.-J. Wang, S.-P. Chen, W.-L. Hu, and X.-Y. Chen. 2007. Gene expression and metabolite profiles of cotton fiber during cell elongation and secondary cell wall synthesis. Cell Res. 17:422–434.

Graves, D., and J. Stewart. 1988. Chronology of the differentiation of cotton (*Gossypium hirsutum* L.) fiber cells. Planta 175:254–258. doi:10.1007/BF00392435

Guan, X., and Z.J. Chen. 2013. Cotton fiber genomics. In: P.W. Becraft, editor, Seed genomics. Wiley-Blackwell, New York. p. 203–216.

Guan, X., J.J. Lee, M. Pang, X. Shi, D.M. Stelly, and Z.J. Chen. 2011. Activation of *Arabidopsis* seed hair development by cotton fiber-related genes. PLoS ONE 6:e21301. doi:10.1371/journal.pone.0021301

Haigler, C.H., L. Betancur, M.R. Stiff, and J.R. Tuttle. 2012. Cotton fiber: A powerful single-cell model for cell wall and cellulose research. Front. Plant Sci. 3:104. doi:10.3389/fpls.2012.00104

Haigler, C.H., N.R. Rao, E.M. Roberts, J.-Y. Huang, D.R. Upchurch, and N.L. Trolinder. 1991. Cultured ovules as models for cotton fiber development under low temperatures. Plant Physiol. 95:88–96. doi:10.1104/pp.95.1.88

Haigler, C.H., B. Singh, G. Wang, and D. Zhang. 2009. Genomics of cotton fiber secondary wall deposition and cellulose biogenesis. In: A.H. Paterson, editor, Genetics and genomics of cotton. Springer, New York. p. 385–417.

Haigler, C.H., D. Zhang, and C.G. Wilkerson. 2005. Biotechnological improvement of cotton fibre maturity. Physiol. Plant. 124:285–294. doi:10.1111/j.1399-3054.2005.00480.x

Han, J., J. Tan, L. Tu, and X. Zhang. 2014. A peptide hormone gene, GhPSK promotes fibre elongation and contributes to longer and finer cotton fibre. Plant Biotechnol. J. 12:861–871. doi: 10.1111/pbi.12187

Han, L.B., Y.B. Li, H.Y. Wang, X.M. Wu, C.L. Li, M. Luo, S.J. Wu, Z.S. Kong, Y. Pei, G.L. Jiao, and G.X. Xia. 2013. the dual functions of WLIM1a in cell elongation and secondary wall formation in developing cotton fibers. Plant Cell 25:4421–4438. doi:10.1105/tpc.113.116970

Hao, J., S. Chen, L. Tu, H. Hu, and X. Zhang. 2014. GhH2A12, a replication-dependent histone H2A gene from *Gossypium hirsutum*, is negatively involved in the development of cotton fiber cells. Plant Cell Rep. 33:1711–1721.

Hao, J., L. Tu, H. Hu, J. Tan, F. Deng, W. Tang, Y. Nie, and X. Zhang. 2012. GbTCP, a cotton TCP transcription factor, confers fibre elongation and root hair development by a complex regulating system. J. Exp. Bot. 63:6267–6281. doi:10.1093/jxb/ers278

Hinchliffe, D.J., W.R. Meredith, K.M. Yeater, H.J. Kim, A.W. Woodward, Z.J. Chen, and B.A. Triplett. 2010. Near-isogenic cotton germplasm lines that differ in fiber-bundle strength have temporal differences in fiber gene expression patterns as revealed by comparative high-throughput profiling. Theor. Appl. Genet. 120:1347–1366. doi:10.1007/s00122-010-1260-6

Hinchliffe, D.J., R.B. Turley, M. Naoumkina, H.J. Kim, Y. Tang, K.M. Yeater, P. Li, and D.D. Fang. 2011. A combined functional and structural genomics approach identified an EST-SSR marker with complete linkage to the Ligon lintless-2 genetic locus in cotton (*Gossypium hirsutum* L.). BMC Genomics 12:445. doi:10.1186/1471-2164-12-445

Hovav, R., J.A. Udall, B. Chaudhary, E. Hovav, L. Flagel, G. Hu, and J.F. Wendel. 2008a. The evolution of spinnable cotton fiber entailed prolonged development and a novel metabolism. PLoS Genet. 4:e25. doi:10.1371/journal.pgen.0040025

Hovav, R., J.A. Udall, B. Chaudhary, R. Rapp, L. Flagel, and J.F. Wendel. 2008b. Partitioned expression of duplicated genes during development and evolution of a single cell in a polyploid plant. Proc. Natl. Acad. Sci. USA 105:6191–6195. doi:10.1073/pnas.0711569105

Hsieh, Y.-L., E. Honik, and M. Hartzell. 1995. A developmental study of single fiber strength: Greenhouse grown SJ-2 Acala cotton. Text. Res. J. 65:101–112. doi:10.1177/004051759506500206

Hu, M.-Y., M. Luo, Y.-H. Xiao, X.-B. Li, K.-L. Tan, L. Hou, J. Dong, D.-M. Li, S.-Q. Song, and J. Zhao. 2011. Brassinosteroids and auxin down-regulate DELLA genes in fiber initiation and elongation of cotton. Agric. Sci. China 10:1168–1176. doi:10.1016/S1671-2927(11)60107-7

Huang, G.Q., S.Y. Gong, W.L. Xu, W. Li, P. Li, C.J. Zhang, D.D. Li, Y. Zheng, F.G. Li, and X.B. Li. 2013a. A fasciclin-like arabinogalactan protein, GhFLA1, is involved in fiber initiation and elongation of cotton. Plant Physiol. 161:1278–1290. doi:10.1104/pp.112.203760

Huang, Q.-S., H.-Y. Wang, P. Gao, G.-Y. Wang, and G.-X. Xia. 2008. Cloning and characterization of a calcium dependent protein kinase gene associated with cotton fiber development. Plant Cell Rep. 27:1869–1875. doi:10.1007/s00299-008-0603-0

Huang, Y., J. Wang, L. Zhang, and K. Zuo. 2013b. A cotton annexin protein AnxGb6 regulates fiber elongation through its interaction with actin 1. PLoS ONE 8:e66160. doi:10.1371/journal.pone.0066160

Islam, M.S., L. Zeng, C.D. Delhom, X. Song, H.J. Kim, P. Li, and D.D. Fang. 2014. Identification of cotton fiber quality quantitative trait loci using intraspecific crosses derived from two near-isogenic lines differing in fiber bundle strength. Mol. Breed. 34:373–384.

Jacob-Wilk, D., I. Kurek, P. Hogan, and D. P. Delmer. 2006. The cotton fiber zinc-binding domain of cellulose synthase A1 from *Gossypium hirsutum* displays rapid turnover in vitro and in vivo. Proc. Natl. Acad. Sci. USA. 103:12191–12196.

Ji, S.J., Y.C. Lu, J.X. Feng, G. Wei, J. Li, Y.H. Shi, Q. Fu, D. Liu, J.C. Luo, and Y.X. Zhu. 2003. Isolation and analyses of genes preferentially expressed during early cotton fiber development by subtractive PCR and cDNA array. Nucleic Acids Res. 31:2534–2543. doi:10.1093/nar/gkg358

Jiang, Y., W. Guo, H. Zhu, Y.-L. Ruan, and T. Zhang. 2012. Overexpression of GhSusA1 increases plant biomass and improves cotton fiber yield and quality. Plant Biotechnol. J. 10:301–312. doi:10.1111/j.1467-7652.2011.00662.x

Jin, X., Q. Li, G. Xiao, and Y.X. Zhu. 2013. Using genome-referenced expressed sequence tag assembly to analyze the origin and expression patterns of *Gossypium hirsutum* transcripts. J. Integr. Plant Biol. 55:576–585. doi:10.1111/jipb.12066

Kearney, T.H., and G.J. Harrison. 1927. Inheritance of smooth seeds in cotton. J. Agric. Res. 35:193–217.

Kim, H.J. 2013. *Agrobacterium rhizogenes*-induced cotton hairy root culture as an alternative tool for cotton functional genomics. Methods Mol. Biol. 958:179–187. doi:10.1007/978-1-62703-212-4_15

Kim, H.J., N. Kato, S. Kim, and B. Triplett. 2008. Cu/Zn superoxide dismutases in developing cotton fibers: Evidence for an extracellular form. Planta 228:281–292. doi:10.1007/s00425-008-0734-0

Kim, H.J., H.S. Moon, C.D. Delhom, L. Zeng, and D.D. Fang. 2013a. Molecular markers associated with the immature fiber (*im*) gene affecting the degree of fiber cell wall thickening in cotton (*Gossypium hirsutum* L.). Theor. Appl. Genet. 126:23–31. doi:10.1007/s00122-012-1956-x

Kim, H.J., N. Murai, D.D. Fang, and B.A. Triplett. 2011. Functional analysis of *Gossypium hirsutum cellulose synthase catalytic subunit 4* promoter in transgenic *Arabidopsis* and cotton tissues. Plant Sci. 180:323–332. doi:10.1016/j.plantsci.2010.10.003

Kim, H.J., J. Rodgers, C. Delhom, and X. Cui. 2014. Comparisons of methods measuring fiber maturity and fineness of Upland cotton fibers containing different degree of fiber cell wall development. Text. Res. J. 84:1622–1633. doi:10.1177/0040517514525878

Kim, H.J., and B.A. Triplett. 2001. Cotton fiber growth in planta and in vitro. Models for plant cell elongation and cell wall biogenesis. Plant Physiol. 127:1361–1366. doi:10.1104/pp.010724

Kim, H.J., and B.A. Triplett. 2004. Characterization of GhRac1 GTPase expressed in developing cotton (*Gossypium hirsutum* L.) fibers. Biochim. Biophys. Acta, Gene Struct. Expression. 1679:214–221. doi:10.1016/j.bbaexp.2004.06.008

Kim, H.J., and B. Triplett. 2008. Involvement of extracellular Cu/Zn superoxide dismutase in cotton fiber primary and secondary cell wall biosynthesis. Plant Signal. Behav. 3:1119–1121. doi:10.4161/psb.3.12.7039

Kim, H.J., Y. Tang, H.S. Moon, C.D. Delhom, and D.D. Fang. 2013b. Functional analyses of cotton (*Gossypium hirsutum* L.) immature fiber (*im*) mutant infer that fiber cell wall development is associated with stress responses. BMC Genomics 14:889. doi:10.1186/1471-2164-14-889

Kohel, R. 1973. Genetic nomenclature in cotton. J. Hered. 64:291–295.

Kohel, R.J., and S.C. McMichael. 1990. Immature fiber mutant of Upland cotton. Crop Sci. 30:419–421. doi:10.2135/cropsci1990.0011183X003000020038x

Kwak, P.B., Q.Q. Wang, X.S. Chen, C.X. Qiu, and Z.M. Yang. 2009. Enrichment of a set of microRNAs during the cotton fiber development. BMC Genomics 10:457. doi:10.1186/1471-2164-10-457

Lang, A.G. 1938. The origin of lint and fuzz hairs of cotton. J. Agric. Res. 56:507–521.

Lee, J., T.H. Burns, G. Light, Y. Sun, M. Fokar, Y. Kasukabe, K. Fujisawa, Y. Maekawa, and R.D. Allen. 2010. Xyloglucan endotransglycosylase/hydrolase genes in cotton and their role in fiber elongation. Planta 232:1191–1205. doi:10.1007/s00425-010-1246-2

Lee, J.J., O.S. Hassan, W. Gao, N.E. Wei, R.J. Kohel, X.-Y. Chen, P. Payton, S.-H. Sze, D.M. Stelly, and Z.J. Chen. 2006. Developmental and gene expression analyses of a cotton naked seed mutant. Planta 223:418–432. doi:10.1007/s00425-005-0098-7

Lee, J.J., A.W. Woodward, and Z.J. Chen. 2007. Gene expression changes and early events in cotton fibre development. Ann. Bot. (Lond.) 100:1391–1401. doi:10.1093/aob/mcm232

Li, C., W. Guo, and T. Zhang. 2009. Fiber initiation development in Upland cotton (*Gossypium hirsutum* L.) cultivars varying in lint percentage. Euphytica 165:223–230. doi:10.1007/s10681-008-9740-3

Li, F., G. Fan, K. Wang, F. Sun, Y. Yuan, G. Song, Q. Li, Z. Ma, C. Lu, and C. Zou. 2014. Genome sequence of the cultivated cotton *Gossypium arboreum*. Nat. Genet. 46:567–572. doi:10.1038/ng.2987

Li, X.-B., X.-P. Fan, X.-L. Wang, L. Cai, and W.-C. Yang. 2005. The cotton ACTIN1 gene is functionally expressed in fibers and participates in fiber elongation. Plant Cell. 17:859–875. doi:10.1105/tpc.104.029629

Li, Y., J. Jiang, L. Li, X.-L. Wang, N.-N. Wang, D.-D. Li, and X.-B. Li. 2013. A cotton LIM domain-containing protein (GhWLIM5) is involved in bundling actin filaments. Plant Physiol. Biochem. 66:34–40. doi:10.1016/j.plaphy.2013.01.018

Li, Y., D. Liu, L. Tu, X. Zhang, L. Wang, L. Zhu, J. Tan, and F. Deng. 2010. Suppression of GhAGP4 gene expression repressed the initiation and elongation of cotton fiber. Plant Cell Rep. 29:193–202. doi:10.1007/s00299-009-0812-1

Liao, W.-b., M.-b. Ruan, B.-m. Cui, N.-f. Xu, J.-j. Lu, and M. Peng. 2009. Isolation and characterization of a *GAI/RGA*-like gene from *Gossypium hirsutum*. Plant Growth Regul. 58:35–45. doi:10.1007/s10725-008-9350-z

Liu, D., L. Tu, L. Wang, Y. Li, L. Zhu, and X. Zhang. 2008. Characterization and expression of plasma and tonoplast membrane aquaporins in elongating cotton fibers. Plant Cell Rep. 27:1385–1394. doi:10.1007/s00299-008-0545-6

Liu, K., J. Sun, L. Yao, and Y. Yuan. 2012. Transcriptome analysis reveals critical genes and key pathways for early cotton fiber elongation in Ligon lintless-1 mutant. Genomics 100:42–50. doi:10.1016/j.ygeno.2012.04.007

Liu, Y. 2013. Recent progress in Fourier transform infrared (FTIR) spectroscopy study of compositional, structural and physical attributes of developmental cotton fibers. Materials (Basel) 6:299–313. doi:10.3390/ma6010299

Luo, M., Y. Xiao, X. Li, X. Lu, W. Deng, D. Li, L. Hou, M. Hu, Y. Li, and Y. Pei. 2007. GhDET2, a steroid 5a-reductase, plays an important role in cotton fiber cell initiation and elongation. Plant J. 51:419–430. doi:10.1111/j.1365-313X.2007.03144.x

Machado, A., Y. Wu, Y. Yang, D.J. Llewellyn, and E.S. Dennis. 2009. The MYB transcription factor GhMYB25 regulates early fibre and trichome development. Plant J. 59:52–62. doi:10.1111/j.1365-313X.2009.03847.x

Macmillan, C., H. Birke, F. Bedon, and F. Pettolino. 2013. Lignin deposition in cotton cells—Where is the lignin. J. Plant Biochem. Physiol. 1:e106.

Martin, L.K., and C.H. Haigler. 2004. Cool temperature hinders flux from glucose to sucrose during cellulose synthesis in secondary wall stage cotton fibers. Cellulose 11:339–349. doi:10.1023/B:CELL.0000046420.10403.15

Marx-Figini, M. 1982. The control of molecular weight and molecular-weight distribution in the biogenesis of cellulose. In: R.M. Brown Jr., editor, Cellulose and other natural polymer systems. Plenum Press, New York. p. 243–271.

McFarlane, H.E., A. Döring, and S. Persson. 2014. The cell biology of cellulose synthesis. Annu. Rev. Plant Biol. 65:69–94. doi:10.1146/annurev-arplant-050213-040240

Mei, W., Y. Qin, W. Song, J. Li, and Y. Zhu. 2009. Cotton *GhPOX1* encoding plant class III peroxidase may be responsible for the high level of reactive oxygen species production that is related to cotton fiber elongation. J. Genet. Genomics 36:141–150. doi:10.1016/S1673-8527(08)60101-0

Meinert, M.C., and D.P. Delmer. 1977. Changes in biochemical composition of the cell wall of the cotton fiber during development. Plant Physiol. 59:1088–1097. doi:10.1104/pp.59.6.1088

Naithani, S., N.R. Rao, and Y. Singh. 1982. Physiological and biochemical changes associated with cotton fibre development. Physiol. Plant. 54:225–229. doi:10.1111/j.1399-3054.1982.tb06330.x

Naoumkina, M., D.J. Hinchliffe, R.B. Turley, J.M. Bland, and D.D. Fang. 2013. Integrated metabolomics and genomics analysis provides new insights into the fiber elongation process in Ligon lintless-2 mutant cotton (*Gossypium hirsutum* L.). BMC Genomics 14:155. doi:10.1186/1471-2164-14-155

Padmalatha, K.V., G. Dhandapani, M. Kanakachari, S. Kumar, A. Dass, D.P. Patil, V. Rajamani, K. Kumar, R. Pathak, B. Rawat, S. Leelavathi, P.S. Reddy, N. Jain, K.N. Powar, V. Hiremath, I.S. Katageri, M.K. Reddy, A.U. Solanke, V.S. Reddy, and P.A. Kumar. 2012. Genome-wide transcriptomic analysis of cotton under drought stress reveal significant down-regulation of genes and pathways involved in fibre elongation and up-regulation of defense responsive genes. Plant Mol. Biol. 78:223–246. doi:10.1007/s11103-011-9857-y

Padmanabhan, M., and S.P. Dinesh-Kumar. 2009. Virus-induced gene silencing as a tool for delivery of dsRNA into plants. Cold Spring Harb. Protoc. 4:1–4. doi:10.1101/pdb.prot5139

Pang, M., A.W. Woodward, V. Agarwal, X. Guan, M. Ha, V. Ramachandran, X. Chen, B.A. Triplett, D.M. Stelly, and Z.J. Chen. 2009. Genome-wide analysis reveals rapid and dynamic changes in miRNA and siRNA sequence and expression during ovule and fiber development in allotetraploid cotton (*Gossypium hirsutum* L.). Genome Biol. 10:R122. doi:10.1186/gb-2009-10-11-r122

Park, W., B.E. Scheffler, P.J. Bauer, and B.T. Campbell. 2010. Identification of the family of aquaporin genes and their expression in upland cotton (*Gossypium hirsutum* L.). BMC Plant Biol. 10:142. doi:10.1186/1471-2229-10-142

Park, W., B.E. Scheffler, P.J. Bauer, and B.T. Campbell. 2012. Genome-wide identification of differentially expressed genes under water deficit stress in upland cotton (*Gossypium hirsutum* L.). BMC Plant Biol. 12:90. doi:10.1186/1471-2229-12-90

Paterson, A.H., Y. Saranga, M. Menz, C.X. Jiang, and R.J. Wright. 2003. QTL analysis of genotype x environment interactions affecting cotton fiber quality. Theor. Appl. Genet. 106:384–396. doi:10.1007/s00122-002-1025-y

Paterson, A.H., J.F. Wendel, H. Gundlach, H. Guo, J. Jenkins, D. Jin, D. Llewellyn, K.C. Showmaker, S. Shu, and J. Udall. 2012. Repeated polyploidization of *Gossypium* genomes and the evolution of spinnable cotton fibres. Nature 492:423–427. doi:10.1038/nature11798

Paudel, D., E. Hequet, and A. Noureddine. 2013. Evaluation of cotton fiber maturity measurements. Ind. Crops Prod. 45:435–441. doi:10.1016/j.indcrop.2012.12.041

Pear, J.R., Y. Kawagoe, W.E. Schreckengost, D.P. Delmer, and D.M. Stalker. 1996. Higher plants contain homologs of the bacterial celA genes encoding the catalytic subunit of cellulose synthase. Proc. Natl. Acad. Sci. USA 93:12637–12642. doi:10.1073/pnas.93.22.12637

Potikha, T.S., C.C. Collins, D.I. Johnson, D.P. Delmer, and A. Levine. 1999. The involvement of hydrogen peroxide in the differentiation of secondary walls in cotton fibers. Plant Physiol. 119:849–858. doi:10.1104/pp.119.3.849

Pu, L., Q. Li, X. Fan, W. Yang, and Y. Xue. 2008. The R2R3 MYB transcription factor GhMYB109 is required for cotton fiber development. Genetics 180:811–820. doi:10.1534/genetics.108.093070

Qin, Y.-M., C.-Y. Hu, Y. Pang, A.J. Kastaniotis, J.K. Hiltunen, and Y.-X. Zhu. 2007. Saturated very-long-chain fatty acids promote cotton fiber and Arabidopsis cell elongation by activating ethylene biosynthesis. Plant Cell. 19:3692–3704. doi:10.1105/tpc.107.054437

Qin, Y.M., F.M. Pujol, C.Y. Hu, J.X. Feng, A.J. Kastaniotis, J.K. Hiltunen, and Y.X. Zhu. 2006. Genetic and biochemical studies in yeast reveal that the cotton fibre-specific *GhCER6* gene functions in fatty acid elongation. J. Exp. Bot. 58:473–481. doi:10.1093/jxb/erl218

Qin, Y.-M., and Y.-X. Zhu. 2011. How cotton fibers elongate: A tale of linear cell-growth mode. Curr. Opin. Plant Biol. 14:106–111. doi:10.1016/j.pbi.2010.09.010

Qu, J., J. Ye, Y.-F. Geng, Y.-W. Sun, S.-Q. Gao, B.-P. Zhang, W. Chen, and N.-H. Chua. 2012. Dissecting functions of KATANIN and WRINKLED1 in cotton fiber development by virus-induced gene silencing. Plant Physiol. 160:738–748. doi:10.1104/pp.112.198564

Ramsey, J., and J. Berlin. 1976. Ultrastructural aspects of early stages in cotton fiber elongation. Am. J. Bot. 63:868–876. doi:10.2307/2442048

Rodgers, J., C. Delhom, D. Hinchliffe, H.J. Kim, and X. Cui. 2013. A rapid measurement for cotton breeders of maturity and fineness from developing and mature fibers. Text. Res. J. 83:1439–1451. doi:10.1177/0040517512471744

Rodgers, J., J. Montalvo, G. Davidonis, and T. VonHoven. 2010. Near infrared measurement of cotton fiber micronaire, maturity and fineness—A comparative investigation. Text. Res. J. 80:780–793. doi:10.1177/0040517509343780

Romano, G.B., E.W. Taliercio, R.B. Turley, and J.A. Scheffler. 2011. Fiber initiation in 18 cultivars and experimental lines of three *Gossypium* species. J. Cotton Sci. 15:61–72.

Ruan, Y.-L. 2014. Regulation of cotton fiber development by sugar and water. Poster presented at: Plant and Animal Genome XXII Conference, Plant and Animal Genome, San Diego, CA. 11–15 Jan. 2014. Poster W225.

Ruan, Y.-L., and P.S. Chourey. 1998. A fiberless seed mutation in cotton is associated with lack of fiber cell initiation in ovule epidermis and alterations in sucrose synthase expression and carbon partitioning in developing seeds. Plant Physiol. 118:399–406. doi:10.1104/pp.118.2.399

Ruan, Y.-L., D.J. Llewellyn, and R.T. Furbank. 2001. The control of single-celled cotton fiber elongation by developmentally reversible gating of plasmodesmata and coordinated expression of sucrose and K+ transporters and expansin. Plant Cell. 13:47–60.

Ruan, Y.-L., D.J. Llewellyn, and R.T. Furbank. 2003. Suppression of sucrose synthase gene expression represses cotton fiber cell initiation, elongation, and seed development. Plant Cell. 15:952–964. doi:10.1105/tpc.010108

Ruan, Y.-L., S.-M. Xu, R. White, and R.T. Furbank. 2004. Genotypic and developmental evidence for the role of plasmodesmatal regulation in cotton fiber elongation mediated by callose turnover. Plant Physiol. 136:4104–4113. doi:10.1104/pp.104.051540

Ryser, U. 1999. Cotton fiber initiation and histodifferentiation. In: A.S. Basra, editor, Cotton fibers: developmental biology, quality improvement, and textile processing. The Haworth Press, New York. p. 1–46.

Salnikov, V.V., M.J. Grimson, R.W. Seagull, and C.H. Haigler. 2003. Localization of sucrose synthase and callose in freeze-substituted secondary-wall-stage cotton fibers. Protoplasma 221:175–184.

Schubert, A., C. Benedict, J. Berlin, and R. Kohel. 1973. Cotton fiber development-kinetics of cell elongation and secondary wall thickening. Crop Sci. 13:704–709. doi:10.2135/cropsci1973.0011183X001300060035x

Seagull, R.W. 1986. Changes in microtubule organization and wall microfibril orientation during in vitro cotton fiber development: An immunofluorescent study. Can. J. Bot. 64:1373–1381. doi:10.1139/b86-188

Seagull R.W. 1990. Tip growth and transition to secondary wall synthesis in developing cotton hairs. In: I.B. Heath, editor, Tip growth in plant and fungal cells. Academic Press, San Diego, CA. p. 261–284.

Seagull, R.W. 1993. Cytoskeletal involvement in cotton fiber growth and development. Micron 24:643–660. doi:10.1016/0968-4328(93)90042-Y

Seagull, R.W., V. Oliveri, K. Murphy, A. Binder, and S. Kothari. 2000. Cotton fiber growth and development 2. Changes in cell diameter and wall birefringence. J. Cotton Sci. 4:97–104.

Seki, M., M. Narusaka, J. Ishida, T. Nanjo, M. Fujita, Y. Oono, A. Kamiya, M. Nakajima, A. Enju, and T. Sakurai. 2002. Monitoring the expression profiles of 7000 *Arabidopsis* genes under drought, cold and high-salinity stresses using a full-length cDNA microarray. Plant J. 31:279–292. doi:10.1046/j.1365-313X.2002.01359.x

Senthil-Kumar, M., and K.S. Mysore. 2011. New dimensions for VIGS in plant functional genomics. Trends Plant Sci. 16:656–665. doi:10.1016/j.tplants.2011.08.006

Shi, Y.-H., S.-W. Zhu, X.-Z. Mao, J.-X. Feng, Y.-M. Qin, L. Zhang, J. Cheng, L.-P. Wei, Z.-Y. Wang, and Y.-X. Zhu. 2006. Transcriptome profiling, molecular biological, and physiological studies reveal a major role for ethylene in cotton fiber cell elongation. Plant Cell 18:651–664. doi:10.1105/tpc.105.040303

Shofner, F.M., G.F. Williams, C.K. Bragg, and P.E. Sasser. 1988. Advanced fiber information system: A new technology for evaluating cotton. Paper presented at: Conference of the Textile Institute, Coventry, UK. 7–8 Dec. 1988.

Singh, B., H.D. Cheek, and C.H. Haigler. 2009. A synthetic auxin (NAA) suppresses secondary wall cellulose synthesis and enhances elongation in cultured cotton fiber. Plant Cell Rep. 28:1023–1032. doi:10.1007/s00299-009-0714-2

Stewart, J.M. 1975. Fiber initiation on the cotton ovule (*Gossypium hirsutum*). Am. J. Bot. 62:723–730. doi:10.2307/2442061

Stiff, M.R., and C.H. Haigler. 2012. Recent advances in cotton fiber development. In: Flowering and fruiting in cotton. The Cotton Foundation, Cordova, TN. p. 163–192.

Sun, Y., S. Veerabomma, H.A. Abdel-Mageed, M. Fokar, T. Asami, S. Yoshida, and R.D. Allen. 2005. Brassinosteroid regulates fiber development on cultured cotton ovules. Plant Cell Physiol. 46:1384–1391.

Suo, J., X. Liang, L. Pu, Y. Zhang, and Y. Xue. 2003. Identification of *GhMYB109* encoding a R2R3 MYB transcription factor that expressed specifically in fiber initials and elongating fibers of cotton (*Gossypium hirsutum* L.). Biochim. Biophys. Acta, Gene Struct. Expression. 1630:25–34. doi:10.1016/j.bbaexp.2003.08.009

Taliercio, E.W., and D. Boykin. 2007. Analysis of gene expression in cotton fiber initials. BMC Plant Biol. 7:22. doi:10.1186/1471-2229-7-22

Tan, J., L. Tu, F. Deng, H. Hu, Y. Nie, and X. Zhang. 2013. A genetic and metabolic analysis revealed that cotton fiber cell development was retarded by flavonoid naringenin. Plant Physiol. 162:86–95. doi:10.1104/pp.112.212142

Tan, J., L. Tu, F. Deng, R. Wu, and X. Zhang. 2012. Exogenous jasmonic acid inhibits cotton fiber elongation. J. Plant Growth Regul. 31:599–605. doi:10.1007/s00344-012-9260-1

Tang, W., Y. He, L. Tu, M. Wang, Y. Li, Y. Ruan, and X. Zhang. 2014a. Down-regulating annexin gene GhAnn2 inhibits cotton fiber elongation and decreases Ca (2+) influx at the cell apex. Plant Mol. Biol. 85:613–25. doi:10.1007/s11103-014-0208-7

Tang, W., L. Tu, X. Yang, J. Tan, F. Deng, J. Hao, K. Guo, K. Lindsey, and X. Zhang. 2014b. The calcium sensor GhCaM7 promotes cotton fiber elongation by modulating reactive oxygen species (ROS) production. New Phytol. 202:509–520. doi:10.1111/nph.12676

Thibodeaux, D.P., and J.P. Evans. 1986. Cotton fiber maturity by image analysis. Text. Res. J. 56:130–139. doi:10.1177/004051758605600211

Timpa, J.D., and B.A. Triplett. 1993. Analysis of cell-wall polymers during cotton fiber development. Planta 189:101–108. doi:10.1007/BF00201350

Tiwari, S.C., and T.A. Wilkins. 1995. Cotton (*Gossypium hirsutum*) seed trichomes expand via diffuse growing mechanism. Can. J. Bot. 73:746–757. doi:10.1139/b95-081

Triplett, B.A. 2000. Cotton ovule culture: A tool for basic biology, biotechnology and cotton improvement. In Vitro Cell. Dev. Biol. Plant 36:93–101. doi:10.1007/s11627-000-0020-8

Trolinder, N.L. 2009. Genetic engineering of cotton. In: A.H. Paterson, editor, Genetics and genomics of cotton. Springer, New York. p. 187–207.

Turley, R., and R. Kloth. 2002. Identification of a third fuzzless seed locus in upland cotton (*Gossypium hirsutum* L.). J. Hered. 93:359–364. doi:10.1093/jhered/93.5.359

Turley, R.B., and R.H. Kloth. 2008. The inheritance model for the fiberless trait in upland cotton (*Gossypium hirsutum* L.) line SL1-7-1: Variation on a theme. Euphytica 164:123–132. doi:10.1007/s10681-008-9670-0

Tuttle, J. 2011. Development of a virus-induced gene silencing system for cotton and its application for functional genomics in fiber. Doctoral diss., NC State Univ., Raleigh, NC.

Tuttle, J.R., A. Idris, J.K. Brown, C.H. Haigler, and D. Robertson. 2008. Geminivirus-mediated gene silencing from Cotton leaf crumple virus is enhanced by low temperature in cotton. Plant Physiol. 148:41–50. doi:10.1104/pp.108.123869

Udall, J.A. 2009. The *Gossypium* transcriptome. In: A.H. Paterson, editor, Genetics and genomics of cotton. Springer, New York. p. 157–185.

Udall, J.A., J.M. Swanson, K. Haller, R.A. Rapp, M.E. Sparks, J. Hatfield, Y. Yu, Y. Wu, C. Dowd, and A.B. Arpat. 2006. A global assembly of cotton ESTs. Genome Res. 16:441–450. doi:10.1101/gr.4602906

Wakelyn, P.J., N.R. Bertoniere, A.D. French, D.P. Thibodeaux, B.A. Triplett, M.-A. Rousselle, W.R. Goynes Jr, J.V. Edwards, L. Hunter, and D.D. McAlister. 2010. Cotton fiber chemistry and technology. CRC Press, Boca Raton, FL.

Walford, S.-A., Y. Wu, D.J. Llewellyn, and E.S. Dennis. 2011. GhMYB25-like: A key factor in early cotton fibre development. Plant J. 65:785–797. doi:10.1111/j.1365-313X.2010.04464.x

Walford S.-A., Y. Wu, D.J. Llewellyn, and E.S. Dennis. 2012. Epidermal cell differentiation in cotton mediated by the homeodomain leucine zipper gene, GhHD-1. Plant J. 71:464–478. doi:10.1111/j.1365-313X.2012.05003.x

Wang, C., Y. Lv, W. Xu, T. Zhang, and W. Guo. 2014a. Aberrant phenotype and transcriptome expression during fiber cell wall thickening caused by the mutation of the Im gene in immature fiber (*im*) mutant in *Gossypium hirsutum* L. BMC Genomics 15:94. doi:10.1186/1471-2164-15-94

Wang, C., T. Zhang, and W. Guo. 2013a. The *im* mutant gene negatively affects many aspects of fiber quality traits and lint percentage in cotton. Crop Sci. 53:27–37. doi:10.2135/cropsci2012.05.0304

Wang, H., Y. Guo, F. Lv, H. Zhu, S. Wu, Y. Jiang, F. Li, B. Zhou, W. Guo, and T. Zhang. 2010a. The essential role of GhPEL gene, encoding a pectate lyase, in cell wall loosening by depolymerization of the de-esterified pectin during fiber elongation in cotton. Plant Mol. Biol. 72:397–406. doi:10.1007/s11103-009-9578-7

Wang, H., W. Mei, Y. Qin, and Y. Zhu. 2011. 1-Aminocyclopropane-1-carboxylic acid synthase 2 is phosphorylated by calcium-dependent protein kinase 1 during cotton fiber elongation. Acta Biochim. Biophys. Sin. (Shanghai) 43:654–661. doi:10.1093/abbs/gmr056

Wang, H.Y., J. Wang, P. Gao, G.L. Jiao, P.M. Zhao, Y. Li, G.L. Wang, and G.X. Xia. 2009. Down-regulation of *GhADF1* gene expression affects cotton fibre properties. Plant Biotechnol. J. 7:13–23. doi:10.1111/j.1467-7652.2008.00367.x

Wang, J., H.-Y. Wang, P.-M. Zhao, L.-B. Han, G.-L. Jiao, Y.-Y. Zheng, S.-J. Huang, and G.-X. Xia. 2010b. Overexpression of a profilin (GhPFN$_2$) promotes the progression of developmental phases in cotton fibers. Plant Cell Physiol. 51:1276–1290. doi:10.1093/pcp/pcq086

Wang, K., Z. Wang, F. Li, W. Ye, J. Wang, G. Song, Z. Yue, L. Cong, H. Shang, and S. Zhu. 2012. The draft genome of a diploid cotton *Gossypium raimondii*. Nat. Genet. 44:1098–1103. doi:10.1038/ng.2371

Wang, L., A. Cook, J.W. Patrick, X.-Y. Chen, and Y.-L. Ruan. 2014b. Silencing the vacuolar invertase geneGhVIN1blocks cotton fiber initiation from the ovule epidermis, probably by suppressing a cohort of regulatory genes via sugar signaling. Plant J. 78:686–696. doi:10.1111/tpj.12512

Wang, L., X.-R. Li, H. Lian, D.-A. Ni, Y.-k. He, X.-Y. Chen, and Y.-L. Ruan. 2010c. Evidence that high activity of vacuolar invertase is required for cotton fiber and Arabidopsis root elongation through osmotic dependent and independent pathways, respectively. Plant Physiol. 154:744–756. doi:10.1104/pp.110.162487

Wang, M.Y., P.M. Zhao, H.Q. Cheng, L.B. Han, X.M. Wu, P. Gao, H.Y. Wang, C.L. Yang, N.Q. Zhong, J.R. Zuo, and G.X. Xia. 2013b. The cotton transcription factor TCP14 functions in auxin-mediated epidermal cell differentiation and elongation. Plant Physiol. 162:1669–1680. doi:10.1104/pp.113.215673

Wang, Q.Q., F. Liu, X.S. Chen, X.J. Ma, H.Q. Zeng, and Z.M. Yang. 2010d. Transcriptome profiling of early developing cotton fiber by deep-sequencing reveals significantly differential expression of genes in a fuzzless/lintless mutant. Genomics 96:369–376. doi:10.1016/j.ygeno.2010.08.009

Wang, S., J.-W. Wang, N. Yu, C.-H. Li, B. Luo, J.-Y. Gou, L.-J. Wang, and X.-Y. Chen. 2004. Control of plant trichome development by a cotton fiber MYB gene. Plant Cell 16:2323–2334. doi:10.1105/tpc.104.024844

Wang, Z.M., W. Xue, C.J. Dong, L.G. Jin, S.M. Bian, C. Wang, X.Y. Wu, and J.Y. Liu. 2012. A comparative miRNAome analysis reveals seven fiber initiation-related and 36 novel miRNAs in developing cotton ovules. Mol. Plant 5:889–900. doi:10.1093/mp/ssr094

Ware, J.O., L.I. Benedict, and W.H. Rolfe. 1947. A recessive naked seed character in Upland cotton. J. Hered. 38:313–319.

Weis, K.G., K.R. Jacobsen, and J.A. Jernstedt. 1999. Cytochemistry of developing cotton fibers: A hypothesized relationship between motes and non-dyeing fibers. Field Crops Res. 62:107–117. doi:10.1016/S0378-4290(99)00004-0

Wendel, J.F., and R.C. Cronn. 2003. Polyploidy and the evolutionary history of cotton. Adv. Agron. 78:139–186. doi:10.1016/S0065-2113(02)78004-8

Wendel, J.F., L.E. Flagel, and K.L. Adams. 2012. Jeans, genes, and genomes: Cotton as a model for studying polyploidy. In: P.S. Soltis and D.E. Soltis, editors, Polyploidy and genome evolution. Springer-Verlag, Heidelberg. p. 181–207.

Wu, Y., D.J. Llewellyn, R. White, K. Ruggiero, Y. Al-Ghazi, and E.S. Dennis. 2007. Laser capture microdissection and cDNA microarrays used to generate gene expression profiles of the rapidly expanding fibre initial cells on the surface of cotton ovules. Planta 226:1475–1490. doi:10.1007/s00425-007-0580-5

Wu, Y., A.C. Machado, R.G. White, D.J. Llewellyn, and E.S. Dennis. 2006. Expression profiling identifies genes expressed early during lint fibre initiation in cotton. Plant Cell Physiol. 47:107–127. doi:10.1093/pcp/pci228

Wuzi, X., N.L. Trolinder, and C.H. Haigler. 1993. Cool temparature effects on cotton fiber initiation and elongation clarified using in vitro cultures. Crop Sci. 33:1258–1264. doi:10.2135/cropsci1993.0011183X003300060029x

Xiao, Y.-H., D.-M. Li, M.-H. Yin, X.-B. Li, M. Zhang, Y.-J. Wang, J. Dong, J. Zhao, M. Luo, and X.-Y. Luo. 2010. Gibberellin 20-oxidase promotes initiation and elongation of cotton fibers by regulating gibberellin synthesis. J. Plant Physiol. 167:829–837. doi:10.1016/j.jplph.2010.01.003

Xue, W., Z. Wang, M. Du, Y. Liu, and J.-Y. Liu. 2013. Genome-wide analysis of small RNAs reveals eight fiber elongation-related and 257 novel microRNAs in elongating cotton fiber cells. BMC Genomics 14:629. doi:10.1186/1471-2164-14-629

Yang, S.S., F. Cheung, J.J. Lee, M. Ha, N.E. Wei, S.H. Sze, D.M. Stelly, P. Thaxton, B. Triplett, and C.D. Town. 2006. Accumulation of genome-specific transcripts, transcription factors and phytohormonal regulators during early stages of fiber cell development in allotetraploid cotton. Plant J. 47:761–775. doi:10.1111/j.1365-313X.2006.02829.x

Yang, Z., C. Zhang, X. Yang, K. Liu, Z. Wu, X. Zhang, W. Zheng, Q. Xun, C. Liu, L. Lu, Z. Yang, Y. Qian, Z. Xu, C. Li, J. Li, and F. Li. 2014. PAG1, a cotton brassinosteroid catabolism gene, modulates fiber elongation. New Phytol. 203:437–448. doi:10.1111/nph.12824

Yoo, M.-J., and J.F. Wendel. 2014. Comparative evolutionary and developmental dynamics of the cotton (*Gossypium hirsutum*) fiber transcriptome. PLoS Genet. 10:e1004073. doi:10.1371/journal.pgen.1004073

Zhang, B. 2013. Transgenic cotton: From biotransformation methods to agricultural application. Methods Mol. Biol. 958:3–15. doi:10.1007/978-1-62703-212-4_1

Zhang, D., M. Hrmova, C.H. Wan, C. Wu, J. Balzen, W. Cai, J. Wang, L.D. Densmore, G.B. Fincher, H. Zhang, and C.H. Haigler. 2004. Members of a new group of chitinase-like genes are expressed preferentially in cotton cells with secondary walls. Plant Mol. Biol. 54:353–372. doi:10.1023/B:PLAN.0000036369.55253.dd

Zhang, D., T. Zhang, and W. Guo. 2010a. Effect of H_2O_2 on fiber initiation using fiber retardation initiation mutants in cotton (*Gossypium hirsutum*). J. Plant Physiol. 167:393–399. doi:10.1016/j.jplph.2009.10.005

Zhang, D.Y., T.Z. Zhang, Z.Q. Sang, and W.Z. Guo. 2007. Comparative development of lint and fuzz using different cotton fiber-specific developmental mutants in *Gossypium hirsutum*. J. Integr. Plant Biol. 49:1038–1046. doi:10.1111/j.1672-9072.2007.00454.x

Zhang, F., K. Zuo, J. Zhang, X. Liu, L. Zhang, X. Sun, and K. Tang. 2010b. An L_1 box binding protein, $GbML_1$, interacts with $GbMYB_{25}$ to control cotton fibre development. J. Exp. Bot. 61:3599–3613. doi:10.1093/jxb/erq173

Zhang, M., X. Zheng, S. Song, Q. Zeng, L. Hou, D. Li, J. Zhao, Y. Wei, X. Li, M. Luo, Y. Xiao, X. Luo, J. Zhang, C. Xiang, and Y. Pei. 2011. Spatiotemporal manipulation of auxin biosynthesis in cotton ovule epidermal cells enhances fiber yield and quality. Nat. Biotechnol. 29:453–458. doi:10.1038/nbt.1843

Zhang, T., and J. Pan. 1991. Genetic analysis of a fuzzless-lintless mutant in *Gossypium hirsutum* L. Jiangsu J. Agric. Sci. 7:13–16.

Zhao, P.-M., L.-L. Wang, L.-B. Han, J. Wang, Y. Yao, H.-Y. Wang, X.-M. Du, Y.-M. Luo, and G.-X. Xia. 2010. Proteomic identification of differentially expressed proteins in the *Ligon lintless* mutant of upland cotton (*Gossypium hirsutum* L.). J. Proteome Res. 9:1076–1087. doi:10.1021/pr900975t

Zhou, G.-Y., J. Weng, Y. Zeng, J. Huang, S. Qian, and G. Liu. 1983. [29] Introduction of exogenous DNA into cotton embryos. Methods Enzymol. 101:433–481. doi:10.1016/0076-6879(83)01032-0

Cytology and Cytogenetics

Tianzhen Zhang* and John E. Endrizzi

Abstract

The cotton genus *Gossypium* is composed of about 45 diploid ($2n = 2x = 26$) and 5 allotetraploid ($2n = 4x = 52$) species. Two allotetraploid species, *Gossypium hirsutum* L. and *G. barbadense* L., consisted of A and D subgenomes, were domesticated after polyploidization and are cultivated worldwide. The identification, morphology, and number of the chromosome; the genomic differentiation and origin of the allotetraploid; and the structure and number variations of chromosome in cotton are reviewed. *Gossypium herbaceum* L. and *G. raimondii* Ulbr. are generally regarded as the best explants of the A and D subgenome progenitors. Diploid *Gossypium* species fall into eight different genome types designated A to G, and K. Cotton chromosomes are numbered in the order of their distinguishing features related to translocations and aneuploids. On the basis of the present molecular marker, the 13 homeologous chromosome pairs have been established. The 62 translocations identify 25 of the 26 chromosomes in the allotetraploid. D12 or chromosome 26 has not been isolated in a translocation, but it is identified as a telocentric chromosome. Therefore, the translocation stocks constitute the only complete set of cytogenetic markers for the allotetraploid cottons. Monosomes for 15 of the 26 chromosomes of *G. hirsutum* have been identified. They have been used extensively to assign genetic factors to specific chromosomes.

The cotton genus *Gossypium* is composed of about 45 diploid and 5 allotetraploid species (Fryxell, 1992, Wendel et al., 2009). Diploid *Gossypium* species fall into eight different genome types designated as A through G and K, based on meiotic pairing behavior (Beasley, 1942; Endrizzi et al., 1985; Percival et al., 1999). Only four are cultivated species: *G. hirsutum* [$n = 2x = 26$, $(AD)_1$], *G. barbadense* L. [$n = 2x = 26$, $(AD)_2$], *G. herbaceum* ($n = x = 13$, A_1), and *G. arboreum* L. ($n = x = 13$, A_2). Of these, the New World allotetraploid species, *G. hirsutum* and *G. barbadense,* dominate worldwide cotton production. Both allotetraploids originated in the New World from interspecific hybridization between diploid species closely related to *G. herbaceum* (A_1) and *G. raimondii* (D_5) (Wendel, 1989) or *G. gossypioides* (Ulbr.) Standl. (D_6) (Wendel et al., 1995). *Gossypium herbaceum* and *G. raimondii* are generally

Abbreviations: BAC, bacterial artificial chromosome; FISH, fluorescence in situ hybridization; mya, million years ago; SSR, simple-sequence repeat.

T.Z. Zhang, Nanjing Agricultural University, Weigang No. 1, Nanjing, Jiangsu 210095, China. *Corresponding author (cotton@njau.edu.cn).

J.E. Endrizzi (retired), Dep. of Plant Sciences, University of Arizona, Tucson, AZ 85721.

doi:10.2134/agronmonogr57.2013.0023

Cotton. 2nd. ed. David D. Fang and Richard G. Percy, editors. Agronomy Monograph 57.

regarded as the best explants of the A and D subgenome progenitors (Endrizzi et al., 1985; Wendel et al., 1995; Zhao et al., 1998), respectively.

Identification of the Chromosome

During the 1920s and 1930s, chromosome numbers of $2n = 26$ and 52 were reported for many *Gossypium* species, including the wild and cultivated forms occurring in America, Asia, Africa, and Australia (Youngman and Pande, 1927; Harland, 1928; Banerji, 1929; Baranov, 1930; Kearney, 1930a; Longley, 1933; Skovsted, 1933, 1934a, 1934b, 1935b, 1935c; Webber, 1934a, 1934b, 1935, 1939). Skovsted's cytological observations of chromosomal differences among the species were extended by Beasley (1940b, 1942), who established a cytological classification of genomes that is closely related to taxonomic affinities. Symbols were devised by Beasley (1940b) to designate the genomes of species. Similar genomes are designated by the same capital letter and closely related genomes are distinguished by a numerical subscript after each letter of that class. Eight diploid and one tetraploid genome groups are now recognized in *Gossypium* (Table 1) (Wendel et al., 2009).

The chromosomes of the tetraploid species and the diploid species *G. herbaceum* were noted by early cotton cytologists to vary in size at metaphase I (Beal, 1928; Davie, 1933). Skovsted (1934a, 1934b, 1935a) reported that the Old World and Australian diploids—*G. arboreum*, *G. herbaceum*, *G. stocksii* Mast., and *G. sturtianum* J.H. Willis—had 13 large chromosomes, whereas the New World diploids *G. thurberi* Tod., *G. davidsonii* Kellogg, *G. aridum* (Rose & Standl.) Skovst., *G. harknessii* Brandegee, and *G. armourianum* Kearney (Skovsted, 1937) had 13 small chromosomes.

Stephens (1947) and Katterman and Ergle (1970) have classified the relative chromosome size of the genome groups as follows: (i) the C genomes have very large chromosomes; (ii) the E and F genomes have large chromosomes that are slightly larger than those of the A genomes; (iii) the B genomes have large chromosomes, some of which are slightly larger than any in the A genomes (in Stephens's classification, the B and A genome groups are switched in relation to one another); (iv) the A genomes have moderately large chromosomes; (v) the G genome has chromosomes that are moderately large but smaller than those of A genomes (Wilson and Fryxell, 1970; Kadir, 1976; Edwards and Mirza, 1979); and (vi) the D genomes have small chromosomes.

Chromosomes are numbered in the order of discovery on the basis of some distinguishing cytological feature associated with a particular chromosome (Kohel, 1973). Numbers 1 through 13 are reserved for the A subgenome and numbers 14 through 26 are reserved for the D subgenome. Kohel (1973) suggested that when chromosome homology has been established, the A subgenome chromosomes should be identified as A01 through A13, and the homeologous D subgenome chromosomes should be designated as D01 through D13 in the cotton genetic nomenclature. Chromosome identification is important for karyotyping, cytogenetics, and physical mapping of genetic regions of agricultural importance. The large genome of cotton is distributed over 26 pairs of small chromosomes without suitable cytogenetic markers such as bands, and it is almost impossible for the routine and unambiguous identification of each chromosomes based only on their morphology. Additionally, polyploid plant species are not similar to diploid species, as their genomes contain two or more subgenomes

Table 1. Diversity and geographic distribution of the major lineages of *Gossypium*. Genomic placements of species enclosed by parentheses are yet to be determined (Wendel et al., 2009).

Genome group	Number of species	Recognized species	Geographic distribution
A	2	*G. arboreum* L.	Africa, Asia
		G. herbaceum L.	
B	3 (4)	*G. anomalum* Wawra	Africa, Cape Verde Islands
		G. triphyllum (Harv.) Hochr.	
		G. capitis-viridis Mauer	
		(*G. trifurcatum* Vollesen)	
C	2	*G. sturtianum* J.H. Willis	Australia
		G. robinsonii F. Muell.	
D	13 (14)	*G. thurberi* Tod.	primarily Mexico; also Peru, Galapagos Islands, Arizona
		G. armourianum Kearney	
		G. harknessii Brandegee	
		G. davidsonii Kellogg	
		G. klotschianum Andersson	
		G. aridum (Rose & Standl.) Skovst.	
		G. raimondii Ulbr.	
		G. gossypioides (Ulbr.) Standl.	
		G. lobatum Gentry	
		G. trilobatum (DC.) Skovst.	
		G. laxum L. Ll. Phillips	
		G. turneri Fryxell	
		G. schwendimanii Fryxell & S.D. Koch	
		(*Gossypium* sp. nov.)	
E	5 (9)	*G. stocksii* Mast.	Arabian Peninsula, Northeast Africa, Southwest Asia
		G. somalense (Gurke) J.B. Hutch.	
		G. areysianum Deflers	
		G. incanum (O. Schwartz) Hillc.	
		G. trifurcatum	
		(*G. benadirense* Mattei)	
		(*G. bricchettii* (Ulbr.) Vollesen)	
		(*G. vollesenii* Fryxell)	
		(G. trifurcatum)	
F	1	*G. longicalyx* J.B. Hutch. & B.J.S. Lee	East Africa
G	3	*G. bickii* Prokh.	Australia
		G. australe F. Muell.	
		G. nelsonii Fryxell	
K	12	*G. anapoides* J.M. Stewart et al., nom. inval.	Northwest Australia, Cobourg Peninsula, Northern Territory
		G. costulatum Tod.	
		G. cunninghamii Tod.	
		G. enthyle Fryxell et al.	
		G. exiguum Fryxell et al.	
		G. londonderriense Fryxell et al.	
		G. marchantii Fryxell et al.	
		G. nobile Fryxell et al.	
		G. pilosum Fryxell	
		G. populifolium (Benth.) F. Muell. ex Tod.	
		G. pulchellum (C.A. Gardner) Fryxell	
		G. rotundifolium Fryxell et al.	
AD	5	*G. hirsutum* L.	New World tropics and subtropics, including Hawaii
		G. barbadense L.	
		G. tomentosum Nutt. ex Seem.	
		G. mustelinum Miers ex G. Watt	
		G. darwinii G. Watt	

that originated from the same ancestor (Wendel, 1989, 2000; Wendel and Cronn, 2002). Specific bacterial artificial chromosome (BAC) clones constructed in *G. hirsutum* were identified by screening the BAC library using linkage group-specific simple-sequence repeat (SSR) markers. A whole set of 26 chromosome-specific BAC clones with corresponding SSR markers and their cytological locations have been developed and applied as fluorescence in situ hybridization (FISH) markers for the cytogenetic and physical mapping in tetraploid cotton (Wang et al., 2007b). These BAC clones were hybridized to 10 translocation heterozygotes of *G. hirsutum* as BAC–FISH probes, which allowed complete assignment of the 6 linkage groups A01, A02, A03, D02, D03, and D08 to chromosomes 13, 8, 11, 21, 24, and 19, respectively. Therefore, the 13 homeologous chromosome pairs and a new chromosome nomenclature for tetraploid cotton have been established (Table 2) (Wang et al., 2006).

Combining 45S and 5S rDNA and a probe cocktail of 20 BAC clones and FISH, all the 13 *G. arboreum* chromosomes are identified simultaneously (Fig. 1). According to their homology with tetraploid cotton, the *G. arboreum* chromosomes were designated as A01 to A13, and a standard karyotype for *G. arboreum* was presented (Wang et al., 2007a) (Table 3). Meiotic FISH methods have also been used to physically map rDNA sites to specific chromosomes or chromosome arms (Hanson et al., 1996; Ji et al., 1999).

Integration of genetic loci into physical localizations of homeologous chromosomes A12 and D12 showed considerable variation of genome organization, structure, and size between A12 and D12 homeologous chromosomes (Wang et al., 2010). The distal regions of the chromosomes displayed relatively lower levels of structural and size variation than other regions of the chromosomes. The highest level of variation was found in the pericentric regions in the long arms of the

Table 2. Length and arm ratio of the mitotic chromosomes in *Gossypium arboreum* cv. jianglinzhngmian.

Chromosome	Total length	Relative length†	DNA content‡	Location of NOR§	Arm ratio¶
	µm	%	Mb		
A01	3.44 ± 0.35	8.02 ± 0.81	94.0 ± 13.91	–	1.30 ± 0.15
A02	3.04 ± 0.26	7.10 ± 0.61	83.2 ± 10.57	–	1.25 ± 0.22
A03	3.60 ± 0.34	8.40 ± 0.79	98.4 ± 13.60	–	1.06 ± 0.06
A04	3.59 ± 0.48	8.37 ± 1.11	98.0 ± 19.07	–	1.20 ± 0.16
A05	3.79 ± 0.39	8.85 ± 0.92	103.6 ± 15.73	short arm	1.19 ± 0.12
A06	3.57 ± 0.48	8.33 ± 1.12	97.6 ± 19.31	–	1.12 ± 0.06
A07	3.69 ± 0.47	8.61 ± 1.09	100.8 ± 18.74	short arm	1.32 ± 0.24
A08	3.01 ± 0.31	7.03 ± 0.72	82.3 ± 12.39	–	1.13 ± 0.10
A09	2.80 ± 0.31	6.55 ± 0.71	76.6 ± 12.21	–	1.29 ± 0.19
A10	3.02 ± 0.31	7.04 ± 0.71	82.4 ± 12.19	–	1.20 ± 0.14
A11	2.91 ± 0.32	6.79 ± 0.74	79.5 ± 12.82	–	1.37 ± 0.21
A12	3.41 ± 0.38	7.96 ± 0.89	93.3 ± 15.17	–	1.86 ± 0.23
A13	2.97 ± 0.32	6.94 ± 0.74	81.2 ± 12.71	–	1.27 ± 0.28

† Relative length = chromosome length/genome length.

‡ NOR, nucleolar organizer region.

§ Estimation of the DNA content of each chromosome is based on the assumption that the haploid genome size of *G. arboreum* is 1171 Mb (Hendrix and Stewart, 2005). Thus, the DNA content will be relative length × estimated genome size, that is, 1171 Mbp.

¶ Arm ratio = length of the long arm/the length of the short arm.

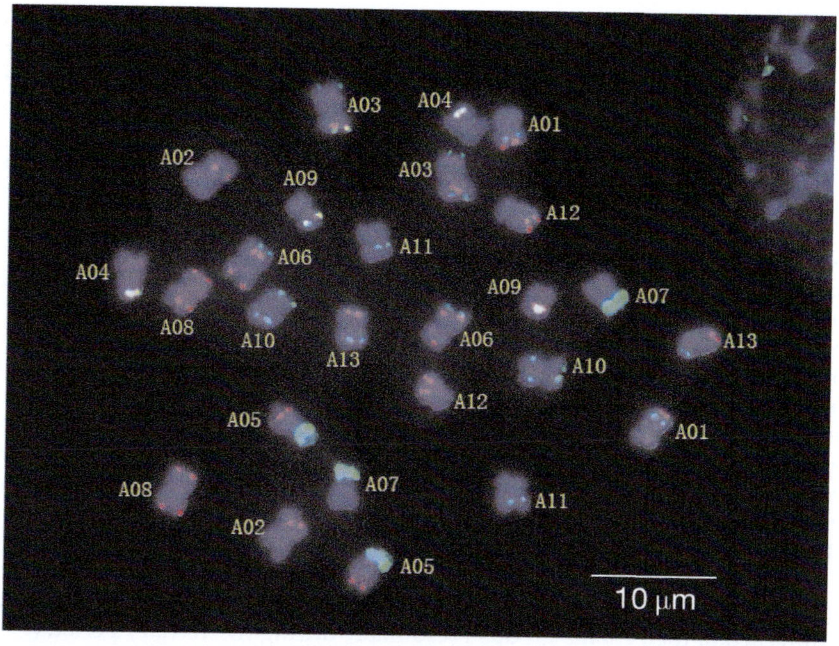

Fig. 1. Fluorescence in situ hybridization results of a 22-probe cocktail hybridized to *Gossypium arboreum* L. mitotic chromosomes simultaneously (Wang et al., 2008). The 5s rDNA and bacterial artificial chromosome 84C03 on chromosome A04 were labeled by both biotin and digoxigenin, thus the white or yellow signals on chromosomes A04 and A09 were found. The patterns of signals enable recognition of each chromosome pair according to the homologous relationship with the A subgenome chromosomes of *G. hirsutum* L.

Table 3. Chromosome identification in tetraploid cotton.

Proposed assignment of A genome chromosome	Former assignment of A genome chromosome	Proposed assignment of D genome chromosome	Former assignment of D genome chromosome
A01	chromosome 1	D01	chromosome 15
A02	chromosome 2	D02	chromosome 14
A03	chromosome 3	D03	chromosome 17
A04	chromosome 4	D04	chromosome 22
A05	chromosome 5	D05	chromosome 19
A06	chromosome 6	D06	chromosome 25
A07	chromosome 7	D07	chromosome 16
A08	chromosome 8	D08	chromosome 24
A09	chromosome 9	D09	chromosome 23
A10	chromosome 10	D10	chromosome 20
A11	chromosome 11	D11	chromosome 21
A12	chromosome 12	D12	chromosome 26
A13	chromosome 13	D13	chromosome 18

two homeologous chromosomes. The genome size difference between A and D subgenomes in allotetraploid cotton is mainly associated with uneven expansion or contraction between different regions of homeologous chromosomes (Wang et al., 2012).

Genome Differentiation and Origin of the Allotetraploids

Genome Differentiation in *Gossypium*

Two groups of diploid species, designated as A genome and D genome, diverged from a common ancestor about 5 to 10 million years ago (mya). These diploid species were combined to form new allotetraploid species through a polyploidization event that occurred 1 to 2 mya, giving rise to five divergent allotetraploid species (Seelanan et al., 1997; Wendel, 1989). Two of the allotetraploid species, *G. hirsutum* and *G. barbadense*, account for ~95% and ~5% of worldwide cotton cultivation, respectively. Saunders (1961) speculates that the center of origin of the genus *Gossypium* is in central Africa because four of the seven diploid genome groups occur on the African continent. Divergence of the genus into the different genome groups is generally assumed to have occurred before or during the separation of the continents in the Cretaceous period (Skovsted, 1934a; Hutchinson et al., 1947; Prokhanov, 1947; Mauer, 1954; Saunders, 1961; Phillips, 1963; Hawkes and Smith, 1965; Fryxell, 1965, 1979). This event was followed by differentiation within each of the genome groups presumably during the Tertiary period (Fryxell, 1965; Phillips, 1966).

Chromosome homologies (Phillips, 1966), plant morphological characters (Fryxell, 1971; Valicek, 1978), and seed protein patterns (Johnson and Thein, 1970) have been utilized to deduce which one of the genomes is likely to be most representative of the initial genome of the genus.

Data on chromosome pairing in intragenomic and intergenomic hybrids of the diploid species were summarized by Phillips (1966, 1974). Phillips hypothesized that reduction of chromosome pairing and chiasma frequency between two genomes can be attributed to structural changes, the degree of which is directly related to the time elapsed since their divergence from a common ancestor. The data showed that the intergenomic hybrids involving the E genome species had the highest frequency of univalents, indicating that the E genome is ancient in origin and therefore closest to the ancestral genome of the genus (Tables 4 and 5). Univalent frequency in genomic hybrids of A and B species was very low, suggesting that these two genomes represent a rather recent divergence in which A evolved from the B genome. Univalent frequency in hybrids involving C or D species indicated that the C and D genomes represent evolutionary lines of intermediate age. Divergence of the D genome was assumed to have occurred somewhat before the divergence of the C genome (Phillips, 1963, 1966).

Genome origin based on intergenomic chromosome affinities as interpreted by Phillips (1966, 1974) suggests that the original ancestral generic genome was composed of large chromosomes similar to the E genome. The later divergence of genomes with smaller and larger chromosomes would require a downward and upward shift in total DNA content.

A comparative study of plant morphological characters among the diploid species by the Wagner Groundplan/Divergence Method was reported by Fryxell (1971). He found that species with the D genome had the least amount of evolutionary

Table 4. Chromosome pairing in intergenomic hybrids of diploid *Gossypium* species.

Intergenomic hybrids	Univalent/cell	Chiasma/bivalent	Intergenomic hybrids	Univalent/cell	Chiasma/bivalent
A × B	2.82	1.44	B × E	22.35	
A × C	8.50	1.19	C × D	13.10	1.10
A × D	13.98	1.10	C × E	24.68	1.01
A × E	17.13	1.08	D × E	25.15	1.00
A × G	16.00	1.13	D × F	21.60	1.02
B × C	11.17	1.11	G × C	3.84	1.48
B × D	18.19	1.07			

Table 5. Chromosome pairing in intergenomic hybrids of tetraploid and diploid *Gossypium species.*

Intergenomic hybrids	Univalent	Bivalent	Multivalent	Chiasma/bivalent
AD × A	13.20	8.63	2.0	1.89
AD × B	26.91	5.41	0.4	1.07
AD × C	26.81	5.33	0.51	1.03
AD × D	13.28	12.21	0.41	1.82
AD × E	37.90	0.55	0	1.00
AD × F	22.25	6.78	1.03	1.03

advancement, suggesting that the small D genomes are closest to the ancestral or prototype genome. This method of analysis indicates that the other six genomes must have arisen by an upward shift in total nuclear DNA content.

Utilizing numerical taxonomy in evaluating the phenetic relationships of the species of the A through F genomes, Valicek (1978) presented a genome phylogeny similar to that of Johnson and Thein (1970). Valicek's analysis indicates that the B genome is the prototype of the diploid species of the genus. The low correlation of the C and F genomes with other genomic groups indicated that they may be of more recent origin.

Origin of the Allotetraploids

Skovsted (1934a) was the first to point out that the cytological data indicate that the New World allotetraploids consisted of A and D subgenomes. This proposal was followed by efforts to produce a synthetic hybrid between an A and D species. This feat was accomplished independently in 1940 by Beasley (1940a) and Harland (1937), who crossed *G. arboreum* with *G. thurberi* and doubled the chromosomes of the F_1 hybrid. Beasley (1940a) observed less than one-half of the chromosomes paired in the diploid F_1, and regular bivalent pairing along with a high degree of fertility in the synthetic amphidiploid, which is what one would expect according to the allotetraploid hypothesis of Skovsted. Harland (1937), Harland and Atteck (1941), and Silow (1946) provided genetic support for the allopolyploid origin by demonstrating that several loci of the allotetraploids were related (homologous) to those of the diploids by common origin.

The next task was to determine which genomes of the two Asiatic cottons and of the several American wild diploids were most similar to the progenitor forms of the allotetraploids. Stephens (1944) studied leaf developmental patterns in hybrids of Asiatics with various alleles for leaf shape and *G. raimondii*, one of the four American D species, with entire leaves. The results suggested that the D subgenome of the allotetraploids is most closely related to either *G. raimondii* or *G. klotzschianum* Andersson and its closely related form *G. davidsonii*.

Hutchinson et al. (1945) discussed a number of factors that should be considered in formation of a natural allotetraploid, which included the kinds of seed hairs occurring in the diploid species and the vigor of the hybrid forms. Among the then-known six D genome species, *G. raimondii* was the most likely candidate as donor of the D subgenome of the allotetraploids, which in combination with the genome of Asiatic cotton would provide the genetic system, including convoluted seed hairs, characteristic of the allotetraploids.

Gerstel (1953a) demonstrated that the two Asiatic diploid species, *G. arboreum* and *G. herbaceum* differ by one reciprocal interchange. The pairing data of the hybrids with *G. hirsutum* showed that the genome of *G. herbaceum* rather than that of *G. arboreum* is less structurally differentiated from that of the allotetraploids. In addition, the chromosomes of *G. herbaceum* were shown to be structurally similar to that of the closely related wild diploid *G. anomalum* Wawra, whereas *G. arboreum* differed from *G. anomalum* by a simple translocation.

Amphidiploids and allohexaploids involving Asiatic, New World diploids, and the natural allotetraploid species were synthesized and analyzed cytologically for multivalent frequencies and genetically for segregation of mutant genes located in the A and D genomes (Gerstel, 1953b, 1956, 1963; Sarvella, 1958; Gerstel and Phillips, 1957, 1958; Phillips and Gerstel, 1959; Phillips, 1960, 1963, 1964). The pairing relationships and genetic ratios observed in the analysis of the hexaploid hybrids involving the D species showed that those involving *G. raimondii* most closely approached autotetraploid behavior, confirming that its genome is indeed most closely related to the contributor of the D subgenome of the tetraploids. Thus, it was concluded that the progenitor genomes of the allotetraploid species were most similar to the genomes extant in *G. herbaceum* and *G. raimondii*. Additional new diploid species have been identified, but they afford no evidence for an origin of the amphidiploids different from that now concluded.

Skovsted (1934a), Saunders (1961), and Valicek (1978) speculated that the origin of the allotetraploids goes back to the time when the A and D diploid parental forms overlapped, before the origin of the rifting of the continents. Others have also suggested the allotetraploids originated in the area of overlap of the two progenitor genomes, but after the rifting of the continents. Harland (1935) assumed that the allotetraploids originated in the Polynesian Islands by migration of *G. arboreum* across a trans-Pacific land bridge where it met with a D species in the Late Cretaceous or Early Tertiary. Stebbins (1947) proposed that an A genome species arrived in North America before the Eocene by way of Behringia. Neither of these hypotheses is now tenable (Stephens, 1947; Gerstel, 1953a; Hutchinson, 1959; Fryxell, 1965, 1979).

Phillips (1974) pointed out that the earliest archeological specimens of cotton in the New World are dated at 4000 to 3000 BC, which is 2000 yr before known archeological cotton in Africa, 1000 yr before agriculture in Africa, and 3000 yr before humans made long-distance voyages in the African area.

Phillips (1963) interpreted the cytogenetic data as establishing that the allotetraploids are neither ancient nor recent in origin. Cytogenetic data showed that the A and D subgenomes of the allotetraploids, primarily the former, had diverged little during their evolutionary history from the progenitor genomes represented in the Asiatics and *G. raimondii*, which was assumed to be incompatible with ancient origin. The fact that the allotetraploids have differentiated into distinct species, which are widely distributed on islands of the Caribbean and the Pacific, suggested that they are not recent in origin. Thus, Phillips (1963) concluded as did Fryxell (1965, 1979), who in addition stressed the importance of the several pairs of sibling species and the association of the allotetraploids with littoral habitats, that the allotetraploids arose most likely in the Pleistocene following an Atlantic transfer of the progenitor A genome species to the New World.

From the extensive cytological data, most studies of *Gossypium* evolution postulate a monophyletic origin for the allotetraploids. However, several individuals have favored a polyphyletic origin (Davie, 1933, 1935; Mauer, 1938, 1954; Kammacher, 1959, 1960; Sherwin, 1970; Parks et al., 1975; Johnson, 1975).

Solely on the basis of chromosome pairing behavior in the F_1 hybrids of *G. hirsutum* and *G. barbadense* with *G. raimondii*, Kammacher (1959, 1960) concluded that *G. hirsutum* is less "diploidized" and therefore of more recent origin than *G. barbadense*. The two allotetraploids are assumed to have arisen by separate hybridization events.

Sherwin (1970) speculated that a *G. herbaceum*–like cotton was transported to northern South America, either by rafting or by humans, where it was cultivated and hybridized, possibly with *G. raimondii*, to produce *G. barbadense*. The diploid A genome species was then carried north to Mexico where it hybridized with either *G. gossypioides* or *G. trilobum* (DC.) Skovst. to produce *G. hirsutum*. However, cytological and taxonomic data exclude the involvement of the D genomes of *G. gossypioides* or *G. trilobum* in the origin of the allotetraploids.

On the basis of studies of flavonoid compounds of flowers, Parks et al. (1975) assumed that the allotetraploids had separate origins. They speculated that *G. barbadense* and *G. tomentosum* Nutt. ex Seem. could have arisen possibly from a combination of *G. klotzschianum* and an A species, whereas *G. hirsutum* and *G. mustelinum* Miers ex G. Watt originated from a combination of *G. raimondii* and an A species. The cytological data do not support their interpretation since the chromosomes of *G. tomentosum* rather than those of *G. mustelinum* are more similar to the chromosomes of *G. hirsutum* as was demonstrated by Hasenkampf and Menzel (1980). Johnson's (1975) polyphyletic hypothesis was based primarily on the electrophoretic banding patterns of seed proteins, from which he concluded that the allotetraploids arose from possibly three independent hybridization events.

The banding patterns of the two A genome species, *G. herbaceum* and *G. arboreum*, were identical, whereas the D genome species were divided into two groups: one group, designated as D_β, consisted of *G. raimondii*, *G. lobatum* Gentry, *G. aridum*, *G. laxum* L. Ll. Phillips, and *G. gossypioides*, and the other group, D_p consisted of *G. thurberi*, *G. trilobum*, *G. davidsonii*, *G. klotzschianum*, *G. armourianum*, and *G. harknessii* (Johnson and Thein, 1970).

Except for some accessions of *G. hirsutum* "race" palmeri, all accessions of *G. barbadense*, *G. hirsutum*, *G. tomentosum*, *G. darwinii* G. Watt, and *G. mustelinum* (*G. caicoense* Aranha et al. in Johnson, 1975) that were examined had identical banding patterns. Their banding patterns were similar to the banding pattern simulating a 1:1 mixture of the protein extract of *G. herbaceum* and anyone of the

D genome species in the D_{β} genome group. The exceptional palmeri accessions had a different banding pattern that simulated a 1:1 mixture of protein extract of *G. herbaceum* and a species of the D genome group. Because of these differences, Johnson (1975) concluded that the "race" palmeri is a distinct species, *G. palmeri* G. Watt (*G. lanceolatum* Tod. in Fryxell, 1979), and that the allotetraploids arose from possibly three independent hybridization events in a center of early American civilization. He presumes that *G. herbaceum* combined with *G. trilobum* of the D group to produce *G. palmeri*. This event was followed by the origin of *G. hirsutum* in Southern Mexico and Central America from a palmeri prototype with the AAD_pD_p genome and from one or more prototypes with the $AAD_{\beta}D_{\beta}$ genomes. *Gossypium barbadense* with the $AAD_{\beta}D_{\beta}$ genomes is assumed to have originated in South America between *G. herbaceum* and *G. raimondii*. From the Mesoamerican center of origin, *G. tomentosum* was carried by humans or ocean currents to Hawaii.

Phillips (1963, 1964) points out that if *G. hirsutum* and *G. barbadense* have different D subgenomes due to a biphyletic origin, then the frequency of multivalents and the segregation for a given locus should be different in hexaploid hybrids involving these two species and the same American diploid species. However, both hexaploid hybrids have quite similar multivalent frequencies and segregation ratios, which establish the sameness or unity of their genomes, and therefore a monophyletic origin.

Normally in constructing a hypothesis for the origin of the allotetraploids, it has been assumed that the following sequence of evolutionary events must be considered if one accepts the modern synthesis concept of speciation (Stephens, 1947): (i) long-period geographical isolation of the American and Asiatic groups, during which their genomes became cytologically differentiated, (ii) geographical reunion of the two, followed by interspecific hybridization and polyploidy, and (iii) geographical reisolation of the two parental species and genetic differentiation of the allotetraploids. The first two are "double-event" rather than "single-event" processes (Stephens, 1947). However, according to our hypotheses, the origin of the allotetraploids could possibly be nearly like a single-event process.

If that is the order of speciation events, it is apparent that the two species would satisfy the three primary requirements for the origin of an allotetraploid: chromosome differentiation, propinquity, and cross compatibility (Hutchinson et al., 1945). Thus, within a relatively short geological time period, a progenitor and the hybrid allotetraploid species could become established as independent, true breeding taxa.

Even though fossil records of *Gossypium* are unknown (Fryxell, 1979), it is presumed that the genus is very old and that continental drift separated the two major genomic chromosome size groups, large chromosomes in the Old World and small chromosomes in the New World, during the Cretaceous period. Thus, according to the present hypothesis, the allotetraploids are ancient in origin occurring perhaps during Cretaceous times.

The events in the origin of the allotetraploid species can be summarized as follows: (i) the origin of an A genome species occurred by macroevolution from a progenitor D genome, (ii) hybridization and polyploidy occurred within a short geological time period following the formation of the A genome parent, (iii) the origin of the amphidiploid probably occurred during the Cretaceous period, (iv) diploidization of the gene regulatory system of the hybrids led to the formation of at least five distinct amphidiploid species and, finally, (v) the generation of genetic diversity that exists primarily in *G. barbadense* and *G. hirsutum.*

Translocations

Origin and Identification

Before the 1950s, cotton cytologists had observed multivalent chromosome configurations in Pollen Mother Cells of species hybrids. Beasley (1942) referred to the multivalents as translocations without determining the number involved. Later, Gerstel (1953a) reported that the multivalents in hybrids of *G. herbaceum, G. arboreum, G. anomalum,* and the allotetraploids were due to chromosomal interchanges.

During this time, cotton cytologists at the Beasley Laboratory in Texas noticed a high frequency of multivalents at metaphase I in trispecies hybrids of *G. hirsutum* × Asiatic–New World diploid F_1. Upon reanalysis of hybrids between *G. hirsutum* and the Asiatic and New World diploid species, they found that all the viable *G. hirsutum*–New World diploid hybrids almost regularly formed 13 small bivalents and 13 large univalents. However, a different modal pairing behavior was observed in the *G. hirsutum* Asiatic hybrids. Like Gerstel (1953a), they observed chromosome structural changes that differentiate the A genomes of *G. arboreum* and *G. hirsutum* from *G. herbaceum,* and they assigned chromosome end arrangements to the first five chromosomes of the three A genomes that would account for the multivalent formation in each hybrid combination. Since only five chromosomes of the three A genomes are involved in the interchanges, this indicates that their remaining eight A genome chromosomes have identical end arrangements (Menzel and Brown, 1954a).

According to Menzel and Brown (1954a), *G. herbaceum* has the standard or primitive chromosome end arrangement for all of its chromosomes, in which the first five are numbered respectively 1–2, 3–4, 5–6, 7–8, and 9–10. *Gossypium arboreum* differs from *G. herbaceum* by a single interchange between chromosomes 1 and 2, thus its end arrangement for the five chromosomes is 1–3, 2–4, 5–6, 7–8 and 9–10. The A subgenome of the allotetraploids differs from the genome of *G. herbaceum* by two independent reciprocal interchanges involving chromosomes 2 and 3, and 4 and 5. Thus, its end arrangement is 1–2, 3–5, 4–6, 7–9, and 8–10. The interchanges in this hybrid were designated, respectively, as VI_2 and 1 IV_2. The A genomes of the allotetraploids and *G. arboreum* differ by three interchanges, which are apparent in their hybrids by the occurrence of a ring-of six (VI_1) and a ring-of-four (IV_2). The ring-of-six forms from combining the interchange between chromosomes 1 and 2 of *G. arboreum* with the interchange between chromosomes 2 and 3 of the allotetraploids.

The IV_1, IV_2, and VI_1 chromosome translocations were transferred to *G. hirsutum,* providing the first set of chromosomally identified translocations in *Gossypium* (Brown, 1980). The first induced translocations reported in cotton were from seed exposed in the atom bomb test at Bikini (Brown, 1950). Later, seeds of various stocks were exposed to radiation (X-rays, γ, and fast neutron) to induce translocations for isolation and identification in *G. hirsutum* (Menzel and Brown, 1954b; Kammacher, 1958; Brown, 1980). Sixty-two translocations, most of which came from irradiation, have been isolated, made homozygous, and identified as to the chromosomes involved (Brown, 1980; Table 6). In Table 6, the A subgenome chromosomes are numbered 1 through 13 and those of the D subgenome, 14 through 26. The 62 translocations identify chromosomes 1 through 25 of the 26 chromosomes in the allotetraploid. Chromosome 26 has not been isolated in a translocation, but it is identified as a telocentric chromosome (Table 7). Four of the 62 interchanges consist of multiple associations involving interchanges between three or four nonhomologous chromosomes. A tester set selected from the group can be used to identify anyone of the 26 chromosomes occurring in a structural or numerical chromosome change (Ray, 1981). Menzel (1955) has described the cytological procedure for assigning chromosomes in translocations to their respective genomes. When crossed with a D diploid species, AA translocations give a model configuration of 13 AA bivalents and 13 univalents (13II + 13I). AD translocations give model configurations of 12 bivalents, 1 trivalent, and 12 univalents (12II + 1III + 12I). Triploid plants involving DD translocations have a model configuration of 11 bivalents, 1 DD quadrivalent, and 13 univalents (11II + 1IV + 13I). This procedure consists of crossing the homozygous translocations to diploid species with the D genome. The identity of the chromosomes in each of the 62 translocations was accomplished in most cases by intercrossing the homozygous translocations and determining whether the same or different chromosomes were involved (Table 8) (Brown, 1980). In some cases, stocks that were monosomic for known chromosomes were used to identify chromosomes in the translocations and vice versa.

Although more 191 translocations were recovered as heterozygotes following combined treatment of seeds by colchicine and c-rays, irradiation of seeds by thermal neutrons, and c-irradiation of pollen, and 12 new reciprocal homozygous translocation lines were obtained, their chromosomes involved in the interchanges remained to be identified (Sanamyan and Rakhmatullina, 2003). The location of breakpoints in the translocated chromosomes of many translocations has been determined by Menzel and Brown (1978a, 1978b) and Brown et al. (1981). These are given provisionally in Table 8 as "right" and "left" arms, in which "right" is arbitrarily assigned to the arm involved in the larger number of breaks.

Uses of Translocations

Most of cotton chromosomes are numbered in the order of their distinguishing features related to translocation (Table 7). Because *G. hirsutum* is an allotetraploid, viable duplications and deficiencies for segments of chromosomes can be recovered as a result of adjacent disjunction in a heterozygous interchange (Table 8) (Brown, 1950; Brown et al., 1981; Menzel and Brown, 1952, 1954b, 1978b). It can be seen in Table 8 that duplication-deficiency types have been recovered from all the simple interchanges in cotton. Therefore, the translocations probably can be used to assign genes to specific chromosomes.

The translocations are also very useful for detecting incipient genome differentiation between the subgenomes of *G. hirsutum* and the A and D genomes of the diploid species and within the A and D subgenomes of the allotetraploid species (Menzel et al., 1978; Hasenkampf and Menzel, 1980; Menzel et al., 1982). For example, plants homozygous for the AD interchanges are crossed to the New World diploid species to produce F_1's for cytological analysis. The (AD)D hybrid will show a trivalent consisting of the two AD interchanged chromosomes with the normal D chromosome between the two. Therefore, in crosses with two or more D genome species, the chiasma frequencies observed in the Dx chromosome

Table 6. Sequence and method of chromosome identification in *Gossypium hirsutum* L. (Brown, 1980).

Chromosome	First assignment	Year	Basis for assignment	Other translocations involving this chromosome (in order of identification)
			A genome	
01	VI1	1953	Assignment to differential chromosome of Ah, A02, and A01 (Menzel and Brown, 1954a); chromosome 1 was designated the chromosome monosomic in M17 (haplo 1) (1960).	Iva, DP-30, 5-4C, 2780, 2775b, 2770, 4669, 4672, 2935
	VI1, IV1, and 2B-1	1953	Assignment to Ah-A02 differential chromosome common to IV1 and 2B-1 (Menzel, 1955).	1509, 1039, 1058b, DP-30, IVa, 7-2b, 8B-3; also DP-6 VI, 4655 VIII
03	VI1 and IV1	1953	As for chromosome 1; chromosome 3 was designated the chromosome common to IV1 and IV1, but not 2B-1.	1059, 8-30-5, 8-5Gb, 2935, E20-7, 4010; also dp-df Z121
04	IV2	1953	As for chromosome 1; chromosome monosomic in M6 was designated chromosome 4 (Endrizzi and Brown, 1964). The same chromosome was associated with duplication deficiency 692 in 1956 and shown to be A chromosome of 1040 in 1960, confirmed 1961–1962.	1040, 10-5Ka
05	IV2	1953	As for chromosome 1; the chromosome of IV2 not associated with 1040.	SL-18, 2775a, 8-5Gb, C14-3; also 2779 (VI)
06	haplo 6 (M1 and M5)	1958	Independent of translocations involving chromosomes 1–5; 7-2B was shown to involve chromosome 6 by association with haplo 6 M1 and M5 in 1962.	1048, Z9-9, AZ-7, 4010; also with aberrant knobbed chromosome in 3008
07	1052	1962	Independent of chromosomes 1–6, thus called A07. Later (1968) verified by association with haplo 7 (M07) (Endrizzi and Taylor, 1968).	1048, 5-4C, 1043, 2790
08	1039	1963	Presence of A genome chromosome independent of chromosomes 1–7.	1058b, 2775b, 5-5B, 2778; also duplication deficiency Z121
09	1036	1963	Presence of A genome chromosome independent of chromosomes 1–8.	8-30-5, 8B-3, 2870, 2772, 6340, C14-3
10	1626 and Z9-9	1968	Presence of A genome chromosome independent of chromosomes 1–9.	2785, 4675
11	1052	1962	Presence of A genome chromosome independent of haplo 6 and of chromosomes 1–5 and 7–10 in appropriate translocations; early confusion between 7 and 11 in associated translocations cleared by tests with haplo 7 (1968).	6-5M, 10-5Kb, 1058a, 2785, 1316; also 2779 VI and 4665 VIII
12	1043, 6-5M, SL-18, and 9-5H simultaneously	1968	Presence of A genome chromosome independent of chromosomes 1–11.	2778
13	10-5Kb	1970	Presence of A genome chromosome independent of chromosomes 1–12.	2925; also 4655 VIII

Continued.

Table 6. Continued.

Chromosome	First assignment	Year	Basis for assignment	Other translocations involving this chromosome (in order of identification)
D genome				
14	2B-1	1953	First marked chromosome demonstrated to be in D genome (Menzel and Brown, 1954b).	AZ-7, 2780, 2777, 2781; also 4655
15	1040	1956	D genome chromosome independent of 2B-1; associated with LGII (Menzel and Brown, 1954b; Wilson and Kohel, 1970).	1058a, SL-15, 8-5Ga, 2767
16	linkage group III (Endrizzi and Brown, 1964)	1964	Spurious association with 1052 (Menzel and Brown, 1954b); later turned out to involve chromosome 7, which carries homoelogous LG II instead. Monosome M25 (1961) and M39 (1962) proved to be associated with LG III (White and Endrizzi, 1965) and hence were chromosome 16; 8-5Ga was the first translocation shown to involve chromosome 16 (1969).	2770, 2767, 4672
17	haplo 17 (M-2)	1963	D genome chromosome independent of haplo 14 and 15 and LGIII (16) (Endrizzi and Brown, 1964); 1036 was associated with chromosome 17 in 1963	1316, 6340
18	haplo 18 (M-10)	1964	D genome chromosome independent of 2B-1(1956); independent of LG II (H15) and III (H16) (Endrizzi, 1963); independent of chromosome 17 in 1036 (1963). D genome assignment demonstrated in hybrid with *G. raimondii* Ulbr. (D05); 4659 is the only translocation involving this chromosome (1974).	
19	10-5Ka 10-5Ka, 9-5H, 1626	1963	(Tentative) D genome chromosome independent of translocations with chromosomes 14–17 and haplo 18 (M10).	2786, 2925, E20-7, 5-5B, E22-13; also DP-6 VI
20	SL-15, DP-4	1968–1969	D genome chromosome independent of translocations with chromosomes 14–17 and haplo 18 (M10).	2772, 4669, 2791 7-3F
21	7-3F	1968	D genome chromosome independent of translocations with chromosomes 14–17, 19 and 20, and haplo 18; chromosome 21 tentatively assigned to 7-3F in 1968 and verified in 1969.	2790, 4675, E22-13; also 2779 VI
22	DP-4	1968	D genome chromosome independent of translocations with chromosomes 14–17, 19–21, and haplo 18 (M10).	none
23	2775a	1970	D genome chromosome independent of translocations with chromosomes 14–17, 19–22, and haplo 18 (M10).	2777; also DP-6 VI
24	2781, 2786 (VIII) simultaneously	1973	D genome chromosome independent of translocations with chromosomes 14–17, 19–23, and haplo 18 (M10).	none
25	2870	1971	D genome chromosome independent of translocations with chromosomes 14–17, 19–24, and haplo 18 (M10).	2791
26	none			

Table 7. Monosomes and telosomes for chromosomes of the A and D subgenomes of *Gossypium hirsutum*.†

Chromosome A subgenome	Chromosome D subgenome
Mono 1	Telo 14L
Telo 1L	
Telo 1S	Telo 15L
Mono 2	Mono 16
Telo 2L	Telo 16L
Telo 2S	
Mono 3	Mono 17
Telo 3L	Telo 17S
Telo3S	
Mono 4	Mono 18
Telo 4L	Telo 18L
Telo 4S	Telo 18S
Telo 5L	Mono 20
	Telo 20L
Mono 6	Telo 20S
Telo 6L	
Telo 6S	Mono 22
	Telo 22L
Mono 7	Telo 22S
Telo 7L	
Telo 7S	Mono 25
	Telo 25L
	Telo 26S
Mono 9	
Telo 9L	
Mono 10	
Telo 10L	
Telo 10S	
Mono 12	
Telo 12L	

† Mono I, monosome for chromosome I; Telo IL and IS, telocentric chromosomes for the long and short arms, respectively, of chromosome I.

of the different hybrids can be compared and any significant differences would indicate that chromosome differentiation had taken place (Menzel et al., 1978).

Chromosome Number Variations

Monosomes

Origin and Identification

Plants that are deficient for one chromosome of a pair are referred to as monosomics. Unlike diploids, allopolyploids such as *G. hirsutum* can transmit the haplo deficiency to subsequent generations. Chromosome pairing in monosomic plants of allopolyploid cotton will normally consist of 25 bivalents plus 1 univalent. Twenty-six different monosomics are theoretically possible.

Monosomes for 15 of the 26 chromosomes of *G. hirsutum* have been identified (Endrizzi and Ramsay, 1979, 1980). These involve chromosomes 1, 2, 3, 4, 6, 7, 9, 10, and 12 of the A subgenome and 16, 17, 18, 20, 22, and 25 of the D subgenome (Table 7). The translocations were used to identify the monosomes. All of

Table 8. Homozygous translocation lines in Gossypium hirsutum.†

Translocation line‡		Frequency of dp-df§	Translocation line‡		Frequency of dp-df§
	AA translocations			AA translocations	
Iva	T1L-2Ra	0.36	8-30-5	T3R-9R	0.47
DP30	T1L-2Lb	0.37	IV,	T4L-5	0.34
2935	T1L-3L	0.41	C14-3	T5-9	0.42
5-4c	T1L-7L	0.57	SL18	T5-12R	0.39
2775b	T1L-8L	0.30	1048	T6-7L	
IV,	T2-3a	0.39	Z9-9	T6L-10R	0.51
1059	T2-3b	0.37	1052	T7R-11R	0.42
7-2b	T2-6	0.33	1043	T7L-12R	0.55
1039	T2-8Ra	0.35	2778	T8-12	0.43
1058b	T2-8Rb	0.30	2785	T10R-11R	0.52
8B-3	T2L-9R	0.46	6-5M	T11R-12L	0.45
8-5Gb	T3R-5	0.51	10-5Kb	T11R-13L	0.50
4010	T3L-6L	0.43			
	AD translocations			AD translocations	
9-5H	T12R-19R	0.55	2780	T1L-14L	0.40
2925	T13R-19R	0.49	2B-1	T2R-14R	0.40
4669	T1L-20R	0.59	AZ-7	T6L-14L	0.51
2772	T9R-20L	0.47	1040	T4R-15L	0.48
2790	T7R-21R	0.50	1058a	T11L-15L	0.43
4675	T10L-2IL	0.39	2770	T1R-16Ra	0.49
2775a	T5-23	0.42	4672	T1R-16Rb	0.43
2870	T9R-25	0.43	1036	T9R-17Ra	0.45
6340	T9R-17Rb	0.54	10-5ka	T4L-19R	0.43
1316	T11R-17R	0.58	5-5B	T8R-19R	0.46
4659	T7-18R	0.53	1626	T10R-19R	0.59
E20-7	T3L-19L				
	DD translocations			DD translocations	
2777	T14L-23	0.48	E22-13	T19-21R	0.52
2781	T14R-24R	0.53	2786	T19R-24R	0.57
8-5Ga	T15R-16La	0.50	7-3F	T20R-21L	0.42
2767	T14L-16Lb	0.55	DP4	T20L-22R	0.54
SU5	T15R-20R	0.54	2791	T20R-25	0.68
	Complex translocations			Complex translocations	
VI	T1-2-3		2779	T5-11-21	
DP6	T2-19-23		4655	T2-11-13-24	

† Data from Seyam and Brown (1973); Menzel and Brown (1978a, 1978b); Brown (1980); Ray (1981); Brown et al. (1981). R, right arm; L, left arm. R arbitrarily assigned to arm with largest number of breaks, not necessarily the longer arm cytologically.

‡ Chromosome designation 1 to 13, A subgenome; chromosome designation 14 to 26, D subgenome.

§ dp-df, duplication deficiency.

the monosomic lines, including the monotelodisomics to be discussed below, are being transferred to the genetic background of Texas Marker 1 (TM-1), a highly inbred line of 'Deltapine 14,' which serves as the standard reference for plant morphological characters in genetic and cytogenetic studies (Kohel et al., 1970).

Because of the chromosome imbalance, a syndrome of plant morphological characters is associated with each monosomic type, which can be used to identify easily the monosomic plants from their disomic sibs. Several of the monosomics can be recognized in the seedling stage with a high degree of success (Endrizzi and Ramsay, 1979).

Breeding Behavior

Two kinds of cytological progenies are recovered when monosomic plants are self-pollinated or outcrossed as female: disomics and monosomics. The frequency of the monosomic progeny is dependent on the rate of transmission of the haplo-deficient gametes.

Normally, the frequency of monosomic plants obtained from unselected seed will range between 20 and 50%. However, the monosomics for chromosomes 3, 9, and 16 usually occur in much lower frequencies of 5 to 10%.

Douglas (1972) and Malek-Hedayat (1981) have found that within a single boll, the seeds having the lowest weight were predominantly monosomic. Malek-Hedayat (1981) also found that the frequency of monosomic progeny can be increased greatly by selecting within a sample of seed those seeds with weights ranging from the minimum weight to minus one standard deviation of the mean weight. This procedure is recommended especially for monosomics 3, 9, and 16, which are normally transmitted in low frequencies.

With the exception of the monosomic for chromosome 9, most of the monosomic chromosomes are fairly stable and misdivide infrequently to give telocentric or isochromosomes (Brown, 1958; Endrizzi and Ramsay, 1979). The monosomic condition of chromosome 9 induces abnormal chromosome separation, apparently during embryo sac development. Monosome 9 plants produce plants that are aneuploid for one or more chromosomes in addition to normal disomes and a few monosomic 9 plants.

The evidence indicates that male transmission of $n - 1$ gametes is rare, if it occurs at all, although controlled studies have not been made. Of eight monosomes tested, chromosomes 4 and 6 were male transmitted, but in a very low frequency.

Use of Monosomes

Monosomes provide a quick and simple method of identifying the chromosomes in translocations. They have been used extensively to assign genetic factors to specific chromosomes. The protocol for locating recessive and dominant marker genes on monosomic chromosomes was outlined by Endrizzi (1963). In such tests, the monosomic plant must be used as the female parent for their recovery in the F_1 generation, which will consist of two kinds, the disomic and monosomic F_1. Linkage groups were assigned to chromosomes mainly using monosomic and mono-telosomic stocks of cotton (Stelly, 1993; Reinisch et al., 1994).

With the development of the monosomic series, more than one alien chromosome may be substituted into one or the other species (White et al., 1967). Monosomics can also be used in the synthesis of chromosome substitution lines as outlined by Endrizzi et al. (1963b) and White et al. (1967). Several of the initially discovered *G. hirsutum* monosomics were used to substitute *G. barbadense* chromosomes into *G. hirsutum* (White et al., 1967; Kohel et al., 1977; Endrizzi et al., 1985). These lines had been used for the study of the genetic effects of individual chromosomes on plant traits and for estimating the number of genes and their interactions and linkage relationships controlling plant characters of economic importance.

As described methods for development of interspecific chromosome substitution in *G. hirsutum* by Endrizzi (1963) and confirmation of the cytogenetic and genetic constitution of the disomic lines by 34 cytological analysis and chromosome-specific SSR markers, 14 disomic *G. barbadense* chromosome substitution

(CS)-B lines have been developed through hypoaneuploid-based backcross chromosome substitution, using as recurrent parents previously developed monosomic or monotelodisomic near-isogenic backcross derivatives of TM-1. In each CS-B line, a pair of chromosomes (or chromosome arms) of *G. hirsutum* inbred TM-1 was replaced by the respective pair from *G. barbadense* doubled-haploid 3-79 lines (Saha et al., 2004). These substitution lines are nearly isogenic to the common parent TM-1 for 25 chromosome. The dominant or epistatic effects on boll size, lint percentage, micronaire, 2.5% span length, elongation, strength, and yield in 14 cotton lines (CS-B) with specific chromosomes or chromosome arms from *G. barbadense* substituted into *G. hirsutum* and chromosome-specific F_2 families were measured (Saha et al., 2006, 2008, 2010, 2011a; Stelly et al., 2005). Alien chromosome substitution lines for *G. tomentosum* and *G. mustelinum* chromosomes or chromosome segment(s) are in development. By crossing Sub16 or CS-B-16 with the recurrent parent TM-1 and producing F_2, Ren et al. (2002) identified QTLs affecting yield and fiber properties in chromosome 16.

Monotelodisomes

Telocentric chromosomes arise by misdivision of univalents at the first or second meiotic division and are recovered as monotelodisomics. A monotelodisomic plant has 25 bivalents plus a heteromorphic bivalent, in which the heteromorphic pair consists of a telocentric chromosome plus a complete homologous chromosome.

Twenty-nine monotelodisomic stocks have been identified (Table 8) that mark one or both arms of 19 chromosomes. With the exception of monotelodisomic 16L, all other monotelodisomics exhibit a syndrome of morphological characters closely paralleling in one form or another that which is associated with their monosomic counterparts. When either self or cross-pollinated, monotelodisomics produce two kinds of progenies: disomes and monotelodisomes. For most, the female transmission frequency of the telocentric chromosome is generally about 50%. In studies in which the monotelodisomics were used as the male parent in crosses, the telocentrics were transmitted through the pollen, ranging from a few percent for most to a high of 33% for Telo 16L. Only the ditelocentric for the long arm of chromosome 16 has been recovered from self-pollination of monotelodisomics. Following the association of a marker gene and a specific chromosome by the monosome test, the monotelodisomics can then be used for determining the arm location of the marker as well as its linkage relationship with the centromere (Endrizzi and Kohel, 1966). The monotelodisomics can be employed to identify the individual chromosomes in translocations and, in some cases, to determine the arm in which the translocation had occurred.

Isochromosomes are chromosomes with two homologous or identical arms. They arise by misdivision of univalent and telocentric chromosomes. Plants with an isochromosome are referred to as monoisodisomics because they have 25 normal bivalents plus a heteromorphic bivalent consisting of a normal chromosome and an isochromosome.

Isochromosomes can be used for positioning genes in arms of the chromosomes in the same manner as the telocentric chromosomes. However, they may give rates of recombination different from that of monotelodisomics because of exchanges in the trisomic arms, and thus they may not be as useful as the telocentrics for mapping studies (Endrizzi and Bray, 1980).

Trisomes and Alien Chromosome Additions

Many primary and a few secondary and tertiary trisomics, either spontaneous in origin or in several cases from unequal disjunction of heterozygous translocations, have been found. Primary trisomic individuals occur occasionally in cultivars of cotton (Endrizzi et al., 1963a).

By γ-ray irradiation of Pollen Mother Cells, Hu et al. (1996a,b) isolated and identified three primary, one secondary, and one tertiary trisomics. The evidence indicates that male transmission of $n + 1$ gametes is rare. By crossing with one set of translocations, one line is identified to be trisomics of chromosome 19 (Guo et al., 1999)

Alien chromosome addition lines have been bred and studied for their practical use (Kammacher and Poisson, 1964; Poisson, 1967; Kammacher and Schwendiman, 1969; Sharif and Islam, 1970). Only one complete set of *hirsutum–australe* monosomic alien chromosome addition lines have been bred using a combination of morphological survey, microsatellite marker assisted selection, and molecular cytogenetic analysis (Chen et al., 2014).

Disomic addition lines involving alien chromosomes provide a method for identifying homeologous chromosomes of different species (Schwendiman, 1974). Schwendiman reported that several *G. anomalum* disomic chromosome addition lines had modified morphological characters that were similar to that of an equal number of *G. stocksii* disomic chromosome addition lines. This finding showed that in each case the *G. anomalum* and *G. stocksii* disomic additions were homeologs.

Haploid

It was not been successfully to develop haploids from pollen culture because of certain hindrance in the procedure of cotton pollen culture. Semigamy is a type of reproductive mechanism in which a sperm nucleus penetrates the egg cell but does not fuse with the egg nucleus (Zhang et al., 1996). The egg and sperm nuclei divide independently and result in a heterogeneous haploid embryo. Turcotte and Feaster (1963) identified a high-frequency haploid production line from the sea island elite cultivars Pima S-1. It was demonstrated that semigamy was responsible for haploid producing (Turcotte and Feaster, 1967). In 1969, they obtained semigamy induced haploids and successfully doubled them with colchicines and transferred the recessive virescent gene (v_7) as a marker into the semigamy line and developed the virescently marked semigamy (VSg) (Turcotte and Feaster, 1969). The frequency of paternal sectors suitable for chromosome doubling from chimeral plants is relatively low, slightly less than 1% of germinated seed. However, the semigamy technique offers cotton researchers another tool for obtaining materials for research and evaluation in genetic studies, plant breeding, and other related disciplines. Using this technique, many doubled haploids have been developed in *G. hirsurum* and *G. barbadense* for uniformity and stability studies (Feaster and Turcotte, 1973, Meredith et al., 1970). VSg were also used to develop a haploid mapping population to construct a molecular linkage map of allotetraploid cotton (Zhang et al., 2002).

References

Banerji, I. 1929. The chromosome numbers of Indian cottons. Ann. Bot. (Lond.) 43:603–607.

Baranov, P. 1930. Work of the cytoanatomical laboratory of N.I.Kh.I. during the growing period of 1930. (In Russian.) Bull. Sci. Res. Cotton Inst. Tashkent 5:7–17.

Beal, J.M. 1928. A study of the heterotypic prophases in the microsporogenesis of cotton. Cellule 38:247–268.

Beasley, J.O. 1940a. The production of polyploids in *Gossypium*. J. Hered. 31:39–48.

Beasley, J.O. 1940b. The origin of American tetraploid *Gossypium* species. Am. Nat. 74:285–286. doi:10.1086/280895

Beasley, J.O. 1942. Meiotic chromosome behavior in species, species hybrids, haploids and induced polyploids of *Gossypium*. Genetics 27:25–54.

Brown, M.S. 1950. Cotton from Bikini. J. Hered. 41:115–121.

Brown, M.S. 1958. The division of univalent chromosomes in *Gossypium*. Am. J. Bot. 45:24–32. doi:10.2307/2438317

Brown, M.S. 1980. The identification of chromosomes of *Gossypium hirsutum* L. by means of translocations. J. Hered. 71:266–274.

Brown, M.S., M.Y. Menzel, C.A. Hasenkampf, and S. Naqi. 1981. Chromosome configurations and orientations in 58 heterozygous translocations in *Gossypium hirsutum*. J. Hered. 72:161–168.

Chen, Y., Y.Y. Wang, K. Wang, X.F. Zhu, W.W. Guo, T.Z. Zhang, and B.L. Zhou. 2014. Construction of a complete set of alien chromosome addition lines from *Gossypium australe* in *Gossypium hirsutum*: Morphological, cytological, and genotypic characterization. Theor. Appl. Genet. 127:1105–1121. doi:10.1007/s00122-014-2283-1

Davie, I.H. 1933. Cytological studies in the Malvaceae and certain related families. J. Genet. 28:33–67. doi:10.1007/BF02981768

Davie, I.H. 1935. Chromosome studies in the Malvaceae and certain related families. II. Genetica (The Hague) 17:487–498.

Douglas, C.R. 1972. Relationship of seed weight to cytotype of monosomic progeny in cotton. Crop Sci. 12:530–531. doi:10.2135/cropsci1972.0011183X001200040045x

Edwards, G.A., and M.A. Mirza. 1979. Genomes of the Australian wild species of cotton. II. The designation of a new G genome for *Gossypium bickii*. Can. J. Genet. Cytol. 21:367–372. doi:10.1139/g79-040

Endrizzi, J.E. 1963. Genetic analysis of six primary monosomes and one tertiary monosome in *Gossypium hirsutum*. Genetics 48:1625–1633.

Endrizzi, J.E., and R. Bray. 1980. Cytogenetics of disomics, monotelo- and monoiso-disomics and ml_1 *and* st_1 mutants of chromosome 4 of cotton. Genetics 94:979–988.

Endrizzi, J.E., and M.S. Brown. 1964. Identification of monosomes for six chromosomes in *Gossypium hirsutum*. Am. J. Bot. 51:117–120.

Endrizzi, J.E., and R.J. Kohel. 1966. Use of telosomes in mapping three chromosomes in cotton. Genetics 54:535–550.

Endrizzi, J.E., S.C. McMichael, and M.S. Brown. 1963a. Chromosomal constitution of "Stag" plants in *Gossypium hirsutum* 'Acala4-42'. Crop Sci. 3:1–3. doi:10.2135/cropsci1963.0011183X000300010001x

Endrizzi, J.E., and G. Ramsay. 1979. Monosomes and telosomes for 18 of the 26 chromosomes of *Gossypium hirsutum*. Can. J. Genet. Cytol. 21:531–536. doi:10.1139/g79-058

Endrizzi, J.E., and G. Ramsay. 1980. Identification of ten chromosome deficiencies in cotton. J. Hered. 71:45–48.

Endrizzi, J.E., T.R. Richmond, R.J. Kohel, and M.S. Brown. 1963b. Monosomes—A tool for developing better cottons. Texas Agric. Prog. 9:9–11.

Endrizzi, J.E., and T. Taylor. 1968. Cytogenetic studies of N LC_1 yg_2 R, marker genes and chromosome deficiencies in cotton. Genet. Res. 12:295–304. doi:10.1017/S0016672300011885

Endrizzi, J.E., E.L. Turcotte, and R.J. Kohel. 1985. Genetics, cytology, and evolution of *Gossypium*. Adv. Genet. 23:271–375. doi:10.1016/S0065-2660(08)60515-5

Feaster, C.V., and E.L. Turcotte. 1973. Yield stability in doubled haploids of American Pima cotton. Crop Sci. 13:232–233. doi:10.2135/cropsci1973.0011183X001300020024x

Fryxell, P.A. 1965. Stages in the evolution of *Gossypium*. Adv. Front. Plant Sci. 10:31–56.

Fryxell, P.A. 1971. Phenetic analysis and the phylogeny of the diploid species of *Gossypium* 1. (Malvaceae). Evolution 25:554–562. doi:10.2307/2407355

Fryxell, P.A. 1979. The natural history of the cotton tribe. Texas A&M Univ. Press, College Station.

Fryxell, P.A. 1992. A revised taxonomic interpretation of *Gossypium* L. (Malvaceae). Rheedea 2:108–165.

Gerstel, D.U. 1953a. Chromosome translocations in interspecific hybrids of the genus *Gossypium*. Evolution 7:234–244. doi:10.2307/2405734

Gerstel, D.U. 1953b. Genetic segregation of allopolyploids in the genus *Gossypium*. Genetics 38:664–665.

Gerstel, D.U. 1956. Segregation in new allopolyploids of *Gossypium*. I. The R_1 locus in certain New World wild American hexaploids. Genetics 41:31–44.

Gerstel, D.U. 1963. Evolutionary problems in some polyploid crop plants. In: J. MacKey, editor, Proceedings of the Second International Wheat Genetics Symposium, Genetics Institute, Lund, Sweden. 18–24 Aug. 1963. Hereditas Suppl. 2. Mendelian Society of Lund, Lund, Sweden. p. 481–504.

Gerstel, D.U., and L.L. Phillips. 1957. Segregation of new allopolyploids of *Gossypium*. II. Tetraploid combinations. Genetics 42:783–797.

Gerstel, D.U., and L.L. Phillips. 1958. Segregation of synthetic amphidiploids in *Gossypium* and *Nicotiana*. Cold Spring Harb. Symp. Quant. Biol. 23:225–237. doi:10.1101/SQB.1958.023.01.025

Guo, W.Z., C.X. Yi, C.M. Tang, J. Sun, J.J. Pan, and T.Z. Zhang. 1999. Studies of trisomic plants in Upland cotton IV. Origin and identification of chromosome 19 trisomic. Acta Agromonica Sin. 25(2):145–149.

Hanson, R.E., M.N. Islam-Faridi, E.A. Percival, C.F. Crane, Y.F. Ji, T.D. McKnight, D.M. Stelly, and H.J. Price. 1996. Distribution of 5S and 18S rDNA loci in a tetraploid cotton (*Gossypium hirsutum* L.) and its putative diploid ancestors. Chromosoma 105:55–61. doi:10.1007/BF02510039

Harland, S.C. 1928. Cotton notes. Tropic. Agric. 5:116–117.

Harland, S.C. 1935. The genetics of cotton. XII. Homologous genes for anthocyanin pigmentation in New and Old World cottons. J. Genet. 30:465–476.

Harland, S.C. 1937. Homologous loci of wild and cultivated American cottons. Nature 140:467–468. doi:10.1038/140467c0

Harland, S.C., and O.M. Atteck. 1941. The genetics of cotton. XVIII. Transference of genes from diploid North American wild cottons (*Gossypium thurberi* Tod., *G. armourianum* Kearney, *G. aridum* comb. nov. Skovsted) to tetraploid New World cottons (*G. barbadense* L. and *G. hirsutum* L.). J. Genet. 42:1–19. doi:10.1007/BF02982508

Hasenkampf, C.A., and M.Y. Menzel. 1980. Incipient genome differentiation in *Gossypium*. II. Comparison of 12 chromosomes of *G. hirsutum, G. mustelinum* and *G. tomentosum* using heterozygous translocations. Genetics 95:971–983.

Hawkes, J.G., and P. Smith. 1965. Continental drift and the age of angiosperm genera. Nature 207:48–50. doi:10.1038/207048a0

Hendrix, B., and J.M. Stewart. 2005. Estimation of the nuclear DNA content of *Gossypium* species. Ann. Bot. (Lond.) 95:789–797. doi:10.1093/aob/mci078

Hu, B.M., T.Z. Zhang, and J.J. Pan. 1996a. Studies of trisomic plants in Upland cotton. I. Origin, cytological identification and their phenotypes. Acta Agromonica Sin. 22:147–151.

Hu, B.M., T.Z. Zhang, and J.J. Pan. 1996b. Studies of trisomic plants in Upland cotton II. Transmission of trisomic chromosome and production of tetrasomes. Acta Agromonica Sinica 22:283–287.

Hutchinson, J.B. 1959. The application of genetics to cotton improvement. Cambridge Univ. Press, London.

Hutchinson, J.B., R.A. Silow, and S.G. Stephens. 1947. The evolution of *Gossypium*. Oxford Univ. Press, London.

Hutchinson, J.B., S.G. Stephens, and K.S. Dodd. 1945. The seed hairs of *Gossypium*. Ann. Bot. (Lond.) 9:361–367.

Ji, Y.F., M. De Donato, C.F. Cranel, W.A. Raska, M.N. Islam-Faridi, T.D. McKnight, H.J. Price, and D.M. Stelly. 1999. New ribosomal RNA gene locations in *Gossypium hirsutum* mapped by meiotic FISH. Chromosoma 108:200–207. doi:10.1007/s004120050369

Johnson, B.L. 1975. *Gossypium palmeri* and a polyphyletic origin of the New World cottons. Bull. Torrey Bot. Club 102:340–349. doi:10.2307/2484760

Johnson, B.L., and M.M. Thein. 1970. Assessment of evolutionary affinities in *Gossypium* by protein electrophoresis. Am. J. Bot. 57:1081–1092. doi:10.2307/2441274

Kadir, Z.B.Z. 1976. DNA evolution in the genus *Gossypium*. Chromosoma 56:85–95. doi:10.1007/BF00293732

Kammacher, P. 1958. La production artificielle d'aberrations chromosomiques chez *Gossypium hirsutum* par les rayons X. Cotton Fib. Trop. 13:1–22.

Kammacher, P. 1959. Relations cytologiques entre l'espece sauvage *Gossypium raimondii* Ulb. et les especes cultivees de cotonniers *G. hirsutum* L. er *G. barbadense* L. Cotton Fib. Trop. 14:1–4.

Kammacher, P. 1960. Observations cytologiques sur deux hybrides F, entre especes cultivees tetraploids de cotonniers et l'espece diploid sauvage *Gossypium raimondii* Ulb. Rev. Cytol. Bil. Veg. 22:1–32.

Kammacher, P., and C. Poisson. 1964. Sur les possibilities de transferer du material genetique du cotonnier sauvage *Gossypium anomalum* Waw. et Peyr. A l'espece cultivee *G. hirsutum* L. Cotton Fib. Trop. 19:243–264.

Kammacher, P., and J. Schwendiman. 1969. Addition au genome de l'espece du cotonnier *Gossypium hirsutum* de deux chromosomes de l'espece sauvage *G. stocksii*. Can. J. Genet. Cytol. 11:169–183. doi:10.1139/g69-021

Katterman, F.R.H., and D.R. Ergle. 1970. A study of quantitative variations of nucleic acids in *Gossypium*. Phytochemistry 9:2007–2010. doi:10.1016/S0031-9422(00)85353-4

Kearney, T.H. 1930a. Cotton plants, tame and wild and genetics of cotton. J. Hered. 21(325336):375–415.

Kohel, R.J. 1973. Genetic nomenclature in cotton. J. Hered. 64:291–295.

Kohel, R.J., C.F. Lewis, and M.N. Christiansen. 1977. The identification of a new mutant and linkage group in cotton. J. Hered. 68:65–66.

Kohel, R.J., T.R. Richmond, and C.F. Lewis. 1970. Texas Marker-I. Description of a genetic standard for *Gossypium hirsutum* L. Crop Sci. 10:670–671. doi:10.2135/cropsci1970.0011183X001000060019x

Longley, A.E. 1933. Chromosomes of *Gossypium* and related genera. J. Agric. Res. 46:217–227.

Malek-Hedayat, S. 1981. The relationship between seed weight and 13 monosomic and two monotelodisomic chromosomes in *Gossypium hirsutum*. M.S. thesis. Univ. of Arizona, Tucson.

Mauer, F.M. 1938. On the origin of cultivated species of cotton. A highly fertile triple hybrid [(*Gossypium barbadense* × *G. thurber*) × *G. arboretum*]. Bull. Acad. Sci. USSR Ser. Biol. p. 695–709. (From Plant Breed. Abstr. 9:318–319, 1939.)

Mauer, F.M. 1954. Origin and systematics of cotton. (In Russian.) Izv. Akad. Nauk. Uzb. SSR, Tashkent. (From English summary by S.S. Kanash, Tashkent Cotton Station, 1956.)

Menzel, M.Y. 1955. A cytological method for genome analysis in *Gossypium*. Genetics 40:214–223.

Menzel, M.Y., and M.S. Brown. 1952. Viable deficiency-duplications from a translocation in *Gossypium hirsutum*. Genetics 37:678–692.

Menzel, M.Y., and M.S. Brown. 1954a. The significance of multivalent formation in three-species *Gossypium* hybrids. Genetics 39:546–557.

Menzel, M.Y., and M.S. Brown. 1954b. The tolerance of *Gossypium hirsutum* for deficiencies and duplications. Am. Nat. 88:407–418. doi:10.1086/281851

Menzel, M.Y., and M.S. Brown. 1978a. Genetic length and breakpoints in twelve chromosomes of *Gossypium hirsutum* involved in 10 reciprocal translocations. Genetics 88:541–558.

Menzel, M.Y., and M.S. Brown. 1978b. Reciprocal chromosome translocations in *Gossypium hirsutum*. Arm location of breakpoints and recovery of duplications and deficiencies. J. Hered. 69:383–390.

Menzel, M.Y., M.S. Brown, and S. Naqi. 1978. Incipient genome differentiation in *Gossypium*. I. Chromosomes 14, 15, 16, 19 and 20 assessed in *G. hirsutum*, *G. raimondii*, and *G. lobatum* by means of seven A-D translocations. Genetics 90:133–149.

Menzel, M.Y., C.A. Hasenkampf, and J.McD. Stewart. 1982. Incipient genome differentiation in *Gossypium*. III. Comparison of chromosomes of *G. hirsutum* and Asiatic diploids using heterozygous translocations. Genetics 100:89–103.

Meredith, W.R., Jr., R.R. Bridge, and J.F. Chism. 1970. Relative performance of F_1 and F_2 hybrids from doubled haploidc and their parent varieties in Upland cotton, *Gossypium hirsutum* L. Crop Sci. 10:295–298. doi:10.2135/cropsci1970.0011183X001000030025x

Parks, C.R., W.L. Ezell, D.E. Williams, and D.L. Dreyer. 1975. VII. The application of flavonoid distribution to taxonomic problems in the genus *Gossypium*. Bull. Torrey Bot. Club 102:350–361. doi:10.2307/2484761

Percival, A.E., J.F. Wendel, and J.M. Stewart. 1999. Taxonomy and germplasm resources. In: C.W. Smith and J.T. Cothren, editors, Cotton: Origin, history, technology, and production. John Wiley & Sons, New York. p. 33–63.

Phillips, L.L. 1960. The cytogenetics of speciation in Asiatic cotton. Genetics 46:77–83.

Phillips, L.L. 1963. The cytogenetics of *Gossypium* and the origin of New World cottons. Evolution 17:460–469. doi:10.2307/2407096

Phillips, L.L. 1964. Segregation in new allopolyploids of *Gossypium*. V. Multivalent formation in New World x Asiatic and New World × wild American hexaploids. Am. J. Bot. 51:324–329. doi:10.2307/2440304

Phillips, L.L. 1966. The cytology and phylogenetics of the diploid species of *Gossypium*. Am. J. Bot. 53:328–335. doi:10.2307/2439872

Phillips, L.L. 1974. Cotton *(Gossypium)*. In: R.C. King, editor, Handbook of genetics 2: Plants, plant viruses, and protists. Plenum Press, New York. p. 111–133.

Phillips, L.L., and D.U. Gerstel. 1959. Segregation in new allopolyploids of *Gossypium*. III. Leaf shape segregation in hexaploid hybrids of New World cottons. J. Hered. 50:103–108.

Poisson, C. 1967. Sur les possibilities de transfert de material genetique du cotonnier sauvage *Gossypium anomalum* (Wav. et Peyr.) A l'espece cultivee *G. hirsutum* L. II. Creation de lignees d'addition a 27 paires de chromosomes. Cotton Fib. Trop. 22:401–415.

Prokhanov, Y.I. 1947. The conspectus of a new system of cottons *(Gossypium* L.). Bot Zhur. SSRR 32:61–78. (Cited from Fryxell, P.A. 1965. Stages in the evolution of *Gossypium*. Adv. Frontiers Plant Sci. 10:31–56.)

Ray, D.T. 1981. Identification of the chromosomes in a set of reciprocal translocations in *Gossypium hirsutum* L. Ph.D. thesis. Univ. of Arizona, Tucson. Available from University Microfilms, Ann Arbor, Mich. Publ. No. 8116695 (Dis. Abstr. 42:499B).

Reinisch, A.J., J.M. Dong, C.L. Brubaker, D.M. Stelly, J.F. Wendel, and A.H. Paterson. 1994. A detailed RFLP map of cotton *Gossypium hirsutum* × *Gossypium barbadense*: Chromosome organization and evolution in a disomic polyploid genome. Genetics 138:829–847.

Ren, L.H., W.Z. Guo, and T.Z. Zhang. 2002. Identification of QTLs affecting yield and fiber properties in chromosome 16 in cotton using substitution line. Acta Bot. Sin. 44:815–820.

Saha, S., J. Jenkins, J. Wu, J. McCarty, O. Gutierrez, R. Percy, R. Cantrell, and D. Stelly. 2006. Effect of chromosome specific introgression in Upland cotton on fiber and agronomic traits. Genetics 172:1927–1938. doi:10.1534/genetics.105.053371

Saha, S., J. Jenkins, J. Wu, J. McCarty, and D. Stelly. 2008. Genetic analysis of agronomic and fiber traits using four interspecific chromosome substitution lines in cotton. Plant Breed. 127:612–618. doi:10.1111/j.1439-0523.2008.01532.x

Saha, S., J. Wu, J. Jenkins, J. McCarty, R. Hayes, and D. Stelly. 2010. Genetic dissection of chromosome substitution lines discovered novel alleles in *Gossypium barbadense* L. with potential for improving agronomic traits including yield. Theor. Appl. Genet. 120:1193–1205. doi:10.1007/s00122-009-1247-3

Saha, S., J. Wu, J.N. Jenkins, J.C. McCarty, O. Gutierrez, D.M. Stelly, R.G. Percy, and D. Raska. 2004. Effect of chromosome substitutions from *Gossypium barbadense* L. 3–79 into *G. hirsutum* L. TM-1 on agronomic and fiber traits. J. Cotton Sci. 8:162–169.

Saha, S., J. Wu, J. Jenkins, J. McCarty, R. Hayes, and D. Stelly. 2011a. Delineation of interspecific epistasis on fiber quality traits in *Gossypium hirsutum* by ADAA analysis of intermated *G. barbadense* chromosome substitution lines. Theor. Appl. Genet. 122:1351–1361. doi:10.1007/s00122-011-1536-5

Sanamyan, M.F., and E.M. Rakhmatullina. 2003. Cytogenetic analysis of translocations in cotton. Plant Breed. 122:511–516. doi:10.1111/j.1439-0523.2003.00887.x

Sarvella, P. 1958. Multivalent formation and genetic segregation in some allopolyploid *Gossypium* hybrids. Genetics 43:601–619.

Saunders, J.H. 1961. The wild species of *Gossypium* and their evolutionary history. Oxford Univ. Press, London.

Schwendiman, A. 1974. Mise en evidence de trois nouvelles homeologies chromosomique entre *Gossypium anomalum* et *G. stocksii*. Can. J. Genet. Cytol. 16:871–881.

Seelanan, T., A. Schnabel, and J.F. Wendel. 1997. Congruence and consensus in the cotton tribe. Syst. Bot. 22:259–290. doi:10.2307/2419457

Seyam, S.M., and M.S. Brown. 1973. Types of disjunction in interchange heterozygotes of *Gossypium hirsutum* L. Egypt. J. Genet. Cytol. 2:322–330.

Sharif, A., and A.S. Islam. 1970. Cytogenetical studies of some backcross derivatives of *Gossypium hirsutum* × *G. anomalum*. Can. J. Genet. Cytol. 12:454–460. doi:10.1139/g70-064

Sherwin, K.H. 1970. Winds across the Atlantic—Possible African origins for some Pre-Columbian New World cultigens. Research Records of the University Museum, Mesoamerican Studies No. 6. Southern Illinois Univ., Carbondale.

Silow, R.A. 1946. Evidence on chromosome homology and gene homology in the amphidiploid New World cottons. J. Genet. 47:213–221. doi:10.1007/BF02986788

Skovsted, A. 1933. Cytological studies in cotton. I. The mitosis and the meiosis in diploid and triploid Asiatic cotton. Ann. Bot. (Lond.) 47:227–258.

Skovsted, A. 1934a. Cytological studies in cotton. II. Two interspecific hybrids between Asiatic and New World cottons. J. Genet. 28:407–424. doi:10.1007/BF02981763

Skovsted, A. 1934b. Cytogenetics in relation to plant breeding in cotton. Emp. Cotton Grow. Corp., Conf. Cotton Grow. Probl., Rep. Summ. Proc., 2nd. p. 46–49.

Skovsted, A. 1935a. Cytological studies in cotton. III. A hybrid between *Gossypium davidsonii* Kell. and *G. sturtii* F. Muell. J. Genet. 30:397–405. doi:10.1007/BF02982248

Skovsted, A. 1935b. Some new interspecific hybrids in the genus *Gossypium* L. J. Genet. 30:447–463.

Skovsted, A. 1935c. Chromosome numbers in the Malvaceae I. J. Genet. 31:263–296. doi:10.1007/BF02982344

Skovsted, A. 1937. Cytological studies in cotton. IV. Chromosome conjugation in interspecific hybrids. J. Genetics 34:97–134. doi:10.1007/BF02982258

Stebbins, G.L., Jr. 1947. Evidence on rates of evolution from the distribution of existing and fossil plant species. Ecol. Monogr. 17:149–158. doi:10.2307/1943261

Stelly, D.M. 1993. Interfacing cytogenetics with the cotton genome mapping effort. In: Proceedings of the Beltwide Cotton Improvement Conferences, New Orleans, LA. 10–14 Jan. 1993. Natl. Cotton Counc. Am., Memphis, TN. p. 1545–1550.

Stelly, D.M., S. Saha, D.A. Raska, J.N. Jenkins, J.C. McCarty, and O.A. Gutiérrez. 2005. Registration of 17 Upland (*Gossypium hirsutum*) germplasm lines disomic for different *G. barbadense* chromosome or arm substitutions. Crop Sci. 45:2663–2665. doi:10.2135/cropsci2004.0642

Stephens, S.G. 1944. Phenogenetic evidence for amphidiploid origin of New World cottons. Nature 153:53–54. doi:10.1038/153053a0

Stephens, S.G. 1947. Cytogenetics of *Gossypium* and the problem of the origin of New World cottons. Adv. Genet. 1:431–442. doi:10.1016/S0065-2660(08)60491-5

Turcotte, E.L., and C.V. Feaster. 1963. Haploids: High-frequency production from single-embryo seeds in a line of Pima cotton. Science 140:1407–1408. doi:10.1126/science.140.3574.1407

Turcotte, E.L., and C.V. Feaster. 1967. Semigamy in Pima cotton. J. Hered. 58:55–57.

Turcotte, E.L., and C.V. Feaster. 1969. Semigametic production of haploids in Pima cotton. Crop Sci. 9:653–655. doi:10.2135/cropsci1969.0011183X000900050045x

Valicek, P. 1978. Wild and cultivated cottons. Cotton Fib. Trop. 33:363–387.

Wang, K., B. Guan, W. Guo, B. Zhou, Y. Hu, Y. Zhu, and T. Zhang. 2008. Completely distinguishing individual A-genome chromosomes and their karyotyping analysis by multiple bacterial artificial chromosome-fluorescence in situ hybridization. Genetics 178:1117–1122. doi:10.1534/genetics.107.083576

Wang, K., W.Z. Guo, and T.Z. Zhang. 2007a. Detection and mapping of homologous and homoeologous segments in homoeologous groups of allotetraploid cotton by BAC-FISH. BMC Genomics 8:178. doi:10.1186/1471-2164-8-178

Wang, K., W.Z. Guo, and T.Z. Zhang. 2007b. Development of one set of chromosome-specific microsatellite-containing BACs and their physical mapping in *Gossypium hirsutum* L. Theor. Appl. Genet. 115:675–682. doi:10.1007/s00122-007-0598-x

Wang, K., X.L. Song, Z.G. Han, W.Z. Guo, J.Z. Yu, J. Sun, J.J. Pan, R.J. Kohel, and T.Z. Zhang. 2006. Complete assignment of the chromosomes of *Gossypium hirsutum* L. by translocation and fluorescence in situ hybridization mapping. Theor. Appl. Genet. 113:73–80. doi:10.1007/s00122-006-0273-7

Wang, K., Z.J. Yang, C.S. Shu, J. Hu, Q.Y. Lin, W.P. Zhang, W.Z. Guo, and T.Z. Zhang. 2010. Variation of genome structure, and size between A12 and D12 homoeologous chromosomes in allotetraploid cotton. Chromosoma 119:255–266. doi:10.1007/s00412-009-0254-0

Wang, K., W.P. Zhang, Y.J. Cao, Z.X. Zhang, D.W. Zheng, X.W. Niu, Y.J. Jiang, X.D. Chen, B.L. Zhou, W.Z. Guo, and T.Z. Zhang. 2012. Localization of high level of sequence conservation and divergence regions in cotton. Theor. Appl. Genet. 124:1173–1182. doi:10.1007/s00122-011-1777-3

Webber, J.M. 1934a. Chromosome number and meiotic behavior in *Gossypium*. J. Agric. Res. 49:223–237.

Webber, J.M. 1934b. Cytogenetic notes on cotton and cotton relatives. Science 80:268–269. doi:10.1126/science.80.2073.268

Webber, J.M. 1935. Interspecific hybridization in *Gossypium* and the behavior of F_1 plants. J. Agric. Res. 51:1047–1070.

Webber, J.M. 1939. Relationships in the genus *Gossypium* as indicated by cytological data. J. Agric. Res. 58:237–261.

Wendel, J.F. 1989. New world cottons contain Old World cytoplasm. Proc. Natl. Acad. Sci. USA 86:4132–4136. doi:10.1073/pnas.86.11.4132

Wendel, J.F. 2000. Genome evolution in polyploids. Plant Mol. Biol. 42:225–249. doi:10.1023/A:1006392424384

Wendel, J.F., C. Brubaker, I. Alvarez, R. Cronn, and J.M. Stewart. 2009. Evolution and natural history of the cotton genus. In: A.H. Paterson, editor, Genetics and genomics of cotton. Springer, New York. p. 3–22.

Wendel, J.F., and R.C. Cronn. 2002. Polyploidy and the evolutionary history of cotton. Adv. Agron. 78:139–186. doi:10.1016/S0065-2113(02)78004-8

Wendel, J.F., A. Schnabel, and T. Seleman. 1995. An unusual ribosomal DNA sequence from *Gossypium gossypioides* reveals ancient, cryptic, intergenomic introgression. Mol. Phylogenet. Evol. 4:298–313. doi:10.1006/mpev.1995.1027

White, T.G., and J.E. Endrizzi. 1965. Tests for the association of marker loci with chromosomes in *Gossypium hirsutum* by the use of aneuploids. Genetics 51:605–612.

White, T.G., T.R. Richmond, and C.F. Lewis. 1967. Use of cotton monosomes in developing interspecific substitution lines. Crops Res. USDA Rep. ARS 34-91. Washington, DC.

Wilson, F.D., and P.A. Fryxell. 1970. Meiotic chromosomes of *Cienjuegosia* species and hybrids and *Hampea* species (Malvaceae). Bull. Torrey Bot. Club 97:367–376. doi:10.2307/2483857

Wilson, F.D., and R.J. Kohel. 1970. Linkage of green-lint and okra-leaf genes in a reciprocal translocation stock of Upland cotton. Can. J. Genet. Cytol. 12:100–104. doi:10.1139/g70-016

Youngman, W., and S.C. Pande. 1927. Occurrence of branched hairs in cotton and upon *Gossypium stocksii*. Nature 119:745. doi:10.1038/119745a0

Zhang, H.Y., T.Z. Zhang, and J.J. Pan. 1996. Embryological investigation of sexual process of semigametic line in *Gossypium barbadense* L. Acta Agromonica Sin. 22:156–160.

Zhang, J., W.Z. Guo, and T.Z. Zhang. 2002. Molecular linkage map of allotetraploid cotton (*Gossypium hirsutum* L.× *Gossypium barbadense* L.) with a haploid population. Theor. Appl. Genet. 105:1166–1174. doi:10.1007/s00122-002-1100-4

Zhao, X.P., Y. Si, R.E. Hanson, C.F. Crane, H.J. Price, D.M. Stelly, J.F. Wendel, and A.H. Paterson. 1998. Dispersed repetitive DNA has spread to new genomes since polyploid formation in cotton. Genome Res. 5:479–492.

Qualitative Genetics and Utilization of Mutants

Richard Percy,* Bralie Hendon, Efrem Bechere, and Dick Auld

Abstract

Qualitatively inherited traits and mutants have provided the first markers used to map the cotton genome (*Gossypium* L.), and they have contributed significantly to cotton improvement. In this chapter, recently described morphological markers are reviewed, a comprehensive listing of markers is updated, the integration of morphological markers with molecular marker maps is summarized, and the use of qualitative traits in breeding is discussed. The role and contributions of induced mutants to genome mapping and cotton improvement efforts is reviewed. The induction of mutants through ionizing radiation and chemical mutagenesis is discussed, and mutation breeding methods are reviewed. A summary of germplasm created through mutation breeding and the traits that have been improved is given.

The discovery and scientific description of mutant traits in cotton coincides with the rediscovery of Mendelian principles. Initially, mutants in cotton were often regarded as interesting oddities, but some were recognized for their potential as heritable traits in crop improvement. Among the first mutants to be described was the colored lint trait, reported in 1906 (Balls), followed by the okra leaf trait in 1908 (Shoemaker, 1908), and the red leaf in 1912 (McLendon, 1912). Before 1955, few efforts were made to use mutants as markers in linkage mapping of the cotton genome, and only four linkage associations had been identified. The earliest of these were associations between the Red (R_1) and cluster (cl_1) loci and the green lint (*Lg*) and crinkle (*cr*) loci (Harland, 1939). Two independent linkages were reported between Petal spot (R_2) and the short branch (cl_2) and yellow-green (yg_2) traits (Silow, 1946; Rhyne, 1955). In 1955, S.G. Stephens identified four linkage groups consisting of 12 loci (Stephens, 1955). The following 25 yr, until 1980, was the most productive era in finding, describing, and mapping mutant loci

Abbreviations: EMS, ethylmethane sulfonate; QTL, quantitative trait locus; TA, terpenoid aldehyde; TILLING, Targeting Induced Local Lesions IN Genomes.

Richard Percy, Crop Germplasm Research Unit, 2881 F&B Rd., College Station, TX 77845. *Corresponding author (Richard.Percy@ars.usda.gov).

Bralie Hendon (bralie.hendon@ttu.edu) and Dick Auld (dick.auld@ttu.edu), Plant and Soil Science Department, Texas Tech University, MS 2122, Lubbock, TX 79409-2122.

Efrem Bechere, USDA-ARS, Crop Genetics Research Unit, 141 Experiment Station Road, Stoneville, MS 38776 (Efrem.bechere@ars.usda.gov).

doi:10.2134/agronmonogr57.2013.0042

Cotton. 2nd. ed. David D. Fang and Richard G. Percy, editors. Agronomy Monograph 57.

in cotton. The rate of discovery and reporting of morphological mutants slowed greatly with the advent of molecular markers. Over the past 15 yr, 12 new mutants have been described and their linkage associations investigated. To date, 157 morphological trait loci have been found in or transferred to allotetraploid cottons. Of these traits, 73 have been assigned to linkage groups. Nineteen linkage groups have been mapped in the allotetraploid species, and 17 of these have been localized on specific chromosomes. One linkage group has been assigned to a specific subgenome, but remains unassigned to a chromosome. Seventeen mutant markers have been fine mapped to locations within molecular marker maps.

Although morphological mutants have been largely superseded by molecular markers in genome mapping in the past 15 yr, qualitatively inherited mutant traits continue to be used extensively as descriptors in cataloging accessions of germplasm collections, in obtaining plant variety protection status, and in germplasm registration. Numerous traits such as plant color, pubescense, leaf shape, fiber color, etc. appear as descriptors in the National Collection of *Gossypium* Germplasm. Many traits have been evaluated for their potential in improving agronomic performance, disease and insect resistance, and product quality of cotton. Traits that have proven useful have been adopted in cultivar development programs and appear in registrations.

Genetic Nomenclature and Symbol Assignment

The first attempt to standardize gene symbols and nomenclature in cotton appears to have occurred in 1939 with the publication of Hutchinson and Silow (1939). A technical committee of the first US Regional Project S-1 took up the responsibility of standardizing genetic nomenclature and symbol assignment beginning in 1948, and its successor projects S-77 and S-258 continued this practice. In 1972, the S-77 Project Technical Committee adopted nomenclature standards based on guidelines of the International Committee on Genetic Symbols and Nomenclature (Anonymous, 1957). The adopted nomenclature rules and standards were published in 1973 to provide a permanent guide for future research (Kohel, 1973b). Today, there is no active committee revising or enforcing nomenclature and symbol assignment rules. There also is an apparent lack of awareness of the established conventions that has resulted in recent deviations in gene symbol assignment. It is incumbent on the cotton research community to reestablish nomenclature standards for mutant markers and the newer and much more numerous molecular markers.

Recently Described Markers and Mutants

Previous comprehensive reviews of morphological marker phenotypes, inheritance, and genetic linkages appear in the literature, and the current review does not attempt to cover the same ground as previously covered by these reviews (Endrizzi et al., 1984; Percy and Kohel, 1999). Significant effort has been made to provide an updated summary of marker names, designations, and chromosomal locations, shown in Table 1. A brief summary of recently described mutants, their inheritance, their map positions, and their relation to previously described markers follow. The convention of grouping markers into broad categories such as foliar, floral, glandular, reproductive, chlorophyll, pigment, or fiber variants has been followed.

Table 1. Symbols, names, chromosome and linkage associations, species origins, and references for allotetraploid cotton genes.

Gene symbol	Name	Chromosome	Linkage group	Origin	Reference
ac	curley cotton	16 (D7)	III	*Gossypium hirsutum* L.	Tiranti (1967)
an_1, an_2	annulated leaf			*G. hirsutum*	Lee (1985a)
$as_1 as_2$	asynapsis			*G. hirsutum*	Beasley and Brown (1942)
AS98	super-dwarf			*G. barbadense* L. × *G. hirsutum* hybrid	Zhang et al. (2011)
av_1	albivirescent			*G. barbadense*	Turcotte and Feaster (1978)
av_2	albivirescent			*G. barbadense*	Turcotte and Feaster (1978)
B_1	blight resistance			*G. hirsutum*	Knight (1947); Knight and Clouston (1939)
B_2	blight resistance			*G. hirsutum*	Knight and Clouston (1939)
B_3	blight resistance			*G. hirsutum*	Knight (1944)
B_4	blight resistance	5 (A5)	XI	*G. arboreum* L.	Knight (1944,1948)
B_5	blight resistance			*G. barbadense*	Knight (1950)
B_6	blight resistance			*G. arboreum*	Knight (1953a)
B_7	blight resistance			*G. hirsutum*	Knight (1953b)
B_8	blight resistance			*G. anomalum* Wawra	Knight (1954b)
B_{9K}	blight resistance			*G. herbaceum* L.	Knight (1963)
B_{9L}	blight resistance				Lagiere (1959)
B_{10K}	blight resistance				Knight (1957)
B_{10L}	blight resistance				Lagiere (1959)
B_{11}	blight resistance			*G. herbaceum*	Knight (1963)
B_{In}	blight resistance			*G. hirsutum*	Green and Brinkerhoff (1956)
B_N	blight resistance			*G. hirsutum*	Green and Brinkerhoff (1956)
B_S	blight resistance			*G. hirsutum*	Green and Brinkerhoff (1956)
bp	smooth boll			*G. barbadense*	Knight (1954a); Smith (1942)
bw_1, bw_2	withering bracts	12 (A12), 26 (D12)	V, IX	*G. hirsutum*	Knight (1952a); Rhyne (1965)
C_1, C_2	cotton leaf crumple virus resistance			*G. hirsutum*	Wilson and Brown (1991)
Cbd	cotton blue disease resistance	10 (A10)		*G. hirsutum*	Fang et al. (2010)
cg_1, cg_2	cleistogamy			*G. hirsutum*, *G. barbadense*	Hau et al. (1980a)
chl_1, chl_2	chlorophyll deficient			*G. hirsutum*, *G. barbadense*	Harland (1932,1934); Hutchinson (1946b)
ck^x, ck^y, ck^o	corky			*G. barbadense*, *G. hirsutum*	Stephens (1950)
cl_1	cluster	16 (D7)	III	*G. hirsutum*	Hutchinson and Silow (1939)
cl_2	cluster	7 (A7)	I	*G. barbadense*	Silow (1946)
cl_3	cluster	16 (D7)	III	*G. hirsutum*	Hau et al. (1980b)

Continued.

Table 1. Continued.

Gene symbol	Name	Chromosome	Linkage group	Origin	Reference
cr	crinkled leaf	15 (D1)	II	*G. barbadense*	Harland (1917); Hutchinson (1946a)
Crp	crumpled			*G. hirsutum*	Kohel (1973d)
cu	cup leaf			*G. hirsutum*	Lewis (1954)
cy	crinkle yellow			*G. hirsutum*	Kohel (1982)
de	depauperate			*G. hirsutum*	Kohel (1973e)
dg	dense glanding			*G. barbadense*	Percy (1999)
du	ultra-dwarf	6 (A6)		*G. hirsutum*	Chen et al. (2007)
Dw	dirty white lint	16 (D7)	III	*G. raimondii* Ulbr.	Rhyne (1957)
E	enhancer, fertility			*G. barbadense*	Sheetz and Weaver (1980)
F	flowering			*G. barbadense*	Lewis and Richmond (1960)
Fbl	fiberless	12 (A12)		*G. hirsutum*	Kearney and Harrison (1927); Rong et al. (2005)
fg, Fg^h	frego bract	3 (A3)	VI	*G. hirsutum*	Green (1955)
fng-a, *fng-b*	fringeless nectaries			*G. barbadense*	Stephens (1974b)
fs_1	female sterile			*G. hirsutum*	Justus and Meyer (1963)
gl_1, gl_6, gl_1^y	glandless boll			*G. hirsutum*, *G. barbadense*	McMichael (1954, 1970); Murray (1965)
gl_2, Gl_2^s, Gl_2^e, gl_3, Gl_3^r, Gl_3^{dav}	glandless and glanded plant	12 (A12), 26 26 (D12)	V, XIII IX	*G. hirsutum*, *G. barbadense*	Barrow and Davis (1974); Kohel and Lee (1984); Lee (1962, 1978); McMichael (1960)
(gl_4, gl_5)	glandless, glanded modifiers			*G. hirsutum*, *G. barbadense*	Lee (1962)
G^o, G^x, G^y	tumorigenesis			*G. hirsutm*, *G. gossypioides* (Ulbr.) Standl.	Phillips (1976)
Gv	golden veins			*barbadense*	Turcotte and Feaster (1983)
hb_1	hairy boll			*G. hirsutum*, *G. barbadense*	Hau et al. (1980a)
ia	accessory involucre	3 (A3)	VI	*G. hirsutum*	Kohel (1965)
im	immature fiber	3 (A3)	VI	*G. hirsutum*	Kohel and McMichael (1990); Kohel et al. (2002)
K	kidney seed			*G. barbadense*	Turcotte and Percy (1990)
L_1^L	laciniate leaf	1 (A1)	VII	*G. arboreum*	Hutchinson (1934)
L_2^o, L_2^s, L_2^u	okra leaf	15 (D1)	II	*G. barbadense*, *G. hirsutum*	Shoemaker (1908); Stephens (1945)
Lc_1	brown lint	7 (A7)	I	*G. hirsutum*	Harland (1935b)
Lc_2	brown lint	6 (A6)	IV	*G. barbadense*, *G. darwinii* G. Watt, *G. tomentosum* Nutt. ex Seem.	Harland (1935b)
Lc_3	brown lint			*G. hirsutum*	Kohel (1985)
Lc_4	brown lint			*G. hirsutum*	Kohel (1985)
Lc_5	brown lint			*G. hirsutum*	Kohel (1985)
Lc_6	brown lint			*G. hirsutum*	Kohel (1985)

Continued.

Table 1. Continued.

Gene symbol	Name	Chromosome	Linkage group	Origin	Reference
Le_1, Le_2, Le_2^{dav}	lethal	12 (A12), 26 (D12)	V, IX	*G. barbadense*, *G. davidsonii* Kellogg	Lee (1981)
Lf	leaf fleck	12 (A12)	V	*G. hirsutum*	Kohel et al. (1977)
L^f	filiform			G. hirsutum	Hau et al. (1981)
Lg, Lg^f	green lint, white lint, and green fuzz	15 (D1)	II	*G. hirsutum*	Kohel (1985)
Li_1	ligon lintless	22 (D4)	XV	*G. hirsutum*	Kohel (1972); Karaca et al. (2002); Rong et al. (2005)
Li_2	ligon lintless	18 (D13)		*G. hirsutum*	Narbuth and Kohel (1990); Kohel et al. (2002)
Lix	ligon lintless	4 (A4)		*G. hirsutum*	Cai et al. (2013)
Lk	leaky glands			*G. darwinii*	Stephens (1974a)
lp_1, lp_2	abnormal palisades	1(A1), 15(D1)	VII, II	*G. hirsutum*	Kohel (1964)
ltg	light green			G. barbadense	Turcotte and Feaster (1978)
me	methylation	9 (A9)		*G. barbadense*	Kohel and Bell (1999); Kohel et al. (2002)
Mi_1, Mi_2	*Meloidogyne incognita* resistance			*G. hirsutum*	McPherson (1993)
ml	mosaic leaf	4(A4)	VIII	*G. hirsutum*	Lewis (1958)
ms_1	male sterile			*G. hirsutum*	Justus and Leinweber (1960)
ms_2	male sterile			*G. hirsutum*	Richmond and Kohel (1961)
ms_3	male sterile	16 (D7)	III	*G. hirsutum*	Justus et al. (1963)
ms_5, ms_6	male sterile			*G. hirsutum*	Weaver (1968)
Ms_7	male sterile			*G. hirsutum*	Weaver and Ashley (1971)
ms_8, ms_9	male sterile	12 (A12), 26 (D12)	V, IX	*G. hirsutum*	Rhyne (1971)
Ms_{10}	male sterile			*G. hirsutum*	Bowman and Weaver (1979)
Ms_{11}	male sterile	12 (A12)	V	*G. barbadense*	Turcotte and Feaster (1979)
Ms_{12}	male sterile			*G. barbadense*	Turcotte and Feaster (1985)
ms_{13}	male sterile			*G. barbadense*	Percy and Turcotte (1991)
mt	mottled leaf			*G. hirsutum*	Lewis (1960)
N_1	naked seed	12 (A12)	V	*G. barbadense*, *G. hirsutum*	Carver (1929); Griffee and Ligon (1929); Harland (1929c); Kearney and Harrison (1927); Thadani (1923); Ware (1929)
n_2	naked seed	12 (A12)		*G. barbadense*, *G. hirsutum*	Harland (1929c); Ware et al. (1947); Rong et al. (2005)
ne_1, ne_2	nectariless	12 (A12), 26 (D12)	V, IX	*G. tomentosum*	Meyer and Meyer (1961)

Continued.

Table 1. Continued.

Gene symbol	Name	Chromosome	Linkage group	Origin	Reference
np_1, np_2	nonpinking				Rhyne and Carter (1991)
ob_1	open bud	18 (D13)	XVI	*G. hirsutum*	Kohel (1973a)
ob_2	open bud	13 (A13)		*G. hirsutum*	Endrizzi and Ray (1991); Zhang (1992); Qian et al. (2009)
ov_1, ov_2	ovate leaf			*G. barbadense*	Lee (1985a)
P_1	pollen color	5 (A5)	XI	*G. hirsutum*	Harland (1929b); Stephens (1954a)
P_2	pollen color			*G. barbadense*	Turcotte and Feaster (1966)
P_3	pollen color			*G. barbadense*	Rhyne and Carter (1992)
Pf_1, Pf_2	pink filament			*G. hirsutum*	Wilson (1987)
pg	pale green			G. hirsutum	Murray and Brinkerhoff (1966)
Pu	southwestern cotton rust resistance			*G. hirsutum*	Percy and Bird (1985)
Ra_1, Ra_2	raimondal			*G. raimondii*	Kohel and Bell (1999)
R_1, R_1^{dar}	red plant	16 (D7)	III	*G. hirsutum*	McLendon (1912); Harland (1935a); Stephens (1974a)
R_2, R_2^{v}	petal spot	7 (A7)	I	*G. barbadense*, *G. darwinii*, *G. hirsutum*	Harland (1929a); Stephens (1974a)
Rs	sub-red	7 (A7)		*G. hirsutum*	Chen et al. (2009)
Rc	cracked root	4 (A4)	XVIII	*G. harknessii* Brandegee	Weaver and Weaver (1979)
Rd	dwarf red		XIV	*G. hirsutum*	McMichael (1942); Johnson (1949)
Ren_1	reniform nematode resistance	11 (A11)		*G. longicalyx* J.B. Hutch. & B.J.S. Lee	Dighe et al. (2009)
Ren_2	reniform nematode resistance	21 (D11)		*G. aridum* (Rose & Standl.) Skovst.	Romano et al. (2009)
Ren_3	reniform nematode resistance	21 (D11)		*G. barbadense*	Gutierrez et al. (2011)
Ren_4	reniform nematode resistance	21 (D11)		*G. barbadense*	Gutierrez et al. (2011)
Ren_5	reniform nematode resistance	18 (D13)		*G. barbadense*	Gutierrez et al. (2011)
Rf_1	fertility restorer	4 (A4)	XVIII	*G. harknessii*	Weaver and Weaver (1977); DaSilva et al. (1981)
Rf_2	fertility restorer			*G. trilobum* (DC.) Skovst.	Stewart (1992); Zhang and Stewart (2001)
Rg	ragged leaf		X	*G. hirsutum*	Kohel and Lewis (1962a)
rl_1	round leaf		X	*G. hirsutum*	Brown and Cotton (1937)
Rl_2	round leaf			*G. hirsutum*	Percival et al. (1976)
Rl_3	round leaf			*G. barbadense*	Turcotte (1986)
rs	rudimentary stigma			*G. barbadense*	Turcotte and Feaster (1964)

Continued.

Table 1. Continued.

Gene symbol	Name	Chromosome	Linkage group	Origin	Reference
Ru	rugate	20 (D10)	XVII	*G. barbadense*	Turcotte and Feaster (1965)
rx	rex		X	*G. hirsutum*	Kohel and Percival (1989)
s, s^r, s^1, S^h	strap leaf	15 (D1)	II	*G. hirsutum*	Dilday and Waddle (1974); Dilday et al. (1975)
Se	semigamy			*G. barbadense*	Turcotte and Feaster (1975)
sma-$4(h_a)$	fiberless	6 (A6)		*G. arboreum*	Beasley and Egli (1977); Rong et al. (2005)
st_1	club stigma-style	4 (A4)	VIII	*G. hirsutum*	McMichael (1965)
st_2	mini stigma-style			*G. hirsutum*	McMichael (1965)
st_3	invert stigma-style		XIV	*G. hirsutum*	McMichael (1965)
T_1, T_1^h, T_1^{sm}, t_1, T_1^{an}, T_1^{rai}	trichome density	6 (A6)	IV	*G. hirsutum*, *G. tomentosum*, *G. anomalum*, *G. raimondii*	Harland (1944); Saunders (1961, 1963, 1965a, 1965b, 1965c); Simpson (1947); Lee (1968, 1985b)
T_2, T_2^{arm}, T^b_2, T_2^{to}, t_2	trichome density	25 (D6)		*G. hirsutum*, *G. armourianum* Kearney, *G. barbadense*, *G. tomentosum*, *G. raimondii*	Lee (1966, 1968, 1976, 1985b); Meyer (1957); Saunders (1965c)
T_3, t_3^h, t_3	trichome density			*G. hirsutum*	Lee (1968, 1985b)
T_4, t_4	trichome density			*G. barbadense*	Lee (1985b)
T_5, t_5	trichome density			*G. barbadense*	Lee (1985b)
ul	undulate leaf			*G. hirsutum*	Wilson (1989)
v_1	virescent	20 (D10)	XVII	*G. hirsutum*	Killough and Horlacher (1933)
v_2	virescent			*G. hirsutum*	Duncan and Pate (1967)
v_3	virescent	22 (D4)	XV	*G. hirsutum*	Percival and Kohel (1974)
v_4	virescent			*G. hirsutum*	Quisenberry and Kohel (1970)
v_5, v_6	virescent	1(A1), 15(D1)	VII, II	*G. hirsutum* (irradiation)	Kohel (1973c)
v_7	virescent			*G. barbadense*	Turcotte and Feaster (1973)
v_8	virescent		XIV	*G. hirsutum*	Kohel (1974)
v_9	virescent			*G. hirsutum*	Percival and Kohel (1974)
v_{10}	virescent	13 (A13)	XII	*G. hirsutum*	Percival and Kohel (1976)
v_{11}	virescent	5 (A5)	XI	*G. hirsutum*	Percival and Kohel (1974)
v_{12}	virescent			*G. hirsutum*	Kohel (1983a)
v_{13}	virescent			*G. hirsutum*	Kohel (1983a)
v_{14}	virescent			*G. hirsutum*	Kohel (1983a)
v_{15}	virescent			*G. hirsutum*	Kohel (1983a)
v_{16}, v_{17}	virescent	1(A1), 15(D1)	VII, II	*G. hirsutum*	Kohel (1983a)
v_{18}	virescent				
v_{19}	virescent				
v_{20}	virescent				

Continued.

Table 1. Continued.

Gene symbol	Name	Chromosome	Linkage group	Origin	Reference
v_{21}	virescent			*G. barbadense*	Turcotte and Percy (1988)
v_{22}	virescent			*G. hirsutum*	Zhang and Pan (1990)
vsp	virescent			*G. hirsutum* (space environment induced)	Song et al. (2012)
vf, *Vf*[n]	veins-fused	15 (D1)	II	*G. hirsutum*	Kohel and Lewis (1962b); Dilday et al. (1975)
wht_1, wht_2	white			*G. hirsutum*	Kohel (1988)
wr_1	wrinkled leaf			*G. barbadense*	Turcotte and Feaster (1980)
wr_2	wrinkled leaf			*G. barbadense*	Turcotte (1987)
Y_1	yellow petals	13 (A13)	XII	*G. barbadense*, *G. hirsutum*	Harland (1920,1936); Turcotte and Feaster (1963); Stephens (1954b)
Y_2	yellow petals	18 (D13)	XVI	*G. darwinii*	Harland (1936); Hutchinson and Silow (1939); Stephens (1954b)
yg_1, yg_2	yellow-green	16 (D7), 7 (A7)	III, I	*G. barbadense*, *hirsutum*	Rhyne (1955)
yp	yellow pulvinus	16 (D7)	III	*G. hirsutum*	Kohel and Bird (2002)
yv	yellow veins	20 (D10)	XVII	*G. hirsutum*	Kohel (1983b)

Fiber—Ligon Lintless (*Lix*), Fiberless—*Fbl* and *sma-4(h$_a$)*

Numerous mutants conditioning fiber color and quality traits have been found, described, and mapped. Among the many fiber color variants described, almost all have had a dominant expression over the white fiber color of commercial cottons (Harland, 1935b; Ware, 1932; Rhyne, 1960; Kohel, 1985). Among the mutants affecting fiber quality, the Ligon lintless-1 and Ligon lintless-2 mutants affect fiber length (Kohel, 1972; Narbuth and Kohel, 1990). Ligon lintless-1 (Li_1) is characterized by contorted leaf laminae, twisted branches and limbs, and extremely short fiber less than 6 mm long. Ligon lintless-2 (Li_2) produces a fiber similar to that of Li_1 but exhibits normal vegetative growth. In 2013, a new Ligon lintless mutant was found and described that has plant morphology similar to Li_1 (Cai et al., 2013). Unlike Li_1, the new mutant, *Lix*, produces a viable but non-productive plant in the homozygous state. The fiber of the heterozygous *Lixlix* mutant, while shorter than the normal fibered *lixlix* homozygote, does not display the extreme shortness of the Li_1 mutant. Cai et al. (2013) determined that *Lix* occurs on chromosome 4, which is the homeolog of chromosome 22, on which Li_1 resides. Another fiber mutant, immature fiber (*im*), produces a phenotype typical of fiber that has been arrested or retarded in development (Kohel and McMichael, 1990; Kohel et al., 2002). The *im* mutant fiber has a reduced secondary cell wall and a reduced weight, which is characteristic of fiber that has undergone severe environmental stress. Two described fiber mutants not included in previous reviews are both characterized by producing a naked or fiberless seed, *Fbl* and *sma-4(h$_a$)*.

The dominant *Fbl* mutant maps to a location that is indistinguishable from that occupied by a third gene conditioning fiber production, naked seed (N_1). The recessive *sma-4*(h_a) mutant, first found in *G. arboreum* L., mapped to a region with apparent homeology to chromosomes 6 and 25 in the tetraploid cottons.

Gossypol Glands—Dense Glanding (*dg*)

The genus *Gossypium* is characterized by the presence of darkly pigmented lysigenous glands that occur throughout the plant. Of six previously described loci conditioning the expression of glanding, dominant alleles at three (Gl_1, Gl_2, and Gl_3) are considered to produce the "normal" glanding pattern in *G. hirsutum* L. Two additional loci, gl_4 and gl_5, modify the expression of the above loci. A sixth locus, gl_6, has been determined to be a duplicate factor of gl_1 (Murray, 1965). A seventh pattern has been added to the six identified loci conditioning glanding patterns that conditions gland density in juvenile tissue, which has been given the name dense glanding and the gene symbol *dg* (Percy, 1999). Expression of *dg* is strongest in the first and second expanding leaves, weakens in the third apical leaf, and disappears by the fourth or fifth leaf. Coloration is also apparent in the first and second apical internodes and on the petioles of the first through third apical leaves. Expression persists in juvenile tissue through the growing season, but becomes less distinct as plant growth slows. The dense glanding trait is inherited as a monogenic, recessively expressed, nuclear gene. To date, no conclusive linkage associations of *dg* with other mutant marker genes have been reported.

Foliar—Yellow Pulvinus (*yp*)

A foliar mutant has been described that is associated with general depauperate growth, reduced glands on the stem, no or reduced normal reddening of the pulvinus, a gathering of veins at the base of the leaf and bract, and a distinctive yellowing of the pulvinus area (Kohel and Bird, 2002). Inheritance studies have established that the mutant is controlled by recessive alleles at a single locus.

Plant Pigment—Sub-red (R_s)

In cotton, anthocyanin pigmentation (imparting a red color) occurs in foliar tissue and as petal spots at the base of flower petals. Although a complicated series of multiple alleles for anthocyanin expression has been described in Asiatic species (Silow and Yu, 1942), within the cultivated tetraploid species, only three loci have been identified that condition anthocyanin expression. Two of these loci (R_1 and R_2) are homeologous, and their dominant alleles condition the expression of the red color in plant tissues. The third previously described locus, *Rd*, produces a red dwarf plant, is incompletely dominant, and maps to the D genome where it is expressed independently of R_1 (McMichael, 1942; Gerstel and Phillips, 1958; Phillips, 1962). Sub-red, R_s, has been added to the above loci and is found on chromosome 7 in close proximity to R_2 (Chen et al., 2009; Yin et al., 2009). The sub-red trait differs in expression from R_2, is incompletely dominant, and is characterized by a lighter pigmented red foliage and darker red flower than other red loci.

Plant Architecture—Ultra-dwarf (*du*) and Super-dwarf (*AS98*)

There are few qualitative mutant markers for plant height described in cotton, one being dwarf red, *Rd* (McMichael, 1942). The crinkled leaf mutant *cr* has also been characterized as being a dwarf and appears to have been reported several times in the literature (Harland, 1917; Hutchinson, 1946a). A recent addition has been the ultra-dwarf (*du*) trait, described in 2007 (Chen et al., 2007; Jing et al., 2011). The trait was conditioned by a single recessive gene, producing a mature plant only 10.5 cm tall that did not flower or produce seed. The mutant could be induced to grow to a height of 56 cm and flower by the addition of exogenous GA_3 hormone. Another recently described dwarf is super-dwarf, irregularly designated as *AS98* (Zhang et al., 2011). This mutant, derived from an interspecific hybrid, is described as having a normal phenotype until the fourth true leaf stage, after which it exhibits dwarfism. Unlike several other dwarf mutants, *AS98* does not exhibit the crinkle leaf phenotype. Super-dwarf also exhibits a single gene inheritance, but unlike several other dwarf mutants, it is expressed as an incompletely dominant trait. Both *du* and *AS98* are reported to have the same number of internodes as their normal counterparts at maturity and to respond positively to GA_3.

Chlorophyll—Virescent (*vsp*)

A series of mutants affecting chloroplast pigment development have been found in both *G. barbadense* L. and *G. hirsutum*. Easily identified, at least 22 loci conditioning the virescent trait have been reported. These mutants are expressed predominantly in the seedling stage; however, they vary widely in intensity and duration of expression during growth of the plant, and each is influenced by the environment. To date, all virescent mutants described have been recessive in expression. Recently, a putatively new virescent locus of novel origin has been described (Song et al., 2012). The newly described mutant *vsp* was identified from seed that had been exposed to a space environment. Genetic analysis of *vsp* indicated control by a single recessive gene in the nucleus. Tests for allelism with 12 other virescent loci (v_1, v_3, v_5v_6, v_8, v_9, v_{10}, v_{11}, v_{13}, v_{14}, v_{18}, v_{19}, and v_8) indicated that *vsp* was non-allelic.

Biochemical—Terpenoid Variants *me* and Ra_1, Ra_2

Terpenoid aldehyde (TA) compounds occur in the lysigenous "gossypol" glands that are common to most structures of the cotton plant. The TA compounds of glands vary in composition between organs of the plant and between species of *Gossypium*. Methylated TA is common in glands of *G. barbadense*, but infrequent in *G. hirsutum*. Raimonal is a TA variant that is unique to the species *G. raimondii* Ulbr. Both traits have been transferred to *G. hirustum* by interspecific hybridization of *G. hirsutum* and *G. barbadense* in the case of enhanced methylation, or by the crossing of *G. arboreum* L. with *G. raimondii*, followed by chromosome doubling to produce a synthetic amphidiploid in the case of raimondol (Kohel and Bell, 1999). In both cases backcrossing to *G. hirsutum* occurred. Inheritance studies indicate that expression of high levels of methylation is conditioned by a recessive allele at a single locus (designated *me*) and that the production of raimondal is conditioned by dominant alleles at two loci (designated Ra_1 and Ra_2).

Resistance—Cotton Blue Disease (*Cbd*) and Reniform Nematode (*Ren*)

Resistance to cotton blue disease, a serious viral disease occurring in the tropics and subtropics, was first described in upland cotton of African origin in 2003 (Royo et al., 2003). Subsequent research has determined that this resistance is determined by a single dominant gene (Junior et al., 2008; Fang et al., 2010). The resistance gene was determined to occur on chromosome 10 and was given the symbol *Cbd* (Fang et al., 2010). Cotton losses due to the reniform nematode (*Rotylenchulus reniformis* Linford & Oliveira) were first reported in 1940 (Smith, 1940), and the nematode has since become a major cause of crop losses in much of the southeastern and south-central United States (Robinson, 2007). To date, five resistance genes to the reniform nematode have been identified. One gene, initially designated as Ren^{lon} and since renamed Ren_1 (Fang and Stetina, 2011), was originally found in the African diploid species *G. longicalyx* J.B. Hutch. & B.J.S. Lee and subsequently transferred to *G. hirsutum*, where it has been localized on chromosome 11 (Dighe et al., 2009). A second reniform nematode resistance gene, originally designated Ren^{ari} and subsequently designated Ren_2, has been identified in *G. aridum* (Rose & Standl.) Skovst. and transferred to *G. hirsutum* (Romano et al., 2009). Romano et al. mapped Ren_2 to chromosome 21 and speculated that it and Ren_1 may be homeologous loci on homeologous chromosomes, a speculation supported by Fang and Stetina (2011). A third source of reniform nematode resistance has been reported in the species *G. barbadense* (Gutierrez et al., 2011). The discoverers of three quantitative trait loci (QTLs) for resistance in the above species correctly referred to the resistance factors as QTLs but then followed an unfortunate trend of treating QTLs as if they were discrete loci and gave the three QTLs gene designations. In any case, qualitative inheritance and segregation into discrete classes are often hard to demonstrate with disease resistance. Following an incorrect precedence set by Dighe et al. (2009), the investigators designated the three *G. barbadense* QTLs as Ren^{barb1}, Ren^{barb2}, and Ren^{barb3}, using the superscript to designate species origin. In their article correcting the designation of Ren_1 and Ren_2, Fang and Stetina (2011) point out that by convention, gene symbol subscripts denote discrete loci when more than one locus is known, and superscripts denote alleles. Using nomenclature conventions, perhaps the correct designation for the three resistant QTLs found in *G. barbadense* should be Ren_3, Ren_4, and Ren_5, once their true gene locus status is demonstrated. Gutierrez et al. (2011) note that "Ren_4" occurs on chromosome 21, shares a flanking molecular marker with Ren_1 and Ren_2, and may be their homeolog.

Advances in Mutant Marker Mapping

Although molecular markers have been the tools of choice for mapping efforts in recent years, incremental advances have been made in mapping and assigning morphological and physiological mutants to chromosomes and marker linkage groups. The mutant trait genes *im*, Li_2, and *me* have been assigned to chromosomes 3, 18, and 9, respectively (Kohel et al., 2002). These results were obtained from association tests with various aneuploid cottons, and therefore the assignment of these markers to positions within linkage groups on these chromosomes was not possible. An investigation of seven fiber mutants in 2005 produced map locations for Li_1, Li_2, and N_1 that were concordant with previous investigations, and it placed the previously unmapped mutant *Fbl* on chromosome 12 (Rong et al., 2005). The more recently described ultra-dwarf (*du*) and sub-red (*Rs*) mutants have been mapped to chromosomes 6 and 7, respectively (Jing et al., 2011; Chen et al., 2009). *Rs* has been mapped to linkage group I, occurring between genes Lc_1 and R_2 with the genetic distance to Lc_1 being 42.2 cM and the distance to R_2 being 1.68 cM (Yin et al., 2009). The yellow pulvinus (*yp*) marker described in 2002 produced a weak linkage with cluster-1 (cl_1) when submitted to linkage analyses with 14 markers representing 9 linage groups on 9 different chromosomes. This association would place yellow pulvinus in linkage group III on chromosome 16 (Kohel and Bird, 2002).

Ongoing mapping efforts also have allowed previously unassigned linkage groups to be assigned to chromosomes and some misassigned linkages to be corrected. Linkage group XII, containing the markers ob_2, Y_1, and v_{10}, previously could only be assigned to the A genome but has now been established to be resident on chromosome 13 of the A genome (Qian et al., 2009). The lintless fiber mutant Li_1, and therefore the linked trait v_3 of linkage group XV, have been assigned to chromosome 22 or D4 (Karaca et al., 2002; Rong et al., 2005). Mapping of the Li_x mutant marker has placed it on chromosome 4 (A4) and established that Li_1 and Li_x are homeologs. Another linkage, XVIII, between the Cracked Root (*Rc*) and Fertility Restorer (Rf_1) markers, has been placed on the long arm of chromosome 4 (A4) on the basis of analyses of monosomic and telesomic lines (Liu et al., 2003). The previous assignment of the naked seed n_2 marker to chromosome 26 using aneuploids (Endrizzi and Ramsay, 1980) has been brought into question by subsequent mapping that failed to detect any linkages between n_2 and other markers found on chromosome 26 (Rong et al., 2005). However, n_2 was linked to several markers of the homeologous chromosome 12.

Many of the recent linkage and chromosomal assignments of morphological mutants have been made with the assistance of molecular marker mapping. However, only a very few morphological traits have been incorporated into molecular marker maps and only a very few have been "fine mapped." Most of the morphological markers that have been incorporated into molecular maps have been traits of potential agronomic or economic interest, or mutants that may be useful in dissecting expression pathways of agronomic traits. Table 2 lists morphological markers, their chromosomal assignments, and their flanking molecular markers. Of the 24 markers listed, 10 are related to fiber quality, development, or color (*Lix*, Li_1, Li_2, n_2, N_1, *Fbl*, *im*, *sma-4*(h_a), Lc_1, and Lc_2). Six markers are associated with cotton blue disease and reniform nematode resistance and have been discussed previously. The remaining mutants of the table are associated with traits having utility in plant breeding methodology or to improving agronomic attributes of the plant. The open bud (ob_1 and ob_2) and fertility restoration (Rf_1) traits have been cited for their potential in the production of hybrid cottons (Qian et al., 2009; Liu et al., 2003). The ultra-dwarf, sub-red, frego bract, and okra leaf traits were candidates for fine mapping because of their perceived agronomic potential. As the name implies, the ultra-dwarf trait confers a very short stature and early maturation to the plant that may be advantageous under high density planting or short season situations. The okra leaf, conferring an open canopy and greater light penetration, and frego bract traits have both been associated with insect resistance but also confer undesirable attributes to the plant. They are discussed below. Red plant color has been associated with non-preference by certain insect pests (Jones and Brand, 1981), but its effectiveness in a monoculture is uncertain. The benefit of the sub-red leaf trait in insect control has not been reported. The glandless condition is of particular interest to breeding efforts aimed at producing a high protein food source from cotton seed and is also discussed below. To date, morphological markers not perceived to confer immediate agronomic benefits have not been candidates for fine mapping and therefore have not been integrated into molecular marker maps.

Table 2. Morphological mutant markers mapped to molecular marker maps and their nearest established flanking markers.

Gene symbol	Name	Chromosome	Flanking markers	Marker type†	Reference
ob_1	open bud	18 (D13)	BNL2652 (2.9 cM) ob_1 (0.5 cM) NAU624	EST-SSR	Qian et al. (2009)
ob_2	open bud	13 (A13)	BNL2449 (1.1 cM) ob_2 (1.7 cM) TMP01	EST-SSR	Qian et al. (2009)
Lix	ligon lintless	4 (A4)	NAU8376 (6.4 cM) *Lix* (24.5 cM) NAU3469	SSR	Cai et al. (2013)
Li_1	ligon lintless	22 (D4)	TMB2500 (0.8 cM) Li_1 (9.6 cM) DPL0489	SSR	Gilbert et al. (2013)
			Gate4CA09 (2.7 cM) Li_1 (1.3 cM) Coau1J04	RFLP	Rong et al. (2005)
			MP4030 (12.5 cM) Li_1	SSR	Karaca et al. (2002)
Li_2	ligon lintless	18 (D13)	A1552 (0.5 cM) Li_2 (1.8 cM) Gate4BF10	RFLP	Rong et al. (2005)
			DPL0547 (0.87 cM) Li_2 (0.52 cM) DPL0922	SSR	Hinchliffe et al. (2011)
n_2	naked seed	12 (A12)	pAR0155 (13.3 cM) n_2 (14.6 cM) Gate2BB08	RFLP	Rong et al. (2005)
Cbd	blue disease resistance	10 (A10)	NG0204310 (0.05 cM) *Cbd* (0.75cM) NG0211495	SNP	Fang et al. (2010)
Ren_1	reniform nematode resistance	11 (A11)	Ren_1 (0.8 cM) BNL3279	SSR	Dighe et al. (2009)
Ren_2	reniform nematode resistance	21 (D11)	TMB1871(5.3 cM) Ren_2 (1 cM) BNL3279	SSR	Romano et al. (2009)
Ren_3	reniform nematode resistance	21 (D11)	BNL1551 (14.4 cM) Ren_3 (8.7 cM) GH132	SSR	Gutierrez et al. (2011)
Ren_4	reniform nematode resistance	21 (D11)	BNL4011 (2.1 cM) Ren_4 (1.8 cM) BNL3279	SSR	Gutierrez et al. (2011)
Ren_5	reniform nematode resistance	18 (D13)	BNL1721 (12.0 cM) Ren_5 (3.3cM) BNL569	SSR	Gutierrez et al. (2011)
N_1	naked seed	12 (A12)	Gafb28I12 (6.5 cM) N_1 (9.0 cM) Gafb29C08	RFLP	Rong et al. (2005)
Fbl	fiberless	12 (A12)	Gafb29C08 (5 cM) *Fbl* (1.9 cM) pAR0244	RFLP	Rong et al. (2005)
du	ultra-dwarf	6 (A6)	NAU2238 (3.3 cM) *du* (1.4 cM) NAU4946	SSR	Jing et al. (2011)
Rs	sub-red	7 (A7)	NAU3735 (0.1 cM) *Rs* (0.2 cM) NAU1048	SSR	Chen et al. (2009)
Gl_2^e	glandless	12 (A12)	CIR362 (9.27 cM) GL_2^e (0.96 cM) NAU2251b	SSR	Dong et al. (2007)
Rf_1	fertility restoration	4 (A4)	NAU/SSR/$Rf_1 1_{170}$ (0.3 cM) Rf_1	SSR	Liu et al. (2003)
im	immature fiber	3 (A3)	DPL1071 (0.32 cM) *im* (0.13 cM) TMB1898	SSR	Kim et al. (2013)
L_2^o	okra leaf	15 (D1)	Gh565 (2.6 cM) L_2^o (2.8 cM) NAU2343	SSR	Andres et al. (2014)
sma-4(h_a)	fiberless	6 (A6)	Gate1BB03 (2.3 cM) *sma-4*(h_a) (0.7 cM) A1691b	RFLP	Rong et al. (2005)
Lc_1	brown lint	7 (A7)	NAU2862 (7.8 cM) Lc_1 (3.8 cM) NAU1043	SSR	Wang et al. (2014)
Lc_2	brown lint	6 (A6)	NAU5433 (4.4 cM) Lc_2 (7.4 cM) NAU2968	SSR	Wang et al. (2014)
fg	frego bract	3 (A3)	NAU3016 (0.3 cM) *fg* (4.7 cM) NAU3172	SSR	Li et al. (2010)

† EST, expressed sequence tag; SSR, simple sequence repeat; RFLP, restriction fragment length polymorphism; SNP, single nucleotide polymorphism.

Utilization of Genetic Mutants

Probably one of the most investigated aspects of qualitatively inherited morphological traits has been their potential for conferring insect resistance. A large body of research on this subject exists, much of it originating in the 1970s and 1980s. Trichome density and distribution conferred by the gene series T_1 through t_5 has received extensive attention. One of the first reports of a simply inherited morphological trait of cotton conferring resistance to an insect was the hirsute trait providing a level of resistance to the jassid (Knight, 1952b). The glabrous condition has been found to alleviate damage caused by the bollworm and reduce populations of fleahopper, banded-wing whitefly, and sweetpotato whitefly (Lukefahr et al., 1969, 1970, 1971; Mound, 1965). Although the pilose trait (T_1) has been found to reduce boll weevil feeding (Stephens and Lee, 1961), it has not been used in cotton improvement efforts because of its various deleterious effects associated with it (Kohel et al., 1967; Lee, 1984, 1985c). The smooth leaf trait (T_3) has been incorporated into commercial cottons, not for its insect resistance potential but primarily because of its potential to reduce trash content at harvest (Lee, 1977; Meredith, 1993; Novick et al., 1990).

As with the case of the pilose trait, a number of morphological traits of cotton have been determined to provide levels of resistance to specific insects, but have associated agronomic or quality costs that must be factored into their adoption and use. The mutant frego bract trait (*fg*) has been determined to confer resistance to the boll weevil, but also has been associated with later plant maturity, smaller boll size, and increased susceptibility to the plantbug (Jenkins and Parrott, 1971; Percival and Kohel, 1978; McGowen et al., 1995). The okra leaf (L_2^o) trait, besides conferring some resistance to boll rotting fungi because of the open canopy mentioned above, also has been reported to confer levels of resistance to boll weevil and whitefly (Pieters and Bird, 1977; Percy et al., 1997) and increased earliness of maturity. Negative associations reported for the okra leaf trait include lower photosynthetic rates and yields, and increased plant determinacy (Andries et al., 1969). Whereas the frego trait has not been incorporated into commercial cultivars, the okra leaf trait has been used in several successful cultivars. Although many qualitatively inherited morphological traits have displayed a blend of beneficial and detrimental attributes, resistance traits to diseases and pests such bacterial blight, cotton rust, cotton blue disease, root knot nematode, and reniform nematode have been of unmixed benefit to cotton production.

Gossypol, a sesquiterpenoid aldehyde found in pigment glands of cotton, has been demonstrated to be an effective antibiosis factor against the bollworm, tobacco budworm, and pink bollworm (Lukefahr and Houghtaling, 1969; Shaver and Parrott, 1970; Shaver et al., 1970). Levels of gossypol can be increased within the cotton plant through breeding, and high gossypol germplasm has been demonstrated to possess higher levels of insect resistance. However, increased gossypol from highly glanded cottonseed is a detrimental factor in the use of seed as a protein source in livestock feeding (Berardi and Goldblatt, 1980; Calhoun and Holmberg, 1991). Conversely, gossypol and gossypol glands can be removed from the cotton plant through the manipulation of a couple of marker loci, but only at the cost of creating a cotton plant that is highly susceptible to predation. A far greater literature on morphological traits as host plant resistance factors exists than is covered here. The above gives a sampling of the extensive work in this field.

Creation and Utilization of Induced Cotton Mutants

Mutagenesis, the act of inducing mutations within an organism's genome, has been used in plant breeding since Muller's discovery of the mutagenic effects of x-rays on *Drosophila* flies to study the genetics of these flies and the mechanisms of heredity (Muller, 1927). Techniques to predictably induce mutations as a method of plant improvement emerged in the middle of the last century (Smart and Simmonds, 1995). Mutation breeding in conjunction with conventional breeding methods has been credited with the development of over 3200 crop varieties, including wheat (*Triticum* L.), barley (*Hordeum vulgare* L.), rice (*Oryza sativa* L.), sunflower (*Helianthus annuus* L.), grapefruit (*Citrus paradisi* Macfad.), and cotton that have been grown worldwide (FAO–IAEA, 2011). The FAO–IAEA database lists 47 varieties of cotton developed with mutagenesis during the period of 1961 to 2009. China (13 varieties), Pakistan (10 varieties), Uzbekistan (9 varieties), and India (9 varieties) were the most active nations using mutagenesis in cotton. Badnawar-1, the first cotton cultivar to be listed as having been created through mutagenesis, was created in 1961 by treating seeds with x-rays and crossing a resulting mutant line ('Indore 2') with 'Dhar Kambodia'. The recent literature would indicate mutation breeding of cotton is currently active in Pakistan, India, Uzbekistan, and Egypt, with at least a few programs in the United States (Ahloowalia and Maluszynski, 2001; Chopra, 2005; Mba, 2013).

The evolution of cotton (*Gossypium* sp.) along with its domestication and intensive breeding efforts focused primarily on the development of elite cultivars has caused a series of bottlenecks that seriously restricts the amount of genetic diversity available in the cotton genome (Wendel, 1989). This lack of genetic diversity especially limits future improvement of the allotetraploid species *G. barbadense* and *G. hirsutum*. However, continued advances in lint yield, fiber quality, pest resistance, and herbicide tolerance are necessary to keep cotton as a competitive crop commodity worldwide. One way to potentially mitigate the narrow germplasm base in cotton is by using mutagenesis as a breeding and genetics tool.

Induced Mutations

Induced mutants have been used to create novel alleles necessary to breed unique and often superior cotton cultivars. The common types of mutations in DNA sequences include the following:

- genome mutations where there are changes in chromosome number;
- structural changes in the chromosomes including chromosal aberrations and translocations;
- gene mutations including single base substitutions or point mutations, in which one base is replaced by another—these could be either transitions (A or G) or transversions (C or T); and
- extra nuclear mutations, which involve plastids and mitochondria (Brown and Caligari, 2008).

Mutagenesis has been shown to be an effective tool in creating a wide range of phenotypic variations in both diploid and tetraploid *Gossypium* populations (Auld et al., 2009; Mba, 2013). Both physical and chemical mutagens have been used successfully in cotton mutagenesis.

Most common physical mutagens use ionizing radiation, which includes γ rays, x-rays, neutrons, α particles, β particles, ultraviolet radiation, and even cosmic radiation experienced in space exploration (Brown and Caligari, 2008). Radiation mutagenesis has been shown to be an effective tool in the creation of a wide range of phenotypic variants in tetraploid (*G. hirsutum*) populations. Mutants were identified with improved earliness (using γ rays as a mutagen) (Kandhro et al., 2002; Xanthopoulos and Kechagia, 2001, 2003); enhanced phosphorous uptake (radio phosphorous) (Nazirov et al., 1981); improved drought tolerance (radio phosphorous) (Nazirov and Satipov, 1979); and increased lint percent (x-rays) (Cornelius et al., 1970); as well as improvements in both earliness and lint yield (γ rays) (Shamsuzzaman et al., 2003). Radiation has also been used to create variation in the electron transport and the Hill reactions (γ rays) (Yakabova et al., 1975); photoperiod insensitivity (γ rays) (Raut et al., 1971); nonspecific esterases (γ ray) (Gulin et al., 1985); fiber properties (x-rays) (Mehetre and Thombre, 1983); cytoplasmic sterility (γ rays) (Negmatov et al., 1975); and pollen size (γ rays) (Savaskan, 2002). However, five cycles of seed irradiation at only 10 kR of γ rays failed to induce high levels of variation in both lint yield and fiber quality characteristics (Fotiadis and Miller.1973).

Mutation breeding programs using physical mutants have been effective and productive. A mutant cotton cultivar, NIAB78, that was released in 1987 and grown widely in Pakistan has accounted for over US$20 million in additional income to farmers annually (Ahloowalia et al., 2004). This cultivar is heat tolerant, has a determinate growth habitat, and escapes bollworm attacks because of its early maturity (Ahloowalia and Maluszynski, 2001). Mutation breeding was also successfully used to improve seed oil and protein content in cotton (Muthusamy and Jayabalan, 2013; Singh and Narayanan, 1991). Waghmare and Koranne (2000) reported plant mutants in cotton with morphological changes that included dwarf and stunted (15–20 cm), dwarf bushy, and compact types (45–80 cm) and very tall types (200–260 cm). Bhat and Dani (1993) observed higher boll weight in cotton with γ irradiation at different doses and ethylmethane sulfonate (EMS) treatment at 0.05%. They also reported higher gin turnout (39%) in *G. hirsutum* mutants with γ irradiation with 5 kR and treatments with EMS independently in M_5 generation. Cornelius et al. (1970) also reported increased lint percent by using x-rays. Bazhanova and Mamedov (1991) reported *G. barbadense* mutants with higher lint index of 6.7 g and seed index of 10.4 g when γ was irradiated with 5 kR in the M_5 generation. Improved mutants have also played an important role in generating materials with resistance to Verticillium wilt (Gaibullaev et al., 1975), bacterial blight (Salanki and Parameswarappa, 1968), and jassids (Chopra, 2005).

Chemical mutagens can affect the DNA molecule by using agents that mimic one of the four nitrogenous bases found in DNA or agents that induce polyploidy such as colchicine. They have gained popularity since they are easy to use, do not require any specialized equipment, and can provide a very high mutation frequency. They cause single bp changes or single nucleotide polymorphisms rather than deletions and translocations common in radiological methods (MacLeod, 1994). Chemical mutagenesis has identified mutants in both *G. barbadense* and *G. hirsutum*. Hussein et al. (1982), working with *G. barbadense*, identified glandless cotton mutants using sodium azide as a mutagenic agent. Mutation frequency in *G. hirsutum* was shown to have a linear response to increasing concentrations of sodium azide (Larik et al., 1983). Ethylmethane sulfonate is probably the most widely used chemical mutagen. Ethylmethane sulfonate selectively alkylates guanine bases causes the DNA polymerase to preferentially pairing a thymine residue instead of a cytosine residue opposite to the guanine residue during DNA replication. This type of base substitution results in a random point mutation. A majority of the changes (70–99%) in EMS-mutated populations are GC to AT bp transitions (Till et al., 2004, 2007; Sikora et al., 2011).

Additional traits identified by researchers in chemically mutated populations of *G. hirsutum* include resistance to Verticillium wilt (dimethyl sulfate) (Gaibullaev et al., 1975) and improved fiber characteristics (ethyl methanesulfonate) (Auld et al., 2000; Brown et al., 2012; Patel et al., 2014). The oil content of cotton has also been shown to be successfully manipulated with mutagenesis (Bhat and Dani, 1993; Singh and Narayanan, 1991). Shattuck and Katterman (1982) also found that cotyledonary tissue of *G. barbadense* showed an unscheduled and enhanced DNA synthesis after treatment with ethyl methanesulfonate. Colchicine has also been used to create genetic variation in lint yield, lint percent, and fiber quality (Luckett, 1989); partially naked seed (Salanki and Parameswarappa, 1968); and parthenogenesis (Shi-Qi et al., 1991). According to Badigannavar (1999), EMS treatment at 0.2% was very effective in inducing a higher magnitude of genetic variance for number of bolls/plant and boll weight in cotton. The same author also reported a wide range of lint percent for EMS mutants (30– 38.5%). The highest turnout for a mutant was 15.3% higher than the control. Katterman (1973) found a decrease in the 5-methyl cytosine content in the DNA and spontaneously induced polyhaploids in both *G. hirsutum* and *G. barbadense* treated with colchicine but did not report any morphological or agronomic mutants. The treatment of cotton seed with colchicine in combination with γ radiation (Sanamyan and Rakhmatullina, 2003) or with γ radiation alone (Sheidai et al., 2002) resulted in increased chromosomal aberrations.

Mutagenic Breeding

Historically, nearly all of the mutant populations developed in cotton have been screened phenotypically for one or more specific trait(s) using a forward genetics approach to plant improvement (Bechere et al., 2010; Herring et al., 2004). Individual plants phenotypically are selected and progeny tested to ensure they are homozygous and breed true for the selected trait. While this process has been moderately effective in increasing the genetic variation in cotton, it has been limited to enhancement of simply inherited traits that can be selected directly from the phenotype. Cotton mutants that have proven useful in crop improvement have fallen into six main categories: disease resistance, pest resistance, herbicide resistance, enhanced fiber quality, reduction in gossypol, and plant canopy architecture (Gaibullaev et al., 1975; Singh and Narayanan, 1991; Auld et al., 2000; Bechere et al., 2010). Induced mutations provide a proven, safe, robust, highly portable, and inexpensive plant breeding strategy. However, some of the characteristics that have retarded the use of mutants are the requirement for generating and evaluating large populations of putative mutants, the incidence of chimeras, and the recessive nature of mutations (Ahloowalia and Maluszynski, 2001; Mba, 2013).

Screening mutant populations for changes in specific enzymes using reverse genetics techniques such as Targeting Induced Local Lesions IN Genomes (TILLING) holds much promise (Till et al., 2004, 2007). The power of TILLING was first demonstrated in model systems such as *Arabidopsis* and *Drosophila* (McCallum et al., 2000; Winkler et al., 2005; Sikora et al., 2011). It has also been shown that single mutations within a specific DNA could be identified in higher plants (Caldwell et al., 2004; Chawade et al., 2010; Cooper et al., 2008; Galeano et al., 2009). However, there are limited reports of TILLING mutagenized cotton populations. The application of reverse genetics could lead to advances in conventional genetics and the development of functional genomics in cotton (Mba, 2013; Henikoff et al., 2004). By adopting the deductive process of identifying which traits have been modified and relating the modifications to changes in genomic regions of the induced mutants, as compared with the normal types, the gene(s) controlling a trait may be identified (Mba, 2013; Tadele et al., 2009; Auld et al., 2009).

USDA–ARS has a number of cotton germplasm and cultivar mutant collections that it holds publicly available under the National Plant Germplasm System (GRIN) (http://www.ars-grin.gov/npgs/holdings.html, accessed 31 Mar. 2015). Mutants have been identified for their superior fiber quality, lint yield, herbicide tolerance, and tufted-naked seed traits (Table 3).

Table 3. *Gossypium hirsutum* L. mutants developed through chemical mutagenesis and stored in the USDA-ARS National Germplasm System (http://www.ars-grin.gov/npgs/holdings.html, accessed 31 Mar. 2015).

Mutant institutions	Plant introduction number	Year of release	Improved traits	Reference
TTU 202-1107-B†,¶,‡‡	613162	1999	fiber length/strength	Auld et al. (2000)
TTU 271-2155-C†,¶,‡‡	613163	1999	fiber length/strength	Auld et al. (2000)
TTU 0808-1-6-1†,‡,¶,§,‡‡	643916	2006	fiber length/strength/uniformity/short fiber content	Bechere et al. (2007)
TTU 0774–3-3†,‡,¶,§,‡‡	643915	2006	fiber length/strength/uniformity/short fiber content	Bechere et al. (2007)
RAIDER 276†,¶,§	645568	2006	yield/fiber length/strength/uniformity/yarn tenacity	Auld et al. (2007)
RM(3)-8-1†,¶,§	657943	2009	imazamox (herbicide) tolerance	Bechere et al. (2010)
SCM(3)-4-3†,¶,§	657941	2009	imazamox (herbicide) tolerance	Bechere et al. (2010)
SCM(3)-7-3†,¶,§	657942	2009	imazamox (herbicide) tolerance	Bechere et al. (2010)
EM(4)-3-1†,¶,§	657944	2009	imazamox (herbicide) tolerance	Bechere et al. (2010)
TTU 0782†,‡,§,‡‡	659504	2010	fiber length/strength/uniformity	Bechere et al. (2011)
TAM 94L-25-M24†,‡,¶,#	664553	2010	upper-half mean length	Brown et al. (2012)
TAM 94L-25-M25†,‡,¶,#	664554	2010	upper-half mean length	Brown et al. (2012)
TAM 94L-25-M30†,‡,¶,#	664555	2010	upper-half mean length	Brown et al. (2012)
MD 15 M4§	665638	2012	fiber strength	Bechere et al. (2013)
9023n_4^t†,§	667553	2013	tufted-naked seed	Bechere and Auld (2013)

† Texas Tech University.

‡ Texas A&M University.

¶ Texas A&M AgriLife Research, Lubbock, TX.

§ USDA-ARS, Stoneville, MS.

Cotton Incorporated.

‡‡ New Mexico State University.

References

Ahloowalia, B.S., and M. Maluszynski. 2001. Induced mutations—A new paradigm in plant breeding. Euphytica 118:167–173. doi:10.1023/A:1004162323428

Ahloowalia, B.S., M. Maluszynski, and K. Nichterlein. 2004. Global impact of mutation-derived varieties. Euphytica 135:187–204. doi:10.1023/B:EUPH.0000014914.85465.4f

Andres, R.J., D.T. Bowman, B. Kaur, and V. Kuraparthy. 2014. Mapping and genomic targeting of the major leaf shape gene (*L*) in Upland cotton (*Gossypium hirsutum* L.). Theor. Appl. Genet. 127:167–177. doi:10.1007/s00122-013-2208-4

Andries, J.A., J.E. Jones, L.W. Sloane, and J.G. Marshall. 1969. Effects of okra leaf shape on boll rot, yield, and other important characters of Upland cotton, *Gossypium hirsutum* L. Crop Sci. 9:705–710. doi:10.2135/cropsci1969.0011183X000900060009x

Anonymous. 1957. Report of the International Committee on Genetic Symbols and Nomenclature. Int. Union Biol. Sci., Series B 30:1-6.

Auld, D.L., E. Bechere, M.D. Ethridge, W.D. Becker, E. Hequet, and R.G. Cantrell. 2000. Registration of TTU 202-1107B and TTU 271-2155C mutant germplasm lines of upland cotton with improved fiber quality. Crop Sci. 40:1835–1836.

Auld, D.L., E. Bechere, M. Krifa, H. Kebede, E. Hequet, R. Wright, and S. Misra. 2007. Registration of 'Raider 276' a high-yielding improved quality upland mutant cotton cultivar. J. Plant Reg. 1:115–116. doi:10.3198/jpr2007.01.0059crc

Auld, D.L., G.G. Light, M. Fokar, E. Bechere, and R.D. Allen. 2009. Mutagenesis systems for genetic analysis of *Gossypium*. In: A.H. Paterson, editor, Genetics and genomics of cotton. Springer, New York. p. 209–226.

Badigannavar, A.M. 1999. Induction of genetic variability through mutations for quantitative characters in cotton. M.Sc. (Agri.) thesis, Univ. of Agric. Sci., Dharwad, India.

Barrow, J.R., and D.D. Davis. 1974. Gl_2^s—A new allele for pigment glands in cotton. Crop Sci. 14:325–326. doi:10.2135/cropsci1974.0011183X001400020047x

Bazhanova, A.P., and K. Mamedov. 1991. Effect of gamma rays from ^{60}Co on changes on the amount of seed fuzz in *Gossypium barbadense*. Ref. Zh. 5Ya3129.

Beasley, C.A., and E. Egli. 1977. Fiber production in vitro from a conditional fiberless mutant of cotton. Dev. Biol. 57:234–237. doi:10.1016/0012-1606(77)90371-2

Beasley, J.O., and M.S. Brown. 1942. Asynaptic Gossypium plants and their polyploids. J. Agric. Res. 65:421–427.

Bechere, E., and D.L. Auld. 2013. Registration of a tufted-naked seed upland cotton germplasm, 9023n_4^t. J. Plant Reg. 8(1):63–67. doi:10.3198/jpr2013.06.0025crg

Bechere, E., D.L. Auld, R.G. Cantrell, E. Hequet, M. Krifa, S. Misra, and C. Wayne Smith. 2007. Registration of TTU 0774-3-3 and TTU 0808-1-6-1 Upland cotton germplasm lines with improved fiber length and strength. J. Plant Reg. 1:58–59. doi:10.3198/jpr2006.09.0587crg

Bechere, E., D.L. Auld, P. Dotray, and H. Kebede. 2010. Registration of four upland cotton (*Gossypium hirsutum* L.) genetic stock mutants with tolerance to Imazamox. J. Plant Reg. 4:155–158. doi:10.3198/jpr2009.08.0446crgs

Bechere, E., D.L. Auld, M. Krifa, C. Wayne Smith, and R.G. Cantrell. 2011. Registration of TTU 0782, an upland cotton germplasm line with superior fiber quality. J. Plant Reg. 5:207–210. doi:10.3198/jpr2010.07.0404crg

Bechere, E., W.R. Meredith, and J.C. Boykin. 2013. Registration of mutant population MD 15M4 *Gossypium hirsutum* L. with enhanced fiber quality. J. Plant Reg. 7(2):216–219. doi:10.3198/jpr2012.07.0006crg

Berardi, L.C., and L.A. Goldblatt. 1980. Gossypol. In: I.E. Liener, editor, Toxic constituents of plant foodstuffs. Academic Press, New York. p. 211–266.

Bhat, M.G., and R.G. Dani. 1993. Improvement in productivity and oil content of cotton (*G. hirsutum*) cultivars through induced polygenic mutations. J. Cotton Res. Dev. 7(1):9–18.

Bowman, D.T., and J.B. Weaver, Jr. 1979. Analyses of a dominant male-sterile character in Upland cotton. II. Genetic studies. Crop Sci. 19:628–630. doi:10.2135/cropsci1979.0011183X001900050020x

Brown, H.B., and J.R. Cotton. 1937. "Round Leaf" cotton. Notes on the appearance and behavior of a peculiar new strain. J. Hered. 28:45–48.

Brown, I.N., C.W. Smith, D.L. Auld, S. Hague, E.F. Hequet, and D. Jones. 2012. Registration of TAM 94L-25-M24, TAM 94L-M25, and TAM 94L-25-M30 mutant Upland cotton germplasm with improved fiber length and strength. J. Plant Reg. 6:195–199. doi:10.3198/jpr2011.05.0245crg

Brown, J., and P.D.S. Caligari. 2008. An introduction to plant breeding. Blackwell Publ., Oxford, UK.

Cai, C., X. Tong, F. Liu, F. Lv, H. Wang, T. Zhang, and W. Guo. 2013. Discovery and identification of a novel Ligon lintless-like mutant (*Lix*) similar to the Ligon lintless (Li_1) in allotetraploid cotton. Theor. Appl. Genet. 126:963–970. doi:10.1007/s00122-012-2029-x

Caldwell, D.G., N. McCallum, P. Shaw, G.J. Muchlbaver, D.F. Marshall, and R. Waugh. 2004. A structural mutant population for forward and reverse genetics in barley (*Hordeum vulgare* L.). Plant J. 40(1):143–150. doi:10.1111/j.1365-313X.2004.02190.x

Calhoun, M.C., and C. Holmberg. 1991. Safe use of cotton by-products as feed ingredients for ruminants: A review. In: L.A. Jones, et al., editors, Cattle research with gossypol-containing feeds. Natl. Cottonseed Prod. Assoc., Memphis, TN. p. 97.

Carver, W.A. 1929. The inheritance of certain seed, leaf, and flower characters in *Gossypium hirsutum* and their genetic interrelations. J. Am. Soc. Agron. 21:467–480. doi:10.2134/agronj1929.00021962002100040012x

Chawade, A., P. Sikora, and M. Brautigam. 2010. Development and characterization of an oat TILLING population and identifications of mutations in lignin and beta-glucan biosynthesis genes. BMC Plant Biol. 10:86. doi:10.1186/1471-2229-10-86

Chen, X.S., J.C. Di, X.J. Ma, and X.M. Liu. 2009. Analysis on linkage inheritance of sub-red plant mutant from upland cotton. Yichuan 31(8):855–858.

Chen, X.S., J.C. Di, N.Y. Xu, S.H. Xiao, and J.G. Liu. 2007. The inheritance of an ultra-dwarf plant mutant from upland cotton. (In Chinese.) Yichuan 29(4):471–474. doi:10.1360/yc-007-0471

Chopra, V.L. 2005. Mutagenesis: Investigating the process and processing the outcome for crop improvement. Curr. Sci. 89(2):353–359.

Cooper, J.L., B.J. Till, R.G. Laport, M.C. Darlow, J.M. Kleffner, A. Jamai, T. El-Mellouki, S. Liu, R. Ritchie, N. Nelson, K.D. Bilyeu, K. Meksem, L. Comai, and S. Henikoff. 2008. TILLING to detect induced mutations in soybean. BMC Plant Biol. 8:9. doi:10.1186/1471-2229-8-9

Cornelius, T.J., S.D. Peter, S.S. Narayanan, and S. Kamalanathan. 1970. Useful mutants in Cambodia cotton (*G. hirsutum*) by ionizing radiation of seeds. Madras Agric. J. 57:594–597.

DaSilva, F.P., J.E. Endrizzi, and L.S. Stith. 1981. Genetic study of restoration of pollen fertility of cytoplasmic male-sterile cotton. Rev. Bras. Genet. 4:411–426.

Dighe, N.D., A.F. Robinson, A.A. Bell, M.A. Menz, R.G. Cantrell, and D.M. Stelly. 2009. Linkage mapping of resistance to reniform nematode in cotton following introgression from *Gossypium longicalyx* (Hutch. & Lee). Crop Sci. 49:1151–1164. doi:10.2135/cropsci2008.03.0129

Dilday, R.H., R.J. Kohel, and T.R. Richmond. 1975. Genetic analysis of leaf differentiation mutants in Upland cotton. Crop Sci. 15:393–397. doi:10.2135/cropsci1975.0011183X001500030033x

Dilday, R.H., and B.A. Waddle. 1974. Inheritance and linkage relatioship of strap leaf mutants in cotton. Crop Sci. 14:387–390. doi:10.2135/cropsci1974.0011183X001400030014x

Dong, C., Y. Ding, W. Guo, and T. Zhang. 2007. Fine mapping of the dominant glandless gene *Gl2e* in Sea-island cotton (*Gossypium barbadense* L.). Chin. Sci. Bull. 52:3105–3109. doi:10.1007/s11434-007-0468-6

Duncan, D.N., and J.B. Pate. 1967. Inheritance and uses of golden crown virescent in cotton. J. Hered. 58:237–239.

Endrizzi, J.E., and G. Ramsay. 1980. Identification of ten chromosome deficiencies in cotton. J. Hered. 71:45–48.

Endrizzi, J.E., and D.T. Ray. 1991. Linkage analysis of open bud (ob_2) and yellow petal (Y_1) in cotton. Genome 34:461–463. doi:10.1139/g91-069

Endrizzi, J.E., E.L. Turcotte, and R.J. Kohel. 1984. Qualitative genetics, cytology, and cytogenetics. In: R.J. Kohel and C.F. Lewis, editors, Cotton. ASA, Madison, WI. p. 81–129.

Fang, D.D., and S.R. Stetina. 2011. Improving cotton (*Gossypium hirsutum* L.) plant resistance to reniform nematodes by pyramiding Ren_1 and Ren_2. Plant Breed. 130:673–678. doi:10.1111/j.1439-0523.2011.01889.x

Fang, D.D., J. Xiao, P.C. Canci, and R.G. Cantrell. 2010. A new SNP haplotype associated with blue disease resistance gene in cotton (*Gossypium hirsutum* L.). Theor. Appl. Genet. 120:943–953. doi:10.1007/s00122-009-1223-y

FAO–IAEA. 2011. Mutant variety database. http://mvgs.iaea.org/AboutMutantVarities.aspx (accessed 26 June 2014).

Fotiadis, N.A., and P.A. Miller. 1973. Effects of recurrent seed irradiation on genetic variability and recombination in cotton (*Gossypium hirsutum* L.). Crop Sci. 13:40–44. doi:10.2135/cropsci1973.0011183X001300010013x

Gaibullaev, I.K., M.K. Avazkhodzhaev, and A.E. Egamberdiev. 1975. Reaction of chemo mutants of cotton to infection of the pathogen of Verticillium wilt. Sov. Genet. 11:34–36.

Galeano, C.H., M. Gomez, L.M. Rodriguez, and M.W. Blair. 2009. CEL I nuclease digestion for SNP discovery and marker development in common bean (*Phaseolus vulgaris* L.). Crop Sci. 49:381–394. doi:10.2135/cropsci2008.07.0413

Gerstel, D.U., and L.L. Phillips. 1958. Segregation of synthetic amphidiploids in *Gossypium* and *Nicotiana*. Cold Spring Harbor Symp. Quant. Biol. 23:225–237.

Gilbert, M.K., R.B. Turley, H.J. Kim, P. Li, G. Thyssen, Y. Tang, C.D. Delhom, M. Naoumkina, and D.D. Fang. 2013. Transcript profiling by microarray and marker analysis of the short cotton (*Gossypium hirsutum* L.) fiber mutant Ligon lintless-1 (Li_1). BMC Genomics 14:403–420. doi:10.1186/1471-2164-14-403

Green, J.M. 1955. Frego bract, a genetic marker in Upland cotton. J. Hered. 46:232.

Green, J.M., and L.A. Brinkerhoff. 1956. Inheritance of three genes for bacterial blight resistance in Upland cotton. Agron. J. 48:481–485. doi:10.2134/agronj1956.00021962004800110001x

Griffee, F., and L.L. Ligon. 1929. Occurrence of "lintless" cotton plants and the inheritance of the character "lintless". J. Am. Soc. Agron. 21:711–717. doi:10.2134/agronj1929.00021962002100070001x

Gulin, V.V., V.A. Kal'chenko, S. Turavekov, and D.A. Musaev. 1985. Polymorphism and mutagenesis of nonspecific esterases in cotton (*Gossypium hirsutum* L.). Dokl. Biol. Sci. 285:700–703.

Gutierrez, O.A., A.F. Robinson, J.N. Jenkins, J.C. McCarty, M.J. Wubben, F.E. Callahan, and R.L. Nichols. 2011. Identification of QTL regions and SSR markers associated with resistance to reniform nematode in *Gossypium barbadense* L. accession GB713. Theor. Appl. Genet. 122:271–280. doi:10.1007/s00122-010-1442-2

Harland, S.C. 1917. On the genetics of crinkled dwarf rogues in Sea Island cotton. West Indian Bull. 16:82–84.

Harland, S.C. 1920. Studies of inheritance in cotton. I. The inheritance of corolla colour. West Indian Bull. 18:13–19.

Harland, S.C. 1929a. The genetics of cotton. I. The inheritance of petal spot in New World cottons. J. Genet. 20:365–385. doi:10.1007/BF02983155

Harland, S.C. 1929b. The genetics of cotton. II. The inheritance of pollen colour in New World cottons. J. Genet. 20:387–399. doi:10.1007/BF02983156

Harland, S.C. 1929c. The work of the Genetics Department of the Cotton Research Station, Trinidad. Empire Cotton Grow. Rev. 6:304–314.

Harland, S.C. 1932. The genetics of cotton. VI. The inheritance of chlorophyll deficiency in New World cottons. J. Genet. 25:271–281. doi:10.1007/BF02984591

Harland, S.C. 1934. The genetics of cotton. XI. Further experiments on the inheritance of chlorophyll deficiency in New World cottons. J. Genet. 29:181–195. doi:10.1007/BF02982195

Harland, S.C. 1935a. The genetics of cotton. XII. Homologous genes for anthocyanin pigmentation in New and Old World cottons. J. Genet. 30:465–476.

Harland, S.C. 1935b. The genetics of cotton. XIV. The inheritance of brown lint in New World cottons. J. Genet. 31:27–37. doi:10.1007/BF02982277

Harland, S.C. 1936. Duplicate genes for corolla colour in *Gossypium barbadense* L. and *G. darwinii* Watt. Z. Indukt. Abstamm. Vererbungsl. 71:417–419.

Harland, S.C. 1939. The genetics of cotton. Jonathan Cape, London, UK.

Harland, S.C. 1944. The selection experiment with Peruvian Tanguis cotton. Bull. No. 1. Soc. Nac. Agrar., Inst. Cotton Genet., Lima, Peru.

Hau, B., E. Koto, and J. Schwendiman. 1980a. Determinisme genetique de deux mutants du cotonnier capsule pileuse et fleur cleistogame. Coton Fibres Trop. 35:355–357.

Hau, B., E. Koto, and J. Schwendiman. 1980b. Examen du groupe de liaison III du cotonnier *Gossypium hirsutum*: Description d'un nouveau phenotype cluster et localisation des genes. Coton Fibres Trop. 35:359–367.

Hau, B., E. Koto, and J. Schwendiman. 1981. Description d'une mutation induisant une feuille filiforme chez le cotonnier *G. hirsutum* L. Coton Fibres Trop. 36:205–208.

Henikoff, S., B.J. Till, and L. Comai. 2004. TILLING. Traditional mutagenesis meets functional genomics. Plant Physiol. 135:630–636. doi:10.1104/pp.104.041061

Herring, A.D., D.L. Auld, M.D. Ethridge, E.F. Hequet, E. Bechere, C.J. Green, and R.G. Cantrell. 2004. Inheritance of fiber quality and lint yield in a chemically mutated population of cotton. Euphytica 136:333–339. doi:10.1023/B:EUPH.0000032747.97343.54

Hinchliffe, D.J., R.B. Turley, M. Naoumkina, H.J. Kim, Y. Tang, K.M. Yeater, P. Li, and D.D. Fang. 2011. A combined functional and structural genomics approach identified an EST-SSR marker with complete linkage to the Ligon lintless-2 genetic locus in cotton (*Gossypium hirsutum* L.). BMC Genomics 12:445. doi:10.1186/1471-2164-12-445

Hussein, H.A.S., A. Al-Enani, and M. El-Moghazi. 1982. Histological and morphological characteristics of glandless cotton mutant induced with sodium azide. Egypt. J. Genet. Cytol. 11:167–173.

Hutchinson, J.B. 1934. The genetics of cotton. X. The inheritance of leaf shape in Asiatic *Gossypiums*. J. Genet. 28:437–513. doi:10.1007/BF02981765

Hutchinson, J.B. 1946a. The crinkled dwarf allelomorph series in New World cottons. J. Genet. 47:178–207. doi:10.1007/BF02986786

Hutchinson, J.B. 1946b. On the occurrence and significance of deleterious genes in cotton. J. Genet. 47:272–289. doi:10.1007/BF02986252

Hutchinson, J.B., and R.A. Silow. 1939. Gene symbols for use in cotton genetics. J. Hered. 30:461–464. doi:10.1093/jhered/30.10.461

Jenkins, J.N., and W.L. Parrott. 1971. Effectiveness of frego bract as a boll weevil resistance character in cotton. Crop Sci. 11:739–743. doi:10.2135/cropsci1971.0011183X0011000500 41x

Jing, C., X.J. Ma, J.C. Di, and X.S Chen. 2011. Gene mapping of an ultra-dwarf mutant in Upland cotton. Yichuan 33:1393–1397.

Johnson, B.L. 1949. Complementary factors for dark-red plant color in Upland cotton. J. Agric. Res. 78:535–543.

Jones, J.E., and J.W. Brand. 1981. Influence of red plant color on boll weevil non-preference and agronomic performance. In: J.M. Brown, editor, Proceedings of the Beltwide Cotton Production Research Conference, Natl. Cotton Counc. Am., New Orleans, LA. 4–8 Jan. 1981. Natl. Cotton Counc. Am., Memphis, TN. p. 76–77.

Junior, O.P., I. Schuster, R.B. Pinto, E. Pires, J.L. Belot, P. Silvie, L.G. Chitarra, L.V. Hoffmann, and P. Barroso. 2008. Inheritance of resistance to cotton blue disease. Pesqui. Agropecu. Bras. 43(5):661–665.

Justus, N.J., and C.L. Leinweber. 1960. A heritable partial male-sterile character in cotton. J. Hered. 51:191–192.

Justus, N.J., and J.R. Meyer. 1963. A female sterile mutant in a double haploid of Upland cotton. J. Hered. 54:65–66.

Justus, N.J., J.R. Meyer, and J.B. Roux. 1963. A partially male-sterile character in Upland cotton. Crop Sci. 3:428–429. doi:10.2135/cropsci1963.0011183X000300050018x

Kandhro, M.M., S. Laghari, M.A. Sial, and G.S. Nizamani. 2002. Performance of early maturing strains of cotton (*Gossypium hirsutum* L.) developed through induced mutation and hybridization. Asian J. Plant Sci. 5:581–582.

Karaca, M., S. Saha, J.N. Jenkins, A. Zipf, R. Kohel, and D.M. Stelly. 2002. Simple sequence repeat (SSR) markers linked to the Ligon lintless (Li_1) mutant in cotton. J. Hered. 93:221–224. doi:10.1093/jhered/93.3.221

Katterman, F.R.H. 1973. 5-methyl cytosine content in the DNA of colchicine and spontaneous induced polyploids of *Gossypium*. Phytochemistry 12:1887–1889. doi:10.1016/S0031-9422(00)91506-1

Kearney, T.H., and G.J. Harrison. 1927. Inheritance of smooth seeds in cotton. J. Agric. Res. 35:193–217.

Killough, D.T., and W.R. Horlacher. 1933. The inheritance of virescent yellow and red plant characters in cotton. Genetics 18:329–334.

Kim, H.J., H.S. Moon, C.D. Delhom, L. Zeng, and D.D. Fang. 2013. Molecular markers associated with the immature fiber (*im*) gene affecting the degree of fiber cell wall thickening in cotton (*Gossypium hirsutum* L.). Theor. Appl. Genet. 126:23–31. doi:10.1007/s00122-012-1956-x

Knight, R.L. 1944. The genetics of blackarm resistance. IV. *G. punctatum* (Sch. and Thon.) crosses. J. Genet. 46:1–27. doi:10.1007/BF02986691

Knight, R.L. 1947. The genetics of blackarm resistance. V. Dwarf-bunched and its relationship to B_1. J. Genet. 48:43–50. doi:10.1007/BF02986096

Knight, R.L. 1948. The genetics of blackarm resistance. VII. *Gossypium arboreum* L. J. Genet. 49:109–116. doi:10.1007/BF02986827

Knight, R.L. 1950. The genetics of blackarm resistance. VIII. *Gossypium barbadense*. J. Genet. 50:67–76. doi:10.1007/BF02986795

Knight, R.L. 1952a. The genetics of withering or deciduous bracteoles in cotton. J. Genet. 50:392–395. doi:10.1007/BF02986836

Knight, R.L. 1952b. The genetics of jassid resistance in cotton. I. The genes H_1 and H_2. J. Genet. 51:47–66. doi:10.1007/BF02986704

Knight, R.L. 1953a. The genetics of blackarm resistance. IX. The gene B_{6m} from *G. arboreum*. J. Genet. 51:270–275. doi:10.1007/BF03023298

Knight, R.L. 1953b. The genetics of blackarm resistance. X. The gene B_7 from Stoneville 20. J. Genet. 51:515–519. doi:10.1007/BF02982941

Knight, R.L. 1954a. Abstract bibliography of cotton breeding and genetics, 1900–1950. Commonw. Agric. Bur., Farnham Royal, Cambridge, UK.

Knight, R.L. 1954b. The genetics of blackarm resistance. XI. *Gossypium anomalum*. J. Genet. 52:466–472. doi:10.1007/BF02981538

Knight, R.L. 1957. Blackarm disease of cotton and its control. In: Plant Protection Conference: Proceedings of the Second International Conference at Fernhurst Research Station, Butterworths, London. p. 53–59.

Knight, R.L. 1963. The genetics of blackarm resistance. XII. Transference of resistance from *Gossypium herbaceum* to *G. barbadense*. J. Genet. 58:328–346. doi:10.1007/BF02986304

Knight, R.L., and T.W. Clouston. 1939. The genetics of blackarm resistance. I. Factors B_1 and B_2. J. Genet. 38:133–159. doi:10.1007/BF02982168

Kohel, R.J. 1964. Inheritance of abnormal palisade mutant in American Upland cotton, *Gossypium hirsutum* L. Crop Sci. 4:112–113. doi:10.2135/cropsci1964.0011183X000400010039x

Kohel, R.J. 1965. Inheritance of accessory involucre mutant in American Upland cotton, *Gossypium hirsutum* L. Crop Sci. 5:119–120. doi:10.2135/cropsci1965.0011183X000500020004x

Kohel, R.J. 1972. Linkage tests in Upland cotton, *Gossypium hirsutum* L. II. Crop Sci. 12:66–69. doi:10.2135/cropsci1972.0011183X001200010022x

Kohel, R.J. 1973a. Genetic analysis of the open bud mutant in cotton. J. Hered. 64:237–238.

Kohel, R.J. 1973b. Genetic nomenclature in cotton. J. Hered. 64:291–295.

Kohel, R.J. 1973c. Analysis of irradiation induced virescent mutants and the identification of a new virescent mutant ($v_5v_5v_6v_6$) in *Gossypium hirsutum* L. Crop Sci. 13:86–88. doi:10.2135/cropsci1973.0011183X001300010027x

Kohel, R.J. 1973d. Genetic analysis of the Crumpled mutant in cotton, *Gossypium hirsutum* L. Crop Sci. 13:384–386. doi:10.2135/cropsci1973.0011183X001300030028x

Kohel, R.J. 1973e. Genetic analysis of the depauperate mutant in cotton, *Gossypium hirsutum* L. Crop Sci. 13:427–428. doi:10.2135/cropsci1973.0011183X001300040011x

Kohel, R.J. 1974. Genetic analysis of a new virescent mutant in cotton. Crop Sci. 14:525–527. doi:10.2135/cropsci1974.0011183X001400040009x

Kohel, R.J. 1982. Crinkle yellow, a new mutant in *Gossypium hirsutum* L. cotton. J. Hered. 73:382–383.

Kohel, R.J. 1983a. Genetic analysis of virescent mutants and the identification of virescents v_{12}, v_{13}, v_{14}, v_{15}, and $v_{16}v_{17}$ in Upland cotton. Crop Sci. 23:289–291. doi:10.2135/cropsci1983.0011183X002300020026x

Kohel, R.J. 1983b. Genetic analysis of the yellow-veins mutant in cotton. Crop Sci. 23:291–293. doi:10.2135/cropsci1983.0011183X002300020027x

Kohel, R.J. 1985. Genetic analysis of fiber color variants in cotton. Crop Sci. 25:793–797. doi:10.2135/cropsci1985.0011183X002500050017x

Kohel, R.J. 1988. Genetic analysis of a white mutant in cotton. Crop Sci. 28:1016–1018. doi:10.2135/cropsci1988.0011183X002800060033x

Kohel, R.J., and A.A. Bell. 1999. Genetic analysis of two terpenoid variants in cotton (*Gossypium hirsutum* L.). J. Hered. 90:249–251. doi:10.1093/jhered/90.1.249

Kohel, R.J., and L.S. Bird. 2002. Inheritance and linkage analysis of the yellow pulvinus mutant of cotton. J. Cotton Sci. 6:115–118.

Kohel, R.J., and J.A. Lee. 1984. Genetic analysis of Egyptian glandless cotton. Crop Sci. 24:1119–1121. doi:10.2135/cropsci1984.0011183X002400060027x

Kohel, R.J., and C.F. Lewis. 1962a. Inheritance of ragged mutant in American Upland cotton, *G. hirsutum* L. Crop Sci. 2:61–62. doi:10.2135/cropsci1962.0011183X000200010020x

Kohel, R.J., and C.F. Lewis. 1962b. Inheritance of veins-fused mutant in American Upland cotton, *G. hirsutum* L. Crop Sci. 2:174–175. doi:10.2135/cropsci1962.0011183X000200020031x

Kohel, R.J., C.F. Lewis, and M.N. Christiansen. 1977. The identification of a new mutant and linkage group in cotton. J. Hered. 68:65–66.

Kohel, R.J., C.F. Lewis, and T.R. Richmond. 1967. Isogenic lines in American Upland cotton, *Gossypium hirsutum* L.: Preliminary evaluation of lint measurements. Crop Sci. 7:67–70. doi:10.2135/cropsci1967.0011183X000700010023x

Kohel, R.J., and S.C. McMichael. 1990. Immature fiber mutant of upland cotton. Crop Sci. 30:419–421. doi:10.2135/cropsci1990.0011183X003000020038x

Kohel, R.J., and A.E. Percival. 1989. Genetic analysis of the Rex mutant in cotton. J. Hered. 80:78–80.

Kohel, R.J., D.M. Stelly, and J. Yu. 2002. Tests of six cotton (*Gossypium hirsutum* L.) mutants for association with aneuploids. J. Hered. 93:130–132. doi:10.1093/jhered/93.2.130

Lagiere, R. 1959. La bacteriose du cotonnier, *Xanthomonas malvacearum* (E.F. Smith) Dowson, dans le monde et en Republique Centrafricaine. Inst. Rech. Cott. Rex. Exot., Paris.

Larik, A.S., H.M.I. Hafiz, and Y.A. Al-Saheal. 1983. Azide mutagenesis in cotton (*Gossypium hirsutum*). Sci. Environ. 5:33–42.

Lee, J.A. 1962. Genetical studies concerning the distribution of pigment glands in the cotyledons and leaves of Upland cotton. Genetics 47:131–142.

Lee, J.A. 1966. Genetics of the smooth stem character in *Gossypium hirsutum* L. Crop Sci. 6:497–498. doi:10.2135/cropsci1966.0011183X000600050035x

Lee, J.A. 1968. Genetical studies concerning the distribution of trichomes on the leaves of *Gossypium hirsutum* L. Genetics 60:567–575.

Lee, J.A. 1976. Transfer of a smooth leaf allele from *Gossypium barbadense* L. to *Gossypium hirsutum* L. Crop Sci. 16:601–602. doi:10.2135/cropsci1976.0011183X001600040047x

Lee, J.A. 1977. Registration of North Carolina 1 and 2 smoothleaf cotton germplasm. Crop Sci. 17:826. doi:10.2135/cropsci1977.0011183X001700050048x

Lee, J.A. 1978. Allele determining rugate fruit surface in cotton. Crop Sci. 18:251–254. doi:10.2135/cropsci1978.0011183X001800020014x

Lee, J.A. 1981. Genetics of D_3 complementary lethality in *Gossypium hirsutum* and *G. barbadense*. J. Hered. 72:299–300.

Lee, J.A. 1984. Effects of two pilosity alleles on agronomic and fiber traits in Upland cotton. Crop Sci. 24:127–129. doi:10.2135/cropsci1984.0011183X002400010029x

Lee, J.A. 1985a. Inheritance of two new mutations in cotton. J. Hered. 76:479–480.

Lee, J.A. 1985b. Revision of the genetics of the hairiness-smoothness system of *Gossypium*. J. Hered. 76:123–126.

Lee, J.A. 1985c. Effects of the density of pubescence on some traits of extra-long-stapled cotton. Crop Sci. 25:517–520. doi:10.2135/cropsci1985.0011183X002500030020x

Lewis, C.F. 1954. The inheritance of cup leaf in cotton. J. Hered. 45:127–128.

Lewis, C.F. 1958. Genetic studies of a leaf mosaic mutant causing somatic instability in cotton. J. Hered. 49:267–271.

Lewis, C.F. 1960. The inheritance of mottled leaf in cotton. J. Hered. 51:209–212.

Lewis, C.F., and T.R. Richmond. 1960. The genetics of flowering response in cotton. Genetics 45:79–85.

Li, F., X. Hu, X. Qiu, and L. Wang. 2010. Fine mapping of the fg gene using the interspecific cross of *Gossypium hirsutum x Gossypium barbadense*. Euphytica 176:303–308. doi:10.1007/s10681-010-0197-9

Liu, L., W. Guo, X. Zhu, and T. Zhang. 2003. Inheritance and fine mapping of fertility restoration for cytoplasmic male sterility in *Gossypium hirsutum* L. Theor. Appl. Genet. 106:461–469.

Luckett, D.J. 1989. Colchicine mutagenesis is associated with substantial heritable variation in cotton. Euphytica 42:177–182. doi:10.1007/BF00042630

Lukefahr, M.J., C.B. Cowan, and J.E. Houghtaling. 1970. Field evaluations of improved cotton strains resistant to the cotton fleahopper. J. Econ. Entomol. 63:1101–1103. doi:10.1093/jee/63.4.1101

Lukefahr, M.J., and J.E. Houghtaling. 1969. Resistance of cotton strains with high gossypol content to *Heliothis* spp. J. Econ. Entomol. 62:588–591. doi:10.1093/jee/62.3.588

Lukefahr, M.J., J.E. Houghtaling, and H.M. Graham. 1971. Suppression of *Heliothis* populations with glabrous cotton strains. J. Econ. Entomol. 64:486–488. doi:10.1093/jee/64.2.486

Lukefahr, M.J., T.N. Shaver, and W.L. Parrot. 1969. Sources and nature of resistance in *Gossypium hirsutum* to bollworm and tobacco budworms. In: Proceedings of the Beltwide Cotton Production Research Conference, Natl. Cotton Counc. Am., New Orleans, LA. 7–8 Jan. 1969. Natl. Cotton Counc. Am., Memphis, TN. p. 81–82.

MacLeod, M.R. 1994. Analysis of an allelic series of mutants at the *r* locus of pea. Ph.D. thesis, Univ. of East Anglia, Norfolk, UK.

Mba, C. 2013. Induced mutations unleash the potentials of plant genetic resources for food and agriculture. Agronomy 3(1):200–231. doi:10.3390/agronomy3010200

McCallum, C.M., L. Comai, E.A. Greene, and S. Henikoff. 2000. Targeted screening for induced mutations. Nat. Biotechnol. 18(4):455–457. doi:10.1038/74542

McGowen, R.E., F.M. Bourland, and N.P. Tugwell. 1995. Evaluation of resistance to tarnished plant bug in cotton. In: D.A. Richter and J. Armour, editors, Proceedings of the Beltwide Cotton Conference, Natl. Cotton Counc. Am., San Antonio, TX. 4–7 Jan. 1995. Natl. Cotton Counc. Am., Memphis, TN. p. 536.

McLendon, C.A. 1912. Mendelian inheritance in cotton hybrids. Ga. Exp. Stn. Bull. 89:141–228.

McMichael, S.C. 1942. Occurrence of the dwarf-red character in Upland cotton. J. Agric. Res. 64:477–481.

McMichael, S.C. 1954. Glandless boll in Upland cotton and its use in the study of natural crossing. Agron. J. 46:527–528. doi:10.2134/agronj1954.00021962004600110016x

McMichael, S.C. 1960. Combined effects of the glandless genes gl_2 and gl_3 on pigment glands in the cotton plant. Agron. J. 52:385–386. doi:10.2134/agronj1960.00021962005200070005x

McMichael, S.C. 1965. Inheritance of aberrant stigmas in the flowers of Upland cotton. J. Hered. 56:21–22.

McMichael, S.C. 1970. Yuma glandless, an allelomorph of glandless-one in cotton, *Gossypium hirsutum* L. Crop Sci. 10:202–203. doi:10.2135/cropsci1970.0011183X001000020027x

McPherson, G.R. 1993. Inheritance of root-knot nematode resistance in cotton as determined by combining ability, generation mean, and mendelian analysis. Ph.D. diss., Mississippi State Univ., Mississippi State.

Mehetre, S.S., and M.V. Thombre. 1983. Fibre properties of x-ray induced glandless mutants in American cotton. J. Maharashtra Agric. Univ. 8:189–190.

Meredith, W.R. 1993. Registration of 'MD51ne' cotton. Crop Sci. 33:1415. doi:10.2135/cropsci1993.0011183X003300060074x

Meyer, J.R. 1957. Origin and inheritance of D_2 smoothness in Upland cotton. J. Hered. 48:249–250.

Meyer, J.R., and V.G. Meyer. 1961. Origin and inheritance of nectariless cotton. Crop Sci. 1:167–169. doi:10.2135/cropsci1961.0011183X000100030004x

Mound, L.A. 1965. Effect of leaf hair on cotton whitefly populations in the Sudan Gezira. Empire Cotton Grow. Rev. 42:33–40.

Muller, H.J. 1927. Artificial transmutation of the gene. Science 66:84–87. doi:10.1126/science.66.1699.84

Murray, J.C. 1965. A new locus for glanded stem in tetraploid cotton. J. Hered. 56:42–44.

Murray, J.C., and L.A. Brinkerhoff. 1966. Inheritance of pale-green color mutant in cotton. Crop Sci. 6:375–376. doi:10.2135/cropsci1966.0011183X000600040026x

Muthusamy, A., and N. Jayabalan. 2013. Variations in seed protein content of cotton (*Gossypium hirsutum* L.) mutant lines by in vivo and in vitro mutagenesis. J. Environ. Biol. 34:11–16.

Narbuth, E.V., and R.J. Kohel. 1990. Inheritance and linkage analysis of a new fiber mutant in cotton. J. Hered. 81:131–133.

Nazirov, N.N., and G. Satipov. 1979. Reaction of radiation induced mutants to different water regimes. Sov. Agric. Sci. 7:12–15.

Nazirov, N.N., N.T. Tashmatov, A. Bakhabov, and A.G. Nabiev. 1981. Respiration rate and rate of 32P incorporation into organophosphorus compounds in radiation-induced cotton mutants and their initial forms on saline soil. Sov. Agric. Sci. 5:27–29.

Negmatov, M., V.I. Kovalenko, V.K. Shumnyi, and K.A. Asrorov. 1975. Induction of cytoplasmic male sterility in cotton by the method of radiation mutagenesis. Sov. Genet. 11:1593–1595.

Novick, R.G., J.E. Jones, W.S. Anthony, W. Aguillard, and J.I. Dickson. 1990. Seedcotton cleanability and non-lint trash at the gin as affected by morphological traits. In: J. Brown and D. Richter, editors, Proceedings of the Beltwide Cotton Production Research Conference, Natl. Cotton Counc. Am., Nashville, TN. 2–7 Jan. 1990. Natl. Cotton Counc. Am., Memphis, TN. p. 80–81.

Patel, J.D., R.J. Wright, D.L. Auld, R. Chandnani, V.H. Goff, J. Ingles, G.J. Pierce, M.J. Torres, and A.H. Paterson. 2014. Alleles conferring improved fiber quality from EMS mutagenesis of elite cotton genotypes. Theor. Appl. Genet. 127:821–830. doi:10.1007/s00122-013-2259-6

Percival, A.E., and R.J. Kohel. 1974. Genetic analysis of virescent mutants in cotton, *Gossypium hirsutum* L. Crop Sci. 14:439–440. doi:10.2135/cropsci1974.0011183X001400030030x

Percival, A.E., and R.J. Kohel. 1976. New virescent cotton mutant linked with the marker gene Yellow petals. Crop Sci. 16:503–505. doi:10.2135/cropsci1976.0011183X001600040015x

Percival, A.E., and R.J. Kohel. 1978. Characterization and performance evaluation of frego bract isolines of cotton. Crop Sci. 18:1080–1082. doi:10.2135/cropsci1978.0011183X001800060045x

Percival, A.E., R.J. Kohel, and R.H. Dilday. 1976. A dominant round leaf mutant in cotton. Crop Sci. 16:794–796. doi:10.2135/cropsci1976.0011183X001600060014x

Percy, R.G. 1999. Inheritance of cytoplasmic-virescent *cyt-V* and dense-glanding *dg* mutants in American Pima cotton. Crop Sci. 39:372–374. doi:10.2135/cropsci1999.0011183X0039000200012x

Percy, R.G., and L.S. Bird. 1985. Rust resistance expression in cotyledons, petioles, and stems of *Gossypium hirsutum* L. J. Hered. 76:202–204.

Percy, R.G., P.C. Ellsworth, and H.S. Moser. 1997. Silverleaf whitefly resistance screening in Pima cotton genotypes. In: Silverleaf Whitefly: 1997 Supplement to the Five-year National Research and Action Plan. USDA–ARS, Washington, DC. p. 186.

Percy, R.G., and R.J. Kohel. 1999. Qualitative genetics. In: C.W. Smith and J.T. Cothren, editors, Cotton: Origin, history, technology, and production. John Wiley & Sons, New York. p. 319–360.

Percy, R.G., and E.L. Turcotte. 1991. Inheritance of male-sterile mutant ms_{13} in American Pima cotton. Crop Sci. 31:1520–1521. doi:10.2135/cropsci1991.0011183X003100060026x

Phillips, L.L. 1962. Segregation in new allopolyploids of *Gossypium*. IV. Segregation in New World × Asiatic and New World × wild American hexaploids. Am. J. Bot. 49:51–57.

Phillips, L.L. 1976. Interspecific incompatibility in *Gossypium*. III. The genetics of tumorigenesis in hybrids of *G. gossypioides*. Can. J. Genet. Cytol. 18:365–369. doi:10.1139/g76-044

Pieters, E.P., and L.S. Bird. 1977. Field studies of boll weevil resistant cotton lines possessing the okra leaf-frego bract characters. Crop Sci. 17:431–433. doi:10.2135/cropsci1977.0011183X001700030022x

Qian, N., X.-W. Zhang, W.-Z. Guo, and T.-Z. Zhang. 2009. Fine mapping of open-bud duplicate genes in homoelogous chromosomes of tetraploid cotton. Euphytica 165:325–331. doi:10.1007/s10681-008-9790-6

Quisenberry, J.E., and R.J. Kohel. 1970. Genetics of the virescent-4 mutant in cotton. J. Hered. 61:212–214.

Raut, R.N., H.K. Jain, and R.S. Panwar. 1971. Radiation-induced photo insensitive mutants in cotton. Curr. Sci. 40:383–384.

Rhyne, C.L. 1955. The inheritance of yellow-green, a possible mutation in cotton. Genetics 40:235–245.

Rhyne, C.L. 1957. Duplicated linkage groups in cotton. J. Hered. 48:59–62.

Rhyne, C.L. 1960. Linkage studies in Gossypium. II. Altered recombination values in a linkage group of allotetraploid *G. hirsutum* L. as a result of transferred diploid species genes. Genetics 45:673–681.

Rhyne, C.L. 1965. Duplicate linkage blocks in glandless-leaf cotton. J. Hered. 56:247–252.

Rhyne, C.L. 1971. Indehiscent anther in cotton. Cotton Grow. Rev. 48:194–199.

Rhyne, C.L., and J.C. Carter. 1991. Anthocyaninless cotton. In: D.J. Herber and D.A. Richter, editors, Proceedings of the Beltwide Cotton Conference, Natl. Cotton Counc. Am., San Antonio, TX. 6–10 Jan. 1991. Natl. Cotton Counc. Am., Memphis, TN. p. 540–541.

Rhyne, C.L., and J.C. Carter. 1992. True orange pollen in cotton. In: D.J. Herber and D.A. Richter, editors, Proceedings of the Beltwide Cotton Conference, Natl. Cotton Counc. Am., Nashville, TN. 6–10 Jan. 1992. Natl. Cotton Counc. Am., Memphis, TN. p. 591.

Richmond, T.R., and R.J. Kohel. 1961. Analysis of a completely male-sterile character in American Upland cotton. Crop Sci. 1:397–401. doi:10.2135/cropsci1961.0011183X000100060005x

Robinson, A.F. 2007. Reniform in U.S. cotton: When, where, why and some remedies. Annu. Rev. Phytopathol. 45:263–288. doi:10.1146/annurev.phyto.45.011107.143949

Romano, G.B., E.J. Sacks, S.R. Stetina, A.F. Robinson, D.D. Fang, O.A. Gutierrez, and J.A. Scheffler. 2009. Identification and genomic location of a reniform nematode (*Rotylenchulus reniformis*) resistance locus (*Ren*^{ari}) introgressed from *Gossypium aridum* into upland cotton (*G. hirsutum*). Theor. Appl. Genet. 120:139–150. doi:10.1007/s00122-009-1165-4

Rong, J., G.J. Pierce, V.N. Waghmare, C.J. Rogers, A. Desai, P.W. Chee, O.L. May, J.R. Gannaway, J.F. Wendel, T.A. Wilkins, and A.H. Paterson. 2005. Genetic mapping and comparative analysis of seven mutants related to seed fiber development in cotton. Theor. Appl. Genet. 111:1137–1146. doi:10.1007/s00122-005-0041-0

Royo, O.M., L Erazzu, I. Bonacic, J. Poisson, and A. Montenegro. 2003. Screening of cotton germplasm for "blue disease: under natural field infestation. In: A. Swanepoel, editor, Proceedings of the World Cotton Research Conference 3: Cotton production for the new millennium, Cape Town, South Africa. 9–13 Mar. 2003. Agricultural Research Council, Institute for Industrial Crops, Pretoria, South Africa. p. 305–316.

Salanki, M.S., and R. Parameswarappa. 1968. Colchine-induced mutant in cotton (*Gossypium hirsutum* L.). Curr. Sci. 12:356–357.

Sanamyan, M.F., and R.M. Rakhmatullina. 2003. Cytogenetic analysis of translocations in cotton. Plant Breed. 122:511–516. doi:10.1111/j.1439-0523.2003.00887.x

Saunders, J.H. 1961. The mechanism of hairiness in *Gossypium*. I. *Gossypium hirsutum*. Heredity 16:331–348. doi:10.1038/hdy.1961.38

Saunders, J.H. 1963. The mechanism of hairiness in *Gossypium*. II. *Gossypium barbadense*—The inheritance of stem hair. Empire Cotton Grow. Rev. 40:104–106.

Saunders, J.H. 1965a. The mechanism of hairiness in *Gossypium*. III. *Gossypium barbadense*–the inheritance of upper leaf laminar hair. Empire Cotton Grow. Rev. 42:15–25.

Saunders, J.H. 1965b. The mechanism of hairiness in *Gossypium*. IV. The inheritance of plant hair length. Empire Cotton Grow. Rev. 42:26–32.

Saunders, J.H. 1965c. Genetics of hairiness transferred from *Gossypium raimondii* to *G. hirsutum*. Euphytica 14:276–282. doi:10.1007/BF00149511

Savaskan, C. 2002. The effects of gamma radiation of the pollen size of *Gossypium hirsutum* L. Turk. J. Biol. 26:477–480.

Shamsuzzaman, K.M., M.A. Hamid, M.A.K. Azad, M. Hussain, and M.A. Majid. 2003. Varietal improvement of cotton (*Gossypium hirsutum*) through mutation breeding. In: Improvement of new and traditional industrial crops by induced mutations and related biotechnology. International Atomic Energy Agency, Vienna. p. 81–94.

Shattuck, V., and F.R. Katterman. 1982. Enhanced unscheduled DNA synthesis in the cotyledons of *Gossypium barbadense* L. by ethyl methanesulfonate (EMS). Biochem. Biophys. Res. Commun. 109:1017–1025. doi:10.1016/0006-291X(82)92041-1

Shaver, T.N., M.J. Lukefahr, and J.A. Garcia. 1970. Food utilization, ingestion, and growth of larvae of the bollworm and tobacco budworm on diets containing gossypol. J. Econ. Entomol. 63:1544–1546. doi:10.1093/jee/63.5.1544

Shaver, T.N., and W.L. Parrott. 1970. Relationship of larval age to toxicity of gossypol to bollworms, tobacco budworms, and pink bollworms. J. Econ. Entomol. 63:1802–1804. doi:10.1093/jee/63.6.1802

Sheetz, R.H., and J.B. Weaver, Jr. 1980. Inheritance of a fertility enhancer factor from Pima cotton when transferred into Upland cotton with *Gossypium harknessii* Brandegee cytoplasm. Crop Sci. 20:272–275. doi:10.2135/cropsci1980.0011183X002000020033x

Sheidai, M., H. Azarani, and Z. Hosseininejad. 2002. Cytogenetic study of gamma irradiated lines of cotton (*Gossypium hirsutum* L.). J. Sci. Islamic Repub. Iran 13:311–322.

Shi-Qi, A., Q. De-Qi, and C. Xiu-Yun. 1991. Induction of parthenogenesis and chromosome behavior in plants of parthenogenesis origin in cotton (*Gossypium hirsutum*). Genome 34:255–260. doi:10.1139/g91-040

Shoemaker, D.N. 1908. A study of leaf characters in cotton hybrids. Rep. Am. Breeders Assoc. 5:116–119.

Sikora, P., C. Aakash, L. Mikael, O. Johanna, and O. Olof. 2011. Mutagenesis as a tool in plant genetics, functional genomics and breeding. Int. J. Plant Genomics. 2011:314829.

Silow, R.A. 1946. Evidence on chromosome homology and gene homology in the amphidiploid New World cottons. J. Genet. 47:213–221. doi:10.1007/BF02986788

Silow, R.A., and C.P. Yu. 1942. Anthocyanin pattern in Asiatic cottons. J. Genet. 43:249–284.

Simpson, D.M. 1947. Fuzzy leaf in cotton and its association with short lint. J. Hered. 38:153–156.

Singh, P., and S.S. Narayanan. 1991. Seed oil improvement in upland cotton (*Gossypium hirsutum*) through pedigree selection. Indian J. Agric. Sci. 61:43–44.

Smart, J., and N.W. Simmonds. 1995. Evolution of crop plants. 2nd ed. Longman, New York.

Smith, A.L. 1940. Distribution and relation of the meadow nematode *Protylenchus pratensis* to Fusarium wilt of cotton in Georgia. Phytopathology 30:710.

Smith, E.G. 1942. Inheritance of smooth and pitted bolls in Pima cotton. J. Agric. Res. 64:101–103.

Song, M., Z. Yang, S. Fan, H. Zhu, C. Pang, M. Tian, and S. Yu. 2012. Cytological and genetic analysis of a virescent mutant in upland cotton (*Gossypium hirsutum* L.). Euphytica 187:235–245. doi:10.1007/s10681-012-0666-4

Stephens, S.G. 1945. A genetic survey of leaf shape in New World cottons—A problem in critical identification of alleles. J. Genet. 46:313–330. doi:10.1007/BF02989273

Stephens, S.G. 1950. The genetics of "corky." II. Further studies on its genetic basis in relation to the general problems of interspecific isolating mechanisms. J. Genet. 50:9–20. doi:10.1007/BF02986790

Stephens, S.G. 1954a. Interspecific homologies between gene loci in *Gossypium*. I. Pollen color. Genetics 39:701–711.

Stephens, S.G. 1954b. Interspecific homologies between gene loci in *Gossypium*. II. Corolla color. Genetics 39:712–723.

Stephens, S.G. 1955. Linkage in Upland cotton. Genetics 40:903–917.

Stephens, S.G. 1974a. Geographic and taxonomic distribution of anthocyanin genes in New World cottons. J. Genet. 61:128–141. doi:10.1007/BF02984077

Stephens, S.G. 1974b. The use of two polymorphic systems, nectary fringe hairs and corky alleles, as indicators of phylogenetic relationships in New World cottons. Biotropica 6:194–201. doi:10.2307/2989652

Stephens, S.G., and H.S. Lee. 1961. Further studies on the feeding and ovipositioning preferences of the boll weevil (*Anthonomus grandis*). J. Econ. Entomol. 54:1085–1090. doi:10.1093/jee/54.6.1085

Stewart, J.M. 1992. A new cytoplasmic male sterile and restorer. In: D.J. Herber and D.A. Richter, editors, Proceedings of the Beltwide Cotton Conference, Natl. Cotton Counc. Am., Nashville, TN. 6–10 Jan. 1992. Natl. Cotton Counc. Am., Memphis, TN. p. 610.

Tadele, Z., C. Mba, and B.J. Till. 2009. Tilling for mutations in model plants and crops. In: S.M. Jain and D.S. Brar, editors, Molecular techniques in crop improvement. 2nd ed. Springer, Dordrecht, the Netherlands. p. 307–322.

Thadani, K.I. 1923. Linkage relations in the cotton plant. Agric. J. Ind. 18:572–579.

Till, B.J., J. Cooper, and T.H. Tai. 2007. Discovery of chemically induced mutations in rice by TILLING. BMC Plant Biol. 7:19. doi:10.1186/1471-2229-7-19

Till, B.J., S.H. Reynolds, C. Weil, N. Springer, C. Burtner, K. Young, E. Bowers, C.A. Codomo, L.C. Enns, A.R. Odden, E.A. Greene, L. Comai, and S. Henikoff. 2004. Discovery of induced point mutations in maize genes by TILLING. BMC Plant Biol. 4:12. doi:10.1186/1471-2229-4-12

Tiranti, I. 1967. "Algodon Crespo", a new mutation in *Gossypium hirsutum* L. Bol. Genet. Inst. Fitotec. Castelar. 3:29–33.

Turcotte, E.L. 1986. Round leaf-3 mutant in American Pima cotton. J. Hered. 77:364–366.

Turcotte, E.L. 1987. Inheritance of a second wrinkled leaf mutant in American Pima cotton. Crop Sci. 27:702–704. doi:10.2135/cropsci1987.0011183X002700040019x

Turcotte, E.L., and C.V. Feaster. 1963. Inheritance of a cream petal mutant in Pima cotton, *Gossypium barbadense* L. Crop Sci. 3:563–564. doi:10.2135/cropsci1963.0011183X000300060035x

Turcotte, E.L., and C.V. Feaster. 1964. Inheritance of a mutant with rudimentary stigma and style in Pima cotton, *Gossypium barbadense* L. Crop Sci. 4:377–378. doi:10.2135/cropsci1964.0011183X000400040013x

Turcotte, E.L., and C.V. Feaster. 1965. The inheritance of rugate leaf in Pima cotton. J. Hered. 56:234–236.

Turcotte, E.L., and C.V. Feaster. 1966. A second locus for pollen color in Pima cotton, *Gossypium barbadense* L. Crop Sci. 6:117–119. doi:10.2135/cropsci1966.0011183X000600020003x

Turcotte, E.L., and C.V. Feaster. 1973. The interaction of two genes for yellow foliage in cotton. J. Hered. 64:231–232.

Turcotte, E.L., and C.V. Feaster. 1975. Inheritance of semigamy in American Pima cotton, *Gossypium barbadense* L. Agron. Abstr. 1975:65.

Turcotte, E.L., and C.V. Feaster. 1978. Inheritance of three genes for plant color in American Pima cotton. Crop Sci. 18:149–150. doi:10.2135/cropsci1978.0011183X001800010040x

Turcotte, E.L., and C.V. Feaster. 1979. Linkage tests in American Pima cotton. Crop Sci. 19:119–120. doi:10.2135/cropsci1979.0011183X001900010030x

Turcotte, E.L., and C.V. Feaster. 1980. Inheritance of a leaf mutant in American Pima cotton. J. Hered. 71:134–135.

Turcotte, E.L., and C.V. Feaster. 1983. Inheritance of a golden-veins mutant in American Pima cotton. J. Hered. 74:213–214.

Turcotte, E.L., and C.V. Feaster. 1985. Inheritance of male-sterile mutant Ms_{12} in American Pima cotton. Crop Sci. 25:688–690. doi:10.2135/cropsci1985.0011183X002500040026x

Turcotte, E.L., and R.G. Percy. 1988. Inheritance of a second virescent mutant in American Pima cotton. Crop Sci. 28:1018–1019. doi:10.2135/cropsci1988.0011183X002800060034x

Turcotte, E.L., and R.G. Percy. 1990. Genetics of kidney seed in *Gossypium barbadense* L. Crop Sci. 30:384–386. doi:10.2135/cropsci1990.0011183X003000020029x

Waghmare, V.N., and K.D. Koranne. 2000. Induced mutation in Asiatic cotton. J. Indian Soc. Cotton Improv. 25:128–135.

Wang, L., H. Liu, X. Li, X. Xiao, X. Ai, C. Luo, L. Zhu, and X. Li. 2014. Genetic mapping of fiber color genes on two brown cotton cultivars in Xinjiang. SpringerPlus 3:480. doi:10.1186/2193-1801-3-480

Ware, J.O. 1929. Inheritance of lint percentage in cotton. J. Am. Soc. Agron. 21:876–894. doi:10.2134/agronj1929.00021962002100090002x

Ware, J.O. 1932. Inheritance of lint colors in Upland cotton. J. Am. Soc. Agron. 24:550–562.

Ware, J.O., L.I. Benedict, and W.H. Rolfe. 1947. A recessive naked seed character in Upland cotton. J. Hered. 38:313–319.

Weaver, J.B., Jr. 1968. Analysis of a double recessive completely male-sterile cotton. Crop Sci. 8:597–600. doi:10.2135/cropsci1968.0011183X000800050027x

Weaver, J.B., Jr., and T. Ashley. 1971. Analysis of dominant gene for male sterility in Upland cotton, *Gossypium hirsutum*. Crop Sci. 11:596–598. doi:10.2135/cropsci1971.0011183X001100040048x

Weaver, D.B., and J.B. Weaver, Jr. 1977. Inheritance of pollen fertility restoration in cytoplasmic male-sterile cotton. Crop Sci. 17:497–499. doi:10.2135/cropsci1977.0011183X001700040003x

Weaver, J.B., Jr., and D.B. Weaver. 1979. Cracked root mutant in cotton. Inheritance and linkage with fertility restoration. Crop Sci. 19:307–309. doi:10.2135/cropsci1979.0011183X001900030006x

Wendel, J.F. 1989. New world tetraploid cottons contain Old World cytoplasm. Proc. Natl. Acad. Sci. USA 86:4132–4136. doi:10.1073/pnas.86.11.4132

Wilson, F.D. 1987. Inheritance of pink filament in cotton. J. Hered. 78:223–224.

Wilson, F.D. 1989. Inheritance of undulate leaf mutant of Upland cotton. J. Hered. 80:482–483.

Wilson, F.D., and J.K. Brown. 1991. Inheritance of response to cotton leaf crumple virus infection in cotton. J. Hered. 82:508–509.

Winkler, S., A. Schawabedissen, and D. Backasch. 2005. Targeted-selected mutant screen by TILLING in *Drosophila*. Genome Res. 15(5):718–723. doi:10.1101/gr.3721805

Xanthopoulos, F.P., and U.E. Kechagia. 2001. Improvement of two locally adapted cultivars in earliness by induced mutations. Aust. J. Agric. Res. 52:523–527. doi:10.1071/AR00027

Xanthopoulos, F.P., and U.E. Kechagia. 2003. Improvement of two locally adapted cultivars by induced mutations In: Improvement of new and traditional industrial crops by induced mutations and related biotechnology. International Atomic Energy Agency, Vienna. p. 61–68.

Yakabova, M.M., A.B. Rubin, G.A. Khramova, and D.N. Matorin. 1975. Hill reaction and delayed florescence in mutants of *Gossypium hirsutum*. In: Y.S. Nasyrov et al., editors, Genetic aspects of photosynthesis. Junk, The Hague, the Netherlands. p. 263–269.

Yin, J.M., X.S. Chen, S. H. Xiao, N.Y. Xu, J.C. Die, J.G. Liu, and Q.J. Wu. 2009. Molecular mapping of a new red mutant gene (R_s) in Upland cotton. Plant Breed. 128:416–419.

Zhang, C., J.L. Sun, Y.H. Jia, J. Wang, Z.J. Xu, and X.M. Du. 2011. Morphological characters, inheritance and response to exogenous hormones of a cotton super-dwarf mutant of *Gossypium hirsutum*. Plant Breed. 130:67–72. doi:10.1111/j.1439-0523.2009.01759.x

Zhang, J.F., and J.M. Stewart. 2001. Inheritance and genetic relationships of the D8 and D2-2 restorer genes for cotton cytoplasmic male sterility. Crop Sci. 41:289–294. doi:10.2135/cropsci2001.412289x

Zhang, T.Z. 1992. Genetic analysis of open bud in tetraploid cotton. China Cottons 19:7.

Zhang, T.Z., and J.J. Pan. 1990. Allelic tests of 11 virescent mutants and genetic identification of virescent v_{22} in Upland cotton. Jiangsu Agric. Sci. 6:24–29.

Quantitative Genetics

B. Todd Campbell* and Gerald O. Myers

Abstract

The majority of economically important traits targeted for cotton improvement are quantitatively inherited. In this chapter, the current state of cotton quantitative genetics is described and separated into four components. These components include the following: (i) traditional quantitative inheritance analysis, (ii) genotype × environment interactions, (iii) genetic association of traits, and (iv) selection progress. An effort is made to discuss quantitative genetics in the context of a contemporary cotton breeding program. In an effort to offer a future perspective of quantitative genetics, contemporary and emerging technologies for quantitative trait analyses are discussed.

The majority of characteristics targeted for cotton improvement are complex and behave as quantitatively inherited traits. Such traits include yield, earliness, disease and pest resistance, stress tolerance, and fiber quality. Although these quantitatively inherited traits are the most economically important for cotton production, their complex nature makes them extremely difficult to manipulate and improve. Simply, the measurement or phenotype of a quantitative trait can be expressed by the following mathematical equation:

$$\text{Phenotype} = \text{Genetics} + \text{Environment} + (\text{Genetics} \times \text{Environment}) + \text{Error}$$

Quantitative genetics theory uses logical mathematical models to put order to the phenotype of complex traits. Partitioning the components of this simple mathematical equation can be extremely difficult as each component of the phenotype is difficult to measure and control.

In addition to being difficult because they are controlled by numerous genes, quantitative traits are also difficult because of the following: (i) they are a composite of numerous, interacting component traits, (ii) lack of suitable quantitative trait measurement tools, (iii) uncontrollable effects of the environment, (iv) environmental sensitivity caused by genotype × environment

---

Abbreviations: AMMI, additive main effects and multiplicative interactions; GGE, genotype main effect plus genotype ´ environments interaction; QTL, quantitative trait locus.

B. Todd Campbell, USDA-ARS Coastal Plains Soil, Water, and Plant Research Center, 2611 W. Lucas St., Florence, SC 29501. *Corresponding author (todd.campbell@ars.usda.gov).

Gerald O. Myers, Louisiana State University, School of Plant, Environmental and Soil Sciences, 104 M.B. Sturgis Hall, Baton Rouge, LA 70803 (GMyers@agcenter.lsu.edu).

doi:10.2134/agronmonogr57.2013.0024

Cotton. 2nd. ed. David D. Fang and Richard G. Percy, editors. Agronomy Monograph 57.

interactions, (v) gene-to-gene interactions, and (vi) negative correlations among quantitative traits.

As described by Meredith (1984), plant breeders must ask a series of questions to develop effective strategies to simultaneously improve multiple quantitatively inherited traits. First posed over 40 yr ago by Dudley and Moll (1969), these questions are still appropriate starting points for contemporary breeding programs. They include the following:

1. Is genetic variability for the desired characteristic sufficient to warrant a selection program?
2. Which parents or genetic populations are most promising as sources of improved breeding material?
3. How should environments be partitioned into regions or subregions for areas of cultivar adaptation?
4. What are the most efficient methods of testing germplasm (in terms of years, specific locations, management systems, replications, etc.)?
5. What are the genetic associations among various characteristics?
6. Which breeding and selection procedures will most rapidly and efficiently produce an acceptable level of breeding progress?
7. What type of cultivar (hybrid, synthetic, pure line, etc.) is the most appropriate goal?
8. What is the most efficient selection procedure for combining several characteristics (such as yield and fiber quality)?

In this chapter, an updated assessment of studies involving quantitatively inherited traits will be provided. Efforts will be made to discuss quantitative genetics in the context of a plant breeding program focused on addressing the eight questions listed above. Although there are numerous examples of quantitatively inherited traits in cotton improvement, the emphasis here will be on yield and fiber quality.

Traditional Quantitative Inheritance Analyses

Genetic variability is necessary for plant improvement. Without variation, there can be no progress unless it is induced (e.g., mutagenesis) or introduced (genetic engineering). Furthermore, if variation is present it must be useful variation, able to impart some advantage. In the context of Mendelian inheritance, this implies allelic diversity at one or a few loci for qualitative traits and at multiple loci for quantitative traits.

Quantitative traits are typically measured on a metric scale since the effect of any particular allele over multiple loci is small. Unlike qualitative traits, which present discrete classes and display non-normal distributions, quantitative trait distributions approximate a normal curve. Under the much simpler qualitative trait systems, the mechanisms of gene action are discernible: additive, dominance, incomplete dominance, overdominance, and epistasis. However, with quantitative traits, the determination of genetic effects is not as obvious and requires the use of secondary statistics such as variances. Secondary statistics are typically estimated within a defined mating scheme or population structure in the organism under study.

Quantitative genetics provides the information necessary to design and implement an effective approach for genetic improvement. For a given quantitative trait, there are typically three types of desirable breeding populations (Bernardo, 2002). These populations are based on the mean and genetic variance and include the following: (i) high mean and small variance, (ii) low mean and high variance, and (iii) high mean and high variance. A high population mean is desirable, because it provides a boost in breeding for the quantitative trait. A high population genetic variance is desirable, because it provides potential for genetic improvement. Effective genetic improvement programs must consider these simple population statistics when designing genetic improvement strategies.

An excellent historical review of research conducted in the first half of the 20th century into cotton genetics and the inheritance of yield, yield component, and fiber quality traits is provided by Breaux (1951). Meredith (1984) more recently reviewed quantitative genetics in cotton. Modern analytical techniques and research results are refining but essentially validating prior investigations in this area. This review seeks to highlight more contemporary research (as of 2014) and compare it to prior results.

Gene Action, Heritability, and Combining Ability

Within the context of the simple mathematical equation presented earlier describing the phenotype of a quantitative trait [phenotype = genetics + environment + (genetics × environment) + error], gene action estimates further partition the genetic variance. Genetic variance is often partitioned into additive variance and dominance deviations. Additive genetic variance is the sum of the average effects of alleles at multiple loci. Dominance deviations can be represented by complete dominance, incomplete dominance, codominance, overdominance, or epistasis. Overall, the extent and type of gene action responsible for the genetic variance observed for a given trait determines the breeding and selection approach used for trait improvement. Estimation of gene action requires four steps, which include the following: (i) one or more types of populations are developed, (ii) the populations are evaluated in a set of environments, (iii) variance components are estimated from mean squares in the analysis of variance, and (iv) variance components are interpreted on the basis of the covariances between relatives (Bernardo, 2002). Typically, a number of different mating designs are used to estimate gene action. For a thorough description of mating design details including computational procedures, readers should consult Bernardo (2002).

A number of studies have been conducted to estimate gene action in cotton (Campbell et al., 2013, 2014b; Cheatham et al., 2003; Jenkins et al., 2009, 2012; McCarty et al., 2004a, 2004b; Saha et al., 2013; Tang et al., 1996; Tang and Xiao, 2013; Zeng and Wu, 2012). Historically, these studies have focused on gene action as it relates to yield and fiber quality traits. Table 1 provides a summary of additive and dominance genetic variance components for a number of recent studies. On average, this summary suggests adequate additive genetic variance for lint yield, yield components, and fiber quality traits. In terms of agronomic traits, on average lint percent displays the greatest amount of additive genetic variation, whereas boll number unit area^{-1} has the lowest. In terms of fiber quality traits, length and strength display substantive additive genetic variance. Overall, each of the traits displays nearly equivalent additive and dominance genetic variation

Table 1. A summary of percent additive and dominance genetic variance for agronomic and fiber quality traits. Data derived from a divergent sample of genetic populations, various selection units, and test instruments.

	Additive		Dominance		Additive/Dominance ratio	
Trait	Mean	Range	Mean	Range	Mean	Range
	%		%			
Lint yield	17.8	6–45	23.7	12–44	0.8	0.2–1.7
Lint percent	36.1	2– 81	33.6	6–92	1.1	0.0–7.4
Boll weight	26.0	12–45	34.0	23–44	0.8	0.4–1.3
Boll number	14.8	0–35	18.5	3–33	0.8	0.0–3.3
Length	26.6	0–60	33.4	14–55	0.8	0.0–3.9
Strength	30.7	0–62	14.9	1–32	2.1	0.0–47.0
Micronaire	19.6	4–63	24.4	2–60	0.8	0.1–7.0

with the exception of strength, which on average displays two times more additive variance as opposed to dominance. It is important to mention that the wide range of variance components estimated in these studies highlights their population and/or evaluation environment dependence.

The estimation of gene action also allows for the calculation of heritability. Ultimately, the heritability of a specific quantitative trait determines the potential for its genetic improvement. There are two types of heritability, which include broad sense (H) and narrow sense (h^2). Broad sense heritability is the amount of genetic variance expressed as a proportion of the total phenotypic variance. Narrow sense heritability is the amount of additive genetic variance expressed as a proportion of the total phenotypic variance. It is important to note that heritability is a function of alleles (not genotypes) being passed along from a parent to offspring (Bernardo, 2002). Heritability estimates have been calculated in cotton, and Table 2 provides a summary of recent estimates (Ali et al., 2008; Basal and Turut, 2005; Batool and Khan, 2012; Cheatham et al., 2003; Hague et al., 2011; Herring et al., 2004; Jia et al., 2014; Jenkins et al., 2009; Lu and Myers, 2011; May and Jividen, 1999; McCarty et al., 2004a, 2004b; Nassar, 2013; Percy et al., 2006; Stiller et al., 2005; Tang et al., 1996; Ulloa, 2006; Zeng and Wu, 2012). Overall, the mean H estimated in these studies is quite high for each of the agronomic and fiber quality traits summarized. On average, lint percent displays the highest H and micronaire the lowest. Lint percent, length, and strength display the highest h^2. Lint yield displays the lowest h^2 among agronomic traits, and micronaire displays the lowest h^2 among fiber quality traits. Similar to estimates of additive and dominance genetic variance percentages, these studies report a wide range of heritability estimates depending on the population being sampled and the environment for which they are evaluated. Overall, this summary of heritability indicates adequate genetic variation for these core agronomic and fiber quality traits suggesting the opportunity for future genetic gain from selection on these traits.

Although most cotton genetic improvement has focused on maximizing additive genetic variation to develop pure-line cultivars, a number of studies have demonstrated significant nonadditive or heterotic effects. Overall, studies have estimated significant midparent heterosis primarily for yield (Davis, 1978; Meredith, 1997; Tang et al., 1993a, 1993b; Meredith and Brown, 1998; Campbell et al., 2008). These studies report lint yield midparent heterosis ranging from 11

Table 2. A summary of studies estimating heritability for agronomic and fiber quality traits. Data were derived from a divergent sample of genetic populations, various selection units, and test instruments.

	Broad-sense		Narrow-sense	
Trait	Mean	Range	Mean	Range
Lint yield	0.58	0.35–0.69	0.22	0.03–0.42
Lint percent	0.72	0.53–0.98	0.50	0.33–0.81
Boll weight	0.63	0.34–0.87	0.30	0.04–0.42
Boll number	0.50	0.21–0.74	0.27	0.10–0.49
Length	0.54	0.25–0.92	0.38	0.23–0.54
Strength	0.53	0.25–0.90	0.40	0.30–0.57
Micronaire	0.45	0.08–0.90	0.29	0.17–0.69

to 31%. However, the same studies indicate minimal heterotic effects for fiber quality traits. Although heterosis is being exploited as a mechanism to increase cotton yield in India (up to 45% of total production) and China (up to 20% of total production), the lack of an economic hybrid seed production system has limited the exploitation of heterosis in US cotton production (Tang et al., 1993a, 1993b; Chaudhry, 1997; Meredith, 1997; Campbell et al., 2008; Zumba et al., 2013). Campbell et al. (2008) found that crosses involving obsolete cultivars as opposed to modern cultivars produced higher heterotic effects for yield. This would suggest that obsolete cultivars have a greater ability to transmit nonadditive genetic effects. As expected, it would also suggest that breeding procedures used over the last 100 yr have resulted in the accumulation of additive genetic effects.

Another application for quantitative trait analyses involves the estimation of combining ability. Estimates of combining ability provide information on the performance of an individual, a family, an inbred, or a population in terms of its average performance when crossed to other individuals (Bernardo, 2002). General combining ability is based on the prediction of additive genetic effects and specific combining is based on the prediction of heterozygous dominance genetic effects (Jenkins et al., 2009). General combining ability estimates provide information most relevant to pure-line breeding programs, whereas specific combining ability estimates provide information most relevant to hybrid breeding programs. A number of combining ability studies have been conducted in cotton based on a wide range of germplasm and unique genetic stocks (Campbell et al., 2013, 2014b; Cheatham et al., 2003; Hinze et al., 2011; Jenkins et al., 2006, 2007, 2009, 2012; McCarty et al., 2004a, 2004b, 2007; Meredith and Brown, 1998; Zeng et al., 2011; Zeng and Wu, 2012). Overall, estimates of combining ability provide useful information that aid breeders in parental line selection.

Genotype × Environment Interactions

Genotype × environment interaction (G × E) occurs when genotypes differ in their relative performance across environments (Bernardo, 2002). Next to yield, G × E interactions have been identified as an important trait when selecting parents in a breeding program (Bowman, 2000). The calculation, interpretation, and exploitation of this interaction have been the subject of considerable research in cotton and breeding in general as it affects resource allocation in an improvement program (Campbell and Jones, 2005; Paterson et al., 2003).

Highlighting the importance of G × E in cotton, Liu et al. (2013) provided an analysis of the effect of the interaction between genetics and management on cotton lint yield gains. The observed gain in yield from 1980 to 2009 was estimated to be due proportionately to genetics (48%), management (28%), and genotype × management (24%). The more recent 18.3 kg ha^{-1} yield gain was attributed to a reduction of G × E interactions over time.

Numerous other studies have estimated the effect of G × E interactions on yield and fiber quality traits in cotton. Typically, these studies compare relative proportions of components of the total phenotypic variance (E, G, and G × E). Table 3 provides a summary of a number of recent studies estimating G × E for yield and fiber quality traits (Blanche et al., 2006; Campbell and Jones, 2005; Campbell et al., 2011; DeLacy et al., 1990; Kerby et al., 2000; Meredith, 2003). Overall, these studies highlight the large effect of E on lint yield and micronaire. The largest G effects are shown for yield component traits lint percent, boll weight, and seed weight. Moderate G effects are typical for fiber length and strength. The ratio of G × E to G provides an understanding of the relative importance of G × E in these studies. Clearly, the mean 1.8 and range of 0.8 to 9.0 ratio of G × E to G indicates the significant impact of G × E on yield. Overall, G × E for fiber quality traits is minimal and micronaire appears to be most impacted by G × E as the mean G × E to G ratio was 0.9 and ranged from 0.5 to 2.9.

It is generally accepted that there are two types of G × E that describe the performance of genotypes across environments (Bernardo, 2002). The first pattern occurs when the magnitude of the difference between genotypes differs across environments (noncrossover interaction). The second pattern occurs when the best genotype differs across environments (crossover interaction). Crossover interactions result in a change in genotype rank across environments, whereas a noncrossover interaction does not result in a rank change. Most reports in cotton identify G × E interactions of the noncrossover type (Campbell and Jones, 2005). This is especially true for studies documenting significant G × E for fiber quality traits. However, there have been reports of crossover G × E interactions for yield and/or yield component traits (Campbell and Jones, 2005).

Overall, given its relative importance economically, the most common subject of G × E analyses in cotton has been yield. To overcome the large influence of E and G × E in concert with the relatively low heritability for yield, Meredith (1984) proposed increasing yield by selecting for yield component traits that have a higher heritability and are under simpler genetic control—at least compared with yield. On the basis of correlation and path analysis, Bechere and Zeng (2014) found that yield components could collectively account for 71% of lint yield. Similarly, Zeng et al. (2014a) used commonality analysis (which may better determine cause–effect relationships than path analysis) and found that a combination of four traits—lint percentage, lint weight $seed^{-1}$, lint weight unit seed surface $area^{-1}$, and lint number unit seed surface $area^{-1}$—collectively accounted for 67 and 44% of total variation in lint yield in two consecutive years.

In addition to path and commonality analysis, other sophisticated G × E analyses have been developed using various mixed models and biplot methods that allow for the estimation of genotype stability across environments and a visual representation of G and G × E. Most notable among these include the additive main effects and multiplicative interactions (AMMI) model (Zobel et al., 1988) and the genotype main effect plus genotype × environments interaction (GGE)

Table 3. A summary of studies partitioning the total variance into components including environment (E), genotype (G), and genotype × environment interaction (GE). Data were derived from a divergent sample of genetic populations, various selection units, and test instruments.

	% of total variance							
	E		G		GE		GE/G ratio	
Trait	Mean	Range	Mean	Range	Mean	Range	Mean	Range
Lint yield	86	72–94	5	1–11	9	5–19	1.8	0.8–9.0
Lint percent	52	23–82	29	6–57	19	12–34	0.7	0.3–2.0
Boll weight	45	36–55	25	10–38	30	19–46	1.2	0.7–4.6
Seed weight	39	26–47	38	29–48	24	18–28	0.6	0.5–1.0
Length	56	32–85	27	6–49	17	8–35	0.6	0.3–1.5
Strength	40	14–66	44	15–65	16	8–22	0.4	0.1–1.3
Micronaire	66	56–82	18	6–33	16	9–20	0.9	0.5–2.9

model (Yan et al., 2000; Yan and Kang, 2003). Using 3 yr of official cotton variety trial data balanced for entries across a total of 18 environments, Blanche et al. (2007) found that GGE biplots provided stability statistics comparable to traditional stability measures.

In an effort to manage G × E and more effectively utilize resources in breeding programs, both AMMI and GGE biplot analyses have been used to characterize testing environments, identify mega-environments, and identify specific areas of adaptation (Gauch and Zobel, 1997; Yan et al., 2000). Geng et al. (1990) used 6 yr of yield data from the national cotton variety testing program and identified four mega-environments in the United States on the basis of discrimination power, consistency of discrimination power, relative potential productivity, and its repeatability. Myers (2002) combined G and G × E criterion using biplots on 3 yr of yield data from the national cotton variety trial and identified three mega-environments and their most discriminating environments. On a more local scale, Campbell and Jones (2005) used AMMI analyses to detect two mega-environments in South Carolina appropriate for discriminating yield potential. In the same analysis, due to noncrossover type G × E for fiber quality, efficient discrimination of fiber quality potential among genotypes could be achieved in a single environment within the target production area. Blanche and Myers (2006) used multiple traits in a weighted selection index to identify the most discriminating locations for the simultaneous selection of lint yield and fiber quality in Louisiana. Zeng et al. (2014b) used GGE biplot analyses and several climatic variables to discriminate testing sites across the US cotton production area. They determined Plains and Western US testing environments were similar to one another but distinct from testing environments east of the Plains region. Low temperatures during early and late season contributed to the separation of the Plains and Western US testing environments. Overall, discriminating testing sites overcome G × E by ranking genotypes similarly, improving testing efficiency, and allowing for greater breeding progress by maximizing genetic variance (Blanche and Myers, 2006; Blanche et al., 2008; Brown et al., 1983; Geng et al., 1990; Lubbers, 2003).

Genetic Association of Traits

Just as partitioning the phenotype and genotype of a quantitative trait into components is essential to better understanding the trait, when designing effective breeding strategies it is also important for plant breeders to consider the associations among two or more traits. Associations among traits are generally calculated using analyses of covariance that estimate either phenotypic or genotypic correlations. Phenotypic correlations consider the combined effects of genotype, environment, G × E, and error and provide a basic glimpse at trait associations. On the contrary, genotypic correlations consider only the genetic component of the phenotype, thereby providing a more focused estimation of trait association. To demonstrate the importance of trait associations, consider an instance where there are two associated traits (either positive or negative association). In the case there is a positive association between two traits, basing selection in one direction on one trait will result in selection for the associated trait in the same direction. In the case there is a negative association between two traits, basing selection in one direction on one trait will result in selection for the associated trait in the opposite direction. Unfortunately, all too often cotton breeders are faced with negatively associated traits.

Historically, a number of studies have been conducted to estimate the associations present among agronomic and fiber quality traits. Such associations are particularly important, as a major breeding objective is to simultaneously improve both lint yield and fiber quality. Table 4 provides a summary of genotypic correlation estimates between lint yield, yield component traits, and fiber quality traits estimated from several studies. In terms of yield component traits, lint yield is positively correlated with lint percent and bolls m^{-2} while negatively correlated with boll weight and seed index. In terms of fiber quality traits, generally lint yield is negatively correlated with fiber length and fiber strength while being positively correlated with micronaire and fineness.

Historically, efforts to increase fiber quality in Upland cotton breeding programs relied on interspecific hybridization. It is widely believed that the two primary sources of increased fiber quality in Upland cotton have been hybridization with two interspecific sources: (i) the triple hybrid (*G. thurberi* Tod. × *G. arboreum* L.) × *G. hirsutum* L. and (ii) *G. barbadense* L. (Bowman and Gutierrez, 2003; Campbell et al., 2009). In addition to being an excellent source of useful genetic variation, interspecific hybridization is known to result in the transfer of blocks of chromatin containing deleterious genes due to linkage drag (Young and Tanksley, 1989). Although efforts were employed in the 1960s to combat the negative relationship among agronomic and fiber quality traits, its persistence today highlights a continued importance in cotton breeding programs.

Studies were initiated in the 1960s to develop strategies to overcome the negative relationship. Generally, these studies suggest the negative relationship is due to linkage rather than pleiotropy (Meredith, 1984). On the basis of the theoretical principles outlined by Hanson (1959), random intercrossing (e.g., recurrent selection) strategies were deployed by several cotton breeding programs in an attempt to break up unfavorable linkage blocks. Miller and Rawlings (1967), Meredith and Bridge (1971), and Culp et al. (1979) employed various recurrent selection strategies and documented a reduction in the negative relationship between fiber quality and yield from the original base population to populations developed following several cycles of recurrent selection. More recently, Clement et al. (2012) and Campbell et al. (2012) conducted studies aimed to retrospectively evaluate progress from efforts to overcome the negative relationship between fiber quality and yield. Evaluating elite breeding and cultivar field evaluation trial datasets

Table 4. A summary of genotypic correlations and their standard errors between lint yield, other agronomic traits, and fiber quality traits.

Lint yield vs. other traits	Miller and Rawlings (1967)	Meredith and Bridge (1971)	Fotiadis and Miller (1973)	Scholl and Miller (1976)	Zeng and Meredith (2009)	Campbell et al. (2012)
Lint percent	0.90 (0.12)	0.70 (0.14)	0.79 (0.09)	0.84 (0.08)	0.81	0.82 (0.06)
Bolls m^{-2}	–	–	–	–	0.93	0.84 (0.05)
Boll weight	0.14 (0.23)	–0.43 (0.22)	–0.14 (0.15)	–0.04 (0.14)	–0.13	–0.30 (0.16)
Seed index	–	–0.45 (0.15)	–0.62 (0.12)	–0.28 (0.15)	–0.64	–0.80 (0.13)
Length	0.02 (0.20)	–0.47 (0.18)	–0.18 (0.14)	–0.36 (0.13)	–0.26	–0.60 (0.10)
Strength	–0.69 (0.14)	–0.54 (0.17)	–0.46 (0.15)	–0.36 (0.14)	–0.48	–0.48 (0.11)
Elongation	0.71 (0.19)	0.03 (0.20)	0.02 (0.18)	0.38 (0.16)	–0.17	–
Micronaire	0.42 (0.17)	0.42 (0.19)	0.62 (0.12)	0.54 (0.12)	–	0.48 (0.12)
Fineness	–	–	–	–	0.57	0.44 (0.13)
Maturity ratio	–	–	–	–	0.38	–

from Australia and the United States, Clement et al. (2012) showed that the negative relationship still persists but can be overcome by (i) using recurrent selection to break up negative linkage blocks, (ii) using adequate early generation breeding population sizes, and (iii) setting breeding population target regression coefficients for the association between fiber quality and yield. Evaluating the high fiber quality Pee Dee cotton germplasm enhancement program, Campbell et al. (2012) found that although the negative relationship persists following nearly 80 yr of breeding, the extent of the relationship declined following eight cycles of recurrent selection. Independently, both Clement et al. (2012) and Campbell et al. (2012) support the hypothesis that (i) linkage is the cause of the negative relationship and (ii) rare recombination events combining high fiber quality and yield can be identified. Hence it is likely that long-term breeding efforts focused on combining fiber quality and yield potential will likely continue resulting in incremental decreases in the negative relationship.

Selection Progress

The self-pollinated nature of cotton reproductive biology has resulted in the vast majority of cotton breeding programs practicing selection based on modifications of a pedigree breeding system aimed to develop pure line cultivars. Significant heterotic effects have been measured for lint yield and used to develop hybrid cultivars (Campbell et al., 2008). However, utilization of heterosis has been slow because of the lack of a cost-effective and efficient hybrid seed production and delivery system (Campbell et al., 2008). Hence, a large majority of the selection progress in improving quantitatively inherited traits such as yield and fiber quality properties has focused on exploiting additive genetic variation.

Numerous studies have estimated genetic gain to evaluate genetic progress in improving lint yield potential. Table 5 shows a summary of yield genetic gain estimates obtained from a number of studies. Collectively, these studies support three primary conclusions. First, steady genetic gain for yield was achieved up to the late 1970s and early 1980s. Second, genetic gain for yield decreased in the 1980s up to the mid-1990s. Third, genetic gain for yield increased in the late 1990s to present. Since the large majority of commercial cotton cultivars represent pure lines derived from pedigree breeding selection schemes, it can be assumed that

Table 5. A summary of genetic gain estimates for yield productivity.

Years evaluated	Genetic gain estimate	Reference
	kg ha^{-1} yr^{-1}	
1910–1978	9.5	Bridge and Meredith (1983)
1922–1962	10.2	Bridge et al. (1971)
1945–1978	10.5	Culp and Green (1992)
1961–1980	−0.9	Meredith (1984)
1980–1994	7.0	Liu et al. (2013)
1981–1995	5.0	Campbell et al. (2014a)
1995–2009	18.3	Liu et al. (2013)
1996–2011	21.6	Campbell et al. (2014a)

the majority of selection progress has capitalized on improving additive genetic variance for lint yield.

Most studies reporting estimates of genetic progress for improved fiber quality potential have demonstrated negligible trends over time (Bridge and Meredith, 1983; Bridge et al., 1971; Culp and Green, 1992; Schwartz and Smith, 2008). Although heritability estimates are generally higher for fiber properties (especially fiber strength) than for lint yield, gains through selection for fiber quality properties have not kept pace with the realized gains for lint yield. As suggested by Schwartz and Smith (2008), this may be due to the relatively recent development of fiber property measurement techniques and/or the lack of fiber quality being a primary breeding objective. It is also plausible that fiber quality genetic gains not matching those for yield potential is due to the continued negative relationship between the two quantitative traits. In support of this notion, Campbell et al. (2011) estimated in the Pee Dee cotton germplasm enhancement program 3% per breeding cycle gains for yield potential associated with less than 1% decreases in fiber quality properties.

One recent advance in selecting for quantitatively inherited fiber quality properties has been the development of a fiber quality selection index or Q-score (Bourland et al., 2010). The Q-score combines a weighted index of several different fiber quality properties. Although the Q-score selection index does not include lint yield or other measures of agronomic productivity, it has already been used to efficiently select for high fiber quality (Bourland and Jones, 2012). In terms of lint yield, selection indices based on components of lint yield have been developed (Manning, 1956). It is unclear how frequently selection indices are used to improve lint yield potential. To aid breeding efforts aimed at simultaneously improving yield potential and fiber quality, Clement et al. (2012) proposed a quasi selection index. This proposal advocates using a selection criteria that sets breeding population target regression coefficients for the association between fiber quality and yield in breeding programs. This and other selection methods that focus on simultaneously improving yield and fiber quality potential may enhance efforts to improve both quantitative traits.

Contemporary and Emerging Technologies for Quantitative Trait Analysis

During the last 25 yr of tremendous advances in molecular biology and the concurrent development of genome-based enabling technologies and tools, our methods of quantitative trait analysis have evolved and advanced. The basic principles of phenotypes and genotypes discussed in this chapter still provide the foundation for quantitative trait analysis. However, the development of genome marker systems beginning with isozymes then moving to restriction length fragment polymorphisms, and finally to various polymerase chain reaction based systems, have revolutionized our ability to measure genotypes at the chromosome level. In turn, this revolution has invoked the development of methods to link DNA-based genotypes with quantitative trait phenotypes.

Linking DNA-based genotypes to quantitative trait phenotypes has allowed for quantitative trait analysis to go beyond simply estimating the proportion of variance attributed to environment, genetics, and G × E. It has allowed for the development of quantitative trait locus (QTL) mapping techniques to associate specific chromosomal regions (or QTL) to quantitative trait genetic variation (Paterson et al., 1988). The most basic of QTL mapping requires a discrete, biparental genetic population such as an F_2, backcross, or recombinant inbred population. Generally speaking, there are two general requirements to accomplish basic QTL mapping using an appropriate genetic population, which include the following: (i) a genetic map with sufficient polympomorphic marker loci that cover the entire nuclear genome and (ii) an accurate estimation of the quantitative trait phenotype. The presence of a chromosomal region or QTL associated with the quantitative trait phenotype is declared when significant differences are observed between the means of phenotypic values of all individuals within a biparental population carrying different parental marker alleles at a genetic locus (Chee and Campbell, 2009). This basic QTL mapping approach has been used to dissect the genetic basis of a number of quantitative traits in cotton. For example, Chee and Campbell (2009) summarized the QTL mapping based approaches used to study the genetic basis of a number of quantitatively inherited fiber quality properties and found (i) 107 QTL for fiber length, (ii) 13 QTL for short fiber content, 83 QTL for fiber strength, 33 QTL for micronaire, and 112 QTL for fiber fineness. Traditional quantitative trait analysis methods have been extended to estimate and dissect genetic variance (additive and dominance) of a specific QTL, epistasis between two or more QTL, and QTL × E interactions.

In addition to basic QTL analyses using biparental genetic populations, more recently association mapping approaches are being used to dissect quantitative trait variation without restricting analysis to a limited number of genetic backgrounds (Yu and Buckler, 2006). Therefore, association mapping approaches offer tremendous potential to associate chromosomal regions with quantitative trait variation, because they are based on associating historical and evolutionary recombination events at the population level without restriction to recombination events occurring during biparental population development (Holland, 2007). In cotton quantitative trait dissection, association mapping approaches are still in their infancy and only a few studies have been reported. To date, Abdurakhmonov et al. (2008) provided one of the most comprehensive association mapping studies in cotton. The study estimated linkage disequilibrium blocks of ~30 cM for improved and ~10 cM for exotic germplasm populations. This suggests that improved germplasm populations would require large population sizes, but fewer polymorphic markers. Exotic germplasm populations would require smaller population sizes, but greater polymorphic markers. These will be important considerations for future cotton association mapping studies.

Two additional emerging quantitative trait analysis tools and approaches worthy of mention for the future of quantitative trait analysis in cotton include nested association mapping and genomic selection. Nested association mapping combines the power of biparental and association analysis of quantitative trait variation such that the analysis accounts for both recent and historical recombination events, thereby allowing for high allele richness, high mapping resolution, and high statistical power (McMullen et al., 2009). Nested association mapping has been successfully applied in maize to unravel the genetic architecture of the quantitative trait flowering time (Buckler et al., 2009). Genomic selection is another emerging technology that capitalizes on the abundance of low cost genome marker data currently being generated in many crop plants including cotton. Rather than relying on linkage or association mapping techniques to first identify genetic loci associated with quantitative trait variation, genomic selection directly uses all available marker data as predictors of performance to deliver accurate phenotype predictions. These phenotype predictions, often for quantitatively inherited traits, are then used as the basis for selection and advance in breeding programs. Simulation studies in crops other than cotton have shown that genomic selection offers tremendous potential for improving quantitatively inherited traits (Jannink et al., 2010).

In total, the merging of rapidly advancing genomic technologies with traditional principles of quantitative genetics are leading to the development of novel statistical genetic approaches to better understand and exploit quantitative trait variation. Although novel techniques such as nested association mapping and genomic selection have yet to be fully evaluated in cotton, they have shown tremendous promise in other crop plants. It is anticipated that these new genomic based quantitative trait analysis tools and others unforeseen will impact our understanding of quantitatively inherited traits and how to improve them in the near future.

References

Abdurakhmonov, I.Y., R.J. Kohel, J.Z. Yu, A.E. Pepper, A.A. Abdullaev, F.N. Kushanov, I.B. Salakhutdinov, Z.T. Buriev, S. Saha, B.E. Scheffler, J.N. Jenkins, and A. Abdukarimov. 2008. Molecular diversity and association mapping of fiber quality traits in exotic *G. hirsitum* L. germplasm. Genomics 92:478–487. doi:10.1016/j.ygeno.2008.07.013

Ali, M.A., I.A. Kahn, S.I. Awan, S. Ali, and S. Niaz. 2008. Genetics of fibre quality traits in cotton (*Gossypium hirsutum* L.). Australian J. Crop Sci. 2:10–17.

Basal, J., and I. Turut. 2005. Genetic analysis of yield components and fiber strength in Upland cotton (*Gossypium hirsutum* L.). Asian J. Plant Sci. 4:293–298. doi:10.3923/ajps.2005.293.298

Batool, S., and N.U. Khan. 2012. Diallel studies and heritability estimates using Hayman's approach in Upland cotton. SABRAO J. Breed. Genet. 44:322–338.

Bechere, E., L. Zeng, and D. Boykin. 2014. Correlation and path-coefficient analyses of lint yield and other traits in Upland cotton (*Gossypium hirsutum* L.). J. Crop Improv. 28:852–870.

Bernardo, R. 2002. Breeding for quantitative traits in plants. Stemma Press, Woodberry, MN.

Blanche, S.B., and G.O. Myers. 2006. Identifying discriminating locations for cultivar selection in Louisiana. Crop Sci. 46:946–949. doi:10.2135/cropsci2005.0279

Blanche, S.B., G.O. Myers, W.D. Caldwell, and T. Wallace. 2008. Determining selection gains and discriminating environments via GGE Biplots. J. Crop Improv. 21:13–25. doi:10.1300/J411v21n01_02

Blanche, S.B., G.O. Myers, and M.S. Kang. 2007. GGE biplots and traditional stability measures for interpreting genotype by environment interactions. J. Crop Improv. 20:123–135. doi:10.1300/J411v20n01_07

Blanche, S.B., G.O. Myers, J.Z. Zumba, D.W. Caldwell, and J. Hayes. 2006. Stability comparisons between conventional and near-isogenic transgenic cotton cultivars. J. Cotton Sci. 10:17–28.

Bourland, F.M., R. Hogan, D.C. Jones, and E. Barnes. 2010. Development and utility of Q-score for characterizing cotton fiber quality. J. Cotton Sci. 14:53–63.

Bourland, F.M., and D.C. Jones. 2012. Registration of 'UA48' cotton cultivar. J. Plant Reg. 6:15–18. doi:10.3198/jpr2011.06.0309crc

Bowman, D.T. 2000. Attributes of public and private cotton breeding programs. J. Cotton Sci. 4:130–136.

Bowman, D.T., and O.A. Gutierrez. 2003. Sources of fiber strength in the U.S. Upland cotton crop from 1980–2000. J. Cotton Sci. 7:164–169.

Breaux, R.D. 1951. A genetic analysis of the major components of yield in American Upland cotton. M.Sc. thesis. Louisiana State Univ., Baton Rouge.

Bridge, R.R., and W.R. Meredith. 1983. Comparative performance of obsolete and current cotton cultivars. Crop Sci. 23:949–952. doi:10.2135/cropsci1983.0011183X002300050032x

Bridge, R.R., W.R. Meredith, and J.F. Chism. 1971. Comparative performance of obsolete varieties and current varieties of Upland cotton. Crop Sci. 11:29–32. doi:10.2135/cropsci1971.0011183X001100010010x

Brown, K.D., M.E. Sorrells, and W.R. Coffman. 1983. A method for classification and evaluation of testing environments. Crop Sci. 23:889–893. doi:10.2135/cropsci1983.0011183X002300050018x

Buckler, E.S., J.B. Holland, P.J. Bradbury, C.B. Acharya, P.J. Brown, C. Browne, E. Ersoz, S. Flint-Garcia, A. Garcia, J.C. Glaubitz, M.M. Goodman, C. Harjes, K. Guill, D.E. Kroon, S. Larsson, N.K. Lepak, H. Li, S.E. Mitchell, G. Pressoir, J.A. Peiffer, M.O. Rosas, T.R. Rocheford, M.C. Romay, S. Romero, S. Salvo, H.S. Villeda, H.S. da Silva, Q. Sun, F. Tian, N. Upadyayula, D. Ware, H. Yates, J. Yu, Z. Zhang, S. Kresovich, and M.D. McMullen. 2009. The genetic architecture of maize flowering time. Science 325:714–718. doi:10.1126/science.1174276

Campbell, B.T., D.T. Bowman, and D.B. Weaver. 2008. Heterotic effects in topcrosses of modern and obsolete cotton cultivars. Crop Sci. 48:593–600. doi:10.2135/cropsci2007.06.0362

Campbell, B.T., D. Boykin, Z. Abdo, and W.R. Meredith, Jr. 2014a. Yield gains in US cotton, contributing factors and future prospects. In: S. Smith, B. Diers, B. Carver, and J. Specht, editors, Yield gains in major U.S. field crops. CSSA, Madison, WI.

Campbell, B.T., P.W. Chee, E.L. Lubbers, D.T. Bowman, W.R. Meredith, J. Johnson, and D.E. Fraser. 2011. Genetic improvement of the Pee Dee cotton germplasm collection following seventy years of plant breeding. Crop Sci. 51:955–968. doi:10.2135/cropsci2010.09.0545

Campbell, B.T., P.W. Chee, E. Lubbers, D.T. Bowman, W.R. Meredith, Jr., J. Johnson, D. Fraser, W. Bridges, and D.C. Jones. 2012. Dissecting genotype × environment interactions and trait correlations present in the Pee Dee cotton germplasm collection following seventy years of plant breeding. Crop Sci. 52:690–699. doi:10.2135/cropsci2011.07.0380

Campbell, B.T., J.K. Greene, J. Wu, and D.C. Jones. 2014b. Assessing the breeding potential of day-neutral converted racestock germplasm in the Pee Dee cotton germplasm enhancement program. Euphytica 195:453–465. doi:10.1007/s10681-013-1007-y

Campbell, B.T., and M.A. Jones. 2005. Assessment of genotype × environment interactions for yield and fiber quality in cotton performance trials. Euphytica 144:69–78. doi:10.1007/s10681-005-4336-7

Campbell, B.T., D.B. Weaver, R. Sharpe, J. Wu, and D.C. Jones. 2013. Breeding potential of elite Pee Dee germplasm in Upland cotton breeding programs. Crop Sci. 53:894–905.

Campbell, B.T., V. Williams, and W. Park. 2009. Using molecular markers and field performance data to characterize the Pee Dee cotton germplasm resources. Euphytica 169:285–301. doi:10.1007/s10681-009-9917-4

Chaudhry, M.R. 1997. Commercial cotton hybrids. Int. Cotton Advis. Comm. Rec. 15(2):3–14.

Cheatham, C.C., J.N. Jenkins, J.C. McCarty, Jr., C.E. Watson, and J. Wu. 2003. Genetic variances and combining ability of crosses of American cultivars, Australian cultivars, and wild cottons. J. Cotton. Sci. 7:16–22.

Chee, P.W., and B.T. Campbell. 2009. Bridging classical and molecular genetics of cotton fiber quality and development. In: A.H. Paterson, editor, Genetics and genomics of cotton. Springer, New York. p. 283–311.

Clement, J.D., G.A. Constable, W.N. Stiller, and S.M. Liu. 2012. Negative associations still exist between yield and fibre quality in cotton breeding programs in Australia and USA. Field Crops Res. 128:1–7. doi:10.1016/j.fcr.2011.12.002

Culp, T.W., and C.C. Green. 1992. Performance of obsolete and current cultivars and Pee Dee germplasm lines of cotton. Crop Sci. 32:35–41. doi:10.2135/cropsci1992.0011183X003200010008x

Culp, T.W., D.C. Harrell, and T. Kerr. 1979. Some genetic implications in the transfer of high strength genes to Upland cotton. Crop Sci. 19:481–484. doi:10.2135/cropsci1979.0011183X001900040013x

Davis, D.D. 1978. Hybrid cotton: Specific problems and potentials. In: N.C. Brady, editor, Advances in agronomy. Vol. 30. Academic Press, New York. p. 129–158.

DeLacy, I.H., R.L. Eisemann, and M. Cooper. 1990. The importance of genotype-by environment interactions in Regional Variety Trials. In: M.S. Kang, editor, Genotype-by-environment interaction and plant breeding. Louisiana State Univ., Baton Rouge. p. 287–301.

Dudley, J.W., and R.H. Moll. 1969. Interpretations and use of estimates of heritability and genetic variances in plant breeding. Crop Sci. 9:257–262. doi:10.2135/cropsci1969.0011183X000900030001x

Fotiadis, N.A., and P.A. Miller. 1973. Effects of recurrent seed irradiation on genetic variability and recombination in cotton (*Gossypium hirsutum* L.). Crop Sci. 13:40–44. doi:10.2135/cropsci1973.0011183X001300010013x

Gauch, J.H.G., and R.W. Zobel. 1997. Identifying mega-environments and targeting genotypes. Crop Sci. 37:311–326. doi:10.2135/cropsci1997.0011183X003700020002x

Geng, S., B.M. Hu, and D.M. Bassett. 1990. Quantification and classification of locational effects on cotton cultivar testing programs. Agron. J. 82:514–518. doi:10.2134/agronj1990.00021962008200030015x

Hague, S.S., C.W. Smith, G. Berger, J. Clement, and D.C. Jones. 2011. Variation in an extra-long staple Upland × medium staple Upland cotton F2 population. J. Cotton Sci. 15:265–270.

Hanson, W.D. 1959. The breakup of initial linkage blocks under selected mating systems. Genetics 44:857–868.

Herring, A.D., D.L. Auld, M.D. Ethridge, E.F. Hequet, E. Bechere, C.J. Green, and R.G. Cantrell. 2004. Inheritance of fiber quality and lint yield in a chemically mutated population of cotton. Euphytica 136:333–339. doi:10.1023/B:EUPH.0000032747.97343.54

Hinze, L.L., B.T. Campbell, and R.J. Kohel. 2011. Performance and combining ability in cotton (*Gossypium hirsutum* L.) populations with diverse parents. Euphytica 181:115–125. doi:10.1007/s10681-011-0442-x

Holland, J.B. 2007. Genetic architecture of complex traits in plants. Curr. Opin. Plant Biol. 10:156–161. doi:10.1016/j.pbi.2007.01.003

Jannink, J.-L., A.J. Lorenz, and H. Iwata. 2010. Genomic selection in plant breeding: From theory to practice. Briefings Funct. Genomics. 9:166–177. doi:10.1093/bfgp/elq001

Jenkins, J.N., J.C. McCarty, S. Saha, O. Gutierrez, R. Hayes, and D.M. Stelly. 2006. Genetic effects of thirteen *Gossypium barbadense* L. chromosome substitution lines in topcrosses with Upland cotton cultivars: I. Yield and yield components. Crop Sci. 46:1169–1178. doi:10.2135/cropsci2005.08-0269

Jenkins, J.N., J.C. McCarty, Jr., J. Wu, and O.A. Gutierrez. 2009. Genetic variance components and genetic effects among eleven diverse Upland cotton lines and their F2 hyrbrids. Euphytica 167:397–408. doi:10.1007/s10681-009-9902-y

Jenkins, J.N., J.C. McCarty, J. Wu, O.A. Gutierrez, R. Hayes, and D.M. Stelly. 2007. Genetic effects of thirteen *Gossypium barbadense* L. chromosome substitution lines in topcrosses with Upland cotton cultivars: II. Fiber quality traits. Crop Sci. 47:561–571. doi:10.2135/cropsci2006.06.0396

Jenkins, J.N., J.C. McCarty, Jr., J. Wu, R. Hayes, and D. Stelly. 2012. Genetic effects of nine *Gossypium barbadense* L. chromosome substitution lines in top crosses with five elite Upland cotton *G. hirsutum* L. cultivars. Euphytica 187:161–173. doi:10.1007/s10681-011-0580-1

Jia, Y., X. Sun, J. Sun, Z. Pan, X. Wang, S. He, S. Xiao, W. Shi, Z. Zhou, B. Pang, L. Wang, J. Liu, J. Ma, X. Du, and J. Zhu. 2014. Association mapping for epistasis and environmental interaction of yield traits in 323 cotton cultivars under 9 different environments. PLoS ONE 9(5):E95882. doi:10.1371/journal.pone.0095882

Kerby, T.A., J. Burgess, M. Bates, D. Albers, and K. Lege. 2000. Partitioning variety and environmental contribution to variation in yield, plant growth, and fiber quality. In: Proceedings of the National Cotton Council Beltwide Cotton Conference, Natl. Cotton Counc. Am., San Antonio, TX. 4–8 Jan. 2000. Natl. Cotton Counc. Am., Memphis, TN. p. 528–532.

Liu, S.M., G.A. Constable, P.E. Reid, W.N. Stiller, and B.R. Cullis. 2013. The interaction between breeding and crop management in improved cotton yield. Field Crops Res. 148:49–60. doi:10.1016/j.fcr.2013.04.006

Lu, H., and G.O. Myers. 2011. Combining abilities and inheritance of yield components in influential Upland cotton varieties. Aust. J. Crop Sci. 5(4):384–390.

Lubbers, E.L. 2003. Evaluation of cotton breeding test environments in the Southeast United States. In: Proceedings of the National Cotton Council Beltwide Cotton Conference, Natl. Cotton Counc. Am., Nashville, TN. 6–10 Jan. 2003. Natl. Cotton Counc. Am., Memphis, TN. p. 791–807.

Manning, H.L. 1956. Yield improvement from a selection index technique with cotton. Heredity 10:303–322. doi:10.1038/hdy.1956.29

May, O.L., and G.M. Jividen. 1999. Genetic modification of cotton fiber properties as measured by single- and high-volume instruments. Crop Sci. 39:328–333. doi:10.2135/cropsci1999.0011183X003900020004x

McCarty, J.C., J.N. Jenkins, and J. Wu. 2004a. Primitive accession derived germplasm by cultivar crosses as sources for cotton improvement: I. Phenotypic values and variance components. Crop Sci. 44:1226–1230. doi:10.2135/cropsci2004.1226

McCarty, J.C., J.N. Jenkins, and J. Wu. 2004b. Primitive accession derived germplasm by cultivar crosses as sources for cotton improvement: II. Genetic effects and genotypic values. Crop Sci. 44:1231–1235. doi:10.2135/cropsci2004.1231

McCarty, J.C., J. Wu, and J.N. Jenkins. 2007. Use of primitive derived cotton accessions for agronomic and fiber trait improvement: Variance components and genetic effects. Crop Sci. 47:100–110. doi:10.2135/cropsci2006.06.0404

McMullen, M.D., S. Kresovich, H.S. Villeda, P. Bradbury, H. Li, Q. Sun, S. Flint-Garcia, J. Thornsberry, C. Acharya, C. Bottoms, P. Brown, C. Browne, M. Eller, K. Guill, C. Harjes, D. Kroon, N. Lepak, S.E. Mitchell, B. Peterson, G. Pressoir, S. Romero, M.O. Rosas, S. Salvo, H. Yates, M. Hanson, E. Jones, S. Smith, J.C. Glaubitz, M. Goodman, D. Ware, J.B. Holland, and E.S. Buckler. 2009. Genetic properties of the maize nested association mapping population. Science 325:737–740. doi:10.1126/science.1174320

Meredith, W.R., Jr. 1984. Quantitative genetics. In: R.J. Kohel and D.F. Lewis, editors, Cotton. Agron. Monogr. 24. ASA, CSSA, and SSSA, Madison, WI. p. 131–150.

Meredith, W.R., Jr. 1997. Heterosis in cotton. In: Book of abstracts. The genetics and exploitation of heterosis in crops. An international symposium. 17–22 Aug. 1997. CIMMYT, Mexico City, Mexico. p. 282–284.

Meredith, W.R., Jr. 2003. Thirty-six years of Regional High Quality Variety Tests. In: Proceedings of the National Cotton Council Beltwide Cotton Conference, Natl. Cotton Counc. Am., Nashville, TN. 6–10 Jan. 2003. Natl. Cotton Counc. Am., Memphis, TN. p. 2561–2563.

Meredith, W.R., Jr., and R.R. Bridge. 1971. Breakup of linkage blocks in cotton, *Gossypium hirsutum* L. Crop Sci. 11:695–698. doi:10.2135/cropsci1971.0011183X001100050027x

Meredith, W.R., Jr., and J.S. Brown. 1998. Heterosis and combining ability of cottons originating from different regions of the United States. J. Cotton Sci. 2:77–84.

Miller, P.A., and J.O. Rawlings. 1967. Breakup of initial linkage blocks through intermating in a cotton breeding program. Crop Sci. 7:199–204. doi:10.2135/cropsci1967.0011183X000700030009x

Myers, G.O. 2002. Biplot analysis of the National Cotton Variety Test. In: Proceedings of the National Cotton Council Beltwide Cotton Conference, Natl. Cotton Counc. Am., Atlanta, GA. 8–12 Jan. 2002. Natl. Cotton Counc. Am., Memphis, TN.

Nassar, M.A.A. 2013. Some genetic parameters and heterosis in two crosses of Egyptian cotton. J. Appl. Sci. Res. 9(1):548–553.

Paterson, A.H., E.S. Lander, J.D. Hewitt, S. Peterson, S.E. Lincoln, and S.D. Tanksley. 1988. Resolution of quantitative traits into Mendelian factors by using a complete linkage map of restriction length polymorphisms. Nature 335:721–726. doi:10.1038/335721a0

Paterson, A.H., Y. Saranga, M. Menz, C.X. Jiang, and R.J. Wright. 2003. QTL analysis of genotype × environment interaction affecting cotton fiber quality. Theor. Appl. Genet. 106:384–396.

Percy, R.G., R.G. Cantrell, and J. Zhang. 2006. Genetic variation for agronomic and fiber properties in an introgressed recombinant inbred population of cotton. Crop Sci. 46:1311–1317. doi:10.2135/cropsci2005.08-0284

Saha, S., J. Wu, J.N. Jenkins, J.C. McCarty, and D.M. Stelly. 2013. Interspecific chromosomal effects on agronomic traits in *Gossypium hirsutum* by AD analysis using intermated *G. barbadense* chromosome substitution lines. Theor. Appl. Genet. 126:109–117. doi:10.1007/s00122-012-1965-9

Scholl, R.L., and P.A. Miller. 1976. Genetic associations between yield and fiber strength in Upland cotton. Crop Sci. 16:780–783. doi:10.2135/cropsci1976.0011183X001600060010x

Schwartz, B.M., and C.W. Smith. 2008. Genetic gain in yield potential of Upland cotton under varying plant densities. Crop Sci. 48:601–605. doi:10.2135/cropsci2007.01.0049

Stiller, W.N., J.J. Read, G.A. Constable, and P.E. Reid. 2005. Selection for water use efficiency traits in a cotton breeding program: Cultivar differences. Crop Sci. 45:1107–1113. doi:10.2135/cropsci2004.0545

Tang, B., J.N. Jenkins, J.C. McCarty, and C.E. Watson. 1993a. F2 hybrids of host plant germplasm and cotton cultivars: I. Heterosis and combining ability for lint yield and yield components. Crop Sci. 33:700–705. doi:10.2135/cropsci1993.0011183X003300040012x

Tang, B., J.N. Jenkins, J.C. McCarty, and C.E. Watson. 1993b. F2 hybrids of host plant germplasm and cotton cultivars: II. Heterosis and combining ability for fiber properties. Crop Sci. 33:706–710. doi:10.2135/cropsci1993.0011183X003300040013x

Tang, B., J.N. Jenkins, C.E. Watson, J.C. McCarty, and R.G. Creech. 1996. Evaluation of genetic variances, heritabilities, and correlations for yield and fiber traits among cotton F2 hybrid populations. Euphytica 91:315–322. doi:10.1007/BF00033093

Tang, F., and W. Xiao. 2013. Genetic effects and heterosis of within-boll yield components in Upland cotton (*Gossypium hirsutum* L.). Euphytica 194:41–51. doi:10.1007/s10681-013-0958-3

Ulloa, M. 2006. Heritability and correlations of agronomic and fiber traits in an okra-leaf Upland cotton population. Crop Sci. 46:1508–1514. doi:10.2135/cropsci2005.08-0271

Yan, W., L.A. Hunt, Q. Sheng, and Z. Szlavnics. 2000. Cultivar evaluation and mega-environment investigation based on the GGE biplot. Crop Sci. 40:597–605. doi:10.2135/cropsci2000.403597x

Yan, W., and M.S. Kang. 2003. GGE Biplot analysis: A graphical tool for breeders, geneticists, and agronomists. CRC Press, Boca Raton, FL.

Young, N.D., and S.D. Tanksley. 1989. RFLP analysis of the size of chromosomal segments retained around the Tm-2 locus of tomato during backcross breeding. Theor. Appl. Genet. 77:353–359. doi:10.1007/BF00305828

Yu, J., and E.S. Buckler. 2006. Genetic association mapping and genome organization of maize. Curr. Opin. Biotechnol. 17:155–160. doi:10.1016/j.copbio.2006.02.003

Zeng, L., E. Bechere, and D.L. Boykin. 2014a. Commonality analysis and selection of parents for within-boll yield components in Upland cotton. Euphytica 199:339–348.

Zeng, L., and W.R. Meredith, Jr. 2009. Associations among lint yield, yield components, and fiber properties in an introgressed population of cotton. Crop Sci. 49:1647–1654. doi:10.2135/cropsci2008.09.0547

Zeng, L., W.R. Meredith, Jr., and D.L. Boykin. 2011. Germplasm potential for continuing improvement of fiber quality in Upland cotton: Combining ability for lint yield and fiber quality. Crop Sci. 51:60–68. doi:10.2135/cropsci2010.07.0413

Zeng, L., W.R. Meredith, Jr., B.T. Campbell, J.K. Dever, J. Zhang, K.M. Glass, A.S. Jones, G.O. Myers, and F.M. Bourland. 2014b. Genotype-by-environment interaction effects on lint yield of cotton cultivars across major regions in the U.S. cotton belt. J. Cotton Sci. 18:75–84.

Zeng, L., and J. Wu. 2012. Germplasm for genetic improvement of lint yield in Upland cotton: Genetic analysis of lint yield with yield components. Euphytica 187:247–261. doi:10.1007/s10681-012-0708-y

Zobel, R.W., M.G. Wright, and H.G. Gauch. 1988. Statistical analysis of a yield trial. Agron. J. 80:388–393. doi:10.2134/agronj1988.00021962008000030002x

Zumba, J., G.O. Myers, E.L. Clawson, D.K. Miller, R.G. Danka, S.B. Blanche. 2013. Developing hybrid cotton (*Gossypium* spp.) using honeybees as pollinators and the roundup ready phenotype as the selection trait. J. Cotton Sci. 17:293–301.

Conventional Cotton Breeding

Fred Bourland* and Gerald O. Myers

Abstract

Most of the current principles used in conventional cotton (*Gossypium hirsutum* L.) breeding programs are similar to those used over the past half century, but techniques and methodology has changed considerably. The objective of this chapter was to chronicle major changes that have occurred in conventional cotton breeding since about 1980 and to review typical methodology and selection criteria presently being used. Today, computers and further mechanization of planting and harvest have enabled cotton breeders to increase the number of plots and samples they evaluate, thus increasing their probability of success. Development of cottons that mature early (short-season cottons) and possess enhanced host plant resistance received much attention from about 1970 until the mid-1990s, but emphases on these traits have subsided with the advent of transgenic Bt cottons and boll weevil eradication. Improvements in host plant resistance, yield selection, and fiber quality are now frequently being facilitated by DNA markers, test site selection via genotype × environment (G × E), and advances in fiber testing equipment, respectively. Basic methodology of conventional cotton breeding still involves crossing diverse parents, making early-generation selections, and then evaluating subsequently developed strains, but now adventitious presence of transgenes must be carefully avoided. Increased lint yield is still the major selection criterion used in conventional breeding programs. Selection and evaluation of yield components, growth habit, maturity, and host plant resistance traits are often employed to enhance yield and/or yield stability. New fiber testing methods have permitted breeders to place more attention on improving fiber quality and to make progress in overcoming the historical negative relationships between yield and fiber quality. Continued genetic improvements in conventional cottons will occur as conventional cotton breeders continue to develop and employ enhanced methods and approaches.

The challenges faced by cotton breeding research have remained much the same over the last century, but methods and approaches have drastically changed. The overall goal continues to be to enhance the sustainability of the cotton industry. The future of cotton is dependent upon cotton prices relative to costs per unit

---

Abbreviations: AFIS, Advanced Fiber Information System; AMMI, additive main effects and multiplicative interaction; GEBV, genomic estimated breeding value; HVI, high volume instrument; IPS, individual plant selection; NACB, nodes above cracked boll; NAWF, nodes above white flower; PVP, Plant Variety Protection; SSD, single seed descend; TPB, tarnished plant bug.

Fred Bourland, University of Arkansas Northeast Research and Extension Center, 1241 West County Road 780, Keiser, AR 72351. *Corresponding author (bourland@uark.edu).

Gerald O. Myers, Louisiana State University, School of Plant, Environmental, and Soil Sciences, 104 M.B. Sturgis Hall, Baton Rouge, LA 70803 (gmyers@agcenter.lsu.edu).

doi:10.2134/agronmonogr57.2013.0025

Cotton. 2nd. ed. David D. Fang and Richard G. Percy, editors. Agronomy Monograph 57.

of production. For cotton to remain competitive, cotton cultivars must be bred to produce more cotton at lower costs and to produce a high fiber quality to meet industry needs. Niles and Feaster (1984) chronicled how breeding efforts had affected the cotton plant from initial cultivation until about 1980. In this chapter, we will primarily chronicle changes that have occurred since 1980 and summarize methods used in conventional breeding programs.

Major Changes in Cotton Breeding Since 1980

Since the mid-20th century, US cotton production has transformed from heavy reliance on hand labor to highly mechanized operations. Because of the scale of operations, mechanization in cotton breeding programs often lags behind mechanization in commercial production. Nevertheless, breeding programs have now become more mechanized and utilize modern fertilization, plant growth control, and pest control.

Methods to Handle Samples

Personal computers and plot pickers with onboard weigh systems were not available in 1980 but are now standard equipment in most cotton breeding programs. Additionally, cone seeder planting units were just becoming widely used in 1980. Personal computers and associated equipment have facilitated rapid accomplishment of many time-consuming, tedious tasks including randomizing plots, printing (labels, field books, etc.), analyzing data, writing reports, and generating presentations. Simple data analyses that might require several hours with a calculator are now completed with a touch of a computer button. Bar coding is often used to label seed cotton samples and/or subsequent fiber and seed samples. Bar coding helps to reduce errors, enhances speed of operation, and facilitates data entry.

Prior to the development of cone seeder planting units, plots were normally seeded by an individual "dribbling" seed through a funnel and pipe on each planter row. The planter was stopped at each alley to allow riders to pour seed for the next plot into their hands. Cone seeder planting units allow users to plant plots without stopping at each alley, thus greatly decreasing the time required to plant a test. By reducing the time required to plant, breeders can plant more plots with the same amount of resources. Vacuum planters that precisely distribute seed within a plot are now available and can provide additional precision to breeding programs.

Mechanical plot pickers with onboard weigh systems have greatly reduced the harshness of work and the number of people needed to harvest plots. Presently available systems require the machine to stop, weigh, and dump samples between plots. Thus, the amount of time required to harvest plots with a weigh system is similar to using a picker modified to catch harvest in mesh bags. However, the latter system usually required three to six persons on a two-row picker plus an additional crew to weigh and empty sacks. Thus, 1 person operating an onboard weigh system can replace 5 to 10 persons using the old system. Since harvesting with an onboard weigh system does not physically tax the operator, the machine can operate more hours per day. This increased efficiency in harvesting permits cotton breeders to expand testing and to obtain harvest weights in tests (e.g., progeny rows) where only a visual estimation of yield was previously practical. Work is underway to develop onboard weigh systems that do not have to stop between plots. Such a system will greatly decrease the amount of time required to harvest plots.

Advent of Short-Season Cottons

Early maturing, short-season cottons became highly desired as the boll weevil (*Anthonomus grandis* Boheman) advanced through the US cotton belt in the early 20th century. Insecticide treatments for boll weevils would often exacerbate other pests, particularly the cotton bollworm, *Helicoverpa zea* Boddie, and the tobacco budworm, *Heliothis virescens* Fabricius, complex (Bradley, 1996). Removing natural predation incited heavy infestation of this complex. The Heliothine complex quickly developed resistance to new insecticides and classes of insecticides as they became available.

In the Mississippi River Delta, the damages and increased control costs associated with the boll weevil and Heliothine complex was hindering the profitability of cotton production. In response to this problem, R.R. Bridge was employed by the Mississippi Agriculture and Forestry Experiment Station and given the challenge to breed short-season cottons for the Mississippi River Delta—even if such lines had to suffer 10% loss in yield (R.R. Bridge, personal communication, 1978). Bridge exceeded that challenge by developing and releasing 'DES 24' (Plant Variety Protection [PVP] 7800040; Bridge and Chism, 1978a) and 'DES 56' (PVP 7800041; Bridge and Chism, 1978b). Compared with 'Deltapine 16' (a widely grown standard), DES 24 and DES 56 matured about 10 d earlier, but also produced nearly 10% higher yields. Additional high yielding, early-maturity cultivars, for example, 'Deltapine 41' (PVP 7900102) and 'Stoneville 506' (PVP 8100059), soon became available from private companies. The subsequent release of 'DES 119', which was derived from a cross of DES 24 and a sister line of DES 56, combined early maturity into a much more widely adapted cultivar (PVP 7800040; Bridge, 1986).

The importance of the shift to short-season cotton was evidenced by a symposium of 16 papers presented in a special session at the 1987 Beltwide Cotton Production Conference entitled "Short-Season Cotton Production Systems." In that symposium, Bridge and McDonald (1987) documented a major shift from full-season to short season cultivars in most states. The percentage of acreage planted to short-season cultivars (defined as earlier than Stoneville 213) shifted in Mississippi, Arkansas, Tennessee, and Missouri from <20% in 1978 to >90% in 1986. Substantial shifts also occurred in Louisiana, Alabama, North Carolina, South Carolina, Georgia, and Arizona. The shift to short-season cultivars occurred earlier in some parts of Texas, and more than 98% of Texas acreage was planted to short-season cultivars in 1986. Thus, an unprecedented shift in cotton cultivars occurred in a span of about 10 yr. This shift was primarily driven by economics and insect pest management. Parvin et al. (1987) indicated that an increase of 1 wk in earliness was accompanied by a 7% increase in yield, an 8% increase in net revenue, and a 27-d shorter harvest season. A 2-wk enhancement of earliness increased yield by 12% and net revenue by 13%. These benefits drove the shift from full-season to short-season cultivars. Short-season cultivars substantially decreased pest control, harvest, and economic risks associated with cotton production.

Morphological Traits for Host Plant Resistance

Jenkins and Wilson (1996) included morphological traits along with earliness, biochemical mutants, and exotic cytoplasms as sources of pest resistance in cultivated cottons. Each morphological trait was typically evaluated in multiple isoline studies to determine their relative merit in response to various pests and their effect on agronomic performance. Most of the morphological traits were found to infer increased resistance to one pest but increased susceptibility to another. For example, Frego (or rolled) bract was associated with increased resistance to boll weevil and the bollworm/tobacco budworm complex, but Frego lines were highly susceptible to tarnished plant bugs (*Lygus lineolaris* Palisot de Beauvois).

Several germplasm lines having single or combinations of the resistance traits (10 morphological traits and 6 cytoplasms listed) were released (Jenkins and Wilson, 1996). Of the 10 listed morphological traits, Upland cultivars having the pilose, Frego bract, red plant, yellow pollen, orange pollen, or male sterile traits have not been released. Cultivars possessing glabrous, okra-leaf, nectariless, and high gossypol have been developed and made commercially available. Of these, the glabrous (smoothleaf) trait has been most widely utilized in commercial cultivars. The glabrous trait was reported to be associated with reduced bollworm/tobacco budworm oviposition, number of whiteflies (*Trialeurodes abutilonea* Haldeman), and cotton leafperforator (*Bucculatrix thurberiella* Busck.), but increased susceptibility to tarnished plant bugs, tobacco thrips (*Frankliniella fusca* Hinds), and leaf hoppers (*Empoasca* spp.). The wide acceptance of the glabrous trait is due to its relation to improved cotton grades—less trash, less lint cleaning required (Rayburn and Libous, 1983; Anthony and Rayburn, 1989)—rather than its relation to host plant resistance.

Jenkins and Wilson (1996) summarized data showing that the nectariless trait is associated with increased resistance to bollworm/tobacco budworm complex, tarnished plant bugs, cotton leafperforator, and pink bollworm (*Pectinophora gossypiella* Saunders) and does not adversely affect any agronomic or fiber quality property. Consequently, much effort has been made to develop nectariless cultivars, but only three have been released: 'Stoneville 731N' (PVP 7600048) in 1976, 'Stoneville 825' (PVP 7900024) in 1978, and 'DP 0935 B2RF (PVP 200900062) in 2008.

Advent of Transgenic Cotton Cultivars

The introduction of transgenic cultivars in the 1990s incited a revolutionary change in cotton cultivars. Since the introduction of commercial transgenic cultivars to Arkansas in 1995, conventional cultivars declined from over 99% of Arkansas cotton acreage in 1995 to less than 1% in 2004. Similar shifts have occurred in most areas of the United States. The first transgenic cultivars were simply popular cultivars (recurrent parents) with an additional transgene inserted. Several studies showed that agronomic performances of the transgenic cultivars were similar to their recurrent parent (e.g., Bourland et al. 1997; Bryant et al. 2003; May et al., 2003). To some degree, this facilitated quick acceptance of the new transgenic cultivars, but also hindered improvement of agronomic performance since the genes were backcrossed into older cultivars.

Many of the first transgenic cultivars were subsequently marketed on the basis of the performance of their recurrent parent. Although transgenic cultivars typically resembled their recurrent parent, small gene shifts during selection sometimes altered the adaptation and performance of the transgenic cultivar.

The transgenic cultivar H1560 BG (PVP 9700130) illustrates this point. In contrast to its recurrent parent 'H1560' (PVP 9600280), H1560 BG matured much earlier and was susceptible to root-knot nematode.

Most transgenic cultivars are still derived from backcrossing transgenes into conventional lines rather than by forward crossing. Thus, conventional cotton breeding techniques are still widely used throughout the industry. As soon as a promising conventional line is identified, specific transgenes are inserted using an aggressive backcross program.

The advent of transgenic cultivars has presented challenges to conventional breeding programs. An initial question is whether conventional and transgenic lines can be fairly compared in adjacent plots. Agronomic traits of herbicide-resistant and conventional cultivars are usually compared by simply using a conventional herbicide program on all cultivars. However, some herbicide-resistant cultivars may possibly be at a disadvantage under a conventional herbicide program. Transgenic cultivars that possess insect-resistant genes present additional testing concerns. Determination of whether treatment thresholds have been exceeded may be biased if insect-resistant lines dominate the test. Insect pests may then congregate on conventional lines and consequently not be accurately quantified by routine scouting. The insect-resistant lines may also incur less subtreatment level damage by pests than adjacent conventional lines in small-plot tests.

Perhaps the greatest effect of the advent of transgenic cultivars on conventional cotton breeding has been the restrictions on available parents to the breeder. Previously, breeders were able to make crosses with any PVP cultivar. Now, crossing is restricted because all transgenic cultivars (and some conventional lines) are patented. Consequently, most breeders are required to make relatively narrow crosses within their own germplasm and may not have the best genetic lines available to them.

DNA Marker-Assisted Selection

Selection efficiency is a fundamental component of trait improvement. This can be easy with such traits as leaf shape or nectary presence/absence for which there are visual clues or done with reasonable accuracy on traits such as yield components and fiber quality parameters. Selection efficiency is, however, reduced in situations where accuracy and/or precision of trait measurement is affected by the environment or is difficult to measure because of time, cost, or subjectivity in scoring.

Markers based upon polymorphism at the DNA level are increasingly being used in cotton breeding to follow specific traits, monitor transgenic trait introgression, construct genetic maps, maintain genetic purity via fingerprinting, and study the evolution and diversity within the genus *Gossypium*. The status of marker availability and their use has recently been reviewed by Rahman et al. (2009) and markers associated with numerous yield, fiber quality, and biotic/abiotic traits have been identified (Mei et al., 2014). Hindering many of these studies is the low level of intraspecific polymorphism in cotton, but the advent of new-high throughput sequencing technology is overcoming this hurdle. A concern with many of the original mapping studies identifying trait-marker associations was the lack of marker portability (Mei et al., 2014). Association mapping approaches including ongoing nested association mapping projects are identifying more robust markers that are usable in a breeding program. These have been aided by the rapid reduction in cost of marker technology, and the advent of next-generation sequencing will add further robustness to crop association mapping studies (Zhu et al., 2008). This is bringing to the forefront the expense and difficulty of phenotyping large populations. High-throughput phenotyping systems have recently been initiated in cotton (Andrade-Sanchez et al., 2013).

Genomic selection is an emerging area for the use of DNA markers in crop improvement. Originally proposed by Meuwissen et al. (2001), it is a form of marker-assisted selection that considers all markers in a population to explain genetic variance. Genotypic and phenotypic data are jointly considered in the development of a prediction model in calculating genomic estimated breeding values (GEBVs) within a training set. These GEBVs are then applied to the breeding material using the marker data only. Genomic selection is an iterative process that has the potential to overcome issues associated with genotype × environment interactions as well (Heffner et al., 2009), especially if additional information such as pedigree is available (Burgueño et al., 2012).

Efficiency of Test Site Selection via G × E

The performance (yield and fiber quality) of a cotton cultivar is dependent upon its genetics (G), the environment (E) in which it is grown, management (M) practices, and interactions between all three (G × E, G × M, G × E × M). This adds several dimensions to breeding program decisions beyond yield per se, for example, stability of performance, adaptation (broad or narrow), and on a programmatic level with program resource allocation. This is less of a concern with qualitative traits (morphological and many biotic stress traits), but in cotton, most traits of interest (yield, fiber quality) are quantitative or semi-quantitative (three or more genes) in nature. As such, the effect of environment is significant on trait expression, with an average of 86% of total variance for yield, 53% for lint percentage, and 54% for the fiber traits of length, strength, and micronaire (see Campbell and Myers, 2015). The average amount of G × E interaction for yield, lint percentage, and these same three fiber traits across several studies is 9, 19, and 16%, respectively (Meredith et al., 2012).

Methodology to partition trait performance into G, E, and G × E (also applicable to G × E × M) components has been extensively documented (Cooper and Hammer, 1996; Kang and Gauch, 1996). Of particular concern are the situations in which the rankings of genotypes change over a range of environments (locations × years). These crossover interactions are typically the norm for cotton yield and to a lesser extent for yield components and fiber traits. This lack of broad adaptation across environments makes it difficult to make blanket cultivar recommendations and has led to attempts to identify environments that elicit similar responses. Pattern analysis based upon ordination (principal components), grouping (cluster), and combination methods such as the additive main effects and multiplicative interaction (AMMI) model have been applied to identify so called mega-environments.

Geng et al. (1990) used indices based upon analyses of variance and clustering to identify four distinct zones across the entire US cotton belt. Interestingly, the productivity of a zone was correlated with the ability to discriminate between the cultivars. This is a general observation across subsequent studies as well: that the highest yielding environment within a region (mega-environment) is the best in which to select since it will maximize the genotype contribution to phenotype.

Within the context of most breeding programs, a more regional analysis is desirable to increase breeding program efficiency, through both identifying homogeneous environments and identifying optimal selection environments (Brown et al., 1983). To this end, Campbell and Jones (2005) used regression based stability estimates and AMMI analysis to assess cultivar stability in South Carolina and to identify similar test environments to allow more effective resource allocation. Environmental effects were largest for yield (90% of variance) and micronaire (81%) and least for fiber strength (14%) and fiber elongation (13%). Genotype contributions to total trait variance were smallest for yield (2%) and micronaire (10%). Genotype × environment interaction effects were largest for gin turnout (20%). The five test sites were able to be assigned to two groups based on yield. For certain traits (e.g. fiber strength), all locations behaved similarly implying that selection results at any one location were extendable to other locations.

Yan et al. (2000) proposed the (G + GE) biplot method to identify mega-environments, perform cultivar stability and adaptation analysis, and identify discriminating environments for selection. In a national context, Lubbers (2003) used both AMMI and GGE biplot on cotton lint yield data to identify similar testing environments within the context of identifying location(s) for establishing a breeding station. Blanche and Myers (2006) used GGE biplot methodology to identify testing environments within Louisiana that were both discriminating and representative. They also extended the use of biplots to multi-trait selection indices. In research on maximizing genetic gain using specifically designed single and multiple-trait segregating populations in select environments meant to be either highly or lowly discriminating, Blanche et al. (2008) demonstrated that the use of highly discriminating selection environments maximized expressed genetic variance and enhanced selection accuracy.

Fiber Testing—HVI and AFIS

Evaluation of cotton fiber quality has evolved from subjective determination of fiber quality by a cotton classer to sophisticated machine classing used throughout the industry today. The cotton classer would visually determine a grade for a cotton lint sample on the basis of color and trash (Perkins et al., 1984). Fiber length was estimated by touch rather than sight. Micronaire, a measure of fineness of fiber based on resistance to airflow through a specified sample, became a part of the USDA Official Classification procedure in 1966. Carefully prepared standards for grade, length, and micronaire were prepared and approved by the Cotton Division of USDA's Agricultural Marketing Service.

Cooperative efforts by USDA and equipment manufacturers began in the mid-1960s to develop high volume instrument (HVI) systems for classing cotton (Perkins et al., 1984; Ramey, 1999). By 1991, USDA began classing all cotton samples provided to the department with the HVI system. Today, HVI class data are accepted throughout the world and are the foundation on which cotton is traded. With this broad use of HVI, genetic improvements of HVI-measured traits became important to cotton breeding programs. The HVI-measured traits normally include micronaire, fiber length, length uniformity, strength, elongation, and short-fiber index.

The Advanced Fiber Information System (AFIS) was developed by the cooperative efforts of USDA–ARS at Clemson, SC, and Schaffner Technologies, with research beginning in 1982 (Bragg and Shofner, 1993). The Advanced Fiber Information System provides data regarding about 20 fiber properties based on individual fiber analysis, but does not include measurement of tensile properties of the fiber. The Advanced Fiber Information System requires extensive and careful preparation of samples and considerable time to evaluate a sample. Thus, turnover and cost of obtaining AFIS data are much greater than obtaining HVI data.

The rapid, relatively inexpensive measurement of fiber quality by HVI has increased the priority of high fiber quality in most cotton breeding programs. The Advanced Fiber Information System provides more detailed information regarding fiber length distribution, but it is generally not practical to evaluate a high number of samples by AFIS. Kelly et al. (2012) found that improvement in fiber length could effectively be done using either HVI or AFIS data and that differences in fiber quality improvement were minimal between the two fiber testing methods.

Basic Cotton Breeding Methodology

Conventional cotton breeding techniques relative to crossing and advancement of genetic lines have not greatly changed over the years. Available resources and specific breeding objectives dictate the exact procedures used within various conventional breeding programs. However, the overall approach is similar across programs—crosses are made among parents differing for specific traits, segregating populations are grown, individual plants are selected from the segregating populations, seed from the individual plants are evaluated in progeny rows, and seed from superior progeny rows are harvested and evaluated in replicated tests over multiple locations and years.

Crossing Methods

A conventional breeding program begins by crossing two parents that vary for one or more distinct traits. Selection and parents should be based on relative performance and on genetic background of the lines. The parents are normally planted in close proximity, and plants are thinned to 1 to 2 plant m^{-2}. Prior to crossing, the parental plants are examined for specific morphological and/or host plant resistance traits and off-types are rogued.

Crossing may be initiated after both parents have commenced flowering. A small "pinhead" square is the first visual evidence of the development of the cotton flower (Fig. 1). About 25 d are required for the square to become a white flower (anthesis). The developing flower forms an elongated "candle" on the day prior to anthesis. Both male and female flowers used in a cross must be at the candle stage on the same day. On day of anthesis (day after candle), the male will shed pollen and the female will be receptive to pollen. The flower petals turn pink on the day after anthesis. Additional flowers occur on the successive sympodial positions at about 6-d intervals (horizontal flowering interval) and on the same horizontal position of successive sympodia at about 3-d intervals (vertical flowering interval).

The female flower for crossing is prepared by removing the male parts (stamen) from the candle (Fig. 2). This is usually accomplished using a "thumbnail"

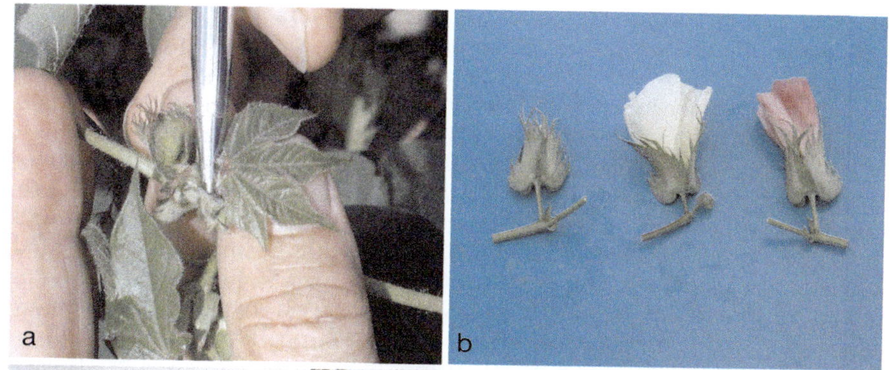

Fig. 1. Development of cotton flower: (a) pinhead and 3-d-old squares in plant terminal; (b) flower relative to anthesis—day before (candle), day of (white flower), and day after (pink flower); and (c) square, white flower and two bolls at 6-d intervals on sympodial branch.

Fig. 2. Crossing cotton flowers using thumbnail emasculation technique: (a) flower candle during afternoon before anthesis; (b) removal of petals and staminal column using thumbnail; and (c) transferring pollen from male flower to pistil on morning of anthesis.

emasculation technique. The petals and sepals are removed, leaving the pistol surrounded by the bracts. Scratching or damaging the membranous structure that surrounds the ovary will cause the young developing boll to shed. Breaking of the style at the junction of the style and ovary is another common problem when emasculating a flower.

Natural cross-pollination in cotton is primarily due to insects, usually bees. In areas where cross-pollination is likely, the emasculated flower (particularly the stigma) should be covered to prevent unintentional cross-pollinating (Fig. 2). The stigma may be covered using a soda straw (with the upper end closed and a strip of wet cotton to secure the straw to the ovary) or some type of bagging or netting to eliminate cross-pollinating insects. To prevent contamination of pollen from other genotypes, the male candle should be self-pollinated using the same type of bagging to cover the female flower or by securing the closed petals with a wire tag or with glue. In areas with none or very few cross-pollinating insects, the emasculated flower may be left uncovered and pollen may be obtained from open white flowers on the day of crossing.

Male and female flowers should be prepared for crossing on the afternoon prior to crossing. Crossing commences in the morning as soon as pollen begins to shed, and should be completed before pollen desiccates. Specific timing of crossing depends on ambient conditions. As temperatures become hotter and relative humidity decreases, crossing must generally be completed earlier in the day.

The cross is made by applying pollen from the male flower to the stigma of the female flower (Fig. 2). This may be accomplished by gently rubbing the anthers of one or more male flowers around the stigma or by using a small camel-hair brush to transfer pollen from a male flower (or a collection of pollen from one or more flowers) to the stigma. After the female flower is pollinated, it should be marked with an appropriate tag. By connecting the labeling tag around the peduncle, the tag will fall from the plant if the developing boll is aborted. Because of mechanical handling and poor pollination, manually crossed bolls often produce about half the number of seed as open-pollinated bolls. From 40 to 80 d are needed from pollination to open boll. Because sample size is usually small, harvested crosses are often ginned on a roller gin.

Early-Generation Procedures

Generally, F_1 seed are advanced to the F_2 generation without any selection. Some breeders will begin individual plant selection (IPS) in the segregating F_2 generation. Bourland (2004, 2013) utilizes a modified single seed descend (SSD) in the F_2 and F_3 generations. The modifications to SSD are as follows: (i) a single boll (instead of a single seed) is taken from each plant; (ii) bolls are not taken from plants deemed to not meet specified selection criteria; (iii) a set number of bolls (100 or 150) is harvested from each population; and (iv) if needed to obtain the set number of bolls, more than one boll may be taken from visually superior plants. After 2 yr of modified SSD, IPS are made in the F_4 generation. Genotypes from the individual plants are evaluated for 2 yr in nonreplicated $F_{4:5}$ and $F_{4:6}$ progeny rows at multiple locations. Seed cotton yields for progenies are measured at each location using a cotton picker modified with a weigh system. Priority is given to progenies that display specified morphological, host plant resistance, and fiber quality traits and have yields that equal or exceed those of check cultivars.

Strain Evaluation

After progenies from individual plants are evaluated, seed from selected progenies are saved and evaluated in subsequent replicated tests that are usually duplicated at multiple locations. Some form of seed increase must be established to have sufficient seed for these tests. Maintaining purity of lines is particularly difficult in areas where high cross-pollination occurs. Bourland (2004, 2013) evaluates lines in nonreplicated progeny rows for 2 yr. After that, selected progenies are promoted to strain status and evaluated in replicated strain tests for up to 4 yr at four locations each year, with only superior strains carried forward each year. To better estimate specific adaptation, some advanced strains are typically entered into regional strain tests (Regional Breeders Testing Network and/or Regional High Quality) during the third and fourth year of replicated testing. After 4 yr of replicated tests, strains are typically released (as a germplasm line or cultivar) or are discarded.

Adventitious Presence of Transgenes

Over the years, accidental crossing of genotypes has increased diversity and probably led to many developed, improved cultivars. Such accidental crosses have sometimes provided unexpected combinations of traits, but were of little concern to the cotton breeding community. The issue of accidental crosses and subsequent adventitious presence of protected transgenes in conventional lines has now become a major concern.

Maintaining pure conventional lines requires careful attention from the initial cross through seed increases of selected lines. To ensure absence of transgenes in F_1 seed, parental plants must be tested for presence of transgenes. This can be done by testing leaf discs from individual plants for the presence of transgenes using immunostrips. Crosses are then only made between plants that do not test positive for transgenes. Alternatively, one seed from each F_1 boll may be sampled and tested for transgenes. This can be done by removing the seed coat, crushing the embryo in an aliquot of buffer, then inserting a test strip. If the one seed is contaminated, the entire boll should be discarded. If the one seed is free of transgenes, the other seeds in that boll are also likely to be free.

The F_1 and subsequent generations should be adequately separated from any transgenic cotton. The space required depends upon the quantity of bees in the area. Since cotton pollen is not effectively carried by wind, natural outcrossing levels are primarily dependent on pollinator (bee) activity (McGregor, 1976; Van Deynze et al., 2005). In an area with low bee population and natural outcrossing of less than 2% (similar to an area described by Meredith and Bridge, 1973), Bourland (2004, 2013) establishes about a 450-m border of no cotton between his breeding nursery and transgenic cotton. In addition, a 15-m border planted to a morphologically distinct cotton genotype is maintained around the breeding nursery. Bees transferring pollen from a distantly planted transgenic cotton are most likely to first visit plants in the border area. Plants derived from outcrosses with the border plants are easily identified and may be rogued from subsequent generations. These procedures have lessened, but not eliminated, adventitious presence of transgenes from this program.

Selection Criteria Used in Conventional Cotton Breeding

Conventional cotton breeding does not have to confirm the presence and activity of specific transgenes (but sometimes must confirm the absence of transgenes). Otherwise, selection criteria used in conventional and transgenic cotton breeding programs are similar. Primary emphasis is always given to developing lines that produce high, stable yields and high fiber quality. Priorities on how lines produce high yields differ among programs, depending on prevalent pest and environmental conditions. Specific morphological traits may be used to enhance pest resistance or to improve fiber quality. Early maturity and plant structure may be emphasized in programs that address areas where the growing season may be shortened by weather or certain pests.

Selecting for Lint Yield and Yield Components

Most of the early progress is cotton breeding was accomplished by visual selection for yield. A successful cotton breeder was often said to have "a good eye for cotton." Visual ratings for yield are still sometimes used to advance lines, particularly in early generations. Using data from cotton breeders in four states, Bowman et al. (2004) found correlations between visual ratings and actual yields ranged from −0.22 to 0.70. Breeders differed in their ability to visually select high yielding genotypes. Three individuals with cotton breeding experience (ranging from >30 yr to <1 yr) visually rated progeny rows in Arkansas. Correlations with harvested yield increased slightly with years of experience. Considering only visual ratings, all three individuals would have discarded the highest yielding progeny row in the test.

Breeding progress based upon visual ratings of yield may be limited to the cosmetic appearances of plants and bolls. Plant structure may mask the yield performance of a line. As plant size increases, the ability to visually estimate yield becomes increasing difficult. Similarly, boll conformation (size and degree of openness of boll) and orientation within the plant canopy can bias visual ratings. Selection based on harvested yields removes those biases.

Harvested yield data have long been used to evaluate and select advanced lines in breeding programs. To genetically improve yields, cotton breeders evaluate yield of lines over multiple locations and years. However, breeding progress based on yield is hindered by high genotype × environment interactions. Modifying the basic components of cotton yield may provide a window to effectively select for improved yield efficiency. Most crop models define yield as the product of some determination of "number of seed per area" times the "weight per seed." For cotton yield, the "weight of fiber per seed" must also be considered. Lewis et al. (2000) modeled cotton yield simply as the "number of seed produced per area" multiplied by the "weight of fiber per seed." Obviously, a high number of seed per area is required for high yields, but heavy reliance on producing yield by increasing this component can lead to less stable yields. Reliance on seed production for yield leads to less stable yields because seed production requires more weight (seed makes up about 60% of seed cotton by weight) and energy (oil compared with cellulose production) than does lint production. Slight changes in the partitioning can result in significant lint yield increases. An increase of only 5 mg of fiber per seed produces about 95 lb ac^{-1} (84 kg ha^{-1}) in lint yield. Furthermore, the "weight of fiber per seed" component eliminates many of the possible interactions that normally affect yield (Groves, 2009).

In addition to the ginning and fiber quality measurements normally available in cotton testing, the only other parameter needed to calculate these yield components is seed index (weight of 100 fuzzy seed). Seed surface area (SSA) can then be estimated by the regression equation suggested by Groves and Bourland (2010): SSA = 35.94 + 6.59SI, where SI is equal to seed index associated with the sample. "Weight of fiber per seed" and its related parameters "number of fibers per seed" and "fiber density" (number of fibers per unit area of seed coat) were much less affected by environment than were lint yield or number of seed per area (Groves, 2009). Selection for high "weight of fiber per seed" alone results in lines with higher lint per seed, but also larger seed. In contrast, selection for high lint percentage will result in lines with higher gin turnout, but also smaller seed. Selecting for high fiber density may be an effective means to increase both yield and yield stability. In Australia, selecting for high fiber density has been an effective way to decrease fineness and micronaire while maintaining yield (Clement et al., 2014).

Selecting for Plant Growth Types

Niles and Feaster (1984) separated Upland cotton cultivars in the United States into four distinctive types—Eastern, Delta, Plains, and Acala types—and suggested different selection criteria for each type. The types formed the basis for separation of regions in the National Cotton Variety Test (Zeng et al., 2014), and specific types were accepted to be best adapted to specific regions. Types were generally based on cultivar background, but also on plant conformation.

Plant conformation (or shape) was a major consideration in cotton breeding programs and was often associated with particular brands of cottonseed. Bourland (1974) evaluated four specific plant conformations: Delta, inverted Delta, diamond, and stovepipe. Variation in these conformations was primarily related to dominance of monopodial (vegetative) branches and length of internodes on sympodial (fruiting) branches. Except for how they relate to maturity, plant conformation is given little attention. Lessened emphasis on plant growth types has likely been associated with increased reliance on harvested yield rather than visual ratings (Bowman et al., 2004). Also, intercrossing of growth types have generated more intermediate types and have masked many of the distinct types.

Selecting for Early Crop Maturity

Early maturing, short-season cottons continue to be a breeding priority in some conventional breeding programs. Because of its indeterminate growth habit and compensation ability, accurate and easy measurement of maturity in cotton is elusive. Prior to the widespread use of ethephon as a boll opener, cotton was usually mechanically picked two times, and maturity was often measured as the percentage of total harvest collected in the first harvest (Richmond and Radwan, 1962). Use of this earliness measurement was limited because it could not be standardized over years or sites, did not express the entire maturity period of the crop, and was not expressed in days. Mean maturity date was found to overcome these drawbacks, but required additional harvests to satisfy accuracy requirements (Richmond and Ray, 1966; Bilbro and Quisenberry, 1973).

Determination of the proportion of open bolls either by visual ratings or by counts of green and open bolls can provide an estimation of maturity (Bourland, 2004). These counts are usually sequenced with the applications of defoliants, so that over 60% of bolls are open. Like percentage in the first harvest, these measures provide a comparison of lines in a particular test, but do not provide a time-based measurement of maturity. Both counts and visual ratings of open bolls depend upon subjective determination of whether specific green bolls are harvestable. Counts of bolls are further restricted by relative small sample sizes and the difficulty of accurately counting bolls in a confined space. Visual ratings may include evaluation of whole plots (large sample size), but are highly subjective. By concentrating on the position of open bolls relative to the height of the plant, trained individuals can effectively estimate maturity by visual ratings of open bolls. Work is underway to develop a method to estimate percentage of open bolls using electronic sensors.

Maturity in cotton may also be determined by counts of nodes above white flower (NAWF) (Bourland et al., 1992; Kerby et al., 2010). Nodes above white flower measures the maturity of plants by the progress of first-position white flowers to the plant terminal and is an accumulative measure of boll load in relation to vegetative growth of the plant. Waddle (1974) was perhaps the first to use this parameter. Among several mid-summer measurements of earliness, he noted that "number of nodes above the last white bloom" was the "quickest and easiest count to make." With only one seasonal count of NAWF (during the third or fourth week of flowering), he determined that it was a good earliness indicator and that it had a degree of genetic control. Bourland et al. (2001) determined the number of days to physiological cutout (NAWF = 5.0) in several cultivar and strain tests by regressing NAWF counts over multiple dates by the number of days from planting. Days to NAWF = 5.0 provides a time-based measure of maturity that has a physiological base (Oosterhuis et al., 1993). The major drawback for determining maturity by days to NAWF = 5.0 is that much labor is required to obtain multiple counts of NAWF for the many entries in breeding and cultivar tests.

Nodes above cracked boll (NACB) is a late-season tool for determining maturity, which like NAWF, monitors the progress of fruit maturity up the plant (Kerby et al., 1992, 2010). Nodes above cracked boll has primarily been used to time crop defoliation. Fields with NACB of 4.0 or less can be defoliated without yield or fiber quality loss. The measurement may also be used to assess maturity variation of genetic lines by a one-time measurement of NACB or by determining the number of days to NACB = 4.0. This later measurement requires multiple counts of NACB.

Plant height measured after physiological cutout may also be used as an indicator of relative maturity of lines. If the plant growth regulator mepiquat chloride is used in a test, it should be applied early (prior to first flower) and at low dosages so that the genetic expression of height is more fully expressed. Plant height does not provide a temporal measure of maturity and is not always related to the progressive maturation of the plant.

Selecting for Pest Resistance

A primary means to enhance yield stability is to improve resistance to insects, diseases, and environmental stress. By decreasing the effects of these factors, optimum yields for specific soil and environmental conditions can be realized. Transgenes have been and are being developed to combat some pests, but conventional plant host resistance is the only available genetic option for combating most of these pests.

Selecting for Increased Resistance to Insects

Because of their sheer numbers and diversity, insect pests are dynamic and will often shift with different cropping systems, control tactics, and crop management strategies. Eliminating one pest species or neutralizing it with insecticides often leads to enhanced damage by another pest species. Recent changes in cotton insect control in the United States illustrate this scenario. By the autumn of 2009, boll weevils were practically eradicated in all US cotton regions with the exception of some areas in Texas. With boll weevils eradicated and Heliothine species (worms) largely controlled with Bt cotton, insecticide applications were greatly reduced. Unfortunately, the elimination of insecticide applications frequently led to increased problems with plant bugs (*Lygus* spp.). Plant bug populations now have documented resistance to organophospate and pyrethroid insecticides. Neonicotinoid insecticides are applied as a seed treatment and as foliar treatments to combat plant bugs in cotton and are frequently used in other crops. With widespread and frequent use, plant bug resistance to this group of insecticides will likely evolve. Transgenes affecting partial resistance to plant bugs may soon be available.

Cotton cultivars that are resistant to plant bugs have long been sought. Nectariless (no glands on leaves or bolls that exude nectar, which attracts insects) cottons have been shown to be effective for lowering plant bug populations—a degree of partial resistance. Bourland et al. (2014) annually evaluate cultivars and breeding lines for resistance to tarnished plant bug (TPB) using a modification of the method developed by Maredia et al (1994). The lines are planted in short 1-row plots (replicated 8 to 12 times) in tests where populations of TPB are encouraged. A highly susceptible Frego bract line is planted (about 1 mo prior to planting tests) in four-row strips between tests and is included as a check in each test. After building up in the early fruiting Frego bract strips, TPB populations move to the tests as the Frego bract strips mature. When TPB damage is observed in the Frego bract checks, examination for discoloration of the anthers in white flowers is initiated in each test. In the absence of boll weevils, this damage (called "dirty flowers") is associated with plant bugs. About six flowers per plot are examined for about five sample dates (near consecutive days). Data are accumulated over sample dates to establish "% dirty flowers" for each plot.

Variation in "dirty flowers" among cotton lines tends to be similar over years, suggesting genetic control of this trait. Studebaker et al. (2014) found that TPB treatment thresholds were reached at a slower rate (thus reduced insect control costs) on lines characterized as resistant by our "dirty flowers" measurement. In addition, we have found "dirty flowers" of some nectaried lines to be as low as "dirty flowers" of nectariless lines. This suggests that sources of resistance to TPB, other than nectariless, are present. A combination of other sources with nectariless may further improve resistance to TPB.

Selecting for Increased Resistance to Diseases

Disease problems in cotton are often localized and generally cause less damage than insect pests. Cotton diseases that commonly evaluated in US cotton breeding programs include bacterial blight, Verticillium wilt, and nematodes.

Bacterial blight (also known as angular leaf spot) is not a major disease, but resistance to the disease is well-defined and easily established. Bird (1982) indicated that resistance to bacterial blight is based on multiple genes and may be related to some degree of resistance to other pests. If a cross includes at least one resistant parent, plants in subsequent generations in the breeding nursery and seed increases may be screened for resistance. Inoculation is usually accomplished using with multiple races (always including race 18) of *Xanthomonas axonopodis* pv. *malvacearum* (Smith) Dye, the causal agent of bacterial blight, using the field inoculation procedures described by Bird and Blank (1951). Plants resistant to race 18 are resistant to all known races occurring in the United States. Addition of an organosilicone surfactant (Agrisolutions Silkin) greatly improves the effectiveness of the inoculum. Plants (about four-leaf stage) are inoculated early in the morning (stomata open) with nozzles pointed upward to the abaxil side of leaves using about 50 psi pressure. Symptoms are evaluated about 2 to 3 wk after inoculation. Susceptible plants may be removed from the early-generation populations and seed increases. Progeny rows having susceptible plants are normally discarded.

Verticillium wilt tends to be localized in areas with specific soils and certain temperature regimes during boll development. Since premature boll opening is one of the symptoms of Verticillium wilt, visual rating of disease symptoms is often confounded with variation in plant maturity. Incidence and severity of the disease may vary greatly from year to year at test sites that are naturally infested with the Verticillium wilt pathogen. Cotton lines that produce high yields over years at a location that has high natural infestation of Verticillium wilt can be assumed to be expressing tolerance to the disease. Severity of the disease may be enhanced by low plant populations, high nitrogen fertility, frequent irrigations, and relatively cool temperatures during boll fill. Because of cool temperatures of irrigation water, disease symptoms are typically most severe in the area nearest to the irrigation pipe.

The most widespread occurring nematode problems on cotton are the root-knot and reniform nematodes. Columbian, lance, and lesion nematode are important in some areas. Genetic resistance to root-knot nematode is presently available, but it has been difficult to combine resistance with good agronomic performance. Marker DNA technology may facilitate this.

Selecting for Increased Stress Tolerance

Lack of water and excessive heat are frequently the most limiting factors in cotton production. When supplemental irrigation is not available, the effects of drought may be lessened by improved rooting systems. Because of the difficulty in observing roots, research on rooting systems is limited. About 30 yr ago, we were able to increase the number of lateral roots on cotton seedlings by selection (Bourland and White, 1984; Bourland et al., 1985). However, simply increasing the number of seedling roots may not necessarily be related to final root mass or effectiveness of the roots. Factors other than root mass determine the plant's ability to absorb water and nutrients. To make progress in this area, a better understanding of the physiology of roots is needed to define the optimum rooting system.

Excessive heat often accompanies drought. In humid areas, high night temperatures are particularly harmful. The ability to tolerate high day temperatures in arid regions may not necessarily be related to the ability to tolerate high night temperatures in humid regions. A simple, quick method to accurately quantify heat stress is needed. Methods such as membrane leakage and fluorescence before and after heat stress, as well as examining certain pre-stress antioxidant levels, have been identified for quantifying the effect of heat stress and measuring innate stress tolerance (Bibi et al., 2008; Snider et al., 2010). The latter is particularly exciting because it can be done relatively quickly in the field. Such a measurement may provide us a method to genetically improve heat tolerance.

Selecting for High Fiber quality

Although the first priority for evaluating a cultivar must be its ability to produce competitive high yields, high yields have less value if the cotton is difficult to market. Marketing of low quality cotton may be subject to high price discounts and delayed cash flow. Adverse environmental conditions will always provide an ample supply of low quality cotton. Historically, marketing of low quality cotton has often been supported by governmental marketing loans. Restructuring or elimination of these loan programs will further weaken marketing opportunities for low quality fibers. Consequently, the development of cotton cultivars possessing enhanced cotton fiber quality is essential for sustaining long-term cotton production in any region. Cultivars possessing a genetic capacity for higher fiber quality can build and sustain greater marketability and price. Even with harsh growing conditions, such cultivars will maintain a better quality than cultivars without a high capacity for quality.

Assessing fiber quality rapidly, precisely, and inexpensively has allowed for significant improvement in cotton fiber quality. Kuraparthy and Bowman (2013) analyzed fiber data from the US cotton crop from 1980 to 2010 and found that fiber strength increased by 25%, fiber length (upper half mean length) by 4%, and micronaire by 5%. Key to these improvements was the utilization of HVI fiber measurements by cotton breeding programs (Kelly et al., 2012). While more expensive per sample and with a slower processing speed, the AFIS has been investigated as another tool for improving cotton fiber quality. Kelly et al. (2012) found few fiber quality differences in a selection experiment comparing HVI and AFIS selection tools.

Selecting for High Fiber Quality Using Q-score

Ideally, breeders would like to have one parameter to characterize fiber quality. To address this desire, Bourland et al. (2010) developed "Q-score," a simple numerical index based on up to six HVI fiber parameters. Fiber properties and their relative contributions to Q-score calculations initially used in Q-score included fiber length (50%), micronaire (25%), fiber length uniformity (15%), and fiber strength (10%). These weights were based upon perceived demands of the current cotton market and were particularly weighted in favor of fibers desirable for ring-spinning technology. Users of Q-score may change the relative weights of these four HVI parameters and add weights for elongation and short fiber content.

Q-score can be effectively used throughout the breeding process beginning with evaluation of nonreplicated data from IPS and progenies and ending with replicated data used in the release of a genotype. A primary benefit of Q-score with regard to IPS and progeny is the reduced time and effort required to make selections. Breeders typically make hundreds or thousands of IPS each year. Discarding IPS on the basis of relative values for multiple fiber property traits is a daunting task. Without the use of Q-score, a breeder will usually examine each fiber parameter value for an IPS and determine whether each value is within some arbitrary tolerance limit. Frequently, the breeder must then mentally assign weights to the different parameters to determine which IPS to discard. Sorting the data by Q-score facilitates rapid discard of lower quality lines and recognition of high quality lines. Once Q-score is calculated in a spreadsheet, the IPS can be sorted by their relative Q-score. Lines having low Q-score values can be quickly and painlessly discarded.

The HVI parameters of elongation and short fiber content may be employed in Q-score calculations, but have received little attention. In most tests, genotypes display significant variation for elongation. Since standard values have not been established, elongation values often vary greatly among years and among testing laboratories. Without established premium and discount values for elongation, its use in Q-score is limited. In contrast to the measure of elongation, genotypes seldom display meaningful variation for short fiber content. This may be related to the small laboratory gins employed by breeders. Without meaningful differences in a trait, breeders can make little progress.

Since Q-score does not include consideration of trash or color, it does not completely define fiber quality. Meaningful measures of trash and color are typically not available from small samples that are taken and processed by breeders. Color is primarily affected by field conditions after boll opening and prior to harvest, and by conditions during storage and ginning. Since color has little genetic basis, breeders have little opportunity to improve color grades of cotton.

Selecting for Reduced Leaf Trash

By reducing plant hairiness, breeders can effectively reduce trash content in ginned fiber. Breeders may consider both leaf and bract pubescence ("hairiness") and give preference to breeding lines and cultivars having lower pubescence on leaves and bracts. Reduced pubescence on cotton leaves has been associated with improved seed cotton cleaning efficiency and low foreign matter levels in harvested lint, and thus higher leaf grades in ginned cotton (Novick et al., 1991). To assist with characterizing leaf pubescence, Bourland et al. (2003) developed a rating system that could be easily used to identify less pubescent genotypes. Using this rating system, six random plants from every progeny row and replicated plots are characterized for leaf pubescence by examining the abaxial side of a fully expanded main stem leaf (approximately five nodes from the plant apex). By making these ratings near the time of physiological cutout, work fatigue (due to bending over to examine younger plants) is lessened and sampling of partially senescent leaves is avoided. Since genotype × location interaction is not strong, these evaluations are performed at only one location. On the basis of these characterizations, off-type plants can be removed from seed increases. We also characterize stem pubescence using this rating system. After plots are defoliated and harvested, the uppermost expanded internodes of six random plants per plot are rated from 1 (glabrous) to 9 (pilose, very hairy).

Pubescence on cotton bracts has received relatively little attention. Bracts are modified leaves surrounding the flower buds and bolls of the cotton plant. Morey et al. (1976) found that bracts are a major contributor to "leaf trash" in harvested cotton. This seems reasonable since bracts are in closer proximity to the cotton fibers than are plant leaves, and most leaves are removed from the plant prior to harvest if defoliation is successful. By examining variation in marginal bract trichomes ("hairs") on different canopy positions and cultivars as well as over time and environments, sampling methods were established by Bourland and Hornbeck (2007). They found that glabrous leaf genotypes tended to have lower marginal bract trichome density than did hairy leaf genotypes, but there was some overlap of bract trichome density among glabrous and hairy leaf genotypes. Of all the Upland cotton genotypes that we have examined, none were found to have glabrous marginal bract surfaces. Hornbeck and Bourland (2007) found significant, but low magnitude ($r = 0.33–0.35$), correlations between trichome density on abaxial leaf and marginal bract surfaces. This suggests some degree of independence of the two traits. Preliminary results from a ginning study of contrasting cultivars appear to verify that lower marginal bract trichomes are related to lower trash in ginned cotton (Boykin et al., 2013). We determine bract trichome density for all replicated tests at one location. After physiological cutout, bracts from a mid-canopy, first-position boll are taken from six random plants per plot. The bracts are labeled and placed in a freezer. At our convenience, the bracts are removed from the freezer and thawed on wetted germination paper. Marginal trichomes are counted in two areas of each bract using a viewing telescope and expressed as number per centimeter. Selection preference is given to lines with low bract trichome density, regardless of their leaf pubescence rating.

Selecting for High Yarn Quality

In the end, how an individual cotton genotype's fiber spins into yarn and performs in fabric is the ultimate test. In the textile industry, yarn quality is a vital component, which determines the quality of fabric and clothes (Zhu and Ethridge, 1996). Attempts to relate fiber quality to yarn quality include the Q-score (detailed above) and various other calculations based upon fiber test results, for example, Spinning Consistency Index (Majumdar et al., 2004). Though HVI provides the overall properties of fiber, textile professionals are also interested in variability of individual fibers properties in a sample because uniform individual fiber properties enhance the spinning efficiency and control the quality of fabric products (Morton and Hearle, 2008). In 1990, the AFIS was developed to measure individual fiber properties and distribution of properties in a sample used for processing. In contrast to HVI, AFIS measures some 20 properties of individual fibers (Kelly et al., 2012). Although various fiber properties are determined by using HVI and AFIS, it is still challenging to give priority to a parameter or group of parameters to select the best fibers for industrial uses (Majumdar, 2010). Since various properties of cotton fibers largely influence the final quality of yarn, researchers have developed mathematical, regression, and other computation models using various properties of fibers obtained from HVI and AFIS technology to predict the yarn properties.

Recently, miniature-spinning equipment has been developed to evaluate cotton genotypes for their yarn quality and fabric potential. Interest in developing spinning tests suitable for breeding dates back to the 1920s (Landstreet et al., 1959). Recent efforts have been made, including those of van der Sluijs et al. (2009), using sample sizes of 170 g of lint. A methodology that is capable of handling breeding program sample sizes as small as 30 to 60 g and that has a much faster processing time has recently been reported (Delhom, 2014).

Overcoming Negative Association

Once a trait is defined and a selection method is established, progress can usually be made to improve that trait. However, improvement in one trait can result in a corresponding negative expression in another trait. The negative relationship between yield and fiber quality provides a good example. If improved yield and fiber quality were positively related, fiber quality would be improved as breeders have selected higher yielding cultivars. This obviously has not been the case. Strong negative associations have long been found between lint yield and many fiber traits (Al-Jibouri et al., 1958; Meredith and Bridge, 1971).

Using data sets from both Australia and the United States, Clement et al. (2012) showed that negative associations still exist between yield and fiber quality parameters. In each set of data, they found that fiber length and strength had significant negative associations with yield; fiber maturity had a positive association with yield, while associations of micronaire and fineness with yield were inconsistent. Progress toward weakening the strong negative associations appears to be occurring in the Australian cotton breeding program.

Poor relationships that are not genetically bound together can be broken, but considerable effort and focus is usually required. Historically, one of the strongest negative relationships was between yield and fiber strength. Culp et al. (1979) were successful in breaking the negative association between fiber strength and yield. Their findings and subsequent germplasm releases have led to improved fiber strength among high yielding cultivars and demonstrate that many negative relationships can be broken.

Placing a high priority on fiber quality traits in early generations is an approach that appears to help break the negative relationships of fiber quality and yield. As noted above, Q-score greatly facilitates the process of discarding IPS and progeny on the basis of fiber quality. Evaluation based on Q-score can be accomplished with little prejudice since other limited data are available and relatively little time and effort have been invested in the genotype. In addition, discarding individual plants prior to planting decreases the time and space required for field evaluation of progeny. Using high selection pressure for fiber quality in early generations ensures that only high fiber quality lines will be advanced in a breeding program. The goal then is to find the best yielding line among the selected high fiber quality lines.

The relative yield and fiber quality of 'UA48' documents the success of this approach. Over years, UA48 produced lint yields equal to two standard conventional cotton cultivars. Its fiber quality greatly exceeded either check cultivar. Moreover, UA48 matures earlier than either DP 393 or SG 105, both of which are considered to be early-maturing cultivars. This combination of early maturation, competitive yields, and exceptional fiber quality is unprecedented. Additional information on this cultivar is available in its registration publication (Bourland and Jones, 2012).

References

Al-Jibouri, H.A., P.A. Miller, and H.F. Robinson. 1958. Genotypic and environmental variances and covariances in an Upland cotton cross of interspecific origin. Agron. J. 50:633–636.

Andrade-Sanchez, P., M.A. Gore, J.T. Heun, K.R. Thorp, A.E. Carmo-Silva, A.N. French, M. Salvucci, and J.W. White. 2013. Development and evaluation of a field-based high-throughput phenotyping platform. Funct. Plant Biol. 41:68–79.

Anthony, W.S., and S.T. Rayburn. 1989. Cleanability of smooth- and hairy-leaf cottons—Quality effects. Trans. ASAE. 32:1127–1130.

Bibi, A.B., D.M. Oosterhuis, and E.D. Gonias. 2008. Photosynthesis, quantum yield of photosystem II and membrane leakage as affected by high temperatures in cotton genotypes J. Cotton Sci. 12:150–159.

Bilbro, J.D., and J.E. Quisenberry. 1973. A yield related measure of earliness for cotton, *Gossypium hirsutum* L. Crop Sci. 13:392–393.

Bird, L.S. 1982. The MAR (Multi-Adversity Resistance) system for genetic improvement of cotton. Plant Dis. 66:172–176.

Bird, L.S., and L.M. Blank. 1951. Breeding strains of cotton resistant to bacterial blight. Bull. 736. Texas Agric. Exp. Stn., College Station.

Blanche, S.B. and G.O. Myers. 2006. Identifying discriminating locations for cultivar selection in Louisiana. Crop Sci. 46:946–949.

Blanche, S.B., G.O. Myers, W.D. Caldwell, and T. Wallace. 2008. Determining selection gains and discriminating environments via GGE biplots. J. Crop Improv. 21:13–25.

Bourland, F.M. 1974. Growth type classification of cotton for genetic studies. MS thesis. Univ. of Arkansas, Fayetteville, AR.

Bourland, F.M. 2004. Overview of the University of Arkansas cotton breeding program. In: Proceedings of the Beltwide Cotton Conference, Natl. Cotton Counc. Am., San Antonio, TX. 5–9 Jan. 2004. Natl. Cotton Counc. Am., Memphis, TN. p. 1093–1097.

Bourland, F.M. 2013. Novel approaches used in the University of Arkansas cotton breeding program. In: Proceedings of the Beltwide Cotton Production Conference, Natl. Cotton Counc. Am., San Antonio, TX. 7–10 Jan. 2013. Natl. Cotton Counc. Am., Memphis, TN. p. 409–418.

Bourland, F.M., N.R. Benson, E.D. Vories, N.P. Tugwell, and D.M. Danforth. 2001. Measuring maturity of cotton using nodes above white flower. J. Cotton Sci. 5:1–8.

Bourland, F.M. D.S. Calhoun, and W.D. Caldwell. 1997. Cultivar evaluation of Bt cottons in the Mid-South. Spec. Rep. 183. Arkansas Agric. Exp. Stn., Fayetteville.

Bourland, F.M., R. Hogan, D.C. Jones, and E. Barnes. 2010. Development and utility of Q-score for characterizing cotton fiber quality. J. Cotton Sci. 14:53–63.

Bourland, F.M., and J.M. Hornbeck. 2007. Variation in marginal bract trichomes on Upland cotton. J. Cotton Sci. 11:242–251.

Bourland, F.M., J.M. Hornbeck, A.B. McFall, and S.D. Calhoun. 2003. A rating system for leaf pubescence of cotton. J. Cotton Sci. 7:8–15.

Bourland, F.M., and D.C. Jones. 2012. Registration of 'UA48' cotton cultivar. J. Plant Regist. 6:15–18.

Bourland, F.M., G.A. Mitchell, and B.W. White. 1985. Effects of selection for tolerance to trifluralin in cotton. In: Proceedings of the Beltwide Cotton Production Research Conference, Natl. Cotton Counc. Am., New Orleans, LA. 6–11 Jan. 1985. Natl. Cotton Counc. Am., Memphis, TN. p. 200–202.

Bourland, F.M., D.M. Oosterhuis, and N.P. Tugwell, 1992. Concept for monitoring the growth and development of cotton plants using main-stem node counts. J. Prod. Agric. 5:532–538.

Bourland, F., G. Studebaker, and T.G. Teague. 2014. Host plant resistance to tarnished plant bug in Arkansas: I. Variation among cotton genotypes in small plots. In: Proceedings of the Beltwide Cotton Production Research Conference, Natl. Cotton Counc. Am., New Orleans, LA. 6–8 Jan. 2014. Natl. Cotton Counc. Am., Memphis, TN. p. 668–678.

Bourland, F.M., and B.W. White. 1984. Techniques for selecting and evaluating tolerance to trifluralin in cotton. In: Proceedings of the Beltwide Cotton Production Research Conference, Natl. Cotton Counc. Am., Atlanta, GA. 8–12 Jan. 1984. Natl. Cotton Counc. Am., Memphis, TN. p. 99–102.

Bowman, D.T., F.M. Bourland, G.O. Myers, T.P. Wallace, and D. Caldwell. 2004. Visual selection for yield in cotton breeding programs. J. Cotton Sci. 8:62–68.

Boykin, J.C., F. Bourland, and D.M. Dobbs. 2013. Relationship of leaf and bract trichomes to trash content of ginned lint. In: Proceedings of the Beltwide Cotton Production

Research Conference, Natl. Cotton Counc. Am., San Antonio, TX. 7–10 Jan. 2013. Natl. Cotton Counc. Am., Memphis, TN. p. 774–782.

Bradley, J.R., Jr. 1996. Major developments in management of insect and mite pests in cotton. In: E.G. King, J.R. Phillips, and R.J. Coleman, editors, Cotton insects and mites: Characterization and management. The Cotton Foundation, Memphis, TN. p. 1–13.

Bragg, C.K., and F.M. Shofner. 1993. A rapid, direct measurement of short fiber content. Text. Res. J. 63:171–176.

Bridge, R.R. 1986. Registration of 'DES 119' cotton. Crop Sci. 26:646–647.

Bridge, R.R., and J.F. Chism. 1978a. Release of DES 24 cotton. Crop Sci. 18:523.

Bridge, R.R., and J.F. Chism. 1978b. Release of DES 56 cotton. Crop Sci. 18:524.

Bridge, R.R., and L.D. McDonald. 1987. Beltwide efforts and trends in development of varieties for short-season production systems. In: Proceedings of the Beltwide Cotton Conference, Natl. Cotton Counc. Am., Dallas, TX. 4–8 Jan. 1987. Natl. Cotton Counc. Am., Memphis, TN. p. 81–85.

Brown, K.D., M.E. Sorrells, and W.R. Coffman. 1983. A method for classification and evaluation of testing environments. Crop Sci. 23:889–894.

Bryant, K.J., R.L. Nichols, C.T. Allen, N.R. Benson, F.M. Bourland, L.D. Earnest, M.S. Kharboutli, K.L. Smith, and E.P. Webster. 2003. Transgenic cotton cultivars: An economic comparison in Arkansas. J. Cotton Sci. 7:194–204.

Burgueño, J., G. de los Campos, K. Weigel, and J. Crossa. 2012. Genomic prediction of breeding values when modeling genotype × environment interaction using pedigree and dense molecular markers. Crop Sci. 52:707–719.

Campbell, B.T., and M.A. Jones. 2005. Assessment of genotype × environment interactions for yield and fiber quality in cotton performance trials. Euphytica. 144:69–78.

Campbell, B.T., and G.O. Myers. 2015. Quantitative genetics. In: D. Fang and R. Percy, editors, Cotton. 2nd ed. Agron. Monogr. 57. ASA, CSSA, and SSSA, Madison, WI. doi:10.2134/agronmonogr57.2013.0024

Clement, J.D., G.A. Constable, and S.M. Liu. 2014. Increasing cotton seed fibre density as a breeding strategy to improve fibre fineness. Field Crops Res. 160:81–89.

Clement, J.D., G.A. Constable, W.N. Stiller, and S.M. Liu. 2012. Negative associations still exist between yield and fibre quality in cotton breeding programs in Australia and USA. Field Crops Res. 128:1–7.

Cooper, M., and G.L. Hammer. 1996. Plant adaptation and crop improvement. CAB International, Wallingford, UK.

Culp, T.W., D.C. Harrell, and T. Kerr. 1979. Some genetic implications in the transfer of high fiber strength genes to Upland cotton. Crop Sci. 19:481–484.

Delhom, C. 2014. Assessing cotton fiber quality from a tiny sample. Biotechnology. http://phys.org/news/2014-10-cotton-fiber-quality-tiny-sample.html (accessed 27 Oct. 2014).

Geng, S., B. Hu, and D.M. Basset. 1990. Quantification and classification of locational effects on cotton cultivar testing programs. Agron. J. 82:514–518.

Groves, F.E. 2009. Relationships of yield component variables to yield and fiber quality parameters in cotton (*Gossypium hirsutum* L.). Ph.D. diss. Univ. of Arkansas, Fayetteville, AR.

Groves, F.E., and F.M. Bourland. 2010. Estimating seed surface area of cottonseed. J. Cotton Sci. 14:74–81.

Heffner, E.L., M.E. Sorrells, and J-L. Jannink. 2009. Genomic selection for crop improvement. Crop Sci. 49:1–12.

Hornbeck, J.M., and F.M. Bourland. 2007. Visual ratings and relationships of trichomes on bracts, leaves, and stems of Upland cotton. J. Cotton Sci. 11:252–258.

Jenkins, J.N., and F.D. Wilson. 1996. Host plant resistance. In: E.G. King, J.R. Phillips, and R.J. Coleman, editors, Cotton insects and mites: Characterization and management. The Cotton Foundation, Memphis, TN. p. 563–597.

Kang, M.S., and H.G. Gauch. 1996. Genotype-by-environment interaction: New perspectives. CRC Press, Boca Raton, FL.

Kelly, C.M., E.F. Hequet, and J.K. Dever. 2012. Interpretation of AFIS and HVI fiber property measurements in breeding for cotton fiber quality improvement. J. Cotton Sci. 16:1–16.

Kerby, T.A., F.M. Bourland, and K.D. Hake. 2010. Physiological rationale in plant monitoring and mapping. In: J.M. Stewart, D.M. Oosterhuis, J.J. Heitholt, and J.R. Mauney, editors, Physiology of cotton. Springer, Dordrecht, the Netherlands. p. 304–317.

Kerby, T.A., J. Supak, J.C. Banks, and C. Snipes 1992. Timing defoliations using nodes above cracked boll. In: Proceedings of the Beltwide Cotton Conference, Natl. Cotton Counc. Am., Nashville, TN. 6–10 Jan. 1992. Natl. Cotton Counc. Am., Memphis, TN. p. 155–156.

Kuraparthy, V., and D.T. Bowman. 2013. Gains in breeding Upland cotton for fiber quality. J. Cotton Sci. 17:157–162.

Landstreet, C.B., P.R. Ewald, and T. Kerr. 1959. A miniature spinning test for cotton. Text. Res. J. 29:699–706.

Lewis, H., L. May, and F.M. Bourland. 2000. Cotton yield components and yield stability. In: Proceedings of the Beltwide Cotton Conference, Natl. Cotton Counc. Am., Dallas, TX. 4–8 Jan. 2000. Natl. Cotton Counc. Am., Memphis, TN. p. 532–537.

Lubbers, E. 2003. Evaluation of cotton breeding test environments in the Southeast United States. In: Proceedings of the Beltwide Cotton Conference, Natl. Cotton Counc. Am., Nashville, TN. 7–9 Jan. 2003. Natl. Cotton Counc. Am., Memphis, TN. p. 791–807.

Majumdar, A. 2010. Selection of raw materials in textile spinning industry using fuzzy multi-criteria decision making approach. Fibers Polym. 11:121–127.

Majumdar, A., P.K. Majumdar, and B. Sarkar. 2004 Selecting cotton bales by spinning consistency index and micronaire using artificial neural networks. AUTEX Res. J. 4:1–8.

Maredia, K.M., N.P. Tugwell, B A. Waddle, and F.M. Bourland. 1994. A simple and rapid technique for screening cotton germplasm for resistance to tarnished plant bug, *Lygus lineolaris* (Palisot de Beauvois). Southwest. Entomol. 19:63–70.

May, O.L., F.M. Bourland, and R.L. Nichols. 2003. Challenges in testing transgenic and nontransgenic cotton cultivars. Crop Sci. 43:1594–1601.

McGregor, S.E. 1976. Insect pollination of cultivated crop plants. Agric. Handb. 496. US Gov. Print. Office, Washington, DC.

Mei, H., X. Zhu, W. Guo, C. Cai, and T. Zhang. 2014. Exploitation of Chinese Upland cotton cultivar germplasm resources to mine favorable QTL alleles using association mapping. In: I.Y. Abdurakhmonov, editor, World cotton germplasm resources. Intech, Rijeka, Croatia. p. 55–85.

Meredith, W.R., Jr., D.L. Boykin, F.M. Bourland, W.D. Caldwell, B.T. Campbell, J.R. Gannaway, K. Glass, A.P. Jones, L. M. May, C.W. Smith, and J. Zhang. 2012. Genotype × environment interactions over seven years for yield, yield components, fiber quality, and gossypol traits in the Regional High Quality Tests. J. Cotton Sci. 16:160–169.

Meredith, W.R., Jr., and R.R. Bridge. 1971. Breakup of linkage blocks in cotton, *Gossypium hirsutum* L. Crop Sci. 11:695–698.

Meredith, W.R., Jr., and R.R. Bridge. 1973. Natural crossing in cotton (*Gossypium hirsutum* L.) in the delta of Mississippi. Crop Sci. 13:551–552.

Meuwissen, T.H.E., B.J. Hayes, and M.E. Goddard. 2001. Prediction of total genetic value using genome-wide dense marker maps. Genetics 157:1819–1829.

Morey, P.R., P.E. Sasser, R.M. Bethea, and M.T. Kopetzky. 1976. Variation in trash composition in raw cottons. Am. Ind. Hyg. Assoc. J. 37:407–412.

Morton, W.E., and J.W.S. Hearle. 2008. Physical properties of textile fibres. 4th ed. Woodhead Publishing, Cambridge, UK.

Niles, G.A., and C.V. Feaster. 1984. Breeding. In: R.J. Kohel and C.F. Lewis, editors, Cotton. ASA, CSSA, and SSA, Madison, WI. p. 201–231.

Novick, R.G., J.E. Jones, W.S. Anthony, W. Aguillard, and J.I. Dickson. 1991. Genetic trait effects on non-lint trash of cotton. Crop Sci. 31:1029–1034.

Oosterhuis, D.M., F.M. Bourland, and N.P. Tugwell. 1993. Physiological basis for the nodes-above-white-flower cotton monitoring system. In: Proceedings of the Beltwide Cotton Production Research Conference, Natl. Cotton Counc. Am., New Orleans, LA. 10–14 Jan. 1993. Natl. Cotton Counc. Am., Memphis, TN. p. 1181–1183.

Parvin, D.W., Jr., J.W. Smith, and F.T. Cooke, Jr. 1987. The economics of cotton harvesting in the midsouth as it relates to shorter season production systems. In: Proceedings of the Beltwide Cotton Production Conference, Natl. Cotton Counc. Am., Dallas, TX. 4–8 Jan. 1987. Natl. Cotton Counc. Am., Memphis, TN. p. 79–81.

Perkins, H.H., Jr., D.E. Ethridge, and C.K. Bragg. 1984. Fiber. In: R.J. Kohel and C.F. Lewis, editors, Cotton. ASA, CSSA, and SSA, Madison, WI. p. 437–510.

Rahman, M., Y. Zafar, and A.H. Paterson. 2009. *Gossypium* DNA markers: types, numbers, and uses. In: A.H. Paterson, editor, Genetics and genomics of cotton. Springer, New York. p. 101–139.

Ramey, H.H., Jr. 1999. Classing of fiber. In: C.W. Smith and J.T. Cochran, editors, Cotton: Origin, history, technology, production. John Wiley & Sons, New York. p. 709–727.

Rayburn, S.T., Jr., and L. Libous. 1983. Preliminary investigation of cleanability of cotton with varying degrees of plant hairiness. In: Proceedings of the Beltwide Cotton Production Research Conference, Natl. Cotton Counc. Am., San Antonio, TX. 2–6 Jan. 1983. Natl. Cotton Counc. Am., Memphis, TN. p. 152–153.

Richmond, T.R., and S.R.H. Radwan. 1962. Comparative study of seven methods of measuring earliness of crop maturity in cotton. Crop Sci. 2:397–400.

Richmond, T.R., and L.L. Ray. 1966. Product-quantity measures of earliness of crop maturity in cotton. Crop Sci. 6:235–239.

Snider, J.L., D.M. Oosterhuis, and E.M. Kawakami. 2010. Genotypic differences in thermotolerance are dependent upon prestress capacity for antioxidant protection of the photosynthetic apparatus in *Gossypium hirsutum*. Physiol. Plant. 138:268–277.

Studebaker, G., F. Bourland, and T.G. Teague. 2014. Host plant resistance to tarnished plant bug in Arkansas: II. Confirmation of small plot data. In: Proceedings of the Beltwide Cotton Production Research Conference, Natl. Cotton Counc. Am., New Orleans, LA. 6–8 Jan. 2014. Natl. Cotton Counc. Am., Memphis, TN. p. 679–684.

van der Sluijs, M., R. Long, and S. Gordon. 2009. An alternative miniature cotton spinning system. In: Proceedings of the Beltwide Cotton Conference, Natl. Cotton Counc. Am., San Antonio, TX. 5–8 Jan. 2009. Natl. Cotton Counc. Am., Memphis, TN. p. 1446–1452.

Van Deynze, A.E., F.J. Sundstrom, and K.J. Bradford. 2005. Pollen-mediated gene flow in California depends on pollinator activity. Crop Sci. 45:1565–1570.

Waddle, B.A. 1974. Using mid-summer measurement as earliness indicators. In: Proceedings of the Beltwide Cotton Production Research Conference, Natl. Cotton Counc. Am., Dallas, TX. 7–9 Jan. 1974. Natl. Cotton Counc. Am., Memphis, TN. p. 78–79.

Yan, W., L.A. Hung, Q. Sheng, and Z. Sxlavnics. 2000. Cultivar evaluation and mega-environment investigation based on the GGE biplot. Crop Sci. 40: 597–605.

Zeng, L., W.R. Meredith, Jr., B.T. Campbell, J.K. Dever, J. Zhang, K.M. Glass, A.S. Jones, G.O. Myers, and F.M. Bourland. 2014. Genotype-by-environment interaction effects on lint yield of cotton cultivars across major regions in the US cotton belt. J. Cotton Sci. 18:75–84.

Zhu, C., M. Gore, E.S. Buckler, and J. Yu. 2008. Status and prospects of association mapping in plants. Plant Genome. 1(1): 5–20.

Zhu, R., and M.D. Ethridge. 1996. The prediction of cotton yarn irregularity based on the 'AFIS' measurement. J. Text. Inst. 87:509–512.

Transgenic Cotton Breeding

Jinfa Zhang

Abstract

Natural variation exists within the genus *Gossypium* L. that affords continuous genetic improvement of cotton in lint yield, fiber quality, and resistance to biotic and abiotic stresses. However, genetic engineering has provided a complementary avenue to introduce novel or desirable genes and traits into cotton. Since 1996, several insect-resistant biotech traits conferred by various *Bacillus thuringiensis* (Bt) genes or their combinations and herbicide-resistant traits have been commercialized in the United States and major cotton-growing countries, accounting for 90 and 60% of the cotton acreages in the United States and the world, respectively. The sources of transgenes used in commercialization are predominantly from bacteria. With an understanding of the biochemical and molecular basis of plant growth and development based on genomic research, genes from other plants or cotton can also be used in transformation. In this chapter, major sources of transgenes used to develop commercial genetically engineered (GE) cotton, selectable marker genes and their promoters, and transformation methods are first described. Methods used in assessments of transgenic cotton for regulatory approval are then discussed. These include genetic and molecular characterization and requirements for addressing issues related to agronomic performance, nutrient analysis, and environmental safety. Finally, breeding methods in developing GE cultivars from approved transgenic events are discussed in detail.

Cotton improvement through classical genetics and breeding targets high yields, better fiber and seed quality, resistance or tolerance to abiotic and biotic stresses, and adaption to production systems where cotton is grown. Conventional cotton breeding takes advantage of desirable alleles that are existent in the cultivated cotton of the same or a different *Gossypium* species. Intraspecific breeding within tetraploid Upland (*G. hirsutum* L.) or Pima cotton (*G. barbadense* L.) does not have any reproductive barriers between parental lines, while interspecific breeding between Upland and Pima cotton is faced with the hybrid breakdown problem in F_2 and advanced generations, even though heterotic

---

Abbreviations: 2,4-D, 2,4-dichlorophenoxyacetic acid ; Bt, Bacillus thuringiensis; DMA, dicamba monooxygenase; ELISA, enzyme-linked immunosorbent assay; EPSP, 5-enolpyruvylshikimate-3-phosphate; EPSPS, enolpyruvylshikimate 3-phosphate synthase; GE, genetically engineered; miRNA, microRNA; PAT, phosphinothricin acetyltransferase; PCR, polymerase chain reaction; PEP, phosphoenolpyruvate; QTL, quantitative trait locus; RNAi, RNA interference; TALEN, transcription activator-like effector nuclease; ZFN, zinc-finger nuclease.

Jinfa Zhang, Dep. of Plant and Environmental Sciences, New Mexico State University, MSC 3Q, Skeen Hall, 945 College Drive, Las Cruces, NM 88003 (jinzhang@nmsu.edu).

doi:10.2134/agronmonogr57.2013.0026

Cotton. 2nd. ed. David D. Fang and Richard G. Percy, editors. Agronomy Monograph 57.

fertile F_1 hybrids are produced (Zhang et al., 2014b). Introducing desirable genes from diploid species to Upland cotton has only limited success due to difficulties in producing F_1 hybrids, chromosome doubling, and recovery of the chromosome number through repeated backcrossing. Stewart (1994) described several avenues to overcome obstacles in interspecific cotton breeding. Since the rediscovery of Mendel's law of heredity, conventional cotton breeding has been making progress in steadily improving lint yield across the US Cotton Belt (Meredith and Bridge, 1984; Zhang et al., 2005). Fiber quality and resistance to certain diseases and insects have also been improved in some commercial cotton cultivars. However, desirable traits from other *Gossypium* species are seldom carried in commercial cotton except for genes for bacterial blight resistance and nectarilessness.

Although improvement of lint yield remains the top priority in cotton breeding, weed control in cotton is more difficult than in other row crops such as maize (*Zea mays* L.), and cotton yield losses due to pests, including insects, are severe. Although genetic traits within cotton can be used to improve insect resistance, farmers still have to rely heavily on multiple applications of insecticides and herbicides to control major insects and weeds. This has created a demand for novel traits that can be introduced from other species to cotton for effective pest control through genetic engineering. In 1996, the first GE herbicide (bromoxynil)–resistant BXN cotton was commercially planted in the United States, but was short lived because of various reasons including its limited weed control spectrum. In the same year, Monsanto commercially released the first GE insect-resistant Bollgard (BG or B) cotton carrying the Bt gene expressing the Cry1Ac endotoxin of Bt in the United States. This was followed by the commercialization of Roundup Ready (RR or R) cotton expressing the CP4-EPSPS protein of *Agrobacterium tumefaciens* that was herbicide (glyphosate)–resistant by the same company in the following year. Transgenic cultivars combining both Bt and RR (a stacked trait BR) were soon introduced to the market. These two biotech traits were later gradually replaced by their improved version, when Monsanto commercially launched GE cotton with Bollgard II (B2 or BGII) carrying two Bt genes (*cry1Ac* and *cry2Ab*) in 2004 and Roundup Ready Flex (RF) cotton with high levels of expression of the *cp4-epsps* gene in 2006. Also in 2004, LibertyLink (LL) GE cotton cultivars, resistant to glufosinate herbicide, were commercially available in the United States through Bayer CropScience. In 2005, WideStrike cotton carrying two gene constructs (with *cry1Ac* and *cry1F*) encoding the endotoxin of Bt was commercialized by Dow AgroSciences. The above commercialized GE traits and other new biotech traits were announced in the annual Beltwide Cotton Conferences, which can be found in their proceedings at www.cotton.org.

Since the first GE cotton was commercialized, transgenic cotton production has grown exponentially in the United States and has also been rapidly adopted by at least 15 other countries, including major cotton-producing ones such as China, India, Pakistan, Australia, and Brazil (James, 2013). Other countries include Argentina, Paraguay, Columbia, Costa Rica, Mexico, Greece, Burkina Faso, Sudan, South Africa, and Myanmar. Since 2010, transgenic cotton has reached more than 90% cotton acreage in the United States (USDA-ERS, 2014) and 60% in the world (Table 1). The other three major biotech crops are maize, soybean [*Glycine max* (L.) Merr.], and canola (*Brassica napus* L.). (Table 2). In 2013, biotech cultivars of these crops accounted for 79, 32, and 22% of the respective crop acreages in the world (James, 2013). The economic, ecological, and social benefits of growing biotech

Table 1. Percentage of cotton acreages accounted by genetically engineered cotton with Bt traits only, herbicide tolerant (HT) traits only, or stacked with both traits in the United States and the world, 1996–2014.

Year	USA†				World‡
	Bt only	HT only	Stacked	Total	
	%				
1996	15	13		28	
1997	15	15		30	
1998	17	19		36	
1999	34	29		63	
2000	15	26	20	61	16
2001	13	32	24	69	20
2002	13	36	22	71	20
2003	14	32	27	73	21
2004	16	30	30	76	28
2005	18	27	34	79	28
2006	18	26	39	83	38
2007	17	28	42	87	43
2008	18	23	45	86	46
2009	17	23	48	88	49
2010	15	20	58	93	64
2011	17	15	58	90	82
2012	14	17	63	94	81
2013	8	15	67	90	70
2014	5	12	79	96	na

† USDA-ERS.
‡ James (2013).

Table 2. Global planting area and major countries growing genetic engineered crops in 2013.†

Rank	Country	Area	Crops‡										
			1	2	3	4	5	6	7	8	9	10	11
		million ha											
1	USA	70.1	x	x	x	x	x	x	x	x			
2	Brazil	40.3	x	x	x								
3	Argentina	24.4	x	x	x								
4	India	11			x								
5	Canada	10.8	x	x			x						
6	China	4.2			x				x		x	x	x
7	Paraguay	3.6	x	x	x								
8	South Africa	2.9	x	x	x								
9	Pakistan	2.9			x								
10	Uruguay	1.5	x	x									
11	Bolivia	1		x									
12	Philippines	0.8	x										
13	Australia	0.6			x	x							
14	Burkina Faso	0.5			x								
15	Myanmar	0.3			x								
16	Spain	0.1	x										
17	Mexico	0.1		x	x								
18	Columbia	0.1	x		x								
19	Sudan	0.1			x								
20	Chile	<0.1	x	x		x							
21	Honduras	<0.1	x										
22	Portugal	<0.1	x										
23	Cuba	<0.1	x										
24	Czech Republic	<0.1	x										
25	Costa Rica	<0.1		x	x								
26	Romania	<0.1	x										
27	Slovakia	<0.1	x										

† Adopted from James (2013).

‡ 1, maize (*Zea mays* L.); 2, soybean [*Glycine max* (L.) Merr.]; 3, cotton (*Gossypium* L.); 4, canola (*Brassica napus* L.); 5, sugarbeet (*Beta vulgaris* L. subsp. *vulgaris*); 6, alfalfa (*Medicago sativa* subsp. *sativa*); 7, papaya (*Carica papaya* L.); 8, squash (*Cucurbita* L.); 9, poplar (*Populus* L.); 10, tomato (*Solanum lycopersicum* L.); 11, sweet pepper (*Capsicum annuum* L.).

crops including cotton have been well documented (Edge et al., 2001; Perlak et al., 2001; Vitale et al., 2008; Lemaux, 2008, 2009; James, 2013; Fernandez-Cornejo et al., 2014; Klümper and Qaim, 2014). These benefits include reduced use of pesticides and/or herbicides and increased crop productivity, among others. It is expected that GE or biotech cotton will continue to gain more acreage in world cotton production. A complete updated list of GE events and traits that have been approved for commercialization and planting can be found on the ISAAA website, http://www.isaaa.org/gmapprovaldatabase/crop/default.asp?CropID=7&Crop=Cotton (accessed 15 Apr. 2015). There are other biotechnology traits that are being tested in various stages including field tests; however, their commercialization has not been approved.

This chapter will briefly describes sources of transgenes that have been introduced and tested in cotton, techniques in introducing foreign genes into cotton, and breeding methods with transgenes.

Sources of Transgenes

Sources of genes for transgenic cotton breeding can be isolated or modified from the same or other *Gossypium* species, or any other living organisms including viruses, bacteria, and other plant species. Here, the focus is only on those genes that have been commercially available or have gone through extensive field tests.

Insect Resistance

A Bt gene is carried in a naturally occurring soil bacterium, *B. thuringiensis* var. *kurstaki*, and encodes the protein called crystal (Cry) δ-endotoxin produced during the sporulation phase. The endotoxin has been used as an organic pesticide to control several lepidopterous pests, since it binds to a receptor site located on the stomach lining, causing cell lysis and death of these insects (Federici et al., 2006). The crystal protein is highly insoluble below pH 8.0, making it safe for humans, animals, and most insects. However, it is highly soluble at a pH of 8.0 or above in specific lepidopteran species (Federici et al., 2006). There are many different strains of Bt bacteria, each of which affects a specific group of insects. In contrast to many chemical insecticides, the Bt toxin does not directly affect many natural enemies of insects. Transgenic cottons carrying the Bt gene encoding *cry1Ac* alone (called Bollgard) or in combination with *cry2Ab* (called Bollgard II) or *cry1F* (called WideStrike) have been transferred into commercially grown cotton. The first transgenic Bt cotton (Bollgard) carrying *cry1Ac* has achieved excellent control of tobacco budworm, *Heliothis virescens* F., American bollworm, *Helicoverpa armigera* (Hubner), and pink bollworms, *Pectinophora gossypiella* (Saund.), but it is less effective in controlling bollworm, *Helicoverpa zea* (Boddie), and various other lepidopterous pests. The second generation of Bt cotton (Bollgard II or WideStrike) provides more effective control on these cotton pests.

As it grows, Bt also secretes another type of protein, that is, vegetative insecticidal protein (VIP), classified as an exotoxin. VIP is structurally, functionally, and biochemically different from Bt endotoxins. Syngenta obtained a final approval of its VipCot trait carrying *Vip3* and *cry1Ab* genes for commercialization in 2011 (Syngenta, 2003). Monsanto has developed a three-gene cotton, Bollgard III combining BGII (*cry1Ac* and *cry2Ab*) and *Vip3A*, and it is expected to be commercially available soon.

Herbicide Tolerance

Bromoxynil (3,5-dibromo-4-hydroxybenzonitrile) is a nitrile herbicide used for post-emergence control of annual broadleaf weeds including morning glory, cocklebur, and velvetleaf. All but 11 bp of the 5′ and 96 bp of the 3′ untranslated region from the BXN gene of 1.2 kb (encoding a bromoxynil-degrading nitrilase), which was isolated from a *Klebsiella* species obtained in a soil sample, was used for cotton transformation using the *kanr* gene coding for the enzyme aminoglycoside 3′-phosphotransferase II as the selective marker. The nitrilase enzyme, which is composed of two identical subunits (a 37-kDa polypeptide each), confers resistance to bromoxynil in GE cotton carrying the BXN gene, as it catalyzes the conversion of bromoxynil to the nonphytotoxic 3,5-dibromo-4-hydroxybenzoic acid (Calgene, 1994).

Glyphosate is a nonselective herbicide and widely used to control weeds in crop production and landscape systems. It is the active ingredient in one of the most popular household products, Roundup, and other generic brands. Since the glyphosate molecule mimics phosphoenolpyruvate (PEP), it specifically binds and therefore blocks the activity of the enzyme called 5-enolpyruvylshikimate 3-phosphate synthase (EPSPS) catalyzing the conversion of shikimate-3-phosphate (SP3) and PEP to 5-enolpyruvylshikimate-3-phosphate (EPSP). The EPSP is involved in the shikimate pathway, critical for many plant functions including biosynthesis of aromatic amino acids such as phenylalanine, tryptophan, and tyrosine. Therefore, glyphosate stops EPSP production, leading to reduction of the essential amino acids and protein production, which in turn causes plant growth shutdown and eventually leads to plant death (Franz et al., 1997). An EPSPS enzyme variant, CP4-EPSPS, identified in the Agrobacterium strain CP4, binds glyphosate less efficiently. The *cp4-epsps* gene was isolated and inserted into the cotton genome to express the CP4-EPSPS protein, which has low levels of affinity to the glyphosate molecule, thus rendering glyphosate resistance in GE cotton. However, because of the weak expression of the *cp4-epsps* gene (driven by the cauliflower mosaic virus CaMV 35S promoter) in male flower tissues, glyphosate can only be applied over the top of cotton from seedling emergence to the four–true leaf stage (Pline et al., 2003). Over-the-top application of glyphosate beyond the growth stage will cause abnormality of reproductive growth including square and boll shedding and misshapen bolls in lower fruiting branches, and yield loss. To overcome this limitation, a construct with two *cp4-epsps* gene cassettes driven by stronger promoters was introduced to cotton, and the resulting Roundup Ready Flex (RF) cotton can tolerate glyphosate in both vegetative and reproductive organs during the entire cotton growth cycle, providing farmers with more options in applying the herbicide (Monsanto, 2004).

The plant *epsps* gene can also be engineered to reduce its binding affinity with glyphosate and then introduced back to plants. For example, Bayer CropScience (2006) introduced two mutations into the wild-type maize *epsps* gene, and the resulting *2mepsps* gene encodes a double mutant EPSPS protein with two amino acid substitutions (2mEPSPS), which showed a decreased binding affinity for glyphosate. This allowed the GE cotton GlyTol to maintain a sufficient enzymatic activity in the presence of glyphosate, conferring tolerance to the herbicide.

Glufosinate is a nonselective herbicide, as it inhibits glutamine synthetase, an enzyme important for the conversion of glutamate and ammonia to the amino acid, glutamine. This results in an accumulation of toxic ammonia, which causes membrane disruption, inhibition of photosynthesis, chloroplast disruption, and eventual plant death. Glufosinate can be effective in controlling glyphosate-resistant weeds such as Palmer amaranth (*Amaranthus palmeri* S.Wats.). Therefore, it is a good alternative to a glyphosate-based weed management system in areas where glyphosate-resistant weeds are a problem. A *bar* gene, isolated from *Streptomyces hygroscopicus* (Jensen) Waksman & Henrici, encodes the phosphinothricin acetyltransferase (PAT) enzyme that acetylates L-phosphinothricin and confers tolerance to glufosinate. Glufosinate-resistant LibertyLink (LL) cotton was developed through insertion of the *bar* gene to cotton (USDA-APHIS, 2003). However, LibertyLink cotton cultivars have not gained a wide adoption because of unavailability of the glyphosate-resistant trait in the same cultivars with desirable agronomic performance. It is interesting to note that WideStrike (W) cotton containing two Bt genes (*cry1Ac* and *cry1F*) was developed using a *pat* gene as a selectable gene marker for plant transformation. The *pat* gene encoding a homologous PAT enzyme was originally isolated from another common soil microorganism, *Streptomyces viridochromogenes*. Therefore, WideStrike cotton has lower levels of PAT activity and lower levels of tolerance to glufosinate than the LibertyLink cotton. Although commercial cotton cultivars carrying both WideStrike and RF traits (designated as WRF) are resistant to both glyphosate and glufosinate, glufosinate is not recommended to use on WideStrike cotton because crop injury may occur.

The herbicide dicamba (3,6-dichloro-2-methoxybenzoic acid) kills broadleaf weeds before and after their emergence and is an organochloride, a derivative of benzoic acid. At sufficient concentrations, dicamba accelerates plant growth at a faster rate than available nutrients can supply, resulting in plant death. Dicamba can be converted to 3,6-dichlorosalicylic acid (DCSA) that lacks herbicidal activity by the three-component enzyme dicamba O-demethylase isolated from the soil bacterium *Pseudomonas maltophilia* (strain DI-6). The three components include a monooxygenase, a reductase, and a ferredoxin, which serve as an electron transfer chain to shuttle electrons from reduced nicotinamide adenine dinucleotide (NADH) through the reductase to the ferredoxin and finally to the terminal component, the dicamba monooxygenase (DMO). The ferredoxin component of the enzyme resembles the ferredoxin found in chloroplasts. Monsanto has recently developed dicamba-resistant cotton by incorporating the gene coding for DMO into the cotton genome (Behrens et al., 2007), which targets the monooxygenase for expression in the chloroplasts with a steady stream of electrons from reduced ferredoxin generated by photosynthesis.

2,4-dichlorophenoxyacetic acid (2,4-D) mimics the plant hormone auxin and was the first synthetic herbicide to be commercially developed for controlling a wide spectrum of broadleaf weeds. Although 2,4-D remains one of the most widely used herbicides globally, only isolated cases of resistant weed species have been reported because of its complex, plant-specific mode of action. A transgenic cotton with 100-fold more tolerance to 2,4-D was obtained by introducing the *tfdA* gene from the bacterium *Alcaligenes eutrophus* (Bayley et al., 1992). The *tfdA* gene encodes the enzyme dioxygenase catalyzing the degradation of 2,4-D to the much less phytotoxic compound 2,4-dichlorophenol (2,4-DCP). Recently, Wright et al. (2010) isolated two genes encoding aryloxyalkanoate dioxygenase (ADD), that is, ADD-1 (*RdpA*) from *Sphingobium herbicidivorans* and AAD-12 (*SdpA*) from *Delftia acidovorans* that have 28 and 31% amino acid sequence identity to TfdA, respectively. Both AADs can effectively degrade 2,4-D in that AAD-1 cleaves the aryloxyphenoxypropionate family of grass-active herbicides, while AAD-12 acts on pyridyloxyacetate auxin herbicides. To improve production of the recombinant protein in plants, the codons of both AAD DNA sequences were optimized for expression in plants. The resulting synthetic gene sequences encode identical protein sequences, except for the addition of an alanine residue in the second position resulting from the creation of an *Nco*I site (CCATGG) across the ATG start codon to enable subsequent cloning operations. Dow AgroSciences is seeking deregulation of its GE 2,4-D–resistant (Enlist) cotton for commercialization.

The rapid evolution of herbicide-resistant weeds requires the use of herbicide-resistant cotton with multiple modes of actions in an integrated weed management system. For example, XtendFlex cotton, which was developed and marketed by Monsanto in 2015, contained the triple-stack trait, that is, glyphosate-tolerant Genuity Roundup Ready Flex stacked with dicamba and glufosinate tolerance (USDA-APHIS, 2014). This provides farmers with more choices and flexibilities in selecting pre- and post-emergence herbicides.

Other Genes of Interest

The recent advances in functional genomics including transcriptomics and proteomics have contributed to the discovery and characterization of numerous genes active during cotton growth and development. For example, numerous stress-responsive plant genes have been tested through transgenic technology by overexpression for abiotic stress tolerance (Allen, 2010). Many of these genes encode protection proteins involved in specific biochemical pathways, such as antioxidants and reactive oxygen species (ROS) scavenging enzymes including ascorbate peroxidase (APX), superoxide dismutase (SOD), catalase, glutathione reductase (GR), and metallothioneins; enzymes in biosynthetic pathways for production of osmoprotectants such as proline, glycine-betaine (choline monooxygenase, *betA* gene encoding choline dehydrogenase), polyamines (e.g., S-adenosyl methionine decarboxylase) and trehalose; dehydrins; late embryogenesis abundant (LEA) proteins; heat shock proteins such as GHSP26; water channel proteins; and membrane transporters such as AtNHX-1, AtAVP1 encoding a vacuolar pyrophosphatase, and AtSOS1. However, transgenic cotton plants transformed with the above single-action genes usually do not show sufficient levels of field abiotic stress tolerance to justify investment from a commercial breeding standpoint (Pasapula et al., 2011). However, Shen et al. (2015) recently showed that transgenic cotton co-overexpressing *AVP1* and *AtNHX1* had enhanced drought and salt tolerance, a higher photosynthetic rate, larger roots, and increased lint yield under field drought conditions than those with single-gene overexpression and the null wild-type plants. More recent efforts have also been focused on genes encoding regulatory proteins or RNA, such as transcription factors including AtABF3, AtCBF3/DREB1A, AtRAV1/2, AtABI5 GhWRKY17, and OsSNAC1; zinc finger proteins such as AtSAP5, GhDi19-1, and GhDi19-2; protein kinases including GhMPK17, GhCIPK6, GbRLK, and GhZFP1; and microRNAs (miRNAs). Other genes have been tested in cotton such as *ipt* encoding isopentenyltransferase, a rate limiting enzyme in cytokinin biosynthesis (Kuppu et al., 2013), *LOS5/*

ABA3 (*LOS5*) encoding a molybdenum co-factor and essential for activating aldehyde oxidase, which is involved in abscisic acid (ABA) biosynthesis, and *AtGF14* lambda encoding for a 14-3-3 protein. Many stress-responsive genes have been isolated from cotton and tested in other model plants such as *Arabidopsis* Heynh. and tobacco (*Nicotiana tabacum* L.).

There are several examples of improved host resistance to fungal pathogens created by the transgenic expression of plant defense molecules. These include proteins from several classes, for example, hevein-like, glucanases and chitinases, protease inhibitors, thionins, defensins, and genes involved in disease response pathways such as those encoding for Ser/Thr protein kinase, GhMKK1 and GhMKK2, mitogen-activated protein kinase (MAPK), WRKY gene GhWRKY39-1, ERF transcription factors (GbERF2), GhHbd1, GhDIR1, GhDIR2, and AtNPR1 (e.g., Gao et al., 2011; Kumar et al., 2013).

Cotton fibers, important natural raw materials for the textile industry, are trichomes elongated from epidermal cells of cotton ovules. To date, a number of genes critical for fiber development have been introduced into cotton through genetic engineering (e.g., Walford et al., 2011; Jiang et al., 2012; Xu et al., 2013). The genes include those coding for expansins, annexins, phytochrome B (PHYB), transcription factors such as GhMYB25 and GhMYB109, sucrose synthase (GhSusA1), sucrose-phosphate synthase (SPS), proline-rich proteins (GhPRP5), fasciclin-like arabinogalactan proteins (FLAs), and xyloglucan endotransglycosylases/hydrolases (XTH), and many other proteins. Specifically, transgenic cotton plants have been developed to overexpress genes responsible for plant hormone biosynthesis such as indole-3-acetic acids (IAA), gibberellins (GA), brassinosteroids (BR), and a novel peptide hormone. A notable example is the development of GE cotton with targeted expression of the IAA biosynthetic gene *iaaM*, which was driven by the promoter of the petunia (*Petunia hybrida*) MADS box gene floral binding protein 7 (FBP7). The IAA level in the epidermis of cotton ovules at the fiber initiation stage was increased by 29%, leading to the increase in the number of lint fibers (Zhang et al., 2011b). In a 4-yr field trial, the lint percentage of the transgenic cotton was consistently higher by 20% than its nontransgenic null control, resulting in a 15 to 24% increase in lint yield. Interestingly, fiber fineness of the transgenic plants was also improved.

Cottonseed, as a byproduct of cotton production, is mainly used for vegetable oil and feed for ruminant animals because of the presence of toxic gossypol. Seed free of gossypol can be used as food and feed for nonruminant animals. Although glandless cotton free of gossypol in both plant body and seed exists, its high susceptibility to insects prohibits its commercial production (Zhang et al., 2014a). Genetically engineered Upland cotton with glanded plant body and ultra-low gossypol cottonseed was developed using a gene silencing technology called RNA interference (RNAi) to knock down δ-cadinene synthase gene(s) involved in the biosynthetic pathway of gossypol (Sunilkumar et al., 2006). The glanded plant and glandless seed trait was stable under field conditions, and it had no deleterious effects on lint yield and fiber quality (Palle et al., 2013). In comparison with the nontransgenic parental plants, the GE plants had significantly higher (4–8%) oil content. In addition, genes involved in cottonseed oil production and biosynthesis of fatty acids can also be introduced to cotton to improve cottonseed oil content and quality.

Transformation Methods

Once the gene(s) of interest for GE is identified, proper vectors carrying suitable promoters and terminators for the gene(s) and selectable gene markers should be constructed. Next, high efficient transformation techniques are employed to introduce the gene construct cassette(s) into a cotton genotype. After the transformed plants are regenerated, molecular and genetic analysis and characterization of the transgenic plants and their progeny should be performed. In this section, only those vectors used to develop commercial GE traits and established transformation methods will be discussed.

Vectors

The expression of gene(s) of interest in transformation is regulated by promoters, enhancers, or other *cis*-acting regulatory elements. The CaMV 35S promoter is the most widely used promoter to construct a DNA cassette for the constitutive expression of the gene(s) of interest in higher plants including cotton (Odell et al., 1985). Another constitutive promoter, the *Agrobaterium tumefaciens* nopaline synthase gene (*nos*) drives the constitutive expression of selectable markers. Therefore, the traditional design of plant transformation vectors includes the CaMV 35S promoter, the *nos* gene, and a terminator sequence. The vector should also have multiple cloning sites in the expression cassettes, allowing easy cloning and expression of one or more genes of interest, choosing from various promoter and terminator sequences and selectable and reporter genes (Chung et al., 2005). For plant transformation, one can use a binary vector system that employs a disarmed Ti plasmid (helper plasmid) in combination with a binary plasmid carrying the T-DNA border sequences, transgenes, and selectable markers that is replication-competent in both *Escherichia coli* and *A. tumefaciens*. In the case of Roundup Ready cotton (Monsanto, 2002a), the coding sequences from three genes (*cp4-epsps*, *nptII*, and *aad*) with the CoMVb promoter region and CaMV 35S promoter were introduced into Coker 312 cotton using an *A. tumefaciens* binary transformation vector PV-GHGT07 that is 12,032 bp long. The *nptII* gene encodes a selectable marker enzyme, neomycin phosphotransferase II (NPTII), which was used to identify transformed cotton cells containing the CP4-EPSPS protein. The *nptII* coding sequence was driven by a CaMV 35S promoter and was followed by a *nos* 3′ region that directed polyadenylation of the mRNA. The transcriptional termination sequence was derived from the T-E9 DNA sequences of *Pisum sativum* L., containing the 3′ nontranslated region of the ribulose-1,5-biphosphate carboxylase (Rubisco) small subunit *E9* gene. The *aad* gene encodes the bacterial selectable marker enzyme 3′′(9)-O-aminoglycoside adenylyltransferase (AAD), which allowed for the selection of bacteria containing the plasmid on culture media containing spectinomycin or streptomycin. Since the *aad* gene was under the control of a bacterial promoter, the encoded protein was not expressed in the Roundup Ready cotton. Similarly, the *bar* gene encoding for PAT enzyme was also driven by the 35S promoter, when LibertyLink cotton was developed by Bayer CropScience through transformation of Coker 312 hypocotyls with the plasmid vector pGSV71 (USFDA, 2002).

To develop Bollgard cotton, a similar vector PV-GHBK04 of 11,407 bp carrying *cry1Ac*, *nptII*, and *aad* genes was used to transform Coker 312 cotton (Monsanto, 2002b). But unlike PV-GHBK07, this *cry1Ac* gene cassette contained an enhanced

CaMV e35S promoter and a 7S 3′ transcriptional termination sequence from soybean. To develop Bollgard II cotton carrying both *cry1Ac* and *cry2Ab* (Monsanto, 2003), Bollgard cotton cultivar DP 50B containing *Cry1Ac* was transformed with the 8.7-kb vector PV-GHBK11 containing two gene expression cassettes (one with *cry2Ab* and another with *uidA*) through a particle bombardment. The *uidA* gene encoding the β-D-glucuronidase (GUS) marker protein served to facilitate selection for transgenic cotton containing *cry2Ab*. Both *cry2Ab* and *uidA* were under the regulation of the enhanced CaMV 35S promoter (e35S) and 3′ untranslated region of the *nos* gene for mRNA polyadenylation. The e35S driving *cry2Ab* was also fused with the 5′ untranslated sequence for the petunia heat shock protein 70 (HSP70) and chloroplast transit peptide from the *A. thaliana* 5-enolpyruvyl shikimate-3-phosphate synthase gene (*cpt2*), which directed the protein to the chloroplast. Therefore, Bollgard II cotton was developed through two separate transgenic events.

As more plant promoters have been isolated and characterized, the use of plant promoters has become more important in plant biotechnology, because transgene(s) can be better regulated to achieve the intended agronomic performance. For example, in the case of Roundup Ready Flex (RF) trait (Monsanto, 2004), the hypocotyl tissue of Coker 312 was transformed with the binary plasmid PV-GHGT35 containing two tandem *cp4-epsps* gene expression cassettes by Agrobacterium-mediated transformation, and plants were regenerated using glyphosate as the selection pressure in tissue culture. The first *cp4-epsps* gene cassette had a chimeric promoter P-FMV/TSF1, containing the *A. thaliana tsf1* gene promoter, the encoding elongation factor EF-1 α, enhancer sequences from the figwort mosaic virus 35S promoter, and the leader and intron sequences from the *tsf1* gene of *A. thaliana*. The second *cp4-epsps* gene cassette contained another chimeric promoter P-35S/ACT8 from the *act8* gene of *A. thaliana* combined with enhancer sequences from the CaMV 35S promoter, and the leader and intron sequences from the *act8* gene of *A. thaliana*. The two cassettes were joined by a chloroplast transit peptide (*cpt2*) sequence, derived from the *A. thaliana epsps* gene. As with the vector PV-GHGT07 used for developing Roundup Ready cotton, the transcriptional termination and polyadenylation sequences were derived from the T-E9 DNA sequences of *P. sativum*, containing the 3′ nontranslated region of the Rubisco small subunit *E9* gene.

The GlyTol cotton was developed by transforming Coker 312 with the vector pTEM2 (derived from pGSC1700) containing the glyphosate-tolerant gene *2mepsps* and *aadA* and *nptI* genes (Bayer CropScience, 2006). The expression of the *2mepsps* gene was constitutively regulated by the Ph4a748At promoter sequence derived from the histone *H4* gene of *A. thaliana* in combination with the first intron of gene II of the histone *H3.III* variant of *A. thaliana*. The 3′ untranslated region of the histone *H4* gene was the termination and polyadenylation signal. The mature protein was targeted to chloroplasts by TPotp C, an optimized transit peptide containing sequences from the Rubisco small subunit genes of maize and sunflower (*Helianthus annuus* L.).

Also in the case of WideStrike cotton (Dow, 2012), plant promoters were used to regulate the two Bt genes *cry1Ac* and *cry1F*, which were introduced to cotton through two independent transgenic events and stacked together through cross breeding. The *cry1Ac* cassette contained a Cry1Ac protein coding region, the maize ubiquitin-1 promoter, and *orf25* trailing regulator element, which defined the length of the DNA to be expressed. The *cry1F* cassette contained a Cry1F protein coding region, the 4ocs Δmas promoter, and *orf25* trailing regulatory element. The *pat* gene, which is located next to both Bt genes, was controlled by a different promoter and the same bi-directional trailing regulatory element. The *pat* gene provides a certain level of tolerance to glufosinate herbicide and is used as a selectable marker for regenerated plants containing *cry1Ac* and *cry1F*.

Another GE cotton with two Bt genes, TwinLink, was recently developed by Bayer CropScience through two independent transgenic events each carrying *cry1Ab* (or *cry2Ae*) and the *bar* gene (for PAT protein). The two gene expression cassettes contained the 35S promoter and other virus promoters, and plant promoters (USDA-APHIS, 2011). The *cry2Ae* + PAT cassette contained *5′cab22L* sequence including the leader sequence of the chlorophyll a/b binding protein gene from petunia (*P. hybrida*) and *TPssuAt* coding sequence of the transit peptide of the Rubisco small subunit gene *ats1A* from *A. thaliana*. The *cry1Ab* + PAT cassette contained the *3′me1* sequence that included the 3′ untranslated region of the NADP-malic enzyme gene from yellowtop (*Flaveria bidentis*) and *5′e1* sequence that included the leader sequence of the tapetum-specific *E1* gene (*GE1*) of rice (*Oryza sativa* L.).

The selection of promoters in vector cassette construction is very important, because a promoter determines the constitutive or spatial/temporal expression of the gene(s) it drives. A constitutive expression of the transgene(s) may be unnecessary or a waste of energy, for example, expression of Bt gene(s) in all cotton tissues, and insufficient expression or expression at the wrong time may not achieve the desired agronomic effects, for example, Roundup Ready cotton. Therefore, tissue-specific promoters should be characterized to control the spatial/temporal expression of a transgene. One of the most successful examples is the test of 16 promoters to drive the bacterial *iaaM* gene in cotton, resulting in the selection of the promoter for the petunia MADS box gene encoding floral binding protein 7 (FBP7) to increase IAA synthesis during fiber initiation, which in turn increased lint percentage and yield (Zhang et al., 2011b).

In crops including cotton, desirable traits may be achieved through reduction or elimination of a native gene expression. One successful example in cotton is the generation of GE Upland cotton with glanded plant body and ultra-low gossypol cottonseed developed using RNAi to knock down the expression of δ-cadinene synthase gene(s) involved in the biosynthetic pathway of gossypol (Sunilkumar et al., 2006). A 604-bp-long internal fragment of the *G. arboreum* L. *cad1-C1* (XC1) gene was used to make an intron-containing hairpin (ihp) construct with the pHANNIBAL/pART27 system. The RNAi transgene cassette was driven by a seed-specific promoter from the cotton globulin B gene, and it silenced the expression of the δ-cadinene synthase gene in the GE seed in a dominant fashion because of the formation of the double-stranded RNA (dsRNA) between the RNA from the transgene and mRNA from the target gene in the cotton cell, which prevents the production of the enzyme.

However, transformation methods, whether biological or direct, insert genes onto the plant genome randomly. Other challenges include multiple transgene copies and unpredictable transgene expression. Homologous recombination and DNA recombinase-mediated site-specific integration (called genome editing) have rapidly emerged as promising technologies to address these challenges for placing a single copy of transgenes into a targeted site (Li et al., 2009). These targeted genome editing strategies make it possible to directly target genes or DNA sequences in a site-specific manner. Zinc-finger nucleases (ZFNs) and transcription activator-like effector nucleases (TALENs) (Urnov et al., 2010; Joung and Sander, 2013), based on protein-DNA interactions, introduce site-specific double-strand DNA breaks. For the ZFN technology, plants are first engineered with ZFN constructs to produce ZFNs that target loci specified in the construction of the zinc finger. The zinc fingers then cut both strands of DNA at the locus and then insert the transgene through recombination (Urnov et al., 2010). However, ZFNs and TALENs have not been widely adopted by the plant research community because of their complicated designs and laborious assembly of specific DNA binding proteins for each target gene. Most recently, an easier method, based on the bacterial type II CRISPR (clustered regularly interspaced short palindromic repeats)/Cas endonuclease, is an RNA-guided DNA endonuclease system and has been successfully tested in plants (Jiang et al., 2013). The CRISPR/Cas system allows targeted cleavage of genomic DNA guided by a customizable small noncoding RNA, resulting in gene modifications by both nonhomologous end joining and homology-directed repair mechanisms. Because of its advantages in the straightforward construct design and assembly as compared with ZFNs and TALENs, it is anticipated that the CRISPR/Cas technology will gain a wide range of applications in plants.

Transformation Techniques

Several transformation methods have been tested and modified in cotton by numerous researchers. However, there are two main methods for transformation in cotton, that is, *A. tumefaciens*–mediated transformation and particle bombardment–mediated transformation. The most often used method in introducing a gene to the cotton genome is the former, pioneered by Umbeck et al. (1987) and Firoozabady et al. (1987) using hypocotyls of Coker 310 or cotyledons of Coker 210. *A. tumefaciens* naturally infects plants at wounded sites where it inserts a specific region of the tumor-inducing (Ti) plasmid carrying genes for opine and plant hormone synthesis into the host plant cell, resulting in the creation of a crown gall. The gall produces opines that can be metabolized by the bacterium, but not the plant. The Ti plasmid is coated by proteins to protect it from plant endonucleases and to direct it to the plant nucleus where one of the coating proteins cleaves the plant DNA, and the Ti plasmid is inserted into the plant genome by employing the plant DNA repair machinery. The plant hormones encoded by Ti plasmid DNA stimulate the infected cells to divide and enlarge, forming galls. A disarmed Ti plasmid by eliminating virulent genes from the original bacterial Ti plasmid can be further engineered to carry gene(s) of interest and transferred onto the host plant genome without the production of galls. The transformed cells or tissues can then be regenerated into whole transgenic plants.

Cotton regeneration is usually performed on the basis of tissue culture through somatic embryogenesis, where somatic cells produce an embryo similar to what is produced by zygotic embryogenesis. Cotton is known to be one of the most recalcitrant species to plant regeneration in higher plants, as only a small number of genotypes can be successfully in vitro induced to produce embryos for regeneration. Through various modifications in culture media and cultural conditions for callus and embryo induction, cotton plants can be successfully regenerated from explants from many different tissues including leaves, cotyledons, petioles, hypocotyls, and apical or embryonic meristems. Detailed technical information regarding cotton regeneration and transformation through somatic embryogenesis can be found in Wilkins et al. (2000), Kumria et al. (2003), and more recently Duncan (2010). Using seedlings of Coker 312 or Coker 315 as explants, this somatic embryogenesis–based transformation developed several commercial transgenic cotton traits including insect-resistant Bollgard cotton, WideStrike cotton, TwinLink cotton and Vip3A cotton, 2,4-D tolerant cotton, bromoxynil tolerant BXN cotton, glyphosate-tolerant Roundup Ready (RR) cotton and Roundup Ready Flex (RF) cotton, and glufosinate-resistant LibertyLink cotton. However, several limitations exist for its use in cotton transformation, including a limited number of regenerable genotypes (Coker cotton cultivars and a few others sharing Coker 100W in the pedigrees), lengthy tissue culture (about 10–12 mo), intensive labor needs, somaclonal variations, and sterility of regenerated plants.

Various Agrobacterium-mediated methods exist to transform plant cells. To circumvent the above problems, especially genotype dependency of regeneration, apical meristems from elite commercial cotton can be excised, wounded or unwounded, and directly transformed with *A. tumefaciens* through tissue culture. Shoot tips can also be inoculated with a bacterial suspension through vacuum infiltration or by being submerged in the bacterial suspension. The meristems can be grafted if necessary and are allowed to grow to maturity. To simplify the process, apical meristems can be transformed with a bacterial suspension in planta or through wounding. However, the successful rate in transformation is very low with these methods. Furthermore, the meristem transformation produces chimeric regenerated plants, requiring a progeny test.

Another genotype independent method is transformation of apical meristems from mature seeds with DNA-coated gold particles via particle bombardment (McCabe and Martinell, 1993; Altpeter et al., 2005; Rech et al., 2008). Although inefficient, this method can achieve 0.2 to 0.55% regenerated transgenic plants. In fact, it was used to introduce *cry2Ab* to Deltapine 50B carrying *cry1Ac* to develop Bollgard II cotton by Monsanto. Similar to the above Agrobacterium-mediated transformation via embryogenesis, embryogenic calli can be also transformed using the ballistic bombardment method or silicon carbide (whisker)-mediated transformation method through vigorous agitation (Duncan, 2010).

For particle bombardment, equipment such as the Biolistic PDS-1000/He Particle Delivery System (PSD-1000) (Bio-Rad, Hercules, CA), also known as the gene gun (Kikkert, 1993), can be used. The biolistic transformation uses a burst of helium air within a vacuum to physically introduce millions of DNA-coated gold particles, known as microcarriers, into the target plant cells or tissues. The biolistic transformation has several advantages such as avoidance of hypersensitive response to Agrobacterium in cell cultures as well as the need to eliminate the Agrobacterium after transformation with antibiotics, because false positives in initial polymerase chain reaction (PCR) screening may occur because of the possible presence of residual Agrobacterium cells. However, it is more likely to produce high copy numbers of transgenes than the Agrobacterium-mediated method. However, improvements in transformation efficiency and producing single copy of transgene(s) can be made through the choice of cell type and its physiological state, and adjustment of the concentration of linearized DNA cassettes and the particle size.

To avoid chimeric tissues produced from meristem transformation, cotton gametes or zygotes can be transformed through the pollen tube pathway (Zhou et al., 1983; Huang et al., 1999). In this method, foreign DNA is injected to cotton bolls and delivered to the embryo sac through the pollen tube pathway formed 10 to 24 h after pollination, where it is inserted onto the genome of the egg and zygote. Once mature bolls are harvested, seeds or seedlings can then be tested for transgene expression. The transformation efficiency is relatively higher (0.5–1% per treated flower). This method is versatile in that it is not only genotype independent, but it also does not need any delicate equipment. It can be used for transfer of any DNA fragments or gene constructs by individuals who do not need any special technical training. This pollen tube pathway technique has been widely used by numerous scientists in China and was successful in developing its first commercial Bt cotton (Cui and Guo, 1995). Unfortunately, this method is internationally controversial and not widely accepted. A similar direct transformation method is a modified floral dip called pistil dip (Clough and Bent, 1998) in an Agrobacterium solution 24 h after pollination of cotton flowers (Chen et al., 2010). When immature ovules are transformed and mature, the resulting seed may be transgenic but heterozygous. This simplest method without a tissue culture step was highly successful in *Arabidopsis* transformation, but it is less successful in other crops including cotton.

Transgenic plants generated from one of the above transformation processes are referred to as the T0 generation. Each independent transgenic plant derived from a single transformed cell is commonly referred to as a transgenic event. Seed harvested from the first transgenic event and its plant is referred to as the T1 generation, which gives rise to T2 generation after self-pollination, etc. Multiple transgenic events are often necessary in transgenic research and breeding for several reasons. Transgenic events with single copy insertions of the gene(s) of interest are preferred, while multiple copies of insertions are usually selected out. The transgene(s) should be stable in different genetic backgrounds under various environmental conditions and follow a traditional Mendelian segregation in each generation. The segregation should be verified at the T2 generation. Similar to traditional plant breeding, a transgenic event must be tested and selected on the basis of high levels of gene and trait expression (i.e., high efficacy) and stability in multiple environments (locations and years) and multiple genetic backgrounds. Furthermore, if generation of "marker-free" transgenic plants is the target, plants are first transformed with two construct cassettes: one carrying the transgene(s) responsible for the trait of interest and the other used for the transgene selection during the tissue culture process leading to regeneration of the transgenic plants. The selectable marker can be removed during segregation subsequent to T0 generation if the two cassettes are integrated at separate genome locations, while the transgene(s) of interest will be transferred into elite or commercial cotton backgrounds through breeding. On the other hand, a transgenic event with multiple copies of the transgene(s) can cause illegitimate recombination or instability of transgenes, and difficulties in characterizing genome locations and expression of transgene(s) and in transgene transfer through cross breeding. Multiple transgenic events are necessary because position effects of transgene(s) may exist (Matzke and Matzke, 1998). Therefore, both upstream and downstream cotton genome sequences of the transgene(s) should be molecularly characterized to ensure that the transgene(s) were not inserted into a known endogenous gene.

Otherwise, it could destroy or alter key biochemical pathways, resulting in unintended deleterious effects on agronomic traits. These transgenic events and others with low levels of transgene expression should be eliminated in transgenic breeding. However, it should be recognized that systemic biosynthesis of transgene protein(s) such as Bt endotoxin and CP4-EPSPS in the cotton plant may compete for photosynthate that otherwise is used for boll, seed, and fiber development.

Molecular and Genetic Characterization of Transgenes

Not only different transgenic events but also different individuals of the same event from the same gene construct may vary in transgene expression, which complicates the analysis of transgene effects. In fact, a high frequency of low-expressing, undesired transgenic plants is often observed. As a result, a large number of transgenic plants should be screened for those with acceptable levels of transgene expression and a desirable phenotype. Furthermore, the stability of the transgene expression at the RNA and protein levels and its phenotype should be monitored over several generations, as selected GE plants might lose the desired transgene expression and trait in subsequent generations. Factors such as copy number of transgenes, homologous gene silencing, and genome position of the transgenes contribute to differences in transgene expression among different events. Multiple transgene copies can increase the transgene expression, but low levels of expression are often the case because of interaction between the transgene and homologous sequences in the plant genome. A transgene may be integrated in different regions of the plant genome such as gene-rich regions with high levels of transcriptional activities or gene islands or highly repeated sequences with low transcriptional activities. Therefore, variable transgene expression, somaclonal variation, and sterility make it necessary to screen a number of plants from multiple transgenic events over several generations.

There are different methods used for the analysis of transgenic plants. Transgenic plants are initially selected on the basis of antibiotic resistance, a selectable gene marker, or herbicide resistance. Subsequently, the Southern blot analysis should be used to screen the GE plants to show genome integration of the transgene in transformed plants and their progeny, which also allows the estimation of transgene copy numbers. Polymerase chain reactions with primers designed from the selectable gene marker and/or the transgene(s) can also be performed as the indication of the existence of the transgenes; however, false positives may appear from the amplification of nonintegrated plasmid DNA from Agrobacterium or from insertion of partial plasmid DNA sequences. Although the exact genome location of the transgene insertion can be revealed from molecular methods such as DNA or genome walking, whole genome sequencing provides the genome-wide sequence analysis of transgenes with regard to their structural organization and possible rearrangements. For example, in the case of Bollgard cotton, the molecular characterization identified two T-DNA inserts: one containing a single copy of the full-length *cry1Ac* gene, the *nptII* gene, and the *aad* antibiotic resistance gene and the other containing 242 bp of a portion of the 7S 3′ polyadenylation sequence from the terminus of the *cry1Ac* gene (Monsanto, 2002b). A transcript was detected by reverse transcription PCR (RT-PCR) from the former, while the latter did not produce a transcript. On the basis of an 8-yr study of numerous locations by Monsanto, Southern blot analyses of numerous generations of the Bollgard cotton showed an identical Southern blot pattern, indicating a stable integration of the functional *cry1Ac* gene into the cotton genome. Although quantitative PCR can be further performed to assess the expression levels of transgene(s) in various tissues and developmental stages and possibly the copy number of the transgene(s), the expression of the transgene at the protein level is more important, as it is directly related to the desired phenotype. An enzyme-linked immunosorbent assay (ELISA) of Bollgard cotton seed obtained from multi-site trials over years showed similar levels of the Cry1Ac and NPTII proteins, and the production of the Cry1Ac protein was further confirmed by immuno-detection and/or efficacy data under different environmental conditions and in numerous Bollgard cotton cultivars. The stability was further confirmed by the Mendelian inheritance of the Bollgard trait after self-pollination or backcrossing with other cotton cultivars. As a result, the insecticidal efficacy had been maintained in a total production in over 17 million acres planted in the United States since 1996 (Monsanto, 2002b).

In the case of Roundup Ready Flex cotton, a Southern blot analysis indicated a single insertion event at a single cotton genome location with an intact integration of the two *cp4-epsps* cassettes and no other insertion of plasmid backbone sequences (Cerny et al., 2010). The Southern blot analysis further assessed the genetic stability of the glyphosate tolerance trait within and across generations. The segregation of T1 progeny from T0 generation was a 3 resistant to 1 nonresistant ratio, and the progeny test in T2 derived from individual resistant T1 plants showed 1 homozygous resistant to 2 hemizygous segregating progeny ratio, as expected for a single locus insertion. More advanced generations of self-pollinated homozygous-resistant plants did not show segregation of any nonresistant plants. A zygosity assay by TaqMan distinguished an F_2 population into 1 homozygous:2 hemizygous:1 null ratio. A molecular linkage analysis further mapped the insertion on chromosome D07, that is, c16 (Cerny et al., 2010). An ELISA analysis showed that the levels of CP4-EPSPS were higher in leaves and seed, lower in roots, and very low in pollen obtained from different environments. However, the expression levels of CP4-EPSPS are higher in Roundup Ready Flex cotton than in Roundup Ready cotton.

Agronomic and Nutrient Analysis and Assessment of Environmental Safety

Before a transgenic event with a desired phenotype is submitted to USDA, USEPA, and/or USFDA for an approval, extensive field tests on agronomic, fiber, and other traits should be performed to compare it with its nontransgenic counterpart in multiple genetic backgrounds and environmental and crop management conditions. In addition to the intended effects such as insect resistance or herbicide resistance, any significant effects of the transgene(s), either positive or negative, should be well documented. Significant deleterious effects of transgene(s) on yield, yield components, fiber and seed quality, and responses to abiotic and biotic stresses may impact the prospect of commercialization of a biotech trait. In addition, effects of transgene(s) on targeted and nontargeted insects or weeds should be monitored and addressed. In a comprehensive report on the first transgenic Bt cotton (Perlak et al., 1990), Perlak et al. (2001) stated that Bollgard cotton reduced cotton production costs and insecticide use by providing an effective alternative to chemical insecticides for the control of tobacco budworm (*H. virescens*), cotton bollworm (*H. zea*), and pink bollworm (*P. gossypiella*). Agronomic traits, fiber quality, and seed composition remain unchanged in the transgenic cotton. However, negative effects from a transgene may exist. For example, in the case of Roundup Ready Flex cotton, a 14-location field study in 2002 showed that Coker 312 with the transgene had reduced overall boll (4.56 vs. 4.70 g boll^{-1}) and seed size (9.56 vs. 9.83 g 100 seed^{-1}), and micronaire (3.76 vs. 3.88 units), as compared with its nontransgenic Coker 312, although these differences were within the variation common for commercial cotton. Other fiber quality traits including length, strength, and elongation were not changed (Horak et al., 2007).

Environmental safety issues should be well addressed, because one of the major concerns of releasing transgene(s) into the environment is the possibility of genetic contamination of herbicide- or insect-resistant transgenes to nontransgenic cultivars or wild or weedy relatives of the GE crop. This could greatly affect the coexistence between GE cotton and conventional or organic cotton and genetic compositions of wild species, although cultivated cotton does not have any weedy relative to cross naturally. While natural outcrossing between cultivated tetraploid cotton and a diploid cotton species is extremely rare if not impossible, it can produce natural fertile F_1 hybrids with three other wild tetraploid species (*G. tomentosum* Nutt. ex Seem., *G. mustelinum* Meers ex Watt, and *G. darwinii* Watt). The potential gene flow from transgenic cotton is only limited to regions where these three wild species and race stocks of *G. hirsutum* and *G. barbadense* originated, such as Mexico, Peru, Brazil, Hawaii, and Galapagos Islands. Therefore, for most cotton-producing countries, this is not a problem.

Besides the gene flow from transgenic cultivars to wild or semi-wild tetraploid cotton species, the major concern is the gene flow from transgenic cultivars to nontransgenic cultivars, because it could hurt farmers when they sell their cotton products to their target markets desiring nontransgenic cotton lint or seed. In addition to cross-pollination, other seed contaminations exist including improper cleaning of equipment during planting, harvesting, ginning and seed processing, accidental seed mixing, volunteer crop or human error during planting, harvesting, and ginning. Cotton is classified as an often cross-pollinated crop, as cotton seed is predominantly produced by self-pollination, but outcrossing can be mediated by pollinating insects such as bees including honey bees and bumble bees and not by wind. A cotton field with more abundant bees will have a higher outcrossing rate than one with few bees. Therefore, physical isolation plays an important role in reducing gene flow. In cotton, outcrossing rate decreases with distance between 0 and 6 m, but it rarely exceeded 1% of seeds at a distance of 7 to 10 m in a nontransgenic field (Umbeck et al., 1991; Van Deynze et al., 2005). Pollen-mediated gene flow was below 1% beyond 1 m (Van Deynze et al., 2005), especially when pollinator activities were low. Even though outcrossing is still detectable beyond 25 m, even 1625 m (Van Deynze et al., 2005), 55 m of nonplanted area may be sufficient to maintain seed purity. In Arizona, Heuberger et al. (2010) monitored 15 non-Bt Upland cotton seed production fields for gene flow of the Bt cotton carrying the *cry1Ac* gene. Adventitious Bt cotton plants from seed-mediated gene flow, due to seed bags and planting error, accounted for over 15% of plants sampled from the edges of three seed production fields. However, the pollen-mediated gene flow affected less than 1% of the seed sampled from field edges. In areas where outcrossing is low, therefore, careful planting and screening of seeds could be more important than field spacing to limit gene flow. It should be pointed out that, in the United States, nontransgenic crops do not require separation from transgenic crops that have received government approval, unless they are labeled as "GE-free" or "organic," and it also does not have strict labeling thresholds for adventitious presence of a biotech trait in seed. Therefore, the purity of conventional cotton seed may be problematic, and accessibility of "GE-free" or "organic" seed may be difficult.

Since cottonseed, as the byproduct in cotton production, is used for oil and animal feed, reliable data on seed composition, feed and food nutrients, toxicity, and allergenicity should also be obtained. Any organization seeking deregulation of a transgenic event should provide the above information in their petition to USDA and USEPA. A number of petition letters for deregulation and approval notices of transgenic events can be found in their websites. However, the cost associated with all the experiments and the process leading to a governmental regulatory approval for commercial release of a transgenic event was estimated at US$7 to $10 million (Kalaitzandonakes et al., 2007). Therefore, it is understandable that only a few large biotech companies can currently afford the investment in developing commercial biotech traits.

Breeding Methods of Transgenes

Once a transgenic event has shown promise, breeding for commercial transgenic cultivars commences even before its governmental regulatory approval. Since cotton transformation can only be successfully performed in a few regenerable, obsolete Coker genotypes, introgression of the transgene into elite genetic backgrounds through backcrossing is necessary for a commercial launch of the GE trait. In fact, the first generations of Bt and herbicide-resistant traits in Coker 312 were backcrossed to the then commercial cultivars, such as ST 4892 BR and ST 4793 R in ST 474 background, FM 989 R and FM 989 BR in FM 989 background, and PM 1560 BR in PM 1560 background. Both transgenic and nontransgenic cotton types coexisted for several years until the replacement by the transgenic versions in cotton production. Therefore, backcross breeding has played an important role in releasing commercially available transgenic cultivars in cotton. The individual biotech traits in the same genetic backgrounds can be further crossed to pyramid in the same cultivars with a double or triple stacked trait, such as Bollgard and Roundup Ready, and Bollgard II and Roundup Ready Flex. Once these biotech traits are carried in multiple genetic backgrounds, forward breeding can be practiced to select transgressive segregants with higher yield and better fiber quality to develop GE cotton cultivars carrying one or more transgenic traits. During the breeding process, generation acceleration techniques in the greenhouse and marker-assisted selection can be used to speed the recovery of the genetic backgrounds of the recurrent parents used in backcrossing. Because these transgenic traits are owned and patented by different biotechnology companies, licensing for using these GE traits in research and commercial breeding is legally required. For example, Monsanto started a Cotton States program in the early 2000s to serve as a foundation cottonseed licensing business, which was used to license conventional germplasm from other cotton breeders to develop commercial GE cultivars.

Backcross Breeding

Backcross breeding is the most important breeding method used to incorporate transgenes into elite genotypes or commercial cultivars. Successful introgression breeding for transgenic cultivars through backcrossing follows the same general guidelines as described by Fehr (1987). Depending on the genetic closeness, three generations of backcrossing are usually required within Upland cotton. For example, Acala 1517–99W (Zhang et al., 2008) and Acala 1517–09R (Zhang et al., 2011a) were developed through three backcrosses between their transgenic donors and the recurrent parent Acala 1517–99. However, the number of backcrosses depends mainly on the following factors, that is, (i) the strength of the linkage between unwanted chromosomal fragments and the transgene; (ii) the genetic and phenotypic differences between the donor and the recurrent parent; and (iii) the population size and selection pressure imposed during the recurrent backcrossing process. The closer the linkage between the transgene and deleterious chromosomal fragments, the more backcrosses are needed. The greater the genetic and phenotypic difference between the donor and recurrent parent, the more backcrosses are also needed. For example, because of great genetic differences and a reproductive barrier between Upland and Pima cotton (Zhang et al., 2014b), seven to nine backcross generations are needed to transfer a transgene from Upland to Pima cotton. This resulted in the development of the current herbicide-resistant Pima cotton cultivars such as PHY 802 RF, PHY 805 RF, PHY 811 RF, and DP 358 RF. However, higher selection pressure in a large backcross population will result in a greater recovery of the recurrent parent genome.

Although backcrossing in developing transgenic cultivars follows the same traditional backcrossing scheme, several strategies can be used to accelerate the recovery of the recurrent parent genotype. First, a generation acceleration method such as an off-season winter nursery or a greenhouse can be used for growing two to three generations a year. With high temperatures in the greenhouse, three generations of cotton can be easily produced on an annual basis. Since the recurrent parent does not carry the transgene, an F_1 between it and the transgene donor is hemizygous for the transgene locus, and its backcross to the recurrent parent results in hemizygous transgene and homozygous null (nontransgene) plants in a 1:1 ratio. The transgene or selectable marker or targeted herbicide or insects can be used to select the hemizygous BC_1F_1 for further backcrossing leading to BC_2F_1 and advanced backcross generations. As transgenes or their protein products per se are perfect markers for marker-assisted backcrossing (MAB), primers designed from the transgenes can be used for PCR followed by electrophoresis. Commercial strips are also developed to detect cotton plants or seed for transgenes on the basis of ELISA. Currently, there are more than 11 biotech traits with 14 transgenes carried in commercial cotton (BXN—1, BG—1, RR—1, BG II—2, RF—1, WS—2, LL—1, GlyTol—1, TwinLink—2, Xtend—3, and Vip3—1). As with multiplex PCR, a strip can be used to test for the existence of multiple transgenes in the same individuals. To avoid the transfer of any possible induced cytoplasmic mutations, the transgene donor should be used as the male parent during crossing and backcrossing.

Although the average percentage of the recurrent parent genome recovery in BC_1, BC_2, and BC_3 is 75.0, 87.5, and 93.75%, respectively, individual backcross plants vary in their genetic compositions. Furthermore, phenotypic selection in the greenhouse for the recurrent parent genotype is impractical because of a small growing space, the lack of the appropriate condition for phenotype evaluation, and the unnecessity of growing plants to maturity. Therefore, marker-assisted background selection should be used to accelerate the recovery of the recurrent parent genome. This is even more important in transferring transgene(s) from Upland to Pima cotton. In background selection, markers from the recurrent parent are used in BC_1F_1, BC_2F_1, and BC_3F_1 to select heterozygous transgenic individual plants with the most number of alleles from the recurrent parent for further backcrossing or selfing if it is the last backcross generation. For better genome coverage, 5 to 10 markers are needed for each chromosome or linkage group. To reduce the linkage drag, closely linked markers from both sides of the transgene locus should be used to select for recombinants because of double crossover events at around the transgene locus. Therefore, possible DNA fragments with deleterious effects from the transgene donor will be eliminated. Chapter 11 in this book has details on molecular breeding in cotton (Fang et al., 2015). Xu (2010) also detailed the theory and practice in marker-assisted selection. Of course, if a transgene donor is known to carry other desirable genes or quantitative trait loci (QTL) in addition to the transgene, foreground selection should also be simultaneously employed with the background selection. In foreground selection, at least two closely linked anchoring marker alleles (one on either side) of the gene or QTL in the donor should be chosen to reduce false positives in selection. For example, the use of one marker with a genetic distance of 5 cM from the target gene or QTL results in 5% false positives in selecting for the gene or QTL on the basis of the marker. However, the simultaneous use of another anchoring marker at 5 cM on the other side of the gene or QTL will reduce the false positives to 0.25%.

Population size and selection pressure in each backcross generation are also important in the recovery of the recurrent parent genome. A small number of backcross plants would not provide a high possibility of segregating an individual with a high percentage of the recurrent parent genome, no matter how high the selection pressure. Therefore, genotyping with genome-wide markers to screen an adequate number of plants is necessary.

Pyramiding of Multiple Transgenes

Two or more transgenes can be pyramided into the same genotype through transformation of a genotype with a construct containing expression cassettes for all the transgenes (Agrawal et al., 2005; Naqvi et al., 2010) or sequential transformation of the genotype with two or more transgenes. For example, Bollgard II, WideStrike, and TwinLink cotton each carrying two Bt genes were all developed using the sequential transformation approach. However, to introduce multiple transgenes into elite genetic backgrounds or commercial cultivars, two approaches can be taken, that is, sequential backcrossing and simultaneous backcrossing. With the sequential backcrossing approach, a transgene is transferred from its donor to a recurrent parent first, followed by crossing and backcrossing the backcrossed progeny carrying the first transgene with the second transgene donor. This approach is a time-consuming process. However, when both transgenes are available, their donors can be crossed and backcrossed with the same recurrent parent at the same time and then intercrossed between the backcrossed progeny carrying separate transgenic traits. This simultaneous backcrossing approach will save time in developing a cultivar with a stacked trait. Of course, marker-assisted backcrossing as described above can be used to pyramid multiple transgenic traits.

Whether it is the transfer of one transgene or multiple transgenes, once homozygous F_2 plants with the maximum recovery of the recurrent parent genome in the last backcross generation (e.g., BC_3) are identified and self-pollinated in the greenhouse, the resulting $BC_3F_{2:3}$ progeny should be moved to the field for a progeny test, followed by tests in multiple environments for a few selected lines. In the end, a commercial cultivar with the transgene(s) and acceptable agronomic traits will be released for production.

Forward Breeding

Once the same transgenic traits are transferred into multiple elite genetic backgrounds or commercial cultivars, forward breeding including pedigree selection can be used to develop improved GE cultivars with the same transgenic traits. During the process, a traditional pedigree method can be used in field conditions because selection for the transgenic traits is unnecessary. Therefore, a large segregating breeding population can be grown for phenotypic selection on yield, fiber quality, and other agronomic traits. Of course, if desirable genes or QTL are known to be carried in these parents, tightly linked markers should be used to select for them during the forward breeding process.

Forward breeding can also be used to cross between an elite GE line and a non-GE line. Therefore, new GE lines or cultivars with better agronomic traits can be developed. As with backcross breeding, gene markers (DNA or protein) for the transgenes are used to select for homozygous individuals with the GE traits for further assessment of agronomic and other traits. Marker-assisted forward breeding can be certainly deployed to select for these traits.

References

Agrawal, P.K., A. Kohli, R.M. Twyman, and P. Christou. 2005. Transformation of plants with multiple cassettes generates simple transgene integration patterns and high expression levels. Mol. Breed. 16:247–260. doi:10.1007/s11032-005-0239-5

Allen, R.D. 2010. Opportunities for engineering abiotic stress tolerance in cotton plants. In: U.B. Zehr, editor, Cotton: Biotechnological advances. Springer, Berlin. p. 127–160.

Altpeter, F., N. Baisakh, R. Beachy, R. Bock, T. Capell, P. Christou, H. Daniell, K. Datta, S. Datta, P.J. Dix, C. Fauquet, N. Huang, A. Kohli, H. Mooibroek, L. Nicholson, T.T. Nguyen, G. Nugent, K. Raemaker, A. Romano, D.A. Somers, E. Stoger, N. Taylor, and R. Visser. 2005. Particle bombardment and the genetic enhancement of crops: Myths and realities. Mol. Breed. 15:305–327. doi:10.1007/s11032-004-8001-y

Bayer CropScience. 2006. Petition for determination of nonregulated status for glyphosate-tolerant cotton: GlyTol™ cotton event GHB614. USDA-APHIS. http://www.aphis.usda.gov/brs/aphisdocs/06_33201p.pdf (accessed 13 Apr. 2015).

Bayley, C., N. Trolinder, C. Ray, M. Morgan, J.E. Quisenberry, and D.W. Ow. 1992. Engineering 2,4-D resistance into cotton. Theor. Appl. Genet. 83:645–649. doi:10.1007/BF00226910

Behrens, M.R., N. Mutlu, S. Chakraborty, R. Dumitru, W.Z. Jiang, B.J. LaVallee, P.L. Herman, T.E. Clemente, and D.P. Weeks. 2007. Dicamba resistance: Enlarging and preserving biotechnology-based weed management strategies. Science 316:1185–1188. doi:10.1126/science.1141596

Calgene. 1994. Petition of determination of non-regulated status: BXN™ cotton. USDA-APHIS. http://www.aphis.usda.gov/brs/aphisdocs/93_19601p.pdf (accessed 13 Apr. 2015).

Cerny, R.E., J. T. Bookout, C.A. CaJacob, J.R. Groat, J.L. Hart, G.R. Heck, S.A. Huber, J. Listello, A.B. Martens, M.E. Oppenhuizen, B. Sammons, N.K. Scanlon, Z.W. Shappley,

J.X. Yang, and J.H. Xiao. 2010. Development and characterization of a cotton (*Gossypium hirsutum* L.) event with enhanced reproductive resistance to glyphosate. Crop Sci. 50:1375–1384. doi:10.2135/cropsci2009.06.0286

Chen, T., S.J. Wu, J. Zhao, W.Z. Guo, and T.Z. Zhang. 2010. Pistil drip following pollination: A simple in planta Agrobacterium-mediated transformation in cotton. Biotechnol. Lett. 32:547–555. doi:10.1007/s10529-009-0179-y

Chung, S.-M., E.L. Frankman, and T. Tzfira. 2005. A versatile vector system for multiple gene expression in plants. Trends Plant Sci. 10:357–361. doi:10.1016/j.tplants.2005.06.001

Clough, S.J., and A.F. Bent. 1998. Floral dip: A simplified method for Agrobacterium-mediated transformation of *Arabidopsis thaliana*. Plant J. 16:735–743. doi:10.1046/j.1365-313x.1998.00343.x

Cui, H.Z., and S.D. Guo. 1995. Advances in studies of transgenic cottons in China. Sci. Agric. Sin. 29(1):93–95.

Dow. 2012. Product safety assessment (PSA): WideStrike™ insect protection. Dow. http://www.dow.com/productsafety/finder/ws.htm (accessed 13 Apr. 2015).

Duncan, D.R. 2010. Cotton transformation. In: U.B. Zehr, editor, Cotton: Biotechnological advances. Springer, Berlin. p. 65–77.

Edge, J.M., J.H. Benedict, J.P. Carroll, and H.K. Reding. 2001. Bollgard cotton: An assessment of global economic, environmental, and social benefits. J. Cotton Sci. 5:121–136.

Fang, D. 2015. Molecular breeding. In: D.D. Fang and R.G. Percy, editors, Cotton, 2nd ed. Agron. Monogr. 57. ASA, CSSA, and SSSA, Madison, WI. doi:10.2134/agronmonogr57.2013.0027

Federici, B.A., H.W. Park, and Y. Sakano. 2006. Insecticidal protein crystals of *Bacillus thuringiensis*. Microbiol. Monogr. 1:195–236. doi:10.1007/3-540-33774-1_8

Fehr, W.R. 1987. Principles of cultivar development: Theory and technique. Macmillan, New York.

Fernandez-Cornejo, J., S. Wechsler, M. Livingston, and L. Mitchell. 2014. Genetically engineered crops in the United States. ERR-162. USDA-ERS. http://www.ers.usda.gov/media/1282246/err162.pdf (accessed 13 Apr. 2015).

Firoozabady, E., D.L. DeBoer, D.J. Merlo, E.L. Halk, L.N. Amerson, K.E. Rashka, and E.E. Murray. 1987. Transformation of cotton (*Gossypium hirsutum* L.) by *Agrobacterium tumefaciens* and regeneration of transgenic plants. Plant Mol. Biol. 10:105–116. doi:10.1007/BF00016148

Franz, J.E., M.K. Mao, and J.A. Sikorski. 1997. Glyphosate: A unique global herbicide. Am. Chem. Soc., Washington, DC.

Gao, X., T. Wheeler, Z. Li, C.M. Kenerley, P. He, and L. Shan. 2011. Silencing *GhNDR1* and *GhMKK2* compromises cotton resistance to Verticillium wilt. Plant J. 66:293–305. doi:10.1111/j.1365-313X.2011.04491.x

Heuberger, S., C. Ellers-Kirk, B.E. Tabashnik, and Y. Carriere. 2010. Pollen- and seed-mediated transgene flow in commercial cotton seed production fields. PLoS ONE 5:e14128.

Horak, J.M., E.W. Rosenbaum, C.L. Woodrum, A.B. Martens, R.F. Mery, J.T. Cothren, J.A. Burns, T.E. Nickson, T.A. Pester, C. Jiang, J.L. Hart, and B. Sammons. 2007. Characterization of Roundup Ready Flex cotton, 'MON 88913' for use in ecological risk assessment: Evaluation of seed germination, vegetative and reproductive growth, and ecological interactions. Crop Sci. 47:268–277. doi:10.2135/cropsci2006.02.0063

Huang, G.C., Y.M. Dong, and J.S. Sun. 1999. Introduction of exogenous DNA into cotton via the pollen-tube pathway with GFP as a reporter. Chin. Sci. Bull. 44:698–701. doi:10.1007/BF02909705

James, C. 2013. Global status of commercialized biotech/GM crops. ISAAA Brief No. 46. ISAAA, Ithaca, NY.

Jiang, W., H. Zhou, H. Bi, M. Fromm, B. Yang, and D.P. Weeks. 2013. Demonstration of CRISPR/Cas9/sgRNA-mediated targeted gene modification in Arabidopsis, tobacco, sorghum and rice. Nucl. Acids Res. 41:e188. doi:10.1093/nar/gkt780

Jiang, Y., W. Guo, H. Zhu, Y.L. Ruan, and T. Zhang. 2012. Overexpression of *GhSusA1* increases plant biomass and improves cotton fiber yield and quality. Plant Biotechnol. J. 10:301–312. doi:10.1111/j.1467-7652.2011.00662.x

Joung, J.K., and J.D. Sander. 2013. TALENs: A widely applicable technology for targeted genome editing. Nat. Rev. Mol. Cell Biol. 14:49–55. doi:10.1038/nrm3486

Kalaitzandonakes, N., J.M. Alston, and K.J. Bradford. 2007. Compliance costs for regulatory approval of new biotech crops. Nat. Biotechnol. 25:509–511. doi:10.1038/nbt0507-509

Kikkert, J.R. 1993. The Biolistic PDS-1000 He device. Plant Cell Tissue Organ Cult. 33:221–226. doi:10.1007/BF02319005

Klümper, W., and M. Qaim. 2014. A meta-analysis of the impacts of genetically modified crops. PLoS ONE 9:e111629. doi:10.1371/journal.pone.0111629

Kumar, V., S.G. Joshi, A.A. Bell, and K.S. Rathore. 2013. Enhanced resistance against *Thielaviopsis basicola* in transgenic cotton plants expressing Arabidopsis NPR1 gene. Transgenic Res. 22:359–368. doi:10.1007/s11248-012-9652-9

Kumria, R., S. Leelavathi, R.K. Bhatnagar, and V.S. Reddy. 2003. Regeneration and genetic transformation of cotton: Present status and future perspectives. Plant Tissue Cult. 13:211–225.

Kuppu, S., N. Mishra, R. Hu, L. Sun, X. Zhu, G. Shen, E. Blumwald, P. Payton, and H. Zhang. 2013. Water-deficit inducible expression of a cytokinin biosynthetic gene *IPT* in cotton improves drought tolerance under controlled environment growth conditions. PLoS ONE 8:e64190. doi:10.1371/journal.pone.0064190

Lemaux, P.G. 2008. Genetically engineered plants and foods: A scientist's analysis of the issues (Part I). Annu. Rev. Plant Biol. 59:771–812. doi:10.1146/annurev.arplant.58.032806.103840

Lemaux, P.G. 2009. Genetically engineered plants and foods: A scientist's analysis of the issues (Part II). Annu. Rev. Plant Biol. 60:511–559. doi:10.1146/annurev.arplant.043008.092013

Li, Z., A. Xing, B.P. Moo, R.P. McCardell, K. Mills, and S.C. Falco. 2009. Site-specific integration of transgenes in soybean via recombinase-mediated DNA cassette exchange. Plant Physiol. 151:1087–1095. doi:10.1104/pp.109.137612

Matzke, A.J.M., and M.A. Matzke. 1998. Position effects and epigenetic silencing of plant transgenes. Curr. Opin. Plant Biol. 1:142–148. doi:10.1016/S1369-5266(98)80016-2

McCabe, D.E., and B.J. Martinell. 1993. Transformation of elite cotton cultivars via particle bombardment of meristem. Nat. Biotechnol. 11:596–598. doi:10.1038/nbt0593-596

Meredith, W.R., Jr., and R.R. Bridge. 1984. Genetic contributions to yield changes in Upland cotton. In: W.R. Fehr, editor, Genetic contributions to yield gains of five major crop plants. CSSA Spec. Publ. 7. ASA, CSSA, and SSSA, Madison, WI. p. 75–87.

Monsanto. 2002a. Safety assessment of Roundup Ready® cotton, event 1445. Monsanto. http://www.monsanto.com/products/documents/safety-summaries/cotton_pss.pdf (accessed 13 Apr. 2015).

Monsanto. 2002b. Safety assessment of Bollgard® cotton event 531. Monsanto. http://www.monsanto.com/products/documents/safety-summaries/bollgard_pss.pdf (accessed 13 Apr. 2015).

Monsanto. 2003. Safety assessment of Bollgard II® cotton MON 15985. Monsanto. http://www.monsanto.com/products/documents/safety-summaries/bollgard_ii_pss.pdf (accessed 13 Apr. 2015).

Monsanto. 2004. Petition for the determination of non-regulated status for Roundup Ready® Flex cotton, MON 88913. Monsanto. http://www.aphis.usda.gov/brs/aphisdocs/04_08601p.pdf (accessed 13 Apr. 2015).

Naqvi, S., G. Farre, G. Sanahuja, T. Capell, C.F. Zhu, and P. Christou. 2010. When more is better: Multigene engineering in plants. Trends Plant Sci. 15:48–56. doi:10.1016/j.tplants.2009.09.010

Odell, J.T., F. Nagy, and N.H. Chua. 1985. Identification of DNA sequences required for activity of the cauliflower mosaic virus 35S promoter. Nature 313:810–812. doi:10.1038/313810a0

Palle, S.R., L.M. Campbell, D. Pandeya, L. Puckhaber, L.K. Tollack, S. Marcel, S. Sundaram, R.D. Stipanovic, T.C. Wedegaertner, L. Hinze, and K.S. Rathore. 2013. RNAi-mediated ultra-low gossypol cottonseed trait: Performance of transgenic lines under field conditions. Plant Biotechnol. J. 11:296–304. doi:10.1111/pbi.12013

Pasapula, V., G. Shen, S. Kuppu, J. Paez-Valencia, M. Mendoza, P. Hou, J. Chen, X. Qiu, L. Zhu, X. Zhang, D. Auld, E. Blumwald, H. Zhang, R. Gaxiola, and P. Payton. 2011. Expression of an Arabidopsis vacuolar H^+-pyrophosphatase gene (AVP1) in cotton improves drought- and salt tolerance and increases fibre yield in the field conditions. Plant Biotechnol. J. 9:88–99. doi:10.1111/j.1467-7652.2010.00535.x

Perlak, F.J., R.W. Deaton, T.A. Armstrong, R.L. Fuchs, S.R. Sims, J.T. Greenplate, and D.A. Fischhoff. 1990. Insect resistant cotton plants. BioTechnology 8:939–943. doi:10.1038/nbt1090-939

Perlak, F.J., M. Oppenhuizen, K. Gustafson, R. Voth, S. Sivasupramaniam, D. Heering, B. Carey, R.A. Ihrig, and J.K. Roberts. 2001. Development and commercial use of Bollgard cotton in the USA—Early promises versus today's reality. Plant J. 27:489–501. doi:10.1046/j.1365-313X.2001.01120.x

Pline, W.A., K.L. Edmisten, J.W. Wilcut, R. Wells, and J. Thomas. 2003. Glyphosate-induced reductions in pollen viability and seed set in glyphosate-resistant cotton and attempted remediation by gibberellic acid (GA_3). Weed Sci. 51:19–27. doi:10.1614/0043-1745(2003)051[0019:GIRIPV]2.0.CO;2

Rech, E.L., G.R. Vianna, and F.J. Aragão. 2008. High-efficiency transformation by biolistics of soybean, common bean and cotton transgenic plants. Nat. Protoc. 3:410–418. doi:10.1038/nprot.2008.9

Shen, G., J. Wei, X. Qiu, R. Hu, S. Kuppu, D. Auld, E. Blumwald, R. Gaxiola, P. Payton, and H. Zhang. 2015. Co-overexpression of AVP1 and AtNHX1 in cotton further improves drought and salt tolerance in transgenic cotton plants. Plant Mol. Biol. Rep. 33:167–177. doi:10.1007/s11105-014-0739-8.

Stewart, J.M. 1994. Potential for crop improvement with exotic germplasm and genetic engineering. In: G.A. Constable and N.W. Forrester, editors, Challenging the future. Proceedings of the World Cotton Research Conference 1. CSIRO, Melbourne, Australia. p. 313–327.

Sunilkumar, G., L.M. Campbell, L. Puckhaber, R.D. Stipanovic, and K.D. Rathore. 2006. Engineering cottonseed for use in human nutrition by tissue-specific reduction of toxic gossypol. Proc. Natl. Acad. Sci. USA 103:18054–18059. doi:10.1073/pnas.0605389103

Syngenta. 2003. Application for determination of non-regulated status for lepidopteran insect protected VIP3A cotton transformation event COT102. USDA-APHIS. http://www.aphis.usda.gov/brs/aphisdocs/03_15501p.pdf (accessed 13 Apr. 2014).

Umbeck, P.F., K.A. Barton, E.V. Nordheim, J.C. McCarty, W.L. Parrott, and J.N. Jenkins. 1991. Degree of pollen dispersal by insects from a field test of genetically engineered cotton. J. Econ. Entomol. 84:1943–1950. doi:10.1093/jee/84.6.1943

Umbeck, P., G. Johnson, K. Barton, and W. Swain. 1987. Genetically transformed cotton (*Gossypium hirsutum* L.) plants. Nat. Biotechnol. 5:263–266. doi:10.1038/nbt0387-263

Urnov, F.D., E.J. Rebar, M.C. Holmes, H.S. Zhang, and P.D. Gregory. 2010. Genome editing with engineered zinc finger nucleases. Nat. Rev. Genet. 11:636–646. doi:10.1038/nrg2842

USDA-APHIS. 2003. Environmental assessment of LibertyLink cotton transformation event LLCotton25. USDA-APHIS. http://www.aphis.usda.gov/brs/aphisdocs2/02_04201p_com.pdf (accessed 13 Apr. 2015).

USDA-APHIS. 2011. Determination of nonregulated status of insect resistant and glufosinate ammonium-tolerant (TwinLink™) cotton, *Gossypium hirsutum*, events T304-40 x GHB119. USDA-APHIS http://www.aphis.usda.gov/brs/aphisdocs/08_34001p_dea.pdf (accessed 13 Apr. 2015).

USDA-APHIS. 2014. Draft environmental impact statement—2014. Monsanto petitions (10-188-01p and12-185-01p) for determinations of nonregulated status for Dicamba resistant soybean and cotton varieties. USDA-APHIS. http://www.aphis.usda.gov/brs/aphisdocs/dicamba_deis.pdf (accessed 13 Apr. 2015).

USDA-ERS. 2014. Genetically engineered varieties of corn, upland cotton, and soybeans, by state and for the United States, 2000–14. USDA-ERS. http://www.ers.usda.gov/data-products/adoption-of-genetically-engineered-crops-in-the-us.aspx (accessed 13 Apr. 2015).

USFDA. 2002. Biotechnology consultation note to the file BNF No. 000086. USFDA. http://www.fda.gov/Food/FoodScienceResearch/Biotechnology/Submissions/ucm155782.htm (accessed 13 Apr. 2015).

Van Deynze, A.E., F.J. Sundstrom, and K.J. Bradford. 2005. Pollen-mediated gene flow in California cotton depends on pollinator activity. Crop Sci. 45:1565–1570. doi:10.2135/cropsci2004.0463

Vitale, J., H. Glick, J. Greenplate, M. Abdennadher, and O. Traoré. 2008. Second-generation Bt cotton field trials in Burkina Faso: Analyzing the potential benefits to west African farmers. Crop Sci. 48:1958–1966. doi:10.2135/cropsci2008.01.0024

Walford, S.A., Y. Wu, D.J. Llewellyn, and E.S. Dennis. 2011. GhMYB25-like: A key factor in early cotton fibre development. Plant J. 65:785–797. doi:10.1111/j.1365-313X.2010.04464.x

Wilkins, T.A., K. Rajasekaran, and D.M. Anderson. 2000. Cotton biotechnology. Crit. Rev. Plant Sci. 19:511–550. doi:10.1016/S0735-2689(01)80007-1

Wright, T.R., G. Shan, T.A. Walsh, J.M. Lira, C. Cui, P. Song, M. Zhuang, N.L. Arnold, G. Lin, K. Yau, S.M. Russell, R.M. Cicchillo, M.A. Peterson, D.M. Simpson, N. Zhou, J. Ponsamuel, and Z. Zhang. 2010. Robust crop resistance to broadleaf and grass herbicides provided by aryloxyalkanoate dioxygenase transgenes. Proc. Natl. Acad. Sci. USA 107:20240–20245. doi:10.1073/pnas.1013154107

Xu, W.L., D.J. Zhang, Y.F. Wu, L.X. Qin, G.Q. Huang, J. Li, L. Li, and X.B. Li. 2013. Cotton *PRP5* gene encoding a proline-rich protein is involved in fiber development. Plant Mol. Biol. 82:353–365. doi:10.1007/s11103-013-0066-8

Xu, Y.B. 2010. Molecular plant breeding. CAB International, Wallingford, UK.

Zhang, J.F., R.G. Cantrell, C. Waddell, R. Flynn, and E. Hughs. 2008. Release of the first Bt Acala cotton cultivar, Acala 1517–99W. In: Proceedings of the Beltwide Cotton Conferences, Natl. Cotton Counc. Am., Nashville, TN. 8–11 Jan. 2008. Natl. Cotton Counc. Am., Memphis, TN. p. 906–912.

Zhang, J.F., R. Flynn, S.E. Hughs, S. Bajaj, and D.C. Jones. 2011a. Registration of 'Acala 1517–09R' cotton. J. Plant Regist. 5:164–169. doi:10.3198/jpr2010.05.0268crc

Zhang, J.F., O.J. Idowu, T. Wedegaertner, and S.E. Hughs. 2014a. Genetic variation and comparative analysis of thrips resistance in glandless and glanded cotton under field conditions. Euphytica 199:373–383. doi:10.1007/s10681-014-1137-x

Zhang, J.F., Y. Lu, H. Adragna, and E. Hughs. 2005. Genetic improvement of New Mexico Acala cotton germplasm and their genetic diversity. Crop Sci. 45:2363–2373. doi:10.2135/cropsci2005.0140

Zhang, J.F., R.G. Percy, and J.C. McCarty, Jr. 2014b. Introgression genetics and breeding between Upland and Pima cotton—A review. Euphytica 198:1–12. doi:10.1007/s10681-014-1094-4

Zhang, M., X. Zheng, S. Song, Q. Zeng, L. Hou, D. Li, J. Zhao, Y. Wei, X. Li, M. Luo, Y. Xiao, X. Luo, J.F. Zhang, C. Xiang, and Y. Pei. 2011b. Spatiotemporal manipulation of auxin biosynthesis in cotton ovule epidermal cells enhances fiber yield and quality. Nat. Biotechnol. 29:453–458. doi:10.1038/nbt.1843

Zhou, G., J. Weng, Y. Zeng, J. Huang, S. Qian, and G. Liu. 1983. Introduction of exogenous DNA into cotton embryos. Methods Enzymol. 101:433–481.

Molecular Breeding

David D. Fang

Abstract

The use of molecular and genomic tools to assist selection of parents or progeny has become an integral part of modern cotton breeding. In this chapter, the basic components of molecular cotton breeding are described: molecular marker development, genetic and physical map construction, identification of markers associated with qualitative (major) genes, quantitative trait loci (QTL) mapping, and the utilization of markers in practical breeding. Major interspecific and *Gossypium hirsutum* intraspecific genetic maps are listed. In addition, previous research reports in mapping qualitative and quantitative traits are summarized, current efforts in using molecular markers in practical breeding are described, and perspectives on the future of molecular cotton breeding are discussed.

In concept, plant breeding is simple: choose parents, cross them, and select progeny that perform better than the parents. In practice, plant breeding consists of three steps: create or assemble populations with useful genetic variations, identify and select individuals with superior phenotypes, and develop varieties from the selected individuals (Moose and Mumm, 2008). Of these three steps, selection of superior individuals is the most difficult and time consuming. In conventional plant breeding, selection is usually based on the phenotypes, but with the advent of molecular biology, selection can be done at the molecular level on the basis of DNA and its effects on the phenotypes, which provides the foundation of molecular plant breeding. Molecular plant breeding may be defined in a broad sense as the use of genetic manipulation performed at the DNA level to improve traits of interest in plants (Jiang, 2013). This includes gene manipulation through transgenic approaches (transgenic breeding), marker-assisted selection (MAS), genome-wide selection (GWS), genomic selection (GS), etc. However,

---

Abbreviations: AFLP, amplified fragment length polymorphism; BAC, bacterial artificial chromosome; BIBAC, binary BAC; CBD, cotton blue disease; E, environment; CBT, cotton bunchy top; CSIL, chromosome-segment introgression line; DP, donor parent; EST, expressed sequence tag; GS, genomic selection; GWS, genome-wide selection; HDC, high-density consensus; MABS, marker-assisted backcrossing selection; MARS, marker-assisted recurrent selection; MAS, marker-assisted selection; PCR, polymerase chain reaction; QTL, quantitative trait loci; RAPD, random amplified polymorphic DNA; RFLP, restriction fragment length polymorphism; RP, recurrent parent; SNP, single-nucleotide polymorphism; SSR, simple sequence repeat.

David D. Fang, Cotton Fiber Bioscience Research Unit, USDA-ARS-SRRC, 1100 Robert E Lee Blvd., New Orleans, LA 70124 (David.Fang@ars.usda.gov).

doi:10.2134/agronmonogr57.2013.0027

Cotton. 2nd. ed. David D. Fang and Richard G. Percy, editors. Agronomy Monograph 57.

molecular plant breeding more often refers to using DNA markers, in combination with trait analysis based on genetic or physical maps, to assist in improving phenotypic traits. This definition has been used to describe several molecular breeding strategies, including MAS, marker-assisted backcrossing selection (MABS), marker-assisted recurrent selection (MARS), GWS, and GS (Xu, 2010). During the past two decades, cotton researchers have made tremendous progress in the development of molecular markers, the construction of genetic maps (interspecific and intraspecific), sequencing of genomes, and identification of markers associated with qualitative (major) and quantitative traits of agronomic importance. Application of molecular markers to improve traits of interest such as resistance to the root-knot nematode has been successfully implemented in both public and private breeding programs (Jenkins et al., 2012). In this chapter, I will discuss the progress that has been made in research and applications as related to molecular cotton breeding in following topics: molecular marker types; genetic and physical map construction; qualitative (major) trait analysis; quantitative trait analysis; and molecular cotton breeding in practice, along with perspectives on future of molecular cotton breeding.

Molecular Marker Types

Genetic markers (or molecular markers) are determined by the allelic forms of genes or genetic loci and can be transmitted between generations; thus, they can be used as genetic tags to keep track of an individual, a cell, a chromosome, or a gene (Jiang, 2013). Genetic markers include both DNA markers and non-DNA ones, which consist of morphological, biochemical and cytological markers. Due to the inherent drawbacks of non-DNA markers, such as limited numbers and occasional changes caused by environmental factors , very few, if any, were used for tagging traits of interest in cotton. Thus, molecular cotton breeding *in essence* is exclusively limited to using DNA markers to assist in breeding.

A DNA marker is a small region of DNA sequence showing polymorphism (a base deletion, insertion, or substitution) between different individuals. DNA markers that are ideal for molecular plant breeding should be abundant and well distributed across the genome, have a high level of polymorphism, and be codominant, genome-specific, low cost, and easy to detect. In addition, analysis of markers can be automated for high throughput application in practical breeding. Five major DNA marker types have been widely used in cotton: restriction fragment length polymorphisms (RFLPs), random amplified polymorphic DNAs (RAPDs), amplified fragment length polymorphisms (AFLPs), simple sequence repeats (SSRs) and single-nucleotide polymorphisms (SNPs). Their characteristics are shown in Table 1.

RFLP Markers

RFLP markers were the first generation of DNA markers developed and utilized in cotton (Rahman et al., 2009). In RFLP analysis, organisms are differentiated by fragments that are obtained after cleaving the genomic DNA with restriction enzymes. The polymorphism is based on the fragment length difference between individuals (Botstein et al., 1980). These markers are highly reproducible and codominant and were widely used in cotton genetics studies between the 1990s and early 2000s, when they were used to construct the first cotton

Table 1. Characteristics of five major DNA marker types in cotton.†

	RFLP	RAPD	AFLP	SSR	SNP
Genomic abundance	high	high	high	high	very high
Inheritance	codominant	dominant	dominant	codominant	codominant
Level of polymorphism	medium	low	medium-high	medium-high	medium-high
Type of polymorphism	restriction site, InDels	base change, InDels	base change, InDels	length of repeats	base change
Amount of DNA required	5–10 μg	10–100 ng	10–100 ng	10–100 ng	10–100 ng
Quality of DNA required	high	low	high	medium	high
PCR based	no	yes	yes	yes	yes
Reproducibility	high	low	high	high	high
Technically demanding	high	low	medium	medium	medium-high
Easiness of automation	low	low	low	medium	high
Cost of development‡	high	very low	very low	medium	medium
Cost of utilization	high	low	medium	low	low-medium
Number of markers available to public	<3000	NA	NA	>20,000	>100,000
Number of polymorphic loci per assay	1–3	1–5	5–50	1–2	1
Suitability for molecular breeding	low	low	low	medium-high	high

† Adapted from Rahman et al., 2009; Xu, 2010; and Jiang, 2013.

‡ As technology develops, the cost to develop new SSR and SNP markers will decrease.

genetic map (Reinisch et al., 1994); two high-density genetic maps (Rong et al., 2004; Lacape et al., 2009) also include many RFLP markers. RFLP markers were also used to tag agronomic traits (Ulloa and Meredith, 2000; Chee et al., 2005; Rong et al., 2005). RFLP analysis usually requires radioisotope detection, which has been a major limitation for the use of RFLP markers. In addition, high demand for DNA quality and quantity further impedes the application of RFLP markers in cotton breeding. Because DNA marker systems based on the polymerase chain reaction (PCR) demonstrated significant advantages over RFLP in early 2000s, the use of RFLP markers gradually diminished in cotton genetics and molecular breeding.

RAPD Markers

RAPD markers are a PCR-based marker system. In this system, the total genomic DNA of an individual is amplified by PCR using a single, short (usually about 10 bases), random primer (Welsh and McClelland, 1990; Williams et al., 1990). Polymorphisms resulting from base insertions, deletions, substitutions, or rearrangements either at or between the primer-binding sites are visible with electrophoresis as the presence or absence of a particular DNA fragment. RAPD marker analysis does not require prior sequence information for primer design. This technique was widely used in many other organisms for genetic studies and germplasm characterization before it was used in cotton. In cotton, RAPD markers were used to construct genetic maps (Lin et al., 2009), tag agronomic traits (Zhang et al., 2003), and distinguish elite cultivars (Iqbal et al., 2001). Although analysis of RAPD markers is simple and easy, a great majority of RAPD markers are dominant, which makes RAPD a less ideal marker type in cotton breeding. Furthermore, the polymorphism of RAPD markers is relatively low in cotton compared with other marker systems. Another drawback of RAPD markers is the low reproducibility. Because of these disadvantages, RAPD markers gradually disappeared from the cotton research beginning in the early 2000s, and, indeed, no single scientific paper using RAPD markers has been published in a mainstream international journal since 2010.

AFLP Markers

AFLP markers are PCR-based markers. Technically, AFLP combines the reliability of RFLP markers with the convenience of PCR (Vos et al., 1995). The former is based on selective PCR amplification of restriction fragments resulting from a double digestion of the genomic DNA under high stringency conditions; that is, the combination of polymorphisms at restriction sites and hybridization of arbitrary primers. AFLP analysis involves three steps: (i) digestion of total genomic DNA with two restriction enzymes, usually a frequent cutter (4-bp recognition site, *Mse*I or *Taq*I) and a rare cutter (6-bp recognition site, *Eco*RI or *Hind*III); (ii) ligation of adaptors to the ends of the restricted fragments; and (iii) PCR and electrophoresis of fragments. AFLP markers are very reliable and can produce many bands per analysis. In cotton, AFLP markers were used for constructing genetic maps, tagging agronomic traits, and characterizing cultivars (Lacape et al., 2003; Mei et al., 2004; Lu et al., 2008). However, like RAPD markers, a great majority of AFLP markers are dominant. In addition, their complex banding pattern makes AFLP markers very difficult to use in high throughput automation. Since 2010, the use of AFLP markers has gradually diminished in cotton research.

SSR Markers

SSR markers, also called microsatellites or short tandem repeats, are PCR-based markers. They are random tandem repeats of short nucleotide motifs (usually 2–6 bp long). SSR markers are characterized as hypervariable, highly reproducible, codominant, locus specific, and randomly distributed across the whole genome in most cases. Furthermore, SSR markers can be multiplexed and automated for high throughput genotyping. Since 2000, SSR markers gradually replaced RFLP, RAPD, and AFLP markers in cotton and have become the current marker of choice. In cotton, SSR markers were developed mainly from two sources: microsatellite enriched genomic DNA libraries and expressed sequence tags (ESTs) (Blenda et al., 2006; Han et al., 2006). With more genome sequences available recently, SSR markers have also been developed directly from genome sequences (K. Wang et al., 2012; Z. Wang et al., 2013). As of today, the CottonGen database (www.cottongen.org) contains more than 20,000 SSR markers, which are usually named after their original developers. Table 2 shows the cotton SSR markers currently in the CottonGen database. SSR markers have been extensively used in the construction of genetic linkage maps, QTL mapping, marker-assisted selection, and germplasm characterization in cotton (Guo et al., 2007; J.Z. Yu et al., 2012a; J. Yu et al., 2013b). The first high-density consensus genetic map of tetraploid cotton was constructed on the basis of many SSR markers that were commonly mapped in individual maps (Blenda et al., 2012). The first comprehensive genetic map of cotton (Yu et al., 2010) was made possible largely as a result of the common SSR markers between maps.

Table 2. Names of cotton SSR markers, their developers, and their origins.

Prefix of SSR marker	Developer	Marker DNA origin†
BNL	Brookhaven National Laboratory	genomic
C2	Delta and Pine Land Company (Monsanto)	genomic
CER	Monsanto	EST
CGR	Monsanto	genomic
CIR	CIRAD, France	genomic
CM	Texas A&M University	genomic
COT	Delta and Pine Land Company (Monsanto)	genomic
DC	Delta and Pine Land Company (Monsanto)	genomic
DOW	Dow AgroScience	genomic
DPL	Delta and Pine Land Company (Monsanto)	genomic
GH	Texas A&M University and USDA-ARS, College Station, TX	genomic
HAU	Huazhong Agricultural University, China	EST‡
JESPR	Texas A&M University and USDA-ARS, Mississippi State, MS	genomic
MGHES	USDA-ARS, Mississippi State, MS	EST§
MUCS	USDA-ARS, Shafter, CA	EST§
MUSB	USDA-ARS, Shafter, CA	genomic (Bacterial artificial chromosome)
MUSS	USDA-ARS, Shafter, CA	EST§
NAU	Nanjing Agricultural University, China	EST§ and genomic (Bacterial artificial chromosome)
NBRI	National Botanical Research Institute, India	genomic and EST
SHIN	Delta and Pine Land Company (Monsanto)	EST§
STV	USDA-ARS, Stoneville, MS	EST
TMB	USDA-ARS, College Station, TX	genomic (Bacterial artificial chromosome)
UCD	Delta and Pine Land Company (Monsanto)	EST§

† The species is *G. hirsutum* if not specially indicated.

‡ Including both *G. hirsutum* and *G. barbadense*.

§ Multiple species including *G. arboreum*, *G. hirsutum* and *G. barbadense*.

SNP Markers

An SNP is a single nucleotide base difference between two DNA sequences or individuals. In general, a few base insertions or deletions (indels) are also considered to be SNPs; however, in this chapter, indels are not included as SNPs. The latter can be categorized according to nucleotide substitutions as either transitions (C/T or G/A) or transversions (C/G, A/T, C/A or T/G) (Byers et al., 2012). SNPs provide the simplest form of molecular marker because a single nucleotide base is the smallest unit of inheritance and thus can provide the maximum number of markers. In general, SNPs are codominant and often present in genes. Thus SNPs have become a very attractive DNA marker type in genetic studies and breeding. SNP markers are highly abundant in living organisms. Moreover, SNP marker genotyping can be easily automated. Due to the significant advantages of SNP markers over other marker types currently available, SNP markers have rapidly replaced the latter and become the marker of choice for genetic studies in humans and in molecular breeding of cattle, maize, and soybean (Xu, 2010). Realizing the potential importance in cotton breeding, large corporations, such as Delta and Pine Land Company, Monsanto, and Bayer CropScience, started large-scale SNP discovery research in the early 2000s and have used

SNPs to tag disease-resistance genes (Fang et al., 2010; Xiao et al., 2010) and other important traits. On the public research side, Van Deynze et al. (2009) conducted the first large-scale discovery of SNPs by sequencing 270 loci of 24 cotton genotypes. Later, Byers et al. (2012) developed more than 10,000 SNPs using 454 sequencing technology. With the dramatic decrease in sequencing expenses due to next-generation sequencing technologies, tens of thousands of cotton SNP markers were developed recently with different approaches (Lacape et al., 2012; Gore et al., 2014; Islam et al., 2015; Zhu et al., 2014). In 2014, the first cotton SNP chip, which was developed by the International Cotton SNP Consortium and manufactured by Illumina, was made available to the public. This SNP chip contains 70,000 SNPs. Of them, 50,000 are from *G. hirsutum* cotton lines and intraspecific crosses, and 20,000 are from interspecific comparisons (mainly between *G. hirsutum* and *G. barbadense*).

Genetic and Physical Maps

A high-density genetic or physical map plays an important role in dissecting economically important traits, identifying molecular markers associated with agronomic traits, and cloning genes of interest through map-based cloning strategies. In cotton, researchers have been constructing genetic maps with multiple types of DNA markers using different populations since the 1990s (Reinisch et al., 1994; Rong et al., 2004; Guo et al., 2007; Blenda et al., 2012; J.Z. Yu et al., 2012b). Rapid advancements in sequencing technologies enabled the completion of the whole genome sequence of *G. raimondii* (Paterson et al., 2012; K. Wang et al., 2012) and *G. arboreum* (Li et al., 2014). More genome sequencing projects are expected in the near future. Cultivated Upland cotton has very low genetic diversity as revealed by different molecular marker types (Iqbal et al., 2001; Hinze et al., 2012; D. Fang et al., 2013). Construction of a moderate density *G. hirsutum* intraspecific genetic map will require several thousand polymorphic markers. However, it remains a big challenge to identify such large number of intraspecific polymorphic markers, and currently, there is no high-density *G. hirsutum* intraspecific genetic map that covers the entire genome. Nonetheless, significant progress has been made in constructing intraspecific genetic maps that have been used to map important agronomic traits in Upland cotton.

Interspecific Genetic Maps

Because of the very low polymorphism within *G. hirsutum*, most genetic maps of tetraploid cotton were constructed with interspecific populations, mainly between *G. hirsutum* and *G. barbadense*. Many laboratories around the world invested heavily in this endeavor. Table 3 is a summary of the major cotton interspecific maps that have been constructed. In 2012, a group of scientists from the United States and France constructed the first high-density consensus (HDC) genetic map of tetraploid cotton by combining the six interspecific maps (Blenda et al., 2012). Before constructing this HDC map, cotton SSR markers (17,343 in total) were checked for sequence redundancy with 90% as a similarity cutoff. After the redundant markers were removed, the HDC map was constructed on the basis of a sufficient number of collinear markers in common among the component maps. The HDC map consists of 8254 unique loci, originating from 6669 non-redundant markers, and spans 4070 cM, with an average of 2 loci per cM.

Table 3. Major interspecific genetic maps of tetraploid cotton.

Population code	Parents†	Generation	Population size	Genetic distance‡	No. of mapped markers						Reference
					Total	RFLP	AFLP	SSR	SNP	Other§	
				cM							
T3	TM-1 / 3-79	RIL	186	3430	2527			2280	247		(Fang and Yu, 2012)
GV¶	(Guazuncho 2 / VH8-4602) / Guazuncho 2, and Guazuncho 2 / VH8-4602	BC_1–RIL	215	3637	1745	190	715	781		59	(Lacape et al., 2009)
PK	Palmeri / K101	F_2	57	4448	2584	2459		124		1	(Rong et al., 2004)
TH	(TM-1 / Hai 7124) / TM-1	BC_1	138	3678	3414		80	2742	73	517	(Zhao et al., 2012)
E3	(Emian22 / 3-79) / Emian22	BC_1	141	4419	2316			2311		5	(Yu et al., 2011)
CH	CRI 36 / Hai 7124	F_2	186	4418	1080		93	690		297	(Yu et al., 2007)
HDC#.	Consensus interspecific map			4070	8254	2607	869	4421	205	152	(Blenda et al., 2012)

† All female and recurrent parents belong to *G. hirsutum*, and all male parents belong to *G. barbadense*.

‡ T3, GV, TH and E3 maps were constructed with JoinMap program. PK and CH maps were constructed using MapMaker program.

§ Includes target region amplification polymorphism (TRAP), sequence-related amplified polymorphism (SRAP), isozyme, gene-derived, and morphological markers.

¶ A consensus map constructed using 75 BC_1 and 140 RIL populations.

TRAP, SRAP, morphological and some AFLP markers could not be easily traced back to their description and were not included in consensus map integration.

Among the 6669 mapped markers, 4744 had sequence information, with a majority representing gene-based markers (EST-SSRs, STS-RFLPs, and EST-derived SNPs). Sequence alignment against the *G. raimondii* genome indicated a high level of homology between the diploid D genome and the tetraploid cotton genetic map (13 D_T chromosomes), with only a few possible minor structural rearrangements. Overall, the HDC map serves as a valuable resource for trait QTL comparative mapping, map-based cloning of important genes, and better understanding of the genome structure and evolution of tetraploid cotton.

In a separate effort, Yu et al. (2010) integrated map data from 28 mapping reports published before 2009 and proposed a consensus order of markers for a tetraploid cotton genome reference map. This comprehensive map also included traits or QTLs.

Besides the above efforts, interspecific genetic maps were constructed using a cross either between *G. hirsutum* and *G. tomentosum* (Waghmare et al., 2005) or between *G. hirsutum* and *G. darwinii* (B. Wang et al., 2012). However, these maps only covered a fraction of the entire genome.

Intraspecific Genetic Maps

A *G. hirsutum* intraspecific genetic map could improve the accuracy of QTL analysis compared with an interspecific genetic map if the QTL is intended to be used in Upland cotton breeding. However, low polymorphism within Upland cotton, in addition to the relatively limited number of informative markers available to researchers, has been a great challenge in constructing a moderate-to high-density intraspecific genetic map. To overcome this challenge, cotton scientists used either a four-way-cross population (Qin et al., 2008) or a three-parent composite population (K. Zhang et al., 2012) to construct maps. The approach using populations involving more than two parents ensured greater genetic diversity and an increased frequency of polymorphisms in the mapping populations and an improved possibility of QTL analysis. Representative *G. hirsutum* intraspecific maps are shown in Table 4. In addition, intraspecific genetic maps of *G. arboreum* (Ma et al., 2008), and of *G. raimondii* (Rong et al., 2004) were also constructed.

As shown in Table 4, most intraspecific maps were constructed with SSR markers. The map constructed using a three-parent composite population (no. 17 in Table 4) contains the most marker loci and covers about 94.1% of tetraploid cotton genome. However, this map consists of 69 linkage groups, and many gaps are larger than 20 cM, thus the 94.1% of coverage is still in question. The TM-1/NM24016 map (no. 13 in Table 4) contains the second highest number of loci (841), and about half of them are SNP loci. So far, this is the only intraspecific map that contains SNP markers. In summary, although many intraspecific cotton genetic maps have been constructed, most cover only a fraction of the genome. Hence, cotton researchers need to develop more markers, especially SNPs, to increase the map density. Furthermore, some genomic regions in *G. hirsutum* are monomorphic due to identity-by-descent (D. Fang et al., 2013). These regions might indicate fixation (homozygosity) in modern Upland cotton and may contain key adaptive genes that have been fixed through breeding and selection processes (D. Fang et al., 2014). To increase marker coverage in these regions will be very difficult, if not impossible.

Table 4. A representative sample of *G. hirsutum* intraspecific maps.

No.	Map name	Population size	Population type	Markers	No. of linkage groups	No. loci	Reference
1	(Simian-3/Sumian-12)// (Zhong-4133/8891), 4WC	280	4WC	SSR	56	286	(Qin et al., 2008)
2	0-153/sGK9708	250	F_2	SSR	26	155	(Sun et al., 2012)
3	0-153/sGK9708	196	RIL	SSR	34	190	(Sun et al., 2012)
4	7235/TM-1	258	RIL	SSR	31	156	(Shen et al., 2007)
5	Deltapine-61/Texas-701	251	F_2	SSR	17	73	(Guo et al., 2008)
6	DH962/Jimian5	137	F_2	SSR, RGA, RAPD, SRAP	51	471	(Lin et al., 2009)
7	HQ95-6/MD51ne	199	$F_{2:3}$	RFLP	24	82	(Ulloa et al., 2002)
8	HS-46/MARCABUCAG8US 1-88	96	$F_{2:3}$	RFLP	31	119	(Shappley et al., 1998a)
9	HS-46/MARKCBUCAG8US 1-88	188	RIL	SSR	30	388	(H. Liu et al., 2012)
10	MCU-5/Siokra1-4	244	$F_{3:4}$	SSR, AFLP	18	151	(Lopez-Lavalle et al., 2012)
11	MD-5678ne/Prema	119	$F_{2:3}$	RFLP	17	81	(Ulloa and Meredith, 2000)
12	Prema/86-1	180	RIL	SSR	54	279	(Ning et al., 2014)
13	TM-1/NM24016	95	RIL	SSR, SNP	117	841	(Gore et al., 2014)
14	Yumian-1/T586	270	RIL	SSR, SRAP, IT-ISJ	60	618	(Zhang et al., 2009)
15	Yumian-1/AcalaMaxxa	180	F_2	SSR	62	323	(Shao et al., 2014)
16	Yumian-1/CA3084	180	F_2	SSR	57	302	(Shao et al., 2014)
17	Yumian-1/CRI35//Yumian-1/7235	172	$F_{2:3}$	SSR	69	978	(K. Zhang et al., 2012)
18	Yumian-1/TAM94L-25	180	F_2	SSR	49	262	(Shao et al., 2014)
19	Zhongmiansuo-12/J8891	180	RIL	SSR, AFLP, RAPD, SRAP	35	181	(R. Liu et al., 2012)
20	Zhongmiansuo-12/J8891	171	RIL	cDNA-AFLP	26	302	(Liu et al., 2011)

Physical Maps and Genome Sequences

Genetically anchored physical maps have proven useful both for their intrinsic merit and as adjuncts to genome sequencing (Lin et al., 2010). In cotton, researchers have been trying to construct physical maps for the cultivated tetraploid cottons—*G. hirsutum* and *G. barbadense*—and their diploid progenitor, *G. raimondii*. Although progress toward constructing a physical map for tetraploid cotton has been very slow, a physical map of *G. raimondii* was published, and whole-genome sequences of this diploid species were also made available (Paterson et al., 2012; K. Wang et al., 2012). Following is a brief description of the *G. raimondii* physical map and its genome sequences.

In 2004, Rong et al. published a 763-loci genetic map of *G. raimondii*, of which most of the markers are RFLPs. The same research group constructed bacterial artificial chromosome (BAC) libraries of *G. raimondii*. A whole-genome physical map was assembled. A total of 13,662 BAC-end sequences and 2828 DNA probes were used in genetically anchoring 1585 contigs to a cotton consensus genetic map (Lin et al., 2010). The contigs were assembled to 13 chromosomes. At the same time, the Joint Genome Institute of the US Department of Energy Sanger-sequenced the very same BAC library. Using the physical map and genetic map described, a reference genome sequence of *G. raimondii* was published in 2012 (Paterson et al., 2012). It has 13 pseudomolecules (corresponding to 13 chromosomes) totaling 737.8 Mb and is oriented to 98.3% of the genome. Aligning the markers mapped on the high-density consensus tetraploid map (Blenda et al., 2012) against the reference *G. raimondii* genome sequence revealed high sequence homology between *G. raimondii* and tetraploid cotton. Recently, Z. Wang et al. (2013) placed 18,597 DNA markers corresponding to 48,958 loci along this reference sequence. This reference genome is being widely used for many research purposes, such as cloning a gene of interest, enriching marker density in a specific region, identifying putative (potential) genes, etc. In a separate effort, a research group at the Cotton Research Institute in China published a draft genome sequence of *G. raimondii* (K. Wang et al., 2012).

Recently, Li et al. (2014) sequenced the genome of the diploid A genome species *G. arboreum*. The authors anchored and oriented 90.4% of the assembly on 13 pseudo-chromosomes and found that 68.5% of the genome comprises repetitive DNA sequences. *G. arboreum* was predicted to have 41,330 protein-coding genes. Compared with the whole-genome sequences of *G. raimondii*, insertions of long terminal repeats in the past 5 million years are responsible for the twofold difference in the sizes of these two species' genomes.

Construction of the physical map of tetraploid cotton has been progressing slowly but steadily, with several laboratories around the world working toward this goal. In 2012, M. Zhang et al. published a physical map of *G. hirsutum* that consisted of 3450 binary BAC (BIBAC) contigs with an N50 contig size of 863 kb, collectively spanning 2244 Mb. They also sorted the map contigs according to their origin of subgenomes, showing that the contigs were assembled into the A and D subgenomes, separately. Moreover, the physical map was marked with nearly 10,000 BIBAC end sequences, making one BIBAC end sequence in approximately 250 kb. This physical map should be useful for fine-mapping and cloning the cotton agronomic genes and QTLs and for assembling the cotton genome with next-generation sequencing technologies. So far, genome sequences of tetraploid cotton species are not yet available.

Qualitative Trait Analysis

In cotton, many agronomical important traits are controlled by major genes. Such traits include resistance to diseases and nematodes, male sterility, fiber development, and others related to shape, color of whole plants and/or plant parts. These traits are often of monogenic inheritance in nature. For some traits, such as disease resistance or fiber quality, one or a few major QTLs or genes can account for a very high proportion of the phenotypic variation of the trait. Incorporation of such a gene or major QTL into a specific line can lead to great improvement of the trait in the variety under development. DNA markers that are tightly associated with the major genes can be used for selection of favorable alleles and are often more efficient and more effective than direct selection for the phenotypic traits. In cotton, great success has been achieved in transferring the genes for resistance to nematodes, cotton blue disease (CBD), and bacterial blight into Upland cotton cultivars through MAS. More examples will be described below. Table 5 lists the major cotton genes that have been mapped with molecular markers. Most are genes for disease or nematode resistance, because of their economic importance. Several traits, such as Li_1, Li_2, im, Rf_1, T_1, t_2, were mapped independently by different research groups, and the results were similar or same. These are naturally occurring mutants that affect fiber development or hair morphology. For the reniform-nematode resistance genes, three sources of resistance were identified: Ren^{lon} (also called Ren_1), from *G. longicalyx* and mapped on chromosome 11(Dighe et al., 2009); $Reni^{ari}$ (renamed as Ren_2), from *G. aridum* and mapped on chromosome 21(Fang and Stetina, 2011); and Ren^{barb1}, Ren^{barb2}, and Ren^{barb3}, from *G. barbadense* accession 713 (GB713) (Gutierrez et al., 2011). It appears that Ren^{lon}, $Reni^{ari}$, and Ren^{barb2} are duplicate genes residing on homeologous chromosomes 11 or 21 (Fang and Stetina, 2011; Table 5).

A biparental mapping strategy was mainly used to identify markers associated with a major gene in cotton. Almost all the research referenced in Table 5 was conducted with that approach. Taking a single cross as an example, the general procedure to map a major gene can be described as follows:

1. Select parents, at least one of which possesses the gene of interest, and make a cross.
2. Analyze the parents with markers with or without chromosome positions to identify polymorphic markers.
3. Plant F_1 plants; self-pollinate to obtain F_2 seeds.
4. Plant the segregating F_2 population. The population size is usually determined by the complexity of the trait. For a single-gene trait, 100 to 200 individual F_2 plants are usually enough. However, for a trait like root-knot nematode resistance, the population size should be larger than 200 if possible and feasible. Isolate DNAs from each F_2 plant. Phenotypes of F_2 plants should be obtained when possible.
5. Plant $F_{2:3}$ progeny rows, and phenotype the progeny if needed. A bulk of $F_{2:3}$ individuals within a row may be used for further marker confirmation when necessary if the preceding F_2 plant is homozygous for the markers.

Table 5. A representative sample of qualitative (major) genes that have been mapped in cotton.

Trait	Abbrev.	Population	Generation	Population size	Cross type	Markers	Location	Interval	Reference
Bacterial blight resistance (race 18)	B_{12}	Delta Opal/DP388	$F_{4:5}$	285	GH × GH	SSR, SNP	c14	CIR246–NG0207069	(Wright et al., 1998; Xiao et al., 2010)
Blue disease resistance	*Cbd*	Delta Opal/DP388	$F_{2:3}$	363	GH × GH	SSR, SNP	c10	DC20027–NG0204310	(Fang et al., 2010)
Envoke herbicide susceptibility	*ehs*	HS26/ST474	F_2	450	GH × GH	SSR	c20	BNL3071–JESPR171	(Thyssen et al., 2014)
Fiberless	N_1	Pima S-7/N1	F_2	159	GB × GH	RFLP	c12	Unig24G11–pAR10F10	(Rong et al., 2005)
Fiberless	n_2	Pima S-7/n2	F_2	123	GB × GH	RFLP	c12/c26		(Rong et al., 2005)
frego	*fg*	T582/Hai7124	F_2	290	GH × GB	SSR	long arm of c3	NAU3016–NAU3172	(Li et al., 2010)
Hybrid lethality	*Le*	multiple	F_2	850	GH × GB	SSR	c24, c21	NAU2306–JESPR157, BNL3279–BNL1154	(Song et al., 2009)
immature mutant	*im*	TM-1/im mutant	F_2	366	GH × GH	SSR	c3	DPL1071–SHIN-151	(Kim et al., 2013; C. Wang et al., 2013)
Ligon-lintless 1	Li_1	Li1 mutant/DP5690	F_2	2553	GH × GH	SSR	c22	TMB2500–DPL0489	(Gilbert et al., 2013)
Ligon-lintless like	*Lix*	Lix/Li1, Lix/TM-1, Hai7124/Lix	F_2	4389	GH × GH, GB × GH	SSR	c04	NAU8376–NAU3469	(Cai et al., 2013)
Ligon-lintless 2	Li_2	Li2 mutant/DP5690	F_2	136	GH × GH	SSR	c18	DL0547–DPL0922	(Hinchliffe et al., 2011b)
Okra leaf shape	*L*	NC05AZ21/NC11–2100	F_2	236	GH × GH	SSR	short arm of c15	Gh565–NAU2343	(Andres et al., 2014)
Open bud	ob_1, ob_2	ob mutant/TM-1, ob mutant/Hai7124, 3–79	F_2	595	GH × GH and GH × GB	SSR	c13, c18	BNL2652–NAU0624, BNL249–TMP001	(Qian et al., 2009)
Pilose	T_1	Yumian 1/T586	RIL	270	GH × GH	SSR	c6	BNL3650–BNL4108	(Wan et al., 2007)
Leaf hairness	t_2	Pima S-7/Empire B2b6	F_2	150	GB × GH	RFLP	c25	pAR026a–P1–34a	(Wright et al., 1999)
Reniform nematode resistance	Ren^{lon}	*G. hirsutum/G. Longicalyx/G. armourianum/*	multiple		Trispecies	SSR	c11	BNL0836–BNL3279	(Dighe et al., 2009)
Reniform nematode resistance	Ren^{ari}	*G. arboreum/(G. hirsutum/G. aridum)*	BC	76	Trispecies	SSR	c21	BNL3279	(Romano et al., 2009)
Reniform nematode resistance	Ren^{barb1}	*G. barbadense* 713/Nem X	multiple	300	GB × GH	SSR	c21	BNL1551–GH132	(Gutierrez et al., 2011)

Trait	Abbrev.	Population	Generation	Population size	Cross type	Markers	Location	Interval	Reference
Reniform nematode resistance	*Ren*^{barb2}	*G. barbadense* 713/Nem X	multiple	300	GB × GH	SSR	c21	BNL4011–BNL3279	(Gutierrez et al., 2011)
Reniform nematode resistance	Ren^{barb3}	*G. barbadense* 713/Nem X	multiple	300	GB × GH	SSR	c18	BNL1721–BNL0569	(Gutierrez et al., 2011)
Restoration of cytoplasmic male sterile (CMS-D2)	Rf_1	(B416R/Ark8518)/Ark8518	BC_1F_1	114	GH × GH	RAPD, STS		UBC147–UBC607	(Feng et al., 2005)
Restoration of cytoplasmic male sterile (CMS-D2)	Rf_1	multiple	F_2	8024	GH × GH	RAPD, SSR	long arm of c4		(Liu et al., 2003)
Restoration of cytoplasmic male sterile (CMS-D8)	Rf_2	multiple	F_2	679	GH × GH	RAPD		UBC188–UBC169	(Zhang and Stewart, 2004)
Root-knot nematode resistance	rkn_1	Nem X/SJ-2	$F_{2:3}$, RIL	96, 60	GH × GH	AFLP	c11	GHACC1–CIR316	(Wang and Roberts, 2006)
Root-knot nematode resistance	None	Alburn 634RNR/ST213	F_2	86	GH × GH	SSR	c14	BNL3661–BNL3545	(Ynturi et al., 2006)
Root-knot nematode resistance	*Mi-C11*	M120RNR/Pima S-6	F_2	1252	GH × GB	SSR, AFLP	c11	CIR069–CIR316	(Shen et al., 2010)
Root-knot nematode resistance	None	CLW/ST213, M240/CLW6, M240/ST213	RILs	289	GH × GH	SSR	c11, c14	CIR316, BNL3661	(Gutierrez et al., 2010)

6. Conduct bulked segregant analysis (Michelmore et al., 1991) to identify molecular markers associated with the trait of interest. Alternatively, analyze all polymorphic markers among the entire F_2 population.
7. Construct a linkage map using software such as JoinMap (Van Ooijen, 2006) to establish a linkage between markers and the genetic locus controlling the trait.

Quantitative Trait Analysis

Most of the important agronomic traits, such as yield and fiber quality, are polygenic or controlled by QTLs. Marker-assisted selection for the improvement of such traits is a complex and difficult task because it is related to many genes or QTLs, QTL × environment (E) interaction, and epistasis. Usually, each of these genes has a small effect on the phenotypic expression, which also is influenced by environmental conditions. The QTL × E interaction reduces the efficiency of MAS, and epistasis can result in a skewed QTL effect on the trait. Furthermore, unexpected epistatic interactions with other genes can also affect phenotypic expression. Consequently, phenotyping quantitative traits is a complex endeavor, and determining marker-phenotype association becomes difficult as well. Therefore, repeated field tests are required to accurately characterize the effects of the QTLs and to evaluate the stability across environments. In cotton, quantitative traits that were analyzed and mapped mainly include three groups: fiber traits, yield and yield components, and disease resistance. At present, more than 2000 QTLs have been reported (Table 6). Despite a tremendous number of QTL mapping experiments during the past decade, application and utilization of the QTL mapping information in cotton breeding has been constrained by a number of factors: (i) a lack of universally valid QTL-marker associations that are applicable across populations. Because QTL mapping is needed to identify new QTL markers whenever a new germplasm is used, some people become discouraged and lose interest in MAS; (ii) a strong QTL × E interaction, which makes phenotyping difficult since expression may vary from one location or year to another; (iii) deficiencies in QTL statistical analysis, which lead to either overestimation or underestimation of the number of QTLs involved and their effect on the trait; (iv) a frequent lack of QTLs with major effects on the trait, which necessitates identification of a large number of QTLs, an often impossible or difficult-to-achieve goal that further complicates identification of marker-QTL associations; (v) QTLs identified in an interspecific population that are not present in *G. hirsutum*. Markers showing polymorphism between two species may be monomorphic within *G. hirsutum*, which in turn makes the markers useless in Upland cotton breeding; (vi) difficulty and limited success in transferring favorable traits from *G. barbadense* to *G. hirsutum* via intermating (Lacape et al., 2005; Gore et al., 2012). Nonetheless, QTLs identified in earlier studies have laid the foundation for further dissection of quantitative traits and will help confirm and validate new QTLs in any future research.

A large majority of cotton QTL analysis has focused on fiber quality and yield, followed by disease resistance (Table 6). Three meta-analyses of fiber QTLs (Rong et al., 2007; Lacape et al., 2010; Said et al., 2013) have revealed QTL clusters and hotspots. Some of the clusters or hotspots were also confirmed by more recent studies (D. Fang et al., 2014). A few of the repeatedly identified fiber QTLs are being used in breeding (see below). Compared with fiber QTLs, QTL analysis

Table 6. A representative sample of quantitative trait loci and their characteristics in cotton.

Trait	Abbrev.	Population	Type	Generation	Size	Markers	No. QTLs	Major	Reference
Fiber		TM-1/3-79	GH × GB	F_2	171	RFLP, RAPD	14		(Kohel et al., 2001)
Fiber		Acala-44/Pima S-7	GH × GB	F_2	94	AFLP, SSR, RFLP	7		(Mei et al., 2004)
Fiber		Random mated population	GH × GH	$C_{11}S_{12}$	260	SSR			(Zeng et al., 2009)
Fiber		Random mated population	GH × GH	C_5S_6	275	SSR	131		(D. Fang et al., 2014)
Fiber		0-153/sGK9708	GH × GH	F_2, F_3, RIL	250	SSR	50		(Sun et al., 2012)
Fiber		Luyuan/Lumianyan22	(GH × GB) × GH	F_2, F_3	209	SSR	24		(F. Wang et al., 2013)
Fiber		Yumian-1/T586	GH × GH	RIL	270	SSR, SRAP, IT-ISJ	13		(Zhang et al., 2009)
Fiber		Yumian-1/CRI35//Yumian-1/7235	GH × GH	$F_{1:3}$	172	SSR	63		(K. Zhang et al., 2012)
Fiber		multiple	GH × GH	$F_{2:4}$	540	SSR	77		(Shao et al., 2014)
Fiber		SG747/Giza 75	GH × GB	BC_2F_1	146	SSR	28		(J. Yu et al., 2013b)
Fiber		CRI36/Hai7124	GH × GB	$F_{2:3}$	180	SSR, AFLP, others	106		(J. Yu et al., 2013a)
Fiber		7235/TM-1	GH × GH	F_2	2517	SSR	5	24	(Chen et al., 2009)
Fiber		MD52ne/MD90ne	GH × GH	F_2	768	SSR	7		(Islam et al., 2014)
Fiber		(Guazuncho 2 / VH8-4602) / Guazuncho 2	(GH × GB) × GH	BC		SSR, AFLP, RFLP	80		(Lacape et al., 2005)
Fiber and yield		(Simian-3/Sumian-12)//(Zhong-4133/8891)	GH × GH	4WC	280	SSR	31		(Qin et al., 2008)
Fiber and yield		HS-46/MARCABUCAG8US 1-88	GH × GH	$F_{2:3}$	96	RFLP	100		(Shappley et al., 1998b)
Fiber and yield		MD-5678ne/Prema	GH × GH	$F_{2:3}$	119	RFLP	26		(Ulloa and Meredith, 2000)
Fiber and yield		7235/TM-1	GH × GH	RIL	258	SSR	16		(Shen et al., 2006b)
Fiber strength		7235/TM-1	GH × GH	F_2, BC		SSR, RAPD, SCAR	1	10	(Guo et al., 2003)

Continued.

Table 6. Continued.

Trait	Abbrev.	Population	Type	Generation	Size	Markers	No. QTLs	Major	Reference
Fusarium wilt resistance (FOV4)	FWR	MCU-5/Siokra 1-4	GH × GH	F_3, F_4	244	SSR, AFLP	4		(Lopez-Lavalle et al., 2012)
Fusarium wilt resistance (FOV4)	FOV	multiple	GH × GH, GH × GB	F_2, RIL	680	SSR	4	14	(Ulloa et al., 2013)
Fusarium wilt	FW^R	ZMS35/Junmian 1	GH × GH	$F_{2:3}$	317	SSR	1 major and 4 minor	17	(Wang et al., 2009)
Gossypol		HS26/MARKCBU-CAG8US 1-88	GH × GH	RIL	188	SSR	10		(H. Liu et al., 2012)
Gossypol		SG747/Giza 75	GH × GB	BC_2F_1	146	SSR	3		(J. Yu et al., 2012)
Height of NFFB	HNFFB	Baimian2/TM-1 and Baimian2/CIR12	GH × GH	$F_{2:3}$	428	SSR	6	17–2	(Li et al., 2012)
Leaf morphology		DP61/Seaberry	GH × GB	F_2	180	RFLP	62		(Jiang et al., 2000)
Leaf pubescence		Pima S-7/Empire B2b6	GB × GH	F_2	150	RFLP	2		(Wright et al., 1999)
Leaf pubescence		Guazuncho 2/VH8-4602	GH × GB	BC_1, BC_2	275	RFLP, SSR, AFLP	9	6	(Lacape and Nguyen, 2005)
Lint percentage	Lint%	MD17/181, MD17/FM966	GH × GH	F_2	264	SSR	2		(An et al., 2010)
Node of first fruit branch	NFFB	Baimian2/TM-1 and Baimian2/CIR12	GH × GH	$F_{2:3}$	428	SSR	8	17–2	(Li et al., 2012)
Node of first fruit branch	NFFB	DP61/Texas 701	GH × GH	F_2	251	SSR	3	21	(Guo et al., 2008)
Oil content		SG747/Giza 75	GH × GB	BC_2F_1	146	SSR	17	21	(J. Yu et al., 2012)
Protein		HS26/MARKCBU-CAG8US 1-88	GH × GH	RIL	188	SSR	12		(H. Liu et al., 2012)
protein content		SG747/Giza 75	GH × GB	BC_2F_1	146	SSR	22		(J. Yu et al., 2012)
Stomatal conductance		TM-1/NM24016	GH × GH	F_2	118	SSR, RAPD	2		(Ulloa and Meredith, 2000)
Verticillium wilt resistance	VWR	verification using mass crosses derived from 17 different hybrid cotton lines							(Li et al., 2013)
Verticillium wilt resistance	VWR	TM-1/NM24016	GH × GH	RIL	95	SSR, SNP, RGA-AFLP	21		(H. Fang et al., 2013)
Verticillium wilt resistance	VWR	SG747/Pima S-7	GH × GB	BC_2F_4	146	RGA-AFLP	3	4	(H. Fang et al., 2014)
Verticillium wilt resistance	VWR	Hai7124/Junmian 1	GB × GH	F_2, BC_1	128, 96	SSR, RGA, DDRT	10		(Yang et al., 2008)

Trait	Abbrev.	Population	Type	Generation	Size	Markers	No. QTLs	Major	Reference
Yield		SG747/Giza 75	GH × GB	BC_2F_1	146	SSR	39		(J. Yu et al., 2013b)
Yield		CRI36/Hai7124	GH × GB	$F_{2:3}$	180	SSR, AFLP, others	71		(J. Yu et al., 2013a)
Yield		CRI12/J8891	GH × GH	IF_2, RIL	180	SSR	111		(R. Liu et al., 2012)
Yield		7235/TM-1	GH × GH	F_2	2517	SSR	6	24	(Chen et al., 2010)
Yield and yield components		ZMS12/8891	GH × GH	RIL	180	SSR	34		(Wang et al., 2007)
Yield and fiber traits		Handan 208/Pima 90	GH × GB	$F_{2:3}$	69	SSR, RAPD, ARAP, REMAP	52		(He et al., 2007)
Yield and fiber traits		HS-46/MARKCBU-CAG8US 1-88	GH × GH	RIL	188	SSR	56		(Wu et al., 2009)
Yield and fiber traits		TM-1/NM24016	GH × GH	RIL	95	SSR, SNP	6		(Gore et al., 2014)
Yield and fiber traits		Acala Prema/86-1	GH × GH	RIL	180	SSR	86		(Ning et al., 2014)
Yield, phenology, morphology		Guazuncho 2 / VH8-4602	GH × GB	RIL	140	SSR, AFLP, RFLP	166		(Lacape et al., 2013)

on yield or yield components lag behind: although many yield QTLs have been identified, very few show congruence between reports. The low inheritability of yield traits is a major reason, which means that more field tests involving different environments are required for more accurately mapping yield QTLs. Due to significant losses resulting from Verticillium wilt, Fusarium wilt, and nematodes (root-knot and reniform) in China, Australia, and the United States, significant efforts have been made to identify QTLs resistant to these diseases. Although cotton leaf curl virus is detrimental to cotton in India and Pakistan, no QTLs resistant (or tolerant) to the virus have been reported yet.

Methods to Map QTLs

Mapping QTLs is more complicated and difficult than mapping major genes. In cotton, several mapping methodologies, including biparental mapping, multiple-parent mapping and association mapping, have been used to identify trait QTLs.

Biparental Mapping

Biparental mapping has been the most frequently used method in cotton, as shown in Table 6. The method is similar to that for mapping major genes, but some differences exist. First, the population size for QTL mapping is usually larger. Second, the segregating population should be repeatedly evaluated for the trait of interest in multiple environments (years or locations) to mitigate the environmental effects on QTL mapping. Third, bulked segregant analysis is not useful. All progeny should be genotyped with all polymorphic markers to construct a linkage map that is useful for QTL identification. Fourth, the phenotypic data may need to be analyzed with appropriate statistical tools such as SAS or ASReml (Gilmour et al., 2009) before QTL mapping. Fifth, an appropriate mapping method, such as composite interval mapping, is required to identify QTLs. This step is usually conducted with software such as MapQTLs (Van Ooijen, 2009), QTL Cartographer, or others.

Multiple-Parent Mapping

As indicated by its name, this method uses more than two parents to develop a population for QTL identification. Because of the very low genetic diversity within Upland cotton, it is very difficult or impossible to obtain a high-resolution genetic map with a biparental population. The approach using populations involving more than two parents ensures greater genetic diversity, an increased polymorphism frequency in the mapping populations, and an improved possibility of QTL analysis. Qin et al. (2008) first used a four-way cross to map fiber and yield QTLs and developed a statistical model to construct the linkage map. Zhang et al. (2012a) used a three-parent composite population to identify fiber QTLs. A linkage map was also constructed using a method similar to that employed by Qin et al. (2008). To further increase the polymorphism, a random-mated population involving more than 10 parents was used to map trait QTLs (Zeng et al., 2009). Recently, D. Fang et al. (2014) used a random-mated recombinant inbred population derived from 11 diverse Upland cotton cultivars as parents for fiber QTL analysis. They genotyped the recombinant inbred lines with 1548 SSR markers. The mapping positions of markers were based on the HDC map (Blenda et al., 2012) and the *G. raimondii* reference genome sequence. The marker-trait associations were analyzed with the software program TASSEL (Bradbury et al., 2007).

Association Mapping

Association mapping based on linkage disequilibrium has become a powerful tool to identify marker-trait QTL associations in plants (Hamblin et al., 2011). It utilizes all the alleles contained in a panel of varieties and phenotypic data that may be obtained through experiments or historical variety tests. In cotton, a few association mapping reports have been published, most focusing on fiber or yield traits (Abdurakhmonov et al., 2009; Mei et al., 2013; Zhang et al., 2013; Cai et al., 2014). In collaboration with cotton researchers in the USDA-ARS, the author and his research group in New Orleans are conducting association mapping of fiber quality traits, yield, and seed composition using the historical data from National Cotton Variety tests that began in 1960. In addition, the cotton community is developing a nested association mapping population for more rigorous trait QTL identification in cotton.

Molecular Cotton Breeding in Practice

Molecular markers and other genomic tools have been used in practical cotton breeding by both public institutions and private companies. Great success has been achieved in transferring root-knot resistance gene(s) into elite cultivars and in introgressing reniform nematode resistance genes from *Gossypium* species into Upland cotton. In this section, I will discuss a few examples of the utilization of molecular markers and other genomic tools in cotton breeding.

MAS for Cotton Blue Disease and Bacterial Blight Resistance Breeding

Cotton blue disease is caused by a virus that belongs to the genus *Polerovirus* of the family Luteoviridae (Correa et al., 2005). Symptoms include leaf rolling, vein yellowing, moderate to severe stunting due to shortening of internodes, and a dramatic dark-green to bluish color of leaves, from which the name "blue disease" originated. In Brazil and other South American countries, CBD is a serious problem for cotton production. Resistance to CBD is present in Upland cotton. The cotton cultivar 'DeltaOpal' is resistant to CBD, and this resistance is controlled by a single dominant gene, *Cbd* (Fang et al., 2010), which was mapped to chromosome 10 and is flanked by the SSR marker DC20027 and the SNP marker NG0204310. The SSR marker DC20027 was converted to the SNP marker NG0211495 (Fang et al., 2010). Bacterial blight, caused by *Xanthomonas axonopodis* pv. *malvacearum*, is a major cotton disease occurring in most cotton-producing countries. In the United States, yield losses caused by this disease have been as high as 3.4% in some years (Brown, 2001). In Australia, resistance to bacterial blight is a mandate for all commercial cotton varieties. In Brazil, bacterial blight is managed by planting resistant cultivars. There are many isolates, but race 18 is the most prevalent. The resistance to race 18 is present in many cotton varieties, including 'DeltaOpal', and is controlled by a single dominant gene, B_{12}. This gene was mapped to chromosome 14 and is flanked by SNP markers NG0207069 and NG0207155. SSR markers CIR246 and BNL3545 are also at adjacent locations (Xiao et al., 2010). Since 2004, at Delta and Pine Land Company (now part of Monsanto), more than 3000 germplasms, cultivars, and breeding lines have been surveyed with multiplex SSR or SNPs via the TaqMan assay. The marker information and the predicted genotypes at the *Cbd* and B_{12} loci of these lines were provided to breeders, who used this information when selecting a parent for a cross.

For the breeding program in Brazil, the Deltapine cotton breeder screened all F_2 plants with markers associated with the *Cbd* and B_{12} genes. Plants without these genes were eliminated. At the F_4 or F_5 stage, progeny rows were screened again with the markers. The breeder made decisions based on marker information and other objectives, such as yield or fiber quality. Then the potential lines were evaluated in a CBD disease nursery or inoculated with bacterial blight by spraying in a field. Almost all cotton lines selected via MAS were resistant to both diseases in field tests. By using markers, the number of progeny rows that went to field trials was significantly reduced, and the breeder focused his efforts on selecting high-yielding, good-quality lines. The same strategy was also applied to the Deltapine breeding program in Australia. Although there is no CBD in Australia, there is a similar viral disease, cotton bunchy top (CBT). Most of the cotton lines that have the CBD markers are also resistant to CBT (unpublished data, 2007). It remains unclear whether CBD and CBT are the same disease occurring in different countries.

MAS for Root-Knot Nematode Resistance Breeding

The southern root-knot nematode, *Meloidogyne incognita* Kofoid and White (Chitwood), is the most important endoparasitic pest of cotton. Found worldwide, the root-knot nematode causes root damage and extensive galling, resulting in significant yield losses. Worldwide cotton yield losses due to this pest were estimated to be 10.7% (Shen et al., 2006a). In addition to direct losses from this nematode, infection by it greatly increases the incidence and severity of Fusarium wilt (caused by *Fusarium oxysporum* f.sp. *vasinfectum*), which results in an estimated loss of another 54.5 million pounds of lint annually in the USA.

Three major sources of resistant germplasm are present in *G. hirsutum*: Auburn 623RNR, Clevewilt 6, and Acala NemX (Shen et al., 2006a). Wang et al. (2006) first identified a recessive root-knot nematode resistance gene, *rkn1*, in Acala NemX. They mapped this gene to the telomeric region of chromosome 11 and tagged it with SSR marker CIR316. Shen et al. (2006a, 2010) mapped the resistance gene (*Mi-C11*) and a major QTL from the source Auburn 623RNR to chromosome 11, and the flanking markers are CIR069 and CIR316. They also identified a second resistance locus (or QTL) on chromosome 7, and the flanking markers were the SSR marker NAU474b and the RFLP marker G1158b. On the basis of the associated marker CIR316, the locus *rkn1* may be the same as *Mi-C11*. In separate research, Gutierrez et al. (2010) identified two QTLs responsible for nematode resistance. One is located on chromosome 11, is tagged by SSR marker CIR316, and is responsible for galling index, and the other one is on chromosome 14, is tagged by the SSR markers BNL3545 and BNL3661, and is responsible for egg production. These three independent studies all confirmed that the gene on chromosome 11 tagged by CIR316 is a major source of resistance. Using the SSR marker CIR316, Fang (unpublished data, 2007) screened more than 3000 progeny lines derived from crosses involving LA887 as one parent. Out of these lines, one resistant line became the commercial cultivar DP174RF, the first root-knot nematode resistant commercial cotton cultivar that was bred partially with the assistance of molecular markers. Monsanto continued to fine-map the resistance gene regions with SNP markers. In 2014, the company launched a new root-knot resistant variety, DP1454NRB2RF, which was bred via MAS (http://southeastfarmpress.com/cotton/cotton-resistant-root-knot-nematodes-unveiled, accessed 18 Jan. 2015).

In public breeding, both CIR316 and BNL3661 were used to track the resistance QTLs on chromosomes 11 and 14, respectively, in the USDA-ARS breeding program at Mississippi State, MS (Jenkins et al., 2012). The researchers concluded that these markers are effective and efficient enough to assist in the development of varieties resistant to root-knot nematode. Roberts and Ulloa (2010) screened 56 known resistant and susceptible genotypes with SSR markers on chromosomes 7, 11, and 14. Except for the marker CIR316 on chromosome 11, which showed consistent differences between the resistant and susceptible genotypes, no such consistency was observed between the resistant and susceptible genotypes with the other two chromosomes. The marker sequence homology between tetraploid and diploid cotton species also indicates that root-knot nematode resistance may be present in other species.

MAS for Reniform Nematode Resistance Breeding

The reniform nematode (*Rotylenchulus reniformis* Linford & Oliveira) is an obligate plant parasite that feeds on roots. In recent years, it has been expanding its geographic distribution in the United States, and has become a major pest of cotton (Robinson et al., 2007). Currently, US cotton production losses from this pest are estimated to exceed $100 million annually. In Mississippi, Louisiana, and Alabama, it has emerged as a very important pathogen that caused about 7% yield loss. No sources of host-plant resistance have been found in cultivated Upland cotton despite extensive evaluation of more than 2000 *G. hirsutum* accessions by various researchers (Robinson et al., 1999; Weaver et al., 2007). To date, control of reniform nematodes has been largely limited to crop rotation and application of nematicides. Breeding Upland cotton varieties with resistance to reniform nematodes would be a great advantage to the US cotton farmers and to the environment. Toward this goal, three research groups in the United States have identified resistance genes from different sources and mapped them with molecular markers, as mentioned above. Utilization of the LonRen germplasm that contains the Ren^{lon} gene for cotton variety development has been in progress in both public institutions and private companies. Breeders use the associated markers, such as BNL3279, to track the introgression of the Ren^{lon} gene in each generation. Weaver et al. (2013) selected 20 homogeneous resistant and 20 homogeneous susceptible $F_{2:4}$ and $F_{2:5}$ lines from the cross LonRen-1/ Fibermax 966 on the basis of the marker genotypes and tested them in two fields, one infested with the reniform nematode and one not infested. There were no yield differences between lines with and without Ren^{lon} in the nematode-free field. Fiber quality was affected by Ren^{lon}, with lines carrying the gene tending to have greater fiber strength, better length uniformity, and lower short-fiber content. The Ren^{lon} gene is effective in reducing nematode populations and may improve fiber quality; however, the LonRen-1 progenies did not reduce yield losses from reniform nematodes. The researchers also noted that cotton plants carrying the Ren^{lon} gene were usually stunted during early season growth and yielded less than those without the gene. This indicates that more research is required to determine the reason (linkage drag or pleiotropic effect) and to fully utilize this resistance gene in variety improvement. The gene Ren^{ari} appears to have similar effects as Ren^{lon}.

Realizing the deleterious effects of *Ren*^{lon} or *Ren*^{ari}, more recent effort is focusing on the introgression of resistance genes from *G. barbadense* accession 713 into Upland cotton (Gutierrez et al., 2011). US cotton breeders are using the markers to assist the introgression process.

MAS for Improving Fiber Quality

Zhang et al. (2003) identified a major fiber strength QTL on chromosome 24. This QTL originally came from the accession 'Suyuan 7235'. Later the same group fine-mapped this QTL using three backcross populations (Chen et al., 2009) and confirmed the effects of this QTL on fiber strength in different genetic backgrounds. A research group in the United States made crosses between 'Suyuan 7235' and two US germplasm lines with different fiber strengths (Kumar et al., 2012). This QTL was consistently expressed across generations and years in both populations. Up to 40% of the total phenotypic variation was explained by the 'Suyuan 7235' allele. DNA marker analysis on a collection of elite cotton lines indicated that this QTL is not present in the elite US public germplasm surveyed.

The Upland cotton lines MD52ne and MD90ne are near-isogenic lines with significant difference in fiber bundle strength. Both lines were developed through backcross breeding in the USDA-ARS cotton breeding program at Stoneville, MS. MD52ne is a BC_6 high bundle strength selection, with MD90ne as the recurrent parent (Meredith, 2005b; Hinchliffe et al., 2010). In these studies, MD52ne was shown to have approximately 10 to 25% higher fiber bundle strength than MD90ne (Meredith, 2005a). This difference in fiber bundle strength between the near-isogenic lines was confirmed across multiple years (2005–2012) and environmental locations (Stoneville, MS, and New Orleans, LA) with no significant genotype × E interaction, implying that the observed difference in fiber bundle strength was highly inheritable and possibly under the control of a major gene (Hinchliffe et al., 2011a). Using MD52ne and MD90ne as parents, crosses were made, and a major fiber strength QTL was identified on chromosome 3 (Islam et al., 2014). This QTL is being validated in different genetic backgrounds, and the markers are being used to assist transfer of this QTL to other germplasm lines.

Shen et al. (2011) used markers to assist transfer of the fiber length QTL on chromosome 1 (*qFL-chr1*) into other cotton lines. The superior allele of this QTL came from *G. barbadense,* which contributed to longer fibers and explained up to 24% of the phenotypic variance. To substantiate the association of this genomic region with fiber length, three BC_3F_2 plants heterozygous for the genetic markers that delineate the *qFL-chr1* QTL were selected to construct three independent populations of near-isogenic introgression lines. The results support the positive effect of *qFL-chr1* on fiber length. A single near-isogenic introgression line, R01-40-08, had about 94.3% of recurrent genome composition and significantly longer fibers than the recurrent parent. In addition to confirming the efficacy of *qFL-chr1* to enhance fiber length, this work provides a valuable genetic resource for the improvement of cotton.

MAS for Improving Fiber Quality by Developing *G. barbadense* Introgressions in *G. hirsutum* Backgrounds

G. barbadense has better fiber quality than *G. hirsutum*. However, transfer of these better fiber-quality traits into *G. hirsutum* through intermating resulted in very limited success despite many decades of continuous effort by many cotton geneticists and breeders. The reasons include hybrid breakdown, segregation toward either parents, linkage drag etc. (Gore et al., 2012, 2014). To overcome some of these problems, cotton scientists developed chromosome substitution lines (Saha et al., 2010) and chromosome-segment introgression lines (CSILs) (P. Wang et al., 2012), both of which were used to dissect the genetic control of fiber quality traits and for QTL analysis. Recently, Cao et al. (2014) demonstrated the first practical use of CSILs for the transfer of fiber quality QTLs into Upland cotton cultivars using SSR markers without detrimentally affecting desirable agronomic characteristics. Of many CSILs, four CSILs, IL040-A4-1, IL080-D6-1, IL088-A7-3, and IL019-A2-6, which were found to be associated with superior fiber qualities including fiber length, strength, and fineness QTL, were selected, backcrossed, and transferred into three commercial Upland cotton cultivars: Xinluzao 26, Xinluzao 41, and Xinluzao 42, grown in Xinjiang, China. By backcrossing and self-pollinating twice, five improved lines (3262-4, 3389-2, 3326-3, 3380-4, and 3426-5) were developed via MAS of background and introgressed segments. In diverse field trials, these QTLs consistently and significantly had additive effects on the target phenotype. Furthermore, the authors also pyramided two segments from different CSILs (IL080-D6-1 and IL019-A2-6) into cultivar 0768 to accelerate the breeding process via MAS. The improved lines pyramided by these two introgressed segments showed significant additive epistatic effects in four separate field trials. In summary, this research showed that the CSILs have great potential to improve fiber qualities in Upland cotton MAS breeding programs.

Marker-Assisted Backcross Selection

Marker-assisted backcross selection is considered to be the simplest form of marker-assisted selection, and at present it has been successfully used in assisting the introgression of transgenes from a donor parent to elite cultivars. Currently, 90% of cotton planted in the United States contain one or more transgenes that are tolerant primarily to herbicides such as Round-up Ready and the insecticidal toxin producing BollGuard. The transgenes were originally transformed into a few often obsolete cotton lines such as Coker 312. Cotton seed companies had to introgress the transgenes into better elite cultivars through breeding. At Delta and Pine Land Company, MABS was widely used to expedite the introgression of transgenes into many elite cultivars. The general procedure of MABS is as follows:

1. Select an elite cultivar as the recurrent parent (RP), and make a cross with a donor parent (DP) containing the transgene(s). The DNA sequences of the transgene are known, and the presence of the transgene can be easily detected with a regular method such as PCR/gel electrophoresis or the TaqMan assay.
2. Identify markers that are polymorphic between the RP and DP. Select 10 markers or more per chromosome whenever possible. The markers should be well distributed along the chromosome.
3. Plant the F_1 population and detect the presence of the transgene at early stages of growth to eliminate false hybrids. False F_1's can also be eliminated by spraying an herbicide if the transgene is an herbicide-tolerant gene. Cross the true F_1 plants back to the RP.
4. Plant the BC_1F_1 population, screen individuals for the transgene at early growth stages, and cross the individuals carrying the transgene in heterozygous status back to the RP.

5. Plant the BC_2F_1. Screen individual plants for the transgene and discard the individuals carrying no transgene. Genotype each individual progeny with 260 markers.
6. Plant the progenies that have the greatest part of the genome of the RP based on the markers. Backcross it to RP to obtain the BC_3F_1. Screen individual plants for the transgene and discard the individuals carrying no transgene. Genotype each individual progeny with markers that are still polymorphic.
7. Plant the progenies that have the greatest part of the genome of the RP based on the markers. Backcross it to RP to obtain the BC_4F_1. Screen individual plants for the transgene and discard the individuals carrying no transgene. Self-pollinate to obtain the BC_4F_2. Harvest homozygous individuals at the transgene locus for further evaluation and release.

MABS can expedite the cultivar development that is essential for a seed company to stay competitive. Early on (mid-2000s), the major hurdle for MABS was the shortage of DNA markers that showed polymorphisms between the DP and RP, in addition to the uneven marker distribution across the genome. With more and more SNP markers becoming available, MABS is expected to play a bigger role in transgenic breeding.

Future Perspectives

Molecular cotton breeding has gained more and more attention from both private and public sectors since 1990s. Compared with conventional breeding, molecular breeding has several advantages, as outlined by Jiang (2013). In cotton, molecular breeding has gone through the following three stages: (i) development of tool boxes, which includes developing large number of easy-to-use molecular markers, constructing genetic and physical maps, and generating genomic resources such as EST databases and *G. raimondii* and *G. arboreum* genome sequences; (ii) identification of molecular markers associated with traits of interest, which includes mapping qualitative and quantitative traits; and (iii) use of molecular breeding methods in practical breeding, which includes introgression of transgenes, major disease resistance genes, and major fiber QTLs into elite cultivars.

Although tremendous progress has been achieved in the application of molecular markers in cotton breeding, the results still have not met expectations in terms of the extent and success compared with other major crops such as rice, soybean, and maize. For the time being, application of molecular breeding methods is still limited to the traits controlled by qualitative genes or major QTLs. Improvement of most agronomic traits that are of complicated inheritance and economic importance, like yield and fiber quality, is still a great challenge for marker-assisted breeding. To fully and successfully implement molecular-marker technologies in practical breeding, the future of molecular breeding research should focus on following:

- Validating already-identified QTLs (mainly fiber QTLs) in more genetic backgrounds, and identify the QTLs that have moderate to high effects. These QTLs in turn will gain priority for use in practical breeding.
- Identifying stable-yield and yield component QTLs. Mapping yield QTLs requires repeated field tests in multiple environments to ensure that the QTLs identified are stable and useful in breeding. Accurate phenotyping

often remains the principal limiting factor for achieving this goal. Semi- or fully automated, high-throughput phenotyping under both controlled conditions and field tests in multiple environments promises to narrow the genotype-phenotype gap and ensure that the QTLs identified are stable and useful in breeding.

- Conducting more investigations on GS, GWS, or family-based association mapping, and their applicability in cotton breeding. GS is a relatively new member of the family of molecular breeding methods and has been used in animal as well as crop breeding. In cotton, however, few, if any, reports on GS have been published. With the significantly reduced cost of DNA sequencing and more SNP markers becoming available, the time has come for cotton geneticists and breeders to investigate and implement GS in practical breeding.
- Placing priority on developing markers related to functional genes. Such markers will improve the accuracy and efficiency of MAS.
- Developing an SNP chip that contains informative SNP markers. Although many trait QTLs have been identified so far, it is very difficult for breeders to use them in a breeding program due to lack of a centralized marker-trait platform. Cotton breeders are very busy with field experiments and selection and rarely have time to navigate through the sea of markers and QTLs. The cotton community needs to develop an SNP chip that can be conveniently used by breeders and that contains SNP markers associated with agronomic traits, preferably distributed across the entire genome.

Acknowledgments

I thank Drs. Wangzhen Guo, Xinlian Shen, and two anonymous reviewers for their constructive comments, which helped improve the manuscript.

References

Abdurakhmonov, I.Y., S. Saha, J.N. Jenkins, Z.T. Buriev, S.E. Shermatov, B.E. Scheffler, A.E. Pepper, J.Z. Yu, R.J. Kohel, and A. Abdukarimov. 2009. Linkage disequilibrium based association mapping of fiber quality traits in *G. hirsutum* L. variety germplasm. Genetica 136:401–417. doi:10.1007/s10709-008-9337-8.

An, C., J.N. Jenkins, J. Wu, Y. Guo, and J.A. McCartin. 2010. Use of fiber and fuzz mutants to detect QTL for yield components, seed and fiber traits of Upland cotton. Euphytica 172:21–34. doi:10.1007/s10681-009-0009-2

Andres, R.J., D.T. Bowman, B. Kaur, and V. Kuraparthy. 2014. Mapping and genomic targeting of the major leaf shape gene (*L*) in Upland cotton (*Gossypium hirsutum* L.). Theor. Appl. Genet. 127:167–177. doi:10.1007/s00122-013-2208-4

Blenda, A., D.D. Fang, J.F. Rami, O. Garsmeur, F. Luo, and J.M. Lacape. 2012. A high density consensus genetic map of tetraploid cotton that integrates multiple component maps through molecular marker redundancy check. PLoS ONE 7:e45739. doi:10.1371/journal.pone.0045739

Blenda, A., J. Scheffler, B. Scheffler, M. Palmer, J.M. Lacape, J.Z Yu, C. Jesudurai, S. Jung, S. Muthukumar, P. Yellambalase, S. Ficklin, M. Staton, R. Eshelman, M. Ulloa, S. Saha, B. Burr, S. Liu, T. Zhang, D. Fang, A. Pepper, S. Kumpatla, J. Jacobs, J. Tomkins, R. Cantrell, and D. Main. 2006. CMD: A cotton microsatellite database resource for *Gossypium* genomics. BMC Genomics 7:132. doi:10.1186/1471-2164-7-132.

Botstein, D., R.L. White, M. Skolnick, and R.W. Davis. 1980. Construction of a genetic linkage map in man using restriction fragment length polymorphisms. Am. J. Hum. Genet. 32:314–331.

Bradbury P.J., Z. Zhang, D.E. Kroon, T.M. Casstevens, Y. Ramdoss, and E.S. Buckler. 2007. TASSEL: Software for association mapping of complex traits in diverse samples. Bioinformatics 23:2633–2635. doi:10.1093/bioinformatics/btm308.

Brown, J.K. 2001. Bacterial blight. In: T.L. Kirkpatrick and C.S. Rothrock, editors, Compedium of cotton diseases. 2nd ed. APS Press, St. Paul, Minnesota. p. 34–35.

Byers, R.L., D.B. Harker, S.M. Yourstone, P.J. Maughan, and J.A. Udall. 2012. Development and mapping of SNP assays in allotetraploid cotton. Theor. Appl. Genet. 124:1201–1214. doi:10.1007/s00122-011-1780-8

Cai, C., X. Tong, F. Liu, F. Lv, H. Wang, T. Zhang, and W. Guo. 2013. Discovery and identification of a novel Ligon lintless-like mutant (Lix) similar to the Ligon lintless (Li1) in allotetraploid cotton. Theor. Appl. Genet. 126:963–970. doi:10.1007/s00122-012-2029-x

Cai, C., W. Ye, T. Zhang, and W. Guo. 2014. Association analysis of fiber quality traits and exploration of elite alleles in Upland cotton cultivars/accessions (*Gossypium hirsutum* L.). J. Integr. Plant Biol. 56:51–62. doi:10.1111/jipb.12124

Cao, Z., P. Wang, X. Zhu, H. Chen, and T. Zhang. 2014. SSR marker-assisted improvement of fiber qualities in *Gossypium hirsutum* using *G. barbadense* introgression lines. Theor. Appl. Genet. 127:587–594. doi:10.1007/s00122-013-2241-3

Chee, P., X. Draye, C.X. Jiang, L. Decanini, T.A. Delmonte, R. Bredhauer, C.W. Smith, and A.H. Paterson. 2005. Molecular dissection of interspecific variation between *Gossypium hirsutum* and *Gossypium barbadense* (cotton) by a backcross-self approach: I. Fiber elongation. Theor. Appl. Genet. 111:757–763. doi:10.1007/s00122-005-2063-z

Chen, H., N. Qian, W. Guo, Q. Song, B. Li, F. Deng, C. Dong, and T. Zhang. 2009. Using three overlapped RILs to dissect genetically clustered QTL for fiber strength on Chro.D8 in Upland cotton. Theor. Appl. Genet. 119:605–612. doi:10.1007/s00122-009-1070-x

Chen, H., N. Qian, W. Guo, Q. Song, B. Li, F.D. Deng, C. Dong, and T. Zhang. 2010. Using three selected overlapping RILs to fine-map the yield component QTL on Chro.D8 in Upland cotton. Euphytica 176:321–329. doi:10.1007/s10681-010-0204-1

Correa, R.L., T.F. Silva, J.L. Simoes-Araujo, P.A.V. Barroso, M.S. Vidal, and M.F.S. Vaslin. 2005. Molecular characterization of a virus from the family *Luteoviridae* associated with cotton blue disease. Arch. Virol. 150:1357–1367. doi:10.1007/s00705-004-0475-8

Dighe, N.D., A.F. Robinson, A.A. Bell, M.A. Menz, R.G. Cantrell, and D.M. Stelly. 2009. Linkage mapping of resistance to reniform nematode in cotton following introgression from *Gossypium longicalyx* (Hutch. & Lee). Crop Sci. 49:1151–1164. doi:10.2135/cropsci2008.03.0129

Fang, D.D., L.L. Hinze, R.G. Percy, P. Li, D. Deng, and G. Thyssen. 2013. A microsatellite-based genome-wide analysis of genetic diversity and linkage disequilibrium in Upland cotton (*Gossypium hirsutum* L.) cultivars from major cotton-growing countries. Euphytica 191:391–401. doi:10.1007/s10681-013-0886-2

Fang, D.D., J.N. Jenkins, D.D. Deng, J.C. McCarty, P. Li, and J. Wu. 2014. Quantitative trait loci analysis of fiber quality traits using a random-mated recombinant inbred population in Upland cotton (*Gossypium hirsutum* L.). BMC Genomics 15:397. doi:10.1186/1471-2164-15-397

Fang, D.D., and S.R. Stetina. 2011. Improving cotton (*Gossypium hirsutum* L.) plant resistance to reniform nematodes by pyramiding Ren_1 and Ren_2. Plant Breed. 130:673–678. doi:10.1111/j.1439-0523.2011.01889.x

Fang, D.D., J. Xiao, P.C. Canci, and R.G. Cantrell. 2010. A new SNP haplotype associated with blue disease resistance gene in cotton (*Gossypium hirsutum* L.). Theor. Appl. Genet. 120:943–953. doi:10.1007/s00122-009-1223-y

Fang, D.D., and J.Z. Yu. 2012. Addition of 455 microsatellite marker loci to the high-density *Gossypium hirsutum* TM-1 x *G. barbadense* 3-79 genetic map. J. Cotton Sci. 16:229–248.

Fang, H., H. Zhou, S. Sanogo, R. Flynn, R.G. Percy, S.E. Hughs, M. Ulloa, D.C. Jones, and J. Zhang. 2013. Quantitative trait locus mapping for Verticillium wilt resistance in a backcross inbred line population of cotton (*Gossypium hirsutum* × *Gossypium barbadense*) based on RGA-AFLP analysis. Euphytica 194:79–91. doi:10.1007/s10681-013-0965-4

Fang, H., H. Zhou, S. Sanogo, A.L. Lipka, D.D. Fang, R.G. Percy, S.E. Hughs, D.C. Jones, and J. Zhang. 2014. Quantitative trait locus analysis of Verticillium wilt resistance in an introgressed recombinant inbred population of Upland cotton. Mol. Breed. 33:709–720. doi:10.1007/s11032-013-9987-9

Feng, C.D., J.M. Stewart, and J.F. Zhang. 2005. STS markers linked to the Rf1 fertility restorer gene of cotton. Theor. Appl. Genet. 110:237–243. doi:10.1007/s00122-004-1817-3

Gilbert, M.K., R.B. Turley, H.J. Kim, P. Li, G. Thyssen, Y. Tang, C.D. Delhom, M. Naoumkina, and D.D. Fang. 2013. Transcript profiling by microarray and marker analysis of the short cotton (*Gossypium hirsutum* L.) fiber mutant Ligon lintless-1 (Li1). BMC Genomics 14:403. doi:10.1186/1471-2164-14-403

Gilmour, A.R., B. Gogel, B. Cullis, R. Thompson, D. Butler, M. Cherry, D. Collins, D. Dutkowski, S. Harding, and K. Haskard. 2009. ASReml user guide release 3.0. VSN International, Hemel Hempstead, UK.

Gore, M.A., D.D. Fang, J.A. Poland, J. Zhang, R.G. Percy, R.G. Cantrell, G. Thyssen, and A.E. Lipka. 2014. Linkage map construction and quantitative trait locus analysis of agronomic and fiber quality traits in cotton. Plant Gen. 7(1). doi:10.3835/plantgenome2013.07.0023

Gore, M.A., R.G. Percy, J. Zhang, D.D. Fang, and R.G. Cantrell. 2012. Registration of the TM-1/NM24016 cotton recombinant inbred mapping population. J. Plant Reg. 6:124–127. doi:10.3198/jpr2011.06.0334crmp

Guo, W., T. Zhang, X. Shen, J.Z. Yu, and R.J. Kohel. 2003. Development of SCAR marker linked to a major QTL for high fiber strength and its usage in molecular-marker assisted selection in Upland cotton. Crop Sci. 43:2252–2256. doi:10.2135/cropsci2003.2252

Guo, W.Z., C.P. Cai, C.B. Wang, Z.G. Han, X.L. Song, K. Wang, X.W. Niu, C. Wang, K.Y. Lu, B. Shi, and T.Z. Zhang. 2007. A microsatellite-based, gene-rich linkage map reveals genome structure, function and evolution in *Gossypium*. Genetics 176:527–541. doi:10.1534/genetics.107.070375

Guo, Y., J.C. Mccarty, J. Jenkins, and S. Saha. 2008. QTLs for node of first fruiting branch in a cross of an Upland cotton, *Gossypium hirsutum* L., cultivar with primitive accession Texas 701. Euphytica 163:113–122. doi:10.1007/s10681-007-9613-1

Gutierrez, O.A., J.N. Jenkins, J.C. McCarty, M.J. Wubben, R.W. Hayes, and F.E. Callahan. 2010. SSR markers closely associated with genes for resistance to root-knot nematode on chromosomes 11 and 14 of Upland cotton. Theor. Appl. Genet. 121:1323–1337. doi:10.1007/s00122-010-1391-9

Gutierrez, O.A., A.F. Robinson, J.N. Jenkins, J.C. McCarty, M.J. Wubben, F.E. Callahan, and R.L. Nichols. 2011. Identification of QTL regions and SSR markers associated with resistance to reniform nematode in *Gossypium barbadense* L. accession GB713. Theor. Appl. Genet. 122:271–280. doi:10.1007/s00122-010-1442-2

Hamblin, M.T., E.S. Buckler, and J.L. Jannink. 2011. Population genetics of genomics-based crop improvement methods. Trends Genet 27:98–106. doi:10.1016/j.tig.2010.12.003

Han, Z., C. Wang, X. Song, W. Guo, J. Gou, C. Li, X. Chen, and T. Zhang. 2006. Characteristics, development and mapping of *Gossypium hirsutum* derived EST-SSRs in allotetraploid cotton. Theor. Appl. Genet. 112:430–439. doi:10.1007/s00122-005-0142-9

He, D., Z. Lin, X. Zhang, Y. Nie, X. Guo, Y. Zhang, and W. Li. 2007. QTL mapping for economic traits based on a dense genetic map of cotton with PCR-based markers using the interspecific cross of *Gossypium hirsutum* × *Gossypium barbadense*. Euphytica 153:181–197. doi:10.1007/s10681-006-9254-9

Hinchliffe, D.J., W.R. Meredith, C.D. Delhom, D.P. Thibodeaux, and D.D. Fang. 2011a. Elevated growing degree days influence transition stage timing during cotton (*Gossypium hirsutum* L.) fiber development and result in increased fiber strength. Crop Sci. 51:1683–1692. doi:10.2135/cropsci2010.10.0569

Hinchliffe, D.J., W.R. Meredith, K.M. Yeater, H.J. Kim, A.W. Woodward, Z.J. Chen, and B.A. Triplett. 2010. Near-isogenic cotton germplasm lines that differ in fiber-bundle strength have temporal differences in fiber gene expression patterns as revealed by comparative high-throughput profiling. Theor. Appl. Genet. 120:1347–1366. doi:10.1007/s00122-010-1260-6

Hinchliffe, D.J., R.B. Turley, M. Naoumkina, H.J. Kim, Y. Tang, K.M. Yeater, P. Li, and D.D. Fang. 2011b. A combined functional and structural genomics approach identified an

EST-SSR marker with complete linkage to the Ligon lintless-2 genetic locus in cotton (*Gossypium hirsutum* L.). BMC Genomics 12:445. doi:10.1186/1471-2164-12-445.

Hinze, L.L., J.K. Dever, and R.G. Percy. 2012. Molecular variation among and within improved cultivars in the U.S. cotton germplasm collection. Crop Sci. 52:222–230. doi:10.2135/cropsci2011.04.0202

Iqbal, A.E., O.U.K. Reddy, K.M. El-Zik, and A.E. Pepper. 2001. A genetic bottleneck in the "evolution under domestication" of upland cotton *Gossypium hirsutum* L. examined using DNA fingerprinting. Theor. Appl. Genet. 103:547–554. doi:10.1007/PL00002908

Islam, M.S., G.N. Thyssen, J.N. Jenkins, and D.D. Fang. 2015. Detection, validation and application of genotyping-by-sequencing based single nucleotide polymorphisms in Upland cotton (*Gossypium hirsutum* L.). Plant Gen. 8(1):1-10. 10.1007/s00438-014-0930-5.

Islam, M.S., L. Zeng, C.D. Delhom, X. Song, H. Kim, P. Li, and D.D. Fang. 2014. Identification of cotton fiber quality quantitative trait loci using intraspecific crosses derived from two near-isogenic lines differing in fiber bundle strength. Mol. Breed. 34:373–384. doi:10.1007/s11032-014-0040-4

Jenkins, J.N., J.C. McCarty, M.J. Wubben, R. Hayes, O.A. Guitierrez, F.E. Callahan, and D.D. Deng. 2012. SSR markers for marker assisted selection of root-knot nematode (*Meloidogyne incognita*) resistant plants in cotton (*Gossypium hirsutum* L). Euphytica 183:49–54. doi:10.1007/s10681-011-0470-6

Jiang, C., R.J. Wright, S.S. Woo, T.A. DelMonte, and A.H. Paterson. 2000. QTL analysis of leaf morphology in tetraploid *Gossypium* (cotton). Theor. Appl. Genet. 100:409–418. doi:10.1007/s001220050054

Jiang, G.L. 2013. Molecular markers and marker-assisted breeding in plants. In: S.B. Anderson, editor, Plant breeding from laboratories to fields. InTech, Rijeka, Croatia, p. 45–83.

Kim, H.J., H.S. Moon, C.D. Delhom, L. Zeng, and D.D. Fang. 2013. Molecular markers associated with the immature fiber (*im*) gene affecting the degree of fiber cell wall thickening in cotton (*Gossypium hirsutum* L.). Theor. Appl. Genet. 126:23–31. doi:10.1007/s00122-012-1956-x

Kohel, R., J. Yu, Y.-H. Park, and G. Lazo. 2001. Molecular mapping and characterization of traits controlling fiber quality in cotton. Euphytica 121:163–172. doi:10.1023/A:1012263413418

Kumar, P., R. Singh, E.L. Lubbers, X. Shen, A.H. Paterson, B.T. Campbell, D.C. Jones, and P.W. Chee. 2012. Mapping and validation of fiber strength quantitative trait loci on chromosome 24 in Upland cotton. Crop Sci. 52:1115–1122. doi:10.2135/cropsci2011.09.0524

Lacape, J.M., M. Claverie, R.O. Vidal, M.F. Carazzolle, G.A. Guimaraes Pereira, M. Ruiz, M. Pre, D. Llewellyn, Y. Al-Ghazi, J. Jacobs, A. Dereeper, S. Huguet, M. Giband, and C. Lanaud. 2012. Deep sequencing reveals differences in the transcriptional landscapes of fibers from two cultivated species of cotton. PLoS ONE 7:e48855. doi:10.1371/journal.pone.0048855.

Lacape, J.M., G. Gawrysiak, T.V. Cao, C. Viot, D. Llewellyn, S. Liu, J. Jacobs, D. Becker, P.A. Vianna Barroso, J.H. Assunção, O. Palaï, S. Georges, J. Jean, and M. Gibanda. 2013. Mapping QTLs for traits related to phenology, morphology and yield components in an inter-specific *Gossypium hirsutum* × *G. barbadense* cotton RIL population. Field Crops Res. 144:256–267. doi:10.1016/j.fcr.2013.01.001

Lacape, J.M., J. Jacobs, T. Arioli, R. Derijcker, N. Forestier-Chiron, D. Llewellyn, J. Jean, E. Thomas, and C. Viot. 2009. A new interspecific, *Gossypium hirsutum* × *G. barbadense*, RIL population: Towards a unified consensus linkage map of tetraploid cotton. Theor. Appl. Genet. 119:281–292. doi:10.1007/s00122-009-1037-y

Lacape, J.M., D. Llewellyn, J. Jacobs, T. Arioli, D. Becker, S. Calhoun, Y. Al-Ghazi, S. Liu, O. Palai, S. Georges, M. Giband, H. de Assuncao, P.A. Barroso, M. Claverie, G. Gawryziak, J. Jean, M. Vialle, and C. Viot. 2010. Meta-analysis of cotton fiber quality QTLs across diverse environments in a *Gossypium hirsutum* × *G. barbadense* RIL population. BMC Plant Biol. 10:132. doi:10.1186/1471-2229-10-132

Lacape, J.M., and T.B. Nguyen. 2005. Mapping quantitative trait loci associated with leaf and stem pubescence in cotton. J. Hered. 96:441–444. doi:10.1093/jhered/esi052

Lacape, J.M., T.B. Nguyen, B. Courtois, J.L. Belot, M. Giband, J.P. Gourlot, G. Gawryziak, S. Roques, and B. Hau. 2005. QTL analysis of cotton fiber quality using multiple *G.*

hirsutum × *G. barbadense* backcross generations. Crop Sci. 45:123–140. doi:10.2135/cropsci2005.0123a

Lacape, J.M., T.B. Nguyen, S. Thibivilliers, B. Bojinov, B. Courtois, R.G. Cantrell, B. Burr, and B. Hau. 2003. A combined RFLP-SSR-AFLP map of tetraploid cotton based on a *Gossypium hirsutum* × *Gossypium barbadense* backcross population. Genome 46:612–626. doi:10.1139/g03-050

Li, C., C. Wang, N. Dong, X. Wang, H. Zhao, R. Converse, Z. Xia, R. Wang, and Q. Wang. 2012. QTL detection for node of first fruiting branch and its height in upland cotton (*Gossypium hirsutum* L.). Euphytica 188:441–451. doi:10.1007/s10681-012-0720-2

Li, C.Q., Q.S. Liu, H.H. Zhao, L.J. Wang, X.F. Zhang, Y. Liu, W.Y. Zhou, L.L. Yang, P.B. Li, and Q.L. Wang. 2013. Marker-assisted selection of *Verticillium* wilt resistance in progeny populations of upland cotton derived from mass selection-mass crossing. Euphytica 191:469–480. doi:10.1007/s10681-013-0917-z

Li, F., G. Fan, K. Wang, F. Sun, Y. Yuan, G. Song, Q. Li, Z. Ma, C. Lu, C. Zou, W. Chen, X. Liang, H. Shang, W. Liu, C. Shi, G. Xiao, C. Gou, W. Ye, X. Xu, X. Zhang, H. Wei, Z. Li, G. Zhang, J. Wang, K. Liu, R.J. Kohel, R.G. Percy, J.Z. Yu, Y.X. Zhu, J. Wang, and S. Yu. 2014. Genome sequence of the cultivated cotton *Gossypium arboreum*. Nat. Genet. 46:567–572. doi:10.1038/ng.2987

Li, F., X. Hu, X. Qiu, and L. Wang. 2010. Fine mapping of the *fg* gene using the interspecific cross of *Gossypium hirsutum* × *Gossypium barbadense*. Euphytica 176:303–308. doi:10.1007/s10681-010-0197-9

Lin, L., G.J. Pierce, J.E. Bowers, J.C. Estill, R.O. Compton, L.K. Rainville, C. Kim, C. Lemke, J. Rong, H. Tang, X. Wang, M. Braidotti, A.H. Chen, K. Chicola, K. Collura, E. Epps, W. Golser, C. Grover, J. Ingles, S. Karunakaran, D. Kudrna, J. Olive, N. Tabassum, E. Um, M. Wissotski, Y. Yu, A. Zuccolo, M. ur Rahman, D.G. Peterson, R.A. Wing, J.F. Wendel, and A.H. Paterson. 2010. A draft physical map of a D-genome cotton species (*Gossypium raimondii*). BMC Genomics 11:395. doi:10.1186/1471-2164-11-395.

Lin, Z., Y. Zhang, X. Zhang, and X. Guo. 2009. A high-density integrative linkage map for *Gossypium hirsutum*. Euphytica 166:35–45. doi:10.1007/s10681-008-9822-2

Liu, H.Y., A. Quampah, J.H. Chen, R. Li, Z.R. Huang, Q.L. He, C.H. Shi, and S.J. Zhu. 2012. QTL analysis for gossypol and protein contents in upland cottonseeds with two different genetic systems across environments. Euphytica 188:453–463. doi:10.1007/s10681-012-0733-x

Liu, L., W. Guo, X. Zhu, and T. Zhang. 2003. Inheritance and fine mapping of fertility restoration for cytoplasmic male sterility in *Gossypium hirsutum* L. Theor. Appl. Genet. 106:461–469. doi:10.1007/s00122-002-1084-0.

Liu, R., B. Wang, W. Guo, Y. Qin, L. Wang, Y. Zhang, and T. Zhang. 2012. Quantitative trait loci mapping for yield and its components by using two immortalized populations of a heterotic hybrid in *Gossypium hirsutum* L. Mol. Breed. 29:297–311. doi:10.1007/s11032-011-9547-0

Liu, R., B. Wang, W. Guo, L. Wang, and T. Zhang. 2011. Differential gene expression and associated QTL mapping for cotton yield based on a cDNA-AFLP transcriptome map in an immortalized F2. Theor. Appl. Genet. 123:439–454. doi:10.1007/s00122-011-1597-5

Lopez-Lavalle, L.A., V.J. Gillespie, W.A. Tate, M.H. Ellis, W.N. Stiller, D.L. Llewellyn, and I.W. Wilson. 2012. Molecular mapping of a new source of Fusarium wilt resistance in tetraploid cotton (*Gossypium hirsutum* L.). Mol. Breed. 30:1181–1191. doi:10.1007/s11032-012-9705-z

Lu, Y., J. Curtiss, D. Miranda, E. Hughs, and J. Zhang. 2008. ATG-anchored AFLP (ATG-AFLP) analysis in cotton. Plant Cell Rep. 27:1645–1653. doi:10.1007/s00299-008-0568-z

Ma, X.X., B.L. Zhou, Y.H. Lu, W.Z. Guo, and T.Z. Zhang. 2008. Simple sequence repeat genetic linkage maps of A-genome diploid cotton (*Gossypium arboreum*). J. Integr. Plant Biol. 50:491–502. doi:10.1111/j.1744-7909.2008.00636.x

Mei, H., X. Zhu, and T. Zhang. 2013. Favorable QTL alleles for yield and its components identified by association mapping in Chinese Upland cotton cultivars. PLoS ONE 8:e82193. doi:10.1371/journal.pone.0082193

Mei, M., N.H. Syed, W. Gao, P.M. Thaxton, C.W. Smith, D.M. Stelly, and Z.J. Chen. 2004. Genetic mapping and QTL analysis of fiber-related traits in cotton (*Gossypium*). Theor. Appl. Genet. 108:280–291. doi:10.1007/s00122-003-1433-7

Meredith, W.R. 2005a. Minimum number of genes controlling cotton fiber strength in a backcross population. Crop Sci. 45:1114–1119. doi:10.2135/cropsci2003.0425

Meredith, W.R. 2005b. Registration of MD 52ne high fiber quality cotton germplasm and recurrent parent MD 90ne. Crop Sci. 45:806–807. doi:10.2135/cropsci2005.0807

Michelmore, R.W., I. Paran, and R.V. Kesseli. 1991. Identification of markers linked to disease-resistance genes by bulked segregant analysis: A rapid method to detect markers in specific genomic regions by using segregating populations. Proc. Natl. Acad. Sci. USA 88:9828–9832. doi:10.1073/pnas.88.21.9828

Moose, S.P., and R.H. Mumm. 2008. Molecular plant breeding as the foundation for 21st century crop improvement. Plant Physiol. 147:969–977. doi:10.1104/pp.108.118232

Ning, Z., H. Chen, H. Mei, and T. Zhang. 2014. Molecular tagging of QTLs for fiber quality and yield in the upland cotton cultivar Acala-Prema. Euphytica 195:143–156. doi:10.1007/s10681-013-0990-3

Paterson, A.H., J.F. Wendel, H. Gundlach, H. Guo, J. Jenkins, et al. 2012. Repeated polyploidization of *Gossypium* genomes and the evolution of spinnable cotton fibres. Nature 492:423–427. doi:10.1038/nature11798

Qian, N., X.W. Zhang, W. Guo, and T. Zhang. 2009. Fine mapping of open-bud duplicate genes in homoelogous chromosomes of tetraploid cotton. Euphytica 165:325–331. doi:10.1007/s10681-008-9790-6

Qin, H., W. Guo, Y.M. Zhang, and T. Zhang. 2008. QTL mapping of yield and fiber traits based on a four-way cross population in *Gossypium hirsutum* L. Theor. Appl. Genet. 117:883–894. doi:10.1007/s00122-008-0828-x

Rahman, M., Y. Zafar, and A.H. Paterson. 2009. *Gossypium* DNA markers: Types, numbers and uses. In: A.H. Paterson, editor, Genetics and genomics of cotton. Springer, New York. p. 101–139.

Reinisch, A.J., J. Dong, C.L. Brubaker, D.M. Stelly, J.F. Wendel, and A.H. Paterson. 1994. A detailed RFLP map of cotton, *Gossypium hirsutum* × *Gossypium barbadense:* Chromosome organization and evolution in a disomic polyploid genome. Genetics 138:829–847.

Roberts, P.A., and M. Ulloa. 2010. Introgression of root-knot nematode resistance into tetraploid cottons. Crop Sci. 50:940–951. doi:10.2135/cropsci2009.05.0281

Robinson, A.F., A.A. Bell, N.D. Dighe, M.A. Menz, R.L. Nichols, and D.M. Stelly. 2007. Introgression of resistance to nematode Rotylenchulus reniformis into Upland cotton (*Gossypium hirsutum*) from *Gossypium longicalyx*. Crop Sci. 47:1865–1877. doi:10.2135/cropsci2006.12.0776

Robinson, A.F., C.G. Cook, and A.E. Percival. 1999. Resistance to *Rotylenchulus reniformis* and *Meloidogyne incognita* race 3 in the major cotton cultivars planted since 1950. Crop Sci. 39:850–858. doi:10.2135/cropsci1999.0011183X003900030039x

Romano, G.B., E.J. Sacks, S.R. Stetina, A.F. Robinson, D.D. Fang, O.A. Gutierrez, and J.A. Scheffler. 2009. Identification and genomic location of a reniform nematode (*Rotylenchulus reniformis*) resistance locus (*Ren ari*) introgressed from *Gossypium aridum* into upland cotton (*G. hirsutum*). Theor. Appl. Genet. 120:139–150. doi:10.1007/s00122-009-1165-4

Rong, J., C. Abbey, J.E. Bowers, C.L. Brubaker, C. Chang, P.W. Chee, T.A. Delmonte, X. Ding, J.J. Garza, B.S. Marler, C.H. Park, G.J. Pierce, K.M. Rainey, V.K. Rastogi, S.R. Schulze, N.L. Trolinder, J.F. Wendel, T.A. Wilkins, T.D. Williams-Coplin, R.A. Wing, R.J. Wright, X. Zhao, L. Zhu, and A.H. Paterson. 2004. A 3347-locus genetic recombination map of sequence-tagged sites reveals features of genome organization, transmission and evolution of cotton (*Gossypium*). Genetics 166:389–417. doi:10.1534/genetics.166.1.389

Rong, J., F.A. Feltus, V.N. Waghmare, G.J. Pierce, P.W. Chee, X. Draye, Y. Saranga, R.J. Wright, T.A. Wilkins, O.L. May, C.W. Smith, J.R. Gannaway, J.F. Wendel, and A.H. Paterson. 2007. Meta-analysis of polyploid cotton QTL shows unequal contributions of subgenomes to a complex network of genes and gene clusters implicated in lint fiber development. Genetics 176:2577–2588. doi:10.1534/genetics.107.074518

Rong, J., G.J. Pierce, V.N. Waghmare, C.J. Rogers, A. Desai, P.W. Chee, O.L. May, J.R. Gannaway, J.F. Wendel, T.A. Wilkins, and A.H. Paterson. 2005. Genetic mapping and comparative analysis of seven mutants related to seed fiber development in cotton. Theor. Appl. Genet. 111:1137–1146. doi:10.1007/s00122-005-0041-0

Saha, S., J. Wu, J.N. Jenkins, J.C. McCarty, R. Hayes, and D.M. Stelly. 2010. Genetic dissection of chromosome substitution lines of cotton to discover novel *Gossypium barbadense* L. alleles for improvement of agronomic traits. Theor. Appl. Genet. 120:1193–1205. doi:10.1007/s00122-009-1247-3

Said, J.I., Z. Lin, X. Zhang, M. Song, and J. Zhang. 2013. A comprehensive meta QTL analysis for fiber quality, yield, yield related and morphological traits, drought tolerance, and disease resistance in tetraploid cotton. BMC Genomics 14:776. doi:10.1186/1471-2164-14-776

Shao, Q., F. Zhang, S. Tang, Y. Liu, X. Fang, D. Liu, D.J. Liu, J. Zhang, Z. Teng, A.H. Paterson, and Z. Zhang. 2014. Identifying QTL for fiber quality traits with three upland cotton (*Gossypium hirsutum* L.) populations. Euphytica 198:43–58. doi:10.1007/s10681-014-1082-8

Shappley, Z.W., J.N. Jenkins, W.R. Meredith, and J.C. McCarty. 1998a. An RFLP linkage map of Upland cotton, *Gossypium hirsutum* L. Theor. Appl. Genet. 97:756–761. doi:10.1007/s001220050952

Shappley, Z.W., J.N. Jenkins, J. Zhu, and J.C. McCarty. 1998b. Quantitative trait loci associated with agronomic and fiber traits of upland cotton. J. Cotton Sci. 2:153–163.

Shen, X., Z. Cao, R. Singh, E.L. Lubbers, P. Xu, C.W. Smith, A.H. Paterson, and P.W. Chee. 2011. Efficacy of qFL-chr1, a quantitative trait locus for fiber length in cotton (*Gossypium* spp.). Crop Sci. 51:2005–2010. doi:10.2135/cropsci2010.11.0653

Shen, X., W. Guo, Q. Lu, X. Zhu, Y. Yuan, and T. Zhang. 2007. Genetic mapping of quantitative trait loci for fiber quality and yield trait by RIL approach in Upland cotton. Euphytica 155:371–380. doi:10.1007/s10681-006-9338-6

Shen, X., Y. He, E.L. Lubbers, R.F. Davis, R.L. Nichols, and P.W. Chee. 2010. Fine mapping QMi-C11 a major QTL controlling root-knot nematodes resistance in Upland cotton. Theor. Appl. Genet. 121:1623–1631. doi:10.1007/s00122-010-1415-5

Shen, X., G. Van Becelaere, P. Kumar, R.F. Davis, O.L. May, and P. Chee. 2006a. QTL mapping for resistance to root-knot nematodes in the M-120 RNR Upland cotton line (*Gossypium hirsutum* L.) of the Auburn 623 RNR source. Theor. Appl. Genet. 113:1539–1549. doi:10.1007/s00122-006-0401-4

Shen, X., T. Zhang, W. Guo, X. Zhu, and X.Y. Zhang. 2006b. Mapping fiber and yield QTLs with main, epistatic, and QTL × environment interaction effects in recombinant inbred lines of Upland cotton. Crop Sci. 46:61–66. doi:10.2135/cropsci2005.0056

Song, L., W. Guo, and T. Zhang. 2009. Interaction of novel Dobzhansky-Muller type genes for the induction of hybrid lethality between *Gossypium hirsutum* and *G. barbadense* cv. Coastland R4-4. Theor. Appl. Genet. 119:33–41. doi:10.1007/s00122-009-1014-5

Sun, F.D., J.H. Zhang, S.F. Wang, W.K. Gong, Y.Z. Shi, A.Y. Liu, J.W. Li, J.W. Gong, H.H. Shang, and Y.L. Yuan. 2012. QTL mapping for fiber quality traits across multiple generations and environments in upland cotton. Mol. Breed. 30:569–582. doi:10.1007/s11032-011-9645-z

Thyssen, G., J.C. McCarty, P. Li, J.N. Jenkins, and D.D. Fang. 2014. Genetic mapping of non-target-site resistance to a sulfonylurea herbicide (Envoke) in Upland cotton (*Gossypium hirsutum* L.). Mol. Breed. 33:341–348. doi:10.1007/s11032-013-9953-6

Ulloa, M., R.B. Hutmacher, P.A. Roberts, S.D. Wright, R.L. Nichols, and R. Michael Davis. 2013. Inheritance and QTL mapping of Fusarium wilt race 4 resistance in cotton. Theor. Appl. Genet. 126:1405–1418. doi:10.1007/s00122-013-2061-5

Ulloa, M., and W.R. Meredith. 2000. Genetic linkage map and QTL analysis of agronomic and fiber quality traits in an intraspecific population. J. Cotton Sci. 4:161–170.

Ulloa, M., W.R. Meredith, Z.W. Shappley, and A.L. Kahler. 2002. RFLP genetic linkage maps from four F2.3 populations and a joinmap of *Gossypium hirsutum* L. Theor. Appl. Genet. 104:200–208. doi:10.1007/s001220100739

Van Deynze, A., K. Stoffel, M. Lee, T.A. Wilkins, A. Kozik, R.G. Cantrell, J.Z. Yu, R.J. Kohel, and D.M. Stelly. 2009. Sampling nucleotide diversity in cotton. BMC Plant Biol 9:125. doi:10.1186/1471-2229-9-125.

Van Ooijen, J.W. 2006. JoinMap 4.0: Software for the calculation of genetic linkage maps in experimental populations. Kyazma, Wageningen, The Netherlands.

Van Ooijen, J.W. 2009. MapQTL 6, software for the mapping of quantitative trait loci in experimental populations of diploid species. Kyazma, Wageningen, The Netherlands.

Vos, P., R. Hogers, M. Bleeker, M. Reijans, T. van de Lee, M. Hornes, A. Frijters, J. Pot, J. Peleman, M. Kuiper, and M. Zabeau. 1995. AFLP: A new technique for DNA fingerprinting. Nucleic Acids Res. 23:4407–4414. doi:10.1093/nar/23.21.4407

Waghmare, V.N., J. Rong, C.J. Rogers, G.J. Pierce, J.F. Wendel, and A.H. Paterson. 2005. Genetic mapping of a cross between *Gossypium hirsutum* (cotton) and the Hawaiian endemic, *Gossypium tomentosum*. Theor. Appl. Genet. 111:665–676. doi:10.1007/s00122-005-2032-6

Wan, Q., Z. Zhang, M. Hu, L. Chen, D. Liu, X. Chen, W. Wang, and J. Zheng. 2007. T_1 locus in cotton is the candidate gene affecting lint percentage, fiber quality and spiny bollworm (*Earias* spp.) resistance. Euphytica 158:241–247. doi:10.1007/s10681-007-9446-y

Wang, B., W. Guo, X. Zhu, Y. Wu, N. Huang, and T. Zhang. 2007. QTL mapping of yield and yield components for elite hybrid derived-RILs in Upland cotton. J. Genet. Genomics 34:35–45. doi:10.1016/S1673-8527(07)60005-8

Wang, B., Y. Nie, Z. Lin, X. Zhang, J. Liu, and J. Bai. 2012. Molecular diversity, genomic constitution, and QTL mapping of fiber quality by mapped SSRs in introgression lines derived from *Gossypium hirsutum* × *G. darwinii* Watt. Theor. Appl. Genet. 125:1263–1274. doi:10.1007/s00122-012-1911-x

Wang, C., and P.A. Roberts. 2006. Development of AFLP and derived CAPS markers for root-knot nematode resistance in cotton. Euphytica 152:185–196. doi:10.1007/s10681-006-9197-1

Wang, C., M. Ulloa, and P.A. Roberts. 2006. Identification and mapping of microsatellite markers linked to a root-knot nematode resistance gene (*rkn1*) in Acala NemX cotton (*Gossypium hirsutum* L.). Theor. Appl. Genet. 112:770–777. doi:10.1007/s00122-005-0183-0

Wang, C., T.Z. Zhang, and W.Z. Guo. 2013. The *im* mutant gene negatively affects many aspects of fiber quality traits and lint percentage in cotton. Crop Sci. 53:27–37. doi:10.2135/cropsci2012.05.0304

Wang, F., Z. Xu, R. Sun, Y. Gong, G. Liu, J. Zhang, L. Wang, C. Zhang, S. Fan, and J. Zhang. 2013. Genetic dissection of the introgressive genomic components from *Gossypium barbadense* L. that contribute to improved fiber quality in *Gossypium hirsutum* L. Mol. Breed. 32:547–562. doi:10.1007/s11032-013-9888-y

Wang, K., Z. Wang, F. Li, W. Ye, J. Wang, G. Song, Z. Yue, L. Cong, H. Shang, S. Zhu, C. Zou, Q. Li, Y. Yuan, C. Lu, H. Wei, C Gou, Z. Zheng, Y. Yin, X. Zhang, K. Liu, B. Wang, C. Song, N. Shi, R.J. Kohel, R.G. Percy, J.Z. Yu, Y.X. Zhu, and S. Yu. 2012. The draft genome of a diploid cotton *Gossypium raimondii*. Nat Genet 44:1098–1103. doi:10.1038/ng.2371.

Wang, P., L. Su, L. Qin, B. Hu, W. Guo, and T. Zhang. 2009. Identification and molecular mapping of a Fusarium wilt resistant gene in upland cotton. Theor. Appl. Genet. 119:733–739. doi:10.1007/s00122-009-1084-4

Wang, P., Y. Zhu, X. Song, Z. Cao, Y. Ding, B. Liu, X. Zhu, S. Wang, W. Guo, and T. Zhang. 2012. Inheritance of long staple fiber quality traits of *Gossypium barbadense* in *G. hirsutum* background using CSILs. Theor. Appl. Genet. 124:1415–1428. doi:10.1007/s00122-012-1797-7

Wang Z., Zhang D., Wang X., Tan X., Guo H., Paterson A.H. (2013) A whole-genome DNA marker map for cotton based on the D-genome sequence of *Gossypium raimondii* L. G3, Genes|Genomes|Genetics 3:1759-1767. doi:10.1534/g3.113.006890.

Weaver, D.B., K.S. Lawrence, and E. van Santen. 2007. Reniform nematode resistance in Upland cotton germplasm. Crop Sci. 47:19–24. doi:10.2135/cropsci2006.02.0130

Weaver, D.B., R.B. Sikkens, K.S. Lawrence, C. Sürmelioglu, E. van Santen, and R.L. Nichols. 2013. REN^{lon} and its effects on agronomic and fiber quality traits in upland cotton. Crop Sci. 53:913–920. doi:10.2135/cropsci2012.08.0504

Welsh, J., and M. McClelland. 1990. Fingerprinting genomes using PCR with arbitrary primers. Nucleic Acids Res. 18:7213–7218. doi:10.1093/nar/18.24.7213

Williams, J.G., A.R. Kubelik, K.J. Livak, J.A. Rafalski, and S.V. Tingey. 1990. DNA polymorphisms amplified by arbitrary primers are useful as genetic markers. Nucleic Acids Res. 18:6531–6535. doi:10.1093/nar/18.22.6531

Wright, R.J., P.M. Thaxton, K.M. El-Zik, and A.H. Paterson. 1998. D-subgenome bias of *Xcm* resistance genes in tetraploid *Gossypium* (cotton) suggests that polyploid formation has created novel avenues for evolution. Genetics 149:1987–1996.

Wright, R.J., P.M. Thaxton, K.M. El-Zik, and A.H. Paterson. 1999. Molecular mapping of genes affecting pubescence of cotton. J. Hered. 90:215–219. doi:10.1093/jhered/90.1.215

Wu, J., O.A. Gutierrez, J.N. Jenkins, J.C. McCarty, and J. Zhu. 2009. Quantitative analysis and QTL mapping for agronomic and fiber traits in an RI population of upland cotton. Euphytica 165:231–245. doi:10.1007/s10681-008-9748-8

Xiao, J., D.D. Fang, M. Bhatti, B. Hendrix, and R. Cantrell. 2010. A SNP haplotype associated with a gene resistant to *Xanthomonas axonopodis* pv. *malvacearum* in upland cotton (*Gossypium hirsutum* L.). Mol. Breed. 25:593–602. doi:10.1007/s11032-009-9355-y

Xu, Y. 2010. Molecular plant breeding. CABI, Cambridge, MA.

Yang, C., W. Guo, G. Li, F. Gao, S. Lin, and T. Zhang. 2008. QTLs mapping for *Verticillium* wilt resistance at seedling and maturity stages in *Gossypium barbadense* L. Plant Sci. 174:290–298. doi:10.1016/j.plantsci.2007.11.016

Ynturi, P., J.N. Jenkins, J.C. McCarty, O.A. Gutierrez, and S. Saha. 2006. Association of root-knot nematode resistance genes with simple sequence repeat markers on two chromosomes in cotton. Crop Sci. 46:2670–2674. doi:10.2135/cropsci2006.05.0319

Yu, J., R.J. Kohel, and C.W. Smith. 2010. The construction of a tetraploid cotton genome wide comprehensive reference map. Genomics 95:230–240. doi:10.1016/j.ygeno.2010.02.001

Yu, J., S. Yu, S. Fan, M. Song, H. Zhai, X. Li, and J. Zhang. 2012. Mapping quantitative trait loci for cottonseed oil, protein and gossypol content in a *Gossypium hirsutum* × *Gossypium barbadense* backcross inbred line population. Euphytica 187:191–201. doi:10.1007/s10681-012-0630-3

Yu, J., S. Yu, M. Gore, M. Wu, H. Zhai, X. Li, S. Fan, M. Song, and J. Zhang. 2013a. Identification of quantitative trait loci across interspecific F2, F2:3 and testcross populations for agronomic and fiber traits in tetraploid cotton. Euphytica 191:375–389. doi:10.1007/s10681-013-0875-5

Yu, J., S. Yu, C.A. Lu, W. Wang, S.L. Fan, M.Z. Song, Z. Lin, X.D. Zhang, and J. Zhang. 2007. High-density linkage map of cultivated allotetraploid cotton based on SSR, TRAP, SRAP and AFLP markers. J. Integr. Plant Biol. 49:716–724. doi:10.1111/j.1744-7909.2007.00459.x

Yu, J., K. Zhang, S. Li, S. Yu, H. Zhai, M. Wu, X. Li, S. Fan, M. Song, D. Yang, Y. Li, and J. Zhang. 2013b. Mapping quantitative trait loci for lint yield and fiber quality across environments in a *Gossypium hirsutum* × *Gossypium barbadense* backcross inbred line population. Theor. Appl. Genet. 126:275–287. doi:10.1007/s00122-012-1980-x

Yu, J.Z., D.D. Fang, R.J. Kohel, M. Ulloa, L.L. Hinze, R.G. Percy, J. Zhang, P. Chee, B.E. Scheffler, and D.C. Jones. 2012a. Development of a core set of SSR markers for the characterization of *Gossypium* germplasm. Euphytica 187:203–213. doi:10.1007/s10681-012-0643-y

Yu, J.Z., R.J. Kohel, D.D. Fang, J. Cho, A. Van Deynze, M. Ulloa, S.M. Hoffman, A.E. Pepper, D.M. Stelly, J.N. Jenkins, S. Saha, S.P. Kumpatla, M.R. Shah, W.V. Hugie, and R.G. Percy. 2012b. A high-density simple sequence repeat and single nucleotide polymorphism genetic map of the tetraploid cotton genome. G3, Gene|Genome|Genetics 2:43–58. doi:10.1534/g3.111.001552.

Yu, Y., D.J. Yuan, S.G. Liang, X.M. Li, X.Q. Wang, Z.X. Lin, and X.L. Zhang. 2011. Genome structure of cotton revealed by a genome-wide SSR genetic map constructed from a BC1 population between *Gossypium hirsutum* and *G. barbadense*. BMC Genomics 12:15. doi:10.1186/1471-2164-12-15

Zeng, L., W.R. Meredith, Jr., O.A. Gutierrez, and D.L. Boykin. 2009. Identification of associations between SSR markers and fiber traits in an exotic germplasm derived from multiple crosses among *Gossypium* tetraploid species. Theor. Appl. Genet. 119:93–103. doi:10.1007/s00122-009-1020-7

Zhang, J., and J.M. Stewart. 2004. Identification of molecular markers linked to the fertility restorer genes for CMS-D8 in cotton. Crop Sci. 44:1209–1217. doi:10.2135/cropsci2004.1209

Zhang, K., J. Zhang, J. Ma, S. Tang, D. Liu, Z. Teng, D. Liu, and Z. Zhang. 2012. Genetic mapping and quantitative trait locus analysis of fiber quality traits using a three-parent composite population in upland cotton (*Gossypium hirsutum* L.). Mol. Breed. 29:335–348. doi:10.1007/s11032-011-9549-y

Zhang, M., Y. Zhang, J.J. Huang, X. Zhang, M.K. Lee, D.M. Stelly, and H.B. Zhang. 2012. Genome physical mapping of polyploids: A BIBAC physical map of cultivated tetraploid cotton, *Gossypium hirsutum* L. PLoS ONE 7:e33644. doi:10.1371/journal.pone.0033644

Zhang, T., N. Qian, X. Zhu, H. Chen, S. Wang, H. Mei, and Y. Zhang. 2013. Variations and transmission of QTL alleles for yield and fiber qualities in Upland cotton cultivars developed in China. PLoS ONE 8:e57220. doi:10.1371/journal.pone.0057220

Zhang, T., Y. Yuan, J. Yu, W. Guo, and R.J. Kohel. 2003. Molecular tagging of a major QTL for fiber strength in Upland cotton and its marker-assisted selection. Theor. Appl. Genet. 106:262–268. doi:10.1007/s00122-002-1101-3.

Zhang, Z.S., M.C. Hu, J. Zhang, D.J. Liu, J. Zheng, K. Zhang, W. Wang, and Q. Wan. 2009. Construction of a comprehensive PCR-based marker linkage map and QTL mapping for fiber quality traits in upland cotton (*Gossypium hirsutum* L.). Mol. Breed. 24:49–61. doi:10.1007/s11032-009-9271-1

Zhao, L., Y. Lv, C. Cai, X. Tong, X. Chen, W. Zhang, H. Du, X. Guo, and W. Guo. 2012. Toward allotetraploid cotton genome assembly: Integration of a high-density molecular genetic linkage map with DNA sequence information. BMC Genomics 13:539. doi:10.1186/1471-2164-13-539.

Zhu, Q.H., A. Spriggs, J.M. Taylor, D. Llewellyn, and I. Wilson. 2014. Transcriptome and complexity-reduced, DNA-based identification of intraspecies single-nucleotide polymorphisms in the polyploid *Gossypium hirsutum* L. G3, Gene|Genome|Genetics 4:1893–1905. doi:10.1534/g3.114.012542.

Genome Sequencing

Maojun Wang, Daojun Yuan, and Xianlong Zhang*

Abstract

As one of the major polyploid species crops, cotton (*Gossypium* spp.) is widely cultivated throughout the world. With the long-term accumulation of knowledge on genome research and advent of high-throughput sequencing technologies, scientists have released the genome sequences of two putative diploid progenitors of tetraploid cotton. Direct sequencing and analysis of tetraploid cotton genomes are in progress and are expected to be released soon. The breakthrough of genome sequencing makes it easier for researchers to exploit functional genes, uncover the mysteries of genome ploidy, and analyze genomic variations. New sequencing technologies and approaches should be adopted to assist decoding the complex genome sequences. Completion of tetraploid cotton genome sequences will facilitate the improvement of this important crop.

With the advent of high-throughput sequencing technologies, the genomes of many crops, such as rice, maize, and soybean, have been sequenced (Goff et al., 2002; Yu et al., 2002; Schnable et al., 2009; Schmutz et al., 2010). Recently, the genome sequences of two important polyploidy crops, bread wheat and rapeseed, have also been released (Chalhoub et al., 2014; IWGSC, 2014). A high-quality reference genome sequence not only provides a complete gene catalog for a species, but also paves the way for exploiting functional genes that contribute to agronomic traits (Edwards and Batley, 2010). Moreover, knowledge of genome sequences will make it easier for understanding and exploiting the genomic variations that are the basis of crop improvement (Bolger et al., 2014). Sequencing of the genomes of cultivated tetraploid cotton has been ongoing through the efforts of many scientists around the world (Chen et al., 2007) and is nearing completion.

The cotton genus, *Gossypium*, is categorized into 45 diploid species that fall into 8 genomes (A–G and K, $2n = 2x = 26$) and 6 tetraploid species (AADD, $2n = 4x = 52$), with genome sizes ranging from approximately 880 Mb in the D-genome species *Gossypium raimondii* to about 2.5 Gb in tetraploid species (Hawkins et al., 2006; Wendel et al., 2010; Grover et al., 2014; see Fig. 2 in Wendel and Grover, 2015).

Abbreviations: ATP, adenosine triphosphate; BAC, bacterial artificial chromosome; BIBAC, binary BAC; FISH, fluorescence in situ hybridization; LTR, long terminal repeat; NGS, next-generation sequencing; PCR, polymerase chain reaction; SNP, single-nucleotide polymorphism.

Maojun Wang (mjwang1234@gmail.com), Daojun Yuan (robert@mail.hzau.edu.cn), and Xianlong Zhang, National Key Laboratory of Crop Genetic Improvement, Huazhong Agricultural University, Wuhan, Hubei 430070, China. *Corresponding author (xlzhang@mail.hzau.edu.cn).

doi:10.2134/agronmonogr57.2013.0028

Cotton. 2nd. ed. David D. Fang and Richard G. Percy, editors. Agronomy Monograph 57.

The tetraploid *G. hirsutum* (Upland cotton, AADD, AD1 genome) is widely cultivated around the world because of its good adaptation and high yield. Another tetraploid species, *G. barbadense* (Sea-Island cotton, or Pima cotton, AADD, AD2 genome), is cultivated in some areas for its superior fiber quality. However, due to the inherent characteristics of cotton genomes, such as large sizes, high content of repetitive sequences, and high sequence similarity between the At and Dt subgenomes in tetraploids, sequencing the tetraploid cotton genome has been a big challenge for cotton scientists. To face and overcome this challenge, in 2007, cotton scientists developed a strategy for sequencing the cotton genomes: first, sequencing the smallest *G. raimondii* genome (the proposed extant progenitor of the Dt subgenome in tetraploid cotton), then sequencing the *G. arboreum* or *G. herbaceum* genome (the proposed extant progenitors of the At subgenome in tetraploid cotton), and finally, sequencing the tetraploid genomes (*G. hirsutum* and *G. barbadense*) (Chen et al., 2007).

In this chapter, we will focus on the DNA sequencing technologies and genome sequencing strategy to illustrate the current status of cotton genome sequencing.

DNA Sequencing Methods

DNA sequencing technologies have advanced rapidly during the past three decades. Genome sequencing began with the application of the Sanger sequencing technology (Sanger et al., 1977), which played a very important role in sequencing the first human genome (Venter et al., 2001). In plants, the Sanger sequencing method contributed to the completion of the *Arabidopsis* genome sequence (AGI, 2000) and later to the rice genome sequence (Goff et al., 2002; Yu et al., 2002). These model species were sequenced using a bacterial artificial chromosome (BAC) by BAC strategy. However, because of its high cost, low throughput, and time cost, the use of the Sanger method for sequencing crop genomes has been very limited, even it was significantly improved upon in the 1990s.

The advent of next-generation sequencing (NGS) has revolutionized biology (Schuster 2008) and genome sequencing. Next-generation sequencing technologies eliminate the multiple steps of library construction, clone selection, and transformation, which can be substituted for by the direct ligation of adaptors to the fragmented DNA or RNA and then amplified by emulsion polymerase chain reaction (PCR) or bridge PCR (Fig. 1). Due to these advantages, the NGS can be conducted in parallel on a large scale. More importantly, the cost of NGS is dramatically lower than the Sanger sequencing.

Next-generation sequencing technologies fall into following three categories according to their original developers: 454 technology, SOLiD technology, and Illumina technology.

454 Technology

The 454 platform developed by the Roche Company adopts the technology of sequencing by synthesis chemistry (Rothberg and Leamon, 2008). This platform differs from other NGS platforms in that it employs a two-step reaction to detect the added nucleotides. First, the triphosphate nucleotide is cleaved to release pyrophosphate, then the pyrophosphate is converted into adenosine triphosphate (ATP) via the enzyme ATP sulfurylase. The following steps include the conversion of luciferin into oxyluciferin and generation of a quantum of light,

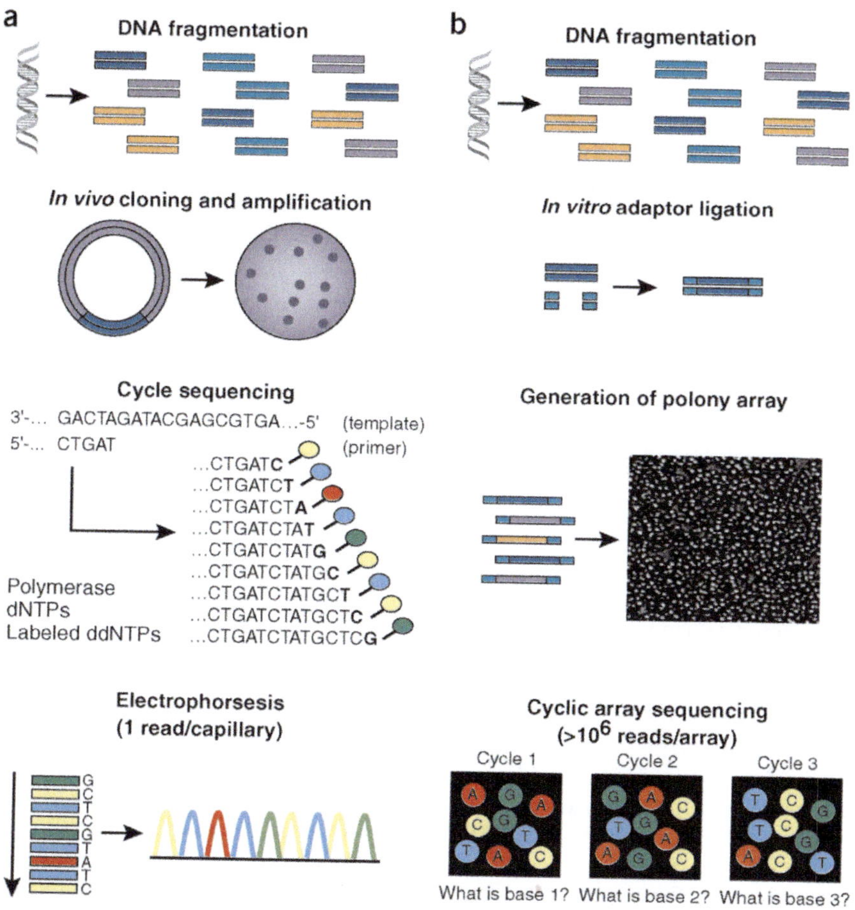

Fig. 1. Comparison of the Sanger sequencing (a) and next-generation sequencing (b) methods. (From Shendure and Ji, 2008).

which is captured from the picotiter plate by a charge-coupled camera. The 454 sequencing technology is superior to other platforms in regard to read lengths, which can be up to 1000 bases. Each run of this platform can generate up to 1 million reads with an average of 700 bp in less than 1 d of run time (http://454.com/products/gs-flx-system/index.asp).

SOLiD Technology

SOLiD technology is a platform that was commercialized by Applied Biosystems (now part of Thermo-Fisher Scientific, Carlsbad, CA) in 2007. The main features of this platform are emulsion PCR, sequence-by-ligation, and two-base encoding technologies (Mardis, 2008). Like 454 sequencing, this platform ligates the adapters to the fragmented DNA and binds them to magnetic beads that are smaller than the 454 platform. After the PCR reaction, the small magnetic beads are fixed to the solid plates. The 8-bp probes will be paired with the single-stranded DNA template using DNA ligase. The ligation step will continue until the full read is completed. The unique ligation step allows for two checks of a nucleotide position and thus provides a great sequencing accuracy level (up to 99.99%). This advantage makes this platform perform well when the sequences are used to locate single-nucleotide polymorphisms (SNPs).

Illumina Technology

The most widely used NGS platform is the technology developed by Illumina. (Quail et al., 2008). This method is based on the sequencing by synthesis (SBS) technology. The sequencing instrument is designed to use DNA cluster and reversible terminator technologies for automated and massively parallel sequencing. Double-strand DNA in a library is passed through a flow cell that will hybridize the individual molecules with adaptor sequences. Hybridized sequences at both ends of the adaptor will be amplified as a bridge (bridge-PCR), and then elongated with modified DNA polymerase and deoxy-ribonucleoside triphosphate. Each nucleotide is detected by a fluorescent signal. After the imaging signal of each reaction, the 3′-terminal fluorescence will be eliminated via chemical methods and then another reaction will continue. The Illumina sequencing method is widely used for its high accuracy, high throughput, high sensitivity, and low cost. Currently, this method is the one most used and has contributed to the complete sequencing of many plant and animal genomes. The sequence read length has been increased from 35 bp to 300 bp (MiSeq sequencer) with dramatically higher throughputs and flexible running times. The latest sequencer, HiSeq4000, can generate 1.5 Tb data for a run, and the NextSeq 500 can provide 20–100 Gb of data within 1 to 2 d (http://www.illumina.com/systems/sequencing.html).

Compared with Sanger sequencing, the NGS methods are superior and have much higher throughputs and lower cost; however, the sequencing reads are shorter, especially for Illumina and SOLiD sequencing (Liu et al., 2012; Grada and Weinbrecht, 2013) (Table 1). In addition, the sequencing accuracy of NGS methods is less than for Sanger sequencing. Different NGS technologies have their own advantages and disadvantages. The 454 sequencing method can generate longer sequence reads, but at higher cost, and plays a vital role in de novo sequencing and meta-genomics studies but results in numerous sequencing errors for single-base repetition regions. The Illumina sequencing method has the advantages of high throughput and lower costs, which is beneficial for resequencing projects. The main problem with the Illumina sequencing method is nucleotide substitution. The SOLiD sequencing technology can generate sequencing accuracy similar to that with 454 sequencing and is also a high-throughput method with higher costs than Illumina sequencing. Table 1 shows a comparison of the different sequencing technologies.

Third-Generation Sequencing

The advent of the Pacific Biosciences sequencing platform represents third-generation sequencing. This platform also adopts the sequencing-by-synthesis system used in NGS, but it reads only single molecules, which is also known as single-cell sequencing (Schadt et al., 2010). The core technology of this platform is single-molecule real time (SMAT). The length of the sequencing read can reach 5000 nucleotides, which is significantly longer than with NGS. Another representative platform of third-generation sequencing is Nanopore technology,

Table 1. Comparison of the main sequencing technologies.

Sequencing method	Sanger 3730xl	454 GS FLX	HiSeq 4000	SOLiDv4	PacBio
Core technology	Dideoxy chain termination	Pyrosequencing	Sequencing by synthesis	Ligation and two-base coding	Sequencing by synthesis
Read length	400–900 bp	700 bp	150 bp	50 + 35 bp or 50 + 50 bp	2700 mean bp, ≤5000 bp
Throughput	1.9–84 Kb	0.7 Gb	1500 Gb	120 Gb	90 Mb per cell
Time per run	20 min–3h	24 h	3.5 d	7–14 d	<1 d
Cost/million bases	$2,400	$10	<$0.07	$0.13	$11–340
Primary error	amplification error	InDel	substitution	A-T bias	CG deletion
Base error rate	0.001%	0.10%	~1–2% over 100 bp	0.05%	13%

which is also a single-molecule technique. So far, this technology is still in development but has been utilized in an RNA-seq study (Li et al., 2014b). Third-generation sequencing is still in its infancy and has many unsolved technical challenges (Reid et al., 2012), but it is quite promising and may revolutionize biological research as the NGS technologies have done (Flusberg et al., 2010; Sharon et al., 2013).

Tools for Genome Assembly

Rapid advances in sequencing technology have made genome sequencing faster, cheaper, and easier. However, constructing a complete genome is still a long-term, complex project that is different from cDNA assembly. Thus, special methods and software, assisted by good physical and genetic maps, are needed to perform genome assembly (Lewin et al., 2009).

BAC Library Construction and Physical Maps

BAC libraries have been playing a fundamental role in genome sequencing, especially for complex plant genomes. These libraries can be used for constructing physical maps, assisting in genome sequencing, map-based cloning of important functional genes, and long-term preservation of genome resources.

In cotton, many BAC libraries have been constructed for a variety of reasons. The first BAC library of the Upland cotton 'Maxxa' was constructed in 2001 (Tomkins et al., 2001). This library contains 129,024 clones, totaling 8.3 times the haploid genome size. Analyses of 435 randomly selected BAC clones showed that the average insertion size ranged from 80 to 275 kb, among which the lengths of 99% of the clones were larger than 100 kb. Hu et al. (2009) constructed a BAC library of Upland cotton TM-1. The library contains147,456 clones, and the average length was 122.8 kb, ranging from 97 to 240 kb, among which the insertion sizes of more than 96% of the clones were larger than 100 kb. Wang et al. (2006) constructed a BAC library of Sea Island cotton Pima 90-53 containing 167,424 clones with an average insertion size of 130 kb, of which more than 94% of the clones were larger than 100 kb. The estimated total length of this library was 6.5 times that of the genome. Hu et al. (2010) constructed a BAC library of *G. arboreum*, which contained 123,648 clones and the average insertion size was 100.2 kb, ranging from 30 to 190 kb. The estimated total size of this library was 7.2 times that of the diploid A genome.

Toward the long-term goal of characterizing cultivated tetraploid genomes, Lin et al. (2010) first constructed a physical map of *G. raimondii* by integrating genetically anchored overgo hybridization probes, agarose-based fingerprints, and high information-content fingerprinting. This map used a total of 13,662 BAC-end sequences and 2828 DNA probes and anchored 1585 contigs to a cotton consensus genetic map. Importantly, this study also anchored 370 and 438 contigs to *Arabidopsis thaliana* and *Vitis vinifera* genome sequences, respectively, suggesting its promise for using translational genomic approaches to understand cotton genomes and exploit important resident genes.

For physical maps of tetraploid cottons, Lee et al. (2013) constructed the first plant-transformation-competent binary BAC (BIBAC) of Upland cotton. This library contains 76,800 clones with an average insertion size of 135 kb. Zhang et al. (2012) constructed a BIBAC physical map consisting of 3450 BIBAC contigs with an N50 contig size of 863 kb, collectively spanning 2244 Mb. They identified 1063 A- and 141 D-subgenome-specific contigs through screening the source BIBAC library with the probes derived from three A-/D-subgenome-specific, interspersed repetitive elements. These contigs covered 1192 Mb, accounting for 50% of the Upland cotton genome. Xu et al. (2008) constructed a physical map of homeologous chromosomes 12 and 26 in Upland cotton accession TM-1. The assigned At subgenome contained approximately 110 contigs, while the Dt subgenome contained 48 contigs. This physical map of the At and Dt subgenomes covered 73.49 Mb and 34.23 Mb, respectively, in physical length. Approximately 67 contigs between the At and Dt subgenomes were highly similar. These physical maps are helpful for the direct analysis of tetraploid cotton genomes.

Using a multiprobe fluorescence in situ hybridization (FISH) system, Gan et al. (2012) conducted individual chromosome identification in *G. darwinii* (A_dD_d), *G. klotzschianum* (D_{3k}), *G. davidsonii* (D_{3d}), *G. armourianum* ($D_{2\text{-}1}$) and *G. aridum* (D_4). Two sets of BAC clones were used as the FISH probes, one is specific to 26 individual chromosomes from the A and D subgenomes of *G. hirsutum* (A_h and D_h) while the other is a D-genome, centromere-specific BAC clone. They found that all the A chromosomes were marked with the 13 At chromosome-specific BAC clones and all the D chromosomes and chromosomal arms were identified with the 13 Dt chromosome-specific BAC clones and the D-genome, centromere-specific BAC. These studies are helpful to establish a gene bank to exploit more genes from cotton relatives for cotton breeding and theoretically even for the direct sequencing of tetraploid genomes.

Cotton Genetic Maps

A genetic map, also called a linkage map, is constructed based on the recombination among the progeny plants of a population. A genetic map is widely used for quantitative trait loci mapping, map-based cloning, and molecular marker assisted selection. For direct whole-genome sequencing, the genetic map can be used to anchor scaffolds to chromosomes (Mascher and Stein, 2014). With the rapid progress of biotechnology, different molecular markers have been employed in constructing cotton genetic maps from traditional restriction fragment length polymorphisms, random amplified polymorphic DNA, amplified fragment length polymorphisms, and sequence-related amplified polymorphisms to the widely used simple sequence repeats and SNPs. The primary mapping population—including F_2 and F_3, backcrosses, recombinant inbred lines, and doubled haploid lines—and secondary mapping population—including near isogenic lines and chromosome segment substitution lines—are used in cotton.

As detailed by Fan (2015), many cotton genetic maps have been constructed and serve as valuable tools for assembling genome sequences. Due to differences in the population sizes, generation, structures, and the parents used in constructing these maps, many discrepancies exist between the published maps. To overcome these problems, Yu et al. (2010) collected 28 published genetic maps containing a total of 7424 markers and obtained a comprehensive reference map. Blenda et al. (2012) integrated six representative maps and constructed a high-density consensus map. That map contained 8254 marker loci and covered 4070 cM with 2 markers per cM. Using this map, they found that the genetic map of tetraploid cotton was highly syntenic with the diploid D genome by sequence alignment. Recently, Wang et al. (2013) constructed a whole-genome DNA marker map for cotton based on the D-genome sequence of *G. raimondii.*

Methods of Cotton Genome Sequencing and Challenges

Before sequencing a genome, biologists should take many factors into consideration (Baker, 2012). De novo assembly of a genome should start with the estimation of the genome size, which can be conducted with experimental approaches such as flow cytometry (Doležel and Bartoš, 2005). After obtaining the sequencing data, biologists can also consider some computational methods. The K-mer algorithm is widely used to verify the experimental result and assess the sequence quality (Compeau et al., 2011). Usually, the low-heterozygosity genome is the first choice for de novo assembly, which requires a series of genome libraries with different insertion sizes. Importantly, for highly heterozygous or highly repetitive genomes, a fosmid-by-fosmid or fosmid pooling strategy is usually required to improve the continuity of assembly (Williams et al., 2012). Recently, an efficient software program, PLATANUS, was developed and provides a novel and efficient approach for the assembly of gigabase-sized, highly heterozygous genomes (Kajitani et al., 2014).

De novo assembly of large complex genomes needs powerful software. Actually, most assemblers employ de Bruijn graph algorithms, such as Velvet, ABySS, ALLPATH-LG, and SOAPdenovo, for whole-genome sequence assembly (Simpson et al., 2009; Zerbino, 2010; Gnerre et al., 2011; Luo et al., 2012). These programs were intended to recover sequences for genomes of a known (estimated) size. SOAPdenovo and ALLPATH-LG are two of the most popular for assembling large genomes. Generally, the first possesses a fast run speed and does not have a strict requirement for computational hardware resources. Comparatively, the second can generate better continuity of assembly but requires high-configuration computational hardware resources. Because of the current prevalence of Illumina short-read data, most assemblers are optimized for this data type. Software that can integrate long-read data, such as 454 data and PacBio reads, is urgently needed to meet the need for genome assembly in the near future, especially for complex genomes.

For polyploid genomes, Udall's laboratory developed a promising tool, PolyCat, to assign sequencing data to each subgenome (Page et al., 2013). Users can utilize GSNAP software for read mapping because of its tolerance of SNPs, and the sequencing reads will be categorized by these SNPs among subgenomes (Wu and Nacu, 2010).

Assembly of cotton genomes is undoubtedly a challenge because of their large genome size, polyploidy, and high repeat content. Although scientists have been developing more genetic makers, the density of genetic maps is still low because of the large genome size. For the assembly of a high-quality cotton genome, the whole-genome shotgun strategy may be insufficient. The traditional BAC-by-BAC method can be a good alternative for such complex genomes, and many plant genomes are being sequenced with a combination of these two strategies, thus providing high-quality genome sequences. Among the three NGS platforms, Illumina can generate abundant sequences for a comparatively low cost; however, it is undoubtedly very difficult to assemble these highly repetitive regions using the short sequence reads. Assembly using only this platform will leave many gaps and miss much data, resulting in discontinuous genome assembly. The 454 long reads and PacBio assemblies could partially improve the quality of these regions (Huddleston et al., 2014). With the advent of third-generation long-read sequencing techniques, especially their improvement of the error rate, researchers can use them to generate low-coverage sequence data, which could be regarded as scaffolds and helpful for assembling highly repetitive regions.

Assembly of the tetraploid cotton genome should be done cautiously. If possible, assignment of sequencing data to subgenomes is a prerequisite for assembly of a polyploid cotton genome, because the At and Dt subgenomes in tetraploid cotton are highly similar, especially in the regions of conserved genes. Previously, cotton scientists managed to assemble the putative progenitors (A and D genomes) first and used these genes as a guide to finish the assembly of tetraploid genomes. Undoubtedly, this strategy is feasible for assembling gene regions, which has been shown with the bread wheat genome, along with machine-learning algorithms that can be used to discriminate a large number of subassemblies between the wheat A, B, or D subgenomes trained by diploid genomes (Brenchley et al., 2012). However, this strategy is less efficient for assembling genome regions with abundant rearrangements and translocations resulting from genome ploidization, thus it cannot provide high-quality genome sequences for analyzing genome evolution. Therefore, a BAC-by-BAC strategy might be a good way to directly sequence tetraploid cotton genomes. If possible, researchers can separate each chromosome and conduct its sequencing one by one, which may be much easier for complex genomes.

Methods of Genome Annotation

After genome assembly, the next challenge is genome annotation. The first step toward the successful annotation of any genome is determining whether its assembly is ready for annotation. The N50 statistic is one of the most important statistics for assessing genome completeness and continuity. The assembly with an N50 scaffold length that is gene-sized is necessary for annotation (Yandell and Ence, 2012). Because repetitive sequences usually mislead gene prediction, biologists will first identify and mask these sequences in genomes. Predictions of protein-coding genes usually require evidence from alignment with proteins of closely related species, transcription supported by expressed sequence tags or RNA-seq data (Liang et al., 2009), and ab initio gene prediction, which includes many software programs modeled by some gene characteristics, such as intron lengths, codon usage, and GC content, to identify genes and to determine their intron-exon structures. Much software for gene prediction has been developed and applied. Nowadays, gene annotation has moved beyond merely identifying protein-coding genes and also included predictions of regulatory regions, pseudogenes, and ncRNA genes. More importantly, coordination of these different prediction approaches requires more effective computational tools to improve the accuracy of prediction. Researchers need to be careful with gene prediction in polyploid genomes because most software was developed and trained using data from diploids.

Progress in Cotton Genome Sequencing

Compared with the model plants, research in sequencing the cotton genome has lagged behind. As mentioned above, in 2007, a coalition of cotton genome scientists developed a strategy for sequencing cotton genomes: sequencing two putative progenitors first, and then sequencing tetraploid cottons. With this strategy, significant accomplishments have been achieved in recent years.

Wang et al. (2012) published the draft genome sequence of *G. raimondii*. Using the Illumina platform, they generated shotgun reads covering 103.6-fold of the genome and assembled them into 775.2 Mb. Using a set of 1369 molecular markers from a consensus genetic linkage map (Lin et al., 2010), they anchored 73.2% of the assembled scaffolds (567.2 Mb) on the *G. raimondii* chromosomes. The *G. raimondii* genome contains at least 57.0% transposable elements and encodes 40,976 protein-coding genes. Comparative genomic analysis showed that *G. raimondii* and *T. cacao* probably diverged from a common ancestor approximately 33.7 million years ago. Through comparative expression analysis of elite fiber-related genes, such as *Sus* (sucrose synthase), *KCS* (3-ketoacyl-CoA synthase), and *ACO* (1-aminocyclopropane-1-carboxylic acid oxidase), they found that expression of these genes were essential for spinnable cotton fiber development. Phylogenetic analysis of the *CDN* [(+)-δ-cadinene synthase] genes indicated their conservation in cotton and *T. cacao*, which might suggest that gossypol production evolved after their divergence from other plant species. Comparative transcriptome analysis of 3-d post-anthesis ovules showed that transcription levels of *ACO1*, *ACO2*, and *ACO3* in *G. raimondii* were 3000-, 30-, and 10-fold those in *G. hirsutum*, respectively. Although previous studies have shown that *ACO* transcription is essential for fiber development, it was suggested that its excessive transcription might be harmful.

Paterson et al. (2012) aimed at assembling a high-quality reference genome of *G. raimondii*, which was published after the publication of D draft genome. They sequenced the *G. raimondii* genome by combining BAC-clone sequencing and the Roche 454 and Illumina sequencing platforms. This reference genome sequence is of high continuity and completeness and obviously of higher quality than the first reported draft genome sequence (Table 2). The assembly was 761.4 Mb with N50 18.8 Mb and oriented 98.3% of the sequences to the genome; both the N50 and the ratio of the orientation of the sequences were significantly higher than for the first D draft genome. The annotated genome contains 61% repetitive sequences, of which 53% sequences are long terminal repeat (LTR) transposons, with only 3% full-length LTR transposons. The *G. raimondii* genome encodes 37,505 protein-coding genes with 77,267 transcript isoforms. Comparative genomic analysis revealed that *G. raimondii* genome underwent a five- to sixfold ploidy increase approximately 60 million years ago, which contributed to

Table 2. Comparison of the sequenced cotton genomes.

Categories	*G. arboreum*	*G. raimondii* (Wang et al., 2012)	*G. ramondii* (Paterson et al., 2012)
Total assembly size, Mb	1,694	775.2	761.4
Total scaffold number	7,914	4,715	1,084
N50, Mb	0.66	2.3	18.8
Longest scaffold, Mb	5.9	12.8	52.1
Anchored and oriented scaffolds, Mb	1,532	406.3	748.7
Number of protein-coding genes	41,330	40,976	37,505
Percentage of repeat sequences	68.5%	57.0%	61.0%

the speciation between cotton and cacao. The tetraploid genomes were formed by the reunion between A-genome and D-genome diploids approximately 1 to 2 million years ago. By sequencing the A genome and tetraploid cotton genomes, they also observed many nonreciprocal DNA exchanges in the At and Dt subgenomes of tetraploid cottons, which might contribute to phenotypic innovation. Analysis of the *G. raimondii* chromosome 1 helped identify a sequence block that includes many genes closely homologous to mitochondrial genes, and the genes in this block might have experienced a coordinated change in expression associated with fiber domestication in *G. barbadense*.

The genome of the diploid *G. arboreum* (cultivar Shixiya1) was released in 2014 (Li et al., 2014a). To assemble this genome, a total of 193.6 Gb of clean Illumina sequence reads were generated by constructing libraries with different insertion lengths, which ranged from 180 bp to 40 kb. They also generated 33,454 BAC end sequences to improve the assembly. Using a high-density genetic map based on 24,569 codominant SNP markers, they anchored the 90.4% scaffolds onto the 13 pseudochromosomes. The *G. arboreum* encoded 41,330 protein-coding genes and contained 68.5% repetitive sequences, among which the LTR retrotransposons occupied 95.12%. Comparative genome analysis showed that the A and D diploid genomes shared two whole-genome duplications before speciation. Most protein-coding genes were conserved in these two genomes and are syntenic. Though the size of A genome was twice that of the D genome, no specific genome duplication event was observed. It was the abundant transposable element expansion that resulted in the twofold change in genome size. The NBS-related gene family was rapidly expanded through tandem duplication in *G. raimondii* but contracted in *G. arboreum*, which may have altered their resistance to *Verticillium dahliae*. The deletion of the myeloblastosis-binding site in the promoter regions of *AGO1* and *AGO3* decreased the level of ethylene, which might contribute to cotton fiber elongation in *G. arboreum*. However, very high levels of *ACO* transcripts in *G. raimondii* ovules in conjunction with an ethylene burst might lead to the senescence of early fibers in *G. raimondii*. These observations suggested that inactivation or overproduction of *ACO* transcripts might suppress fiber development.

The sequences of the two putative diploid progenitors laid the foundation for elucidating cotton biology and also accelerated basic and applied research on the crop. These sequences will help exploit elite functional genes and provide an opportunity for genome-based breeding. They also may be important for assembly and evolutionary analysis of the cultivated tetraploid cotton genomes.

For tetraploid cotton, Rai et al. (2013) conducted the 454 sequencing of six genotypes using a hypomethylated partial restriction method and generated more than 18,360,000 sequences amounting to 5.2 Gb covering 1.27 Gb of genome. The gene prediction result showed that tetraploid genome encoded about 93,363 putative protein-coding genes, including 21,399 full-length genes, 1093 transcription factors, 3135 putative promoter sequences, and 78 miRNA loci.

Prospects

The new sequencing technologies have revolutionized the way we conduct biological research, and research on cotton is no exception. Due to inherent nature of tetraploid cotton genomes, it is not easy to assemble high-quality cotton genome sequences. However, the accomplishments of the diploid D reference genome and A draft genome have dramatically facilitated the whole-genome sequencing projects in tetraploid cotton. Although no tetraploid cotton genome sequences have been publicly released, many scientists around the world are currently working toward this goal. As the cost of sequencing continues to drop and sequencing techniques advance, completion of a draft or reference tetraploid cotton genome is imminent.

In fact, sequencing a genome is much easier than understanding it. To deeply decode the cotton genomes, much more effort needs to be made to understand them. With whole-genome sequences, biologists could make an in-depth study of genome polyploidization. The genome sequences should make it easier to distinguish multiple-copy homeologous genes in the post-genome era and exploit key functional genes in each subgenome that contribute to the development of tetraploid cotton fiber. Moreover, the genome sequences should offer an opportunity for analyzing genome variations and can be utilized for genome-based breeding. Combined with promising high-throughput phenotyping platforms, we can expect genome-wide association studies to make significant breakthroughs in breeding cotton for desirable agronomic traits and elucidating fundamental biological issues.

Acknowledgments

The funding was provided by the Fundamental Research Funds for the Central Universities (NO. 2013JC011).

References

Arabidopsis Genome Initiative (AGI). 2000. Analysis of the genome sequence of the flowering plant *Arabidopsis thaliana*. Nature 408(6814):796–815. doi:10.1038/35048692

Baker, M. 2012. De novo genome assembly: What every biologist should know. Nat. Methods 9(4):333–337. doi:10.1038/nmeth.1935

Blenda, A., D.D. Fang, J.F. Rami, O. Garsmeur, F. Luo, and J.M. Lacape. 2012. A high density consensus genetic map of tetraploid cotton that integrates multiple component maps through molecular marker redundancy check. PLoS ONE 7(9):E45739. doi:10.1371/journal.pone.0045739

Bolger, M.E., B. Weisshaar, U. Scholz, N. Stein, B. Usadel, and K.F. Mayer. 2014. Plant genome sequencing- applications for crop improvement. Curr. Opin. Biotechnol. 26:31–37. doi:10.1016/j.copbio.2013.08.019

Brenchley, R., M. Spannagl, M. Pfeifer, G.L.A. Barker, R. D'Amore, A.M. Allen, et al. 2012. Analysis of the bread wheat genome using whole-genome shotgun sequencing. Nature 491(7426):705–710. doi:10.1038/nature11650

Chalhoub, B., F. Denoeud, S. Liu, I.A.P. Parkin, H. Tang, X. Wang, et al. 2014. Early allopolyploid evolution in the post-Neolithic *Brassica napus* oilseed genome. Science 345(6199):950–953. doi:10.1126/science.1253435

Chen, Z.J., B.E. Scheffler, E. Dennis, B.A. Triplett, T. Zhang, W. Guo, et al. 2007. Toward sequencing cotton (*Gossypium*) genomes. Plant Physiol. 145(4):1303–1310. doi:10.1104/pp.107.107672

Compeau, P.E., P.A. Pevzner, and G. Tesler. 2011. How to apply de Bruijn graphs to genome assembly. Nat. Biotechnol. 29(11):987–991. doi:10.1038/nbt.2023

Doležel, J., and J. Bartoš. 2005. Plant DNA flow cytometry and estimation of nuclear genome size. Ann. Bot. (Lond.) 95(1):99–110. doi:10.1093/aob/mci005

Edwards, D., and J. Batley. 2010. Plant genome sequencing: Applications for crop improvement. Plant Biotechnol. J. 8(1):2–9. doi:10.1111/j.1467-7652.2009.00459.x

Fang, D. 2015. Molecular breeding. In: D. Fang and R. Percy, editors, Cotton. 2nd ed. Agron. Monogr. 57. ASA, CSSA, and SSSA, Madison, WI. doi:10.2134/agronmonogr57.2013.0027

Flusberg, B.A., D.R. Webster, J.H. Lee, K.J. Travers, E.C. Olivares, T.A. Clark, et al. 2010. Direct detection of DNA methylation during single-molecule, real-time sequencing. Nat. Methods 7(6):461–465. doi:10.1038/nmeth.1459

Gan, Y., F. Liu, R. Peng, C. Wang, S. Li, X. Zhang, et al. 2012. Individual chromosome identification, chromosomal collinearity and genetic-physical integrated map in *Gossypium darwinii* and four D genome cotton species revealed by BAC-FISH. Genes Genet. Syst. 87(4):233–241. doi:10.1266/ggs.87.233

Gnerre, S., I. MacCallum, D. Przybylski, F.J. Ribeiro, J.N. Burton, B.J. Walker, et al. 2011. High-quality draft assemblies of mammalian genomes from massively parallel sequence data. Proc. Natl. Acad. Sci. USA 108(4):1513–1518. doi:10.1073/pnas.1017351108

Goff, S.A., D. Ricke, T.-H. Lan, G. Presting, R. Wang, M. Dunn, et al. 2002. A draft sequence of the rice genome (*Oryza sativa* L. ssp. *japonica*). Science 296(5565):92–100. doi:10.1126/science.1068275

Grada, A., and K. Weinbrecht. 2013. Next-generation sequencing: Methodology and application. J. Invest. Dermatol. 133(8):E11. doi:10.1038/jid.2013.248

Grover, C.E., X. Zhu, K.K. Grupp, J.J. Jareczek, J.P. Gallagher, E. Szadkowski, et al. 2014. Molecular confirmation of species status for the allopolyploid cotton species, *Gossypium ekmanianum* Wittmack. Genet. Resour. Crop Evol. 62(1):1–12.

Hawkins, J.S., H. Kim, J.D. Nason, R.A. Wing, and J.F. Wendel. 2006. Differential lineage-specific amplification of transposable elements is responsible for genome size variation in *Gossypium*. Genome Res. 16(10):1252–1261. doi:10.1101/gr.5282906

Hu, Y., W.Z. Guo, and T.Z. Zhang. 2009. Construction of a bacterial artificial chromosome library of TM-1, a standard line for genetics and genomics in Upland cotton. J. Integr. Plant Biol. 51(1):107–112. doi:10.1111/j.1744-7909.2008.00773.x

Hu, Y., Y. Lu, D. Ma, W. Guo, and T. Zhang. 2010. Construction and characterization of a bacterial artificial chromosome library for the A-genome of cotton (*G. arboreum* L.). J. Biomed. Biotechnol. 2011:457137.

Huddleston, J., S. Ranade, M. Malig, F. Antonacci, M. Chaisson, L. Hon, et al. 2014. Reconstructing complex regions of genomes using long-read sequencing technology. Genome Res. 24(4):688–696. doi:10.1101/gr.168450.113

International Wheat Genome Sequencing Consortium (IWGSC). 2014. A chromosome-based draft sequence of the hexaploid bread wheat (*Triticum aestivum*) genome. Science 345(6194):1251788. doi:10.1126/science.1251788

Kajitani, R., K. Toshimoto, H. Noguchi, A. Toyoda, Y. Ogura, M. Okuno, et al. 2014. Efficient de novo assembly of highly heterozygous genomes from whole-genome shotgun short reads. Genome Res. 24(8):1384–1395. doi:10.1101/gr.170720.113

Lee, M.K., Y. Zhang, M. Zhang, M. Goebel, H.J. Kim, B.A. Triplett, et al. 2013. Construction of a plant-transformation-competent BIBAC library and genome sequence analysis of polyploid Upland cotton (*Gossypium hirsutum* L.). BMC Genomics 14:208. doi:10.1186/1471-2164-14-208

Lewin, H.A., D.M. Larkin, J. Pontius, and S.J. O'Brien. 2009. Every genome sequence needs a good map. Genome Res. 19(11):1925–1928. doi:10.1101/gr.094557.109

Li, F., G. Fan, K. Wang, F. Sun, Y. Yuan, G. Song, et al. 2014a. Genome sequence of the cultivated cotton *Gossypium arboreum*. Nat. Genet. 46(6):567–572. doi:10.1038/ng.2987

Li, S., S.W. Tighe, C.M. Nicolet, D. Grove, S. Levy, W. Farmerie, et al. 2014b. Multi-platform assessment of transcriptome profiling using RNA-seq in the ABRF next-generation sequencing study. Nat. Biotechnol. 32(9):915–925. doi:10.1038/nbt.2972

Liang, C., L. Mao, D. Ware, and L. Stein. 2009. Evidence-based gene predictions in plant genomes. Genome Res. 19(10):1912–1923. doi:10.1101/gr.088997.108

Lin, L., G.J. Pierce, J.E. Bowers, J.C. Estill, R.O. Compton, L.K. Rainville, et al. 2010. A draft physical map of a D-genome cotton species (*Gossypium raimondii*). BMC Genomics 11:395. doi:10.1186/1471-2164-11-395

Liu, L., Y. Li, S. Li, N. Hu, Y. He, R. Pong, et al. 2012. Comparison of next-generation sequencing systems. J. Biomed. Biotechnol. 2012:251364.

Luo, R., B. Liu, Y. Xie, Z. Li, W. Huang, J. Yuan, et al. 2012. SOAPdenovo2: An empirically improved memory-efficient short-read de novo assembler. GigaScience 1(1):18. doi:10.1186/2047-217X-1-18

Mardis, E.R. 2008. Next-generation DNA sequencing methods. Annu. Rev. Genomics Hum. Genet. 9(1):387–402. doi:10.1146/annurev.genom.9.081307.164359

Mascher, M., and N. Stein. 2014. Genetic anchoring of whole-genome shotgun assemblies. Front. Genet. 5:208. doi:10.3389/fgene.2014.00208

Page, J.T., A.R. Gingle, and J.A. Udall. 2013. PolyCat: A resource for genome categorization of sequencing reads from allopolyploid organisms. G3: Genes, Genomes, Genet. 3(3):517–525</jrn>

Paterson, A.H., J.F. Wendel, H. Gundlach, H. Guo, J. Jenkins, D. Jin, et al. 2012. Repeated polyploidization of *Gossypium* genomes and the evolution of spinnable cotton fibres. Nature 492(7429):423–427. doi:10.1038/nature11798

Quail, M.A., I. Kozarewa, F. Smith, A. Scally, P.J. Stephens, R. Durbin, et al. 2008. A large genome centre's improvements to the Illumina sequencing system. Nat. Methods 5(12):1005–1010. doi:10.1038/nmeth.1270

Rai, K.M., S.K. Singh, A. Bhardwaj, V. Kumar, D. Lakhwani, A. Srivastava, et al. 2013. Large-scale resource development in *Gossypium hirsutum* L. by 454 sequencing of genic-enriched libraries from six diverse genotypes. Plant Biotechnol. J. 11(8):953–963. doi:10.1111/pbi.12088

Reid, A.J., S.J. Vermont, J.A. Cotton, D. Harris, G.A. Hill-Cawthorne, S. Konen-Waisman, et al. 2012. Comparative genomics of the apicomplexan parasites *Toxoplasma gondii* and *Neospora caninum*: Coccidia differing in host range and transmission strategy. PLoS Pathog. 8(3):E1002567. doi:10.1371/journal.ppat.1002567

Rothberg, J.M., and J.H. Leamon. 2008. The development and impact of 454 sequencing. Nat. Biotechnol. 26(10):1117–1124. doi:10.1038/nbt1485

Sanger, F., S. Nicklen, and A.R. Coulson. 1977. DNA sequencing with chain-terminating inhibitors. Proc. Natl. Acad. Sci. USA 74(12):5463–5467. doi:10.1073/pnas.74.12.5463

Schadt, E.E., S. Turner, and A. Kasarskis. 2010. A window into third-generation sequencing. Hum. Mol. Genet. 19(R2):R227–R240. doi:10.1093/hmg/ddq416

Schmutz, J., S.B. Cannon, J. Schlueter, J. Ma, T. Mitros, W. Nelson, et al. 2010. Genome sequence of the palaeopolyploid soybean. Nature 463(7278):178–183. doi:10.1038/nature08670

Schnable, P.S., D. Ware, R.S. Fulton, J.C. Stein, F. Wei, S. Pasternak, et al. 2009. The B73 maize genome: Complexity, diversity, and dynamics. Science 326(5956):1112–1115. doi:10.1126/science.1178534

Schuster, S.C. 2008. Next-generation sequencing transforms today's biology. Nat. Methods 5(1):16–18. doi:10.1038/nmeth1156

Sharon, D., H. Tilgner, F. Grubert, and M. Snyder. 2013. A single-molecule long-read survey of the human transcriptome. Nat. Biotechnol. 31(11):1009–1014. doi:10.1038/nbt.2705

Shendure, J., and H. Ji. 2008. Next-generation DNA sequencing. Nat. Biotechnol. 26(10):1135–1145. doi:10.1038/nbt1486

Simpson, J.T., K. Wong, S.D. Jackman, J.E. Schein, S.J.M. Jones, and İ. Birol. 2009. ABySS: A parallel assembler for short read sequence data. Genome Res. 19(6):1117–1123. doi:10.1101/gr.089532.108

Tomkins, J.P., D.G. Peterson, T.J. Yang, D. Main, T.A. Wilkins, A.H. Paterson, and R.A. Wing. 2001. Development of genomic resources for cotton (*Gossypium hirsutum* L.): BAC library construction, preliminary STC analysis, and identification of clones associated with fiber development. Mol. Breed. 8(3):255–261. doi:10.1023/A:1013798716098

Venter, J.C., M.D. Adams, E.W. Myers, P.W. Li, R.J. Mural, G.G. Sutton, et al. 2001. The sequence of the human genome. Science 291(5507):1304–1351. doi:10.1126/science.1058040

Wang, K., Z. Wang, F. Li, W. Ye, J. Wang, G. Song, et al. 2012. The draft genome of a diploid cotton *Gossypium raimondii*. Nat. Genet. 44(10):1098–1103. doi:10.1038/ng.2371

Wang, X.F., J. Ma, W.S. Wang, Y.M. Zheng, G.Y. Zhang, C.J. Liu, and Z.Y. Ma. 2006. Construction and characterization of the first bacterial artificial chromosome library for the cotton species *Gossypium barbadense* L. Genome 49(11):1393–1398. doi:10.1139/g06-113

Wang, Z., D. Zhang, X. Wang, X. Tan, H. Guo, and A.H. Paterson. 2013. A whole-genome DNA marker map for cotton based on the D-genome sequence of *Gossypium raimondii* L. G3: Genes, Genomes, Genet. 3(10): 1759–1767. doi:10.1534/g3.113.006890

Wendel, J.F., C.L. Brubaker, and T. Seelanan. 2010. The origin and evolution of *Gossypium*. In: J.M. Stewart et al., editors, The physiology of cotton. Springer, New York. p. 1–18.

Wendel, J.F., and C.E. Grover. 2015. Taxonomy and evolution of the cotton genus *Gossypium*. In: D. Fang and R. Percy, editors, Cotton. 2nd ed. Agron. Monogr. 57. ASA, CSSA, and SSSA, Madison, WI. doi:10.2134/agronmonogr57.2013.0020

Williams, L.J.S., D.G. Tabbaa, N. Li, A.M. Berlin, T.P. Shea, I. MacCallum, et al. 2012. Paired-end sequencing of Fosmid libraries by Illumina. Genome Res. 22(11):2241–2249. doi:10.1101/gr.138925.112

Wu, T.D., and S. Nacu. 2010. Fast and SNP-tolerant detection of complex variants and splicing in short reads. Bioinformatics 26(7):873–881. doi:10.1093/bioinformatics/btq057

Xu, Z., R.J. Kohel, G. Song, J. Cho, J. Yu, S. Yu, J. Tomkins, and J.Z. Yu. 2008. An integrated genetic and physical map of homoeologous chromosomes 12 and 26 in Upland cotton (*G. hirsutum* L.). BMC Genomics 9:108. doi:10.1186/1471-2164-9-108

Yandell, M., and D. Ence. 2012. A beginner's guide to eukaryotic genome annotation. Nat. Rev. Genet. 13(5):329–342. doi:10.1038/nrg3174

Yu, J., S. Hu, J. Wang, G.K.-S. Wong, S. Li, B. Liu, et al. 2002. A draft sequence of the rice genome (*Oryza sativa* L. ssp. *indica*). Science 296(5565):79–92. doi:10.1126/science.1068037

Yu, J., R.J. Kohel, and C.W. Smith. 2010. The construction of a tetraploid cotton genome wide comprehensive reference map. Genomics 95(4):230–240. doi:10.1016/j.ygeno.2010.02.001

Zerbino, D.R. 2010. Using the Velvet de novo assembler for short-read sequencing technologies. Curr. Protoc. Bioinf. 31:11.5.1–11.5.12. doi:10.1002/0471250953.bi1105s31

Zhang, M., Y. Zhang, J.J. Huang, X. Zhang, M.K. Lee, D.M. Stelly, and H.B. Zhang. 2012. Genome physical mapping of polyploids: A BIBAC physical map of cultivated tetraploid cotton, *Gossypium hirsutum* L. PLoS ONE 7(3):E33644. doi:10.1371/journal.pone.0033644

Role of Bioinformatic Tools and Databases in Cotton Research

Jing Yu* and Dorrie Main*

Abstract

Rapid advances in high-throughput genomic and phenomic technologies have led to an explosion of research data for cotton (*Gossypium* L.; Malvaceae) over the last decade. These include genetic and physical maps, genotypic and phenotypic data, whole-genome sequence annotations, transcriptomes and expression data, and proteomic and metabolomic data. Storage, analysis, integration, and dissemination of this vast abundance of data has led to the development of bioinformatic tools and advanced databases. Databases range from large generic ones that hold specific data for a broad range of species to curated databases with carefully integrated data for single species. This chapter discusses the different types of data available to the cotton community, such as genome sequences, molecular markers, and germplasm information. In addition, generic databases, such as GenBank, and cotton specific databases, such as CottonGen, along with tools for data access, analysis, and online visualization are discussed.

Advances during the last 20 yr in high-throughput sequencing and biotechnology have revolutionized the life sciences and heralded an age of information-driven biological research, as manifested by generation of large amounts of data. The advent of these large-scale biological data sets led to development of bioinformatic resources to manage, analyze, store, and access these data sets. Combining the disciplines of biology, statistics, computer science, and

---

Abbreviations: AFLP, amplified fragment length polymorphism; BAC, bacterial artificial chromosome; BGI, Beijing Genomics Institute; BLAST, Basic Local Alignment Search Tool; cDNA, complementary DNA; CDS, coding sequence; CGP, Cotton Genome Project; CoGe, Comparative Genomics; DDBJ, DNA Data Bank of Japan; EBI, European Bioinformatics Institute; ENA, European Nucleotide Archive; EMBL, European Molecular Biology Laboratory; EST, expressed sequence tag; FPC, fingerprinted contig; GEO, Gene Expression Omnibus; GO, Gene Ontology; GRIN, Germplasm Resources Information Network; GSS, genomic survey sequence; ICGI, International Cotton Genome Initiative; JGI, Joint Genome Institute; KEGG, Kyoto Encyclopedia of Genes and Genomes; LD, linkage disequilibrium; NCBI, National Center for Biotechnology Information; NCVT, National Cotton Variety Test; PCR, polymerase chain reaction; PLEXdb, Plant Expression Database; PUT, predicted unique transcript; QTL, quantitative trait locus; RFLP, restriction fragment length polymorphism; SCVT, State Cotton Variety Tests; SNP, single nucleotide polymorphism; SRAP, sequence related amplified polymorphism; SSR, simple sequence repeat; TAIR, the Arabidopsis Information Resource; TC, tentative consensus; UniProt, Universal Protein Resource; UniProtKB, UniProt KnowledgeBase; wwPDB, Worldwide Protein Data Bank.

Jing Yu and Dorrie Main, Dep. of Horticulture, Washington State University, 45 Johnson Hall, Pullman, WA 99164-6414. *Corresponding authors (jing.yu@wsu.edu and dorrie@wsu.edu).

doi:10.2134/agronmonogr57.2013.0043

Cotton. 2nd. ed. David D. Fang and Richard G. Percy, editors. Agronomy Monograph 57.

information technology, bioinformatics has become a fundamental part of many biological experiments. Online databases are an integral component of the bioinformatic tool kits available to researchers. Providing access to raw, analyzed, and integrated data sets and tools, these databases have become fundamental resources for researchers, enabling interrogation of publicly available data in a fully targeted manner to help address research questions.

This chapter will discuss the various types of genetic, genomic, and breeding data that are currently available to the cotton community; general bioinformatics resources relevant for cotton species (*Gossypium* spp.); publicly available databases, including CottonGen (Yu et al., 2014), the cotton community database; and how these tools and this information can be efficiently used to aid cotton research.

Available Data Types

Genetic Information

Molecular Markers

A molecular or genetic marker is a short sequence of DNA uniquely associated with a known location in the genome. Advances in the development and deployment of DNA markers associated with key agronomic traits are enabling marker-assisted selection and the streamlining of many crop-improvement efforts (Burr et al., 1983; Tanksley et al., 1988). Robust DNA-based markers are by definition phenotype neutral, free of epistatic effects, and show simple Mendelian inheritance. There are two types of molecular markers: those based on DNA-DNA hybridization, such as restriction fragment length polymorphism (RFLP), and markers based on the amplification of DNA sequences using polymerase chain reaction (PCR), such as amplified fragment length polymorphisms (AFLPs), simple sequence repeats (SSRs), sequence-related amplified polymorphisms (SRAPs), and single nucleotide polymorphisms (SNPs). As with most other crops, RFLPs were the first DNA markers used in cotton and demonstrated reliability in linkage analysis and breeding and in determining whether a linked trait was present in a homozygous or heterozygous state in an individual. PCR-based DNA markers superseded RFLPs because the former are highly polymorphic, inexpensive, and do not require large amounts of DNA. Among PCR-based DNA markers, SSRs, or microsatellite markers, and SNPs are now the two most prevalent marker types used in cotton research. SSRs are ubiquitous in most eukaryote genomes and tend to be multiallelic, reproducible, and highly informative (Senior and Heun, 1993; Vos et al., 1995). The development of a large number of expressed sequence tags (ESTs) in cotton has provided a rich source of gene-based, highly polymorphic SSRs (EST-SSRs). SNPs, created by a single base change, insertion, or deletion (indels) in homologous DNA fragments are the most abundant source of polymorphisms encountered in genomes (Cho et al., 1999; Griffin and Smith, 2000). SNP-based markers are codominant with normal distributions and are often associated with morphological changes (Lindblad-Toh et al., 2000), making them superior to other marker systems. SNPs are useful in characterizing allelic variation, mapping quantitative trait loci (QTLs), and being implemented as markers for marker-assisted selection. For cotton, however, a narrow germplasm base and a complex allotetraploid genome have made the discovery of SNP markers more difficult (Byers et al., 2012). Only a few cotton SNPs have been discovered using earlier SNP detection techniques (Shah et al., 2004; Lu et al., 2005; Ahmad et al., 2007; An et al., 2008). More recently, the advent of low-cost, massive-throughput, next-generation sequencing technology and advanced computational pipelines has facilitated the rapid discovery of large numbers of SNP markers in cotton (Van Deynze et al., 2009; Byers et al., 2012; Lacape et al., 2012; Rai et al., 2013). As of June 2014, more than 230,000 molecular markers (146,404 SNPs; 78,340 SSR and EST-SSRs; 3541 RFLPs; and 2146 AFLPs) have been developed and are being used in cotton research. Information associated with these markers includes characteristics such as primer sequence, marker sequence, repeat motif (if SSRs), marker allele size and polymorphism information, chromosome location and linkage to other markers or QTLs, and gel images, as well as literature citations. Marker-associated project information includes marker source germplasm, clone, DNA library, etc. Marker-derived information includes genetic maps, physical maps, consensus maps, comparative maps, and genetic diversity studies.

Quantitative Trait Loci

Many agronomic traits of cotton, such as yield and fiber quality, are controlled by multiple genes and are quantitatively inherited. As the gene number controlling trait expression increases, the probability of finding an individual with beneficial alleles at all of the genetic loci decreases. Such traits do not consistently fall into discrete classes because environmental conditions greatly modify their performance. The identification of genetic factors responsible for these complex traits has been difficult, limiting the efficiency of breeding efforts by the traditional approach, where only well-defined characters, usually controlled by a single, dominant gene, can be identified (Kohel, 1999; Tanksley and McCouch, 1997). QTLs are essentially a statistical creation that identifies a particular region of the genome (containing one or more genes) that is associated with the expression of a trait that is being measured. QTLs are used in linkage mapping, and markers are identified that are associated with the QTL. Markers derived from QTL regions are then used in breeding programs to screen for individual plants that have the trait(s) of interest. As the world's leading natural textile fiber crop, fiber yield and quality have been the focus of most cotton QTL studies, followed by disease, pest resistance, and abiotic stress responses. Currently, more than 2000 QTLs representing more than 80 agronomically important traits have been mapped in cotton using different populations, including F_2, recombinant inbred lines (RILs), advanced backcross populations, and chromosome segment introgression lines (CSILs). Among these QTLs, 60% are related to various aspects of fiber quality and yield (Table 1). QTL mapping usually involves developing populations for genotyping and phenotyping, generating genetic linkage maps, and establishing marker-trait associations. Information associated with QTLs includes types of mapping population, sets of molecular markers, descriptions of traits of interest, genotyping and phenotyping data, environment information, and genetic-mapping data.

DNA Sequences

New technologies that facilitate the rapid and relatively inexpensive generation of large amounts of DNA sequences have led to the availability of a significant number of sequences for cotton species. Common information associated with DNA sequences includes the molecule type, such as genomic DNA or mRNA; the source organism; gene and/or gene product names and coding sequence

Table 1. Summary of quantitative trait loci (QTL) from published journal articles.†

Year of publication	No. QTL in publication	QTL categories: Fiber	Yield	Seed	Early maturity	Resistant and tolerant	Morphological	Other	Reference
2013	1712	1245		89	153	88		137	Said et al., 2013
2013	166		61		17		66	22	Lacape et al., 2013
2013	54				54				Li et al., 2013
2013	37				37				Ai et al., 2013
2014	28	22	6						Gore et al., 2014
2014	2					2			He et al., 2014
2014	37						37		Liang et al., 2014
2014	7					7			Mei et al., 2014
2014	86	66	20						Ning et al., 2014
2014	77	77							Shao et al., 2014
2014	59	59							Tan et al., 2014
2013	24	24							Wang et al., 2013
2014	50		50						Xia et al., 2014
2014	2					2			Zhang et al., 2014
2014	6		6						Wang M et al., 2014
Total	2347	1493	143	89	261	99	103	159	

† This table expands the summary of Said et al. (2013), which contains 1998–2013 cotton QTL publications, by adding new publications between 2013 and 30 June 2014.

(CDS); corresponding protein name and amino acid translation; source DNA library; and publication information. DNA sequences can be divided into three types: (i) core nucleotide sequences, which are genomic sequences that include sequences from whole-genome sequencing projects, sequences of complete or partial genes, and sequences of transcripts data and provide the foundation for biomedical research and discovery; (ii) expressed sequence tags (ESTs), which are short single-read transcript sequences; and (iii) genome survey sequences (GSSs), which are unannotated short single-read genomic sequences or random genomic survey sequences, such as BAC (bacterial artificial chromosome)_end sequences and transposon-tagged sequences. Nucleotide sequences derived from ESTs are typically assembled into longer gene fragments, creating unigene data sets that provide one sequence (often a consensus sequence) representing a single gene. As of 30 June 2014, nearly 700,000 DNA sequences were available for cotton. These include 133,280 core nucleotide sequences, 464,568 EST sequences, and 94,046 GSS sequences in the US National Center for Biotechnology Information (NCBI) GenBank (Benson et al., 2013). Figure 1 shows that the number of nucleotide sequences for cotton species grew from only 5 sequences in 1990 to nearly 700,000 in 2014. From Fig. 1, we can see that the number of sequences for different sequence types grew at different rates as technology has evolved. For instance, EST sequences were widely available and used in cotton research since 2003, while many cotton BAC end sequences became available beginning in 2006. The rapid increase in the number of nucleotide sequences for cotton partially reflects the whole genome sequencing projects that were recently completed.

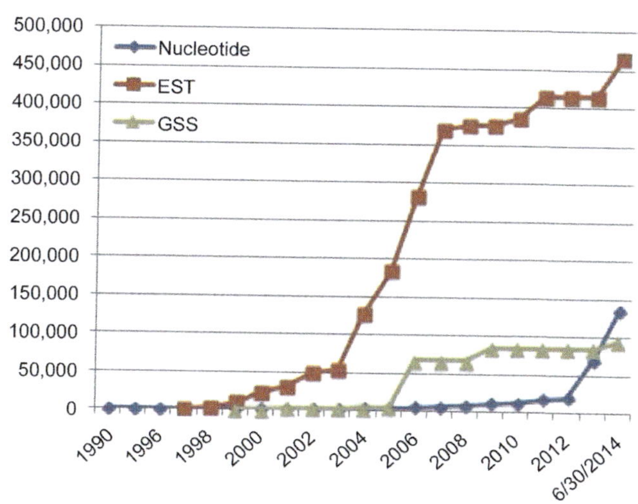

Fig. 1. The growth of the three types of DNA sequences for cotton species, core nucleotide, expressed sequence tag (EST) sequences and genomic survey sequences (GSS) from 1990 to 30 June 2014. As of 30 June 2014, there are nearly 700,000 cotton species sequences, including 133,280 core nucleotide sequences, 464,568 EST sequences, and 94,046 GSSs in the NCBI GenBank. Data source: National Center for Biotechnology Information GenBank.

Genetic Resources

Plant genetic resources refer to the use of plants as a source of genetic variation. Plant breeding uses specific traits to develop improved commercial cultivars and frequently focuses on the same genetic strains to combine the most desired traits. Out of 50 *Gossypium* species, 4 have been domesticated—two diploids (*G. arboreum, G. herbaceum*) and two tetraploids (*G. hirsutum, G. barbadense*). A suite of traits involved in domestication have included elimination of the hard seed coat, improvement in fiber length and the percentage of lint, development of larger boll size, and development of a day-neutral reproductive habit. *Gossypium hirsutum* has come to dominate world cotton production, representing over 90% of the annual cotton crop. The major drawback with domestication is the lack of genetic diversity (Mehboob-ur-Rahman et al., 2012). The current genetic base of US commercial cotton cultivars represents less than 1% of US cotton germplasm (Esbroeck and Bowman, 1998). The resulting homogeneity and uniformity can have substantial advantages in both the quantity and quality of cotton harvested, but this same genetic homogeneity can also reflect greater susceptibility to pests and pathogens. To mitigate against future disaster, a coordinated effort to collect and maintain cotton genetic resources has been ongoing in nine countries—Australia, Brazil, China, France, India, Pakistan, Russia, United States, and Uzbekistan (Campbell et al., 2010; FAO, 2010; GRIN, 2014). Table 2 describes the cotton species present in the worldwide cotton germplasm collections and shows that the nine countries have provided more than half of the accessions in the world cotton GeneBank. Overall, 46% of the GeneBank accessions are unknown or a mixture of two or more germplasm types, with the remainder composed of 4% wild species, 14% landrace/old cultivars, 19% research material/breeding lines, and 17% advanced cultivars. A germplasm collection represents preservation of genetic diversity (as plants, seeds, and/or tissues) and usually contains current and heirloom crop varieties, breeding lines used to develop hybrids and new varieties, landraces of crops developed over time by farmer selection, wild plants involved in domestication, and mutants maintained for research. Information for a specific germplasm accession can include passport data (genus and species, genotype or cultivar name, collection information, country of origin, donor, and seed availability), pedigree and genealogy information, measurements at a genotypic or phenotypic level, and images of physical traits, such as boll shape, lint color, seed fuzz, etc. Other information associated with a specific accession can include associated DNA libraries, sequences, and derived molecular markers.

Genomic Information

Linkage Mapping

Genetic mapping provides observed positions of genetic markers along chromosome(s) in genetic map units or centimorgans calculated on the basis of recombination frequency. Recombination frequencies reveal the genetic distance between traits and markers, and trait-linked markers can be anchored, allowing determination of the location in the genome that likely contains the trait and gene(s) of interest. Recombination is a biological phenomenon and can occur at very different rates in different genomic regions, populations, and environments. Nonetheless, it directly measures the transmission of genes (or groups of genes) from parent to progeny, a central component of crop improvement. Advances in molecular biology have provided a broad spectrum of technologies to assess genetics at

Table 2. Description of worldwide cotton germplasm collections.†

FAO GenBank				Percent of all within collection‡				
Country	Instcode	No. of acc	% or total	WS	LR	BL	AC	OT
		no.	%	%				
Uzbekistan	UZB036	12,048	11					100
USA‡	USA049	10,501	10	21	2	8	4	64
India	IND512	9000	8		100			
China	CHN001	7226	7	7				93
Russia	RUS001	6205	6		23	16	58	3
France	FRA002	4116	4	12	38			50
Brazil	BRA003	3179	3					100
Pakistan	PAK009	1830	2	2		98		
Australia§		1622	2		1		39	60
Viet Nam	VNM013	1400	1			100		
Azerbaijan	AZE015	1370	1			<1	100	
	Others (98)	47,397	45	5	6	7	5	77
Total		105,894	100	4	14	19	17	46

† All information was obtained from the FAO GenBank (Food and Agriculture Organization of the United Nations, 2010), except that for USA is from GRIN, 2014 and that for Australia is from Campbell et al., 2010.

‡ WS, wild species; LR, landraces/old cultivars; BL, research materials/breeding lines; AC, advanced cultivars; OT, the type is unknown or a mixture of two or more types.

the DNA level. Since the first RFLP-marker-based cotton genetic map reported in 1994 (Reinisch et al., 1994), more than 50 cotton genetic maps accompanied by over 34,000 loci have been reported (Yu et al., 2014). These maps were derived from different types of populations such as F_2 plants, recombinant inbred lines, backcross populations, and doubled haploid lines, and they were developed in the AD, A, D, G, and C cotton genomes. General information associated with genetic maps includes markers, mapping populations, environments, and QTLs.

Association Mapping

Association mapping is a high-resolution method used for mapping QTLs based on linkage disequilibrium (LD), a useful technique for dissection of complex genetic traits. In contrast to QTL mapping (or linkage mapping) which uses biparental-cross populations, association mapping identifies QTLs by identifying marker-trait associations that can be attributed to the strength of LD between markers and functional polymorphisms across a wide range of individuals. Because the unobserved ancestry among individuals can extend over thousands of generations, shared inheritance will only persist for neighboring loci after many generations of recombination. Effectively, association mapping exploits evolutionary and historical recombination at the population level (Remington et al., 2001; Thornsberry et al., 2001). Association mapping offers several advantages over pedigree-based linkage analysis, namely, much higher mapping resolution, greater allele number, a broader reference population, and less research time in establishing an association (Buckler and Thornsberry, 2002; Flint-Garcia et al., 2003). A general understanding of the value of association mapping has increased significantly since its inception in plants. In cotton, association mapping (or LD mapping) has been used in fiber-quality and yield QTL mapping in tetraploid species (Abdurakhmonov et al., 2009; Zeng et al., 2009; Mei et al., 2013; Jia et al., 2014) and diploid species (Kantartzi and Stewart, 2008), in disease-resistance QTL mapping (Mei et al., 2014; Zhao et al., 2014), and in studies of genetic diversity and population structure (Jena et al., 2011; Fang et al., 2013).

Physical Mapping

Genetic linkage mapping is a powerful method but suffers from some limitations: only polymorphic markers can be mapped, the resolution that can be achieved is often limited, the distances are based on recombination frequencies rather than physical distance, and the recombination frequencies are not equally distributed across the genome. Fingerprinted contigs (FPCs; Nelson and Soderlund, 2009) and cytogenetic maps provide complementary information because they are based on the fine physical structure of chromosomes. FPC uses the restriction fragments (bands) generated for a set of BAC clones and assembles them into contigs. If at least some of the clones contain markers that have been independently mapped by other means, then the position of the entire contig in the genome is also known. Selecting overlapping clones from a BAC and sequencing them provides the DNA sequence for the BAC (typically 100– 120 Kb), and software can be used to predict the genes in that region. Cytomolecular mapping uses microscopy to directly show DNA landmarks on chromosomes, and the distance between landmarks is measured directly. In cotton, physical maps have been used to complete assignment of the chromosomes of *G. hirsutum* (Wang et al., 2006), to develop chromosome-specific microsatellite-containing BACs in *G. hirsutum* (Wang et al., 2007), to search for the homeologous regions between the A subgenome and D subgenome of tetraploid species *G. hirsutum* (Xu et al., 2008), to investigate the organization and evolution of fiber genes (Xu et al., 2010), to construct whole-genome physical maps (Lin et al., 2010; Zhang et al., 2012), and to perform genome comparative analysis (Lin et al., 2011). Physical-mapping-associated information includes clones, DNA libraries, germplasm, and markers.

Whole Genome Sequencing

The availability of whole genome cotton sequences provides a fundamental resource for understanding the functional and agronomic significance of genome size variation and polyploidy within *Gossypium*. With falling costs and faster sequencing technologies available for large-scale DNA sequencing, the cotton community started to build and take advantage of comprehensive cotton genomic resources. *Gossypium raimondii* (D_5 genome) was chosen as the first species for whole genome sequencing. This species has the smaller genome of the two diploid progenitors (A and D genomes) of the domesticated allotetraploid cotton (*G. hirsutum*, or AD_1 genome) that dominates world cotton commerce (Wendel and Cronn, 2003). In 2012, two independent research teams released whole genome sequences of *G. raimondii*, the Joint Genome Institute (JGI) version (Paterson et al., 2012) and Beijing Genomics Institute (BGI) version (Wang et al., 2012). Soon after, one of the two A genome species, *G. arboreum* (A_2 genome), was reported to be fully sequenced (Li et al., 2014). The two versions of the D_5 genome sequencing projects generated 761,405,269 bp sequences with 77,627 CDSs (Paterson et al., 2012) and 773,767,704 bp sequences with 40,976 CDSs (Wang et al., 2012). In comparison, the A_2 genome sequencing project generated 1,560,831,891 bp sequences with 43,110 CDSs. Simultaneously, data of annotations for these sequenced genomes, such as CDSs and the corresponding proteins, and alignments of cotton unigenes and markers were made available to the community. Table 3 shows the numbers of

Table 3. Assembly information and putative genes of the fully sequenced *Gossypium arboreum* and the two versions of the *Gossypium raimondii* genome sequence.

Genome	Total contigs	N50(L50) contigs	Total scaffolds	N50(L50) scaffolds	Coding sequences	Total length	Reference
BGI-D_5	41,307	4918 (44.8 kb)	4434	95 (2.28 Mb)	40,976	773,767,704	Wang et al., 2012
JGI-D_5	19,735	1596 (135.6 kb)	1033	6 (62.2 Mb)	77,267	761,405,269	Paterson et al., 2012
BGI-A_2	40,381	N/A (72.0 kb)	7914	N/A (665.8 kb)	43,110	1,560,831,891	Li et al., 2014

whole genome sequences and putative genes that are derived from these three genome sequencing projects.

Microarrays

Microarrays provide high-density in situ synthesis of oligonucleotides or ex situ spotting of target molecules (complementary DNA, cDNA) for conducting genome-wide comparative studies. Although they have now been largely replaced by more accurate RNA-Seq technology for gene expression profiling, several microarray platforms were developed and expression data were generated for plant response under different abiotic and biotic stresses. The use of microarrays was reported for the first time in cotton in 2004 (Arpat et al., 2004). In the study, an array of 12,227 nonredundant cDNA probes from a *G. arboreum* cDNA library was used for genetic characterization of rapid cell elongation in cotton fibers. Since then, many other cotton research projects have used microarrays. The microarray platforms developed in some of these projects include a cotton-fiber-specific cDNA array of 11,692 probes representing more than 8000 cotton genes and being used to identify the role for ethylene in cotton-fiber cell elongation (Shi et al., 2006); an array of 22,787 probes generated from a global assembly of 211,397 cotton ESTs derived from more than 50 cDNA libraries representing different tissue types and tissue treatments, being used to target transcriptome diversity and previously studied cotton genes, transcription factors, and genes with homology to *Arabidopsis* (Udall et al., 2007); a cDNA array that includes 9126 plasmids randomly selected from a normalized fiber cDNA library (from −2 to 25 d post anthesis) of *G. barbadense* to study gene expression during fiber development (Tu et al., 2007); a cotton community instigated Affymetrix GeneChip Consortia Program array composed of 23,977 probe sets representing 21,854 cotton transcripts (Affymetrix, 2009) based on EST sequences from four species of cotton (*G. hirsutum, G. raimondii, G. arboreum,* and *G. barbadense*); a 24K Affymetrix cotton genome array developed to generate transcription profiles of the response to sodium chloride treatment and used to study genes that regulate salt stress in cotton (Yao et al., 2011); and a 29K cotton genome array used in functional analysis of a esterase/lipase gene from *G. hirsutum* L. (Tong et al., 2013).

RNA Sequencing

RNA sequencing is a relatively new next-generation sequencing technology that provides a snapshot of the presence and quantity of RNA from a genome at a given moment in time. Compared with other older transcriptome profiling technologies, RNA sequencing is a much more accurate technique for detecting transcript abundance; identifying transcript sequence polymorphisms, novel *trans*-splicing, and splice isoforms; and distinguishing paralagous genes (Lister et al., 2009; Wang et al., 2009). RNA sequencing has been used extensively for humans, mammals, and yeast, and it is now being adopted widely in plants. *Arabidopsis* was the first plant species to be studied with this technique (Weber et al., 2007). A comparison of deep transcriptome sequencing with the publicly available ESTs confirmed most of the annotated introns and identified thousands of novel, alternatively spliced mRNA isoforms, suggesting that at least 42% of the intron-containing genes of *Arabidopsis* are alternatively spliced (Filichkin et al., 2010). These RNA sequencing data (Lister et al., 2008; Filichkin et al., 2010), combined with proteomics data (Baerenfaller et al., 2008; Castellana et al., 2008), were used to revise *Arabidopsis* gene models and were incorporated into release 10 of the Arabidopsis Information Resource (TAIR) (Lamesch et al., 2010). In cotton, RNA sequencing experiments were performed for the annotation of the *G. raimondii* (D_5) genome sequencing project (JGI v2.1; Paterson et al., 2012); in the identification of the role of lignin that contributes to the resistance of cotton to verticillium wilt (Xu et al., 2011); to identify differentially expressed transcripts under water-deficit stress (Bowman et al., 2013); and in the study of comparative evolutionary and developmental dynamics of the cotton (*G. hirsutum*) fiber transcriptome (Yoo and Wendel, 2014).

Breeding Information

Germplasm Characterization and Evaluation

Germplasm characterization involves describing the diversity contained in a germplasm collection and is the basic description of morphological traits that are generally qualitative and do not change in different environments. Providing information about the traits of individual accessions helps ensure more optimal utilization of the germplasm collection by breeders. Recording and compiling data on the significant characteristics that differentiate accessions within a species enables easy and quick discrimination among phenotypes. It allows simple grouping of accessions, classifying of core collections, retrieval of valuable germplasm for breeding, and identification of germplasm gaps, resulting in improved knowledge about the composition and genetic diversity of the collection. Characterization of the US Cotton Germplasm Collection indicated a need to conduct more morphological, agronomic, molecular, and priority trait assessments to improve the use of the collection. To this end phenotypical (or morphological) characterization or recharacterization of germplasm in the collection has been conducted, and digital imaging has been adopted. To establish clear standards and reduce the subjectivity that occurs when scoring descriptors on such a spectrum of diversity in the collection, a pictorial reference of the descriptor scores has been generated (Percy et al., 2014). To date, more than 3268 germplasm have been characterized and more than 22,000 digital images have been produced.

Germplasm evaluation, in the broad sense and in the context of genetic resources, is the process of describing quantitative traits that can be influenced by environmental conditions. Evaluation of a germplasm collection therefore uses the genetic resources in the collection. In the US Cotton Germplasm Collection, 7000 germplasm accessions were evaluated for 49 traits; and several ongoing projects are investigating germplasm in the national collection for different economically important traits, such as resistance to pests or disease and drought tolerance. Similarly, evaluation data for 34 traits from nearly 3000 germplasm in the China Cotton Germplasm Collection and data on 49 traits evaluated in 847 germplasm in the Uzbekistan Cotton Germplasm Collection are also available to the community. These data are of great value in basic and translational cotton research as well as in breeding efforts.

Variety Testing

The US National Cotton Variety Test (NCVT) program was initiated in 1960 to standardize collection and analysis of the field data necessary to enable objective evaluation of new Upland and Pima varieties. It started with tests in the 1950s of Pima cotton species in six distinct growing regions of the US Cotton Belt (which extends from Virginia south to Georgia and west to southern California) to today's arrangement of testing four or five national standards for 3-yr cycles; at the end of the cycle, two or three of these varieties will be replaced by new national standards. The NCVT program has enabled researchers to systematically test new, region-specific varieties and establish national standards for each test, serving as a link between regions during the last 60 years (Suszkiw, 2010). More than 1300 varieties, germplasm, and strains have been tested (for fiber quality and yield) by the NCVT during this period. In addition, the US States Cotton Variety Tests (SCVT) have been conducted to evaluate commercial cultivars in every cotton-growing region. Both of these cotton variety testing programs have produced valuable information for the cotton community. The NCVT and SCVT data have not yet been stored in any database system. They are currently maintained as either pdf files or website tables created by different states in different years. For users' convenience, the cotton community database, CottonGen (Yu et al., 2014), has gathered all those pdf files and websites into one web page— http://www.cottongen.org/VarietyTest—and have started the process of converting the data into a format for uploading to the database and enabling targeted comparison, querying, and downloading of the data.

Data Repositories and Analysis Tools

Publicly available databases house data that is readily available for download and use. Listed below are many general databases that are useful for cotton researchers. The URLs are listed in Table 4. Data from many of these resources have been integrated or hyperlinked in CottonGen.

DNA-Based Sequence Repositories

GenBank, European Nucleotide Archive, and DNA Data Bank of Japan

The availability of extensive sequence data provides essential genomics resources help understand the biology of crops and apply this knowledge for the improvement of varieties. The International Nucleotide Sequence Database Collaboration is composed of the three primary sequence databases, which work together to catalog all available sequence data: the US NCBI public genome database GenBank (Benson et al., 2013), the European Nucleotide Archive (ENA) of the European Molecular Biology Laboratory–European Bioinformatics Institute (EMBL-EBI; Leinonen et al., 2010), and the DNA Data Bank of Japan (DDBJ) (Kosuge et al., 2014). This collaboration, which includes the daily exchange of newly submitted data between

Table 4. Publicly available databases used in cotton research.

Type	Name	URL	Reference
Sequence	National Center for Biotechnology Information (NCBI) GenBank	http://www.ncbi.nlm.nih.gov/genbank	Benson et al., 2013
	European Nucleotide Archive (ENR)	http://www.ebi.ac.uk/ena/	Leinonen et al., 2010
	DNA Data Bank of Japan (DDBJ)	http://www.ddbj.nig. ac.jp/	Kosuge et al., 2014
	Phytozome	http://phytozome.jgi.doe.gov	Goodstein et al., 2012
	Phytozome (*G. raimondii*)	http://phytozome.jgi.doe.gov/pz/portal.html#!info?alias=Org_Graimondii	Paterson et al., 2012
	BGI Cotton Genome Project (CGP)	http://cgp.genomics.org.cn/page/species/index.jsp	Wang et al., 2012; Li et al., 2014
	PlantGDB	http://www.plantgdb.org	Duvick et al., 2008
	TIGR/DFCI Gene Indices	http://compbio.dfci.harvard.edu/tgi/plant.html	Lee et al., 2006
Gene expression	Gene Expression Omnibus (GEO)	http://www.ncbi.nlm.nih.gov/geo/	Barrett et al., 2013
	ArrayExpress	http://www.ebi.ac.uk/arrayexpress	Rustici et al., 2013
	Omics	http://trace.ddbj.nig.ac.jp/dor/index_e.html	Ogasawara et al., 2013
	Plant Expression Database (PLEXdb)	http://www.plexdb.org	Dash et al., 2012
Gene sequence and structure	UniProt	http://www.uniprot.org	The UniProt Consortium, 2014
	Worldwide Protein Data Bank (wwPDB)	http://www.wwpdb.org	Berman et al., 2007
	Classification of Protein Structure (CATH)	http://www.cathdb.info/	Sillitoe et al., 2013
Gene classification	Gene Ontology (GO)	http://geneontology.org/	Gene Ontology Consortium, 2013
	InterPro	http://www.ebi.ac.uk/interpro/	Hunter et al., 2012
	GreenPhylDB	http://www.greenphyl.org/cgi-bin/index.cgi	Rouard et al., 2012
Comparative genomics	Comparative Genomics (CoGe)	https://genomevolution.org/CoGe/	Lyons and Freeling, 2008
	PLAZA	http://bioinformatics.psb.ugent.be/plaza/	Van Bel et al., 2012
	VISTA	http://genome.lbl.gov/vista/index.shtml	Frazer et al., 2004
Metabolic pathway	Kyoto Encyclopedia of Genes and Genomes (KEGG)	http://www.genome.jp/kegg/	Kanehisa et al., 2014
	BioCyc	http://biocyc.org/	Caspi et al., 2014
Germplasm database	Germplasm Resources Information Network (GRIN)	http://www.ars-grin.gov	GRIN
	GRIN National Plant Germplasm System	http://www.ars-grin.gov/npgs/index.html	GRIN-NPGS
Community database	CottonGen	http://www.cottongen.org/	Yu et al., 2014

the databases, ensures that a comprehensive, up-to-date, and secure collection of sequence data is available worldwide (Benson et al., 2013). The GenBank database is a primary source for DNA and protein sequences of all organisms and is organized into different sub-databases, such as whole-genome sequences and annotations, dbEST for short single-read transcripts, and UniGene for gene-oriented clusters of transcripts. Other divisions include three-dimensional macromolecular structures, taxonomy, single nucleotide polymorphisms, chemical molecules and substances, protein domains, microarray data, the Gene Expression Omnibus (GEO) data sets, and the Sequence Read Archive to store high-throughput sequence raw data and alignments. These databases provide comprehensive information about a gene or protein of interest. Selected available maps connect directly to markers with known sequences. Various analysis tools are available, such as the Basic Local Alignment Search Tool (BLAST), which is a sequence-similarity search engine that can be used to compare nucleotide or protein sequences and find the closest matches to other organisms. The Entrez search site on the home page of the NCBI (www.ncbi.nih.gov) allows researchers to batch-download all the different data that it houses. Searching for information on cotton species by typing "Gossypium[ORGN]" in the search box, returns a results page categorized by the different NCBI databases. As of 30 June 2014, this search returned three genome assemblies; 2723 genes; 464,568 ESTs; 24,580 clusters of expressed transcripts; 10,143 proteins; four three-dimensional structures; 36,113 genomic and cDNA clones; 94,046 GSSs; 133,396 nucleotides; 14,357 probes; and 39,862 SNPs. NCBI is, therefore, an excellent source of sequence data for cotton species. Researchers can keep up to date on new data submitted to the NCBI for their species of interest by creating a "My NCBI" account, which saves the search history and filters data, among other features. EMBL-EBI is Europe's primary collection of molecular biology/genomics data and also contains copies of known DNA and protein sequences. This site provides access to an excellent source of online analysis tools that can be used for cotton research. DDBJ is the primary DNA sequence data collection repository for Japanese researchers but also accepts data from researchers in other countries. More information about NCBI GenBank, DDBJ, and EMBL ENA can be found from their online tutorials and the Nucleic Acids Research Database Addition, which is published every January.

Phytozome

Phytozome (http://phytozome.jgi.doe.gov), a joint project of the Department of Energy's JGI and the Center for Integrative Genomics, is a tool to facilitate comparative genomic studies of green plants (Goodstein et al., 2012). The modern descendants of ancestral genes are constructed at key phylogenetic nodes to identify families of orthologous and paralogous genes. These groupings allow easy access to clade-specific genes and gene expansions as well as to clade-specific orthology/paralogy relationships. As of release v10.0, Phytozome provides access to 46 sequenced and annotated green-plant genomes, which have been clustered into gene families at 13 evolutionarily significant nodes. Genes are annotated with Gene Ontology (GO; Gene Ontology Consortium, 2013), Kyoto Encyclopedia of Genes and Genomes (KEGG; Aoki and Kanehisa, 2005), Protein Families Database (Finn et al., 2010), Protein Analysis through Evolutionary Relationships (Mi et al., 2010), and Eukaryotic Orthologous Group (Tatusov et al., 2003) assignments where possible. Query-based data access is provided by Phytozome's InterMine and BioMart instances, while bulk data sets can be accessed via JGI's Genome

Portal (registration required). JBrowse genome browsers are available for all genomes. Phytozome houses the annotated *G. raimondii* genome sequence (Paterson et al., 2012) and contains tools for closer investigation. Data can also be easily downloaded after registering for a free account.

China Cotton Genome Project

The Cotton Genome Project (CGP; http://cgp.genomics.org.cn/page/species/index.jsp) website, has been developed by the Cotton Research Institute of the Chinese Academy of Agricultural Sciences. In collaboration with BGI, the CGP is mainly focused on cotton sequencing and functional analysis, and currently houses BGI versions of the *G. raimondii* and *G. arboreum* whole-genome sequences and related annotation data (Wang et al., 2012; Li et al., 2014). The genome data (sequence and annotation) can be downloaded and viewed in a genome browser or pairwise analysis can be conducted using the BLAST algorithm.

PlantGDB

PlantGDB (http://www.plantgdb.org/; Duvick et al., 2008) provides annotated transcript assemblies for more than 100 plant species, with transcripts mapped to their respective genomes where possible, and various analysis tools and computational access to the data. For plant species with emerging or complete genome sequence, PlantGDB's genome browsers (xGDB) provide graphical interfaces for viewing, evaluating, and annotating transcript and protein alignments to the genomic sequence. Community curation of the gene models are facilitated by access to PlantGDB's web-based gene-structure annotation tool. Direct computational access to PlantGDB includes the BioExtract Server, a web-based server for executing custom sequence analysis workflows, and Tracembler, an interactive alignment tool that generates contigs from GenBank trace file data. A research outreach portal at PlantGDB facilitates access to a large number of resources for research and training (Duvick et al., 2008). PlantGDB, using its custom assembly pipeline, generated and stores cotton predicted unique transcripts (PUTs) based on sequences from GenBank and UniProt (uploaded and parsed every 4 mo). These transcripts include 24,597 *Gossypium arboreum* PUTs, 6712 *Gossypium barbadense* PUTs, 98,420 *Gossypium hirsutum* PUTs and 27,355 *Gossypium raimondii* PUTs. PlantGDB provides a genome browser based on 150 *G. hirsutum* BAC sequences (from GenBank version 154) with gene annotations and alignments of ESTs, cDNAs, PUTs, and proteins.

The Institute for Genome Research Gene Indices

The Institute for Genome Research was formerly the major institute participating in the Human Genome Project. They also collected publicly available EST sequences (including cotton ESTs) and developed bioinformatics tools to assemble ESTs into tentative consensus (TC) sequences, representing genes which they also computationally annotated (Lee et al., 2006). No longer updated, the current Cotton Gene Index (CGI) v11 consists of 50,873 TC sequences from 351,954 EST sequences. EST and TC information can be retrieved from their website, which has moved to the Dana-Farber Cancer Institute and the Harvard School of Public Health (http://compbio.dfci.harvard.edu/tgi/plant.html).

Gene Expression Resources

Gene Expression Omnibus, ArrayExpress, and Omics

The Gene Expression Ominibus (GEO, http://www.ncbi.nlm.nih.gov/geo/) at NCBI is one of the three international public repositories that archives and freely distributes microarray, next-generation sequencing, and other forms of high-throughput functional genomic data sets (Barrett et al., 2013). The other two public repositories include the ArrayExpress Archive of Functional Genomics Data (Rustici et al., 2013) at EMBL-EBI (http://www.ebi.ac.uk/arrayexpress/) and the DDBJ Omics Archive (http://trace.ddbj.nig.ac.jp/dor/index_e.html, Ogasawara et al., 2013). Approximately 90% of the data in GEO are gene-expression studies that investigate a broad range of biological themes including disease, development, evolution, immunity, ecology, toxicology, metabolism, and more. The nonexpression data in GEO represent other categories of functional genomic and epigenomic studies including those that examine genome methylation, chromatin structure, genome copy-number variations, and genome-protein interactions. The data in GEO represent original research submitted by the scientific community and made available for journal- and grant-compliance purposes to facilitate independent evaluation of results, reanalysis, and full access to all parts of each study. This resource supports archiving of all parts of a study, including raw data files, processed data, and descriptive metadata, which are indexed, cross-linked, and searchable. While the principal role of GEO is to serve as a primary data archive, this resource also offers several tools and features that allow users to explore, analyze, and display expression data from both gene-centric and study-centric perspectives. As of 30 June 2014, a search for "Gossypium" in NCBI GEO returned 1226 GEO data sets of nine species: 1019 for *Gossypium hirsutum*, 109 for *Gossypium barbadense*, 63 for *Gossypium herbaceum*, 48 for *Gossypium arboreum*, 38 for *Gossypium raimondii*, 37 for *Gossypium longicalyx*, 15 for *Gossypium tomontosum*, 13 for *Gossypium mustelinum*, and 11 for *Gossypium darwinii*.

Plant Expression Database

The Plant Expression Database (PLEXdb; http://www.plexdb.org) is a gene-expression (transcript) resource developed to help correlate genotype with phenotype through transcript profiling. It integrates multiple data sets from a wide variety of plant and plant-pathogen microarray experiments and provides a single site to access, analyze, and disseminate expression data for plant comparative functional genomics studies (Dash et al., 2012). Comparison analyses using PLEXdb data sets and tools facilitate interpretation of structure, function, and regulation of genes in economically important plants. Gene Atlas lists of experiments highlight data sets that give responses across different developmental stages, conditions, and tissues. These gene lists can be mapped onto corresponding model plant genomes with the PlEXdb Model Genome Interrogator tool, which also displays gene structures and supporting information for annotated genes and full-length cDNAs and predicts gene homology. Other tools in PLEXdb include a gene list-processing wizard guide for creating, analyzing, annotating, and managing gene lists. Higher-level analyses, such as ANOVA and clustering, are also available in PLEXdb, enabling identification of up-regulation under diverse conditions. To date, there are 12 publicly available cotton experiments in PLEXdb (CottonPLEX: http://www.plexdb.org/plex.php?database=Cotton).

Gene Sequence and Structure Resources

UniProt

The Universal Protein Resource (UniProt) is a resource developed by the Swiss Institute of Bioinformatics, the EBI, and the Protein Information Resource. It provides a comprehensive, accurately annotated protein sequence knowledgebase comprised of four major components (UniProt Consortium, 2014). First, the UniProt KnowledgeBase (UniProtKB) is the central hub for the collection of functional information on proteins, with accurate, consistent, and extensive annotation. It consists of a section containing manually annotated records with information extracted from the literature and curator-evaluated computational analysis and a section with computationally analyzed records that await full manual annotation. For the sake of continuity and name recognition, the two sections are referred to as UniProtKB/Swiss-Prot (reviewed, manually annotated) and UniProtKB/TrEMBL (unreviewed, automatically annotated), respectively. Second, the UniProt Reference Clusters contains protein sequence clusters, used to accelerate sequence similarity searches. Third, the UniProt Archive is a comprehensive and nonredundant database that contains most of the publicly available protein sequences in the world. Fourth, the UniProt Metagenomic and Environmental Sequence Database contains sequence data for metagenomics research. UniProt is updated and distributed every 4 wk and can be accessed online for searching or download (UniProt Consortium, 2014). As of 30 June 2014, there were 239 cotton proteins in UniProtKB/Swiss-Prot and 5393 in UniProtKB/TrEMBL. UniProt is also routinely used as a resource to identify closely related homologs of cotton genes in structural and functional annotation studies.

Worldwide Protein Data Bank

The Worldwide Protein Data Bank (wwPDB; http://www.wwpdb.org) is a repository for the three-dimensional structural data of large biological molecules, such as proteins and nucleic acids. It provides a single archive of macromolecular structural data that is freely and publicly available to the global community. Currently wwPDB contains the coordinates and related information of more than 90,000 structures of proteins, nucleic acids, and protein/nucleic acid complexes. As of 30 June 2014, only two *Gossypium hirsutum* and two *Gossypium arboreum* protein structures were found in the wwPDB, but the resource can be used to help predict three-dimensional structure for unknown cotton proteins and nucleic acids.

Protein Structure Classification Database

While the content of the wwPDB is considered primary data, there are many derived (i.e., secondary) databases that categorize the data differently, such as CATH (Class, Architecture, Topology, Homology) and the Protein Structure Classification Database (Sillitoe et al., 2013), which categorize structures according to type of structure and assumed evolutionary relations. The latest released CATH v4.0 identified 277,687 CATH domains and 2738 CATH superfamilies from 69,058 protein data banks. A search for "Gossypium" in CATH v4.0 returns structures of three CATH superfamilies and 16 CATH domains that contain proteins from *Gossypium hirsutum* and *Gossypium arboreum*.

Gene and Gene Product Classification Resources

Gene Ontology

Gene and protein sequences require annotation to describe their function. GO provides a standardized terminology for describing the activities of gene products to facilitate the integration of data from multiple species and different databases (Gene Ontology Consortium, 2013). GO annotation enables identification of gene products in one species with similar functions in other species. It is composed of three ontologies (controlled vocabularies) that each describe the gene products. The molecular function ontology describes the biochemical functions; the biological process ontology describes the cell physiological, regulatory, developmental, or metabolic processes; and the cellular component ontology describes the intracellular location. Each ontology is hierarchical; for example, a protein that functions in the G1 phase of the cell cycle would be associated with the biological process ontology terms "cell cycle," "interphase," and "G1 phase" (Gene Ontology Consortium, 2013). GO has grown to become an integrated resource containing essential functional information for 347,778 species (including strains) covering plants, animals, and the microbial world, and it is routinely used in publications involving any functional annotation analysis. The GO site includes tools developed both within and outside the GO Consortium for searching, browsing, and editing the GO ontologies and for using the ontologies to annotate data or analyze microarray data sets.

InterPro

InterPro (http://www.ebi.ac.uk/interpro/) is a database of protein families, domains, and functional sites in which information from known proteins can be used to infer function for new protein sequences (Hunter et al., 2012). It is an integrated, searchable resource compiled from multiple databases that reduces data redundancy and provides tools to help users conduct and interpret sequence analysis results. It is commonly used in large-scale analysis of whole proteomes, genomes, and metagenomes, as well as to characterize individual protein sequences. InterPro has been used within EBI to annotate protein sequences in UniProKB. It is used by the GO annotation group to automatically assign GO terms to protein sequences. It is routinely used to identify functional domains in cotton transcripts and putative genes from transcriptome- and genome-sequencing projects.

GreenPhylDB

With an increasing number of genomes being sequenced, a major goal is to transfer accurate annotation from characterized proteins to uncharacterized sequences. Consequently, sequenced-based comparative genomics has become a common and efficient strategy in functional genomics. GreenPhylDB (Rouard et al., 2011) is a web resource designed for comparative and functional genomics in plants. It includes a catalog of gene families based on gene predictions of sequenced genomes and covering a broad taxonomy of plants. Computational clustering in GreenPhylDB is followed by manual annotation and analysis by phylogenetics to predict homologous relationships. It supports functional and evolutionary research to discover candidate genes affecting agronomic traits in crops. The GreenPhylDB v4 comprises protein sequences of 37 plant species fully sequenced that were grouped into 5545 homeomorphic families out of 8347 clusters. The *Gossypium raimondii* sequence (JGI v2.1) is one of the 37 plant species grouped into gene families.

Comparative Genomics Database Resources

Comparative genomics compares the structure and function of genomes from different species or varieties. It is often used to obtain insight into how genomes have evolved through detection of duplicated regions within species and syntenic regions among species. The comparison of genomes and their contents among species enables identification of similar genes and other sequence features that are conserved among species and/or genes that are specific to certain clades. Comparative genomics also enables transfer of knowledge from well-studied species to less-studied species. With an increasing number of sequenced genomes, several web-based databases are available to compare this data. A few, described below, contain the whole genome sequences of cotton species.

Accelerating Comparative Genomics

Accelerating Comparative Genomics (CoGe; https://genomevolution.org/CoGe/; Lyons and Freeling, 2008) is an online system for enabling retrieval and comparison of genomic information and sequence data. As of 30 June 2014, CoGe contains the whole-genome data of 22,486 genomes from 16,489 organisms across all domains of life, including *Gossypium raimondii* and several cotton chloroplast genome sequences, such as those from *Gossypium hirsutum* and *Gossypium barbadense*. CoGe contains a series of tools with which researchers can select, view, and compare genomes of interest. It includes OrganismView for searching organisms, CoGeBlast for blasting sequences against the genomes of interest, FeatView for searching genomic features by name, SynMap for whole genome syntenic dotplot analyses, and GEvo for high-resolution sequence analysis of genomic regions. SynMap is a tool for comparative genomics which uses DAGChainer (Haas et al., 2004) as the underlying software for synteny detection.

PLAZA

PLAZA (http://bioinformatics.psb.ugent.be/plaza/; Van Bel et al., 2012) is a platform for plant genome comparisons. Data in PLAZA includes structural and functional gene annotations, homologous gene families, multiple sequence alignments, phylogenetic trees, and colinear regions between and within species. An orthology viewer combines information from different orthology prediction methodologies enabling investigation of complex orthology relationships. As of 30 June 30, PLAZA housed 15 monocot (v3.0) and 31 dicot (v3.0) whole-genome sequences including that for *Gossypium raimondii* (JGI v2.1).

VISTA

VISTA (http://genome.lbl.gov/vista/index.shtml; Frazer et al., 2004) is a comprehensive suite of programs and databases for comparative analysis of genomic sequences. Researchers can view precomputed whole-genome alignments of different species and submit sequences and alignments for analysis. VISTA uses global alignment strategies and a curve-based visualization technique for rapid identification of conserved sequences in these long alignments. As of 30 June 2014, VISTA contained plant genome sequences, including the whole-genome sequence of *Gossypium raimondii* (JGI v2.0).

Metabolic Pathway Resources

Kyoto Encyclopedia of Genes and Genomes

KEGG (http://www.genome.jp/kegg/; Kanehisa et al., 2014) is a long-established knowledge base containing metabolic function information on gene products that is routinely used for biological interpretation of genome and transcript sequences. Genomes are linked to pathways through the KEGG Orthology system, a collection of manually defined ortholog groups identified by K numbers (i.e., KEGG Ortholog identifiers). Once annotated with K numbers, the genes in the genome are further evaluated through other KEGG modules to show metabolic capacities and other phenotypic features. Representing chemical units of reactions, the reaction modules are used to analyze the design principles of metabolic networks and also improve the definition of K numbers and associated annotations. As of 30 June 2014, KEGG v71.0 houses 462 KEGG Pathway Maps that were derived from knowledge based on 299,248 references. KEGG analysis is routinely used in structural and functional annotations studies in cotton.

BioCyc

BioCyc (http://biocyc.org) is a collection of more than 3000 organism-specific pathway/genome databases housing metabolites, enzymes, reactions, metabolic pathways, predicted operons, transport systems, and pathway-hole fillers for each organism (Caspi et al., 2014). BioCyc offers a variety of online and desktop tools for querying and analysis of pathway/genome databases, including omics viewers and tools for comparative analysis. As of 30 June 2014, a CottonCyc Pathway Database (http://pathways.cottongen.org/) obtained from the JGI v2.1 *G. raimondii* genome sequences has been made publicly available through CottonGen.

Germplasm Information Repositories

Germplasm Resources Information Network

The Germplasm Resources Information Network (GRIN; http://www.ars-grin.gov/; Wiersema, 1994) is a program of the USDA Agricultural Research Service (USDA-ARS) that provides germplasm information on various organisms important for US food and agricultural production. GRIN provides access to passport, characterization, evaluation, inventory, and distribution data important for the use of national germplasm collections. The website also provides information on registered crop cultivars, parental lines, and germplasm and genetic stocks, and records of plants protected under the Plant Variety Protection Act (to provide legal intellectual property rights protection to breeders of new varieties of plants). It also provides access to online ordering of seeds from the ARS germplasm collections, including the Cotton Germplasm Repository Site, located in College Station, Texas. As of 30 June 2014, the GRIN National Plant Germplasm System contained 566,895 accessions of 2373 genera collected from 245 countries, including 15,501 accessions of 41 cotton species collected from 123 countries (GRIN, 2014).

Other countries, such as Australia, Brazil, China, France, India, Pakistan, Russia, and Uzbekistan, also have cotton germplasm collections, but we are unable to discuss these in this chapter due to a lack of detailed information and online access for them.

CottonGen: A Community Database for Cotton

Plant community databases combine expertly curated data repositories for a single or closely related species with data mining and analysis tools. Long-established examples of such databases include TAIR (Lamesch et al., 2012), Gramene (Monaco et al., 2014), MaizeGDB (Schaeffer et al., 2011), the Genome Database for Rosaceae (Jung et al., 2014), the Solanaceae Genomics Network (Bombarely et al., 2011), and SoyBase (Grant et al., 2010). Such databases generally store comprehensively integrated data such as annotated sequences of genomes and transcriptomes, genetic data, molecular and phenotypic diversity data, pathway data, and cultivar performance data. As a result of data integration, the value of the individual types of data increase exponentially, providing essential resources to accelerate the molecular understanding of phenotypic traits and the use of DNA information in crop improvement. The CottonGen database (http://www.cottongen.org) serves as the central data repository and analysis resource for the cotton community. Figure 2 depicts the overall organization of the CottonGen database system. The following description and contents of CottonGen are current as of 30 June 2014.

CottonGen Background

CottonGen is funded by Cotton Incorporated, the USDA-ARS Crop Germplasm Research Unit at College Station, TX; Bayer CropScience, Dow/Phytogen, Monsanto, the Southern Association of Agricultural Experiment Station Directors,

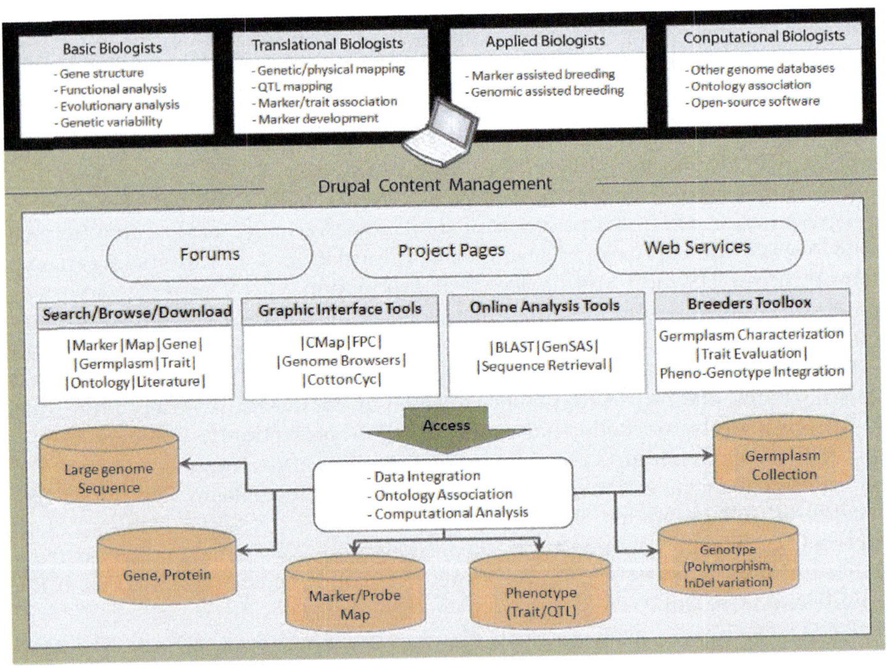

Fig. 2. Schema of CottonGen database system. Scientists with different research interests, (top), access the system. Data from different sources are integrated and organized in the system (bottom). The data are visible, searchable, and downloadable via Drupal content management that uses Tripal as infrastructure (middle).

Washington State University, and USDA National Research Support Project 10 (NRSP10). Aiming to provide centralized access to an integrated, comprehensive, online information system of genomics, genetics, and breeding data for cotton, the database was initiated in 2012 as the successor to both CottonDB (Yu et al., 2012)—a USDA-ARS database established in 1995 for genomic, genetic, germplasm, and taxonomic information for cotton—and the Cotton Marker Database (Blenda et al., 2006), a cotton marker database established in 2005 and funded by Cotton Incorporated. CottonGen was expanded to include annotated transcriptome, genome sequence, pathway, and breeding data and enhanced tools for easy querying and viewing of integrated research data. CottonGen also hosts the website for the International Cotton Genome Initiative (ICGI), which was formerly included in CottonDB. ICGI is a nonprofit organization that was created in 2000 to facilitate global communication and collaboration, and to increase knowledge of the structure and function of the cotton genome for the benefit of the community. Since its release in 2012, CottonGen usage has been doubling every year, and in 2014, it had 15,666 visits by 8091 users from 130 countries, with 90,994 pages viewed.

CottonGen was developed with Tripal (Ficklin et al., 2011; Sanderson et al., 2013), an open-source standards-based platform for construction of online genomic and genetic databases. Tripal integrates a community-derived database schema named Chado (Mungall et al., 2011, Jung et al., 2014) with Drupal (http://www.drupal.org), a popular web-content management system. Chado is designed for the storage of such common modern classes of data as genotypes, phenotypes, sequence and sequence comparisons, publications, and phylogeny. Chado employs the use of community-developed controlled vocabularies, such as GO (Gene Ontology Consortium, 2013), Sequence Ontology (Eilbeck et al., 2005), and others, to ensure standardization of data storage. Drupal is an open-source, free software package that allows for construction, maintenance, and an endless variety of customizations without needing to know any coding languages. The Drupal Application Programming Interface contains thousands of freely shared extensions created by third-party developers that enable computer programmers to easily create new functionalities. Using Chado as the underlying database schema and a suite of Drupal interface extension modules, Tripal facilitates development of scalable, efficient, community-driven databases. Tripal has been used to construct several genome databases (Genome Database for Rosaceae, Jung et al., 2014; the Banana Genome Hub, Droc et al., 2013; KnowPulse, Sanderson et al., 2012; the Citrus Genome Database; Main et al., 2013), with many of the larger commodity and model databases now converting to this more resource-efficient system for ease of management and use.

Data and Web Interfaces

CottonGen contains data including whole-genome sequences and annotations, EST sequences, genes, molecular markers, genetic and QTL maps, traits and trait evaluations, taxonomy, germplasm information, and relevant bibliographic citations. For easy access, a common navigation menu is displayed on every CottonGen web page to provide links for general information, help, data, search, tools, ICGI, and community resources. The data section lists an overview of the major classes of data and associated data entries within CottonGen. Each data class page has a summary of the data and various links to access the data. The data section also provides links for researchers to download major data sets and templates to submit data. Additionally, the data section contains a list of people who contributed data directly to the database. The search section lists query tools for searching for genes, germplasm, markers, QTLs, publications, and trait-evaluation data. Customized options are provided in each query tool to filter the data search results. From the search results site, researchers can access the various data pages or download the entire or partial query result. Major CottonGen data and web interfaces to the data are described in detail below.

Whole-Genome Sequence Data

CottonGen contains genome sequences of *Gossypium raimondii*, the first cotton species with its entire genome sequenced. The whole genome was sequenced, assembled, and annotated by two independent research teams (JGI v2.1, Paterson et al. 2012; and BGI v1, Wang et al., 2012), and CottonGen includes both versions. The CottonGen team further analyzed the putative genes from these assemblies to include homologies to proteins in closely related or other well-annotated species, and in silico prediction of GO terms, KEGG pathway terms, and InterPro protein associate domains. Additional analyses by the CottonGen team include the alignment of cotton transcripts—such as CottonGen Unigene version v1, Udall cotton Unigene contigs (Udall et al., 2006), and cotton unigenes from PlantGDB and GenBank—to both the BGI and JGI versions of *G. raimondii* genome sequences. In addition, SNPs that were identified between the two diploid A_2 and D_5 genomes (i.e., A_2:D_5) and between each of the two diploid genomes (A_2 or D_5) and their related tetraploid subgenomes (i.e., A_2:A_t, A_2:D_t, D_5:A_t, D_5:D_t, where t in A_t or D_t represents tetraploid) (Page et al., 2013a, 2013b) were mapped to the JGI version of the *G. raimondii* reference genome. All cotton genetic marker sequences also were aligned to both the JGI and BGI versions of the *G. raimondii* reference genome. The assembled and annotated sequence data for the genome can be viewed or downloaded from CottonGen via the genome sequencing projects' home pages, the sequence retrieval tools, and the genome browsers GBrowse (Stein et al., 2002) and JBrowse (Skinner et al., 2009). Both the genome sequences and the data-download page (listed as "Data Download") can be found under the data navigation menu, and both contain FASTA-formatted sequences and general-feature-format alignments of various downloadable files include CDSs, predicted gene transcripts, and putative gene peptides. Excel-formatted homologs between cotton proteins and proteins of other well-annotated species or those closely related to cotton that are found in UniProt and GenBank are also available via hyperlinks to those databases. Other downloadable files include ESTs and genetic markers in FASTA and Excel formats that map to the whole-genome sequences and functional annotation files from protein, InterPro, and KEGG alignments. In the gene and sequence search tools, whole-genome data can be found by filtering by name, GO terms, InterPro domains, or KEGG pathway terms (Fig. 3).

Annotated EST UniGene

All GenBank *Gossypium* ESTs are available in CottonGen and in a CottonGen unigene data set developed on the basis of 442,954 ESTs downloaded from GenBank on 12 Sept. 2012. Creation of a unigene data set reduces inherent redundancy among the ESTs, providing a nonredundant expressed gene list for *Gossypium* based on these EST data. To create the EST unigene data set, the ESTs are first filtered to remove any contaminating vectors, species-specific chloroplasts, mitochondrial RNA, transfer RNA, and ribosomal RNA by means of the BLAST

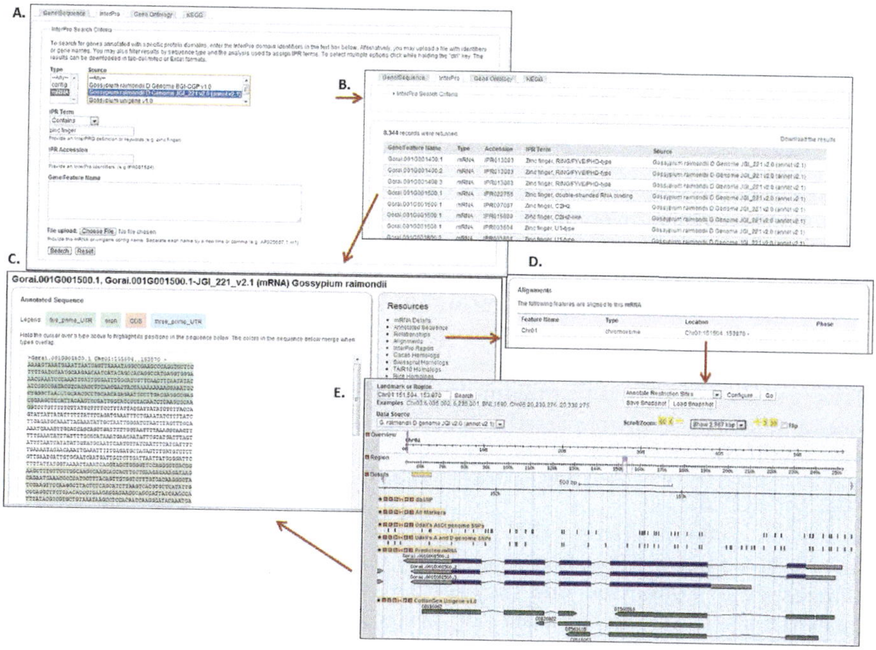

Fig. 3. Gene and sequence search site in CottonGen. (A) Genes and sequences can be searched according to various categories, such as by name, Gene Ontology (GO) terms, InterPro protein domain name, or KEGG (Kyoto Encyclopedia of Genes and Genomes) pathway terms. The example shows an InterPro search site. (B) The search results page has links for downloads, gene/sequence detail pages, and external databases. (C) The gene detail page has various tabs to show the data. Shown is the annotated sequence page. (D) The alignment tab of the gene detail page shows the position in the whole genome and a link to GBrowse. E. Gbrowse page linked from the alignment tab of the gene detail page. Users can return to the gene detail page from Gbrowse.

algorithm with NCBI UniVec-recommended parameters. The contaminant-free ESTs are then trimmed to remove low-quality sequences and assembled into longer gene fragments, known as contigs, with the CAP3 (Huang and Madan, 1999) algorithm having an overlap percentage parameter of 90% (–p 90), followed by computational annotation of gene function. By means of this protocol, 437,185 filtered sequences were assembled into 21,698 contigs and 128,218 singletons to make a unigene set of 149,916 sequences. Because the contigs comprise two or more overlapping ESTs, the contig data set should be considered the more accurate component of the unigene data set. Annotation includes comparison of both the filtered ESTs and the EST contig consensus sequences using the BLASTX algorithm against SWISS-PROT, TrEMBL, InterPro, TAIR (Lamesch et al., 2012), and other well-annotated species' protein databases. For each EST and contig, the top 10 matches with an expectation value of less than $1e^{-9}$ are recorded and stored in the database for each EST and contig. The results of computational assignment of GO terms and functional classification by pathways from KEGG are also stored in the database. The 21,698 contigs from the v1.0 unigene are searchable by name, Interpro domain, GO term, or KEGG term or gene in the gene and sequence search sites, and the results can be downloaded as Excel files from these search pages. The entire annotations from the unigenes can also be downloaded from the CottonGen download page (http://www.cottongen.org/data/download).

GenBank Sequences and Genes

All GenBank *Gossypium* sequences were parsed for genes, mRNA, CDS, and 3′ and 5′ untranslated regions and made available in CottonGen. Further annotations of these sequences include homology to genes in other species and identification of InterPro protein domains and KEGG and GO terms. A community-driven gene database for cotton has been built by storing separately in CottonGen the distinct gene names for *Gossypium*. Each gene unique in the *Gossypium* genus is currently linked to all NCBI genes from various species and serves as a base entity to link to other associated data, such as predicted genes from whole-genome sequences, QTLs, genetic markers, and mutant phenotypes, as annotation progresses. All genes and mRNAs that are parsed out from NCBI sequences are searchable in the gene search site.

Map and QTL Data

CottonGen contains genetic, QTL, and physical maps of cotton including the underlying mapping populations, molecular markers, QTLs, and FPCs. Currently, 50 maps are in CottonGen, and they represent five kinds of maps (genetic map, bin map, consensus map, associate map, and in silico map); 45 of the 50 are genetic maps. These maps cover both tetraploid and diploid species of five *Gossypium* genome groups (AD, A, C, D, and G), are derived from various types of populations (such as F_2, backcross, recombinant inbred line, and doubled haploid), and consist of approximately 34,000 marker loci and a thousand QTLs. CottonGen houses each QTL with such associated information as the CottonGen-assigned QTL label, published symbols, alias, trait name, populations, map positions, associated markers, and statistical values. QTLs can be searched via the CottonGen QTL search interface that is listed under the search section on the navigation menu or at http://www.cottongen.org/search/qtl, where QTLs can be queried by any combination of trait name, published symbol, or CottonGen label. Mapped marker sequences or nearby loci can be queried via marker search interfaces listed under the search section on the navigation menu or http://www.cottongen.org/search/markers. Search results are downloadable in Excel format and mapped markers or QTLs are hyperlinked to CMap (Youens-Clark et al., 2009). Maps and mapped QTLs can be visualized via CottonGen CMap, a CottonGen-implemented free browser-based tool that enables users to view and compare genetic and physical maps between and among species. Maps and mapped QTLs can be batch downloaded from CottonGen interfaces of CMap (under the tool section on the CottonGen navigation menu) or Data Download (under the section "Data" on the CottonGen navigation menu).

Genetic Markers and Traits

More than 230,000 genetic markers are contained in CottonGen, including nearly 140,000 SNPs and more than 78,000 SSRs. An extensively annotated marker database in CottonGen provides detailed information about markers and traits that have been used in genetic map development or genetic diversity studies. Such information includes marker aliases, source cultivars and descriptions, motif type or repeat patterns, map positions, associated EST or BAC sequences, and published references. For sequence-based markers such as SSRs, AFLPs, SRAPs, and cDNA-rapid RFLPs, CottonGen provides details on experimental conditions, such as the marker primer(s), amplicon sequence information, and the PCR amplification conditions. Annotation of trait data in CottonGen is at an initial stage and currently includes trait aliases, published symbol(s), trait descriptions, trait category curations, taxons, trait-screening methods, map positions, and associated publications.

The CottonGen marker search site allows both a simple search by name and an advanced search with various categories. The search categories include marker type, the species from which the marker was developed, which chromosome or genome group the marker was mapped with, map position, markers with associated BAC clones, and markers with associated ESTs. Users can also upload a file of names to get the detailed data. In the trait-search site, users can search traits by name, symbol, taxon, or curated trait category.

Germplasm and Trait-Evaluation Data

Species information is stored in CottonGen for each of the 50 *Gossypium* species. This information includes genome groups, geographic origins, interspecies compatibilities, and germplasm. As of 30 June 2014 more than 15,200 germplasm accessions were available in CottonGen, identified from more than 48,700 entries mainly from the USDA-ARS GRIN (Wiersema, 1994) cotton collection, the cotton germplasm collection of the China Cotton Research Institute, the Chinese Academy of Agricultural Sciences, and the cotton germplasm collection of the Uzbekistan Center of Genomics and Bioinformatics, Academy of Sciences of Uzbekistan. The germplasm data include aliases, pedigrees, passport information, stock collection center, associated maps, libraries, and sequences. In addition, trait-evaluation data is available for approximately 9000 germplasm accessions with more than 118,000 trait scores. The *Gossypium* species summary page, which is found under data/species in the navigation menu, provides a list of species along with available information including genome group, haploid chromosome number, and geographic origin. More summary data available includes the number of germplasms, sequences, and DNA libraries. Accessing the species page through the species hyperlink in the summary table provides more detailed information, such as common name, images, and additional data. The species page also shows the results of functional analysis (KEGG and GO reports) of species genes, both from any available whole-genome sequences and NCBI sequences. The germplasm search pages provide access to different types of data (Fig. 4). "Search by collection" provides a list of the selected germplasms along with information about the stock collection center. These data can be filtered by collection center name, and germplasm name and/or accession name. The "search by pedigree" page allows searching the germplasms by pedigree, and the "search germplasm by country" page searches by the country of origin. From the germplasm search results page, researchers can access the page with details for the individual germplasm, which shows known pedigree, passport, collection center, image, and associated genotypic and phenotypic data. The trait-evaluation data available for the germplasm data can also be searched. Search sites for both qualitative and quantitative trait-evaluation data allow the values of up to three trait descriptors to be specified. The results can be viewed or downloaded as Excel files.

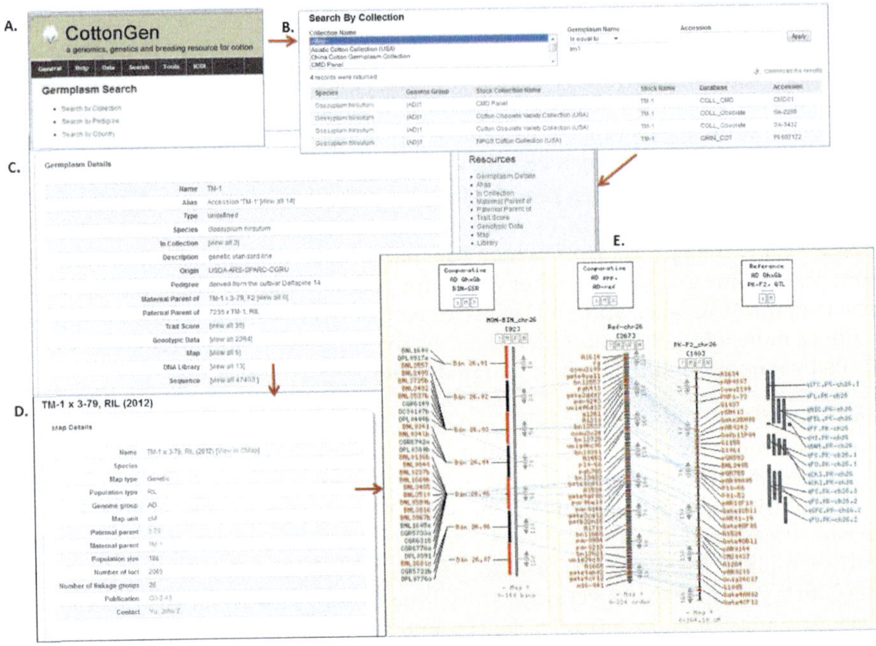

Fig. 4. Germplasm search site in CottonGen. (A) Multiple germplasm search sites are available for the type of information users are interested in. (B) An example search interface in which users can view and search for germplasm and all the collections that contains the germplasm. (C) Germplasm detail page with various tabs to show the information. (D) Map tab of the germplasm page shows all the maps for which the germplasm has been used. (E) From the map page, users can go to the CMap for further exploration.

Cotton Metabolic Pathway Database

CottonGen currently houses the first version of the CottonCyc *G. raimondii* metabolic pathway database (http://pathways.cottongen.org/). The data in CottonCyc were inferred by PathwayTools (Karp et al., 2010) version 18.0 from the JGI v2.0 genome assembly sequence and pathway predictions and in silico v2.1 annotations provided by JGI. The GraimondiiCyc contains 382 pathways, 12,468 enzymes, 2057 reactions, 1473 compounds, and 77,269 polypeptides. To date there has been no manual curation of this GraimondiiCyc database v1.0.

Bibliographic Data

CottonGen contains numerous bibliographic data from diverse sources that are used by cotton genetic, genomic, and breeding researchers. Relevant publications from NCBI PubMed (http://www.ncbi.nlm.nih.gov/pubmed) and the USDA National Agricultural Library (http://agricola. nal.usda.gov/) databases were automatically imported to CottonGen via the Tripal pipeline. Additionally, references from other journals not included in those two databases were manually imported to CottonGen; such publications included conference proceedings, book chapters, theses, and patents. Currently, CottonGen houses more than 14,000 references. Bibliographic data can be searched via a combination of keywords (in the abstract or title), all or partial titles, authors, journal or conference names, years, and other categories. For ease of use and familiarity, the CottonGen publications search page mirrors the PubMed search interface. Search results link to pages that contain such publication details as the abstract, citation, and external link to the full article, among others.

Future Directions

CottonGen will continue to provide a comprehensive, accurate, up-to-date, convenient medium for researchers to easily exchange and analyze genetic, genomic, and breeding data of cotton. To achieve this goal, the future directions of CottonGen are to

1. Continue collecting, consolidating, and integrating data generated from ongoing cotton research efforts and making these data available to the international cotton scientific community. With the expected availability of whole-genome sequences of *G. hirsutum* and *G. barbadense*, the two commercial important tetraploid cotton species, CottonGen will integrate and further annotate these sequences to help researches understand the functional and agronomic significance and structure of polyploidy. Large data sets that include data from genotyping by sequencing, RNA-sequencing of complex transcriptomes, and genome-wide association studies will also be consolidated where possible and integrated into the database. Other large data sets that will be obtained, curated, analyzed, and uploaded to the database include updated map and QTL data, as well as the US NCVT data, which will be enhanced with online searchable features through the database.
2. Improve database functionality and utilize existing database tools or adopt open-source software to enhance data interaction and visualization capabilities. With the availability of whole-genome sequences of more cotton species, GBrowse-syn, a GBrowse-based synteny browser (McKay et al., 2010), will be added to view synteny or colinearity data from closely related or useful species such as cacao and *Arabidopsis*. Meanwhile, an online genome sequence annotation server, GenSAS (Lee et al., 2011), will be made available. GenSAS provides a customizable pipeline for users to upload, view, and edit genome sequences for further annotation and curation. To integrate the rapidly expanding genetic and genomic data with traditional breeding data, a comprehensive breeder's toolbox similar to that developed for the Rosaceae community as part of the RosBREED project funded by the USDA National Institute of Food and Agriculture Specialty Crop Research Initiative will be developed. This toolbox will provide access for breeders to make predictive analyses by mapping genotypes to phenotypes and enabling breeders to select on marker data alone with easy-to-use and breeder-friendly relevant tools. Other tools to be implemented in CottonGen include PolyCat (Page et al., 2013a), a pipeline for mapping and categorizing allopolyploid sequences, and a functional SNP index for sequences of both diploid and polyploid cotton species. With funding from the USDA NRSP10 Project and NSF DIBBS Tripal Project, web services will be implemented to enable computational access to public data within CottonGen. These services will facilitate seamless access to analysis capabilities on high-performance computing and software workflows within iPlant, from within CottonGen, or using researcher-selected or researcher-uploaded data. The results of the analysis will be returned to the researcher within CottonGen with the option to store the results in private user accounts.
3. Continue to closely connect and interact with the cotton research community and work together with specialists to generate trait ontology and plant ontology

for cotton and develop easier data submission procedures. Standardized nomenclatures on genes, QTLs, trait descriptors, and markers will be set and expanded to use within the entire community as well as in future publications. Data submission will include providing access to the FieldBook App to facilitate more efficient electronic recording of field and lab data and upload them to CottonGen.

4. Continue maintaining extensive interaction between CottonGen and the cotton data in other databases (such as GRIN, GenBank, PlantGDB, and PLEXdb) as well as with corresponding data in closely related species, so that CottonGen can support data inquiry from the broader scientific community.

Conclusions

It is clear there are many useful database resources that house excellent data and tools that can facilitate cotton research. This chapter has highlighted many of them while also focusing in more depth on CottonGen, the crop community database for cotton research. The wealth of genomics, genetics, and breeding data generated by the cotton community that is now stored, integrated, and made easily available in CottonGen provides an invaluable resource for facilitating basic, translational, and applied research. With cotton, as with other important crops, the exponential growth in the generation of large-scale genomic, transcriptomic, phenomic, metabolomic, and proteomic data sets ("big data") will further accelerate information-driven cotton research if it is integrated and made accessible through user friendly database resources. Imperative to the success of these efforts in data generation, integration, and access will be the continued development of standards and ontologies for data collection and submission to databases by the community. Providing credit for these efforts to data submitters and developing the necessary analytical tools from resource-efficient, scalable databases will also be central to delivering on the promise of information-driven basic discovery and its translation to crop improvement.

References

Abdurakhmonov, I.Y., S. Saha, J.N. Jenkins, Z.T. Buriev, S.E. Shermatov, B.E. Scheffler, et al. 2009. Linkage disequilibrium based association mapping of fiber quality traits in *G. hirsutum* L. variety germplasm. Genetica (The Hague) 136(3):401–417.

Affymetrix. 2009. GeneChip Cotton Genome Array: Product description. http://www.affymetrix.com/estore/browse/products.jsp?productId=131430&navMode=34000&navAction=jump&aId=productsNav#1_1 (accessed 9 June 2014).

Ahmad, S., T.Z. Zhang, and T. Shaheen. 2007. Identifying genetic variation in *Gossypium* based on single nucleotide polymorphism. Pakistan Journal of Botany 39(4):1245–1250.

Ai, N.J., R.Z. Liu, T.Q. Zhao, J.H. Qin, and T.Z. Zhang. 2013. Analysis of early maturity gene sources in upland cotton using molecular markers. Acta Agron. Sin. 39(9):1548–1561. doi:10.3724/SP.J.1006.2013.01548

An, C.F., Saha S, J.N. Jenkins, D.P. Ma, B.E. Scheffler, R.J. Kohel, et al. 2008. Cotton (*Gossypium* spp.) R2R3-MYB transcription factors SNP identification, phylogenomic characterization, chromosome localization, and linkage mapping. Theoretical and Applied Genetics 116(7):1015–1026.

Aoki, K.F., and M. Kanehisa. 2005. Using the KEGG Database resource. Curr. Protoc. Bioinf. 11: 1.12.1–1.12.43. doi:10.1002/0471250953.bi0112s38

Arpat, A., M. Waugh, J.P. Sullivan, M. Gonzales, D. Frisch, D. Main, et al. 2004. Functional genomics of cell elongation in developing cotton fibers. Plant Mol. Biol. 54:911–929. doi:10.1007/s11103-004-0392-y

Baerenfaller, K., J. Grossmann, M.A. Grobei, R. Hull, M. Hirsch-Hoffmann, S. Yalovsky, et al. 2008. Genome-scale proteomics reveals Arabidopsis thaliana gene models and proteome dynamics. Science 320:938–941. doi:10.1126/science.1157956

Barrett, T., S.E. Wilhite, P. Ledoux, C. Evangelista, I.F. Kim, M. Tomashevsky, et al. 2013. NCBI GEO: Archive for functional genomics data sets–update. Nucleic Acids Res. 41(D1):D991–D995. doi:10.1093/nar/gks1193

Benson, D.A., M. Cavanaugh, K. Clark, I. Karsch-Mizrachi, D.J. Lipman, J. Ostell, and E.W. Sayers. 2013. GenBank. Nucleic Acids Res. 41(D1):D36–D42. doi:10.1093/nar/gks1195

Berman, H., K. Henrick, H. Nakamura, and J.I. Markley. 2007. The worldwide Protein Data Bank (wwPDB): Ensuring a single, uniform archive of PDB data. Nucleic Acids Res. 35:D301–D303. doi:10.1093/nar/gkl971

Blenda, A., J. Scheffler, B. Scheffler, M. Palmer, J. Lacape, J.Z. Yu, et al. 2006. CMD: A Cotton Microsatellite Database resource for *Gossypium* genomics. BMC Genomics 7:132. doi:10.1186/1471-2164-7-132

Bombarely, A., N. Menda, I.Y. Tecle, R.M. Buels, S. Strickler, T. Fischer-York, A. Pujar, et al. 2011. The Sol Genomics Network (solgenomics.net): Growing tomatoes using Perl. Nucleic Acids Res. 39(suppl. 1):D1149–D1155. doi:10.1093/nar/gkq866

Bowman, M.J., W. Park, P.J. Bauer, J.A. Udall, J.T. Page, J. Raney, B.E. Scheffler, et al. 2013. RNA-Seq transcriptome profiling of upland cotton (*Gossypium hirsutum* L.) root tissue under water-deficit stress. PLoS ONE 8(12):e82634. doi:10.1371/journal.pone.0082634

Buckler, E.S., and J.M. Thornsberry. 2002. Plant molecular diversity and applications to genomics. Curr. Opin. Plant Biol. 5:107–111. doi:10.1016/S1369-5266(02)00238-8

Burr, B., S.V. Evola, F.A. Burr, and J.S. Beckmann. 1983. The application of restriction fragment length polymorphism to plant breeding. In: J.K. Setlow and A. Hollaender, editors, Genetic engineering. Vol. 5. Plenum Press, New York. p. 45–59.

Byers, R.L., D.B. Harker, S.M. Yourstone, P.J. Maughan, and J.A. Udall. 2012. Development and mapping of SNP assays in allotetraploid cotton. Theor. Appl. Genet. 124:1201–1214. doi:10.1007/s00122-011-1780-8

Campbell, B.T., S. Saha, R. Percy, J. Frelichowski, J.N. Jenkins, W. Park, et al. 2010. Status of the global cotton germplasm resources. Crop Sci. 50:1161–1179. doi:10.2135/cropsci2009.09.0551

Caspi, R., T. Altman, R. Billington, K. Dreher, H. Foerster, C.A. Fulcher, et al. 2014. The MetaCyc database of metabolic pathways and enzymes and the BioCyc collection of Pathway/Genome Databases. Nucleic Acids Res. 42(D1):D459–D471. doi:10.1093/nar/gkt1103

Castellana, N.E., S.H. Payne, Z. Shen, M. Stanke, V. Bafna, and S.P. Briggs. 2008. Discovery and revision of *Arabidopsis* genes by proteogenomics. Proc. Natl. Acad. Sci. USA 105:21034–21038. doi:10.1073/pnas.0811066106

Cho, R.J., M. Mindrinos, D.R. Richards, R.J. Sapolsky, M. Anderson, E. Drenkard, et al. 1999. Genome-wide mapping with biallelic markers in *Arabidopsis thaliana*. Nat. Genet. 23:203–207. doi:10.1038/13833

Dash, S., J.V. Hemert, L. Hong, R.P. Wise, and J.A. Dickerson. 2012. PLEXdb: Gene expression resources for plants and plant pathogens. Nucleic Acids Res. 40(D1):D1194–D1201. doi:10.1093/nar/gkr938

Droc, G., D. Larivière, V. Guignon, N. Yahiaoui, D. This, O. Garsmeur, et al. 2013. The banana genome hub. Database 2013:bat035. doi:10.1093/database/bat035

Duvick, J., A. Fu, U. Muppirala, M. Sabharwal, M.D. Wilkerson, C.J. Lawrence, et al. 2008. PlantGDB: A resource for comparative plant genomics. Nucleic Acids Res. 36(suppl. 1):D959–D965. doi:10.1093/nar/gkm1041.

Eilbeck, K., S.E. Lewis, C.J. Mungall, M. Yandell, L. Stein, R. Durbin, and M. Ashburner. 2005. The Sequence Ontology: A tool for unification of genome annotations. Genome Biol. 6:R44. doi:10.1186/gb-2005-6-5-r44

Esbroeck, G.V., and D.T. Bowman. 1998. Cotton germplasm diversity and its importance to cultivar development. J. Cotton Sci. 2:121–129.

Fang, D.D., L.L. Hinze, R.G. Percy, P. Li, D. Deng, and G. Thyssen. 2013. A microsatellite-based genome-wide analysis of genetic diversity and linkage disequilibrium in

Upland cotton (*Gossypium hirsutum* L.) cultivars from major cotton-growing countries. Euphytica 191(3):391–401. doi:10.1007/s10681-013-0886-2

Ficklin, S.P. L-A Sanderson, C-H Cheng, M. Staton, T. Lee, I.H. Cho, et al. 2011. Tripal: A construction toolkit for online genome databases. Database 2011:bar044. doi:10.1093/database/bar044.

Filichkin, S.A., H.D. Priest, S.A. Givan, R.K. Shen, D.W. Bryant, S.E. Fox, et al. 2010. Genome-wide mapping of alternative splicing in *Arabidopsis thaliana*. Genet. Res. 20:45–58. doi:10.1101/gr.093302.109

Finn, R.D., J. Mistry, J. Tate, P. Coggill, A. Heger, J.E. Pollington, et al. 2010. The Pfam protein families database. Nucleic Acids Res. 38:D211–D222. doi:10.1093/nar/gkp985

Flint-Garcia, S.A., J.M. Thornsberry, and E.S. Buckler. 2003. Structure of linkage disequilibrium in plants. Annu. Rev. Plant Biol. 54:357–374. doi:10.1146/annurev.arplant.54.031902.134907

Food and Agriculture Organization of the United Nations (FAO). 2010. The second report on the state of the world's plant genetic resources for food and agriculture. Commission on Genetic Resources for Food and Agriculture, FAO, United Nations, Rome. http://www.fao.org/agriculture/crops/thematic-sitemap/theme/seeds-pgr/sow/sow2/en/ (accessed 9 May 2015).

Frazer, K.A., L. Pachter, A. Poliakov, E.M. Rubin, and I. Dubchak. 2004. VISTA: Computational tools for comparative genomics. Nucleic Acids Res. 32(suppl 2):W273–W279. doi:10.1093/nar/gkh458

Gene Ontology Consortium. 2013. Gene Ontology annotations and resources. Nucleic Acids Res. 41:D530–D535. doi:10.1093/nar/gks1050

Germplasm Resources Information Network (GRIN). 2014. Summary statistics for GRIN. http://www.ars-grin.gov/npgs/stats/ (accessed on 30 June 2014).

Goodstein, D.M., S.Q. Shu, R. Howson, R. Neupane, R.D. Hayes, J. Fazo, et al. 2012. Phytozome: A comparative platform for green plant genomics. Nucleic Acids Res. 40(D1):D1178–D1186. doi:10.1093/nar/gkr944

Gore, M.A., D.D. Fang, J.A. Poland, J. Zhang, R.G. Percy, R.G. Cantrell, et al. 2014. Linkage map construction and quantitative trait locus analysis of agronomic and fiber quality traits in cotton. Plant Gen. 7(1):1–10. doi:10.3835/plantgenome2013.07.0023

Grant, D., R.T. Nelson, S.B. Cannon, and R.C. Shoemaker. 2010. SoyBase, the USDA-ARS soybean genetics and genomics database. Nucleic Acids Res. 38:D843–D846. doi:10.1093/nar/gkp798.

Griffin, T.J., and L.M. Smith. 2000. Single-nucleotide polymorphism analysis by MALDI-TOF mass spectrometry. Trends Biotechnol. 18:77–84. doi:10.1016/S0167-7799(99)01401-8

Haas, B.J., A.L. Delcher, J.R. Wortman, and S.L. Salzberg. 2004. DAGchainer: A tool for mining segmental genome duplications and synteny. Bioinformatics 20:3643–3646. doi:10.1093/bioinformatics/bth397

He, Y., P. Kumar, X. Shen, R.F. Davis, G. Van Becelaere, O.L. May, et al. 2014. Re-evaluation of the inheritance for root-knot nematode resistance in the Upland cotton germplasm line M-120 RNR revealed two epistatic QTLs conferring resistance. Theor. Appl. Genet. 127(6):1343–1351. doi:10.1007/s00122-014-2302-2

Huang, X., and A. Madan. 1999. CAP3: A DNA sequence assembly program. Genome Res. 9:868–877. doi:10.1101/gr.9.9.868

Hunter, S., P. Jones, A. Mitchell, R. Apweiler, T.K. Attwood, A. Bateman, et al. 2012. InterPro in 2011: New developments in the family and domain prediction database. Nucleic Acids Res. 40(D1):D306–D312. doi:10.1093/nar/gkr948.

Jena, S.N., A. Srivastava, U.M. Singh, S. Roy, N. Banerjee, K.M. Rai, et al. 2011. Analysis of genetic diversity, population structure and linkage disequilibrium in elite cotton (*Gossypium* L.) germplasm in India. Crop Pasture Sci. 62(10):859–875. doi:10.1071/CP11161

Jia, Y., X. Sun, J. Sun, Z. Pan, X. Wang, S. He, et al. 2014. Association mapping for epistasis and environmental interaction of yield traits in 323 cotton cultivars under 9 different environments. PLoS ONE 9(5):e95882. doi:10.1371/journal.pone.0095882

Jung, S., S.P. Ficklin, T. Lee, C. Cheng, A. Blenda, P. Zheng, et al. 2014. The Genome Database for Rosaceae (GDR): Year 10 update. Nucleic Acids Res. 42:D1237–D1244. doi:10.1093/nar/gkt1012

Kanehisa, M., S. Goto, Y. Sato, M. Kawashima, M. Furumichi, and M. Tanabe. 2014. Data, information, knowledge and principle: Back to metabolism in KEGG. Nucleic Acids Res. 42(D1):D199–D205. doi:10.1093/nar/gkt1076

Kantartzi, S.K., and J.M. Stewart. 2008. Association analysis of fibre traits in *Gossypium arboreum* accessions. Plant Breed. 127(2):173–179. doi:10.1111/j.1439-0523.2008.01490.x

Karp, P.D., S.M. Paley, M. Krummenacker, M. Latendresse, J.M. Dale, T.J. Kee, et al. 2010. Pathway Tools version 13.0: Integrated software for pathway/genome informatics and systems biology. Brief. Bioinform. 11(1):40–79. doi:10.1093/bib/bbp043

Kohel, R.J. 1999. Cotton germplasm resources and the potential for improved fiber production and quality. In: A.S. Basra, editor, Cotton fibers. Haworth Press, New York. p. 167–182.

Kosuge, T., J. Mashima, Y. Kodama, T. Fujisawa, E. Kaminuma, O. Ogasawara, et al. 2014. DDBJ progress report: A new submission system for leading to a correct annotation. Nucleic Acids Res. 42(D1):D44–D49. doi:10.1093/nar/gkt1066

Lacape, J.-M., M. Claverie, R.O. Vidal, M.F. Carazzolle, G.A. Guimaraes, M. Ruiz, et al. 2012. Deep sequencing reveals differences in the transcriptional landscapes of fibers from two cultivated species of cotton. PLoS ONE 7(11):e48855. doi:10.1371/journal.pone.0048855

Lacape, J.M., G. Gawrysiak, T.V. Cao, C. Viot, D. Llewellyn, S. Liu, et al. 2013. Mapping QTLs for traits related to phenology, morphology and yield components in an inter-specific *Gossypium hirsutum* × *G. barbadense* cotton RIL population. Field Crops Res. 144:256–267. doi:10.1016/j.fcr.2013.01.001

Lamesch, P., K. Dreher, D. Swarbreck, R. Sasidharan, L. Reiser, and F. Huala. 2010. Using The Arabidopsis Information Resource (TAIR) to find information about Arabidopsis genes. Curr. Protoc. Bioinformatics 2010:1–11.

Lamesch, P., T.Z. Berardini, D. Li, D. Swarbreck, C. Wilks, R. Sasidharan, R. Muller, et al. 2012. The *Arabidopsis* Information Resource (TAIR): Improved gene annotation and new tools. Nucleic Acids Res. 40(D1):D1202–D1210. doi:10.1093/nar/gkr1090

Lee, J.J., O.S.S. Hassan, W. Gao, J. Wang, E.N. Wei, R.J. Kohel, et al. 2006. Developmental and gene expression analyses of a cotton naked seed mutant. Planta 223:418–432. doi:10.1007/s00425-005-0098-7

Lee, T., C. Peace, S. Jung, P. Zheng, D. Main, and I. Cho. 2011. GenSAS—An online integrated genome sequence annotation pipeline. In: Y. Ding, et al., editors, 2011 4th International Conference on Biomedical Engineering and Informatics. Shanghai, China. 15–17 Oct. 2011. IEEE, Piscataway, NJ. Vol. 4, p. 1967–1973.

Leinonen, R., R. Akhtar, E. Birney, L. Bower, A. Cerdeno-Tarraga, Y. Cheng, et al. 2010. The European Nucleotide Archive. Nucleic Acids Res. 39:D28–D31: doi:10.1093/nar/gkq967.

Li, C.Q., X.Y. Wang, N. Dong, H.H. Zhao, X. Zhe, R. Wang, R.L. Converse, et al. 2013. QTL analysis for early-maturing traits in cotton using two upland cotton (*Gossypium hirsutum* L.) crosses. Breed. Sci. 63(2):154. doi:10.1270/jsbbs.63.154

Li, F.G., G.Y. Fan, K.B. Wang, F.M. Sun, Y.Y. Yuan, G.L. Song, et al. 2014. Genome sequence of the cultivated cotton *Gossypium arboreum*. Nat. Genet. 46:567–572. doi:10.1038/ng.2987

Liang, Q., P. Li, C. Hu, H. Hua, Z. Li, Y. Rong, et al. 2014. Dynamic QTL and epistasis analysis on seedling root traits in upland cotton. J. Genet. 93(1):63–78. doi:10.1007/s12041-014-0341-8

Lin, L.F., G.J. Pierce, J.E. Bowers, J.C. Estill, R.O. Compton, L.K. Rainville, C. Kim, et al. 2010. A draft physical map of a D-genome cotton species (*Gossypium raimondii*). BMC Genomics 11(1):395.

Lin, L.F., H. Tang, R.O. Compton, C. Lemke, L.K. Rainville, X. Wang, et al. 2011. Comparative analysis of *Gossypium* and *Vitis* genomes indicates genome duplication specific to the *Gossypium* lineage. Genomics 97(5):313–320. doi:10.1016/j.ygeno.2011.02.007

Lindblad-Toh, K., E. Wibchester, M.J. Daly, D.G. Wang, J.N. Hirchhof, J.P. Laviolette, et al. 2000. Large-scale discovery and genotyping of single nucleotide polymorphism in mouse. Nat. Genet. 24:381–386. doi:10.1038/74215

Lister, R., B.D. Gregory, and J.R. Ecker. 2009. Next is now: New technologies for sequencing of genomes, transcriptomes, and beyond. Curr. Opin. Plant Biol. 12(2):107–118. doi:10.1016/j.pbi.2008.11.004

Lister, R., R.C. O'Malley, J. Tonti-Filippini, B.D. Gregory, C.C. Berry, A.H. Millar, and J.R. Ecker. 2008. Highly integrated single-base resolution maps of the epigenome in *Arabidopsis*. Cell 133:523–536. doi:10.1016/j.cell.2008.03.029

Lu, Y., J. Curtiss, J. Zhang, R.G. Percy, and R.G. Cantrell. 2005. Discovery of single nucleotide polymorphisms in selected fiber genes in cultivated tetraploid cotton. In: Proceedings—Beltwide Cotton Conference, New Orleans, LA. 4–7 Jan. 2005. Natl. Cotton Counc. Am., Memphis, TN. p. 946.

Lyons, E., and M. Freeling. 2008. How to usefully compare homologous plant genes and chromosomes as DNA sequences. Plant J. 53:661–673. doi:10.1111/j.1365-313X.2007.03326.x

Main, D., S. Jung, S.P. Ficklin, P. Zheng, C.-H. Cheng, M. Olmstead, et al. 2013. Citrus genome database: Updates and new functionality. Paper presented at: Plant and Animal Genome 21, San Diego, CA. 12–16 Jan. 2013. Paper 6584. https://pag.confex.com/pag/xxi/webprogram/Paper6584.html (accessed 9 May 2015).

McKay, S.J., I.A. Vergara, and J.E. Stajich. 2010. Using the Generic Synteny Browser (GBrowse_syn). Curr. Protoc. Bioinformatics 39:9.12.1–9.12.25. doi:10.1002/0471250953.bi0912s31

Mehboob-ur-Rahman, T. Shaheen, N. Tabbasam, M.A. Iqbal, M. Ashraf, Y. Zafar, and A.H. Paterson. 2012. Cotton genetic resources. A review. Agron. Sustainable Dev. 32(2):419–432. doi:10.1007/s13593-011-0051-z

Mei, H., N. Ai, X. Zhang, Z. Ning, and T. Zhang. 2014. QTLs conferring FOV 7 resistance detected by linkage and association mapping in Upland cotton. Euphytica 197(2):237–249. doi:10.1007/s10681-014-1063-y

Mei, H., X. Zhu, and T. Zhang. 2013. Favorable QTL alleles for yield and its components identified by association mapping in Chinese Upland cotton cultivars. PLoS ONE 8(12):e82193. doi:10.1371/journal.pone.0082193

Mi, H., Q. Dong, A. Muruganujan, P. Gaudet, S. Lewis, and P.D. Thomas. 2010. PANTHER version 7: Improved phylogenetic trees, orthologs and collaboration with the Gene Ontology Consortium. Nucleic Acids Res. 38:D204–D210. doi:10.1093/nar/gkp1019

Monaco, M.K., J. Stein, S. Naithani, S. Wei, P. Dharmawardhana, S. Kumari, et al. 2014. Gramene 2013: Comparative plant genomics resources. Nucleic Acids Res. 42(D1):D1193–D1199. doi:10.1093/nar/gkt1110

Mungall, C.J., C. Batchelor, and K. Eilbeck. 2011. Evolution of the Sequence Ontology terms and relationships. J. Biomed. Inform. 44:87–93. doi:10.1016/j.jbi.2010.03.002

Nelson, W., and C. Soderlund. 2009. Integrating sequence with FPC fingerprint maps. Nucleic Acids Res. 37(5):e36. doi:10.1093/nar/gkp034

Ning, Z.Y., H. Chen, H. Mei, and T. Zhang. 2014. Molecular tagging of QTLs for fiber quality and yield in the upland cotton cultivar Acala-Prema. Euphytica 195(1):143–156. doi:10.1007/s10681-013-0990-3

Ogasawara, O., J. Mashima, Y. Kodama, E. Kaminuma, Y. Nakamura, K. Okubo, and T. Takagi. 2013. DDBJ new system and service refactoring. Nucleic Acids Res. 41(D1):D25–D29. doi:10.1093/nar/gks1152

Page, J.T., A.R. Gingle, and J.A. Udall. 2013a. PolyCat: A resource for genome categorization of sequencing reads from allopolyploid organisms. G3: Genes, Genomes, Genet. 3:517–525.

Page, J.T., M.D. Huynh, Z.S. Liechty, K. Grupp, D. Stelly, A.M. Hulse, et al. 2013b. Insights into the evolution of cotton diploids and polyploids from whole-genome re-sequencing. G3: Genes, Genomes, Genet. 3:1809–1818.

Paterson, A.H., J.F. Wendel, H. Gundlach, H. Guo, J. Jenkins, D. Jin, et al. 2012. Repeated polyploidization of *Gossypium* genomes and the evolution of spinnable cotton fibres. Nature 492:423–427. doi:10.1038/nature11798

Percy, R.G., J.E. Frelichowski, M.D. Arnold, T.B. Campbell, J.K. Dever, D.D. Fang, et al. 2014. The U.S. National Cotton Germplasm Collection––Its contents, preservation, characterization, and evaluation. In: I. Abdurakhmonov, editor, World cotton germplasm resources. InTech, Rijeka, Croatia. p. 167–201.

Rai, K.M., S.K. Singh, A. Bhardwaj, V. Kumar, D. Lakhwani, A. Srivastava, et al. 2013. Large-scale resource development in *Gossypium hirsutum* L. by 454 sequencing of genic-enriched libraries from six diverse genotypes. Plant Biotechnol. J. 11:953–963. doi:10.1111/pbi.12088

Reinisch, A.J., J.M. Dong, C.L. Brubaker, D.M. Stelly, J.F. Wendel, and A.H. Paterson. 1994. A detailed RFLP map of cotton, *Gossypium hirsutum* ′ *Gossypium barbadense*: Chromosome organization and evolution in a disomic polyploidy genome. Genetics 138:829–847.

Remington, D.L., J.M. Thornsberry, Y. Matsuoka, L.M. Wilson, S.R. Whitt, J. Doebley, et al. 2001. Structure of linkage disequilibrium and phenotypic associations in the maize genome. Proc. Natl. Acad. Sci. USA 98:11479–11484. doi:10.1073/pnas.201394398

Rouard, M., V. Guignon, C. Aluome, M.A. Laporte, G. Droc, C. Walde, et al. 2011. GreenPhylDB v2.0: Comparative and functional genomics in plants. Nucleic Acids Res. 39:D1095–D1102. doi:10.1093/nar/gkq811

Rustici, G., N. Kolesnikov, M. Brandizi, T. Burdett, M. Dylag, I. Emam, et al. 2013. ArrayExpress update—Trends in database growth and links to data analysis tools. Nucleic Acids Res. 41(D1):D987–D990. doi:10.1093/nar/gks1174

Said, J.I., Z. Lin, X. Zhang, M. Song, and J. Zhang. 2013. A comprehensive meta QTL analysis for fiber quality, yield, yield related and morphological traits, drought tolerance, and disease resistance in tetraploid cotton. BMC Genomics 14(1):776. doi:10.1186/1471-2164-14-776

Sanderson, L.A., S.P. Ficklin, C.-H. Cheng, S. Jung, F.A. Feltus, K.E. Bett, and D. Main. 2013. Tripal v1.1: A standards-based toolkit for construction of online genetic and genomic databases. Database 2013:bat075. doi:10.1093/database/bat075

Sanderson, L.A., C. Krilow, B. Vandenberg, W. Tom, B. Taran, and K. Bett. 2012. KnowPulse: A breeder-focused web portal that integrates genetics and genomics of pulse crops with model genomes. Paper presented at: Plant and Animal Genome 23, San Diego, CA. 14–18 Jan. Paper 17190. https://pag.confex.com/pag/xxiii/webprogram/Paper17190.html (accessed 9 May2015).

Schaeffer, M. L., L.C. Harper, J.M. Gardiner, C.M. Andorf, D.A. Campbell, E.K. Cannon, E. K. et al. 2011. MaizeGDB: Curation and outreach go hand-in-hand. Database 2011:bar022. doi:10.1093/database/bar022

Senior, M.L., and M. Heun. 1993. Mapping maize microsatellites and polymerase chain reaction confirmation of the targeted repeats using a CT primer. Genome 36:884–889. doi:10.1139/g93-116

Shah, M.R., S.P. Kumpatla, M. Gupta, and S.A. Thompson. 2004. SSR-Based SNP discovery: An economical approach for the development of SNP markers in cotton. Paper presented at: Plant and Animal Genome 12, San Diego, CA. 10–14 Jan.

Shao, Q.S., F.J. Zhang, S.Y. Tang, Y. Liu, X.M. Fang, D.X. Liu, et al. 2014. Identifying QTL for fiber quality traits with three upland cotton (*Gossypium hirsutum* L.) populations. Euphytica 198(1):43–58. doi:10.1007/s10681-014-1082-8

Shi, Y.H., S.W. Zhu, X.Z. Mao, J.X. Feng, Y.M. Qin, L. Zhang, et al. 2006. Transcriptome profiling, molecular biological, and physiological studies reveal a major role for ethylene in cotton fiber cell elongation. Plant Cell 18(3):651–664. doi:10.1105/tpc.105.040303

Sillitoe, I., A.L. Cuff, B.H. Dessailly, N.L. Dawson, N. Furnham, D. Lee, et al. 2013. New functional families (FunFams) in CATH to improve the mapping of conserved functional sites to 3D structures. Nucleic Acids Res. 41(D1):D490–D498. doi:10.1093/nar/gks1211

Skinner, M.E., A.V. Uzilov, L.D. Stein, C.J. Mungall, and I.H. Holmes. 2009. JBrowse: A next-generation genome browser. Genome Res. 19(9):1630–1638. doi:10.1101/gr.094607.109

Stein, L.D., C. Mungall, S.Q. Shu, M. Caudy, M. Mangone, A. Day, et al. 2002. The Generic Genome Browser: A building block for a model organism system database. Genome Res. 12:1599–1610. doi:10.1101/gr.403602

Suszkiw, J. 2010. The National Cotton Variety Test: 50 years old and growing strong. Agric. Res. 58(1):22.

Tan, Z.Y., X.M. Fang, S.Y. Tang, J. Zhang, D.J. Liu, Z.H. Teng, et al. 2014. Genetic map and QTL controlling fiber quality traits in upland cotton (*Gossypium hirsutum* L.). Euphytica. doi:10.1007/s10681-014-1288-9

Tanksley, S.D., and S.R. McCouch. 1997. Seed bank and molecular maps: Unlocking genetic potential from the wild. Science 277:1063–1066. doi:10.1126/science.277.5329.1063

Tanksley, S.D., J.C. Miller, A.H. Paterson, and R. Bernatzky. 1988. Molecular mapping of plant chromosomes. In: J. Gustafson and R. Appels, editors, Chromosome structure and function. Plenum Press, New York. p. 157–172.

Tatusov, R.L., N.D. Fedorova, J.D. Jackson, A.R. Jacobs, B. Kiryutin B, E.V. Koonin, et al. 2003. The COG database: An updated version includes eukaryotes. BMC Bioinformatics 4:41. doi:10.1186/1471-2105-4-41

Thornsberry, J.M., M.M. Goodman, J. Doebley, S. Kresovich, D. Nielsen, and E.S. Buckler. 2001. Dwarf8 polymorphisms associate with variation in flowering time. Nat. Genet. 28:286–289. doi:10.1038/90135

Tong, X.C., L.M. Wang, W.J. Hu, X.Y. Zhang, T.Z. Zhang, and W.Z. Guo. 2013. Molecular cloning and functional analysis of a GDSL lipase gene from *Gossypium hirsutum* L. Acta Agron. Sin. 39(7):1164–1171. doi:10.3724/SP.J.1006.2013.01164

Tu, L.L., X.L. Zhang, S.G. Liang, D.Q. Liu, L.F. Zhu, F.C. Zeng, et al. 2007. Genes expression analyses of sea-island cotton (*Gossypium barbadense* L.) during fiber development. Plant Cell Rep. 26(8):1309–1320. doi:10.1007/s00299-007-0337-4

Udall, J.A., J.M. Swanson, K. Haller, R.A. Rapp, M.E. Sparks, J. Hatfield, et al. 2006. A global assembly of cotton ESTs. Genome Res. 16:441–450. doi:10.1101/gr.4602906

Udall, J.A., L.E. Flagel, F. Cheung, A.W. Woodward, R. Hovav, R.A. Rapp, et al. 2007. Spotted cotton oligonucleotide microarrays for gene expression analysis. BMC Genomics 8:81. doi:10.1186/1471-2164-8-81

UniProt Consortium. 2014. Update on activities at the Universal Protein Resource (UniProt) in 2013. Nucleic Acids Res. 41(D1):D43–D47.

Van Bel, M., S. Proost, E. Wischnitzki, S. Movahedi, C. Scheerlinck, Y. Van de Peer, and K. Vandepoele. 2012. Dissecting plant genomes with the PLAZA comparative genomics platform. Plant Physiol. 158:590–600. doi:10.1104/pp.111.189514

Van Deynze, A., K. Stoffel, M. Lee, T.A. Wilkins, A. Kozik, R.G. Cantrell, et al. 2009. Sampling nucleotide diversity in cotton. BMC Plant Biol. 9:125. doi:10.1186/1471-2229-9-125

Vos, P., R. Hogers, M. Bleeker, M. Reijans, T. Van de Lee, M. Hornes, et al. 1995. AFLP: A new technique for DNA finger printing. Nucleic Acids Res. 23:4407–4414. doi:10.1093/nar/23.21.4407

Wang, F.R., Z.Z. Xu, R. Sun, Y.C. Gong, G.D. Liu, J.X. Zhang, et al. 2013. Genetic dissection of the introgressive genomic components from *Gossypium barbadense* L. that contribute to improved fiber quality in *Gossypium hirsutum* L. Mol. Breed. 32(3):547–562. doi:10.1007/s11032-013-9888-y

Wang, K., W. Guo, and T. Zhang. 2007. Development of one set of chromosome-specific microsatellite-containing BACs and their physical mapping in *Gossypium hirsutum* L. Theor. Appl. Genet. 115(5):675–682. doi:10.1007/s00122-007-0598-x

Wang, K., X. Song, Z. Han, W. Guo, J.Z. Yu, J. Sun, et al. 2006. Complete assignment of the chromosomes of Gossypium hirsutum L. by translocation and fluorescence in situ hybridization mapping. Theor. Appl. Genet. 113(1):73–80. doi:10.1007/s00122-006-0273-7

Wang, K.B., Z.W. Wang, F.G. Li, W.W. Ye, J.Y. Wang, G.L. Song, et al. 2012. The draft genome of a diploid cotton *Gossypium raimondii*. Nat. Genet. 44:1098–1103. doi:10.1038/ng.2371

Wang, M., C. Li, and Q. Wang. 2014. Quantitative trait loci mapping and genetic dissection for lint percentage in upland cotton (*Gossypium hirsutum*). J. Genet. 93(2):371–378. doi:10.1007/s12041-014-0385-9

Wang, Z., M. Gerstein, and M. Snyder. 2009. RNA-Seq: A revolutionary tool for transcriptomics. Nat. Rev. Genet. 10(1):57–63. doi:10.1038/nrg2484

Weber, A.P.M., K.L. Weber, K. Carr, C. Wilkerson, and J.B. Ohlrogge. 2007. Sampling the Arabidopsis transcriptome with massively parallel pyrosequencing. Plant Physiol. 144:32–42. doi:10.1104/pp.107.096677

Wendel, J.F., and R.C. Cronn. 2003. Polyploidy and the evolutionary history of cotton. Adv. Agron. 78:139–186. doi:10.1016/S0065-2113(02)78004-8

Wiersema, J.H. 1994. Taxonomic information on cultivated plants in the USDA/ARS Germplasm Resources Information Network (GRIN). Acta Hortic. (ISHS) 413:109–116.

Xia, Z., X. Zhang, Y.Y. Liu, Z.F. Jia, H.H. Zhao, C.Q. Li, and Q.L. Wang. 2014. Major gene identification and quantitative trait locus mapping for yield-related traits in upland cotton (*Gossypium hirsutum* L.). J. Integr. Agric. 13(2):299–309. doi:10.1016/S2095-3119(13)60508-0

Xu, L., L. Zhu, L. Tu, D. Yuan, J. Li, L. Long, and X.L. Zhang. 2011. Lignin metabolism has a central role in the resistance of cotton to the wilt fungus *Verticillium dahliae* as revealed by RNA-Seq-dependent transcriptional analysis and histochemistry. J. Exp. Bot. 62(15):5607–5621. doi:10.1093/jxb/err245

Xu, Z., R.J. Kohel, G. Song, J. Cho, J. Yu, S. Yu, J. Tomkins, et al. 2008. An integrated genetic and physical map of homoeologous chromosomes 12 and 26 in Upland cotton (*G. hirsutum* L.). BMC Genomics 9:108.

Xu, Z., J.Z. Yu, J. Cho, J. Yu, R.J. Kohel, and R.G. Percy. 2010. Polyploidization altered gene functions in cotton (*Gossypium spp.*). PLoS ONE 5(12):e14351. doi:10.1371/journal.pone.0014351

Yao, D., X. Zhang, X. Zhao, C. Liu, C. Wang, Z.H. Zhang, et al. 2011. Transcriptome analysis reveals salt-stress-regulated biological processes and key pathways in roots of cotton (*Gossypium hirsutum* L.). Genomics 98(1):47–55. doi:10.1016/j.ygeno.2011.04.007

Yoo, M.J., and J.F. Wendel. 2014. Comparative evolutionary and developmental dynamics of the cotton (*Gossypium hirsutum*) fiber transcriptome. PLoS Genet. 10(1):e1004073. doi:10.1371/journal.pgen.1004073

Youens-Clark, K., B. Faga, I.V. Yap, L. Stein, and D. Ware. 2009. CMap 1.01: A comparative mapping application for the Internet. Bioinformatics 25:3040–3042. doi:10.1093/bioinformatics/btp458

Yu, J., S. Jung, C. Cheng, S.P. Ficklin, T. Lee, P. Zheng, et al. 2014. CottonGen: A genomics, genetics and breeding database for cotton research. Nucleic Acids Res. 42(D1):D1229–D1236. doi:10.1093/nar/gkt1064

Yu, J., R. Kohel, L. Hinze, J.Z. Yu, J. Frelichowski, S.P. Ficklin, et al. 2012. CottonDB. Paper presented at: Plant and Animal Genome 20, 14–18 Jan. 2012, San Diego, CA. Paper C16.

Zeng, L., W.R. Meredith, O.A. Gutiérrez, and D.L. Boykin. 2009. Identification of associations between SSR markers and fiber traits in an exotic germplasm derived from multiple crosses among *Gossypium* tetraploid species. Theor. Appl. Genet. 119(1):93–103. doi:10.1007/s00122-009-1020-7

Zhang, M., Y. Zhang, J.J. Huang, X. Zhang, M.K. Lee, D.M. Stelly, and H.B. Zhang. 2012. Genome physical mapping of polyploids: A BIBAC physical map of cultivated tetraploid cotton, *Gossypium hirsutum* L. PLoS ONE 7(3):e33644. doi:10.1371/journal.pone.0033644

Zhang, X., Y. Yuan, Z. Wei, X. Guo, Y. Guo, S. Zhang, et al. 2014. Molecular mapping and validation of a major QTL conferring resistance to a defoliating isolate of verticillium wilt in cotton (*Gossypium hirsutum* L.). PLoS ONE 9(4):e96226. doi:10.1371/journal.pone.0096226

Zhao, Y., H. Wang, W. Chen, and Y. Li. 2014. Genetic structure, linkage disequilibrium and association mapping of verticillium wilt resistance in elite cotton (*Gossypium hirsutum* L.) germplasm population. PLoS ONE 9(1):e86308. doi:10.1371/journal.pone.0086308

Physiology

John L. Snider* and Derrick M. Oosterhuis

Abstract

Since the last physiology chapter was written by Benedict (1984) for Agronomy Monograph 24, *Cotton* three decades ago, numerous advances have been made in understanding the function of this complicated row crop, and numerous reviews have been written. Given the daunting task of providing a single chapter that is a thorough yet concise review of a topic about which volumes of information have been written, the authors have chosen the following approach. Rather than describe fundamental metabolic processes that have been described extensively elsewhere, we have chosen a phenology approach that considers three overarching phases of plant development: seed germination and stand establishment, vegetative development, and reproductive development. Within each phase, typical developmental patterns are considered first, followed by a discussion of factors that influence the physiology of the cotton (*Gossypium hirsutum* L.) plant and cause deviations from normal developmental trends.

Seed Germination and Stand Establishment

Seed Germination and Early Seedling Development

Uniform and vigorous seed germination and early seedling growth is important for stand establishment and can, in many situations, be the difference between a record crop yield and total crop loss. Even in situations where stand remediation can be accomplished through replanting, the additional input costs incurred to replant can substantially limit economic productivity. Wanjura et al. (1969) demonstrated that time to emergence, a measure of seedling vigor, could largely account for variation in end-of-season lint yields. That is, seedlings that emerged more rapidly than others also produced the highest yields. Although most breeding programs have not selected for seedling vigor as a primary objective, it is important to note that uniform stand establishment and rapid early seedling growth (i) maximizes solar radiation interception by photosynthetic plant tissues

Abbreviations: ATP, adenosine triphosphate; DPA, days past anthesis; ETR, electron transport rates; NPQ, non-photochemical quenching; PAR, photosynthetically active radiation; PSII, photosystem II; ROS, reactive oxygen species.

John L. Snider, University of Georgia, Department of Crop and Soil Sciences, 115 Coastal Way, Tifton, GA 31794. *Corresponding author (jlsnider@uga.edu).

Derrick M. Oosterhuis, University of Arkansas, Department of Crop, Soil, and Environmental Sciences, 1366 Altheimer Drive, Fayetteville, AR 72704 (oosterhu@uark.edu).

doi:10.2134/agronmonogr57.2013.0044

Cotton. 2nd. ed. David D. Fang and Richard G. Percy, editors. Agronomy Monograph 57.

at a time during the season when daily insolation is at its peak, (ii) improves competitiveness with weedy species early in the season, and (iii) lessens the long-term damage that can be caused by early-season insect herbivory (e.g., thrips). Simply put, the longer a seedling is in the seedling phase, the more pronounced will be the impact of biotic and abiotic stresses on crop development. First, we consider the process of seed germination in cotton. This process begins with imbibition, entry of water into the quiescent seed, through the chalazal aperture (Benedict, 1984; Christiansen and Rowland, 1986; Simpson et al., 1940). When exposed to adequate moisture, cotton seeds will reach full hydration within 4 to 5 h (Cothren, 1999). Imbibition initiates a host of physiological responses that promote, repair, and resume active growth of the newly hydrated cotton embryo (Turley and Chapman, 2010). During the first day of germination, metabolic activity and oxygen uptake increase substantially, with energy demands during this early growth phase being fueled by preexistent carbohydrate reserves (raffinose, stachyose, and sucrose) of the cotyledons (Kuo et al., 1988; Turley and Chapman, 2010). Following germination, lipid mobilization and the production of lipid-derived precursors for gluconeogenesis (production of glucose from non-sugar precursors) occurs, which ultimately drives post-germinative growth before the conversion of the cotton seedling to photosynthetic autotrophy. The process of gluconeogenesis via the glyoxylate cycle in cotton has been discussed by numerous authors (Benedict, 1984; Christiansen and Rowland, 1986; Cothren, 1999; Trelease, 1984; Trelease and Doman, 1984; Turley and Chapman, 2010) and requires coordination between lipid bodies, glyoxysomes, and mitochondria. The first visible structure to protrude beyond the seed coat is the radicle, which becomes the primary root or the taproot of the cotton plant. Root development is discussed in later sections. Between 5 and 10 d after sowing, elongation of the hypocotyl pulls the cotyledons above the soil surface during emergence, and the greening of the newly expanded cotyledons signals the transition from dependence on seed reserves to photosynthetic autotrophy (Oosterhuis, 1990). Innate seed and seedling characteristics along with environmental factors will strongly influence seed germination, seedling vigor, and stand establishment; therefore, these factors will be considered next.

Factors Affecting Seed Germination and Seedling Vigor

For cotton, seed germination and early seedling growth are clearly influenced by the physical and chemical characteristics of the quiescent seed. As a general rule, seed dormancy, the inability of a viable seed to germinate in an environment conducive to germination (Bewley, 1997), is an undesirable trait for an annual row crop and has been selected against during domestication (Graeber et al., 2012). Despite this, some cotton species exhibit a hard-seeded characteristic, which is a type of physical dormancy in which seeds do not readily imbibe due to the presence of water-impermeable parenchyma tissue that prevents water entry through the chalazal aperture (Christiansen and Rowland, 1986; Bradow and Bauer, 2010). Seedling vigor is well recognized as a desirable trait, and early studies indicated that seedling vigor could be indicative of final yield (Wanjura et al., 1969). Bourland et al. (1992) defined cotton seedling vigor as rapid true-leaf differentiation, well-developed seedling roots, and low incidence of seedling disease. There have been additional metrics used to assess vigor, including seedling fresh weight, dry weight, emergence time, plant height, rate of nodal development, and root

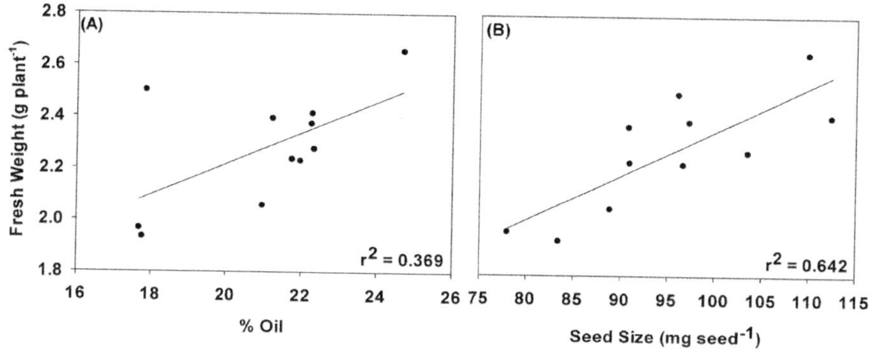

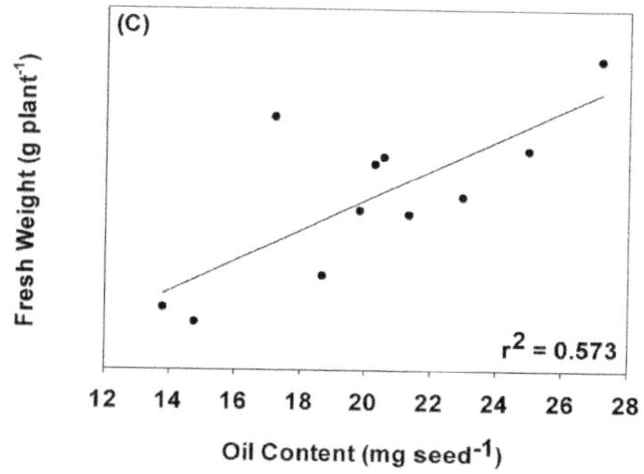

Fig. 1. Linear regression of seedling fresh weight at the 2 to 3 leaf stage versus (A) percent oil, (B) seed size, and (C) total seed oil content for 11 commercially available cotton cultivars. Fresh weight data were averaged from five locations, four replicate plots at each location, and 20 plants per plot (each data point represents the average weight of 400 seedlings) (from Snider et al., 2014a).

growth (Ferguson and Turner, 1971; Krieg and Bartee, 1975; Leffler and Williams, 1983; Snider et al., 2014a). Regardless of definition, a number of authors have tested for correlations between seed physical or chemical characteristics and seedling vigor. For example, Krieg and Bartee (1975) and Leffler and Williams (1983) found that seedling vigor was more closely associated with seed density (dense, large seeds were most vigorous) than seed size alone, and Ferguson and Turner (1971) suggested that the degree of seed filling was a strong indicator of seedling vigor. Other authors have suggested that the total lipid content of the cotton seed could be a strong predictor of seedling vigor because this represents the energy reserves available to promote seedling growth before the seedling becoming self-sufficient (Bartee and Krieg, 1974). Because substantial genotypic variation exists with regard to the nutrient content (lipids, proteins, carbohydrates) of quiescent cotton seeds (Dowd et al., 2010; Pettigrew and Dowd, 2012), cultivar differences in seedling vigor and stand establishment could potentially be attributed to inherent differences in seed nutrient composition. Recently, Snider et al. (2014a) evaluated relationships between seed size, oil concentration, total oil content per seed and seedling fresh weight for 11 commercially available cotton cultivars grown in five different cotton-producing counties throughout Georgia. In their study, seed oil content was inversely related to protein content and seedling fresh weight but was closely associated with seed size and total oil content per seed (Fig. 1). Because seed quality can also be influenced by environment during seed maturation and subsequent storage (Cothren, 1999), current methods that allow for rapid nondestructive measurements of oil and protein content of whole cottonseed (Horn et al., 2011) may prove useful in predicting seedling vigor for individual cultivars and individual seed lots.

During germination and early seedling growth, temperature, oxygen availability, and soil moisture content will strongly influence stand establishment. For example, it has been argued that temperature is most important among the abiotic factors impacting stand establishment (Bradow and Bauer, 2010). As with a number of physiological processes in cotton, the optimal temperature for cottonseed germination and emergence is 30°C, and the lower threshold is 16°C (Bradow and Bauer, 2010; Cothren, 1999; Oosterhuis, 1990). Camp et al. (1997) demonstrated a negative linear relationship between the days after planting that ambient temperatures were <16°C and lint yield, illustrating the inhibitory effect of cool early-season temperatures on final yield. Exposure to chilling temperatures less than 5°C for even short periods of time (4 h) during imbibition will result in the death of the embryo (Christiansen, 1963; Cothren, 1999; Bradow and Bauer, 2010). Regarding the impact of oxygen on seed germination and seedling development, given the high respiration rates of germinating cotton seeds mentioned previously, oxygen availability is essential for successful germination and stand establishment, where waterlogging or other restrictions to free air movement can inhibit germination and completely prevent stand establishment (Christiansen and Rowland, 1986; Wanjura, 1986). Although it is an important factor governing seedling vigor, the impact of low oxygen availability on developing cotton root systems will be discussed in subsequent sections of this chapter. As with many other mature seeds, the cotton seed is extremely dry at maturity (11% moisture, Cothren, 1999). Thus, imbibition generally occurs unimpeded if care is taken to plant cotton seed after a sufficient rainfall event (50 mm of soil-penetrating rainfall; Bradow and Bauer, 2010; Munro, 1987). However, following imbibition, cell expansion processes (root or shoot) require sufficient turgor pressure to continue unabated and are the most sensitive physiological processes to water deficit (Hsiao, 1973). Thus hot, dry conditions soon after planting can limit root and shoot growth of the developing seedling due to decreased pressure potential needed for cellular expansion. Furthermore, soil drying increases mechanical impedance, which limits growth of the radicle and the hypocotyl, thereby limiting subsequent root growth and shoot emergence (Bennie, 1996).

Vegetative Development

Root Growth and Development

The root system plays a critical role in plant growth and productivity, and the development of the root system is affected by numerous factors in the soil environment. The development of the cotton root system has been described by a number of authors (Balls, 1919; Collings and Warner, 1927; Brown and Ware, 1958; Taylor and Klepper, 1978; McMichael, 1986; McMichael et al., 2010). The radicle, or primary root, develops rapidly and is the first organ to emerge from the seed coat during germination. The primary root, or taproot, penetrates the soil rapidly, reaching a depth of up to 25 cm or more by the time the cotyledons unfold aboveground (Oosterhuis, 1990), and penetrating to depths of more than 2 to 3 m (Balls, 1919). On deep alluvial and irrigated soils in California, roots may reach a depth of 1 to 1.5 m when the young plants are only 20 to 25 cm high, with a final depth at maturity of 3 m (Hubbard, 1933). In the humid mid-southern United

States, the higher water tables and compacted soils result in much shallower root systems. Root development in cotton continues rapidly during the early vegetative stage at the rate of up to 6 cm d^{-1}, depending on conditions (Taylor and Radcliff, 1969; McMichael, 1986), such that the roots may be 1.0 m deep when the aboveground portion of the plant is only about 0.35 m (Oosterhuis, 1990; Fig. 2).

Lateral roots generally develop about 12 cm behind the primary root apex (Mauney, 1968) and proliferate outward from the taproot, forming a mat of roots that may extend horizontally more than 2 m from the taproot (Taylor and Klepper, 1978). Lateral roots generally remain fairly shallow (Hayward, 1938), such that the bulk of the root system is located in the upper meter, but this is dependent on soil moisture, soil physical structure, and vigor of the plant itself. Root distribution within the soil profile, expressed as root length density (total root length per unit soil volume, cm root cm^{-3} soil), can vary considerably with soil and plant conditions. Root length densities of cotton decrease with depth in the soil (Rogers and Prior, 1992) but increase with depth as the soil profile dries (Klepper et al., 1973). Zhao et al. (2010) showed root length densities for cotton to be highest near the surface, approximately 20 mm cm^{-3} at 10 cm, decreasing with depth to about 10 mm cm^{-3} at 40 cm (Fig. 3). As the cotton plant develops, total root length continues to increase until maximum plant height is achieved and bolls begin to form at flowering (Taylor and Klepper, 1974; Pearson and Lund, 1968; Zhao et al., 2010), after which root length begins to decline as older roots die (Cappy, 1979; McMichael, 1986). This decline in root activity begins as the developing boll load becomes a stronger sink and carbohydrates are increasingly partitioned toward the fruit (Fig. 4, Oosterhuis and Jernstedt, 1999).

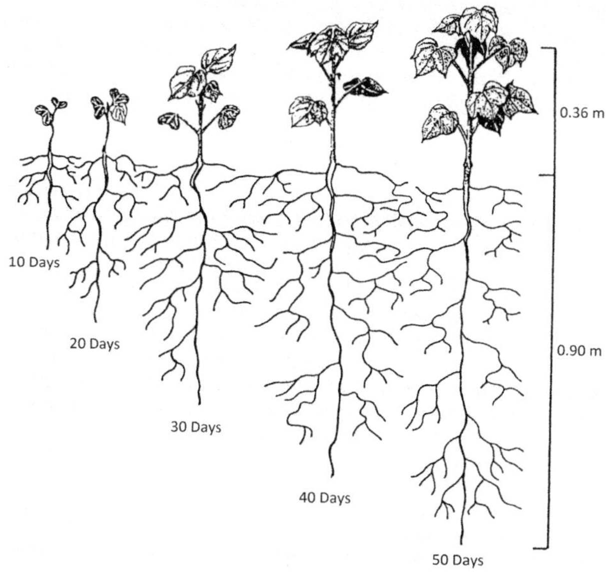

Fig. 2. Root development of the cotton plant during the first 50 d after germination showing the comparative root and shoot growth (from Oosterhuis, 1990).

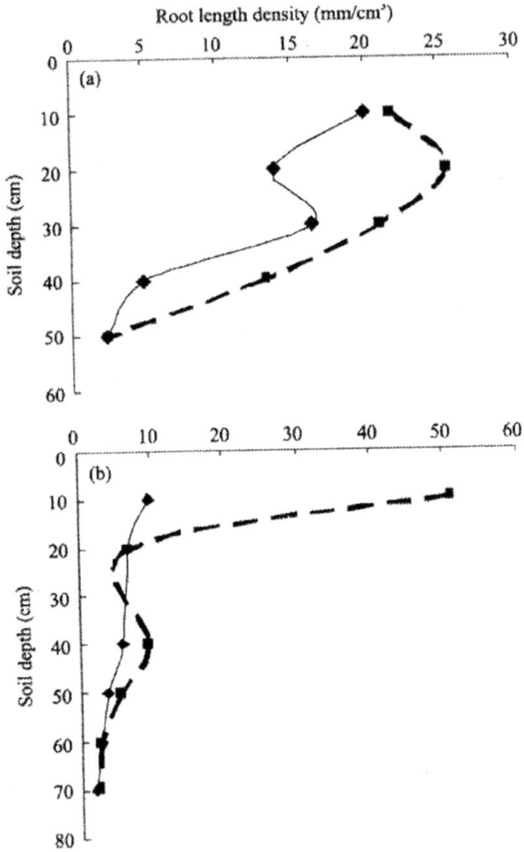

Fig. 3. Vertical distribution of cotton optimal root length density at (A) flowering and (B) boll opening stages (from Zhao et al., 2010, with permission).

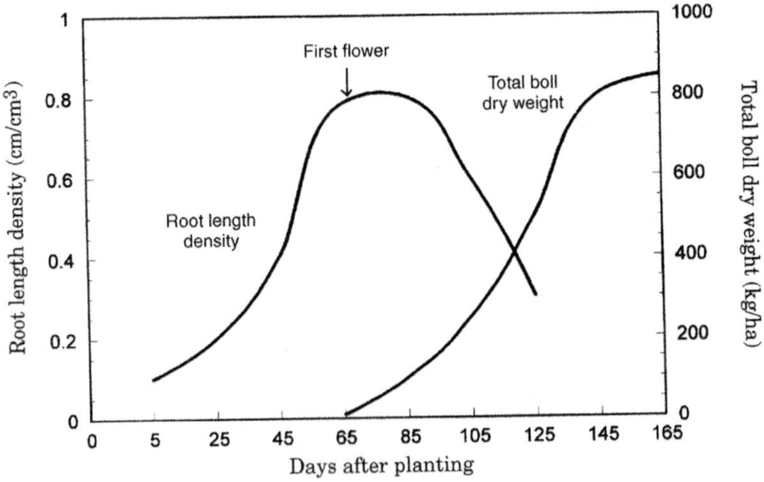

Fig. 4. Typical growth pattern of cotton roots in relation to the developing boll load during a growing season (from Oosterhuis and Jernstedt, 1999).

The total volume of soil explored by a cotton root system for water and mineral nutrients is related to the ability of the root system to produce secondary roots and increase root length density (McMichael, 1986). Lateral roots arise in the pericycle on the poles of the xylem elements and are arranged in rows along the primary root, with the number of rows corresponding to the number of vascular bundles in the stele (Brown and Ware, 1958). The normal vascular arrangement in commercial cultivars of cotton is a tetrarch arrangement with four protoxylem poles in the stele (Hayward, 1938). McMichael et al. (1987) reported a correlation between the number of vascular bundles in the taproot and the production of lateral roots. They showed an increased number of vascular bundles in the stele for a wild-type T25 strain of cotton and suggested this would increase the potential for greater drought tolerance, increased hydraulic conductance, and greater number of lateral roots. However, Oosterhuis and Wullschleger (1987a) showed that this increased number of vascular bundles and lateral roots had no significant effect on axial resistance to water movement or water flow for the primary root.

Factors Affecting Cotton Root Growth

Numerous factors affect cotton root growth including soil water content, soil temperature, compaction, aeration, fertility and soil microorganisms. The impact of the soil environment on the growth of cotton roots was reviewed by McMichael and Quisenberry (1993) and more recently by McMichael et al. (2010).

Soil Water

The water content of the soil has a significant influence on the growth and function of roots. Rooting densities increase with depth and decrease in upper soil layers as the soil profile dries (McMichael et al., 2010; Klepper et al., 1973; Zhao et al., 2010). Water stress increased cotton root length density in deeper soil layers (Taylor and Klepper, 1974; Zhao et al., 2010), due to the disappearance and death of roots in the upper horizons and growth of new roots in the lower horizons to maintain water uptake rates (Klepper et al., 1973). Although growth of the whole root system decreases with water stress (Zhou and Oosterhuis, 2012), the production of dry matter by cotton roots has been observed to be less severely inhibited than shoots under decreasing soil moisture (Carmi and Shalhevet, 1983; Ball et al., 1994), and therefore, the root/shoot ratio increases as the soil water content decreases (Malik et al., 1979). Sharp et al. (2004) showed that the primary root of four agronomic species, including cotton, maintained elongation under water deficit for a longer period of time than shoots, with root elongation being maintained at water potentials lower than −1.5 MPa, whereas shoot growth was completely inhibited (Fig. 5). Furthermore, these authors showed that elongation rates of maize root apices differ with water stress. The elongation rates for well-watered maize (*Zea mays* L.) roots peaked about 4.5 mm from the apex and then gradually declined to zero, whereas the zone of elongation for a water-stressed root was shorter, peaking at 3.0 mm and then declining to zero at 7.0 mm (Fig. 6). Similar studies have not been reported for cotton.

Irrigation practices can also change water distribution in the profile and impact the growth of cotton roots. Long irrigation cycles caused more rapid deterioration of the root system during heavy fruiting, above that normally experienced by cotton (Radin et al., 1989). Drip irrigation resulted in a shallow root system concentrated around the drippers, necessitating a more frequent supply

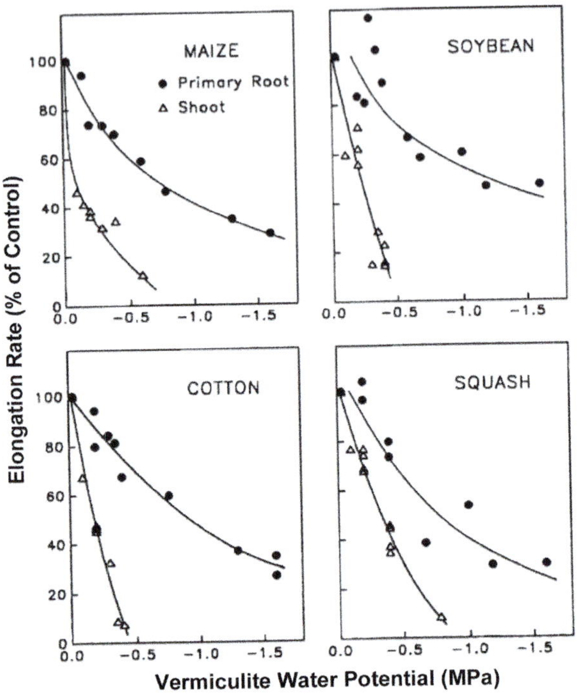

Fig. 5. Elongation rates of primary root (solid circles) and shoot (open circles) of seedlings of four species at various water potentials (from Sharp et al., 2004, J. Exp. Bot. 55:2343–2351, with permission).

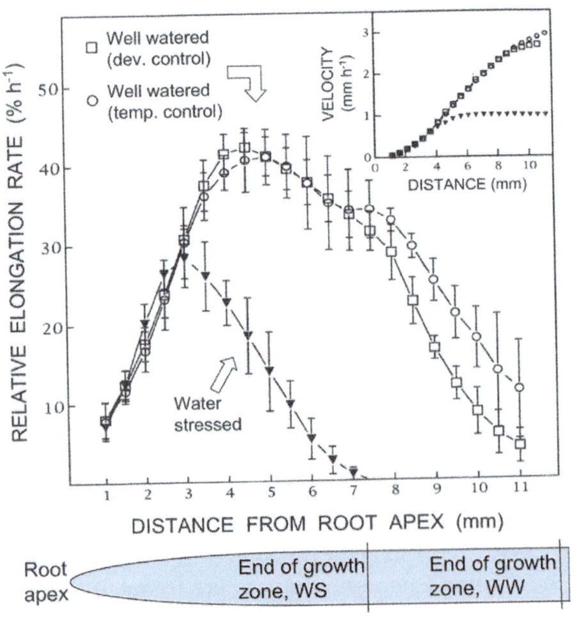

Fig. 6. Relative elongation rate as a function of distance from the apex of the primary root of maize (from Liang et al., 1997, modified by Sharp et al. 2004, with permission).

of water for continued growth (Carmi et al., 1992). Water uptake in cotton has been shown to be proportional to the rooting density as well as the difference in water potential between the root xylem and the bulk soil (Taylor and Klepper, 1975). Taylor and Klepper (1974) showed that root length did not increase when the water content fell below 0.06 cm^3 cm^{-3} or about 0.1 MPa, although Jordan (1983) reported that rooting densities may decrease to as low as 0.2 cm cm^{-3} and still extract water. Cultivar differences in root length density and soil moisture extraction for cotton have been reported (Kennedy et al., 1987). In other work, Taylor and Klepper (1971) observed that water extraction per unit length of root was greater in wet soil and decreased exponentially with decreasing soil water potential. They also found that deep roots were as effective as shallow roots in extracting water. Ball (1992) showed with cotton that the more abundant small-sized roots (medium diameter 0.30 mm) were more susceptible to drought than medium-sized roots (0.62-mm diameter) and suggested that the larger roots may be more important for continued growth in stress and recovery.

Taylor and Klepper (1975) reported that the major resistance to water flow was either in the root or the root–soil interface, although Passioura (1980) found no evidence of any major hydraulic resistance at the root–soil interface. Oosterhuis (1981) showed that radial resistance to water flow in 10-d-old cotton roots accounts for about 76% of total root resistance. Taylor and Klepper (1978) assumed that the radial resistance–water content relations did not vary with root age or soil depth and also that axial resistance was negligible. Root hydraulic conductivity was decreased in previously water-stressed cotton seedling roots in part due to increased radial root resistance (Oosterhuis and Wiebe, 1980).

Numerous mechanisms are involved in maintaining growth of the primary root under water deficits, including turgor maintenance by osmotic adjustment, hormonal involvement, enhanced cell wall loosening, and adjustment of growth zone dimensions (Sharp et al., 2004). Osmotic adjustment, or osmoregulation, is a plant mechanism for drought tolerance and the maintenance of water potential gradients (Wyn Jones and Gorham, 1983) and cell turgor under water stress, which allows a continuation of growth, although at a reduced rate in comparison to optimal conditions (Sharp and Davies, 1979). The role of osmotic adjustment in maintaining root growth, allowing water uptake longer in drying soil, has been emphasized by Acevedo and Hsiao (1974). Osmotic adjustment has been documented in cotton roots (Oosterhuis and Wullschleger, 1987b; Ball, 1992), as well as in leaves (Ackerson and Herbert, 1981; Cutler and Rains, 1977). The magnitude of osmotic adjustment in cotton is greater in leaves (0.41 MPa) than roots (0.19 MPa), although the percentage change in roots (46%) was greater than in leaves (22%) (Oosterhuis and Wullschleger, 1987b). The authors related this to the drought tolerance and survival capabilities of cotton. It has been shown that there is substantial osmotic adjustment in the tips of maize roots (Sharp and Davies, 1979; Westgate and Boyer, 1985), which could contribute to growth maintenance. The primary contributors to the osmotica were hexoses and particularly proline (Voetberg and Sharp, 1991). The increase in proline deposition at low water potentials required an accumulation of abscisic acid (Ober and Sharp, 1994). The premise that osmotic adjustment allows for turgor maintenance and increased root growth at low water potentials implies that the plant will be able to exploit a greater and deeper soil volume for water. Roots at different positions in the soil profile may vary in osmotic adjustment in response to water deficit (Turner et al.,

1987). Liu et al. (2008) reported that the root vigor and drought tolerance of cotton under short-duration water stress was increased with adequate availability of nitrogen, due to changes in antioxidant activities and decreased membrane damage in the roots. The function of cotton roots is also influenced by interactions between soil water status and soil temperature. Hydraulic conductance of cotton roots has been shown to decline at cooler temperatures (Radin, 1990; Bolger et al., 1992), which would affect water uptake. These results suggested that water uptake by cotton roots may decrease as a result of low soil temperatures even when water is not a limiting factor.

Soil Temperature

The growth of cotton root systems can be significantly influenced by both soil and air temperature. Soil temperature is usually lower than air temperature, but seasonal fluctuations can occur with depth depending on soil and aboveground factors (McMichael and Burke, 1998). This is a particular problem in the southeastern and midsouth US Cotton Belt, where soil temperatures early in the growing season are usually cool and well below the optimum for root growth. The rate of cotton root downward growth increases with the rate at which the soil warms (Bland, 1993). Research has generally shown that the growth of cotton roots increases with soil temperature until an optimum of between 28 and 35°C is reached (Bloodworth, 1960; Taylor et al., 1972; Mills et al., 2012), after which growth declines (Glinski and Lipiec, 1990; McMichael et al., 2010; Pearson et al., 1970). McMichael and Burke (1994) reported that root growth in cotton in response to temperature is maximum within the thermal kinetic window defined by Burke et al. (1988). The detrimental effects of high temperature on root growth are probably related to the level of available substrate or stored seed reserves (McMichael and Burke, 1994). Kaspar and Bland (1992) indicated that root systems expanded downward into deeper soil layers as the soil profile warmed progressively downward. The taproot of cotton may be more adapted to adverse soil temperatures than subsequent branch roots, at least until the taproot had developed to approximately 10 cm in length (Arndt, 1945). McMichael et al. (2010) reported similar results with exotic cotton strains grown in hydroponics. The response of the taproot to changes in temperature was different from that of lateral roots, with the taproot able to grow over a broader temperature range than the branch roots (Fig. 7) (McMichael and Quisenberry, 1993). Root branching increases at high soil temperatures approaching the optimum (Nielsen, 1974), while low soil temperature can reduce cotton root branching (Brower and Hoagland, 1964). Mills et al. (2012) showed that genetic diversity exists among cotton genotypes for taproot and lateral root growth under cool soil temperatures, and suggested that this could be exploited to develop cultivars able to tolerate cool temperatures and produce more vigorous cotton seedlings. The uptake of water by roots is reduced at low temperature (Nielsen, 1974), while higher temperatures result in increased uptake. Bolger et al. (1992) demonstrated that the hydraulic conductance of cotton roots declined as the root zone temperature decreased below 30°C such that conductance at 18°C averaged only 43% of that at 30°C. The interaction between soil and air temperature also influences cotton root development, as shown in Table 1 for 10-d-old cotton seedlings (McMichael and Burke, 1994). When the root temperature was low (20°C), root growth was reduced regardless of the temperature of the air. This root–shoot interaction in response to temperature is probably

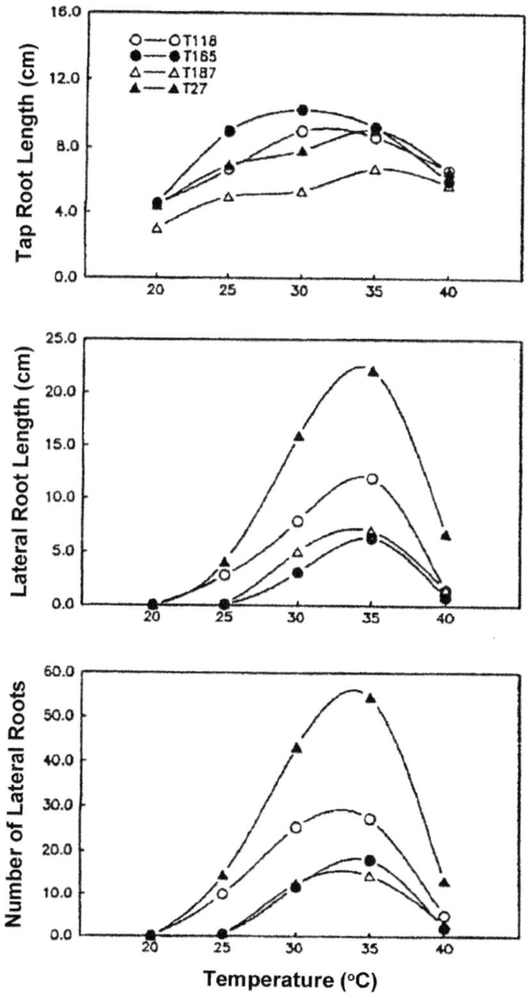

Fig. 7. The influence of temperature on the growth of primary (tap) roots and lateral roots of 10-d-old cotton seedlings. (From McMichael and Quisenberry, 1993).

related to changes in source–sink relationships. For example, Guinn and Hunter (1968) showed changes in carbohydrate levels in shoots and roots in response to temperature with a buildup of sugars occurring at low root temperatures.

Soil Strength

Compacted soil layers, either natural or as a result of wheel traffic, can significantly restrict root growth in cotton and impact productivity. Comprehensive reviews of the effects of mechanical impedance on root growth were provided by Bennie (1996) and Taylor (1983). In general, root growth is restricted as the soil strength increases (Wanjura and Buxton, 1972; Grimes et al., 1975; Gerard et al., 1982). Taylor and Ratliff (1969) observed that once the taproot of cotton entered a compacted zone, the root length varied inversely with soil resistance, as measured by penetrometers. Gerard et al. (1982) indicated that the critical strength at which root growth stopped range from 6.0 to 7.0 MPa in coarse-textured soil

Table 1. Root development of 10-d-old cotton seedlings grown in growth pouch system at four temperature regimes (from McMichael and Burke, 1994).

Treatment					
Shoot temperature	Root temperature	Taproot length	Lateral root length	Total root length	Number of Laterals
C°		cm			
28	20	3.05a†	0.0 a	3.08a	0.0a
20	28	5.06b	1.83b	6.89b	8.00b
28	28	7.71c	5.82c	13.53c	8.38 9c
20	20	4.71	0.0 a	4.71a	0.0a

† Means followed by the same letter are not significantly different at the 0.05 probability level based on a Duncan's Multiple Range Test.

and approximately 2.5 MPa in clay soil. Glinski and Lipiec (1990) noted that the response of most roots grown in mechanically impeded soil were reduced root size, diminished elongation rates, uneven root distribution, thickening of roots, greater lateral branching, and a reduction of water and nutrient uptake. The cells of cotton roots grown in loose soil were elongated with thicker cell walls compared to similar cells of roots grown in compacted soils (Camp and Lund, 1964; Bennie, 1996). Under soil impedance, plant roots exhibit various anatomical and physiological changes that facilitate elongation in hard soil, including shorter, fatter cells, exudation from the root cap to decrease friction, and increased sloughing off of border cells (Bengough et al., 2006). Croser et al. (1999) showed that compacted soils changed the pattern of root elongation rate behind the root meristem of pea (*Pisum sativum* L.) from a bell-shaped curve with a shorter accelerating region behind the apex and a longer-tailed decelerating region in a loose soil, to a truncated shape in a compact soil. This was similar to the root response reported by Sharp et al. (2004) for corn roots under water deficit.

Bennie (1996) suggested that the decrease in root growth in a compacted zone was the same for most species, and that the main differences were related to the ability of plants to produce branch roots in the compacted layers. Similarly, Lowry et al. (1970) concluded that the restricted supply of water due to a limited rooting volume in high strength soil pans was the main factor causing reduced yield for cotton. Interactions between soil strength and soil water status have been shown to be important in the response of cotton root systems to mechanical impedance. Studies have shown that root elongation was more sensitive to soil strength than water content (Taylor and Ratliff, 1969). Taylor and Gardner (1963) concluded that the critical factor controlling root penetration particularly in sandy soils was soil strength and not bulk density changes. However, Bennie (1996) showed that impedance at a specific bulk density increases with soil drying, and Camp and Lund (1964) showed that as moisture decreased, the effects of soil density became more critical, with no root penetration of cotton roots occurring at densities of 1.7 to 1.8 g mL^{-1}. Regardless of the magnitude of the interaction of soil strength and moisture content, mechanical impedance can be a significant factor in reducing aboveground growth of the cotton plant (Carmi, 1986) and cotton yields (Carmi and Shalhevet, 1983), probably due to a restricted water supply as a result of the limited rooting volume (Lowry et al., 1970). The reasons underlying root growth response to mechanical impedance are likely due to changes in carbon fluxes and sink strengths within the roots (Thaler and Pages, 1998). It has been suggested that ethylene may be involved in the response of roots to mechanical impedance, and a higher rate of ethylene evolution by mechanically impeded roots has been reported (Kays et al., 1974). The actual cause of the increase in the rate of ethylene evolution by impeded roots is not clear, and Moss et al. (1988) argued that it may not be a direct result of the mechanical impedance but could be a reaction to physical wounding of the root tissue as the roots grow through the impeding soil.

Soil Aeration

The soil air composition is similar to the aboveground air except for higher CO_2 levels. Soil air has a composition of approximately 79% nitrogen, 20% oxygen, and 0.5 to 1% CO_2 at depths of 15 to 20 cm (Stolzy, 1974). Oxygen levels can vary considerably, as low as 5%, and CO_2 levels can increase to approximately 20% depending on such factors as soil temperature and soil water content (Cannon, 1925). However, it is the lack of oxygen rather than the increase in CO_2 that is important because roots need about 2% oxygen as a minimum to survive (Kirkham, 2011). Cotton roots are very sensitive to changes in oxygen concentrations relative to changes in CO_2 concentrations (Kramer, 1969). Excess water due to temporary saturation and increased water tables can deplete dissolved oxygen from the soil, resulting in depressed root growth (Drew and Stolzy, 1991; Leonard, 1945). Christianson et al. (2010) showed that waterlogging and the resulting lack of oxygen results in significant reductions in root growth and altered expression of genes as early as 4 h after flooding. The genes involved were associated with changes in cell walls, metabolic pathways, mitochondrial function, and nitrogen metabolism. Cotton can survive under low oxygen conditions, but for only a few hours, after which the tap roots dies in anoxic conditions followed by development of new lateral root initiation above the killed portion (Huck, 1970). Whitney (1941) observed that cotton root growth was depressed when exposed to toxic levels of carbon dioxide or reduced levels of oxygen. Camp and Lund (1964) showed that even at low bulk densities, root growth was depressed when either oxygen was decreased below 19% or the CO_2 levels were increased.

Canopy Development

Leaves are a major determinant in crop growth and yield, as they set up the structure and produce most of the requirements for the flower and boll. In addition to the role of leaves in light interception, and as a source for photosynthetic assimilation and subsequent carbon export, they also serve as a sink for mineral nutrient assimilation and provide the surface for evaporative cooling through transpiration. The development of the cotton canopy and description of leaf types has been described by Oosterhuis (1990). The distribution of leaves in the canopy, together with the pattern of photosynthate production of leaves, are key characteristics of cotton that make it very different from other row crops (Constable and Oosterhuis, 2010). The spatial distribution of leaf area within the cotton canopy and its overall involvement in boll development are key ingredients essential for understanding growth and development, but also for obtaining optimal yield.

The development of total leaf area follows a typical sigmoidal pattern, increasing slowly at first during the first 6 to 7 wk, and then more rapidly during early fruiting with canopy closure (Fig. 8). Early in the season, the total leaf area is comprised entirely of main-stem leaves, since sympodial branch formation is

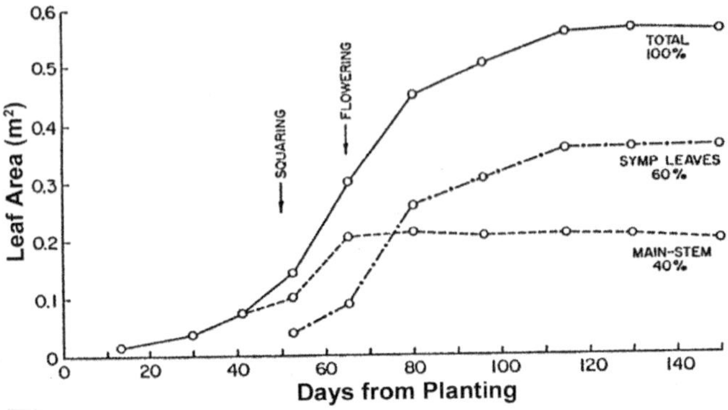

Fig. 8. Development of total leaf area, main-stem leaf area, and sympodial leaf area per plant in field-grown cotton (from Oosterhuis and Wullschleger, 1988).

initiated only after 40 to 45 d of vegetative growth. As the season progresses sympodial leaf area increases rapidly as fruiting branches develop and accounts for approximately 60% of the total leaf area at 120 d after planting. Leaf area reaches a maximum of approximately 0.57 m^2 $plant^{-1}$ at 85 d after planting, equivalent to a leaf area index of about 3.5. During the period of rapid leaf area development (Days 40–80), leaf area increases at a rate of more than 850 m^2 ha^{-1} d^{-1} and accounts for approximately 75% of the maximum leaf area index accumulated (Oosterhuis and Wullschleger, 1988). During the early vegetative period much of the partitioning of assimilate emphasis is on root development, whereas during the later vegetative period the emphasis changes to square and flower development. It is imperative that leaf area development during the early period is not delayed so as to achieve maximum utilization of incoming solar radiation for photosynthesis as soon as possible. Canopy closure also improves weed control and decreases soil evaporative losses by confining evaporation to that occurring by transpiration through the plant, which provides more control over water loss.

Monopodial and Sympodial Leaves

Within the cotton canopy two basic leaf types characteristically develop following expansion of the cotyledons: main-stem leaves and sympodial leaves. Main-stem leaves are formed at regular intervals in a 3/8's phyllotaxy along the main stem or central axis of the plant, while sympodial leaves form in a sequential fashion along the sympodial or fruiting branches (Oosterhuis and Jernstedt, 1999) (Fig. 9). Leaves on fruiting branches are usually one-half the area or less than the main-stem leaf on the same node. The largest leaves are produced in the lower to central regions of the plant (McClelland, 1916; Hearn, 1969; Constable and Rawson, 1980b; Mutsaers, 1983; Constable, 1986).

A typical profile of final leaf size on intensively managed upland cotton plants is shown in Fig. 10 (Constable and Oosterhuis, 2010). In this example, the largest leaves (140 cm^2) are on the main stem at node 8, whereas the largest leaves on the fruiting branch at that position in the canopy are only 77 cm^2. Leaves on fruiting branches are smaller than the corresponding main-stem leaf at the same node position by a factor of about 0.55, 0.4, and 0.3 for the first three nodal

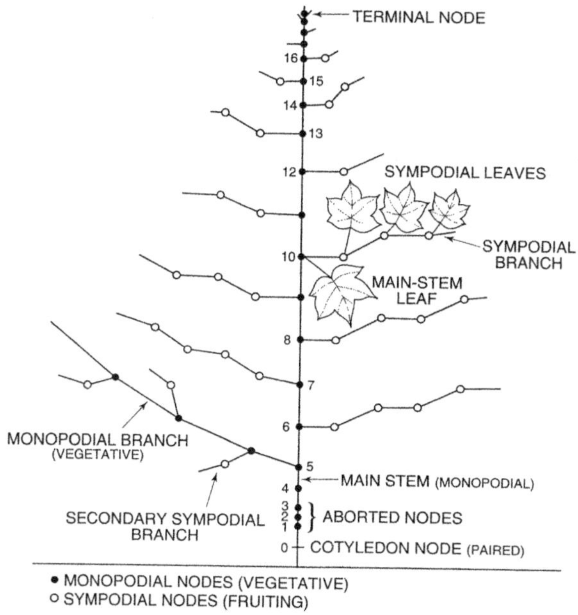

Fig. 9. Diagrammatic cotton plant showing the main stem, monopodial and sympodial branches, and the distribution of main-stem leaves and sympodial leaves (from Oosterhuis, 1990).

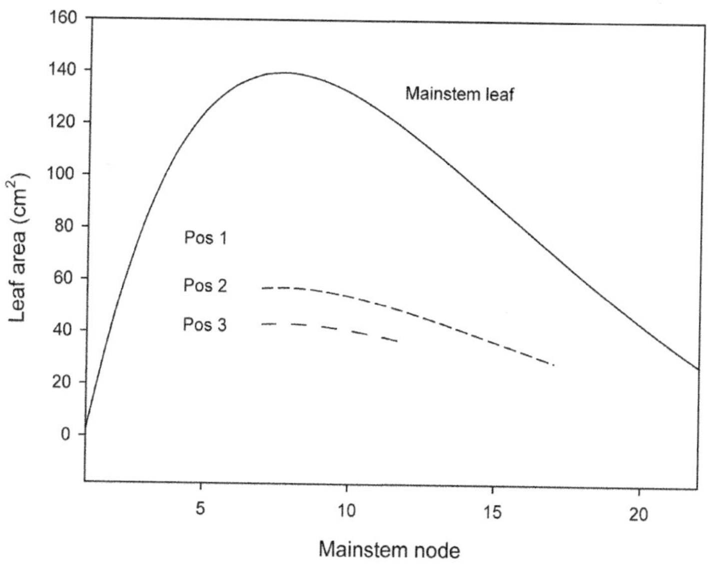

Fig. 10. Typical profile of final area of individual main stem and fruiting branch leaves on a cotton plant (pos 1, pos 2, and pos 3 refer to the first three sympodial leaves on each fruiting branch). This example has no vegetative branches and the first fruiting branch at node 7. The total area of leaves is 3500 cm^2 if plant density was 10 m^{-2} giving a maximum leaf area index of 3.5 (from Constable and Oosterhuis, 2010).

positions on a fruiting branch, respectively (Hearn, 1969; Constable and Rawson, 1980b; Wullschleger and Oosterhuis, 1987). Competition with other leaves and particularly with fruit has been proposed as the mechanisms behind these leaf size patterns. Leaf size is also strongly influenced by position on the plant, plant density, and planting date (Constable, 1986).

Studies using various leaf removal treatments and detailed analysis of carbon allocation patterns (Constable and Rawson, 1980a; Oosterhuis and Urwiler, 1988; Wullschleger and Oosterhuis, 1987) have indicated that these two leaf types may have differing contributions to growth and yield development. Main-stem leaves are associated with providing photosynthate for structural growth (Oosterhuis and Urwiler, 1988), as well as to developing bolls on a sympodia (Peoples and Matthews, 1981), while sympodial leaves, being spatially close to potential fruiting sites, are primarily involved in boll development (Brown, 1968; Constable and Rawson, 1982; Ashley, 1972).

The vertical distribution of main-stem and sympodial leaves within the cotton canopy is shown in Fig. 11. The lower main-stem leaves characteristically senesce and abscise under the heavily shaded conditions of the lower canopy. This is evident in Fig. 4 by the absence of main-stem leaves below main-stem node 7, although sympodial leaves were present. Individual main-stem leaf area increases with position on the plant to a maximum of 170 cm^2 at main-stem node 10 and then decreases with increasing position up the plant. Total leaf area per main-stem node increases with height to a maximum of approximately 475 cm^2 at main-stem node 8, with considerable leaf area between main-stem nodes 7 and 11, where the majority of the boll yield is formed (Jenkins et al., 1990). The first, second, and third fruiting positions along the sympodial branches comprised 31, 22, and 14% of the total leaf area (Fig. 12). There is a similar hierarchy of yields contributed from these three positions, with the first fruiting position producing 66 to 75% of total yield, the second position 18 to 21%, and 2 to 4% for all other positions on a fruiting branch (Jenkins et a., 1990).

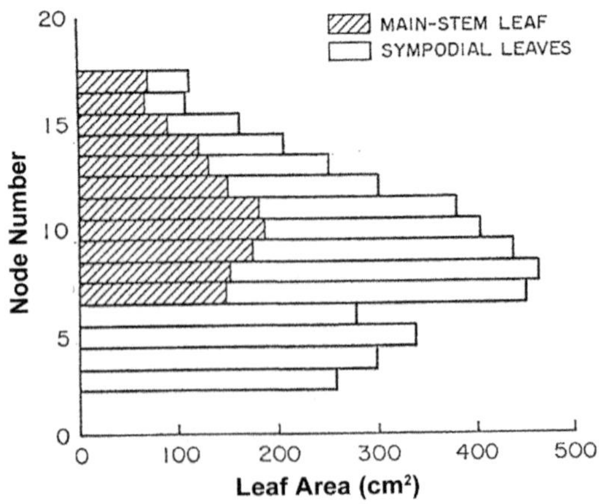

Fig. 11. Contribution of main-stem and sympodial leaves to total leaf area relative to main-stem node number at 120 d after planting (from Oosterhuis and Wullschleger, 1988).

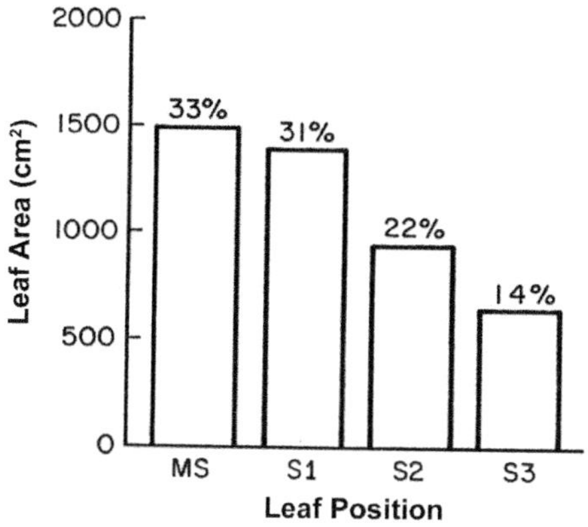

Fig. 12. Relative distribution of leaf area by leaf position along a sympodial branch at 120 d after planting. (Oosterhuis and Wullschleger, 1988).

Leaf Shape, Radiation Interception, and Yield

Upland cotton exhibits considerable variation in leaf shape ranging from the normally broad, palmately lobed leaves to slightly bifurcated (subokra) to highly indented (superokra) (Meredith, 1984). However, most of the cultivars grown commercially possess the normal leaf shape (Heitholt et al., 1992). These okra-leaf shapes can alter the architecture of the canopy and therefore canopy light interception characteristics (Wells et al., 1986). Okra-leaf types have been reported to offer various characteristics not exhibited by normal-leaf types, including higher single-leaf photosynthesis (Pettigrew et al., 1990), early maturity (Jones, 1982), great flower production (Kerby and Buxton, 1978), less boll rot, and increased resistance to some insect pests (Jones, 1982; Wilson, 1986). There have, however, been conflicting results comparing lint yield of normal-leaf and okra-leaf isolines (Heitholt and Meredith, 1998). Earlier reports showed yields sometimes being below those of normal-leaf cotton (Wilson, 1986), but more recent studies have found no difference in yield comparisons (Gonias et al., 2011). Heitholt et al. (1992) reported higher yields of okra-leaf isolines for a given amount of intercepted radiation, indicating that okra-leaf utilized intercepted radiation more efficiently than normal-leaf types. This was confirmed by Gonias et al. (2011), who showed that an okra-leaf isoline utilized intercepted radiation more efficiently for dry matter production than the normal-leaf isoline, with values of radiation use efficiency of 2.636 and 1.897 MJ^{-1} of intercepted photosynthetically active radiation, respectively. Wells and Meredith (1986) concluded that okra-leaf cottons can produce plant canopies competitive in yield with those of normal-leaf cotton.

Cotton Leaf Growth, Gas Exchange, and Carbon Requirements

The pattern of leaf development and gas exchange is shown in Fig. 13. Cotton leaves reach their maximum size about 24 d after unfolding, but achieve maximum photosynthesis in about 16 to 22 d after unfolding; that is, they are about 70 to 90% expanded when they reach maximum efficiency (Constable and Rawson, 1980a; Muramoto et al., 1967; Wullschleger and Oosterhuis, 1990a). For about the first week after unfolding they are net "importers" of carbohydrates, after which the leaf becomes a net carbon exporter increasing to a short "plateau" for maximum photosynthesis in the third week after leaf unfolding (Constable and

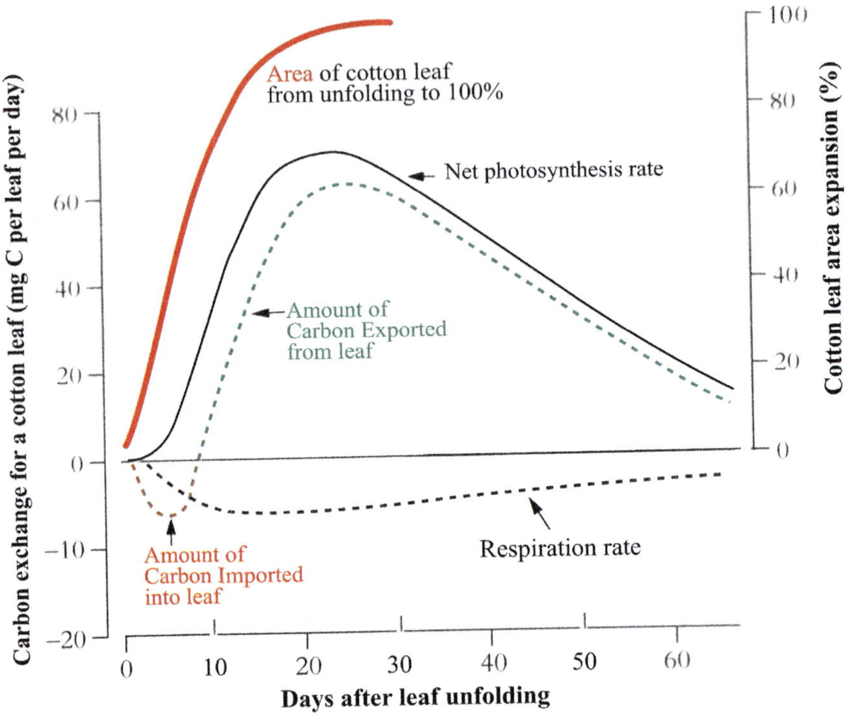

Fig. 13. The typical developmental pattern of daily net photosynthesis during leaf ontogeny from main-stem leaf 7 unfolding to senescence at Day 70. Note carbon is initially imported to help sustain leaf expansion, and the leaf is 70 to 80% expanded when maximum photosynthetic efficiency is achieved (adapted from Atwell et al., 1999, redrawn from Constable and Rawson, 1980a, with permission).

Rawson, 1980a; Wullschleger and Oosterhuis, 1990a). The peak carbon export of 1 mg C d^{-1} occurs at 23 d from unfolding (Constable and Rawson, 1980a), after which photosynthesis declines rapidly to zero at about 70 d.

Researchers have shown that the decline in leaf photosynthetic activity with leaf age is related to a number of factors, including a reduction in stomatal conductance (Holmgren et al., 1965; Ludlow and Wilson, 1971), decreasing light intensity in the lower regions of the canopy as the crop develops (Constable and Rawson, 1980a), nitrogen status (Wullschleger and Oosterhuis, 1990b), water stress, nutrient deficiency, plant growth hormones, and the transport of material out of the leaf to the developing bolls (Cornish, 1988). Constable and Oosterhuis (2010) concluded from a review of the literature, that "this balance in development sequence makes the most sense as there is little evolutionary advantage in a leaf making investment in protein resources in a leaf with low potential/capacity for photosynthesis due to shading." In relation to leaf development and photosynthesis, it is worth noting that there is a lack of synchronization between carbon production and utilization by bolls both at the node and whole plant level, which necessitates considerable assimilate movement within the plant (Constable and Rawson, 1980a; Wullschleger and Oosterhuis, 1990d).

As the cotton crop grows the number of leaves in the canopy increases. At the same time the average leaf age increases such that the developing boll load is being fed by a canopy made up of mostly older leaves past their photosynthetic prime (Wullschleger and Oosterhuis, 1992). Remembering that the activity of an individual leaf changes from a "sink" (0–2 wk), to a strong "source" (3–4 wk), and finally to a declining "sink" (5–10 wk). This means that average leaf photosynthetic activity in the crop changes with age such that at 60 d after planting (late squaring) 39% of the canopy leaves are in their optimum activity (strong sink) phase. This is in contrast with 21% at 90 d (mid-bloom stage) and only 11% at 120 d (boll fill period). When peak leaf area occurs at about 100 d after planting 75% of the canopy leaf area is older than 29 d with an average canopy leaf age of 34 d. At 120 d most of the leaves are old and not very active. Hence the boll load is really being filled by a canopy consisting of mostly older leaves with a rapidly declining photosynthetic capacity. The potential growth activity of a crop is obviously reflected in the sum of the growth activity of the individual leaves.

During the early season canopy photosynthetic activity of a crop parallels the development of leaf area. Crop photosynthetic capabilities are low during the early part of the season because of the slow growth of the crop and the relatively small amount of leaf area. Canopy photosynthesis increases rapidly during the squaring stage, as the leaf area expands greatly, and peaks a few weeks after flowering (Wells et al., 1986; Peng and Krieg, 1991; Bourland et al., 1992). This decline in canopy photosynthesis has been related to the indeterminate nature of cotton, which results in increased shading of lower leaves (Mauney, 1986) and also the increasing average leaf age in the canopy (Constable and Rawson, 1980a; Wullschleger and Oosterhuis, 1992).

Relationship of Leaf Area Distribution to Boll Development and Yield

The leaf subtending the boll is considered to be the major supplier of carbon to the developing boll (Benedict and Kohel, 1975; Wullschleger and Oosterhuis, 1990d; Ashley, 1972) and there is a strong source–sink association between a sympodial leaf and the boll it subtends. Oosterhuis and Wullschleger (1988) reported a relationship between the size of fully matured sympodial leaves and the weight of the boll they subtended (Fig. 14), suggesting that the larger the sympodial leaf, the larger will be the boll at that position. However, there is a lack of synchrony

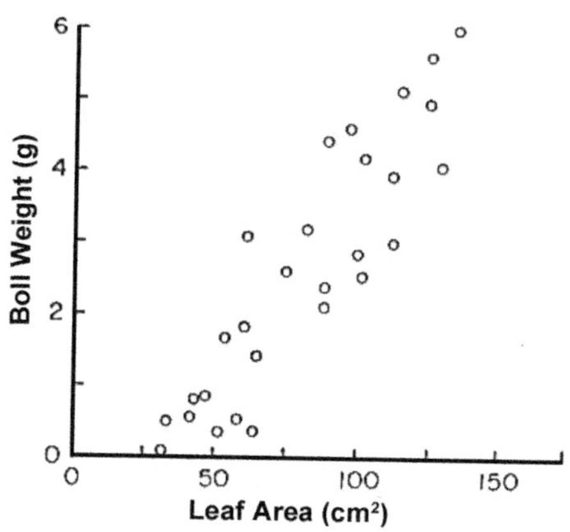

Fig. 14. The relationship between boll weight and area of the subtending leaf (Oosterhuis and Wullschleger, 1988).

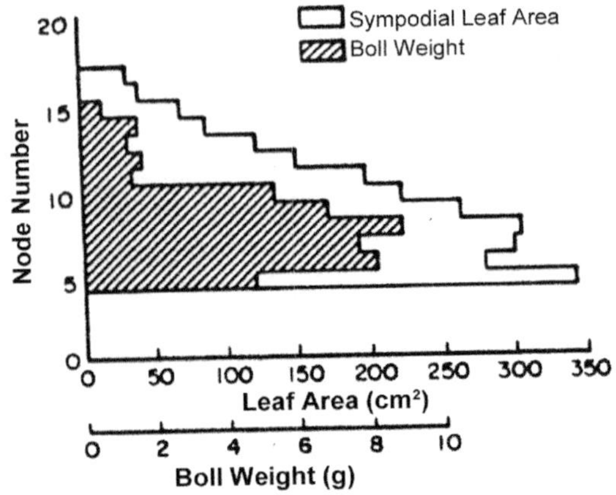

Fig. 15. Spatial distribution of sympodial leaf area and boll dry weight relative to main-stem nodes at 120 d after planting. (Oosterhuis and Wullschleger, 1988).

between the leaf and subtended boll with the leaf being past its maximum photosynthetic output when the boll requirement is increasing exponentially (Constable and Rawson, 1980a; Wullschleger and Oosterhuis, 1990a). The carbon sources for the boll include the leaf subtending the boll, the leaf subtending adjacent fruiting position (in the absence of a developing boll), the main-stem leaf subtending the sympodium (Horrocks et al., 1978; Peoples and Matthews, 1981; Constable and Rawson, 1982), the bracts at the base of the boll (Wullschleger et al., 1990; Morris, 1965), the green capsule wall (Elmore, 1973; Wullschleger and Oosterhuis, 1990c; Constable and Rawson, 1982), and possibly from storage in the stem and root (Fry, 1986; Wells, 2002).

It has been shown that boll size profiles measured in cotton canopies are similar to leaf size profiles (Fig. 15) (Kerby et al., 1987; Constable 1991; Oosterhuis and Wullschleger, 1988), emphasizing the importance of local assimilate supply in determining boll growth rates. The spatial distribution of sympodial leaves and boll weights within the cotton canopy illustrated in Fig. 15 shows that approximately 75% of the total sympodial leaf area was located between main-stem nodes 5 and 10, which coincided with the distribution of 80% of the total yield, expressed as boll dry weight (shaded region in Fig. 15). This is similar to the report of Jenkins et al. (1990) for upland cotton and Kerby et al. (1987) for Acala cotton. The bulk of the yield in commercially grown cotton comes from main-stem nodes 9 to 14 (Jenkins et al., 1990). Generally the largest bolls are found near the bottom of a cotton plant, on about the third fruiting branch and with bolls closest to the main stem on each fruiting branch (Constable, 1991). As mentioned earlier, 80 to 90% of the total yield is produced at fruiting positions one and two on sympodial branches (Mauney, 1986; Jenkins et al., 1990).

Single Leaf Photosynthesis

As stated by Gardner et al. (1985), agriculture is a system that exploits solar energy through photosynthesis. Although there are production scenarios in which cotton yields are sink-limited, productivity is also strongly influenced by total source strength, which is equal to the product of effective leaf area and the average photosynthetic efficiency of all leaves within the crop canopy (Gardner et al., 1985; Krieg and Sung, 1986). Development of photosynthetic area (canopy) and factors affecting canopy development were discussed in the previous section. Regardless of photosynthetic pathway utilized by a given crop species, photosynthesis utilizes the energy derived from solar radiation to convert H_2O and CO_2 into carbohydrate that can be utilized for crop growth and maintenance (Taiz and Zeiger, 2010), and the photosynthetic pathway employed will govern carbon fixation responses to management and environmental factors. It is well established that cotton is a C_3 plant (Baker and Baker, 2010; Benedict, 1984; Benedict et al., 1972; Cothren, 1999; Hesketh, 1967; Krizek, 1986; Whelan et al., 1970). Because the process of photosynthetic carbon fixation by C_3 plants has been described extensively elsewhere (Jensen 1986; Cothren, 1999; Hopkins and Hüner, 2004; Taiz and Zeiger, 2010), it is not the aim of the authors to provide an extensive overview of the photosynthetic process. Rather, our aim is to review recent advances in our understanding of how key environmental factors impact net carbon assimilation (A_N) in cotton and to provide, where possible, an overview of the underlying physiological mechanisms affecting cotton's photosynthetic response to environment.

Light

Because solar radiation, derived from hydrogen fusion reactions in the sun's core, ultimately serves as the energy source that drives photosynthesis (Hopkins and Hüner, 2004), the impact of light on net carbon assimilation (A_N) will be considered first. Not surprisingly, the amount of photosynthetically active radiation (PAR) at a leaf surface strongly impacts A_N. Incident PAR levels depend on the solar constant (1360 W m^{-2} at the outer edge of Earth's atmosphere), atmospheric transmission, and the angle at which the sun's rays strike the leaf surface (Fitter and Hay, 2002). The latter can be influenced by a number of factors, including the time of year, time of day, latitude at which the crop is grown, leaf position within a plant canopy, and leaf movements. Leaves of *G. hirsutum* have been shown to exhibit strong diaheliotropism, where solar tracking allows leaves to be nearly perpendicular to the sun's rays throughout much of the day (Constable, 1986; Ehleringer and Hammond, 1987; Lang, 1973; Sassenrath-Cole, 1995; Thanisawanyangkura et al., 1997; Zhang et al., 2009; Wells 2011). Although it could be expected that solar tracking would provide a photosynthetic advantage throughout the day over leaves that do not exhibit solar tracking, *G. hirsutum* leaves that are allowed to track the sun have slightly higher A_N than horizontally restrained leaves only when PAR levels are very low (early morning or late afternoon) (Ehleringer and Hammond, 1987; Zhang et al., 2009). In contrast with the leaf movements observed for *G. hirsutum*, *G. barbadense* leaves remain in a fixed position throughout the day (Ehleringer and Hammond, 1987; Sassenrath-Cole, 1995) and have extensive leaf cupping, which results in lower cumulative diurnal insolation at the adaxial surface relative to *G. hirsutum*. However, the leaf morphology described for *G. barbadense* also allows for greater light penetration into the crop canopy and may promote whole-canopy photosynthesis (Sassenrath-Cole, 1995; Wise et al., 2000), although this has not been demonstrated.

The pattern of net photosynthesis in response to increasing PAR is referred to as the light response curve and is characteristic of a given plant species and growth environment; light response curves have been evaluated extensively for cotton by a number of authors (reviewed in Cothren, 1999; Constable and Oosterhuis, 2010; Wells, 2011). For example, Constable and Rawson (1980a) observed a light saturation point for single-leaf net photosynthesis in *G. hirsutum* of ~1100 μmol m^{-2} s^{-1} PAR and a light compensation point of ~10 μmol m^{-2} s^{-1}. Similarly, other authors have since indicated that the light saturation point for *G. hirsutum* is at or near 1000 μmol m^{-2} s^{-1} or 50% of full sunlight (Krizek, 1986; Ehleringer and Hammond, 1987; Constable and Oosterhuis, 2010; Wells, 2011). In contrast, other authors have demonstrated that net photosynthesis in *G. hirsutum* does not reach light saturation even at light intensities equal to full sun (2000 μmol m^{-2} s^{-1} PAR) (Sassenrath-Cole et al., 1996; Wise et al., 2000; Zhang et al., 2009; 2011). Although these findings appear somewhat conflicting at face value, it is important to note that *G. hirsutum* exhibits exceptional ability to acclimate to the prevailing light environment during plant development, leading to different light response curves for cotton plants grown in different environments. For example, Patterson et al. (1977) demonstrated that *G. hirsutum* grown in controlled environments under peak irradiance of 600–700 μmol m^{-2} s^{-1} had a lower light saturation point (750 μmol m^{-2} s^{-1}) than plants grown in the field (light saturation point = 1500 μmol m^{-2} s^{-1}) under peak irradiances of ≥ 2000 μmol m^{-2} s^{-1}. In the same study, field-grown plants had higher light-saturated photosynthetic rates, leaf thickness, and chlorophyll content per unit leaf area than chamber-grown plants. In a comparison of photosynthetic responses and leaf anatomy in *G hirsutum* and *G barbadense*, Wise et al. (2000) illustrated that *G. barbadense* leaves were broader, thinner, and had lower light-saturated photosynthetic rates and lower light saturation points than *G. hirsutum* (Fig. 16). In addition to light environment and genetic differences in light response, other authors have demonstrated that leaf senescence (Constable and Rawson, 1980a; Sassenrath-Cole et al., 1996) and abiotic stresses such as drought (Constable and Rawson, 1980a; Zhang et al., 2011) that decrease the light saturated photosynthetic rate can decrease the light saturation point. In general, the light saturation point for net photosynthesis is closely linked to the light saturated rate of photosynthesis. Based on a review of the aforementioned studies, it appears that leaves capable of achieving light saturated photosynthetic rates greater than 30 μmol CO_2 m^{-2} s^{-1} may require light intensities greater than or equal to full sun to attain saturation (Sassenrath-Cole et al., 1996; Wise et al., 2000; Zhang et al., 2009; 2011), whereas light saturated photosynthetic rates lower than

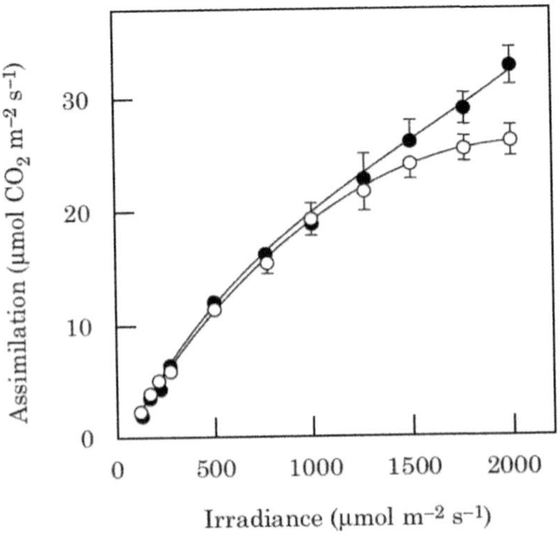

Fig. 16. Light saturation curves of photosynthesis for *G. hirsutum* (filled circles) and *G. barbadense* (open circles) leaves. Mean ± SD, *n* = 5 for each datum point (from Wise et al., 2000).

30 μmol CO_2 m^{-2} s^{-1} require light intensities near 50% of full sun to reach saturation (Constable and Rawson, 1980a; Ehleringer and Hammond, 1987; Constable and Oosterhuis, 2010).

Carbon Dioxide

Plant species differ substantially in their photosynthetic response to atmospheric CO_2 concentration, where A_N in most C_3 species responds positively to increasing CO_2 above current atmospheric levels. Even between species utilizing C_3 photosynthesis, response to elevated CO_2 can be substantially different (Mauney et al., 1978). Cotton (*G. hirsutum*) is highly responsive to increases in atmospheric CO_2, such that El-Sharkawy and Hesketh (1965) reported increases of 56 and 111% for cotton grown at 1600 and 4750 mg kg^{-1} CO_2 when compared with cotton grown at 300 mg kg^{-1}. Furthermore, Mauney et al. (1978) illustrated that cotton grown at 630 mg kg^{-1} CO_2 increased A_N 13 to 15% above plants grown under 330 mg kg^{-1} with nearly doubled lint yields (Mauney et al., 1978). A number of subsequent studies have also demonstrated that cotton increases A_N and aboveground productivity when grown under high CO_2 conditions (Sasek et al., 1985; Radin et al., 1987; Thomas and Strain, 1991; Harley et al., 1992; Mauney et al., 1994; Zhao et al., 2004; Reddy and Zhao, 2005). Furthermore, regardless of the objective of a given experiment, numerous investigators have used A_N responses to short-term increases in CO_2 for individual leaves to characterize photosynthetic characteristics in cotton, and in the absence of other confounding stresses, all have demonstrated increases in A_N when cotton leaves are exposed to short-term increases in CO_2 concentration higher than current ambient levels (Krizek, 1986; Thomas and Strain 1991; Harley et al., 1992; Wise et al., 2004; Ennahli and Earl, 2005; Zhao et al., 2004). All these data taken together indicate that increases in atmospheric CO_2, similar to those projected to result near the end of the 21st century (IPCC, 2013) should greatly increase net carbon assimilation rates and productivity in cotton. However, it is important to note that environmental stressors, such as mineral deficiency, severe drought, and high temperature, all of which can occur concurrently due to the importance of water in maintaining cool foliage temperatures and in moving nutrients throughout the plant, have been shown to abolish any positive effects of CO_2 enrichment on net photosynthesis (Wise et al., 2004; Reddy and Zhao, 2005; Ennahli and Earl, 2005).

High Temperature

Cotton (Pima or Upland) exhibits a broad optimal temperature range from 25 to 35°C for net photosynthesis (Law and Crafts-Brandner, 1999; Crafts-Brandner and Salvucci, 2000; Salvucci and Crafts-Brandner, 2004; Wise et al., 2004; Bibi et al., 2008a), where temperatures in excess of 35°C have consistently been shown to negatively impact net photosynthesis in most cotton genotypes (Law and Crafts-Brandner, 1999; Crafts-Brandner and Salvucci, 2000; Salvucci and Crafts-Brandner, 2004; Schrader et al., 2004; Wise et al., 2004; Bibi et al., 2008a; Snider et al., 2009, 2010; Conaty et al., 2012). Importantly, the optimal temperature range for net photosynthesis is below temperatures often experienced under field conditions in many cotton producing regions throughout the world (Oosterhuis and Snider, 2011; Snider and Oosterhuis, 2012). Three physiological mechanisms could potentially contribute to decreased net photosynthetic rates under high temperature conditions: (i) increased photorespiration rates, (ii) higher dark respiration rates, and

(iii) direct inhibition of photosynthesis. For example, because cotton is a C_3 plant, it is susceptible to increased photorespiratory carbon losses under high temperature conditions due to decreased affinity of Rubisco for CO_2 and low CO_2/O_2 ratios inside the leaf (Salvucci and Crafts-Brandner, 2004; Taiz and Zeiger, 2010). Perry et al. (1983) observed a linear increase in photorespiration from 15 to 50% of gross photosynthesis in *G. hirsutum* as air temperatures increased from 22 to 40°C, and the positive effect of high temperature on photorespiration in cotton has been reiterated by other authors since (Krieg, 1986; Singh et al., 2007). Dark respiration also increases predictably as leaf temperatures increase, where the Q_{10} for dark respiration in cotton is 1.86. Only at leaf temperatures $\geq$ 42.5°C does dark respiration begin to decline (Law and Crafts-Brandner, 1999; Crafts-Brandner and Salvucci, 2000; Salvucci and Crafts-Brandner, 2004). Although not affecting instantaneous measures of net photosynthesis, it is important to note that nighttime respiration rates also increase with elevated nighttime temperatures, thereby consuming a larger portion of the carbohydrate that was synthesized during daylight hours (Loka and Oosterhuis, 2010). This would be expected to decrease overall net assimilation rates when considered over multi-day time periods.

In addition to the carbon loss mechanisms described above, moderately high temperatures (35–40°C) limit net photosynthesis through direct inhibition of the photosynthetic apparatus (Salvucci and Crafts-Brandner, 2004; Schrader et al., 2004; Wise et al., 2004). Whether moderately high temperature primarily limits photosynthesis by negatively impacting the thylakoid reactions or the carbon fixation reactions has been the focus of numerous investigations and is heavily debated. Photosystem II (PSII) was initially considered the most heat-sensitive component of the photosynthetic apparatus in plants (Berry and Bjorkman, 1980), but a number of authors have since demonstrated that F_v/F_m, maximum quantum yield of PSII as estimated using chlorophyll *a* fluorescence, is unaffected by leaf temperatures lower than 40°C and often does not decline appreciably until leaf temperatures reach or exceed 45°C (Law and Crafts-Brandner, 1999; Crafts-Brandner and Law, 2000; Salvucci and Crafts-Brandner 2004; Snider et al., 2013a). Furthermore, the activity of isolated Rubisco responds positively to assay temperatures as high as 50°C, which excludes Rubisco as the target of direct photosynthetic inhibition under moderately elevated temperature (Crafts-Brandner and Salvucci, 2000). However, Rubisco activase (hereafter simply referred to as activase) is exceptionally heat labile, and a number of authors have shown that the kinetic response of activase to temperature can account entirely for direct photosynthetic inhibition under moderately high temperature in cotton leaves (Feller et al., 1998; Crafts-Brandner and Salvucci, 2000; Salvucci and Crafts-Brandner, 2004). Figure 17 neatly sums up the preceding discussion. In contrast, subsequent studies have indicated that thylakoid membrane perturbations, at sites other than PSII, occur under moderately high temperature, and Schrader et al. (2004) and Wise et al. (2004) provided convincing evidence that reduced photosynthetic electron transport is the primary limitation to photosynthesis under high temperature in cotton. Given the aforementioned evidence, it appears that the thylakoid reactions and the carbon fixation reactions are equally likely targets of heat-induced photosynthetic inhibition under moderately high temperature.

It is also important to note that high temperature can result in excessive production of reactive oxygen species (ROS) in a number of plant species (Foyer and Noctor, 2005). Consequently, increases in the activity of ROS scavenging enzymes

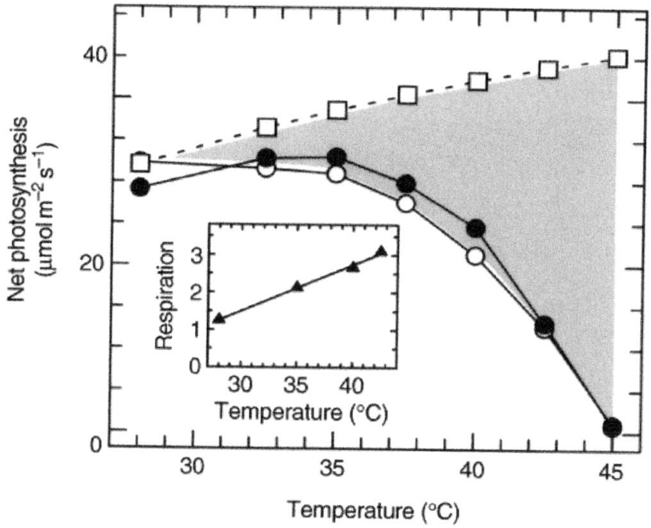

Fig. 17. Response of net photosynthesis and dark respiration to temperature in cotton leaves. Net photosynthesis in air was measured at the indicated temperatures (circles) or predicted from the kinetics of Rubisco with (points) and without (squares) adjustment for the measured changes in activation state (Law and Crafts-Brandner 1999). The gray area indicates the difference between the measured rates and rates predicted without adjust for changes in Rubisco activation. The data for net photosynthesis was from Crafts-Brandner and Salvucci (2000). Calculations of net photosynthesis (Laing et al., 1974) were based on the activity of fully activated Rubisco at 25°C determined after extraction from cotton leaves. These rates were adjusted for temperature based on improved measurements of the temperature response of isolated cotton Rubisco (Fig. 2), the effects of temperature on the other kinetic constants of Rubisco (Jordan and Ogren, 1984) and changes in the solubilities of CO_2 and O_2 with temperature without adjustment for respiration. Inset, dark respiration (filled triangles) was measured at the indicated leaf temperatures. (From Salvucci and Crafts-Brandner, 2004.)

have been observed in different cotton tissues following exposure to high temperature conditions (Bibi et al., 2008b; Snider et al., 2009). Furthermore, Snider et al. (2010) recently demonstrated the importance of ROS scavenging enzymes in promoting innate heat tolerance in cotton. Specifically, the temperature response of actual quantum yield of electron transport (Φ_{PSII}) was evaluated in situ for two cotton cultivars that were both grown under optimal temperature conditions (30/20°C day/night): VH260 (a cotton cultivar grown in Pakistan where midday air temperatures can exceed 45°C) and ST4554 (a widely grown cotton cultivar in the midsouthern region of the United States). VH260 was found to be substantially more heat tolerant than ST4554 (Fig. 18), and heat tolerance was strongly dependent on pre-stress antioxidant enzyme activity.

Low Temperature

Although some plant species are relatively tolerant of low, above freezing temperatures, cotton is well-recognized as a chilling sensitive crop (Wise et al., 1983), and the negative impacts of chilling temperatures on the photosynthetic apparatus

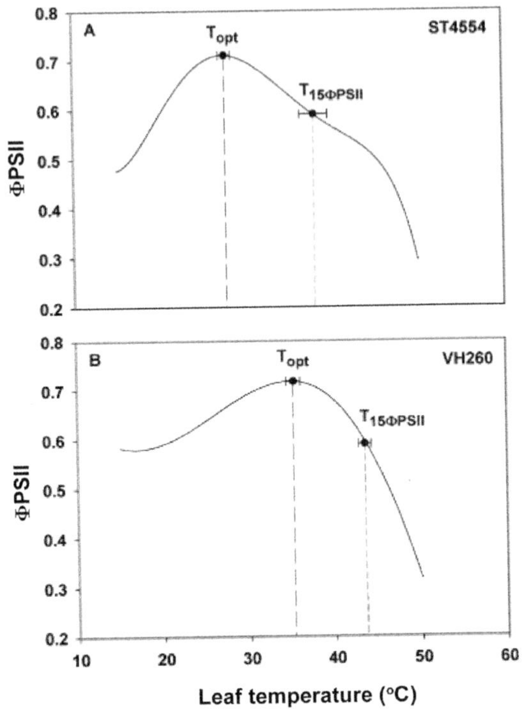

Fig. 18. Effect of temperature on Φ_{PSII} in *G. hirsutum* (A) cultivars ST4554 and (B) VH260. Leaves were illuminated with 500 µmol photons m^{-2} s^{-1} and incubated at temperatures ranging from 15 to 50°C for 5 min at each temperature before Φ_{PSII} determination. In each graph, a representative curve illustrates how T_{opt} (the temperature at which the highest quantum efficiency was obtained for a given leaf) and $T_{15\Phi PSII}$ (the temperature causing a 15% decline in Φ_{PSII} from the value at T_{opt}) were determined for a given cultivar. *G. hirsutum* cultivar ST4554 had a 7.5 and 5.5°C lower (Student's *t* test; $P < 0.05$) mean T_{opt} (27.7°C) and $T_{15\Phi PSII}$ (38°C), respectively, than VH260 (35.2 and 43.5°C, respectively). Horizontal bars = standard error. (From Snider et al., 2010.)

have been well documented for a number of chilling sensitive species, including cotton (Wise et al., 1983; Wise and Naylor, 1987a, 1987b; Warner and Burke 1993; Payton et al., 1997; Kratsch and Wise, 2000; Payton et al., 2001; Kornyeyev et al., 2001, 2003a, 2003b). For example, Wise et al. (1983) reported that leaves of cotton seedlings exposed to chilling temperatures (5°C) became permanently wilted within 24 h and began to exhibit abnormal chloroplast ultrastructure as early as 48 h after exposure to chilling temperatures. Furthermore, a number of authors have demonstrated chilling-induced declines in PSII function (Warner and Burke 1993; Kornyeyev et al., 2001, 2003a, 2003b) and A_N (Payton et al., 1997, 2001) for cotton. The consensus view of the basis for chilling-induced photosynthetic inhibition is that exposure of chilling-sensitive species to cold temperatures under high light intensity (typical of field conditions encountered early in the cotton growing season) more negatively impacts the enzymatic processes of the carbon fixation reactions than the electron transport processes of the thylakoid reactions, thereby favoring the production of reactive oxygen species, which damage the photosynthetic apparatus (reviewed in Kratsch and Wise, 2000). Consequently, efforts have been made to improve chilling tolerance in cotton by producing cotton lines that overexpress chloroplast-localized antioxidant enzymes (Payton et al., 1997, 2001; Kornyeyev et al., 2001, 2003a, 2003b). In general, chilling-induced photosynthetic inhibition can be lessened when the capacity for H_2O_2 scavenging is increased (e.g., through enhanced ascorbate peroxidase activity or glutathione reductase activity) but not when superoxide scavenging capacity (Mn-superoxide dismutase activity) is enhanced (Payton et al., 2001).

Water Deficit

Water deficit is well known to negatively impact net photosynthesis in cotton (Ackerson et al., 1977; McMichael and Hesketh, 1982; Turner et al., 1986; Wise et al., 1992; Pettigrew, 2004; Ennahli and Earl, 2005; Snider et al., 2014b). Whether photosynthesis is primarily limited due to impairment of metabolic processes (nonstomatal limitation) or by limitations to CO_2 diffusion to the carboxylation site (stomatal limitation) is heavily debated and an active area of drought stress research for a number of plant species, including cotton (Medrano et al., 2002; Flexas et al., 2004; Lawlor and Tezara, 2009; Loka et al., 2011). We first consider the impact of drought on the thylakoid reactions. Most of the available literature dealing with water deficit effects on primary photochemical processes in situ have utilized chlorophyll fluorescence methodology. Upon absorbing solar radiation, electrons within the chlorophyll molecule are raised to an excited state, and the energy associated with the excited state can be (i) dissipated as heat via xanthophyll-dependent non-photochemical quenching (NPQ), (ii) utilized in photochemistry, or (iii) emitted as a longer wavelength of light (fluorescence). Importantly, these are competitive processes, and chlorophyll fluorescence can be used to estimate photochemistry and NPQ. An overview of chlorophyll fluorescence theory, methodological considerations, and parameter calculations was provided by Maxwell and Johnson (2000). Importantly, where data are available, maximal quantum yield of PSII (F_v/F_m) is either unaffected or responds positively to water deficit in field-grown cotton (Genty et al., 1987; Pettigrew, 2004; Massacci et al., 2008). Thus, PSII is likely unaffected by mild to moderate drought stress. Furthermore, when assessed during afternoon hours (1200–1500 h) in situ for field-grown cotton, electron transport rates through PSII (ETR) are either unaffected by water deficit or are enhanced by water deficit sufficient to significantly limit net photosynthesis (Chastain et al., 2014; Pettigrew, 2004; Kitao and Lei, 2007; Zhang et al., 2011; Carmo-Silva et al., 2012; Snider et al., 2013b, 2014b), indicating that ETR is also unaffected by mild to moderate water deficit (Fig. 19).

Other proposed nondiffusional limitations to net photosynthesis during drought exposure include inhibition of adenosine triphosphate (ATP) production needed for the carbon fixation reactions and inhibition of Rubisco activity. For example, a number of authors have demonstrated that ATP synthase activity can be rapidly inhibited by water deficit and have proposed that decreased ATP production under water deficit is the primary limitation to photosynthesis (Tezara et al., 1999, 2008; Lawlor and Tezara, 2009); however, the impact of drought on ATP production in field-grown cotton has been largely unexplored. Importantly, a strong, inverse correlation exists between NPQ and ATP synthase activity (Lawlor and Tezara, 2009). That is to say that ATP synthase inhibition coupled with decreased photosynthesis results in the production of a large proton gradient

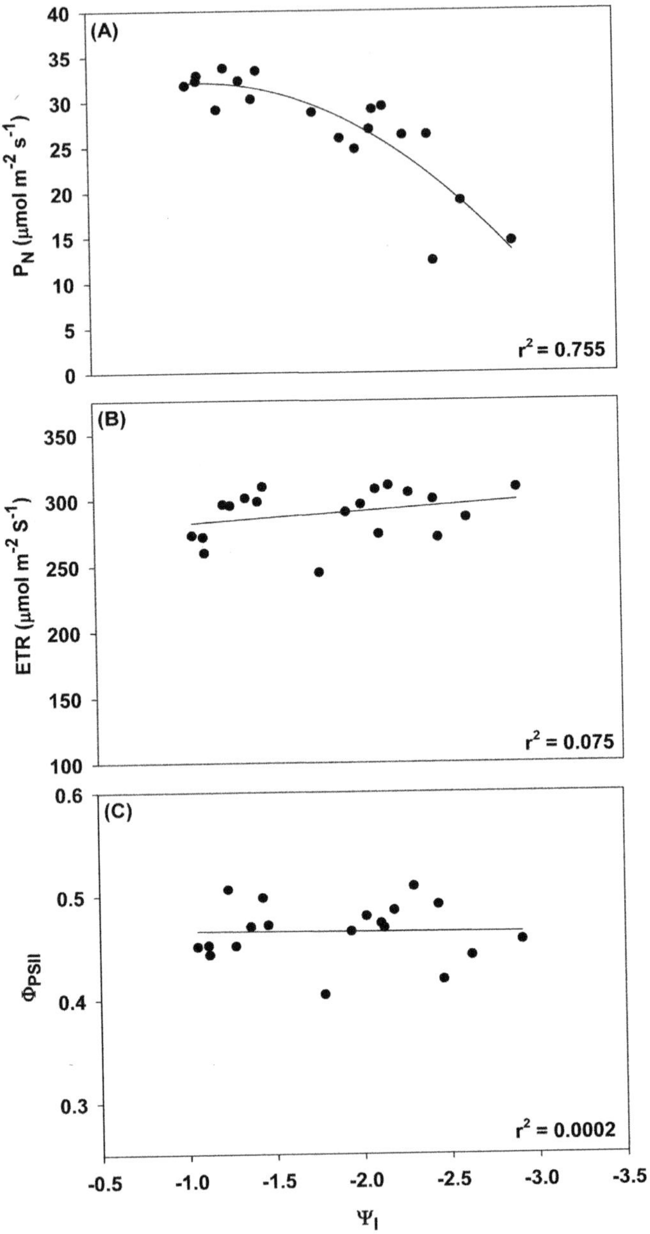

Fig. 19. The relationship between midday (1200–1400 h) leaf water potential (Ψ_l in MPa) and (A) net photosynthesis (P_N), (B) electron transport through photosystem II (ETR), and (C) actual quantum efficiency of photosystem II (Φ_{PSII}). Each data point represents an average of four replicate plots, where three measurements were taken in each replicate plot. The data presented were obtained from two study sites in Georgia on four sample dates from July 9 to July 26. All measurements were conducted on fourth-node, main-stem leaves. (From Snider et al., 2013b.)

between the thylakoid lumen and the stroma, resulting in an increase in xanthophyll-dependent NPQ. Rather than demonstrating an increase in NPQ under drought, studies evaluating NPQ for field-grown cotton exposed to photosynthesis-limiting water deficit have demonstrated decreased NPQ in drought stressed leaves relative to those maintained under well-watered conditions (Pettigrew 2004; Zhang et al., 2011). Although direct measurements of ATP production under drought will certainly be needed to more definitively rule out ATP production as a limitation to photosynthesis in drought-stressed cotton, it can be inferred from the aforementioned studies that photosynthetic ATP production may be unaffected by mild to moderate drought exposure. Regarding the impact of water deficit on Rubisco activity, it is generally accepted that Rubisco activity is not inhibited until water deficit conditions are severe (Bota et al., 2004). Carmo-Silva et al. (2012) demonstrated a decline in Rubisco activity for field-grown cotton exposed to water deficit. However, high leaf temperatures were also observed for drought-stressed plants in their experiment ($T_{leaf} \sim 37–40°C$), and the authors attributed the decline in Rubisco activity under drought to heat-induced inhibition of Rubisco activase.

Based on the evidence provided in the preceding sections, it seems that nondiffusional limitations to photosynthesis are not often observed for field-grown cotton under mild to moderate water-deficit stress. In contrast, an extensive body of knowledge, in cotton and many other plant species, supports the hypothesis that under most drought-stressed conditions, photosynthetic inhibition occurs as the result of decreased CO_2 availability at the carboxylation site of Rubisco due to increases in resistance to CO_2 diffusion from the atmosphere to the stroma (McMichael and Hesketh, 1982; Turner et al., 1986; Wise et al., 1992; Pettigrew, 2004; Ennahli and Earl, 2005; Baker et al., 2007; Snider et al., 2014b). For example, where net photosynthesis and stomatal conductance are measured on cotton leaves exposed to different levels of plant water status, characterized using either predawn or afternoon leaf water potential (Ψ_L), stomatal conductance begins to decline at Ψ_L values that also cause declines in net photosynthesis (McMichael and Hesketh, 1982; Turner et al., 1986; Pettigrew, 2004; Ennahli and Earl, 2005; Kitao and Lei, 2007; Snider et al., 2014b), and A_N has been strongly correlated with g_s in multiple drought stress studies (McMichael and Hesketh, 1982; Baker et al., 2007; Snider et al., 2014b). Moreover, measurements of photosynthetic responses to internal leaf CO_2 concentration (A_N/C_i curves, where C_i is intercellular CO_2 concentration) or CO_2 concentration at the carboxylation (C_c) site of Rubisco (A_N/C_C curves) have been used to assess whether photosynthesis in drought-stressed plants can recover to levels comparable to well-watered plants when C_i is maintained the same for both well watered and drought stressed leaves; this would be evidence for a diffusional limitation to photosynthesis at a given level of stress (Flexas et al., 2002; Ennahli and Earl, 2005). In an elegantly conducted greenhouse study with Upland cotton, Ennahli and Earl (2005) utilized A_N/C_i measurements along A_N/C_C measurements on cotton plants exposed to a range of soil moisture and plant water status to demonstrate that diffusional limitations to photosynthesis dominate at mild to moderate drought stress conditions, but metabolic limitations (nondiffusional limitations) to photosynthesis become apparent only under the most severe stress conditions (Ennahli and Earl, 2005).

In addition to direct photosynthetic inhibition via the diffusional and nondiffusional limitations discussed above, net photosynthesis could potentially be limited through increased respiratory and photorespiratory carbon losses. Dark respiration (R_D) has been found to either increase or decrease under water deficit depending on the severity of stress (Loka et al., 2011). Flexas et al. (2005) proposed a biphasic response to drought stress, where (i) R_D initially decreases as mild drought stress develops then, under more severe stress, R_D increases. Pallas et al. (1967) observed a similar response of R_D to drought progression in greenhouse-grown cotton. However, other investigators have demonstrated an overall decrease in R_D under water deficit (Massacci et al., 2008; Zhang et al., 2011). Although a number of conflicting reports exist on R_D responses to drought when data are expressed on a leaf area basis, Atkin and Macherel (2009) conducted a thorough review on respiratory responses to drought and suggested that the ratio of R_D to gross photosynthesis (A_G) increases under drought stress in nearly all instances, regardless of plant species or severity of stress, and this response was recently observed for field-grown cotton (Chastain et al., 2014). Additionally, the oxygenation reaction of Rubisco and associated photorespiration (R_l) have been shown to increase (Cornic and Fresneau, 2002; Ennahli and Earl, 2005; Massacci et al., 2008) under water deficit although some studies with growth chamber and field-grown cotton have indicated that R_l can decrease (Perry et al., 1983; Zhang et al., 2011) or remain stable under drought in a cultivar-dependent manner (Perry et al., 1983).

Mineral Nutrition

Many of the nutrient deficiencies discussed in mineral nutrition and plant physiology texts limit plant growth and development by directly or indirectly inhibiting photosynthesis (Marschner, 1995; Taiz and Zeiger, 2010). For the sake of brevity, the impacts of N, P, and K on net photosynthetic rates in cotton are addressed here. The importance of N in photosynthesis cannot be overstated because N is a major component of chlorophyll and of the proteins that function as the machinery of the photosynthetic apparatus; furthermore, the N containing compounds, ATP and nicotinamide adenine dinucleotide phosphate, serve as substrates for the carbon fixation reactions (Taiz and Zeiger, 2010). Using a hydroponic system, Longstreth and Nobel (1980) observed significant declines in cotton net assimilation rates, leaf chlorophyll concentration, and mesophyll thickness under N deficiency. Radin and Ackerson (1981) similarly observed declines in net assimilation rate for N-deficient cotton leaves under greenhouse conditions, and Milroy and Bange (2003) demonstrated strong, nonlinear relationships between specific leaf N and light-saturated carbon assimilation rates of individual leaves for field-grown cotton. Similar relationships between leaf N content and photosynthetic rate have been documented for Pima cotton (Reddy et al., 1996). Furthermore, N deficiency can also limit leaf area development before a negative impact on net assimilation rate is observed (Pettigrew and Gerik, 2007; Wullschleger and Oosterhuis, 1990b), thereby limiting whole canopy photosynthesis, which can ultimately hinder productivity. Constable and Rawson (1980a) and Wullschleger and Oosterhuis (1990a) described the pattern of a single leaf photosynthesis peaking at 18 to 20 d after leaf unfurling and then declining to leaf senescence at about 70 d, and Bondada and Oosterhuis (1998) showed that the decline in photosynthesis during leaf aging was associated with stomatal closure and removal of N during leaf ontogeny.

Potassium plays a key role in regulating osmotic potential in plant cells and is an important activator for a number of enzymes involved in photosynthesis

(Taiz and Zeiger, 2010; Oosterhuis et al., 2013). Thus, a number of studies have demonstrated significant declines in net photosynthetic rates per unit leaf area under K-deficient conditions (Bednarz et al., 1998; Pervez et al., 2004; Longstreth and Nobel, 1980; Zhao et al., 2001; Reddy and Zhao, 2005). Although K deficiency has been demonstrated to limit net assimilation rate by decreasing stomatal conductance under acute K-deficiency stress (Bednarz et al., 1998), work conducted by Zhao et al. (2001) illustrated that severe and chronic K-deficiency stress (35 d without K added to the nutrient solution) was capable of decreasing A_N to levels 23% of K-sufficient plants and more severely limited chlorophyll content, chloroplast ultrastructure, and carbohydrate translocation out of the leaf than stomatal conductance. Phosphorous is a key element required for photophosphorylation, the light-driven production of ATP during the thylakoid reactions (Plaxton and Carswell, 1999; Taiz and Zeiger, 2010). Therefore, declines in A_N under P deficiency (Barrett and Gifford, 1995; Longstreth and Nobel, 1980) might be expected due to a direct limitation to photosynthesis proper, causing limited RuBP regeneration and decreased triose phosphate production (Hodges and Constable, 2010; Pettigrew and Gerik, 2007). Additionally, P deficiency results in carbohydrate accumulation in cotton leaves, limits carbon export (Radin and Eidenbock, 1986), and may limit ATP production required for phloem loading (Hodges and Constable, 2010).

Reproduction

Cotton, like many other row crops, is grown for its economically important reproductive organs. Specifically, fiber yield is the product of bolls per hectare, seeds per boll, and fiber weight per seed. Therefore, knowledge of the typical pattern of reproductive development in cotton and of factors that impact any of the aforementioned yield components will be essential for researchers and producers alike to make gains in productivity or to improve production efficiency. Because cotton is a flowering plant, reproductive development occurs in three general phases: (i) gametophyte development—this is the stage of reproductive development beginning at meiosis and ending at pollination and for cotton, occurring during squaring; (ii) progamic phase—this stage begins at pollination and ends at fertilization and for cotton is a 12- to 24-h period occurring on the day of flowering; (iii) embryo development—this is the stage of reproduction beginning at zygote formation and ending with the production of a mature seed and, for cotton, seed coat associated fibers (in subsequent sections we simply refer to this stage as boll development). Herein, the authors briefly characterize the typical pattern of development for each reproductive phase and discuss the impact of environmental and management factors on each of the aforementioned reproductive stages.

Squaring

Approximately 4 to 5 wk after planting, the first floral buds, referred to as squares, are visible near the top of the plant as small pyramidal structures consisting of developing floral parts encased in three triangular bracts. Although the floral bud initially grows slowly, 1 to 2 d before anthesis, a period of rapid expansion ultimately culminates in the opening of the flower approximately 3 wk after squares are first visible. Though squares are only visible during the 3 wk preceding flowering, squares actually begin to differentiate in the apex of the plant approximately 6 wk before anthesis. (This description of square development draws from Beasley, 1975; Oosterhuis, 1990; Oosterhuis and Jernstedt, 1999; and Stewart, 1986.) Between 2 and 3 wk before flowering, micro- and megaspore meiosis occurs, and external stresses that limit meiosis will also limit the amount of viable pollen and functional ovules available at flowering (Mauney, 2012; Snider and Oosterhuis, 2011, 2012; Stewart, 1986). Furthermore, young squares are particularly vulnerable to shedding when exposed to environmental stresses (Cothren, 1999). Thus, in subsequent sections it is our aim to characterize some of the key factors that alter normal reproductive development during the squaring phase of reproductive development.

Water

The available reports on water-deficit effects during squaring indicate that drought stress during this time period primarily decreases the number of fruiting sites initiated (Loka and Oosterhuis, 2012; Krieg, 2000). In addition, young squares (≤1 wk after first becoming visible to the naked eye) are very likely to abscise after experiencing water-deficit conditions (Loka and Oosterhuis, 2012; McMichael and Hesketh, 1982). Thus, of those squares produced, fewer of them are retained under water deficit, either due to hormonal imbalance across the abscission zone or due to decreased carbohydrate availability to developing floral buds (Guinn, 1982). Furthermore, water deficit has also been shown to decrease the individual mass of floral buds when imposed during early squaring (Zhao and Oosterhuis, 1997). Furthermore, excessive irrigation during squaring can also limit yield by accelerating vegetative growth such that fruiting sites at lower positions on the plant are not retained due to excessive shading by leaves at higher positions in the canopy (Munk and Farahani, 2012). The impact of shading on square development (Zhao and Oosterhuis, 1998) is discussed in a subsequent section of this chapter.

Temperature

High temperature during squaring predominantly limits yield by (i) enhancing young square abscission and (ii) negatively impacting meiosis. For example, Guinn (1974) reported that high night temperature (30°C) enhanced square abscission, and Ehlig and LeMert (1973) observed that maximum daily temperatures in excess of 42°C during the growing season limited flower production 3 wk after the high temperature event, likely due to high rates of square abscission (Stewart, 1986). Similarly, Reddy et al. (1992b) observed that *G. hirsutum* plants exposed to maximum daily temperatures of 40°C in a controlled environment for 12 h d^{-1}, beginning at the seedling stage, exhibited 100% square abscission. As a general rule, day/night temperatures in excess of the 30/20°C optimum day/night temperature regime for cotton are well known to limit square retention (Reddy et al., 1992a,b, 1999; Zhao et al., 2005). An examination of the physiological causes of square abscission under high temperature illustrated that heat stress-induced abscission of young squares was closely correlated with low nonstructural carbohydrate contents of the developing floral buds (Zhao et al., 2005).

Because megaspore and microspore meiosis along with development of parental tissues of the gynoecium and androecium occur during the early squaring period, exposure to above-optimal temperatures during this time frame will substantially limit subsequent success during flowering. For example, Meyer (1966, 1969) noted a positive correlation between high temperatures 15 to 16 d before anthesis and percent anther sterility in cotton. Furthermore, Brown (2001) reported that high temperature exposure during this same time frame resulted in a number of floral abnormalities, including short filaments (resulting in the stigmatic surface being perched well above the anthers), anther indehiscence, and flowers that would not open. A number of elegantly conducted studies on other angiosperm species have demonstrated that microspore meiosis is the most sensitive reproductive process to high temperature exposure during gametophyte development, where high temperature exposure during this time frame decreases the amount of pollen produced and the viability of mature pollen (Ahmed et al., 1992; Porch and Jahn, 2001). It also important to note that very few studies have evaluated the impact of high temperature on megagametophyte (embryo sac) development, but a study conducted on wheat (*Triticum aestivum* L.) by Saini et al. (1983) illustrated that high temperature exposure during megaspore meiosis increased the proportion of defective ovules, which would be expected to limit fertilization and seed set.

Light and Carbon Dioxide

We will consider the impact of light and atmospheric CO_2 on square development. Regarding the impacts of light on square development, most of the work conducted on the impacts of shading has been directed at boll development (reported in later sections of this chapter). However, investigations conducted by Zhao and Oosterhuis (1998, 2000) demonstrated that shading (8-d period of 63% reduction in photosynthetically active radiation) significantly reduced net photosynthetic rates, resulting in lower nonstructural carbohydrate contents of 20-d-old squares and lower square dry weight per plant, particularly when shade was induced during the boll development period.

Regarding the impact of CO_2 on square development, the primary impact of elevated atmospheric CO_2 concentration is to increase the number of squares produced and the number of squares retained, relative to ambient CO_2 levels (Yoon et al., 2009; Reddy et al., 1995a) due to higher canopy-level carbon assimilation rates (Reddy et al., 1995b).

Flowering

Approximately 3 wk after squares are first observed, the first cotton flower of the season opens as a white flower at dawn, and pollination occurs between 0700 and 1100 h that same day, initiating the start of the progamic phase (Cothren, 1999; Snider et al., 2011a; Oosterhuis and Snider, 2011; Stewart, 1986; Snider and Oosterhuis, 2011, 2012). Within ~30 min after pollination of a receptive stigma, pollen grains germinate, and the newly elongating pollen tubes begin to penetrate the stigmatic surface. Thereafter, under favorable conditions, pollen tubes can grow through the transmitting tissue of the style at rates between 3 and 3.4 mm h^{-1} (Snider and Oosterhuis, 2011; Snider and Oosterhuis, 2012; Stewart, 1986). The progamic phase concludes between 12 and 24 h after pollen germination. At this time, individual pollen tubes enter the ovule via the micropyle, and one of the two sperm cells housed in the growing pollen tube fuses with the egg cell to produce the zygote, whereas the other sperm nucleus fuses with the polar cell to produce the endosperm nucleus in the process of double fertilization, which is typical of flowering plants (Oosterhuis and Jernstedt, 1999; Stewart, 1986). Although the timing of events during the progamic phase is somewhat predictable in cotton, it is important to note that environmental factors (e.g., humidity and temperature) will strongly influence the timing and rate at which the aforementioned processes occur, as well as how successfully fertilization occurs (Burke, 2011; Burke and Wanjura, 2010; Oosterhuis and Snider, 2011; Snider and Oosterhuis, 2011, 2012; Snider et al., 2009, 2011a,b,c; Stewart, 1986). Although flowers on the day of anthesis are not susceptible to abscission, limitations to fertilization during this time frame will ultimately limit seed set, and if fertilization is severely inhibited, young bolls are very likely to abscise (Stewart, 1986). Thus, any factor that substantially limits fertilization would be expected to limit final yield.

Temperature

Although cold temperatures are a common concern early in the growing season and could potentially limit reproductive success during the progamic phase if they occurred at that time, our discussion of temperature effects will be limited to a discussion of heat stress since this is the most likely temperature stress to occur during flowering in most cotton production regions. Because fertilization is required for seed production and yield development in cotton, it is important to understand how high temperature limits reproductive processes during the progamic phase. Assuming proper preanthesis pollen development has occurred, high temperatures on the day of anthesis have been shown to limit pollen germination, pollen tube growth, and ultimately fertilization of the ovules. For example, recent reports have demonstrated that the optimal temperature for pollen germination in cotton is 28 to 37°C, and percent germination declines substantially thereafter (Burke et al., 2004; Kakani et al., 2005). Pollen tube growth responses to temperature have demonstrated even greater sensitivity to high temperature than germination, where pollen tube growth is maximal between 28 and 32°C, declining substantially at higher temperatures (Fig. 20) (Burke et al., 2004; Kakani et al., 2005; Liu et al., 2006; Snider et al., 2011a). Importantly, pollen tube growth in vivo is heterotrophic during the linear growth phase as the pollen tubes traverse the transmitting tissue of the style. That is, pollen tube growth is not only dependent on the nutritive reserves of the mature pollen grain but also on the carbohydrate and mineral nutrient reserves of the style. For field-grown cotton,

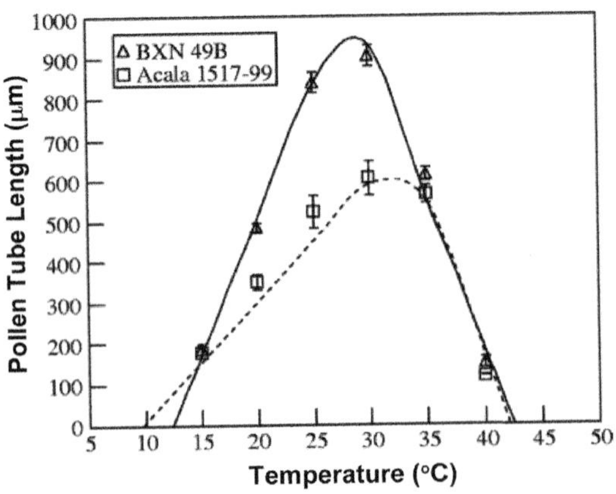

Fig. 20. Pollen tube length responses to temperature for two cotton cultivars (BXN 49B and Acala 1517–99). Error bars indicate ± SE. (From Kakani et al., 2005.)

Snider et al. (2011c) demonstrated a strong relationship between in vivo pollen tube growth rates and soluble carbohydrate content of the pistil during pollen tube growth, where pistils exposed to a maximum diurnal air temperature of 34.6°C had lower pollen tube growth rates and soluble carbohydrate contents than those at 29.9°C (Fig. 21 and 22).

As noted previously, high temperature exposure, typical of that experienced during the summer months in many cotton growing regions of the United States, has been shown to limit fertilization. For example, Barrow (1983) demonstrated that ovule penetration by growing pollen tubes was severely inhibited when pollen was previously exposed to temperatures ≥35°C, and Snider et al. (2009, 2011a) observed significant declines in fertilization efficiency for cotton plants exposed to 38°C (day temperature) relative to plants grown under optimal temperature conditions (30°C) beginning 1 wk before flowering. Concomitant with declines in fertilization efficiency under high temperature, Snider et al. (2009) reported an increase in oxidative stress (evidenced by increases in ROS scavenging enzyme activity), a decline in ATP content, and a decline in soluble carbohydrate content

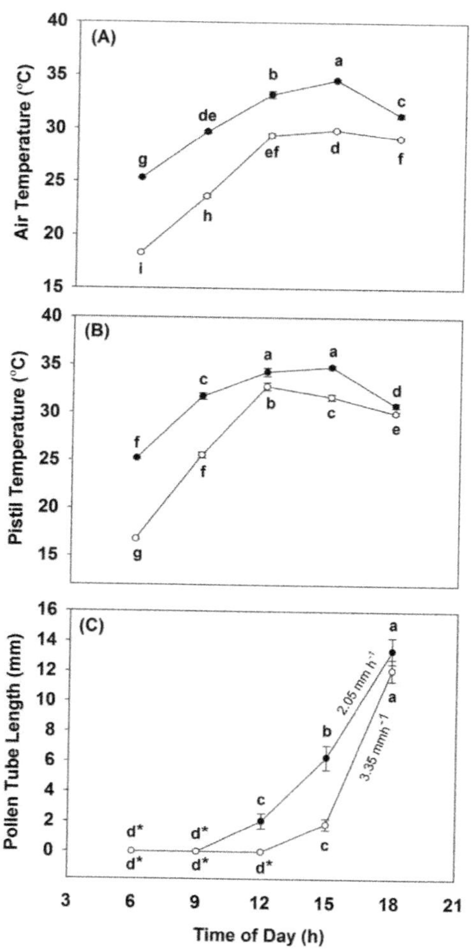

Fig. 21. (A) Diurnal air temperature, (B) pistil temperature, and (C) in vivo pollen tube growth under optimal (T_{max} = 29.9°C; open circles) and high (T_{max} = 34.6°C; closed circles) air temperature conditions from 0600 to 1800 h in 3-h increments. An asterisk next to a data point indicates that no pollen grains were present on the stigmatic surface at that time of day (pollen tube length = 0). All values are means ± standard error (*n* = 6), and values not sharing a common letter are significantly different (LSD; *P* < 0.05). Pollen tube growth rates (mm h^{-1}) under optimal and high temperature conditions are shown adjacent to the corresponding line. (From Snider et al., 2011a, modified by Snider and Oosterhuis, 2011.)

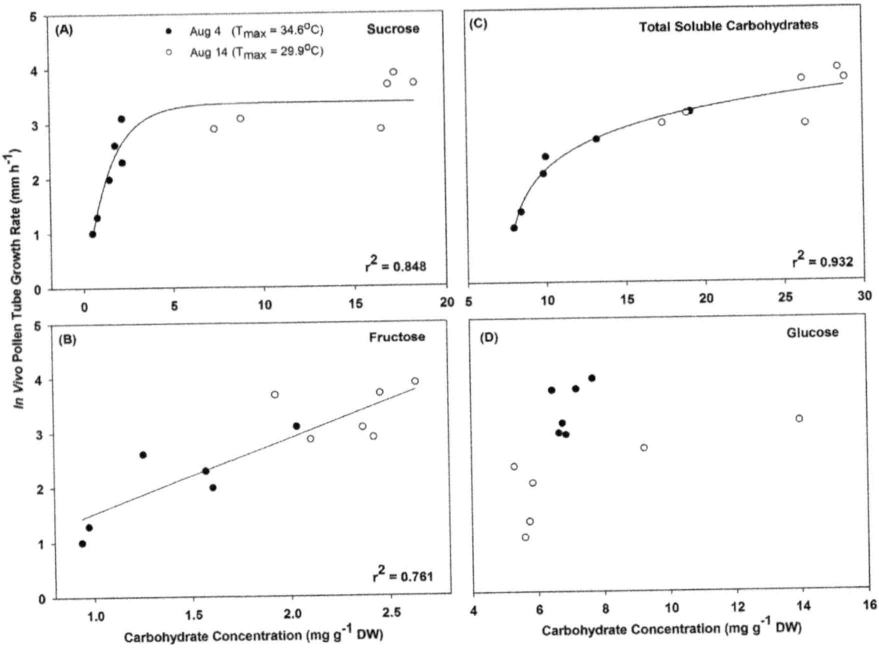

Fig. 22. The relationship between (A) pistil sucrose, (B) fructose, (C) total soluble carbohydrate, and (D) glucose concentrations and pollen tube growth rate in *Gossypium hirsutum* pistils sampled on 4 August (closed circles) and 14 Aug. 2009 (open circles) between 1500 and 1800 h (from Snider et al., 2011c).

of the pistil on the day of anthesis, suggesting that oxidative stress and altered carbohydrate/energy balance in the pistil can negatively impact the fertilization process. A subsequent study (Snider et al., 2011b) comparing the Upland cotton cultivars VH260 and ST4554 (compared for their contrasting heat tolerance as described in a previous section) demonstrated that fertilization efficiency was unaffected under high day temperature conditions (38°C) for VH260, whereas fertilization efficiency declined for ST4554 at 38°C relative to optimal temperature conditions (30°C). Along with greater reproductive thermotolerance, VH260 also had higher pistil antioxidant enzyme activity than ST4554 under optimal temperature conditions but comparable antioxidant enzyme activity under high temperature conditions. These findings indicate that elevated pre-stress levels of antioxidant enzyme activity may promote reproductive thermotolerance by minimizing oxidative stress in the pistil.

Water

Although it is well established that water deficit during flowering can cause significant declines in yield (Loka et al., 2011), the cotton flower on the day of anthesis is widely considered buffered from water deficit stress (Trolinder et al., 1993; Loka and Oosterhuis, 2012). That is, even under extreme water-deficit stress conditions (evidenced by low leaf water potential), cotton flowers are able to expand unhindered and maintain petal, style, and ovary water potentials that are substantially higher than the surrounding leaves (Loka and Oosterhuis, 2012). However, detailed information is limited on the effect of water deficit stress on progamic processes in cotton. One factor that is well documented to have a major impact on reproductive processes during flowering is relative humidity. Under conditions of high relative humidity (100%), pollen grains are susceptible to rupturing (Burke, 2011; Burke et al., 2004). Furthermore, when mature pollen grains come into contact with water, they rupture within 2 min following exposure, and pollen losses and yield reductions due to sprinkler irrigation into open flowers have been observed (Burke, 2003). Under low relative humidity (35%) mature pollen grains will germinate, but soon thereafter, pollen tube growth ceases, producing only very short pollen tubes (Burke, 2011; Burke et al., 2004). It is important to note that cultivar differences exist in humidity sensitivity of mature pollen grains, where cultivars exhibiting rapid water loss and greater sensitivity to dry environments will germinate more readily under high humidity environments than pollen grains that tend to be resistant to water loss (Burke, 2011).

Mineral Nutrition

It has been argued that nutrient imbalances in the flower could limit lint yields and contribute to yield variability, and alterations in floral nutrient concentrations have been reported for cotton in response to N fertilizer application rate, where N concentration increased with increasing N rate while the concentrations of P, K, and B decreased (Oosterhuis et al., 2008; Rochester et al., 2012). However, studies describing how floral nutrient concentrations or imbalances affect reproductive processes during the progamic phase are limited. Studies with a number of plant species, including cotton, have demonstrated a major role for Ca during flowering. For example, Ca promotes pollen germination (Brewbaker and Kwack, 1963); the presence of loosely bound Ca in the transmitting tissue of the style is essential for the elongation of growing pollen tubes (Zhang et al., 1997. In plant species investigated, Ca promotes successful fertilization (Ge et al., 2007). Also, B deficiency during flowering can limit lint yields and negatively impact fiber quality (Miley et al., 1969; Rosolem and Costa, 2000; Zhao and Oosterhuis, 2003; Rosolem and Bogiani, 2011), and soil or foliar applications of B fertilizers are commonly made to field-grown cotton to prevent deficiency. In many other plant species, it is well established that B is a key nutrient involved in pollen germination and pollen tube growth (Blevins and Lukaszewski, 1998), which may partially account for the sensitivity of cotton to B deficiency during flowering.

Boll Development

The cotton boll (the ovary after the progamic phase has occurred) expands rapidly after anthesis, reaching maximum expansion between 3 and 4 wk after flowering; thereafter, boll dry mass continues to increase until 45 to 50 d past anthesis (DPA) (Stewart, 1986). Fiber development is a four-phase process: (i) fiber cell initiation, (ii) fiber elongation, (iii) fiber thickening, and (iv) fiber desiccation and death. The preceding characterization of fiber development is from Haigler (2010), although the fourth phase is sometimes also referred to as maturation. Whether an ovule is fertilized or not, fiber initiation (the bulbous enlargement of epidermal fiber cell initials on the ovule surface) begins near the time of anthesis, although the initiation phase can also begin a few days before or after anthesis (Oosterhuis and Jernstedt, 1999; Stewart, 1986). After the first few DPA, fiber cell elongation has already begun and will continue until approximately 20 DPA (Oosterhuis and Jernstedt, 1999; Lee et al., 2007). Near the end of the fiber elongation phase, 15 to 20 DPA, secondary cell wall thickening begins and will continue for 40 to 45 DPA, at which time the boll opens, and fiber cells desiccate and become twisted (Oosterhuis and Jernstedt, 1999). Concurrent with fiber development, embryogenesis occurs and consists of: (i) morphogenesis, (ii) maturation, and (iii) desiccation. The following general description of events occurring during embryogenesis will draw heavily from Turley and Chapman (2010). Morphogenesis begins with the formation of the zygote and continues until maximum embryonic length is reached at 25 DPA. The maturation period encompasses the period between 20 and 45 DPA and involves the accumulation of nutritive reserves (lipids and proteins) in the cotyledons. Embryo desiccation begins at 35 DPA, thereby overlapping with the maturation phase. Thus, during this period of overlap, seed fresh weight will decrease concomitant with increases in abscisic acid content of the desiccating cotton seed while dry weight increases. Subsequently, abscission of the funiculus at 45 to 47 DPA signals the end of nutrient transport to the embryo, and fully desiccated, quiescent cotton seeds can be obtained at approximately 50 DPA. As will be discussed in subsequent sections, adverse environmental conditions encountered either during early boll development (within the first 2 wk after anthesis) or sometime thereafter can strongly influence boll retention, fiber quality, and final lint yield. It is, therefore, our aim in successive sections to address the current state of knowledge regarding environmental impacts on boll and fiber development.

Water

The cotton boll is considered less sensitive to water deficit than subtending leaves because it is essentially a nontranspiring organ (reviewed in Loka and Oosterhuis, 2012). Furthermore, Van Iersel and Oosterhuis (1995) showed that water potentials of subtending leaves were approximately 0.3 MPa lower than the boll and that xylem transport of water to young fruit was unlikely (due to poorly developed xylem connections), and these authors further demonstrated that most water entered the boll via the phloem. Nonetheless, water deficit does have a pronounced negative effect on boll development. For example, one of the major ways water deficit limits yield in cotton is to limit boll retention (Guinn and Mauney, 1984; Turner et al., 1986; Guinn and Brummett, 1988; Krieg and Sung, 1986; Lokhande and Reddy, 2014; Pettigrew 2004) (Table 2). Young cotton bolls (<14 d post anthesis) are particularly sensitive to drought-induced abscission (McMichael et al., 1973), and a number of authors have demonstrated strong relationships between leaf water potential (predawn or afternoon Ψ_l) and boll retention in both greenhouse and field-grown *G. hirsutum* plants, where afternoon $\Psi_l \leq -1.9$ MPa (Guinn and Mauney 1984) or predawn $\Psi_l \leq -1.0$ MPa (McMichael et al., 1973) severely limited boll retention. Boll abscission due to drought stress is associated with enhanced ethylene production (recognized as the direct hormonal regulator of abscission; Guinn, 1986; Taiz and Zeiger, 2010) along with decreased free indoleacetic acid levels and higher concentrations of free and conjugated ABA (Guinn, 1976; Guinn and Brummett, 1988; Loka and Oosterhuis, 2012). It is important to note that enhanced boll abortion under drought can also be attributed to reduced leaf area and photosynthetic efficiency relative to well-watered

Table 2. Plant biomass (total dry weight, total bolls, open bolls, seedcotton, and seed weight per plant) and boll components (boll, seedcotton, lint, and seed weight per boll) of cotton plants grown at 100, 80, 60, and 40% evapotranspiration (ET) treatments between 44 and 117 d after emergence. Final harvest was carried out at 80% boll opening stage (117 DAE) in all treatments with 20 plants per treatment. (From Lokhande and Reddy 2014a.)

	Water stress treatments, %ET			
Parameters	100	80	60	40
Plant components				
Total biomass, g plant^{-1}	241.4a†	228.4a	180.1b	160.6b
Total bolls, no. plant^{-1}	20.9a	18.8a	13.2b	11.8b
Open bolls, no. plant^{-1}	13.2a	13.0a	11.6ab	10.5bc
Seed cotton weight, g plant^{-1}	63.9a	62.1a	55.9b	51.7b
Seed weight, g plant^{-1}	43.4a	42.2a	37.6b	35.1b
Boll components				
Boll weight, g boll^{-1}	63.1ns	6.46ns	6.44ns	6.42ns
Seed cotton weight, g boll^{-1}	4.69a	4.67a	4.60ab	4.53b
Lint weight, g boll^{-1}	1.43ns	1.52ns	1.51ns	1.50ns
Seed weight, g boll^{-1}	3.22a	3.15a	3.09a	3.03b

† Values in each row followed by the same letter are not significantly different ($P < 0.05$) according to Fisher's LSD. ns, not significant.

conditions, which results in decreased total source strength and an inability to support a large number of bolls per plant (Krieg and Sung, 1986).

The impacts of water deficit on individual boll development have been well documented. For example, Grimes and Yamada (1982) defined afternoon Ψ_l thresholds for fiber growth parameters and lint yield. Fiber length and weight were shown to be negatively affected by mid-day leaf water potentials below −2.8 MPa, but fiber growth proceeded unabated at higher water potentials (Grimes and Yamada, 1982). In the same study, yield was negatively affected when irrigations were withheld until water potential dropped below −1.9 MPa, indicating that yield was more sensitive to water deficit than fiber growth. This is also consistent with Krieg and Sung (1986), who demonstrated that significant declines in retained boll number per plant could be obtained under water deficit conditions without any decline in boll size or carbohydrate translocation to developing bolls. However, alterations in fiber quality parameters have been observed as drought stress increases in severity. For example, Lokhande and Reddy (2014a; Fig. 23) found that fiber length, strength, and uniformity declined linearly as the average Ψ_l for each drought stress treatment declined, whereas a negative linear relationship was observed between micronaire and Ψ_l. This is not surprising because turgor pressure is required for fiber elongation (Dhindsa et al., 1975); therefore, drought during boll development can be expected to decrease fiber length.

Temperature

Daily temperatures in excess of 32°C during flowering and boll production limit lint yield (Balls, 1919; Oosterhuis and Snider, 2011). Depending on the timing, duration, and severity, high temperature can limit yields by decreasing the number of bolls produced per plant as well as by decreasing individual boll size. For Upland cotton, Reddy et al. (1992b) demonstrated that total boll weight per plant was maximized at 47% of total aboveground dry matter under a day/night temperature regime of 30/22°C. When plants were exposed to higher day night/

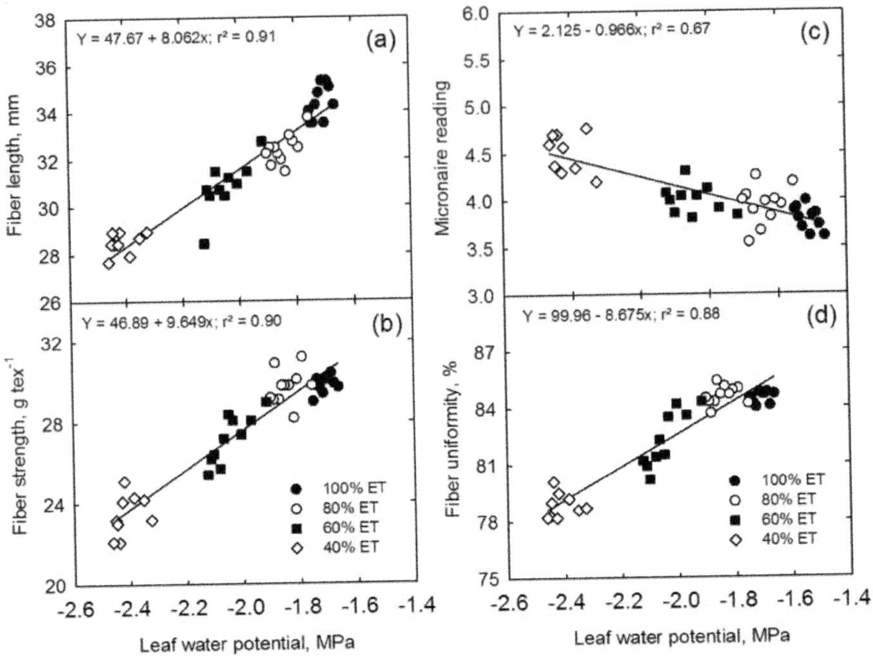

Fig. 23. Water stress effects on cotton (a) fiber length, (b) fiber strength, (c) micronaire reading, and (d) fiber uniformity as a function of midday leaf water potential measured with high volume instrumentation method. The leaf water potentials were averaged from flowering to open bolls. Cotton lint samples were collected at the final harvest performed at 80% of boll opening in each treatment, 117 d after sowing, and 73 d after water stress treatments. (From Lokhande and Reddy 2014a.)

night temperature regimes of 35/25 and 40/30°C, the proportion of aboveground dry matter attributed to bolls was 5.7 and <1%, respectively. This decline in boll mass was primarily attributed to enhanced boll abscission under high temperature as the number of fruiting sites per plant increased with high temperature, but boll retention declined with higher temperature, being nearly abolished at the 40/30°C temperature regime. Similarly, boll retention was shown to decline precipitously in Pima cotton when daytime growth temperatures were in excess of 30°C (Reddy et al., 1992b). Subsequent studies have consistently demonstrated substantial declines in boll retention at high temperatures and have indicated that boll abscission is a major limitation to lint yield at temperatures in excess of the 30°C day temperature optimum (Lokhande and Reddy, 2014b; Reddy et al., 1999; Zhao et al., 2005) (Table 3). Zhao et al. (2005) demonstrated that high temperature-induced abscission of young cotton bolls was strongly and negatively correlated with the total nonstructural carbohydrate content (starches, soluble sugars) of those bolls. Regarding individual boll size, high temperatures (35–40°C) have been shown to limit the final size of the mature boll by decreasing the number of seeds per boll (Pettigrew, 2008a), likely by limiting fertilization efficiency (reviewed in Oosterhuis and Snider, 2011; Snider and Oosterhuis, 2011, 2012). Furthermore, the boll maturation period is shortened with increasing

Table 3. Treatment means for all plant and boll biomass attributes studied; plant parameters (Plant height, main stem nodes, total dry weight, total bolls, open bolls and reproductive potential); boll components (Boll, seedcotton, lint and seed weight per boll). Final harvest was carried out at physiological maturity (20 plants per treatment) (Lokhande and Reddy, 2014b).

Parameters	Mean day/night temperature, °C			
	18.1	21.5	25.5	29.5
Plant components				
Plant height, cm plant^{-1}	205.0c†	227a	222a	212b
Mainstem nodes, no. plant^{-1}	20.1c	21.8b	21.2b	22.6a
Total biomass, g plant^{-1}	265.0a	268a	241ab	217bc
Total bolls, no. plant^{-1}	21.6a	19.8a	20.9a	11.6b
Open bolls, no. plant^{-1}	4.6b	12.8a	13.2a	4.7b
Reproductive potential, g kg^{-1}	390.0	458	496	229
Boll components				
Boll weight, g boll^{-1}	6.50a	6.57a	6.31a	5.02b
Seed cotton weight, g boll^{-1}	4.88a	4.95a	4.69a	3.43b
Lint weight, g boll^{-1}	1.51a	1.55a	1.43	0.86b
Seed weight, g boll^{-1}	3.27a	3.34a	3.22a	2.51b
Seed number, number boll^{-1}	27.3a	28.1a	28.4a	21.2b

† Values in each row followed by the same letter are not significantly different ($P < 0.05$) according to Fisher's LSD.

temperature, which lessens the amount of time dry matter that can accumulate in developing bolls, thereby decreasing boll component weights (seedcotton, lint, and seed weight per boll) and further limiting final boll size (Reddy et al., 1999; Lokhande and Reddy, 2014b).

In studies where high temperature has a quantifiable effect on fiber quality, it has been demonstrated that fiber length is a particularly sensitive fiber quality parameter to above-optimal temperature (Reddy et al., 1999; Lokhande and Reddy, 2014b). Obviously, because the fiber elongation phase occurs during the first 20 d postanthesis (Lee et al., 2007), high temperature exposure that occurs during this time frame will have the greatest impact on fiber length. In a recent study Lokhande and Reddy (2014b) (Fig. 24) demonstrated a quadratic relationship between temperature and fiber length, where temperatures in excess of an average daily temperature of 21.5°C (26/18°C day/night temperature regime) decreased fiber length. Similar relationships were observed for fiber uniformity and micronaire, but these parameters were maximal at an average daily temperature of 25.5°C (30/22°C day/night temperature regime). In the aforementioned study and in other studies evaluating fiber quality in response to high temperature, fiber strength increases as growth temperatures exceed that required to optimize yield (Pettigrew, 2008a; Lokhande and Reddy, 2014b), most likely because the shorter fibers produced under high temperature have greater secondary cell wall thickening, which positively affects fiber strength (Lokhande and Reddy, 2014b). Although we have placed substantial emphasis on high temperatures, cold temperatures during late-season boll development can also have a pronounced effect on fiber development. For example, early fiber elongation can be delayed and final fiber length decreased when night temperatures are less than 19°C (Bradow and Davidonis, 2000; Gipson and Joham, 1968; Gipson and Ray, 1970; Lokhande and Reddy, 2014b). As noted previously, Lokhande and

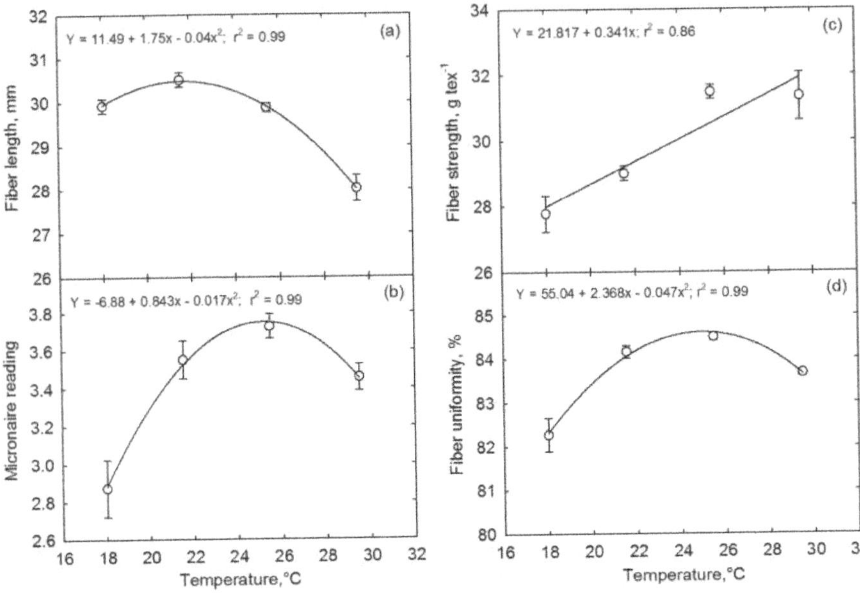

Fig. 24. Temperature effects on cotton (a) fiber length, (b) micronaire reading, (c) fiber strength, and (d) fiber uniformity measured with high volume instrumentation (HVI) method. The temperatures were averaged from flowering to open bolls. Lint samples were collected at final the harvest performed in each treatment. The values are mean ± standard error pooled from all groups within each temperature treatment for each quality parameter. (From Lokhande and Reddy, 2014b.)

Reddy (2014b) reported nonlinear relationships between fiber length, micronaire, and uniformity; these parameters were maximal under 26/18°C (fiber length) and 30/22°C (fiber micronaire and fiber uniformity) day/night temperature regimes. In this same study, a negative, linear relationship between fiber strength and temperature was observed as growth temperatures declined below the highest (34/26°C) day/night temperature regime.

Light and Carbon Dioxide

Light and CO_2 effects on overall boll development and lint yield will be considered together due to their importance as raw energy and carbon sources for photosynthesis. Shading studies with cotton have shown a negative impact of low light on lint yield and fiber quality (Eaton and Ergle, 1954; Zhao and Oosterhuis, 1998; Pettigrew, 1994, 1995, 2001). This is important because extended periods of cloudy, overcast conditions commonly occur in humid areas of the US cotton belt. As expected, low light environments severely limit net photosynthetic rates in cotton (Zhao and Oosterhuis, 1998), thereby limiting carbohydrate availability to developing reproductive units (Eaton and Ergle, 1954; Pettigrew, 2001). For example, an early study by Eaton and Ergle (1954) demonstrated that cotton plants grown under light conditions equal to 32% of full sun produced approximately one-half the seedcotton yield and number of bolls as control plants. Furthermore, these authors demonstrated declines in seedcotton per boll and reduced fiber strength and percent fiber maturity in shaded versus full sun growth conditions. A subsequent study by Pettigrew (1994) demonstrated that shading to 70% of full sunlight reduced lint yield 20% and boll number 21%, further confirming that boll number was the primary yield component to be limited under shaded conditions. Regarding fiber quality, it was demonstrated that fiber strength and micronaire were negatively impacted under the environment. The timing of shading is also important. For example, Zhao and Oosterhuis (2000) reported that an 8-d period of shading (37% of full sun) had no effect on yield or fiber quality when shading was imposed during early squaring; however, when shading was imposed during flowering and boll development, significant declines in boll retention, lint yield, fiber micronaire, and fiber strength were observed. Regarding the mechanism of reduced fiber quality under shade, Pettigrew (2001) noted significant reductions in fiber starch and sucrose concentrations during the fiber development period (14 and 21 DPA) that were associated with lower fiber micronaire and strength at the end of the season, indicating that shading reduces fiber quality by limiting carbohydrate availability to developing cotton fibers.

As noted briefly in previous sections of this chapter, elevated CO_2 concentrations positively impact net photosynthesis in cotton (Mauney et al., 1978), and studies by numerous authors have demonstrated that in the absence of other abiotic stresses, substantial increases in atmospheric CO_2 concentrations positively impact aboveground dry matter accumulation and final lint yield (Mauney et al., 1978, 1994; Reddy et al., 2004; Reddy and Zhao, 2005; Zhao et al., 2003). For example, Mauney et al. (1978) showed that growing cotton under greenhouse conditions at 630 mg kg^{-1} atmospheric CO_2 resulted in nearly double the lint yield when compared with plants grown under 330 mg kg^{-1} CO_2. A subsequent study conducted under a free air CO_2 enrichment environment (FACE) similarly demonstrated yield enhancements for cotton grown under elevated CO_2 conditions. It is now well established high CO_2 does not alter dry matter partitioning to reproductive units (Mauney et al., 1994; Reddy and Zhao, 2005; Zhao et al., 2003). Instead, total boll load per plant is increased under elevated CO_2 conditions primarily by increasing boll number per plant and in some instances, increasing boll size (Mauney et al., 1978, 1994; Reddy et al., 1999, 2004). Studies that include information on fiber growth and quality in response to atmospheric CO_2 have not demonstrated an effect of atmospheric CO_2 level on fiber characteristics (Reddy et al., 1999, 2004).

Mineral Nutrition

By definition, deficiencies in any mineral nutrient will limit productivity in cotton. For brevity, we will focus on N, P, and K nutrition effects on yield, yield components, and fiber quality. The importance of N for overall growth and yield is well-established, with positive lint yield responses to N fertilization being observed by numerous authors (Fritschi et al., 2003; Kawakami et al., 2012; Meredith et al., 1997; Mullins and Burmester, 1990; Read et al., 2006), where Mullins and Burmester (1990) reported that cotton N uptake averaged 19.9 kg N for every 100 kg of lint produced. The positive effect of N fertility on lint yield is due to an overall increase in total aboveground dry matter production rather than an increase in harvest index (Boquet and Breitenbeck, 2000; Mullins and Burmester, 1990). Nitrogen deficiency negatively impacts yield primarily by decreasing the number of bolls per plant and by extension, the number of bolls ha^{-1} (Boquet and Breitenbeck, 2000; Read et al., 2006). Regarding fiber quality, Read et al. (2006) observed declines in fiber length and strength for N-deficit cotton plants relative to adequately fertilized control plants. Fritschi et al. (2003) demonstrated fiber length, strength, and micronaire responses to N fertility rate, but the magnitude and direction of response for each quality parameter was highly dependent on study site and year. Reddy et al. (2004) evaluated the impact of N nutrition and CO_2 growth environment on lint yield and fiber quality and observed strong quadratic relationships between leaf N concentration during the boll maturation period for cotton and number of fiber parameters, including fiber length and diameter (Fig. 25). Importantly, these authors observed no interaction between atmospheric CO_2 and N for any recorded fiber characteristic, indicating that fiber quality could be maintained under elevated CO_2 if N fertility was optimized. As with N fertility studies, positive yield responses to K fertility have been observed by a number of authors (Cassman et al., 1990; Mullins and Burmester, 1990; Pettigrew, 2003, 2008b; Pettigrew et al., 2005), where 15.3 kg K are taken up by the cotton crop for every 100 kg lint weight produced (Mullins and Burmester, 1990). Yield declines due to Kdeficiency are due to declines in total biomass production and declines in the fraction of total biomass accounted for by reproductive dry matter (Reddy and Zhao, 2005). Furthermore, positive lint yield responses to increased K fertility are often associated with an increase in individual boll mass (Read et al., 2006; Pettigrew, 2008b).

Fiber elongation, micronaire, and uniformity have been shown to respond positively to enhanced K fertility relative to fiber from plants receiving 0 kg ha^{-1} K during the growing season (Pettigrew, 2008b). However, fiber quality responses to K are highly variable and dependent on environment (above and below ground) and cultivar (Bradow and Davidonis, 2000). Phosphorus is well recognized as a requirement for obtaining maximal lint yields (Nelson, 1949), where 2.5 kg ha^{-1} P are removed from the soil for every 100 kg lint produced (Mullins and Burmester, 1990). Nelson (1949) reported a positive yield response to 56 kg ha^{-1} P_2O_5 relative to unfertilized plots and observed that the increase in lint yield was also associated with an increase in individual boll size. However, lint yield responses to soil-applied P fertilizer are highly variable and very difficult to predict (Bronson et al., 2001, 2003). Furthermore, fiber quality is not affected by P fertilization in either Upland or Pima cotton (Nelson, 1949; Tewolde and Fernandez, 2003).

Conclusions

While it is difficult to summarize and keep up with the ever-changing and constant volume of physiological research, we have reviewed and summarized the current literature, concentrating on three phases of cotton plant development, seed germination and stand establishment, vegetative development, and reproductive development, with a discussion of factors that influence the physiology of the cotton plant at each phase and cause deviations from normal developmental trends. Specifically, we have discussed how physiological processes and the environment interact at each stage to affect the growth and yield of the cotton plant. It is the authors' hope that this chapter will serve as valuable reference for the experienced and new cotton researcher alike.

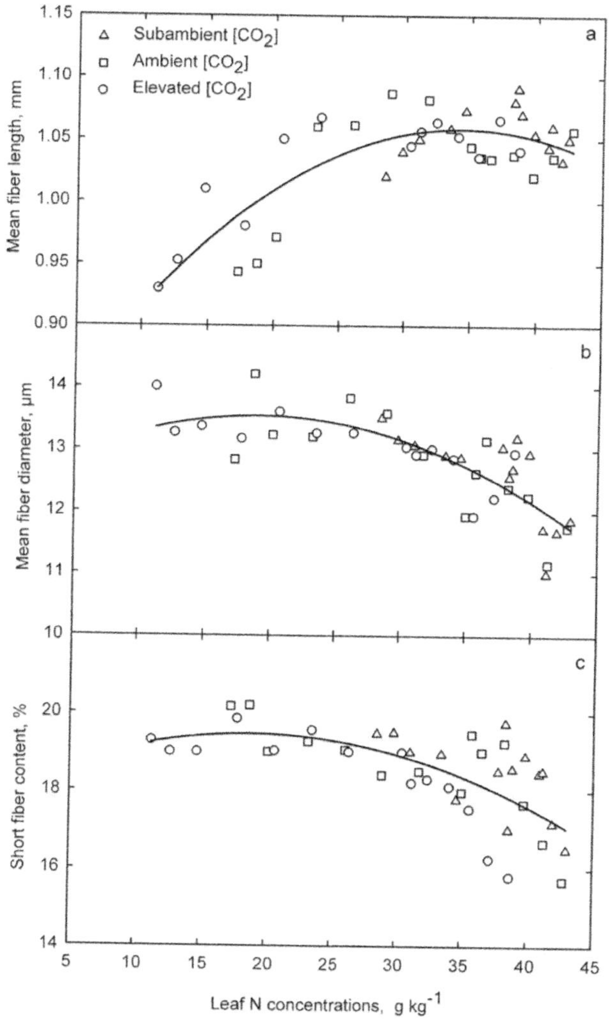

Fig. 25. Fiber length, diameter, and short fiber content as a function of leaf N concentrations (g kg^{-1}) during the boll maturation period in cotton. Observations are from plants that were grown under three levels of carbon dioxide concentrations ([CO_2]), two levels of N, and seven groups of bolls grouped based on flowering dates. Fitted regressions are (a) $y = -0.0003x^2 + 0.0224x + 0.6968$, $r^2 = 0.63$, $n = 42$ for mean fiber length (mm); (b) $y = -0.0031x^2 + 0.118x + 12.394$, $r^2 = 0.61$, $n = 42$ for mean fiber diameter (μ); and(c) $y = -0.005x^2 + 0.189x + 17.647$, $r^2 = 0.50$, $n = 42$ for short fiber content (%). (From Reddy et al., 2004.)

References

Acevedo, E., and T.C. Hsiao. 1974. Plant responses to water deficit, water use efficiency and drought resistance. Agric. Meteorol. 1:458–484.

Ackerson, R.C., and R.R. Herbert. 1981. Osmoregulation in cotton in response to water stress: 1. Alterations in photosynthesis, leaf conductance, translocation, and ultrastructure. Plant Physiol. 67:484–488. doi:10.1104/pp.67.3.484

Ackerson, R.C., D.R. Krieg, C.L. Haring, and N. Chang. 1977. Effects of plant water status on stomatal activity, photosynthesis, and nitrate reductase activity of field-grown cotton. Crop Sci. 17:81–84. doi:10.2135/cropsci1977.0011183X001700010023x

Ahmed, F.E., A.E. Hall, and D.A. DeMason. 1992. Heat injury during floral development in cowpea (*Vigna unguiculata*, Fabaceae). Am. J. Bot. 79:784–791. doi:10.2307/2444945

Arndt, C.H. 1945. Temperature-growth relations of the roots and hypocotyls of cotton seedlings. Plant Physiol. 20:200–219. doi:10.1104/pp.20.2.200

Ashley, D.A. 1972. 14C-labelled photosynthate translocation and utilization in cotton plants. Crop Sci. 12:69–74. doi:10.2135/cropsci1972.0011183X001200010023x

Atkin, O.K., and D. Macherel. 2009. The crucial role of plant mitochondria in orchestrating drought tolerance. Ann. Bot. (Lond.) 103:581–597. doi:10.1093/aob/mcn094

Atwell, B.J., P.E. Kriedemann, and C. Turnbull. 1999. Plants in action: Adaptation in nature, performance in cultivation. Macmillan Education, Australia. South Yarra, VIC, Australia.

Baker, D.N., and J.T. Baker. 2010. Cotton source/sink relationships. In: J.M. Stewart, D.M. Oosterhuis, J.J. Heitholt, and J.R. Mauney, editors, Physiology of cotton. Springer, New York. p. 80–96.

Baker, J.T., D.C. Gitz, P. Payton, D.F. Wanjura, and D.R. Upchurch. 2007. Using leaf gas exchange to quantify drought in cotton irrigated based on canopy temperature measurements. Agron. J. 99:637–644. doi:10.2134/agronj2006.0062

Ball, R.A. 1992. Root dynamics, shoot growth, and solute accumulation in cotton (*Gossypium hirsutum* L.) during water deficit stress and recovery. M.S. thesis. University of Arkansas, Fayetteville.

Ball, R.A., D.M. Oosterhuis, and A. Mauromoustakos. 1994. Growth dynamics of the cotton plant during water stress. Agron. J. 86:788–795. doi:10.2134/agronj1994.000219620 08600050008x

Balls, W.L. 1919. The cotton plant in Egypt. MacMillan and Co., London.

Barrett, D.J., and R.M. Gifford. 1995. Acclimation of photosynthesis and growth by cotton to elevated CO_2: Interactions with severe phosphate deficiency and restricted rooting volume. Aust. J. Plant Physiol. 22:955–963. doi:10.1071/PP9950955

Barrow, J.R. 1983. Comparisons among pollen viability measurement methods in cotton. Crop Sci. 23:734–736. doi:10.2135/cropsci1983.0011183X002300040031x

Bartee, S.N., and D.R. Krieg. 1974. Cottonseed density: Associated physical and chemical properties of 10 cultivars. Agron. J. 66:433–435. doi:10.2134/agronj1974.0002196200660 0030028x

Beasley, C.A. 1975. Developmental morphology of cotton flowers and seed as seen with the scanning electron microscope. Am. J. Bot. 62:584–592. doi:10.2307/2441936

Bednarz, C.W., D.M. Oosterhuis, and R.D. Evans. 1998. Leaf photosynthesis and carbon isotope discrimination of cotton in response to potassium deficiency. Environ. Exp. Bot. 39:131–139. doi:10.1016/S0098-8472(97)00039-7

Benedict, C.R. 1984. Physiology. In: R.J. Kohel and C.F. Lewis, editors, Cotton, Agron. Monogr. 24. ASA, CSSA, and SSSA, Madison, WI. p. 151–200. doi:10.2134/agronmonogr24.c6

Benedict, C.R., and R.J. Kohel. 1975. Export of 14C-assimilates in cotton leaves. Crop Sci. 15:367–372. doi:10.2135/cropsci1975.0011183X001500030026x

Benedict, C.R., K.J. McCree, and R.J. Kohel. 1972. High photosynthetic rate of chlorophyll mutant of cotton. Plant Physiol. 49:968–971. doi:10.1104/pp.49.6.968

Bengough, A.G., M.F. Bransby, J. Hans, S.J. McKenna, T.J. Roberts, and T.A. Valentine. 2006. Root responses to soil physical conditions; Growth dynamics from field to cell. J. Exp. Bot. 57(2):437–447. doi:10.1093/jxb/erj003

Bennie, A.T.P. 1996. Growth and mechanical impedance. In: Y. Waisel, A. Eshel, and U. Kafkafi, editors, Plant roots: The hidden half. Marcel Dekker, New York. p. 453–470.

Berry, J., and O. Bjorkman. 1980. Photosynthetic response and adaptation to temperature in higher plants. Annu. Rev. Plant Physiol. 31:491–543. doi:10.1146/annurev.pp.31.060180.002423

Bewley, J.D. 1997. Seed germination and dormancy. Plant Cell 9:1055–1066. doi:10.1105/tpc.9.7.1055

Bibi, A.C., D.M. Oosterhuis, and E.D. Gonias. 2008a. Photosynthesis, quantum yield of photosystem II and membrane leakage as affected by high temperatures in cotton genotypes. J. Cotton Sci. 12:150–159.

Bibi, A.C., D.M. Oosterhuis, and E.D. Gonias. 2008b. Changes in the antioxidant enzymes activity of cotton genotypes during high temperature stress. Life Sci. Int. J. 2:621–627.

Bland, W.L. 1993. Cotton and soybean root system growth in three soil temperature regimes. Agron. J. 85:906–911. doi:10.2134/agronj1993.00021962008500040023x

Blevins, D.G., and K.M. Lukaszewski. 1998. Boron in plant structure and function. Annu. Rev. Plant Physiol. Plant Mol. Biol. 49:481–500. doi:10.1146/annurev.arplant.49.1.481

Bloodworth, M.E. 1960. Effect of soil temperature on water use by plants. Trans. 7th Intl. Cong. Soil Sci. 1:153–163.

Bolger, T.P., D.P. Upchurch, and B.L. McMichael. 1992. Temperature effects on cotton root hydraulic conductance. Environ. Exp. Bot. 32:49–54. doi:10.1016/0098-8472(92)90029-2

Bondada, B.R., and D.M. Oosterhuis. 1998. Relationship between nitrogen content and net gas exchange components of a cotton leaf during ontogeny. Photosynthetica 35:631–635. doi:10.1023/A:1006999511907

Boquet, D.J., and G.A. Breitenbeck. 2000. Nitrogen rate effect on partitioning of nitrogen and dry matter by cotton. Crop Sci. 40:1685–1693. doi:10.2135/cropsci2000.4061685x

Bota, J., H. Medrano, and J. Flexas. 2004. Is photosynthesis limited by decreased Rubisco activity and RuBP content under progressive water stress? New Phytol. 162:671–681. doi:10.1111/j.1469-8137.2004.01056.x

Bourland, F.M., D.M. Oosterhuis, and N.P. Tugwell. 1992. Concept for monitoring the growth and development of cotton plants using main-stem node counts. J. Prod. Agric. 5:532–538. doi:10.2134/jpa1992.0532

Bradow, J.M., and P.J. Bauer. 2010. Germination and seedling development. In: J.M. Stewart, D.M. Oosterhuis, J.J. Heitholt, and J.R. Mauney, editors, Physiology of cotton. Springer, New York. p. 48–56.

Bradow, J.M., and G.H. Davidonis. 2000. Quantification of fiber quality and the cotton production-processing interface: A physiologist's perspective. J. Cotton Sci. 4:34–64.

Brewbaker, J.L., and B.H. Kwack. 1963. The essential role of calcium ion in pollen germination and pollen tube growth. Am. J. Bot. 50:859–865. doi:10.2307/2439772

Bronson, K.F., J.W. Keeling, J.D. Booker, T.T. Chua, T.A. Wheeler, R.K. Boman, and R. Lascano. 2003. Influence of landscape position, soil series, and phosphorus fertilizer on cotton lint yield. Agron. J. 95:949–957. doi:10.2134/agronj2003.0949

Bronson, K.F., A.B. Onken, J.D. Booker, R.J. Lascano, T.L. Provin, and H.A. Torbert. 2001. Irrigated cotton lint yields as affected by phosphorus fertilizer and landscape position. Commun. Soil Sci. Plant Anal. 32:1959–1967. doi:10.1081/CSS-120000261

Brower, R., and A. Hoagland. 1964. Responses of bean plants to root temperatures. II. Anatomical aspects. Meded. Inst. Biol. Scheik. Onderz. Land. Gewass. 236:23–31.

Brown, K.J. 1968. Translocation of carbohydrate in cotton: Movement to the fruiting bodies. Ann. Bot. (Lond.) 32:703–713.

Brown, P.W. 2001. Heat stress and cotton yields in Arizona. In: J. Silvertooth, editor, Cotton: A College of Agriculture Report for 2001. University of Arizona, Tucson.

Brown, H.B., and J.O. Ware. 1958. Cotton. 3rd ed. McGraw-Hill, New York.

Burke, J.J. 2003. Sprinkler-induced flower losses and yield reductions in cotton (*Gossypium hirsutum* L.). Agron. J. 95:709–714. doi:10.2134/agronj2003.0709

Burke, J.J. 2011. Cotton flowers: Pollen and petal humidity sensitivities determine reproductive competitiveness in diverse environments. In: D.M. Oosterhuis, editor, Stress physiology in cotton. The Cotton Foundation, Cordova, TN. p. 24–36.

Burke, J.J., J.R. Mahan, and J.L. Hatfield. 1988. Crop-specific thermal kinetic windows in relation to wheat and cotton biomass production. Agron. J. 80:553–556. doi:10.2134/agronj1988.00021962008000040001x

Burke, J.J., J. Velten, and M.J. Oliver. 2004. In vitro analysis of cotton pollen germination. Agron. J. 96:359–368. doi:10.2134/agronj2004.0359

Burke, J.J., and D.F. Wanjura. 2010. Plant responses to temperature extremes. In: J.M. Stewart, D.M. Oosterhuis, J.J. Heitholt, and J.R. Mauney, editors, Physiology of cotton. Springer, New York. p. 123–128.

Camp, C.R., P.J. Bauer, and P.G. Hunt. 1997. Subsurface drip irrigation lateral spacing and management for cotton in the southeastern coastal plain. Trans. ASAE 40:993–999. doi:10.13031/2013.21351

Camp, C.R., and Z.F. Lund. 1964. Effect of soil compaction on cotton roots. Crops and Soils, November, p. 8–9.

Cannon, W.A. 1925. Physiological features of roots with special reference to the relation of roots to the aeration of the soil. Publ. 368. Carnegie Institution, Washington, DC.

Cappy, J.J. 1979. The rooting patterns of soybeans and cotton throughout the growing season. Ph.D. diss., Diss. Abstr. 1015. University of Arkansas, Fayetteville.

Carmi, A. 1986. Effect of root zone volume and plant density on the vegetative and reproductive development of cotton. Field Crops Res. 13:25–32. doi:10.1016/0378-4290(86)90004-3

Carmi, A., Z. Plaut, and A. Grava. 1992. Establishment of shallow and restricted root systems in cotton and its impact on plant response to irrigation. Irrig. Sci. 13:87–91. doi:10.1007/BF00193985

Carmi, A., and J. Shalhevet. 1983. Root effects on cotton growth and yield. Crop Sci. 23:875–878. doi:10.2135/cropsci1983.0011183X002300050014x

Carmo-Silva, A.E., M.A. Gore, P. Andrade-Sanchez, A.N. French, D.J. Hunsaker, and M.E. Salvucci. 2012. Decreased CO_2 availability and inactivation of Rubisco limit photosynthesis in cotton plants under heat and drought stress in the field. Environ. Exp. Bot. 83:1–11. doi:10.1016/j.envexpbot.2012.04.001

Cassman, K.G., T.A. Kerby, B.A. Roberts, D.C. Bryant, and S.L. Higashi. 1990. Potassium nutrition effects on lint yield and fiber quality of Acala cotton. Crop Sci. 30:672–677. doi:10.2135/cropsci1990.0011183X003000030039x

Chastain, D.R., J.L. Snider, G.D. Collins, C.D. Perry, J. Whitaker, and S.A. Byrd. 2014. Water deficit in field-grown Gossypium hirsutum primarily limits net photosynthesis by decreasing stomatal conductance, increasing photorespiration, and increasing the ratio of dark respiration to gross photosynthesis. J. Plant Physiol. 171:1576–1585. doi:10.1016/j.jplph.2014.07.014

Christiansen, M.N., and R.A. Rowland. 1986. Germination and stand establishment. In: J.R. Mauney and J.M. Stewart, editors, Cotton physiology. The Cotton Foundation, Memphis, TN. p. 535–542.

Christiansen, M.N. 1963. Influence of chilling upon seedling development of cotton. Plant Physiol. 38:520–522. doi:10.1104/pp.38.5.520

Christianson, J.A., D.J. Llewellyn, E.S. Dennis, and L.W. Wilson. 2010. Global gene expression responses to waterlogging in roots and leaves of cotton (*Gossypium hirsutum* L.). Plant Cell Physiol. 51:21–37. doi:10.1093/pcp/pcp163

Collings, G.H., and J.D. Warner. 1927. Root development of cotton on Cecil sandy loam during 1926. Agron. J. 19:839–842. doi:10.2134/agronj1927.00021962001900090008x

Conaty, W.C., J.J. Burke, J.R. Mahan, J.E. Neilsen, and B.G. Sutton. 2012. Determining the optimum plant temperature of cotton physiology and yield to improve plant-based irrigation scheduling. Crop Sci. 52:1828–1836. doi:10.2135/cropsci2011.11.0581

Constable, G.A. 1986. Growth and light receipt by mainstem cotton leaves in relation to plant density in the field. Agric. For. Meteorol. 37:279–292. doi:10.1016/0168-1923(86)90066-3

Constable, G.A. 1991. Mapping the production and survival of fruit on field-grown cotton. Agron. J. 83:374–378. doi:10.2134/agronj1991.00021962008300020022x

Constable, G.A., and H.M. Rawson. 1980a. Effect of leaf position, expansion and age on photosynthesis, transpiration and water use efficiency of cotton. Aust. J. Plant Physiol. 7:89–100. doi:10.1071/PP9800089

Constable, G.A., and H.M. Rawson. 1980b. Carbon production and utilisation in cotton: Inferences from a carbon budget. Aust. J. Plant Physiol. 7:539–553. doi:10.1071/PP9800539

Constable, G.A., and H.M. Rawson. 1982. Distribution of a 14-C label from cotton leaves: Consequences of changed water and nitrogen status. Aust. J. Plant Physiol. 9:735–747. doi:10.1071/PP9820735

Constable, G.A., and D.M. Oosterhuis. 2010. Temporal dynamics of cotton leaves and canopies. In: J.M. Stewart, D.M. Oosterhuis, J.J. Heitholt, and J.R. Mauney, editors, Physiology of cotton. Springer, New York. p. 72–79.

Cornic, G., and C. Fresneau. 2002. Photosynthetic carbon reduction and carbon oxidation are the main electron sinks for photosystem II activity during a mild drought. Ann. Bot. (Lond.) 89:887–894. doi:10.1093/aob/mcf064

Cornish, K. 1988. Why does photosynthesis decline with leaf age? Proceedings Beltwide Cotton Production Research Conferences. p. 50–55. National Cotton Council of America, Memphis, TN.

Cothren, J.T. 1999. Physiology of the cotton plant. In: C.W. Smith and J.T. Cothren, editors, Cotton: Origin, history, technology, and production. John Wiley and Sons, New York.

Crafts-Brandner, S.J., and R.D. Law. 2000. Effect of heat stress on the inhibition and recovery of the ribulose-1,5-bisphosphate carboxylase/oxygenase activation state. Planta 212:67–74. doi:10.1007/s004250000364

Crafts-Brandner, S.J., and M.E. Salvucci. 2000. Rubisco activase constrains the photosynthetic potential of leaves at high temperature and CO_2. Proc. Natl. Acad. Sci. USA 97:13430–13435. doi:10.1073/pnas.230451497

Croser, C., A.G. Bengough, and J. Pritchard. 1999. The effect of mechanical impedance on root growth in pea (*Pisum sativum*). 1. Rates of cell flux, mitosis, and strain during recovery. Physiol. Plant. 107:277–286. doi:10.1034/j.1399-3054.1999.100304.x

Cutler, J.M., and D.W. Rains. 1977. Effects of irrigation history on responses of cotton to subsequent water stress. Crop Sci. 17:329–335. doi:10.2135/cropsci1977.0011183X001700020025x

Dhindsa, R.S., C.A. Beasley, and I.P. Ting. 1975. Osmoregulation in cotton fiber. Plant Physiol. 56:394–398.

Drew, M.C., and L.H. Stolzy. 1991. Growth under oxygen stress. In: Y. Waisel, A. Eshel, and U. Kafkafi, editors, Plant roots: The hidden half. Marcel Dekker, New York, NY. p. 397–414.

Dowd, M.K., D.L. Boykin, W.R. Meredith, Jr., B.T. Campbell, F.M. Bourland, J.R. Gannaway, K.M. Glass, and J. Zhang. 2010. Fatty acid profiles of cottonseed genotypes from the national cotton variety trials. J. Cotton Sci. 14:64–73.

Eaton, F.M., and D.R. Ergle. 1954. Effects of shade and partial defoliation on carbohydrate levels and the growth, fruiting and fiber properties of cotton plants. Plant Physiol. 29:39–49. doi:10.1104/pp.29.1.39

Ehleringer, J.R., and S.D. Hammond. 1987. Solar tracking and photosynthesis in cotton leaves. Agric. For. Meteorol. 39:25–35. doi:10.1016/0168-1923(87)90013-X

Ehlig, C.F., and R.D. LeMert. 1973 Effects of fruit load, temperature, and relative humidity on boll retention of cotton. Crop Sci. 13:168–171. doi:10.2135/cropsci1973.0011183X001300020006x

Elmore, C.D. 1973. Contributions of the capsule wall and bracts to the developing cotton fruit. Crop Sci. 13:751–752. doi:10.2135/cropsci1973.0011183X001300060049x

El-Sharkawy, M., and J. Hesketh. 1965. Photosynthesis among species in relation to characteristics of leaf anatomy and CO_2 diffusion resistances. Crop Sci. 5:517–521. doi:10.2135/cropsci1965.0011183X000500060010x

Ennahli, S., and H.J. Earl. 2005. Physiological limitations to photosynthetic carbon assimilation in cotton under water stress. Crop Sci. 45:2374–2382. doi:10.2135/cropsci2005.0147

Feller, U., S.J. Crafts-Brandner, and M.E. Salvucci. 1998. Moderately high temperatures inhibit Ribulose-1,5-bisphosphate carboxylase/oxygenase (Rubisco) activase-mediated activation of Rubisco. Plant Physiol. 116:539–546. doi:10.1104/pp.116.2.539

Ferguson, D., and J.L. Turner. 1971. Influence of unfilled cotton seed upon emergence and vigor. Crop Sci. 11:713–715. doi:10.2135/cropsci1971.0011183X001100050033x

Fitter, A., and R. Hay. 2002. Environmental physiology of plants. 3rd ed. Academic Press, San Diego, CA.

Flexas, J., J. Bota, J.M. Escalona, B. Sampol, and H. Medrano. 2002. Effects of drought on photosynthesis in grapevines under field conditions: An evaluation of stomatal and mesophyll limitations. Funct. Plant Biol. 29:461–471. doi:10.1071/PP01119

Flexas, J., J. Bota, F. Loreto, G. Cornic, and T.D. Sharkey. 2004. Diffusive and metabolic limitations to photosynthesis under drought and salinity in C_3 plants. Plant Biol. 6:269–279. doi:10.1055/s-2004-820867

Flexas, J., J. Galmes, M. Ribas-Carbo, and H. Medrano. 2005. The effects of water stress on plant respiration. In: H. Lambers and M. Ribas-Carbo, editors, Plant respiration. Advances in photosynthesis and respiration. Springer, Netherlands. p. 85–94.

Foyer, C.H., and G. Noctor. 2005. Oxidant and antioxidant signaling in plants: A re-evaluation of the concept of oxidative stress in a physiological context. Plant Cell Environ. 28:1056–1071. doi:10.1111/j.1365-3040.2005.01327.x

Fritschi, F., B.A. Roberts, R.L. Travis, D.W. Rains, and R.B. Hutmacher. 2003. Response of irrigated Acala and Pima cotton to nitrogen fertilization. Agron. J. 95:133–146. doi:10.2134/agronj2003.0133

Fry, K.E. 1986. Reserve starch in Pima cotton plants as influenced by boll load. Cotton: A College of Agriculture Report. Series Rep. 370063. University of Arizona, Tucson.

Gardner, F.P., R.B. Pearce, and R.L. Mitchell. 1985. Physiology of crop plants. Iowa State Univ. Press, Ames.

Ge, L.L., H.Q. Tian, and S.D. Russell. 2007. Calcium function and distribution during fertilization in angiosperms. Am. J. Bot. 94:1046–1060. doi:10.3732/ajb.94.6.1046

Genty, B., J.M. Briantais, and J.B.V. Da Silva. 1987. Effects of drought on primary photosynthetic processes of cotton leaves. Plant Physiol. 83:360–364. doi:10.1104/pp.83.2.360

Gerard, C.J., P. Sexton, and G. Shaw. 1982. Physical factors influencing soil strength and root growth. Agron. J. 74:875–879. doi:10.2134/agronj1982.00021962007400050025x

Gipson, J.R., and H.E. Joham. 1968. Influence of night temperature on growth and development of cotton (*Gossypium hirsutum* L.). II. Fiber properties. Agron. J. 60:296–298. doi:10.2134/agronj1968.00021962006000030015x

Gipson, J.R., and L.L. Ray. 1970. Temperature-variety interrelationships in cotton. I. Boll and fibre development. Cotton Growing Rev. 47:257–263.

Glinski, J., and J. Lipiec. 1990. Soil physical conditions and plant roots. CRC Press, Inc. Boca Raton, Fla.

Gonias, D., D.M. Oosterhuis, and A.C. Bibi. 2011. Light interception and radiation use efficiency of okra and normal leaf cotton isolines. J. Agric. Sci. 72:217–222.

Grimes, D.W., R.J. Miller, and P.L. Wiley. 1975. Cotton and corn root development in two field soils of different strength characteristics. Agron. J. 67:519–523. doi:10.2134/agronj1975.00021962006700040016x

Grimes, D., and H. Yamada. 1982. Relation of cotton growth and yield to minimum leaf water potential. Crop Sci. 22:134–139. doi:10.2135/cropsci1982.0011183X002200010031x

Graeber, K., K. Nakabayashi, E. Miatton, G. Leubner-Metzger, and W.J.J. Soppe. 2012. Molecular mechanisms of seed dormancy. Plant Cell Environ. 35:1769–1786. doi:10.1111/j.1365-3040.2012.02542.x

Guinn, G. 1974. Abscission of cotton floral buds and bolls as influenced by factors affecting photosynthesis and respiration. Crop Sci. 14:291–293. doi:10.2135/cropsci1974.0011183X001400020036x

Guinn, G. 1976. Water deficit and ethylene evolution by young cotton bolls. Plant Physiol. 57:403–405. doi:10.1104/pp.57.3.403

Guinn, G. 1982. Causes of square and boll shedding in cotton. USDA Tech. Bull. 1672. US Gov. Print. Office, Washington, DC.

Guinn, G. 1986. Hormonal relations during reproduction. In: J.R. Mauney and J.M. Stewart, editors, Cotton physiology. The Cotton Foundation, Memphis, TN. p. 113–136.

Guinn, G., and D.L. Brummett. 1988. Changes in free and conjugated indole-3-acetic acid and abscisic acid in young cotton fruits and their abscission zones in relation to fruit retention during and after moisture stress. Plant Physiol. 86:28–31. doi:10.1104/pp.86.1.28

Guinn, G., and R.E. Hunter. 1968. Root temperature and carbohydrate status of young cotton plants. Crop Sci. 8:67–70. doi:10.2135/cropsci1968.0011183X000800010020x

Guinn, G., and J.R. Mauney. 1984. Fruiting of cotton. II. Effects of plant moisture status and active boll load on boll retention. Agron. J. 76:94–98. doi:10.2134/agronj1984.00021962007600010024x

Haigler, C.H. 2010. Physiological and anatomical factors determining fiber structure and utility. In: J.M. Stewart, D.M. Oosterhuis, J.J. Heitholt, and J.R. Mauney, editors, Physiology of cotton. Springer, New York. p. 33–47.

Harley, P.C., R.B. Thomas, J.F. Reynolds, and B.R. Strain. 1992. Modelling photosynthesis of cotton grown in elevated CO_2. Plant Cell Environ. 15:271–282. doi:10.1111/j.1365-3040.1992.tb00974.x

Hayward, H.E. 1938. The structure of economic plants. MacMillan, New York.

Hearn, A.B. 1969. The growth and performance of cotton in a desert environment. II. Dry matter production. J. Agric. Sci. 73:75–86. doi:10.1017/S0021859600024151

Heitholt, J.J., and W.R. Meredith, Jr. 1998. Yield, flowering, and leaf area of okra-leaf and normal-leaf cotton isolines. Crop Sci. 38:643–648. doi:10.2135/cropsci1998.0011183X003800030003x

Heitholt, J.J., W.T. Pettigrew, and W.R. Meredith, Jr. 1992. Light interception and lint yield of narrow-row cotton. Crop Sci. 32:728–733. doi:10.2135/cropsci1992.0011183X003200030030x

Hesketh, J. 1967. Enhancement of photosynthetic CO_2 assimilation in the absence of oxygen, as dependent upon species and temperature. Planta 76:371–374. doi:10.1007/BF00387543

Hodges, S.C., and G. Constable. 2010. Plant responses to mineral deficiencies and toxicities. In: J.M. Stewart, D.M. Oosterhuis, J.J. Heitholt, and J.R. Mauney, editors, Physiology of cotton. Springer, New York. p. 72–79.

Holmgren, P., P.G. Jarvis, and M.S. Jarvis. 1965. Resistance to carbon dioxide and water vapour transfer in leaves of different plant species. Physiol. Plant. 18:557–573. doi:10.1111/j.1399-3054.1965.tb06917.x

Hopkins, W.G., and N.P.A. Hüner. 2004. Introduction to plant physiology. 3rd ed. John Wiley and Sons, Hoboken, NJ.

Horn, P.J., P. Neogi, X. Tombokan, S. Ghosh, B.T. Campbell, and K.D. Chapman. 2011. Simultaneous quantification of oil and protein in cottonseed by low-field time-domain nuclear magnetic resonance. J. Am. Oil Chem. Soc. 88:1521–1529. doi:10.1007/s11746-011-1829-5

Horrocks, R.D., T.A. Kerby, and D.R. Buxton. 1978. Carbon source for developing bolls in normal and superokra cotton. New Phytol. 80:335–340. doi:10.1111/j.1469-8137.1978.tb01566.x

Hsiao, T.C. 1973. Plant responses to water stress. Annu. Rev. Plant Physiol. 24:519–570. doi:10.1146/annurev.pp.24.060173.002511

Hubbard, J.W. 1933. Root development of cotton plants in the San Joaquin valley of California. USDA Circular 262. https://archive.org/stream/rootdevelopmento262hubb#page/n1/mode/2up (accessed 19 Jan. 2015).

Huck, M.G. 1970. Variation in taproot elongation rate as influenced by composition of the soil air. Agron. J. 62:815–818. doi:10.2134/agronj1970.00021962006200060042x

IPCC. 2013. Climate change 2013: The physical science basis. Contribution of working group I to the fifth assessment report to the intergovernmental panel on climate change. Cambridge Univ. Press, New York.

Jenkins, J.N., J.C. McCarty, and W.L. Parrott. 1990. Effectiveness of fruiting sites in cotton: Yield. Crop Sci. 30:365–369. doi:10.2135/cropsci1990.0011183X003000020024x

Jensen, R. 1986. The biochemistry of photosynthesis. In: J.R. Mauney and J.M. Stewart, editors, Cotton physiology. The Cotton Foundation, Memphis, TN. p. 157–182.

Jones, J.E. 1982. The present state of the art and science of cotton breeding for leaf-morphological shapes. p. 93-99. In: J.M. Brown, editor, Proc. Beltwide Cotton prod. Conf., Las Vegas, NV. 3–7 Jan. National Cotton Council of America, Memphis, TN.

Jordan, W.R. 1983. Cotton. In: J.D. Teare and M.M. Peet, editors, Crop water relations. John Wiley and Sons, New York, NY. p. 213–254.

Jordan, D.B., and W.L. Ogren. 1984. Tge CO_2/O_2 specificity of ribulose 1,5-bisphosphate carboxylase/oxygenase. Planta 161:308–313.

Kakani, V.G., K.R. Reddy, S. Koti, T.P. Wallace, P.V.V. Prasad, V.R. Reddy, and D. Zhao. 2005. Differences in in vitro pollen germination and pollen tube growth of cotton cultivars in response to high temperature. Ann. Bot. (Lond.) 96:59–67. doi:10.1093/aob/mci149

Kaspar, T.C., and W.L. Bland. 1992. Soil temperature and root growth. Soil Sci. 154:290–299. doi:10.1097/00010694-199210000-00005

Kawakami, E.M., D.M. Oosterhuis, J.L. Snider, and M. Mozaffari. 2012. Physiological and yield responses of field grown cotton to application of urea with the urease inhibitor NBPT and the nitrification inhibitor DCD. Eur. J. Agron. 43:147–154. doi:10.1016/j.eja.2012.06.005

Kays, S.J., C.W. Nicklow, and D.H. Simons. 1974. Ethylene in relation to the response of roots to physical impedance. Plant Soil 40:565–571. doi:10.1007/BF00010513

Kennedy, C.W., M.T. Ba, A.G. Caldwell, R.L. Hutchinson, and J.E. Jones. 1987. Differences in root and shoot growth and soil moisture extraction between cotton cultivars in an acid subsoil. Plant Soil 101:241–246. doi:10.1007/BF02370651

Kerby, T.A., and D.R. Buxton. 1978. Effect of leaf shape and plant population on rate of fruiting position appearance in cotton. Agron. J. 70:535–538. doi:10.2134/agronj1978.00021962007000040004x

Kerby, T.A., M. Keeley, and S. Johnson. 1987. Growth and development of Acala cotton. California Agric. Exp. Stn. Bull. 1921.

Kirkham, M.B. 2011. Elevated carbon dioxide in the soil variable oxygen concentration and root growth. In: Elevated carbon dioxide: Impacts on soil and plant water relations. CRC Press, Boca Raton, FL. p. 69–79.

Kitao, M., and T.T. Lei. 2007. Circumvention of over-excitation of PSII by maintaining electron transport rate in leaves of four cotton genotypes developed under long-term drought. Plant Biol. 9:69–76. doi:10.1055/s-2006-924280

Klepper, B., H.M. Taylor, M.G. Huck, and E.L. Fiscus. 1973. Water relations and growth of cotton in drying soil. Agron. J. 65:307–310. doi:10.2134/agronj1973.00021962006500020036x

Kornyeyev, D., B.A. Logan, P. Payton, R.D. Allen, and A.S. Holaday. 2001. Enhanced photochemical light utilization and decreased chilling-induced photoinhibition of photosystem II in cotton overexpressing genes encoding chloroplast-targeted antioxidant enzymes. Physiol. Plant. 113:323–331. doi:10.1034/j.1399-3054.2001.1130304.x

Kornyeyev, D., B.A. Logan, R.D. Allen, and A.S. Holaday. 2003a. Effect of chloroplastic overproduction of ascorbate peroxidase on photosynthesis and photoprotection in cotton leaves subjected to low temperature photoinhibition. Plant Sci. 165:1033–1041. doi:10.1016/S0168-9452(03)00294-2

Kornyeyev, D., B.A. Logan, P. Payton, R.D. Allen, and A.S. Holaday. 2003b. Elevated chloroplastic glutathione reductase activities decrease chilling-induced photoinhibition by increasing rates of photochemistry, but not thermal energy dissipation, in transgenic cotton. Funct. Plant Biol. 30:101–110. doi:10.1071/FP02144

Kramer, P.J. 1969. Plant and soil relationships: A modern synthesis. McGraw-Hill, New York.

Kratsch, H.A., and R.R. Wise. 2000. The ultrastructure of chilling stress. Plant Cell Environ. 23:337–350. doi:10.1046/j.1365-3040.2000.00560.x

Krieg, D.R. 1986. Feedback control and stress effects on photosynthesis. In: J.R. Mauney and J.M. Stewart, editors, Cotton physiology. The Cotton Foundation, Memphis, TN. p. 227–244.

Krieg, D.R. 2000. Cotton water relations. In: D.M. Oosterhuis, editor, Summaries of cotton research in 2000. Special Rep. 198. Univ. of Arkansas Agric. Exp. Stn., Fayetteville. p. 7–15.

Krieg, D.R., and S.N. Bartee. 1975. Cottonseed density: Associated germination and seedling emergence properties. Agron. J. 67:343–347. doi:10.2134/agronj1975.00021962006700030015x

Krieg, D.R., and J.F.M. Sung. 1986. Source-sink relations as affected by water stress during boll development. In: J.R. Mauney and J.M. Stewart, editors, Cotton physiology. The Cotton Foundation, Memphis, TN. p. 227–244.

Krizek, D.T. 1986. Photosynthesis, dry matter production and growth in CO_2 enriched atmospheres. In: J.R. Mauney and J.M. Stewart, editors, Cotton physiology. The Cotton Foundation, Memphis, TN. p. 193–226.

Kuo, T.M., J.F. VanMiddlesworth, and W.J. Wolf. 1988. Content of raffinose oligosaccharides and sucrose in various plant seeds. J. Agric. Food Chem. 36:32–36. doi:10.1021/jf00079a008

Lang, A.R.G. 1973. Leaf orientation of a cotton plant. Agric. Meteorol. 11:37–51. doi:10.1016/0002-1571(73)90049-6

Laing, W.A., W.L. Ogren, and R.H. Hageman. 1974. Regulation of soybean net photosynthetic CO_2 fixation by the interaction of CO_2, O_2, and ribulose 1,5-diphosphate carboxylase. Plant Physiol. 54:678–685.

Law, R.D., and S.J. Crafts-Brandner. 1999. Inhibition and acclimation of photosynthesis to heat stress is closely correlated with activation of ribulose-1,5-bisphosphate carboxylase/oxygenase. Plant Physiol. 120:173–181. doi:10.1104/pp.120.1.173

Lawlor, D.W., and W. Tezara. 2009. Causes of decreased photosynthetic rate and metabolic capacity in water-deficient leaf cells: A critical evaluation of mechanisms and integration of processes. Ann. Bot. (Lond.) 103:561–579. doi:10.1093/aob/mcn244

Lee, J.J., A.W. Woodward, and Z.J. Chen. 2007. Gene expression changes and early events in cotton fibre development. Ann. Bot. (Lond.) 100:1391–1401. doi:10.1093/aob/mcm232

Leffler, H.R., and R.D. Williams. 1983. Seed density classification influences germination and seedling growth of cotton. Crop Sci. 23:161–165. doi:10.2135/cropsci1983.0011183X002300010044x

Leonard, O.A. 1945. Cotton root development in relation to natural aeration of some Mississippi blackbelt and delta soils. Agron. J. 37:55–71. doi:10.2134/agronj1945.00021962003700010005x

Liang, B.M., R.E. Sharp, and T.I. Baskin. 1997. Regulation of growth anisotropy in well-watered and water-stressed maize roots. I. Spatial distribution of longitudinal, radial and tangential expansion rates. Plant Physiol. 115:101–111.

Liu, Z., Y.L. Yuan, S.Q. Liu, X.N. Yu, and L.Q. Rao. 2006. Screening for high temperature tolerant cotton cultivars by testing in vitro pollen germination, pollen tube growth, and boll retention. J. Integr. Plant Biol. 48:706–714. doi:10.1111/j.1744-7909.2006.00276.x

Liu, Z., Z.G. Zhou, W.Q. Guo, B.L. Chen, and D.M. Oosterhuis. 2008. Effects of N fertilization on root development and activity of water-stressed cotton (*Gossypium hirsutum* L.) plants. Agric. Water Manage. 95:1261–1270. doi:10.1016/j.agwat.2008.05.002

Loka, D.A., and D.M. Oosterhuis. 2010. Effect of high night temperatures on cotton respiration, ATP levels and carbohydrate content. Environ. Exp. Bot. 68:258–263. doi:10.1016/j.envexpbot.2010.01.006

Loka, D.A., and D.M. Oosterhuis. 2012. Water stress and reproductive development in cotton. In: D.M. Oosterhuis and J.T. Cothren, editors, Flowering and fruiting in cotton. The Cotton Foundation, Cordova, TN. p. 51–58.

Loka, D.A., D.M. Oosterhuis, and G.L. Ritchie. 2011. Water-deficit stress in cotton. In: D.M. Oosterhuis, editor, Stress physiology in cotton. The Cotton Foundation, Cordova, TN. p. 37–72.

Lokhande, S., and K.R. Reddy. 2014a. Reproductive and fiber quality responses of upland cotton to moisture deficiency. Agron. J. 106:1060–1069. doi:10.2134/agronj13.0537

Lokhande, S., and K.R. Reddy. 2014b. Quantifying temperature effects on cotton reproductive efficiency and fiber quality. Agron. J. 106:1275–1282. doi:10.2134/agronj13.0531

Longstreth, D.J., and P.S. Nobel. 1980. Effects of nitrogen, phosphorus, and potassium for *Gossypium hirsutum* L. Plant Physiol. 65:541–543. doi:10.1104/pp.65.3.541

Lowry, F.E., H.M. Taylor, and M.G. Huck. 1970. Growth rate and yield of cotton as influenced by depth and bulk density of soil pans. Soil Sci. Soc. Am. J. 34:306–309. doi:10.2136/sssaj1970.03615995003400020034x

Ludlow, M.M., and G.L. Wilson. 1971. Photosynthesis of tropical pasture plants. III. Leaf age. Aust. J. Biol. Sci. 24:1077–1087.

Malik, R.S., J.S. Dhankar, and N.C. Turner. 1979. Influence of soil water deficits on root growth of cotton seedlings. Plant Physiol. 53:109–115.

Marschner, H. 1995. Mineral nutrition of higher plants. 2nd ed. Academic Press, San Diego, CA.

Massacci, A., S.M. Nabiev, L. Pietrosanti, S.K. Nematov, T.N. Chernikova, K. Thor, and J. Leipner. 2008. Plant Physiol. Biochem. 46:189–195. doi:10.1016/j.plaphy.2007.10.006

Mauney, J.R. 1968. Morphology of the cotton plant In: F.C. Elliot, M. Hoover, and W.K. Porter, editors, Cotton: Principles and practices. Iowa State Press, Ames. p. 23–40.

Mauney, J.R. 1986. Vegetative growth and development of fruiting sites. In: J.R. Mauney and J.M. Stewart, editors, Cotton physiology. Cotton Foundation, Memphis, TN. p. 11–28.

Mauney, J.R. 2012. Anatomy and morphology of fruiting forms. In: D.M. Oosterhuis and J.T. Cothren, editors, Flowering and fruiting in cotton. The Cotton Foundation, Cordova, TN. p. 1–12.

Mauney, J.R., K.E. Fry, and G. Guinn. 1978. Relationship of photosynthetic rate to growth and fruiting of cotton, soybean, sorghum, and sunflower. Crop Sci. 18:259–263. doi:10.2135/cropsci1978.0011183X001800020016x

Mauney, J.R., B.A. Kimball, P.J. Pinter, Jr., R.L. LaMorte, K.F. Lewin, J. Nagy, and G.R. Hendrey. 1994. Growth and yield of cotton in response to a free-air carbon dioxide enrichment (FACE) environment. Agric. For. Meteorol. 70:49–67. doi:10.1016/0168-1923(94)90047-7

Maxwell, K., and G.N. Johnson. 2000. Chlorophyll fluorescence—A practical guide. J. Exp. Bot. 51:659–668. doi:10.1093/jexbot/51.345.659

McClelland, C.K. 1916. On the regularity of blooming in the cotton plant. Science 44:578–581. doi:10.1126/science.44.1138.578

McMichael, B.L. 1986. Growth of roots. In: J.R. Mauney and J.McD. Stewart, editors, Cotton physiology. The Cotton Foundation, Memphis, TN. p. 29–38.

McMichael, B.L., and J.J. Burke. 1994. Metabolic activity of cotton roots in response to temperature. Environ. Exp. Bot. 34:201–206. doi:10.1016/0098-8472(94)90039-6

McMichael, B.L., and J.J. Burke. 1998. Soil temperature and root growth. HortScience 33:947–958.

McMichael, B.L., and J.D. Hesketh. 1982. Field investigations of the response of cotton to water deficits. Field Crops Res. 5:319–333. doi:10.1016/0378-4290(82)90034-X

McMichael, B.L., W.R. Jordan, and R.D. Powell. 1973. Abscission processes in cotton: Induction by plant water deficit. Agron. J. 65:202–204. doi:10.2134/agronj1973.00021962006500020005x

McMichael, B.L., D.M. Oosterhuis, J.C. Zak, and C.A. Beyrouty. 2010. Growth and development of root systems. In: J.M. Stewart, D.M. Oosterhuis, J. Heitholt, and J.R. Mauney, editors, Physiology of cotton. Springer, Dordrecht. p. 57–71.

McMichael, B.L., and J.E. Quisenberry. 1993. The impact of the soil environment on the growth of root systems. Environ. Exp. Bot. 33:53–61. doi:10.1016/0098-8472(93)90055-K

McMichael, B.L., J.E. Quisenberry, and D.R. Upchurch. 1987. Lateral root development in exotic cottons. Environ. Exp. Bot. 27:499–502. doi:10.1016/0098-8472(87)90031-1

Medrano, H., J.M. Escalona, J. Bota, J. Gulias, and J. Flexas. 2002. Regulation of photosynthesis of C_3 plants in response to progressive drought: Stomatal conductance as a reference parameter. Ann. Bot. (Lond.) 89:895–905. doi:10.1093/aob/mcf079

Meredith, W.R., Jr. 1984. Influence of leaf morphology on lint yield of cotton: Enhancement by sub okra trait. Crop Sci. 24:855–857. doi:10.2135/cropsci1984.0011183X002400050007x

Meredith, W.R., Jr., J.J. Heitholt, W.T. Pettigrew, and S.T. Rayburn, Jr. 1997. Comparison of obsolete and modern cotton cultivars at two nitrogen levels. Crop Sci. 37:1453–1457.

Meyer, V.G. 1966. Environmental effects on the differentiation of abnormal cotton flowers. Am. J. Bot. 53:976–980. doi:10.2307/2440674

Meyer, V.G. 1969. Some effects of genes, cytoplasm, and environment on male sterility of cotton (*Gossypium*). Crop Sci. 9:237–242. doi:10.2135/cropsci1969.0011183X000900020039x

Miley, W.N., G.W. Hardy, M.B. Sturgis, and F.E. Sedberry, Jr. 1969. Influence of boron, nitrogen, and potassium on yield, nutrient uptake and abnormalities of cotton. Agron. J. 61:9–13. doi:10.2134/agronj1969.00021962006100010004x

Mills, N., D.M. Oosterhuis, and B.L. McMichael. 2012. Seedling root growth response to cool environmental conditions of diverse cotton genotypes. Am. J. Agric. Sci. 2:269–287.

Milroy, S.P., and M.P. Bange. 2003. Nitrogen and light responses of cotton photosynthesis and implications for crop growth. Crop Sci. 43:904–913. doi:10.2135/cropsci2003.9040

Morris, D.A. 1965. Photosynthesis by the capsule wall and bracteoles of the cotton plant. Cotton Grow. Rev. 42:49–51.

Moss, G.I., K.C. Hall, and M.B. Jackson. 1988. Ethylene and the response of roots of maize to physical impedance. New Phytol. 109:303–311. doi:10.1111/j.1469-8137.1988.tb04199.x

Mullins, G.L., and C.H. Burmester. 1990. Dry matter, nitrogen, phosphorus, and potassium accumulation by four cotton varieties. Agron. J. 82:729–736. doi:10.2134/agronj1990.00021962008200040017x

Munk, D., and H. Farahani. 2012. Why schedule irrigation? In: C.D. Perry and E.M. Barnes, editors, Cotton irrigation management for humid regions. Cotton Inc., Cary, NC.

Munro, J.M. 1987. Cotton. 2nd. ed. Longman–John Wiley and Sons, New York.

Muramoto, H., J.D. Hesketh, and C.D. Elmore. 1967. Leaf growth, leaf aging and leaf photosynthetic rates of cotton plants. Proc. Cotton Production Research Conference. National Cotton Council, Memphis, TN. p. 161–165.

Mutsaers, H.J.W. 1983. Leaf growth in cotton (*Gossypium hirsutum* L.). 1. Growth in area of mainstem and sympodial leaves. Ann. Bot. (Lond.) 51:503–520.

Nelson, W.L. 1949. The effect of nitrogen, phosphorus, and potash on certain lint and seed properties of cotton. Agron. J. 41:289–293. doi:10.2134/agronj1949.00021962004100070003x

Nielsen, K.F. 1974. Roots and root temperature. In: E.W. Carson, editor, The plant and its environment. Univ. Press of Virginia, Charlottesville. p. 293–335.

Ober, E.S., and R.E. Sharp. 1994. Proline accumulation in maize (*Zea mays* L.) primary roots at low water potentials. I. Requirement for increased levels of ABA. Plant Physiol. 105:981–987.

Oosterhuis, D.M. 1981. Hydraulic conductivity and drought acclimation of cotton root systems. Ph.D. diss. Diss. Abst. Int.42(6). Utah State Univ., Logan, Utah. Univ. Microfilm, Ann Arbor, MI.

Oosterhuis, D.M. 1990. Growth and development of the cotton plant. p. 1–24. In: W.N. Miley and D.M. Oosterhuis (eds.) Nitrogen nutrition in cotton: Practical ossues. Proc. Southern Branch Workshop for Practicing Agronomists. Little Rock, AR. 3–7 Feb. 1990. Publ. American Society of Agronomy., Madison, WI.

Oosterhuis, D.M., and J. Jernstedt. 1999. Morphology and anatomy of the cotton plant. In: W. Smith and J.S. Cothren, editors, Cotton: Origin, history, technology and production. John Wiley and Sons, New York. p. 175–206.

Oosterhuis, D.M., D. Loka, and T. Raper. 2013. Potassium and stress alleviation: Physiological functions and management. J. Plant Nutr. Soil Sci. 176:331–343. doi:10.1002/jpln.201200414

Oosterhuis, D.M., M.A. Okuba, and M. Mozaffari. 2008. Effect of soil-applied nitrogen fertilizer on the nitrogen content of cotton flowers. In: N.A. Slaton, editor, Arkansas cotton fertility studies 2007. Ark. Agric. Exp. Stn., University of Arkansas, Fayetteville, AR. p. 43–44.

Oosterhuis, D.M., and J.L. Snider. 2011. High temperature stress on floral development and yield of cotton. In: D.M. Oosterhuis, editor, Stress physiology in cotton. The Cotton Foundation, Cordova, TN. p. 1–24.

Oosterhuis, D.M., and M.J. Urwiler. 1988. Cotton main-stem leaves in relation to vegetative development and yield. Agron. J. 80:65–67. doi:10.2134/agronj1988.00021962008000010014x

Oosterhuis, O.M., and H.H. Wiebe. 1980. Hydraulic conductivity and osmotic adjustment in drought acclimated cotton. Plant Physiol. 65:5.

Oosterhuis, D.M., and S.D. Wullschleger. 1987a. Osmotic adjustment in cotton (*Gossypium hirsutum* L.) leaves and roots in response to water stress. Plant Physiol. 84:1154–1157. doi:10.1104/pp.84.4.1154

Oosterhuis, D.M., and S. Wullschleger. 1987b. Water flow through cotton roots in relation to xylem anatomy. J. Exp. Bot. 38:1866–1874. doi:10.1093/jxb/38.11.1866

Oosterhuis, D.M., and S.D. Wullschleger. 1988. Carbon partitioning and photosynthetic efficiency during boll development. In: D.A. Richter, editors, 1988 Proceedings Beltwide Cotton Conferences, New Orleans, LA. 3–8 Jan. National Cotton Council, Memphis, TN. p. 57–60.

Pallas, J.E., B.E. Michel, and D.G. Harris. 1967. Photosynthesis, transpiration, leaf temperature, and stomatal activity of cotton plants under varying water potentials. Plant Physiol. 42:76–88. doi:10.1104/pp.42.1.76

Passioura, J.B. 1980. Transport of water from soil to shoot in wheat seedlings. J. Exp. Bot. 31:333–345. doi:10.1093/jxb/31.1.333

Patterson, D.T., J.A. Bunce, R.S. Alberte, and E. Van Volkenburgh. 1977. Photosynthesis in relation to leaf characteristics of cotton from controlled and field environments. Plant Physiol. 59:384–387. doi:10.1104/pp.59.3.384

Payton, P., R. Webb, D. Kornyeyev, R. Allen, and S. Holaday. 2001. Protecting cotton photosynthesis during moderate chilling at high light intensity by increasing chloroplastic antioxidant enzyme activity. J. Exp. Bot. 52:2345–2354. doi:10.1093/jexbot/52.365.2345

Payton, P., R.D. Allen, N. Trolinder, and A.S. Holaday. 1997. Over-expression of chloroplast-targeted Mn superoxide dismutase in cotton (*Gossypium hirsutum* L., cv. Coker 312) does not alter the reduction of photosynthesis after short exposures to low temperature and high light intensity. Photosynth. Res. 52:233–244. doi:10.1023/A:1005873105596

Pearson, R.W., and Z.F. Lund. 1968. Direct observation of cotton root growth under field conditions. Agron. J. 60:442–443. doi:10.2134/agronj1968.00021962006000040036x

Pearson, R.W., L.F. Ratliff, and H.M. Taylor. 1970. Effect of soil temperature, strength and pH on cotton seedling root elongation. Agron. J. 62:243–246. doi:10.2134/agronj1970.00021962006200020019x

Peng, S., and D.R. Krieg. 1991. Single leaf and canopy photosynthesis response to plant age in cotton. Agron. J. 83:704–708. doi:10.2134/agronj1991.00021962008300040011x

Peoples, T.R., and M.A. Matthews. 1981. Influence of boll removal on assimilate partitioning in cotton. Crop Sci. 21:283–286. doi:10.2135/cropsci1981.0011183X002100020019x

Perry, S.W., D.R. Krieg, and R.B. Hutmacher. 1983. Photosynthetic rate control in cotton. Plant Physiol. 73:662–665. doi:10.1104/pp.73.3.662

Pervez, H., M. Ashraf, and M.I. Makhdm. 2004. Influence of potassium nutrition on gas exchange characteristics and water relations in cotton (*Gossypium hirsutum* L.). Photosynthetica 42:251–255. doi:10.1023/B:PHOT.0000040597.62743.5b

Pettigrew, W.T. 1994. Source-to-sink manipulation effects on cotton lint yield and yield components. Agron. J. 86:731–735. doi:10.2134/agronj1994.00021962008600040027x

Pettigrew, W.T. 1995. Source-to-sink manipulation effects on cotton fiber quality. Agron. J. 87:947–952. doi:10.2134/agronj1995.00021962008700050029x

Pettigrew, W.T. 2001. Environmental effects on cotton fiber carbohydrate concentration and quality. Crop Sci. 41:1108–1113. doi:10.2135/cropsci2001.4141108x

Pettigrew, W.T. 2003. Relationships between insufficient potassium and crop maturity in cotton. Agron. J. 95:1323–1329. doi:10.2134/agronj2003.1323

Pettigrew, W.T. 2004. Physiological consequences of moisture deficit stress in cotton. Crop Sci. 44:1265–1272. doi:10.2135/cropsci2004.1265

Pettigrew, W.T. 2008a. The effect of higher temperatures on cotton lint yield production and fiber quality. Crop Sci. 48:278–285. doi:10.2135/cropsci2007.05.0261

Pettigrew, W.T. 2008b. Potassium influences on yield and quality production for maize, wheat, soybean, and cotton. Physiol. Plant. 133:670–681. doi:10.1111/j.1399-3054.2008.01073.x

Pettigrew, W.T., and M.K. Dowd. 2012. Interactions between irrigation regimes and varieties result in altered cottonseed composition. J. Cotton Sci. 16:42–52.

Pettigrew, W.T., and T.J. Gerik. 2007. Cotton leaf photosynthesis and carbon metabolism. Adv. Agron. 94:209–236. doi:10.1016/S0065-2113(06)94005-X

Pettigrew, W.T., J.J. Heitholt, and W.R. Meredith. 1990. Early season fruit removal effects on cotton photosynthesis. Agronomy Abstracts, p. 128.

Pettigrew, W.T., W.R. Meredith, and L.D. Young. 2005. Potassium fertilization effects on cotton lint yield, yield components, and reniform nematode populations. Agron. J. 97:1245–1251. doi:10.2134/agronj2004.0321

Plaxton, W.C., and M.C. Carswell. 1999. Metabolic aspects of phosphate starvation response in plants. In: H.R. Lerner, editor, Plant responses to environmental stresses: From phytohormones to genome reorganization. Marcel Dekker, New York. p. 349–372.

Porch, T.G., and M. Jahn. 2001. Effects of high-temperature stress on microsporogenesis in heat-sensitive and heat-tolerant genotypes of *Phaseolus vulgaris*. Plant Cell Environ. 24:723–731. doi:10.1046/j.1365-3040.2001.00716.x

Radin, J.W. 1990. Response of transpiration and hydraulic conductance to root temperature in nitrogen and phosphorus deficient cotton seedlings. Plant Physiol. 92:855–857. doi:10.1104/pp.92.3.855

Radin, J.W., and R.C. Ackerson. 1981. Water relations of cotton plants under nitrogen deficiency III. Stomatal conductance, photosynthesis, and abscisic acid accumulation during drought. Plant Physiol. 67:115–119. doi:10.1104/pp.67.1.115

Radin, J.W., and M.P. Eidenbock. 1986. Carbon accumulation during photosynthesis in leaves of nitrogen and phosphorus-stressed cotton. Plant Physiol. 82:869–871. doi:10.1104/pp.82.3.869

Radin, J.W., B.A. Kimball, D.L. Hendrix, and J.R. Mauney. 1987. Photosynthesis of cotton plants exposed to elevated levels of carbon dioxide in the field. Photosynth. Res. 12:191–203. doi:10.1007/BF00055120

Radin, J.W., J.R. Mauney, and P.C. Kerridge. 1989. Water uptake of cotton roots during fruit filling in relation to irrigation frequency. Crop Sci. 29:1000–1005. doi:10.2135/cropsci1989.0011183X002900040035x

Read, J.J., K.R. Reddy, and J.N. Jenkins. 2006. Yield and fiber quality of upland cotton as influenced by nitrogen and potassium nutrition. Eur. J. Agron. 24:282–290. doi:10.1016/j.eja.2005.10.004

Reddy, K.R., G.H. Davidonis, A.S. Johnson, and B.T. Vinyard. 1999. Temperature regime and carbon dioxide enrichment alter cotton boll development and fiber properties. Agron. J. 91:851–858. doi:10.2134/agronj1999.915851x

Reddy, K.R., H.F. Hodges, and J.M. McKinion. 1995a. Carbon dioxide and temperature effects on pima cotton development. Agron. J. 87:820–826. doi:10.2134/agronj1995.00021962008700050007x

Reddy, K.R., H.F. Hodges, J.M. McKinion, and G.W. Wall. 1992a. Temperature effects on Pima cotton growth and development. Agron. J. 84:237–243. doi:10.2134/agronj1992.00021962008400020022x

Reddy, K.R., H.F. Hodges, and V.R. Reddy. 1992b. Temperature effects on cotton fruit retention. Agron. J. 84:26–30. doi:10.2134/agronj1992.00021962008400010006x

Reddy, K.R., S. Koti, G.H. Davidonis, and V.R. Reddy. 2004. Interactive effects of carbon dioxide and nitrogen nutrition on cotton growth, development, yield, and fiber quality. Agron. J. 96:1148–1157. doi:10.2134/agronj2004.1148

Reddy, V.R., K.R. Reddy, and H.F. Hodges. 1995b. Carbon dioxide enrichment and temperature effects on cotton canopy photosynthesis, transpiration, and water use efficiency. Field Crops Res. 41:13–23. doi:10.1016/0378-4290(94)00104-K

Reddy, A.R., K.R. Reddy, R. Padjung, and H.F. Hodges. 1996. Nitrogen nutrition and photosynthesis in leaves of Pima cotton. J. Plant Nutr. 19:755–770. doi:10.1080/01904169609365158

Reddy, K.R., and D. Zhao. 2005. Interactive effects of elevated CO2 and potassium deficiency on photosynthesis, growth, and biomass partitioning of cotton. Field Crops Res. 94:201–213. doi:10.1016/j.fcr.2005.01.004

Rochester, I.J., G.A. Constable, D.M. Oosterhuis, and M. Errington. 2012. Nutritional requirements of cotton during flowering and fruiting. In: D.M. Oosterhuis and J.T. Cothren, editors, Flowering and fruiting in cotton. The Cotton Foundation, Cordova, TN. p. 35–50.

Rogers, H.H., and S.A. Prior. 1992. Cotton root and rhizosphere responses to free-air CO_2 enrichment. Crit. Rev. Plant Sci. 11:251–263.

Rosolem, C.A., and J.C. Bogiani. 2011. Physiology of boron stress in cotton. In: D.M. Oosterhuis, editor, Stress physiology in cotton. The Cotton Foundation, Cordova, TN. p. 113–124.

Rosolem, C.A., and A. Costa. 2000. Cotton growth and boron distribution in plants as affected by a temporary deficiency of boron. J. Plant Nutr. 23:815–825. doi:10.1080/01904160009382062

Saini, H.S., M. Sedgley, and D. Aspinall. 1983. Effect of heat stress during floral development on pollen tube growth and ovary anatomy in wheat (*Triticum aestivum* L.). Aust. J. Plant Physiol. 10:137–144. doi:10.1071/PP9830137

Salvucci, M.E., and S.J. Crafts-Brandner. 2004. Inhibition of photosynthesis by heat stress: The activation state of rubisco as a limiting factor in photosynthesis. Physiol. Plant. 120:179–186. doi:10.1111/j.0031-9317.2004.0173.x

Sasek, T.W., E.H. Delucia, and B.R. Strain. 1985. Reversibility of photosynthetic inhibition in cotton after long-term exposure to elevated CO_2 concentrations. Plant Physiol. 78:619–622. doi:10.1104/pp.78.3.619

Sassenrath-Cole, G.F. 1995. Dependence of canopy light distribution on leaf and canopy structure for two cotton (*Gossypium*) species. Agric. For. Meteorol. 77:55–72.

Sassenrath-Cole, G.F., G. Lu, H.F. Hodges, and J.M. McKinion. 1996. Photon flux density versus leaf senescence in determining photosynthetic efficiency and capacity of *Gossypium hirsutum* L. leaves. Environ. Exp. Bot. 36:439–446. doi:10.1016/S0098-8472(96)01019-2

Schrader, S.M., R.R. Wise, W.F. Wacholtz, D.R. Ort, and T.D. Sharkey. 2004. Thylakoid membrane responses to moderately high leaf temperature in Pima cotton. Plant Cell Environ. 27:725–735. doi:10.1111/j.1365-3040.2004.01172.x

Simpson, D.M., C.L. Adams, and G.L. Stone. 1940. Anatomical structure of the cottonseed coat as related to problems of germination. USDA Tech. Bull. 734. US Gov. Print. Office, Washington, DC.

Singh, R.P., P.V.V. Prasad, K. Sunita, S.N. Giri, and K.R. Reddy. 2007. Influence of high temperature and breeding for heat tolerance in cotton: A review. Adv. Agron. 93:313–385. doi:10.1016/S0065-2113(06)93006-5

Sharp, R.E., and W.I. Davies. 1979. Solute regulation and growth by roots and shoots of water-stressed maize plants. Plant 147:43–49. doi:10.1007/BF00384589

Sharp, R.E., V. Poroyko, L.G. Hejlek, W.G. Spollen, G.K. Springer, H.J. Bohnert, and H.T. Nguyen. 2004. Root growth maintenance during water deficits: Physiology to functional genomics. J. Exp. Bot. 55:2343–2351. doi:10.1093/jxb/erh276

Snider, J.L., G.D. Collins, J. Whitaker, K.D. Chapman, P. Horn, and T.L. Grey. 2014a. Seed size and oil content are key determinants of seedling vigor in *Gossypium hirsutum*. J. Cotton Sci. 18:1–9.

Snider, J.L., G.D. Collins, J. Whitaker, C.D. Perry, and D.R. Chastain. 2014b. Electron transport through photosystem II is not limited by a wide range of water deficit conditions in field-grown *Gossypium hirsutum*. J. Agron. Crop Sci. 200:77–82. doi:10.1111/jac.12044

Snider, J.L., G.D. Collins J. Whitaker, C.D. Perry, and D.R. Chastain. 2013b. The effect of water deficit on photosynthetic electron transport and net CO_2 assimilation rates in field-grown cotton. In: D. Shurley, J. Snider, and J. Whitaker, editors, 2012 Georgia Cotton Research and Extension Report. University of Georgia, Athens. p. 51–55.

Snider, J.L., and D.M. Oosterhuis. 2011. How does timing, duration, and severity of heat stress influence pollen–pistil interactions in angiosperms? Plant Signal. Behav. 6:930–933. doi:10.4161/psb.6.7.15315

Snider, J.L., and D.M. Oosterhuis. 2012. Heat stress and pollen–pistil interactions. In: D.M. Oosterhuis and J.T. Cothren, editors, Flowering and fruiting in cotton. The Cotton Foundation, Cordova, TN. p. 59–78.

Snider, J.L., D.M. Oosterhuis, G.D. Collins, C. Pilon, and T.R. FitzSimons. 2013a. Field-acclimated Gossypium hirsutum cultivars exhibit genotypic and seasonal differences in photosystem II thermostability. J. Plant Physiol. 170:489–496. doi:10.1016/j.jplph.2012.11.004

Snider, J.L., D.M. Oosterhuis, and E.M. Kawakami. 2010. Genotypic differences in thermotolerance are dependent upon prestress capacity for antioxidant protection of the photosynthetic apparatus in *Gossypium hirsutum*. Physiol. Plant. 138:268–277. doi:10.1111/j.1399-3054.2009.01325.x

Snider, J.L., D.M. Oosterhuis, and E.M. Kawakami. 2011a. Pollen tube growth rate is slowed by high temperature in field-grown *Gossypium hirsutum* pistils. J. Plant Physiol. 168:441–448. doi:10.1016/j.jplph.2010.08.003

Snider, J.L., D.M. Oosterhuis, and E.M. Kawakami. 2011b. Mechanisms of reproductive thermotolerance in *Gossypium hirsutum*: The effect of genotype and exogenous calcium application. J. Agron. Crop Sci. 197:228–236. doi:10.1111/j.1439-037X.2010.00457.x

Snider, J.L., D.M. Oosterhuis, D.A. Loka, and E.M. Kawakami. 2011c. High temperature limits in vivo pollen tube growth rates by altering diurnal carbohydrate balance in field-grown *Gossypium hirsutum* pistils. J. Plant Physiol. 168:1168–1175. doi:10.1016/j.jplph.2010.12.011

Snider, J.L., D.M. Oosterhuis, B.W. Skulman, and E.M. Kawakami. 2009. Heat stress-induced limitations to reproductive success in *Gossypium hirsutum*. Physiol. Plant. 137:125–138. doi:10.1111/j.1399-3054.2009.01266.x

Stewart, J.M. 1986. Integrated events in flower and fruit. In: J.R. Mauney and J.M. Stewart, editors, Cotton physiology. The Cotton Foundation, Memphis, TN. p. 261–300.

Stolzy, L.H. 1974. Soil atmosphere. In: E.W. Carson, editor, The plant root and its environment. Univ. Press of Virginia, Charlottesville. p. 335–361.

Taiz, L., and E. Zeiger. 2010. Plant physiology. 5th ed. Sinauer Associates, Sunderland, MA.

Taylor, H.M. 1983. Managing root systems for efficient water use: An overview. In: H.M. Taylor, W.R. Jordan, and T.R. Sinclair, editors, Limitations to efficient water use in crop production. ASA, CSSA, and SSSA, Madison, WI. p. 87–113. doi:10.2134/1983.limitationstoefficientwateruse.c6

Taylor, H.M. and H.R. Gardner. 1963. Penetration of cotton seedling taproots as influenced by bulk density, moisture content, and strength of soil. Soil Sci. 96:153–1 56.

Taylor, H.M., M.G. Huck, and B. Klepper. 1972. Root development in relation to soil physical conditions. In: D.I. Hillel, editor, Optimizing the soil physical environment toward greater crop yields. Academic Press, New York. p. 57–77.

Taylor, H.M., and B. Klepper. 1971. Water uptake by cotton roots during an irrigation cycle. Aust. J. Biol. Sci. 24:853–859.

Taylor, H.M., and B. Klepper. 1975. Water uptake by cotton root systems: An examination of assumptions in the single root model. Soil Sci. 120:57–67. doi:10.1097/00010694-197507000-00009

Taylor, H.M., and B. Klepper. 1978. The role of rooting characteristics in the supply of water to plants. Adv. Agron. 30:99–128. doi:10.1016/S0065-2113(08)60704-X

Taylor, H.M., and B. Klepper. 1974. Water relations of cotton. 1. Root growth and water use as related to top growth and soil water content. Agron. J. 66:584–588. doi:10.2134/agronj1974.00021962006600040031x

Taylor, H.M., and L.F. Ratliff. 1969. Root elongation rates of cotton and peanuts as a function of soil strength and soil water content. Soil Sci. 108:113–119. doi:10.1097/00010694-196908000-00006

Tewolde, H., and C.J. Fernandez. 2003. Fiber quality response of Pima cotton to nitrogen and phosphorus deficiency. J. Plant Nutr. 26:223–235. doi:10.1081/PLN-120016506

Tezara, W., V.J. Mitchell, S.D. Driscoll, and D.W. Lawlor. 1999. Water stress inhibits plant photosynthesis by decreasing coupling factor and ATP. Nature 401:914–917. doi:10.1038/44842

Tezara, W., S.D. Driscoll, and D.W. Lawlor. 2008. Partitioning of photosynthetic electron flow between CO_2 assimilation and O_2 reduction in sunflower plants under water deficit. Photosynthetica 46:127–134. doi:10.1007/s11099-008-0020-1

Thaler, P., and L. Pages. 1998. Modelling the influence of assimilate availability on root growth and architecture. Plant Soil 201:307–320. doi:10.1023/A:1004380021699

Thanisawanyangkura, S., H. Sinoquet, P. Rivet, M. Cretenet, and E. Jallas. 1997. Leaf orientation and sunlight leaf area distribution in cotton. Agric. For. Meteorol. 86:1–15. doi:10.1016/S0168-1923(96)02417-3

Thomas, R.B., and B.R. Strain. 1991. Root restriction as a factor in photosynthetic acclimation of cotton seedlings grown in elevated carbon dioxide. Plant Physiol. 96:627–634. doi:10.1104/pp.96.2.627

Trelease, R.N. 1984. Biogenesis of glyoxysomes. Annu. Rev. Plant Physiol. 35:321–347. doi:10.1146/annurev.pp.35.060184.001541

Trelease, R.N., and D.C. Doman. 1984. Mobilization of oil and wax reserves. In: D.R. Murray, editor, Seed physiology. II. Germination and reserve mobilization. Academic Press Australia, Sydney. p. 201–245.

Trolinder, N.L., B.L. McMichael, and D.R. Upchurch. 1993. Water relations of cotton flower petals and fruits. Plant Cell Environ. 16:755–760. doi:10.1111/j.1365-3040.1993.tb00496.x

Turley, R.B., and K.D. Chapman. 2010. Ontogeny of cotton seeds: Gametogenesis, embryogenesis, germination, and seedling growth. In: J.M. Stewart, D.M. Oosterhuis, J.J. Heitholt, and J.R. Mauney, editors, Physiology of cotton. Springer, New York. p. 332–341.

Turner, N., A. Hearn, J. Begg, and G. Constable. 1986. Cotton (*Gossypium hirsutum, L*).: Physiological and morphological responses to water deficits and their relationship to yield. Field Crops Res. 14:153–170. doi:10.1016/0378-4290(86)90054-7

Turner, N.C., W.R. Stern, and P. Evans. 1987. Water relations and osmotic adjustment of leaves and roots of lupins in response to water deficits. Crop Sci. 27:977–983. doi:10.2135/cropsci1987.0011183X002700050032x

Van Iersel, M.W., and D.M. Oosterhuis. 1995. Diurnal water relations of expanding and full-sized cotton fruits and subtending leaves. Plant Cell Environ. 18:807–812. doi:10.1111/j.1365-3040.1995.tb00585.x

Voetberg, G.S., and R.E. Sharp. 1991. Growth of the maize primary root at low water potentials. III. Role of increased proline deposition in osmotic adjustment. Plant Physiol. 96:1125–1130. doi:10.1104/pp.96.4.1125

Wanjura, D.F. 1986. Field environment and stand establishment. In: J.R. Mauney and J.M. Stewart, editors, Cotton physiology. The Cotton Foundation, Memphis, TN. p. 551–556.

Wanjura, D.F., and D.R. Buxton. 1972. Hypocotyl and radicle elongation of cotton as affected by soil environment. Agron. J. 64:431–434. doi:10.2134/agronj1972.00021962006400040006x

Wanjura, D.F., E.B. Hudspeth, Jr., and J.D. Bilbro, Jr. 1969. Emergence time, seed quality, and planting depth effects on yield and survival of cotton (*Gossypium hirsutum* L.). Agron. J. 61:63–65. doi:10.2134/agronj1969.00021962006100010021x

Warner, D.A., and J.J. Burke. 1993. Cool night temperatures alter leaf starch and photosystem II chlorophyll fluorescence in cotton. Agron. J. 85:836–840. doi:10.2134/agronj1993.00021962008500040011x

Wells, R. 2002. Stem and root carbohydrate dynamics of two cotton cultivars bred fifty years apart. Agron. J. 94:876–882. doi:10.2134/agronj2002.8760

Wells, R. 2011. Light and the cotton plant. In: D.M. Oosterhuis, editor, Stress physiology in cotton. The Cotton Foundation, Cordova, TN. p. 73–84.

Wells, R., and W.R. Meredith, Jr. 1986. Normal vs okra leaf yield interactions in cotton. II. Analysis of vegetative and reproductive growth. Crop Sci. 26:223–228. doi:10.2135/cropsci1986.0011183X002600020002x

Wells, R., W.R. Meredith, Jr., and J.R. Williford. 1986. Canopy photosynthesis and its relationship to plant productivity in near-isogenic cotton lines differing in leaf morphology. Plant Physiol. 82:635–640. doi:10.1104/pp.82.3.635

Westgate, M.E., and J.S. Boyer. 1985. Osmotic adjustment and the inhibition of leaf, root, stem and silk growth at low water potentials in maize. Planta 164:540–549. doi:10.1007/BF00395973

Whelan, T., W.M. Sackett, and C.R. Benedict. 1970. Carbon isotope discrimination of a plant possessing the C_4 dicarboxylic acid pathway. Biochem. Biophys. Res. Commun. 41:1205–1210. doi:10.1016/0006-291X(70)90214-7

Whitney, J.B. 1941. Effects of the composition of the soil atmosphere on the absorption of water by plants. Am. J. Bot. 28:14.

Wilson, F.D. 1986. Pink bollworm resistance, lint yield, and lint yield components of okra-leaf cotton in different genetic backgrounds. Crop Sci. 26:1164–1167. doi:10.2135/cropsci1986.0011183X002600060018x

Wise, R.R., J.R. McWilliam, and A.W. Naylor. 1983. A comparative study of low-temperature-induced ultrastructural alterations of three species with differing chilling sensitivities. Plant Cell Environ. 6:525–535. doi:10.1111/j.1365-3040.1983.tb01165.x

Wise, R.R., and A.W. Naylor. 1987a. Chilling-enhanced photooxidation: Evidence for the role of singlet oxygen and superoxide in the breakdown of pigments and endogenous antioxidants. Plant Physiol. 83:278–282. doi:10.1104/pp.83.2.278

Wise, R.R., and A.W. Naylor. 1987b. Chilling-enhanced photooxidation: The peroxidative destruction of lipids during chilling injury to photosynthesis and ultrastructure. Plant Physiol. 83:272–277. doi:10.1104/pp.83.2.272

Wise, R.R., A.J. Olson, S.M. Schrader, and T.D. Sharkey. 2004. Electron transport is the functional limitation of photosynthesis in field-grown Pima cotton plants at high temperature. Plant Cell Environ. 27:717–724. doi:10.1111/j.1365-3040.2004.01171.x

Wise, R.R., A. Ortiz-Lopez, and D.R. Ort. 1992. Spatial distribution of photosynthesis during drought in field-grown and acclimated and nonacclimated growth chamber-grown cotton. Plant Physiol. 100:26–32. doi:10.1104/pp.100.1.26

Wise, R.R., G.F. Sassenrath-Cole, and R.G. Percy. 2000. A comparison of leaf anatomy in field-grown *Gossypium hirsutum* and *G. barbadense*. Ann. Bot. (Lond.) 86:731–738. doi:10.1006/anbo.2000.1235

Wullschleger, S.D., and D.M. Oosterhuis. 1987. Carbon partitioning within a sympodial branch during cotton boll development. Ark. Farm Res. 36:4.

Wullschleger, S.D., and D.M. Oosterhuis. 1990a. Photosynthesis of individual field-grown cotton leaves during ontogeny. Photosynth. Res. 23:163–170. doi:10.1007/BF00035007

Wullschleger, S.D., and D.M. Oosterhuis. 1990b. Canopy development and photosynthesis of cotton as influenced by nitrogen nutrition. J. Plant Nutr. 13:1141–1154. doi:10.1080/01904169009364140

Wullschleger, S.D., and D.M. Oosterhuis. 1990d. Photosynthetic carbon production and use by developing cotton leaves and bolls. Crop Sci. 30:1259–1264. doi:10.2135/cropsci1990.0011183X003000060021x

Wullschleger, S.D., and D.M. Oosterhuis. 1990c. Photosynthetic and respiratory activity of fruiting forms within the cotton canopy. Plant Physiol. 94:463–469. doi:10.1104/pp.94.2.463

Wullschleger, S.D., and D.M. Oosterhuis. 1992. Canopy leaf area development and age-class dynamics in cotton. Crop Sci. 32:451–456. doi:10.2135/cropsci1992.0011183X003200020033x

Wullschleger, S.D., D.M. Oosterhuis, and S. Rurtherford. 1990. Importance of bracts in the carbon economy of cotton. Arkansas Farm Res. 39(3):4.

Wyn Jones, R.G., and J. Gorham. 1983. Osmoregulation. In: O.L. Lange, P.S. Nobel, C.B. Osmond, and H. Ziegler, editors, Encyclopedia of plant physiology. New Series 12, Physiology, Physiological Plant Ecology III. Springer Verlag, Berlin. p. 35–38.

Yoon, S.T., G. Hoogenboom, I. Flitcroft, and M. Bannayan. 2009. Growth and development of cotton (*Gossypium hirsutum* L.) in response to CO_2 enrichment under two different temperature regimes. Environ. Exp. Bot. 67:178–187. doi:10.1016/j.envexpbot.2009.06.015

Zhang, J.S., H.Y. Yang, L. Zhu, and H. Tong. 1997. Ultracytochemical localization of calcium in the pollen tube track of the cotton gynoecium. Acta Bot. Sin. 39:121–125.

Zhang, Y.L., Y.Y. Hu, H.H. Luo, W.S. Chow, and W.F. Zhang. 2011. Two distinct strategies of cotton and soybean differing in leaf movement to perform photosynthesis under drought in the field. Funct. Plant Biol. 38:567–575. doi:10.1071/FP11065

Zhang, Y.L., H.Z. Zhang, G.Y. Feng, J.S. Tian, and W.F. Zhang. 2009. Leaf diaheliotropic movement can improve carbon gain and water use efficiency and not intensify photoinhibition in upland cotton (*Gossypium hirsutum*). Photosynthetica 47:609–615. doi:10.1007/s11099-009-0087-3

Zhao, D., and D.M. Oosterhuis. 1997. Physiological response of growth chamber-grown cotton plants to the plant growth regulator PGR IV under water-deficit stress. Environ. Exp. Bot. 38:7–14. doi:10.1016/S0098-8472(97)00002-6

Zhao, D., and D.M. Oosterhuis. 1998. Cotton responses to shade at different growth stages: Nonstructural carbohydrate composition. Crop Sci. 38:1196–1203. doi:10.2135/cropsci1998.0011183X003800050014x

Zhao, D., and D.M. Oosterhuis. 2000. Cotton responses to shade at different growth stages: Growth, lint yield and fiber quality. Exp. Agric. 36:27–39. doi:10.1017/S0014479700361014

Zhao, D., and D.M. Oosterhuis. 2003. Cotton growth and physiological responses to boron deficiency. J. Plant Nutr. 26:855–865. doi:10.1081/PLN-120018570

Zhao, D., D.M. Oosterhuis, and C.W. Bednarz. 2001. Influence of potassium deficiency on photosynthesis, chlorophyll content, and chloroplast ultrastructure of cotton plants. Photosynthetica 39:103–109. doi:10.1023/A:1012404204910

Zhao, D., K.R. Reddy, V.G. Kakani, S. Koti, and W. Gao. 2005. Physiological causes of cotton fruit abscission under conditions of high temperature and enhanced ultraviolet-B radiation. Physiol. Plant. 124:189–199. doi:10.1111/j.1399-3054.2005.00491.x

Zhao, D., K.R. Reddy, V.G. Kakani, A.R. Mohammed, J.J. Read, and W. Gao. 2004. Leaf and canopy photosynthetic characteristics of cotton (*Gossypium hirsutum*) under elevated CO2 concentration, and UV-B radiation. J. Plant Physiol. 161:581–590. doi:10.1078/0176-1617-01229

Zhao, D., K.R. Reddy, V.G. Kakani, J.J. Read, and J.H. Sullivan. 2003. Growth and physiological responses of cotton (*Gossypium hirsutum* L.) to elevated carbon dioxide and ultraviolet-B radiation under controlled environmental conditions. Plant Cell Environ. 26:771–782. doi:10.1046/j.1365-3040.2003.01019.x

Zhao, C.Y., Y.Y. Yan, Y. Yimamu, J.Y. Li, Z. Zhao, and L. Wu. 2010. Effects of soil moisture on cotton root length density and yield under drip irrigation with plastic mulch in Aksu Oasis farmland. J. Arid Land. 2:243–249.

Zhou, Z., and D.M. Oosterhuis. 2012. Physiological mechanism of nitrogen mediating cotton (*Gossypium hirsutum* l.) seedlings growth under water-stress conditions. Amer. J. Plant Sci. 3:721–730. doi:10.4236/ajps.2012.36087

Climate and Its Interaction with Cotton Morphology

Warren Conaty, Rose Brodrick, James Mahan, and Paxton Payton*

Abstract

This chapter is an extension of previous work by Jack R. Mauney and is written to extend basic information regarding the morphological adaptations to the wide diversity of cotton production regions. Our goal, similar to the previous edition of this book, is to outline the fundamental vegetative framework of the cotton plant and discuss its phenotypic plasticity in terms of production regions and management schemes. Additionally, we attempt to highlight the key limiting factors across broad regions of production and primary mitigation strategies employed by growers across those regions, focusing specifically on season length, water requirements and availability, and key climate characteristics. Not included in this chapter is detailed information about soil types, which can vary significantly within each production region.

Cotton is an indeterminate, xerophytic, deciduous plant of the Malvaceae family and *Gossypium* genus. Cultivated species are native to subtropical climates, with two Old World diploid species (*G. arboreum* L. and *G. herbaceum* L.) and two New World tetraploid species (*G. hirsutum* L. and *G. barbadense* L.) independently domesticated. Ninety-five percent of global cotton production is of *G. hirsutum*, or Upland cotton, and thus this review will discuss only this species, especially its fundamental vegetative framework and phenotypic plasticity in terms of production regions and management schemes. A complete discussion of the developmental patterns in cotton is given by Hector (1936) and Baranov and Maltzev (1937), as well as in published reports by Mauney and Ball (1959), Quintanilha et al. (1962), Mauney (1984, 1986), and Oosterhuis and Jernstedt (1999).

The geographic center of origin of Upland cotton is Central America, and cotton was probably first domesticated by pre-Columbian people of the Yucatan Peninsula (Brubaker and Wendel, 1994). Today cotton is indigenous to Central and Southern America, the Caribbean, and some Pacific Islands. However, cotton

Warren Conaty (warren.conaty@csiro.au) and Rose Brodrick (Rose.Brodrick@csiro.au), CSIRO Agriculture, 21888 Kamilaroi Highway, Narrabri, NSW, Australia 2390.

James Mahan (james.mahan@ars.usda.gov) and Paxton Payton, USDA-ARS Cropping Systems Research Laboratory, 3810 4th Street, Lubbock, TX. *Corresponding author (paxton.payton@ars.usda.gov).

doi:10.2134/agronmonogr57.2013.0021

Cotton. 2nd. ed. David D. Fang and Richard G. Percy, editors. Agronomy Monograph 57.

is successfully cultivated as an annual in a wide variety of tropical (long-season, humid), subtropical (short-season, humid), and temperate (long-season, arid; short-season, semiarid) locations between 36°S and 46°N (Fig. 1). Cotton grown in Mediterranean and arid climates requires irrigation.

The morphological plasticity of the cotton plant enables it to be produced in a wide variety of agro-ecological regions (Oosterhuis and Jernstedt, 1999). This plasticity essentially translates to the lengthening, shortening, or interruption of its effective flowering period in response to the length of the season and availability of water (Hake and Kerby, 1996). As a result, cotton plants generally produce 60% more squares (floral buds) than they can physiologically retain. This response is a survival strategy whereby the plant can compensate in the event of weather- and pest-related losses. Because climate has a significant effect on season length and water availability, the growth and morphology of the cotton plant in the context of agro-ecological zones is important. The range in global cotton production climatic zones can be broadly classed into four separate groupings based on season length and humidity: (i) long season, arid; (ii) long season, humid; (iii) short season, arid; and (iv) short season, humid. The length of the season ultimately controls the ability to accumulate degree days (heat units) and is affected primarily by latitude and elevation. Here we define long season as more than 150 d from planting to harvest, and short season as production of a crop in less than 150 d. The length of the growing season is a climatic factor that is often a function of the interval between lethal low temperatures at each end of the season. At a given location (climate), the environmental conditions within a given growing season can be quite variable: a harvestable, economically viable yield can often be obtained on the Texas High Plains in as little as 100 d in rainfed crops, while 160 d might be required for irrigated crops. The aridity of the location not only affects the water supply (via rainfall) and the atmospheric humidity,

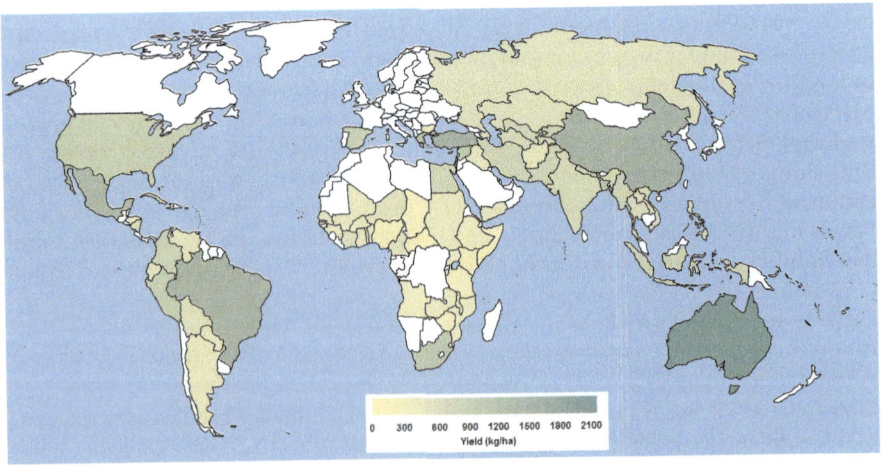

Fig. 1. The range of environments, season lengths and rainfall suitable for cotton production is highlighted by the latitude of the 73 cotton producing nations and the average lint yield (kg/ha) obtained in each of these countries in 2014. Data sourced from ICAC World Cotton Database, https://icac.generation10.net/.

but also influences the amount and quality of solar radiation the crop can capture for gas exchange. The characteristics of these climatic zones are listed in Table 1. However, it should be noted that in reality, most global production areas would be best represented by an intermediate classification of medium season length and moderate humidity.

Factors Driving Season Length

Latitude and elevation in different cotton-growing regions have the biggest effect on season length as well as day and night temperatures (Hake and Kerby, 1996). Low latitude and elevation generally result in a longer growing season. High yield potential can be achieved in these areas; however, adequate water supply and radiation are still required. Furthermore, as insect populations can breed more generations in a longer cropping cycle, a long season has a greater potential for yield loss due to heavy insect pressure.

Day length is also controlled by latitude. Lower latitudes tend to have shorter summer day lengths than higher latitudes. However, at these higher latitudes, the season length is truncated by suboptimal temperatures. As a result, lower latitudes may have up to 600 extra day length hours per season (Table 1). However, the quality of radiation in these daylight hours must also be taken into consideration.

Shorter day lengths and cooler, less predictable temperatures are more likely to occur at latitudes further from the equator. Cotton planted at these more extreme latitudes are subject to a much shorter growing season and a higher risk of chilling injury. Regions at the extent of these latitudes are often coastal or adjacent to large bodies of water, which moderate the temperature and minimize the risk of early- and late-season frosts (e.g., the eastern USA). Elevation also controls season length through its temperature effects (Hake and Kerby, 1996). At higher elevations, night cooling occurs because heat is not trapped by the thinner, drier atmosphere, thus shortening the potential season length (e.g., the Texas High Plains, USA). Night temperatures usually drop by about 1°C for each increase in elevation of 300 m. Hence, cotton grown at high elevations has an increased risk of seedling and late-boll damage due to frost injury. Conversely, higher elevations can be beneficial for mid-season cotton growth and development, because cooler night temperatures allow the plant to approach the optimum night temperature. This is due to limited transpiration at leaf temperatures less than 27.5°C, thus ensuring the synchrony of leaf and ambient air temperature (Hatfield, et al., 1987). Warm night conditions (>25°C) can lead to altered flowering times, decreased pollen viability, and loss of fruiting positions (Mauney 1966; Hesketh and Low 1968; Gipson and Ray 1969; Zeiher et al., 1994; Hake and Kerby, 1996; Pettigrew 2008; Echer et al., 2014). Furthermore, night-time transpiration and maintenance respiration occurs under supraoptimal night temperatures. These processes consume stored water and energy resources, thereby reducing the supply available for new daytime growth and boll filling (reviewed in Snider and Oosterhuis, 2011).

Cotton growth and development are also affected by atmospheric humidity and cloud cover, and hence the amount and quality of solar radiation. When reduced solar radiation is combined with a reduction in season length, the effect of season length on the availability of solar radiation for gas exchange can be significant. For example, a crop growing at Al Kawd in Yemen will receive

Table 1. Examples of major cotton-growing regions and production environments. Data sourced from Weatherbase, www.weatherbase.com.

	Long season, arid	Full season, humid	Short season, arid	Short season, humid
		Location and season		
Name	Al Kawd, Yemen	Primavera do Leste, Mato Grosso, Brazil	Lubbock, TX, USA	Franklin, VA,USA
Region	Abyan Delta, Yemen	Cerrado, Mato Grosso, Brazil	Texas High Plains, USA	Northeastern Cotton Belt, USA
Latitude (°)	13.1	−15.6	33.6	36.7
Elevation (m)	20	636	980	11
Growing season (d)	200	210	150	150
Planting to harvest	September–March	December–July	May–September	May–October
		In-crop climate averages		
Daily global solar radiation (MJ m^{-2})	21.0	17.8	23.2	17.0
Day length (h)	12.1	12.4	14.0	13.4
Maximum temperature (°C)	28.9	29.5	31.3	24.8
Minimum temperature (°C)	25.7	16.1	16.6	17.0
Growing degree days (base temp = 10°C)	3707	3190	2138	2417
Rainfall (mm)	24	790	303	590
Rainfall days (d $season^{-1}$)	7	39	33	37
Relative humidity (%)	70.6	80.2	45.2	72.9
Other similar locations	Indus Valley, Pakistan. San Joaquin Valley, CA, USA. Western New South Wales, Australia.	Yangtze River region, China. The Burdekin, Queensland, Australia.	Xinjian, northwestern China. Southern Greece.	

approximately 4200 MJ m^{-2} of solar radiation during the season, while one growing at Franklin, VA, in the USA will only receive 2500 MJ m^{-2} (Table 1). This has an obvious effect on yield potential and the crop's ability to compensate following periods of unfavorable conditions.

Cotton is diverse in terms of morphology and phenology, having a wide variety of characteristics even among commercial varieties. Leaf shape, canopy density, plant height, rooting patterns, and bolls size and number are all potentially variable. Given the wide range of environments within which cotton is cultivated, it might be expected that certain morphological types might be better suited to some environments than others. In the mid-2000s, cotton varieties developed for high-input, high-yield production systems in New South Wales, Australia, (long season, arid) were widely adopted in lower-input, lower-yield production systems in the Lubbock, Texas, region (short season, arid). The suitability of these varieties in Texas surprised many in the field. In 2010 and 2011, five Brazilian varieties were compared with varieties from Chad, Mali, Burkina Faso, and Benin at two locations in Burkina Faso in 2010 and 2011 (Bourgou and Sanfo, 2012). While the varieties varied in terms of morphological characteristics, the differences did not translate into significant differences in agronomic performance. Similarly, Hague et al. (2014) compared the fruiting morphology of five commercial cultivars for 2 yr at four locations. Although they found significant responses of the cultivars due to variations in location and year, the effects were consistent among the cultivars. They concluded: "Breeders can select plant size and fruiting habits at one location with confidence that such parameters will be stable across a large geographical range." These limited results underscore the general ability of cotton to respond to changes in growth conditions that is the foundation of its profitable cultivation across a wide range of environments and management schemes.

Case Studies of Climate Influences on the Morphological Development of Cotton

Long-Season Arid Regions: Abyan Delta, Yemen

Given adequate water supply and control of insect pressure, crops grown in arid conditions such as the Middle East, Australia, and the southwestern USA have some of the highest yields. This is generally due to high amounts of solar radiation and non-limiting season lengths. Crop development in such environments is accurately described by degree day accumulation (Constable, 1976), the switch from vegetative to reproductive growth, and the associated source-sink dynamics that reliably predict the cessation of vegetative growth as the developing fruit (sinks) deplete the supply of photosynthates (source) from vegetative material.

The Abyan Delta in Yemen has a hot desert climate with largely unbroken sunshine and little rainfall. Cotton is grown on deep alluvial soils on stored flood water from seasonal rivers, with some water supply supplemented by irrigation. The key climatic challenge to growing cotton in long-season arid environments is the water supply. Given an adequate water supply, very high yield and fiber quality can be achieved in these environments. In long-season arid environments, cotton growth and development will be very predictable, with a new main stem node being produced every 2 to 3 d and adjacent floral buds (squares) being produced on fruiting branches above the fifth to eighth main stem node every 3 to 7 d (Hesketh et al., 1975; Oosterhuis, 1990; Godoy and Palomo, 1999; Bednarz and Nichols, 2005). Because radiation is not limiting, the number of nodes, fruiting organs, and leaf biomass accumulate exponentially, provided other abiotic (water and nutrition) and biotic stresses do not occur (Hearn, 1969). This pattern continues until the internal demand for photoassimilates by the fruiting bodies cannot be met by the supply. As a result, the young fruiting bodies are shed until the production of squares is stopped. When this limit is reached, the crop is said to have physiologically cut out, and photoassimilates are used to develop set fruit. Provided there is sufficient water and nutrition, once carbohydrates are again in ample supply, a second fruiting cycle can occur. In these areas, planting dates can be more flexible, and the loss of early fruit is less significant because compensation will occur, with new fruit being produced and retained through the growth of larger bolls (Wells and Meredith, 1984; Constable, 1991; Sadras, 1995). Such early-season fruit loss can lead to an increased yield potential because it produces a larger plant structure before the effective fruit period is initiated, allowing for an increase in photosynthate supply for use in boll development

In arid environments where water limits production, vegetative growth occurs at a rate independent of the water supply, until approximately three-quarters of the available water supply is diminished (Hearn, 1969). The remaining water is used to mature the bolls already set on the plant. Thus the amount of preplanting moisture (or the ability to supply supplementary irrigation water) will determine the size and yield potential of the plant.

Cotton grown in arid climates has a higher risk of experiencing high temperatures and undergoing heat stress, particularly if it is grown in inland temperate regions. Due to the Abyan Delta's proximity to the Arabian Sea with its moderating effects on temperature, cotton grown in the delta usually does not experience heat stress. However, other long-season, arid regions, such as the Indus Valley in Pakistan, Arizona, and the San Joaquin Valley in the USA, and western New South Wales in Australia, can all potentially experience heat waves. In general, higher average air temperatures accelerate crop growth but can negatively impact reproductive development. This is particularly evident when transpirational cooling is unable to maintain optimum crop temperature. For a 1°C increase in air temperature the time required to produce squares, flowers, and mature bolls is reduced by an average of 1.6, 3.1, and 6.9 degree days, respectively (Reddy et al., 1997).

Reducing the developmental time for bolls results in smaller bolls, lower lint quality, and reduced yield. Heat stress and associated fruit shedding begins to occur when ambient daytime temperatures exceed 30°C (Reddy et al., 1992), and controlled-environment growth chamber studies showed that at temperatures in excess of 40°C, all flowers can be shed (Reddy et al., 1991b). Additionally, in controlled-environment chambers, an increase in nighttime air temperature from 28°C to 32°C resulted in increased abortion of bolls less than 10 d after anthesis in chronological order (Zeiher et al., 1995). If this increased air temperature is coupled with increased night temperatures, further increases in boll abortion are observed, because there is an increase in the demand for photoassimilate for maintenance respiration. High night temperatures can also reduce boll set through effects on square development, either by suppressing the development of the reproductive meristem or by increased shedding and abortion of young squares (Singh et al., 2007).

Long-Season Humid Regions: The Cerrado, Mato Grosso, Brazil

The Brazilian Cerrado has a tropical climate with distinct wet and dry seasons. Cotton in this environment is unlikely to experience temperature extremes: no cold shocks (<12°C) and rarely heat stress (>35°C). The key climatic challenge to growing cotton in this region is the variability of cloud cover and rainfall during early and mid-fruiting. In this environment, cotton is sown during the wet season, and matures and is harvested in the dry season. The practice ensures that the crop grows vegetatively during the period with the most rainfall, and bolls mature during periods of high quality radiation. This is important because cotton is least sensitive to the low amounts of radiation associated with rainfall during its vegetative growth phase. However, adequate root development is important because late-season water stress has the potential to limit crop growth.

In temperate regions, crop development is reliably predicted by degree-day accumulation. However in the tropics the degree-day relationship is less clear due to temperature extremes before squaring and the possibility of delayed flowering due to shedding (Yeates et al., 2010a, 2010b). Days after sowing can be a better predictor of phenology (Grundy et al., 2012).

Cloudy conditions combined with high night temperatures can reduce yield potential (Grundy et al., 2012; Yeates et al., 2010b). This is probably due to the combination of an increase in aborted fruit as well as a decrease in the production of fruiting sites as a result of the reduction in the supply of carbon assimilates, as well as of an increase in respiration rates. The abscission of squares, flowers, and young bolls is a survival mechanism employed by the cotton crop. In the tropics, this response is associated with periods of low radiation, where the plant rebalances its internal demand for assimilates by shedding developing fruit. Doing so allows for continued vegetative expansion and production of fruiting sites that have the potential to develop into bolls when conditions are more favorable. If radiation is limiting assimilate supply, the retention of fruit can be counterproductive because bolls set during these periods are smaller and then compete with future canopy growth and fruit production. Grundy et al. (2012) outlined the two types of flower shedding that occur in response to reduced radiation:

1. Shedding as a result of cloudy conditions that extend beyond 4 or 5 d. This type of shedding is in response to reduced photoassimilate supply and, in a manner similar to heat stress fruit abortion, progresses from the most mature fruiting sites (small bolls >10 d old) to the least developed. As a result, first-position fruit will be shed before second-position fruit. This is an adaptive response whereby the younger fruit will have more access to radiation when conditions are more favorable because they will be less susceptible to self-shading.
2. A sudden change in weather from extended cloudy conditions to hot, dry conditions. Small squares are generally most susceptible to this form of fruit shedding.

Because the season length is sufficiently long, some monsoon-influenced fruit shedding leading to loss of lower fruit (due to lack of radiation or boll rots associated with humid weather conditions) is acceptable (Fig. 2). Such shedding results

Fig. 2. Due to excess moisture, low radiation, and fungal diseases (boll rots), cotton in tropical regions often sheds most of its lower and first-position bolls developing in the early flowering period (left). Fruit infested with boll rot (right).

in taller crops with a larger proportion of fruit at the top of the plant. Shedding of the lower fruit during early flowering in this environment can result in a yield advantage because plants will have more resources to set additional, and larger, fruit when high radiation occurs later in the season. Flower abortion can also occur when rainfall on newly opened flowers kills pollen. This is only an issue for midmorning showers when pollen development and pollination is susceptible to rainfall (Pennington et al., 1989; Burke, 2003).

Although tropical areas can receive up to 20% less radiation than more arid environments, this loss can be partially compensated for by the combination of access to a long season and more favorable temperatures during boll development, ensuring more efficient photosynthesis (Yeates et al., 2010a; Grundy, et al., 2012). Cotton further compensates for low radiation by growing larger leaves with decreased cuticle thickness much earlier in the season, thus enabling more light interception (Oosterhuis et al., 1991; Grundy et al., 2012).

Short-Season Humid Regions: Northeastern US Cotton Belt

The northeastern part of the US Cotton Belt includes the coastal states of Virginia and North Carolina. These regions have a humid, temperate environment in which crops are predominantly rainfed and irrigation is required only to supplement in-crop rainfall. The key climatic challenge to growing cotton in such an environment is the shortened season combined with reduced solar radiation levels and less reliability of consistent optimal temperatures. Yield reductions resulting from low amounts of radiation are primarily due to fewer numbers of bolls produced per plant (Pettigrew, 1994). Fiber quality is impacted by cloudy weather and cooler conditions because an associated decrease in photosynthates reduces the carbohydrates available for fiber development. Cotton grown in such regions must be grown under a short-season management strategy based on maximizing early-season growth, stimulating early flowering, preventing rank growth with chemicals such as mepiquat chloride, and avoiding boll losses due to insect pressure. These management approaches are designed to retain as many fruit as possible produced by the plant, resulting in earlier crop maturation (Edmiston, 2013b). Unlike long-season humid environments, these factors are of great importance in short-season environments because the plants' ability to grow compensatory bolls later in the season is reduced.

Early and midseason rank cotton growth is one of the biggest challenges in cotton production in short-season humid environments. Rank cotton growth results in an imbalance between vegetative and reproductive plant growth, favoring vegetative growth (Hearn, 1975, 1979). It can be caused by abundant water and nitrogen accompanied by warm temperatures and is more likely to occur if there is a delay in the initiation of fruiting branches, particularly if the early squares are not retained. As such, the effect of rank growth can be exacerbated in short-season humid regions, where cool, cloudy conditions can lead to square shedding as a result of decreased photosynthesis as well as to harbor pests that may damage squares. Furthermore, rank growth can result in a negative feedback loop in which the developing fruit and their subtending leaf (the principal carbohydrate supplier) are shaded by the rapidly expanding canopy. The shade results in reduced photosynthates for the developing fruit, which can then be shed due to lack of resources. As cotton evolved as a perennial, indeterminate plant native to the subtropics, rank growth allowed the plant to use resources when available to provide enough plant structure to support increased fruit development later in the season (Hearn, 1994). However, unlike in long-season environments, plants grown in short-season environments do not have the ability to compensate for early and midseason fruit losses. As a result, producers in short-season environments tend to grow early-maturing cultivars.

Earliness in cotton is a complex polygenic trait influenced by a number of factors, such as morphology, phenology, physiology, and environmental attributes. Plant features including the initiation of fruiting branches at lower nodes, high retention of early fruit, and rapid boll maturity and opening contribute to earliness (Edmisten, 2013b). The most important factor in earliness is high retention of early fruit. Therefore short-season crops are smaller in stature with more fruit at lower nodes, and consequently a lower yield potential. Additionally, Bednarz and Nichols (2005) showed that shorter horizontal flowering intervals and boll maturation times are prevalent traits in higher-yielding short-season varieties. The morphology of a well-managed cotton plant grown in a short-season humid region tends to be shorter than in full-season locations. Plants will typically have up to 12 or 15 fruiting branches and bolls that have a higher retention (50–70%) on lower nodes (6–10). Bolls produced on node 18 have a less than 10% chance of full maturation. Second-position boll set also peaks at 20 to 30% between nodes 6 and 10 (Edmisten, 2013a, 2013b).

Short-Season Arid Regions: Texas High Plains, USA

Plants in short-season arid areas have growth patterns similar to those in long-season arid regions; however, the truncated season length results in shorter plants with reduced fruiting sites. Furthermore, because bolls can mature in cooler conditions, the fiber quality can be reduced. The southern High Plains of Texas represents the largest contiguous cotton production area in the USA. In this region, as in all arid cotton-production regions, water supply is the principal driver of yield, which is primarily a result of limiting the number of fruiting sites (Morrow and Krieg 1990). In years with ample rain or under supplemental irrigation, season length is the major limitation on yield, and earliness is a desirable trait or management scheme (Peng et al., 1989; Morrow and Krieg 1990; Gwathmey et al., 2011). In general, short-season arid environments result in retarded leaf growth rates and stem elongation, leading to shorter, lower-yielding plants with fewer nodes (Wells and Meredith, 1984; Reddy, Baker, and Hodges, 1991; Reddy et al., 1997; Howell et al., 2004; Bednarz and Nichols, 2005; Snowden et al., 2013a, 2013b). Although plants with fewer bolls can compensate to some degree producing larger bolls, the number of bolls per unit area is the most significant component that explains crop production (Wells and Meredith, 1984). In addition, drought and associated heat stress can adversely affect fiber length and other fiber-quality characteristics (Antony and Kutty, 1975; Guinn and Mauney, 1984a, 1984b; Bradow and Bauer, 1997; Liakatas et al., 1998; El-Zik et al., 2000).

The Response of Root Morphology to Environment

Root morphology can vary due to genetic or environmental factors. Environmental differences that impede the development of a vigorous and healthy root system can significantly impact aboveground productivity, because the roots are the plant's source of water and nutrients.

Differences in the soil environment are the major factors influencing root growth; however, aboveground processes in the plant that impact photosynthesis, fruiting, and leaf area interact to influence root development. Although during the past four to five decades there has been progress in understanding cotton root systems, much remains unknown or without overwhelming consensus. McMichael et al. (2010, 2011) provide a comprehensive discussion of the knowledge of how cotton root growth responds to environmental factors.

As with aboveground growth and development, root growth is responsive to temperature, with root growth increasing as the soil temperature increases. The optimum temperature for cotton root growth has been reported to range between 28 and 35 °C (McMichael and Quinsenberry, 1993). In short-season environments where temperatures are cooler at the beginning of the season, soil temperatures can be significantly lower than optimum for root development (Bolger et al., 1992). In these environments, early root growth, which is critical in the establishment of healthy and vigorous plants, is needed for growth and crop maturity within the short season environment. In response to low temperatures ($<20°C$), root growth is reduced, and less branching can occur. Water and nutrient uptake can also be reduced under low temperatures and have possible impacts on above- and belowground growth (Bolger et al., 1992).

In long-season regions, high temperatures can lead to soil temperatures exceeding the optimum for root development. At optimum temperatures, the branching of roots is greater and water uptake increases; however, these processes then decline at higher temperatures, with the number of lateral roots declining rapidly after the optimum temperature is exceeded. Tap root growth may have a slightly higher resilience with McMichael et al. (2010) reporting that tap root length was maintained in cotton grown at high temperatures, whereas lateral branching and lateral root length declined.

Cotton roots are sensitive to changes in the soil moisture level and plant water potential, altering their structure accordingly in an attempt to maintain water uptake as the soil dries (Taylor and Klepper 1974). Patterns of soil wetting and drying can also influence the distribution of roots within the soil profile. In high-rainfall environments or with irrigation systems providing a frequent water supply (e.g., drip), root growth is often concentrated in the upper part of the soil profile. Carmi et al. (1992) found that there was a higher percentage of small roots around drippers, indicating adaptation and dependence on the frequent supply of water. Root morphology of cotton grown in the wet season and harvested in the dry season in the tropics can also develop a shallow root system in response to a readily available water supply from excessive rainfall. The plant will not invest assimilates in root growth if water and nutrient supplies can be adequately met. However, once dry conditions develop, water stress and premature cut-out can occur because the roots are not able to adapt to the dramatic drop in water supply. In a study in India by Sampathkumar et al. (2012), cotton exposed to water stress had a greater rooting depth in comparison with various irrigation treatments. Similar responses were described by Klepper et al. (1991), where a reduction in soil moisture resulted in death of early roots in the upper part of the soil profile and subsequent root development occured deeper in the soil profile. Additionally, water-stressed cotton root systems have reduced lateral root lengths and diameters. Both commercial and exotic cottons have been shown to increase their root biomass when grown in environments having evaporative stress and water limitation, thus resulting in an increased root-to-shoot biomass ratio (McMichael and Quisenberry, 1991).

The Influence of Soil Type on Root Morphology

Across the globe, cotton is grown in a number of different soil types ranging from deep sandy soils (Arizona, USA) to silt loams (Tennessee, USA) and heavy clay soils (Australia). Regions that might have similar atmospheric climatic conditions can vary widely in soil type, with heavy clays in the Mississippi Delta, fine sandy loams in the southeast Coastal Plains, and erodible silt loams in western Tennessee. Similar soil types can be found in very different climatic zones, for example, heavy, clay soils in the Mississippi Delta and in areas of India and northwestern New South Wales.

The characteristics of the soil environment have the most influence on root development, and within different environments, differences in soil type will influence root morphology. In general, soils with a low water-holding capacity have deeper roots, whereas those with a higher capacity have shallower roots.

Cotton roots can reach a depth of 3 m (McMichael, 1986) and extend out laterally more than 2 m (Taylor and Klepper, 1974). The root system continues to grow and increase in length until fruiting begins, at which time the total root length begins to decline as older roots die and fewer new roots are produced. The size and distribution of the root systems depends on the ability of the primary root to penetrate the soil and the ability of the root system to produce secondary (lateral) roots. The structure of the soil will determine how easily roots can penetrate, and the latter in turn will influence their depth and lateral root growth. Compacted layers or layers of high density will impede root growth, and root elongation will decrease. Comparing root growth in two different soil types in California,

Grimes et al. (1975) found that root growth was rapid to depth in the clay loam but impeded by a high-density layer at 30 cm in the profile of a sandy loam soil.

In their study conducted on silt loam soil in the southeastern United States, Schwab et al. (2000) estimated that 50% of the cotton root system is within the upper 15 cm of the soil profile during vegetative growth. As the plant matures, the root system no longer grows exponentially and will redistribute, with as much as 70% of the biomass existing within the top 30 cm of the soil following flowering initiation and early boll-filling stages. This overall architecture has been suggested to result in complete water extraction in the upper soil profile, though approximately 85 to 120 mm of soil water considered to be available to the plant is not taken up due to minimal root investment at depth. Some sandy clay loam soils in the southeastern USA have limiting layers in the subsoil that restrict root growth and water movement (Bauer and Busscher, 1993; Bauer et al., 1995). Root-limiting layers of this type may arise from compaction due to machinery traffic on the soil but may also occur naturally. Plants growing in a silty clay loam in Alabama and dug up 2 wk after emergence had developed lateral taproots that ran along the top of a compacted zone 5 to 10 cm below the soil surface (Burmester et al., 1995).

In some soil types, particularly heavy clays, waterlogging can have adverse impacts on root growth. In Australia, the production of cotton is concentrated on cracking gray clay soils (vertisols), which have inherently low drainage rates (Hodgson and Chan, 1982; McKenzie et al., 1984; Triantafilis et al., 2003). Combined with the almost exclusive use of furrow irrigation and a summer-dominant rainfall pattern, these heavy clay soils pose a significant risk of intermittent waterlogging. The problem can be exacerbated by poor land forming, such as excessive field length and inadequate slope; by poor soil structure, poor leveling, or bed formation; or by substantial rainfall after an irrigation. The low tolerance of cotton roots to poor soil aeration leads to reduced root elongation associated with flooding and water-logging of the soil. The consequences of prolonged waterlogging for the plant may include reduction or cessation of growth, the death of root apices, boll shedding, and changes in the patterns of nutrient accumulation. Leonard (1945) noted an abundance of lateral roots in cotton near the soil surface where waterlogging was a problem in cotton grown in Houston clay soil in Mississippi.

Although we have some knowledge of cotton root morphology in response to the soil environment, many studies of roots have been in controlled environments, so there is a need for research aimed at understanding root growth in a wide range of different environments and soil conditions.

Summary

Cotton is an indeterminate, subtropical plant that is grown across a wide geographical range and is able to adapt its morphology in response to the environment. Cotton yield is influenced by the number of bolls, their size, and the retention of bolls and is dependent on an adequate carbohydrate supply to support the initiation and growth of the bolls. Optimizing yield depends on managing plant structure and fruiting patterns to ensure that the fruiting cycle and maturation of the fruit match the length of the season. Management also needs to account for the constraints of the environment (temperature, radiation, humidity, and water availability) to maintain the balance of assimilate supply and the demand for it by the developing bolls. Plant mapping can be used to evaluate plant structure and to monitor vegetative and reproductive development and the distribution and retention of developing fruit. Optimum management practices vary widely across the cotton production environments, and understanding cotton's morphological response to the environment is critical.

The information presented in this review demonstrates, if nothing else, the complexity of cotton production. Genotype × environment × management interactions, particularly when not well controlled, documented, or understood, result in many apparently contradictory relationships. The contradictions can lead to the conclusion that the physiology of cotton is particularly complex, and possibly intractable. Alternatively, perhaps as the environment and management aspects are more fully quantified and understood, the underlying physiological mechanisms may prove to be much more straightforward, and thus cotton improvement more easily achieved.

References

Antony, A., and K. Kutty. 1975. Effects of soil and climatic conditions on yield and fiber properties of improved strains of upland cotton. Ind. J. Agric. Sci. 45:199–203.

Baranov, P.A. and A.M. Maltzev. 1937. The structure and development of the cotton plant: An atlas. Ogis-Isogis, Mosow-Leningrad.

Bauer, P.J., and W.J. Busscher. 1993. Effect of winter cover on soil moisture content in conventional and strip tillage cotton. In: P.K. Bollich, editor, The evolution of conservation tillage systems. Proceedings of the 1993 Southern Conservation Tillage Conference for Sustainable Agriculture, Monroe, LA. 15–17 June 1993. Louisiana State Univ. Agric. Center, Baton Rouge, LA. p. 8–10.

Bauer, P.J., W.J. Busscher, and J.M. Bradow. 1995. Cotton response to reduced tillage and cover crops in the southeastern Coastal Plain. In: W.L. Kingery and N. Buehring, editors, Conservation farming: A focus on water quality. Proceedings of the 1995 Southern Conservation Tillage Conference for Sustainable Agriculture, Jackson, MS. 26–28 June 1995. Mississippi State Univ., Mississippi State. p. 100–102.

Bednarz, C.W., and R.L. Nichols. 2005. Phenological and morphological components of cotton crop maturity. Crop Sci. 45:1497–1503. doi:10.2135/cropsci2004.0321

Burmester, C.H., M.G. Patterson, and D.W. Reeves. 1995. No-till cotton response to cover crop system and starter fertilizer placement in northern Alabama. In: Proc. Beltwide Cotton Conferences, San Antonio TX. 4–7 Jan. 1995. Natl. Cotton Counc. Am., Memphis, TN. p. 1310–1313.

Bolger, T.P., D.R. Upchurch, and B.L. McMichael. 1992. Temperature effects on cotton root hydraulic conductance. Environ. Exp. Bot. 32:49–54. doi:10.1016/0098-8472(92)90029-2

Bourgou, L., and D. Sanfo. 2012. Potential genetic benefits of using Brazilian cotton varieties to improve those cultivated in the C4 countries: 1. Analysis of major architectural and agronomic characteristics. Tropicultura 30 (4): 243–248.

Bradow, J.M., and P.J. Bauer. 1997. Fiber-quality variations related to cotton planting date and temperature. In: Proceedings of the Beltwide Cotton Conferences, New Orleans, LA. 6–10 Jan. 1997. Natl. Cotton Counc. Am., Memphis, TN. p. 1491–1495.

Brubaker, C.L., and J.F. Wendel. 1994. Re-evaluating the origin of domesticated cotton (*Gossypium hirsutum*; Malvaceae) using nuclear restriction fragment length polymorphisms (RFLPs). Am. J. Bot. 81:1309–1326. doi:10.2307/2445407

Burke, J.J. 2003. Sprinkler-induced flower losses and yield reductions in cotton (*Gossypium hirsutum* L.). Agron. J. 95:709–714. doi:10.2134/agronj2003.0709

Carmi, A., Z. Plaut, B. Heuer, and A. Grava. 1992. Establishment of shallow and restricted root systems in cotton and its impact on plant response to irrigation. Irrig. Sci. 13:87–91. doi:10.1007/BF00193985

Constable, G.A. 1976. Temperature effects on the early field development of cotton. Aus. J. Exp. Agr. Anim. Husb. 16:905–910.

Constable, G.A. 1991. Mapping the survival of fruit on field-grown cotton. Agron. J. 83:374–378. doi:10.2134/agronj1991.00021962008300020022x

Echer, F.R., D.M. Oosterhuis, D.A. Loka, and C.A. Rosolem. 2014. High night temperatures during the floral bud stage increase the abscission of reproductive structures in cotton. J. Agron. Crop Sci. 200:191–198. doi:10.1111/jac.12056

Edmisten, K.L. 2013a. The cotton plant. In: 2013 Cotton Information. North Carolina Coop. Ext. Serv., North Carolina State Univ., Raleigh, NC. p. 5–12.

Edmisten, K.L. 2013b. Developing a management strategy: Short-season timeliness. In: 2013 Cotton Information. North Carolina Coop. Ext. Serv., North Carolina State Univ., Raleigh. p. 16–22.

El-Zik, K.M., P.M. Thaxton, T.F. Dusek, T.D. Brooks, K. Schaefer, P. Dugger, and P. Richterx, 2000. Effect of planting date, moisture and heat units on cotton yield and fiber quality under dryland production. Proceedings of the Beltwide Cotton Conferences, San Antonio, TX. 4–8 Jan. 2000. p. 525–528.

Gipson, J.R., and L.L. Ray. 1969. Fiber elongation rates in five varieties of cotton (*Gossypium hirsutum* L.) as influenced by night temperature. Crop Sci. 9:339–341. doi:10.2135/cropsci1969.0011183X000900030027x

Godoy, A.S., and G.A. Palomo. 1999. Genetic analysis of earliness in upland cotton (*Gossypium hirsutum* L.). I. Morphological and phenological variables. Euphytica 105:155–160. doi:10.1023/A:1003490016166

Grimes, D.W., R.J. Miller, and P.L. Wiley. 1975. Cotton and corn root development in two field soils of different strength characteristics. Agron. J. 67:519–523. doi:10.2134/agronj1975.00021962006700040016x

Grundy, P., S. Yeates, and T. Grundy. 2012. NORpak: Cotton production and management guidelines for the Burdekin and north Queensland coastal dry tropics region. Cotton Catchments Communities CRC, Narrabri, NSW, Australia.

Guinn, G., and J. Mauney. 1984a. Fruiting of cotton. I. Effects of plant moisture stress on flowering. Agron. J. 76:90–94. doi:10.2134/agronj1984.00021962007600010023x

Guinn, G., and J. Mauney. 1984b. Fruiting of cotton II. Effects of plant moisture stress on active boll load and boll retention. Agron. J. 76:94–98. doi:10.2134/agronj1984.00021962007600010024x

Gwathmey, C.O., B.G. Leib, and C.L. Main. 2011. Lint yield and crop maturity responses to irrigation in a short-season environment. J. Cotton Sci. 15:1–10.

Hague S.S., W.C. Smith, J.C. Faircloth, J. Dever, G. , J. Clement, and D.C. Jones. 2014. Evaluation of cotton productivity by fruiting zone in diverse growing locations. J. Cotton Sci. 18:166–172

Hake, K.D., and T.A. Kerby. 1996. Cotton and the environment. In: S.J. Hake, T.A. Kerby, and K.D. Hake, editors, Cotton production manual. Agric. Nat. Resources, University of California, Richmond, CA. p. 324–331.

Hatfield, J.L., J.J. Burke, J.R. Mahan, and D.E. Wanjura. 1987. Foliage temperature measurements: A link between the biological and physical environments, In: R. J. Hanks, editor, Proceedings of international conference on measurement of soil plant and water status, Logan, UT. 6–10 July 1987. Utah State Univ., Logan. p. 99–102.

Hearn, A.B. 1969. The growth and performance of cotton in a desert environment. J. Agric. Sci. 73:75–97. doi:10.1017/S0021859600024151

Hearn, A.B. 1975. Response of cotton to water and nitrogen in a tropical environment.1. Frequency of watering and method of application of nitrogen. J. Agric. Sci. 84:407–417.

Hearn, A.B. 1979. Water relationships in cotton. Outlook Agric. 10:159–166.

Hearn, A.B. 1994. The principles of cotton water relations and their application in management. In: G.A. Constable and N.W. Forrester, editors, Challenging the future. Proceedings of the World Cotton Research Conference—1, Brisbane, Australia. 13–17 Feb. 1994. Intl. Cotton Advisory Committee, Washington, DC. p. 66–92.

Hector, J.M. 1936. Introduction to the botany of field crops. Vol. 2. Non-cereals. Central News Agency, Johannesburg, South Africa.

Hesketh, J.D., and A. Low. 1968. The effect of temperature on components of yield and fibre quality of cotton varieties of diverse origin. Cotton Grower Rev. 45:243–257.

Hesketh, J.D., H.C. Lane, R.S. Alberte, and S. Fox. 1975. Earliness factors in cotton: New comparisons among genotypes. Cotton Grower Rev. 52:126–133.

Hodgson, A.S., and K.Y. Chan. 1982. The effect of short-term waterlogging during furrow irrigation of cotton in a cracking grey clay. Aust. J. Agric. Res. 33:109–116. doi:10.1071/AR9820109

Howell, T.A., S.R. Evett, J. Tolk, and A.D. Schneider. 2004. Evapotranspiration of full-, or deficit-irrigated, and dryland cotton on the northern Texas High Plains. J. Irrig. Drain. Eng. 130:277–285. doi:10.1061/(ASCE)0733-9437(2004)130:4(277)

Klepper, B. 1991. Crop root system response to irrigation. Irrig. Sci. 12: 105–108.

Leonard, O.A. 1945. Cotton root development in relation to natural aeration of some Mississippi Blackbelt and Delta soils. J. Am. Soc. Agron. 37:55–71. doi:10.2134/agronj1945.00021962003700010005x

Liakatas, A., D. Roussopoulos, and W.J. Whittington. 1998. Controlled-temperature effects on cotton yield and fiber properties. J. Agric. Sci. 130:463–471. doi:10.1017/S0021859698005413

McKenzie, D.C., T.S. Abbott, D.A. Anthony, D.J.M. Hall, and F.R. Higginson. 1984. Soil management strategies for improving the yields of irrigated crops in the lower Macquarie valley. In: W.A. Muirhead and E. Humphries, editors, Root zone limitations to crop production on clay soils. Proceedings of the Symposium of the Australian Society of Soil Science. Griffith, New South Wales. 25–27 Sept. 1984. CSIRO Publishing, Melbourne, Australia. p. 149–161.

Mauney, J.R. 1966. Floral initiation of upland cotton *(Gossypium hirsutum* L.) in response to temperature. J. Exp. Bot. 17:452–459. doi:10.1093/jxb/17.3.452

Mauney, J.R. 1984. Anatomy and morphology of cultivated cottons. In: R.J. Kohel and C.F. Lewis, editors, Cotton. Agron. Monogr. 24. ASA, CSSA, and SSSA, Madison, WI. p. 25–79.

Mauney, J.R. 1986. Vegetative growth and development of fruiting sites. In: J.R. Mauney and J.M. Stewart, editors, Cotton physiology. Cotton Foundation, Memphis, TN. p. 11–28.

Mauney, J.R. and E. Ball. 1959. The axillary buds of cotton. Bull. Torr. Bot. Club. 86:236–244.

McMichael, B.L. 1986. Growth of roots. In: J.R. Mauney and J.M. Stewart, editors, Cotton physiology. Cotton Foundation Reference Book 1. Cotton Foundation, Memphis, TN. p. 29–38.

McMichael, B.L., D.M. Oosterhuis, and J.C. Zak. 2011. Stress response in cotton root systems. In: D.M. Oosterhuis, editor, Stress physiology in cotton. Cotton Foundation, Cordova, TN. p. 97–112.

McMichael, B.L., D.M. Oosterhuis, J.C. Zak, and C. Beyrouty. 2010. Growth and development of root systems. In: J.M. Stewart, D.M. Oosterhuis, J.J. Heitholt, and J.R. Mauney, editors, Physiology of cotton. Springer, London. p. 57–71.

McMichael, B.L., and J.E. Quisenberry. 1991. Genetic variation for root-shoot relationships among cotton germplasm. Environ. Exp. Bot. 31:461–470. doi:10.1016/0098-8472(91)90045-P

McMichael, B.L., and J.E. Quisenberry. 1993. The impact of the soil environment on the growth of root systems. Environ. Exp. Bot. 33:53–61. doi:10.1016/0098-8472(93)90055-K

Morrow, M., and D. Krieg. 1990. Cotton management strategies for a short growing season environment: Water-nitrogen consideration. Agron. J. 82:52–56. doi:10.2134/agronj1990.00021962008200010011x

Oosterhuis, D.M. 1990. Growth and development of the cotton plant. In: W.N. Miley and D.M. Oosterhuis, editors, Nitrogen nutrition in cotton: Practical issues. Proceedings of the Southern Branch Workshop for Practicing Agronomists. ASA, Madison, WI. p. 1–24.

Oosterhuis, D.M., R.E. Hampton, and K.S. Kim. 1991. Characteristics of the cotton leaf cuticle. Arkansas Farm Res. 40:12–14.

Oosterhuis, D.M., and J. Jernstedt. 1999. Morphology and anatomy of the cotton plant. In: W. Smith and J.S. Cothren, editors, Cotton: Origin, history, technology and production. John Wiley & Sons, New York. p. 175–206.

Peng, S., D.R. Kreig, and S.K. Hicks. 1989. Cotton lint yield response to accumulated heat units and soil water supply. Field Crops Res. 19:253–262. doi:10.1016/0378-4290(89)90097-X

Pennington, D.A., H.C. Pringle, and R. Wells. 1989. Effect of sprinkler irrigation on open cotton flowers. Bulletin 960/ Mississippi Agric. Forestry Exp. Sta., Mississippi State Univ., Mississippi State..

Pettigrew, W.T. 1994. Source-to-sink manipulation effects on cotton lint yield and yield components. Agron. J. 86:731–735. doi:10.2134/agronj1994.00021962008600040027x

Pettigrew, W.T. 2008. The effect of higher temperatures on cotton lint yield production and fiber quality. Crop Sci. 48:278–285. doi:10.2135/cropsci2007.05.0261

Quintanilha, A., D.L. Salazar, and A. Cabral. 1962. Desenvolimeto do butao flora algodoerio em funcao do tempo. Bot. Soc. Broteriana 36:189–215.

Reddy, V.R., D.N. Baker, and H.F. Hodges. 1991a. The effects of temperature on cotton canopy growth, photosynthesis, and respiration. Agron. J. 83:699–704. doi:10.2134/agronj1991.00021962008300040010x

Reddy, K.R., H.F. Hodges, and J.M. McKinion. 1997. A comparison of scenarios for the effect of global climate change on cotton growth and yield. Aust. J. Plant Physiol. 24:707–713.

Reddy, K.R., H.F. Hodges, and V.R. Reddy. 1992. Temperature effects on cotton fruit retention. Agron. J. 84:26–30. doi:10.2134/agronj1992.00021962008400010006x

Reddy, V.R., K.R. Reddy, and D.N. Baker. 1991b. Temperature effect on growth and development of cotton during the fruiting period. Agron. J. 83:211–217. doi:10.2134/agronj1991.00021962008300010050x

Sadras, V.O. 1995. Compensatory growth in cotton after loss of reproductive organs. Field Crops Res. 40:1–18.

Sampathkumar, T., B.J. Pandian, and S. Mahimairaja. 2012. Soil moisture distribution and root characters as influenced by deficit irrigation through drip system in cotton–maize cropping sequence. Agric. Water Manage. 103:43–53. doi:10.1016/j.agwat.2011.10.016

Schwab, G.J., G.L. Mullins, and C.H. Burmester. 2000. Growth and nutrient uptake by cotton roots under field conditions. Commun. Soil Sci. Plant Anal. 31:149–164. doi:10.1080/00103620009370426

Singh, R.P., P.V.V. Prasad, K. Sunita, S.N. Giri, and K.R. Reddy. 2007. Influence of high temperature and breeding for heat tolerance in cotton: A review. Adv. Agron. 93:313–385. doi:10.1016/S0065-2113(06)93006-5

Snider, J.L., and D.M. Oosterhuis. 2011. How does timing, duration and severity of heat stress influence pollen-pistil interactions in angiosperms? Plant Signal. Behav. 6:930–933. doi:10.4161/psb.6.7.15315

Snowden, M.C., G.L. Ritchie, J. Cave, W. Keeling, and N. Rajan. 2013a. Multiple irrigation levels affect boll distribution, yield, and fiber micronaire in cotton. Agron. J. 105:1536–1544. doi:10.2134/agronj2013.0084

Snowden, M.C., G.L. Ritchie, and T.L. Thompson. 2013b. Water use efficiency and irrigation response of cotton cultivars on subsurface drip in West Texas. J. Cotton Sci. 17:1–9.

Taylor, H.M., and B. Klepper. 1974. Water relations of cotton. I. root growth and water use as related to top growth and soil water content. Agron. J. 66:584–588. doi:10.2134/agronj1974.00021962006600040031x

Triantafilis, J., A.I. Huckel, and I.O.A. Odeh. 2003. Field-scale assessment of deep drainage risk. Irrig. Sci. 21:183–192.

Wells, R., and W.R. Meredith. 1984. Comparative growth of obsolete and modern cultivars: Relationship of yield to observed growth characteristics. Crop Sci. 24:868–872. doi:10.2135/cropsci1984.0011183X002400050010x

Yeates, S.J., G.A. Constable, and T. McCumstie. 2010a. Irrigated cotton in the tropical dry season. I: Yield, its components and crop development. Field Crops Res. 116:278–289. doi:10.1016/j.fcr.2010.01.005

Yeates, S.J., G.A. Constable, and T. McCumstie. 2010b. Irrigated cotton in the tropical dry season. II: Biomass accumulation, partitioning and RUE. Field Crops Res. 116:290–299. doi:10.1016/j.fcr.2010.01.007

Zeiher, C.A., P.W. Brown, J.C. Silvertooth, N. Matumba, and N. Mitton. 1994. The effect of night temperature on cotton reproductive development In: Cotton: A College of Agriculture Rep. Series P-96. University of Arizona, Tucson, AZ. p. 26–32.

Zeiher, C., N. Matumba, P. Brown, and J. Silvertooth. 1995. Response of upland cotton to elevated night temperatures. II. Results of controlled environment studies. In: Proceedings of the Beltwide Cotton Conferences, San Antonio, TX. 4–7 Jan. 1995. Natl. Cotton Counc. Am., Memphis, TN. p. 1129–1130.

Crop Growing Practices

Philip J. Bauer

Abstract

Cotton (*Gossypium hirsutum* L.) production involves managing the crop and the soil. Crop management includes making decisions on planting date, plant population, irrigation, and the use of plant growth regulators for managing canopy size and preparing the crop for harvest. In-season monitoring of the growth and nutrient status of the crop is used to help make some of these decisions. Sustainability of soils for cotton production includes minimizing losses to erosion and maintenance and enhancement of soil productivity. Conservation tillage is a proven technique to reduce erosion. Recent literature and grower use suggests management with conservation tillage can provide improved soil productivity, especially in locales where soil water limitations exist during the season. Winter cover crops with conservation tillage systems provide the potential to enhance productivity. Economic and environmental concerns are driving the development of improved nutrient management techniques for crop production. Site-specific application of nutrients is being developed for improved nutrient use efficiency in cotton. In addition, considerable research has been conducted recently on the use of animal manures as nutrient sources for the crop.

Numerous changes to the cotton industry have occurred since the first edition of this monograph was published in 1984. Regarding growing practices, notable changes include development of transgenic cultivars with herbicide and insect resistance, eradication of the boll weevil from most of the US cotton growing areas, more interaction between producers and processors about the needs for high-quality fiber to enhance spinning efficiency and fabric quality, and increased awareness of how production practices influence the environment. Though much has changed, many of the principles outlined by Waddle (1984) are still very applicable.

Growing practices around the globe for cotton are diverse, depending on environmental, economic, and resource availability conditions. The International Cotton Advisory Committee (ICAC, 2011) survey reported that there were 12,486 growers producing cotton on approximately 4.32 million hectares in the USA in 2011, while in China, 90,000,000 producers were growing cotton on approximately 5.6 million hectares. Not only are there major differences between countries,

---

Abbreviations: ICAC, International Cotton Advisory Committee; NDVI, normalized difference vegetative index.

Philip J. Bauer, USDA-ARS, Coastal Plains Soil, Water, and Plant Research Center, 2611 West Lucas St., Florence, SC, 29501 (Phil.bauer@ars.usda.gov).

doi:10.2134/agronmonogr57.2013.0029

Cotton. 2nd. ed. David D. Fang and Richard G. Percy, editors. Agronomy Monograph 57.

differences within countries can also be dramatic. In Brazil, for example, the ICAC estimated 384 growers grew cotton on 1.3 million hectares in the Brazilian savannas region, while 9750 growers grew cotton on 117,000 ha in the northeast region.

Even though production systems are diverse, the goal of every cotton producer is to attain high yield. Cotton yield is primarily a function of the number of bolls per unit land area, the size of the bolls, and the ratio of lint to seed in the bolls. Hake and Grimes (2010), writing about water management in cotton, stated that "the most important factor to consider when designing optimal plant–water relations is the impact of water management decisions on the ability of the plant to initiate, retain, and mature harvestable bolls." It seems similar goals are appropriate for any component of the management system. When agronomists are developing goals for their research programs, whether they are for near-term economic gain or long-term production sustainability, a major portion of the focus needs to remain on maximizing the chances of producing harvestable bolls.

It is not the purpose of this chapter to provide a guide for growing cotton. Excellent presentations on this are available, including Silvertooth et al. (1999). Online sources and local and regional specialists in the public and private sectors can provide details for specific locales. Cotton grower guides in the United States are updated annually by the extension services of land grant universities to provide information and recommendations on all facets of cotton production from field and soil selection through harvest management. The ICAC surveys cotton producing countries and provides a summary every 3 yr of specific practices used in countries and different regions within some countries with cotton production. *Cotton Production Practices* (ICAC, 2011) is most recent version at this writing.

A viable cotton industry depends on cotton growers providing the textile industry cotton of sufficient quantity and quality so that it can provide consumers cotton apparel and other products that are price competitive with products of other fibers, primarily synthetic fibers. Cotton growers have remained competitive by producing higher yield and improving quality with increased crop production efficiency. Improving nutrient management and preserving and enhancing soil productivity are critical to sustaining cotton production into the future. The remainder of this chapter contains a brief discussion of in-season management practices, followed by a discussion of recent research conducted in conservation tillage management and nutrient management in cotton.

In-Season Management

Planting Date and Plant Population

Numerous studies have been conducted in cotton growing locales to provide growers planting date windows that provide the best chance for optimal yield and quality. Within those windows, recommendations for when to begin planting are generally based on soil temperature (generally at or above 16°C at a certain depth in the profile for consecutive days) and temperature forecasts for the next few days. Cotton is a chill-sensitive species, and growth ceases at temperatures below 16°C (Tharp, 1960; Munro, 1987). Exposing imbibing seeds and young seedlings to chilling temperatures can impact stands and ultimately may impact yield and quality (reviewed by Bradow and Bauer, 2010).

Efforts have been made to develop production systems with planting occurring outside of the planting date window. Improvements in seed fungicides and seed quality that may reduce risk of stand and seedling damage when temperatures are cooler prompted Pettigrew (2002) to evaluate very early planting in Mississippi. He hypothesized that very early planting would provide increased yield because reproductive growth would occur nearer the summer solstice, providing more fixed carbon from photosynthesis to developing bolls. He tested a very early planting date compared to normal under irrigated conditions and found a 10% yield increase in four of five years with this system. In a subsequent 4-yr study (Pettigrew, 2010) to evaluate very early planting under irrigated and rainfed conditions, low amounts of precipitation during June resulted in higher yield with the system only under irrigated conditions.

Optimal seeding rates for a given field are determined by climate, production practices, and soil. Current recommended seeding rates are lower than previous recommendations. Waddle (1984) suggested planting rates should provide 100,000 to 140,000 plants/ha under irrigated conditions in temperate areas. Under nonirrigated conditions, 200,000 to 250,000 plants/ha were recommended in temperate areas, and 100,000 to 150,000 plants/ha were recommended in tropical areas. Current recommendations in South Carolina are 70,000 to 100,000 plants/ha under nonirrigated conditions, but only 34,000 to 70,000 plants/ha under irrigated conditions, where severe soil water deficit stress is likely, and where boll rot may be a problem (Jones et al., 2013).

Irrigation

Methods used to irrigate cotton include surface (furrow or flood), overhead sprinklers, and surface and buried drip irrigation. The determination of the best method for a particular field depends on many factors, including field size, field slope, field geometry, soil type, and water source. Variable rate overhead sprinkler irrigation is a new technology that offers promise for maximizing water use in fields with variable soils (Sadler et al., 2007). This technology allows for differing rates of irrigation along the length of the system and as the system travels through the field. Irrigation use for cotton is increasing in humid regions because of the need to eliminate the risk of low yields due to water deficit stress as the cost of other production inputs rises and because of an increase in rotations with crops that require irrigation for successful production.

Cotton seasonal water use differs due to differences in climate and other factors. Estimates of seasonal water use range from 550 to 1050 mm (Martin et al., 2007). Within the season, crop water use increases with crop growth and leaf area development and begins declining at about the time of the first open boll.

The sensitivity of cotton to water deficit stress differs depending on the stage of growth. The most sensitive growth stage is from early squaring into bloom. Plant growth and leaf area development is rapid during this time, and water deficit stress will limit yield potential through reduced leaf area for photosynthesis and reduced numbers of potential fruiting sites. Plant water deficit stress at this time can also result in shedding of squares and bolls. Severe water deficit during the first 30 d of a cotton boll's development lowers fruit retention, reduces fiber length, and causes high micronaire (Hake and Grimes, 2010).

Knowing the pattern of crop water use and the effect water deficit stress have on the crop is useful for irrigation scheduling. Current techniques used to schedule irrigation are soil based (e.g., tensiometers and electrical resistance sensors), plant based (e.g., pressure chambers to measure leaf water potential and infrared thermometers to measure canopy temperature), and computer based (Sadler et al., 2007). Modern communication technologies are allowing some soil and plant sensors to be remotely monitored via satellite, cell modem, and radio telemetry which greatly reduces the labor needed to use them for scheduling irrigation since data do not have to be collected at each sampling site.

Plant Growth Regulators

Plant growth regulators are natural or synthetic organic compounds that modify or control physiological processes within plants. Hormones are plant growth regulators that are produced within the plant. In general, hormones either promote growth (auxins, gibberellins, and cytokinins) or inhibit growth (ethylene and abscisic acid). Manipulating physiological processes can be accomplished through altering hormone levels within the plant or by changing the ability of the plant to respond to hormones (Cothren and Oosterhuis, 2010).

Plant growth regulator products are commonly used in cotton production throughout the world. Products containing cytokinins, cytokinin-like compounds, auxins, and gibberellins alone or in mixtures have been used. Research results on yield improvements with application of these products have been inconsistent (Cothren and Oosterhuis, 2010).

The product most widely used on cotton during the growing season is mepiquat chloride (*N,N*-dimethypiperidinum chloride). Mepiquat chloride is a growth inhibitor that reduces the amount of gibberellins in plant tissues. Recent evidence (Wang et al., 2014) suggests mepiquat chloride downregulates biosynthetic and metabolic genes of gibberellins in cotton. The lower amount of gibberellins in the plant tissues reduces internode elongation of both main stem and fruiting branches, resulting in shorter and more compact plants.

Although the response of cotton to mepiquat chloride for plant height has been consistent across many trials, the response of cotton to mepiquat chloride application for yield has been inconsistent, with reports of increased yield (Cathey and Meredith, 1988; Kerby, 1985; York, 1983), decreased yield (Cathey and Meredith, 1988; York, 1983; Zhao and Oosterhuis, 2000), and no effect on yield (Boman and Westerman, 1994; Gwathmey and Craig, 2003; Kerby et al., 1986; Zhao and Oosterhuis, 2000). A recent 2-yr study across 22 locations in 12 US cotton growing states found no influence of mepiquat chloride or a similar product, mepiquat pentaborate, on yield (Dodds et al., 2010). Kerby (1985) conducted 35 experiments over 5 yr in California and found a yield response to mepiquat chloride only when untreated plants exceeded 1.10 m at maturity or when growing season length was short. In a plant mapping study, Kerby et al. (1986) found mepiquat chloride increased the number of harvestable bolls at nodes 12 or lower but decreased the number of harvestable bolls at nodes 14 or above.

As a growth inhibitor, mepiquat chloride is applied to prevent excessive vegetative growth. Excessive vegetative growth can cause a delay in maturity, boll rot, and reduced yields. Thus, recommendations for its use are primarily for conditions that result in an undesirable ratio of vegetative to reproductive growth. These conditions include late planting, excessive rainfall or irrigation, high nitrogen fertility, high plant density, and other biotic or abiotic factors affecting the crop that may delay maturity.

Plant growth regulators are used at the end of the season to chemically terminate the crop, hasten boll opening, and precondition plants for defoliation.

Ethephon (2-chloroethyl phosphonic acid) is a plant growth regulator commonly applied at this time. Ethephon breaks down in the plant cells into ethylene, which stimulates cellulase and other hydrolase enzymes. These enzymes weaken and dissolve cell walls (Abeles, 1969).

In-Season Plant Monitoring

Waddle (1984) described the petiole nitrate monitoring system that was developed by Maples et al. (1977) for determining the need for additional N fertilizer during the growing season. In this system, petioles from uppermost fully expanded leaves are collected beginning at about 2 wk after the first flower buds appear and then at regular intervals. Analysis of the nitrate content in the petioles is then compared to experimentally determined sufficiency levels.

The petiole nitrate system is less dependable under rainfed conditions than under irrigated conditions because soil water content has an influence on petiole nitrate concentration. For rainfed conditions, leaf blade N concentration is measured (Bell et al., 2003).

In-season plant measurements are used to assist in making management decisions. Vigor indexes during squaring and early bloom are used to determine the need for application of mepiquat chloride. Measurements used include plant height, height/node ratio, average length of the top five internodes, and nodes above white bloom. Defoliation timing is critical for optimum yield and quality. Methods used to trigger defoliant application are percent open bolls and nodes above cracked boll.

Tillage and Nutrient Management

Historically, the soil has been thought of as the medium in which cotton plants obtain water and mineral nutrients, and it serve to anchor plants. Management practices and inputs that increased yield, reduced costs, and/or improved quality were often designed and implemented based on this view of soils. In recent years, "soil quality" and "soil health" have become parts of the lexicon of agricultural production, reflecting the balance needed between current production economics and providing future generations the ability to continue producing crops. Sustaining and improving soils and protecting air and water resources have become higher priorities as the world population, and thus the demand for food and fiber, grow.

Tillage

Conventional Tillage

Cotton produced with conventional tillage continues to be the norm in many of the cotton growing regions. Although specifics vary with production area, these systems usually consist of primary and secondary tillage tools, as well as between-row cultivators that bury nearly all previous crop residue and disturb all of the soil (see USDA-NRCS definition, http://ecat.sc.egov.usda.gov/Help.aspx, accessed 2 June 2015). Conventional tillage is practiced to prepare seedbeds, improve the physical condition of the soil, and eliminate weed competition. Tillage is also practiced (and in some areas mandated) for control of other pests, including pink bollworm.

Tillage buries crop residues, breaks up soil clods, and loosens soil so that planters can accurately place and firm the soil around seeds. Tillage also loosens compacted soils that can inhibit and modify root growth (McMichael et al., 2010). Cotton seed germination and early seedling growth are influenced by temperature (Wanjura, 1986; Bradow and Bauer, 2010), soil water availability, and oxygen levels (Christiansen and Rowland, 1986). To help warm soil and to help moderate soil water availability, often the last tillage practice before cotton is planted in humid regions is to construct raised beds. Low beds are used in well-drained soils while higher beds are used in wet soils to facilitate drainage from the soil around planted seeds.

Compared to most other agronomic crops, using tillage for weed control has been an especially important aspect in cotton production. Cotton is less able to compete with weeds during the early season because canopy closure is slower in cotton than in other crops. Slower canopy closure results from both production practices and from inherent characteristics of the species. In regions where cotton is spindle-picked, cotton has traditionally been planted in wide (usually ~1 m) rows to facilitate mechanical harvest. Slow canopy closure also occurs because cotton seeds have no true leaves, only cotyledons, so all differentiation of vegetative tissues above the cotyledons occurs after germination, causing cotton to remain a seedling longer than other crop species (Mauney, 1986). For corn (*Zea mays* L.) and soybean [*Glycine max* (L.) Merr.], weed control is necessary for 5 or 6 wk after planting while for cotton control may be necessary for 10 to 12 wk (Triplett and Dick, 2008).

Conservation Tillage

Although conventional tillage is used on most cotton fields, soils planted to cotton with conventional tillage are quite susceptible to erosion (Mutchler et al., 1985; Soileau et al., 1994; Truman et al., 2003). Soil erosion is a serious threat to worldwide sustainable crop production (Pimentel et al., 1995). Montgomery (2007), using estimates of soil erosion from the literature, compared soil erosion rates for conventional tillage agriculture, no-till agriculture, native vegetation, and geological erosion to natural rates of soil production. This analysis (which indicated erosion rates for conventional agriculture are >1 mm yr^{-1}) provided evidence that the historic 500 to several thousand year lifespan of major civilizations around the world could be due to loss of soil productivity due to erosion. As part of this analysis, data from 39 studies that had direct comparisons between conventional and no-till production indicate erosion is reduced with no-till 2.5 to >1000 times from conventional tillage rates and that the reduction is enough so that losses are of the same magnitude as natural rates of soil production.

The loss of soil through wind and water erosion drove the development of conservation tillage. Conservation tillage systems are any system where 30% or more of the soil is covered by plant residues after planting. These systems include mulch tillage, ridge-tillage, strip-tillage, and no-tillage (http://www.ctic.purdue.edu/resourcedisplay/322/, accessed 4 June 2015). These systems differ in the amount of soil that is disturbed before planting and the amount of residues remaining on the soil surface to protect the soil. No-tillage is defined as a planting system where one-third or less of the row area is disturbed.

Cotton production with conservation tillage has increased substantially since the first edition of this monograph. In 2008, it was estimated that no-tillage was used on 17.5% of US cotton hectares (http://www.ctic.purdue.edu/crm_results/, accessed 4 June 2015). Cotton yield response to conservation tillage systems was reviewed by Gwathmey et al. (2010). Over the years, researchers have found yields with conservation tillage to be lower, the same, and higher than conventional tillage systems. Lower yields with conservation tillage under rainfed conditions were especially apparent during the first growing season of conservation tillage. Including winter cover crops could overcome the yield decrease in the humid southeastern United States (Brown et al., 1985; Bauer and Busscher, 1996; Raper et al., 2000). In a large study that was conducted with grower cooperators across seven US states between 1998 and 2002, Buman et al. (2005) found no difference in cotton lint yield between conventional and conservation tillage.

The development and practice of conservation tillage on farms was made possible by modern herbicides (Triplett and Dick, 2008). In the southeastern United States, rapid growth in the number of hectares of cotton produced with conservation tillage occurred in the mid-1990s after the release of transgenic cotton cultivars that were resistant to glyphosate [*N*-(phosphonomethyl)glycine] herbicide (Buman et al., 2005). The widespread use of transgenic crops with herbicide resistance, along with the lack of rotation of crop species and herbicides with different modes of action, has resulted in the development of herbicide-resistant weeds. Herbicide-resistant weeds are a threat to the sustainability and future increase of hectares managed with conservation tillage.

For cotton production, residues covering the soil surface following cotton harvest are often not enough for sufficient soil protection from water erosion (Mutchler et al., 1985). Planting winter cover crops to increase surface residues can significantly reduce erosion losses (Mutchler and McDowell, 1990; Yoo and Touchton, 1988). Truman et al. (2009) found that the reduction in erosion with conservation tillage and a winter cover crop was due to reducing interrill erodibilities. To protect soil from wind erosion, keeping the soil protected with surface residues is the basic method used (Fryrear and Skidmore, 1985) and high residue rotation or winter crops are needed in cotton because of the low amounts of cotton residues.

Continuous no-tillage production increases the amount of soil organic matter near the soil surface. Potential benefits of soil organic matter are improved soil aggregation, enhanced structural stability, and increased cation exchange capacity. On the "Old Rotation" plots at Auburn, Alabama, where cover crops and crop rotations in cotton production have been studied for more than 100 yr (Entry et al., 1996), Mitchell and Entry (1998) reported a quadratic response of cotton yield with soil organic carbon (soil organic carbon ranged from $\sim$3 to 16 g kg^{-1} or 0.5 to 2.75% soil organic matter using 1.72 as the conversion from soil organic C to soil organic matter).

A benefit often associated with increasing soil organic matter is the increase in soil available water holding capacity. Hudson (1994) showed that as soil organic matter increased from 0.5 to 3.0% for sand, silt loam, and silty clay loam soil textures, available water content more than doubled. Temporarily increasing soil organic matter can be done with physical additions of organic materials, green manures, and through rotation (Waddle, 1984). Building soil organic matter by sequestering carbon in the soil with conservation tillage takes time, even with cover crops and rotations. Franzluebbers et al. (2012) estimated the potential of soils planted to cotton to sequester carbon using the soil conditioning index in the Revised Universal Soil Loss Equation (RUSLE2) model. In that analysis, which was done across 10 US cotton production regions, soil organic C losses averaged −0.31 (±0.19) Mg C ha^{-1} yr^{-1} for conventional tillage cotton using a moldboard plow. For conservation tillage, an average increase of 0.12 (±0.06) Mg C ha^{-1} yr^{-1} of sequestered C was estimated for cotton grown under a number of different rotations.

Improved water relations of cotton grown with conservation tillage where residues are left on the surface, especially during the first years of adoption, appear to be more related to changes in soil physical characteristics other than increased organic matter. Surface residues act as mulch in crop production systems to enhance rainfall (or irrigation) infiltration and reduce soil water evaporation, which can result in increased crop yield (Gallaher, 1977). Bosch et al. (2005) measured surface runoff and lateral subsurface flow in conservation (strip tillage) and conventional tillage plots for 5 yr in Georgia, USA. Cotton was grown on the plots for four of the years and peanut (*Arachis hypogaea* L.) in the other. A rye (*Secale cereale* L.) winter cover crop was grown every year. Surface runoff from the conservation tillage plots was 81% less than the runoff in the conventional tillage plots, but subsurface lateral flow was 73% higher with conservation tillage. The researchers estimated that net annual water gain in the conservation tillage plots was 60 mm.

Besides increasing infiltration, conservation tillage reduces water loss by soil evaporation. Lascano et al. (1994) quantified the value surface mulches provide to cotton productivity by reducing water evaporation from the soil surface. In their study, cotton was grown following a wheat (*Triticum aestivum* L.) cover crop. The wheat was shredded and incorporated in half of the plots and left as stubble in the other half. They found that total evapotranspiration (E_t) was similar between the two management practices but that the ratio of transpired water to total E_t was higher in cotton grown with wheat stubble than in cotton grown in conventional tillage. The higher amount of water lost through transpiration was associated with 35% higher cotton yields, increasing water use efficiency from 1.9 to 2.6 g lint kg^{-1} water. Both increased rainfall infiltration and decreased soil evaporation likely explain the results of an experiment in the southeastern United States, where cotton yields were higher with conservation tillage than with conventional tillage during an extended drought from 1998 through 2002 (Fig. 1; Bauer et al., 2010).

In addition to the physical changes in soils caused by tillage and residue management, management practices can influence the soil microbial communities in cotton fields. In long-term tillage studies, microbial biomass C (Wright et al., 2005; Feng et al., 2003) and microbial biomass N (Wright et al., 2005) were higher in soils with cotton grown with conservation tillage than conventional tillage. The timing and perhaps the intensity of tillage seem to be important, though, as Locke et al. (2013) found that minimal disruption by a one-time tillage operation in the fall did not affect microbial community structure. Particular interest has recently been given to the impacts of production systems on arbuscular mycorrhizae. These fungi are seen as an important component of sustainable agricultural systems because of their well-known ability to enhance host-plant uptake of P. Conventional tillage reduces arbuscular mycorrhizae hyphae survival and proliferation and affects fungal community composition (reviewed by Kabir, 2005). Zak et al. (1998) found higher amounts of root infection by arbuscular mycorrhizae fungi in young seedlings when cotton was grown with conservation tillage

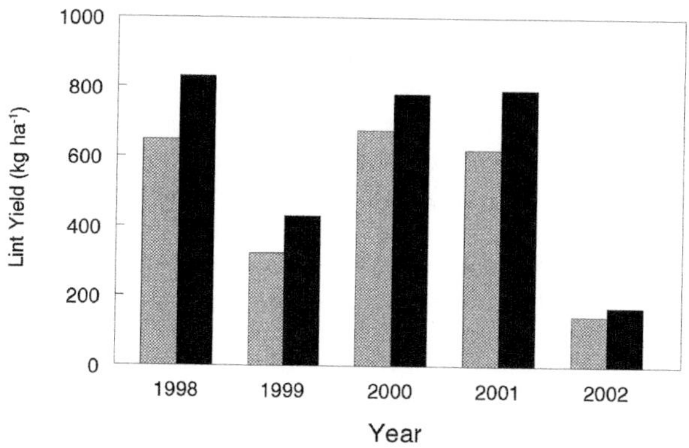

Fig. 1. Cotton lint yields during a multiyear drought period in the southeastern United States. Data are from Bauer et al. (2010).

following a terminated wheat crop than when cotton was grown with conventional tillage and no winter cover.

Winter Cover Crops in Conservation Systems

Use of cover crops and green manures is an ancient practice. Cover crops protect soil from erosion, improve soil physical and chemical properties by supplying organic matter, recycle nutrients, and in the case of legumes, provide N for a subsequent cash crop (Reeves, 1994). Waddle (1984) stated that government subsidies in the United States at various times have encouraged growers to use green manure crops (cover crops that are tilled into the soil), but growers often quit growing them when the subsidy ended. Recent emphasis on improving soil quality, the need to reduce N and P losses to surface and groundwaters, and research results showing cotton production benefits with abundant surface residue in conservation tillage systems have renewed interest in their use. Reeves (1994) suggested a cover crop should be easy to establish, provide ground cover quickly with rapid growth, produce sufficient biomass for maintaining residues, be disease resistant and not serve as a host for pests of cash crops, be easy to terminate, and be economically viable.

Many legume and cereal species have been evaluated as cover crops for cotton. In comparisons among cover crop species, cereals have been found to provide cotton yields with conservation tillage management that are equal or superior to legumes (Bauer and Busscher, 1996; Schomberg et al., 2006; Brown et al., 1985; Boquet et al., 2004) provided adequate N fertilizer was applied to the cotton following the cereal. Since cereals have a higher C/N ratio than legumes, their decomposition is slower, and residues from cereals keep soils covered longer than residues from legumes. Cereals scavenge residual fertilizer N. Legumes can scavenge N, but N inputs into the soil system through fixation is the predominant benefit of using legumes. The impact of the cover crop on the amount of N scavenged by cereals or fixed by legumes in the system is dependent on a number of factors, including cover crop seeding date, growing conditions, tolerance of cold temperature, and innate ability of the species being grown (reviewed by Reeves, 1994).

Strategies to increase farm profits in cotton production systems by growing cash crops during the fallow period are receiving renewed interest, particularly for small and medium-sized producers. Cotton can be double cropped with winter small grains in areas with sufficient water for production of both crops (Smith and Varvil, 1982) using conservation tillage (Baker, 1987; Hunt et al., 1997), but the risk associated with an early freeze due to late planting is considerable in many areas. Harvesting of cover crop biomass earlier in the season by grazing stocker cattle during the winter shows promise to increase farm income (Franzluebbers, 2007). Compaction of the soil by the cattle can reduce cotton yields (Mullins and Burmester, 1997) but can be alleviated noninversion deep tillage (Siri-Prieto et al., 2007) in conservation tillage systems. In addition to grazing, using rye cover crops for cellulosic biofuel feedstocks has been proposed (Feyereisen et al., 2013). Research is just beginning in harvesting rye cover crops for this purpose in cotton rotations. Raper et al. (2011) found cotton and peanut yields were reduced by an average of 9%, and rye biomass crops following the first year of study were reduced by 17% when rye was harvested compared to leaving the residues on the soil.

Soil Fertility and Nutrient Management

Within the industry, there is wide agreement that better management of applied fertilizers is needed. Fertilizer prices are associated with energy prices and their application is becoming a larger part of producer budgets. Environmental concerns of nonpoint pollution into water bodies from nutrients leaving fields are continuing to escalate. World manufactured N fertilizer use since the 1960s has increased nine-fold and P fertilizer use has tripled (Sutton et al., 2013). Agricultural fields are a major source of the N and P nonpoint-source pollution that are causing eutrophication in lakes, rivers, estuaries, and coastal oceans (Carpenter et al., 1998). In 2008, there were more than 400 scientifically reported accounts of coastal systems being eutrophication-associated dead zones in the world (Diaz and Rosenberg, 2008).

The 16 elements considered essential for plant growth and development are C, H, O, N, P, K, Ca, Mg, S, B, Fe, Mn, Cu, Zn, Mo, and Cl. Of these, C, H, and O are taken up as water through the roots and from photosynthesis in the leaves. The other elements are primarily taken up from the soil. Nitrogen, P, and K are the main nutrients in fertilizers across almost all soils. Most soils growing cotton require the addition of supplemental nutrients for optimal yield.

The amount of nutrients estimated to be taken up by two cotton crops is shown in Table 1. These data are from two yield potentials—rainfed cotton in

Table 1. Nutrient uptake by a rainfed cotton crop in the southeastern United States and an irrigated crop in Australia. Data from Mullins and Burmester (1990, 1992, 1993) and Rochester et al. (2012).

Nutrient	840 kg lint ha^{-1} yield crop	2250 kg lint ha^{-1} yield crop
N, kg ha^{-1}	128	232
P, kg ha^{-1}	17	49
K, kg ha^{-1}	106	312
Ca, kg ha^{-1}	64	289
Mg, kg ha^{-1}	18	72
Fe, g ha^{-1}	626	2592
Mn, g ha^{-1}	388	829
Cu, g ha^{-1}	28	77
Zn, g ha^{-1}	103	272

the southeastern United States (Mullins and Burmester, 1990, 1992, 1993) and irrigated cotton in Narrabri, Australia (Rochester et al., 2012). Values from the high yield crop were generated from regression equations from nutrient uptake curves. These data demonstrate that optimal fertilizer management requires realistic estimates of yield potential so that nutrients are not under- or overapplied. Knowledge of nutrient removal at harvest is useful for planning crop rotations and fertilizer applications for succeeding crops. Removal of nutrients in the harvested cotton is closely related to yield, although actual concentrations of nutrients in seeds and lint depend on growing conditions and nutrient levels in the soil. Estimates of the removal of nutrients in a harvest of 2900 kg ha^{-1} of seed and lint of cotton 71 kg N, 12.3 kg P, 29 kg K, 4.5 kg Ca, 7.8 kg Mg, 5.6 kg S, 0.2 kg Cu, 0.4 kg Mn, and 1.1 kg Zn (Osmond and Kang, 2008).

Fertilizer application rate recommendations for some nutrients are usually based on soil test results. Tests and nutrient extraction methods can vary in different growing regions due to differences in soils, climates, and production conditions. Silvertooth et al. (1999) stated that reliable recommendations from a laboratory for nutrients consist of the laboratory relating the soil test results, the fertilization rate, and crop yield.

In humid areas, blanket application rates for the mobile elements in the soil are generally made, as these can be leached from the surface soils. Soil tests for N are useful in arid areas (Silvertooth et al., 1999) but are not generally used in humid areas because of nitrate loss due to leaching. In these areas, fertilizer N application rates are based on historical yield responses in N rate experiments and projected yield potential of the crop. For example, Clemson University prescribes 78 kg N ha^{-1} for rainfed cotton and 112 kg N ha^{-1} for irrigated cotton based on the higher yield potential with irrigation (Jones et al., 2013). In-season plant tissue tests of petioles and leaf blades can be made to monitor and prescribe sidedress applications. More recently, on-the-go sensors that can measure normalized difference vegetative index (NDVI) of the canopy have become commercially available and algorithms to provide site-specific sidedress N prescriptions based on NDVI have been developed. More information of this technology is presented below in "Precision Nutrient Management."

Historically, animal manures have not been widely used for cotton in the United States (Waddle, 1984). Mitchell and Tu (2005) suggested there were several reasons cotton producers in the southeastern United States have not used manures as a fertilizer source. These included past experience with overapplication of manure that produced cotton with excessive vegetation that was late maturing, suspicion of weed seed in the manure, cotton fields being remote from the poultry growing areas, and availability of the manure not coinciding with optimum time for fertilizing. A considerable amount of research beginning in the 1990s demonstrated the value of litter as a fertilizer source. This, along with the large amounts of litter available as the poultry industry grew, experience by cotton producers using litter for fertilizer in other crops, and the high cost of commercial fertilizers in the 2000s has resulted in litter being used for fertilizing cotton in the region.

Animal manures contain significant amounts of nutrients, but unlike commercial fertilizers, they are variable in nutrient concentrations, and the ratio of N/P/K does not match cotton fertility needs. Wilkinson (1979) presented representative amounts of nutrients from different livestock operations at typical moisture contents ranging from 0.6% N for dairy and swine to 3.0% N for broiler litter. Ranges for P_2O_5 were from 0.3 to 3.3%, and ranges for K_2O were from 0.4 to 1.8%. Cotton uptake of N and K are more than six times higher than that of P (Table 1); the ratio of these three nutrients in manure is much closer. Mitchell and Tu (2005) conducted a 13-yr study with broiler litter, and the average fertilizer grade of the litter used was about 2.6:3.5:2.5 ($N/P_2O_5/K_2O$). Manures should be tested for nutrient concentration before application. Application of manures to cotton at rates high enough to supply all N will increase soil P and will likely increase soil levels of Cu and Zn. Conversely, application at rates for optimal P fertility will require additional N fertilizer.

Precision Nutrient Management

Precision agriculture is agricultural management that combines information and technology to manage site-specific variability within fields. It began in the mid-1980s to improve the application of fertilizers within fields (Robert, 2002). Precision agriculture arose from the convergence of several trends in the agricultural industry, including access to differential global positioning satellites to precisely determine location, controllers for variable rate application, on-the-go yield monitors, and computer hardware and software to analyze the large amounts of data (Sadler et al., 1998).

Precision nutrient management offers increased environmental protection from overapplication of nutrients. This management practice also has the potential to increase farmer income by avoiding over- or underapplying nutrients in portions of variable fields when uniform applications are made. Along with variation in soil physical properties, spatial variability in soil nutrient status has been reported to be related to both yield and fiber property variability. For example, Johnson et al. (2002) found that pH, P, and organic matter content were correlated with yield and the fiber properties associated with fiber secondary wall thickness (i.e., micronaire, immature fiber content, and cross-sectional area) in a US Atlantic Coastal Plain production field.

Precision nutrient and lime management combines site-specific sampling with variable rate soil amendment application technology. Grid sampling is commonly used; it is necessary when application equipment is controlled by software that is based on grid sampling. Grid size is a concern because too large of a grid may miss variability in the field, which reduces the effectiveness of the precision application. On the other hand, costs for sampling and analysis increase as grid size gets smaller. Thus, grid size chosen depends on both the amount of variability expected and resources of the producer. Also, US government programs that provide financial assistance for precision nutrient application for environmental protection specify a minimal grid size. An alternative to grid mapping is zone sampling. Zone sampling consists of using information to classify areas as homogenous zones within the field. Examples of information used include topography, multi-year yield maps, aerial images, remotely sensed images and data, and apparent soil electrical conductivity where salinity is present (Corwin et al., 2003) or when soils are coarse-textured. An advantage of zone management over grid sampling is that it can result in fewer sampling areas and reduced expense compared to grid sampling.

Remote sensing of plant canopies is just beginning to be used for site-specific N application. The remotely sensed data currently being used by commercial companies is the light reflectance from the canopy in the red (r) and far-red (fr) portions of the spectrum. From these, the NDVI is calculated as NDVI = (ir − r)/(ir + r). The NDVI has been correlated with plant biomass and can be used to detect the presence of N deficiency in cotton experiments (Plant et al., 2000). Early research on NDVI was done with satellite and aerial platform images. More recently, ground-based sensors have been developed and are commercially available.

Two types of ground-based sensors have been developed for measuring NDVI. Passive sensors are those that measure reflected natural sunlight from the canopy, and active sensors measure reflectance of modulated light from a light source on the sensor. Active sensors require less calibration and can be used under all lighting conditions. Application of sidedress N based on NDVI can be accomplished in two ways. Maps of NDVI in cotton fields can be made by collecting the data just before the N application. Variable rate controllers on applicators can then be programmed to apply the N based on these maps. Alternatively, the sensors can be mounted on the applicator in front of the nozzles. The remotely sensed data are then immediately translated into N application rates for site-specific application.

Different approaches have been used in developing algorithms to convert sensor measurements into sidedress N application rates. Raun et al. (2005) used a potential yield approach. This includes using NDVI to estimate yield potential at the time of the measurement and then estimating the response to applied N by comparing NDVI of the plants to be fertilized to the NDVI of plants that received sufficient N at planting so that yield would not be N limited. From those, an N rate is calculated. Holland and Schepers (2010) developed an N application model based on user inputs (including management zones, recommended optimal N, and N from other sources) and a generalized response curve to N. Comparison of NDVI to plants that have been supplied sufficient N is also used in their model. Oliveira et al. (2013) developed instrument specific regression equations based on economically optimal N rates for calculating applied N rates to cotton from mid-season NDVI measurements.

Timing of the sensing within the growing season appears to be important. Rutto et al. (2013) suggested that NDVI-based sidedress N applications can be made up until the crop is at the first white flower growth stage. Oliveira et al. (2013) found that equations were more accurate when measurements were made during the mid-squaring or early flower stages than during early squaring.

Fertilizer Management in Conservation Tillage

Repeated surface application of lime and nutrients without tillage to mix them with the soil can result in the surface soil to have a higher pH and relatively immobile nutrients to remain near the soil surface. Vertical stratification of Mehlich-1 extractable P and K on three silt loam soils was found by Howard et al. (1999) after 6 yr of no-tillage in cotton fields in Tennessee (Fig. 2). Because of this, soil sampling procedures may need to be modified (Triplet and Dick, 2008) for conservation systems. Kovar (1994) proposed sampling sampling only the surface 7.5 cm when no-tillage is used. Even if no fertilizers are applied, stratification can occur due to nutrients being removed by roots deep in the profile and deposited on the surface when crop residues decompose. Wright et al. (2007) found differences in stratification of nutrients when only N fertilizer was applied between tillage systems in a 5-yr study on a silty clay loam soil in Texas. In that

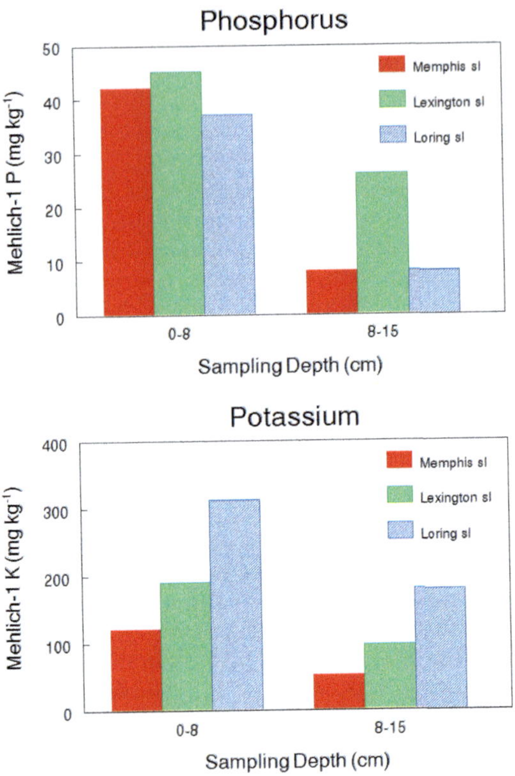

Fig. 2. Vertical stratification of Mehlich-1 P and K in silt loam soils in Tennessee after 6 yr of continuous cotton produced with no-tillage. Phosphorus and K are from the in-row sampling position and plots that were fertilized with 56 kg K ha^{-1} annually. Data are from Howard et al. (1999).

study, the conservation tillage treatment included reforming beds in the fall and two cultivations during the season for weed control. They suggested that the increased available P and NO_3 in the surface soil with the conservation tillage treatment may have contributed to increased yield.

More fertilizer N than is recommended when cotton is grown with conventional tillage may be needed for cotton that is grown with conservation tillage into winter cereal residues. This is because of the high C/N ratio of the residues and the potential for N losses due to denitrification, volatilization, and immobilization. On the basis of research in Alabama, Reeves et al. (1993) recommended 34 kg ha^{-1} more N for cotton grown following cereals compared to conventional tillage grown cotton. They also stated that starter N (28–34 kg ha^{-1}) must be applied in this system. Similarly, Bronson et al. (2001) found that on the Texas High Plains 19 to 38 kg ha^{-1} more N was needed for cotton grown with conservation tillage into residues from a winter wheat cover crop than for cotton grown with conventional tillage. Boquet et al. (2004) compared conservation tillage to no-tillage following native cover, wheat, and hairy vetch [*Vicia hirsuta* (L.) Gray] with N rates of 0 to 159 kg ha^{-1} in Louisiana. They found optimal N (in terms of lint yield) following native cover depended on tillage; a lower N rate (78 kg ha^{-1}) was needed in conventional tillage than was needed for conservation tillage (118 kg ha^{-1}). Following wheat, both tillage systems were optimum at 118 kg ha^{-1} in both tillage systems. No fertilizer N was needed in either tillage system following hairy vetch.

Research has been conducted investigating the use of animal manures as a fertilizer source with conservation tillage on cotton. Gascho et al. (2001) found broiler litter to be valuable as an amendment for cotton grown with conservation tillage and recommended an application rate of 4.5 Mg ha^{-1}. Mitchell and Tu (2005) found that N availability from litter was similar between conventional tillage where the litter was incorporated and conservation tillage when the litter remained on the soil surface. Surface application of litter in conservation tillage systems can result in some nutrient losses through volatilization and in runoff. For example, Adeli et al. (2008) found less NO_3–N and P in the soil when litter was left on the soil surface than when it was incorporated. Use of poultry litter as a fertilizer source in combination with conservation tillage resulted in a substantial increase in cotton yield over conventional management with tillage and commercial fertilizer on a US Southern Piedmont soil (Endale et al., 2010).

Summary

Cotton yields are increasing in modern production systems due to a number of factors including higher genetic yield potential and improvements in pest management. Higher yielding cotton production systems will require more nutrients for optimal production. Improved soil and plant nutrient management will be important to grow cotton successfully in the future. As temperatures rise due to global climate change, evaporative demand will increase. Global climate change is also predicted to result in changes in precipitation patterns causing more water deficit stress periods in some cotton growing areas. Cotton growers with access to irrigation will be able to adapt more easily to these stresses. However, groundwater supplies for irrigation are declining in many areas, and competition for water with municipal, industrial, and recreational uses is growing. Conservation techniques must continue to be developed and improved to reduce losses of surface soil, increase soil water holding capacity, and increase crop water use efficiency.

References

Abeles, F.B. 1969. Abscission: Role of cellulase. Plant Physiol. 44:447–452. doi:10.1104/pp.44.3.447

Adeli, A., M.W. Shankle, H. Tewolde, K.R. Sistani, and D.E. Rowe. 2008. Nutrient dynamics from broiler litter applied to no-till cotton in an upland soil. Agron. J. 100:564–570. doi:10.2134/agronj2007.0224

Baker, S.H. 1987. Effects of tillage practices on cotton double cropped with wheat. Agron. J. 79:513–516. doi:10.2134/agronj1987.00021962007900030022x

Bauer, P.J., and W.J. Busscher. 1996. Winter cover and tillage influences on Coastal Plain cotton production. J. Prod. Agric. 9:50–54. doi:10.2134/jpa1996.0050

Bauer, P.J., B.F. Fortnum, and J.R. Frederick. 2010. Cotton response to tillage during the turn of the century drought. Agron. J. 102:1145–1148. doi:10.2134/agronj2010.0037

Bell, P.F., D.J. Boquet, E. Millhollon, S. Moore, W. Ebelhar, C.C. Mitchell, J. Varco, E.R. Funderburg, C. Kennedy, G.A. Breitenbeck, C. Craig, M. Holman, W. Baker, and J.S. McConnell. 2003. Relationship between leaf-blade nitrogen and relative seed cotton yields. Crop Sci. 43:1367–1374. doi:10.2135/cropsci2003.1367

Boman, R.K., and R.L. Westerman. 1994. Nitrogen and mepiquat chloride effects on the production of nonrank, irrigated, short-season cotton. J. Prod. Agric. 7:70–75. doi:10.2134/jpa1994.0070

Boquet, D.J., R.L. Hutchinson, and G.A. Breitenbeck. 2004. Long-term tillage, cover crop, and nitrogen rate effects on cotton: Yield and fiber properties. Agron. J. 96:1436–1442. doi:10.2134/agronj2004.1436

Bosch, D.D., T.L. Potter, C.C. Truman, C.W. Bednarz, and T.C. Strickland. 2005. Surface runoff and lateral subsurface flow as a response to conservation tillage and soil-water conditions. Trans. ASABE 48:2137–2144. doi:10.13031/2013.20099

Bradow, J.M., and P.J. Bauer. 2010. Germination and seedling development. In: J.McD. Stewart, D.M. Oosterhuis, J.J. Heitholt, and J.R. Mauney, editors, Physiology of cotton. Springer, New York. p. 48–56.

Bronson, K.F., A.B. Onken, J.W. Keeling, J.D. Booker, and H.D. Torbert. 2001. Nitrogen response in cotton as affected by tillage system and irrigation level. Soil Sci. Soc. Am. J. 65:1153–1163. doi:10.2136/sssaj2001.6541153x

Brown, S.M., T. Whitwell, J.T. Touchton, and C.H. Burmester. 1985. Conservation tillage systems for cotton production. Soil Sci. Soc. Am. J. 49:1256–1260. doi:10.2136/sssaj1985.03615995004900050037x

Buman, R.A., B.A. Alesii, J.F. Bradley, J.L. Hatfield, and D.L. Karlen. 2005. Profit and yield of tillage in cotton production systems. J. Soil Water Conserv. 60:235–242.

Carpenter, S.R., N.F. Caraco, D.L. Correll, R.W. Howarth, A.N. Sharpley, and V.H. Smith. 1998. Nonpoint pollution of surface waters with phosphorus and nitrogen. Ecol. Appl. 8:559–568. doi:10.1890/1051-0761(1998)008[0559:NPOSWW]2.0.CO;2

Cathey, G.W., and W.R. Meredith, Jr. 1988. Cotton response to planting date and mepiquat chloride. Agron. J. 80:463–466. doi:10.2134/agronj1988.00021962008000030014x

Christiansen, M.N., and R. Rowland. 1986. Germination and stand establishment. In: J.R. Mauney and J.McD. Stewart, editors, Cotton physiology. The Cotton Foundation, Memphis, Tenn. p. 535–541.

Corwin, D.L., S.M. Lesch, P.J. Shouse, R. Soppe, and J.E. Ayar. 2003. Identifying soil properties that influence cotton yield using soil sampling directed by apparent soil electrical conductivity. Agron. J. 95:352–364. doi:10.2134/agronj2003.0352

Cothren, J.T., and D.M. Oosterhuis. 2010. Use of growth regulators in cotton production. In: J.McD. Stewart, D.M. Oosterhuis, J.J. Heitholt, and J.R. Mauney, editors, Physiology of cotton. Springer, New York. p. 289–303.

Diaz, R.J., and R. Rosenberg. 2008. Spreading dead zones and consequences for marine ecosystems. Science 321:926–929. doi:10.1126/science.1156401

Dodds, D.M., J.C. Banks, L.T. Barber, R.K. Boman, S.M. Brown, K.L. Edmisten, J.C. Faircloth, M.A. Jones, R.G. Lemon, C.L. Main, C. D. Monks, E.R. Norton, A.M. Stewart, and R.L. Nichols. 2010. Beltwide evaluation of commercially available plant growth regulators. J. Cotton Sci. 14:119–130.

Endale, D.M., H.H. Schomberg, M.B. Jenkins, D.H. Franklin, and D.S. Fisher. 2010. Management implications of conservation tillage and poultry litter use for Southern Piedmont USA cropping systems. Nutr. Cycling Agroecosyst. 88:299–313. doi:10.1007/s10705-009-9318-z

Entry, J.A., C.C. Mitchell, and C.B. Backman. 1996. Influence of management practices on soil organic matter, microbial biomass and cotton yield in Alabama's "Old Rotation." Biol. Fertil. Soils 23:353–358. doi:10.1007/BF00335906

Feng, Y., A.C. Motta, D.W. Reeves, C.H. Burmester, E. van Santen, and J.A. Osborne. 2003. Soil microbial communities under conventional-till and no-till continuous cotton systems. Soil Biol. Biochem. 35:1693–1703. doi:10.1016/j.soilbio.2003.08.016

Feyereisen, G.W., G.G.T. Camargo, R.E. Baxter, J.M. Baker, and T.L. Richard. 2013. Cellulosic biofuel potential of a winter rye double crop across the U.S. corn–soybean belt. Agron. J. 105:631–642. doi:10.2134/agronj2012.0282

Franzluebbers, A.J. 2007. Integrated crop–livestock systems in the southeastern USA. Agron. J. 99:361–372. doi:10.2134/agronj2006.0076

Franzluebbers, A.J., M.D. Hubbs, and M.L. Norfleet. 2012. Evaluating soil organic carbon sequestration potential in the Cotton Belt with the soil conditioning index. J. Soil Water Conserv. 67:378–389. doi:10.2489/jswc.67.5.378

Fryrear, D.W., and E.L. Skidmore. 1985. Methods for controlling wind erosion. In: R.F. Follett and B.A. Stewart, editors, Soil erosion and crop productivity. ASA, CSSA, and SSSA, Madison, WI. p. 443–457. doi:10.2134/1985.soilerosionandcrop.c24

Gascho, G.J., R.K. Hubbard, T.B. Brenneman, A.W. Johnson, D.R. Sumner, and G.H. Harris. 2001. Effects of broiler litter in an irrigated, double-cropped, conservation-tilled rotation. Agron. J. 93:1315–1320. doi:10.2134/agronj2001.1315

Gallaher, R.N. 1977. Soil moisture conservation and yield of crops no-till planted in rye. Soil Sci. Soc. Am. J. 41:145–147. doi:10.2136/sssaj1977.03615995004100010040x

Gwathmey, C.O., J.F. Bradley, A.Y. Chambers, D.D. Howard, and D.D. Tyler. 2010. Physiological responses to tillage systems, cover crops, and residue management. In: J.McD. Stewart, D.M. Oosterhuis, J.J. Heitholt, and J.R. Mauney, editors, Physiology of cotton. Springer, New York. p. 246–254.

Gwathmey, C.O., and C.C. Craig, Jr. 2003. Managing earliness in cotton with mepiquat-type growth regulators. Crop Manage. 2(1). doi: 10.1094/CM-2003-1222-01-RS

Hake, K.D., and D.W. Grimes. 2010. Crop water management to optimize growth and yield. In: J.McD. Stewart, D.M. Oosterhuis, J.J. Heitholt, and J.R. Mauney, editors, Physiology of cotton. Springer, New York. p. 255–264.

Holland, K.H., and J.S. Schepers. 2010. Derivation of a variable rate nitrogen application model for in-season fertilization of corn. Agron. J. 102:1415–1424. doi:10.2134/agronj2010.0015

Howard, D.D., M.E. Essington, and D.D. Tyler. 1999. Vertical phosphorus and potassium stratification in no-till cotton soils. Agron. J. 91:266–269. doi:10.2134/agronj1999.00021962009100020014x

Hudson, B.D. 1994. Soil organic matter and available water capacity. J. Soil Water Conserv. 49:189–194.

Hunt, P.G., P.J. Bauer, and T.A. Matheny. 1997. Crop production in a wheat doublecrop rotation with conservation tillage. J. Prod. Agric. 10:462–465. doi:10.2134/jpa1997.0462

ICAC. 2011. Cotton production practices. International Cotton Advisory Committee, Washington, DC.

Johnson, R.M., R.G. Downer, J.M. Bradow, P.J. Bauer, and E.J. Sadler. 2002. Variability in cotton fiber yield, fiber quality, and soil properties in a Southeastern Coastal Plain. Agron. J. 94:1305–1316. doi:10.2134/agronj2002.1305

Jones, M.A., J. Greene, M. Marshall, and J.D. Mueller. 2013. South Carolina cotton growers' guide. Coop. Ext. Serv. Publ. EC 589. Clemson Univ., Clemson, SC.

Kabir, Z. 2005. Tillage or no-tillage: Impact on mycorrhizae. Can. J. Plant Sci. 85:23–29. doi:10.4141/P03-160

Kerby, T.A. 1985. Cotton response to mepiquat chloride. Agron. J. 77:515–518. doi:10.2134/agronj1985.00021962007700040003x

Kerby, T.A., K. Hake, and M. Keeley. 1986. Cotton fruiting modification with mepiquat chloride. Agron. J. 78:907–912. doi:10.2134/agronj1986.00021962007800050032x

Kovar, J.L. 1994. Sampling soils under reduced row crop tillage. In: W.O. Thom and W. Sabbe, editors, Soil sampling procedures for the southern region of the United States. Southern Coop. Series Bull. 377. Univ. of Kentucky, Lexington. p. 19–23.

Lascano, R.J., R.L. Baumhardt, S.K. Hicks, and J.L. Heilman. 1994. Soil and plant water evaporation from strip-tilled cotton: Measurement and simulation. Agron. J. 86:987–994. doi:10.2134/agronj1994.00021962008600060011x

Locke, M.A., R.M. Zablotowicz, R.W. Steinriede, Jr., S. Testa, and K.N. Reddy. 2013. Conservation management in cotton production: Long-term soil biological, chemical, and physical changes. Soil Sci. Soc. Am. J. 77(3):974–984. doi:10.2136/sssaj2012.0325

Maples, R., J.G. Keogh, and W.E. Sabbe. 1977. Nitrate monitoring for cotton production in Loring–Calloway Silt Loam. Arkansas Agric. Exp. Stn. Bull. 825. Univ. Arkansas, Fayetteville.

Martin, E.C., W. Stephens, R. Wiedenfeld, H.C. Bittenbender, J.P. Beasley, Jr., J.M. Moore, H. Neibling, and J.H. Gallian. 2007. Sugar, oil, and fiber. In: R.J. Lascano and R.E. Sojka, editors, Irrigation of agricultural crops. Agronomy Monogr. 30. 2nd ed. ASA, CSSA, and SSSA, Madison, WI. p. 279–335. doi:10.2134/agronmonogr30.2ed.c9

Mauney, J.R. 1986. Vegetative growth and development of fruiting sites. In: J.R. Mauney and J.McD. Stewart, editors, Cotton physiology. The Cotton Foundation, Memphis, TN. p. 11–28.

McMichael, B.L., D.M. Oosterhuis, J.C. Zak, and C.A. Beyrouty. 2010. Growth and development of root systems. In: J.McD. Stewart, D.M. Oosterhuis, J.J. Heitholt, and J.R. Mauney, editors, Physiology of cotton. Springer, New York. p. 57–71.

Mitchell, C.C., and J.A. Entry. 1998. Soil C, N, and crop yields in Alabama's long-term 'Old Rotation' cotton experiment. Soil Tillage Res. 47:331–338. doi:10.1016/S0167-1987(98)00121-4

Mitchell, C.C., and S. Tu. 2005. Long-term evaluation of poultry litter as a source of nitrogen for cotton and corn. Agron. J. 97:399–407. doi:10.2134/agronj2005.0399

Montgomery, D.R. 2007. Soil erosion and agricultural sustainability. Proc. Natl. Acad. Sci. USA 104:13268–13272. doi:10.1073/pnas.0611508104

Mullins, G.L., and C.H. Burmester. 1990. Dry matter, nitrogen, phosphorus, and potassium accumulation by four cotton varieties. Agron. J. 82:729–736. doi:10.2134/agronj1990.00021962008200040017x

Mullins, G.L., and C.H. Burmester. 1992. Uptake of calcium and magnesium by cotton grown under dryland conditions. Agron. J. 84:564–569. doi:10.2134/agronj1992.00021962008400040004x

Mullins, G.L., and C.H. Burmester. 1993. Accumulation of copper, iron, manganese and zinc by four cotton cultivars. Field Crops Res. 32:129–140. doi:10.1016/0378-4290(93)90025-I

Mullins, G.L., and C.H. Burmester. 1997. Starter fertilizer and the method and rate of potassium fertilizer effects on cotton grown on soils with and without winter grazing by cattle. Commun. Soil Sci. Plant Anal. 28:739–746. doi:10.1080/00103629709369826

Munro, J.M. 1987. Cotton. 2nd ed. Longman–John Wiley and Sons, New York.

Mutchler, C.K., and L.L. McDowell. 1990. Soil loss from cotton with winter cover crops. Trans. ASAE 33:432–436. doi:10.13031/2013.31347

Mutchler, C.K., L.L. McDowell, and J.D. Greer. 1985. Soil loss from cotton with conservation tillage. Trans. ASAE 28:160–163. doi:10.13031/2013.32220

Oliveira, L.R., P.C. Scharf, E.D. Vories, S.T. Drummond, D. Dunn, W.G. Stevens, K.F. Bronson, N.R. Benson, V.C. Hubbard, and A.S. Jones. 2013. Calibrating canopy reflectance sensors to predict optimal mid-season nitrogen rate for cotton. Soil Sci. Soc. Am. J. 77:173–183. doi:10.2136/sssaj2012.0154

Osmond, D.L., and J. Kang. 2008. Nutrient removal by crops in North Carolina. North Carolina Coop. Ext. Serv. North Carolina State Univ., Raleigh, NC. Available at http://content.ces.ncsu.edu/20688.pdf (accessed 2 June 2015).

Pettigrew, W.T. 2002. Improved yield potential with an early planting cotton production system. Agron. J. 94:997–1003. doi:10.2134/agronj2002.0997

Pettigrew, W.T. 2010. Impact of varying planting dates and irrigation regimes on cotton growth and lint yield production. Agron. J. 102:1379–1387. doi:10.2134/agronj2010.0172

Pimentel, D., C. Harvey, P. Resosudarmo, K. Sinclair, D. Kurz, M. McNair, S. Crist, L. Shpritz, L. Fitton, R. Saffouri, and R. Blair. 1995. Environmental and economic costs of soil erosion and conservation benefits. Science 267:1117–1123. doi:10.1126/science.267.5201.1117

Plant, R.E., D.S. Munk, B.R. Roberts, R.L. Vargas, D.W. Rains, R.L. Travis, and R.B. Hutmacher. 2000. Relationships between remotely sensed reflectance data and cotton growth and yield. Trans. ASAE 43:535–546. doi:10.13031/2013.2733

Raper, R.L., D.W. Reeves, C.H. Burmester, and E.B. Schwab. 2000. Tillage depth, tillage timing, and cover crop effects on cotton yield, soil strength, and tillage energy requirements. Appl. Eng. Agric. 16(4):379–385. doi:10.13031/2013.5363

Raper, R.L., E.B. Schwab, F.J. Arriga, K.S. Balkcom, A.J. Price, and T.S. Kornecki. 2011. Effects of cover crop removal on a cotton-peanut rotation. Trans. ASABE 54:1213–1218. doi:10.13031/2013.39019

Raun, W.R., J.B. Solie, M.L. Stone, K.L. Martin, K.W. Freeman, R.W. Mullen, H. Zhang, J.S. Schepers, and G.V. Johnson. 2005. Optical sensor-based algorithm for crop nitrogen fertilization. Comm. Soil Sci. Plant Nutr. 36:2759–2781. doi:10.1080/00103620500303988

Reeves, D.W. 1994. Cover crops and rotations. In: J.L. Hatfield and B.A. Stewart, editors, Advances in Soil Science– Crops Residue Management. Lewis Publ., CRC Press, Boca Raton, FL. p. 125–172.

Reeves, D.W., C. Mitchell, G. Mullins, and J. Touchton. 1993. Nutrient management for conservation-tillage cotton in the southeast. In: M.R. McClelland, et al., editors, Conservation-tillage systems for cotton: A review of research and demonstration results from across the cotton belt. Special Rep. 160. Arkansas Agric. Exp. Stn. Univ. of Arkansas, Fayetteville, AR. p. 23–28.

Rochester, I.J., G.A. Constable, D.M. Oosterhuis, and M. Errington. 2012. Nutritional requirements of cotton during flowering and fruiting. In: D.M. Oosterhuis and J.T. Cothren, editors, Flowering and fruiting in cotton. The Cotton Foundation, Cordova, TN. p. 35–50.

Robert, P.C. 2002. Precision agriculture: A challenge for crop nutrition management. Plant Soil 247:143–149. doi:10.1023/A:1021171514148

Rutto, E., B.D. Arnall, J.L. May, K. Butchee, and W.R. Raun. 2013. Ability of cotton (*Gossypium hirsutum* L.) to recover from early season nitrogen deficiency. J. Cotton Sci. 17:70–79.

Sadler, E.J., W.J. Busscher, P.J. Bauer, and D.L. Karlen. 1998. Spatial scale requirements for precision farming: A case study in the Southeastern USA. Agron. J. 90:191–197. doi:10.2134/agronj1998.00021962009000020012x

Sadler, E.J., C.R. Camp, and R.G. Evans. 2007. New and future technology. In: R.J. Lascano and R.E. Sojka, editors, Irrigation of agricultural crops. Agron. Monogr. 30. 2nd ed. ASA, CSSA, and SSSA, Madison, WI. p. 609–626. doi:10.2134/agronmonogr30.2ed.c16

Schomberg, H.H., R.G. McDaniel, E. Mallard, D.M. Endale, D.S. Fisher, and K.L. Cabrera. 2006. Conservation tillage and cover crop influences on cotton production on a Southeastern U.S. Coastal Plain soil. Agron. J. 98:1247–1256. doi:10.2134/agronj2005.0335

Silvertooth, J.C., K.L. Edmisten, and W.H. McCarty. 1999. Production practices. In: W.C. Smith and J.T. Cothren, editors, Cotton: Origin, history, technology, and production. John Wiley and Sons, New York. p. 451–488.

Siri-Prieto, G., D.W. Reeves, and R.L. Raper. 2007. Tillage requirements for integrating winter-annual grazing in cotton production: Plant water status and productivity. Soil Sci. Soc. Am. J. 71:197–205. doi:10.2136/sssaj2004.0258

Soileau, J.M., J.T. Touchton, B.B. Hajek, and K.H. Yoo. 1994. Sediment, nitrogen, and phosphorus runoff with conventional- and conservation-tillage cotton in a small watershed. J. Soil Water Conserv. 49:82–89.

Smith, C.W., and J.J. Varvil. 1982. Double cropping cotton and wheat. Agron. J. 74:862–865. doi:10.2134/agronj1982.00021962007400050022x

Sutton, M.A., A. Bleeker, C.M. Howard, M. Bekunda, B. Grizzetti, W. de Vries, H.J.M. van Grinsven, Y.P. Abrol, T.K. Adhya, G. Billen, E.A. Davidson, A. Datta, R. Diaz, J.W. Erisman, X.J. Liu, O. Oenema, C. Palm, N. Raghuram, S. Reis, R.W. Scholz, T. Sims, H. Westhoek, and F.S. Zhang with contributions from S. Ayyappan, A.F. Bouwman, M. Bustamante, D. Fowler, J.N. Galloway, M.E. Gavito, J. Garnier, S. Greenwood, D.T. Hellums, M. Holland, C. Hoysall, V.J. Jarramillo, Z. Klimont, J.P. Ometto, H. Pathak, V. Plocq Fichelet, D. Powlson, K. Ramakrishna, A. Roy, K. Sanders, C. Sharma, B. Singh, U. Singh, X.Y. Yan, and Y. Zhang. 2013. Our nutrient world: The challenge to produce more food and energy with less pollution. Global overview of nutrient management. Centre for Ecology and Hydrology, Edinburgh on behalf of the Global Partnership on Nutrient Management and the International Nitrogen Initiative. Available at http://initrogen.org/sites/default/files/documents/files/ONW.pdf (accessed 2 June 2015).

Tharp, W.H. 1960. The cotton plant. USDA Agric. Handb. 178. US Gov. Print. Office, Washington, DC.

Triplett, G.B., Jr., and W.A. Dick. 2008. No-tillage crop production: A revolution in agriculture! Agron. J. 100:S-153–S-165. doi:10.2134/agronj2007.0005c

Truman, C.C., W.D. Reeves, J.N. Shaw, A. Motta, C.H. Burmester, R. Raper, and E. Schwab. 2003. Tillage impacts on soil property, runoff, and soil loss variations from a Rhodic Paleudult under simulated rainfall. J. Soil Water Conserv. 58:258–267.

Truman, C.C., J.N. Shaw, D.C. Flanagan, D.W. Reeves, and J.C. Ascough, II. 2009. Conservation tillage to effectively reduce interrill erodibility of highly-weathered Ultisols. J. Soil Water Conserv. 64:265–275. doi:10.2489/jswc.64.4.265

Waddle, B.A. 1984. Crop growing practices. In: R.J. Kohel and C.F. Lewis, editors, Cotton. Agron. Monogr. 24. ASA, CSSA, and SSSA, Madison, WI. p. 233–263. doi:10.2134/agronmonogr24.c8

Wang, L., C. Mu, M. Du, Y. Chen, X. Tian, M. Zhang, and S. Li. 2014. The effect of mepiquat chloride on elongation of cotton (*Gossypium hirsutum* L.) internode is associated with low concentration of gibberellic acid. Plant Sci. 225:15–23. doi:10.1016/j.plantsci.2014.05.005

Wanjura, D.L. 1986. Field environment and stand establishment. In: J.R. Mauney and J.McD. Stewart, editors, Cotton physiology. The Cotton Foundation, Memphis, Tenn. p. 551–554.

Wilkinson, S.R. 1979. Plant nutrient and economic value of animal manures. J. Anim. Sci. 48:121–133.

Wright, A.L., F.M. Hons, and J.E. Matocha, Jr. 2005. Tillage impacts on microbial biomass and soil carbon and nitrogen dynamics of corn and cotton rotations. App. Soil Ecol. 29:85–92. doi:10.1016/j.apsoil.2004.09.006

Wright, A.L., F.M. Hons, R.G. Lemon, M.L. McFarland, and R.L. Nichols. 2007. Stratification of nutrients in soil for different tillage regimes and cotton rotations. Soil Tillage Res. 96:19–27. doi:10.1016/j.still.2007.02.005

Yoo, K.H., and J.T. Touchton. 1988. Surface runoff and sediment yield from various tillage systems of cotton. Trans. ASAE 31:1154–1158. doi:10.13031/2013.30837

York, A.C. 1983. Cotton cultivar response to mepiquat chloride. Agron. J. 75:663–667. doi:10.2134/agronj1983.00021962007500040020x

Zhao, D., and D.M. Oosterhuis. 2000. Pix plus and mepiquat chloride effects on physiology, growth, and yield of field-grown cotton. J. Plant Growth Regul. 19:415–422. doi:10.1007/s003440000018

Zak, J.C., B. McMichael, S. Dhillion, and C. Friese. 1998. Arbuscular-mycorrhizal colonization dynamics of cotton (*Gossypium hirsutum* L.) growing under several production systems on the Southern High Plains, Texas. Agric. Ecosyst. Environ. 68:245–254. doi:10.1016/S0167-8809(97)00151-5

Remote Sensing for Cotton Farming

Yanbo Huang* and Steven J. Thomson

Abstract

Application of remote sensing technologies in agriculture began with the use of aerial photography to identify cotton root rot in the late 1920s. From then on, agricultural remote sensing has developed gradually until the introduction of precision farming technologies in the late 1980s and biotechnological innovations in the early 21st century. Remote sensing technologies for crop production management have advanced rapidly since then and are being developed for cotton growth monitoring, estimation of evapotranspiration and irrigation scheduling, nitrogen efficiency analysis, pest management (weeds, insects, and diseases), application of harvesting aids, and prediction of yield. In the next 10 yr or so, more specialized remote sensing analysis techniques for cotton management will be developed with improvements in unmanned aerial technologies and dual uses for agricultural aircraft. This chapter begins with the history of agricultural remote sensing, then discusses precision farming and the biotechnology interface for remote sensing on cotton, presents several research aspects of remote sensing for cotton, presents two research cases for remote sensing of cotton, discusses limitations of remote sensing for support of cotton management, and presents a perspective on remote sensing for cotton farming in the future.

Remote sensing, in the sense of a traditional description, is a technique that acquires data and information from an object without contacting the object. Remote sensing's beginnings can be traced to the invention of the camera about 180 yr ago, but the term "remote sensing" was not used until the 1960s. The purpose of modern remote sensing is earth observation, which had been conducted via aerial photography from balloons soon after the invention of the camera,

---

Abbreviations: AVHRR, Advanced Very High Resolution Radiometer; AVNIR, airborne visible and near infrared; AWIFS, advanced wide field sensor; Bt, Bacillus thuringiensis; CCD, charge-coupled device; CIR, color infrared; CLCV, cotton leaf curl virus; DVI, difference vegetation index; EC, electrical conductivity; ET, evapotranspiration; FCA, first canonical axis; FOSI, first order sensitivity index; GM, genetically modified; GR, glyphosate-resistant; HAT, hours after treatment; HR, herbicide-resistant; HS, herbicide-sensitive; IRT, infrared transducer; LAI, leaf area index; LISA, local indicator of spatial association; LISS, linear imaging self-scanning system: MCARI, modified chlorophyll absorption index; MODIS, Moderate Resolution Imaging Spectroradiometer; MTVI, modified triangular vegetation index; NDVI, normalized difference vegetation index; NIR, near infrared; OSAVI, optimized soil-adjusted vegetation index; PROSPECT, leaf optical property spectra; RDVI, renormalized difference vegetation index: RR, radiance ratio; RVI, ratio vegetation index; SAR, synthetic aperture radar; SAVI, soil adjusted vegetation index; SOM, self organized map; SR, simple ratio; TCARI, transformed chlorophyll absorption index; TIR, thermal infrared; TM, Thematic Mapper; TSI, total sensitivity index; TVI, transformed vegetation index; UAV, unmanned aerial vehicle; WAGT, weeks after glyphosate treatment.

Yanbo Huang and Steven J. Thomson (steve.thomson@ars.usda.gov), USDA-ARS, Crop Production Systems Research Unit, 141 Experiment Station Rd., P.O. Box 350, Stoneville, MS 38776. *Corresponding author (yanbo.huang@ars.usda.gov).

doi:10.2134/agronmonogr57.2013.0030

Cotton. 2nd. ed. David D. Fang and Richard G. Percy, editors. Agronomy Monograph 57.

airplanes in World Wars I and II for military reconnaissance, and airplanes for earth surface monitoring until the early 1960s when satellite remote sensing began in earnest.

Remote sensing science is a field of study devoted to sensing technologies that detect radiation at different wavelengths reflected or emitted from the earth's surface. Remote sensing science further develops methods that analyze the energy balance relationship and identifies, categorizes, and maps the spatial distribution of earth surface attributes. Applications of remote sensing began with topographic mapping in the middle of the 1880s and then military reconnaissance in World Wars I and II. Presently, remote sensing is widely applied in meteorological observation, ocean monitoring, hazard assessment, and natural resource management (http://oceanservice.noaa.gov/facts/remotesensing.html).

Agricultural remote sensing can be traced back to the late 1920s when aerial photography was used to identify cotton root rot at College Station, TX (Neblette, 1927; Taubenhaus et al., 1929; Steddom et al., 2005). Soon after, Bawden (1933) found that aerial infrared film could capture some virus symptoms on potato, which might not be easily distinguished on panchromatic film that is sensitive to visible light. In the mid 1950s, experiments were conducted using aerial color infrared (CIR) photography to classify vegetation types and detect diseases and stresses of certain cereal crops (Colwell, 1956).

Modern agriculture has seen two major changes: precision farming in the late 1980s (Yao and Huang, 2013) and biotechnological innovations in the early 2000s (Brookes and Barfoot, 2011). Precision farming implements site-specific crop management by integrating agricultural mechanization and information technology in GIS, GPS, and remote sensing. Remote sensing plays important roles in crop monitoring and control (Pinter et al., 2003; Yao and Huang, 2013). Biotechnological innovations basically provide genetically modified (GM) crops to allow for the suppression of weed pests with herbicide-tolerant varieties and insect pests with insect-protected *Bacillus thuringiensis* (*Bt*) varieties. Remote sensing is a rapid and cost-effective tool for managing GM crop adoption (K. Huang, Louisiana State University, personal communication, 2010) and for detecting crop injury from off-target herbicide spray (Huang et al., 2010b; Ortiz et al., 2011). Farming of cotton, as well as other commodity crops, has been studied and applied using remote sensing as a tool to supplement modern agricultural practices, such as precision management and biotechnological innovations. This chapter will be organized to include the following:

- "Precision Cotton Farming" to compare conventional cotton farming and precision cotton farming;
- "Genetically modified cotton" to indicate the benefits of the adoption of the biotechnologically innovative crops;
- "Remote cotton sensing" to describe available platforms, systems, and methods for remotely sensing cotton fields;
- "Cases" to present details of two successful projects involving remote cotton sensing;
- "Limitations of remote sensing" to point out the limits of agricultural remote sensing, especially in cotton farming; and
- "Future outlook" to describe the new areas and the outlook for the development of and application of remote sensing for cotton farming.

Precision Cotton Farming

In the past two decades, agriculture has been experiencing a significant change from traditional farming to modern precision farming (Stafford, 2000; Zhang et al., 2002). Traditional agriculture counts on mechanization to treat crop fields uniformly from crop seeding/planting, fertilizer and pesticide application, and irrigation, to harvesting. Precision agriculture is realized in through site-specific management for crop production. Instead of treating a crop field as a uniform unit as in traditional agriculture, precision agriculture considers variability within the crop field, including variability in the soil conditions, crop stress caused by pests, water, and nutrients, and yield. Accordingly, precision agriculture can handle in-field variability site-specifically based on the local needs of the field. Thus, the concept behind precision agriculture/precision farming is to manage the in-field variability. Modern precision agriculture is enabled by the integration of agricultural mechanization and information technology in GIS, GPS, and remote sensing (Yao and Huang, 2013; Mulla, 2013). With these technologies, farmers can decide when, where, and how to take precision farming actions such as variable-rate seeding, variable-rate application of fertilizers and pesticides, and yield variability monitoring.

Cotton is a high-value crop that requires the extensive use of fertilizers and chemicals (Brooks, 2001). Cotton is often grown in fields that have heterogeneous soil, such as in the southern United States (Mallarino and Wittry, 2004). Remote sensing is well suited as a tool for site-specific cotton production, and it has been used to target the in-season, site-specific management of the crop with new sensors that provide farming services (Larson et al., 2008; Huang et al., 2010a).

Genetically Modified Cotton

In the past 20 years, agriculture has experienced another major change due to wide adoption of GM crops (Brookes and Barfoot, 2011). Up to 2010, 148 million ha were planted to transgenic crops globally (referring to single or stacked genes for herbicide, insect, drought, and salinity resistance, and value-added traits), of which 61% of the total field area was planted to herbicide resistant crops [soybean (*Glycine max* Merr.), corn (maize; *Zea mays* L.), cotton (*Gossypium hirsutum* L.), canola (*Brassica napus* L.), sugar beet (*Beta vulgaris* L.), and alfalfa (*Medicago sativa* L.)] and 22% was planted to stacked herbicide and insect traits (Reddy and Nandula, 2012). A transgenic crop is created by the stable integration of a foreign gene to introduce a new trait for the plant to provide resistance to certain pests, diseases, environmental conditions, or resistance to chemical treatments. For herbicide resistance, glyphosate-resistant (GR) canola, soybean, and cotton were commercially released in 1996–1997, and the new generation of GR cotton was developed in 2006 to enhance tolerance to glyphosate and allow for in-season glyphosate application (Reddy and Nandula, 2012). Also, most transgenic cotton and corn on the market carry insect-resistant traits (*Bt* trait; Reddy and Nandula, 2012).

The area planted to transgenic crops (both herbicide and insect resistant) in the world has doubled from 67.7 million ha to 148 million ha from 2003 (Dill, 2005) to 2010 (James, 2010). In 2010, there are 29 countries in the world growing transgenic crops, among which the USA has dominated with 45% of the total planting area. By 2015, about 40 countries are expected to grow transgenic crops (Reddy and Nandula, 2012).

On the basis of the 2013 data of the National Agricultural Statistics Service of the USDA, 90% of upland cotton that was planted was genetically engineered, including herbicide-tolerant varieties, insect-resistant (*Bt*) varieties, and stacked gene varieties (Economic Research Service, 2014) compared to 61% in 2000.

With the rapid development and application of transgenic cotton, issues have to be handled to smooth the process. For example, management of transgenic crop planting may require planning, monitoring, and evaluation of the planting location, area, time, and benefits. Also, with the adoption of transgenic crops, chemical application may require evaluation of crop injury due to the off-target drift of herbicides and their environmental impact. These effects are spatially variable, and remote sensing is well suited to provide useful information for support when making a decision on how to handle these issues.

Remote Cotton Sensing

Since agricultural remote sensing began with the aerial monitoring of cotton in the late 1920s, and especially since the last two decades of the 20th century, remote sensing in spaceborne, airborne, and ground-based measuring platforms has been developed to help analyze and support the decisions that are required for the different aspects of cotton production (Table 1). With the advancement of the technologies, more remote sensors will be used to monitor cotton, such as the Operational Land Imager and the Thermal Infrared Sensor on Landsat 8 from NASA and the new radiometric thermal cameras (640 × 480 pixel array) used on agricultural aircraft. Here is an explanation of how the sensors and systems in different platforms were used to remotely sense cotton fields.

Growth Monitoring

Growth monitoring of cotton is an important remote sensing application in cotton production. Bai et al. (2012) determined cotton growth conditions at the flower-boll stages by using the multitemporal reflectance values of Landsat Thematic Mapper 5 images. The study showed that the productivity of cotton fields was closely related to soil salinity, texture, and topography. Previously, Bai et al. (2011a) conducted a field study on cotton growth conditions based on leaf area index (LAI) inversed from remotely sensed spectral measurements in a cotton field. Qi et al. (2005) surveyed the research on the extraction of characteristic information of a crop canopy from hyperspectral data, and Qi et al. (2011) sensed the cotton canopy at different growth stages in Xinjiang, China, using a FieldSpec Pro spectroradiometer (ASD Inc., Boulder, CO). The hyperspectral data analysis indicated that the spectral reflectance of the cotton canopy was closed in relation to the cotton leaf chlorophyll content, leaf area, biomass, growth stages, and plant health conditions.

Vegetation indices derived from various data sources were used in cotton growth monitoring as well (Ansari et al., 1999; Leon et al., 2003; Sudbrink et al., 2003; Read et al., 2004; Ritchie et al., 2005; Willers et al., 2005; Zarco-Tejada et al., 2005; Mukherjee et al., 2008, 2009; Huang et al. 2010b, 2010c; Tubana, 2010; Bai et al., 2011a; Kumar et al., 2013; Ortiz et al., 2011; Musande et al., 2012; Ahmed et al.,

Table 1. Sensor systems and platforms that have been used for remote cotton sensing.

System	Sensor	Platform	Bands	Resolution
Landsat 1	Multispectral scanner	Satellite	Blue, green, orange, red, and near-infrared	80 m
Landsat 5	TM	Satellite	Green, blue, red, near-infrared, short infrared, and thermal infrared	30 m except 120 m for thermal infrared
Landsat 7	Enhanced Thematic Mapper Plus	Satellite	Green, blue, red, near infrared, short infrared, thermal infrared, and panchromatic	30 m except 60 m for thermal infrared and15 m for panchromatic
Terra and Aqua	Moderate Resolution Imaging Spectroradiometer	Satellite	36 bands from 0.4 to 14.5 μm	250, 500, and 1000 m
NOAA	Advanced Very High Resolution Radiometer	Satellite	Visible, near-infrared, and thermal infrared	1100 m
SPOT	VEGETATION	Satellite	Green, red, near-infrared, short infrared, and panchromatic	5 or 2.5 m for panchromatic and 10 m for multispectral
QuickBird	QuickBird	Satellite	Green, blue, red, near-infrared, and panchromatic	0.61 m for panchromatic and 2.44 m for multispectral
IRS	Linear imaging self-scanning system (LISS) I, II, and III, advanced wide field sensor (AWIFS)	Satellite	Green, red, near-infrared, and short infrared	23.5 m for LISS III and 56 m for AWIFS
HJ-1	Charge-coupled device (CCD) and hyperspectrometer	Satellite	CCD: blue, green, red, near-infrared, panchromatic, and thermal infrared; hyperspectrometer: 128 bands from 0.45 to 0.95 μm	30 or 150 m for CCD multispectral, 300 m for CCD thermal, and 100 m for hyperspectrometer
Airborne visible and near infrared	airborne visible and near infrared hyperspectral sensor	Airborne	430 nm to 1030 nm for 60 bands at 10 nm nominal	1.25 nm/pixel * spatial resolution: 654 pixels
DuncanTech	MS4100 3-CCD multispectral camera	Airborne	Blue, green, red, and near-infrared	Image size 1920 × 1080
Long wave-length infrared	TVS-700 thermal camera	Airborne	8 to 14 μm broadband	Image size 320 × 240
Electrophysics	PV-320T thermal camera	Airborne	8 to 14 μm broadband	Image size 320 × 240
ADC	FieldSpec handheld spectroradiometer	Ground	325 to 1075 nm	< 3.0 nm @ 700 nm of spectral resolution
ADC	FieldSpec Pro spectroradiometer	Ground	325 to 2500 nm	3 nm @700 nm and 30 nm @ 1400/2100 nm for spectral resolution
GER	GER 1500 spectroradiometer	Ground	350 to 1050 nm	1.5 nm sampling interval
Crop Circle	Crop Circle Handheld Crop Canopy Sensor	Ground	670, 730, and 780 nm	
GreenSeeker	GreenSeeker HandHeld Data Collection and Mapping Unit	Ground		
CropScan	CropScan Radiometer	Ground	Visible and near-infrared narrow bands	

2013). In later parts of the paper, we will provide the detail of the works using vegetation indices to remotely sense cotton for different purposes.

Evapotranspiration Estimation and Irrigation Scheduling

Evapotranspiration (ET) is an important parameter in determining crop water requirements for irrigation system design and operation. Accurate estimation of ET from different data sources has been a topic of research for many years. Remote sensing has provided a technology and a source of data for the estimation. Garatuza-Payan and Watts (2005) estimated ET of irrigated wheat and cotton in northwest Mexico using GOES-10 data from NOAA's satellite system to determine crop water requirements of irrigated crops. The NOAA GOES satellite system provides continuous monitoring of locations on the surface on the Earth with a 1-km spatial resolution while systems such as Moderate Resolution Imaging Spectroradiometer (MODIS) and Landsat cannot provide the capability of such continuous earth observations. Dalezios et al. (2011) estimated cotton ET for irrigation management in central Greece by determining water-limited growth environment zones from the normalized difference vegetation index (NDVI) and brightness temperature derived from NOAA's Advanced Very High Resolution Radiometer (AVHRR) satellite imagery for the vegetation health index to represent overall vegetation health. The method developed in this study can be applied in large-scale areas for the calculation of cotton crop coefficients using satellite data. Singh et al. (2013) performed a field study in Haryana, India, for the estimation of cotton crop coefficients using the soil adjusted vegetation index (SAVI) derived from Landsat TM 5 images. The estimated crop coefficients were used along with the crop reference ET values to estimate cotton crop ET. The results showed that the spatial distribution of crop ET varied in the growth season with farming conditions. Spatially varied cotton ET presented the spatially varied water demand in the cotton field, which can be used for site-specific irrigation.

For irrigation scheduling, Ritchie et al. (2005) studied cotton irrigation timing in Camilla, GA, using aerial imagery collected using a two-Nikon digital camera system. One camera was modified to be near infrared (NIR) sensitive and was mounted and operated remotely on a 15-ft-long tethered blimp. From the imagery, NDVI was calculated to provide information for remote sensing–based irrigation. The results were compared with those from irrigation triggered by soil water sensor readings, and they indicated that remote sensing–based irrigation scheduling can produce cotton yields comparable to those from standard sensor triggered irrigation scheduling.

Accurate determination of crop water stress is critical for the effective operation of irrigation. Maas et al. (2000) presented a 2-yr field study in Shafter, CA, to estimate the temperature of a cotton canopy with incomplete ground cover, which might indicate the water status of the crop. The estimation was started from acquiring visible, NIR, and thermal images through an airborne multispectral remote sensing system. The acquired images were then calibrated and used to estimate field vegetation ground cover (from red and NIR bands) and field-surface temperature (thermal band). With all the information and estimation of soil temperature, a linear mixture model was built to estimate the temperature of the cotton canopy. Falkenberg et al. (2007) also conducted a 2-yr cotton field study in Uvalde, TX, to evaluate the applicability of a TVS-700 long wave-length infrared camera (Indigo Systems, Dallas, TX) mounted in a helicopter for site-specific management of abiotic (drought) and biotic (root rot) stress of cotton under a center-pivot, low energy precision application irrigation system. In the evaluation, the ability of canopy temperature difference detection in the 8- to 14-μm band using a thermal camera was compared with infrared transducers (IRTs) mounted on the center pivot. The results indicated that pivot-mounted IRTs and the thermal camera were both able to differentiate water stress among different irrigation regimes, and the thermal imagery differences between biotic and abiotic stresses could be distinguished well.

Fulton et al. (2007) studied the use of thermal infrared (TIR) imagery for early detection of crop stress. The TIR data were acquired using a TIR sensor (L3 Communications Infrared Products, Garland, TX) mounted on an unmanned aerial vehicle (UAV). The TIR emittance correlated well with cotton canopy cover and stomatal conductance.

Thomson et al. (2012) used a Sofradir Electrophysics PV-320T thermal imaging camera mounted in an agricultural aircraft to detect spatial and temporal changes in canopy temperature that can be related to crop water stress. They developed a preliminary linear model indicating the effect of altitude ($r^2 = 0.58$) on canopy temperature represented at the sensor. When the variable solar radiation was added, its combined effect with altitude accounted for 73% of the variability in canopy temperature. It was indicated that a model such as this could be used to rescale thermal images spectrally so that accurate comparisons could be made between images on a temporal basis. However, difficulty would still exist under low-vapor-pressure deficits, as canopy-air temperature differences are quite small for a given difference in stress (Pinter and Reginato, 1981). Spatial relationships between yield, canopy cover change, and canopy temperature were also developed for cotton, indicating consistently sub-yielding field areas.

Multifrequency microwave has been used to sense cotton field moisture. Moran et al. (1997) combined the synthetic aperture radar (SAR) data of wavelengths in the Ku (14.85 GHz) and C (5.3 GHz) bands and optical remote sensing measurement in visible, NIR and thermal bands to measure the moisture in cotton field. The results indicated that SAR C backscatter coefficient (C-band $\sigma°$) was sensitive to soil moisture for cotton fields with green LAI less than 1.0. For low-green LAI cotton, C-band $\sigma°$ was correlated with measurements of the cotton field surface temperature.

Nitrogen Efficiency Analysis

Nitrogen deficiency is an important issue, and many studies have been conducted on nitrogen use and efficiency in cotton growth. Tarpley et al. (2000) predicted nitrogen concentration in cotton leaf using reflected radiance, measured using a GER 1500 spectroradiometer (Geophysical and Environmental Research Corp., Millbrook, NY) in the spectral range of 350 to 1050 nm with a 1.5-nm sampling interval. Results indicated that the ratios of red-edge to NIR reflectance provided satisfactory precision and accuracy for the prediction. Read et al. (2002) used the same type of spectroradiometer to measure cotton leaf and canopy reflectance, identifying reflectance ratios that were sensitive to leaf chlorophyll, which is associated with the nitrogen status of the crop. The results indicated that narrow-waveband reflectance ratios related to violet or blue and red-edge wavebands were sensitive to remotely sensing cotton's nitrogen status. Zhao et al. (2005) conducted a field study for determining the relationship between the cotton leaf's spectral reflectance and its chlorophyll/nitrogen concentration. The cotton leaf hyperspectral reflectances were measured using the portable 350 to 2500 nm ASD FieldSpec FR spectroradiometer. Results provided optimal reflectance ratios for determining cotton leaf chlorophyll and nitrogen concentrations and suggested that the method had potential for monitoring the status of cotton plant nitrogen in real-time. Sui and Thomasson (2006) developed a ground on-the-go sensing system with a multispectral optical sensor that could measure plant reflectance in the red, green, blue, and NIR wavebands. An ultrasonic sensor also on the system was used to measure plant height simultaneously with spectral measurement. The system was tested in the field for 2 yr, and the results indicated that the measured spectral information and the plant height were highly correlated with the cotton leaf nitrogen concentration.

Huang et al. (2013) estimated cotton yield with varied irrigation and nitrogen treatments by using vegetation indices that were derived from aerial CIR imagery acquired by a MS4100 3–charge-coupled device (CCD) camera mounted on an Air Tractor 402B airplane (Air Tractor, Inc., Olney, TX). The correlation between the yield and nitrogen treatments was not strong, possibly due to high residual nitrogen in the soil.

Farias de Oliveira (2008) measured reflectance of a cotton canopy in the field at different growth stages to provide rate recommendations for on-the-go side-dress nitrogen application. The measurements were conducted using Crop Circle (Holland Scientific, Lincoln, NE), GreenSeeker (NTech Industries, Inc., Ukiah, CA), and Cropscan (CropScan, Inc., Rochester, MN) crop spectral sensors. The results suggested that the reflectance measurements can assist in real-time variable-rate nitrogen application in the mid-square and early flower growth stages. Tubana (2010) measured the NDVI of a cotton canopy using the GreenSeeker handheld optical sensor at different growth stages, and for each stage, a relationship between cotton lint yield and NDVI was determined. The results indicated that an increase in lint yield was closely related to nitrogen fertilizer, and that the ratio vegetation index (RVI) had a close relationship with yield under the irrigation and nitrogen treatments. The relationship could be improved if the soil electrical conductivity (EC) of the field was incorporated as an additional explanatory variable.

Insect Control

Cotton insect control is a critical issue for plant protection. Remote sensing has been used at different scales to help with the control of cotton insects. At a large scale, Coleman et al. (1974) evaluated satellite imagery to help control a pink bollworm infestation in cotton in the southern deserts of California. The imagery was acquired by the multispectral scanner installed on the Earth Resources Technology Satellite, i.e., Landsat 1, launched by NASA in 1972. The study was an early one that used satellite imagery to investigate insects in cotton and indicated that the cotton could be identified by sequential photography from the satellite. Yones et al. (2012) studied the timing of insecticide application against cotton leaf worm (*Spodoptera littoralis*) based on the calculation of the average of thermal units from air temperature and soil temperature derived from NOAA satellite images. This study determined the best timing for applications and provided a method for quick prediction of a cotton insect infestation. Ahmed et al. (2013) correlated NDVI, derived from the images acquired from the SPOT satellite (Spot Image, Toulouse,

France), with cotton leaf curl virus (CLCV). Compared to NDVI, the results showed that weather data, air temperature, and pan evaporation were more significantly correlated with CLCV. Reisig and Godfrey (2006) developed satellite [QuickBird (DigitalGlobe, Inc., Longmont, CO)], airborne (customized and proprietary multispectral and hyperspectral imaging), and ground-based (ASD FieldSpec Pro FR spectroradiometer) remote sensing to distinguish aphid (*Aphis gossypii* Glover) and spider mite (*Tetranychus* spp.) infested cotton from uninfested cotton in Shafter, CA. The same satellite and airborne data were later used to detect aphid-infested cotton (Reisig and Godfrey, 2010), but this time, the focus was on distinguishing cotton crop stress caused by nitrogen deficiency and aphids.

At the field scale, Sudbrink et al. (2003) conducted three field studies in the Mississippi Delta to evaluate airborne and ground-based (hand-held) remote sensors to map the variability in cotton fields and monitor larval densities of beet armyworm and cabbage looper, relating NDVI to pest infestation. Willers et al. (2005) used NDVI derived from CIR images that were acquired from an airborne camera system (ITD-Spectral Visions remote data acquisition camera system; ITD Inc., Stennis Space Center, MS) to map the within-field variability of cotton fields for insect sampling. Willers et al. (2006) continued to estimate the boundaries of different pest densities by mapping habitat variability within cotton fields using the high-resolution remotely sensed imagery for analysis of insect dispersion and abundance in the plants. The boundaries of various densities of the pest provided information for site-specific prescriptions to reduce management costs and amounts of pesticide used. Lawrence et al. (2007) conducted a multiyear study using the 350 to 2500 nm ASD Fieldspec Pro spectroradiometer to measure hyperspectral reflectance of cotton plants at different reniform nematode population levels in controlled microplot tests and in naturally infested fields. The hyperspectral signatures and reniform populations were classified using an unsupervised artificial neural network, self-organized map (SOM). Nematode contour maps based on the classification were developed to provide zones for different nematicide rates. Doshi et al. (2010) continued this study using ground-measured hyperspectral reflectance of cotton plants at different nematode population numbers. Classification of the reflectance using SOM indicated that the spectral regions of the red edge, NIR, and mid-IR were correlated with the nematode numbers in the cotton plant's rhizosphere. Prabhakar et al. (2011) used the 350 to 2500 nm ASD FieldSpec 3 Hi-Res spectroradiometer to measure the reflectance of a cotton canopy for detection of crop stress caused by leafhopper (Hemiptera: Cicadellidae). From the reflectance measurements, six sensitive bands were selected, and a regression analysis of the ratio indices derived from two or more of these sensitive bands resulted in a good estimation of the leafhopper indices, which have potential to detect leafhoppers over a cotton canopy. Kumar et al. (2013) studied the relationship between spectral indices, such as the simple ratio (SR), NDVI, and transformed vegetation index (TVI) as derived from the reflectance measured with a spectroradiometer in the field, and the effect of various growth environments on sucking pest population dynamics in cotton. The regression built with the relationship could predict the sucking pest population of leafhopper and whitefly on cotton.

Insect-resistant (*Bt*) cotton has been widely used in cotton farming. Mukherjee et al. (2008, 2009) used NDVI derived from IRS-1D linear imaging self-scanning system (LISS) III satellite data [Indian Space Research Organization, Bangalore,

India] to identify the *Bt* cotton area cover. The NDVI had a fairly good correlation with the yield of *Bt* cotton; however, the authors did not differentiate *Bt* cotton areas from non-*Bt* cotton areas.

Plant Disease Infestation

Identification of root rot spots in a cotton field is a starting point for the application of remote sensing. Cotton root rot is caused by the soil-borne fungus *Phymatotrichopsis omnivore*, one of the most destructive plant diseases. Neblette (1927) and Taubenhaus et al. (1929) used aerial photography to survey infection of cotton root rot in Texas. Nixon et al. (1975, 1987) evaluated the distribution of cotton root rot damage in cotton fields using color infrared photography and multispectral false-color video, respectively. Pinter et al. (1979), using infrared thermography, found that when cotton roots were infected by *Phymatotrichum omnivorum*, leaf temperatures increased by 3 to 4°C above those of healthy plants. Yang et al. (2005a) classified airborne multispectral imagery to map root rot infections within cotton fields close to the end of the growing season when cotton root rot was fully developed in south Texas. Yang et al. (2010) compared airborne multispectral and hyperspectral imagery for mapping root rot infestations in cotton fields close to the end of the growing season, and later, Yang et al. (2014) used temporal airborne multispectral images to monitor the progress of cotton root rot over the growing season.

Crop Field Weed Infestation and Herbicide Usage

Insect and weed control are both important components of pest management. Remote sensing has been playing a significant role in detecting these pests or pest environments for effective control. For weed management, Iqbal et al. (2006) conducted a study in central Mississippi to discriminate weeds (morning glory and grasses) from cotton canopy based on the spectral analysis of aerial CIR imagery and field hyperspectral measurements [GER 1500 spectroradiometer (SVC, Poughkeepsie, NY)]. Results showed that the NIR band of the CIR imagery and the wavelength region of 750 to 1000 nm in hyperspectral responses demonstrated good discrimination between the cotton canopy and the cotton canopy when infested with morning glory.

Richardson et al. (1992) determined that 20-m SPOT satellite imagery could be used to monitor the cotton stalk destruction program in the Rio Grande Valley area. Richardson et al. (1993) further found that satellite multispectral image classification techniques could not reliably distinguish cotton from sorghum [*Sorghum bicolor* (L.) Moench] and greenness screening for cotton plant stalks after the stalk destruction deadline was possible only where ground observations locating cotton fields were available. Yang et al. (2005b) evaluated the control of postharvest regrowth with herbicides in cotton fields by using cotton canopy reflectance measurements from an ASD FieldSpec handheld 350 to 1050 nm spectroradiometer, airborne CIR imagery of a customized three-camera system for the green, red, and NIR bands, and the physical data of the plants. Yang et al. (2006) later used a similar method to assess the destruction of cotton stalks with herbicide to detect the difference in regrowth among treatments.

Huang et al. (2010b, 2010c) conducted a field study to examine the effect of aerially applied glyphosate drift on non-GR cotton using spray drift sampling and aerial multispectral remote sensing. NDVI and SAVI were derived from the aerial CIR imagery and were used to characterize crop injury from the glyphosate drift downwind of the aerial application. Ortiz et al. (2011) evaluated seven vegetation indices generated from aerial CIR imagery and determined that the chlorophyll vegetation index (Vincini et al., 2008) was the best index for correlation with field-measured plant injury data. To detect the onset of cotton plant injury from glyphosate application before the injury symptom became visible, Zhao et al. (2014a) created a new first canonical axis (FCA) feature from the band selection through a biochemical parameter sensitivity analysis and a physically based leaf radiation transfer model. Zhao et al. (2014b) conducted an inversion of the leaf biochemical parameters through the model for the early detection of glyphosate injury to cotton plants as well.

Yield Prediction

Cotton yield is the result of cotton growth management and is a measure of the performance of various management methods. So, the accurate prediction or estimation of the cotton yield is important. Ansari et al. (1999) predicted the yield of cotton species (American and *Desi* cotton) under different nitrogen treatments. The prediction was based on the integrated radiance ratio (RR) and NDVI that were derived from radiance measurements of the cotton canopy using a handheld multiband ground truth radiometer [GTR 12-0.4 (Optomech Engineers Pvt. Ltd., Hyderabad, Andhra Pradesh, India)]. Results indicated that RR and NDVI were significantly correlated quadratically with yield in several growth stages, especially 81 to 110 d after sowing. Leon et al. (2003) established a grid coordinate system to sample cotton fields to determine the relationship between spectral radiance, soil parameters, and cotton yield. The aerial multispectral imagery were acquired using a real-time digital airborne camera system (Spectral Visions, Inc., Stennis Space Center, MS), which is equipped with narrow band-pass filters in the regions of green, red, and NIR. From the imagery, NDVI, SAVI, RVI, and the difference vegetation index (DVI; the NIR and red difference vegetation index) were generated. A stepwise regression was performed on the green, red, and NIR bands and the RVI, DVI, NDVI, and SAVI indices to select the best band combinations with the cotton plant's height and yield. Read et al. (2004) found plant cotton height, LAI, and lint yield were closely associated with NDVI- and NIR-band values from the spectral measurement from either an aircraft or a handheld GER-1500 spectroradiometer during the peak bloom stage of the cotton plants. To estimate the within-field cotton yield variability, Zarco-Tejada et al. (2005) acquired airborne hyperspectral imagery using the airborne visible and near infrared (AVNIR) hyperspectral sensor (OKSI, Inc., Torrance, CA). With the hyperspectral imagery, cotton canopy structure indices of the renormalized difference vegetation index (RDVI), modified triangular vegetation index (MTVI), optimized soil-adjusted vegetation index (OSAVI), modified chlorophyll absorption index (MCARI), and transformed chlorophyll absorption index (TCARI) were obtained. The results suggested that the indices related to LAI, such as RDVI, MTVI, and OSAVI, provided the best relationships with yield and field segmentation at early growing stages, and the indices related to the crop's physiological status, such as MCARI and TCARI, were good predictors at the later growing stages when the plants were close to harvest.

Plant Discrimination and Estimation of the Crop Field Area

Discrimination of cotton plants from other plants (crops and weeds) in the field and the estimation of cotton planting area are important for the management of cotton production. Shifa et al. (2011) differentiated cotton from sugarcane based on the classification of airborne hyperspectral imagery using the AVNIR hyperspectral sensor (OKSI, Inc., Torrance, CA). The classification accuracy was 98%. Musande et al. (2012) discriminated the cotton crop from weeds through a fuzzy classification approach that used SR, NDVI, the transformed normalized difference vegetation index, SAVI, and TVI, which were derived from the temporal advanced wide field sensor (AWIFS) and LISS II imagery of IRS-P6 Resourcesat-I satellite. Among the vegetation indices, SAVI achieved the best accuracy, at 93%, for cotton crop discrimination. Zhang et al. (2013) identified cotton plants by the fusion of data from airborne CIR imagery that was collected using a customized multicamera system and a ground crop canopy reflectance measurement with an ASD FieldSpec handheld spectroradiometer. Results suggested that the data fusion greatly enhanced the detection of cotton from other plants.

Dutta et al. (1994) estimated cotton acreage based on the land cover classification of IRS-1B LISS I and II satellite imagery. Dong et al. (2008) estimated the cotton area portions along with other crops, such as maize, sorghum, and rice (*Oryza sativa* L.), in the North China Plain on the basis of the classification of imagery from the Landsat satellite Thematic Mapper (TM) sensor and SPOT-Vegetation (VGT-S10; European Commission Joint Research Centre, Ispra, Italy). Zhang et al. (2012) conducted a similar study in the Heilonggang area in the north of Huang-Hui-Hai Plain in China. For enhancing land cover classification, Landsat TM/Enhanced Thematic Mapper Plus imagery and MODIS MOD13Q1 vegetation index images were combined. The result provided a 1.3% relative error in the extraction of cotton plant information. Bai et al. (2011b) used an enhanced vegetation index and a differenced enhanced vegetation index derived from Landsat TM 5 imagery to estimate the density of cotton plants in Xinjiang, China. Wang et al. (2012) extracted the cotton planting area from multitemporal HJ satellite images (China Center for Resources Satellite Data and Application, Beijing, China) also in Xinjiang, China. The accuracy of the extraction was 87.7%.

Precision Cotton Farming

Precision cotton farming has been implemented with the development and integration of agricultural mechanization and information technology. Campanella (2000) evaluated remote sensing and precision agricultural technologies to solve problems in decision making for cotton production. Plant et al. (2001) discussed remote sensing data uses in strategic and tactical cotton management. They found that uncalibrated remote sensing data could be used to answer strategic questions related to the spatial variability of the cotton crop status such that the costs of remote sensing data collection and processing could be lowered considerably in site-specific management. Robertson et al. (2002) evaluated multispectral aerial imagery from a Duncan-Tech camera (formerly Duncan Technologies, Inc., Auburn, CA; now part of Optech Inc., Vaughan, Ontario, Canada) and Veris 2000 XA Soil EC mapping system (Veris Technologies, Salina, KS) as tools to identify cotton field variability. Ramarao (2003) conducted a conformity analysis of a cotton crop using IRS 1D LISS III remotely sensed imagery and GIS in India. Pinter et al. (2003) reviewed all aspects of crop management, including that of cotton crops, using remote sensing technologies from the practices of scientists in the ARS USDA. Barnes (2004) reviewed the implementation of site-specific technologies, including remote sensing, in cotton production in the US Midwest. Larson et al. (2008) evaluated the factors that influence cotton producers to use remote sensing technologies for implementing variable-rate application and found that younger, more educated farmers with large farms tended to use remote sensing and agricultural mapping services. Roboson and Medway (2009) compared classified NDVI images of cotton crops derived from both low- and high-resolution satellite imagery to determine the most accurate and feasible option for Australian cotton growers. This work also demonstrated a rapid, automated processing and internet delivery system for distributing satellite SPOT-2 imagery. Li et al. (2012) created management zones in a cotton field for site-specific management based on information extraction from HJ-1 satellite imagery on a county-wide scale in Shandong, China.

Cases

Crop Herbicide Injury Determination

Glyphosate [*N*-(phosphonomethyl) glycine] is a non-selective, broad-spectrum systemic herbicide, which is applied directly onto the plant and absorbed through plant leaves. Glyphosate is the most widely used herbicide in the world, largely because of the wide use of transgenic, GR cropping systems. Aerial application of glyphosate can cover large areas rapidly to provide timely and effective control of weeds in crop fields. However, off-target drift of aerially applied glyphosate is highly active on sensitive plant species, even at low rates. Scientists characterized glyphosate drift in crop fields using spray sampling (Payne, 1993; Kirk, 2000) and also characterized crop injury based on the plant's biological responses (Reddy et al., 2010; Ding et al., 2011). Measurement and analysis of the plant's biological responses for evaluating crop injury are laborious and time consuming. Remote sensing provides an alternative for the rapid assessment of crop injury in order to provide timely information on the effects of glyphosate drift. Rowland (2000) illustrated how low rates of glyphosate could reduce the yield of maize. Henry et al. (2004) determined that hyperspectral remote sensing could be used to identify and quantify herbicide injury to soybean and maize.

To investigate the effects of glyphosate drift from aerial application, a field study was conducted on non-GR cotton using spray drift sampling and aerial multispectral remote sensing (Huang et al., 2010b, 2010c). In the study, non-GR cotton (cultivar "FM955LL") was planted at a rate of 100,000 seed ha^{-1} in eight rows spaced 1.02-m apart and 80-m long with four replicated blocks (Fig. 1). Aerial application of glyphosate was made over the cotton field by the Air Tractor 402B aircraft equipped with 54 equally spaced CP-09 spray nozzles (CP Products, Tempe, AZ) on two 4.93-m booms installed on each side of the aircraft. The sprayed liquid was a glyphosate solution of Roundup WeatherMax (Monsanto Co., St. Louis, MO) applied at a rate of 0.866 kg acid equivalent ha^{-1}. Weather conditions were recorded during the flight while applying the liquid over the field. The average wind speed was 11.2 km h^{-1} with an average direction of 64° from true north. The average air temperature was 28.5°C, and the relative humidity was 72%.

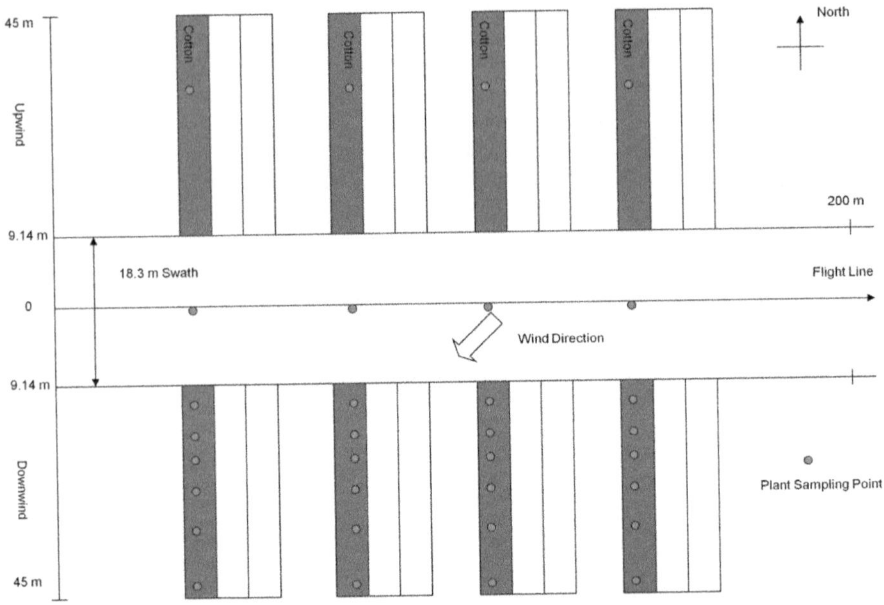

Fig. 1. Field layout for the spray test. The small round circles are the points of measurement for the plant's biological response. (Modified from Fig. 1 from Biosystems Engineering, 107, Y. Huang et al., Airborne remote sensing assessment of the damage to cotton caused by spray drift from aerially applied glyphosate through spray deposition measurements, 2010, with permission from Elsevier.)

An MS 4100 camera was mounted on the Air Tractor aircraft to acquire images over the cotton field on a weekly basis for 3 wk after glyphosate treatment (WAGT). The MS4100 camera was used to provide CIR imagery of the field with an image quality of 1920 (horizontal) × 1080 (vertical) pixels and a 60° field of view. Therefore, with a flight altitude of approximately 366-m above ground level, the ground spatial resolution of the images was 11 × 20 cm pixel^{-1}. The acquired images were georeferenced and transformed to vegetation indices of NDVI and SAVI to illustrate crop injury downwind glyphosate spray (Fig. 2). The images on the day of glyphosate treatment did not show visible crop damage. The images at 1 WAGT indicate clear crop damage, which was more pronounced in 2 and 3 WAGT.

NDVIs were regressed with the field visual grading (0–100%) of crop injury at 1, 2, and 3 WAGT, respectively (Table 2). The results indicated that NDVI explained crop injury well at 2 and 3 WAGT. At 1 WAGT, NDVI values were relatively low, with much interference from the soil. From 2 to 3 WAGT, the crop continued to grow, and the canopy developed gradually in the uninjured crop areas. The continuing crop growth and gradually developing canopy enhanced the contrast between the injured, regrown, and newly grown crop in both the injured and uninjured crop areas, which had relatively large NDVI values. This indicates that NDVI can be used to describe the cotton injury gradient downwind from the glyphosate treatment.

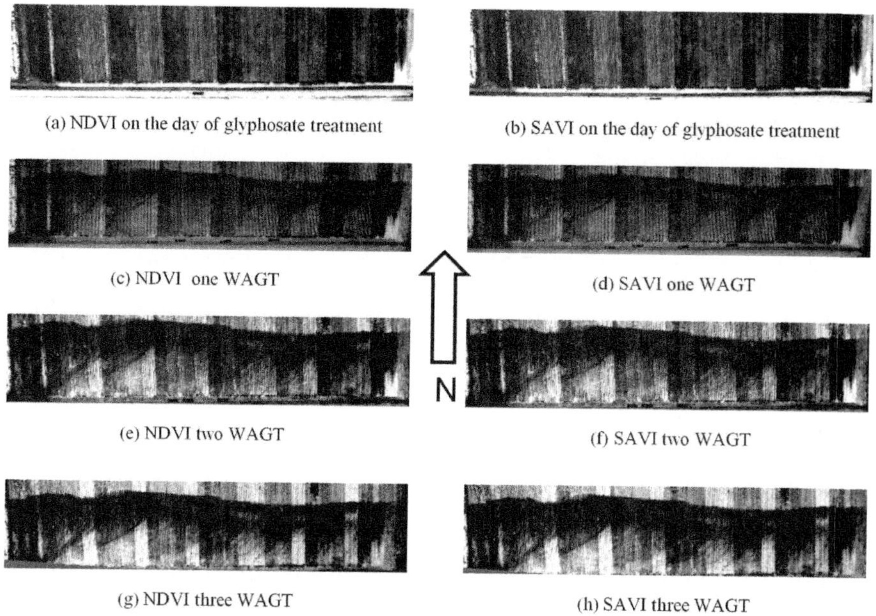

Fig. 2. NDVI and soil adjusted vegetation index images on the day of glyphosate treatment, 1 wk after glyphosate treatment (WAGT), 2 WAGT, and 3 WAGT (modified and expanded from Fig. 4 from Biosystems Engineering, 107, Y. Huang et al., Airborne remote sensing assessment of the damage to cotton caused by spray drift from aerially applied glyphosate through spray deposition measurements, 2010, with permission from Elsevier.).

In practice, farmers want to know the coverage and extent of the glyphosate drift and be able to project the severity of the injury as early as possible. With this information farmers can take appropriate remedial measures to best protect their interests. Such actions may include replanting or seeking insurance compensation. Therefore, an early warning system for herbicide drift detection in weed management would greatly benefit farmers. To develop such a system, we applied hyperspectral remote sensing to extract features from the narrow-band reflectance to detect subtle changes in the plant reflectance for the onset of crop injury.

This research began with a greenhouse experiment. Four weeks after planting, cotton plants were treated and measured for spectral reflectance. In the experiment, 36 pots of non-GR cotton (cultivar FM955LL) were used to obtain leaf reflectance spectra and the plant's biochemical properties. The 36 pots of plants were randomly divided into three treatment groups: (1) 12 pots were treated with a 0.5X (X = 0.866 kg acid equivalent ha^{-1}) solution of glyphosate; (2) 12 pots were treated with a 0.25X glyphosate solution; and (3) 12 pots were used as controls with no glyphosate treatment (CTRL group). A glyphosate solution was prepared using the commercial formulation of glyphosate, Roundup WeatherMax. After the glyphosate treatment, the cotton leaf reflectance and plant biochemical parameters, chlorophyll content (C_{a+b}), equivalent water thickness (C_w), and leaf mass per area (C_m) of three plants for each group were measured at 6, 24, 48, and 72 h after treatment (HAT) to analyze the plant's response to the glyphosate.

Table 2. Regressions of normalized difference vegetation index (NDVI) with field visual grading of cotton plant injury.†

Equation	R^2
VG_1 = 42.99 – 106.57 × NDVI_1	0.21
VG_2 = 44.45 – 163.02 × NDVI_2	0.85
VG_3 = 104.91 – 348.51 × NDVI_3	0.90

† VG_1, VG_2 and VG_3 are the visual grading at 1 wk after glyphosate treatment (WAGT), 2 WAGT, and 3 WAGT, respectively; NDVI_1, NDVI_2, and NDVI_3 are the NDVI values extracted from the aerial color infrared images at 1 WAGT, 2 WAGT, and 3 WAGT, respectively (extracted from Table 3 of Huang et al., 2010b).

Leaf reflectance measurements were conducted using an ASD integrating sphere (ASD Inc., Boulder, CO). This apparatus was coupled with the ASD FieldSpec 3 Hi-Res spectroradiometer which has a 1-nm spectral sampling interval in the spectral range of 400 to 2500 nm. Two lamps (Osram no. 64225, 6V, 10W) were connected with the integrating sphere and provided a collimated beam as the light source to illuminate the plant sample and the white reference board.

To determine the sensitive bands of the measured reflectance, a sensitivity analysis was performed through a physically based leaf radiation transfer model: the leaf optical property spectra (PROSPECT) model (Jacquemoud and Baret, 1990). In the sensitivity analysis, the ranges of the plant's biochemical parameters were determined. Then, a thousand combinations of the plant's biochemical parameters were randomly selected as the inputs to the PROSPECT model. With the inputs, 1000 reflectance spectra were produced accordingly from model simulation implementation. Figure 3 shows the results of the PROSPECT sensitivity analysis of the plant leaf's biochemical parameters. Chlorophyll is an important indicator of crop injury. Therefore, two bands, 479 and 508 nm, located in the blue domain were selected that correspond to peaks of the first order sensitivity index (FOSI) and the total sensitivity index (TSI) curves of C_{a+b}, respectively. Similarly, two other bands, 654 and 673 nm, that are located at the peaks in the red domain were selected. The band, 750 nm, which corresponds to the peaks of the FOSI and TSI of N (a parameter that describes the mesophyll structures of a leaf species), was also selected. At 750 nm, the chlorophyll sensitivity was close to zero. This band could potentially provide information about the leaf structure instead of the chlorophyll content. These five bands were then used to formulate new spectral features. Canonical analysis (Lejeune and Caliński, 2000) was used for the generation of transformed spectral features with optimized group separation. With canonical analysis, a new axis can be constructed for which σ_a^2/σ_b^2 (σ_a is the standard deviation among groups and σ_b is the standard deviations within the groups) is driven as large as possible, and the highest possibility will be generated to discriminate injured plant leaves from healthy ones. The FCA feature for cotton was created as follows:

$$\mathrm{FCA} = 0.26R_1 + 0.65R_2 + 0.64R_3 + 0.33R_4 - 0.03R_5$$

where R_1, R_2, R_3, R_4, and R_5 are the reflectance values at 479, 508, 652, 673, and 750 nm, respectively (Zhao et al., 2014a).

Figure 4 shows the temporal variation of FCA for the cotton leaves of all three groups. It could be seen that at 6 HAT, there were no noticeable differences among the FCA values of the three groups. At 24 HAT and later, the

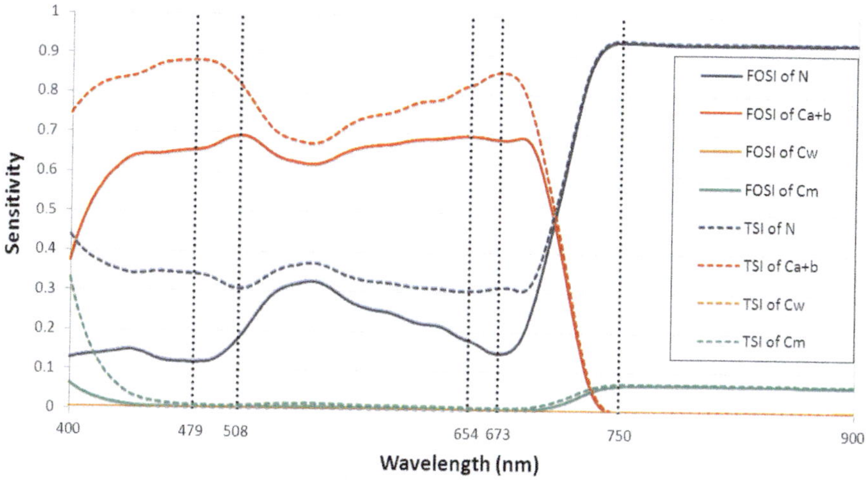

Fig. 3. Sensitivity indices of biochemical parameters through the leaf optical property spectra model (PROSPECT) model and the first order sensitivity index (FOSI), total sensitivity index (TSI), C_{a+b} is chlorophyll content, C_w is equivalent water thickness, C_m is leaf mass per area, and N is a parameter describing the mesophyll structures of he leaf species [modified from Fig. 2(a) Zhao et al., 2014a].

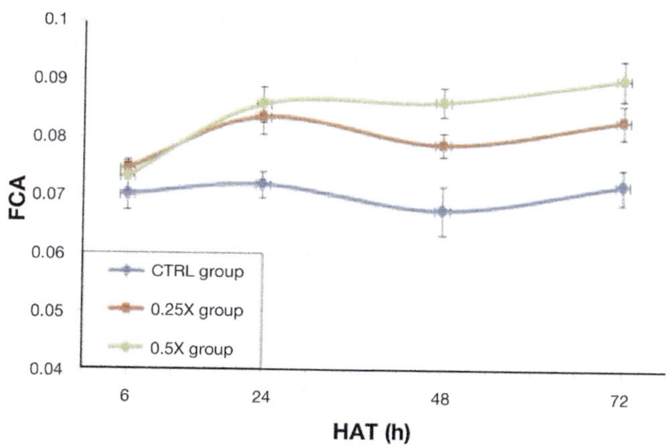

Fig. 4. The first canonical axis variation of cotton leaves of the three groups at 6, 24, 48, and 72 h after treatment [modified from Fig. 3(b) Zhao et al., 2014a].

differences appear more pronounced, with a consistent trend of a higher spray rate with larger FCA values. This indicates that the FCA features can detect the onset of injury to the cotton plant before it becomes visible, which is typically 2 to 3 d after glyphosate treatment.

Water Management in Cotton

Water management is one important aspect of crop production, but the crop production system frequently requires multiple decisions to be made. Nowhere is this more apparent than in cotton production. Cotton management involves many factors that must work together properly to ensure a high-yielding and quality crop. Depending on soil texture, nutrient levels, and other field characteristics, some field areas have a lower yield potential and thus have differing needs with regard to the timing of water application and insecticide management. We thus decided to investigate the fusion of multiple sensor inputs that includes thermal imaging to delineate field areas requiring more attention with regard to water and pest management.

Thermal images acquired over 5 wk in the summer (using a Sofradir Electrophysics PV-320T camera mounted in an agricultural aircraft and flown at an altitude of 457 m) were composited to produce a cumulative thermal map of a cotton canopy. This 2.2-ha field was managed uniformly with regard to nutrient levels and was not irrigated. In addition to the thermal images, two CIR images (12-bit, 0.15-m resolution) were obtained using the Emerge Digital Sensor System flown at an altitude of approximately 609 m on two dates in the summer of 2006. Images were processed using an intensity normalization method followed by the calculation of NDVI. The CIR images were also processed using the Isodata unsupervised classification method (Jensen, 1996) to establish two classes (coded as zero = no vegetation and one = vegetation). The means in the management zones were subsequently calculated for each image and resampled to a 1-m resolution, resulting in an approximation of the percentage of cover. The change in canopy cover was determined by subtracting a classified image from the vegetation mapped 1 mo later. Figure 5 illustrates the description above with the combination of thermal and CIR imaging.

Significant patterns of yield that are linked to the thermal zones were evident, using a bivariate local indicator of spatial association (LISA) map. About 65% of the field extent showed a significant correlation at the 1% level, and 15% of the field area was significant between the 1% and 5% level. Another bivariate LISA map was developed using GeoDa (Spatial Analysis Laboratory, University of Illinois, Urbana-Champaign, IL). Significant autocorrelations can be seen ($p \leq 0.05$) between the change in cotton canopy cover and the cumulative thermal maps (Fig. 5A). A low amount of change in the canopy cover was coupled with the lowest cumulative canopy temperatures, as depicted by category 4 of the legend. Similarly, adjacent areas of the field indicated by category 3 displayed a relatively high change in canopy cover paired with low canopy temperature change. The second bivariate LISA map compared patterns of yield with in-season changes in canopy cover (Fig. 5B). Four field-scale production zones are delineated including the following: (1) high-yielding areas with low changes in canopy cover; (2) a spotty grouping of high-yielding areas paired with medium to high changes in canopy cover; (3) "stressed" portions of the field with a low yield and low change in canopy cover; and (4) low-yielding areas coupled with medium to high changes in canopy cover. Ground scouting demonstrated that from mid to late summer, cotton plants that were subjected to consistently high temperatures had more than 50% open bolls. These results indicate that composited thermal imagery combined with tracking the change in canopy cover at key phenological stages could be useful for the in-season prediction of yield potential as well as early senescence promoted by heat/water stress in highly mixed cotton fields. Termination of irrigation is tied to the phenological stage in cotton. Maps such as these could promote development of site-specific insecticide applications to

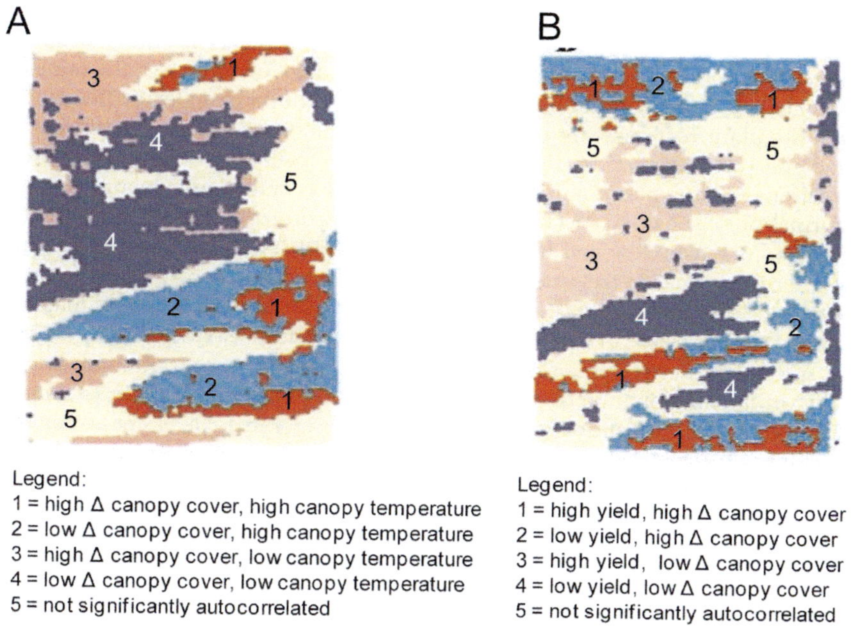

Fig. 5. Spatial relationships between (A) change in canopy cover and canopy temperature, and (B) yield and change in canopy cover. The numbers correspond to relative shading in the image, where Δ means change (Thomson et al., 2012).

protect high-yielding areas and promote the cost-effective application of defoliants/harvesting aids. Assuming an irrigation system can vary water application rates or could be zonally scheduled, spatial differences as indicated by differing canopy temperatures could be useful for spatial water management.

Limitations of Remote Sensing

Remote sensing platforms are available for cotton production. Satellite systems such as Landsat (30-m spatial resolution for multispectral bands) and MODIS (250-m spatial resolution) are only suitable for large-scale observation. Although high-resolution satellite imaging platforms such as WorldView-2 (1.8-m multispectral spatial resolution) and GeoEye-1 (1.65-m multispectral spatial resolution) could be used for cotton field observation, in many cases, sub-meter spatial resolutions are needed to monitor cotton plants in different growth stages. Also, it is difficult to have a satellite visit cover a specific field within a designated time frame and frequency. Agricultural airplanes are more favorable to provide a cost-effective, rapid, and flexible platform to conduct high-resolution (cm level) remote sensing of cotton fields. For example, in the study of crop injury from glyphosate, images of the field were needed 1, 2, and 3 wk after the field treatment. During this time, the cotton plants were only at the 3- to 5-leaf stage and were surrounded by soil. To acquire the images that reliably segment the small cotton plants from the soil background, the airborne imaging system achieved 20/40 cm (horizontal/vertical) per pixel, which is much better than any current satellite data.

Multispectral remote sensing provides broadband spectral information of cotton plants. The broadband spectral information is typically sufficient to produce vegetation indices for mapping the status of a cotton field, e.g., the injury to cotton plants from glyphosate spray. However, if subtle changes of the spectra from the plant response are required to detect the onset of crop stress, narrowband spectral information is required from hyperspectral remote sensing. Early detection of cotton plant injury from glyphosate was determined by narrowband information of sensitive spectral wavelengths to produce features for differentiating the injury from low and high doses of glyphosate spray and control 1, 2, and 3 d after the herbicide treatment. Further, the narrowband information could be converted to a sensor design of multiple sensitive bands, which could lead to a device directly usable in spectral measurement of cotton leaves.

Although thermal imaging can provide an instant representation of "hot spots" in the field, that give an indication of relative water or heat stress, a major challenge still exists for delineating canopy-air temperature differences on a temporal basis under the low vapor pressure deficits that are prevalent in humid climates. A proper accounting is necessary of the micrometeorological conditions and altitude effects on the canopy temperature as represented by the camera. Although some progress has been made, more research is also needed on how to accurately characterize the canopy temperature by using thermal imagery before the complete closure of the canopy.

Future Outlook

In the last decade, remote sensing technologies have seen a rapid development for the use in crop monitoring and production management. In the next 10 yr or so, new systems, technologies, and applications will be developed for specialized remote sensing analysis of cotton fields.

With the so-called genomics revolution, the DNA in crop plant varieties is being investigated to determine how DNA influences the growth of plants under different conditions of soil, weather, nutrition, and pests. Plant breeders want to associate the performance of plants with DNA information; however, genotyping analysis alone cannot make the association. Therefore, an analysis of phenotyping is also necessary. The phenotype is the physical characteristic of an organism that results from its genes and environment and the interaction between them. A phenotype can be any response to environmental factors, such as plant height, chemical composition, and crop yield. Measuring plant phenotypes for the association mentioned above is a challenging task. To discover the subtle effects of genes on plant traits may require multiple repeated measurements. Remote sensing could contribute to such measurements for the inversion of biophysical and biochemical parameters of soil and crops from measured spectral data, and it is a powerful tool for phenomic studies of the routine measurement of plant height, canopy cover, and canopy temperature throughout the growing season for cotton.

UAV technologies are developing very quickly for agricultural operations and management. The Association for Unmanned Vehicle Systems International predicts that agriculture will make up 80% of UAV commercial growth, creating 100,000 jobs and having an economic impact of $82 billion by the year 2025 (https://deltaagtech.com/). Although we typically have the remote sensing platforms of space-borne satellites, airborne (manned) airplanes, and ground on-the-go systems, UAVs provide a unique platform to remotely sense crop fields with a high spatial resolution, rapid coverage, and flexible scheduling and operation. UAVs can fly much lower than manned aircraft and can acquire continuous images over a crop field, while ground-based systems only provide point measurements over the field. UAVs will be an important remote sensing platform for cotton field monitoring and management.

Repeated and intensive use of herbicide has exerted a high selection pressure on weed populations resulting in the evolution of herbicide-resistant (HR), along with herbicide-sensitive (HS), weeds in crop fields. To date, a total of 24 weed species have evolved resistance to glyphosate worldwide. HR weed populations could seriously jeopardize pre-plant burndown options and planting operations in crop production systems. To effectively manage HR weeds, it is necessary to identify and determine the distribution of the HR and HS weed field populations. With this information, site-specific weed management can be conducted using different post-emergence options to treat HR and HS weed field populations. Currently, HR and HS plants are identified by assessing physiological and biochemical changes in plants after glyphosate treatment. The current methods with the whole plant or leaf segment/disc shikimate assays, while undoubtedly reliable in distinguishing HR from HS plants, are tedious and labor-intensive and may cause delay in critical decision-making for weed management. Therefore, alternative approaches for direct differentiation of HR from HS weeds are highly desirable. Hyperspectral remote sensing technology has a great potential for rapid, consistent, and accurate differentiation between HR and HS weed plants in crop fields. Research is being conducted to offer guidelines to crop producers and consultants to deal with HR and HS weeds in crop fields. The ultimate goal is to use the information to develop low-cost sensing systems or devices that can be placed in the hands of producers and crop consultants.

References

Ahmed, A., A. Akhtar, B. Khalid, and A. Shamim. 2013. Correlation of meteorological parameters and remotely sensed normalized difference vegetation index (NDVI) with cotton leaf curl virus (CLCV). J. Phys.: Conf. Ser. 439:012044. doi:10.1088/1742-6596/439/1/012044

Ansari, M.S., R.K. Mahey, and S.S. Sidhu. 1999. Cotton yield prediction through spectral parameters. J. Ind. Soc. Remote Sensing. 27:185–192. doi:10.1007/BF02990831

Bai, J., J. Li, and S. Li. 2011. Monitoring the cotton growth conditions based on LAI from remote sensing and expert knowledge. Adv. Mater. Res. 317-319:1050–1057. doi:10.4028/www.scientific.net/AMR.317-319.1050

Bai, J., J. Li, and S. Li. 2011. Monitoring the plant density of cotton with remotely sensed data. In: IFIP Advances in Information and Communication Technology. Computer and Computing Technologies in Agriculture IV 345. Springer, New York. p. 90–101. doi:10.1007/978-3-642-18336-2_11

Bai, J., J. Li, Q. Liu, X. Wang, and S. Li. 2012. Diagnosing cotton farmland quality using multi-temporal remotely sensed data. Sensor Letters. 10(1-2):475–483.

Barnes, E.M. 2004. Current implementation of site-specific technologies in U.S. cotton production. In: W. Gao, D.R. Shaw, editors, Remote sensing and modeling of ecosystems for sustainability. Proc. SPIE 4:12–19.

Bawden, F.C. 1933. Infra-red photography and plant virus diseases. Nature 132:168. doi:10.1038/132168a0

Brookes, G., and P. Barfoot. 2011. The income and production effects of biotech crops globally 1996-2009. Int. J. Biotechnol. 12:1–49. doi:10.1504/IJBT.2011.042680

Brooks, N.L. 2001. Characteristics and production costs of U.S. cotton farms. Stat. Bull. 974-2. Economic Res. Serv., USDA, Washington, DC.

Campanella, R. 2000. Testing components toward a remote-sensing-based decision support system for cotton production. Photogramm. Eng. Remote Sensing 66(10):1219–1227.

Coleman, V.B., C.W. Johnson, and L.N. Lewis. 1974. Remote sensing in control of pink bollworm infestation in cotton. Calif. Agric. 28:10–12.

Colwell, R.N. 1956. Determining the prevalence of certain cereal crop diseases by means of aerial photography. Hilgardia 26:223–286. doi:10.3733/hilg.v26n05p223

Dalezios, N.R., A. Mplanta, and C. Doemnikiotis. 2011. Remotely sensed cotton evapotranspiration for irrigation water management in vulnerable agriculture of central Greece. J. Inf. Technol. Agric. 4(1):1–14.

Dill, G.M. 2005. Glyphosate-resistant crops: History, status, and future. Pest Manag. Sci. 61(3):219–224. doi:10.1002/ps.1008

Ding, W., K.N. Reddy, L.J. Krutz, S.J. Thomson, Y. Huang, and R.M. Zablotowicz. 2011. Biological response of soybean and cotton to aerial glyphosate drift. J. Crop Improv. 25:291–302. doi:10.1080/15427528.2011.559633

Dong, Q., E. Herman, and Z. Chen. 2008. Crop area assessment using remote sensing on the North China Plain. Int. Arch. Photogramm. Remote Sens. Spatial Inf. Sci. 37(B8):957–962.

Doshi, R.A., R.L. King, and G.W. Lawrence. 2010. Classification of rotylenchulus reniformis numbers in cotton using remotely sensed hyperspectral data on self-organizing maps. J. Nematol. 42(3):179–193.

Dutta, S., S.A. Sharma, A.P. Khera, Ajai, M. Yadav, R.S. Hooda, K.E. Mothikumar, and M.L. Manchanda. 1994. Accuracy assessment in cotton acreage estimation using Indian remote sensing satellite data. ISPRS J. Photogramm. Remote Sens. 49(6):21–26. doi:10.1016/0924-2716(94)90011-6

Economic Research Service. 2014. Adoption of genetically engineered crops in the U.S., USDA, Washington, DC. http://www.ers.usda.gov/data-products/adoption-of-genetically-engineered-crops-in-the-us.aspx (accessed date 25 May 2014)

Falkenberg, N.R., P. Giovanni, J.T. Cothren, D.I. Leskovar, and C.M. Rush. 2007. Remote sensing of biotic and abiotic stress for irrigation management of cotton. Agric. Water Manage. 87(1):23–31. doi:10.1016/j.agwat.2006.05.021

Farias de Oliveira, L. 2008. Reflectance sensors to predict mid-season nitrogen need of cotton. M.S. thesis, Univ. of Missouri-Columbia, Columbia.

Fulton, J., D. Sullivan, J. Shaw, M. Dougherty, and G. Bland. 2007. Site-specific management of cotton production using remote sensed thermal imagery within a conservation tillage system. Southern Conservation Agriculture Systems Conference, Quincy, FL, 25–27 June.

Garatuza-Payan, J., and C.J. Watts. 2005. The use of remote sensing for estimating ET of irrigated wheat and cotton in Northwest Mexico. Irrig. Drain. Syst. 19:301–320. doi:10.1007/s10795-005-5192-2

Henry, W.B., D.R. Shaw, K.R. Reddy, L.M. Bruce, and H.D. Tamhankar. 2004. Remote sensing to detect herbicide drift on crops. Weed Technol. 18:358–368. doi:10.1614/WT-03-098

Huang, Y., S.J. Thomson, Y. Lan, and S.J. Maas. 2010a. Multispectral imaging systems for airborne remote sensing to support agricultural production management. Int. J. Agric. Biol. Eng. 3(1):50–62.

Huang, Y., S.J. Thomson, B.V. Ortiz, K.N. Reddy, W. Ding, R.M. Zablotowicz, and J.R. Bright, Jr. 2010b. Airborne remote sensing assessment of the damage to cotton caused by spray drift from aerially applied glyphosate through spray deposition measurements. Biosystems Eng. 107:212–220. doi:10.1016/j.biosystemseng.2010.08.003

Huang, Y., R. Sui, S.J. Thomson, and D.K. Fisher. 2013. Estimation of cotton yield with varied irrigation and nitrogen treatments using aerial multispectral imagery. Int. J. Agric. Biol. Eng. 6(2):37–41.

Huang, Y., S.J. Thomson, B.V. Ortiz, K.N. Reddy, W. Ding, R.M. Zablotowicz, and J.R. Bright, Jr. 2010c. Determination of cotton plant injury by aerial application of glyphosate using remote sensing and spray drift sampling. Proceedings of 2010 National Cotton Council, Beltwide Cotton Conference, New Orleans, LA, January 4–7.

Iqbal, J., P.R. Owens, and I. Ali. 2006. Application of remote sensing data to assess weed infestation in cotton. Agric. J. 1(4):186–191.

Jacquemoud, S., and F. Baret. 1990. PROSPECT: A model of leaf optical properties spectra. Remote Sens. Environ. 34(2):75–91. doi:10.1016/0034-4257(90)90100-Z

James, C. 2010. Global status of commercialized biotech/GM crops: 2010. Brief 42-2010. International Service for the Acquisition of Agri-Biotech Applications, Ithaca, NY. http://www.isaaa.org/resources/publications/briefs/42/executivesummary/default.asp.

Jensen, J.R. 1996. Introductory digital image processing—A remote sensing perspective. Prentice Hall, Saddle River, NJ. p. 197–256.

Kirk, I.W. 2000. Aerial spray drift from different formulations of glyphosate. Trans. ASABE 43:555–559. doi:10.13031/2013.2735

Kumar, N., R. Niwas, M.L. Khichar, R.K. Saini, and B. Biswas. 2013. Effect of different growing environments on population dynamics of sucking pests in relation to various spectral indices in cotton. J. Indian Soc. Remote Sens. 41(2):309–317. doi:10.1007/s12524-012-0218-3

Larson, J., R.K. Roberts, B.C. English, S.L. Larkin, M.C. Marra, S.W. Martin, K.W. Paxton, and J.M. Reeves. 2008. Factors affecting farmer's adoption of remotely sensed imagery for precision management in cotton production. Precis. Agric. 9:195–208. doi:10.1007/s11119-008-9065-1

Lawrence, G.W., R.A. Doshi, R.L. King, K.S. Lawrence, and J. Caceres. 2007. Nematode management using remote sensing technology, self-organized maps and variable rate nematicide applications. Proceedings of World Cotton Research Conference 4. Lubbock, TX, September 10–14. International Cotton Advisory Committee, Washington, DC.

Lejeune, M., and T. Caliński. 2000. Canonical analysis applied to multivariate analysis of variance. J. Multivariate Anal. 72:100–119. doi:10.1006/jmva.1999.1852

Leon, C.T., D.R. Shaw, M.S. Cox, M.J. Abshire, B. Ward, and M.C. Wardlaw, III. 2003. Utility of remote sensing in predicting crop and soil characteristics. Precis. Agric. 4:359–384. doi:10.1023/A:1026387830942

Li, M., G. Zhao, M. Cai, Q. Zhao, and J. Tang. 2012. Extraction of cotton Information using remote sensing and precision management zoning at county scale. J. Nat. Resour. 27(11):1971–1980 (in Chinese).

Maas, S.J., G.J. Fitzgerald, and W.R. DeTar. 2000. Determining cotton leaf canopy temperature using multispectral remote sensing. Proceedings of Beltwide Cotton Conf. 1, San Antonio, TX. Cotton Foundation Publisher: Memphis, TN. p. 623–626.

Mallarino, A.P., and D.J. Wittry. 2004. Efficacy of grid and zone soil sampling approaches for site-specific assessment of phosphorus, potassium, pH, and organic matter. Precis. Agric. 5:131–144. doi:10.1023/B:PRAG.0000022358.24102.1b

Moran, S., A. Vidal, D. Troufleau, J. Qi, T.R. Clarke, P.J. Pinter, Jr., T.A. Mitchell, Y. Inoue, and C.M.U. Neale. 1997. Combining multifrequency microwave and optical data for crop management. Remote Sens. Environ. 61(1):96–109. doi:10.1016/S0034-4257(96)00243-X

Mukherjee, S., S. Shashtri, C.K. Singh, B. Kumari, R. Kumari, A. Ram, A. Mukherjee, and B. Singh. 2009. Remote sensing techniques to infer Bt cotton in a part of Bhatinda, Punjab, India. Gene Conserve. 643-675.

Mukherjee, S., S. Shashtri, C.K. Singh, B. Kumari, R. Kumari, A. Ram, A. Singh, A. Mukherjee, and B. Singh. 2008. Genetically modified cotton species detection by LISS-III satellite data. Nat. Preceedings http://dx.doi.org/10.1038/npre.2008.2659.2.

Mulla, D.J. 2013. Twenty five years of remote sensing in precision agriculture: Key advances and remaining knowledge gaps. Biosyst. Eng. 114:358–371. doi:10.1016/j.biosystemseng.2012.08.009

Musande, V., K. Anil, and K. Karbhari. 2012. Cotton crop discrimination using fuzzy classification approach. J. Indian Soc. Remote Sens. 40:589–597.

Neblette, C.B. 1927. Aerial photography for the study of plant diseases. Photo-Era Mag. 58:346.

Nixon, P.R., D.E. Escobar, and R.L. Bowen. 1987. A multispectral false-color video imaging system for remote sensing applications. Proceedings of 11th Biennial Workshop on Color Aerial Photography and Videography in the Plant Sciences and Related Fields, Am. Soc. for Photogramm. and Remote Sens., Bethesda, MD. p. 295-305, 340.

Nixon, P.R., S.D. Lyda, M.D. Heilman, and R.L. Bowen. 1975. Incidence and control of cotton root rot observed with color infrared photography. Pub. No. MP1241. Agric. Exp. Station, Texas A&M, College Station, TX.

Ortiz, B.V., S.J. Thomson, Y. Huang, K.N. Reddy, and W. Ding. 2011. Determination of differences in crop injury from aerial application of glyphosate using vegetation indices. Comput. Electron. Agric. 77:204–213. doi:10.1016/j.compag.2011.05.004

Payne, N.J. 1993. Spray dispersal from aerial silvicultural glyphosate applications. Crop Prot. 12(6):463–469. doi:10.1016/0261-2194(93)90009-8

Pinter, P.J., Jr., J.L. Hatfield, J.S. Schepers, E.M. Barnes, M.S. Moran, C.S.T. Daughtry, and D.R. Upchurch. 2003. Remote sensing for crop management. Photogramm. Eng. Remote Sensing 69(6):647–664. doi:10.14358/PERS.69.6.647

Pinter, P.J., Jr., and R.J. Reginato. 1981. Thermal infrared techniques for assessing plant water status. In: Irrigation Scheduling for Water and Energy Conservation in the 80's. Proceedings of the ASAE Irrigation Scheduling Conference. St. Joseph, MI. ASAE.

Pinter, P.J., Jr., M.E. Stanghellini, R.J. Reginato, S.B. Idso, A.D. Jenkins, and R.D. Jackson. 1979. Remote detection of biological stress in plants with infrared thermography. Science 205:585–587. doi:10.1126/science.205.4406.585

Plant, R.E., D.S. Munk, B.R. Roberts, R.N. Vargas, R.L. Travis, D.W. Rains, and R.B. Hutmacher. 2001. Application of remote sensing to strategic questions in cotton management and research. J. Cotton Sci. 5:30–41.

Prabhakar, M., Y.G. Prasad, M. Thirupathi, G. Sreedevi, B. Dharajothi, and B. Venkateswarlu. 2011. Use of ground based hyperspectral remote sensing for detection of stress in cotton caused by leafhopper (Hemiptera: Cicadellidae). Comput. Electron. Agric. 79:189–198. doi:10.1016/j.compag.2011.09.012

Qi, Y., Z. Duan, and X. Lv. 2011. Research of cotton canopy characteristic information by hyperspectral remote sensing data. Cotton Sci. 23(2):167–171.

Qi, Y., D. Wang, G. Chen, C. Huang, Z. Duan, Y. Chen, J. Yuan, and K. Yang. 2005. Research of crop canopy characteristic information using hyperspectral remote sensing data. Cotton Sci. 17(6):371–375.

Read, J.J., L. Tarpley, J.M. McKinion, and K.R. Reddy. 2002. Narrow-waveband reflectance ratios for remote estimation of nitrogen status in cotton. J. Environ. Qual. 31:1442–1452. doi:10.2134/jeq2002.1442

Read, J.J., J. Iqbal, J.A. Thomasson, J.L. Willers, and J.N. Jenkins. 2004. Remote sensing in dryland cotton: Relation to yield potential and soil properties. In: Proceedings of SPIE 5153, ecosystems' dynamics, agricultural remote sensing and modeling, and site-specific agriculture, 5 Jan. p. 61–72. doi:10.1117/12.505875.

Ramarao, N. 2003. Conformity analysis of cotton crop using remote sensing and GIS. In: Proceedings of Map Asia Conference. Geospatial Media and Communications Pvt Ltd, Noida, Uttar Pradesh, India. p. 1–11.

Reddy, K.N., W. Ding, R.M. Zablotowicz, S.J. Thomson, Y. Huang, and L.J. Krutz. 2010. Biological responses to glyphosate drift from aerial application in non-glyphosate-resistant corn. Pest Manag. Sci. 66:1148–1154. doi:10.1002/ps.1996

Reddy, K.N., and V.K. Nandula. 2012. Herbicide resistant crops: History, development and current technologies. Indian J. Agron. 57(1):1–7.

Reisig, D.D., and L.D. Godfrey. 2006. Remote sensing for detection of cotton aphid- (Homoptera: Aphididae) and spider mite- (Acari: Tetranychidae) infested cotton in the San Joaquin Valley. Environ. Entomol. 35:1635–1646. doi:10.1603/0046-225X(2006)35[1635:RSFDOC]2.0.CO;2

Reisig, D.D., and L.D. Godfrey. 2010. Remotely sensing arthropod and nutrient stressed plants: A case study with nitrogen and cotton aphid (Hemiptera: Aphididae). Environ. Entomol. 39:1255–1263. doi:10.1603/EN09218

Richardson, A.J., M.M. Crawford, J.H. Everitt, and K.R. Summy. 1992. Monitoring the Rio Grande Valley cotton stalk destruction program with satellite imagery. Geocarto Int. 7(2):15–20. doi:10.1080/10106049209354362

Richardson, A.J., A.H. Gerbermann, K.R. Summy, and G.L. Anderson. 1993. Remote sensing techniques for monitoring the Rio Grande Valley cotton stalk destruction program. Geocarto Int. 8(3):15–22. doi:10.1080/10106049309354416

Ritchie, G.L., C.W. Bednarz, J. Whitaker, and C. Mills. 2005. Cotton irrigation timing using remote sensing. In: P. Dugger and D. Richter, editors, Proceedings of the Beltwide Cotton Conferences, New Orleans, LA, 4–7 Jan. 2005. National Cotton Counc. Am., Memphis, TN. p. 1956–1960.

Robertson, W., C. Jayroe, W. Baker, D. Plunkett, and T. Kirkpatrick. 2002. Remote sensing technologies used as management tools in cotton production. AAES Res. Ser. 502, Arkansas Soil Fertility Studies. Arkansas Agric. Exp. Station, Univ. of Arkansas, Fayetteville. p. 70–74.

Roboson, A., and J. Medway. 2009. Remote sensing applications for cotton. Australian Cottongrower 30(4):40–43.

Rowland, C. D. (2000). Crop tolerance to non-target and labeled herbicide applications. MS thesis. Department of Plant and Soil Sciences, Mississippi State University, Mississippi State.

Shifa, M.S., M.S. Naweed, M. Omar, M.Z. Jhandir, and T. Ahmed. 2011. Classification of cotton and sugarcane plants on the basis of their spectral behavior. Pak. J. Bot. 43(4):2119–2125.

Singh, S.K., S. Dutta, and N. Dharaiya. 2013. Estimation of crop evapotranspiration of cotton using remote sensing technique. Int. J. Environ. Eng. Manage. 4(5):523–528.

Stafford, J.V. 2000. Implementing precision agriculture in the 21st century. J. Agric. Eng. Res. 76:267–275. doi:10.1006/jaer.2000.0577

Steddom, K., D. Jones, and C. Rush. 2005. A picture is worth a thousand words. Am. Phytopathol. Soc. http://www.apsnet.org/publications/apsnetfeatures/Documents/2005/RemoteSensing.pdf. (accessed date 15 April 2014).

Sudbrink, D.L., Jr., F.A. Harris, J.T. Robbins, P.J. English, and J.L. Willers. 2003. Evaluation of remote sensing to identify variability in cotton plant growth and correlation with larval densities of beet armyworm and cabbage looper (Lepidoptera: Noctuidae). Fla. Entomol. 86(3):290–294. doi:10.1653/0015-4040(2003)086[0290:EORSTI]2.0.CO;2

Sui, R., and J.A. Thomasson. 2006. Ground-based sensing system for cotton nitrogen status determination. Trans. ASABE 49(6):1983–1991. doi:10.13031/2013.22279

Tarpley, L., K.R. Reddy, and G.F. Sassenrath-Cole. 2000. Reflectance indices with precision and accuracy in predicting cotton leaf nitrogen concentration. Crop Sci. 40:1814–1819. doi:10.2135/cropsci2000.4061814x

Taubenhaus, J.J., W.N. Ezekiel, and C.B. Neblette. 1929. Airplane photography in the study of cotton root rot. Phytopathology 19:1025–1029.

Thomson, S.J., C.M. Ouellet-Plamondon, S.L. DeFauw, Y. Huang, D.K. Fisher, and P.J. English. 2012. Potential and challenges in use of thermal imaging for humid region irrigation system management. J. Agric. Sci. 4(4):103–116.

Tubana, B.S. 2010. Improving nitrogen use efficiency in cotton: A remote sensor-based approach. In: Proceedings of the 13th Annual National Conservation Systems Cotton and Rice Conference, Tunica, MS. Cotton Incorporated, Cary, NC. p. 32–34.

Vincini, M., E. Frazzi, and P.D. Alessio. 2008. A broad-band leaf chlorophyll vegetation index at the canopy scale. Precis. Agric. 9:303–319. doi:10.1007/s11119-008-9075-z

Wang, Q., K. Wang, S. Li, C. Xiao, J. Dai, and L. Fang. 2012. Cotton planting area extraction based on multi-temporal HJ remote sensing. Cotton Sci. 24(6):503–510 (in Chinese).

Willers, J.L., J.N. Jenkins, W.L. Ladner, P.D. Gerard, D.L. Boykin, K.B. Hood, P.L. McKibben, S.A. Samson, and M.M. Bethel. 2005. Site-specific approaches to cotton insect control. Sampling and remote sensing analysis techniques. Precis. Agric. 6:431–452. doi:10.1007/s11119-005-3680-x

Willers, J.L., J.M. McKinion, and J.N. Jenkins. 2006. Remote sensing, sampling and simulation applications in analyses of insect dispersion and abundance in cotton. USDA Forest Serv. Proc. RMRS-P-42CD, p. 879–885.

Yang, C., C.J. Fernandez, and J.H. Everitt. 2005a. Mapping Phymatotrichum root rot of cotton using airborne three-band digital imagery. Trans. ASAE 48(4):1619–1626. doi:10.13031/2013.19176

Yang, C., C.J. Fernandez, and J.H. Everitt. 2010. Comparison of airborne multispectral and hyperspectral imagery for mapping cotton root rot. Biosystems Eng. 107:131–139. doi:10.1016/j.biosystemseng.2010.07.011

Yang, C., S.M. Greenberg, and J.H. Everitt. 2005b. Evaluation of herbicide-based cotton regrowth control using remote sensing. Trans. ASAE 48(5):1987–1994. doi:10.13031/2013.19991

Yang, C., S.M. Greenberg, J.H. Everitt, and J.W. Norman, Jr. 2006. Assessing cotton stalk destruction with herbicides using remote sensing technology. J. Cotton Sci. 10:136–145.

Yang, C., G.N. Odvody, C.J. Fernandez, J.A. Landivar, R.R. Minzenmayer, R.L. Nichols, and J.A. Thomasson. 2014. Monitoring cotton root rot progression within a growing season using airborne multispectral imagery. J. Cotton Sci. 18:85–93.

Yao, H., and Y. Huang. 2013. Remote sensing applications to precision farming. In: G. Wang and Q. Weng, editors, Remote sensing of natural resources. Chap 18. CRC Press, Boca Raton, FL. p. 333–352.

Yones, M.S., S. Arafat, A.F. Abou Hadid, H.A. Abd Elrahman, and H.F. Dahi. 2012. Determination of the best timing for control application against cotton leaf worm using remote sensing and geographical information techniques. Egyptian J. Remote Sens. Space Sci. 15:151–160. doi:10.1016/j.ejrs.2012.05.004

Zarco-Tejada, P.J., S.L. Ustin, and M.L. Whiting. 2005. Temporal and spatial relationships between within-field yield variability in cotton and high-spatial hyperspectral remote sensing imagery. Agron. J. 97(3):641–653. doi:10.2134/agronj2003.0257

Zhang, H., Y. Lan, C. Suh, J. Westbrook, W.C. Hoffmann, C. Yang, and Y. Huang. 2013. Fusion of remotely sensed data from airborne and ground-based sensors to enhance detection of cotton plants. Comput. Electron. Agric. 93:55–59. doi:10.1016/j.compag.2013.02.001

Zhang, J., Y. Cheng, F. Zhang, D. Yue, X. Guo, H. Dong, J. Wang, and H. Tang. 2012. Multitemporal remote sensing image-based extraction on the crop planting information. Int. J. Remote Sens. Appl. 2(4):31–40.

Zhang, N., M. Wang, and N. Wang. 2002. Precision agriculture- a worldwide overview. Comput. Electron. Agric. 36:113–132. doi:10.1016/S0168-1699(02)00096-0

Zhao, D., K.R. Reddy, V.G. Kakani, J.J. Read, and S. Koti. 2005. Selection of optimum reflectance ratios for estimating leaf nitrogen and chlorophyll concentrations of field-grown cotton. Agron. J. 97:89–98. doi:10.2134/agronj2005.0089

Zhao, F., Y. Guo, Y. Huang, K.N. Reddy, M.A. Lee, R.S. Fletcher, and S.J. Thomson. 2014b. Early detection of crop injury from herbicide glyphosate by leaf biochemical parameter inversion. Int. J. Appl. Earth Obs. Geoinf. 31:78–85. doi:10.1016/j.jag.2014.03.010

Zhao, F., Y. Huang, Y. Gao, K.N. Reddy, M.A. Lee, R.S. Fletcher, and S.J. Thomson. 2014a. Early detection of crop injury from glyphosate on soybean and cotton using plant leaf hyperspectral data. Remote Sens. 6:1538–1563. doi:10.3390/rs6021538

Diseases

Craig S. Rothrock,* Jason E. Woodward, and Robert C. Kemerait, Jr.

Abstract

Cotton is affected by a number of important diseases that limit production wherever the crop is grown. In the United States, about 8 to 17% of yields were estimated to be lost from diseases. In this chapter, the authors address a number of important diseases of cotton in US and world cotton production, and strategies for managing these diseases.

Cotton (*Gossypium* spp.) is a crop that is affected by a number of important diseases that limit production wherever the crop is grown. Being a subtropical to tropical crop that is grown over a wide range of latitudes, as well as a perennial plant grown as an annual crop, cotton is often under stresses that may exacerbate specific disease problems. In the United States, scientists working with cotton diseases have estimated disease losses annually since 1952 as part of the Cotton Disease Council, http://www.cotton.org/tech/pest/seedling/index.cfm. These estimates show the occurrence and prevalence of diseases over years and locations. Estimated losses are scientists' perception of the importance of diseases in their states and are not quantitative loss estimates. From the disease loss estimates from 1952 to 2013, one can see the consistent occurrence and estimated losses for seedling diseases, boll rots, and Fusarium wilt. The importance of other diseases has decreased. For example, losses from bacterial blight in the United States were greatest in 1958 and dramatically declined after 1973 as a result of a number of control practices. The importance of other diseases and pathogens has increased, for example, the dramatic increase in losses associated with nematodes since the 1990s. Similarly, leaf spot diseases were more important in the 1960s and have resurged in recent years. Other diseases have regional importance. For example, Cotton root rot is restricted to the southwestern United States and northern Mexico as a result of soil factors. Although disease losses vary among years, 8 to 17% of yields in the United States were estimated to be lost from diseases. In Australia, quantitative systematic surveys conducted early in the season and before defoliation from 1984

Abbreviations: AG, anastamosis group.

Craig S. Rothrock, Dep. of Plant Pathology, Plant Science 217, Univ. of Arkansas, Fayetteville, AR 72701. *Corresponding author (Rothrock@uark.edu). Jason E. Woodward, Texas A&M AgriLife Extension Service, 1102 East FM 1294, Lubbock, TX 79403 (jewoodward@ag.tamu.edu); Robert C. Kemerait Jr., Dep. of Plant Pathology, 104 Research Way, Univ. of Georgia, Coastal Plain Experiment Station, Tifton, GA 31793 (kemerait@uga.edu).

doi:10.2134/agronmonogr57.2014.0071

Cotton. 2nd. ed. David D. Fang and Richard G. Percy, editors. Agronomy Monograph 57.

to 2011 have documented the occurrence, relative importance, and spread of cotton diseases that affected the industry and demonstrated the impact of production practices on disease incidence and severity (Kirkby et al., 2013). Surveys in Australia documented the reduction in bacterial blight with reduced seed infestation and the introduction of resistant cultivars to trace levels. These surveys also found an increase in Verticillium wilt with the adoption of permanent beds. The appearance and spread of black root rot and Fusarium wilt also were documented. Quantitative disease losses require an effective preventative fungicide or resistant cultivar and will be presented for specific diseases in each section. Unlike many other crops, the most important diseases of cotton are caused by soilborne pathogens, which survive in fields for extended periods and often affect belowground plant parts. Thus, the incidence or severity of these diseases is only recognized when sufficient stunting, changes in plant appearance, or death allow the problem to be recognized, even though chronic disease levels may be reducing yields.

This chapter will address a number of important diseases of cotton for US and world cotton production. It is not meant to be an exhaustive list of all agents that can cause disease in cotton. A more comprehensive list of diseases can be found at http://www.apsnet.org/publications/commonnames/Pages/Cotton.aspx. More exhaustive discussions of specific diseases can be found in Kirkpatrick and Rothrock (2001) or Hillocks (1992).

Seed and Seedling Diseases

Seedling diseases of cotton refer to a group of diseases that are caused by various seedborne and soilborne pathogens. These diseases affect germination of cottonseed and emergence, survival, and development of seedlings. Estimated losses due to seedling diseases accounted for over 20% of the disease losses in the United States. Cotton seedling diseases are an important factor affecting cotton stand establishment worldwide (Hillocks, 1992; DeVay, 2001).

Symptoms

Symptoms of cotton seedling diseases are reduced stands and stand uniformity from rotted seed or preemergence damping-off of the seedlings prior to emergence. In addition to preemergence mortality, seedlings may die within the first 1 to 4 wk after planting from postemergence damping-off (Fig. 1). Younger seedlings are more susceptible than older seedlings. Surviving seedlings may suffer from chronic seedling disease effects, especially when the environment is not favorable for rapid development of the crop. Symptoms may include poor root development and root rot, as well as hypocotyl lesions, leading to stunting and slower crop development. This delay in development of the crop leads to management problems, such as the timing of pesticide applications or delaying harvest. Stand losses may be great enough in some fields and years to require replanting.

Causal Agents

Hillocks (1992) states that over 40 microorganisms have been isolated from diseased seedlings. The four primary agents of seedling disease worldwide are *Pythium* spp., *Rhizoctonia solani* Kühn [teleomorph *Thanatephorus cucumeris* (A.B. Frank) Donk], *Fusarium* spp., and *Thielaviopsis basicola* (Berk. & Broome) Ferraris (syn. *Chalara elegans* Nag Raj & Kendrick).

Fig. 1. Postemergence damping-off (seedling death), one of the symptoms associated with seedling disease (courtesy of C.S. Rothrock).

Seedling diseases caused by *Pythium* spp. are widespread in cotton-growing areas. The genus *Pythium,* an Oomycete in the kingdom Chromalveolata, is no longer considered to be a fungus. Many species of *Pythium* contribute to the cotton seedling disease complex, including *P. ultimum, P. irregular, P. aphanidermatum,* and *P. sylvaticum* (Johnson and Chambers, 1973; DeVay et al., 1982). The most common symptoms associated with *Pythium* spp. are seed rot and preemergence damping-off resulting in stand losses (Arndt, 1943; DeVay et al., 1982; Ogle et al., 1993). *Pythium* spp. also cause soft wet hypocotyl lesions and a root rot. A significant negative correlation was obtained between the percentage of seedlings from which *Pythium* spp. were isolated and the percent emergence of seedlings (Johnson and Doyle, 1986). The use of the Pythium selective fungicide seed treatment, metalaxyl, has demonstrated the importance of *Pythium* spp. in stand establishment, providing an increase in stand in 40 of 119 trials that had a stand response to a fungicide treatment (Rothrock et al., 2012).

Rhizoctonia solani also has been reported to be the most important pathogen in the seedling disease complex of cotton (Ray and McLaughlin, 1942; Davis, 1975; Rothrock, 1996). The fungus can cause preemergence death of cotton (Fulton and Bollenbacher, 1959; Ogle et al., 1993); however, the symptom most often associated with *R. solani* is postemergence damping-off caused by brown to reddish-brown sunken lesions that girdle the hypocotyl near the soil line, referred to as "sore shin" or "wirestem." *Rhizoctonia solani* is a diverse species with hyphae of genetically similar isolates recognizing and fusing with each other (anastomosis). Isolates in an anastamosis group (AG) generally have similar host ranges and cause similar symptoms on hosts. *Rhizoctonia solani* AG4, which causes seedling diseases on a wide range of crops, is the primarily AG causing seedling disease on cotton (Weinhold, 1977; Rothrock et al., 1995). The use of the selective fungicide treatment PCNB increased plant stands over nontreated seed in 44 of 119 trials where a significant fungicide seed treatment response was found (Rothrock et al., 2012), demonstrating the widespread importance of *R. solani* in stand establishment.

Fusarium spp. often are the most frequently isolated fungi from cotton seedlings (Fulton and Bollenbacher, 1959; Davis, 1975; Johnson et al., 1978; Colyer, 1988; Melero-Vara and Jimenaz-Diaz, 1990). Isolates of several species, most frequently *F. oxysporum* Schlechtend.:Fr. and *F. solani* (Mart.) Sacc., have been demonstrated to be pathogenic (Roy and Bourland, 1982; Colyer, 1988, 2001; Melero-Vara and Jimenaz-Diaz, 1990; Abd-Elsalam et al., 2006). Symptoms caused by *Fusarium* spp. include decayed seed and brown or black lesions on the hypocotyls and roots causing chlorosis and stunting, reduced root systems, and death of seedlings (Colyer, 2001).

Thielaviopsis basicola, the causal agent of black root rot, has become an important seedling disease on cotton throughout the world (Mathre et al., 1966; Roy and Bourland, 1982; Allen, 2001; Nehl et al., 2004). Symptoms on seedlings associated with *T. basicola* include a black cortical rot of the root and the belowground portions of the hypocotyl and stunting of cotton plants (Melero-Vara and Jimenaz-Diaz, 1990; Rothrock, 1992; Allen, 2001). Generally, the discoloration is limited to the cortical layer and does not progress into the endodermis and vascular cylinder (Mathre et al., 1966; Mauk and Hine, 1988; Walker et al., 1999; Allen, 2001). When disease is severe, the development of the lateral roots is restricted or prevented. Incidence of *T. basicola* was negatively correlated with cotton stands and positively correlated with root and hypocotyl disease indices (Roy and Bourland, 1982). In microplot studies, soils infested with *T. basicola* reduced yields 15 to 21% compared with noninfested plots (Jaraba et al., 2014).

Limited research has examined the importance of soil pathogen populations on disease development. Multi-year data from Rothrock et al. (2012) show a small but significant negative correlation between soil populations of *Pythium* spp. and stand, but not the other pathogens examined. Johnson and Doyle (1986) found no relationship between soil populations of *Pythium* spp. and isolation frequency or disease incidence on cotton seedlings in Tennessee. In California, DeVay et al. (1982) observed that survival of cotton was negatively correlated with populations of *P. ultimum*; however, Davis et al. (1997) did not find an association between soil populations of *R. solani* or *Pythium* spp. and stands.

The seedling pathogens discussed are all considered soilborne pathogens allowing them to persist for years in soil in the absence of a cotton crop. These pathogens often persist in soil by long-lived survival structures: *Pythium* spp., oospores (thick-walled sexual spores) or hyphal swellings (asexual globose sporangia); *R. solani*, sclerotia (compact masses of hyphal cells); and *Fusarium* spp. and *T. basicola*, chlamydospores (thick-walled cells). *Pythium* spp., *R. solani*, and *Fusarium* spp. also may rapidly colonize organic material as saprophytes maintaining or increasing inoculum of the pathogen by producing spores or sclerotia. These pathogens also can cause disease on a wide range of plants, as well as parasitize a range of plant species producing no symptoms on non-hosts allowing them to persist in soils (Colyer, 2001). All of these pathogens are ubiquitous in cotton soils, except *T. basicola*.

Epidemiology

Early planting is a common practice for growers, in order to maximize the growing season, limit late-season insect pressure, and allow for favorable weather at harvest. Poor stands are often associated with early planting (Johnson et al., 1969; Colyer et al., 1991). Rothrock et al. (2012) found that cotton stands decreased in trials where the average minimum soil temperature the first 3 d after planting decreased from 24 to 12°C and rainfall increased. Seedling diseases accounted for most of these stand reductions as evidenced by the difference in percent stand between seed having a broad-spectrum fungicide treatment and seed not receiving a fungicide treatment (Fig. 2). A fungicide response was found across all environmental conditions; however, the efficacy of fungicide treatment increased dramatically as temperature after planting decreased, particularly at lower temperatures as rainfall increased after planting. Stand losses from *Pythium* spp. increase as soil temperature decreases and soil moisture increases (Arndt, 1943; Johnson et al., 1969; Rothrock et al., 2012). Black root rot also is more severe under planting temperatures less than 24°C and increased soil water (Blank et al., 1953; Rothrock, 1992; Allen, 2001), while stand losses from *R. solani* occur over a wide range of temperatures and soil water (Walker, 1928; Arndt, 1943; Hunter et al., 1960; Huisman, 1988; Rothrock et al., 2012). Differences in soil texture may affect seedling disease severity, with soils high in clay content being more favorable for disease than sandier soils (Johnson and Doyle, 1986; Hillocks, 1992). Seedling disease pathogens may interact with each other to form a complex or interact with nematodes and thrips to increase losses.

Disease Control

Cotton seedling disease control relies on avoidance. Practices limiting seedling disease severity include delaying planting until soil temperatures become favorable for cotton germination and growth with a favorable weather forecast, along with planting high-quality seed, proper planting depth, and planting on

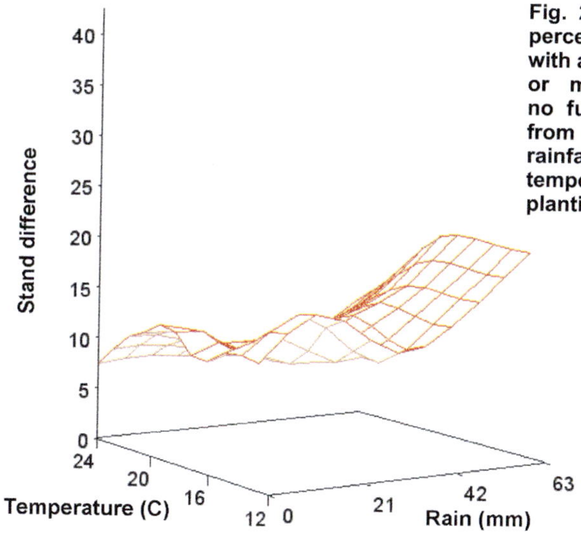

Fig. 2. Difference in mean stand percentages between seed treated with a triazole and either metalaxyl or mefenoxam and seed with no fungicide treatment for trials from 1999 to 2004 in relation to rainfall and average minimal soil temperature for the first 3 d after planting (Rothrock et al., 2012).

a well-prepared raised seedbed in well-drained soils. Cottonseed is frequently treated with a combination of fungicides to target a range of seedling pathogens prior to sale (Minton and Garber, 1983; Hillocks, 1992). Fungicide seed treatments increased stands in 119 of 211 trials across 12 yr of trials and 29 locations clearly demonstrating the importance of seedling diseases and the value of seed treatment fungicides (Rothrock et al., 2012). Numerous other studies have demonstrated consistent stand improvements and increases in yields with the use of fungicide seed treatments (Davis et al., 1997; Kaufman et al., 1998; Wheeler et al., 1997). New chemical classes and combinations of seed treatments have shown significantly greater efficacy over historical standards (Fichtner et al., 2005; Rothrock et al., 2012). Triazole fungicides also have increased the spectrum of seedling diseases being controlled (Kaufman et al., 1998; Toksoz et al., 2009). Increased control of seedling diseases may be achieved by the application of additional fungicides either on the seed before planting (custom seed treatments) or in the planting furrow (in-furrow treatments).

Bacterial Blight

Bacterial blight has been a yield limiting disease in cotton production worldwide if appropriate control practices are not implemented. Bacterial blight is more destructive on *G. barbadense* L. and *G. hirsutum* L. than the diploid cultivated species *G. herbaceum* L. and *G. arboreum* L., which are highly resistant or immune. Since the cotton plant can be affected by the pathogen at all stages of development, the disease was given a number of names based on symptoms including angular leaf spot, blackarm, and bacterial boll rot. These disease symptoms were all shown to be caused by the same bacterium, and thus the disease is now referred to as bacterial blight. Hillocks (1992) provides an extensive review of the history of the disease since its description in 1891 and its occurrence in cotton growing areas throughout the world.

Arnold (1965) reported losses of 7 to 31% from bacterial blight in Uganda using four genetically similar cultivars that differed in resistance to the disease. Seed treatments increased yield from 16 to 19% depending on the year. In India, Mathur et al. (1973) demonstrated 14 to 55% losses associated with bacterial blight with the combination of seed treatments and foliar applications of antibiotics and fungicides. Losses from bacterial blight can be significant even after eliminating seed infestation by acid delinting or seed treatments. Using artificial inoculation, Bird (1959) found a 9 to 34% yield loss across a number of cotton cultivars from leaf symptoms alone. This is similar to work in Uganda with a relative loss of 15% for a susceptible cultivar by foliar inoculation over seed infestation with no foliar bactericidal applications (Wickens, 1961).

Symptoms

Bacterial blight symptoms may appear at any growth stage. For seedling blight, cotyledons may develop lesions that appear as dark green water-soaked areas first becoming visible on the underside of the cotyledons (Hillocks, 1992). This may result in the death of the seedling. Foliar symptoms also start as water-soaked affected areas with the colonized tissue quickly becoming dark brown to black necrotic lesions. These lesions are often delimited by the major veins, resulting in an angular appearance. Lesions may coalesce as they expand. In addition to the angular leaf lesions, localized systemic movement of the bacterium may result in necrotic areas extending along major veins in the leaves. Leaf abscission is common for symptomatic leaves. The bacterium may spread from the cotyledon or leaf into the petiole to the stem, resulting in a black lesion that may girdle the stem (Arnold, 1965). Blackarm symptoms also may occur on stem internodes from infections initiated on the stem (Weindling, 1948). Girdled stems easily break resulting in loss of fruiting branches. The bacterium may infect the boll causing young bolls to shed. Lesions on bolls are initially water-soaked spots, which become dark brown to black circular lesions and may exceed 1 cm in diameter. Lesions on the boll are often associated with the carpel suture or extend up from the base of the boll, the epicalyx. Lesions may penetrate the boll wall, especially when associated with insect damage, for example, from the stainer bug, *Dysdercus* spp. (Hillocks, 1992). Affected bolls also may be colonized by other organisms associated with boll rots to the point of obscuring the bacterial blight symptoms. Internal colonization results in the yellow discoloration and deterioration of the lint and seed.

Causal Agent

The pathogen causing bacterial blight is classified currently as *Xanthomonas citri* subsp. *malvacearum* (ex Smith 1901) Schaad et al. 2007 (Schaad et al., 2006) [syn. *Xanthomonas axonopodis* pv. *malvacearum* (Smith 1901 Vauterin et al. 1995) and *Xanthomonas campestris* pv. *malvacearum* (Smith 1901) Dye 1978]. The pathogen is closely related to *Xanthomonas citri* subsp. *citri*, the citrus canker pathogen. The bacterium is gram negative and in culture produces large amounts of yellow mucoid wet growth on 2% peptone-sucrose agar. Optimum temperature for growth is 25 to 30°C (Hillocks, 1992).

Biological races of the *X. citri* subsp. *malvacearum* were recognized as early as 1950 following the identification and incorporation of genes for resistance (Balasubramanyan and Raghavan, 1950; Hunter and Blank, 1954). Standard differential cultivars or lines have been used to characterize 20 races of the pathogen (Bird, 1986). A diversity of races are commonly found where cotton is grown, with race 18 often being the dominant race (Thaxton et al., 2001; Akello and Hillocks, 2002; Madani et al., 2010). Different races of the pathogen can be isolated from the same leaf or after serial passage. Genomic sequencing of selected races has advanced our knowledge of the taxonomy and genes associated with virulence of this pathogen (Zhai et al., 2013).

Epidemiology

Sources of inoculum for initial primary infections in the spring include cotton residue from the previous season, volunteer plants, and seed. No important alternative hosts are known. In warm climates, volunteer plants should be eliminated to prevent overseasoning of the pathogen. In Tanzania and the United States, the bacterium was able to persist when residue remained on the soil surface for at least 200 d, suggesting the potential for primary inoculum to come from residue from the previous season but not for longer periods of time (Arnold and Arnold, 1961; Brinkerhoff and Fink, 1964). When residue was buried, disease did not develop and infectively was lost within a couple of months. Survival in residue is reduced with increasing rainfall or in wetter areas of the field and at warmer temperatures, which increase decomposition of residue from general microbial activity.

Planting seed is an important method for the introduction of inoculum into fields and new regions. The bacterium is most frequently carried externally on the seed. Acid delinting is very effective at eliminating external contamination of the seed, dramatically reducing the occurrence and the severity of the disease; however, the bacterium may also infect the seed. Brinkerhoff and Hunter (1963) found 0 to 4% infection of seed lots from a number of sources. Although the pathogen is carried at a very low frequency internally in seed, levels of seed infection resulting in disease in field studies have been as low as 1 in 5000 to 6000 seed (Tarr, 1961; Mehta et al. 2005). Schnathorst (1964) demonstrated that seed may also become contaminated during the ginning process.

The bacterium enters the plant through open stomata or wounds, multiplies within intercellular spaces in the sub-stomatal chamber or tissue, and kills cells (Weindling, 1948). The pathogen may become locally systemic. Younger tissue is more susceptible than older tissue. Disease is favored by temperatures above 30°C and relative humidity above 75% (Stoughton, 1931). Brinkerhoff and Presley (1967) reported disease was favored by relatively low night temperatures of 19°C compared with 26.5°C and day temperatures of 25.5 to 36.5°C.

Local spread of the pathogen and secondary infections resulting in the rapid buildup of disease is accomplished by wind-driven rains and rain splashing. Surface and overhead irrigation have been demonstrated to spread the pathogen. Insects also may be important in the spread of the pathogen in a nonspecific manner internally or externally. Research for the same species *X. citri* subsp. *citri* showed that within the first 5 min of rain the maximum levels of bacteria are exuded from citrus leaves (Bock et al., 2005). Using GIS and vector analysis, wind speeds of 8 m s^{-1} (17.9 mph) and rainfall of 0.318 cm h^{-1} (0.125 in h^{-1}) explained 80% or greater of trees infected (Irey et al., 2006). The pathogen was found to spread approximately 579 m or 1900 ft during a 30-d period in southern Florida (Gottwald et al., 2002).

Disease Control

Limiting carryover is important since little evidence exists that the pathogen can survive for more than one growing season. Tillage operations that break up and bury residue increases the rate of decomposition and reduces the risk of spreading inoculum. Moist conditions including irrigation or flooding speeds up this process. The value of these cultural practices along with processing planting seed in the gin first followed by the process of acid delinting successfully eradicated the pathogen in California (Schnathorst, 1966).

Since seed often serves as the primary source of inoculum, acid delinting has been very effective for the elimination of external inoculum of this pathogen. Verma (1986) summarized research on the efficacy of seed treatments including hot water treatment, antibiotics, and fungicides for the eradication of *X. citri* subsp. *malvacearum*. Foliar applications may be used as a method of limiting secondary spread of the pathogen to limit disease development and protect yields in some areas of the world (Verma, 1986).

Over 20 resistance genes have been described since breeding for resistance was started in 1939. Hillocks (1992) summarized breeding efforts around the world, as well as variation in disease evaluations and reactions. Resistance in cultivars is often associated with multiple resistance genes, increasing the effective range against races and the level of resistance. A number of cultivars have a high level of resistance approaching immunity. The D genome, from New World cotton in the tetraploid *G. hirsutum*, is responsible for many of these resistance genes (Wright et al., 1998). The resistance gene B12 confers resistance to a wide range of races including race 18 (Wallace and El-Zik, 1989). Although a number of resistance genes are available, much of the Cotton Belt in the United States where bacterial blight is not a recurring problem does not emphasize the use of resistance in the cultivars planted (Bayles and Verhalen, 2007). Molecular markers including simple sequence repeats (SSR), single nucleotide polymorphisms (SNP), and restriction fragment length polymorphisms (RFLP) have advanced the understanding of the location of these resistance genes (Wright et al, 1998; Xiao et al., 2010). Resistance follows the gene for gene concept with the resistance gene in the host having a corresponding avirulence (*avr*) gene in the pathogen. In the resistant reaction, instead of developing a water-soaked expanding lesion, affected cells in the leaves undergo a rapid hypersensitive response within 2 d collapsing and becoming necrotic. The resistance is associated with peroxidase production of superoxide anions and sesquiterpenoid phytoalexins, which accumulate at the site of infection at levels inhibiting the development of the bacterium, and the salicylic acid and jasmonic acid signaling pathways initiating a systemic acquired resistance response. All *avr* genes described belong to a group known as Xanthomonas avirulence/pathogenicity (avrBs3/pthA) type III secretion genes found widely in this genus (Delannoy et al., 2005; Zhai et al., 2013).

Foliar Diseases

Numerous foliar diseases occur on cotton and are generally caused by fungal pathogens, with the exception of the bacterial pathogen *Xanthomons citri* subsp. *malvacearum*, which is discussed previously in this chapter. Kirkpatrick and Rothrock (2001) and Hillocks (1992) provide an overview of foliar diseases on cotton. Given the importance of the many production and pest management issues during the growing season, for example, insect and weed management, soil fertility and irrigation, use of growth regulators, and harvest decisions, the importance and impact of foliar diseases can easily be overlooked. Foliar diseases caused by fungal pathogens are commonly encountered; however, the impact of these diseases varies by production regions worldwide. In some areas, foliar diseases are considered to be of only minor importance, while in other areas these diseases are considered to be an important threat to cotton yields.

The foliar disease situation is changing dramatically across regions and years with the emergence of new diseases or resurgence of other diseases. For example, Stemphylium leaf spot, which is increasing in importance in some areas of the United States, is not mentioned in Hillocks (1992). Also, an apparently emerging foliar disease of cotton known as "Target spot" or "Corynespora leaf spot" (Jones, 1961) is not mentioned in either Kirkpatrick and Rothrock (2001) or Hillocks (1992). Additionally, the use of fungicides to manage foliar diseases of cotton was limited to a few specific areas of world production, especially India; however, in recent years not only has the potential for use of foliar-applied fungicides increased in many cotton production areas, but the number of fungicides registered and modes of action has expanded as well.

Foliar diseases to be discussed here are all of significance in the United States or elsewhere in the world. The diseases include Alternaria leaf spot, Stemphylium leaf spot, Cercospora leaf spot, Areolete (grey) mildew, Myothecium leaf spot, Target (Corynespora) leaf spot, Southwestern cotton rust, and Tropical cotton rust.

Symptoms

Correct diagnosis of foliar diseases of cotton is necessary in order to determine importance and best management practices; however, diagnosis of foliar diseases of cotton can be a challenge. This is especially true for identification in the field, but diagnosis in the lab can be problematic as well. Though rust and Areolate mildew are easy to differentiate from other diseases affecting the foliage, correctly differentiating spots on cotton leaves, especially early in disease development, can be difficult. This is because leaf spots caused by different pathogens can look similar and diagnosis often requires observing signs of the pathogen. For example, zonate, target-like spots may be associated with Ascochyta blight, Cercospora leaf spot, Target spot, or other diseases. Early identification of a leaf spot, even in the diagnostic lab, can be difficult if no fungal fruiting structures (e.g., spores) are observed. Identification of spots when the disease is well established in the field may be difficult as numerous opportunistic fungi, including *Alternaria* spp., may be present. In addition to descriptive characteristics for diagnosis, one must also consider the growth and development of the crop, canopy development, and the presence or absence of stress factors, such as nutritional issues.

Alternaria leaf spot is one of the most common foliar diseases reported to affect cotton worldwide (Baird, 2001a). Reports from Israel, India, United States, East Africa, and Zimbabwe indicate Alternaria leaf spot is an important disease affecting plants as early as the cotyledonary and seedling stages. Symptoms of Alternaria leaf spot may be highly variable, but begin as small brown spots bordered by a purple margin. When conditions are favorable for disease, the spots can enlarge to approximately 10 mm in diameter. On green leaves, the purple margin may appear more pronounced and the affected tissue within the margin may be zonate with concentric rings. In advanced cases, the spots may coalesce into larger necrotic areas on the leaf (Hillocks, 1991).

Symptoms of Stemphylium leaf spot are characterized as small, circular to irregular, dark brown to black spots, varying between 2 and 5 mm and becoming reddish brown with age. Fully developed lesions are up to 10 mm in diameter with a whitish center and cracks, giving the appearance of a "frog's eye." At a later stage, the cracked area drops out, leaving a hole in the center of the lesion. In severely affected leaves, the lesions may coalesce, and yellowing of the leaves occurs. After a few days the affected leaves desiccate and abscise (Mehta, 1998). Three leaf symptoms have been described from Louisiana and include leaf spots with light centers and red margins, leaf spots with large, brown spreading lesions with concentric rings, and finally, smaller lesions lacking concentric rings (Sinclair and Shatla, 1962). Lesions associated with Stemphylium leaf spot may be found on leaves anywhere in the canopy, but the greatest number are typically found on the upper leaves. Additionally, leaves affected by Stemphylium leaf spot often demonstrate stress, typically nutritional stress, and are reddish and yellowish in color.

Symptoms of Ascochyta blight are reported to develop on all aboveground parts of the cotton plant including cotyledons, leaves, stems, and bolls (Elliott, 1922). Leaf symptoms are variable. Spots occurring on the cotyledons and leaves are characterized initially as white to pale brown circular spots with dark brown to purple borders (Hillocks, 1992; Koenning, 2000). Spots may enlarge from 2 mm in diameter to coalesce and form light brown areas affecting much of the leaf surface. Borders of these necrotic areas are often dark in color, which is characteristic of this disease (Hillocks, 1992; Baird, 2001c). At later stages, black specks (pycnidia) are often visible in the older leaf spots (Koenning, 2000). Lesions on the petioles and stems are often elongated (6–7 mm) sunken cankers varying in color from reddish-purple to black or ashy. As the cankers age, they often become dark with a shredded appearance. The presence of small black fungal structures, pycnidia, is characteristic of tissue affected by Ascochyta blight (Hillocks, 1992; Koenning, 2000; Baird, 2001c).

Symptoms of Cercospora leaf spot include irregular brown lesions, usually surrounded by a considerable area of chlorotic tissue. Symptoms also may include red dots early in the development of the disease that increase in diameter to approximately 2 cm. A narrow red margin encloses a white to light brown center of dead tissue. The variability in symptoms may result from the fact that Cercospora leaf spot typically occurs in conjunction with other diseases (Hillocks, 1992). In the southeastern United States, mature spots are likely to be enclosed by a dark margin and the enclosed affected area may have a zonate, target-like appearance. Cercospora leaf spot and Stemphylium leaf spot often occur on the same leaves.

Symptoms of Areolate mildew, also known as false mildew or grey mildew, first appear on older leaves low in the canopy as the plant reaches maturity soon after boll-set. Initial symptoms are pale, translucent lesions 1 to 10 mm in diameter with an irregular to angular shape with the margin often defined by the veins of the leaf. Similar lesions may develop on the bracts of bolls. Copious sporulation occurs on the underside of the leaf, producing a white mildew-like appearance. As the disease progresses, spots become necrotic and dark brown in color. These necrotic symptoms may be confused with bacterial blight (Ehrlich and Wolf, 1931; Hillocks, 1992; Sharma and Bambawale, 2008).

Symptoms of Myrothecium leaf spot often appear on the leaves of young plants, but may also cause preemergence and postemergence damping-off of young seedlings (Hillocks, 1992). Leaf spots are initially reported to be circular and tan-colored with violet-brown margins and may be surrounded by a translucent area. Spots may expand to 3 cm in diameter. The lesions often include concentric necrotic rings, with salient structures (sporodochia) appearing as dark green to black, pinhead-sized dots irregularly distributed throughout the lesion. Symptoms are reported from petioles, bracts, leaves, and bolls; stem lesions and die-back of cotton have been reported (Munjal, 1960; Hillocks, 1992; Meyer et al., 2006).

Symptoms of Target spot (Corynespora leaf spot) initially develop as brick red dots that lead to the formation of irregular to circular lesions with tan to light brown centers (Fig. 3). Lesions further enlarge and often demonstrate a target-like appearance formed from concentric rings within the spot (Jones, 1961; Fulmer et al., 2012; Conner et al., 2013; Wei et al., 2014). These spots may measure up to 10 mm in diameter. Target spots begin on the lower leaves of the canopy and progress to within approximately the upper 10 to 20% of the canopy. Field observations included estimates of premature defoliation up to 70%, with abundant characteristic spots on the leaves and bracts (Fulmer et al., 2012; Conner et al., 2013).

Symptoms of Southwestern cotton rust begin as pale green lesions approximately 0.5 mm in diameter 4 to 6 d after infection. These lesions develop into

Fig. 3. Foliar symptoms of Target (*Corynespora*) leaf spot having a target-like appearance within the brown lesions (courtesy of R.C. Kemerait Jr.).

bright yellow to orange pustules. The pustules may be enclosed by a dark red or maroon border. These pycnidial pustules can occur on all aboveground parts of the plant; however, they are most abundant on the upper leaf surfaces. In extreme instances, several hundred pustules may crowd the upper leaf surface. The pycnia within the pustules exude a thick yellow fluid. Within 8 to 10 d of formation of pycnia, cuplike aecia (0.2–2 mm in diameter) protrude from the lower surface of the leaf, forming circular clusters that protrude 1 to 2 mm beyond the leaf surface. The aecia are yellow to orange in color. As the aecia on the underside of the leaf age, they fade to a light yellow color; the aging pycnia on the top of the leaf darken to brown. As a result of the pustules on both sides of the leaf, the foliage may curl and prematurely abscise, and yield and quality may be affected by small bolls coupled with premature boll opening (Black, 2001a).

Symptoms of Tropical cotton rust are typically described as clusters of 7 to 20 uredinia (pustules, 0.5–3.0 mm in diameter) on the leaves of the cotton plant. These pustules are variable in color, having been described as yellow, brown, pink-brown, or violet. The pustules have been described as ash-white, corky, or powdery in the center and are positioned in lesions on the leaf (1–5 mm in diameter) that are described as purple, red, or brown in color. When tropical rust is severe, defoliation may occur (Black, 2001b).

Causal Agents

The causal agents for the various foliar diseases will be discussed in some detail since the signs of the pathogens are often essential to the correct identification of the disease being managed.

Causal agents of Alternaria leaf spot are *Alternaria macrospora* A. Zimmern. and *A. alternata* (Fr.:Fr.). *Alternaria macrospora* is more commonly associated with severe outbreaks of Alternaria leaf spot. Conidiophores (specialized fungal hyphae bearing spores) of *A. macrospora* occur solitarily or in clusters. The conidia (asexual spores) are ellipsoidal in shape and taper toward a narrow beak. The conidia have muriform septations, meaning that they have both transverse longitudinal and oblique septa. The conidia of *A. macrospora* typically occur singly on the conidiophores, but may be found in chains of two linked conidia. In contrast, conidia of the less important pathogen of cotton, *A. alternata,* are almost always catenulate, that is, borne in chains of multiple spores. The conidia of *A. alternata* have shorter "beaks" than conidia of *A. macrospora* (Baird, 2001a). *Alternaria* spp. are common saprophytes on diseased or necrotic cotton leaves. In diagnosing Alternaria leaf spot, care must be taken to avoid a misdiagnosis based upon the presence of conidia of *Alternaria* spp. not associated with disease.

The causal agent of Stemphylium leaf spot is *Stemphylium solani* G.F. Weber, which has erect, septate condiophores with prominent apical swellings. The conidia are olive-brown, oblong, and muriform, meaning that they have longitudinal, transverse, and oblique septa. Condiophores and conidia are scattered singularly across the spots associated with Stemphylium leaf spot (Baird, 2001d). At least initially, differentiating conidia of *S. solani* from *Alternaria* spp. may be a bit confusing, but conidia of *S. solani* never have a beak.

The causal agent of Ascochyta blight is *Ascochyta gossypii* Woronichin, (syn. *Phoma exigua* Desmaz). The conidia are hyaline (transparent and colorless) and are two-celled. The conidia are cylindrical to ovoid and may be straight or slightly curved. The conidia have rounded ends and are formed in globose brown to black pycnidia (Baird, 2001c). The presence of these structures and the observation of spores being exuded from them provide strong evidence that the disease observed is Ascochyta blight.

The causal agent of Cercospora leaf spot is *Cercospora gossypina* Cooke. Conidiophores are reported to mononematous and in fascicles of 2 to 10. The conidiophores are not branched and are brown in color with 1 to 3 septations.

Conidia are hyaline (colorless), straight or curved, with 5 to 10 septations and a truncate bases. The perfect (sexual) state of *C. gossypina* has been identified as *Mycosphaerella gossypina* (Hillocks, 1992; Baird, 2001b).

The causal agent of Areolate mildew (false mildew, grey mildew) is *Ramularia gossypii* (Speg.) Cif. The teleomorph (sexual stage) for the fungus is *Mycosphaerella areola* J. Ehrlich & F.A. Wolf. Conidia are hyaline with 1 or 2 septations. Spermagonia can be observed as raised black dots on the underside of fallen leaves. When the spermogonia mature, rod-shaped spermatia are released in a mucilaginous mass. Perithecia later replace the spermogonia and produce fusiform asci with two-celled ascospores (Hillocks, 1992; Baird, 2001e).

The causal agent of Myrothecium leaf spot is *Myrothecium roridum* Tode:Fr. Sporulation occurs in the lesions on the leaves and appears as black dots. Conidiophores of this pathogen arise directly from the epidermal cells of the host and divide to form two to five branches that bear phialides. The conidiophores are arranged in sporodochia and appear as a black mass on the host leaf. Conidia are rod-shaped with rounded ends; the conidia vary from hyaline to olive colored (Hillocks, 1992; Baird 2001e).

The causal agent of Target spot is *Corynespora cassicola* (Berk. & M.A. Curtis.) C.T. Wei. Conidiophores are simple, erect, cylindrical, brown or olivaceous, and unbranched with 2 to 7 septa. Condia are borne singly, ranging from subhyaline to olivaceous, straight to slightly curved, containing 4 to 15 pseudosepta (Fulmer et al., 2012; Conner et al., 2013).

The causal agent of Southwestern cotton rust is *Puccinia cacabata* Arth. & Holw. This rust fungus is macrocylic and heteroecious, meaning that in the course of its life cycle the fungus infects not only cotton but also the alternate host grama grass (*Bouteloua* spp.). In simple terms, the first symptoms observed in the field are small yellow-orange spots on the upper side of the cotton leaf that contain the pycnia. Seven to 10 d after the formation of the pycnia on the top of the leaf, yellow-orange cup-like aecia form on the underside of the leaf. Aeciospores that are produced in these aecia are not capable of infecting cotton plants, but must infect grama grass (Hillocks, 1992).

The causal agent of Tropical cotton rust is *Phakopsora gossypii* (Lagerh.) Hiratsuka, which is an autoecious rust. Unlike in southwestern cotton rust, tropical rust only produces urediniospores in uredinial pustules (Hillocks, 1992; Black, 2001a).

Epidemiology

An understanding of the development of the diseases affecting the leaves of the cotton plant is essential in order to determine the best management practices. Of primary importance in the epidemiology of the diseases discussed here are (i) source of initial inoculum (i.e., the origin of the spores that initially infect the cotton crop), (ii) the importance of secondary inoculum (i.e., the spores that maintain and increase an epidemic in the field), and (iii) the impact of the environment on the development of disease. For this discussion, "environment" includes abundance of rainfall, humidity, temperature, soil types, and soil fertility.

The disease cycle for *M. roridum* is thought to begin when spores from infested soil or from infected weed species (this pathogen has a wide host range) are deposited on the cotton plant and infection occurs. This pathogen is capable of causing symptoms throughout the development of the cotton plant, including boll development. Optimum temperature for spore germination is reported to be 29°C, but this may depend upon the geographic origin of the pathogen (Hillocks, 1992). This general life cycle is repeated for many of the leaf spot diseases of cotton.

On cotton species other than *G. hirsutum*, primary inoculum of *A. macrospora* originates from infested crop residues and from infected seed and can infect cotyledons and seedlings, which support the initial stages of the epidemic. Periods of high humidity and rainfall are favorable for sporulation to occur and for spread of spores. Disease can spread from the cotyledons to the lower leaves. For *G. hirsutum*, Alternaria leaf spot typically does not occur until the canopy is closed and increased humidity favors development of disease. Primary infection seems to occur first on the older leaves low in the canopy and likely is the result of spores splashed from debris on the soil or blown in from other diseased plants. Conidia produced in diseased lesions serve as a source for secondary inoculum and further development of disease. The disease cycle is completed when diseased leaves abscise and drop to the soil, serving as a source for primary inoculum for a future crop (Hillocks, 1992).

When considering *G. hirsutum*, the life cycles for *S. solani*, *C. cassiicola*, and *C. gossypina* are similar in many ways to that described for *A. macrospora*. The source of initial inoculum is the soil, infested crop residue, or other host species. Infection of older leaves and development within the canopy where humidity is elevated and leaf wetness is prolonged are important in the development of disease. Extended periods of leaf wetness are especially important for the development and spread of Target spot.

Ascochyta blight (also referred to as "wet weather blight") can become a problem during periods of prolonged cloudy weather and rainfall early in the season, leading to significant infection of cotyledons and leaves, and can negatively affect yield (Baird, 2001c).

Disease Control

Management decisions throughout the growing season often influence leaves and the canopy of foliage. Though such decisions have historically focused on plant nutrition and management of crop growth with irrigation and growth regulators, it is increasingly apparent that management of diseases is also an important consideration. Yield and quality of the cotton crop benefit when healthy, young leaves are sustained on the plant (Oosterhuis et al., 1990). An important goal in cotton production must be to maintain healthy leaves and to promote earliness in boll retention. The three types of leaves for the cotton plant are cotyledons on the emerging seedling, the main-stem leaves, and the subtending leaves. Cotyledons feed the developing shoot tip and root system, and loss of cotyledons within the first week after emergence can severely delay maturity (Oosterhuis et al., 1990). Main-stem leaves are important in that they provide photosynthates to the developing terminal, branches, and even bolls of the cotton plants. Fruiting branch leaves develop adjacent to squares (the flower bud that appears on the cotton plant when reproductive growth occurs) and are referred to as the "subtending leaf." The subtending leaf, along with the nearest main-stem leaf, provides much of the carbon and nitrogen nutrients to feed the boll during the early stages of growth when it is most vulnerable to shed (Hake et al., 1990).

Considerations in the management of foliar diseases should include the level of damage, as well as what leaves are affected. Thus, an understanding of sources of photosynthates, carbohydrates, and sinks, those plant parts needing carbohydrates for growth and development, is important. The most productive sources are middle-aged, fully expanded, and fully illuminated leaves—those that are approximately 80% of full size (16–18 d of age) until about 25 d of age (Hake et al., 1990). Mid-season, the main-stem leaf that is approximately 5 nodes from the top of the plant is the most photosynthetically active, whereas a leaf 13 nodes from the top is likely inactive and a "sink." The greatest "sink" to be fed by photosynthetically active leaves is reported to be a 20- to 30-d-old boll as it is during this time frame that it is experiencing its greatest gain in dry weight. Approximately 50% of the dry weight is provided to the boll from its subtending leaf, while approximately 35% is provided by the nearest main-stem leaf. As a boll ages, it must rely on leaves farther away and even those illuminated leaves at the top of plant as sources for nutrition. Once bolls are set on the plant, any damage to the leaves is of greater concern, regardless of growing season, and boll growth is limited by the ability of leaves to provide adequate nutrition. Maturation of late-set bolls requires the presence of healthy, late-season young, and well-illuminated leaves to sustain boll production. Plant stress during the growing season may increase disease. For example, the cotton plant becomes more susceptible to Alternaria leaf spot if it suffers periods of adverse growing conditions (Hillocks, 1992). Furthermore, premature aging and senescence of leaves is often associated with water stress and nutrient deficiencies (Hake et al., 1990)

Five basic questions must be answered in deciding on best management practices for control of leaf diseases affecting cotton. The first question is, "What cultural and production practices are available to reduce the risk to and impact of these diseases?" The second question is, "Are cultivars with acceptable yield potential and increased resistance to a disease (or diseases) available?" The third question is, "Are there products (e.g., fungicides or biocontrol agents) available that are effective in reducing disease severity and increasing yield?" The fourth question is, "What is the appropriate time for application of a fungicide or biocontrol product?" The fifth and last question asks, "Is the risk from the disease sufficient enough to even consider management efforts?"

Cultural and production practices can be important in the management of leaf diseases of cotton. For example, the pathogens causing Alternaria leaf spot, Stemphylium leaf spot, Cercospora leaf spot, Areolate mildew, Myrothecium leaf spot, and Target spot all can survive in crop debris. For this reason, growers can take steps to manage these diseases by (i) rotating fields away from cotton production, (ii) destroying crop residue, typically by burying the debris, and (iii) managing "volunteer" cotton plants that may grow between seasons. Such steps should help to reduce the amount of primary inoculum for the next growing season. Additional measures would be to ensure that non-infested seed is planted to help reduce the risk of diseases like Alternaria leaf spot and Tropical cotton rust (Hillocks, 1992). Growers can significantly reduce the risk to Alternaria leaf spot, Stemphylium leaf spot, and Cercospora leaf spot diseases by taking necessary steps to reduce stress on the crop, especially stress resulting in poor fertility, for example, insufficient potassium (Hillocks, 1991, 1992; Howard et al., 1997; Baird, 2001d; Bhuiyan et al., 2007). Finally, growers can use growth regulators to manage rank growth in the cotton field. Management of excessive growth increases airflow in the canopy and reduces humidity and periods of leaf wetness. Preliminary reports suggest that reducing humidity and periods of leaf wetness are useful in the management of Target spot (Hagan, 2014).

Ideally, cotton farmers could plant cultivars that are resistant to the important foliar diseases; however, planting cultivars with resistance to specific foliar diseases is generally not an option for cotton growers. Resistance has been reported to Alternaria leaf spot where Pima cotton cultivars (*G. barbadense*) are more susceptible to the disease than are Upland cultivars (*G. hirsutum*). There also have been reports of different levels of resistance to this disease in commercial Upland cultivars (Hillocks, 1992). Resistance to Areolate mildew is reported in India (Hillocks, 1992).

Applications of fungicides for management of foliar diseases of cotton have not been of general practice among cotton growers in the United States; however, fungicides have been used in other countries with success. Researchers in India have assessed use of the fungicides mancozeb, propineb, propiconazole, copper oxychloride, and tebuconazole for management of Alternaria leaf spot, Areolate mildew, and Myrothecium leaf spot (Govindappa and Chattannavar, 2008; Dighule et al., 2011; Hosagoudar and Chattannavar, 2014). Research efforts in India have also focused on the use of *Pseudomonas fluorescens* as a biological control agent for these same diseases (Bhattiprolu and Prasada Rao, 2009; Singh et al., 2012). The use of fungicides has also been investigated in Australia (Colson-Hanks et al., 2000; Bhuiyan et al., 2007) and Israel (Shtienberg, 1992; Shtienberg et al., 1993) for management of Alternaria leaf spot. Fungicides currently labeled for management of cotton diseases in the United States are part of the strobilurin and triazole classes of chemistry. Until recently, use of fungicides in the United States has been primarily for the management of Southwestern cotton rust; however, use of foliar-applied fungicides could become more widespread in the United States as research efforts continue on the management of Target spot or as additional fungicides become available (Hagan, 2014). An important constraint to the use of fungicides for management of a new disease such as Target spot is that the impact of the disease on yield and the quantity of yield protected with a fungicide program is variable. Growers are concerned about the impact of the disease and are generally willing to use fungicides in the management of target spot; however, they are cautious in adding expense to production without assurance that applications will increase profit. As research on Target spot and other leaf diseases of cotton progresses, basic questions, for example, impact of fungicides on yield, the optimal number of applications for control of the disease and the best timing for those applications, will be answered.

At present, the timing of a fungicide application for management of Target spot is problematic. When a fungicide is applied too early, it may not be effective if and when the disease begins to develop; however, delaying the application of a fungicide may result in missing the critical initiation of the disease when management is most appropriate. Additionally, as the cotton plant ages, it becomes increasingly difficult to penetrate the thick canopy of foliage so that the fungicide is appropriately placed in the lower interior leaves of the plant. Hence, the best timing of a fungicide application for the management of a foliar disease of cotton must consider both the growth and development of the crop and the likely time at which the disease will begin to occur.

As discussed earlier, to achieve optimal yield it is critical to protect those leaves that are essential for development of bolls and to ensure a continued supply of healthy, young leaves to mature the later bolls. Therefore, it is essential to protect these subtending leaves and leaves in the upper canopy from disease. Although cotyledons do not directly feed the bolls, these leaves are important for the early development of the plant, and premature loss can lead to significant delays in maturation of the crop (Oosterhuis et al., 1990). Ascochyta blight may lead to significant infection of cotyledons and early leaves during periods of prolonged cloudy weather and rainfall early in the season (Baird, 2001c). Other diseases, for example, Myrothecium leaf spot and Alternaria leaf spot, also are reported to affect cotyledons. In general, fungicides are most often applied later in the season to protect leaves that will directly affect the growth and development of bolls; however, there is precedent for use of fungicides earlier in the season if young plants are severely affected by disease.

Vascular Wilt Diseases

Fusarium wilt and Verticillium wilt are capable of causing significant yield reductions in cotton. Both fungal pathogens are soilborne and infect plants through the roots, later colonizing the xylem elements of the vascular system (Garber and Houston, 1966; Hall et al., 2013). These diseases have been reported from all of the major cotton growing regions around the world and have been increasing in prevalence the past 10 yr. Subtle differences in the onset, appearance, and severity of symptoms can be used to distinguish Fusarium wilt and Verticillium wilt in the field; however, proper identification often requires isolation of the actual causal agent from infected tissues. This is extremely important in developing management strategies, especially when the two diseases occur in the same field.

Symptoms

Symptoms of Verticillium wilt can vary depending on cultivar and aggressiveness or abundance of the pathogen, as well as environmental conditions. While symptoms can appear at any time, they most commonly appear after first flower. Leaves of infected plants have a lime-green, flaccid appearance with chlorotic and necrotic areas developing on the leaf margins and between the veins (Fig. 4). Stunting, premature defoliation, and death may occur. Stems and branches of diseased plants will have a light to dark brown discoloration of the vascular tissue.

Symptoms of Fusarium wilt also can occur at any stage of crop development; however, the most destructive phase tends to affect younger plants. Cotyledons and leaves of infected seedlings wilt and turn necrotic, similar to what is observed with postemergence damping-off. The general appearance of affected plants is most obvious on warm or windy days and can be characterized by the yellowing of lower leaves. These symptoms are accompanied by a continuous brown discoloration of the vascular tissue that is most apparent in the upper taproot and lower stem.

Causal Agents

Several *Verticillium* spp. have been associated with cotton; however, only *V. dahliae* Kleb. (formerly referred to as *V. albo-atrum*) is known to incite disease under field conditions. The pathogen is widely distributed and is favored by cooler temperature; thus, it is more problematic in temperate areas than in tropical or subtropical regions (Pegg, 1984). Much confusion surrounds the classification of *V. dahliae*

Fig. 4. Characteristic foliar symptoms of Verticillium wilt showing chlorosis (yellowing) and necrosis between major veins (courtesy of J.E. Woodward).

isolates. Several systems including race designations and vegetative compatibility groups (Typas and Heale, 1976; Puhalla and Hummel, 1983) have been used to characterize *V. dahliae* strains, and advantages and disadvantages of each method have been reviewed (Correll et al., 1988). A combination of studies have been used to condense all cotton isolates to two pathotypes or strains (Schnathorst and Mathre, 1966; Schnathorst, 1969; Puhalla and Hummel, 1983; Joaquim and Rowe, 1990)—the defoliating type (formerly designated P_1 or T_1) and the non-defoliating type (P_2 formerly termed the SS_4). Overall, the defoliating isolates are more virulent than non-defoliating isolates (Joaquim and Rowe, 1990). According to Bell (1992), the defoliating type is found naturally in the southern and southwestern United States, but has also been reported in other areas around the world. The non-defoliating type of *V. dahliae* has also been reported from cotton-producing areas of California and Peru (Schnathorst, 1969).

According to Davis et al. (2006), Fusarium wilt, caused by *Fusarium oxysporum* Schlechtend. f. sp. *vasinfectum* (Atk.) Snyd. & Hans, referred to as (*Fov*), has been reported from cotton growing regions around the world and is capable of infecting all four species of domesticated cotton. Numerous pathogenic and saprophytic biotypes of *F. oxysporum* exist and have similar morphological features. Fernandez et al. (1994) found that at least 10 vegetative compatibility groups exist for *Fov*. This grouping encompasses biotypes that have been traditionally divided into various races based on results from pathogenicity tests with host differentials (Fahmy, 1927; Armstrong and Armstrong, 1958, 1960, 1978; Ibrahim, 1966). A total of eight races have been described throughout the world.

Historically, race 1 and race 2 have been the most prevalent races associated with cotton in the United States (Smith et al., 1981) and were found in fields infested with the root-knot nematode (*Meloidogyne incognita* Kofoid & White, Chitwood) (Bell, 1984). Race 5 was initially described in Sudan (Ibrahim, 1966), but was later combined with race 3 on the basis of similarities in pathogenicity and morphological characteristics (Nirenberg et al., 1994). Armstrong and Armstrong (1978) described race 6 from Brazil; and Chen et al. (1985) described races 7 and 8 from cotton in China. More recently, extremely virulent strains of *Fov*, capable of causing extensive damage in the absence of *M. incognita*, were identified in Australia (Davis et al., 1996). Several researchers reported finding race 4 in Asia (Armstrong and Armstrong, 1960; Assigbetse et al., 1994; Fernandez et al., 1994; Skovgaard et al., 2001). Kim et al. (2005) recovered isolates, which caused serious economic losses in the absence of *M. incognita* in California. These isolates were identified as race 4 using a combination of host differentials and molecular techniques. Subsequently, *Fov* race 4 has spread throughout much of the San Joaquin Valley and is now the primary concern for the industry in the state (Davis et al., 2006). Furthermore, recent surveys have identified novel types and new distributions of known races in the United States (Bennett et al., 2011, 2013; Castillo et al., 2010; Holmes et al., 2009).

Epidemiology

Both *V. dahliae* and *Fov* are capable of persisting in the soil for long periods of time via the formation of specialized survival structures, microsclerotia (Wilhelm, 1955) or chlamydospores (Smith and Snyder, 1975), respectively. Microsclerotia are considered the initial inoculum of *V. dahliae* and are generally distributed in an aggregated or clustered pattern within a field. The majority of microsclerotia form in undecomposed crop residue, where they are subsequently released and germinate in response to root exudates (Booth, 1968). Viable microsclerotia have been recovered from depths up to 40 cm (Huisman and Gerik, 1989). Paplomatas et al. (1992) found that densities of microsclerotia prior to planting positively correlated with final disease incidence, which supports reports of other researchers (Ashworth et al., 1979; Pullman and DeVay, 1982). Furthermore, Waker and Onkin (1969) demonstrated that the dissemination of infested crop residue led to an increase in disease incidence. Disease development is regulated greatly by temperature, with Verticillium wilt generally being more severe under cooler conditions than Fusarium wilt.

Fusarium wilt is favored by temperatures above 23°C (Young, 1928). Chlamydospores of *Fov* remain dormant in the soil until plant exudates or leachates stimulate their germination (Mai and Abawi, 1987). Once a field is infested, the fungus can persist for long periods of time because of the pathogen's saprophytic ability, resilient nature of the chlamydospores, and ability to parasitize other plant species (Smith and Snyder, 1975). The emergence of new *Fov* races has renewed interest in the seedborne aspects of the fungus (Bennett et al., 2008, 2012). With the exception of race 4 and newly identified biotypes, Fusarium wilt severity is minor in the absence of *M. incognita*. Researchers have found that disease development will occur with *Fov* alone at high densities (Wang et al., 1999; Davis et al., 2006; Hao et al., 2009). Studies have demonstrated the synergistic effect of nematode populations and *Fov* on cotton yield (Roberts et al., 1985; Starr et al., 1989). While interactions between the *M. incognita* and *Fusarium* spp. have been extensively studied, the overall infection process is poorly defined (Mai and Abawi, 1987).

Disease Control

Because of the ability of both *Fov* and *V. dahliae* to survive in the absence of a host, management of vascular wilt diseases is difficult. Cultivars with partial resistance to either disease have been developed and have been moderately successful in disease management. Since Fusarium wilt is often associated with *M. incognita*, utilization of cultivars with nematode resistance or nematicides generally leads to reduced disease severity (Hyer et al., 1979; Shepherd, 1982). Shepherd (1974) suggested that Fusarium wilt resistance existed independently of nematode resistance. Hillocks (1992) reported that disease resistance differs among *Gossypium* spp. Smith and Dick (1960) suggested that the mechanism of resistance in *G. hirsutum* is monogenic; whereas, resistance in *G. barbadense* is controlled by two dominant genes. Later studies indicated that Fusarium wilt resistance in *G. hirsutum* is quantitative and controlled by several major genes, as well as a number of different minor genes (Kappelman, 1971; Reid et al., 2004). Deployment of Fusarium wilt resistant cultivars may be ineffective as resistance may be specific to certain races of the fungus. For example, some cultivars of *G. barbadense* have high levels of resistance to race 4 (Ulloa et al., 2006).

Partially resistant cultivars have been identified for Verticillium wilt (Bell, 1973; Bassett, 1974; Wilhelm et al., 1974); however, sources of strong resistance are limited. The resurgence of Verticillium wilt in the High Plains of Texas has led to a renewed interest in the development of germplasm with increased resistance (Dever et al., 2013). Resistance to Verticillium wilt has been associated with rank vegetative growth, late maturity, and a small fruit load (Marani and Yaacobi, 1976), all of which negatively affect yield, thus limiting commercial usage. Environmental conditions (primarily temperature) and population densities of the fungus can greatly affect the response of cultivars to Verticillium wilt. Infections fail to develop at temperatures above 30°C (El-Zik, 1985). Paplomatas et al. (1992) found that high population densities of microsclerotia (>10 per gram of soil) were required before differences in yield between susceptible and resistant cultivars were observed; however, other studies report susceptible cultivars being significantly affected at densities as low as 1.8 microsclerotia per gram of soil (Ashworth et al., 1979).

Chemical options for management of vascular wilt diseases are limited. Seed treatments have been evaluated to limit the spread *Fov* into new areas (Allen and Kochman, 2001); however, these treatments have no effect on the pathogen in infested fields. There are currently no fungicides registered for use against Verticillium wilt in cotton. Bell (1992) provides a limited overview of fungicides used internationally. Chemical control for wilt diseases is limited to fumigation with products such as metam sodium or 1,3-dichloropropene; however, effective rates of these products are often cost prohibitive (Williams et al., 2002; Woodward et al., 2011).

For Fusarium wilt, crop rotation is often recommended more so for the impact on *M. incognita* than the impact on the fungus (Smith and Snyder, 1975). Tillage will have little effect on *Fov* because of the persistence of chlamydospores. Colyer et al. (1994) demonstrated that disease severity was similar for conventional and conservation tillage systems, which included various cover crops. Irrigation has a minimal effect on disease development. Grinstein et al. (1983) reported dispersal of the pathogen via in-furrow irrigation, which was greater than movement of the pathogen during cultivation. Solarization and summer flooding have been proven to be effective treatments for *Fov* (Allen and Lonergan, 2000; Bennett et al., 2011), but are rarely used because of cost.

In addition to the use of partially resistant cultivars, management of Verticillium wilt includes integrating various cultural practices (El-Zik, 1985). Rotation with a non-host will limit reproduction of the fungus compared with continuous cotton production; however, the overall effect is inadequate in fields with a history of severe disease (Butterfield et al., 1978). In a recent study conducted by Wheeler et al. (2012), crop rotation was effective at reducing damage caused by Verticillium wilt and limited the buildup of microsclerotia. The difference in the latter study was that rotations were implemented prior to the disease occurring in the field. Other reports have shown the impact of plant density on disease development. Wheeler et al. (2010) found that disease incidence was negatively correlated with seeding rate and differences among resistant and susceptible cultivars were not observed at higher seeding rates.

Numerous water management practices have been shown to affect development of Verticillium wilt. El-Zik (1985) and Wheeler et al. (2012) reported that lower irrigation regimes reduced disease incidence and suppressed reproduction of the fungus. The use of raised beds or drains, as well as selecting soil with good drainage, reduces excessive moisture and limits damage caused by the disease (Ranney, 1973). While cultivars with improved resistance to Verticillium wilt are available, integration of multiple disease management strategies is most effective in minimizing losses due to this disease (El-Zik, 1985; Wheeler et al., 2010, 2012).

Cotton Root Rot and Other Stem Diseases

Cotton root rot or Texas root rot is caused by the soilborne fungus *Phymatotrichopsis omnivorum* (Duggar) Hennebert and is considered one of the most destructive and difficult diseases to manage. The disease was first described in the 1880s by Pammel (1888, 1889) and was referred to as Ozonium or Phymatotrichum root rot (after former names of the causal agent). The pathogen has an extensive host range of more than 2000 broadleaf plant species including many field and ornamental crops (Streets and Bloss, 1973).

Symptoms

Disease onset is highly dependent on environmental conditions and typically occurs during bloom; however, initial infections take place well before symptom expression. Symptoms observed on aboveground tissues resemble other diseases such as Verticillium wilt or other factors such as water stress. Symptoms associated with cotton root rot can be characterized by a slight yellowing of the leaves, which quickly intensifies as young leaves begin to wilt (Rush et al., 1984). The lack of discoloration in the xylem elements can be used to distinguish cotton root rot from symptoms associated with non-defoliating forms of Verticillium wilt. Permanent wilting occurs rapidly and dry necrotic leaves remain attached to diseased plants. Wilted leaves exhibit increased canopy temperatures because of the disruption of transpiration. In general, symptoms occur in localized areas within a field radiating outward from individual foci. As the disease progresses, these areas may coalesce affecting a large areas of the field, or under severe cases the entire field.

Evidence of the disease can also be observed on belowground portions of diseased plants. Physical signs of the fungus consist of intertwined strands of hyphae or rhizomorphs along the taproot, which is most obvious at or just below the soil surface. Examination of the root will reveal a red discoloration of the periderm, which increases in size and intensifies in color as the disease progresses. When adequate moisture is available, cottony mycelia can be observed at or below the soil surface. Occasionally, yellow to brown spore mats can be seen at the base of infected plants; however, the role of these structures in the disease cycle is unknown.

Causal Agent

Confusion has shrouded the taxonomy of the cotton root rot fungus for many years. Currently, *P. omnivorum* is considered the causal agent of the disease; however, previous reports have identified the causal agent as *Ozonium omnivorum* (Shear) *Phymatotrichum omnivorum* (Shear) Duggar. Marek et al. (2009) confirmed placement of the fungus in the genus *Phymatotrichopsis* based on molecular systematics.

Distribution of *P. omnivorum* is limited primarily to the southwestern United States and northern Mexico; however, the disease has also been observed in areas of Arkansas and Louisiana (Smith et al., 1959). Soils that favor cotton root rot development are generally alkaline and calcareous, with low organic matter. In Texas, the disease only occurs at elevations below 1500 m (5000 ft) and has not been identified in the High Plains region where cotton production is concentrated.

The fungus survives in the form of sclerotia, which form in the soil following infection and serve as primary inoculum. These structures have an irregular shape, generally conforming to the shape of space in which they are formed (Lyda, 1978). Viable sclerotia have been recovered from throughout the soil profile and have been shown to survive for up to 12 yr. Despite the resilient nature of sclerotia, the fungus is believed to have a low relative saprophytic capacity (Lyda, 1978). The fungus is only active during summer months when air and soil temperatures are high. Lyda and Burnett (1971) found optimal disease development was observed at 27°C and that foliar symptoms were not observed at soil temperatures below 22.5°C (Rush et al., 1984).

Epidemiology

Prior to the identification of sclerotia, the fungus was thought to persist in soil in the form of mycelia or hyphal strands. Sclerotia are commonly produced close to the soil surface 15 to 20 cm (6–8 in), but have been recovered from depths greater than 2.1 m (7 ft). Following the germination of sclerotia, mycelial strands entwine descending roots and grow toward the soil surface. All cells within a sclerotium are capable of germinating, a process that may continue, as long as the propagule remains viable (Presley, 1939). When the fungus reaches the upper portion of the root system, it secretes cell wall degrading enzymes, which kills the periderm allowing for colonization of the cortex and blockage of the vascular elements (Watkins, 1938a, 1938b). Jeger et al. (1987) suggested that disease development originates from discrete foci and that subsequent spread to adjacent plants is achieved via hyphal strands. Lyda and Kissel (1974) suggested that sodium availability was responsible for the localized distribution of sclerotia within a field. Other studies concluded that various sodium-based compounds adversely affected growth of the fungus and disease development (Curtis, 1892; Taubenhaus et al., 1932).

Disease Control

Because of the nonspecific mode of parasitism utilized by *P. omnivorum*, the development of cultivars with improved resistance has been unsuccessful. Early maturing cultivars may escape the onset of disease, as they mature before optimal conditions for disease development occur. These cultivars typically have lower yield potential and are not widely grown in areas affected by cotton root rot. Most management practices have proven ineffective or are cost prohibitive. Practices such as crop rotation or fallow periods are ineffective because of the long-lived sclerotia (Streets and Bloss, 1973; Rush and Lyda, 1984). Deep tillage practices led to a delay in the onset of disease and reduced incidence compared with conventional practices; however, the efficacy was erratic (Rush and Lyda, 1984). The erratic performance can be attributed to the vertical distribution of sclerotia within the soil and tillage practices necessary to increase efficacy are cost prohibitive. The application of soil amendments have been shown to lessen the severity of the disease in some instances, but such an approach has not been adopted because of inconsistent results (Matocha et al., 1988).

Several soil fumigants and fungicides are known to adversely affect *P. omnivorum* in vitro and in vivo; however, limitations exist for each approach. Lyda et al. (1967) achieved adequate disease control using various rates of 1,3-dichloropropene applied at different depths, but this approach proved to be too expensive. Rush and Lyda (1982) and Riggs and Lyda (1987) found that sclerotia of *P. omnivorum* are extremely sensitive to anhydrous ammonia; however, anhydrous ammonia applied in conjunction with deep tillage did not improve control over deep tillage alone (Rush and Lyda, 1984). Triazole fungicides were found to be lethal to growth and development of *P. omnivorum*; however, delayed emergence, poor stand establishment, and other deleterious effects were observed (Riggs and Lyda, 1988). Field studies examining different application methods and various timings of fungicides have led to improved control, and applications of propiconazole and triadimenol applied through subsurface drip irrigation reduced the number of dead plants (Olsen and George, 1987).

More recently, the fungicide flutriafol has been evaluated for control of cotton root rot in Texas. Initial studies found that flutriafol applied via drip irrigation reduced disease incidence compared with the non-treated control, as well as propiconazole; however, evidence of phytotoxicity was observed on foliage later in the season (Isakeit et al., 2007, 2009). Subsequent studies found that flutriafol regardless of rate applied through the drip tape was more consistent than stem drench applications (Isakeit et al., 2010). Additional studies investigating at-plant or in-furrow applications have shown promising results with regard to managing the disease and increasing yield; however, sporadic phytotoxic effects were reported (Isakeit et al., 2011, 2012). Alternative methods in placing the fungicide at planting are being investigated in order to minimize damage to cotton seedlings (Woodward et al., 2014). The apparent persistence of flutriafol in the soil and activity toward *P. omnivorum* coupled with precision applications hold promise for managing cotton root rot in the future.

Charcoal Rot

Charcoal rot is a disease that occurs worldwide on cotton. It is of importance in warmer tropical semi-arid to arid conditions found in east and central Africa and the Indian subcontinent. Symptoms of charcoal rot may include wilting of leaves, a dry rot of the lower stem and taproot causing a discoloration of the tissue, and plant death. The disease is distinguished from other diseases by the presence of small black microsclerotia in the tissue giving the tissue a gray or ashen appearance. The pathogen causing charcoal rot is *Macrophomina phaseolina* (Tassi) Goid. This pathogen has a very wide host range including field and ornamental crops. The pathogen can survive for long periods as microsclerotia in residue or in the soil. Infection may occur at any time in the growing season. Environmental conditions favoring infection are cooler and wetter than when disease symptoms are expressed. At temperatures greater than 30°C and water stress, rapid colonization of plant tissue occurs resulting in the development of disease symptoms (Ghaffar and Erwin, 1969). As disease progresses, the pathogen forms microsclerotia throughout the colonized tissue. The importance of charcoal rot is difficult to determine, as disease is associated with drought conditions. Disease management for charcoal rot involves limiting stresses on the crop that are known to be associated with disease, including irrigation management to limit water stress and high soil temperatures. Crop rotation or other cultural practices have not been shown to have a large impact on disease.

Virus and Phytoplasma Diseases

To date, 19 viral or virus-like diseases have been described on cotton. Of these, only African cotton mosaic, Cotton anthocyanosis, Cotton blue disease, Cotton leaf crumple, Cotton leaf curl, and Cotton yellow vein have been completely or partially characterized. Brown (1992) summarized the major cotton viruses. In general, viral diseases have been considered of minor importance in cotton with severe epidemics only occurring sporadically, except in Africa, Central Asia, and South America. Environmental conditions in tropical or subtropical climates allow alternate hosts in the Malvaceae and volunteer cotton plants to be present all year, thus serving as reservoirs for inoculum. Furthermore, many of the insect vectors that transmit these infectious agents thrive under these conditions.

Field diagnosis of viral diseases is much more difficult than diagnosing diseases caused by fungal or bacterial pathogens. The viruses that infect cotton consist only of nucleic acids (either RNA or DNA) and a few proteins. Thus, viruses do not produce visible signs as with other causal agents. Virus diseases are typically named on the basis of the symptoms infected plants express. This too is confusing as symptom expression may differ among cultivars or with changes in environmental conditions. Furthermore, different viruses may cause similar symptoms on the same host. Most viruses are transmitted by insects and have developed very specific relationships with their vectors. The two most common vectors of cotton viruses are the sweet potato whitefly [*Bemisia tabaci* (Genn.)] and the cotton aphid [*Aphis gossypii* (Glover)].

Cotton Viruses

Cotton viruses classified as geminiviruses (family: *Geminiviridae*; genus *Begomovirus*), including African cotton mosaic, Cotton leaf crumple, and Cotton leaf curl, are considered the most economically important. As a result, these viruses are the most thoroughly studied as is the vector *B. tabaci*. Symptoms of African cotton mosaic include small chlorotic patches and vein-yellowing on leaves; larger, more angular patches that represent the characteristic "mosaic" pattern; thickening of the veins; and deformation of the leaf, as well as severe stunting. Plants infected with Cotton leaf crumple exhibit an overall stunting of the plant, as well as crumpling, curling, and mosaic of the foliage. These symptoms are most severe on the newest growth. Small outgrowths distort flowers and bolls, resulting in yield loss from reductions in boll number and weight. Reductions in fiber length may also occur. Different types of symptoms are associated with Cotton leaf curl. The most common are mosaic patterns, streaking, vein-thickening, and twisting or curving of the leaves (Farooq et al., 2011). Plants exhibiting these symptoms typically do not appear stunted. A more severe set of symptoms include severe curling and blistering of the leaves, thickening of the main leaf vein, shortened internodes, and stunting. A third type of symptom, referred to as "bunchy top," is characterized by shortened internodes, vein-thickening, and reductions in the size of leaves and flowers.

Cotton anthocyanosis, also referred to as Vermelhão or reddening disease, is transmitted by *A. gossypii* and was first observed on *G. barbadense* in Brazil in 1954 (Costa, 1957). It commonly occurs in Brazil and is considered the most damaging disease of cotton in that country. A similar disease is thought to occur in India, the Philippines, Sudan, and Thailand; however, these reports are based mostly on field observations. Symptoms consisting of chlorosis followed by purpling or reddening generally occur. As the disease progresses, discoloration of the leaves will intensify; however, the veins will not be affected. Lower leaves will eventually abcise, as symptoms become more obvious in younger tissues.

Blue disease is responsible for considerable losses in cotton-producing regions of Africa and is transmitted by *A. gossypii*. Symptoms of Blue disease consist of severe stunting or "dwarfing." Affected plants have thick, dark blue-green leaves with lighter areas adjacent to veins and edges that curl downward (Brown, 2001a). Stems of infected plants are often malformed, having a zig-zag appearance.

Cotton yellow vein, also referred to as Texas cotton vein-clearing, was first described in 1957 in the Lower Rio Grande Valley of Texas (Rosberg, 1957) and is also known to occur in areas of Northern Mexico (Halliwell et al., 1980). The virus is transmitted by the leafhopper *Scaphytopius albifrons* rather than whiteflies, thus it is distinct from other viruses such as Cotton leaf crumple (Brown, 2001b). Symptoms of Cotton yellow vein disease include bright yellow vein-clearing, vein banding, and mosaic (Brown, 2001b). Mild to severe stunting can occur in addition to the crumpling and mottling of leaves. Subtle differences in symptom expression can be used to distinguish Cotton yellow vein from Cotton leaf crumple.

Phytoplasma Diseases

Several diseases of cotton, including flavescence, flower virescence and phyllody, psyllosis, and small leaf, are caused by known or suspected phytoplasmas. Occurrences of these types of diseases appear to be geographically restricted; as a result, information on these diseases is limited. In general, diseases caused by phytoplasmas disrupt plant metabolism, drastically reducing fruit production and retention. Seeds from infected plants fail to mature. Furthermore, fiber of infected plants is immature and of very low quality.

Disease Control

Disease reactions differ among *Gossypium* spp. for viral and phytoplasma diseases. For example, cotton cultivars susceptible to African cotton leaf curl are no longer grown because of the severity of the disease, whereas Egyptian cotton cultivars (*G. barbadense*) appear to be immune to small leaf. Chemical control options are limited to the application of insecticides to reduce populations of insect vectors. Depending on the manner in which a virus or phytoplasma is transmitted, the effectiveness and the chemical mode of action will affect efficacy of the insecticide. For example, virus particles transmitted in a non-persistent manner can be transmitted before or while the insect is acquiring an insecticide. Furthermore, more insect feeding may occur before a lethal dosage of a contact insecticide is obtained, resulting in increased transmission of virus particles. Cultural practices targeting alternate weed hosts or ratooned or volunteer cotton have been partially effective at reducing incidence of certain viruses (Idris, 1990).

Boll Rots of Cotton

Microbiological fiber deterioration or boll rot is an important problem in cotton production. Boll rot can be especially severe when rainfall and wet weather are common during the reproductive growth stages of the cotton crop. Boll rot is a problem that can cause significant losses in yield and quality while undermining other production practices, such a planting date, plant population, and soil fertility (Guthrie et al., 1994). Although such practices can improve yield potential, they can also increase the risk of boll rot. For example, adjusting the planting date may lead to situations where boll opening occurs during periods where rain is likely to occur, thus increasing risk to boll rot. Increased seeding rates and soil fertility may elevate risk to boll rot as denser plant populations and excessive growth can lead to higher humidity and longer wetness periods around the developing bolls, thus causing more severe boll rot. The amount of boll damage is very much related to the weather over a period of a few days during boll opening (Marsh et al., 1950), and thus the amount of boll damage is variable from season to season. Boll rot is not a universal or consistent problem for cotton growers; however, when it occurs, management options are limited. Cotton growers in the southeastern United States must anticipate boll rot every year, and losses as high as 33% have been reported in certain fields. For example, boll rot is reported to be the most destructive disease of cotton in the Mississippi Delta, especially under hot and humid conditions (Bagga, 1968, 1970). Growers in drier production regions, both in the United States and elsewhere, rarely face this level of loss from boll rots (Guthrie et al., 1994).

Symptoms

Although compact and poorly fluffed locules (not discolored) with relatively low fiber strength can and do occur in the absence of microbiological damage to the fiber (Marsh et al., 1950), pathogens are often associated with damaged bolls and fiber in the field. "Cotton boll rot" is a term generally applied to any deterioration or breakdown of the boll tissue and may include dark, sunken lesions on the boll surface (e.g., anthracnose and bacterial blight) and a tight-lock/hardlock condition of partially open bolls that is associated with many pathogens. Boll rot is a generic term referring to numerous diseases caused by bacteria and fungi that damage bolls, lint and seed. In "tight-lock" the locules within the bolls do not fluff but remain hard and rotted. It is speculated that tight-lock may result from late infections that occur as bolls open and the moist, susceptible lint is rapidly colonized (Crawford, 1982). In contrast to "tight-lock," Fusarium boll rot and Diplodia boll rot have been referred to as "basal boll rots." It is believed that the bracts associated with the bolls are infected first, followed by the peduncle, which leads to penetration by the pathogen of the base of boll and subsequent colonization of the developing lint within the boll (Crawford, 1982).

In Diplodia boll rot, also referred to as "black rot," infection by the fungal pathogen *Lasiodiplodia theobromae* (Pat.) Griffon & Mabul. (syn. *Diplodia gossypina* Cooke) progresses initially from the bracts to the peduncles and then into the boll, or the pathogen may simply penetrate directly through the carpel wall or through sutures between carpels. Diplodia boll rot is characterized by a dark mass of filaments and spores leading to a "black rot" (Crawford, 1982; Batson, 2001).

Symptoms of anthracnose or Colletotrichum boll rot (caused by *Colletotrichum* spp., *C. gossypii, Glomerella gossypii*) begin as reddish sunken dots at the top of the boll that enlarge to form black regions covered by a mass of pink spores. For anthracnose/Colletotrichum rot, more than half of the boll may become decayed (Batson, 2001). *Colletotrichum capsici* has been reported to cause a boll rot that is distinct from the anthracnose rot cause by *C. gossypii* on the basis of appearance of the rotted bolls and morphological differences between the fungal pathogens (Roberts and Snow, 1984). Favorable environmental conditions for *C. capsici* lead to rapid expansion of lesions that encompass the entire boll. Lesions associated with other species of *Colletotrichum* tend to remain sunken, restricted, and more anthracnose-like (Roberts and Snow, 1984). Bolls infected with *C. capsici* are often "hardlocked."

In Fusarium boll rot, *Fusarium* spp. (especially *F. oxysporum* and *F. roseum*) infects the bracts surrounding the boll first and then colonizes the boll through the peduncle. Infection progresses to the top of the boll, producing a blue-black rot that is covered by a proliferation of pink spores (Crawford, 1982; Batson, 2001). "Fusarium hardlock" has been used in recent years to describe a rot reported to be different than earlier described for Fusarium boll rot (Mailhot et al., 2007). Fusarium hardlock is described as a failure of cotton fibers to expand outward (fluff) after boll opening. Instead, fibers remain in compressed locules, which often have a faint pink or orange color. Although Fusarium hardlock has not been formally recognized as a disease, some consider it to be a subset of the boll rot complex. Fusarium hardlock is most severe along the Gulf Coast of the southeastern United States, possibly because of the region's high temperatures and humidity (Mailhot et al., 2007).

Causal Agents

It has been reported that at least 170 opportunistic pathogens exploit the nutrition provided by the moist lint either during growth and development of the boll or once the boll begins to crack and open, thus leading to boll rot. The most important of these are the "primary" pathogens capable of infecting and rotting a boll without association with some wound, injury, or insect damage. One of the most important primary pathogens associated with boll rot is the bacterial blight pathogen, *X. citri* subsp. *malvacearum*. Important fungal pathogens include *Lasiodiplodia theobromae* (Diplodia boll rot), *Colletotrichum gossypii* (anthracnose boll rot), *Fusarium* spp., especially *Fusarium oxysporum* (Fusarium boll rot/hard lock), *Phytophthora* spp. (Phytophthora boll rot), *Aspergillus flavus* (Aspergillus boll rot), *Nigrospora oryzae* (Nigrospora lint rot), and *Pantoea agglomerans* (Pantoea boll rot) (Guthrie et al., 1994; Batson, 2001).

Epidemiology

Primary inoculum for pathogens associated with bolls rots may be seedborne, as reported for *X. citri* subsp. *malvacearum*. Spores of opportunistic boll rot pathogens, like *Alternaria* spp., may be produced on weeds and other hosts and later deposited on the cotton bolls through wind and blowing rain. These spores may also be introduced by insects that feed on the bolls. When conditions are favorable (e.g., suitable moisture and entry port), these pathogens will infect and colonize the boll. Last are those pathogens, like *L. theobromae* and others, that survive in the crop residue and debris and are disseminated to cotton bolls and bracts (Batson, 2001), perhaps through rain-splash or other mechanism. Wounds from insects or mechanical wounds allowing entry of pathogens are common ways in which boll rot is initiated; however, some pathogens may also penetrate through stomata, nectaries, or sutures that occur between the carpels forming the boll. Several fungal pathogens are capable of direct penetration through the wall of the boll. The presence of separate locules within a boll may actually help to isolate the colonization of the entire boll (Guthrie et al., 1994).

Development of Diplodia boll rot is favored by warm humid conditions (Wang and Pinkard, 1973). For *L. theobromae*, the causal agent of Diplodia boll rot, maximum spore production occurs on bolls that are over 40 d of age. Young nonseptate spores germinate more quickly than do older spores. The surface of the boll supports an enormous microflora, presumably supported by debris, nectarial residues, pollen, and boll exudates adhering to the surface of the boll and conidia of *L. theobromae* are likely present as well. Diplodia boll rot tends to be more problematic following periods of heavy dew, rather than heavy rain, suggesting that the mechanical action of the rain washes spores present on the surface of the boll away from potential sites for infection (Wang and Pinkard, 1973).

In Fusarium boll rot, *F. oxysporum* and *F. roseum* are capable of infecting noninjured bolls, whereas other *Fusarium* spp. can cause boll rot but require a wound for infection to occur. Initial infection by *F. oysporum* and *F. roseum* is to the bracts that subtend the boll, and later the capsule base of the boll is invaded through the receptacle. The age of the boll is important in the development of Fusarium boll rot; once a boll has reached an age of 33 to 35 d or older, the bolls are resistant to infection. In the field, the pathogen becomes established in bracts soon after bloom, but basal rot, an early symptom of Fusarium boll rot, is typically not observed until bolls are 6 to 10 wk old. In the southeastern portion of cotton production in the United States, those species of *Fusarium* associated with boll rot are often considered to be secondary colonizers following insect injury or wounds to the pericarp by *Colletotrichum* and *Xanthomonas*. Other points of potential entry for *Fusarium* include nectaries, the calyx pit, the pedicel, and sutural gaps associated with the bolls (Sparnicht and Roncadori, 1972).

Aspergillus boll rot (*Aspergillus flavus* Link) and Nigrospora lint rot [*Nigrospora oryzae* (Berk. And Broome) Petch] are both examples of damage that can occur to the boll from pathogens that are secondary invaders (Batson, 2001). Both pathogens can cause significant damage once they become established in the developing lint and aflatoxin is associated with Aspergillus boll rot. Aggressivenss of isolates of *A. flavus* has been correlated with their ability to secrete pectinase(s) that degrade the boll tissue. Isolates with reduced ability to spread through boll tissues and cause rot were found to lack major endopolygalacturonase activity (Cleveland and Cotty, 1991).

Disease Control

To best reduce the impact of boll rots, timely management of crop growth, insect pests, and earliness of harvest maturity of the crop is essential. Growth regulation is critical for producing robust, well-fruited plants on compact plants without excessive vegetation. Management strategies must consider acceptable boll-opening windows that may require deliberate delays in planting to avoid predicted periods of rainfall (Guthrie et al., 1994).

Specific tactics to reduce the risk to boll rot include the following recommendations. Growers should correct any factor that unnecessarily promotes excessive growth in order to reduce humidity and periods of leaf wetness within the canopy. Effective use of a cotton growth regulator (e.g., mepiquat chloride) to manage growth is a critical tool in the management of boll rot. Reducing plant populations and modifying row patterns so as to increase airflow and reduce humidity within the canopy are also helpful (Crawford, 1982; Hillocks, 1992; Batson, 2001). Growers should guard against overfertilization and poor insect control as overfertilization leads to rank growth and insects wound bolls and may facilitate the transmission of pathogens. As failure to set bolls promotes excessive vegetative growth, growers should make every effort to manage the crop for maximum boll retention so as to reduce risk to boll rot. As mentioned earlier, growers may consider adjusting planting dates so that boll opening and maturity are less likely to coincide with wetter periods of the season. Timeliness is not only important when considering the implications of wet periods during the season that are important for growing the crop, but it is also important for defoliation of the crop and harvest in order to reduce the risk of fungal rot of lint in the field (Crawford, 1982).

Efforts have been made to protect a cotton crop against boll rot by use of fungicides. For example, in one study use of the fungicides carbendizim and mancozeb increased yields by 17.3 and 15.1%, respectively (Alagarsamy et al., 1988); however, chemical control of boll rots has generally failed because it is difficult to achieve effective coverage of the bolls with fungicides. Achieving adequate coverage is made more difficult with a rank leaf canopy, wet fields, and the sequential development of bolls (Crawford, 1982). Where boll rots occur largely as a consequence of boll damage from insects, judicious use of insecticides may reduce incidence of boll rot (Hillocks, 1992). In such instances, a combination of fungicide and insecticide for the control of bollworm and the boll rot complex can reduce losses associate with both.

Breeding efforts have focused on selecting for arrangement of leaves within the canopy and also specific leaf shapes (e.g., okra leaf) to allow more airflow and sunlight to penetrate the canopy and reduce overall humidity. Efforts also have focused on reducing the number of nectaries so as to repel insects, selecting for bracts that do not enclose the boll and for bolls with tighter sutures that are more difficult for fungi to penetrate (Wang and Pinkard, 1973; Crawford, 1982; Hillocks, 1992).

Cotton Disease Management

Management of plant diseases involves the adoption of a number of tactics to have economical and effective disease control. Knowledge of the pathogen's disease cycle and life cycle are critical, as well as an understanding of the producer's cultural practices and preferences to integrate successfully into a sustainable crop management system. Essential to the understanding of disease occurrence and development is the "disease triangle," the interactions between the disease causal agent (pathogen), the host (the cotton plant), and the environment. For disease to develop, the pathogen must be virulent, the host susceptible, and the environment conducive for infection and spread by the pathogen.

Disease control recommendations can be divided into a number of principles. Exclusion includes practices where the pathogen is not present in an area and voluntary measures or government agencies limit the introduction of the pathogen or a population of the pathogen. In the United States, the American Plant Health Inspection Service (APHIS) is involved in determining which pathogens are not established in the country and limiting the introduction of pathogens such as Australian biotypes of the Fusarium wilt pathogen. Exclusion can also be practiced by the seed industry by ensuring that seedborne pathogens are not present in planting seed. Eradication involves practices that favor the elimination or reduction of populations of a pathogen that results in the elimination or reduction of the disease in the subsequent crop. This might include practices such a crop rotation, which is effective for bacterial blight, a pathogen with limited ability to survive and limited ability to spread over long distances. Tillage practices to bury residue are effective for bacterial blight and other foliar diseases. Other practices may include soil solarization, flooding, or fumigation. Host resistance is one of the most effective and economical disease control practices. For bacterial blight, resistance genes have been used to provide cultivars with high levels of resistance. For a number of other diseases, there is some level of resistance that has been increased over time from selection of individuals or populations that exhibit less disease than other genotypes examined. This has been successful for vascular wilt diseases and provides some disease reduction.

Protection involves the application of materials that prevent or limit infection of the cotton plant. Seed treatment fungicides are universally used to manage seedling disease pathogens. Protection would also be used with the foliar application of fungicides for control of foliar diseases. Avoidance involves production of cotton using decisions to create an environment that limits disease development. When we look at how cotton is being produced in a region, many cultural practices are adopted to limit diseases including land preparation, planting on beds, planting date, and irrigation practices. Effective disease management includes using a number of strategies to limit disease development and maintain effective controls over long periods. Integrating management strategies for cotton diseases is essential, because few options are available after disease problems arise.

Future Directions

Production agriculture is changing faster than ever including the introduction of new cultivars, deployment of new technologies, and changing cultural practices. As in the past, changing cotton production practices will affect the diseases observed and their severity. Globalization has resulted in increased movement of plant material, equipment, and people, increasing the risk of the movement of pathogens. Thus, it is not unexpected to see the reemergence of certain diseases or changes in pathogen populations or new diseases in a region.

Incorporating resistance to important diseases of cotton should be a priority in the development of new cultivars. In addition, identification of new sources of resistance for important diseases should continue. The development of new molecular breeding techniques has increased the rate of cultivar development and the traits being screened in the breeding process. These techniques when applied to the plant or pathogen allow us to examine at the gene level the factors that allow a pathogen to cause disease or the genes that limit disease development. With the advent of transgenic cultivars, new transgenic traits are being incorporated into cultivars being released. Although not used in cotton for disease management at this time, a number of successes have been realized in other crops, for example, transgenic papaya for resistance to papaya ringspot virus in Hawaii (Gonsalvas et al., 2004).

Fungicide chemistries continue to improve and increase the spectrum of control. With cotton, fungicide resistance has not been a major issue in large part because of their limited use for the control of foliar diseases. In addition, methods of delivery of chemicals have been improved as evidenced by the management of cotton root rot via drip irrigation. Use of GIS technology in the precision application of products, such as flutriafol is going to have a large impact on how cotton is produced in fields that are infested with *P. omnivorum*. Furthermore, the value of field prescriptions for the selection of cultivars, seeding rates, fertility, and irrigation based on the soil environment are being developed. Seedling diseases are a limiting factor in establishing adequate stands and should be a part of the prescription for variable seeding rates in the future. Prescription applications have already been realized with nematode management on cotton. Despite all of the past successes in disease management, disease control will continue to play an important role in the profitability and sustainability of cotton production in the future.

References

Abd-Elsalam, K.A., A.-M. Asran-Amal, M.R. Omar, and A.A. Aly. 2006. Frequency and diversity of *Fusarium* spp. colonizing roots of Egyptian cottons. Arch. Phytopathol. Plant Prot. 39:165–177. doi:10.1080/03235400400007533

Akello, B., and R.J. Hillocks. 2002. Distribution and races of *Xanthomonas axonopodis* pv. *malvacearum* on cotton (*Gossypium hirsutum*) in Uganda. J. Phytopathol. 150:65–69. doi:10.1046/j.1439-0434.2002.00718.x

Alagarsamy, G., G. Krishnamohan, K. Sivaprakasam, and R. Jeyarajan. 1988. Chemical control of boll rot complex of cotton. Madras Agric. J. 75:69–70.

Allen, S.J. 2001. Black root rot. In: T.L. Kirkpatrick and C.S. Rothrock, editors, Compendium of cotton diseases. American Phytopathological Society, St. Paul, MN. p. 16–17.

Allen, S.J., and J.K. Kochman. 2001. Eliminating seed-borne inoculum of *Fusarium oxysporium* f. sp. *vasinfectum* in cotton. In: Proceedings of the Beltwide Cotton Conferences. Anaheim, CA. 9–13 Jan. 2000. Natl. Cotton Counc., Memphis, TN. p. 139–140.

Allen, S.J., and P.A. Lonergan. 2000. Control strategies for Fusarium wilt of cotton in Australia. In: Proceedings of the Beltwide Cotton Conferences. San Antonio, TX. 4–8 Jan. 2000. Natl. Cotton Counc., Memphis, TN. p. 136–138.

Armstrong, G.M., and J.K. Armstrong. 1960. American, Egyptian, and Indian cotton-wilt Fusaria: Their pathogenicity and relationship to other wilt Fusaria. Tech. Bull. 1219. USDA, Washington, DC.

Armstrong, G.M., and J.K. Armstrong. 1978. A new race (race 6) of the cotton-wilt Fusarium from Brazil. Plant Dis. Rep. 62:421–423.

Armstrong, J.K., and G.M. Armstrong. 1958. A race of the cotton-wilt Fusarium causing wilt of Yelredo soybean and flue-cured tobacco. Plant Dis. Rep. 42:147–151.

Arndt, C.H. 1943. *Pythium ultimum* and the damping-off of cotton seedlings. Phytopathology 33:607–611.

Arnold, M.H. 1965. The control of bacterial blight in rain-grown cotton II. Some effects of infection on growth and yield. J. Agric. Sci. 65:29–40. doi:10.1017/S0021859600085336

Arnold, M.H., and K.M. Arnold. 1961. Bacterial blight of cotton: Trash-borne infection. Emp. Cotton Grow. Rev. 38:258–270.

Ashworth, L.J., Jr., O.C. Huisman, D.M. Harper, L.K. Stromberg, and D.M. Bassett. 1979. Verticillium wilt disease of cotton: Influence of inoculum density in the field. Phytopathology 69:483–489. doi:10.1094/Phyto-69-483

Assigbetse, K.B., D. Fernandez, M.P. Dubois, and J.-P. Geiger. 1994. Differentiation of *Fusarium oxysporum* f. sp. *vasinfectum* races on cotton by random amplified polymorphic DNA (RAPD) analysis. Phytopathology 84:622–626. doi:10.1094/Phyto-84-622

Bagga, H.S. 1968. Fungi associated with cotton boll rot and their frequency. Plant Dis. Rep. 52:582–584.

Bagga, H.S. 1970. Pathogenicity studies of organisms involved in the cotton boll-rot complex. Phytopathology 60:158–160. doi:10.1094/Phyto-60-158

Baird, R.E. 2001a. Alternaria leaf spot. In: T.L. Kirkpatrick and C.S. Rothrock, editors, Compendium of cotton diseases. American Phytopathological Society, St. Paul, MN. p. 20.

Baird, R.E. 2001b. Cercospora leaf spot. In: T.L. Kirkpatrick and C.S. Rothrock, editors, Compendium of cotton diseases. American Phytopathological Society, St. Paul, MN. p. 21–22.

Baird, R.E. 2001c. Ascochyta blight. In: T.L. Kirkpatrick and C.S. Rothrock, editors, Compendium of cotton diseases. American Phytopathological Society, St. Paul, MN. p. 22.

Baird, R.E. 2001d. Stemphyllium leaf spot. In: T.L. Kirkpatrick and C.S. Rothrock, editors, Compendium of cotton diseases. American Phytopathological Society, St. Paul, MN. p. 23.

Baird, R.E. 2001e. Other foliar diseases. In: T.L. Kirkpatrick and C.S. Rothrock, editors, Compendium of cotton diseases. American Phytopathological Society, St. Paul, MN. p. 23.

Balasubramanyan, R., and A. Raghavan. 1950. Bacterial blight on cotton in Madras. Indian Cotton Grow. Rev. 4:118–123.

Bassett, D.M. 1974. Resistance of cotton cultivars to Verticillium wilt and its relation to yield. Crop Sci. 14:864–867. doi:10.2135/cropsci1974.0011183X001400060027x

Batson, W.E. 2001. Boll rots. In: T.L. Kirkpatrick and C.S. Rothrock, editors, Compendium of cotton diseases. American Phytopathological Society, St. Paul, MN. p. 36–38.

Bayles, M.B., and L.M. Verhalen. 2007. Bacterial blight reactions of sixty-one upland cotton cultivars. J. Cotton Sci. 11:40–51.

Bell, A.A. 1973. Nature of disease resistance. In: C.D. Ranney, editor, Verticillium wilt of cotton. Publ. ARS-S-19. USDA, Washington, DC. p. 47–62.

Bell, A.A. 1984. Cotton protection practices in the USA and world. Section B: Diseases. In: R.J. Koehl and C.F. Lewis, editors, Cotton. Agron. Monogr. 24. ASA, CSSA, and SSSA, Madison, WI. p. 288–309.

Bell, A.A. 1992. Verticillium wilt. In: R.J. Hillocks, editor, Cotton diseases. CAB International, Wallingford, UK. p. 87–126.

Bennett, R.S. 2012. Survival of *Fusarium oxysporum* f. sp. *vasinfectum chlamydospores* under solarization temperatures. Plant Dis. 96:1564–1568. doi:10.1094/PDIS-09-11-0812-RE

Bennett, R.S., R.B. Hutmacher, and R.M. Davis. 2008. Seed transmission of *Fusarium oxysporum* f. sp. *vasinfectum* race 4 in California. J. Cotton Sci. 12:160–164.

Bennett, R.S., W. O'Neill, L. Smith, and R.B. Hutmacher. 2011. Activity of commercial detergents against conidia and chlamydospores of *Fusarium oxysporum* f. sp. *vasinfectum*. J. Cotton Sci. 15:162–169.

Bennett, R.S., T.Z. Scott, K.S. Lawrence, and G.W. Lawrence. 2013. Characterization of race 4-like isolates of *Fusarium oxysporum* from Alabama and Mississippi. J. Cotton Sci. 17:125–130.

Bhattiprolu, S.L., and M.D. Prasada Rao. 2009. Management of Alternaria leaf spot (*Alternaria macrospora* Zimm) on cotton by biological approach. J. Cotton Res. Dev. 23:135–137.

Bhuiyan, S.A., M.C. Boyd, A.J. Dougall, C. Martin, and M. Hearnden. 2007. Effects of foliar application of potassium nitrate on suppression of Alternaria leaf blight of cotton (*Gossypium hirsutum*) in northern Australia. Australas. Plant Pathol. 36:462–465. doi:10.1071/AP07051

Bird, L.S. 1959. Loss measurements caused by the bacterial blight disease of cotton. Phytopathology 49:315.

Bird, L.S. 1986. Half a century dynamics and control of cotton diseases: Bacterial blight. In: Proceedings of the Beltwide Cotton Conferences, Natl. Cotton Counc., Dallas, TX. 5–9 Jan. 1986. Natl. Cotton Counc., Memphis, TN. p. 41–48.

Black, M.C. 2001a. Southwestern cotton rust. In: T.L. Kirkpatrick and C.S. Rothrock, editors, Compendium of cotton diseases. American Phytopathological Society, St. Paul, MN. p. 24–25.

Black, M.C. 2001b. Tropical cotton rust. In: T.L. Kirkpatrick and C.S. Rothrock, editors, Compendium of cotton diseases. American Phytopathological Society, St. Paul, MN. p. 25–26.

Blank, L.M., P.J. Leyendecker, Jr., and R.M. Nakayama. 1953. Observations on black root rot symptoms on cotton seedlings at different soil temperatures. Plant Dis. Rep. 37:473–476.

Bock, C.H., P.E. Parker, and T.R. Gottwald. 2005. Effect of simulated wind-driven rain on duration and distance of dispersal of *Xanthomonas axonopodis* pv. *citri* from canker-infected citrus trees. Plant Dis. 89:71–80. doi:10.1094/PD-89-0071

Booth, J.A. 1968. *Gossypium hirsutum* tolerance to *Verticillium albo-atrum* infection I. Amino acid exudation from aseptic roots of tolerant and susceptible cotton. Phytopathology 59:43–46.

Brinkerhoff, L.A., and G.B. Fink. 1964. Survival and infectivity of *Xanthomonas malvacearum*, in cotton plant debris and soil. Phytopathology 54:1198–1201.

Brinkerhoff, L.A., and R.E. Hunter. 1963. Internally infected seed as a source of inoculum for the primary cycle of bacterial blight of cotton. Phytopathology 53:1397–1401.

Brinkerhoff, L.A., and J.T. Presley. 1967. Effect of four day and night temperature regimes on bacterial blight reactions of immune, resistant and susceptible strains of Upland cotton. Phytopathology 57:47–51.

Brown, J.K. 1992. Virus diseases. In: R.J. Hillocks, editor, Cotton diseases. CAB International, Wallingford, UK. p. 275–329.

Brown, J.K. 2001a. Cotton blue disease. In: T.L. Kirkpatrick and C.S. Rothrock, editors, Compendium of cotton diseases. American Phytopathological Society, St. Paul, MN. p. 50–51.

Brown, J.K. 2001b. Cotton yellow vein disease. In: T.L. Kirkpatrick and C.S. Rothrock, editors, Compendium of cotton diseases. American Phytopathological Society, St. Paul, MN. p. 54–55.

Butterfield, E.J., J.E. DeVay, and R.H. Garber. 1978. The influence of several crop sequences on the incidence of Verticillium wilt of cotton and on the populations of *Verticillium dahliae* in field soil. Phytopathology 68:1217–1220. doi:10.1094/Phyto-68-1217

Castillo, J.D., K.S. Lawrence, and K. Glass. 2010. Isolation and molecular identification of *Fusarium oxysporum* isolates from cotton crops in Alabama. In: Proceedings of the Beltwide Cotton Conferences, Natl. Cotton Counc., New Orleans, LA. 4–7 Jan. 2010. Natl. Cotton Counc., Memphis, TN. p. 273–277.

Chen, Q., X. Ji, and W. Sun. 1985. Identification of races of cotton wilt *Fusarium* in China. Agric. Sci. China 6:1–6.

Cleveland, T.E., and P.J. Cotty. 1991. Invasiveness of *Aspergillus flavus* isolates in wounded cotton bolls in associated production of a specific fungal polygalacturonase. Phytopathology 81:155–158. doi:10.1094/Phyto-81-155

Colson-Hanks, E.S., S.J. Allen, and B.J. Deverall. 2000. Effect of 2,6-dichloroisonicotonic acid or benzothiadiazole on Alternaria leaf spot, bacterial blight and Verticillium wilt in cotton under field conditions. Australas. Plant Pathol. 29:170–177. doi:10.1071/AP00030

Colyer, P.D. 1988. Frequency and pathogenicity of *Fusarium* spp. associated with seedling diseases of cotton in Louisiana. Plant Dis. 72:400–402. doi:10.1094/PD-72-0400

Colyer, P.D. 2001. Diseases caused by *Fusarium* species. In: T.L. Kirkpatrick and C.S. Rothrock, editors, Compendium of cotton diseases. American Phytopathological Society, St. Paul, MN. p. 18–19.

Colyer, P.D., T.L. Kirkpatrick, P.R. Vernon, J.D. Barman, and R.J. Bateman. 1994. Incidence of Fusarium wilt under conventional and reduced tillage following different cover crops. In: M. McClelland and T. Keisling, editors, Conservation technology in Arkansas agriculture 1994. Res. Ser. 449. Arkansas Agricultural Experiment Station, Fayetteville. p. 42–50.

Colyer, P.D., S. Micinski, and K.T. Nguyen. 1991. Effect of planting date on the efficiency of an in-furrow pesticide and the development of cotton seedling disease. Plant Dis. 75:739–742. doi:10.1094/PD-75-0739

Conner, K.N., A.K. Hagan, and L. Zhang. 2013. First report of *Corynespora cassiicola*-induced target spot on cotton in Alabama. Plant Dis. 97:1379. doi:10.1094/PDIS-02-13-0133-PDN

Correll, J.C., T.R. Gordon, and A.H. McCain. 1988. Vegetative compatibility and pathogenicity of *Verticillium albo-atrum*. Phytopathology 78:1017–1021. doi:10.1094/Phyto-78-1017

Costa, A.S. 1957. Anthocyanosis, a virus disease in cotton in Brazil. Phytopathol. Z. 28:167–186.

Crawford, J.L. 1982. Cotton boll rot. Plant Pathol. 3 Leafl. 143. Coop. Ext. Serv., Univ. of Georgia Coll. of Agric., Athens, GA.

Curtis, G.W. 1892. Alfalfa root rot. Tex. Agric. Exp. Stn. Bull. 22:211–215.

Davis, R.D., N.Y. Moore, and J.K. Kochman. 1996. Characterisation of a population of *Fusarium oxysporum* f. sp. *vasinfectum* causing wilt of cotton in Australia. Aust. J. Agric. Res. 47:1143–1156. doi:10.1071/AR9961143

Davis, R.G. 1975. Microorganisms associated with diseased cotton seedlings in Mississippi. Plant Dis. Rep. 59:277–280.

Davis, R.M., P.D. Colyer, C.S. Rothrock, and J.K. Kochman. 2006. Fusarium wilt of cotton: Population diversity and implications for management. Plant Dis. 90:692–703. doi:10.1094/PD-90-0692

Davis, R.M., J.J. Nunez, and K.V. Subbarao. 1997. Benefits of cotton seed treatments for the control of seedling diseases in relation to inoculum densities of *Pythium* species and *Rhizoctonia solani*. Plant Dis. 81:766–768. doi:10.1094/PDIS.1997.81.7.766

Delannoy, E., B.R. Lyon, P. Marmey, A. Jalloul, J.F. Daniel, J.L. Montillet, M. Essenberg, and M. Nicole. 2005. Resistance of cotton towards *Xanthomonas campestris* pv. *malvacearum*. Annu. Rev. Phytopathol. 43:63–82. doi:10.1146/annurev.phyto.43.040204.140251

DeVay, J.E. 2001. Seedling diseases. In: T.L. Kirkpatrick and C.S. Rothrock, editors, Compendium of cotton diseases. American Phytopathological Society, St. Paul, MN. p. 13–14.

DeVay, J.E., R.H. Garber, and D. Matheron. 1982. Role of *Pythium* species in the seedling disease complex of cotton in California. Plant Dis. 66:151–154. doi:10.1094/PD-66-151

Dever, J.K., T.A. Wheeler, and C.M. Kelly. 2013. Registration of CA 4002 cotton germplasm line partially resistant to Verticillium wilt. J. Plant Reg. 7:209–215. doi:10.3198/jpr2012.04.0261crg

Dighule, S.B., R.R. Perane, P.E. Moore, and K.S. Amle. 2011. Efficacy of chemical fungicides and bio-agents against major cotton fungal diseases. Int. J. Plant Protect. 4:263–266.

Ehrlich, J., and F.A. Wolf. 1931. Areolate mildew of cotton. Phytopathology 22:229–240.

Elliott, J.A. 1922. A new ascochyta disease of cotton. Bull. 178. Univ. of Arkansas Agric. Exp. Stn., Fayetteville, AR.

El-Zik, K.M. 1985. Integrated control of Verticillium wilt of cotton. Plant Dis. 69:1025–1032. doi:10.1094/PD-69-1025

Fahmy, T. 1927. The Fusarium disease (wilt) of cotton and its control. Phytopathology 17:749–767.

Farooq, A., J. Farooq, A. Mahmood. A. Shakeel, A. Rehman, A. Batool, M. Riaz, M.T.H. Shahid, and S. Mehboob. 2011. An overview of cotton leaf curl virus disease (CLCuD) a serious threat to cotton productivity. Aust. J. Crop Sci. 5:1823–1831.

Fernandez, D., K. Assigbetse, M.-P. Dubios, and J.-P. Geiger. 1994. Molecular characterization of races and vegetative compatibility groups in *Fusarium oxysporum* f. sp. *vasinfectum*. Ann. Environ. Microbiol. 60:4039–4046.

Fichtner, S.M., T. Isakeit, T.A. Wheeler, H.W. Kaufman, and J.R. Gannaway. 2005. Evaluation of several approaches to manage *Meloidogyne incognita* and cotton seedling disease complexes in the high plains of Texas. J. Nematol. 37:66–70.

Fulmer, A.M., J.T. Walls, B. Dutta, V. Parkunan, J. Brock, and R.C. Kemerait, Jr. 2012. First report of target spot caused by *Corynespora cassiicola* on cotton in Georgia. Plant Dis. 96:1066. doi:10.1094/PDIS-01-12-0035-PDN

Fulton, N.D., and K. Bollenbacher. 1959. Pathogenicity of fungi isolated from diseased cotton seedlings. Phytopathology 49:684–689.

Garber, R.H., and B.R. Houston. 1966. Penetration and development of *Verticillium albo-atrum* in the cotton plant. Phytopathology 56:1121–1126.

Ghaffar, A., and D.C. Erwin. 1969. Effect of soil water stress on root rot of cotton caused by *Macrophomina phaseoli*. Phytopathology 59:795–797.

Gonsalvas, D., C. Gonsalvas, S. Ferreira, K. Pitz, M. Fitch, R. Manshardt, and J. Slightom. 2004. Trangenic virus resistant papaya: From hope to reality for controlling papaya ringspot virus in Hawaii. American Phytopathological Society. http://www.apsnet.org/publications/apsnetfeatures/Pages/PapayaRingspot.aspx (accessed 20 May 2015).

Gottwald, T.R., X. Sun, T. Riley, J.H. Graham, F. Ferrandino, and E.L. Taylor. 2002. Georeferenced spatiotemporal analysis of the urban citrus canker epidemic in Florida. Phytopathology 92:361–377. doi:10.1094/PHYTO.2002.92.4.361

Govindappa, N.H., and S.N. Chattannavar. 2008. Chemical and biological control of foliar diseases of cotton. J. Cotton Res. Dev. 2:225–228.

Grinstein, A., G. Fishler, J. Katan, and D. Hakohen. 1983. Dispersal of the Fusarium wilt pathogen in furrow-irrigated cotton in Israel. Plant Dis. 67:742–743. doi:10.1094/PD-67-742

Guthrie, D., K. Whitman, B. Batson, J. Crawford, and G. Jividen. 1994. Boll rot. Cotton Physiol. Today. 5(8).

Hake, K., K. El-ZIk, S. Johnson-Hake, and J. Mauney. 1990. Leaves. In: Cotton growth and development for production. Cotton Physiol. Educ. Program, Natl. Cotton Counc., Memphis, TN. p. 11–12.

Hagan, A.K. 2014. Target spot management options in Alabama. In: Proceedings of the Beltwide Cotton Conferences, Natl. Cotton Counc., New Orleans, LA. 6–8 Jan. 2014. Natl. Cotton Counc., Memphis, TN. p. 45–48.

Hall, C., R. Heath, and D. Guest. 2013. The infection process of *Fusarium oxysporum* f. sp. *vasinfectum* in Australian cotton. Australas. Plant Pathol. 42:1–8. doi:10.1007/s13313-012-0169-8

Halliwell, R.S, S.D. Lyda, and M.J. Lukefahr. 1980. A leafhopper-transmitted virus of cotton. Bull. MP-1465. Texas Agric. Exp. Stn., College Station.

Hao, J.J., M.E. Yang, and R.M. Davis. 2009. Effect of soil inoculum density of *Fusarium oxysporum* f. sp. *vasinfectum* race 4 on disease development in cotton. Plant Dis. 93:1324–1328. doi:10.1094/PDIS-93-12-1324

Hillocks, R.J. 1991. *Alternaria* leaf spot of cotton with special reference to Zimbabwe. Trop. Pest Manage. 37:124–128. doi:10.1080/09670879109371557

Hillocks, R.J., editor. 1992. Cotton diseases. CAB International, Oxon, UK.

Holmes, E.A., R.S. Bennett, D.W. Spurgeon, P.D. Colyer, and R.M. Davis. 2009. New genotypes of *Fusarium oxysporum* f. sp. *vasinfectum* from the southeastern United States. Plant Dis. 93:1298–1304. doi:10.1094/PDIS-93-12-1298

Hosagoudar, G.N., and S.N. Chattannavar. 2014. Integrated disease management of Alternaria leaf spot on Bt cotton through fungicides. J. Cotton Res. Dev. 28:120–124.

Howard, D.D., M.A. Newman, and A.Y. Chambers. 1997. Soil and foliar potassium effects on Alternaria leaf spot disease in cotton. Better Crops 81:17–19.

Huisman, O.C. 1988. Colonization of field-grown cotton roots by pathogenic soilborne fungi. Phytopathology 78:716–722. doi:10.1094/Phyto-78-716

Huisman, O.C., and J.S. Gerik. 1989. Dynamics of colonization of plant roots by *Verticillium dahliae* and other fungi. NATO ASI Ser. H. 28:1–17.

Hunter, R.E., and L.M. Blank. 1954. Pathogenicity differences of *Xanthomonas malvacearum* isolates. Phytopathology 44:332.

Hunter, R.E., E.E. Staffeldt, and C.R. Maier. 1960. Effects of soil temperature on the pathogenicity of *Rhizoctonia solani* isolates. Plant Dis. Rep. 44:793–795.

Hyer, A.H., E.C. Jorgenson, R.H. Garber, and S. Smith. 1979. Resistance to root-knot nematode in control of root-knot nematode *Fusarium* wilt disease complex in cotton *Gossypium hirsutum*. Crop Sci. 19:898–901. doi:10.2135/cropsci1979.0011183X001900060036x

Ibrahim, F.M. 1966. A new race of the cotton-wilt *Fusarium* in the Sudan Gezira. Emp. Cotton Grow. Rev. 43:296–299.

Idris, A.M. 1990. Cotton leaf curl virus disease in the Sudan. Meded. Fac. Landbouwwet. Univ. Gent 55:263–267.

Irey, M., T.R. Gottwald, J.H. Graham, T.D. Riley, and G. Carton. 2006. Post-hurricane analysis of citrus canker spread and progress towards the development of a predictive model to estimate disease spread due to catastrophic weather events. Plant Health Prog. doi:10.1094/PHP-2006-0822-01-RS

Isakeit, T., R. Minzenmayer, A. Abrameit, G. Moore, and J.D. Scasta. 2010. Control of Phymatotrichopsis root rot of cotton with flutriafol. In: Proceedings of the Beltwide Cotton Conferences, Natl. Cotton Counc., New Orleans, LA. 4–7 Jan. 2010. Natl. Cotton Counc., Memphis, TN. p. 200–203.

Isakeit, T., R.R. Minzenmayer, D.R. Drake, G. Morgan, D. Mott, D.D. Fromme, W.L. Multer, M. Jungman, and A. Abrameit. 2012. Fungicide management of cotton root rot (*Phymatotrichopsis omnivorum*): 2011 results. In: Proceedings of the Beltwide Cotton Conferences, Natl. Cotton Counc., Orlando, FL. 3–6 Jan. 2012. Natl. Cotton Counc., Memphis, TN. p. 235.

Isakeit, T., R.R. Minzenmayer, D.R. Drake, W. Multer, A. Abrameit, M. Jungman, C. Crossland, D.T. Campion, N. Fryar, G. Morgan, and D. Mott. 2011. Flutriafol control of Phymatotrichopsis root rot of cotton. In: Proceedings of the Beltwide Cotton Conferences, Natl. Cotton Counc., Atlanta, GA. 4–7 Jan. 2011. Natl. Cotton Counc., Memphis, TN. p. 207.

Isakeit, T., R.R. Minzenmayer, and C.G. Sansone. 2007. Evaluation of fungicides for control of cotton root rot caused by *Phymatotrichopsis omnivorum*. In: Proceedings of the Beltwide Cotton Conferences, Natl. Cotton Counc., New Orleans, LA. 9–12 Jan. 2007. Natl. Cotton Counc., Memphis, TN. p. 181–184.

Isakeit, T., R. Minzenmayer, and C. Sansone. 2009. Flutriafol control of cotton root rot caused by *Phymatotrichopsis omnivorum*. In: Proceedings of the Beltwide Cotton Conferences, Natl. Cotton Counc., San Antonio, TX. 5–8 Jan. 2009. Natl. Cotton Counc., Memphis, TN. p. 130–133.

Jaraba, J., C.S. Rothrock, T.L. Kirkpatrick, and K.R. Brye. 2014. Soil texture influence on *Meloidogyne incognita* and *Thielaviopsis basicola* and their interaction on cotton. Plant Dis. 98:336–343. doi:10.1094/PDIS-04-13-0357-RE

Jeger, M.J., C.M. Kenerley, T.J. Gerik, and D.O. Koch. 1987. Spatial dynamics of Phymatotrichum root rot in row crops in the Blackland region of north and central Texas. Phytopathology 77:1647–1656. doi:10.1094/Phyto-77-1647

Joaquim, T.R., and R.C. Rowe. 1990. Reassessment of vegetative compatibility relationships among strains of *Verticillium dahliae* using nitrate-nonutilizing mutants. Phytopathology 80:1160–1166. doi:10.1094/Phyto-80-1160

Johnson, L.F., D.D. Baird, A.Y. Chambers, and N.B. Shamiyeh. 1978. Fungi associated with postemergence seedling disease of cotton in three soils. Phytopathology 68:917–920. doi:10.1094/Phyto-68-917

Johnson, L.F., and A.Y. Chambers. 1973. Isolation and identity of three species of *Pythium* that cause cotton seedling blight. Plant Dis. Rep. 57:848–852.

Johnson, L.F., A.Y. Chambers, and J.W. Measells. 1969. Influence of soil moisture, temperature, and planting date on severity of cotton seedling blight. Bull. 461, Tennessee Agric. Exp. Stn., Knoxville.

Johnson, L.F., and J.H. Doyle. 1986. Relationship of seedling disease of cotton to characteristics of loessial soils in Tennessee. Phytopathology 76:286–290. doi:10.1094/Phyto-76-286

Jones, J.P. 1961. A leaf spot of cotton caused by *Corynespora cassiicola*. Phytopathology 51:305–308.

Kappelman, A.J., Jr. 1971. Inheritance of resistance to Fusarium wilt in cotton. Crop Sci. 11:672–674. doi:10.2135/cropsci1971.0011183X001100050020x

Kaufman, H., T. Wheeler, R. Graves, G. Schuster, P. Kidd, and K. Siders. 1998. Large plot performance of seedling disease seed treatment fungicides. In: Proceedings of the Beltwide Cotton Conferences, Natl. Cotton Counc., Anaheim, CA. 5–8 Jan. 1998. Natl. Cotton Counc., Memphis, TN. p. 149–152.

Kim, Y., R.B. Hutmacher, and R.M. Davis. 2005. Characterization of California isolates of *Fusarium oxysporum* f. sp. *vasinfectum*. Plant Dis. 89:366–372. doi:10.1094/PD-89-0366

Kirkby, K.A., P.A. Lonergan, and S.J. Allen. 2013. Three decades of cotton disease surveys in NSW, Australia. Crop Pasture Sci. 64:774–779.

Kirkpatrick, T.L., and C.S. Rothrock, editors. 2001. Compendium of cotton diseases. American Phytopathological Society, St. Paul, MN.

Koenning, S. 2000. Cotton stem canker, wet weather blight, or Ascochyta blight. Cotton Dis. Info. Note No. 2. North Carolina State Univ. Plant Pathol. Ext., Raleigh.

Lyda, S.D. 1978. Ecology of *Phymatotrichum omnivorum*. Annu. Rev. Phytopathol. 16:193–209. doi:10.1146/annurev.py.16.090178.001205

Lyda, S.D., and E. Burnett. 1971. Influence of temperature on *Phymatotrichum omnivorum* sclerotial formation and disease development. Phytopathology 61:728–730. doi:10.1094/Phyto-61-728

Lyda, S.D., and D.E. Kissel. 1974. Sodium influence on disease development and sclerotial formation by *Phymatotrichum omnivorum*. Proc. Am. Phytopathol. Soc. 1:163–164.

Lyda, S.D., G.D. Robison, and H.W. Lembright. 1967. Soil fumigation control of Phymatotrichum root rot in Nevada. Plant Dis. Rep. 51:331–333.

Madani, A.S., A. Marefat, K. Behboudi, and A. Ghasemi. 2010. Phenotypic and genetic characterisitics of *Xanthomonas citri* subsp. *malvacearum*, causal agent of cotton blight, and identification of races in Iran. Australas. Plant Pathol. 39:440–445. doi:10.1071/AP10098

Mai, W.F., and G.S. Abawi. 1987. Interactions among root-knot nematodes and Fusarium wilt fungi in host plants. Annu. Rev. Phytopathol. 25:317–338. doi:10.1146/annurev.py.25.090187.001533

Mailhot, D.J., J.J. Marois, and D.L. Wright. 2007. Influence of flower thrips on Fusarium hardlock severity. Plant Dis. 91:1423–1429. doi:10.1094/PDIS-91-11-1423

Marani, A., and Y.Z. Yaacobi. 1976. Evaluation of Verticillium wilt tolerance in upland cotton relative to lint yield reduction. Crop Sci. 16:392–395. doi:10.2135/cropsci1976.0011183X001600030019x

Marek, S.M., K. Hansen, M. Romanish, and R.G. Thorn. 2009. Molecular systematics of the cotton root rot pathogen, *Phymatotrichopsis omnivora*. Persoonia 22:63–74. doi:10.3767/003158509X430930

Marsh, P.B., L.R. Guthrie, K. Bollenbacher, and D.C. Harrell. 1950. Observations of microbiological deterioration of cotton fiber during the periods of boll opening in 1949. Plant Dis. Rep. 34:165–175.

Mathre, D.E., A.V. Ravenscroft, and R.H. Garber. 1966. The role of *Thielaviopsis basicola* as a primary cause of yield reduction in cotton in California. Phytopathology 56:1213–1216.

Mathur, R.L., L.N. Daftari, and S.L. Jhamaria. 1973. Controlling blackarm of cotton by fungicides and antibiotics. Indian J. Mycol. Plant Pathol. 3:107–108.

Matocha, J.E., S. Mostaghimi, and C. Crenshaw. 1988. Effect of soil amendments on cotton growth and incidence of *Phymatotrichum* root rot. In: Proceedings of the Beltwide Cotton Conferences, Natl. Cotton Counc., New Orleans, LA. 5–8 Jan. 1988. Natl. Cotton Counc., Memphis, TN. p. 41–45.

Mauk, P.A., and R.B. Hine. 1988. Infection, colonization of *Gossypium hirsutum* and *G. barbadense*, and development of black root rot caused by *Thielaviopsis basicola*. Phytopathology 78:1662–1667. doi:10.1094/Phyto-78-1662

Mehta, Y.R. 1998. Severe outbreak of Stemphylium leaf blight, a new disease of cotton in Brazil. Plant Dis. 82:333–336. doi:10.1094/PDIS.1998.82.3.333

Mehta, Y.R., C. Bomfeti, and V. Bolognini. 2005. A semi-selective agar medium to detect the presence of *Xanthomonas axonopodis* pv. *malvacearum* in naturally infected cotton seed. Fitopatol. Bras. 30:489–496. doi:10.1590/S0100-41582005000500005

Melero-Vara, J.M., and R.M. Jimenaz-Diaz. 1990. Etiology, incidence, and distribution of cotton seedling damping-off in southern Spain. Plant Dis. 74:597–600. doi:10.1094/PD-74-0597

Meyer, M.C., J.C. da Silva, G.L. Maia, C.J. Bueno, and N.L. Souza. 2006. Myrothecium leaf spot of cotton caused by *Myrothecium roridum*. Summa Phytopathol. 32:390–393. doi:10.1590/S0100-54052006000400016

Minton, E.B., and R.H. Garber. 1983. Controlling the seedling disease complex of cotton. Plant Dis. 67:115–118. doi:10.1094/PD-67-115

Munjal, R.L. 1960. A commonly occurring leaf spot disease caused by *Myrothecium roridum* Tode ex Fr. Indian Phytopathol. 8:150–155.

Nehl, D.B., S.J. Allen, A.H. Mondal, and P.A. Lonergan. 2004. Black root rot: A pandemic in Australian cotton. Australas. Plant Pathol. 33:87–95. doi:10.1071/AP03085

Nirenberg, H.I., G. Ibrahim, and S.H. Michail. 1994. Race identity of three isolates of *Fusarium oxysporum* Schlect. f. sp. *vasinfectum* (Atk.) Snyd. & Hans. from Egypt and the Sudan. Z. Pflanzenkr. Pflanzenschutz 101:594–597.

Ogle, H.J., A.M. Stirling, and P.J. Dart. 1993. Pathogenicity of fungi associated with seedling disease of cotton. Aust. J. Exp. Agric. 33:923–929. doi:10.1071/EA9930923

Olsen, M.W., and S. George. 1987. Application of systemic fungicides through subsurface drip irrigation for control of *Phymatotrichum* root rot of cotton. Phytopathology 77:1748.

Oosterhuis, D., T. Kirby, and K. Hake. 1990. Leaf physiology and management. Cotton Physiol. Today. May:1–6.

Pammel, L.H. 1888. Root rot of cotton, or "cotton blight". Tex. Agric. Exp. Stn. Ann. Rep. 1:50–65.

Pammel, L.H. 1889. Cotton root-rot. Tex. Agric. Exp. Stn. Annu. Rep. 2:61–86.

Paplomatas, E.J., D.M. Bassett, J.C. Broome, and J.E. DeVay. 1992. Incidence of Verticillium wilt and yield losses of cotton cultivars (*Gossypium hirsutum*) based on soil inoculum density of *Verticilium dahliae*. Phytopathology 82:1417–1420. doi:10.1094/Phyto-82-1417

Pegg, G.F. 1984. The impact of Verticillium diseases in agriculture. Phytopathol. Mediterr. 23:176–192.

Presley, J.T. 1939. Unusual features in the behavior of sclerotia of *Phymatotrichum omnivorum*. Phytopathology 29:498–502.

Puhalla, J.E., and M. Hummel. 1983. Vegetative compatibility groups within *Verticillium dahliae*. Phytopathology 73:1305–1308. doi:10.1094/Phyto-73-1305

Pullman, G.S., and J.E. DeVay. 1982. Epidemiology of Verticillium wilt of cotton: A relationship between inoculum density and disease progression. Phytopathology 72:549–554. doi:10.1094/Phyto-72-549

Ranney, C.D. 1973. Verticillium wilt of cotton. Publ. ARS-S-19. USDA, Washington, DC.

Ray, W.W., and J.H. McLaughlin. 1942. Isolation and infection tests with seed- and soil-borne pathogens. Phytopathology 32:233–238.

Reid, P., S. Allen, J. Kochman, W. Stiller, G. McNamara, J. Lehane, and G. Constable. 2004. Sicot F-1: A variety with increased resistance to Fusarium wilt. In: Proceedings of the 12th Australian Cotton Conference, Gold Coast. 10–12 Aug. 2004. Aust. Cotton Growers Res. Assoc., Queensland p. 739–740.

Riggs, J.L., and S.D. Lyda. 1987. Response of sclerotia of *Phymatotrichum omnivorum* to five soil fumigants. In: Proceedings of the Beltwide Cotton Conferences, Natl. Cotton Counc., Dallas, TX. 5–8 Jan. 1987. Natl. Cotton Counc., Memphis, TN. p. 49–50.

Riggs, J.L., and S.D. Lyda. 1988. Laboratory and field tests with triazole fungicides to control *Phymatotrichum* root rot of cotton. In: Proceedings of the Beltwide Cotton Conferences, Natl. Cotton Counc., Dallas, TX. 5–8 Jan. 1988. Natl. Cotton Counc., Memphis, TN. p. 45–48.

Roberts, R.G., and J.P. Snow. 1984. Histopathology of cotton boll rot caused by *Colletotrichum capsici*. Phytopathology 74:390–397. doi:10.1094/Phyto-74-390

Roberts, P.A., S.N. Smith, and W.C. Matthews. 1985. Quantitative aspects of the interaction of *Meloidogyne incognita* with Fusarium wilt on Acala cotton in field plots. In: Proceedings of the Beltwide Cotton Conferences, Natl. Cotton Counc., New Orleans, LA. 7–11 Jan. 1985. Natl. Cotton Counc., Memphis, TN. p. 21–22.

Rosberg, D.W. 1957. A new virus disease of cotton in Texas. Plant Dis. Rep. 41:726–729.

Rothrock, C.S. 1992. Influence of soil temperature, water, and texture on *Thielaviopsis basicola* and black root rot on cotton. Phytopathology 82:1202–1206. doi:10.1094/Phyto-82-1202

Rothrock, C.S. 1996. Cotton diseases incited by *Rhizoctonia solani*. In: B. Sneh, S. Jabaji-Hare, S. Neate, and G. Dijst, editors, *Rhizoctonia* species: Taxonomy, molecular biology, ecology, pathology, and disease control. Kluwer Academic Publishers, Boston. p. 269–277.

Rothrock, C.S., T.L. Kirkpatrick, R.E. Frans, and H.D. Scott. 1995. The influence of winter legume cover crops on soilborne plant pathogens and cotton seedling diseases. Plant Dis. 79:167–171. doi:10.1094/PD-79-0167

Rothrock, C.S., S.A. Winters, P.K. Miller, E. Gbur, L.M. Verhalen, B.E. Greenhagen, T.S. Isakeit, W.E. Batson, Jr., F.M. Bourland, P.D. Colyer, T.A. Wheeler, H.W. Kaufman, G.L. Sciumbato, P.M. Thaxton, K.S. Lawrence, W.S. Gazaway, A.Y. Chambers, M.A. Newman, T.L. Kirkpatrick, J.D. Barham, P.M. Phipps, F.M. Shokes, L.J. Littlefield, G.B. Padgett, R.B. Hutmacher, R.M. Davis, R.C. Kemerait, D.R. Sumner, K.W. Seebold, Jr., J.D. Mueller, and R.H. Garber. 2012. Importance of fungicide seed treatment and environment on seedling diseases of cotton. Plant Dis. 96:1805–1817. doi:10.1094/PDIS-01-12-0031-SR

Roy, K.W., and F.B. Bourland. 1982. Epidemiological and mycofloral relationships in cotton seedling disease in Mississippi. Phytopathology 72:868–872. doi:10.1094/Phyto-72-868

Rush, C.M., and S.D. Lyda. 1982. The effects of anhydrous ammonia on membrane stability of *Phymatotrichum omnivorum*. Mycopathologia 79:147–152. doi:10.1007/BF01837194

Rush, C.M., and S.D. Lyda. 1984. Evaluation of deep-chiselled anhydrous ammonia as a control for Phymatotrichum root rot of cotton. Phytopathology 74:1466–1469. doi:10.1094/Phyto-74-1466

Rush, C.M., T.J. Gerik, and S.D. Lyda. 1984. The relationship between the time of cortical senescence and foliar symptom development of Phymatotrichum root rot of cotton. Phytopathology 74:1464–1466. doi:10.1094/Phyto-74-1464

Schaad, N.W., E. Postnikova, G. Lacy, A. Sechler, I. Agarkova, P.E. Stromberg, V.K. Stromberg, and A.K. Vidaver. 2006. Emended classification of xanthomonad pathogens on citrus. Syst. Appl. Microbiol. 29:690–695.

Schnathorst, W.C. 1964. Longevity of *Xanthomonas manlvacearum* in dried cotton plants and its significance in dissemination of the pathogen on seed. Phytopathology 54:1009–1011.

Schnathorst, W.C. 1966. Eradication of *Xanthomonas malvacearum* from California through sanitation Plant Dis. Rep. 50:168–171.

Schnathorst, W.C. 1969. A severe form of *Verticillium albo-atrum* in *Gossypium barbadense* in Peru. Plant Dis. Rep. 53:149–151.

Schnathorst, W.C., and D.E. Mathre. 1966. Host range and differentiation of a severe form of *Verticillium albo-atrum* in cotton. Phytopathology 56:1155–1161.

Sharma, O.P., and O.M. Bambawale. 2008. Integrated management of key diseases of cotton and rice. In: A. Ciancio and K.G. Mukerji, editors, Integrated management of diseases caused by fungi, phytoplasma and bacteria. Springer, Dordrecht, Netherlands. p. 271–302.

Shepherd, R.L. 1974. Transgressive segregation for root-knot nematode resistance in cotton. Crop Sci. 14:872–875.

Shepherd, R.L. 1982. Genetic resistance and its residual effects on for control of the root-knot nematode-Fusarium wilt complex in cotton. Crop Sci. 22:1151–1155. doi:10.2135/cropsci1982.0011183X002200060015x

Sinclair, J.B., and M.N. Shatla. 1962. Stemphylium leaf spot of cotton reported from Louisiana. Plant Dis. Rep. 46:744.

Singh, D., N. Singh, J.S. Gill, and P. Singh. 2012. Management of bacterial blight and leaf spot diseases of upland cotton using *Pseudomonas fluorescens*. J. Cotton Res. Dev. 26:99–100.

Skovgaard, K., H.I. Nirenberg, K. O'Donnell, and S. Rodendahl. 2001. Evolution of *Fusarium oxysporum* f. sp. *vasinfectum* races inferred from multigene genealogies. Phytopathology 91:1231–1237. doi:10.1094/PHYTO.2001.91.12.1231

Smith, A.L., and J.B. Dick. 1960. Inheritance of resistance to Fusarium wilt in upland Sea Island cottons as complicated by nematodes under field conditions. Phytopathology 50:44–48.

Smith, H.E., A.L. Smith, W.E. Cooper, and L. Lett. 1959. Reduction in yield of cotton caused by diseases in 1958. Plant Dis. Rep. 43:368–370.

Smith, S.N., D.L. Ebbels, R.H. Garber, and A.J. Kappelman, Jr. 1981. Fusarium wilt of cotton. In: P.E. Nelson, T.A. Toussoun, and R.J. Cook, editors, Fusarium: diseases, biology, and taxonomy. Pennsylvania State Univ., University Park. p. 29–38.

Smith, S.N., and W.C. Snyder. 1975. Persistence of *Fusarium oxysporum* f. sp. *vasinfectum* in fields in the absence of cotton. Phytopathology 65:190–196. doi:10.1094/Phyto-65-190

Sparnicht, R.H., and R.W. Roncadori. 1972. Fusarium boll rot of cotton: Pathogenicity and histopathology. Phytopathology 62:1381–1386. doi:10.1094/Phyto-62-1381

Starr, J.L., M.J. Jerger, R.D. Martyn, and K. Schilling. 1989. Effects of *Meloidogyne incognita* and *Fusarium oxysporum* f. sp. *vasinfectum* on plant mortality and yield of cotton. Phytopathology 79:640–646. doi:10.1094/Phyto-79-640

Stoughton, R.H. 1931. The influence of environmental conditions on the development of the angular leaf-spot disease of cotton III. The influence of air temperature on infection. Ann. Appl. Biol. 18:524–534. doi:10.1111/j.1744-7348.1931.tb02323.x

Shtienberg, D. 1992. Development and evaluation guidelines for the initiation of chemical control of Alternaria leaf spot in Pima cotton in Israel. Plant Dis. 76:1164–1168. doi:10.1094/PD-76-1164

Shtienberg, D., Y. Kremer, and A. Dinoor. 1993. Influence of physiological age of Pima cotton on the need for fungicide treatment to suppress Alternaria leaf spot. Phytopathology 83:1235–1239.

Streets, R.B., and H.E. Bloss. 1973. Phymatotrichum root rot. Monogr. 8. American Phytopathological Society, St. Paul, MN.

Tarr, S.A.J. 1961. Seed treatment against blackarm disease of cotton in the Sudan. V. Results with seed carrying light and moderate infection. Emp. Cotton Grow. Rev. 38:30–35.

Taubenhaus, J.J., W.N. Ezekiel, and J.F. Fudge. 1932. Incidence of root rot as affected by additions to the soil of fertilizer and other salts to the soil. Tex. Agric. Exp. Stn. Annu. Rep. 45:70.

Thaxton, P.M., T.D. Brooks, and K.M. El-Zik. 2001. Race identification and severity of bacterial blight from natural infestations across the cotton belt. In: Proceedings of the Beltwide Cotton Conferences, Natl. Cotton Counc., Anaheim, CA. 9–13 Jan. 2001. Natl. Cotton Counc., Memphis, TN. p. 137–139.

Toksoz, H., C.S. Rothrock, and T.L. Kirkpatrick. 2009. Efficacy of seed treatment chemicals for black root rot, caused by *Thielaviopsis basicola*, on cotton. Plant Dis. 93:354–362. doi:10.1094/PDIS-93-4-0354

Typas, M.A., and J.B. Heale. 1976. Heterokaryosis and the role of cytoplasmic inheritance in dark resting structure formation in *Verticillium* spp. Mol. Gen. Genet. 146:17–26. doi:10.1007/BF00267978

Ulloa, M., R.B. Hutmacher, R.M. Davis, S.D. Wright, R. Percy, and B. Marsh. 2006. Breeding for Fusarium wilt race 4 resistance in cotton under field and greenhouse conditions. J. Cotton Sci. 10:114–127.

Verma, J.P. 1986. Bacterial blight of cotton. CRC Press, Boca Raton, FL.

Waker, H.J., and A.B. Onkin. 1969. Fertilizing irrigated cotton, Southern High Plains of Texas. Publ. MP-913. Texas Agric. Exp. Stn., College Station.

Walker, M.N. 1928. Soil temperature studies with cotton. III. Relation of soil temperature and soil moisture to the soreshin disease of cotton. Univ. Florida Agric. Exp. Stn. Bull. 197:343–371.

Walker, N.R., T.L. Kirkpatrick, and C.S. Rothrock. 1999. Effects of temperature on histopathology of the interaction between *Meloidogyne incognita* and *Thielaviopsis basicola* on cotton. Phytopathology 89:613–617. doi:10.1094/PHYTO.1999.89.8.613

Wallace, T.P., and K.M. El-Zik. 1989. Inheritance of resistance in three cotton cultivars to the HV_1 isolate of bacterial blight. Crop Sci. 29:1114–1119. doi:10.2135/cropsci1989.0011183X002900050003x

Wang, C., M.L. Dale, and J.K. Kochman. 1999. Studies on a pathogenicity assay for screening cotton germplasms for resistance to *Fusarium oxysporum* f. sp. *vasinfectum* in the glasshouse. Aust. J. Exp. Agric. 39:967–974. doi:10.1071/EA99070

Wang, S.-Y., and J.A. Pinkard. 1973. Spore germination of *Diplodia gossypina* in the presence of carbohydrates and phenolic compounds in relation to boll rot of cotton. Phytopathology 63:1181–1185. doi:10.1094/Phyto-63-1181

Watkins, G.M. 1938a. Cytology of Phymatotrichum root rot of cotton seedlings grown in pure culture. Am. J. Bot. 25:118–124. doi:10.2307/2436858

Watkins, G.M. 1938b. Histology of Phymatotrichum root rot of field grown cotton. Phytopathology 28:195–202.

Wei, Y.-X., H. Zhang, J.-J. Pu, and X.-M. Liu. 2014. First report of target spot on cotton caused by *Corynespora cassiicola* in China. Plant Dis. 98:1006. doi:10.1094/PDIS-12-13-1243-PDN

Weindling, R. 1948. Bacterial blight of cotton under conditions of artificial inoculation. Tech. Bull 956. USDA, Washington, DC.

Weinhold, A.R. 1977. Population of *Rhizoctonia solani* in agriculture soil determined by a screening procedure. Phytopathology 67:566–569. doi:10.1094/Phyto-67-566

Wheeler, T.A., J.P. Bordovsky, J.W. Keeling, B.G. Mullinix, Jr., and J.E. Woodward. 2012. Effects of crop rotation, cultivar, and irrigation and nitrogen rate on Verticillium wilt in cotton. Plant Dis. 96:985–989. doi:10.1094/PDIS-02-11-0111-RE

Wheeler, T.A., J.R. Gannaway, H.W. Kaufman, J.K. Dever, J.C. Mertley, and J.W. Keeling. 1997. Influence of tillage, seed quality, and fungicide seed treatments on cotton emergence and yield. J. Prod. Agric. 10:394–400. doi:10.2134/jpa1997.0394

Wheeler, T.A., J.E. Woodward, and B.G. Mullinix, Jr. 2010. Effect of seeding rate on Verticillium wilt incidence, yield and value for three cotton cultivars. J. Cotton Sci. 14:173–180.

Wickens, G.M. 1961. Bacterial blight of cotton: Its influence on yield. Emp. Cotton Grow. Rev. 38:241–257.

Wilhelm, S. 1955. Longevity of the Verticillium wilt fungus in the laboratory and field. Phytopathology 45:180–181.

Wilhelm, S., J.E. Sagen, and H. Tietz. 1974. Resistance to Verticillium wilt in cotton: Sources, techniques of identification, inheritance trends, and the resistance potential of multiline cultivars. Phytopathology 64:924–931. doi:10.1094/Phyto-64-924

Williams, K.R., T. Kirkpatrick, B. Bond, and J. Jaggers. 2002. Rot-knot nematode (*Meloidogyne incognita*) control and carry-over effects of Telone II (1,3-dichloropropene) in cotton in

Ashley County, Arkansas. In: Proceedings of the Beltwide Cotton Conferences, Natl. Cotton Counc., Atlanta, GA. 8–12 Jan. 2002. Natl. Cotton Counc., Memphis, TN.

Woodward, J.E., J.H. Ramirez, and G.D. Morgan. 2014. Effect of flutriafol rate and application method on cotton root rot and lint yield. In: Proceedings of the Beltwide Cotton Conferences, Natl. Cotton Counc., New Orleans, LA. 6–8 Jan. 2014. Natl. Cotton Counc., Memphis, TN. p. 220.

Woodward, J.E., T.A. Wheeler, M.G. Cattaneo, S.A. Russell, and T.A. Baughman. 2011. Evaluation of soil fumigants for the management of Verticillium wilt in Texas peanut. Plant Health Progress. doi:10.1094/PHP-2011-0323-02-RS.

Wright, R.J., P.M. Thaxton, K.M. El-Zik, and A.H. Paterson. 1998. D-subgenome bias of Xcm resistance genes in tetraploid *Gossypium* (Cotton) suggests that polyploidy formation has created novel avenues for evolution. Genetics 149:1987–1996.

Xiao, J., D.D. Fang, M. Bhatti, B. Hendrix, and R. Cantrell. 2010. A SNP haplotype associated with a gene resistant to *Xanthomonas axonopodis* pv. *malvacearum* in upland cotton (*Gossypium hirsutum* L.). Mol. Breed. 25:593–602. doi:10.1007/s11032-009-9355-y

Young, V.H. 1928. Cotton wilt studies: 1. Relation of soil temperature to the development of cotton wilt. Bull. 226. Arkansas Agricultural Experiment Station, Fayetteville.

Zhai, J., Z. Xia, W. Liu, X. Jiang, and X. Huang. 2013. Genomic sequencing globally identifies functional genes and potential virulence-related effectors of *Xanthomonas axonompodis* pv. *malvacearum*. Eur. J. Plant Pathol. 136:657–663.

Cotton Insect Pest Management

Randall G. Luttrell,* Tina Gray Teague, and Michael J. Brewer

Abstract

Cotton production is challenged worldwide by a diversity of arthropod pests that require management to prevent or reduce crop damage. Advances in arthropod control technologies and improved insect and crop management systems have dramatically reduced levels of arthropod damage and the need for insecticidal control of several key pests in cotton over the past 30 yr. The near-complete eradication of boll weevil (*Anthonomous grandis* Boheman) from the United States and the worldwide deployment of transgenic cottons expressing insecticidal proteins of *Bacillus thuringiensis* (Bt cotton) are among the most notable advances. Insects and mites are highly adaptable, and new pest problems have emerged that continue to challenge or limit cotton production and profitability. In this update of cotton insect pest management, we review evolving technologies for cotton arthropod management, the purposeful deployment of resistance management programs, dynamic responses of arthropod pests to management actions, and improved integration of insect and crop management systems. We emphasize ecological context, including varying spatial scales of management, and the adaptive capacity of the arthropods impacted by management decisions as continuing influences on humankind's effort to limit arthropod damage to cotton.

Extraordinary changes in management of insects and mites attacking cotton (*Gossypium* L.) have transpired since Ridgway et al. (1984) reviewed cotton protection practices in the previous Agronomy Monograph, *Cotton*. Reviews published subsequent to that period (Frisbie and Adkisson, 1985; Frisbie et al., 1989; Luttrell et al., 1994; Luttrell, 1994; Fitt, 1994; Ramalho, 1994; Sugonyaev, 1994; Smith, 1998; Fitt, 2000; Hardee and Harris, 2003; Wu and Guo, 2005; Allen, 2008; Spurgeon, 2008; Wu et al., 2008; Naranjo and Luttrell, 2009) provide detailed information on cotton insects and chronicle changes in insect management strategies in response to new pests, advances in control technologies, and changing management structures associated with insect resistance management and government regulations. Of particular relevance are comprehensive reviews by Matthews and Tunstall (1994), King et al. (1996), and Dickerson et al. (2001).

Abbreviations: IPM, integrated pest management; NAWF, nodes above white flower.

Randall G. Luttrell, USDA-ARS, Southern Insect Management Research Unit, 141 Experiment Station Rd., Stoneville, MS 38776. Tina G. Teague, University of Arkansas, Division of Agriculture and Arkansas State University, 245 Agriculture Bldg., Jonesboro, AR 72467 (tteague@astate.edu). Michael J. Brewer, Texas A&M AgriLife Research, 10345 Hwy. 44, Corpus Christi, TX 78406 (mjbrewer@ag.tamu.edu). *Corresponding author (randy.luttrell@ars.usda.gov).

doi:10.2134/agronmonogr57.2014.0072

Cotton. 2nd. ed. David D. Fang and Richard G. Percy, editors. Agronomy Monograph 57.

Any modification in production practices or insect management approaches may impact the abundance and diversity of insects in an agronomic cropping system. Cotton production has a long history of evolving insect pest problems—for better or worse—often triggered in reaction to insecticides (Luttrell et al., 1994). Ridgway et al. (1984) concluded "that the world cotton crop will continue to be threatened by insect pests, and insecticides will continue to be essential for protecting the crop." They also suggested that insecticides could be used more selectively in conjunction with improved surveillance and decision-making approaches. The most important key pests recognized at that time were the heliothines referred to as *Heliothis* spp. in the original review [*Helicoverpa zea* (Boddie), *Heliothis virescens* (F.), *Helicoverpa armigera* (Hubner)], pink bollworm [*Pectinophera gossypiella* (Saunders)], boll weevil, and Egyptian cotton leafworm [*Spodoptera literoralis* (Boisduval)].

In this current review, we continue to emphasize the ecological context of cotton insect management with a special focus on the adaptive capacity of arthropods targeted directly or indirectly by control activities. We also recognize the intricate ecological connections of insects and mites to cotton and the need to integrate pest management practices into more sustainable crop production systems. The managed cotton production environment affords resources that fuel insect and mite population growth. Monitoring plant development and cotton plant response to insect injury in conjunction with routine assessments of arthropod populations provides the critical information needed to maintain profitable and efficient cotton production. Evolving management practices encourage dynamic pest problems in cotton, and socio-economic influences often define boundaries of management capabilities to address new pest problems. Linkages to private and government research and developments, services, regulations, and incentives have been notable in past advancements of cotton entomology, given cotton's place as a global commodity and its worldwide cultivation.

Major advancements have been made in the management of the key pests listed by Ridgway et al. (1984). The pest status of all six of those specifically referenced has been dramatically reduced through deployment of new technologies and/or coordinated management systems. Transgenic cottons expressing insecticidal proteins derived from Bt cotton have been developed that provide high levels of control of the five lepidopteran key pests listed by Ridgway et al. (1984). The remaining key pest in the list, boll weevil, has been eradicated from much of US cotton and in North America is now only a problem in the Lower Rio Grande Valley and adjoining cotton production areas of Mexico (Allen, 2008). This dramatic shift in the pest status of key pests reflects significant advances in insect control capabilities and an evolution of crop management practices over the past 30 yr. These changes have led to substantial economic and ecological benefits, including reductions in use of traditional broad-spectrum insecticides in many cotton production regions of the world (Naranjo, 2011), but also an ironic increase in other pests released from insecticide control originally targeting key pest species (Lu et al., 2010). In this update of Ridgway et al. (1984), we examine the suggested future management options envisioned in the previous review and provide an overview of additional changes not envisioned in 1984. Just as the original work of Ridgway et al. (1984) has become outdated, we recognize that adaptive insects as well as new technologies and evolving scientific capacity can impact dynamic production systems and will soon render the current review out of date. As a benchmark of current approaches, capacity, and concepts, we highlight key changes in the pest complexes, management approaches, and ecological influences that continue to evolve insect management capabilities.

Evolving Technology, Resistance Management, and Dynamic Pest Problems

A number of major insect control technologies were developed over the last three decades that provide superior levels of insect pest control as compared to those available when Ridgway et al. (1984) assessed cotton insect management options. Cotton yields have consistently increased during this period, at least in part because of the plant protection provided by insect management options that include the use of new insecticides and transgenic plants. Production costs also have risen, but to a large extent, increased yield has offset increasing production costs.

A comparison of 1983 and 2013 cotton insect loss data for the United States is provided in both Table 1 and Fig. 1 as an example of the changing status and influence of insect and mite management on cotton production. On the basis of calculations from these data, insect control costs were about US$0.02 per kilogram of lint produced for the United States in 1983, a year before the previous review. They were $0.04 in 2013. Adjusting for the US inflation rate of about 134% over this 30-yr period, average insect control costs across this time period are about the same or currently slightly lower, but regional differences illustrate the influence of changing insect pest problems and management options on cotton

Table 1. Comparative percent of crop loss, insect control costs, hectares harvested, and yield of US cotton in 1983 and 2013.†

State	Crop loss to insects		Insect control costs		Hectares harvested		Yield	
	1983	2013	1983	2013	1983	2013	1983	2013
	%		US$ ha^{-1}		ha		kg lint ha^{-1}	
AL	13.60	4.08	$24.28	$26.75	87,006	145,369	451	947
AZ	7.40	5.61	$29.95	$52.99	128,485	66,885	1,193	1,707
AR	5.30	7.34	$9.83	$54.94	125,450	128,688	555	1,255
CA	3.30	0.32	$7.61	$36.56	390,514	112,501	1,104	1,773
FL	15.70	6.43	$55.00	$29.80	3,844	50,585	696	841
GA	9.10	3.10	$21.04	$24.72	46,538	540,245	562	947
KS	–‡	0.43	–	$6.47	–	10,522	–	704
LA	9.40	5.34	$23.39	$42.53	165,918	51,799	708	1,513
MS	6.50	8.10	$21.89	$63.59	273,158	129,497	717	1,222
MO	2.30	5.00	$2.83	$45.22	37,635	105,216	423	1,054
NM	20.00	6.92	$6.07	$13.77	24,281	17,101	655	1,401
NC	11.50	3.59	$23.07	$27.69	23,876	203,148	410	807
OK	17.10	0.87	$15.38	$10.15	121,403	69,645	251	673
SC	12.40	3.52	$32.37	$21.81	27,923	103,193	428	785
TN	5.60	8.03	$3.44	$27.23	87,006	95,099	375	984
TX	7.80	0.74	$3.93	$12.84	1,344,664	1,335,438	377	687
VA	9.00	1.25	$10.12	$18.59	121	33,791	359	1,121
United States (all states)	7.67	2.68	$10.52	$24.93	2,887,823	3,198,721	574	902

† Calculated from survey data reported by Head (1984) and Williams (2014) unadjusted for inflation. The cumulative inflation rate during this 30-yr period was approximately +130%, making the United States (all states) insect control cost similar across this 30-yr period.

‡ No cotton production statistics were reported for Kansas in 1983 (Head, 1984).

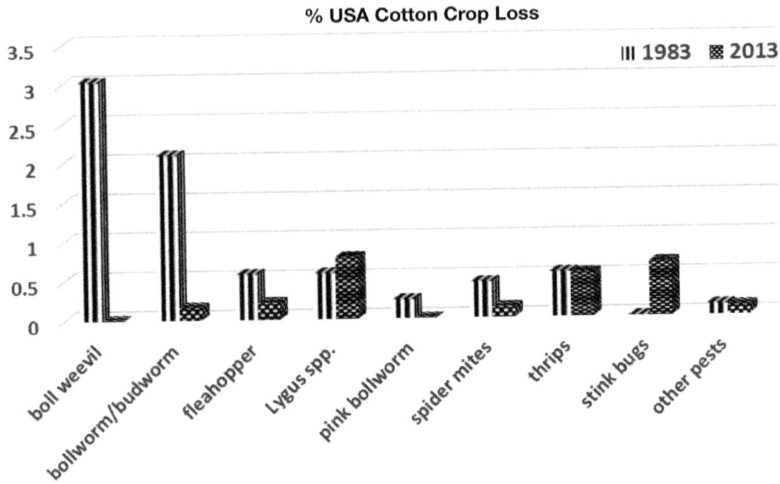

Fig. 1. Relative impacts of different insect pests on % crop loss of US cotton in 1983 and 2013 (based on summaries of cotton insect loss data Head [1984] and Williams [2014]).

profitability. In 1983, insect control costs per kilogram of lint produced were estimated at US$0.01 per kilogram for California and Missouri; however, costs were US$0.10 for South Carolina and Florida, states experiencing the initiation of boll weevil eradication efforts. In 2013, insect control costs per pound of lint produced ranged from a low of US$0.01 in Kansas, New Mexico, and Oklahoma to US$0.05 to $0.06 in Arkansas, Mississippi, and Missouri where insect control costs for tarnished plant bug [*Lygus lineolaris* (Palisot de Beauvois)] were elevated following boll weevil eradication and wide-scale acceptance of Bt cottons. While insect control costs adjusted for inflation are about the same or slightly reduced over a 30-yr period, the amount of cotton produced per managed unit has increased. Across the entire US Cotton Belt, 2013 yields were 328 kg of lint greater per hectare than 1983 yields on similar total harvested area (2,887,823 ha in 1983 and 3,198,721 ha in 2013). Dramatic changes of yield greater than 560 kg of lint per hectare were observed for Arkansas, California, Louisiana, Missouri, New Mexico, Tennessee, and Virginia. Production costs, including those associated with insect and mite management, have also had an influence on regional planting of cotton. Small increases in planted area, based on comparisons of the 1983 and 2013 data (Table 1), were observed in Alabama, Arkansas, California, Kansas, Oklahoma, Tennessee, and Virginia. Corresponding with the eradication of boll weevil and introduction of transgenic cotton, large expansions of area planted to cotton occurred in Florida (13-fold), Georgia (11-fold), North Carolina (8-fold), South Carolina (3-fold), and Missouri (2-fold). Boll weevil was a major limitation to cotton production in these southern states in the 1980s. Reductions in acreage were observed in Arizona, California, Louisiana, Mississippi, New Mexico, Oklahoma, and Texas. The largest reductions were those in Arizona, California, Louisiana, Mississippi, and Oklahoma where 2013 production levels were 28 to 57% of those recorded in 1983.

During the past three decades, there have been major shifts in the relative abundance and importance of major insect pests. Examples of these changes in the US production systems are illustrated in Fig. 1 where the overall crop losses from insects and mites declined from 7.6% of the harvestable crop in 1983 to 2.6% of the harvestable crop in 2013. For the United States, there has been an elimination of 3% crop loss associated with boll weevil and a 10-fold reduction in crop losses associated with bollworm and tobacco budworm (from 2.1% in 1983 to 0.2% in 2013). Crop losses due to pink bollworm, spider mites (*Tetranychus* spp.), and fleahoppers [primarily *Pseudatomoscelis seriatus* (Reuter)] were less in 2013 than in 1983, but still evident and recognized. Increased losses were observed due to *Lygus* spp. [*L. lineolaris* and *L. hesperus* (Knight)] and other plant bugs [*Creontiades signatus* Distant and *Creontiades dilutus* Stal, *Neurocolpus leucopterus* (Say), and other phytophagous Miridae], and stink bugs [several *Euschistus* spp., especially *Euschistus servus* (Say), *Nezara viridula* (L.), *Chinavia halaris* (Say), and a number of other Pentatomidae], which have become the primary pest problems in many production regions. An increase in problems with sucking pest species, especially hemipteran pests, has been observed worldwide. This is discussed in more detail below.

Changes in the diversity and intensity of insect pest problems in the past 30 yr were not new or unexpected, but they serve as a continuing reminder of the interrelationships among organisms in the cotton production system and the importance of ecological context in insect pest management (e.g., Pedigo and Rice, 2009). Perhaps the greatest changes were those associated with successful eradication of boll weevil, historically the most important pest of cotton in the United States, and the development and subsequent wide-scale deployment of Bt cottons that provide high levels of control of key lepidopteran pest species around the world (Naranjo and Luttrell, 2009). The boll weevil was successfully eradicated from US cotton via a structured and coordinated system of region-wide management efforts (Dickerson et al., 2001; Hardee and Harris, 2003; Allen, 2008). Foundational to this approach was area-wide, well-coordinated use of boll weevil pheromone-baited traps and foliar insecticides, joint planning by federal and state governmental entities, and federal and state financial support combined with grower-approved taxing and regulatory authority. The high planning, regulatory, and financial "buy-in" were key to conducting a multiple year effort that led to eradication achieved in progressive fashion from the eastern US cotton growing region (now with eradication lasting for several decades and monitoring continuing) into southern Texas (where eradication sprays are still being conducted and boll weevil has not been detected in some areas for 2 to 6 yr) (Dickerson et al., 2001; Allen, 2008). Boll weevil has now invaded the expanding cotton production system of Brazil and other parts of South America (Ramalho and Wanderly, 1996; Showler, 2009; Oliveira et al., 2013). The coordinated effort of the boll weevil eradication effort of the southern United States may be instructive for other areas, but ecological differences as well as limitations in technical and regulatory governmental services may limit the duplication of this effort in those systems.

The development of Bt cottons and the rapid widespread adoption of these insecticidal plants had a dramatic impact on cotton insect management and cotton production. It was a historic application of biotechnology to agriculture, and it is undoubtedly one of the most important technological advancements in insect control worldwide (Qaim and Zilberman, 2003; Purcell and Perlak, 2004). As of 2013, 12 countries grow Bt cottons, with 7 growing less than 100,000 ha, suggesting the technology's availability is expanding into more local production areas (James, 2013). Other non-lepidopteran insect pests are beginning to be targeted by this approach, such as mirid pests (Chougule and Bonning, 2012), and coleopteran-active proteins are widely used to control rootworms in corn (*Zea mays* L.), but commercial use of Bt transgenes in cotton has been limited to those with activity against Lepidoptera (reviewed below). Bt cotton has dramatically reduced crop damage from major lepidopteran pests in the United States, including the tobacco budworm, the bollworm, beet armyworm, fall armyworm, and the pink bollworm. Similar results have been observed with lepidopteran pests in cotton production systems worldwide (Deutscher et al., 2005; Fitt, 2003; Kathage and Qaim, 2012; Qaim and Zilberman, 2003; Wu and Guo, 2005; Luttrell and Jackson, 2012), especially those impacted by *H. armigera*.

Evolving Insect Control Technologies

Cotton production currently has a diversity of different insect management tools that are highly effective as evidenced by the reduction in insect losses reported for the United States (Fig. 1). Table 2 provides an example of the insecticides currently available and recommended for control of cotton insects and mites. The list includes 37 insecticides or miticides specifically recommended (Catchot, 2014; Studebaker, 2014), with an additional 31 labeled for use in Mississippi but not currently recommended. Recommended insecticides for control of cotton pest species are similar in other states. Arkansas and Mississippi information was used because it represents a major production region with diverse pests and a strong reliance on chemical control measures. The recommended 37 insecticides and miticides represent 16 main groups or primary sites of action materials and 17 different chemical subgroups (or classes) based on mode of action, as categorized by the International Resistance Action Committee (2012). The Insecticide Resistance Advisory Committee is a group representing the global crop protection industry. It was organized in 1984 with the goal of facilitating communications and providing advice on insecticide resistance management. Ridgway et al. (1984) listed 36 different insecticides or miticides from 5 major insecticide classes (chlorinated hydrocarbons, organophosphates, carbamates, pyrethroids, and inorganics). Herzog et al. (1996) provide transitional information between the 1983 review and the current summary, and Parencia et al. (1983) provide additional detail to the historical availability of cotton insecticides summarized by Ridgway et al. (1984). Of special benchmark importance is the current diversity of new materials that provide pest managers with a range of different tools and modes of action. Current insecticide and miticide groups not available in 1983 include the neonicotinoids, sulfoxaflur, spinosyns, avermectins, pyriproxyfen, flonicamid, etoxazole, benzoylureas, diacylhydrazines, METI acaracides, indoxacarb, tetronic and tetramic acid derivatives, and diamides. Government agencies, including extension specialists in most cotton production regions, provide similar lists of available and recommended insecticides.

The development of Bt cotton was not envisioned by Ridgway et al. (1984). Rapid adoption and widespread use of Bt cottons following commercialization of Bollgard (Monsanto Company) cotton in 1995 dramatically changed cotton insect management in the United States (Luttrell and Jackson, 2012) and around the world (Edge et al., 2001; Purcell and Perlak, 2004). Major commercial traits with insecticidal activity include Bollgard that expresses Cry1Ac toxin, Bollgard II (Monsanto Company) that expresses Cry1Ac and Cry2Ab2, WideStrike

Table 2. Insecticides registered and recommended by federal and state sanctioned grower advisory bodies (Extension Services and Ministries of Agriculture) for control of insect and mite pests of cotton in 2014. Refer to recommendation guides for specific guidance and restrictions.†

IRAC main group or primary site of action‡	Chemical sub-group (exemplifying mode of action [class])‡	Insecticide/ acaricide common name	Recommended target pest groups
Group 1A	carbamates	aldicarb	thrips, western flower thrips
Group 1A	carbamates	methomyl	bollworm, fall armyworm, loopers, tobacco budworm
Group 1A	carbamates	oxamyl	fleahoppers, plant bugs, stink bugs
Group 1A	carbamates	thiodicarb	beet armyworm, bollworm, cutworms, fall armyworm, loopers, tobacco budworm
Group 1B	organophosphates	acephate	cutworms, fall armyworm, fleahoppers, plant bugs, stink bugs, thrips, whiteflies
Group 1B	organophosphates	dicrotophos	aphids, fleahoppers, plant bugs, stink bugs, thrips
Group 1B	organophosphates	dimethoate	fleahoppers, plant bugs, thrips
Group 1B	organophosphates	malathion	fleahoppers, plant bugs
Group 1B	organophosphates	methyl parathion	stink bugs
Group 1B	organophosphates	profenofos	bollworm, fall armyworm, tobacco budworm
Group 3A	pyrethroids	beta-cyfluthrin	bollworm, cutworm, tobacco budworm, stink bugs
Group 3A	pyrethroids	bifenthrin	bollworm, cutworm, spider mites, stink bugs, tobacco budworm, banded-winged whitefly
Group 3A	pyrethroids	cyfluthrin	bollworm, cutworm, stink bugs
Group 3A	pyrethroids	cypermethrin	bollworm, cutworm, tobacco budworm
Group 3A	pyrethroids	esfenvalerate	bollworm, cutworm, stink bugs, tobacco budworm
Group 3A	pyrethroids	gamma-cyhalothrin	bollworm, cutworm, stink bugs, tobacco budworm
Group 3A	pyrethroids	lambda-cyhalothrin	bollworm, cutworm, stink bugs, tobacco budworm
Group 3A	pyrethroids	zeta-cypermethrin	bollworm, cutworm, stink bugs, tobacco budworm
Group 4A	neonicotinoids	acetamiprid	aphids, fleahoppers, plant bugs, banded-winged whitefly, silver leaf whitefly
Group 4A	neonicotinoids	clothianidin	fleahoppers, plant bugs
Group 4A	neonicotinoids	imidacloprid	aphids, fleahoppers, plant bugs, thrips
Group 4A	neonicotinoids	thiamethoxam	aphids, fleahoppers, plant bugs, banded-winged whitefly
Group 4C	sulfoxaflur	sulfoxaflor	aphids, fleahoppers, plant bugs
Group 5	spinosyns	spinosad	beet armyworm, bollworm, fall armyworm, loopers, tobacco budworm
Group 6	avermectins, milbemycins	abamectin	spider mites
Group 6	avermectins, milbemycins	emamectin benzoate	beet armyworm, bollworm, fall armyworm, loopers, tobacco budworm
Group 7C	pyriproxyfen	pyriproxyfen	silver leaf whitefly
Group 9C	flonicamid	flonicamid	aphids, fleahoppers, plant bugs
Group 10B	etoxazole	etoxazole	spider mites
Group 12 C	propargite	propargite	spider mites
Group 15	benzoylureas	novaluron	fall armyworm, fleahoppers, plant bugs
Group 18	diacylhydrazines	methoxyfenozide	beet armyworm, fall armyworm, loopers
Group 21A	met1 acaricides and insecticides	fenpyroximate	spider mites
Group 22A	indoxacarb	indoxacarb	beet armyworm, bollworm, fall armyworm, loopers, tobacco budworm
Group 23	tetronic and tetramic acid derivatives	spiromesifen	spider mites, banded-winged whitefly
Group 28	diamides	chlorantraniliprole	beet armyworm, bollworm, fall armyworm, loopers, tobacco budworm
Group 28	diamides	flubendiamide	beet armyworm, bollworm, fall armyworm, loopers, tobacco budworm

† Insecticides/acaricides were identified from a Crop Data Management Systems search of insecticides labeled for use on cotton, using as a gateway of registered products for use in Mississippi during 2014 (http://premeir.cdms.net). The resulting list includes those that correspond to selected cotton insect control recommendations for Arkansas (Studebaker, 2014), Mississippi (Catchot, 2014), and other southern states of the United States and other selected countries. Labeled insecticides for cotton were not included in the recommendations, but with selected special applications such as growing cotton certified as organic by various agencies, include aluminum phosphide, azadirachin, *Beauveria bassiana* strain GHA, chloropicrin, dicofol, diflubenzuron, dinotefuran,disulfoton, fenpropathrin, hydramethylon, kaolin, methidathion, methoprene, mineral oil, naled, paraffinic oil, petroleum oil, phorate, piperonyl butoxide, spinetoram, sulfur, and tebufenozide.

‡ Grouping and subgrouping based on International Resistance Action Committee (2012).

(Dow AgroSciences) that expresses Cry1F and Cry1Ac, WideStrike 3 (Dow AgroSciences) that expresses Cry1F, Cry1Ac, and Vip3A(a), and TwinLink (Bayer CropScience) that expresses Cry1Ab and Cry2Ae insecticidal proteins. Bollgard III cotton expressing Cry1Ac, Cry2Ab2, and Vip3A(a) insecticidal proteins has been registered for use in Australia. Table 3 provides an overview of 40 different transgenic cotton events that have been approved by different governments for cultivation. Decisions to proceed with cultivation of traits may proceed for different reasons including private testing and for commercial development. Regulatory approvals for transgenic cotton events as food or feed components are tracked worldwide and reported by the International Services for Acquisition of Agri-Biotech Applications (http://www.isaa.org).

Although not directly affecting cotton production, European Union regulations on genetically engineered agricultural products presents challenges to implementing the technology (Davison, 2010) and represents the importance of public policy and perception toward evolving technologies in agriculture. International treaties on free trade and government "green payments" versus "commodity payments" will also evolve and likely shift in impact and complexity in the future (Brewer and Goodell, 2012). Also outside the scope of this chapter for in-depth discussion, sanitation standards and border inspection technologies are anticipated to be developed given the worldwide interest in detection of elicit materials, including new potential pest species. This has direct implications to the use of transgenic traits and invasive insect and mite pests (Magarey et al., 2009). With good planning and policy, reduction in the spread of cotton insect pests can benefit from these science and technology advancements. Invasive pest problems in cotton are reviewed below.

Insect Resistance Management

Concepts of insect resistance management and purposeful manipulations of insect resistance traits have become integral components of insect management policy (National Research Council, 1986) and cotton production practices over the past three decades. Insect resistance to insecticides and the impact of insect adaptation on cotton insect management have long been recognized as major influences in our ability to protect cotton from insect damage. Conceptual interest in managing resistance in cotton insects, and in other insect pests, intensified with the widespread use of pyrethroid insecticides in the 1980s. The pyrethroids were highly effective against previously difficult to control insects and generally had longer residual activity than the widely used organophosphate and carbamate insecticides used at that time. The resulting increases in cotton yields were followed by expanded use of the pyrethroids. Selection for resistance genes was intense. Populations of pyrethroid-resistant tobacco budworm were documented in Texas in 1985 and in Mississippi and Louisiana in 1986 (reviewed by Luttrell, 1994). Additional rapid changes were also observed in the spectrum and an abundance of other pest species in the system. Aphids, spider mites, and true bugs became more of a problem, and in some ways the cyclic problem of the "pesticide treadmill" originally associated with overuse of chlorinated hydrocarbon insecticides in the 1950s and 1960s (Metcalf, 1980) was reappearing. However, there were important differences. Pyrethroids were not highly effective against boll weevil, and organophosphate insecticides were still used to control boll weevil and other pests. Pyrethroids, although highly toxic to target insects and of

Table 3. Recorded government approvals for the cultivation of transgenic cotton events expressing insecticidal proteins, based on the September 2014 summary of information from the International Service for the Acquisition of Agri-Biotech Applications.

Event name	Commercial technology (trade name) involved	Developer	Method of gene transfer	Insecticidal genes involved	Country approving cultivation (year)
COT102 × MON15985 × MON88913	Bollgard III × Roundup Ready Flex	Monsanto Company	conventional breeding	*vip3A(a)*, *cry1Ac*, *cry2Ab2*	Australia (2014)
281-24-236	–	Dow AgroSciences	agrobacterium	*cry1F*	Mexico (2004), United States (2004)
281-24-236 × 3006-210-23 (MXB-13)	WideStrike Cotton	Dow AgroSciences	conventional breeding	*cry1F*, *cry1Ac*	Australia (2009), Brazil (2009), Costa Rica (2009), Mexico (2004)
3006-210-23	–	Dow AgroSciences	agrobacterium	*cry1Ac*	Mexico (2004), United States (2004)
3006-210-23 × 281-24-236 × MON1445	WideStrike Roundup Ready Cotton	Monsanto Company and Dow Agrosciences	conventional breeding	*cry1F*, *cry1Ac*	Japan (2006), Mexico (2006)
3006-210-23 × 281-24-236 × MON88913	Widestrike Roundup Ready Flex Cotton	Monsanto Company and Dow Agrosciences	conventional breeding	*cry1F*, *cry1Ac*	Costa Rica (2009), Japan (2006)
3006-210-23 × 281-24-236 × MON88913 × COT102	Widestrike × Roundup Ready Flex × VIPCOT Cotton	Dow AgroSciences	conventional breeding	*cry1Ac*, *vip3A(a)*, *cry1F*	–
31707	BXN Plus Bollgard Cotton	Monsanto Company	agrobacterium	*cry1Ac*	–
31803	BXN Plus Bollgard Cotton	Monsanto Company	agrobacterium	*cry1Ac*	–
31807 × 31808	–	Monsanto Company	conventional breeding	*cry1Ac*	–
31807	BXN Plus Bollgard Cotton	Monsanto Company	agrobacterium	*cry1Ac*	Japan (1998), United States (1997)
31808	Bollgard III × Roundup Ready Flex	Monsanto Company	agrobacterium	*cry1Ac*	Japan (1998), United States (1997)
42317	BXN Plus Bollgard Cotton	Monsanto Company	agrobacterium	*cry1Ac*	–
BNLA-601	–	Central Institute and University of Agricultural Sciences Dharwad (India)	–	*cry1Ac*	India (2008)
COT102 (IR102)	VIPCOT Cotton	Syngenta	agrobacterium	*vip3A(a)*	Japan (2012), Mexico (2010), South Korea (2014), United States (2011)
COT102 × COT67B	–	Syngenta	conventional breeding	*cry1Ab*, *vip3A(a)*	Costa Rica (2009)
COT102 × COT67B × MON88913	VIPCOT Roundup Ready Flex Cotton	Syngenta and Monsanto Company	conventional breeding	*vip3A(a)*, *cry1Ab*	Costa Rica (2009)

Table continued.

Table 3. Continued.

Event name	Commercial technology (trade name) involved	Developer	Method of gene transfer	Insecticidal genes involved	Country approving cultivation (year)
COT102 × MON15985	Bollgard III	Monsanto	conventional breeding	*cry1Ac*, *cry2Ab2*	Australia (2014)
COT67B (IR67B)	–	Syngenta		*cry1Ab*	Japan (2012), United States (2011)
Event1	JK1	JK AgriGenetics (India)	microparticle bombardment	*cry1Ac*	India (2006)
GFM Cry1A	–	Nath Seeds/Global Transgenes Ltd (India)	pollen-tube pathway	*cry1Ab-Ac*	India (2006), Pakistan (2012)
GHB119	–	Bayer CropScience	agrobacterium	*cry2Ae*	Japan (2012), United States (2011)
GHB614 × LLCotton25 × MON15985	–	Bayer CropScience	conventional breeding	*cry1Ac*, *cry2Ab2*	Japan (2011)
GHB614 × MON15985	–	Bayer CropScience	conventional breeding	*cry1AC*, *cry2Ab2*	–
GHB614 × T304-40 × GHB119	Glytol × Twinlink	Bayer CropScience	conventional breeding	*cry1Ab*, *cry2Ae*	Brazil (2012)
GK12	–	Chinese Academy of Agricultural Science	agrobacterium	*cry1Ab-Ac*	China (1997)
LLCotton25 × MON15985	Fibermax Liberty Link Bollgard II	Bayer CropScience	conventional breeding	*cry1Ac*, *cry2Ab2*	Australia (2006), Japan (2007)
MLS 9124		Metahelix Life Sciences Put. Ltd (India)	agrobacterium	*cry1C*	India (2009)
MON1076	Bollgard Cotton	Monsanto Company	agrobacterium	*cry1Ac*	South Africa (1997), United States (1995)
MON15985	Bollgard II Cotton	Monsanto Company	microparticle bombardment	*cry1Ac*, *cry2Ab2*	Australia (2002), Brazil (2009), Barkina Faso (2009), Costa Rica (2009), India (2006), Japan (2004), Mexico (2003), South Africa (2003), United States (2002)
MON15985 × MON1445	Roundup Ready Bollgard II Cotton	Monsanto Company	conventional breeding	*cry2Ab2*, *cry1Ac*	Australia (2002), Costa Rica (2009), Japan (2005)
MON531	Bollgard Cotton, Ingard	Monsanto Company	agrobacterium	*cry1Ac*	Argentina (1998), Australia (2003), Brazil (2005), China (2014), Columbia (2003), Costa Rica (2008), India (2002), Japan (2004), Mexico (1996), Pakistan (2010), Paraguay (2007), South Africa (1997), United States (1995)
MON531 × MON1445	Roundup Ready Bollgard Cotton	Monsanto Company	conventional breeding	*cry1Ac*	Argentina (2009), Australia (2003), Brazil (2009), Columbia (2007), Costa Rica (2009), Japan (2004), Paraguay (2013), South Africa (2005)

Event name	Commercial technology (trade name) involved	Developer	Method of gene transfer	Insecticidal genes involved	Country approving cultivation (year)
MON757	Bollgard Cotton	Monsanto Company	agrobacterium	*cry1Ac*	Japan (2005), South Africa (1997), United States (1995)
MON88913 × MON15985	Roundup Ready Flex Bollgard II Cotton	Monsanto Company	conventional breeding	*cry2Ab2*, *cry1Ac*	Australia (2006), Brazil (2012), Columbia (2007), Costa Rica (2009), Japan (2006), Mexico (2006), South Africa (2007)
Ngwe Chi 6 Bt	–	Cotton and Sericulture Department (Myanmar)	–	–	Myanmar (2006)
SGK321	–	Chinese Academy of Agricultural Science	pollen-tube pathway	*cry1Ac*	China (1999)
T303-3	–	Bayer CropScience		*cry1Ab*	United States (2012)
T304-40	–	Bayer CropScience	agrobacterium	*cry1Ab*	Japan (2012), United States (2011)
T304-40 × GHB119	TwinLink Cotton	Bayer CropScience	conventional breeding	*cry1Ab*, *cry2Ae*	Brazil (2011)

relatively long residual activity, were not as damaging to the environment and higher trophic levels in the ecosystem as previous insecticide classes. Pest managers and agricultural scientists also had knowledge of the previous problems and sought options to lessen the ecological impacts and preserve the effective use of these important management tools via preservation of susceptible genes in targeted pest populations.

Voluntary resistance management programs to manage resistance to pyrethroids were organized around the world during the 1980s with the cooperation of university researchers, extension specialists, government scientists and regulators, and most major industry groups. Organized cooperation within the cotton industry was extremely important, and close collaborations between public sector scientists and industry scientists were critical components of implementing coordinated resistance management efforts. The primary pests targeted were *H. virescens* in North America and *H. armigera* in Australia and Asia. A "window strategy" for use of pyrethroids was widely adopted and based on observations of declining frequencies of resistance in overwintering populations of tobacco budworm. The concept of managing resistance alleles became solidly fixed in cotton management systems. It also became a primary example for an expanding academic and regulatory interest in resistance management systems for other crops and other pesticides. An important practical outcome of the work was the real-time use of the adult vial technique for monitoring resistance levels in adult males in pheromone traps (Plapp et al., 1987, 1990). University and USDA researchers routinely monitored resistance levels and provided real-time information to extension specialists who disseminated information to growers on region-wide levels of resistance being observed. This information was used by consultants and farmers to determine the type or class of insecticide that should be utilized. Related developments in real-time communication of scientific information were the expansion of the adult vial concept to track insecticide resistance in tarnished plant bug (Snodgrass and Scott, 2000; Snodgrass et al., 2008, 2009), projections of fungal infections in cotton aphids (*Aphis gossypii* Glover) that alerted growers and consultants of the possible options to delay insecticide applications (Hollingsworth et al., 1995), and development of the sophisticated mode of action nomenclature by the International Resistance Action Committee to classify and characterize different insecticide modes of action (Table 2). The International Resistance Action Committee continues to be a strong supporter and industry-wide advocate for practical resistance management programs, and the adult vial assay is still widely used to track insecticide resistance in heliothines and other insects (Snodgrass et al., 2009).

Resistance management, largely founded in cotton production by concepts associated with pyrethroid resistance, became a mandated component of Bt cotton management when Bollgard cotton was first registered in 1996 (Table 3). This was the first example of USEPA regulation of a pest management tool for resistance management. Bollgard cotton was the first Bt cotton available on the market, and it was highly effective and widely used. Bollgard cottons expressed a single insecticidal toxin, Cry1Ac, and provided exceptional control of tobacco budworm populations that had become highly resistant to pyrethroid insecticides (Luttrell and Jackson, 2012). Availability of Bt cotton was a much needed and rapidly adopted alternative to foliar sprays of insecticide, and it was rapidly adopted by cotton growers, especially growers located in the southern United

States that were struggling with control of pyrethroid-resistant populations of tobacco budworm. Secondary pest outbreaks following the westward expansion of boll weevil eradication programs were commonly observed because of the ecological disruption of natural biological control associated with area-wide spraying of malathion. Serious pest problems were documented with beet armyworm in Georgia (Ruberson et al., 1994) and Alabama (Smith, 1989; Duffy et al., 1994), tobacco budworm in Mississippi (Williams and Layton, 1996; Luttrell et al., 1997), and beet armyworm in west and south Texas (Summy et al., 1996). High densities of pyrethroid-resistant tobacco budworms were not manageable, and devastating crop loss was observed following outbreaks of tobacco budworm in parts of Mississippi (Luttrell et al., 1997). Bollgard cotton provided an effective solution to this problem. Given the history of insect resistance in cotton pests and the high activity and potential selection for resistance in lepidopteran pest species exposed to these insecticidal plants via a continuous expression of insecticidal activity in all plant tissues, concerns for resistance to Bt cotton were widely recognized (National Research Council, 1986).

The high dose–refuge strategy for managing resistance to insecticidal Bt cottons became the foundation for Bt resistance management programs in cotton worldwide (National Research Council, 1986; USEPA, 2001). This strategy generally assumes that resistance is functionally inherited as a recessive trait, that it is attributed to a major gene effect, and that there is a source of susceptible alleles to dilute the rare resistance alleles concentrated through selection (National Research Council, 1986). Georghiou and Taylor (1977) and Tabashnik and Croft (1982) described genetic and operational factors influencing the high dose–refuge strategy and resistance evolution long before transgenic plants were developed. When Bt cotton became a reality, transgenic plants were envisioned as ideal targets for insect resistance management because of the general consistent expression of a high dose of toxin (i.e., a dose sufficient to kill resistant heterozygotes). This has held true for some species, especially tobacco budworm and pink bollworm, where toxin expression is sufficient to kill most insects exposed to the plants. Unfortunately, insecticidal activity against other targeted pests like bollworm, fall armyworm [*Spodoptera frugiperda* (J. E. Smith)], and beet armyworm [*S. exigua* (Hubner)] has been lower, with survivors that fed on Bt cotton commonly found in field environments. A number of articles have reported insect resistance to Bt cotton. Resistant populations of pink bollworm in India have been studied extensively in recent years (Dhurua and Gujar, 2011). Tabashnik and Carrière (2010) suggest that resistance populations may have resulted from variable or low-expressing plants. Resistance to Bt in fall armyworm appears to be associated with continuous cultivation of Bt corn hybrids in Puerto Rico (Storer et al., 2012). The status of resistance in bollworm has been highly debated. Tabashnik and Carrière (2010) concluded that field resistance to Cry1Ac had evolved. Luttrell and Ali (2009) suggested that the variability observed in assay data was evident before Bt cotton was widely planted. Regardless, bollworm appears to have the genetic capacity to survive exposure to Bt cotton under some situations. Insects collected from these field situations are difficult to maintain in laboratory culture, partially because of reproductive fitness costs (Anilkumar et al., 2008), and mechanisms associated with the surviving field insects have been difficult to discern. Understanding the genetic, or possible polygenic and epigenetic, mechanisms responsible for the segregation of these populations surviving on Bt cotton is an important challenge for contemporary research. Describing resistance evolution in species targeted by Bt technologies that are not high dose is a research need for all agricultural crops utilizing the transgenic insect controls.

Dynamic Pest Problems

Changes in the spectrum of insect and mite pests of cotton are often associated with widespread use of new management tools or changing production systems as referenced above. Sometimes changes are due to the introduction of exotic or invasive species. Boll weevil, pink bollworm, and silverleaf whitefly (*Bemesia argentifolii* Bellows and Perring) are introduced pests of cotton in the United States. All three of these species have been recognized as serious pests in some regions of the United States, requiring dedicated management programs to reduce their pest status. Boll weevil also became a serious introduced pest of cotton in Argentina, Brazil, and Paraguay in the 1990s (Ramalho and Wanderly, 1996), and Brazilian farmers are now concerned with a documented establishment of *H. armigera* that threatens cotton and soybean [*Glycine max* (L.) Merr.] production systems in the region (Tay et al., 2013). Silverleaf whitely is also an important invasive pest in Australian cotton. A number of different mealybug and whitefly species have recently become serious invasive pests of cotton around the world. *Phenacoccus solenopsis* Tinsley is a serious introduced pest of cotton in Asia (Wang et al., 2010).

Cotton production has long had a well-recognized cyclic pattern of high use of new insect control technologies, followed by increased resistance levels in targeted pests and declining performance of the technology, and elevated pest status of secondary pests. Metcalf (1980) referred to this as the "pesticide treadmill" and advocated a stronger ecological approach to insect management. This emphasis on ecological context and recognition of the negative consequences of pesticide use was a major theme in Ridgway et al. (1984) and the preferred integrated management approach to minimize outbreaks of new pest problems. They advocated more research and adoption of alternatives to insecticides, especially the use of biological control, incorporation of host plant resistance traits into commercial cottons, and sophisticated management systems based on detailed crop and pest information. Applications of techniques would ideally proceed in a tiered structure, from insect monitoring and use of economic thresholds as a foundation to the addition of complementary additional cultural, biocontrol, and other tactics (Kogan, 1998), and as specifically proposed for cotton (Ellsworth and Martinez-Carrillo, 2001). The need to develop diverse insect management tools for sustainable cotton production and avoid ecological disruptions of natural controls that trigger outbreaks of pest populations continues. Gross and Rosenheim (2011) used statistical methods to draw formal causal inferences from non-experimental data about the economic impact of secondary pest outbreaks following early-season applications of insecticide for *Lygus* control in the San Joaquin Valley of California. They reported that 20% of late-season pesticide costs were attributed to secondary outbreaks of pests elicited by the early-season applications. This equated to an additional US$2.72 per hectare in management costs. We report a wide diversity of insect control options throughout this chapter, although the simplicity of single-tactic approaches makes transition to multi-tactic approaches challenging (Kogan, 1998; Brewer and Goodell, 2012).

While cotton production currently has a great diversity of insect management tools and major advancements in insect control have been observed through eradication of key pests and wide-scale deployment of Bt cotton, the production environment is less diverse. Large acreages of cotton are planted to the same or similar varieties expressing the same or similar genetics including insecticidal proteins. Area- or region-wide management programs, similar to boll weevil eradication, tend to reduce or even eliminate the spatial and temporal diversity of production fields and pest control options needed to encourage ecological stability envisioned by Ridgway et al. (1984) and many in the integrated pest management (IPM) community (Kogan, 1998). This creates unique challenges for sustainable and ecological management of cotton insects, and it continues to encourage the emergence of new pest problems.

Outbreaks of aphids, whiteflies, and spider mites, common when cotton was sprayed routinely with chlorinated hydrocarbon insecticides in the 1960s and 1970s, reappeared in cotton systems around the world during the 1980s and 1990s (Dittrich et al., 1985; Slosser et al., 1989; Gerson and Cohen, 1989; Bradley, 1996; Wilson et al., 1998; Ellsworth and Jones, 2001; Oliveira et al., 2001) when cotton insect control was heavily reliant on foliar sprays of pyrethroids. This resurgence of secondary pest problems was associated with disruptions of natural control and evolving resistance levels to insecticides targeted at key pests. Widespread use of the neonicotinoid insecticides in the 1990s eliminated some of these problems and introduced a new approach to insecticide delivery through translocation and translaminar movement of insecticides in the plant (Elbert et al., 2008; Zhang et al., 2011). Seed treatments with neonicotinoid insecticides became widely accepted on many agricultural crops. As with all previous classes of insecticide used in cotton, insect resistance has become a concern with the neonicotinoids (Nauen and Denholm, 2005). Issues related to honey bee (*Aphis mellifera* Linnaeus) toxicity and possible associations of honey bee decline with neonicotinoid insecticides (Girolami et al., 2009; Stewart et al., 2014) are a serious environmental concern and another reminder of the importance of ecological context in insect pest management.

With the decreased use of insecticides originally targeting heliothines, due to the success of Bt cotton (and boll weevil eradication within the United States), the sucking bug complex of stink bugs and plant bugs that damage squares and bolls have likely been released from indirect control by these insecticides (Lu et al., 2010). This evolving sucking bug complex is appreciably more spatially and temporally variable in damage severity and in abundance than the more consistently severe damage from heliothines and boll weevil. Cotton damage from stink bugs and plant bugs is attributed to a combination of direct damage from abscission of squares and small bolls and from indirect damage associated with lint and seed deterioration. Indirect damage, which may magnify the overall loss, takes the form of square and boll rots, which occur when pathogens are introduced during insect feeding on the fruiting bodies (Medrano et al., 2009; Brewer et al., 2012b). The feeding sites where economic damage occurs varies, with plant bugs more damaging to squares and young bolls, and stink bugs able to feed on older bolls (Medrano et al., 2009; Brewer et al., 2012a). Severity of injury, both direct and indirect, may vary by pest species. Such species diversity presents a difficult target for work in classical breeding and bioengineered cotton. An additional challenge is the variable abundance in densities of these pests across time and space, attributed partly to their mobility and migration from a wide range of alternative hosts, including non-cultivated plants and other cultivated crops, such as soybeans, corn, alfalfa (*Medicago sativa* L.), and sorghum [*Sorghum bicolor* (L.) Moench] (Carrière et al., 2006; Tillman et al., 2009; Herbert and Toews, 2011).

Pest monitoring combined with plant monitoring is advisable under such situations (Pedigo and Buntin, 1994) to make timely application of effective insecticides. Developing economic thresholds and other management tools to guide decision-making is highly appropriate for this sucking bug complex and other cotton insect pests (Benedict et al., 1989; Musser et al., 2009a, 2009b). Sampling methods, economic thresholds, and insecticide use strategies are available to growers (Benedict et al., 1989; Parajulee et al., 2006; Musser et al., 2009a, 2009b; Reay-Jones et al., 2010; Brewer et al., 2013), and the traditional approach of field-based monitoring by private and public-based services has been effective but also is faced with challenges. Looking to the future, field-based monitoring may become less viable, especially in regions where labor is limiting and production is large scale (Allen, 2014). The same agroecological knowledge base discussed in the plant-based manipulation of this chapter can be used to evaluate pest risk, thereby allowing pest monitoring and decision-making to focus on the areas at highest risk to pests. This process of decision-making is performed often informally by many pest management practitioners, utilizing their local knowledge of the cropping area and field histories. As application of GIS and remote sensing technologies to agriculture and pest management improve (Clay, 2011), evaluation of pest risk should improve regionally and may have smaller-scale applications envisioned in precision agriculture (see next section).

Integration of Insect and Crop Management

New approaches in incorporating insect management into total crop production systems have evolved over the past 30 yr, just as advances have been made in insect control technologies. Expanding capabilities in information management, remote sensing, and management systems capable of integrating agroecological as well as socio-economic influences into crop production have also broadened the scope of insect management options. Pest managers now have many of the management tools Ridgway et al. (1984) envisioned, but there are still challenges. Diverse production systems have different resources, and a wide range of ecological influences associated with different production regions impact pest abundance and pest control options. Implementing advanced systems of management is also associated with socio-economic constraints that vary among production regions. That said, current IPM programs in cotton have the capacity to carefully define plant susceptibility to insect injury, manipulate insect resistance levels, and design landscape level management systems for polyphagous pests attacking multiple crops with different risks of insect crop damage.

Bt and Herbicide-Tolerant Cotton

Cotton cultivars with advanced Bt traits are used throughout the world. Producers can access these highly specific and effective host plant resistance traits for the materials input cost of seed that includes a technology fee. Bt cotton has simplified insect pest control processes across all farm classes regardless of the level of mechanization or farm size (Kathage and Qaim, 2012; Vitale et al., 2010; Naranjo and Ellsworth, 2010). By deploying the most advanced seed technology with insect-resistant germplasm and the most effective seed protectants, managers shift a portion of pest management decision-making to the pre-planting period, thereby lessening reliance on in-season, reactive management decisions. This simplifies some of the previous problems associated with field monitoring and reactionary timing of insecticide applications, but intensifies risks for pest resistance and some impacts on non-target and secondary pests as described above.

Herbicide-tolerant traits typically are included in the seed technology array in mechanized production systems. Input structures for these systems are less reliant on labor and have greater emphasis on capital investments and material inputs purchased off-farm. Producers in capital-intensive systems integrate seed technology with advanced agronomic and engineering innovations in efforts to increase production efficiency. Information management is a critical component in the most progressive systems, particularly those using precision farming technologies where managers use detailed field-level information to optimize production inputs on a spatially variable basis. Ideally, managers exploit their capital investments in machinery and information technologies to optimize profitability while limiting negative environmental impact (Sui and Thomasson, 2013). It is an ongoing challenge for producers and crop advisors, as well as researchers and extension personnel, to sustain the continual adaptation required to integrate the newest machine and information technology into evolving IPM programs.

Farm-Level Planning

Information management innovations allow managers to integrate data from plant, pest, soil, and environmental monitoring into IPM programs that may operate at different scales, from individual management units to region-wide suppression programs. This process of information synthesis includes analysis of farm records and information regarding historical yields (yield and revenue maps) and pest and plant monitoring records (trapping data, survey information, insecticide cost accounts) as well as past and future (short- and long-term) weather data. This information allows examination of temporal and spatial variability in the distribution and abundance of insect pests over contiguous areas. Ultimately this can improve efficiency in defining management units and pest monitoring in fields that tend to have similar pest densities and crop features (Luttrell et al., 2006; Allen et al., 2005, 2007).

Farm-level planning moves IPM approaches beyond a field-based focus. Expansion of pest management tactics among multiple farms allow a population suppression approach to counter high pest risk that may be associated with connected landscapes. This is one basis for managing insect pests on an area-wide basis where the goal is to lower pest population densities. Scales of management can vary from individual farms to communities or large production regions (Carrière et al., 2006, 2012). Spatial distribution of target pests within the managed unit and mobility/dispersal capacity of the managed pest are important considerations impacting the size of area being managed. Agricultural landscapes often are considered well connected, but there may be heterogeneity within the greater landscape (mix of crop types or management practices) or within the field (variable plant vigor or maturity associated with soil or elevation). Such seemingly fragmented landscapes can be connected if a pest species has sufficient dispersal capability to colonize otherwise isolated fields (e.g., Schneider, 1989; Dillon et al., 1996; Showler, 2006; Kennedy and Storer, 2000; Tillman et al., 2009; Sivakoff et al., 2013). Pest dispersal linked to landscape connectivity may exhibit different scales of movement depending on the time of season (e.g., from overwintering habitat or spring hosts to early season cotton, or within-field movement in late season, from cutout plants to patches of plants with second growth, or to new crop or non-crop hosts or overwintering habitat) (Carrière et al., 2006; Dickerson et al., 2001).

Within-field examples of spatial variability also have direct application to insect management in cotton, especially as related to precision agriculture. Precision agriculture technologies include site-specific approaches for seeding rates, fertilizer applications, irrigation practices, or pesticide application regimes at a field level. Availability and access of the global position system (GPS) for agricultural use in the 1990s was key technology to enable site-specific practices. Practical application of this technology for insect pest management in cotton has been limited in comparison with on-farm implementation of site-specific application of fertilizers, plant growth regulators, harvest aids, and nematicides as well as variable rate seeding (Ortiz et al., 2012; Sui and Thomasson, 2013; Oliver et al., 2013). For insect control practices, the use of GPS guidance and control systems with sprayer booms equipped with automatic section controls is recognized as a practical new tool that can improve foliar application efficiency (e.g., eliminate off-target sprays in irregularly shaped fields or sensitive edge-of-field areas) (Luck et al., 2010). Other research examples of spatially variable approaches to insect control have been map-based, spatially variable applications of insecticides triggered on insect abundance, crop biomass, and/or landscape features (Willers et al., 1999).

Prescriptions for variable application of seeds, fertilizer, irrigation, and crop protectants can be associated with yield potential or maturity of management zones based on soil physical properties or landscape features that may affect potential pest risks. Pest managers may adjust and re-direct insect scouting, plant monitoring, or protective insecticide applications in relation to plant maturity, variability, and pest risk in these zones (Willers et al., 2009b). The ability to determine within-field variation of pest densities and plant susceptibility is central to developing reliable decision rules for site-specific insecticide applications. Remote sensing can provide for large-scale collection of image data that can enable users to establish relationships between soil and crop physical and biophysical properties and spectral signatures. For example, spectral reflectance data from both satellite and airborne imagery have been used to establish vegetation indices that have been correlated with crop biomass, nutrient deficiencies, water stress, and yield. Willers et al. (1999) used high-resolution multispectral imagery to create maps to direct sampling for estimating tarnished plant bug densities and then prescribed spatially variable insecticide applications. McKinion et al. (2009) compared spatially variable applications of insecticide for early-season control of tarnished plant bug compared with broadcast uniform application and found a cost savings of up to 60%. Teague et al. (2014) demonstrated zone management for insect control termination decisions in center pivot irrigated fields where plants under the pivot were compared to plants in the non-irrigated, rainfed corners. Rainfed plants with a reduced effective flowering period and earlier cutout did not require extended protection from tarnished plant bug infestations compared with irrigated portions of the field. Eliminating one to two late-season sprays because of early maturity of rainfed plants saved 14 to 18% of insecticide costs in late season depending on field arrangement. As precision practices expand, improvements in productivity as well as reductions in negative environmental impacts should accrue as producers increase adoption of site-specific technology and practices in IPM (Willers et al., 2009a, 2009b).

Precision agriculture is often touted as a means of reducing in-field crop heterogeneity; however, from an insect perspective, increasing variability may aid in pest management efforts. Increased heterogeneity may be helpful in a trap crop approach or in development of a push–pull pest management tactic (Cook et al., 2007). These tactics integrate use of behavioral stimuli that make the crop unattractive while luring pests to another attractive source where they are controlled. Push–pull tactics in cotton have been used most often in smallholder systems (Duraimurugan and Regupathy, 2005). They are designed for population reduction and typically emphasize biological control. Intercropping and trap cropping has been used extensively in Chinese production systems to provide vegetative diversification that disrupts plant host finding or enhances natural enemy abundance for managing bollworms and aphids (Wu and Guo, 2005). Similarly, pheromones may be deployed to concentrate pests into specific areas of a field where insecticides or other control options can be delivered on a site-specific basis. The boll weevil pheromone was used in the attract-and-kill device designed by McKibben et al. (1990) that featured an insecticide-coated tube combined with high concentration of pheromone (Grandlure) to kill responding adult boll weevils. Known as bait sticks, and later commercially developed as BWACT, results from evaluations of field efficacy and performance of the tactic in US cotton systems were mixed (Spurgeon et al., 1999). Bait sticks also were evaluated for use in boll weevil eradication programs in the United States and more recently in development of boll weevil IPM programs in Central and South America (review included in Miranda and Da Silva, 2005).

Even though there is a sound research base with sampling protocols and decision aids that support use of precision agriculture approaches with insect suppression, there has been limited use of site-specific applications by crop managers. This is somewhat puzzling, because historically, site-specific management has always been a critical component of IPM. Technological advances have simply provided opportunities on a larger scale. In observations of boll weevil responses to fruiting characteristics of cotton in small plot variety trials, Isley (1934) noted that late-maturing cotton varieties had highest gains when protected with calcium arsenate dusts, but higher yields were observed from early-maturing varieties without insecticidal protection. He suggested use of early-maturing varieties where managers planned not to use insecticides (dust) and also planting these early maturing varieties proximal to field margins adjacent to overwintering habitat. Isley's recommendations were based on knowledge of insect ecology and plant response to resource availability and were directed at farms with mule-driven planters. Such considerations for cultivar deployment using precision management in spatially variable fields is a focus of 21st century research. Newly available controllers for planters allow variable seeding rates as well as multiple cultivars in the same planter pass. It should be a simple step to implement Isley's 1934 decision guides to use insect behavioral manipulation varying crop characteristics of plant maturity, host plant resistance (resistance and tolerance), and spatially applied insecticides to improve efficiency in crop protection practices.

Linking Insect Management to Crop Development

Cotton is a tropical, perennial shrub, and because of its indeterminate growth habit, the plant is capable of a stop-and-regrow growth pattern that allows for survival and reproduction in variable growing conditions. Long-term efforts in agronomic research have produced management tools and practices that attempt to restrain cotton's indeterminate habit. A more determinant growth pattern allows for improved mechanized production and harvest. Insect herbivory still manages to disrupt these systems. Whether insect feeding results in trivial cosmetic injury or economic damage is dependent not only on the particular pest species and its abundance but also on growth stage and particular cultivar and its level of host plant resistance. Susceptibility to and recovery capacity of plants following herbivory also depend on resource availability (e.g., soil, water, nutrients) and growing environment (Sadras and Felton, 2010).

An interdisciplinary team of researchers at the University of Arkansas, in collaboration with other southern states and with support from Cotton Incorporated, developed COTMAN (Oosterhuis and Bourland, 2008) in the 1990s to help growers and pest managers monitor and summarize plant development, detect plant stress, and define crop sensitivity to insect injury. The goal of the effort was to promote crop earliness and plant vigor and to improve end-of-season decision-making for timing termination of insect control and irrigation and for defoliation (Bourland et al., 2008). This management system is widely recognized as a benchmark of defining final crop development stages vulnerable to insect injury, based on straightforward definitions of crop maturity and cutout (Bourland et al., 2001) that have been widely adopted in the cotton industry (Kerby et al., 2010). Specific use of the software and adoption of recommended sampling schemes have been less widely adopted, especially in recent years with a renewed interest in full-season cotton production, but the importance of the associated management of insects and cotton crop development was advanced through the COTMAN effort. Cotton phenological development shown in Fig. 2 is typical for US upland cotton production systems, particularly those in the southeastern and Midsouth cotton production regions. It is based on developmental timing that was used to generate the standard for crop growth curves in the COTMAN plant monitoring system (Oosterhuis and Bourland, 2008). The four plant development stages are aligned with major insect pests that may have direct and indirect impacts on crop growth and production of seed and fiber. The pre-square stage occurs in the first month after planting and includes germination and seedling emergence with appearance of cotyledonary nodes and monopodial main-stem (vegetative branches). The phenomenon of first flower represents a critical shift in plant development with bolls becoming the dominant physiological sink. During the effective flowering period, there is both vegetative and reproductive growth; however, allocation of assimilates and other plant resources are shifted from the terminal, roots, and other plant parts to the bolls. The rate of squaring node development slows as boll load increases, and it eventually stalls as bolls mature. That temporary hiatus in growth is called cutout (Kerby and Hake, 1996; Guinn, 1985). As plants approach cutout, white flowers appear progressively closer to the plant apex, and sequential counts of squaring nodes can be used as a measure of plant maturity (i.e., nodes above the uppermost first position white flower) (Bourland et al., 1992). Bernhardt et al. (1986) suggested that the measurement be used to identify the last effective flowers, the purpose being to time insect control termination in late season. Crop managers routinely use dates of cutout plus heat unit accumulations to aid in timing endpoints for insect control and irrigation and for scheduling defoliation (reviewed in Oosterhuis and Bourland, 2008).

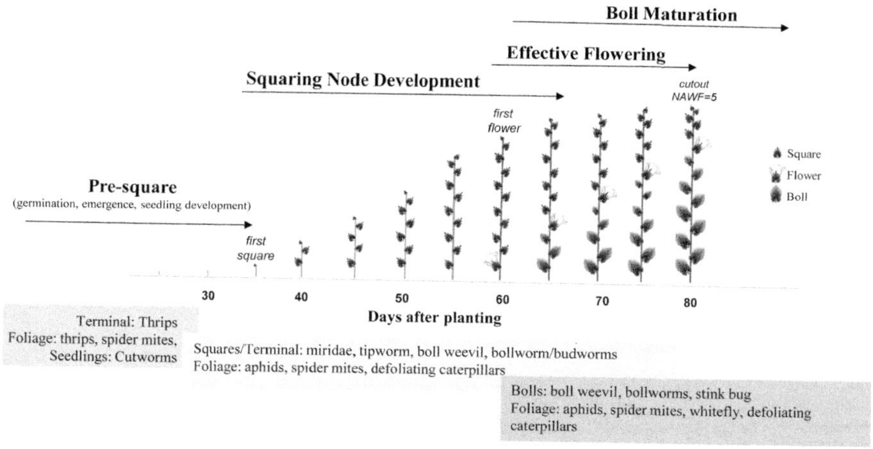

Fig. 2. Cotton main-stem sympodial development is depicted with first position squaring nodes, flowers, and boll nodes. Common insect pests that typically are economically important during the four crop stages are shown below.

Within-season assessment of crop development through regular plant monitoring provides critical context of crop development for within-season insect management decisions. A systematic plant-monitoring program can provide timely information to assure managers that the crop is on target for early and high yields or it can signal if there are problems. With increased system control using modern agronomy and engineering tools, managers may approach crop management similar to process and quality control monitoring practices used in manufacturing (Shewhart, 1939). Crop advisors monitor fields and provide timely information to inform managers if plant growth parameters meet, exceed, or deviate from predefined specifications. Once deviations are detected, for example, in nodal development rates or fruit retention, managers can attempt to identify an assignable cause and try to take action to address the variation. As in manufacturing, process monitoring is essential in an overall program for process improvement.

Plant monitoring requires a standard with which to compare actual plant growth, similar to a quality control chart. In cotton, simple crop growth curves that depict plant fruiting dynamics can be created that consist of squaring nodes plotted against days after planting. These growth curves can be compared to the COTMAN standard curves to represent optimum combinations of early maturity and high yield (Bourland et al., 1992). Termed the Target Development Curve (Fig. 3), it was based on the phenology and main-stem sympodial architecture suggested by Tharp (1960) and represented in Fig. 2. The standard curve shows main stem sympodial nodes that have not yet produced flowers (squaring nodes) through a season, ascending at a pace of 1 node each 2.7 d through first flower at 60 d after planting, and then descending to physiological cutout, defined as 5 nodes above white flower (NAWF = 5), by 80 d. Similar growth curves using squaring nodes and NAWF have been included as crop and pest management tools in Australia. In that system, squaring nodes are plotted against accumulated heat units (developmental degree-days; baseline 15.5°F [60°F] expressed as DD_{60s})

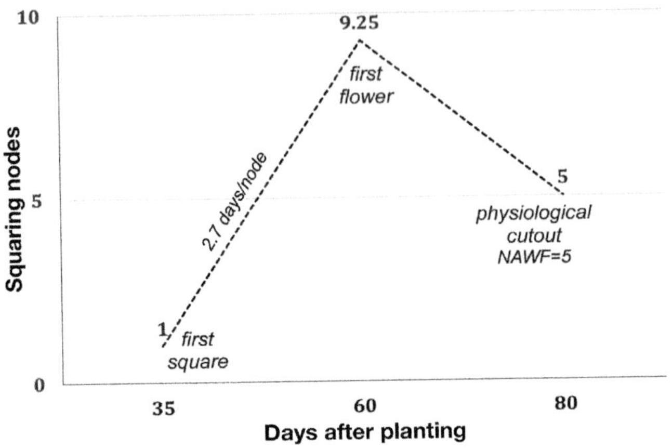

Fig. 3. Target Development Curve used as a standard to compare actual crop growth curves in which squaring nodes are plotted against days after planting. Once flowering begins, squaring nodes are counted above the highest first position white flower, referred to as nodes above white flower (NAWF). Plants attain physiological cutout at NAWF = 5 (Danforth and O'Leary, 1998).

with first squares at approximately 500 DD_{60s}, first flower at 800 DD_{60s}, and cutout at NAWF = 5 at approximately 1200 DD_{60s} (Deutscher et al., 2005).

Growth curves can be used to document plant response to insect herbivory as well as other biotic and abiotic stress. Other production and weather information, such as irrigation timing and rainfall, can be added to the growth chart. They provide a composite picture of plant development through the effective fruiting period, and from a pest management perspective they can show when plants are vulnerable or have matured past a stage susceptible to damage from fruit feeding insect pests. Growth curves can show effects of feeding injury on plant maturity delay resulting from insect induced injury or abiotic stress, either of which can lead to excessive late-season terminal growth (second growth) that is highly attractive to insect pests (Teague, 2011; Teague et al., 2003, 2008, 2012). A post-season retrospective analysis of plant monitoring information and growth curves may reveal important risk factors limiting profitable production. Adjustments in IPM practices for the next crop may include increased use of host plant–resistant cultivars and modifications in seeding rates, use of plant growth regulators, or changes in fertility or irrigation practices.

The most widespread use of crop monitoring information in pest management is the use of NAWF measures to time termination of late-season insect control practices. Cutout represents the flowering date of the last effective boll population, the last cohort of developing bolls that has a high probability to contribute to economic yield (Bourland et al., 1992; Oosterhuis et al., 1996). As the cohort of last effective bolls mature, they eventually reach a stage where specific arthropod pests are no longer of economic importance. This is the final stage of crop susceptibility (Pedigo et al., 1986). Managers use this concept for termination decisions for insect control. Plant monitoring is used to determine the flowering date of the last effective bolls, and then heat unit accumulations are used to gauge maturity. Once the last effective boll population has reached a defined protection endpoint, then insect control and surveillance of specific insect pests can be suspended. The heat unit requirement from anthesis to development of a boll of acceptable size and quality is ≈850 DD_{60s} (Oosterhuis and Bourland, 2008); however, bolls become less susceptible to certain pests much sooner than 850 DD_{60s}. Methods and baselines that use NAWF monitoring as an end-of-season gauge were developed and reported by Bernhardt et al. (1986). Working with boll weevil and bollworm in succeeding studies, Bagwell and Tugwell (1992) examined the relation of boll age and susceptibility to insect feeding and determined that damage to cotton bolls decreased as the boll matured with minimal penetration by ≈350 DD_{60s} after anthesis (Bagwell, 1995). Research by Kim and Oosterhuis (1998) showed that the increased resistance of the boll to penetration by insects at ≈350 DD_{60s} after anthesis was associated with increased lignification and tannin concentration of the boll wall endocarp (Kim and Oosterhuis, 1998; Oosterhuis and Kim, 2004). This was the basis of the COTMAN system for insect control termination (Cochran et al., 1999; Leonard et al., 2008). Insect control termination timing based on NAWF monitoring plus heat unit accumulations is included in Midsouth and Texas state extension recommendations. The broad concept of terminating insect control when harvestable cotton fruit is no longer vulnerable to insect injury is universally recognized, largely because of COTMAN's articulation of the importance of linking insect management to crop development.

Farm to Region-Scale Management Systems

Specific knowledge of plant–pest–natural enemy interactions and the effect of spatial scale and arrangement of crops on these interactions are critical in developing more durable approaches in managing cotton insect pests. Recent advances in this area are based on habitat manipulation and ecological engineering concepts for pest management (e.g., Gurr et al., 2004) and more specific area-wide approaches to pest management (e.g., Elliott et al., 2008) including examples in cotton production systems (e.g., Spurgeon, 2008). Cotton is grown largely as a monoculture. It can be found in expansive plantings as seen in parts of Brazil, China, and the southwestern United States, or cotton plantings can be interspersed with other cultivated crops and non-cropped habitat to varying degrees as seen in many areas of the southern United States and developing countries. Advocates of increasing plant diversification, particularly in large-scale cropping systems, point to a large body of work that indicates pest risk tends to increase in monocultures (Altieri and Nicholls, 2004). In parts of Brazil, for example, wide-area cotton crop plantings, multiple crops per year, and a diverse complex of cotton insect pests has led to very high rates of insecticide application that can only partly be addressed through genetically modified cotton (Oliveira et al., 2013). In contrast, there are many exceptions where ill-structured patchwork of crops increase pests, including instances in cotton when the source of pest immigration into cotton is from a nearby crop or weeds that serve as alternative hosts to the pest insect (i.e., Ellsworth and Martinez-Carrillo, 2001; Tillman et al., 2009; Herbert and Toews, 2011).

Plant-based manipulation for pest management includes crop manipulation such as planting and harvest scheduling, crop remnant destruction, non-crop habitat management, and spatial crop arrangement recommendations. Recommendations to reduce pest risk may be crafted using knowledge of plant–pest–natural enemy interactions and the effect of spatial arrangement and size of plantings. These approaches may require application area-wide to see substantial pest risk decline (Elliott et al., 2008; Brewer et al., 2008), although benefits can be seen with local adoption if ecological interactions are driven primarily at a local scale (Carrière et al., 2006) and even within the field (Stern et al., 1969; Mensah and Sequeira, 2004). Progress applying these concepts to cotton insect pest management has been aided from advances in community and landscape ecology (Forman, 1995), in spatial analysis facilitated by geographic information system technology (Clay, 2011), and the prior work of cotton entomologists working in this field (e.g., Isley, 1934; Sevacherian and Stern, 1974). This knowledge base also has been an essential feature in developing other technologies applied area-wide for cotton pest management, including the use of insecticides and mating and sterile male population disruption techniques seen in boll weevil and pink bollworm eradication programs (Spurgeon, 2008).

Pest risk reduction through plant-based manipulation may well produce a more agroecologically stable system less prone to pest disruptions. Here we focus on the agroecological aspects of utilizing plant-based manipulation. We also briefly address the socio-economic aspects of selected well-documented experiences, because agroecological aspects must be balanced with economics (i.e., markets for and processing of the crop, costs of labor, services, and technology) and social (i.e., the nature of labor force, technology transfer and education structure, and cultural and community structure affecting agriculture) considerations to achieve sustained effort (Brewer and Ishii-Eiteman, 2009). Related to worldwide cotton production, the most sustainable approach will undoubtedly vary by region. These examples are intended to be instructive as a starting point for those formulating best methods of integrating plant-based manipulation with other pest management techniques to achieve a more sustainable pest management cotton system well-tuned to the economic, agroecological, and social structure of the cotton production region of interest.

An arguably effective case of combining techniques of crop manipulation and insecticide technology was exemplified by the case of the silverleaf whitefly in Arizona, the United States, and Mexico, as reviewed in detail by Ellsworth and Martinez-Carrillo (2001). Silverleaf whitefly is a polyphagous insect that attacks both cotton and a variety of vegetable and vine crops grown in the irrigated desert of Arizona, southern California, and northwest Mexico. It readily moves across crops, as triggered by sequential plantings, plant maturation of crops and weedy hosts, and harvesting schedules. Combining use of the insect growth regulators (IGRs) buprofezin and pyriproxyfen in cotton and imidacloprid use in vegetables were key insecticide components of multi-crop regional silverleaf whitefly management (Ellsworth and Martinez-Carrillo, 2001). In commercial-scale experiments in central Arizona, IGR insecticides reduced the number of insecticide applications for whitefly control by 50% (Ellsworth and Jones, 2001). This insecticide technology advance was paired with an agreement within the farmer community on voluntary adherence to guidelines for use of key active ingredients of insecticides and planting date scheduling to sustain crop protection (Palumbo et al., 2001). In nearby growing area of northwestern Mexico, mandatory planting and harvest date schedules, control of volunteer cotton and weedy hosts, host-free periods, and strategic use of insecticides have been implemented area-wide (Ellsworth and Martinez-Carrillo, 2001). Related to the social structure needed to implement these area-wide activities, regional coordination benefitted from the infrastructure originally funded and deployed for eradication of the boll weevil and pink bollworm, grower-supported voluntary or mandated activities enabled by government regulatory bodies, and government- and university-supported education (Ellsworth and Martinez-Carrillo, 2001; Spurgeon, 2008).

For polyphagous cotton pest species, management of pest populations on uncultivated host plants has been explored as a means to reduce risk to cotton. In an area-wide, multi-year suppression experiment, tarnished plant bugs were suppressed in a 23 km^2 region of the Mississippi Delta cotton production region by control of its alternate weedy broadleaf hosts using herbicides. Economic analysis showed that the cost of weed control was more than offset by insecticide use reduction in cotton in most cases (Snodgrass et al., 2006). But long-term stability of the technique is challenged by agroecological and social considerations. There is now a greater diversity of crops grown in the Mississippi Delta that are crop hosts of tarnished plant bug (i.e., soybean and corn), which may serve as source populations (Snodgrass et al., 2006). Also, area-wide application of herbicides requires coordination as best accomplished by a grower-supported management district. Environmental concerns, such as nutrient management, real or perceived pesticide runoff issues (Hardee et al., 1999), and other unintended non-target effects, create challenges to widespread community acceptance as well.

A similar approach was applied in an area-wide effort to management heliothines by applying spray formulation of insect pathogens onto their weedy host plants along roadsides and other uncultivated areas (Hardee et al., 1999). Use of the highly insect host specific baculoviruses and supplemental roadside mowing of weeds addressed environmental concerns in using herbicides for weed control in ditches and other drainages. In an area-wide application, the technique reduced the number of moths emerging from weed hosts (Hayes and Bell, 1994) but did not consistently result in reduced heliothine populations in cotton (Hardee et al., 1999). The scale of movement of the moths, along with local heliothine sources in other crops not sprayed with baculovirus, are significant challenges to the area-wide application of the technique. In both these examples in controlling uncultivated host plants of the target pest, positive benefits of weed control also needs to be balanced with the potential reduced benefits of weeds serving as a habitat to beneficial insects and other organisms (Norris and Kogan, 2000; Bianchi et al., 2006).

More recently, principles of landscape ecology enabled by advances in GIS have been applied where scale and crop arrangement effects may provide guidance for planting strategies that reduces pest risk from *Lygus* spp. in western US cotton (Carrière et al., 2006, 2012; Goodell, 2009; Sivakoff et al., 2013). Carrière et al. (2006) evaluated 50 cotton fields and surrounding forage and seed alfalfa and weeds to serve as sources and sinks of *L. hesperus*. Crop distance and the fruiting cycle of cotton were crucial components, with scales of 375 to 1500 m most relevant to intercrop pest movement and pest risk to cotton at the early fruiting period of growth. The data suggested that spatial arrangements of the crops affected *Lygus* incidence in cotton (Carrière et al., 2006). Educational activities in support of planting strategies, including the spatial arrangement of crops locally and regionally, to reduce pest risk have been initiated (Goodell, 2009; Ellsworth, 2013). In the future, the same concepts and advancements in information technology tied to spatial analysis (GIS) systems (e.g., Clay, 2011) and pest or plant injury-specific sensing systems (e.g., Mustafic et al., 2013) may be appropriate to assess risk of current and proposed cotton-focused cropping systems and gauge pest risk across a cotton production region and possibly in the field. Such pest risk evaluation systems may be particularly relevant to large-scale cotton productions regions that have a limited labor force to conduct traditional pest monitoring and in-field evaluation techniques (Allen, 2014).

Mensah and Sequeira (2004) have recently refreshed the role that in-field plant diversification can play in cotton insect pest management. In Australia, as well as other parts of the world, cotton is grown as a short-lived irrigated crop in regions with low rainfall or prone to drought, resulting in sparse natural vegetation and the irrigated cotton crop serving as the dominant source of shelter and food for pest and beneficial insects. Alfalfa (lucerne) strips intercropped in cotton fields have been used to provide refuge for beneficial insects that prey on *Helicoverpa* spp. or to serve as a trap crop for plant bugs, specifically *Creontiades* spp. (Mensah, 1999). Similar intensive cotton cropping was seen earlier in California, and intercropping of alfalfa in cotton or split harvesting alfalfa adjacent to cotton was proposed to manage *Lygus* spp. (Stern et al., 1969). To be effective, the width of the alfalfa intercrop must permit good agronomic practices but also must provide a sufficient edge-to-area ratio with the cotton crop for movement of beneficial insects between crops and restriction of the pest insect onto the more favored alfalfa (Mensah, 1999). On the basis of Australian research conducted on a commercial scale, significant reductions in the abundance of *Helicoverpa* spp. and insecticide applications in cotton interplanted with lucerne have been seen (Sequeira and Playford, 2001). In previous research in California, cotton interplanted with alfalfa resulted in most *Lygus* spp. reproducing on alfalfa rather than cotton (Sevacherian and Stern, 1974). Although agroecologically sound, this approach was not applied widely in the southwestern United States, as the approach added management complexity and was typically less profitable than chemical control approaches at the time (Goodell, 2009).

As previously discussed in this chapter, boll weevil was largely eradicated at a sub-continent scale in US cotton via a structured and coordinated system of region-wide management efforts. Although the eradication operations focused on monitoring technologies used to trigger coordinated use of foliar insecticides (Allen, 2008), knowledge of key agroecological features of the southern US cotton region aided the eradication program. The host plant range of boll weevil was essentially limited to cotton, and periodic low winter temperatures over most areas of production reduced the boll weevil's overwintering success. End-of-season crop destruction greatly aided management by further eliminating host material for the boll weevil and reduced its spread to more marginal areas (Spurgeon, 2008). Combined with a decades-long coordinated effort on in-season insecticide use, these features led to eradication throughout southern US cotton, with the only substantial population remaining in the most southern tip of Texas. In this remaining area of infestation, the key ecological features of high overwintering survival and cotton regrowth occurring throughout the year may result in a remnant population for the foreseeable future despite concentrated eradication efforts. For this region, anticipated advancements in remotely sensing cotton fields before blooming (Westbrook et al., 2014) may assist coordinating boll weevil monitoring systems as the eradication programs reduces its scale of activities. Boll weevil has now invaded the expanding cotton production system of Brazil and other parts of South America (Oliveira et al., 2013). The coordinated effort of the boll weevil eradication programs of the southern United States may be instructive for others, but ecological consideration should be fully analyzed as well as the needed social structure in the form of coordinated grower- and government-supported action.

Overview and Future Implications

Significant advances have been made in control technologies and management systems for cotton insects since Ridgway et al. (1984) reviewed the topic 30 yr ago. Numerous reviews of cotton insects and cotton insect pest management have been written during this period and are referenced in this current update. The original research summarized in these review articles, typically published in scientific journals, should be consulted for in-depth information on specific topics. Major changes in cotton insect management over the last 30 yr include widely adopted insect management strategies such as boll weevil eradication in the United States, planting of Bt cottons worldwide, and development and use of a more diverse group of insecticides based on new chemistries often with improved environmental and human safety profiles. Information and decision-making management strategies for pest complexes have been well researched and adopted to a varying extent, including insect monitoring and decision-making tools such as dynamic insect thresholds and targeted models of plant development. Perhaps more important for future cotton production systems, many underutilized strategies in insect management have unrealized potential. These include applications of precision agriculture and remote sensing, and area-wide and landscape management programs for polyphagous pests like silverleaf whitely, tobacco budworm, and *Lygus* spp.

While cotton production and associated insect management approaches have dramatically changed over the last 30 yr and production is highly mechanized and investment intense in many production regions of the world, worldwide cotton production is still highly diverse and directly linked to the socio-economic and ecological environments of the local geographic region. This creates differing management needs and variable options for ecological management of cotton insects. Understanding insect pest problems within an ecological context and recognizing the complex interrelationships involved remains as the critical foundation for cotton insect management. Shifts in pest abundance and diversity associated with wide-scale adoption of Bt cotton is the most recent example, but cotton production has a long history of elevated pest problems linked to heavy reliance on one or a few control materials (Luttrell et al., 1994; Ridgway et al., 1984). Bt cotton has dramatically reduced cotton crop loss and insecticide use for major lepidopteran pests around the world (Naranjo and Luttrell, 2009; Qaim and Zilberman, 2003; Purcell and Perlak, 2004), but changes have been observed in pest abundance of species not targeted by the Bt toxins. Shelton et al. (2002), Naranjo (2005), Wu and Guo (2005), Whitehouse et al. (2005), Vitale et al. (2010), and others discuss increased pest status of sucking pests, especially hemipteran pests, in Bt cotton systems. Presumably, much of this increase may be attributed to the reduction in broad-spectrum insecticides previously targeted at Lepidoptera (Cannon, 2000), a positive ecological outcome, but the broad-spectrum insecticides also provided indirect control of hemipteran pests. When the sprays were removed, these pests became more abundant, necessitating additional use of insecticide. The kinds and abundance of insect and mites pests within cotton or other agricultural crops are ecologically linked. Management actions to reduce targeted pests will have impacts on other species.

Challenges to cotton insect management include a continuing reliance on insecticides exacerbated by the relative low cost of insecticides and ignorance or disregard of the ecological consequences of insecticide effects on beneficial natural enemies and selection of insecticide-resistant pest populations. Many of the alternative management tools envisioned by Ridgway et al. (1984), especially biological control agents, have not been widely adopted. Cotton production is still largely reliant on insect management tools that are either seed embedded, seed treated, or foliarly applied. These insect management tools are often deployed unilaterally across large production areas. Sustainability is a common goal of agricultural crop production, but much work remains to understand the socioeconomic factors driving pest management decisions at the farm level. As long as simple insecticide-based tactics remain effective and available, use of complex ecologically based management tactics tend to be reduced (Luttrell, 1994). Pest management practices are more likely to be adopted if they are (i) reliable, (ii) require minimal time and expense, and (iii) can be easily integrated with current crop production practices. Spurgeon (2008) reminds us that cotton management is already a complicated affair, and IPM tactics that require high levels of management are unlikely to be widely adopted. He stressed the need for simplicity.

Even with these challenges, there are great opportunities to further elevate cotton insect management and build more sustainable production systems with the ecological context of landscape management. Genomics and bioinformatics will soon link genetic codes of cotton and pest organisms to biological function, providing insight into the interface between genetics and biological function. Plant phenotyping for resistance breeding and insecticide screening will produce advanced traits and products that will be technologically and ecologically complex but very simple to implement because deployment will be seed based. Both smallholder and highly capitalized, large-scale producers will have access to these pest management tools for input costs (i.e., no capital-intensive equipment necessary).

Crop selection and edge-of-field management with appropriate plant materials and management will be included in overall farm planning to assist with nutrient management, soil conservation, and water quality objectives, but also to provide resources to enhance beneficial natural enemies for conservation biological control as well as for wildlife and pollinators. These well-selected plant materials and management regimes will not aggravate insect pest problems by providing alternate host plants that move into the cotton crop. Such benefits accrue regionally and beyond the arena of the cotton grower. Financial assistance from conservation programs, currently available in the United States, European Union, and selected other regions, may support growers to implement these systems with broad region and conservation benefits (Brewer and Goodell, 2012). Similarly, molecular markers will be deployed to track the evolution of insect resistance traits geographically and temporally. This will provide real-time information for managing insect pest genetics and a foundation for area-wide pest and resistance management programs.

Information management systems will continue to grow requiring creative approaches to data collection. We envision expanded use of remote sensing, greater deployment of unmanned aerial vehicles (drones), and new biological sensors, deployed in precision agriculture systems with defined landscape risk management components. On highly mechanized farms, fully autonomous machines will use advanced GPS receivers to track expanded numbers of satellite systems to enhance location computations for vehicular guidance. Map-based control systems will be linked with location-aware planters equipped with on-the-go sensors that will detect physiochemical soil properties. Planters will deploy seeds of pest-resistant and stress-tolerant cultivars with varying maturity and with varying seed protectants matched with soil properties and pest risks from disease, weeds, and insect pests. All of these technological and management advances will be essential to improve overall system efficiency for profitable and environmentally sound cotton production. Sustainability of future cotton production systems will continue to require ecological approaches to insect management where decisions are made with an understanding of insect–plant–natural enemy interrelationships and an appreciation for risk management as a component of overall crop production.

Acknowledgments

Don Steinkraus, David Kerns, and Katherine Parys are recognized for their review of early drafts of this manuscript and helpful suggestions for revision. The scientific content of this review is directly attributed to the many authors of the original research cited and the numerous scientific reviews of cotton insect management previously published.

References

Allen, C.T. 2008. Boll weevil eradication—An areawide pest management effort. In: O. Koul, G. Cuperus, and N. Elliot, editors, Areawide pest management: Theory and implementation. CAB International, Wallingford, UK. p. 467–559.

Allen, C.T. 2014. Integrated pest management in the southern United States of America: Changing technology and infrastructure—Implications for the future. In: R. Peshin and D. Pimentel, editors, Integrated pest management, experiences with implementation, global overview. Vol. 4. Springer, New York. p. 99–152.

Allen, K.C., R.G. Luttrell, and M.J. Cochran. 2005. Using farm records to explore spatial and temporal patterns of heliothine distributions on cotton in heterogeneous cropping environments in southeast Arkansas. In: Proceedings of the Beltwide Cotton Conferences, Natl. Cotton Counc. Am., San Antonio, TX. 4–7 Jan. 2005. Natl. Cotton Counc. Am., Memphis, TN. p. 1762–1766.

Allen, K.C., R.G. Lutttrell, T.G. Teague, and P. O'Leary. 2007. Using farm records to explain insect management strategies. In: Proceedings of the Beltwide Cotton Conferences, Natl. Cotton Counc., New Orleans, LA. 9–12 Jan. 2007. Natl. Cotton Counc., Memphis, TN. p. 375–381.

Altieri, M.A., and C.I. Nicholls. 2004. Biodiversity and pest management in agroecosystems. 2nd Ed. Food Products Press, New York.

Anilkumar, K.J., M. Pusztai-Carey, and W.J. Moar. 2008. Fitness costs associated with Cry1Ac-resistant *Helicoverpa zea* (Lepidoptera: Noctuidae): A factor countering selection for resistance to Bt cotton? J. Econ. Entomol. 101:1421–1431. doi:10.1603/0022-0493(2008)101[1421:FCAWCH]2.0.CO;2

Bagwell, R.D. 1995. Monitoring the cotton plant for insecticide effects and late-season insecticide termination. Ph.D. diss., Univ. of Arkansas, Fayetteville.

Bagwell, R.D., and N.P. Tugwell. 1992. Defining the period of boll susceptibility to insect damage in heat units from flower. In: Proceedings of the Beltwide Cotton Conferences,

Natl. Cotton Counc., New Orleans, LA. 6–10 Jan. 1992. Natl. Cotton Counc., Memphis, TN. p. 767–768.

Benedict, J.H., K.M. El-Zik, L.R. Oliver, P.A. Roberts, and L.T. Wilson. 1989. Economic injury levels and thresholds for pests of cotton. In: E. Frisbie, K.M. El-Zik, and L.T. Wilson, editors, Integrated pest managements systems and cotton production. John Wiley & Sons, New York. p. 121–153.

Bernhardt, J.L., J.R. Phillips, and N.P. Tugwell. 1986. Positon of the uppermost white flower defined by node counts as an indicator for termination of insecticide treatments in cotton. J. Econ. Entomol. 79:1430–1438.

Bianchi, F.J.J.A., C.H.J. Booij, and T.T. Scharntke. 2006. Sustainable pest regulation in agricultural landscapes: A review on landscape composition, biodiversity and natural pest control. Proc. Biol. Sci. 273:1715–1727. doi:10.1098/rspb.2006.3530

Bourland, F.M., N.R. Benson, E.D. Vories, N.P. Tugwell, and D.M. Danforth. 2001. Measuring maturity of cotton using nodes above white flower. J. Cotton Sci. 5:1–8.

Bourland, F.M., D.M. Oosterhuis, and N.P. Tugwell. 1992. Concept for monitoring the growth and development of cotton plants using the main-stem node counts. J. Prod. Agric. 5:532–538. doi:10.2134/jpa1992.0532

Bourland, F.M., N.P. Tugwell, D.M. Oosterhuis, and M.J. Cochran. 2008. Initial development of COTMAN program. In: D.M. Oosterhuis and F.M. Bourland, editors, COTMAN Crop Management System. Univ. of Arkansas Agric. Exp. Stn, Fayetteville, AR. p. 15–19.

Bradley, J.R. 1996. Major developments in management of insect mite pests in cotton. In: E.G. King, J.R. Phillips, and R. J. Coleman, editors, Cotton insects and mites: Characterization and management. Ch. 1. The Cotton Found., Memphis, TN. p. 1–16.

Brewer, M.J., D.J. Anderson, J.S. Armstrong, and R.T. Villanueva. 2012a. Sampling strategies for square and boll-feeding plant bugs (Hemiptera: Miridae) occurring on cotton. J. Econ. Entomol. 105:896–905. doi:10.1603/EC12109

Brewer, M.J., J.S. Armstrong, E.G. Medrano, and J.F. Esquivel. 2012b. Association of verde plant bug, *Creontiades signatus* (Hemiptera: Miridae), with cotton boll rot. J. Cotton Sci. 16:144–151.

Brewer, M.J., J.S. Armstrong, and R.D. Parker. 2013. Single and multiple in-season measurements as indicators of at-harvest cotton boll damage caused by verde plant bug (Hemiptera: Miridae). J. Econ. Entomol. 106:1310–1316. doi:10.1603/EC12422

Brewer, M.J., and P.B. Goodell. 2012. Approaches and incentives to implement integrated pest management that addresses regional and environmental issues. Annu. Rev. Entomol. 57:41–59. doi:10.1146/annurev-ento-120709-144748

Brewer, M.J., and M. Ishii-Eiteman. 2009. Integrated pest management, with special focus on sustainability and risk: Principles, policy and practice. In: L.E. Phoenix, editor, Critical food issues: Problems and state-of-the-art solutions worldwide. Praeger, Santa Barbara, CA. p. 33–52.

Brewer, M.J., T. Noma, and N.C. Elliott. 2008. A landscape perspective in managing vegetation for beneficial plant-pest-natural enemy interactions: A foundation for area-wide pest management In: O. Koul, G.W. Cuperus, and N.C Elliott. Area-wide pest management: Theory to implementation. CABI Publ., Oxfordshire, UK. p. 81–96.

Cannon, R.J. 2000. Bt transgenic crops: Risks and benefits. Integr. Pest Manage. Rev. 5:151–173. doi:10.1023/A:1011347122894

Carrière, Y., P.C. Ellsworth, P. Dutilleul, C. Ellers-Kirk, V. Barkley, and L. Antilla. 2006. A GIS-based approach for areawide pest management: The scales of *Lygus hesperus* movements to cotton from alfalfa, weeds, and cotton. Entomol. Exp. Appl. 118:203–210. doi:10.1111/j.1570-7458.2006.00384.x

Carrière, Y., P.B. Goodell, C. Ellers-Kirk, G. Larocque, P. Dutilleul, S.E. Naranjo, and P.C. Ellsworth. 2012. Effects of local and landscape factors on population dynamics of a cotton pest. PLoS ONE 7:e39862. doi:10.1371/journal.pone.0039862

Catchot, A. 2014. 2014 insect control guide for agronomic crops. Publ. 2471. Mississippi State Univ. Ext. Serv., Mississippi State, MS.

Chougule, N.P., and B.C. Bonning. 2012. Toxins for transgenic resistance to hemipteran pests. Toxins 4:405–429. doi:10.3390/toxins4060405

Clay, S.A. 2011. GIS applications in agriculture. Invasive species. Vol. 3. CRC Press, Boca Raton, FL.

Cochran, M.J., D.M. Danforth, S. Mi, N.P. Tugwell, N.R. Benson, and K.J. Bryant. 1999. Validation and economic benefits of COTMAN insecticide termination rules: Four years of research. In: D.M. Oosterhuis, editor, Proceedings of the 1999 Cotton Research Meeting and Summaries of Cotton Research in Progress. Spec. Rep. 193, Arkansas Agric. Exp. Stn., Fayetteville.

Cook, S.M., Z.R. Khan, and J.A. Pickett. 2007. The use of push-pull strategies in integrated pest management. Annu. Rev. Entomol. 52:375–400. doi:10.1146/annurev.ento.52.110405.091407

Danforth, D.M., and P.F. O'Leary, editors. 1998. COTMAN expert system 5.0. User's manual. Univ. of Arkansas Agric. Exp. Stn., Fayetteville.

Davison, J. 2010. GM plants: Science, politics and EC regulations. Plant Sci. 178:94–98. doi:10.1016/j.plantsci.2009.12.005

Deutscher, S., L.J. Wilson, and R.K. Mensah. 2005. Integrated pest management guidelines for cotton production systems in Australia. The Australian Cotton CRC, Narrabri, Australia.

Dhurua, S., and G.T. Gujar. 2011. Field-evolved resistance to Bt toxin Cry1Ac in the pink bollworm, *Pectinophora gossypiella* (Saunders) (Lepidoptera: Gelechiidae), from India. Pest Manage. Sci. 67:898–903. doi:10.1002/ps.2127

Dickerson, W.A., A.L. Brashear, J.T. Brumley, F.L. Carter, W.J. Grefenstette, and F.A. Harris. 2001. Boll weevil eradication in the United States through 1999. Ref. Book Ser. No. 6. The Cotton Found., Memphis, TN.

Dillon, M.L., G.P. Fitt, J.G. Hamilton, and W.A. Rochester. 1996. A simulation model of wind-driven dispersal of *Helicoverpa* moths. Ecol. Modell. 86:145–150. doi:10.1016/0304-3800(95)00042-9

Dittrich, V., S.O. Hassan, and G.H. Ernst. 1985. Sudanese cotton and the whitefly: A case study of the emergence of a new primary pest. Crop Prot. 4:161–176. doi:10.1016/0261-2194(85)90015-8

Duffy, P.A., D.L. Cain, G.L. Young, and M.W. Wetzstein. 1994. Effects of the boll weevil eradication program on Alabama cotton farms. J. Agribusiness 12:1–14.

Duraimurugan, P., and A. Regupathy. 2005. Push-pull strategy with trap crops, Neem and nuclear polyhedrosis virus for insecticide resistance management in *Helicoverpa armigera* (Hubner) in cotton. Am. J. Appl. Sci. 2:1042–1048. doi:10.3844/ajassp.2005.1042.1048

Edge, J.M., J.H. Benedict, J.P. Carroll, and H.K. Reding. 2001. Bollgard®cotton: An assessment of global economic, environmental, and social benefits. J. Cotton Sci. 5:121–136.

Elbert, A., M. Huas, B. Springer, W. Thielert, and R. Nauen. 2008. Applied aspects of neonictinoid uses in crop protection. Pest Manage. Sci. 64:1099–1105. doi:10.1002/ps.1616

Elliott, N.C., D.W. Onstad, and M.J. Brewer. 2008. History and ecological basis for areawide pest management. In: O. Koul, G.W. Cuperus, and N.C. Elliott, editors, Area-wide pest management: Theory to implementation. CABI Publ., Oxfordshire, UK. p. 15–33.

Ellsworth, P.C. 2013. Landscape influences on *Lygus* bugs in western cotton. APS crop protection and management collection: Webcasts, focus on cotton. Plant Management Network. http://www.plantmanagementnetwork.org/edcenter/seminars/cotton/LygusBugs/ (accessed 10 Apr. 2015).

Ellsworth, P.C., and J.S. Jones. 2001. Cotton IPM in Arizona: A decade of research, implementation & education. In: Proceedings of the Beltwide Cotton Conferences, Natl. Cotton Counc., Anaheim, CA. 9–13 Jan. 2001. Natl. Cotton Counc., Memphis, TN. p. 1088–1096.

Ellsworth, P.C., and J.L. Martinez-Carrillo. 2001. IPM for *Bemisia tabaci*: A case study from North America. Crop Prot. 20:853–869. doi:10.1016/S0261-2194(01)00116-8

Fitt, G.P. 1994. Cotton pest management: Part 3. An Australian perspective. Annu. Rev. Entomol. 39:543–562. doi:10.1146/annurev.en.39.010194.002551

Fitt, G.P. 2000. An Australian approach to IPM in cotton: Integrating new technologies to minimise insecticide dependence. Crop Prot. 19:793–800. doi:10.1016/S0261-2194(00)00106-X

Fitt, G.P. 2003. Deployment and impact of transgenic Bt cotton in Australia. In: N. Kalaitzandonakes, editor, The economic and environmental impacts of agbiotech. Springer US, New York. p. 141–164.

Forman, R.T.T. 1995. Land mosaics: The ecology of landscapes and regions. Cambridge Univ. Press, Cambridge, UK.

Frisbie, R.E., and P.L. Adkisson. 1985. CIPM: Integrated pest management on major agricultural systems, from a symposium sponsored by The Consortium for Integrated Pest Management and USDA/CSRS. Texas Agric. Exp. Stn. MP-1616, Texas A&M Univ., College Station, TX.

Frisbie, R.E., K.M. El-Zik, and L.T. Wilson, editors. 1989. Integrated pest management systems and cotton production. John Wiley & Sons, New York.

Georghiou, G.P., and C.E. Taylor. 1977. Operational influences in the evaluation of insecticide resistance. J. Econ. Entomol. 70:653–658.

Gerson, U., and E. Cohen. 1989. Resurgences of spider mites (Acari: Tetranychidae) induced by synthetic pyrethroids. Exp. Appl. Acarol. 6:29–46. doi:10.1007/BF01193231

Girolami, V., L. Mazzon, A. Squartini, N. Mori, M. Marzaro, A. DiBernardo, M. Greatti, C. Giorio, and A. Tapparo. 2009. Translocation of neonicotinoid insecticides from seeds to seedling guttation drops: A novel way of intoxication of bees. J. Econ. Entomol. 102:1808–1815. doi:10.1603/029.102.0511

Goodell, P.B. 2009. Fifty years of the integrated control concept: The role of landscape ecology in IPM in San Joaquin valley cotton. Pest Manage. Sci. 65:1293–1297. doi:10.1002/ps.1859

Gross, K., and J.A. Rosenheim. 2011. Quantifying secondary pest outbreaks in cotton and their monetary cost with causal-inference statistics. Ecol. Appl. 21:2770–2780. doi:10.1890/11-0118.1

Guinn, G. 1985. Fruiting of cotton. III. Nutritional stress and cutout. Crop Sci. 25:981–985. doi:10.2135/cropsci1985.0011183X002500060020x

Gurr, G.M., S.D. Wratten, and M.A. Altieri. 2004. Ecological engineering for pest management advances in habitat manipulation for arthropods. CSIRO Publ., Collingwood, Australia.

Hardee, D.D., M.R. Bell, and D.A. Streett. 1999. A review of area-wide management of *Helicoverpa* and *Heliothis* (Lepidoptera: Noctuidae) with pathogens (1987–1997). Southwest. Entomol. 24:62–75.

Hardee, D.D., and F.A. Harris. 2003. Eradicating the boll weevil (Coleoptera: Curculionidae): A clash between a highly successful insect, good scientific achievement, and differing agricultural philosophies. Am. Entomol. 48:82–97.

Hayes, J.L., and M. Bell. 1994. Evaluation of early-season baculovirus treatment for suppression of *Heliothis virescens* and *Helicoverpa zea* (Lepidoptera: Noctuidae) over a wide area. J. Econ. Entomol. 87:58–66.

Head, R.L. 1984. Report of the Cotton Insect Loss Committee of the Thirty-Seventh Annual Conference on Cotton Insect Research and Control. In: Proceedings of the Beltwide Cotton Production Research Conferences, Natl. Cotton Counc., Atlanta, GA. 8–12 Jan. 1984. Natl. Cotton Counc., Memphis, TN. p. 180.

Herbert, J.J., and M.D. Toews. 2011. Seasonal abundance and population structure of the brown stink bug in farmscapes containing corn, cotton, peanut and soybean. Ann. Entomol. Assoc. Amer. 104:909–918. doi:10.1603/AN11060

Herzog, G.A., J.B. Graves, J.T. Reed, W.P. Scott, and T.F. Watson. 1996. Chemical control. In: E.G. King, J. R. Phillips, and R. J. Coleman, editors, Cotton insects and mites: Characterization and management. The Cotton Found., Memphis, TN. p. 447–468.

Hollingsworth, R.G., D.G. Steinkraus, and R.W. McNew. 1995. Sampling to predict fungal epizootics in cotton aphids (Homoptera: Aphididae). Environ. Entomol. 24:414–421.

International Resistance Action Committee. 2012. IRAC MoA Classification Scheme, Version 7.2. International Resistance Action Committee.

Isley, D. 1934. Relationship between early varieties of cotton and boll weevil injury. J. Econ. Entomol. 27:762–766.

James, C. 2013. Global status of commercialized biotech/GM crops: 2013. ISAAA Brief 46-2013. International Service for the Acquisition of Agri-Biotech Applications, Ithaca, NY .

Kathage, J., and M. Qaim. 2012. Economic impacts and impact dynamics of Bt (*Bacillus thuringienis*) cotton in India. Proc. Natl. Acad. Sci. USA 109:11652–11656. doi:10.1073/pnas.1203647109

Kennedy, G.G., and N.P. Storer. 2000. Life systems of polyphagous arthropod pests in temporally unstable cropping systems. Annu. Rev. Entomol. 45:467–493. doi:10.1146/annurev.ento.45.1.467

Kerby, T.A., F.M. Bourland, and K.D. Hake. 2010. Physiological rationales in plant monitoring and mapping. In: J.McD Stewart, D.M. Oosterhuis, J.J. Heitholt, and J.R. Mauney, editors, Physiology of cotton. Springer, New York. p. 304–317.

Kerby, T.A., and K. Hake. 1996. Monitoring cotton's growth. In: T.A. Kerby, K. Hake, and S. Hake, editors, Cotton production manual. Publ. 3352. Div. of Agric. and Nat. Resour., Univ. of California, Oakland, CA. p. 335–355.

Kim, M.J., and D.M. Oosterhuis. 1998. Changes in the cotton boll wall with age in relation to bollworm and boll weevil feeding habits. In: Proceedings of the Beltwide Cotton Conferences, Natl. Cotton Counc., San Diego, CA.5–9 Jan. 1998. Natl. Cotton Counc., Memphis, TN. p. 1367–1368.

King, E.G., J.R. Phillips, and R.J. Coleman, editors. 1996. Cotton insects and mites: Characterization and management. The Cotton Found., Memphis, TN.

Kogan, M. 1998. Integrated pest management: Historical perspectives and contemporary developments—Levels of IPM. Annu. Rev. Entomol. 43:243–270. doi:10.1146/annurev.ento.43.1.243

Leonard, B.R., T.G. Teague, R.G. Luttrell, and S.A. Harris. 2008. Effects of insect-plant interactions on crop development. In: D.M. Oosterhuis and F.M. Bourland, editors, COTMAN Crop Management System. Univ. of Arkansas Agric. Exp. Stn., Fayetteville. p. 27–30.

Lu, Y., K. Wu, Y. Jiang, B. Xia, P. Li, H. Feng, and Y. Guo. 2010. Mirid bug outbreaks in multiple crops correlated with wide-scale adoption of Bt cotton in China. Science 328(5982):1151–1154. doi:10.1126/science.1187881

Luck, J.D., S.K. Pitla, S.A. Shearer, T.G. Mueller, C.R. Dillon, J.P. Fulton, and S.F. Higgins. 2010. Potential for pesticide and nutrient savings via map-based automatic boom section control of spray nozzles. Comput. Electron. Agric. 70:19–26. doi:10.1016/j.compag.2009.08.003

Luttrell, R.G. 1994. Cotton pest management: Part 2. A US perspective. Annu. Rev. Entomol. 39:527–542. doi:10.1146/annurev.en.39.010194.002523

Luttrell, R.G., and M.I. Ali. 2009. Variability in the response of *Helicoverpa zea* and *Heliothis virescens* (Lepidoptera: Noctuidae) to Cry1Ac and Cry2Ab2 in diet incorporation assays. Resist. Pest. Manage. Newsletter 19:33–36.

Luttrell, R.G., K.C. Allen, T.G. Teague, and P.F. O'Leary. 2006. Large farm management of cotton insects via the use of insect samples and COTMAN information. In: Proceedings of the Beltwide Cotton Conferences, Natl. Cotton Counc., San Antonio, TX. 3–6 Jan. 2006. Natl. Cotton Counc., Memphis, TN. p. 1326–1336.

Luttrell, R.G., F.T. Cooke, T.B. Freeland, J. Gibson, F.A. Harris, D.H. Laughlin, J.T. Reed, and C.E. Watson. 1997. A study of cotton insect control problems in Mississippi during 1995 with special emphasis on boll weevil eradication and tobacco budworm. Tech. Bull. 217. Mississippi Agric. For. Exp. Stn., Mississippi State, MS

Luttrell, R.G., G.P. Fitt, F.S. Ramalho, and E.S. Sugonyaev. 1994. Cotton pest management: Part 1: A worldwide perspective. Annu. Rev. Entomol. 39:517–526. doi:10.1146/annurev.en.39.010194.002505

Luttrell, R.G., and R.E. Jackson. 2012. *Helicoverpa zea* and Bt cotton in the United States. GM Crops Food 3(3):213–227. doi:10.4161/gmcr.20742

Magarey, R.D., M. Colunga-Garcia, and D.A. Fieselmann. 2009. Plant biosecurity in the United States: Roles, responsibilities, and information needs. BioScience 59:875–884. doi:10.1525/bio.2009.59.10.9

Matthews, G.A., and J.P. Tunstall, editors. 1994. Insect pests of cotton. CAB Int. Wallingford, UK.

McKibben, G.H., J.W. Smith, and W.L. McGovern. 1990. Design of an attract-and-kill device for the boll weevil (Coleoptera: Curculionidae). J. Entomol. Sci. 25:581–586.

McKinion, J.M., J.N. Jenkins, J.L. Willers, and A. Zumanis. 2009. Spatially variable insecticide applications for early season control of cotton insect pests. Comput. Electron. Agric. 67:71–79. doi:10.1016/j.compag.2009.03.004

Medrano, E.G., J.F. Esquivel, R.L. Nichols, and A.A. Bell. 2009. Temporal analysis of cotton boll symptoms resulting from southern green stink bug feeding and transmission of a bacterial pathogen. J. Econ. Entomol. 102:36–42. doi:10.1603/029.102.0106

Mensah, R.K. 1999. Habitat diversity: Implications for the conservation and use of predatory insects of *Helicoverpa* spp. In: cotton systems in Australia. Int. J. Pest Manage. 45:91–100. doi:10.1080/096708799227879

Mensah, R.K., and R.V. Sequeira. 2004. Habitat manipulation for insect pest management in cotton cropping systems. In: G.M. Gurr, S.D. Wratten, and M.A. Altieri, editors, Ecological engineering for pest management advances in habitat manipulation for arthropods. CSIRO Publ., Collingwood, Australia. p. 187–198.

Metcalf, R.L. 1980. Changing role of insecticides in crop protection. Annu. Rev. Entomol. 25:219–256. doi:10.1146/annurev.en.25.010180.001251

Miranda, J.E., and C.A.D. Da Silva. 2005. Behavioural control of the cotton boll weevil, *Anthonomus grandis* (Coleoptera: Curculionidae) in Northeast Brazil. Bol. Sanid. Veg. Plagas 31:509–515.

Musser, F.R., A.L. Catchot, S.D. Stewart, R.D. Bagwell, G.M. Lorenz, K.V. Tindall, G.E. Studebaker, B.R. Leonard, D.S. Akin, D.R. Cook, and C.A. Daves. 2009a. Tarnished plant bug (Hemiptera:Miridae) thresholds and sampling comparisons for flowering cotton in the midsouthern United States. J. Econ. Entomol. 102:1827–1836. doi:10.1603/029.102.0513

Musser, F.R., G.M. Lorenz, S.D. Stewart, R.D. Bagwell, B.R. Leonard, A.L. Catchot, K.V. Tindall, G.E. Studebaker, D.S. Akin, D.R. Cook, and C.A. Daves. 2009b. Tarnished plant bug (Hemiptera: Miridae) thresholds for cotton before bloom in the Midsouth of the United States. J. Econ. Entomol. 102:2109–2115. doi:10.1603/029.102.0614

Mustafic, A., E.E. Roberts, M.D. Toews, and A. Haidekker. 2013. LED-induced fluorescence and image analysis to detect stink bug damage in cotton bolls. J. Biol. Eng. 7:1–9. doi:10.1186/1754-1611-7-5

Naranjo, S.E. 2005. Long-term assessment of the effect of transgenic Bt cotton on the abundance of nontarget arthropod natural enemies. Environ. Entomol. 34:1193–1210. doi:10.1603/0046-225X(2005)034[1193:LAOTEO]2.0.CO;2

Naranjo, S.E. 2011. Impacts of Bt transgenic cotton on integrated pest management. J. Agric. Food Chem. 59:5842–5851. doi:10.1021/jf102939c

Naranjo, S.E., and P.C. Ellsworth. 2010. Fourteen years of Bt cotton advances IPM in Arizona. Southwest. Entomol. 35:437–444. doi:10.3958/059.035.0329

Naranjo, S.E., and R.G. Luttrell. 2009. Cotton arthropod IPM. In: E.B. Ratcliffe, W.D. Hutchison, and R.E. Cancelado, editors, Integrated pest management: Concepts, tactics, strategies and case studies. Cambridge Univ. Press, Cambridge, UK. p. 324–340.

National Research Council. 1986. Pesticide resistance: Strategies and tactics for management. Natl. Acad. Press, Washington, DC.

Nauen, R., and I. Denholm. 2005. Resistance of insect pests to neonictinoid insecticides: Current status and future prospects. Arch. Insect Biochem. Physiol. 58:200–215. doi:10.1002/arch.20043

Norris, R.F., and M. Kogan. 2000. Interactions between weeds, arthropod pests and their natural enemies in managed ecosystems. Weed Sci. 48:94–158. doi:10.1614/0043-1745(2000)048[0094:IBWAPA]2.0.CO;2

Oliveira, C.M., A.M. Auad, S.M. Mendes, and M.R. Frizzas. 2013. Economic impact of exotic insect pests in Brazilian agriculture. J. Appl. Entomol. 137:1–15. doi:10.1111/jen.12018

Oliveira, M.R.V., T.J. Henneberry, and P. Anderson. 2001. History, current status, and collaborative research projects for *Bemesia tabaci*. Crop Prot. 20:709–723. doi:10.1016/S0261-2194(01)00108-9

Oliver, M., T. Bishop, and B. Marchant. 2013. Precision agriculture for sustainability and environmental protection. Routeledge, New York.

Oosterhuis, D.M., and F.M. Bourland, editors. 2008. COTMAN Crop Management System, Univ. of Arkansas Agric. Exp. Stn., Fayetteville.

Oosterhuis, D.M., F.M. Bourland, N.P. Tugwell, and M.J. Cochran. 1996. Terminology and concepts related to the COTMAN crop monitoring system. Spec. Rep. 174. Univ. of Arkansas Agric. Exp. Stn., Fayetteville.

Oosterhuis, D.M., and M. Kim. 2004. Structural and biochemical changes in the boll wall in support of the NAWF5+350 heat units rule for insecticide termination. In: Proceedings of the Beltwide Cotton Conferences, Natl. Cotton Counc., San Antonio, TX. 5–9 Jan. 2004. Natl. Cotton Counc., Memphis, TN. p. 258–295.

Ortiz, B.V., D.G. Sullivan, C. Perry, P. Lu, R. Kemerait, R.F. Davis, A. Smith, G. Vellidis, and R. Nicholds. 2012. Variable rate application of nematicides on cotton fields: A promising site-specific management strategy. J. Nematol. 44(1):31–39.

Palumbo, J., A. Horowitz, and N. Prabhaker. 2001. Insecticidal control and resistance management for *Bemisia tabaci*. Crop Prot. 20:739–765. doi:10.1016/S0261-2194(01)00117-X

Parajulee, M.N., R.B. Shrestha, and J.F. Leser. 2006. Sampling methods, dispersion patterns, and fixed precision sequential sampling plans for western flower thrips (Thysanoptera: Thripidae) and cotton Fleahoppers (Hemiptera: Miridae) in cotton. J. Econ. Entomol. 99:568–577. doi:10.1603/0022-0493-99.2.568

Parencia, C.R., T.R. Pfrimmer, and A.R. Hopkins. 1983. Insecticides for control of cotton insects. In: R.L. Ridgway, E.P. Lloyd, and W.H. Cross, editors, Cotton insect management with special reference to the boll weevil. Agric. Handb. No. 589. USDA-ARS, Washington, DC. p. 237–264.

Pedigo, L.P., and G.D. Buntin. 1994. Handbook of sampling methods for arthropods in agriculture. CRC, Boca Raton, FL.

Pedigo, L.P., S.H. Hutchins, and L.G. Higley. 1986. Economic injury levels in theory and practice. Annu. Rev. Entomol. 31:341–368. doi:10.1146/annurev.en.31.010186.002013

Pedigo, L.P., and M.E. Rice. 2009. Entomology and pest management. 6th ed. Pearson Prentice Hall, Upper Saddle River, NJ.

Plapp, F.W., G.M. McWhorter, and W.H. Vance. 1987. Monitoring for pyrethroid resistance in the tobacco budworm in Texas—1986, In: Proceedings of the Beltwide Cotton Production Research Conferences, Natl. Cotton Counc., Dallas, TX. 4–8 Jan. 1987. Natl. Cotton Counc., Memphis, TN, p.324–326..

Plapp, F.W., Jr., J.A. Jackman, C. Campanhola, R.E. Frisbie, J.B. Graves, R.G. Luttrell, W.F. Kitten, and M. Wall. 1990. Monitoring and management of pyrethroid resistance in tobacco budworm (Lepidoptera: Noctuidae) in Texas, Mississippi, Louisiana, Arkansas, and Oklahoma. J. Econ. Entomol. 83:335–341.

Purcell, J.P., and F.J. Perlak. 2004. Global impact of insect-resistant (Bt) cotton. AgBioForum 7:27–30.

Qaim, M., and D. Zilberman. 2003. Yield effects of genetically modified crops in developing countries. Science 299:900–902. doi:10.1126/science.1080609

Ramalho, F.S. 1994. Cotton pest management: Part 4. A Brazilian perspective. Annu. Rev. Entomol. 39:563–578. doi:10.1146/annurev.en.39.010194.003023

Ramalho, F.S., and P.A. Wanderly. 1996. Ecology and management of the boll weevil in South American cotton. Am. Entomol. 42:41–47.

Reay-Jones, F.P.F., M.D. Toews, J.K. Greene, and R.B. Reeves. 2010. Development of sampling plans for cotton bolls injured by stink bugs (Hemiptera: Pentatomidae). J. Econ. Entomol. 103:525–532. doi:10.1603/EC09290

Ridgway, R.L., A.A. Bell, J.A. Veech, and J.M. Chandler. 1984. Cotton production practices in the USA and World. In: R.J. Kohel and C.F. Lewis, editors, Cotton. ASA, CSSA, and SSSA, Madison, WI. p. 266–361.

Ruberson, J.R., G.A. Herzog, W.R. Lambert, and W.J. Lewis. 1994. Management of the beet armyworm (Lepidoptera:Noctuidae) in cotton: Role of natural enemies. Fla. Entomol. 77:440–453. doi:10.2307/3495698

Sadras, V., and G. Felton. 2010. Mechanisms of cotton resistance to arthropod herbivory. In: J.McD Stewart, D.M. Oosterhuis, J.J. Heitholt, and J.R. Mauney, editors, Physiology of cotton. Springer, New York. p. 213–228.

Schneider, J.C. 1989. Role of movement in evaluation of area-wide insect pest management. Environ. Entomol. 18:868–874.

Shelton, A.M., J.Z. Zhao, and R.T. Roush. 2002. Economic, ecological, food safety, and social consequences of the deployment of Bt transgenic plants. Annu. Rev. Entomol. 47:845–881. doi:10.1146/annurev.ento.47.091201.145309

Shewhart, W.A. 1939. Statistical methods from the viewpoint of quality control. The Graduate School, USDA, Washington, DC.

Sequeira, R.V., and C.L. Playford. 2001. Abundance of *Helicoverpa* (Lepidoptera: Noctuidae) pupae under cotton and other crops in central Queensland: Implications for resistance management. Aust. J. Entomol. 40:264–269. doi:10.1046/j.1440-6055.2001.00234.x

Sevacherian, V., and V.M. Stern. 1974. Movements of *Lygus* bugs between alfalfa and cotton. Environ. Entomol. 4:163–165.

Showler, A.T. 2006. Short-range dispersal and overwintering habitats of boll weevils (Coleopter: Curculionidae) during and after harvest in the subtropics. J. Econ. Entomol. 99(4):1152–1160. doi:10.1603/0022-0493-99.4.1152

Showler, A.T. 2009. Roles of host plants in boll weevil range expansion beyond tropical Mesoamerica. Am. Entomol. 55:228–236.

Sivakoff, F.S., J.A. Rosenheim, P. Dutilleui, and Y. Carrière. 2013. Influences of the surrounding landscape on crop colonization by a polyphagous insect pest. Entomol. Exp. Appl. 149:11–21. doi:10.1111/eea.12101

Slosser, J.E., W.E. Pinchak, and D.R. Rummel. 1989. A review of known and potential factors affecting the population dynamics of the cotton aphid. Southwest. Entomol. 14:301–313.

Smith, J.W. 1998. Boll weevil eradication: Area-wide management. Ann. Entomol. Soc. Am. 91:239–247.

Smith, R.H. 1989. Experiences with beet armyworm control in cotton in 1988. In: Proceedings of the Beltwide Cotton Production Research Conferences, Natl. Cotton Counc., Nashville, TN. 2–7 Jan. 1989. Natl. Cotton Counc., Memphis, TN. p. 273–275.

Snodgrass, G.L., C. Abel, R. Jackson, and J. Gore. 2008. Bioassay for determining resistance levels in tarnished plant bug populations to neonicotinoid insecticides. Southwest. Entomol. 33:173–180. doi:10.3958/0147-1724-33.3.173

Snodgrass, G.L., J. Gore, C.A. Abel, and R. Jackson. 2009. Acephate resistance in populations of the tarnished plant bug (Heteroptera: Miridae) from the Mississippi River Delta. J. Econ. Entomol. 102:699–707. doi:10.1603/029.102.0231

Snodgrass, G.L., and W.P. Scott. 2000. Seasonal changes in pyrethroid resistance in tarnished plant bug (Heteroptera: Miridae) populations during a three-year period in the Delta area of Arkansas, Louisiana, and Mississippi. J. Econ. Entomol. 93:441–446. doi:10.1603/0022-0493-93.2.441

Snodgrass, G.L., W.P. Scott, C.A. Abel, J.T. Robbins, J. Gore, and D.D. Hardee. 2006. Suppression of tarnished plant bugs (Heteroptera: Miridae) in cotton by control of early season wild host plants with herbicides. Environ. Entomol. 35:1417–1422. doi:10.1603/0046-225X(2006)35[1417:SOTPBH]2.0.CO;2

Spurgeon, D.W. 2008. Ecologically based integrated pest management in cotton. In: O. Koul, G.W. Cuperus, and N.C. Elliott, editors, Area-wide pest management: Theory to implementation. CABI Publ., Oxfordshire, UK. p. 367–405.

Spurgeon, D.W., J.R. Coppedge, J.R. Raulston, and H. Marshall. 1999. Mechanisms of boll weevil (Coleptera: Curculionidae) bait stick activity relative to pheromone traps. J. Econ. Entomol. 92:960–966.

Stern, V.M., A. Muller, V. Sevacherian, and M. Way. 1969. *Lygus* bug control in cotton through alfalfa interplanting. Calif. Agric. 23:8–10.

Stewart, S.D., G.M. Lorenz, A.L. Catchot, J. Gore, D. Cook, J. Skinner, T.C. Mueller, D.R. Johnson, J. Zawislak, and J. Barber. 2014. Potential exposure of pollinators to neonicotinoid insecticide from the use of insecticide seed treatments in the mid-southern United States. Environ. Sci. Technol. 48:9762–9769. doi:10.1021/es501657w

Storer, N.P., M.E. Kubiszak, J.E. King, G.D. Thompson, and A.C. Santos. 2012. Status of resistance to Bt maize in *Spodoptera frugiperda:* Lessons from Puerto Rico. J. Invertebr. Pathol. 110:294–300. doi:10.1016/j.jip.2012.04.007

Studebaker, G. 2014. 2014 insecticide recommendations for Arkansas. MP144. Arkansas Coop. Ext. Serv., Little Rock.

Sugonyaev, E.S. 1994. Cotton pest management: Part 5. A Commonwealth of Independent States perspective. Annu. Rev. Entomol. 39:579–592. doi:10.1146/annurev.en.39.010194.003051

Sui, R., and J.A. Thomasson. 2013. Mechanization, sensing, and control in cotton production. In: Q. Zang and F.J. Pierce, editors, Agricultural automation: Fundamentals and practices. CRC Press, Boca Raton, FL. p. 125–144.

Summy, K.R., J.R. Raulston, D. Spurgeon, and J. Vargus. 1996. An analysis of the beet armyworm outbreak on cotton in the Lower Rio Grande Valley of Texas during the 1995 production season. In: Proceedings of the Beltwide Cotton Conferences, Natl. Cotton Counc., Nashville, TN. 9–12 Jan. 1996. Natl. Cotton Counc., Memphis, TN. p. 837–842.

Tabashnik, B.E., and Y. Carrière. 2010. Field-evolved resistance to Bt cotton: Bollworm in the U.S. and pink bollworm in India. Southwest. Entomol. 35:417–424. doi:10.3958/059.035.0326

Tabashnik, B.E., and B.A. Croft. 1982. Managing pesticide resistance in crop-arthropod complexes: Interactions between biological and operational factors. Environ. Entomol. 11:1137–1144.

Tay, W.T., M.F. Soria, T. Walsh, D. Thomazoni, P. Silvie, G.T. Behere, C. Anderson, and S. Downes. 2013. A brave new world for an Old World pest: *Helicoverpa armigera* (Lepidoptera: Noctuidae) in Brazil. PLoS ONE 8(11): e80134.

Teague, T.G. 2011. Irrigation timing and tarnished plant bug management—Implications for late season susceptibility to tarnished plant bug and crop termination decisions—Year III. In: Proceedings of the Beltwide Cotton Conferences, Natl. Cotton Counc., Atlanta, GA. 4–7 Jan. 2011. Natl. Cotton Counc., Memphis, TN. p. 1186–1198.

Teague, T.G., D.M. Danforth, and E.J. Villavaso. 2003. Changes in cotton fruiting dynamics following pre-flower injury from tarnished plant bug. In: D.M. Oosterhuis, editor, Summaries of Arkansas Cotton Research 2002. Spec. Res. Ser. 507. Arkansas Agric. Exp. Stn., Fayetteville. p. 245–253.

Teague, T.G., E.J. Kelly, D.M. Danforth, and D. Wildy. 2014. Insect control termination decisions across irrigated and rainfed management zones in center pivot irrigated cotton. In: Proceedings of the Beltwide Cotton Conferences, Natl. Cotton Counc., New Orleans, LA. 6–8 Jan. 2014. Natl. Cotton Counc., Memphis, TN. p. 741–753..

Teague, T.G., C. Rothrock, and C.R. Shumway. 2012. Cotton response to irrigation timing and use of enhanced efficiency nitrogen fertilizer and biosolids– Year II. In: D.M. Oosterhuis, editor, Summaries of Arkansas Cotton Research 2011. Spec. Res. Ser. 602. Arkansas Agric. Exp. Stn., Fayetteville. p. 80–91.

Teague, T.G., J. Smith, D.M. Danforth, and P.F. O'Leary. 2008. Manually applied infestations of tarnished plant bug nymphs in late season cotton to identify the final stage of crop susceptibility. In: Proceedings of the Beltwide Cotton Conferences, Natl. Cotton Counc., Nashville, TN. 8–11 Jan. 2008. Natl. Cotton Counc., Memphis, TN. p. 1239–1250.

Tharp, W.H. 1960. The cotton plant, how it grows and why its growth varies. Agric. Handb. 178. USDA-ARS, Washington, DC.

Tillman, P.G., T.D. Northfield, R.F. Mizell, and T.C. Riddle. 2009. Spatiotemporal patterns and dispersal of stink bugs (Heteroptera: Pentatomidae) in peanut-cotton farmscapes. Environ. Entomol. 38:1038–1052. doi:10.1603/022.038.0411

USEPA. 2001. Biopesticides registration action document—*Bacillus thuringiensis* plant-incorporated protectants (10/16/2001). USEPA. http://www.epa.gov/pesticides/biopesticides/pips/bt_brad.htm (accessed 10 Apr. 2015).

Vitale, J.D., G. Vognan, M. Ouattarra, and O. Traore. 2010. The commercial application of GMO crops in Africa: Burkina Faso's decade of experience with Bt Cotton. AgBioForum 13:320–332.

Wang, Y., G.W. Watson, and R. Zhang. 2010. The potential distribution of an invasive mealybug *Phenacoccus solenopsis* and its threat to cotton in Asia. Agric. For. Entomol. 12:403–416. doi:10.1111/j.1461-9563.2010.00490.x

Westbrook, J.K., C.P. Suh, C. Yang, and R.S. Eyster. 2014. Temporal variability of spectral reflectance and estimated canopy cover of cotton plants supports early detection of potential boll weevil infestations, In: Proceedings of the Beltwide Cotton Conferences, Natl. Cotton Counc., New Orleans, LA. 6–8 Jan. 2014. Natl. Cotton Counc., Memphis, TN. p. 863–868.

Whitehouse, M.E.A., L.J. Wilson, and G.P. Fitt. 2005. A comparison of arthropod communities in transgenic Bt and conventional cotton in Australia. Environ. Entomol. 34:1224–1241. doi:10.1603/0046-225X(2005)034[1224:ACOACI]2.0.CO;2

Willers, J.L., E. Jallas, J.M. McKinion, M.R. Seal, and S. Turner. 2009a. Precision farming, Myth or reality: Selected case studies from Mississippi cotton fields. In: P.J. Papajorgji and P.M. Pardalos, editors, Advances in modeling agricultural systems. Springer, New York. p. 243–272.

Willers, J.L., J.N. Jenkins, J.M. McKinion, P. Gerard, K.B. Hood, J.R. Bassie, and M.D. Cauthen. 2009b. Methods of analysis for georeferenced sample counts of tarnished plant bugs in cotton. Precis. Agric. 10:189–212. doi:10.1007/s11119-008-9085-x

Willers, J.L., M.R. Seal, and R.G. Luttrell. 1999. Remote sensing, line-intercept sampling for tarnished plant bugs (Heteroptera:Miridae) in Mid-south cotton. J. Cotton Sci. 3:160–170.

Williams, M.R. 2014. Cotton insect losses—2013. In: Proceedings of the Beltwide Cotton Conferences, Natl. Cotton Counc., New Orleans, LA. 6–8 Jan. 2014. Natl. Cotton Counc., Memphis, TN. p. 798–812.

Williams, M.R., and M.B. Layton. 1996. Taken by storm: A report of the tobacco budworm problem in Mississippi. In: Proceedings of the Beltwide Cotton Conferences, Natl. Cotton Counc., Nashville, TN. 9–12 Jan. 1996. Natl. Cotton Counc., Memphis, TN. p. 823–825.

Wilson, L.J., L.R. Bauer, and G.A. Lully. 1998. Effect of early season insecticide use on predators and outbreaks of spider mites (Acari: Tetramychidae) in cotton. Bull. Entomol. Res. 88:477–488. doi:10.1017/S000748530004222X

Wu, K.M., and Y.Y. Guo. 2005. The evolution of cotton pest management practices in China. Annu. Rev. Entomol. 50:31–52. doi:10.1146/annurev.ento.50.071803.130349

Wu, K.M., Y.H. Lu, H.Q. Feng, Y.Y. Jiang, and J.Z. Zhao. 2008. Suppression of cotton bollworm in multiple crops in China in areas with Bt toxin–containing cotton. Science 321:1676–1678. doi:10.1126/science.1160550

Zhang, L., S.M. Greenburg, Y. Zhang, and T. Liu. 2011. Effectiveness of thiamethoxam and imidacloprid seed treatments against *Bemisia tabaci* (Hemiptera: Aleyrodidae) on cotton. Pest Manag. Sci. 67:226–232. doi:10.1002/ps.2056

Cotton Nematodes

David B. Weaver

Abstract

Cotton (*Gossypium* L.) is susceptible to at least four species of parasitic nematodes. Two major economic pathogens are the southern root-knot [*Meloidogyne incognita* (Kofoid and White) Chitwood] and reniform nematode (*Rotylenchulus reniformis* Linford and Oliveria). Of lesser importance are the Columbia lance (*Hoplolaimus columbus* Sher) and sting (*Belonolaimus longicaudatus* Rau) nematodes. It is estimated that about 4% of total US cotton production is lost each year due to nematode feeding. Fields with high nematode populations can sustain much greater losses. A variety of approaches have been taken to manage nematodes, including chemical and biological nematicides, crop rotation, and genetic resistance. The use of crop rotation is limited by a general lack of suitable rotation crops and economic factors that limit their usefulness. Nematicides, while effective in reducing nematode populations and yield losses, are limited by their availability and environmental concerns. Extensive searches of the germplasm pools in cotton have revealed several sources of resistance to both root-knot and reniform nematodes, with the subsequent release of several germplasms that have resistance to one or both nematode species. Development of molecular markers associated with resistance has greatly facilitated the introgression of these resistance genes into adapted germplasm. For the first time, cultivars with high levels of genetic resistance to root-knot nematode are available, and cultivars with resistance to reniform nematode are being developed and should be on the market soon.

Cotton is subject to attack and economic yield loss by two main species of biotrophic, plant parasitic nematodes, the southern root-knot nematode and the reniform nematode (Koenning et al., 2004; Starr et al., 2007). Yield losses to these two species are estimated to be between 4 and 5% of total cotton yield in the United States (National Cotton Council of America, 2013), more or less equally divided between the two. Estimates have shown a recent upward trend, however, with estimated yield losses increasing steadily from about 2% in the mid-1980s to the current level (Fig. 1). There are several possible reasons for this increase, which include an increased awareness of nematodes due to better grower education, increased sampling, and better assay methods; a general lack of crop rotation as a management tool; and the almost entire lack of cultivars with good genetic resistance (Starr et al., 2007). Either way, with current production levels this 4%

Abbreviations: QTL, quantitative trait locus; SSR, simple sequence repeat.

David B. Weaver, Dep. of Crop, Soil and Environmental Sciences, Auburn University, Auburn University, AL 36849-5412 (weavedb@auburn.edu).

doi:10.2134/agronmonogr57.2013.0045

Cotton. 2nd. ed. David D. Fang and Richard G. Percy, editors. Agronomy Monograph 57.

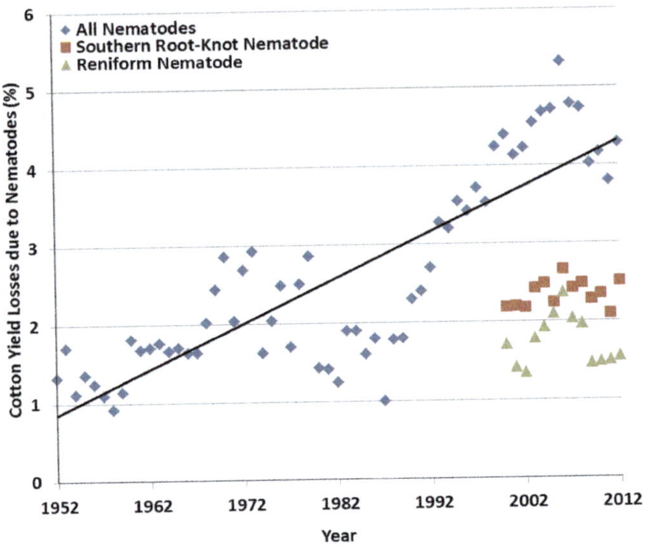

Fig. 1. Cotton yield loss estimates due to nematode damage from all nematodes. All nematodes were separated into southern root knot and reniform nematode beginning in 2000.

yield loss amounts to more than 800,000 bales lost to nematode feeding each year. It is estimated that more than 400,000 kg of aldicarb [2-methyl-2-(methylthio)propionaldehyde *O*-(methylcarbamoyl)oxime] were applied to 16% of the total US Upland cotton (*G. hirsutum* L.) hectarage in 2010, mostly in an effort to manage nematodes (USDA-NASS, 2014). Other species can attack cotton, primarily the sting nematode and Columbia lance nematode, but they are much less widely distributed and their impacts on an economic level are much more limited. Each species will be discussed in turn with regard to economic importance, species characteristics and symptoms, management strategies, and genetic resistance.

Root-Knot Nematode

The root-knot nematode is a ubiquitous apomictic nematode parasitic and pathogenic to a large number of plant species (Trudgill and Blok, 2001). With such a wide host range, it has been described as the world's most damaging crop pathogen (Trudgill and Blok, 2001). The species of root-knot nematode that is responsible for economic damage to cotton is the southern root-knot nematode (*Meloidogyne incognita,* also formerly known as *M. incognita acrita*), which is the only species in the genus *Meloidogyne* that is capable of significant parasitism and reproduction on cotton (Koenning et al., 2004). *Meloidogyne incognita* has been subdivided into races based on a set of host species differentials (Taylor and Sasser, 1978), with races 3 and 4 being able to reproduce on cotton. Southern root-knot nematode is found in every US state where cotton is produced (Koenning et al., 2004; Starr et al., 2007, 2010). Although yield loss estimates for many cotton pathogens have been compiled since 1952 (National Cotton Council of America, 2013), estimates for losses to nematodes were lumped into one category, as *Meloidogyne* spp., until 2000. Other species were not recognized until 2000, when yield losses were subdivided into *Meloidogyne* spp., *Rotylenchulus* spp., and "other" nematode species.

Several excellent reviews have been written in description of the taxonomy, life cycle, and biological attributes of *Meloidogyne* spp. (Taylor and Sasser, 1978). Briefly, the cotton root-knot nematode is a sedentary ectoparasite, with juveniles hatching from eggs laid in a gelatinous mass on roots. They then pass through four metamorphic stages, with the second juvenile stage (J2) being the only stage that infects cotton roots. They enter actively growing cotton roots behind the meristematic region and begin feeding in the zone of differentiation. The nematodes then pass through the remaining two juvenile stages while sedentary in the root. Secretions from the nematode stimulate the production of multinucleated giant cells. The hyperplasia development (giant cells) in root tissue results in the formation of the characteristic galls that give the nematode its name. Aboveground visual symptoms such as poor growth (stunting) and wilting of the plants may occur, but root galls are the most useful disease-specific diagnostic tool. The presence or absence of galling has been widely used as a differentiating phenotype in efforts to discover, evaluate, and utilize genetic resistance (Shepherd, 1974a, 1974b; Robinson et al., 1999, 2004; Zhang et al., 2006). It can also be used as a diagnostic tool in the field (Wrather et al., 2010).

The use of planting time applications of nematicides can reduce soil populations of root-knot nematode. There are several chemicals that have nematicidal properties. Many of the nematicides used in earlier years were the fumigant types, such as EDB (ethylene dibromide), chloropicrin (trichloronitromethane), 1,3-dicloropropene, and DBCP (1,2-dibromo-3-chloropropane) (Jorgenson,1979). These options for root-knot nematode management were proven to be very effective and have come and gone over the years, as fumigants have been withdrawn from agricultural use. Other non-fumigant soil-applied granular materials such as aldicarb have gradually taken their place, but even these chemicals have been recently disallowed by the USEPA (USEPA, 2010). More recently seed treatment nematicides have been compared with soil-applied granular nematicides with variable results. Lawrence and Lawrence (2007) summarized the results of multiple trials comparing thiodicarb (marketed as Aeris) and abamectin (marketed as Avicta) nematicidal seed treatments with aldicarb and non-treated controls. The nematicidal properties of abamectin have been documented (Faske and Starr, 2006). Yield response to abamectin was similar to that for aldicarb treatment in 80% of the trials. Although these experiments targeted root-knot nematode as well as other pathogenic species (reniform and Columbia lance nematode, depending on location), the results indicated the effectiveness of the nematicidal seed treatments in general. Wheeler et al. (2013) compared thiodicarb and abamectin for root-knot nematode management in a range of environments in Texas and Alabama. They found that efficacy of the nematicide treatments was improved when soil moisture was not limiting. When corrected for environmental variables, the nematicidal seed treatments were effective compared with the aldicarb control. The questions surrounding the economics, environmental impacts, and safety of nematicides emphasizes the need for alternative management strategies, with crop rotation and host-plant resistance receiving the most attention.

The value of crop rotation with non-host crops for the management of root-knot nematode has been demonstrated and recommended for many years

(Nusbaum and Ferris, 1973). Unfortunately, the wide host range of the pathogen leaves few suitable rotation crop alternatives. Peanut (*Arachis hypogeae* L.) is immune to southern root-knot nematode (Kirkpatrick and Sasser, 1984) and is recommended and works well as a rotation crop for cotton (Kirkpatrick and Sasser, 1984; Koenning et al., 2004). However, this is by no means the complete solution, as the acreage devoted to cotton in the United States is more than 10 times that for peanut, so there is not enough peanut production to fully satisfy the need for rotation with cotton. Also, peanut, like cotton, requires a specialized infrastructure and specialized equipment for harvesting and processing and does not lend itself to production in some environments in which cotton is currently grown, such as those with heavier soil textures. Although not considered to be a crop rotation system, as it is still a cotton monoculture, the use of root-knot resistant 'NemX' in rotation with susceptible cotton 'Maxxa' was found to reduce nematode population density and improve the yield of the susceptible cultivar (Ogallo et al., 1999).

Because of the limited availability of rotations as a management option, genetic resistance to root-knot nematode has been a long-time goal of cotton breeding. One of the earliest successes in systematic breeding for host-plant resistance to a pathogen was reported by Ware in 1936, when he described efforts to breed cotton for resistance to the disease complex known as the "nematode-wilt" complex, a synergistic pathogen combination of root-knot nematode and Fusarium wilt (caused by *Fusarium oxysporum* f. sp. *vasinfectum* W.C. Snyder & H.N. Hansen) (Ware, 1936; Starr et al., 1989). This was a problem that was prevalent in cotton fields during the early part of the 20th century and continues to cause problems today. Reports of success in developing root-knot nematode resistant cotton and differential reaction to root-knot nematode within the adapted cotton germplasm pool began to appear in the literature as early as 1941 when Smith (1941) attributed an increase in resistance to root-knot nematode in certain cultivars to selection for resistance to Fusarium wilt and suggested that increased resistance to Fusarium wilt could be achieved by increased attention to breeding for root-knot nematode resistance. Some of the cultivar differences, based on severity of galling, were quite distinct. Smith (1941) also noted a connection between the geographic origin of the cultivars and their level of resistance. Those developed in the Coastal Plain area of the southeastern United States, where the Fusarium-root-knot complex was prevalent, were more resistant to root-knot nematode than cultivars developed in the Mid-South area, where root-knot nematode was much less of a problem. 'Auburn 56', released in 1952 (Smith, 1964), was the culmination of more than 50 yr of efforts to breed root-knot nematode resistant cotton and was the first commercially available cotton with any appreciable level of root-knot nematode resistance. However, it was described to be only "moderately tolerant" (Minton, 1962), and thus it was clear that higher levels of resistance were needed.

Shepherd (1974b) developed two germplasm lines of cotton with root-knot nematode resistance, A623 and A61, using the *G. hirsutum* resistance source he referred to as "Wild Mexico," variously also known as "Wild Mexican Jack Jones" or "Mexico Wild," PI 593649 (TX 2516) in the USDA cotton germplasm collection (USDA-ARS, 2014). This accession was collected in Mexico by J.L. Stulb of New Orleans in 1951 and deposited in the germplasm collection in 1963. It was determined to have resistance to root-knot nematode by Jack Jones in 1958 (Jones et al., 1958), hence the name variation. However, Shepherd also observed transgressive segregation for resistance in the populations he studied, finding that the two germplasm lines he developed actually had higher levels of resistance than the original source. Thus, he concluded that the other parent in the cross, 'Clevewilt-6' (PI 528617, SA 235), must have also contributed genes for resistance. A623 was released as the germplasm line Auburn 623 RNR (Shepherd, 1974a). The Clevewilt-6 source of resistance was also utilized in other breeding programs, resulting in the release of germplasm LA434-RKR (Jones et al., 1988), and eventually released in two cultivars, 'Stoneville LA887' and 'Paymaster H1560', which have been described as moderately resistant to root-knot nematode (Robinson et al., 2001) and were widely grown in the South during the 1990s.

Later work by Shepherd (1986) concluded that increased resistance to Fusarium wilt could only be achieved by breeding for resistance to root-knot nematode. Several germplasm lines were released based on the resistance in Auburn 623 RNR. These included Auburn 566 RNR, Auburn 612 RNR, and Auburn 634 RNR (Shepherd, 1982), which were backcrossed-derived lines with different recurrent parents in an effort to improve the agronomic and fiber quality traits of the original Auburn 623 RNR. Additional agronomic improvements were made, resulting in the release of a series of nine germplasm lines, the "M" series of lines (M-92 RNR, M-120 RNR, M-155 RNR, M-240 RNR, M-249 RNR, M-272 RNR, M-315 RNR, M-331 RNR, and M-725 RNR) (Shepherd et al., 1996). The source of resistance in these lines traces back to the resistance developed from the original Wild Mexico × Clevewilt-6 population, with each genotype being a backcross-derived line using a different, widely used and adapted recurrent parent cultivar. The significance of the relatively large number of lines was because of the wide range in recurrent parent genotypes, under the assumption that crosses could be made within M-series germplasm lines with primary selection among the progenies for yield and fiber characteristics, without extensive screening for nematode resistance (Shepherd et al., 1996), as this was considered to be one of the primary barriers to incorporation of resistance in a breeding program and underlies one of the problems in breeding high-yielding, adapted cultivars with good resistance. Shepherd's assessment of resistance was based primarily on a root galling index (Shepherd, 1974a) and nematode reproduction, as measured by determining the number of nematode egg masses per plant (Shepherd, 1979). Methods of resistance assessment are important in comparing studies related to inheritance of resistance and development of marker systems for marker-assisted selection. Most all of the studies cited herein use a gall index, generally on a scale of 1 to 5 (1 = no galling and 5 = extreme galling) or some similar scale, as a measure of plant reaction to the nematode under the assumption that a lower gall index means a higher level of resistance. Many times this is combined with a measure of nematode reproduction, usually by counting the number of egg masses per plant. This assumes that nematode reproduction is higher on more susceptible plants. These quantitative measures are then separated into qualitative classes based on comparison to parental responses in the same environment. Shepherd (1979) determined that measurement of both gall index (plant response to the nematode) and reproduction (nematode response to the plant) was necessary to identify genotypes with high levels of resistance.

A later source of resistance that contributed to breeding efforts was NemX, which was released by a grower cooperative, the California Planting Cottonseed Distributors, in 1995 (Ogallo et al., 1997) and was intended for production by growers in the San Joaquin Valley of California. NemX was derived from several lines including lines collected from the Acala region of Mexico, but the main source of nematode resistance in NemX is thought to be N6072, based on the pedigree of NemX (Bowman et al., 2006) and the reaction of N6072 to root-knot nematode in experiments by Hyer et al. (1979).

Many studies have been conducted on the inheritance, mechanism, and molecular aspects of both the M-series and the NemX sources of resistance. Bezawada et al. (2003) studied the inheritance of resistance in Clevewilt-6 and found it to be determined by a single recessive gene, but could find no significant association with several simple sequence repeat (SSR) markers that were polymorphic in the population studied. McPherson et al. (2004) studied inheritance of resistance in M-315 RNR and found it to be controlled by two genes, a dominant gene at one locus and an additive gene at another. When both genes were present, they contributed equally to the phenotype in an additive fashion. He thus assigned the gene symbols Mi_1 for the dominant gene and Mi_2 for the additive gene and proposed the genotype $Mi_1Mi_1Mi_2Mi_2$ for M-315 RNR. The two-gene model in M-315 RNR had previously been supported by phenotypic observations in studies on post root-penetration and nematode development by Jenkins et al. (1995). McPherson et al. (2004) also hypothesized that the gene in Clevewilt-6 studied by Bezawada et al. (2003) was the same as Mi_2.

Inheritance of resistance in NemX was also found to be conditioned by a single gene (Zhou et al., 1999), but the exact nature of gene action was unclear and could be deduced as either recessive or additive depending on how the data were interpreted.

The evaluation of resistance in inheritance studies and in incorporating resistance into agronomically acceptable cultivars is, at best, a time-consuming, expensive process that is subject to many environmental variables, complicated by the fact that two traits related to the host reaction need to be measured to incorporate high levels of resistance (Shepherd, 1979). Pyramiding different genes based solely on phenotypic selection is even more challenging. Dominant gene action further complicates the process. Even with the effort to develop sources of resistance, evaluate their inheritance, and develop improved germplasm lines with high levels of resistance and improved agronomics, by 1995 there were still no cultivars available with more than moderate levels of resistance to root-knot nematode (Jenkins et al., 1993; Gutiérrez et al., 2010).

Thus, efforts were made beginning in the 1990s to discover molecular markers associated with resistance. Bezawada et al. (2003) screened a limited number of SSR markers for association with the resistance gene in Clevewilt 6-1, but could find no significant linkage. Using a bulk segregant analysis, Wang et al. (2006a) investigated inheritance of the gene in NemX, confirmed it to be controlled by a single gene, and assigned it the gene symbol rkn_1. Furthermore, they identified the rkn_1 locus to be on linkage group A03 and tightly linked to the 221 bp SSR marker CIR316a, with an average across three populations of about 2.6% recombination (2.6 cM) between the two loci. Linkage group A03 was later found to be on the long arm of chromosome 11 (Wang et al., 2006b; Ynturi et al., 2006). An additional marker, BNL1231 was also found to flank rkn_1, but was 17 cM distant. This was the first report of an association between molecular markers and root-knot nematode resistance in cotton. Ynturi et al. (2006) studied the association between SSR markers and root-knot nematode resistance in the Auburn 634 RNR (Shepherd,

1982) source. They confirmed the two-gene model and found four markers to be associated with resistance. Three markers were located on the short arm of chromosome 14—BNL3661, BNL3664, and BNL3545—and explained 52% of the observed phenotypic variation for root galling index. A fourth marker, BNL1231 was not on chromosome 14, but was nonetheless associated with resistance, reinforcing the two-gene model. This same marker was the one found to be associated with the resistance gene in NemX, rkn_1, and led to the conclusion that one of the genes in Auburn 634 RNR was in the same linkage group as rkn_1. Furthermore, Ynturi et al. (2006) concluded that the gene linked to BNL3661 on the short arm of chromosome 14 had additive and dominance genetic effects; the gene linked to BNL1231 on the long arm of chromosome 11 had only additive effects. Using two interspecific hybrid populations (*G. hirsutum* × *G. barbadense* L., 'Pima S-6') Shen et al. (2006) also mapped resistance in an Auburn 634 RNR-derived line, M-120 RNR (Shepherd et al., 1996). Phenotypic data of the parents and F_1 generation, together with skewness toward resistance in the F_2, suggested that resistance was dominant. Restriction fragment length polymorphism (RFLP) markers associated with resistance mapped to chromosome 11. Further investigation revealed a major dominant quantitative trait locus (QTL), flanked by codominant SSR markers CIR069 and CIR316 on chromosome 11. This QTL explained about 64% of the phenotypic variation for reaction to infection by root-knot nematodes. They designated this locus, *Mi-C11*, for resistance to *M. incognita* on chromosome 11. Shen et al. (2006) also found a marker (NAU474b) associated with resistance on chromosome 7, but it had only a minor effect, explaining 7.7% of the phenotypic variation for gall index. Indeed, it turned out that this allele for resistance was derived from the Pima S-6 parent and explains why this allele was discovered only by this group. In comparing the markers found in this study with several different root-knot resistant genotypes, including Clevewilt-6, Wild Mexico Jack Jones, M-92 RNR, M-240 RNR, M-249 RNR, M-272 RNR, M-315 RNR, M-331 RNR, and M-725 RNR, as well as cultivars NemX and Stoneville LA887, Shen et al. (2006) found the marker to be present in all lines except Wild Mexico Jack Jones, suggesting that locus *Mi-C11* was derived from Clevewilt-6. They thus proposed a single dominant gene model for root-knot nematode resistance, with the gene coming from Clevewilt-6. Shen et al. (2010) later fine-mapped this locus to a 3.6 cM interval bounded by the SSR markers CIR069 and amplified fragment length polymorphism (AFLP) marker E14M27-375. Gutiérrez et al. (2010) confirmed the presence of two genes in M-240 (a sister line of M-120 and M-315, both derived from the Auburn 634 RNR source) and placed them on chromosomes 11 and 14. The gene on chromosome 11 was linked to the SSR marker C316-201 and the gene on chromosome 14 was linked to the SSR markers BNL3545-118 and BNL3661-185. The gene on chromosome 11 was present in Clevewilt-6 but not in Mexico Wild, and the gene on chromosome 14 was present in Mexico Wild but not in Clevewilt-6. These findings supported those previously reported by Ynturi et al. (2006) and agreed somewhat with the findings of Shen et al. (2006); however, Gutiérrez et al. (2010) were able to shed new light on the specific phenotypic effects of the two genes. They determined that the gene linked to C316-201 (chromosome 11) was primarily responsible for a reduction in root galling, while the gene linked to BNL3545-118 and BNL3661-185 (chromosome 14) primarily affected nematode reproduction, causing a decrease in egg production. This was the first report of a specific resistance mechanism difference caused by the two genes believed to be in the Auburn 623 RNR source of resistance and explains why the observation of both root galling and nematode reproduction are necessary to achieve high levels of resistance in selected lines.

The work of Shen et al. (2006) was revisited by that group (He et al., 2014), studying the same M-120 RNR × Pima S-6 population using more SSR markers. This time their findings were more in line with those of Ynturi et al. (2006) and Gutiérrez et al. (2010) in that they identified two major QTLs associated with resistance, the one they had previously named on the long arm of chromosome 11, *Mi-C11*, and another on the short arm of chromosome 14, which they called *Mi-C14*. Furthermore, they presented more conclusive evidence on the origin of the two genes; *Mi-C11* was present in Clevewilt-6 and *Mi-C14* was present in Wild Mexico. *Mi-C14* was also absent in the moderately resistant Stoneville LA887 and NemX. They also confirmed that *Mi-C14* was primarily associated with a reduction in egg production, while *Mi-C11* primarily affected root galling. *Mi-C14* was mapped to an interval between markers NAU3584 and CGR5668, and the gene action for this locus was primarily found to be dominance. Further comparisons among the three distinct genotypes that carried the resistance genes (positive for both *Mi-C11* and *Mi-C14*; positive for *Mi-C11* and null for *Mi-C14*; null for *Mi-C11* and positive for *Mi-C14*) suggested an epistatic interaction between the two genes; genotypes carrying both genes showed a greater reduction in galling and egg production than could be explained by the additive effects of the two genes individually (He et al., 2014). Epistasis was of an additive × additive nature, particularly for suppression of egg production. A summary of genes associated with resistance to root-knot nematode in cotton is given in Table 1.

Together, the studies of McPherson et al. (2004), Ynturi et al. (2006), Shen et al. (2006, 2010), Gutiérrez et al. (2010), and He et al. (2014) conclusively demonstrate the inheritance, gene action, and molecular markers associated with root-knot nematode resistance in the Auburn 623 RNR source of resistance. They further validate the observations and work of Jones et al. (1958), Shepherd (1974a, 1974b, 1974c, 1982, 1986), and Shepherd et al. (1988) in bringing this valuable genetic resource to the cotton breeding community. This two-gene source of resistance confers near-immunity to southern root-knot nematode (He et al., 2014) and should result in much better management options for growers in the near future. Several commercial cultivars have been recently released, utilizing the resistance sources in the Auburn 623 RNR source (Keim and Dighe, 2013)

The mechanisms of resistance in the Auburn 623 RNR resistance source have been investigated (Creech et al., 1995; Jenkins et al., 1995). Mechanisms involving both galling and egg production are involved, with resistant types showing reduced development of giant cells 6 d after root penetration compared with susceptible types. Creech et al. (1995) and Jenkins et al. (1995) also found that 24 d after penetration, *M. incognita* females showed reduced fecundity.

Other sources of resistance to root-knot nematode in cotton have been found. Shepherd evaluated 471 primitive race stocks, found 12 to be resistant, and subsequently released day neutral converted race stocks (Shepherd et al., 1988), including M-78 RNR. Robinson and Percival (1997) identified five additional sources within the primary germplasm pool of *G. hirsutum*, TX-1174, TX-1440, TX-2076, TX-2107, and TX-2079, but none of these were as resistant as A623 RNR. Later work by Faske and Starr (2009) and Starr et al. (2010) confirmed the resistance of these accessions, and suggested, based on development of the nematodes

Table 1. Genes conferring resistance to *Meloidogyne incognita* (Kofoid and White) Chitwood in Upland cotton.

Gene	Gene action	Source	Reference
Mi_1†	dominant	Wild Mexico	McPherson et al. (2004)
Mi_2‡	additive	Clevewilt-6	McPherson et al. (2004)
rkn_1	recessive	N6072	Wang et al. (2006a)
Mi-C14†	partial dominant	Wild Mexico	He et al. (2014)
Mi-C11‡	dominant	Clevewilt-6	Shen et al. (2006); He et al. (2014)

† Mi_1 and *Mi-C14* are likely the same gene.

‡ Mi_2 and *Mi-C11* are likely the same gene.

after infection, that two of the accessions (TX-1174 and TX-2079) may have new genes for resistance that are not present in either Clevewilt-6 or Wild Mexican. Shepherd also reported the resistance of a wild accession of *G. barbadense*, but the identification and origin of this particular line was unclear (Shepherd, 1974c).

Genetics of resistance in *G. barbadense* has been less well-studied. Most of the studies cited here (Shepherd, 1974c; Robinson and Percival, 1997) have reported on the use of *G. barbadense* as a genetic resource for root-knot nematode resistance in Upland cotton, *G. hirsutum*. There are no reports in the literature on efforts to develop genotypes or cultivars of *G. barbadense* with resistance to root-knot nematode.

Reniform Nematode

The reniform nematode was first recognized as a pathogen of cotton near Cuthbert, GA in the late 1930s when Smith (1940) reported a heavy infestation of an unknown nematode species that was in the process of being described by Linford and Oliveira (1940). Linford first described the pathogen on cowpea [*Vigna unguiculata* (L.) Walp.] being grown in a field following production of pineapple (*Anasas comosus* L. Merr) in Hawaii and observed that females were only partially embedded in roots and that males did not feed at all. He proposed the name "reniform nematode" because of the kidney-shaped bodies of the adult females. Like southern root-knot nematode, reniform nematode also has a fairly wide host range (Robinson et al., 1997). Unlike southern root-knot nematode, there are no good visible symptoms, such as galling, to indicate the presence of the disease. The only way to ascertain the presence of the pathogen is by sampling the soil. Reniform nematodes affect cotton production primarily through reductions in yield, boll size, and lint percentage (Cook et al., 1997; Jones et al., 1959). An excellent review of the life-cycle, taxonomy, morphology, and host range of *R. reniformis* and other species of this genus has been written by Robinson et al. (1997). Briefly, second stage juveniles (J2) emerge from eggs, but none of the juvenile stages of the nematode are parasitic. The final molt produces about equal numbers of male and female vermiform nematodes, but only the females are parasitic. Unlike the root-knot nematode, which infects only at the root area just behind the meristematic region, the reniform nematode can infect at any root site. Vermiform females penetrate the cortex with the anterior portion of their bodies and establish a feeding site in the stele, becoming sedentary and semiendoparasitic. As the female feeds, the posterior portion of the body protrudes from the root and swells as the reproductive system matures, producing the familiar kidney shape associated with reniform nematode. Eggs are laid in a gelatinous mass outside the root. The life cycle of reniform nematode is rapid and can go from infective adult to infective adult in as little as 21 d under ideal conditions of soil temperature and in the presence of a good host. This allows it to build up to high population levels relatively quickly (Lawrence and McLean, 2001), even in the first year of cotton following a rotation with a non-host crop (Westphal and Smart, 2003). Populations tend to be more uniformly distributed across the field than other nematode species (Koenning et al., 2004).

In infested fields, yield loss estimates in the range of 40 to 60% were reported in early investigations by Birchfield and Jones (1961), but population densities were not given. In a later study, comparisons between the most effective nematicide treatment (1,3-dicloropropene + aldicarb) and a non-treated control revealed a yield reduction of 38.9% at one site and a 66.9% reduction at another (Lawrence et al., 1990). Initial population densities at both sites were just over 10 juveniles and vermiform adults cm^{-3} of soil. Another study based on nematicide treatment comparisons reported yield losses in the 30 to 40% range (Gazaway et al., 1992). Economic thresholds for management of reniform nematode are a function of many factors including cost of treatment, cotton price, and endaphic factors, but generally fall in the range of 2 to 4 nematodes cm^{-3} of soil at the beginning of the growing season and 10 to 20 nematodes cm^{-3} of soil when sampling is done at the end of the growing season (Robinson, 2007). However, there is little consensus among nematologists regarding these levels. On the basis of these results, it is safe to assume that in many situations growers probably experience less extreme yield losses where nematode population densities are lower. Unless plants are actually stunted, yield losses may not be associated with the presence of and feeding damage by the nematode. While reniform nematode can occur throughout the cotton belt, Alabama, Louisiana, and Mississippi are the states where the largest populations of *R. reniformis* are found and where the largest yield losses occur (Blasingame and Patel, 2013).

Management strategies can hinge on several factors. Populations can build up rapidly (Lawrence and McLean, 2001) and exist in substantial numbers throughout the soil profile, at much deeper depths (>1 m) than other soil-borne pathogens (Westphal and Smart, 2003; Robinson et al., 2005a, 2005b), putting it beyond reach of effective depth of most nematicide treatments. Furthermore, reniform nematodes can survive in the soil for extended periods without a suitable host. Reniform nematodes can enter into an anhydrobiotic or desiccated state for up to 2 yr (Birchfield and Martin, 1967) and thus survive periods where fields are left fallow or planted to a non-host crop.

Rotation with non-host crops, where feasible, is recommended for management of reniform nematode. Peanut is a non-host (Robinson et al., 1997), and peanut rotation with cotton is recommended where it fits into the overall crop management system (Koenning et al., 2004; Starr et al., 2007). Peanut, however, much like cotton, is a specialty crop with specialized equipment for harvesting and processing, and the necessary accompanying infrastructure. Corn (*Zea mays* L.) is also a poor host (Windham and Lawrence, 1992) and does not require a specialized infrastructure. It can reduce soil populations of reniform nematode by 90% (Gazaway et al., 2007). Other studies have shown that 2 yr of corn are necessary to reduce reniform nematode populations to below-economic threshold levels (Stetina et al., 2007), primarily due to the ability of the nematode to survive for more than a year in soil without a host (Heald and Inserra, 1988). A corn-cotton rotation was found to produce better cotton yield than continuous cotton at one location but not another (Davis et al., 2003). However, the fact that corn is harvested much earlier than most other warm-season crops in the South makes it very important to effectively manage late-season weeds. Many weed species are susceptible to reniform nematode, and soil populations can build up to damaging economic levels on weeds even after the corn is harvested (Lawrence et al., 2008). Soybean also has been suggested as a rotation crop, but because soybean cultivars can vary in their reaction to reniform nematode, the effectiveness of soybean as a rotation crop can be soybean cultivar-dependent (Davis et al., 2003). The use of winter cover crops in a cotton production system is highly recommended for a variety of reasons, but options for cover crops can be limited, as many leguminous cover crops such as crimson clover (*Trifolium incarnatum* L), subterranean clover (*Trifolium subterraneum* L.), and hairy vetch (*Vicia villosa* Roth) have been found to serve as hosts for reniform nematode and increase populations in the greenhouse. However, they did not increase reniform nematode populations in the field, and their use did not affect cotton yields compared with plots where no cover crops were used (Jones et al., 2006). Other cover crops, mostly monocotyledonous, such as ryegrass (*Lolium* spp.), wheat (*Triticum aestivum* L.), oat (*Avena sativa* L.), rye (*Secale cereale* L.), and other grass cover crops, have been found to be poor hosts at best (Jones et al., 2006). Radish (*Raphanus sativus* L.) also proved to be a poor host (Jones et al., 2006).

As with root-knot nematode, nematicides have been much-used in the past to manage reniform nematode in cotton. Aldicarb and 1,3-dicloropropene have been widely used (Lawrence et al., 1990; Koenning et al., 2004, 2007), but as with nematicides for management of root-knot nematode, these chemical options are no longer available (US Environmental Protection Agency, 2010). Nematicidal seed treatments have also been used for reniform nematode management. Abamectin and thiodicarb have proven to be effective (Lawrence et al., 2006). Yield response to abamectin was comparable to aldicarb treatment at two locations (Lawrence et al., 2006). While nematicides often improve yield, they do little to suppress reniform nematode populations, as populations can rebound quickly during the growing season and are usually the same in nematicide-treated plots as they are in untreated plots by the end of the growing season (Lawrence et al., 1990; Koenning et al., 2007).

The possibility of using biological control agents for management of reniform nematode has been explored. Schmidt et al. (2010) evaluated the use of *Pasteuria* spp., a genus of soil borne bacteria that is known to be parasitic to a number of plant parasitic nematodes. They found a decrease in reniform nematode reproduction on cotton that was dependent on dosage, or amount of bacteria applied as a treatment. The maximum spore loading rate of 2.4×10^5 spores cm^{-3} reduced female nematodes by 82.3% compared with the control. Similarly, Castillo et al. (2013) investigated the biocontrol properties of *Bacillus firmus* strain GB-126 and *Paecilomyces lilacinus* strain 251 for reniform nematode management. *Bacillus firmus* GB-126 is marketed as a seed treatment under the name VOTiVO or as a wettable powder under the name Nortica 5% WP. *Paecilomyces lilacinus* 251 is marketed as a wettable powder under the name NemOut. In the greenhouse *P. lilacinus* 251 alone reduced females and eggs of reniform nematode compared with the untreated control, but was not as effective as the aldicarb control. Similar results were found for *B. firmus* GB-126 when used alone. In microplot and field studies the best results were obtained when both biocontrol agents were used in combination, with *P. lilacinus* 251 observed to reduce females and eggs while *B. firmus* GB-126 had more effect on the vermiform stages. Together, these agents reduced nematode populations as well as the aldicarb control.

The history of the use of genetic resistance or tolerance to reniform nematode with regard to sources, inheritance of resistance, and incorporation of resistance into adapted types is considerably more recent than that for root-knot nematode resistance due to the much later recognition of the nematode as a pathogen of cotton. Although reports of tolerance can be found in the literature (Cook et al., 1997; Koenning et al., 2000; Usery et al., 2005; Stetina et al., 2009; Blessitt et al., 2012), no appreciable levels of resistance have been found in adapted Upland cultivars (Robinson et al., 1999; Usery et al., 2005). While tolerance can be a useful management tool, it does nothing to reduce soil nematode populations. In work to develop cotton with resistance to root-knot nematode (using the Clevewilt resistance source), Jones et al. (1988) observed some experimental lines that had reduced reproduction of reniform nematode in the greenhouse compared with susceptible genotypes. Subsequently, they released four germplasm lines with resistance to both species. In a study to investigate the inheritance of resistance in one of these lines (LA RN-910), they concluded resistance was of a quantitative nature (Muhammad and Jones, 1990), but they also observed transgressive segregation for susceptibility and evidence of epistatic gene action. On the basis of these results, they suggested that breeding for resistance to reniform nematode with these genetic resources would be a very difficult task, and suggested alternative resistance sources be looked for and utilized for future efforts.

Extensive searches of the primary, secondary, and tertiary germplasm pools of cotton for resistance to the reniform nematode have been conducted since the 1960s and have revealed some accessions of *G. hirsutum* with moderate resistance, generally based on reduced reproduction of the nematode compared with a susceptible check genotype (Yik and Birchfield, 1984; Robinson et al., 2004; Weaver et al., 2007), but no high levels of resistance have been reported. Apparent inconsistencies in reports of resistance to reniform nematode within *G. hirsutum* are also a problem. Yik and Birchfield (1984) reported three wild *G. hirsutum* accessions as resistant, but later Robinson and Percival (1997) reported the same accessions as not different in reaction from the susceptible 'Deltapine 16'. Similarly, Weaver et al. (2007) found seven accessions to be moderately resistant following repeated testing, but resistance was not confirmed in later studies (Sürmelioğlu et al., 2010). One of the accessions found to be moderately resistant by Robinson et al. (2004) was TX-2468 (PI 607773), an accession collected from Brazil in 1988. This photoperiodic line was crossed with 'Deltapine 61' with selection in the intervening generations of self-pollination for daylength neutrality and resistance to reniform nematode, resulting in the release of three germplasm lines, MT2468 Ren1, MT2468 Ren2, and MT2468 Ren3 (McCarty et al., 2012), each with moderate resistance to reniform nematode and day neutrality. The genetics of resistance in these lines is unknown.

Searches for resistance in other species of *Gossypium* have resulted in discovery of genotypes with higher levels of resistance. Yik and Birchfield (1984) evaluated 200 genotypes in the family Malvaceae including cultivated *G. hirsutum*,

G. herbaceum L., *G. arboreum* L., and *G. barbadense*, several wild cotton species, and other species of Malvaceae. They reported the wild diploid *G. longicalyx* J.B. Hutch. & B.J.S. Lee (a native species of east Africa) and the related species *Sida rhombifolia* L. to be immune. High levels of resistance were also found within *G. arboreum*, *G. stocksii* Mast., *G. somalense* (Gurke) J.B. Hutch., and *G. barbadense* TX-110 (PI 163608). Utilization of the immunity in *G. longicalyx* initially proved to be problematic, however, because of chromosomal incompatibility between *G. hirsutum* and *G. longicalyx*. *G. hirsutum* is an allotetraploid ($2n = 4x = 52$), comprised of the A and D genomes, while *G. longicalyx* is a diploid ($2n = 2x = 26$) and is cytogenetically distinct from other cottons as the only species with the F genome (Wendel and Cronn, 2003). Through a collaborative effort of USDA-ARS, Texas A&M AgriLife Research, and Cotton Incorporated, a high level of resistance from the diploid *G. longicalyx* was introgressed into a *G. hirsutum* genetic background through a series of bridging crosses using a triple species hybrid (*hirsutum–longicalyx–armourianum* Kearney, or HLA) to overcome the genomic incompatibility between the two species of different genome and ploidy (Robinson et al., 2007). In 2007 two germplasm lines, LONREN-1 and LONREN-2, were developed and made available to research scientists (Bell et al., 2014b). These germplasm lines were developed by backcrossing seven times to *G. hirsutum* as the recurrent parent and included NemX, 'Deltapine 458 B/RR', and 'Stoneville 474', with the final backcross being made to 'Fibermax 958'. The gene for resistance was mapped to chromosome 11 (A genome). It was determined to be dominant (Robinson et al., 2007) and linked to the microsatellite markers BNL1066_156 in LONREN-1 only and BNL3279_114 in both LONREN lines, and it was given the gene symbol REN^{lon} (Dighe et al., 2009). BNL3279_114 was the most closely linked and found to be between 0.5 and 1.4 cM distant from REN^{lon} depending on the method used to calculate the map distance. This gene has also been recently referred to as Ren_1 by Fang and Stetina (2011). REN^{lon} is linked also to a morphological marker, green seed fuzz, with gene symbol Fzg^{lon} (Dighe et al., 2009), with this gene being between 2.8 and 4.5 cM from REN^{lon}. The chromosome segment introgressed from *G. longicalyx* is slightly shorter in LONREN-2 than in LONREN-1 and would account for the lack of association with marker BNL 1066_156 in LONREN-2 (Dighe et al., 2009; Bell et al., 2014b). Konan et al. (2007) were also successful in transferring resistance from *G. longicalyx* into a *G. hirsutum* background using a tri-species hybrid.

The reaction of the original resistance source, *G. longicalyx*, to infection by reniform nematode was investigated by Agudelo et al. (2005). Female nematodes were observed to penetrate the roots of *G. longicalyx*, but remained vermiform during the test period without producing eggs. Cells that would normally become feeding sites were observed to collapse and die, thus disrupting the egg-laying by the female. Thus, the nematodes were able to infect the plant, but were not able to reproduce. Most evaluation for reniform resistance is done based on reproduction, or egg production, as no other symptoms are reliable indicators of resistance or susceptibility (Stetina and Young, 2006). Yik and Birchfield (1984) arbitrarily defined a resistant genotype as one that would reduce egg production by 75% compared with a susceptible control, which in their experiments was Deltapine 16.

During development the LONREN-1 and LONREN-2 germplasms were selected primarily on the basis of the lack of nematode reproduction and selection for presence of the SSR marker BNL3279_114 during the backcross generations.

With the large number of backcrosses to *G. hirsutum*, the yield, agronomic, and fiber quality traits of the LONREN lines and breeding lines developed from them are quite good (Bell et al., 2014b; Weaver et al., 2013). There is no question that the LONREN lines are capable of reducing populations of *R. reniformis* in the field (Bell et al., 2014b; Weaver et al., 2013). However, extensive field testing of the LONREN lines showed that while they performed quite well in fields without high nematode populations, some stunting and plant development issues among lines carrying the *REN*^{lon} gene in fields where reniform nematode is present were observed (Nichols et al., 2010; Weaver et al., 2013). Populations in the field at 25 d after planting were about 3 nematodes cm^{-3} of soil in the research reported by Weaver et al. (2013), exactly intermediate to the 2 to 4 nematodes cm^{-3} of soil economic threshold suggested by Robinson (2007). A greenhouse study by Sikkens et al. (2011) revealed that increasing the initial inoculum level to as high as 300 juveniles and vermiform nematodes cm^{-3} of soil resulted in root necrosis and decline in root mass as a major problem in LONREN-1, LONREN-2, and a LONREN-derived resistant progeny as the inoculum level was gradually increased from 3 to 300 nematodes cm^{-3} of soil. Susceptible genotypes tested under the same conditions showed an increase in root mass in response to increasing inoculum level. Less root necrosis was observed in susceptible genotypes. This has led to speculation as to the reasons for the observed stunting in the LONREN genotypes. Secondary pathogenic fungi, particularly *Thielaviopsis basicola* (Berk. & Broome) Ferraris or *Rhizoctonia solani* J.G. Kühn, have been isolated from necrotic roots and are known to affect lines that carry the *REN*^{lon} gene more severely than lines that do not (Bell et al., 2012). Treatment with fungicides at planting causes the stunting to go away almost entirely (Bell et al., 2011). Another hypothesis, supported by the observations of Agudelo et al. (2005) and Sikkens et al. (2011), is that response to reniform nematode in materials derived from *G. longicalyx* is of a hypersensitive nature. Nematodes can infect and begin to feed, but the plant's reaction (necrosis, cell collapse) prevents the nematode from continuing to feed and reproduce.

This led to further experiments aimed at assessing the effect of the *G. longicalyx* chromosome segment on cotton performance with and without reniform nematodes present. Weaver et al. (2013) compared resistant lines (carrying *REN*^{lon}) and sister susceptible lines (null for *REN*^{lon}) from the same population in adjacent fields, one with a high reniform nematode population and the other with no reniform nematodes. Lines with *REN*^{lon} were stunted during early season growth and yielded less than did their sister lines null for *REN*^{lon} in the nematode-infested field. Even so, the *REN*^{lon} gene reduced nematode populations as at-harvest nematode populations were about one-third of the populations in plots where lines without *REN*^{lon} were planted. There were no yield differences between lines with and without *REN*^{lon} in the nematode-free field. Fiber quality was affected by *REN*^{lon}, with lines carrying the gene tending to have greater fiber strength, better length uniformity, and lower short fiber content. The *REN*^{lon} gene was effective in reducing nematode populations and may improve fiber quality; however, *REN*^{lon} did not reduce yield losses from reniform nematodes. Work is still being conducted to investigate more fully the stunting problem and possible solutions. The *G. longicalyx* chromosomal block that has been transferred into *G. hirsutum* is still relatively large, about 3.68 Mbp in LONREN-2 (Fang and Stetina, 2011). Much effort is being made to reduce the size of this fragment and to further fine-map this important chromosomal region, under the assumption that deleterious genes related to plant stunting and other problems may be contained in this chromosomal segment. Until this problem is solved, it is unlikely that *G. longicalyx*–derived resistance in the form of *REN*^{lon} can be useful commercially.

Other sources of resistance to reniform nematode have been found in *G. barbadense.* Yik and Birchfield (1984) identified *G. barbadense* accession TX-110, a photoperiodic accession from Guatemala, as being resistant. More recently, Robinson et al. (2004) reaffirmed the resistance of TX-110 and identified another *G. barbadense* accession, GB-713 (PI 608139), as being highly resistant, producing only 3% of the eggs as the susceptible control. As a cultivated allotetraploid with the same genome as *G. hirsutum,* this would be considered a primary germplasm pool relative of Upland cotton and resistance should be relatively easy to transfer. Using the TX-110 source of resistance, Starr et al. (2011) released TAM RNRNR-9 and TAM RNRNR-12 germplasms, which combined reniform resistance in TX-110 with root-knot resistance in M315 RNR (Shepherd et al., 1996). Although they did not study inheritance of resistance, their observations on nematode development suggested that resistance was the result of delayed development of mature females. Gutiérrez et al. (2011) investigated the inheritance of resistance derived from GB-713 in populations involving GB-713 and reniform nematode–susceptible parent NemX. Two major QTLs were determined to be on chromosome 21, one at map position 168.2 and flanked by SSR markers BNL1551_162 and GH132_199 and the other at map position 182.7 and flanked by markers BNL4011_155 and BNL3279_106. These two loci were about 14.5 cM apart on chromosome 21. A third locus associated with resistance was found on chromosome 18 at map position 42.0 and flanked by markers BNL1721_178 and BNL569_131. Thus they proposed a three-gene model governing resistance, with the gene symbols *Ren*^{barb1}, *Ren*^{barb2}, and *Ren*^{barb3} for the genes located at positions 168.2 and 182.7 on chromosome 21 and position 42.0 on chromosome 18, respectively. An effort is underway to transfer the GB-713 resistance source into adapted *G. hirsutum* types. Germplasm lines derived from GB-713 have been developed and tested (Bell et al., 2014a; Sikkens et al., 2014), but not yet officially released. Sikkens et al. (2014) compared the performance of GB-713-derived BAR 41, BARBREN 713 (both developed by USDA-ARS and Texas A&M AgriLife Research), and five variations of M713 (developed by USDA-ARS and Mississippi Agriculture and Forestry Experiment Station) (McCarty et al., 2013), along with *G. longicalyx*–derived LONREN-1 and LONREN-2, and MT2468 Ren1, MT2468 Ren2, and MT2468 Ren3 (McCarty et al., 2012) in adjacent fields, one infested with reniform nematode and the other not infested. The performance of all the GB-713-derived lines was very good, with BAR 41 and all of the M713 lines averaging less than 10% lint yield loss due to nematodes while susceptible genotypes and the MT2468 series of lines averaged 65% yield loss and LONREN germplasm lines averaged more than 70% yield loss. In the nematode-free field, the germplasm lines were comparable to the adapted checks in yield and compared favorably in fiber quality. Thus, it appears the *G. barbadense* source of resistance has more promise for commercial utilization than other identified sources. Many of these GB-713-derived resistance sources are currently being used by industry to develop cultivars with resistance to reniform nematode (M. McPherson, personal communication, 2014).

Sources of resistance from other *Gossypium* species have been investigated. Stewart and Robbins (1995) found seven out of nine accessions of the diploid *G.*

arboreum to be resistant to reniform nematodes, and six out of 14 accessions of *G. arboreum* were found to be resistant by Yik and Birchfield (1984). The interaction of the nematode and host plant appeared to be different from that in *G. longicalyx*, as the nematode was observed to attempt to feed, but the host cells reacted by lignifying cells adjacent to the nematode head and lysing the cell walls (Carter, 1981). To investigate the inheritance of resistance in *G. arboreum*, Erpelding and Stetina (2013) crossed the moderately resistant *G. arboreum* accession PI 529728 with a highly susceptible accession of *G. arboreum*, PI 529729. They found resistance to be under the control of a single recessive gene.

Using a chromosomal bridging strategy similar to that used to transfer resistance from *G. longicalyx* into *G. hirsutum*, Sacks and Robinson (2009) attempted to transfer a resistance gene from *G. arboreum* into *G. hirsutum*. They used G 371, a hexaploid ADD genome *G. hirsutum*/*G. aridum* (Rose & Standl.) Skovst. breeding line as the bridge in crossing with *G. arboreum* accession A_2-190 (PI 615699), one of the seven accessions determined by Stewart and Robbins (1995) to be resistant and later confirmed resistant in independent studies (Sacks and Robinson, 2009). They confirmed that the level of resistance in A_2-190 was about equal to that observed in the *G. barbadense* accession GB-713, but the reaction of the bridging hexaploid G 371 was unknown. Surprisingly, the initial evaluation of the parental materials in this study showed G 371 to support about 10% of the nematode reproduction of susceptible controls, and thus it was classified as resistant. On the basis of the general lack of high levels of resistance in *G. hirsutum*, they speculated that the resistance in G 371 must have come from the diploid *G. aridum* parent. Reaction of the progenies showed genetic segregation for resistance, so it was thought to be unlikely the gene in A_2-190 and G 371 were the same gene. Other than the development of the two-gene model, large environmental variations known to plague many of the studies of inheritance of resistance to reniform nematode under controlled conditions (Roberts, 2002), and partial dominant gene action of the resistance genes made it difficult to draw definitive conclusions regarding inheritance of resistance. They concluded that the data fit a 3:1 genetic ratio, which supported the hypothesis that the parents had different genes for resistance and proposed partial dominance as the gene action. Romano et al. (2009) mapped a gene in this population, named *REN*^{ari}, and found two SSR markers, BNL3279_132 and BNL2662_090, to map within 1 cM of *REN*^{ari}. They concluded that because these same SSR fragments were found in the bridging line parent *G. aridum*, the resistance gene must have come from that source. They found this resistance to be under the control of a single dominant gene. They further placed *REN*^{lon} on chromosome 11 (A genome) and *Ren*^{ari} on chromosome 21 (D genome), which are homeologous chromosomes. Fang and Stetina (2011) attempted to pyramid *REN*^{lon} with *Ren*^{ari} (referred to as Ren_1 and Ren_2 in their paper) to help clarify some of these relationships and to see if resistance could be strengthened by combining the two genes. They definitely concluded that the two resistance genes were not allelic, but were duplicate genes residing on homeologous chromosomes, and both could be tagged by slightly different fragment sizes of the SSR marker BNL3279. While they were not able to identify plants carrying both genes on the basis of phenotype with regard to resistance to reniform nematode, the use of markers allowed them to identify plants with both genes. They found that plants with neither gene showed the same level of nematode reproduction as the susceptible controls. Plants with only one gene, regardless of whether it was *REN*^{lon}

Table 2. Genes conferring resistance to *Rotylenchulus reniformis* Linford and Oliveria in Upland cotton.

Gene	Gene action	Source	Reference
REN^{lon}	dominant	*Gossypium longicalyx* J.B. Hutch. & B.J.S. Lee	Robinson et al. (2007)
REN^{ari}	dominant	*G. aridum* (Rose & Standl.) Skovst.	Romano et al. (2009)
Ren^{barb1}	partial dominant	*G. barbadense* L.	Gutiérrez et al. (2011)
Ren^{barb2}	partial dominant	*G. barbadense*	Gutiérrez et al. (2011)
Ren^{barb3}	partial dominant	*G. barbadense*	Gutiérrez et al. (2011)
Unnamed	recessive	*G. arboreum* L.	Erpelding and Stetina (2013)

or Ren^{ari}, were significantly more resistant to the susceptible controls, but were not different from one another. Both genes also behaved in a dominant fashion. Plants with both genes had numerically the lowest nematode reproduction, but could not be statistically separated from the groups carrying the single genes. As the Ren^{ari} gene has not been tested in a field situation, it is unknown whether it may suffer some of the same problems with plant stunting and associated yield loss as the REN^{lon} gene. A summary of genes associated with resistance to reniform nematode in cotton is given in Table 2.

Columbia Lance Nematode

The Columbia lance nematode was originally referred to as "cotton stunt" because of the symptoms and first recognized as a pathogen of cotton in South Carolina (Fassuliotis et al., 1968). Plants are generally stunted and yields are reduced under high infestation levels, with economic thresholds around 1 nematode cm^{-3} of soil (Noe, 1993). It is fairly limited in distribution, occurring mostly in the sandy coastal plains soils of North Carolina, South Carolina, and Georgia (Koenning et al., 2004). Lance nematodes are migratory and can feed as ecto-, endo-, or semi-endoparasites. Besides stunting, symptoms include root necrosis and lesions on the root surface. Stunting is attributed to the lack of deep root penetration caused by feeding in the meristematic areas of growing roots (Koenning et al., 2004). The life cycle is completed in 33 d, with all Columbia lance nematodes remaining as vermiform throughout their life cycle, reproducing parthogenically. This pathogen has a fairly wide host range, including maize (*Zea mays* L.) and soybean [*Glycine max* (L.) Merr.] (Lewis and Smith, 1976), which limits the extent to which crop rotation can be used as a management tool. Nematicides can be effective in reducing yield loss (Mueller and Sullivan, 1988), but again the environmental concerns and other issues surrounding the use of chemical inputs for nematode control make this a limited management option. Subsoiling to allow deeper root penetration was found to be very effective in increasing cotton yield where fields are infested with Columbia lance nematode (Bird et al., 1974; Hussey, 1977), but subsoiling is practiced less frequently now because of efforts to grow cotton under conservation tillage systems. Also, the use of conservation tillage decreases the likelihood of hardpan development, one of the exacerbating factors in Columbia lance nematode as indicated by the response to subsoiling. There is no evidence and no research that conservation tillage systems are less likely to be affected by lance nematode, but it is an idea that is worthy of consideration. There is some evidence that cotton cultivars may differ in level of tolerance (Koenning et al., 2000, 2003), but tolerance seems to be related to maturity, with late maturing cultivars being more tolerant (Koenning et al., 2004).

Sting Nematode

Like Columbia lance nematode, the sting nematode is also migratory and is an ectoparasitic feeder. It remains vermiform throughout its life cycle. Symptoms have been described as "stubby roots" (Crow et al., 1997) and "shrunken lesions along the root axis or at the root tip" (Graham and Holdeman, 1953). Distribution of sting nematode as a pathogen of cotton is limited to very sandy soils, with sand content greater than 80% (Robbins and Barker, 1974). Because little cotton is produced on soils with such a high sand content, the economic damage caused by sting nematode is low compared with root-knot and reniform nematode. However, when cotton is grown in fields with initial sting nematode infestation levels of less than 1 nematode cm^{-3} of soil, yield can be reduced to near zero (Crow et al., 2000). With very limited distribution and economic impact, little research has been conducted on the management of sting nematode in cotton.

References

Agudelo, P., R.T. Robbins, J.M. Stewart, A. Bell, and A.F. Robinson. 2005. Histological observations of *Rotylenchulus reniformis* on *Gossypium longicalyx* and interspecific cotton hybrids. J. Nematol. 37:444–447.

Bell, A.A., A.F. Robinson, J. Quintana, N.D. Dighe, M.A. Menz, D.M. Stelly, X. Zheng, J.E. Jones, C. Overstreet, E. Burris, R.G. Cantrell, and R.L. Nichols. 2014b. Registration of LONREN-1 and LONREN-2 germplasm lines of Upland cotton resistant to reniform nematode. J. Plant Reg. 8:187–190. doi:10.3198/jpr2013.11.0069crg

Bell, A.A., J. Quintana, S. Prom, D.M. Stelly, X. Zheng, V. Saladino, and R.L. Nichols. 2014a. New germplasm lines with high yield and fiber quality combined with nematode resistance. In: Proceedings of the Beltwide Cotton Conferences, New Orleans, LA. 6–8 Jan. 2014. Natl. Cotton Counc. of Am., Memphis TN. p. 660.

Bell, A.A., J. Quintana, X. Zheng, D.M. Stelly, and R.L. Nichols. 2012. A protocol for isolating and enumerating *Thielaviopsis basicola* from cotton roots: Applications to LONREN stunt. In: Proceedings of the Beltwide Cotton Conferences, Orlando, FL. 3–6 Jan. 2012. Natl. Cotton Counc. of Am., Memphis TN. p. 329–340.

Bell, A.A., X. Zheng, D.M. Stelly, and R.L. Nichols. 2011. Importance of fungal pathogens in stunting of LONREN cotton lines grown in reniform nematode-infested fields. In: Proceedings of the Beltwide Cotton Conferences, Atlanta, GA. 4–7 Jan. 2011. Natl. Cotton Counc. of Am., Memphis, TN. p. 250.

Bezawada, C., S. Saha, J.N. Jenkins, R.G. Creech, and J.C. McCarty. 2003. SSR marker(s) associated with root knot nematode resistance gene(s) in cotton. J. Cotton Sci. 7:179–184.

Bird, G.W., O.L. Brooks, C.E. Perry, J.G. Futral, T.D. Canerday, and F.C. Boswell. 1974. Influence of subsoiling and soil fumigation on the cotton stunt disease, *Hoplolaimus columbus*, and *Meloidogyne incognita*. Plant Dis. Rep. 58:541–544.

Birchfield, W., and J.E. Jones. 1961. Distribution of the reniform nematode in relation to crop failure of cotton in Louisiana. Plant Dis. Rep. 45:671–673.

Birchfield, W., and W.J. Martin. 1967. Reniform nematode survival in air-dried soil. Phytopathology 57:804.

Blasingame, D., and M.V. Patel. 2013. Cotton disease loss estimate committee report. In: Proceedings of the Beltwide Cotton Conferences, San Antonio, TX. 7–10 Jan. 2013. Natl. Cotton Counc. of Am., Memphis, TN. p. 1242–1246.

Blessitt, J.A., S.R. Stetina, T.P. Wallace, P.T. Smith, and G.L. Sciumbato. 2012. Cotton (*Gossypium hirsutum*) cultivars exhibiting tolerance to the reniform nematode (*Rotylenchulus reniformis*). Int. J. Agron. 2012:893178. doi:10.1155/2012/893178

Bowman, D.T., O.A. Gutiérrez, R.G. Percy, D.S. Calhoun, and O.L. May. 2006. Pedigrees of Upland and Pima cotton cultivars released between 1970 and 2005. Bull. 1155. Mississippi Agric. and For. Exp. Stn., Mississippi State, MS.

Carter, W.W. 1981. Resistance and resistant reaction of *Gossypium arboreum* to the reniform nematode, *Rotylenchulus reniformis*. J. Nematol. 13:368–374.

Castillo, J.D., K.S. Lawrence, and J.W. Kloepper. 2013. Biocontrol of the reniform nematode by *Bacillus firmus* GB-126 and *Paecilomyces lilacinus* 251 on cotton. Plant Dis. 97:967–976. doi:10.1094/PDIS-10-12-0978-RE

Cook, C.G., A.F. Robinson, and L.N. Namken. 1997. Tolerance to *Rotylenchulus reniformis* and resistance to *Meloidogyne incognita* race 3 in high-yielding breeding lines of Upland cotton. J. Nematol. 29:322–328.

Creech, R.G., J.N. Jenkins, B. Tang, G.W. Lawrence, and J.C. McCarty. 1995. Cotton resistance to root-knot nematode. I. Penetration and reproduction. Crop Sci. 35:365–368. doi:10.2135/cropsci1995.0011183X003500020012x

Crow, W.T., D.W. Dickson, and D.P. Weingartner. 1997. Stubby-root symptoms on cotton induced by *Belonolaimus longicaudatus*. J. Nematol. 29:574.

Crow, W.T., D.W. Dickson, D.P. Weingartner, R. McSorley, and G.L. Miller. 2000. Yield reduction and root damage to cotton induced by *Belonolaimus longicaudatus*. J. Nematol. 32:205–209.

Davis, R.F., S.R. Koenning, R.C. Kamerait, T.D. Cummings, and W.D. Shurley. 2003. *Rotylenchulus reniformis* management in cotton with crop rotation. J. Nematol. 35:58–64.

Dighe, N.D., A.F. Robinson, A.A. Bell, M.A. Menz, R.G. Cantrell, and D.M. Stelly. 2009. Linkage mapping of resistance to reniform nematode in cotton following introgression from *Gossypium longicalyx* (Hutch. & Lee). Crop Sci. 49:1151–1164. doi:10.2135/cropsci2008.03.0129

Erpelding, J.E., and S.R. Stetina. 2013. Genetics of reniform nematode resistance in *Gossypium arboreum* germplasm line PI 529728. World J. Agric. Res. 1:48–53.

Fang, D.D., and S.R. Stetina. 2011. Improving cotton (*Gossypium hirsutum* L.) plant resistance to reniform nematodes by pyramiding Ren_1 and Ren_2. Plant Breed. 130:673–678. doi:10.1111/j.1439-0523.2011.01889.x

Faske, T.R., and J.L. Starr. 2006. Sensitivity of *Meloidogyne incognita* and *Rotylenchulus reniformis* to abamectin. J. Nematol. 38:240–244.

Faske, T.R., and J.L. Starr. 2009. Mechanism of resistance to *Meloidogyne incognita* in resistant cotton genotypes. Nematropica 39:281–288.

Fassuliotis, G., G.J. Rau, and F.H. Smith. 1968. *Hoplolaimus columbus*, a nematode parasite associated with cotton and soybeans in South Carolina. Plant Dis. Rep. 52:571–572.

Gazaway, W.S., K.S. Lawrence, and J.R. Akridge. 2007. Impact of crop rotation and fumigation on cotton production in reniform infested fields. In: Proceedings of the Beltwide Cotton Conferences, New Orleans, LA. 9–12 Jan. 2007. Natl. Cotton Counc. of Am., Memphis, TN. p. 1357–1360.

Gazaway, W., D. Rush, and K. Edmisten. 1992. An evaluation of various Temik and Telone rates for controlling reniform nematodes in cotton. Fungic. Nematicide Tests 47:161.

Graham, T.W., and Q.L. Holdeman. 1953. The sting nematode *Belonolaimus gracilis* Steiner: A parasite on cotton and other crops in South Carolina. Phytopathology 43:434–439.

Gutiérrez, O.A., J.N. Jenkins, J.C. McCarty, M.J. Wubben, R.W. Hayes, and F.E. Callahan. 2010. SSR markers closely associated with genes for resistance to root-knot nematode on chromosomes 11 and 14 of upland cotton. Theor. Appl. Genet. 121:1323–1337. doi:10.1007/s00122-010-1391-9

Gutiérrez, O.A., A.F. Robinson, J.N. Jenkins, J.C. McCarty, M.J. Wubben, F.E. Callahan, and R.L. Nichols. 2011. Identification of QTL regions and SSR markers associated with resistance to reniform nematode in *Gossypium barbadense* L. accession GB713. Theor. Appl. Genet. 122:271–280. doi:10.1007/s00122-010-1442-2

He, Y., P. Kumar, X. Shen, R.F. Davis, G. Van Becelaere, O.L. May, R.L. Nichols, and P.W. Chee. 2014. Re-evaluation of the inheritance for root-knot nematode resistance in the Upland cotton germplasm line M-120 RNR revealed two epistatic QTLs conferring resistance. Theor. Appl. Genet. 127:1343–1351. doi:10.1007/s00122-014-2302-2.

Heald, C.M., and R.N. Inserra. 1988. Effect of temperature on infection and survival of *Rotylenchulus reniformis*. J. Nematol. 20:356–361.

Hussey, R.S. 1977. Effects of subsoiling and nematicides on *Hoplolaimus columbus* populations and cotton yield. J. Nematol. 9:83–86.

Hyer, A.J., E.C. Jorgenson, R.H. Garber, and S. Smith. 1979. Resistance to root-knot nematode in control of root-knot nematode-Fusarium wilt disease complex in cotton. Crop Sci. 19:898–901. doi:10.2135/cropsci1979.0011183X001900060036x

Jenkins, J.N., R.G. Creech, G. Lawrence, J.C. McCarty, and R. McPherson. 1993. Resistance of cotton cultivars to root-knot nematode. Tech. Bull. 994, Mississippi Agriculture and Forestry Experiment Station, Mississippi State, MS.

Jenkins, J.N., R.G. Creech, B. Tang, G.W. Lawrence, and J.C. McCarty. 1995. Cotton resistance to root-knot nematodes: II. Post-penetration development. Crop Sci. 35:369–373. doi:10.2135/cropsci1995.0011183X003500020013x

Jones, J.E., J.P. Beasley, J.I. Dickson, and W.D. Caldwell. 1988. Registration of four cotton germplasm lines with resistance to reniform and root-knot nematodes. Crop Sci. 28:199–200.

Jones, J.E., L.D. Newsom, and E.L. Finley. 1959. Effect of the reniform nematode on yield, plant characters, and fiber properties of Upland cotton. Agron. J. 51:353–356. doi:10.2134/agronj1959.00021962005100060014x

Jones, J.E., S.L. Wright, and L.D. Newsome. 1958. Sources of tolerance to and inheritance of resistance to root-knot nematodes in cotton. In: 11th Annual Cotton Improvement Conference, Houston, TX. 15–16 Dec. 1958. Natl. Cotton Counc. of Am., Memphis, TN. p. 34–39.

Jones, J.R., K.S. Lawrence, and G.W. Lawrence. 2006. Evaluation of winter cover crops in cotton cropping for management of *Rotylenchulus reniformis*. Nematropica 36:53–56.

Jorgenson, E.C. 1979. Granular nematicides as adjuncts to fumigants for control of root-knot nematodes. J. Nematol. 11:144–150.

Keim, D.L., and N. Dighe. 2013. Breeding for high levels of root knot nematode resistance in a commercial cotton breeding program. In: Proceedings of the Beltwide Cotton Conferences, San Antonio, TX. 7–10 Jan. 2013. Natl. Cotton Counc. of Am., Memphis, TN. p. 407.

Kirkpatrick, T.L., and J.N. Sasser. 1984. Crop rotation and races of *Meloidogyne incognita* in cotton root-knot management. J. Nematol. 16:323–329.

Koenning, S.R., K.R. Barker, and D.T. Bowman. 2000. Tolerance of selected cotton lines to *Rotylenchulus reniformis*. J. Nematol. 32:519–523.

Koenning, S.R., K.L. Edmisten, K.R. Barker, and D.E. Morrison. 2003. Impact of cotton production systems on management of *Hoplolaimus columbus*. J. Nematol. 35:73–77.

Koenning, S.R., D.E. Morrison, and K.L. Edminsten. 2007. Relative efficiency of selected nematicides for management of *Rotylenchulus reniformis* in cotton. Nematropica 37:227–235.

Koenning, S.R., J.A. Wrather, T.L. Kirkpatrick, N.R. Walker, J.L. Starr, and J.D. Mueller. 2004. Plant-parasitic nematodes attacking cotton in the United States: Old and emerging production challenges. Plant Dis. 88:100–113. doi:10.1094/PDIS.2004.88.2.100

Konan, N.O., A. D'Hont, M. Rodier-Goud, J.P. Baudoin, and G.G. Mergeai. 2007. Production of new cotton interspecific hybrids highly resistant to the reniform nematode. In: Proceedings of the World Cotton Research Conference-4, Lubbock, TX. 10–14 Sept. 2007. International Cotton Advisory Committee, Washington DC. 1190.

Lawrence, K.S., and G.W. Lawrence. 2007. Performance of the new nematicide seed treatments on cotton. In: Proceedings of the Beltwide Cotton Conferences, New Orleans, LA. 9–12 Jan. 2007. Natl. Cotton Counc. of Am., Memphis, TN. p. 602–605.

Lawrence, G.W., K.S. Lawrence, and J. Caceres. 2006. AVICTA Complete Pak and Temik 15G for reniform nematode management on cotton. J. Nematol. 38:278.

Lawrence, G.W., and K.S. McLean. 2001. Reniform nematodes. In: T.L. Kirkpatrick and C.S. Rothrock, editors, Compendium of cotton diseases. 2nd ed. Am. Phytopathol. Soc, St. Paul, MN. p. 42–44.

Lawrence, G.W., K.S. McLean, W.E. Batson, D. Miller, and J.C. Borbon. 1990. Response of *Rotylenchulus reniformis* to nematicide applications on cotton. J. Nematol. 22:707–711.

Lawrence, K.S., A.J. Price, G.W. Lawrence, J.R. Jones, and J.R. Akridge. 2008. Weed hosts for *Rotylenchulus reniformis* in cotton fields rotated with corn in the southeast of the United States. Nematropica 38:13–22.

Lewis, S.A., and F.H. Smith. 1976. Host plant distribution and ecological associations of *Hoplolaimus columbus*. J. Nematol. 8:264–270.

Linford, M.B., and J.M. Oliveira. 1940. *Rotylenchulus reniformis*, nov. gen., n. sp., a nematode parasite of roots. Proc. Helminthol. Soc. Wash. 7:35–42.

McCarty, J.C., J.N. Jenkins, M.J. Wubben, O.A. Gutiérrez, R.W. Hayes, F.E. Callahan, and D. Deng. 2013. Registration of three germplasm lines of cotton derived from *Gossypium barbadense* L. accession GB713 with resistance to the reniform nematode. J. Plant Reg. 7:220–223.

McCarty, J.C., J.N. Jenkins, M.J. Wubben, R.W. Hayes, and J.M. LaFoe II. 2012. Registration of three germplasm lines of cotton derived from *Gossypium hirsutum* L. accession T2468 with moderate resistance to the reniform nematode. J. Plant Reg. 6:85–87. doi:10.3198/jpr2011.07.0381crg

McPherson, M.G., J.N. Jenkins, C.E. Watson, and J.C. McCarty, Jr. 2004. Inheritance of root-knot nematode resistance in M-315 RNR and M78-RNR cotton. J. Cotton Sci. 8:154–161.

Minton, N. A. 1962. Factors influencing resistance of cotton to root-knot nematodes (*Meloidogyne* spp.). Phytopathology 52:272:279.

Mueller, J.D., and M.J. Sullivan. 1988. Response of cotton to infection by *Hoplolaimus columbus*. J. Nematol. 2:86–89.

Muhammad, N., and J.E. Jones. 1990. Genetics of resistance to reniform nematode in Upland cotton. Crop Sci. 30:13–16. doi:10.2135/cropsci1990.0011183X003000010003x

National Cotton Council of America. 2013. Beltwide yield losses to nematode damage. National Cotton Council of America. https://www.cotton.org/tech/pest/nematode/losses.cfm (accessed 1 Apr. 2014).

Nichols, R.L., A. Bell, D. Stelly, N. Dighe, F. Robinson, M. Menz, J. Starr, P. Agudelo, J. Jones, C. Overstreet, E. Burris, C. Cook, R. Lemon, and D. Fang. 2010. Phenotypic and genetic evaluation of LONREN germplasm. In: Proceedings of the Beltwide Cotton Conferences, New Orleans, LA. 4–7 Jan. 2010. Natl. Cotton Counc. of Am., Memphis, TN. p. 798–799.

Noe, J.P. 1993. Damage functions and population changes of *Hoplolaimus columbus* on cotton and soybean. J. Nematol. 25:440–445.

Nusbaum, C.J., and H. Ferris. 1973. The role of cropping systems in nematode population management. Annu. Rev. Phytopathol. 11:423–440. doi:10.1146/annurev.py.11.090173.002231

Ogallo, J.L., P.B. Goodell, J.W. Eckert, and P.A. Roberts. 1997. Evaluation of NemX, a new cotton cultivar with high resistance to *Meloidogyne incognita*. J. Nematol. 29:531–537.

Ogallo, J.L., P.B. Goodell, J.W. Eckert, and P.A. Roberts. 1999. Management of root-knot nematodes with resistant cotton cv. NemX. Crop Sci. 39:418–421. doi:10.2135/cropsci1999.0011183X0039000200020x

Robbins, R.T., and K.R. Barker. 1974. The effects of soil type, particle size, temperature, and moisture on reproduction of *Belonolaimus longicaudatus*. J. Nematol. 6:1–6.

Roberts, P.A. 2002. Concepts and consequences of resistance. In: J.L. Starr, R. Cook, and J. Bridge, editors, Plant resistance to parasitic nematodes. CABI, NY. p. 23–41.

Robinson, A.F. 2007. Reniform in U.S. cotton: When, where, why, and some remedies. Annu. Rev. Phytopathol. 45:263–268. doi:10.1146/annurev.phyto.45.011107.143949

Robinson, A.F., R. Akridge, J.M. Bradford, C.G. Cook, W.S. Gazaway, T.L. Kirkpatrick, G.W. Lawrence, G. Lee, E.C. McGawley, C. Overstreet, B. Padgett, R. Rodriguez-Kabana, A. Westphal, and L.D. Young. 2005a. Vertical distribution of *Rotylenchulus reniformis* in cotton fields. J. Nematol. 37:265–271.

Robinson, A.F., A.A. Bell, N.D. Dighe, M.A. Menz, R.L. Nichols, and D.M. Stelly. 2007. Introgression of resistance to nematode *Rotylenchulus reniformis* into Upland cotton

(*Gossypium hirsutum*) from *Gossypium longicalyx*. Crop Sci. 47:1865–1877. doi:10.2135/cropsci2006.12.0776

Robinson, A.F., D.T. Bowman, C.G. Cook, J.N. Jenkins, J.E. Jones, O. L. May, S.R. Oakley, M.J. Oliver, P.A. Roberts, M. Robinson, C.W. Smith, J.L. Starr, and J.M. Stewart. 2001. Nematode resistance. In: T.L. Kirkpatrick and C.S. Rothrock, editors, Compendium of cotton diseases. Am. Phytopathol. Soc., St. Paul, MN. p. 68–72.

Robinson, A.F., A.C. Bridges, and A.E. Percival. 2004. New sources of resistance to the reniform (*Rotylenchulus reniformis* Linford and Oliveira) and root-knot (*Meloidogyne incognita* (Kofoid & White) Chitwood) nematode in Upland (*Gossypium hirsutum* L.) and sea island (*G. barbadense* L.) cotton. J. Cotton Sci. 8:191–197.

Robinson, A.F., C.G. Cook, and A.E. Percival. 1999. Resistance to *Rotylenchulus reniformis* and *Meloidogyne incognita* race 3 in the major cotton cultivars planted since 1950. Crop Sci. 39:850–858. doi:10.2135/cropsci1999.0011183X003900030039x

Robinson, A.F., C.G. Cook, A. Westphal, and J.M. Bradford. 2005b. *Rotylenchulus reniformis* below plow depth suppresses cotton yield and root growth. J. Nematol. 37:285–291.

Robinson, A.F., R.N. Inserra, E.P. Caswell-Chen, N. Vovlas, and A. Troccoli. 1997. *Rotylenchulus* species: Identification, distribution, host ranges, and crop plant resistance. Nematropica 27:127–180.

Robinson, A.F., and A.E. Percival. 1997. Resistance to *Meloidogyne incognita* race 3 and *Rotylenchulus reniformis* in wild accession of *Gossypium hirsutum* and *G. barbadense* from Mexico. J. Nematol. 29:746–755.

Romano, G.B., E.J. Sacks, S.R. Stetina, A.F. Robinson, D.D. Fang, O.A. Gutierrez, and J.A. Scheffler. 2009. Identification and genomic location of a reniform nematode (*Rotylenchulus reniformis*) resistance locus (*Ren^ari^*) introgressed from *Gossypium aridum* into Upland cotton (*G. hirsutum*). Theor. Appl. Genet. 120:139–150. doi:10.1007/s00122-009-1165-4

Sacks, E.J., and A.F. Robinson. 2009. Introgression of resistance to reniform nematode (*Rotylenchulus reniformis*) into Upland cotton (*Gossypium hirsutum*) from *Gossypium arboreum* and a *G. hirsutum/Gossypium aridum* bridging line. Field Crops Res. 112:1–6. doi:10.1016/j.fcr.2009.01.006

Schmidt, L.M., T.E. Hewlett, A. Green, L.J. Simmons, K. Kelley, M. Doroh, and S.R. Stetina. 2010. Molecular and morphological characterization and biological control capabilities of a *Pasteuria* ssp. parasitizing *Rotylenchulus reniformis*, the reniform nematode. J. Nematol. 42:207–217.

Shen X., G.V. Becelaere, P. Kumar, R.F. Davis, O.L. May, and P. Chee. 2006. QTL mapping for resistance to root-knot nematodes in the M-120 RNR Upland cotton line (*Gossypium hirsutum* L.) of the Auburn 623 RNR source. Theor. Appl. Genet. 113:1539–1549. doi:10.1007/s00122-006-0401-4

Shen X.L., Y.J. He, E.L. Lubbers, R.F. Davis, R.L. Nichols, and P.W. Chee. 2010. Fine mapping *QMi-C11* a major QTL controlling root-knot nematodes resistance in Upland cotton. Theor. Appl. Genet. 121:1623–1631. doi:10.1007/s00122-010-1415-5

Shepherd, R.L. 1974a. Registration of Auburn 623 RNR cotton germplasm. Crop Sci. 14:911. doi:10.2135/cropsci1974.0011183X001400060058x

Shepherd, R.L. 1974b. Transgressive segregation for root-knot nematode resistance in cotton. Crop Sci. 14:872–875. doi:10.2135/cropsci1974.0011183X001400060029x

Shepherd, R.L. 1974c. Breeding root-knot resistant *Gossypium hirsutum* L. using a resistant wild *G. barbadense* L. Crop Sci. 14:687–691. doi:10.2135/cropsci1974.0011183X001400050023x

Shepherd, R.L. 1979. A quantitative technique for evaluating cotton for root-knot nematode resistance. Phytopathology 69:427–430. doi:10.1094/Phyto-69-427

Shepherd, R.L. 1982. Registration of three germplasm lines of cotton. Crop Sci. 22:692.

Shepherd, R.L. 1986. Cotton resistance to the root knot-Fusarium wilt complex. II. Relation to root-knot resistance and its implications on breeding for resistance. Crop Sci. 26:233–237. doi:10.2135/cropsci1986.0011183X002600020004x

Shepherd, R.L., J.C. McCarty, J.N. Jenkins, and W.L. Parrott. 1988. Registration of twelve nonphotoperiodic lines with root-knot nematode resistant primitive cotton germplasm. Crop Sci. 28:868–869. doi:10.2135/cropsci1988.0011183X002800050036x

Shepherd, R.L., J.C. McCarty, J.N. Jenkins, and W.L. Parrott. 1996. Registration of nine cotton germplasm lines with resistance to root-knot nematode. Crop Sci. 36:820. doi:10.2135/cropsci1996.0011183X003600030071x

Sikkens, R.B., D.B. Weaver, K.S. Lawrence, S.R. Moore, and E. van Santen. 2011. LONREN Upland cotton germplasm response to *Rotylenchulus reniformis* inoculum level. Nematropica 41:68–74.

Sikkens, R.B., D.B. Weaver, K.S. Lawrence, and R.L. Nichols. 2014. Comparative performance of reniform nematode resistant germplasm lines. In: Proceedings of the Beltwide Cotton Conferences, New Orleans, LA. 6–8 Jan. 2014. Natl. Cotton Counc. of Am., Memphis TN. p. 652–659.

Smith, A.L. 1940. Distribution and relation of meadow nematode, *Pratylenchus pratensis*, to Fusarium wilt of cotton in Georgia. Phytopathology 30:710.

Smith, A.L. 1941. The reaction of cotton varieties to Fusarium wilt and root-knot nematode. Phytopathology 31:1099–1107.

Smith, A.L. 1964. Registration of Auburn 56 cotton. Crop Sci. 4:446. doi:10.2135/cropsci1964.0011183X000400040046x

Starr, J.L., M.J. Jeger, R.D. Martyn, and K. Schilling. 1989. Effects of *Meloidogyne incognita* and *Fusarium oxysporum* f. sp. *vasinfectum* on plant mortality and yield of cotton. Phytopathology 79:640–646. doi:10.1094/Phyto-79-640

Starr, J.L., S.R. Koenning, T.L. Kirkpatrick, A.F. Robinson, P.A. Roberts, and R.L. Nichols. 2007. The future of nematode management in cotton. J. Nematol. 39:283–294.

Starr, J.L., E.R. Moresco, C.W. Smith, R.L. Nichols, P.A. Roberts, and P. Chee. 2010. Inheritance of resistance to *Meloidoygne incognita* in primitive cotton accessions from Mexico. J. Nematol. 42:352–358.

Starr, J.L., C.W. Smith, K. Ripple, E. Zhou, R.L. Nichols, and T.R. Faske. 2011. Registration of TAM RKRNR-9 and TAM RKRNR-12 germplasm lines of Upland cotton resistant to reniform and root-knot nematodes. J. Plant Reg. 5:393–396. doi:10.3198/jpr2011.01.0048crg

Stetina, S.R., G.L. Sciumbato, L.D. Young, and J.A. Blessitt. 2009. Cotton cultivars evaluated for tolerance to reniform nematode. Plant Health Prog. doi:10.1094/PHP-2009-0312-01-RS.

Stetina, S.R., and L.D. Young. 2006. Comparisons of female and egg assays to identify *Rotylenchulus reniformis* resistance in cotton. J. Nematol. 38:326–332.

Stetina, S.R., L.D. Young, W.T. Pettigrew, and H.A. Bruns. 2007. Effect of corn-cotton rotations on reniform nematode populations and crop yield. Nematropica 37:237–248.

Stewart, J.M., and R.T. Robbins. 1995. Evaluation of Asiatic cottons for resistance to reniform nematode. In: D.M. Oosterhuis, editor, Proceedings of the 1994 Cotton Research Meeting and 1994 summaries of cotton research in progress. Spec. Rep.166. Arkansas Agric. Exp. Stn., Fayetteville. p. 165–168.

Sürmelioğlu, C., R.B. Sikkens, R.R. Sharpe, S.R. Moore, E. van Santen, K.S. Lawrence, and D.B. Weaver. 2010. Resistance to reniform nematode from *G. hirsutum* and *G. longicalyx* sources: A comparison. In: Proceedings of the Beltwide Cotton Conferences, New Orleans, LA. 4–7 Jan. 2010. Natl. Cotton Counc. of Am., Memphis, TN. p. 787– 788.

Taylor, A.L., and J.N. Sasser. 1978. Biology, identification and control of root-knot nematodes (*Meloidogyne* species). Dep. of Plant Pathol., North Carolina State Univ. and the US Agency for Int. Dev., Raleigh, NC.

Trudgill, D.L., and V.C. Blok. 2001. Apomictic, polyphagous root-knot nematodes: Exceptionally successful and damaging biotrophic root pathogens. Annu. Rev. Phytopathol. 39:53–77. doi:10.1146/annurev.phyto.39.1.53

USDA-ARS. 2014. Germplasm Resources Information Network. USDA-ARS. http://www.ars-grin.gov/cgi-bin/npgs/html/tax_site_acc.pl?COT%20Gossypium%20hirsutum (accessed 21 May 2014).

USDA-NASS. 2014. Upland cotton pesticide use, 2010. USDA-NASS. http://quickstats.nass.usda.gov/results/5F60FCA9-E070-313D-999E-460025E1F0F3#98CBD111-5812-3F69-B903-CCE8E77D7D1D (accessed 2 May 2014).

USEPA. 2010. Agreement to terminate all uses of aldicarb. USEPA. http://www.epa.gov/oppsrrd1/REDs/factsheets/aldicarb_fs.html (accessed 18 June 2014).

Usery, S.R., K.S. Lawrence, G.W. Lawrence, and C.H. Burmester. 2005. Evaluation of cotton cultivars for resistance and tolerance to *Rotylenchulus reniformis*. Nematropica 35:121–134.

Wang, C., M. Ulloa, and P.A. Roberts. 2006a. Identification and mapping of microsatellite markers linked to a root-knot nematode resistance gene (*rkn*$_1$) in Acala NemX cotton (*Gossypium hirsutum* L.). Theor. Appl. Genet. 112:770–777. doi:10.1007/s00122-005-0183-0

Wang, K., X. Song, Z. Han, W. Guo, J. Sun, J. Pan, R.J. Kohel, and T. Zhang. 2006b. Complete assignment of the chromosomes of *Gossypium hirsutum* L. by translocation and fluorescence in situ hybridization mapping. Theor. Appl. Genet. 113:73–80. doi:10.1007/s00122-006-0273-7

Ware, J.O. 1936. Plant breeding and the cotton industry In: Yearbook of Agriculture. US Gov. Print. Office, Washington, DC. p. 657–744.

Weaver, D.B., K. Lawrence, and E. van Santen. 2007. Reniform nematode resistance in Upland cotton germplasm. Crop Sci. 47:19–24. doi:10.2135/cropsci2006.02.0130

Weaver, D.B., R.B. Sikkens, K.S. Lawrence, C. Sürmelioğlu, E. van Santen, and R.L. Nichols. 2013. *REN*^{lon} and its effects on agronomic and fiber quality traits in Upland cotton. Crop Sci. 53:913–920. doi:10.2135/cropsci2012.08.0504

Wendel, J.F., and R.C. Cronn. 2003. Polyploidy and the evolutionary history of cotton. Adv. Agron. 78:139–186. doi:10.1016/S0065-2113(02)78004-8

Westphal, A., and J.R. Smart. 2003. Depth distribution of *Rotylenchulus reniformis* under different tillage and crop sequence systems. Phytopathology 93:1182–1189. doi:10.1094/PHYTO.2003.93.9.1182

Wheeler, T.A., K.S. Lawrence, D.O. Porter, W. Keeling, and B.G. Mullinix, Jr. 2013. The relationship between environmental variables and response of cotton to nematicides. J. Nematol. 45:8–16.

Windham, G.L., and G.W. Lawrence. 1992. Host status of commercial maize hybrids to *Rotylenchulus reniformis*. J. Nematol. 24:745–748.

Wrather, J.A., W.E. Stevens, E.D. Vories, T.L. Kirkpatrick, J.D. Mueller, and A. Mauromoustakos. 2010. The use of root gall ratings to determine high risk zones in cotton fields infested by *Meloidogyne incognita*. Crop Sci. 50:2575–2579. doi:10.2135/cropsci2010.02.0054

Yik, C.P., and W. Birchfield. 1984. Resistant germplasm in *Gossypium* species and related plants to *Rotylenchulus reniformis*. J. Nematol. 16:146–153.

Ynturi, P., J.N. Jenkins, J.C. McCarty, Jr., O.A. Gutiérrez, and S. Saha. 2006. Association of root-knot nematode resistance genes with simple sequence repeat markers on two chromosomes in cotton. Crop Sci. 46:2670–2674. doi:10.2135/cropsci2006.05.0319

Zhang, J., C. Waddell, C. Sengupta-Gopalan, C. Potenza, and R.G. Cantrell. 2006. Relationships between root-knot nematode resistance and plant growth in Upland cotton: Galling index as a criterion. Crop Sci. 46:1581–1586. doi:10.2135/cropsci2005.09-0333

Zhou, E., J.L. Starr, and C.W. Smith. 1999. Inheritance of resistance to *Meloidogyne incognita* in the cotton cultivar Acala NemX. J. Nematol. 31:584–585.

Harvesting

John D. Wanjura,* Edward M. Barnes, Mark S. Kelley, and Randal K. Boman

Abstract

Since the introduction of the first successful mechanical harvester, mechanized harvest has continued to decrease the cost and labor hours required to produce a bale of cotton (*Gossypium* L.). Cotton harvesting in the United States is completely mechanized and is accomplished by two primary machines, the spindle picker and the brush-roll stripper. Significant advances with regard to harvesting and in-field storage systems have occurred over the last 40 yr, which have resulted in substantial increases in harvest productivity. State-of-the-art cotton harvesters are able to harvest at a rate exceeding 4 ha/h (10 ac/h). Harvesting efficiency of pickers and strippers is influenced by a host of production, environmental, design, and operating factors. Thus, adoption of each type of harvester varies by geographic region across the US Cotton Belt. Careful attention to harvester configuration and operation by cotton producers has helped to develop the reputation of the United States as a dependable source of high-quality and contaminant-free cotton.

Since the first patent was awarded in 1850 to S. Rembert and Jebediah Prescott of Memphis, TN (Rembert and Prescott, 1850), there have been hundreds of patents for cotton picking devices ranging from vacuum systems to threshers. Ultimately, two primary systems emerged for cotton harvest: strippers, starting with a horse-drawn "sled" around 1920, and also in the 1920s development of a spindle-type picker by Price–Campbell cotton machinery. Complex social and economic factors limited the adoption of mechanized harvest systems in the United States until the early 1940s (Heinicke and Grove, 2005), but there is no

Abbreviations: NIS, non-ionic surfactant; OMB, onboard module building; PPO, protoporphyrinogen oxidase; RFID, radio frequency identification.

John D. Wanjura, USDA-ARS Cotton Production and Processing Research Unit, 1604 E. FM 1294, Lubbock, TX 79403. *Corresponding author (John.wanjura@ars.usda.gov).

Edward M. Barnes, Cotton Incorporated, 6399 Weston Parkway, Cary, NC 27513 (EBarnes@cottoninc.com).

Mark S. Kelley, Texas A&M AgriLife Extension Service, 1102 E. FM 1294, Lubbock, TX 79403 (m-kelley@tamu.edu).

Randal K. Boman, Southwest Research and Extension Center, Dep. of Plant and Soil Sciences, Oklahoma State University, 16721 U.S. Highway 283, Altus, OK 73521 (randy.boman@okstate.edu).

doi:10.2134/agronmonogr57.2013.0047

Cotton. 2nd. ed. David D. Fang and Richard G. Percy, editors. Agronomy Monograph 57.

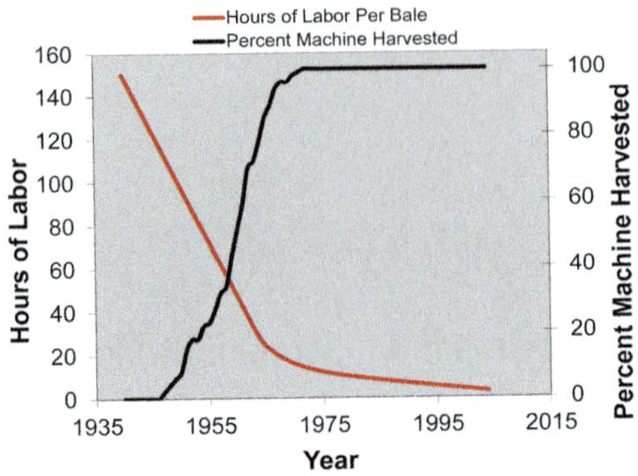

Fig. 1. Hours of labor to create a 218-kg bale of cotton fiber and percent of cotton mechanically harvested from 1940 to 2000 in the United States. Labor data for 1940 to 1960 from USDC (1970), and 1965 to 1987 from USDA-ERS (2000). Percent mechanically harvested compiled from USDA, Economic Research "Cotton Ginning Charges, Harvesting Practices, and Selected Marketing Costs" reports. Similar trends in labor requirements also shown in Ibendahl et al. (2014).

question that the mechanization of cotton harvest drastically decreased cotton's labor requirements as illustrated in Fig. 1.

While mechanical harvest has been the primary method in the United States for over 40 yr, there are many areas in the world that still rely on hand harvest, including the two largest global cotton producing countries of India and China (International Cotton Advisory Committee, 2011). However, in both of those countries there is strong interest in mechanical harvest because of increasing labor costs, and in the case of China, cotton production is moving to less populated areas in the northwest part of the country where available labor is limited.

Part of the decline in US labor requirements that continued into the 1970s after harvest was fully mechanized can be attributed to the development of the cotton module builder. Before the module builder, seed cotton was transported to the cotton gin in trailers. When a producer's trailers were full, harvest had to cease until the gin could process the cotton. With the advent of the module builder, cotton harvest was decoupled from ginning as the modules could be stored on the edge of the field. Ultimately this decreased the number of gins needed and the associated labor required (Hughs et al., 2008).

Machinery

Types of Machines and Principles of Operation

Picker- and stripper-type harvesters are the two primary machines used to mechanically harvest cotton in the United States. Picker-type harvesters utilize a selective harvesting mechanism that collects seed cotton from open bolls, leaving much of the bur and other plant material in the field. Stripper-type harvesters employ a non-selective harvesting mechanism that collects seed cotton along with a significant amount of plant material such as burs, sticks, and leaves. A third type of mechanical harvester is used infrequently in the Southwest and western production areas of the Unites States to glean seed cotton from the ground. Gleaner-type machines are used rarely because of the excessive amount of soil and foreign material they collect with the seed cotton, but provide a means for mitigating production losses from adverse weather that blows cotton out of the bur and onto the ground.

Picker Harvesters

Although early cotton pickers harvested only one row of cotton per field pass, the increase in harvesting productivity that they offered over hand picking was astounding. Modern cotton operations require an even greater level of productivity, and as a result, picking machines produced today often harvest six rows per field pass with first gear (slowest) ground speed in the range of 6.44 to 7.1 km/h (4–4.4 mi/h). Additional harvesting productivity gains have been realized in recent years with the advent of onboard module building (OMB) picker harvesters (Willcutt et al., 2009). The OMB harvesters form seed cotton modules on board the harvester and thus have reduced the amount of labor and additional machinery required to accomplish harvest.

Tapered spindles are used in the harvesting units of modern picker-type harvesters to remove seed cotton from open bolls (Fig. 2). The spindles rotate at high speed (approximately 4000–4200 rpm), and each spindle has three rows of notched teeth or barbs machined along its length at an angle of approximately 30 degrees with respect to the spindle axis. When seed cotton is exposed to a spindle, the barbs engage the lint and wrap the fiber around the spindle as it is pulled from the bur. Approximately 18 to 20 spindles are vertically spaced 41.28 mm (1.625 in) apart on each vertical picker bar, and the picker bars are mounted on rotating drums. When operating in first gear, the drum rotation speed is synchronized to the harvester ground speed so that the speed of the plants moving through the picking zone equals the lateral speed of the spindles. Picking in second gear increases harvester ground speed, but the picking unit drums are not synchronized with second gear ground speed and do not speed up proportionately resulting in a condition where the picking units "lag" the plant row. Although picking in second gear allows more area to be harvested in a given time, the lagging condition generally increases the amount of foreign material collected with the seed cotton.

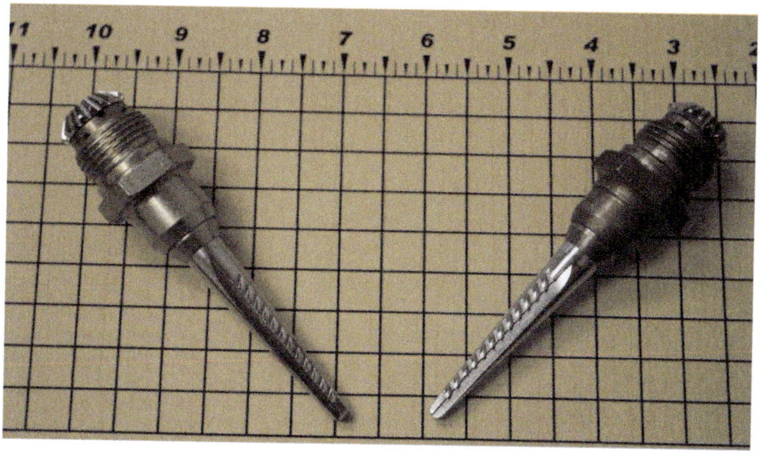

Fig. 2. Left-handed (left) and right-handed (right) tapered spindles.

A cam and follower mechanism at the top of the row unit controls the angular position of the picker bars so that the spindles are properly positioned for picking, doffing, and cleaning. The spindles move into the picking zone pointed slightly toward the rear of the machine and gradually rotate to point slightly toward the front of the machine as they leave the picking zone. As the spindles move through the picking zone, they pass between metal ribs that help to channel the plants through the row unit and prevent other plant material from being harvested with the seed cotton. Compressor plates push the plants against the ribs and help expose seed cotton to the rotating spindles in the picking zone. Scrapping plates can be attached to compressor plates to help increase the exposure of bolls to the spindles under less favorable crop conditions (e.g., low-yielding small plants with closed or tight bolls).

Cotton is removed from the spindles by lugged disks known as doffer pads. Lugs on the bottom side of each doffer pad move in relation to the spindle such that the cotton is unwrapped and pushed off the pointed end of the spindle. The doffed cotton passes into an air stream that conveys the harvested material into the harvester basket. After doffing, the spindles pass under moisture applicator pads that help to wash away residue from green plant material. The moisture applied to each spindle makes it easier for lint to adhere to the spindle surface, thus improving picking efficiency.

All modern picker harvester row units utilize two drums to harvest cotton in either an opposed (Fig. 3) or inline (Fig. 4) drum configuration. The opposed

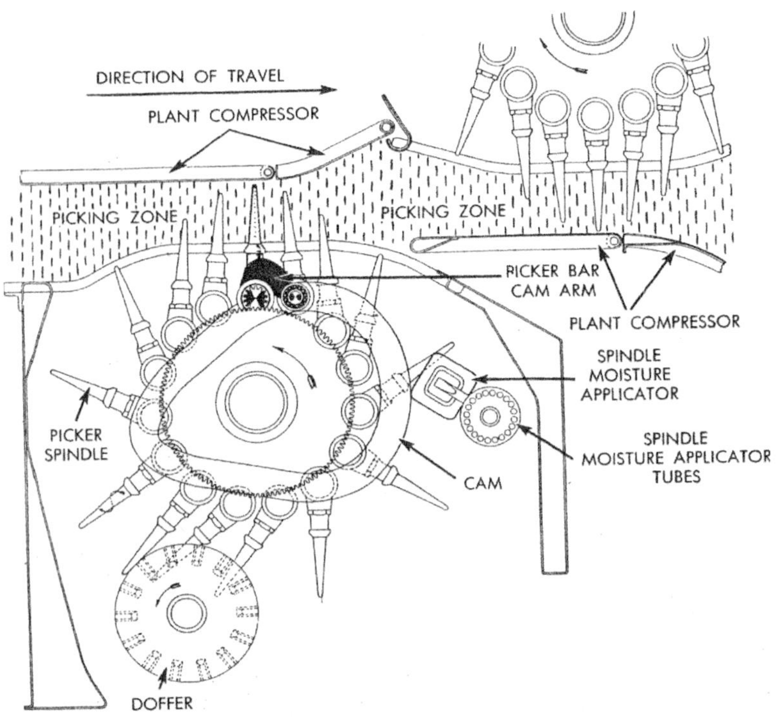

Fig. 3. Opposed drum picking unit schematic.

drum configuration (Fig. 3) positions a picker drum on either side of the plant row so that cotton is picked from both sides of the row. The inline drum configuration positions both picking drums on the right side of the plant row. Row units with opposed drum configurations use both left- and right-handed spindles (Fig. 2), picker bars, and cam tracks. The use of left- and right-handed components can make manufacturing and repair more cumbersome as the number of components is increased and workers must be extra attentive to ensure the installation of correct components. Opposed drum row units require different header frames for narrow (76.2–86.4 cm [30–34 in]) and wide (91.4–101.6 cm [36–40 in]) row spacing patterns due to their width, whereas inline row units can be mounted on a single header frame for all row spacing patterns.

Stripper Harvesters

Cotton stripper harvesters indiscriminately harvest seed cotton along with undesirable plant material such as sticks, burs, and leaves by snapping the material from the plants. Stripping devices used in the early 1900s were horse or mule drawn and consisted of a wooden box on skids with a tapered slot through which the plants were pulled (Fig. 5) (Jones et al., 1928). As the plants passed through the narrow end of the slot, open and unopened bolls along with other plant material too large to pass through were stripped off and collected in the "sled" box. After

Fig. 4. Inline drum picking unit schematic (used with permission, courtesy of John Deere).

Fig. 5. Cotton sled used to harvest cotton in the Southern High Plains in the early 1900s.

sledding, unopened bolls were often piled at the end of the row to allow them to open before being transported to a cotton gin.

Two mechanized systems employing the stripping principal were developed for harvesting cotton, the finger stripper and the brush-roll stripper. Finger strippers (Fig. 6) were developed in response to agronomic research in the Southern High Plains, which showed economic benefits to producing cotton in narrow row or broadcast planting patterns. The finger stripper works well for harvesting short plants with small main stem diameters and minimal vegetative branching; plant growth characteristics common to broadcast or narrow row planting (Kirk et al., 1964). Crop material is stripped off as plants pass between a series of closely spaced angle iron bars or fingers. Paddles mounted on a shaft extending the width of the finger stripper header help to move the harvested material to a large center conveying auger. The auger conveys the material to the center of the header where it is conveyed over a set of cleaning grids by an undershot flighted belt. A final auger is used to carry the harvested material to the side of the header where it is pneumatically conveyed to the harvester basket.

The brush-roll stripper (Fig. 7) is the only mechanical stripper-type cotton harvester commercially produced in the Unites States today. These machines were developed to harvest cotton planted in rows evenly spaced 76.2 to 101.6 cm (30–40 in) apart and utilize two cylinders to remove the material from the plants. The cylinders are positioned on opposite sides of the plant row and are inclined at approximately 30 degrees with respect to the ground surface. The cylinders rotate in opposite directions counter to the movement of the plants through the stripper row unit. Stiff brushes and rubber strips known as "bats" are mounted along the length of each cylinder in an alternating sequence and break bolls and other material off the plants as they pass through the narrow space between the cylinders. The harvested material is conveyed out of the row unit by small augers

Fig. 6. Finger stripper harvester.

Fig. 7. Brush roll stripping unit.

located adjacent to the stripping rolls. The row unit augers carry harvested material to the top of the row unit where it is dropped onto a large auger (cross auger) and conveyed to the center of the machine. The row unit augers and the cross auger convey the harvested material across slotted panels that help to remove soil and other small foreign material contained in the seed cotton. The harvested material is pneumatically conveyed from the header cross auger to an onboard cleaning device. Green bolls and other heavy trash materials (stumps, rocks, etc.) are removed either at the base (intake) or the top (discharge) of the pneumatic conveying duct by gravitational settling (Fig. 8). The onboard cleaner used on modern stripper harvesters is similar in design to a two-saw stick machine used in the cotton ginning process to remove large foreign matter from seed cotton (Fig. 9). Inside the cleaner, cotton is engaged by the teeth of the primary cleaning saw cylinder and foreign material is thrown off the cylinder as the harvested material is exposed to a series of grid bars located around the circumference of the saw cylinder. Cotton remaining on the saw cylinder surface is removed by a doffer brush and carried by air to the harvester basket. The material thrown off by the primary saw passes to the reclaiming saw cylinder where the cotton is engaged by the saw teeth and large foreign material is thrown off by the cleaner.

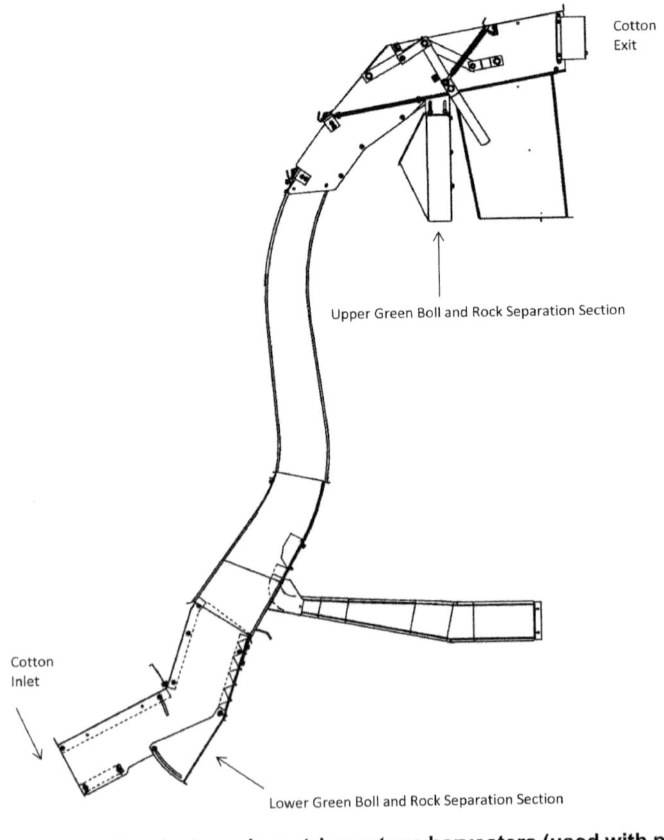

Fig. 8. Cotton conveying duct used on stripper-type harvesters (used with permission, courtesy of John Deere).

The reclaimed cotton is removed by the doffer brush and conveyed by air to the harvester basket.

Application by Location and Culture

Although hand picking has been completely replaced by mechanical harvesting in the Unites States today, the application of mechanical harvest systems varies with regard to geographic location. The brush-roll stripper is used primarily in the semiarid Southern High Plains region of Texas, Oklahoma, New Mexico, and Kansas. Stripping is also used on a more limited basis in the Blackland and coastal bend regions of Texas depending on crop conditions that can vary annually. Stripper harvesters are best suited to harvest cotton with relatively short stalks (i.e., less than 1–1.2 m) and small lateral limbs—characteristics typical of cultivars grown in the Southern High Plains and central and south Texas. Crop desiccation by chemical application or frost is necessary to affect efficient stripper harvesting. Climatic conditions (high winds, rain, etc.) typical to the Southern Great Plains that result in high pre-harvest loss of cotton have promoted the adoption of cultivars that produce bolls with high levels of storm tolerance. Plants with highly storm tolerant boll conformations that hold mature seed cotton tightly inside the

Fig. 9. Schematic of a field cleaner used on cotton stripper harvesters (used with permission, courtesy of John Deere).

open bur can reduce spindle picking efficiency and are better suited for stripper harvesting. The spindle picker is the predominate machine used to harvest cotton in the traditionally rainfed Mid-South, southeast, and irrigated west regions of the United States. Pickers are better suited to harvest high yielding cotton from larger plants similar to those usually produced in these regions. Humid conditions during harvest in the Mid-South and southeastern United States often limit the available daily harvest window for effective stripper harvesting. Pickers can harvest cotton from non-desiccated pliable plants under wet field conditions so long as the leaves have been adequately defoliated and the lint and seed are below 12% w.b. moisture content for safe infield storage.

Machine Performance

Harvesting Efficiency and Foreign Matter Content

The indiscriminate harvesting action used by stripper-type cotton harvesters can remove up to 99% of seed cotton from the plant (Williford et al., 1994). Commonly, stripper harvesting efficiency falls in the 95 to 98% range (Tupper, 1966; Faulkner et al., 2011a) and is influenced by factors such as crop defoliation/desiccation level, harvest date, stripper roll spacing, sequencing of brushes and bats on stripper rolls, and stripper roll speed (Batchelder et al., 1961; Roberson, 1964; Wanjura and Brashears, 1983; Brashears, 1984a, 1986a, 1992; Supak et al., 1992). Picking efficiency is usually in the range of 85 to 90% but can be higher than 95% given favorable crop conditions (Corley, 1966; Corley and Stokes, 1966; Tupper, 1966; Williford et al., 1994; Wanjura et al., 2010a; Faulkner et al., 2011a). Picker harvesters are considerably more mechanically complex than strippers and maximizing picking efficiency requires careful attention to cultural practices (e.g., cultivar selection for proper boll and plant type, plant population, tillage practices, plant growth regulator use, etc.) and harvester setup, adjustment, operation, and maintenance practices (Willcutt et al., 2010).

The amount of foreign material collected by either harvesting system can vary by crop condition and harvester operation. The use of field cleaners on stripper harvesters helps to remove approximately 50 to 60% of the foreign material harvested with the seed cotton, which helps to reduce transportation and ginning charges based on module weight delivered to the gin (Kirk et al., 1972, 1973; Brashears, 1991;Bennett et al., 1995; Baker and Brashears, 2000; Wanjura et al., 2009, 2012a). Wanjura et al. (2012b) compiled lint, seed, and foreign matter content data for several recent studies comparing picker and stripper harvesters in the Southern High Plains and reported average lint turnout values of 35, 30, and 25% for picker and stripper harvested cotton with and without a field cleaner, respectively (Table 1). The values reported by Wanjura et al. (2012b) correspond with previously published data (Baker et al., 1994; Williford et al., 1994).

Field Efficiency, Productivity, and Harvest Capacity

A key factor in maximizing value from harvest equipment is to ensure that the harvester has as little downtime as possible. ASABE (2006) define field efficiency (E_f) as the ratio between the productivity under field conditions and its theoretical maximum productivity, where maximum productivity is equal to the speed of the harvester times the width harvested. The field efficiency relationship can be rearranged to estimate the productivity of a harvester in hectares per hour (C_a) as

Table 1. Lint, seed, and foreign matter components of picker and stripper harvested cotton (Wanjura et al., 2012b).†

	Picker‡	Stripper with field cleaner‡	Stripper without field cleaner‡
Lint turnout, %	34.7	29.7	25.1
Total foreign matter, %	11.1	23.2	39.7
Bur cotton weight, kg/bale (lb/bale)	628 (1384)	733 (1614)	867 (1910)
Seed weight, kg/bale (lb/bale)	344 (758)	354 (780)	341 (750)
Total foreign matter, kg/bale (lb/bale)	70 (154)	173 (381)	368 (810)
Bur weight, kg/bale (lb/bale)	21 (45)	83 (183)	198 (436)
Stick weight, kg/bale (lb/bale)	7 (15)	32 (71)	67 (148)
Leaf/fines/motes, kg/bale (lb/bale)	42 (94)	58 (127)	103 (226)

† Based on a 218-kg lint bale.

‡ Values reported for once over harvest after harvest-aid chemical applications.

$$C_a = \frac{swE_f}{10} \quad [1]$$

where s is field speed while harvesting in kilometers per hour, and w is the width harvested in meters.

For cotton harvest, the following factors impact the field efficiency of Eq. [1]:

- time to empty the harvester;
- time to turn at the end of the row;
- length of row (impacts the number of turns in the field);
- yield (higher yields increase the number of times material needs to be offloaded from the harvester, and higher yields can lower field speed); and
- downtime (time to clean out chokes or blockages; operator breaks, etc.).

Willcutt et al. (2009) measured the field efficiency in the Mississippi Delta region of the United States for three picker types: John Deere model 7760 round module building picker; Case IH 625 OMB picker; and traditional basket six-row pickers (mixture of John Deere and CaseIH models). Similarly, Faulkner et al. (2011a) compared the field efficiency of a basket-type cotton picker with a cotton stripper with and without field cleaning in the Texas High Plains area. Field efficiencies ranged from 75% for conventional basket pickers and strippers to a maximum of 90% with the John Deere 7760. Table 2 lists several scenarios and presents the calculated productivity using Eq. [1]. The scenarios are reflective of the data presented in both Willcutt et al. (2009) and Faulkner et al. (2011a).

From Table 2 and Fig. 10, it is clear that the number of rows harvested has a major impact on harvest productivity. It is also interesting to note the importance of row width—all current cotton harvesters have a fixed number of row units, so decreasing row spacing decreases harvested area per pass. Because harvest costs tends to be a large percentage of annual production costs and cotton yields are typically not as responsive to row spacing relative to crops such as corn (*Zea mays* L.) and soybean [*Glycine max* (L.) Merr.], row spacing less than 1 m are not as desirable from a harvest perspective. In part because of the impact on harvest costs, Buehring et al. (2009) concluded that row patterns that optimized harvest efficiency were the most economical for Mississippi Delta conditions.

Table 2. Cotton harvester productivity levels for different harvest system scenarios.

Scenario	Rows harvested	Row width	Speed	Field efficiency	Productivity
		m	kph		ha/h
Two-row harvester	2	1.0	4.8	0.75	0.7
Four-row harvester	4	1.0	4.8	0.75	1.5
Older six-row spindle harvester	6	1.0	4.8	0.75	2.2
New six-row spindle harvester	6	1.0	6.8	0.75	3.1
New six-row spindle harvester—second gear	6	1.0	8.0	0.75	3.7
Eight-row stripper harvester	8	1.0	6.4	0.75	3.9
New six-row spindle harvester—high efficiency	6	1.0	6.8	0.90	3.7
New six-row picker—high efficiency, second gear	6	1.0	8.0	0.90	4.4
Eight-row stripper—low yield, with field cleaner	8	1.0	8.8	0.78	5.6
New Six-Row Spindle Harvester—Various Row Spacing					
1.5-m row spacing	6	1.5	6.8	0.75	4.6
1.0-m row spacing	6	1.0	6.8	0.75	3.1
0.9-m row spacing	6	0.9	6.8	0.75	2.8
0.8-m row spacing	6	0.8	6.8	0.75	2.3

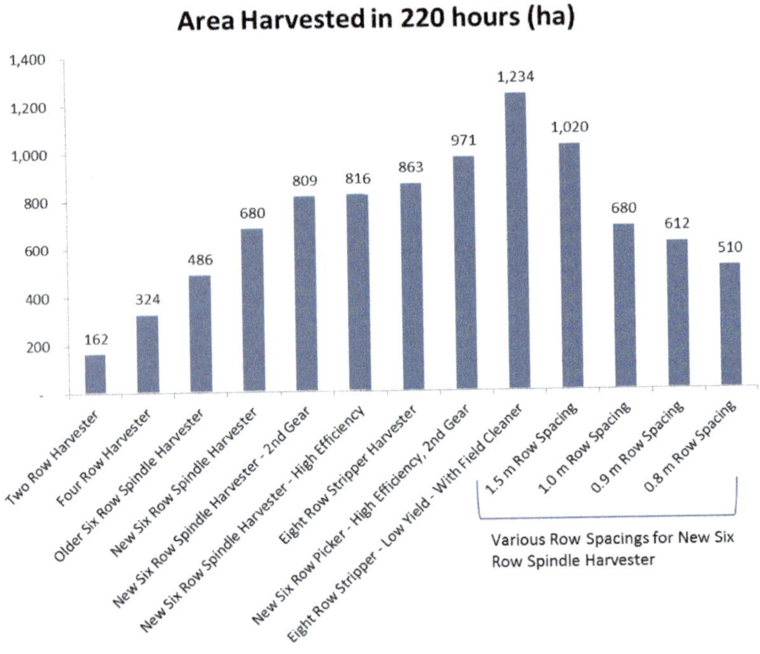

Fig. 10. Area harvested in 220 h (ha) for the scenarios listed in Table 2.

In addition to row spacing, row length can have a large impact on field efficiency. The parameters given in Table 2 for a "new six-row spindle harvester" were used with simulated field efficiencies assuming a turning time of 30 s, a time to unload seed cotton of 90 s, yield of 1120 kg/ha, and a downtime of 10% of operating time for different row lengths. The results of that simulation are presented in Fig. 11. Note that as row length decreases, so does field efficiency as more and more time is spent turning at the end of the row. Row lengths less than 300 m significantly reduce field efficiency, so when planting fields, arranging rows to avoid short distances will optimize harvest efficiency.

Knowledge of the expected field efficiency is important for determining the land area that can be harvested by a single machine to provide sufficient time for harvest. Chen et al. (1992) performed an analysis for conditions in Mississippi and estimated that sufficient cotton harvest capacity should be available to complete harvest in approximately 220 h. A greater number of hours will be available in other regions, as time for harvest is limited in Mississippi because of the humid climate and the need to get the crop harvested before peak hurricane season. Therefore, total time needed for harvest will vary on the basis of climatic conditions and will definitely increase for more arid regions such as the desert Southwest. Having insufficient harvest capacity risks pre-harvest yield loss and lint quality loss due to weathering.

Cost of Harvest

Harvesting cost varies by location because of factors such as annual area harvested, type and age of harvester, crop yield, available harvest period, and production practices. In some areas, custom harvesting is available to producers that do not own harvesting equipment. Custom harvesting rates are usually based on a lint weight basis. Klose et al. (2013) reported the most common custom rates paid in Texas for cotton harvesting with a conventional basket stripper, conventional basket picker, and OMB picker were US$0.22, $0.26, and $0.29 per kg of lint (US$0.10, $0.12, and $0.13 per lb of lint), respectively. These values include machinery, labor, and fuel expenses. In scenarios where a producer owns and operates their own machinery, harvesting costs must consider both fixed and variable costs as well

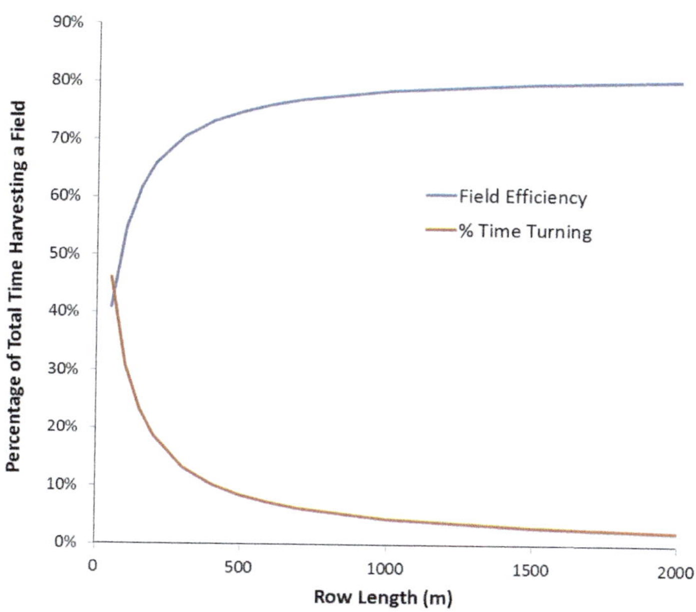

Fig. 11. Impact of field row length on harvest field efficiency and the percent of total harvest time where the machine is turning at the end of the row.

as his specific case in regard to harvested area, available harvest window, and labor. Nelson et al. (2001) demonstrated a method for comparing costs between picker and stripper based harvest systems of various sizes over a range in annually harvested acres. They also compared harvest costs between the machine ownership scenarios to custom harvested scenarios and showed that, given their assumptions, it costs less to own and operate an eight-row stripper if the farm size exceeds 364 ha (900 ac). The same was found true for a six-row picker when the farm size exceeds 324 ha (800 ac). Faulkner et al. (2011b) conducted a net present value analysis to compare picker and stripper based harvest systems in irrigated cotton. Stripper based systems were found more profitable under most yield scenarios when comparing harvest systems with one stripper to those with one picker. However, Faulkner et al. (2011b) found a harvesting system with one picker was more profitable than a harvest system with two strippers. Farrell and Ibendahl (2009) conducted an analysis comparing the optimum utilization rates of conventional and new picking systems with OMB capabilities. They found that optimum harvester utilization occurred where average harvesting costs were minimized for each system and corresponding levels were 647, 809, and 688 ha/yr (1600, 2000, and 1700 ac/yr) for the conventional basket picker, John Deere 7760, and CaseIH ME625, respectively. Falconer et al. (2010) developed a web-based decision aid for helping producers compare the costs associated with implementing various harvesting systems including conventional strippers and pickers and OMB pickers.

Cultural and Production Practices Influencing Mechanical Harvest

Field Optimization for Mechanical Harvest

An efficient harvest system begins at the time of planting. A key element, especially if not using equipment controlled by high-precision GPS steering, is to make sure the number of row units on the planter is an even multiple of the row units on the harvester. For example, if using a four-row spindle picker, the planter should have 8, 12, or 16 row units. This ensures that the distance between the rows is always constant for the harvester and that plants will be positioned in the center of the harvester row unit. Such concerns are less if using a finger-type cotton stripper header.

Previously, the importance of row length and row width was discussed in terms of maximizing harvest productivity. Increasing row width increases the area harvested in each pass, and many studies have found cotton yields are not highly sensitive to row spacing between 20 and 150 cm if a minimum plant population is maintained. For example, Gwathmey et al. (2011) found that if at least 6.1 plants/m^2 were established, row spacing did not impact yields in the northern Mississippi Delta. Similar results have been reported in other environments in the Unites States (Riar et al., 2013; Buehring et al., 2009). Corley and Stokes (1966) recommended plant populations of 49,400 to 148,200 plants/ha (20,000–60,000 plants/ac) for dryland and irrigated production in Alabama to prevent yield declines and produce plant characteristics conducive to mechanical harvesting. Therefore, from a harvest perspective, wider row spacing distances are preferred as the machine can cover more area per pass without significant cotton yield loss. However, rotation with other crops and other agronomic considerations such as better early weed suppression from narrower rows may be overriding factors in determining row spacing as opposed to harvest optimization.

During the season, management practices will also have an impact on ease of crop harvest. If crop height is too great, the increased amount of biomass passing through the row unit can limit the speed of machine travel and increase the amount of foreign matter harvested. Therefore, the use of growth regulators may be necessary for some cultivars. Height control becomes even more important if harvesting ultra-narrow row cotton with a finger-type stripper header. Plant heights above 55 cm are not recommend for harvest by a finger stripper (Bader et al., 1998). Similar problems can be created if adequate weed control is not obtained—large weeds in the field at the time of harvest can slow the harvester (Smith et al., 2000) and contaminate the cotton (especially with grasses, which can be hard to remove during ginning). At the time of harvest, it is particularly important that there not be a significant presence of green weeds as these can stain the cotton fibers.

Another in-season concern is insect pest control, particularly those that excrete sugars such as whiteflies and aphids. The excretions can lead to "sticky cotton" that becomes extremely problematic for textile mills. A well-defined integrated pest management program is essential to control these pests, especially in arid regions where it is less likely that rainfall will remove some of the residues (Ellsworth et al., 1999).

Cotton root rot caused by the fungus *Phymatotrichopsis omnivora* causes production losses in several areas across the Cotton Belt and hinders mechanical harvest especially with stripper-type harvesters. During harvest, plants dead from disease break off at the soil surface and become lodged inside stripping units, which leads to excessive downtime to unchoke row units. Picker harvesters can be used to harvest crops impacted by cotton root rot, but in severe cases low yield and reduced plant population can result in reduced picking efficiency. Recently, the fungicide flutriafol has been tested and sold as an at-planting treatment for cotton root rot in several states. The fungicide has been shown to reduce disease incidence and improve lint yields compared to untreated control plots (Isakeit et al., 2012, 2013; Drake et al., 2013; Goodson et al., 2014; Norton et al., 2014).

Plant Characteristics Impacting Mechanical Cotton Harvest

Several plant attributes impact the ability to efficiently harvest cotton, including plant height; canopy geometry (number and location of branches, and fruiting positions); boll size and shape; time to full crop maturity; cultivar determinacy; stem diameter; and the degree of leaf pubescence. The optimal conditions for many of the listed attributes vary by the type of harvester used (i.e., brush stripper–, finger stripper–, or spindle-type harvesters). In general, stripper harvesters (brush and finger) are most efficient with smaller plants and in moderate to low yielding conditions in part because of the additional material harvested with the seed cotton (Faulkner et al., 2011a). Stripper harvest also has the advantage of being able to collect seed cotton more tightly held in the boll—a characteristic of "storm-resistant" cultivars.

Spindle-type harvesters are more suited to higher yielding environments and ideally need the first boll to be at least 10 to 15 cm above the ground to be efficiently harvested (Husman et al., 2005). In-season cultivation practices that create a small ridge of soil on either side of the plant row can hamper the ability of picker row units to operate low enough on the plant to harvest low set bolls, which can result in reduced picking efficiency. Also, the locks of seed cotton need to be exposed from the carpel so it can be freely presented to the spindle. The disadvantage of exposed locks is that they can be easily removed by heavy rainfall and wind before harvest, so a balance between easy removal and ability to be retained on the plant until harvest is needed.

Despite the harvest system, increased leaf hairiness is not ideal for mechanically harvested cotton as the leaf hairs become enmeshed in the fiber and can be difficult to remove (Eder et al., 2013). Another common attribute across harvester systems is the need for a cotton stalk that does not have an excessive diameter, but is strong enough to endure the upward pulling forces during harvest as well as to prevent lodging during harvest. Excessive stalk diameter (>20 mm measured just above the cotyledon nodes near time of harvest) can increase the potential of bark removal during harvest. Additionally, for stripper harvesters it has been demonstrated that increasing the number of branches in the harvested seed cotton is also a source of bark contamination, especially due to stick breakage during conveying and ginning (Wanjura and Baker, 1979). Therefore, more compact plants with short braches are desirable for stripper harvest; however, for spindle harvest, some elongation of branches is desirable so all bolls have an opportunity to contact a spindle. Finally, a plant characteristic that is beneficial to all mechanical harvest systems are earlier maturing, more determinate cultivars. This limits the probability of harvest of immature bolls.

Pre-Harvest Preparation

Cotton harvest-aid chemicals are widely used across the US Cotton Belt. Varying products are used to prepare the crop for either stripper or picker harvesting by hastening boll dehiscence and opening, inducing leaf abscission and defoliation, desiccating remaining leaves, and inhibiting or removing terminal and/or basal regrowth. Efficacy and profitability of these cotton harvest-aid chemicals are highly dependent on environmental conditions before and after application, crop physiological condition at application, and timing of application. Weather conditions favorable for harvest-aid performance include warm, sunny days with a low probability of precipitation before harvest. Conversely, extended periods of cool (below 15.6°C [60°F]), cloudy, wet, or humid weather can result in decreased performance. Additionally, delayed harvest may lead to additional harvest-aid applications to achieve desired crop conditioning, which results in increased production costs. Soil conditions optimal for performance of these chemicals include moisture levels that are sufficient to sustain plant physiological activities without excessive moisture stress, and relatively low or depleted soil nitrate nitrogen (N) (Main and Hayes, 2010). High moisture and N levels that delay crop maturity can significantly reduce harvest-aid performance in terms of defoliation, can result in low micronaire for stripper harvested cotton, and can increase the likelihood of regrowth following application. In the absence of precipitation, soil moisture levels can be closely managed in irrigated production systems by timely irrigation termination. Soil N levels can also be successfully managed through proper soil sampling practices and fertilizer applications based on realistic lint yield goals. Crop condition at application can also have significant effects, both positive and negative, on harvest-aid performance. A cotton crop that has initiated senescence, has greater than 60% open bolls, and exhibits some natural defoliation will respond more readily to harvest-aid applications (Hake et al., 1996). However, a drought stressed cotton crop will have thicker leaf cuticles and will typically be more difficult to prepare for harvest (Sanders et al., 2009). In general, there are two recognized types or categories of harvest-aid chemicals: those with contact or herbicidal activity and those with hormonal activity. Additionally, harvest-aid products can be divided into groups, which include boll openers, defoliants, and desiccants. Proper selection and application of these harvest-aid chemicals to a relatively mature cotton crop under ideal environmental conditions will help ensure a successful and timely mechanical harvest resulting in optimum yields and quality of cotton fiber and seed.

Boll Openers

Boll openers consist mainly of products containing ethephon [(2-chloroethyl) phosphonic acid]. Ethephon was first registered in the Unites States by Union Carbide Agricultural Products Co. in 1981 (Supak and Snipes, 2001). Since that time, use of ethephon has increased significantly and is now considered common practice in both stripper and picker harvesting systems. Today there are several commercially available ethephon products that are labeled for use in cotton. These products typically contain 0.72 kg of ethephon per L (6 lb/gal). Other, "enhanced" ethephon containing products are also available that include a synergist to improve boll opening performance. One of these products contains cyclanalide and the other contains AMADS, or urea sulfate. These chemicals affect natural plant processes associated with boll opening, but do not increase the rate of boll or fiber maturation. Once inside the plant, ethephon is converted to ethylene, a senescence hormone that increases the rate of abscission layer formation. This response requires physiological activity. Ethephon-based products result in an increased rate of boll opening and defoliation that generally reaches a maximum in about 14 d. Tank mixes of ethephon and defoliants are effective in higher yielding cotton to open bolls and drop leaves. Higher rates of ethephon products usually result in significant defoliation responses, but generally lower rates can be used to obtain effective boll opening. The maximum labeled application rate for ethephon products is 0.367 kg a.i./ha (2 lb a.i./ac). Defoliant chemicals can be tank mixed with ethephon products to enhance defoliation. The response to ethephon requires an active plant and it is generally driven by air temperatures. Logan and Gwathmey (2002) found that ambient day and night temperatures had a significant influence on boll opening response to ethephon treatments. Ethephon product labels typically state that there should be "sufficient mature unopened bolls present to produce desired crop." Mature bolls are defined as "too hard to be dented when squeezed between the thumb and fingers, too hard to be sliced with a sharp knife, and when the seed coat becomes light brown in color." Applications of boll opening products when bolls lack adequate maturity will likely result in reduced lint yield and micronaire (Larson et al., 2002).

Defoliants

Defoliant chemicals result in initiation of an abscission layer at the base of the leaf petiole where it attaches to the stem. The natural abscission layer formation process is enhanced by the defoliant, which results in leaf drop. When considering mode of action, some defoliants are classified as hormonal and others are herbicidal, and some products have mixtures of both. Hormonal defoliants are characterized by enhancing ethylene production and/or inhibiting auxin transport. Response to these products is generally more sensitive to lower temperatures than herbicidal products. Hormonal-type products would include thidiazuron and ethephon. Herbicidal defoliants would include tribufos and related products, protoporphyrinogen oxidase (PPO) inhibitors (carfentrazone-ethyl, fluthiacet-methyl, pyraflufen ethyl, saflufenacil, etc.), and low rates of paraquat or other desiccants (which injure but do not kill the leaves). Some products may have mixtures of both hormonal and herbicidal types of activities. These would include a mixture of thidiazuron and diuron, for example. To obtain maximum leaf drop, defoliants require fairly healthy and active leaves that still properly function and are not severely drought stressed (Supak and Snipes, 2001). Warm air temperatures generally enhance activity. Generally, defoliants can be safely applied when 50 to 60% of the bolls are open and the remaining bolls are of sufficient maturity to obtain desired yield. Although a boll opening response is generally obtained as a result of defoliation, green unopened bolls can still remain a challenge. Many times a follow-up application of paraquat or other products with desiccant activity or a killing freeze is necessary to allow stripper harvest of the crop. Defoliant rates of PPO inhibitor products cause cell membrane disruption, which in turn triggers increased ethylene in leaves. Texas High Plains research trials have indicated that the PPO inhibitor products can be effective defoliants, as well as desiccants in some instances when used at higher rates. These products tend to perform fairly similarly, although under certain crop conditions, some may perform better than others. The PPO inhibitor products can be tank mixed with other products such as paraquat, tribufos, thidiazuron/diuron, and ethephon products. Use of crop oil concentrate (COC) or other adjuvant is suggested for most PPO tank mixtures. Proper adjuvant selection is highly important in obtaining optimum performance from these products, and product labels should be followed closely.

Desiccants

Desiccant products such as paraquat formulations and PPO inhibitors (at higher use rates) desiccate, or "dry down," plant tissue by causing individual cells to rupture. Generally these type of products are more frequently used in stripper harvesting as a final step to properly condition the crop for harvest. Desiccants are herbicidal in action and, when utilized as a stand-alone application, are typically applied when approximately 80% of the productive bolls are open, or when the cotton crop has reached the two to three nodes above cracked boll (NACB) stage. However, if sufficient numbers of bolls are mature, on the basis of the knife test, then these chemicals may be applied to somewhat lower percent open boll fields. Additionally, for stripper harvested cotton, desiccants are used to "finish" the crop for harvest by desiccating any leaves that may remain on the plant following an initial application of a boll opener or boll opener/defoliant tank mixture. Use of paraquat-based desiccants should be discouraged when seedling-stage small grains or other crop species are in close proximity to targeted cotton fields. Paraquat drift can cause severe damage to developing small grains plants grown for cover or harvest (Osborne et al., 2012). Some similar products that have paraquat as the active ingredient differ only in the concentration (0.24 or 0.36 kg a.i./L [2 lb a.i./gal or 3 lb a.i./gal]). Paraquat applications made in the late afternoon before a bright, sunny day appear to enhance the effectiveness of desiccation and tend to increase regrowth control. Use of non-ionic surfactant (NIS) at a minimum rate of 0.125% or 0.25% v/v, depending on the percentage concentration of surface-active agent (see individual product labels), with paraquat is suggested.

It may be necessary to increase the NIS rate to 1% v/v and spray late in the day to effectively desiccate some fields. In some years, PPO inhibitor defoliant/desiccant products, when applied at higher rates, work well to desiccate juvenile growth and regrowth, which many times are difficult to accomplish with paraquat alone. Furthermore, unlike with paraquat, drift from desiccant rates of PPO inhibitor products should not injure small grains (Osborne et al., 2012).

Harvest-Aid Application Timing and Harvest

Timing of harvest-aid applications are critical and may be one of the most important decisions a producer makes. Not only is it important to time applications on the basis of crop conditions and weather but also on availability of harvest equipment. Harvest-aid chemicals applied too early can considerably reduce yield and quality. Larson et al. (2002) observed significantly lower yields, fiber quality, and overall net revenues from premature applications of harvest aids to cotton. Other researchers have presented similar results, thus solidifying the benefits of delaying harvest-aid applications until optimum crop maturity. Harvest-aid applications should also be synchronized with availability of harvest equipment. If left too long in the field following chemical termination, significant reductions in yield of lint and seed, as well as quality of both, can occur. A comprehensive 3-yr, large-plot research project was conducted in the Texas High Plains from 2000 to 2002, which indicated that reductions in yield and quality of lint and seed can occur with delayed stripper harvest of a cotton cultivar with highly storm resistant bolls (Kelley, 2006). On the basis of the results from this study, where a highly storm resistant cultivar was utilized, it is expected that with delayed harvesting, greater physical losses of seed cotton would occur when looser boll-type cultivars are planted. Lower-quality lint, whether the result of premature harvest-aid applications or delayed harvest of a prepared crop, will ultimately result in reduced revenues to the producer. Furthermore, lint processing and spinning can also be negatively affected. Untimely application of harvest aids can reduce seed yields leading to reduced profitability. The quality of seed saved for planting can also be negatively impacted by untimely harvest-aid applications and/or delayed harvesting. Poor planting seed quality can result in reduced seedling vigor and difficulties in stand establishment. As many researchers, past and present, have observed, timely applications of proper harvest-aid materials under good environmental conditions will provide optimum yields and quality of both cotton fiber and seed.

Harvester Setup, Operation, and Maintenance

Proper setup, operation, and maintenance procedures are essential to ensuring maximum harvesting efficiency, field productivity, and cotton quality with any type of cotton harvester. Properly trained personnel should perform pre-season setup procedures on picker and stripper harvesters, and all manufacturer recommendations should be followed. Harvesting equipment is large, heavy, and can be cumbersome to work on. Extreme caution should be taken to ensure worker safety in all aspects of setup, operation, and maintenance. All personnel in charge of setup, operation, and maintenance of cotton harvesting equipment should read and understand the operator's manual. Additional detailed guidance on setup and operating procedures is available from machinery dealers as well as State Extension Service and USDA-ARS engineers. Information on state of the art cotton pickers, strippers, and seed cotton storage systems is documented by Willcutt et al. (2010), Wanjura et al. (2010b), and Searcy et al. (2010), and this information along with other useful videos on cotton harvesting can be found on the Cotton Harvest Systems website developed by Cotton Incorporated (Cotton Incorporated, 2010). Since additional detailed information on machine setup, operation, and maintenance is available in the provided references, this section will briefly cover the main points associated with maximizing harvesting efficiency, productivity, and cotton quality for picker and stripper harvesters.

Pre-Harvest Maintenance and Setup

It is good practice to develop and implement a pre-season checklist of setup and maintenance procedures for all cotton harvesting equipment. Cleanliness is a key aspect to ensuring proper operation and prevention of cotton contamination, and it is a good idea to begin the setup process for either pickers or strippers with a thorough cleaning. Residual seed cotton, trash, and dirt should be removed from basket screens, cab platforms, chassis and frame structures, engine compartments, and row units. High-volume compressed air is recommended to quickly remove material from these locations. Remove any accumulated grease or oil from engine, drivetrain, and row unit components using warm soapy water and a pressure washer. Engine oil, hydraulic oil, transmission/gear case oil, and associated filters should be changed according to recommended service intervals listed in the operator's manual. Check all fuel system components for leaks and repair as necessary. Fluid power systems are used extensively on modern harvesters of either type. Hydraulically powered subsystems should be engaged or actuated before the start of harvest to check for proper operation. Inspection of hydraulic system components such as fluid lines, valves, cylinders, and motors should be completed and any service or repair completed as needed. Engine air filters should be replaced at the beginning of each season and cleaned with compressed air as needed in season. Engine fuel filters should also be replaced according to manufacturer recommendations. Often, power losses experienced during the harvest season can be avoided or remedied by cleaning or replacing air filters and replacing fuel filters. Belt drive systems used to power hydraulic systems, cotton conveying fans, and cooling fans should be inspected and adjusted/repaired as needed. Check to make sure all tires are inflated to the proper pressure. For either type of machine, complete chassis lubrication is recommended before harvest begins and at the manufacturer recommended intervals during the season. On picker harvesters, ensure that the lubrication system tank is full of grease and that the system is applying grease adequately to all row units. Be sure to remove any excess grease from picking units to prevent seed cotton contamination.

Setup of cotton pickers before harvest should begin with a thorough inspection of the row units. The height of each picker bar on each drum should be checked and adjusting shims added as needed. The bars should all operate within the tolerable height difference listed in the operator's manual to ensure proper picking and doffing. Check the spindles on each bar to make sure the spindle teeth are sharp and that there are no broken spindles, dead spindles (spindles that turn freely when twisted by hand), or end play issues between the spindle gear and drive gear inside the picker bar that would lead to accelerated wear. Inspect the row unit grid bars and fix or replace any that are loose, bent, or broken. Row unit compressor plates should be adjusted so that there is even clearance from the top row of spindles to the bottom row of spindles. Generally, the compressor plate to spindle tip clearance should be set between 3.2 and 6.4 mm (0.125 and 0.25 in) at the closest point. Initially adjust the tension on the pressure plates by rotating the torsion rod to apply about 20.3 N-m (15 ft-lb) on the front drum door and about 40.6 N-m (30 ft-lb) on the rear. The pressure plate tension settings should be adjusted on the basis of crop conditions. Scrapping plates can be added to the compressor plates to increase the exposure of bolls to the spindles to help improve picking efficiency when harvesting small or partially opened bolls with immature seed cotton.

Inspect doffer columns for broken or worn doffers and replace or regrind the doffer columns as needed. Properly ground doffers should have a sharp leading edge to ensure proper doffing of cotton from the spindle. The height of the doffers should be checked and adjusted so that each doffer has about 0.08 mm (0.003 in) of clearance with the spindles. The moistener system strainer should be cleaned and the solution tank should be flushed to remove any sediment or residue and filled with fresh solution. Moistener pads should be checked for wear and replaced as needed. Adjust the moistener pad columns so that the spindles lightly depress the moistener pad ribs as they pass underneath. Make certain that each pad is applying solution and that there are no plugged solution lines. Moistener system pressure should be set on the basis of field conditions to provide only the necessary amount of solution to keep the spindles clean. The picker header frame should be adjusted to tilt the row units forward so that the rear of the units are about one to two inches higher than the front when lowered into picking position. Row unit tilt helps to improve picking efficiency and facilitate the shedding of trash out of the units.

By design, cotton strippers are less complex to set up and maintain than picker harvesters, but attention to several main points will help to ensure high harvesting efficiency and field productivity and minimize foreign matter content. Before harvest begins, check the row units for worn components such as stripper roll and row unit auger bearings, bats and brushes on stripper rolls, and unit drive components (belts, pulleys, gears, etc.) and replace as necessary. Stripper row units come from the factory configured with three bats and three brushes in an alternating pattern (Fig. 12). This configuration is more aggressive (i.e., will

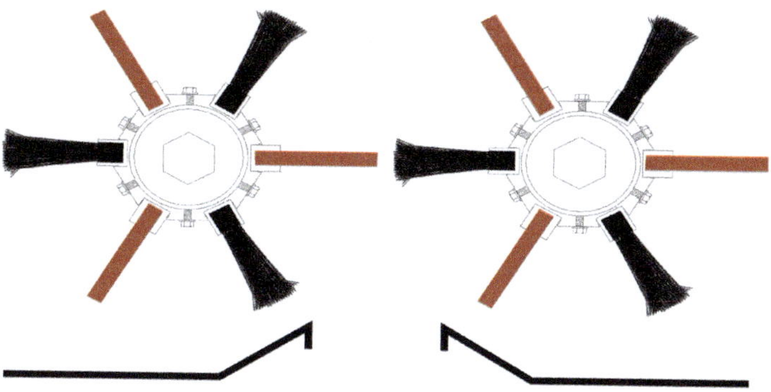

Fig. 12. Factory configuration of stripper rolls with three bats and three brushes.

remove more material from the plants) than several of the configurations tested by Brashears (1992) who found that decreasing the number of bats used in the stripper roll sequence or using half width bats (2.54 cm [1 in] wide vs. standard 5.08 cm [2 in] wide bats) in the factory configuration reduced foreign matter content (especially stick content that leads to bark contamination) with a negligible decrease in harvest efficiency. Bats and brushes should be installed on stripper rolls so that a bat on one roll engages the plant at the same time as a brush on the adjacent roll (i.e., bat to brush timing as shown in Fig. 12) to help reduce aggressiveness. The spacing between stripper rolls at the base of the units should be checked and made consistent across all row units. The closer the spacing between bats and brushes at their inward most position, the more aggressive the harvesting action will be. However, excessively wide spacing will decrease harvesting efficiency. Brashears (1986a) recommended a bottom roll spacing between 0.95 and 1.59 cm (0.375 and 0.625 in) to minimize harvest losses.

Other points on a cotton stripper to inspect before harvest begins are the cross auger bearings and drive components, lower and upper green boll separators on the air duct, and the basket auger drive motors and bearings. Field cleaners should be checked for broken saws, worn bearings, worn doffer brushes, and bent grid bars. The clearance between the field cleaner grid bars and the saw drums should be checked and set according to the operator's manual recommendations.

Harvester Operation

Cotton picking or stripping units should be operated at a height low enough to collect the lowest set bolls, but not so low that they plow the soil as the machine moves through the field. Plant lifters on cotton pickers should be set to just skim the ground surface but not dig in or drag leaves along with the unit. Operating row units of either type too low to the ground will increase the amount of soil and other foreign matter contained in the harvested seed cotton and may lead to accelerated wear of row unit components.

Harvest speed for cotton strippers is dictated most by machine capacity and field/crop conditions. Operators should regulate machine ground speed to ensure that no part of the harvesting system experiences a choke-up due to excess crop throughput. Experienced stripper operators have a "feel" for the optimum harvesting speed based on visual observation of the header and the sound of the field cleaner. Generally, the field cleaner's top saw speed should not slow down by more than 5 to 7% of the no-load set point speed at full engine speed. First picking or once-over picking should be done in first gear to minimize seed cotton loss, and ground speed should be modulated on the basis of field and crop conditions. Picking in second gear causes the row unit drum speed to be slower than or "lag" the speed of the crop row entering the picking unit. When the picking unit drums lag the ground speed of the picker, excess foreign material can be stripped off the plants and increase the foreign matter content of the seed cotton. Picking in second gear is only advisable for second picking or "scrapping." Scrapping refers to the practice of harvesting a field a second time, usually several weeks after the initial harvest, to gather seed cotton not harvested during the initial harvest or from bolls that open after the initial harvest.

An option included on many modern harvesters of either type is a row sensing system that helps to maintain the row units centered on the plant row as the machine moves through the field. These systems help to maximize harvesting efficiency and reduce operator fatigue leading to increased productivity. Additionally, well-trained and experienced operators play an essential role in maximizing productivity. Larson (1977) observed that the picker operator was the single most important factor in maximizing harvest machinery capacity and harvesting efficiency.

In-Season Maintenance

Maintenance of both picker and stripper harvesters should focus on cleaning the machines according to the methods previously discussed. Cleaning accumulated material from the machines helps to prevent fires and facilitate additional maintenance or repairs that should be done before harvest resumes to minimize downtime in the field. High-volume compressed air is a good option for quickly removing accumulated material from harvesters. Be sure to remove lint stringers from inside the basket or accumulator and discard all material cleaned from the machine to prevent contamination of pristine seed cotton. Check engine oil, coolant, and hydraulic oil levels before the beginning of harvest each day, and on picker harvesters re-fill the moistener system tank with fresh solution. On cotton strippers, daily inspection of the row units for worn bearings, bats, and brushes will help reduce downtime in the field. Picker row units should be blown out to remove the buildup of foreign material from interior cabinet walls, moisture pad columns, and around doffer columns. Make sure that the openings in the row unit floor are unobstructed so that trash can easily fall out of the unit during field use. Operators should carefully monitor the condition, setting, and operation of doffers and moisture pads during daily maintenance. Proper lubrication of picker row units is critical to maintaining a highly productive machine. It is good practice to perform row unit lubrication after the units have warmed up so that the grease will flow more easily into the picker bars and onto upper cabinet drive components. Be sure to remove any excess grease to prevent contamination of seed cotton.

Handling Seed Cotton after Harvest

Historically, once cotton was removed from the plants by a harvester it was placed in some type of container (sacks, baskets, trailers, etc.) and taken to a cotton gin for processing. Although ginning capacity could be overwhelmed during the peak of hand harvest, total ginning capacity was usually sufficient to process cotton at the rate of harvesting. To help assuage these periods of inadequate ginning capacity, some gins used seed cotton storage houses to temporarily store cotton so that producers could unload trailers and return to harvest. Some storage houses were equipped with floor ducts to aerate and dry the cotton to help preserve quality. However, seed cotton storage houses were often mismanaged, which led to inefficient use and the generation of ill will from some producers who were delayed in getting their crops harvested and ginned (Ward et al., 1953). The advent of mechanical harvesting exacerbated this problem as the processing capacity of gins did not increase in proportion to the increase in harvesting productivity. Thus, the need for post-harvest storage of seed cotton became a major concern for producers and ginners in the mid-20th century.

Storage Systems

A variety of in-field storage devices and systems have been developed and employed over time to help address the problem of decoupling the harvesting and ginning phases of cotton production. All of these systems integrate the following key functionalities to some degree:

- the ability to accept cotton at the rate it is unloaded by the harvester,
- the ability to protect cotton from losses in regard to quantity and quality during the storage period, and
- the ability to be loaded in the field, transported, and unloaded at the gin in an economical manner.

The ultimate driving factor behind the adoption and use of these storage systems is the economic effect on the production system: which system is the least cost alternative in regard to financial cost as well as labor and management considerations?

Initial efforts to store seed cotton on the farm before ginning included simply piling the cotton and shaping the top and sides to help shed moisture. The limitations to this method were that sand and other foreign material would build up against the side of the piles, limited protection from rainfall or snow, and low compression density resulted in some cotton loss from blowing wind. Additional cost was incurred when cotton had to be handled a second time when it was loaded into trailers for transport to the gin.

The cotton trailer is arguably the first storage system utilized in the Unites States on a large-scale basis. Moreover, trailers remained as a primary means of transporting cotton from the field to the gin for many of the infield storage systems developed in the late 20th century. Large trailers used in the Unites States could hold about 6 bales of seed cotton on average and were constructed with open tops, solid floors, and sides with wire mesh or expanded metal to allow airflow around the cotton. Typically a producer would own enough trailers to allow harvest to continue for several days before having to stop and take the cotton to the gin. Loaded trailers could wait at the gin for several days before they were unloaded and made available so that harvest operations could continue. Inefficiencies in the trailer system lead to the development of the "basket" system (Campbell, 1965), where cotton from trailers would be transferred at the gin into open top baskets for storage. The baskets were essentially trailers constructed without wheels and running gear and required specialized equipment to load and move at the gin. Baskets were typically owned by the gin and could hold about the same amount of cotton as a large cotton trailer.

The "cotton caddy" was the first system developed that formed cotton into a compressed stack for infield storage (McNeal, 1967). Cotton piled onto a wooden pallet inside a specially designed trailer was manually compressed by workers to a final density of about 80 to 112 kg/m^3 (5–7 lb/ft^3). Once the stack was formed, the palletized cotton was unloaded onto the ground and the caddy was moved to a new location where the process was repeated with a new wooden pallet. The pallets helped to protect the cotton from moisture and ground contaminants and facilitated loading back into the caddy trailer for transport to the gin.

The cotton ricking system was developed in a joint project between Cotton Incorporated and Texas Tech University (Smith, 1971). Cotton ricking consisted of forming a continuous stack of compressed seed cotton directly on the ground in the field. The rick forming equipment could accept cotton directly from the harvester without delaying the harvest operation and hydraulically compress the material to a density of about 100 kg/m^3 (6.25 lb/ft^3). The length of each rick varied, and thus the amount of cotton in each rick varied but was normally around 25 bales in the 2.4 m wide by 1.8 m tall (8 ft wide by 6 ft tall) cross-section. After storage, cotton was loaded into conventional cotton trailers by tractor mounted loaders and hauled to a gin for processing. The ricking system was primarily used in arid regions with little rainfall during the storage period. Placement of the ricks on clean high ground was important to promote runoff of precipitation away from the cotton and to prevent contamination.

The cotton module system of storing and handling seed cotton was developed by Wilkes and Jones (1973) in a cooperative effort between Texas A&M University and Cotton Incorporated. Combining aspects of the Cotton Caddy and ricking systems, the original cotton module system mechanically compressed seed cotton into 7.3 m (24 ft) long free-standing stacks built on wooden pallets. The main differences between the module system and the Cotton Caddy system were that modules were mechanically compressed to a much higher density (192–224 kg/m^3 [12–14 lb/ft^3]) than the original manual compression caddy system and the module system required use of different equipment to form (the module builder) and transport the modules (module truck or trailer). Over time, seed cotton modules increased in length to 9.8 m (32 ft) containing approximately 8 to 12 bales of stripper harvested cotton or 10 to 14 picker harvested bales. The cross-sectional shape of the modules is trapezoidal; the base is 2.3 m (7.5 ft) wide, the height varies from 2.9 to 3.4 m (9.5–11 ft), and the sides slope in decreasing the top width by 8.3% (1 in for each 12 in of height). The wooden pallets used with the original module system added to the per bale storage and transportation cost as the pallets had a useful life of about 10 to 15 modules. Parnell (1977) noted that a specially designed module mover was developed in 1974 that could load modules formed directly on the ground, and subsequently the module storage and handling system was rapidly adopted. By 1991 it was estimated that 57% of the US crop was stored in modules (Willcutt et al., 1992), and it has been unofficially estimated that over 95% of the US crop was stored in modules before the release of OMB picker harvesters. Standardization of module forming and transporting equipment was accomplished by the American Society of Agricultural Engineers in 1977 (latest standard update, ASABE, 2010) to help address issues regarding the legal transportation of modules on public roads. The capacity of the module building system to quickly form seed cotton into a package for infield storage has facilitated additional gains in harvesting productivity through the use of boll buggies and high-capacity harvesters.

In 2007 both John Deere and CaseIH introduced cotton pickers with new technology that allowed for the formation of seed cotton modules on board the picker. The John Deere 7760 picker (updated and now named the CP690, Fig. 13) forms modules with a round cross-section, approximately 2.3 to 2.4 m (7.5–8 ft) in diameter and 2.4 m (8 ft) in length. The modules are formed on the picker when harvested cotton is fed out of the accumulator and into the module formation chamber. The round module formation mechanism works similarly to a round hay baler as belts turn the cotton inside the chamber wrapping the cotton in continuous layers around the center. Each round module contains approximately 3.75

Fig. 13. John Deere CP690 cotton picker (used with permission, courtesy of John Deere).

bales of picker harvested cotton. Once completed, the round modules are covered with three layers of polyethylene plastic to protect the cotton and provide the restraining force to hold the cotton inside the modules. The plastic wrap forms a lip around the outer circumference of the modules to help prevent water on the ground from penetrating into the bottom of the modules. After applying the plastic wrap, the modules are ejected onto a cradle located at the rear of the machine as the picker continues to harvest cotton from the field. In theory, the picker does not have to slow down or stop harvesting during the module formation or ejection process since the harvested material is deposited into an accumulator chamber before being formed into a module. Once ejected onto the cradle, the picker can carry the module to the end of the field or other desired location to be placed on the ground for staging. Four round modules can be loaded onto a conventional module truck for transport to the gin. Care must be taken in the field, during transport, and at the gin to prevent damage to the plastic wrap that can lead to contamination of the cotton and/or wrap failure. John Deere (2008, 2013) and Searcy et al. (2010) provide additional guidance on handling, transporting, and storing round modules.

The CaseIH Module Express 625 picker (Fig. 14) forms rectangular modules that are 4.9 m (16 ft) long and trapezoidal in cross-section, similar to the cross-section of conventional modules. The Module Express 625 was updated a few years later and is now named the Module Express 635. The half modules contain 5 to 7 bales of picker harvested cotton and are covered with tarpaulins to protect the cotton from inclement weather. The half-sized modules are formed in the compression chamber at the rear of the picker by a rectangular frame that compresses the cotton as it is raised and lowered on hydraulic cylinders. Augers attached to the frame distribute the cotton evenly across the footprint of the module as it is formed. Once the module is finished, the picker driver moves the machine to

Fig. 14. CaseIH Module Express 635 cotton picker (used with permission, courtesy of CaseIH).

the location where the module is to be stored and begins the unloading cycle. To unload the module, the compression chamber tilts up, the rear door and ramp unfold, and conveyor chains move the completed module out of the chamber and place it on the ground at the same speed at which the picker moves forward. Two half-sized modules formed by a CaseIH Module Express picker can be loaded on a standard module truck for transport to the gin.

The CS690 (Fig. 15), a brush-roll stripper harvester with OMB technology was released by John Deere for commercial use in the 2014/2015 cotton harvest season. The new stripper harvester uses the same round module forming mechanism and plastic module wrap system used on the John Deere CP690 picker. The new stripper utilizes existing technology in regard to the harvesting units (either six or eight row units per header), conveying system, and field cleaner and has a rated material harvesting/processing capacity of 454 kg/min (1000 lb/min). Research is needed to quantify the improvement in harvesting productivity for the CS690 compared with a harvesting system based on a conventional basket stripper harvester with accompanying boll buggies and module builders. John Deere is currently the only manufacturer of cotton stripper harvesters.

Regardless of harvester type or storage system, certain precautions must be followed to ensure the protection and quality of cotton lint and seed. It is generally accepted that seed cotton should not be harvested and stored on a long-term basis if moisture content is greater than 12%. Curley et al. (1987) observed significant module internal heating and decreases in fiber reflectance over a short 5-d period when initial seed cotton moisture levels were over 13.5%. Elevated storage temperatures due to seed cotton moisture above 12% w.b. at harvest lead to degradation of seed quality and germination rates (Metzer, 1976). Foreign material in harvested seed cotton can often contain high moisture content that will migrate into the lint and seed affecting quality losses during storage. Thus, production,

Fig. 15. John Deere CS690 cotton stripper (used with permission, courtesy of John Deere).

harvest-aid application, and harvesting practices to reduce the amount of foreign material in stored seed cotton should be implemented.

Not only is moisture a major concern at harvest but precipitation received during the storage period can cause additional economic losses if modules are not stored in a well-drained location so that they do not wick moisture from the ground. Module shape and cover quality are also essential elements to protecting stored cotton. Modules should be formed with a crowned shape to facilitate moisture shedding across the width and length of the module. Module covers or tarps should be in good condition such that water does not easily penetrate through to the cotton. Searcy et al. (2012) observed that modules formed with a poor shape but covered with a good tarp lost US$200 per module in value, while modules formed with a good shape but covered with a poor tarp lost US$400 per module relative to modules with good shapes and good tarps. The detrimental effects of poor module shape and poor tarp condition are additive and, when combined, could cost a producer US$600 per module (Searcy et al., 2012). ASABE (2012) lists testing protocols and performance specifications for module cover materials.

Protection from fire and vandalism is also a concern with the storage of cotton in modules. The number of seed cotton modules stored in one field location and the required separation distance between groups of modules for fire protection is typically governed by insurance policies that protect the cotton before ginning. Modules should be placed away from normal traffic areas to help prevent vandalism but close enough for easy access for inspection and loading. It is also a good idea to tag or mark the modules and notify the gin as soon as the modules are formed so that they can be covered by insurance.

Transportation

Seed cotton modules are moved on trucks with live chain beds that tilt down at the rear to allow the module to be loaded and unloaded on the ground. The beds used on module trucks have multiple rows of conveyor chains running the length of the bed that pull the module onto the truck at the same speed that the truck moves under the module when loading. Synchronizing the bed chain and truck ground speed is important so that the modules are not pulled apart or otherwise damaged during the loading or unloading process. The bed chains usually contain some type of tooth or lug to help engage and move the module. It is important that smooth lugs (or no lugs) be used on module truck bed chains when handling round modules to ensure the protection of the plastic wrap. Usually, module truck beds are covered to protect the modules from wind and moisture during transport.

State transportation laws govern the maximum size and weight of module trucks. In some states, transport systems have been devised to carry multiple modules on two or three trailers pulled by one tractor unit. This practice increases the productivity and efficiency of the transport system but requires a specialized loading system to load trailers in the field and unload at the gin. This concept has been used to further extend transporter productivity recently as producers switch over to harvesters with OMB capabilities. Depending on the size of the trailer used, six to eight round modules, or two to three half-size modules may be loaded on a semi-tractor trailer.

Fiber Quality

Harvesting Effects on Quality

Cotton fiber quality can be influenced by a number of production and processing factors, including harvesting. Fundamental differences between the harvesting mechanisms used by spindle pickers and stripper-type harvesters affect differences in harvesting efficiency and thereby differences in the population of fibers gathered by each system. Since strippers generally have greater harvesting efficiency, they gather a larger subset of all fibers available for harvest, regardless of maturity. Pickers harvest a smaller subset of all fibers available for harvest as they harvest only the portion of cotton that can be engaged by the spindles. Cotton harvested by pickers is from bolls that open adequately to expose seed cotton for picking, usually from bolls of higher maturity as the spindles cannot efficiently harvest cotton from less mature bolls at the top of the plant that do not open well. In theory, if all fibers were of the same maturity and quality at the time of harvest, there would be no harvest method difference in the average level of quality. However, cotton is an indeterminate plant and continues to add fruiting positions regardless of the amount of time remaining in the growing season available to fully mature bolls. Thus, fiber quality can vary substantially within a particular plant and/or field at harvest time depending on the number of physiologically immature bolls that contribute to yield. It should be noted that not all immature bolls contribute to yield as many will remain closed and either remain in the field or be separated out by the conveying and cleaning systems used on harvesters or at the gin.

The influence of fiber maturity cannot be overstated as it is likely the single most important element in explaining harvest method differences in fiber properties such as micronaire, length distribution, strength, and nep content. Several studies show that fiber quality differences between picker and stripper harvested cotton are exacerbated for crops with low levels of physiological crop maturity (Brashears and Hake, 1995; Vories and Bonner, 1995; Faircloth et al., 2004; Faulkner et al., 2011c; Wanjura et al., 2013). Generally, picker harvested cotton will have improved fiber quality properties because of the influence of enhanced fiber maturity arising from the selective nature of the spindles during harvest.

Fiber color, specifically reflectance and yellowness, can be improved for picker harvested cotton as it is often the case where picker harvesters can begin harvesting a crop earlier than strippers because pickers do not require the crop to be desiccated (either by freeze or chemicals) before harvest. Thus, weathering effects can be avoided somewhat by the use of picker harvesters.

Moreover, lint color can be negatively influenced, especially for stripper harvested cotton, by the presence of excessive high moisture trash during infield storage in modules.

Although gin cleaning recommendations vary by harvest method (Anthony, 1994; Baker, 1994), differences in seed cotton foreign matter content between harvest methods generally result in higher leaf grades for stripper harvested cotton. Interactive effects between trash content, moisture content, and gin cleaning practices can result in greater harvest method differences for other fiber properties. For example, stripper harvested cotton normally contains more foreign material than picked cotton and foreign material is known to reach higher equilibrium moisture content than lint or seed for a given relative humidity level. Additional drying of stripper harvested cotton may be needed to afford efficient seed cotton cleaning. However, excessive drying may make the fibers brittle, causing them to break more during ginning and lint cleaning and thus lowering length, length uniformity, and bundle strength. Although this scenario is not common, it points out the necessity on the part of ginners to pay close attention to process control to maintain fiber quality.

Bark

Bark is a lint contaminant most commonly associated with stripper harvested cotton. In 2013, USDA-AMS classing data indicated that bark contamination levels one and two reduced the value of a 218 kg bale by US$12.72 and $23.28, respectively (USDA-CCC, 2013) for cotton grown in Texas, Oklahoma, New Mexico, and Kansas. Although bark contamination is more rare in other parts of the US Cotton Belt, bark contamination penalties in all other areas (essentially all picker harvested) of the Unites States were more severe as levels one and two docked producers US$19.20 and $33.84 per bale, respectively (USDA-CCC, 2013). Bark contamination is extremely difficult to remove from lint either at the gin or the in the textile mill, and excessive mill cleaning to remove bark adds to production costs by generating additional wasted lint. Although some bark can be tolerated by rotor spinning, ring spinning is much more sensitive to the presence of bark, which leads to decreased spinning efficiency due to increased ends down (Ethridge et al., 1992; Bragg et al., 1995).

Bark is generally considered to originate from the woody outer layer of plant stems and branches, but Morey et al. (1978) indicate that materials called "bark" can arise from the woody core of stems and branches or from the fibrous components of the phloem and the suberized portion of the periderm. Bark content in lint is highly related to the amount of sticks present in harvested seed cotton (Morey et al., 1978; Wanjura and Baker, 1979; Wanjura et al., 1979; Brashears, 1984b; Supak et al., 1989). Wanjura and Baker (1979) showed that bark is generated on the harvester from stick material broken up by the auger conveyors and also by seed cotton cleaning equipment used before the gin stand. Laird and Baker (1975) observed minimal levels of bark contamination in lint from cotton that had less than 1.5% stick content at the extractor feeder apron.

Bark contamination potential can be increased by a number of factors including:

- stripper harvesting of plants with tall, thick stems and long vegetative branches,
- prolonged exposure of plants to field weathering following desiccation by chemicals or freezing conditions,
- a sudden freeze that occurs when plants are in a lush condition,
- extreme seasonal variation in climatic conditions,
- cultivar.

In years with severe bark contamination, bark problems tend to get worse as the harvest season progresses. However, work by Brashears (1986a, 1986b, 1992) showed that decreasing the aggressiveness of stripping rolls via wider bottom spacing, using reduced width bats, and using stripper roll sequences with fewer bats helped reduce both seed cotton stick content and bark contamination. Wanjura et al. (2013) observed that picker harvesting substantially reduced bark contamination levels relative to stripper harvesting in all years of a 3-yr commercially harvested and ginned study on the Texas High Plains. Although picker harvesting helps to reduce the incidence of bark contamination, it does not completely eliminate the problem as bark contamination has increased about twofold over the last few years in the traditionally picker harvested Mid-South and southeastern regions of the Unites States. Possible reasons for this recent increase in bark contamination include cultivar by environment interactions, increased disease/weed pressure, and picking in second gear (i.e., non-synchronized picking drum and ground speeds) to hasten harvest in light of weather shortened harvest periods. Regardless of the harvest method used, producers must be vigilant in their setup and operating procedures to help minimize the incidence of bark contamination.

Contamination

US cotton has developed a reputation for being a reliable source of high-quality, contaminant-free fiber. Cotton bale contaminants are defined simply as anything that is not clean cotton lint. Lint can be contaminated during harvest by materials such as hydraulic oil or gear lubricants, broken pieces of doffer or moisture pads, row unit or field cleaner doffer brush bristles, plastic from litter, plastic bed covering material used in vegetable production, pieces of rope or twine picked up by the harvester, using non-washable module marking paint, and excessive foreign matter from building modules on standing stalks. During ginning, contamination can arise from materials such as machinery lubricants or hydraulic oil, doffer brushes, pieces of worn module tarps or covers, seed coat fragments, and garbage pulled in at the module feeder. With the introduction of the plastic wrapped round module from John Deere, producers and ginners must be diligent to not allow this material to enter the cotton either in the field or at the gin. Close attention should be paid when handling round modules to prevent puncturing or tearing the wrap and proper procedures must be followed when cutting the plastic during wrap removal to keep pieces of the wrap from remaining inside the cotton as it is fed into the gin. Zero tolerance with respect to contamination should be the shared goal of producers and ginners to maintain the reputation of the United States as a reliable source of high-quality and contaminant-free fiber.

Advancements in Harvesting Technology

There are several emerging and new technologies that are impacting cotton harvest systems. These include the use of yield monitors, wireless data transfer, and radio frequency identification (RFID) tags.

Yield Monitoring

Cotton yield monitors have been commercially available since at least 1997 (Durrence et al., 1998), but until recently yield monitors were not a factory installed option. Installation required cutting holes in the conveyance ducts for pickers or modifying the seed cotton handling system in strippers. While several versions of the cotton yield monitor were fairly accurate when properly calibrated (e.g., Wilkerson et al., 2002), the need to install the sensors has limited the adoption rates, with the latest survey indicating 20% of US cotton producers are using yield monitors (Boyer et al., 2014). However, the most recent cotton harvesters from John Deere and Case IH now both come standard with a yield monitor installed. Additionally, Boyer et al. (2014) did find that 73% of US cotton producers have adopted at least one precision agricultural technology and therefore have the ability to take advantage of the data a yield monitor can provide.

There are currently two primary types of sensors used to estimate mass flow of seed cotton: (i) light based sensors that either respond to breaks in a light beam as seed cotton passes by (Wilkerson et al., 2002) or rely on the change in amount of reflected light as the amount cotton in the duct changes; and (ii) a microwave system from John Deere that measures the energy reflected from the seed cotton. Both approaches do not provide a direct measure of mass flow and must be carefully calibrated. Both Wilkerson et al. (2002) and Taylor et al. (2013) have found evidence that cultivar can influence yield monitor calibration, so cultivar specific calibration should be conducted.

Despite some challenges in the calibration of yield monitors, Vellidis and Brannen (2012) illustrate several case studies across the United States where data from a yield monitor have proven useful on farm. One use of a yield map is to make "profit maps" by estimating the gross value of the cotton and subtracting the input costs to identify areas of the field that may be losing money. In many cases if the yield limiting factor cannot be controlled such as the soil type or topography, a good strategy is to minimize all inputs in those areas. If areas of the field continue to show negative returns after a number of years and on the edge of the field, it may be possible to convert the field from row crop production to pasture or some type of conservation easement.

In addition to identifying natural yield variability in the field, yield monitors can also be a good tool to facilitate on farm testing of different production practices such as fertilizer or seeding rates where cultivar is held constant. For example, yield monitor data were used to evaluate the optimal seeding rate for different soil types in on-farm tests in Arkansas (Vellidis and Brannen, 2012). In that particular study it was found that optimizing seeding rate by soil type saved US$8.10/ha ($20/acre).

Moisture Sensing

Sensor technology for measuring seed cotton moisture content has been used in the ginning industry for years to help control drying systems. However, none of these moisture sensing systems have been adapted and commercialized for use on conventional basket picker or stripper harvesters. To date, only one Australian company has adapted their moisture sensor for use on the John Deere OMB picker. The ability to measure seed cotton moisture on board the harvester could help producers objectively gauge when moisture conditions are too wet to begin or to continue harvesting on a daily basis. Additionally, if module moisture information was transmitted to the gin, managers could more properly stage modules for ginning to help optimize the use of gin resources (i.e., labor, dryer fuel, electricity) and help to prevent costly quality losses for producers. Additional work is needed to develop and implement this technology into the next generation of cotton harvesters.

Information Technology for Advanced Harvest Management

Beyond yield monitors, several other technologies are posed to impact the way seed cotton is handled between the field and the gin. The ability of agricultural equipment to collect, store, and receive data continues to increase, but the full value of that data is not always realized. One technology that could help in capturing more value from that data is partly driven by the increased use of round modules that already contain RFID tags. This could easily allow for automated transfer of ownership information, removing the need to manually tag modules; improve efficiency at the gin; and help ensure a producer's cotton is never lost. Standardization of RFID tags for use on cotton modules is needed for this to be fully implemented.

The ability to automate data collection and sharing has additional implications. For example, progress continues in the development of an industry standard leaf hairiness index that will quantify the relative number of leaf hairs of a cultivar that can contribute to poor leaf grades (Morgan et al., 2014). With the automated flow of harvest information, the cultivar contained in a module can easily be transferred to the gin, allowing the ginner to know if they need to add additional seed cotton or lint cleaning if that cultivar is known to be hairy. In addition, through linking the time of harvest with weather data, it is possible to estimate seed cotton moisture content and identify modules that may be too wet for prolonged storage and that require immediate processing. Standardization of harvest information should make it easier for growers to pass yield maps to their consultants, or use that data to support their participation in sustainable programs.

References

Anthony, W.S. 1994. Ginning recommendations for processing machine-picked cotton. In: Cotton ginners handbook. Agric. Handb. 503. USDA-ARS, Washington, DC. p. 240–241.

ASABE. 2006. ASAE EP496.3: Agricultural machinery management. ASABE, St. Joseph, MI.

ASABE. 2010. ANSI/ASAE S392.2: Cotton module builder and transporter standard. ASABE, St. Joseph, MI.

ASABE. 2012. ASABE S615: Cotton module cover material performance. ASABE, St. Joseph, MI.

Bader, M., A. Brashears, T. Deutsch, and C. Hanson. 1998. Finger stripper harvesting technology. In: summary of ultra-narrow row cotton harvest to textiles, Memphis, TN, 26

Feb. 1998. Cotton Inc. http://www.cottoninc.com/fiber/AgriculturalDisciplines/Engineering/Ultra-Narrow-Row-Cotton-Harvest/Finger-Stripper-Harvesting-Technology/ (accessed 26 Mar. 2015).

Baker, R.V. 1994. Ginning recommendations for processing machine-stripped cotton. In: W.S. Anthony and W.D. Mayfield, editors, Cotton ginners handbook. Agric. Handb. 503. USDA-ARS, Washington, DC. p. 242–243.

Baker, R.V., W.S. Anthony, and R.M. Sutton. 1994. Seed cotton cleaning and extracting. In: W.S. Anthony and W.D. Mayfield, editors, Cotton ginners handbook. Agric. Handb. 503. USDA-ARS, Washington, DC. p. 69–90.

Baker, R.V., and A.D. Brashears. 2000. Combined effects of field cleaning and lint cleaning on stripper-harvested cotton. In: Proceedings of the Beltwide Cotton Conferences, San Antonio, TX. 4–8 Jan. 2000. Natl. Cotton Counc. Am., Memphis, TN. p. 1616–1621.

Batchelder, D.G., W.E. Taylor, and J.G. Porterfield. 1961. Tests of roll materials, speeds, arrangements, and designs. Bull B-589. Oklahoma State Univ. Exp. Stn., Stillwater, OK.

Bennett, B.K., S.K. Misra, A.D. Brashears, and T.L. Dowty. 1995. Effect of bur extractor on trash in seed cotton and fiber quality for different harvest dates. In: Proceedings of the Beltwide Cotton Conferences, San Antonio, TX. 4–7 Jan. 1995. Natl. Cotton Counc. Am., Memphis, TN. p. 649–652.

Boyer, C.N., B.C. English, R.K. Roberts, J.A. Larson, D.M. Lambert, M. Velandia, X. Zhou, S. Larkin, M. Marra, R. Rejesus, L.L. Falconer, S.W. Martin, A.K. Mishra, K.P. Paudel, J. Johnson, E. Segarra, C. Wang, and J.M. Reeves. 2014. Results from a cotton precision farming survey across fourteen southern states. In: Proceedings of the Beltwide Cotton Conferences, New Orleans, LA. 6–8 Jan. 2014. Natl. Cotton Counc. Am., Memphis, TN. p. 652–654.

Bragg, G.K., C.L. Simpson, A.D. Brashears, and R.V. Baker. 1995. Bark effect on spinning efficiency of cotton. Trans. ASAE 38(1):57–64. doi:10.13031/2013.27812

Brashears, A.D. 1984a. Modification of stripper rolls to reduce stick content in bur cotton. In: Proceedings of the Beltwide Cotton Production Research Conferences, Atlanta, GA. 8–12 Jan. 1984. Natl. Cotton Counc. Am., Memphis, TN. p. 155.

Brashears, A.D. 1984b. Causes and cures for barky cotton: Stripper adjustment and operation. Western Cotton Prod. Conf. Summ. Proc. 1984:27–35.

Brashears, A.D. 1986a. Variable spacing of cotton stripper rolls. ASAE Paper 861083. ASAE, St. Joseph, MI.

Brashears, A.D. 1986b. Utilization of narrow stripper paddles throughout harvest season. In: Proceedings of the Beltwide Cotton Production Research Conferences, Las Vegas, NV. 4–9 Jan. 1986. Natl. Cotton Counc. Am., Memphis, TN. p. 435.

Brashears, A.D. 1991. Field cleaner efficiency as affected by variety. ASAE Paper 911078. ASAE, St. Joseph, MI.

Brashears, A.D. 1992. Effect of stripper roll configuration on foreign matter and harvesting efficiency. In: Proceedings of the Beltwide Cotton Conferences, Nashville, TN. 6–10 Jan. 1992. Natl. Cotton Counc. Am., Memphis, TN. p. 1369–1370.

Brashears, A.D., and K.D. Hake. 1995. Comparing picking and stripping on the Texas High Plains. In: Proceedings of the Beltwide Cotton Conferences, San Antonio, TX. 4–7 Jan. 1995. Natl. Cotton Counc. Am., Memphis, TN. p. 652–654.

Buehring, N.W., R.R. Dodds, M.P. Harrison, M.H. Willcutt, and S.R. Spurlock. 2009. Nonirrigated spindle picker 15-inch and wide-row cotton production system analysis. Bull. 1178. Office of Agric. Commun., Mississippi State Univ., Starkville.

Campbell, J.D. 1965. Costs of using cotton basket-storage systems, California and Texas, 1964. Mark. Res. Rep. 736. Farm Coop. Serv., USDA, Washington, DC.

Chen, L.H., T. Ahmad, and M.H. Willcutt. 1992. Simulation model for cotton harvesting machinery. Trans. ASAE 35(3):1071–1077. doi:10.13031/2013.28702

Corley, T.E. 1966. Basic factors affecting performance of mechanical cotton pickers. Trans. ASAE 9(3):326–332. doi:10.13031/2013.39967

Corley, T.E., and C.M. Stokes. 1966. Mechanical cotton harvester performance as influenced by plant spacing and varietal characteristics. Trans. ASAE 7(3):281–290.

Cotton Incorporated. 2010. Cotton harvest systems. Cotton Inc. http://www.cottoninc.com/fiber/AgriculturalDisciplines/Engineering/Cotton-Harvest-Systems/ (accessed 21 Aug. 2014).

Curley, R.G., B. Roberts, T. Kerby, C. Brooks, and J. Knutson. 1987. Effect of seed cotton moisture level and storage time on the quality of lint in stored modules. In: Proceedings of the Beltwide Cotton Production Research Conferences, Dallas, TX. 4–8 Jan. 1987. Natl. Cotton Counc. Am., Memphis, TN. p. 504–505.

Drake, D.R., R.R. Minzenmayer, W.L. Multer, T. Isakeit, and G.D. Morgan. 2013. Evaluation of farmer applications of TopGuard (flutriafol) for cotton root rot control in the first section 18 exemption year. In: Proceedings of the Beltwide Cotton Conferences, San Antonio, TX. 7–10 Jan. 2013. [CD-ROM]. Natl. Cotton Counc. Am., Memphis, TN.

Durrence, J.S., C.D. Perry, G. Vellidis, D.L. Thomas, and C.K. Kvien. 1998. Evaluation of commercially-available yield monitors in Georgia field conditions. ASAE Paper 98-3106. ASAE, St. Joseph, MI.

Eder, Z.P., G.D. Morgan, D. Fromme, D.A. Mott, G. Collins, and F. Bourland. 2013. Cotton variety characteristics and their impact on cotton leaf grade. In: Proceedings of the Beltwide Cotton Conferences, San Antonio, TX. 7–10 Jan. 2013. Natl. Cotton Counc. Am., Memphis, TN. p. 456.

Ellsworth, P., R. Tronstad, J. Lesser, P. Goodell, L. Godfrey, T. Henneberry, D. Hendrix, D. Burshwood, S. Naranjo, S. Castle, and R. Nichols. 1999. Sticky cotton sources and solutions. IPM Ser.13. Univ. of Arizona Coop. Ext. Serv., Tucson.

Ethridge, D., J. Brown, J. Price, and C.K. Bragg. 1992. Bark in cotton lint: Effects on processing costs. In: Proceedings of the Beltwide Cotton Conferences, Nashville, TN. 6–10 Jan. 1992. Natl. Cotton Counc. Am., Memphis, TN. p. 1223–1226.

Faircloth, J.C., R. Hutchinson, J. Barnett, K. Paxson, A. Coco, and P. Price. 2004. An evaluation of alternative cotton harvesting methods in Northeast Louisiana—A comparison of the brush stripper and spindle harvester. J. Cotton Sci. 8(2):55–61.

Falconer, L., W.B. Faulkner, and J. Reeves. 2010. Cotton harvest equipment investment analysis decision aid. Cotton Inc. http://www.cottoninc.com/fiber/AgriculturalDisciplines/AgriculturalEconomics/Cotton-Farming-Decision-Aids/EconomicDecisionAid/ (accessed 26 Aug. 2014).

Farrell, M., and G. Ibendahl. 2009. Estimating cotton harvest cost per acre when harvest days are stochastic. Paper presented at: Southern Agricultural Economics Association Meeting, Atlanta, GA. 31 Jan. to 3 Feb. Paper 46813.

Faulkner, W.B., J.D. Wanjura, R.K. Boman, B.W. Shaw, and C.B. Parnell. 2011a. Evaluation of modern cotton harvest systems on irrigated cotton: Harvester performance. Appl. Eng. Agric. 27(4):497–506. doi:10.13031/2013.38196

Faulkner, W.B., J.D. Wanjura, R.K. Boman, B.W. Shaw, and C.B. Parnell. 2011b. Evaluation of modern cotton harvest systems on irrigated cotton: Economic returns. Appl. Eng. Agric. 27(4):515–522. doi:10.13031/2013.38198

Faulkner, W.B., J.D. Wanjura, E.F. Hequet, R.K. Boman, B.W. Shaw, and C.B. Parnell. 2011c. Evaluation of modern cotton harvest systems on irrigated cotton: Fiber quality. Appl. Eng. Agric. 27(4):507–513. doi:10.13031/2013.38197

Goodson, J., R. Boman, S. Osborne, T. Royer, and R. Minzenmayer. 2014. Field evaluation of Topguard (flutriafol) for cotton root rot management in Oklahoma. In: Proceedings of the Beltwide Cotton Conferences, New Orleans, LA. 6–8 Jan. 2014. Natl. Cotton Counc. Am., Memphis, TN. p. 207–211.

Gwathmey, C.O., L.E. Steckel, J.A. Larson, and D.F. Mooney. 2011. Lower limits of cotton seeding rates in alternative row width and patterns. Agron. J. 103(3):584–592. doi:10.2134/agronj2010.0333

Hake, S.J., K.D. Hake, and T.A. Kerby. 1996. Preharvest/harvest decisions. In: S.J. Hake, T.A. Kerby, and K.D. Hake, editors, Cotton production manual. Publ. 3352. Div. of Agric. and Nat. Resour., Univ. of California, Oakland. p. 73–81.

Heinicke, C., and W.A. Grove. 2005. Labor markets, regional diversity and cotton harvest mechanization in the post-World War II United States. Soc. Sci. Hist. 29(2):269–297. doi:10.1215/01455532-29-2-269

Hughs, S.E., T.D. Valco, and J.R. Williford. 2008. 100 years of cotton production, harvesting, and ginning systems engineering: 1907–2007. Trans. ASABE 51(4):1187–1198. doi:10.13031/2013.25234

Husman, S., P. Clay, E. Taylor, and K. White. 2005. Twin line cotton production in a conservation tillage system. Arizona Cotton Rep. P-142. Coll. of Agric., Univ. of Arizona, Tucson.

Ibendahl, G., M. Farrell, S. Spurlock, and J. Tack. 2014. Optimal replacement age of a conventional cotton harvest system. Agric. Finance Rev. 74(1):2–16. doi:10.1108/AFR-02-2013-0008

International Cotton Advisory Committee. 2011. Cotton production practices. Report. Technical Information Section of the International Cotton Advisory Committee, Washington, DC.

Isakeit, T., R.R. Minzenmayer, D.R. Drake, G.D. Morgan, D.A. Mott, D.D. Fromme, W.L. Multer, M. Jungman, and A. Abrameit. 2012. Fungicide management of cotton root rot (*Phymatotrichopsis omnivora*). In: Proceedings of the Beltwide Cotton Conferences, Natl. Cotton Counc. Am., Orlando, FL. 3–6 Jan. 2012. Natl. Cotton Counc. Am., Memphis, TN. p. 235– 237.

Isakeit, T., R.R. Minzenmayer, G.D. Morgan, D.A. Mott, D.R. Drake, D.D. Fromme, W.L. Multer, M. Jungman, R.M. Collett, A. Abrameit, and J. Stapper. 2013. Cotton root rot control with TopGuard in Texas: 2012 Research. In: Proceedings of the Beltwide Cotton Conferences, San Antonio, TX. 7–10 Jan. 2013. [CD-ROM]. Natl. Cotton Counc. Am., Memphis, TN.

John Deere. 2008. Cotton module staging instructions and truck requirements. John Deere Des Moines Works. N390094. Deere and Company. Moline, IL.

John Deere. 2013. Round cotton module ginning recommendations. John Deere Des Moines Works. KK11359. Deere and Company. Moline, IL.

Jones, D.L., W.M. Hurst, and D. Scoates. 1928. Mechanical harvesting of cotton in Northwest Texas. Circ. 52. Texas Agric. Exp. Stn. College Station.

Kelley, M.S. 2006. Field weathering effects on stripper harvested cotton in the Texas High Plains. Ph.D. diss., Texas Tech Univ., Lubbock, TX.

Kirk, I.W., A.D. Brashears, and E.B. Hudspeth. 1972. Field performance of cleaners on cotton stripper harvesters. Trans. ASAE 15(6):1024–1027. doi:10.13031/2013.38063

Kirk, I.W., E.B. Hudspeth, Jr., and A.D. Brashears. 1973. Optimizing saw grid cleaner design for cotton stripper harvesters. Trans. ASAE 16(3):446–449. doi:10.13031/2013.37539

Kirk, I.W., E.B. Hudspeth, Jr., and D.F. Wanjura. 1964. A broadcast and narrow-row cotton harvester. Prog. Rep. 2311. Texas Agric. Exp. Stn. College Station.

Klose, S.L., S. Amosson, J. Smith, S. Bevers, M. Waller, B. Thompson, and M. Young. 2013. 2013 Texas agricultural custom rates. Texas A&M AgriLife Ext. Serv., College Station.

Laird, J.W., and R.V. Baker. 1975. From sticks to bark. Cotton Ginners J. Yearb. 43(1):27–32.

Larson, D.L. 1977. Cotton picker capacity: The operator is the key. Trans. ASAE 20(6):1042–1044. doi:10.13031/2013.35699

Larson, J.A., C.O. Gwathmey, and R.M. Hayes. 2002. Cotton defoliation and harvest timing effects on yields, quality, and net revenues. J. Cotton Sci. 6:13–27.

Logan, L., and C.O. Gwathmey. 2002. Effects of weather on cotton responses to harvest-aid chemicals. J. Cotton Sci. 6(1):1–12.

Main, C.L., and R.M. Hayes. 2010. Cotton harvest aids. Publ. W225. Inst. of Agric., Univ. of Tennessee, Knoxville.

McNeal, X. 1967. Cotton-stacking pallet trailer. Arkansas Farm Res. 16(2):3.

Metzer, R.B. 1976. High quality cotton planting seed—A guide to production and handling. L-1423. Texas Agric. Exp. Stn., Texas A&M Univ., College Station, TX.

Morey, P.R., D.F. Wanjura, and R.V. Baker. 1978. Identification of bark in raw cottons. Agron. J. 70:85–91. doi:10.2134/agronj1978.00021962007000010020x

Morgan, G., F. Bourland, L. Day, D. Fromme, J. Dever, A. Ibrahim, E. Barnes, and D. Jones. 2014. Development of an industry-wide standard for leaf hairiness. In: Proceedings of the Beltwide Cotton Conferences, New Orleans, LA. 6–8 Jan. 2014. Natl. Cotton Counc. Am., Memphis, TN. p. 187.

Nelson, J.M., S.K. Misra, and A.D. Brashears. 2001. Cost comparison of alternative stripper and picker cotton harvest systems. Appl. Eng. Agric. 17(2):137–142. doi:10.13031/2013.5455

Norton, R., M. Olsen, and A. Mostafa. 2014. Efficacy of flutriafol fungicide for control of cotton root rot in Arizona. In: Proceedings of the Beltwide Cotton Conferences, New Orleans, LA. 6–8 Jan. 2014. Natl. Cotton Counc. Am., Memphis, TN. p. 237– 244.

Osborne, S., R. Boman, and G. Strickland. 2012. Performance of Sharpen in Oklahoma. In: Proceedings of the Beltwide Cotton Conferences, Natl. Cotton Counc. Am., Orlando, FL. 3–6 Jan. 2012. Natl. Cotton Counc. Am., Memphis, TN. p. 1508–1512.

Parnell, C.B., Jr. 1977. Module building equipment standards for seed cotton handling and storage. ASAE Paper 77-1542. ASAE, St. Joseph, MI.

Rembert, S.S., and J. Prescott. 1850. Improvement in picking cotton from the bolls in the field. US Patent 7631. Date issued: 10 September.

Riar, R., R. Wells, K. Edmisten, D. Jordan, and J. Bacheler. 2013. Cotton yield and canopy closure in North Carolina as influenced by row width, plant population, and leaf morphology. Crop Sci. 53(4):1704–1711. doi:10.2135/cropsci2012.08.0490

Roberson, P.M. 1964. Design of flexible rolls for cotton harvesters. ASAE Paper 64-631. ASAE, St. Joseph, MI.

Sanders, J.C., J.C. Faircloth, and H.P. Wilson. 2009. Defoliating cotton under adverse conditions: Drought-stress, cool temperatures, and rank growth. Publ. 427-208. Virginia Polytech. Inst. and State Univ. http://pubs.ext.vt.edu/427/427-208/427-208_pdf.pdf (accessed 28 Aug. 2014).

Searcy, S.W., S.L. Simpson, and E.M. Barnes. 2012. Performance of cotton module covers: A summary of testing and standardization efforts. In: Proceedings of the Beltwide Cotton Conferences, Natl. Cotton Counc. Am., Orlando, FL. 3–6 Jan. 2012. Natl. Cotton Counc. Am., Memphis, TN. p. 620–632.

Searcy, S.W., M.H. Willcutt, M.J. Buschermohle, J.D. Wanjura, A.D. Brashears, and E.M. Barnes. 2010. Seed cotton handling and storage. Cotton Inc. http://www.cottoninc.com/Cotton-Harvest-Systems/ (accessed 28 Aug. 2014).

Smith, D.T., R.V. Baker, and G.L. Steele. 2000. Palmer amaranth (*Amaranthus palmeri*) impacts on yield, harvesting, and ginning in dryland cotton (*Gossypium hirsutum*). Weed Technol. 14(1):122–126. doi:10.1614/0890-037X(2000)014[0122:PAAPIO]2.0.CO;2

Smith, M.L. 1971. A feasibility study of field storage and handling of seed cotton. Final Rep. C.P.I. Proj. 70-198. Cotton Inc., Cary, NC.

Supak, J.R., A.D. Brashears, and W.F. Lalor. 1989. Harvesting practices to reduce bark. In: Proceedings of the Beltwide Cotton Production Research Conferences, Nashville, TN. 2–7 Jan. 1989. Natl. Cotton Counc. Am., Memphis, TN. p. 614–615.

Supak, J.R., and C.E. Snipes. 2001. Cotton harvest management: Use and influence of harvest aids. Ref. Book Ser. 5. Cotton Found., Memphis, TN.

Supak, J., R. Stair, A.D. Brashears, and W.F. Lalor. 1992. The influence of a stripper roll modification on harvest efficiency and foreign matter contents of stripped cotton. In: Proceedings of the Beltwide Cotton Conferences, Nashville, TN. 6–10 Jan. 1992. Natl. Cotton Counc. Am., Memphis, TN. p. 578–580.

Taylor, R.K., R.K. Boman, W.M., Porter, S. Osborne, T. Barber, B. McClelland, W. Henderson, M. Buschermohle, J. Wilkerson, C. Main, L. Gibson, J. Fulton, D. Monks, G. Pate, J.D. Wanjura, M. Kelley, C. Ashbrook, A. Thomasson, G. Morgan, Y. Ge, G. Collins, G. Vellidis, and E.M. Barnes. 2013. Using yield monitors to evaluate cotton variety test. In: Proceedings of the Beltwide Cotton Conferences, San Antonio, TX. 7–10 Jan. 2013. Natl. Cotton Counc. Am., Memphis, TN. p. 720–721.

Tupper, G.R. 1966. Stripper harvesting vs. spindle picking of open-boll and experimental stripper varieties of cotton. Trans. ASAE 9(1):110–113. doi:10.13031/2013.39890

USDA-CCC. 2013. 2013 Commodity Credit Corporation (CCC) loan schedule of premiums and discounts for upland and ELS cotton. USDA-CCC. http://www.fsa.usda.gov/FSA/webapp?area = home&subject = prsu&topic = lor (accessed 29 Aug. 2014).

USDA-ERS. 2000. A history of American agriculture, 1607–2000. ERS-POST-12. USDA-ERS, Washington, DC.

USDC. 1970. Historical statistics of the United States, colonial times to 1970. US Govt. Print. Office, Washington, DC.

Vellidis, G., and H.S. Brannen. 2012. Cotton yield maps: Tools for increasing efficiency and profitability. Eight-page, full color booklet summarizing benefits of using cotton yield maps. Booklet funded by Cotton Inc. and distributed at the 2013 Beltwide Cotton Conference, various other workshops, and county extension offices. www.cottoninc.com/fiber/AgriculturalDisciplines/Engineering/Precision-Crop-Management-for-Cotton (accessed 26 Mar. 2015).

Vories, E.D., and C.M. Bonner. 1995. Comparison of harvest methods for dryland cotton. Bull. 945. Arkansas Agric. Exp. Stn., Fayetteville.

Wanjura, D.F., and R.V. Baker. 1979. Stick and bark relationships in the mechanical stripper harvesting and seed cotton cleaning system. Trans. ASAE 22(2):273–278, 282. doi:10.13031/2013.35004

Wanjura, D.F., R.V. Baker, and E.B. Hudspeth. 1979. Characteristics of sticks in mechanically stripped cotton. Trans. ASAE 22(2):233–236, 242. doi:10.13031/2013.34996

Wanjura, D.F., and A.D. Brashears. 1983. Factors influencing cotton stripper performance. Trans. ASAE 26(1):54–58. doi:10.13031/2013.33875

Wanjura, J.D., R.K. Boman, M.S. Kelley, C.W. Ashbrook, W.B. Faulkner, G.A. Holt, and M.G. Pelletier. 2013. Evaluation of commercial cotton harvesting systems in the Southern High Plains. Appl. Eng. Agric. 29(3):321–332.

Wanjura, J.D., G.R. Gamble, G.A. Holt, and M.G. Pelletier. 2012a. Influence of grid bar shape on field cleaner performance—Field testing. J. Cotton Sci. 16:255–267.

Wanjura, J.D., G.A. Holt, R.K. Byler, A.D. Brashears, and R.V. Baker. 2009. Development of a high capacity extractor cleaner for cotton stripper harvesters—Machine design and optimization. Trans. ASABE 52(6):1821–1829. doi:10.13031/2013.29202

Wanjura, J.D., M.S. Kelley, R.K. Boman, and G.A. Holt. 2010a. Harvest timing and techniques to optimize fiber quality—Initial findings. In: Proceedings of the Beltwide Cotton Conferences, Natl. Cotton Counc. Am., New Orleans, LA. 4–7 Jan. 2010. Natl. Cotton Counc. Am., Memphis, TN. p. 539–550.

Wanjura, J.D., W.B. Faulkner, R.K. Boman, M.S. Kelley, A.D. Brashears, E.M. Barnes, S.W. Searcy, M.H. Willcutt, and M.J. Buschermohle. 2010b. Stripper harvesting. Cotton Inc. http://www.cottoninc.com/Cotton-Harvest-Systems/ (accessed 8 Apr. 2015).

Wanjura, J.D., W.B. Faulkner, E.M. Barnes, G.A. Holt, and M.G. Pelletier. 2012b. Crop residue inventory estimates for Texas High Plains Cotton. In: Proceedings of the Beltwide Cotton Conferences, Natl. Cotton Counc. Am., Orlando, FL. 3–6 Jan. 2012. Natl. Cotton Counc. Am., Memphis, TN. p. 542–552.

Ward, J.M., W.E. Paulson, and D.L. Jones. 1953. Storing of seed cotton as an aid to more efficient ginning and marketing. Bull. 765. Texas Agric. Exp. Stn., College Station, TX.

Wilkerson, J.B., F.H. Moody, and W.E. Hart. 2002. Implementation and field evaluation of a cotton yield monitor. Appl. Eng. Agric. 18(2):153–159. doi:10.13031/2013.7784

Wilkes, L.H. and J.K. Jones. 1973. Potentials of a modular system of handling seed cotton. ASAE Paper 73-149. ASABE, St. Joseph, MI.

Willcutt, M.H., M.J. Buschermohle, E. Barnes, F. To, J. Field, and P. Allen. 2009. In field time in motion comparisons of conventional, John Deere 7760, and Case 625 Module Express cotton pickers. In: Proceedings of the Beltwide Cotton Conferences, San Antonio, TX. 5–8 Jan. 2009. Natl. Cotton Counc. Am., Memphis, TN p. 462–476.

Willcutt, M.H., M.J. Buschermohle, G.W. Huitink, E.M. Barnes, J.D. Wanjura, and S.W. Searcy. 2010. The Spindle type cotton harvester. Cotton Inc. http://www.cottoninc.com/Cotton-Harvest-Systems/ (accessed 20 Aug. 2014).

Willcutt, M.H., W.D. Mayfield, W.F. Lalor, and R.G. Curley. 1992. Seed cotton storage and handling in modules. Cotton Inc. Cary, NC.

Williford, J.R., A.D. Brashears, and G.L. Barker. 1994. Harvesting. In: Cotton ginners handbook. USDA-ARS, Washington, DC. p. 11–16.

Ginning

Ed Hughs* and Greg Holt

Abstract

The purpose of the cotton ginning process is to separate a raw field crop into its salable components, a necessary step between the cotton farmer and the textile manufacturer. The original ginning process used a simple manually operated device that took hand-harvested seed cotton and separated the cotton fiber from the cottonseed so that the fiber could then be manually spun into yarn and woven into cloth. Today's modern cotton gin has evolved into a complex mechanical system that takes in mechanically harvested seed cotton to condition, clean, and separate its components, further clean and then package the cotton fiber, and handle the seed and trash for marketing. The modern cotton gin produces four products that have market value: the cotton fiber, the cottonseed, motes (small, immature seed with attached short fiber), and the trash. Currently the two most valuable products are the fiber and the seed, but the motes and trash also have market value. The fiber, also sometimes referred to as lint, is the most valuable product, and the basic design and operation of the ginning process is aimed at producing fiber and maintaining its value. Cottonseed also has significant value and is primarily used either as a feed supplement for cattle or is processed by oil mills into various food products. Some cottonseed is specially ginned and handled to be used as planting seed for the next cotton crop. Motes are separately baled and sold to be used in nonwoven-lint-based products. Gin trash is of relatively small value but is utilized for many purposes that range from compost to manufactured construction materials. The modern cotton gin takes a raw material, mechanically harvested seed cotton, and gives it market value by separating the fibers from the seed, separating foreign material, and packaging and handling all of these for commerce.

Ginning, in its simplest context, refers to the process of separating cotton fibers from the seeds. The first cultivation of cotton and use of cotton fiber are lost in the mists of time, but the first cotton producer also had to remove the fiber from the seed to utilize the fiber. While it is possible to remove the fiber from the seed with the human thumb and forefinger, it is a very slow and laborious process and probably led very quickly to the invention of the first cotton gin (Lakwete, 2003). The modern term "gin" may be a contraction of the word "engine" and is simply

Abbreviations: CBS, combination bur and stick; HVI, High Volume Instrument; JCIBPC, Joint Cotton Industry Bale Packaging Committee; NAAQS, National Ambient Air Quality Standards; OSHA, Occupational Safety and Health Administration; PBI, permanent bale identification; PM, particulate matter; UD, universal density.

Ed Hughs, USDA-ARS Southwestern Cotton Ginning Research Laboratory, 300 E. College Drive, Mesilla Park, NM 88047. *Corresponding author (ed.hughs@ars.usda.gov).

Greg Holt, USDA-ARS Cotton Production and Processing Research, 1204 E FM 1294, Lubbock, TX 79403 (greg.holt@ars.usda.gov).

doi:10.2134/agronmonogr57.2013.0048

Cotton. 2nd. ed. David D. Fang and Richard G. Percy, editors. Agronomy Monograph 57.

applied to any mechanical device that applies force to remove cotton lint from the cottonseed (Vandergriff, 1997).

The first documented cotton gin dates back at least 2000 yr and consisted of a small round bar made of iron or other metal approximately 30 cm (12 in) long and a flat base approximately 13 cm (5 in) wide and 20 to 25 cm (8 to 10 in) long (Lakwete, 2003; Fig. 1A and 1B). The "ginner" grasped both ends of the bar and rolled it over seed cotton lying on the base to pinch the seed apart from the lint. Skill was required to remove the seed from the fiber without crushing the seed when using this single-roller gin.

The next documented development of the gin is generally called the churkha gin. This machine has existed for centuries in various forms in China, India, Southeast Asia, and the Middle East (Bennett, 1959; Lakwete, 2003). This gin consisted of two small-diameter rollers that rotated simultaneously in opposite directions (double-roller gin). These rollers were roughly 30 cm (12 in) long, made of some combination of wood or metal, and powered by either one or two cranks. The gin maker wedged the rollers into a vertical frame (Fig. 2). As the rollers rotated they pulled the cotton fibers between them, pinching the seed off. The seed could not pass through the tightly wedged rotating rollers and so dropped away as the fiber passed between the two rollers, effectively separating fiber from the seed. It has been stated that the churkha gin had an average fiber turnout of five pounds of fiber per day, which was five times that of the single-roller gin (Lakwete, 2003). Thus the faster churkha gin became the dominant method of ginning cotton in medieval cotton production.

The beginnings of modern cotton production came to the New World with the Spanish planting of cotton in Florida in 1556 and the settling of the Virginia Colony in North America by the British in 1607. The Virginia colonists apparently received cottonseed from the British East India Company, probably the Old World species *Gossypium arboretum* or *G. herbaceum*, before departing for the voyage to America (May and Lege, 1999; Smith et al., 1999; Lakwete, 2003). Cotton production slowly spread from Virginia to other colonies but for a long time remained

Fig. 1. (a) Single-roller gin bar. (b) Single-roller gin flat base.

Fig. 2. Churkha gin.

relatively small compared with other cash crops such as tobacco, indigo, and rice. Up to just before the time of the American Revolution, cotton was grown in the colonies mainly for homespun yarn and domestic consumption, but around 1770, American colonists started exporting larger amounts of ginned cotton fiber to Great Britain. Until recently, it was thought that even for this small amount of cotton production, the cotton fiber and seed was separated by hand before Eli Whitney patented a cotton gin in 1794. Such was not the case. Along with the cottonseed from the British East India Company, the colonists also probably brought along some form of the churkha or double-roller gin to separate the cotton fiber from the seed (Lakwete, 2003).

The early history of ginning machines in the American colonies is murky, but there is enough written evidence to say that churkha-type double-roller gins were being used on the North American mainland by the first half of the eighteenth century and probably before (Lakwete, 2003). As cotton production evolved in North America, so did the churkha. In 1762, Denis Diderot, a Frenchman, published the *Encyclopédie*, which depicted an engraving from "an island in America where cotton is grown" and showed both a small, manual and a larger, foot-powered double-roller gin. The more productive foot-powered double-roller gin largely replaced the use of smaller manual gin, but it was superseded in North America in the 1770s by the barrel gin (Fig. 3). The barrel gin was a composite gin made by belting from 2 to 24 roller gin heads to a single drive wheel, or barrel (Lakwete, 2003; Porcher and Fick, 2005). A single ginner could power the smaller units, but animal treadmills or waterwheels were necessary to power larger units. There were other roller-gin innovations during this time that used the double pinch-roller principle, such as the self-feeding gin, which was invented by Joseph Eve in 1788. The Eve gin eliminated the manual feed required for the foot-powered and barrel gins and was safer

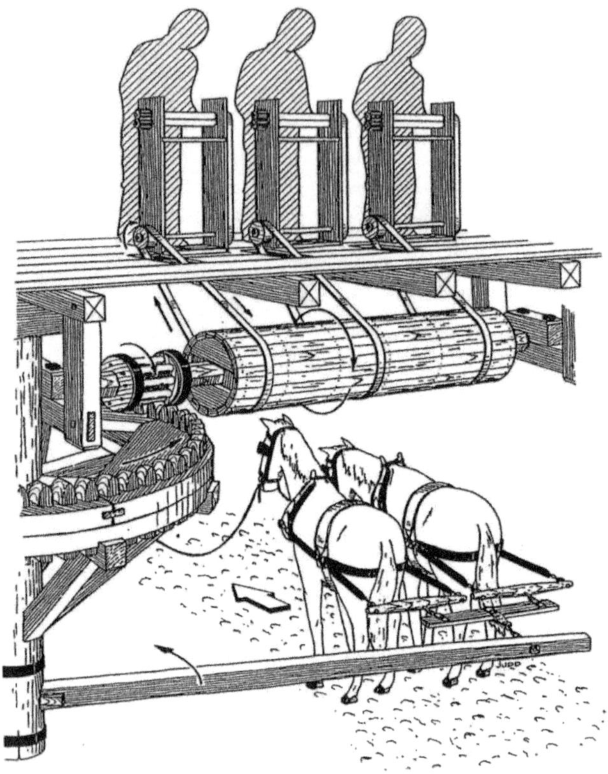

Fig. 3. Barrel gin (From *The Story of Sea Island Cotton*, p. 201, Porcher and Fick, 2005).

to operate. It increased ginning productivity but was more expensive to build and house and required more power to operate (Lakwete, 2003). A 1796 advertisement for Eve's gin promised a daily production of 300 lb for smooth-seed cotton and from 150 to 200 lb of fuzzy-seed varieties (Lakwete, 2003).

All of these various double-roller gins were used in North America to gin the Upland, or "fuzzy" seed, cottons (*G. arboretum*, *G. herbaceum*, or *G. hirsutum*) being grown on the mainland at the time. *Gossypium hirsutum* is a New World "fuzzy" seeded species that probably originated in the Yucatan peninsula and made its way to North America. Another New World species, *G. barbadense* (also known as Sea Island), whose seeds are not "fuzzy" but relatively smooth and whose fiber is longer and finer than that of the Upland cottons, was being grown on the Atlantic Coast by 1785–86 (Porcher and Fick, 2005). The fibers of Sea Island cotton are more easily removed from the seeds than are the fibers of the Upland varieties and thus were more easily ginned on the double-roller gins of the time. This relative ease of removal may have been part of the origin of the story that before Eli Whitney all cotton grown in America except for Sea Island was hand ginned (Bennett, 1959). Even though prerevolutionary cotton shipments from America to Britain were relatively small, in 1768, planters in Virginia, South Carolina, and Georgia alone shipped to Great Britain a total of 46,850 lb of cotton fiber (most all of which would have been from fuzzy-seed, Upland varieties) ginned with the foot gin. Anyone who has tried hand ginning modern varieties of Upland cotton would understand that hand ginning even this relatively small amount of cotton would have been a daunting task.

Several factors in the late 1700s caused cotton production in America to increase greatly, which resulted in the need for more speed in ginning the cotton fiber being produced. The Revolutionary War (1775–1783) changed the relationship of America to Great Britain, from a colony that supplied raw materials to the mother country for finished goods to being a partner with it in world trade. Richard Arkwright invented the water-driven spinning frame in 1769, which greatly improved yarn quality and production speed for British spinners, helping to usher in the industrial revolution at the end of the eighteenth century (May and Lege, 1999; Lakwete, 2003). Other similar textile-processing improvements significantly boosted Great Britain's need for raw cotton for its own industry at a time when America needed trading partners. Finally, as a young country, America needed to build and expand its own textile industry and participate in world trade for the country's economic health and growth.

The Saw Gin

The stage was set for the Eli Whitney's invention of and patent on the spike-tooth cotton gin in 1794 and for Hodgen Holmes's patent on a saw-tooth gin in 1796. It is interesting to note that both Whitney's and Holmes's patents were signed by George Washington on 14 Mar. 1794 and 12 May 1796, respectively (Moore, 1977; Lakwete, 2003). Whitney's gin (Fig. 4) removed the fibers from the seed by means of small spikes driven into a wooden cylinder. The spikes engaged the fibers and pulled them through narrow slots in a metal bar at the back of a seed-cotton roll box. The slots, being too narrow for the seeds to pass through, retained the seeds and allowed the fiber to be pulled free. A brush cylinder behind the slotted bar removed the fibers from the spikes. From patent descriptions, the ginning cylinder was from 15 to 23 cm (6 to 9 in) in diameter and 0.6 to 1.5 m (2 to 5 ft) in length. It was a wood core studded with coarse iron teeth arranged in annular rows spaced by the size of a cotton seed (Lakwete, 2003). Whitney's patent was not

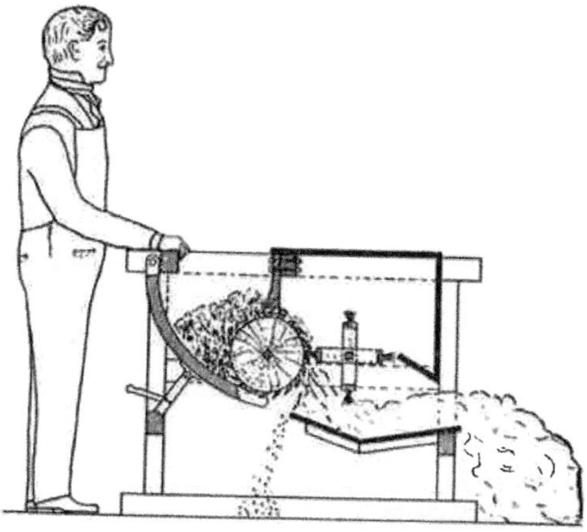

Fig. 4. Whitney's spike-tooth gin.

another improvement on existing double-roll pinch gins but a whole new concept that significantly increased ginning capacity. In a 1793 letter to Thomas Jefferson, Whitney described his machine's performance (Mirsky and Nevins, 1952):

> It is turned by hand, and requires the strength of one man to keep it in constant motion. It is the stated task of one Negro to clean fifty weight (I mean fifty lbs. after it is separated from the seed) of the green-seed cotton per day. This task he usually completes by 1:00 o'clock in the afternoon. He is paid so much per pound for all he cleans over and above his task and to ten or fifteen days successively he had cleaned from sixty to eighty weight per day and left work every day before sunset.

Whitney's gin as patented had two weaknesses: (i) the spiked teeth tended to loosen from the wood core and fall out, and (ii) it was a batch process. The ginner filled the roll box with seed cotton and cranked the cylinder until all the seed was cleaned of fiber. Then the ginner would stop turning the cylinder and dig all of the cleaned seed out of the roll box before filling it again with seed cotton and repeating the process. Holmes's patent, while similar, had three major improvements over Whitney's (Fig. 5). Spikes in a wooden core were replaced with circular saws that were much more durable, and the slotted bar was replaced by flat-iron ribs. Holmes also opened up the bottom of the roll box for the continuous discharge of ginned seed, permitting continuous operation as opposed to Whitney's batch operation (Moore, 1977).

During the next few years following these two gin patents, there were a number of legal actions by Whitney, Holmes, and others to establish legal ownership of what finally became generally known as the saw gin. Whitney and his partner, Phineas Miller, tried early on to monopolize both the ginning industry and the manufacture of gins (May and Lege, 1999; Lakwete, 2003). However, the gin designs patented by Whitney and Holmes were relatively easy to copy, and knockoff copies of the gins were made and used. History probably affords too much credit to Whitney and not enough to Holmes and others for the initiation and development of the saw gin and its impact on the American cotton industry. What is certainly true is that as the American cotton industry grew with the increased production of fuzzy-seed, Upland cottons, the saw gin came to dominate the double-roller gin for the ginning of these cottons. The use of the

Fig. 5. Holmes' saw gin.

older double-roller gin designs for Upland cottons diminished over time, and the gins became primarily used for the higher-quality, longer-staple, Sea Island cottons. The inventions and improvements by Whitney, Holmes, and others were the origin of today's modern saw gin stand, but the older roller ginning technology persisted and adapted as well.

The Roller Gin

As mentioned previously, early roller gins were of the double-roller churkha type that had undergone modification and improvement. The first major modern change in roller gin design occurred in 1840 with the invention of the reciprocating knife principle by Fones McCarthy of Demopolis, Alabama (Alberson and Stedronsky, 1964; Gillum, 1985; Lakwete, 2003, Porcher and Fick, 2005). From the invention of the saw gin up until 1840, American cotton producers had largely segregated into those who produced the shorter-staple, fuzzy-seeded, Upland cottons in the interior of the mainland and those who produced the longer-staple, slick-seeded, Sea Island cotton in the coastal areas. While Upland cotton producers had largely abandoned the use of roller gins for ginning their cottons, Sea Island cotton producers continued to rely on roller gins, primarily the Eve gin (a type of roller gin) and the foot gin, to gin the Sea Island cotton because of fiber-quality concerns. Saw gins tended to shorten and damage the staple (a measure of fiber length) of the longer and finer Sea Island variety, whereas roller gins better preserved the fiber length. However, roller gins had a much slower ginning rate than saw gins and so increased the labor and expense of ginning (Lakwete, 2003). While the foot gin with one operator could produce 11 kg (25 lb) of ginned fiber in a day, the saw gin could produce many times that amount. Sea Island producers needed a faster way to gin cotton without damaging the fiber.

The McCarthy reciprocating-knife roller gin basically consisted of a large single roller ($\geq$4-in diameter) covered with leather, a stationary knife that rode with some pressure against the roller, and a reciprocating knife that pushed up and down on the seed to separate it from the fiber (Fig. 6). In operation, the seed cotton was pushed up against the revolving roll, the fiber was pulled under the stationary knife (the seed could not go under), and the reciprocating knife pushed on the seed until it was entirely separated from its lint and the seed fell away. This design for removing fiber was a great departure from the operating principle of fiber removal used by the older double-roller gins and was not as damaging to the cotton fiber as saw ginning. Paradoxically, McCarthy originally intended his gin for the shorter-staple Upland cotton. However, when growers of Sea Island cotton near Savannah, Georgia, saw the McCarthy gin demonstrated in 1842, they were ready to adopt the new roller gin for long-staple cottons (Lakwete, 2003; Porcher and Fick, 2005). While the foot gin could gin 11 kg (25 lb) of long-staple fiber in a day, the original McCarthy gin could gin 68 to 91 kg (150 to 200 lb) of fiber in a day with only 1 hp (746 W) and one person to feed it, thus replacing several old foot gins and their operators.

The McCarthy reciprocating-knife roller gin and its later modifications were adopted and greatly facilitated the ginning of longer-staple cottons throughout the world. These McCarthy roller gins are no longer used in America, but they continue to be used in many cotton-growing areas of the world to gin both long- and short-staple cotton. Gillum (1985) noted that modern McCarthy-type gins

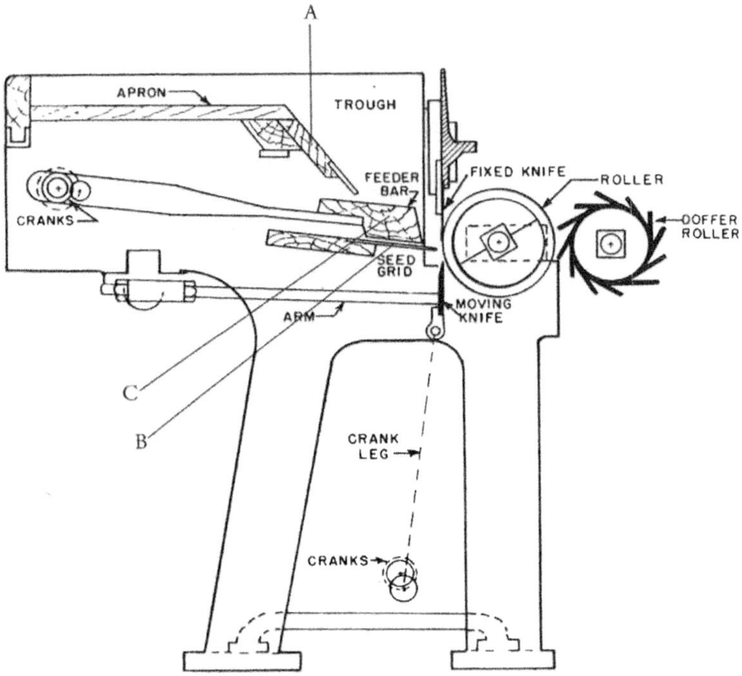

Fig. 6. Reciprocating-knife roller gin.

had a maximum ginning rate of approximately 55 kg h^{-1} (120 lb h^{-1}) of ginned fiber on long-staple Pima cotton (an extra-long-staple cotton grown in the United States that is related to Sea Island cotton) and about 36 kg h^{-1} (80 lb h^{-1}) on Upland cotton.

A significant evolution of the roller gin occurred when the rotary-knife roller gin was developed between 1955 and 1963 through the efforts of the USDA-ARS Southwestern Cotton Ginning Research Laboratory, the gin equipment manufacturers, and private roller ginneries (Fig. 7; Gillum, 1985). As with the McCarthy reciprocating-knife roller gin, the rotary-knife roller gin uses a rotating roll to pull the cotton fiber under a stationary knife. The roll diameter was increased from approximately 20 cm (8 in) in the McCarthy design to 40 cm (16 in) in the rotary-knife design. Also, with the rotary-knife design, the pressure of the stationary knife to the roll surface is greater, and the speed of the roller surface is higher. However, the biggest change, as the name implies, was the replacement of the reciprocating knife with a rotary knife (Fig. 7). The action of the reciprocating knife actually blocked the ginning point approximately 50% of the time, so that no new seed cotton could be applied, whereas the rotary-knife design allowed access of seed cotton to the ginning point 100% of the time (Stedronsky, 1956; Gillum, 1985). All of these changes increased the approximate maximum ginning rate on long-staple, Pima cotton from 55 kg h^{-1} (120 lb h^{-1}) on the McCarthy gin to 341 kg h^{-1} (750 lb h^{-1}) on the rotary-knife gin stand, the ginning rolls of both gin types being 102 cm (40 in) in width. The fiber quality from both gin designs was essentially the same, although somewhat different in appearance, so in a very few

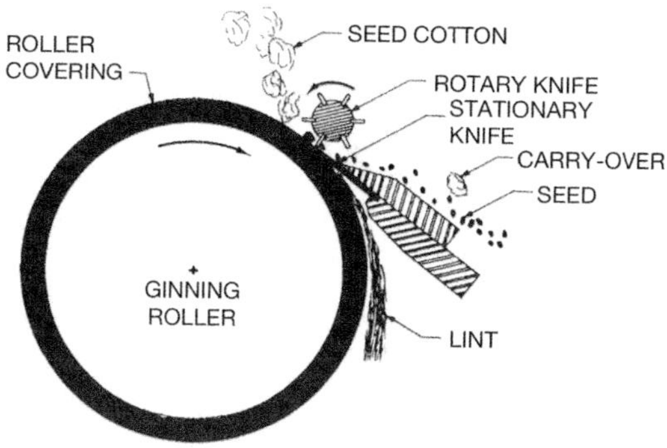

Fig. 7. Rotary-knife roller gin.

years after its development, the rotary-knife roller gin replaced the reciprocating-knife roller gin in the United States.

Even though the rotary-knife roller gin gained 100% acceptance in the United States, the reciprocating-knife roller gin is still widely used overseas because it is significantly cheaper to build and easier to maintain than the more complex rotary-knife roller gin. There is also the cost of labor and a labor-safety factor, because the rotary-knife roller gin must be mechanically fed so that workers do not get caught in the rotary-knife, whereas the reciprocating-knife roller gin can be fed manually with relative safety to the ginner.

Rotary-knife roller gins have recently undergone another significant improvement with the development of the high-speed rotary-knife roller gin (Gillum and Armijo, 1991; Armijo and Gillum, 2007). The most recent development was a design to significantly increase the roller-ginning rate of Pima cotton as well as of Upland cotton. Because the fiber turnout (yield) and fiber-length properties of cotton fiber ginned by rotary-knife roller gins is superior to that same cotton ginned on saw gins, a certain small quantity of the long-staple Acala and other long-staple Upland cottons have been ginned on rotary-knife roller gins for special purposes (Hughs and Leonard, 1986; Hughs and Lalor; 1990). However, the ginning rate of Upland cotton on the rotary-knife roller gins had been only approximately one-half to two-thirds that of Pima cottons, making large-scale roller ginning of Upland cotton not economically feasible in the past. The most recent roller-ginning development was aimed at raising the ginning rate of Upland cotton on rotary-knife roller gins to be competitive with the rates on saw gins.

It had been surmised that significantly raising the revolutions per minute of both the ginning roll and the rotary knife and increasing the pressure of the ginning roll on the stationary knife would greatly increase the roller ginning rate of both Pima and Upland cottons. However, until recently, the rotary-knife roller gins were manually controlled and tended. A ginner might be responsible for manually monitoring and controlling 10 or more roller gin stands and could not safely tend them at rates exceeding more than about 1.5 to 2 cotton bales per stand per hour (a bale is the standard ginned-cotton package weighing approximately 227 kg [500 lb]). At higher ginning rates there were the possibilities of overfeeding seed cotton to the gin stand and causing a chokage, which can mechanically damage the expensive ginning roll or rotary knife, or of not feeding enough cotton to cool the roll (ginned fiber in a roller gin actually acts as a coolant and carries heat away from the ginning point), thus overheating and damaging the ginning roll from lack of adequate seed-cotton feed. These problems were overcome by developing computer-aided feed controls, to feed seed cotton to the ginning point at very high rates, and a reliable air-cooling method for the gin roll (Gillum and Armijo, 1991; Armijo and Gillum, 2007). As a result of these changes, ginning rate of a computer-controlled rotary-knife roller gin for Upland cotton was comparable, for a given-width gin stand, to saw ginning. This improvement, along with the superior fiber-length properties and turnout from the roller gin stands, has caused producers of long-staple Acala cottons in California to install high-speed rotary-knife roller gin stands to gin their Upland cotton.

As of 2014, there are currently approximately 28 cotton gins located in far-west Texas, New Mexico, Arizona, and California that are roller gins. Not all are the newest, high-speed rotary-knife roller gins, and so the number of gin stands per plant and their ginning capacity varies. Some of these roller gins are complete stand-alone gin plants whose entire processing system serves only the rotary-knife roller gin stands. Others are termed "combo gins," where the processing system is shared between roller gin stands and saw gin stands. These combo gins will operate either the saw gin stands or the roller gin stands at any given time, depending on need, but the two types never gin at the same time. Also, some of these roller gins, depending on the needs of their cotton producers, process only Pima cotton, whereas others gin both Upland and Pima cotton during the ginning season.

The Development of Ginning Systems

For many years after 1796, ginning was simply another farm operation performed at the plantation after harvesting had been completed. These plantation ginneries came to be dominated by only one hand-fed saw gin that was part of a ginning system that generally included a two-story building with the saw gin stand on the second floor (Fig. 8; Bennett, 1962; Hughs et al., 2008b). In the first-floor area, harnessed animals or waterwheels provided power through a simple cogwheel/flat-belt drive system. The building provided storage for seed cotton waiting to

Fig. 8. Early plantation gin.

be ginned and a blow room to store the ginned lint before it was pressed into bales. Early presses were animal-powered, single-box, wooden-screw presses located apart from the main building (Fig. 8). The packaging and bagging of the lint was a separate operation and all cotton was manipulated by hand. By the 1880s, mechanical screw presses, gin feeders, and pneumatic cotton-handling systems had eliminated much of the hand labor from ginning and made it practical to operate several gin stands simultaneously in a single system. As cotton ginning systems became larger and more complex, many plantations found that they could no longer afford to buy gins for their individual use. This situation prompted the establishment of custom ginning operations in most local communities by the close of the nineteenth century and marked the end of the plantation gin and the beginning of modern ginning systems.

During the early days of ginning, cotton was carefully handpicked by farm workers who had a direct interest in the crop. This cotton was very clean and carefully handled so that the only machines needed in a ginning system were a gin stand, a bale press, and a manual or pneumatic means of conveying ginned fiber between the two. However, as cotton production increased, it became necessary for cotton producers to employ additional labor for hand harvesting. This practice led to faster and rougher hand snapping and hand stripping, in which the seed cotton was no longer plucked from the boll; the entire boll and all was taken from the plant. These rough hand-harvesting practices greatly increased the foreign matter and moisture content of the harvested seed cotton (Moore and Merkel, 1953), creating the need for additional cleaning at the gin. During the early 1900s, many types of seed-cotton cleaners and extractors were developed to remove the extra trash. Seed-cotton drying equipment began to be installed by the 1930s to improve the performance of seed-cotton cleaners when handling damp cotton and to prevent rough ginning preparation (visual appearance) of the ginned fiber.

Mechanical-stripper harvesting methods developed before the 1940s, and subsequent mechanical picking methods that came into use after World War II spurred the development of other seed-cotton cleaning, drying, and lint-cleaning methods to compensate for the trashier harvesting methods (Moore, 1977; Mayfield et al., 1991; Hughs et al., 2008b). Faster mechanical harvesting also shortened the ginning season significantly, stimulating the development of higher-capacity ginning machinery. More recent increases in the speed of cotton harvesting have resulted from the development of the cotton module (see "Seed-Cotton Handling" below) and the abandonment of slower cotton trailers. This, along with the development of higher-capacity mechanical cotton harvesters, has maintained the pressure for even higher-capacity cotton gins. Historically, ginning has evolved in response to changing production and harvesting practices, and today's modern cotton gin represents the latest development in that long evolutionary process.

Seed-Cotton Handling

Delivery and Handling in the Gin Yard

Harvested seed cotton is stored in modules in the field until it is transported to the gin yard. Moduling is an efficient and economical method of handling and storing large volumes of seed cotton until the cotton can be ginned. Modules are

Fig. 9. Rectangular and round modules.

either rectangular or round (Fig. 9) and typically contain 3 (round module) to 16 (rectangular module) bales worth of lint, depending on the harvesting equipment used. Currently, the most common modules seen in the United States are the large rectangular type, which are 2.2 m wide by 9.7 m long by 2.59 m high (7.25 ft wide by 32 ft long by 8.5 ft high). There are also smaller rectangular modules that are half as long, and thus known as half-modules. Recently, round modules, which are 2.2 m wide by 2.4 m in diameter (7.25 ft wide by 8 ft in diam.) have become more prevalent with the newer harvesting systems.

Most modules are transported to the gin yard by module trucks having live-bottom chain beds that are specifically designed to pick up and transport seed-cotton modules (Fig. 10). With the advent of the round module, semitrailer trucks with a flat-bed trailer are increasingly being used. Once the module is transported to the gin yard, it is weighed and either fed directly into the ginning system or stored in the gin yard. Stored modules are transported to the gin's seed-cotton unloading system as needed by a variety of mobile equipment.

Unloading and Conveying Systems

The term "unloading system" refers to the point at which the seed cotton enters the ginning system and originates from when seed cotton was brought to the cotton gin in trailers, before the advent of modules. Today the unloading system refers not only to the area, but also to the equipment with which the incoming modules of seed cotton are fed into the cotton gin. The primary purpose of the unloading system is to feed cotton into the gin as uniformly and consistently as possible while removing foreign material such as rocks, green bolls, metal, and some sand and dirt. The unloading system is primarily composed of a module feeder, rock-and-green-boll separator, seed-cotton separator, feed control system, and a conveying system. The most common means of conveying seed cotton is pneumatically, which has been done since the late 1800s (Bennett, 1962).

Fig. 10. Module truck loading round modules.

Module Feeder

Module feeders come in two styles: moving head and stationary head. The stationary-head feeder is more commonly used because (i) it has fewer moving parts, (ii) it is easy to operate, and (iii) it has a smaller horsepower requirement compared with the moving-head type (Laird et al., 1994). The stationary-head feeders also do not require repositioning of the feeder head, which allows for a much more consistent gin operation. The primary operating principle of both feeder types is the same, with the exception of which components move: the head or the modules. Both the stationary- and moving-head feeders have spiked dispersing cylinders that break up the dense seed-cotton module to allow loose seed cotton to be fed into the ginning system. The moving-head feeder (Fig. 11) moves across

Fig. 11. Moving-head module feeder.

modules that have been placed one after another in a line end to end for rectangular modules, or side-by-side for round modules, on a concrete pad and disperses the cotton onto a belt conveyor running parallel to the concrete pad. The belt conveyor moves the seed cotton to a pneumatic pickup pipe and into the ginning system. The stationary-head feeder (Fig. 12) moves the staged modules, again one after another end to end or side-by-side, into the feeder-head dispersing cylinders using powered rollers, a wire belt, or alternately moving slats (commonly known as a walking floor). Seed cotton from the dispersing head is conveyed into the rock-and-green-boll separator.

For handling round modules, changes to the operation of the module feeding system have been necessary to accommodate the differences between rectangular

Fig. 12. Stationary-head module feeder shown with rectangular and round modules.

and round modules. The rectangular modules are covered with tarps that typically enclose the top quarter of the module and can easily be removed, folded, and reused. The round modules are wrapped with a plastic that is used once and must be cut off before encountering the module feeder's dispersing cylinders. There are several round-module unwrapping systems, manufactured by various companies, ranging from fully automated to manual. The primary concern is that the entire plastic wrap surrounding the module be removed before the seed cotton enters the dispersing cylinders, since plastic is a contaminant that can be very difficult to remove once it enters the ginning system.

Rock-and-Green-Boll Separator

Seed cotton can contain green bolls (unripened cotton bolls), especially when machine-stripped early in the season. In addition to green bolls, objects such as rocks, metal, and/or dirt can be present. All of these objects need to be removed as early as possible in the ginning process. Rock-and-green-boll separators (labeled "turbulent dryer trap" in Fig. 13) are located in the pneumatic drying system and work via gravity and centrifugal forces. Such separators, called turbulent dryer traps, are the most common means of separating heavy foreign objects from the incoming seed cotton. The two most common types of rock-and-green-boll separators include a device, commonly called a rock trap, mounted in the air line upstream of the seed-cotton separator, and a device mounted on the pneumatic pickup line at the exit of the module feeder and commonly called a hot box. The hot box is a combination drying and rock-and-green-boll-separating system.

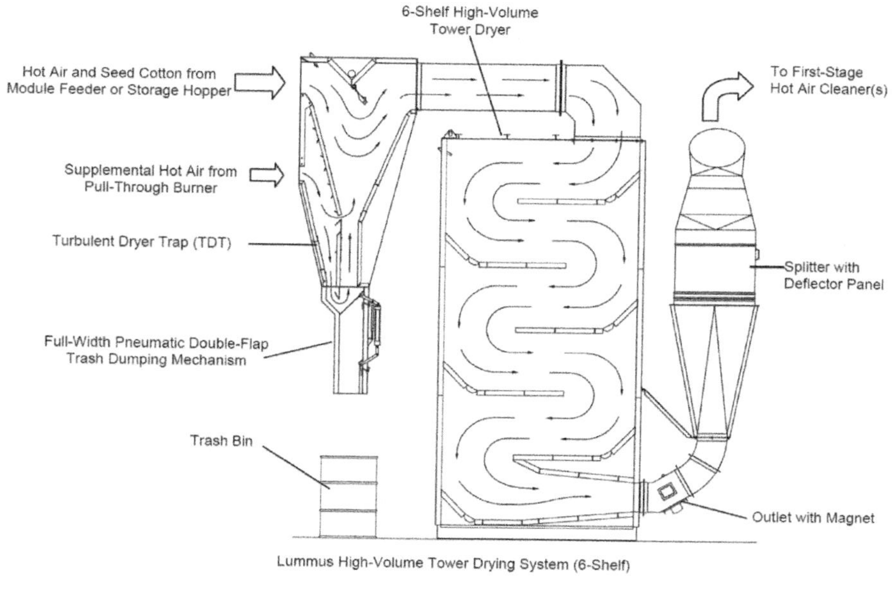

Fig. 13. Lummus drying system with turbulent dryer trap. (Courtesy of Ross Rutherford, Lummus Corp.)

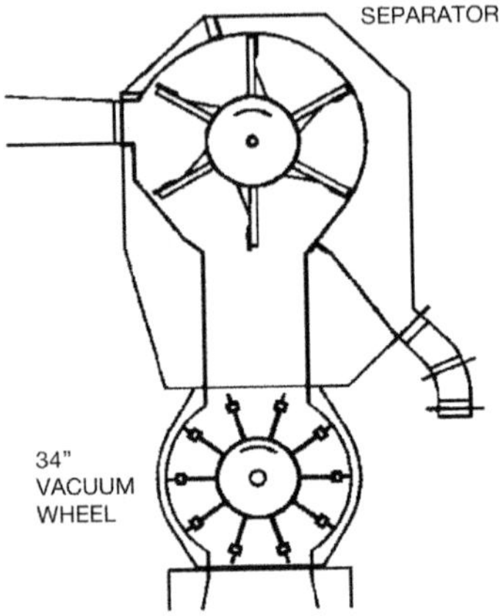

Fig. 14. Separator and vacuum wheel.

Seed-Cotton Separator and Vacuum Dropper

Separation of seed cotton from the conveying air stream is the main function of the seed-cotton separator. Separators generally consist of stationary or revolving screens on which the seed cotton is caught while the air moves through the screen, thus separating the cotton from the air stream. The screen is kept clean by wipers that are attached to rotating reels. The seed cotton falls to a vacuum dropper, which allows the cotton to continue moving to the next stage in the ginning system (Fig. 14). Most separators are operated under negative air pressure. The vacuum dropper is needed to prevent excessive air from flowing with the seed cotton as it continues through the ginning system.

Feed Control Unit

A feed control unit, which is used by most gins, is critical for regulating the flow of seed cotton from the unloading system into the cotton gin. A feed control is basically a surge bin with variable-speed feed rollers and is typically installed in sequence after the seed-cotton separator in the unloading system, although a feed control can be installed directly beneath the module feeder. The variable-speed feed rollers are controlled by the ginner to allow the seed cotton to be more precisely metered into the ginning system. The advantages of a feed control unit are that it (i) provides an even flow of seed cotton to the conditioning and cleaning equipment, (ii) minimizes choke-ups, and (iii) improves the efficiency and performance of the downstream equipment by providing a consistent flow of seed cotton.

Conveying System

In modern ginning systems, the seed cotton is conveyed through a cotton gin by a variety of methods, such as gravity, mechanical conveyors (screws and belts), and pneumatics. The most common means of conveyance is pneumatics. In pneumatic systems, seed cotton is fed into pipes and conveyed at velocities of 1219 to 1524 m min^{-1} (4000 to 5000 ft min^{-1}) from one stage of processing to the next. Upon arriving at the next stage of processing, the seed cotton is removed from the air stream with a seed-cotton separator. Often the conveying air is heated or humidified, and the pneumatic conveyor serves as a drying or moisture-adding

Table 1. Recommended air velocities for conveying seed cotton, cottonseed, gin waste/byproducts, and lint.

Materials	Velocity
	m min^{-1} (ft min^{-1})
Seed cotton	
Suction pipe (from module feeder)	1676–1981 (5500–6500)
Pipes	1067–1676 (3500–5500)
Dryers	610–762 (2000–2500)
Cottonseed	1219–1524 (4000–5000)
Gin waste/byproducts	1219–1524 (4000–5000)
Lint	472–610 (1500–2000)

system (Baker et al., 1994b). Pneumatic conveying systems are designed for a recommended air volume of 12.5 m^3 kg^{-1} seed cotton (20 ft^3 lb^{-1} seed cotton) being conveyed; 31.2 m^3 kg^{-1} (50 ft^3 lb^{-1}) for conveying systems that are also used for drying (12.5 m^3 kg^{-1} [20 ft^3 lb^{-1}]) and up to 31.2 m^3 kg^{-1} (50 ft^3 lb^{-1}) for moving cotton in drying stages of the ginning process. A cotton gin processing 60 bales h^{-1} can easily move more than 73,000 m^3 h^{-1} of air (2.6 million ft^3 h^{-1}), which is why pneumatic systems account for approximately 60% of the energy consumption involved in cotton ginning. Table 1 (Baker et al., 1994b) shows the recommended conveying velocities for materials in the ginning system. Those velocities are designed to make sure that the material stays in the air stream (lower end of the range) without damaging the material and/or pipes and equipment (higher end of the range).

Moisture Conditioning

Importance of Proper Moisture (Fiber–Moisture Relationships)

The moisture content of seed cotton is one of the most important factors affecting cleaning efficiency, ginning performance, and fiber-quality preservation during ginning. Seed cotton that is too wet will not clean and gin properly and will receive low grades due to excessive foreign matter and rough preparation (Leonard et al., 1979). On the other hand, ginning seed cotton that is too dry may adversely affect the fiber quality. The apparent strength of cotton fibers is directly proportional to the fiber moisture content and is, therefore, greater at higher moisture levels. Consequently, as the fiber moisture content is lowered, the apparent strength is reduced and the frequency of fiber breakage during ginning is increased (Griffin and Moore, 1965). Also, a low fiber moisture content contributes to the generation of static electricity, causing chokages and decreased operating efficiency. Early drying recommendations were that drying temperatures should not exceed 93°C (200°F; Bennett, 1932). The effort required to control moisture will pay dividends in the gin-operating efficiency and market value of the baled cotton (Hughs et al., 1994).

There is an optimum fiber moisture content for gin processing. Past research on seed-cotton drying established an optimum range of fiber moisture content within which satisfactory cleaning and ginning can be achieved while maintaining the inherent qualities of the fibers. The currently accepted optimum fiber moisture content lies within the 6.5 to 8.0% wet base range (Griffin and Moore, 1965; Hughs 1985). This optimum range is a compromise between effective cleaning and quality preservation on the one hand, and quality preservation and smooth ginning on the other.

Recent studies have shown that if the fiber moisture is above 8% in a universal-density (UD) bale then, during bale storage, there will be changes in the fiber color and even mold growth and further deterioration of the fiber at higher moisture levels (Anthony, 2003; Hughs et al., 2004; Chun et al., 2007). A UD bale is a package that weighs approximately 227 kg (500 lb) and has dimensions and packaging materials that have been accepted and standardized by the US cotton industry (Joint Cotton Industry Bale Packaging Committee, 2013). These studies also showed that these fiber changes resulted in lower-quality yarn when processed in the textile mill. As a result, the National Cotton Council issued a notice in 2004 recommending that gins carefully monitor bale moisture, that no bale should exceed an average of 7.5% moisture at any point within the bale, and that this recommended level is a maximum ceiling and not a target.

Drying Systems

The moisture in and on the cotton exerts vapor pressure that is a direct function of temperature. If the water vapor pressure in the surrounding air is less than the vapor pressure of the moisture in the cotton, drying will occur. Conversely, if the water vapor pressure in the surrounding air is greater than the vapor pressure of the moisture in the cotton, wetting will occur. Moisture in the air moves from the higher to the lower vapor pressure. Besides vapor pressure differences, other factors that determine rate of drying include (i) temperature of the drying air, (ii) mass of drying air applied per mass of cotton, (iii) openness of the cotton being dried, (iv) exposure time, (v) relative velocity of the drying air to the cotton, and (vi) moisture content of the cotton entering the dryer. Openness (open cotton has few clumps and is fluffed), exposure time, and relative velocity influence how efficiently the drying system uses the potential of the drying air (Hughs et al., 1994). A typical relationship between the cotton fiber moisture content and ambient relative humidity is shown in Fig. 15.

Historically, nearly all systems for drying seed cotton used the shelf-type tower dryer (Fig. 16). It operates on a parallel-flow principle where the drying air is

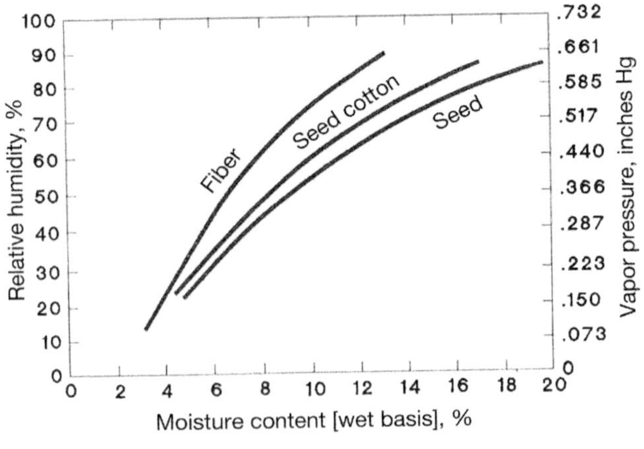

Fig. 15. Relationship between cotton fiber and humidity.

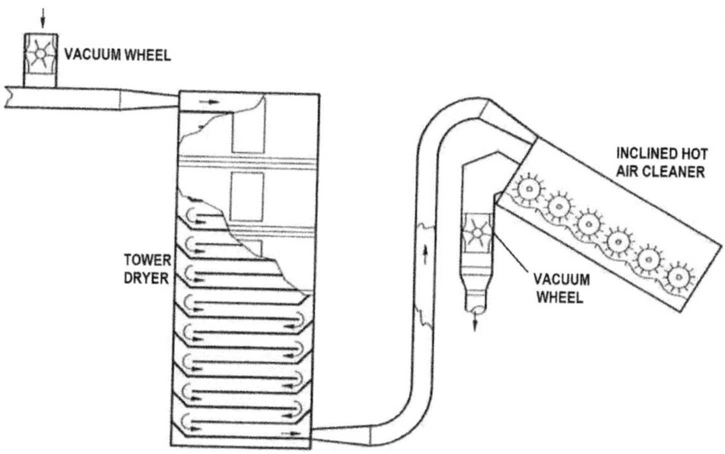

Fig. 16. Tower dryer system. (Courtesy of Ross Rutherford, Lummus Corp.)

also the cotton-conveying medium. Tower dryers vary in width, length, height, and internal shelf spacing depending on the manufacturer and the mass-flow volume of seed cotton that would be dried. Conveying air velocity through the passageways between the shelves is generally in the range of 10 m s^{-1} (2000 ft min^{-1}).

A typical single-stage drying system utilizing a tower dryer is composed of two centrifugal fans, a gas-fired burner, a tower dryer, and connecting piping (Fig. 13). The first fan, called a push fan, takes ambient air and pushes it through the burner and conveying lines to the tower dryer. The seed cotton is introduced via a vacuum dropper into the heated air stream between the burner and tower dryer (Fig. 16). The hot, conveying air transports the seed cotton through serpentine passageways in the tower dryer and delivers it to an air-fed cylinder cleaner or seed-cotton separator while the spent drying air is routed to the second centrifugal fan (pull fan) and discharged back to the atmosphere through particle abatement devices. The push-pull fan arrangement is used to overcome the airflow resistance of the tower dryer and air lines.

The purpose of the design of the tower dryer is to extend seed-cotton residence time, create differential velocities between the conveying air and cotton, and facilitate mixing of the cotton with the heated air as cotton impacts the dryer walls and changes direction between each shelf. All these factors improve the efficiency of the drying process. The dwell time for seed cotton in a drying system is dependent on both the velocity of the conveying air and the aerodynamic characteristics of the seed cotton. Well-opened seed cotton easily becomes airborne and is transported rapidly through the system, whereas damp, heavy seed cotton moves through the system more slowly. While the drying period may vary with the seed cotton's moisture content and from one drying system to another, the total exposure is seldom more than 12 s in a tower-drying system (Baker and Griffin, 1984). In practice the moisture content of the seed cotton does not reach equilibrium with the drying air. For extremely wet seed cotton, it is usually necessary to employ two stages of drying for adequate moisture control. From the standpoint of quality preservation, two stages of drying at low to moderate temperatures are preferable to a single stage at a higher temperature.

While the traditional tower dryer, with 18 or more internal shelves, has long been the industry drying standard, there are a number of other seed-cotton dryer designs that have been developed and are being used by the industry. Table 2 (Baker and Hughs, 2012) lists the common industry names and a brief description of most of the various drying systems that are being used or are available. Some of the dryer types are just variations of the standard tower but vary in complexity

Table 2. Dryer types observed in gin drying systems during the fall 2007–2010 gin dryer study.†

Dryer type	Definition
Standard tower	A device that has 18 or more internal shelves that force the cotton and heated air to frequently change directions and that force the air to pass through the cotton, increasing the exposure time and dispersing the cotton to facilitate drying. Heated airflow is concurrent with seed-cotton flow.
Short (stub) tower	Similar to a standard tower but with less than 18 shelves (typically 11 shelves or fewer). Heated airflow is concurrent with seed-cotton flow.
Hi-volume tower	Similar to a standard tower but with fewer shelves (usually 6) and much larger spacing between shelves. The volume of air is generally greater than in standard tower dryers. Heated airflow is concurrent with seed-cotton flow.
Hot shelf	A tower dryer with a split flow. Conveyed seed cotton flows through a series of shelves; however, these shelves are interspersed with heated air chambers. This heated air is transferred between chambers by U-shaped ducts at the end of the dryer. Heated airflow is separate from the seed-cotton flow.
Belt	A device that has seed cotton conveyed on a perforated belt while heated air is pushed down through the seed cotton. Heated airflow is through seed-cotton flow (cross-flow).
Big reel	A device that has seed cotton fed onto the outside of a rotating perforated drum with heated air flowing from the outside to inside of the drum. Heated airflow is through seed-cotton flow (cross-flow).
Vertical	A device which contains rotating cylinders to break up wads. A stream of heated air and seed cotton enter at the top of the dryer and fall through the dryer hitting baffles and the rotating cylinders until they exit at the bottom of the dryer. Heated airflow is concurrent with seed-cotton flow.
Hot air cleaner	Seed-cotton cleaning machine (i.e., inclined cleaner, impact cleaner) through which heated air with remaining drying potential passes. Heated airflow is across or through the seed-cotton flow.
Pipe	Standard conveying of seed cotton but with heated air. Airflow is concurrent with seed-cotton flow. Pipe length must be sufficient for effective drying, typically greater than 8 m (25 ft).
Cross-flow	A device that uses a jet of high-speed air (in excess of 50.8 m s^{-1} [10,000 ft min^{-1}]) directed horizontally across the width of a vacuum dropper to open and dry seed cotton as it accelerates into the conveying duct.
Hi-slip	A device with a spiked or lugged cylinder that slows down the seed-cotton flow, allowing heated air to pass through the seed cotton, thereby creating a high slip rate (velocity differential) between the air and seed cotton in a concurrent flow.
Hot box	A device that uses heated air to suck up seed cotton from a belt (usually at the module feeder) and convey it into a pipe. Airflow is concurrent with seed cotton flow.
Collider	A rectangular box-shaped structure in which a stream of air-conveyed seed cotton and another stream of heated air enter at the top on opposite sides. The heated air stream and seed cotton collide and tumble down through the dryer until they exit at the bottom of the dryer in a concurrent flow.
Fountain	A rectangular box-shaped structure in which a stream of seed cotton and heated air is injected into the bottom. The force of the airstream suspends the seed cotton for a short time until it falls out through the outlet on the opposite side of the bottom of the dryer in a concurrent flow.
Jet	A cylinder-shaped structure with a cone-shaped bottom in which a stream of seed cotton and heated air is injected into the bottom. The force of the airstream suspends the seed cotton for a short time until it falls back to the bottom and is collected in a concurrent flow out of the dryer.

† Baker and Hughs, 2012.

from the pipe dryer, which is just a length of pipe through which seed cotton is conveyed by heated air, to the vertical dryer, which utilizes rotating cylinders and baffles as part of the system. Regardless of the specific methods used, all systems attempt to open the seed cotton and then use some combination of exposure time, air-to-cotton velocity differential, and air temperature to dry the seed cotton to an appropriate level. Traditionally, cotton gins had at least two stages of tower drying as part of their seed-cotton precleaning system. Current ginning systems typically still have a minimum of two seed-cotton drying systems, but they may be a mix and match of any of the systems listed in Table 2.

Modern drying systems are capable of overdrying and damaging the seed cotton if they are not used properly. Drying-air temperatures in excess of 175°C (350°F) may adversely affect fiber quality, and temperatures above 230°C (450°F) will often scorch the fibers or even set them afire. Drying temperatures in all systems should be monitored by strategically located sensors and regulated by an automatic temperature controller (Baker and Griffin, 1984). The standard for dryer control sensors (ASABE Standards, 2012) recommends one sensor upstream of the seed cotton and heated-air mix point and one sensor at some distance downstream of the mix point. The upstream sensor serves as a maximum air temperature limit not to exceed a preset value, and the downstream serves as the dryer control setting that can be varied depending on incoming seed-cotton moisture conditions. These control systems, when adjusted properly, prevent serious overdrying problems but often leave much to be desired from the standpoint of precise moisture control. Drying control systems based on the measurement of seed-cotton moisture content made either before or after drying are also available. These systems provide improved moisture control during ginning and often reduce the energy requirements for drying.

Moisture Restoration Systems

Cotton harvested under low-humidity conditions may be very dry when delivered to the cotton gin. Since the fiber breakage rate during ginning is inversely related to the fiber moisture content, the average fiber length of the cotton may be maintained by adding moisture before ginning and cleaning the lint (Griffin and Moore, 1965; Leonard et al., 1979). Adding moisture to lint that has already been ginned, however, will not increase average fiber length in the bale (Griffin and Harrell, 1957). Other benefits resulting from moisture restoration include a lower level of static electricity in the seed cotton and reductions in the force required to tramp and press a bale of cotton. Ideally, the fiber moisture content should be increased to the recommended 6.5 to 8.0% level at the bale press.

One of the most common methods of raising the fiber moisture content is to expose the seed cotton to warm, humid air. This vapor method of moisture restoration may be applied at the extractor-feeders, in hoppers above the feeders, in the lint flue, at the battery condenser, or at the lint slide before pressing (these components will be discussed in greater detail later in the chapter). With the vapor system, warm, humid air is supplied by a special air humidifier. The ambient air is first heated and then drawn through a humidifying chamber equipped with water-spray nozzles. The temperature and humidity of the processed air can be varied by adjusting the burner temperature and the water-spray pressure. Typically, humidifiers of this type supply air at temperatures up to 60°C (140°F) and at relative humidity up to no more than 95% to avoid condensation on metal surfaces (Baker and Griffin, 1984). Heated humid air is somewhat self-limiting in the amount of moisture that can be added to the cotton fiber and is usually limited to no more than a 1.5 to 2% moisture increase.

Another approach that is sometimes used is to spray a fine water mist directly onto the surface of the cotton batt, typically on the lint slide that feeds the bale press. This approach has been used at least since the 1950s (Griffin and Harrell, 1957). Spray systems can easily add too much moisture and cause wet spots and fiber degradation in the cotton bale. Any moisture-addition system should be carefully monitored at the gin, but spray systems are especially susceptible to overapplication of moisture to the baled fiber.

Seed-Cotton Cleaning

Foreign Matter or Trash

Foreign matter or trash is any material in the seed cotton other than lint or seed and can consist of organic and inorganic materials. When the terms are used in regard to seed-cotton cleaning, they are, more often than not, referring to vegetative plant material in the seed cotton that needs to be removed prior to the gin stand to allow the gin stand to operate more efficiently and to maximize the market value of the lint. The cotton-plant constituents remaining in the machine-harvested seed cotton can consist of large trash, such as green bolls, burs, sticks, and leaves, and smaller trash, such as soil, pin trash (small vegetative particles from leaf, bract, or sticks that have been ground up into pepper-size particles), and motes.

Even though hand harvesting is still common in many cotton-growing countries, it has not been practiced in the United States since the early 1970s. Essentially all cotton produced there is machine harvested with pickers or strippers. The amount of foreign matter or trash in the seed cotton arriving at the cotton gin will vary significantly depending on the harvest method. Therefore, the cotton gin must be designed to handle cotton with the largest amount of foreign matter expected in the service area. Table 3 shows the typical foreign-matter content of machine-harvested cotton.

Generally, there are two types of machines used to remove foreign matter from seed cotton before ginning: cylinder cleaners and extractors. Extractors can comprise a range of machines, such as stick machines, bur machines, combination bur and stick (CBS) machines, and extractor-feeders. These machines are commonly arranged in series and can include multiple stages of both cylinder cleaners and extractors, depending on the prevalent harvesting practice of the service area.

Table 3. Typical composition of foreign matter in seed cotton harvested by mechanical harvesters.

Harvesting method	Composition of foreign matter		
	Burs	Sticks	Small trash
		%	
Machine picked	2.0	0.7	3.6
Machine stripped without field cleaner†	21.9	5.6	5.7
Machine stripped with field cleaner†	11.2	4.3	6.7

† A field cleaner is also known as a bur extractor.

Cylinder Cleaners

Cylinder cleaners, also known as horizontal cleaners or inclined cleaners, employ rotating spiked cylinders to remove fine particles and to open and prepare the seed cotton for the drying and extraction processes (Mayfield et al., 1983). "Horizontal" and "inclined" refer to how the machines are installed and operated: either horizontally or at an angle. The number of cylinders varies from manufacturer to manufacturer. Cylinder cleaners that are inclined commonly have 6 to 9 cylinders, whereas horizontal machines commonly have 12 to 15 cylinders; however, there have been horizontal cylinder cleaners with 21 cylinders installed in a ginning system. The cylinders can be spiked or fluted or have serrated flighting. The most common is the spiked cylinder (Fig. 17), which can have spikes of varying length and configuration. The cylinders, placed in series, agitate and convey the seed cotton across cleaning surfaces, which have small openings or slots (Fig. 18). The cleaning surfaces may be concave screens, grid-rod sections, or serrated disks. Foreign matter that is displaced from the seed cotton by the agitating action of the cylinders passes through the openings in the cleaning surfaces for collection and disposal.

The most commonly encountered cleaning surface used in cylinder cleaners today is the grid rod. Grid sections are normally constructed of 9.5-mm-diam. (0.375-in-diam.) rods equidistantly spaced (Fig. 18). Variations in grid-rod spacing can alter the performance of the cylinder cleaner. If the grid-rod spacing is too close, less cleaning occurs, and if the grid-rod spacing is too far apart, more cleaning occurs but the seed cotton can drop through the openings of the cleaning surface, resulting in excess waste (Laird et al., 1984).

Cylinder cleaners can also be categorized as air fed or gravity fed, on the basis of how they are used in the ginning system. Air-fed and gravity-fed cylinder

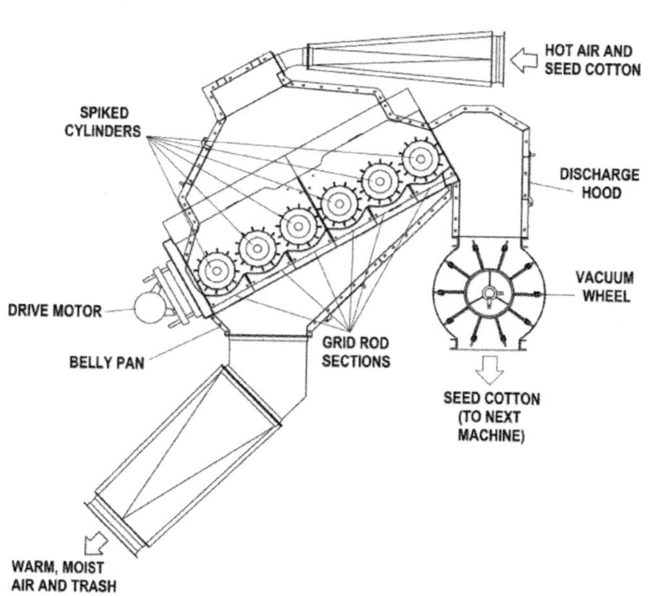

Fig. 17. Inclined-cylinder hot-air cleaner. (Courtesy of Ross Rutherford, Lummus Corp.)

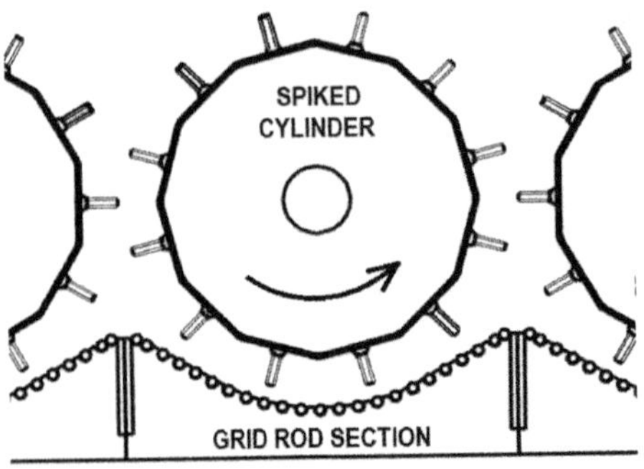

Fig. 18. Detail of inclined-cylinder grid bar. (Courtesy of Ross Rutherford, Lummus Corp.)

cleaners are frequently used immediately ahead of extractors or as finishing cleaners just before the seed cotton is fed into the gin stand by a large auger known as a conveyor-distributor. Air-fed cleaners are commonly connected to the drying system, serve as an extension of the drying system as a means for separating seed cotton from the drying air, and are called hot-air cleaners (Fig. 17). Cylinder cleaners are currently manufactured in widths ranging from 1.8 to 3.7 m (6–12 ft). Historically, rated capacities of cylinder cleaners were listed as 5 to 8 bales h^{-1} m^{-1} of width (1.5–2.5 bales h^{-1} ft^{-1} of width) (Baker et al., 1994a), but machines are currently being operated at 10 to 13 bales h^{-1} m^{-1} of width (3 to 4 bales h^{-1} ft^{-1} of width) and in some instances as high as 16 bales h^{-1} m^{-1} (5 bales h^{-1} ft^{-1}). When higher capacities are needed, machines can be run in parallel, commonly referred to as split-stream, with each machine cleaning half the seed cotton.

Extractors

Extractors are designed to remove larger foreign matter, such as burs (carpel), sticks, and stems from the seed cotton. Machines commonly considered extractors are the bur machine, stick machine, CBS machine, and extractor-feeders.

Bur Machine

The bur machine is not commonly found in modern gins today because of their limited operating capacity of 1.5 to 2.5 bales h^{-1} m^{-1} of length (0.50–0.75 bales h^{-1} ft^{-1} of length), even though some are still in use in older gins. The bur machine was one of the first extractors developed, which occurred in the 1920s. The advent of the bur machine is credited with making it possible for a gin to adequately remove the additional foreign matter that resulted with hand snapping. The bur machine was also a very useful cleaner with the advent of early mechanical strippers that did not use field cleaners to help limit the amount of trash resulting from the stripping process (Pendelton and Moore, 1967). The bur machine was the precursor to the more commonly used extractors in modern cotton gins.

The bur machine (Fig. 19) operates on the principle of dislodging foreign matter (mostly burs) from the seed cotton by knocking or stripping it out with a device known as a stripper roller. The seed cotton fed into the bur machine encounters a kicker-conveyor, which moves the seed cotton onto a large-diameter saw cylinder. The seed cotton adheres to the saw cylinder and is carried underneath a metal-flighted stripper roller rotating in the opposite direction. The metal flights on the stripper roller make contact with the foreign matter in the seed cotton and knock or strip the material out. The dislodged material finds its way

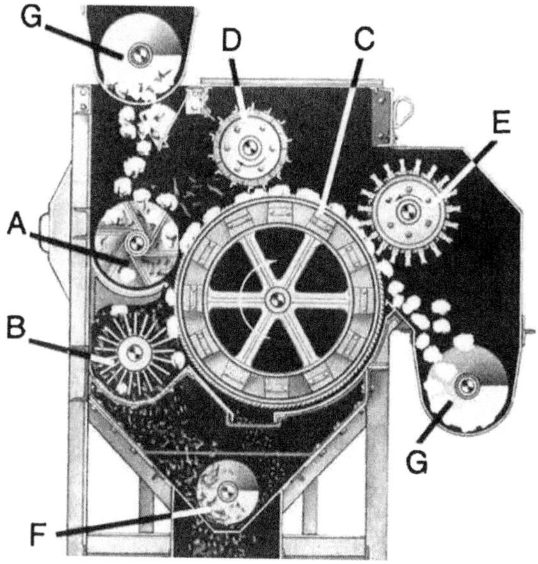

Fig. 19. Bur machine. (A) Kicker conveyor. (B) Spiked conveyor. (C) Saw cylinder. (D) Stripper roller. (E) Doffing brush. (F) Hull conveyor. (G) Cotton auger. (Courtesy of Continental Eagle Corp.)

through the incoming stream of seed cotton to the kicker-conveyer, which moves the material to one end of the machine. At this point, the material falls onto a spiked conveyer and is moved back along the entire length of the saw cylinder so that the seed cotton can be reclaimed from the dislodged material. Fine particles and dirt sift through a screened trough to an auger located underneath the spiked conveyer (Baker et al., 1994a).

Stick Machine

Unlike the bur machine, which knocks or strips the foreign matter out of the seed cotton, the stick machine utilizes centrifugal force to dislodge and throw off burs and sticks from the seed cotton (Fig. 20). The stick machine can be gravity or air fed. By far, the most common method of feeding the stick machine in modern gins is by gravity (either with a separator or cylinder cleaner). The incoming seed cotton is fed onto the main saw cylinder and pressed or wiped onto the saw cylinder teeth by one or more stationary brushes. Foreign matter and some seed cotton are slung off the saw cylinders by centrifugal forces 25 to 50 times the force of gravity (Baker et al., 1994a). Grid bars are spaced strategically about the periphery of the saw cylinder to minimize the loss of seed cotton and to aid in the extraction process. Invariably, some seed cotton will be lost during the extraction process. To recover the lost seed cotton, additional saw cylinders are used to reclaim it. The last saw cylinder is commonly referred to as a reclaiming saw. They are the same as the main saw except that they commonly operate at slower speeds and have more grid bars around the periphery. The number of saws in a stick machine is commonly used to classify the stick machine: thus, a two-saw or three-saw machine. Even though the configurations of stick machines can vary

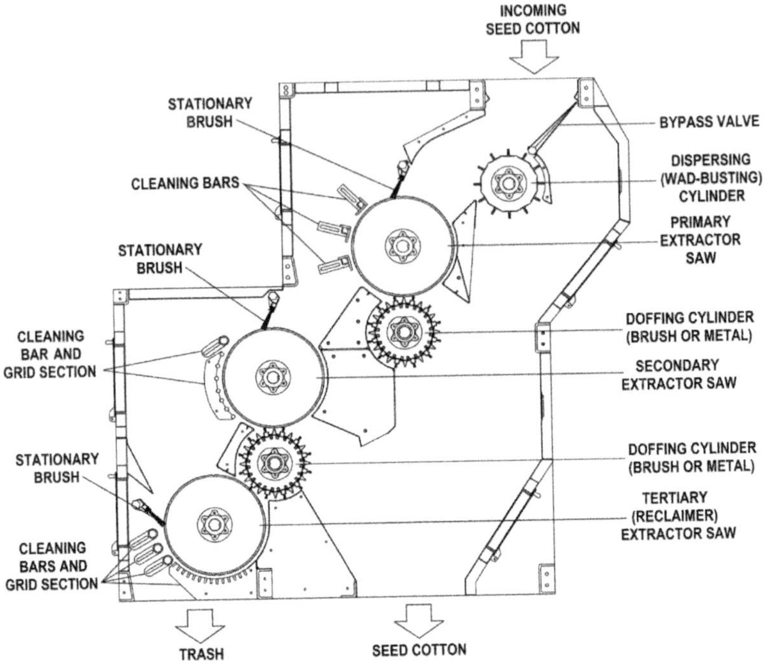

Fig. 20. Three-saw stick machine. (Courtesy of Ross Rutherford, Lummus Corp.)

in the number of saw cylinders, grid-bar configurations, saw speeds, and size and location of the stationary brushes and doffing brushes, all models will have at least one main saw and one reclaimer saw. Stick machines have widths and capacities similar to those of cylinder cleaners: 1.8 m to 3.7 m (6–12 ft) wide with operational capacities of 10 to 13 bales h^{-1} m^{-1} of width (3–4 bales h^{-1} ft^{-1} of width).

Combination Bur and Stick Machine

A commonly encountered extractor in modern gins, especially in stripper-harvested areas, the CBS is a hybrid of a bur machine and stick machine that incorporates the best features of both. Even though various trade names are used to identify these machines, the basic operating principles are the same. The upper section of a CBS machine (Fig. 21) is similar to a bur machine, having a large-diameter saw cylinder. However, the CBS differs from a bur machine in several significant aspects: (i) it has a higher capacity rating; (ii) the seed cotton is commonly fed across the entire width of the machine, as opposed to the end-feeding technique used for bur machines; (iii) it is equipped with a stationary grid bar (stripper plate) instead of a steel-flighted stripper roller; and (iv) it has a foreign-matter removal section not found in conventional bur machines (Baker et al., 1994a). Approximately half of the seed cotton and virtually all of the foreign matter are discharged from the top section of the CBS, by the centrifugal force of the saw cylinder, into the lower section. The lower section of the CBS is virtually identical to a two- or three-saw stick machine and acts as the secondary stage of cleaning, further reducing the foreign matter in the seed cotton. Combination bur and stick machines are available in

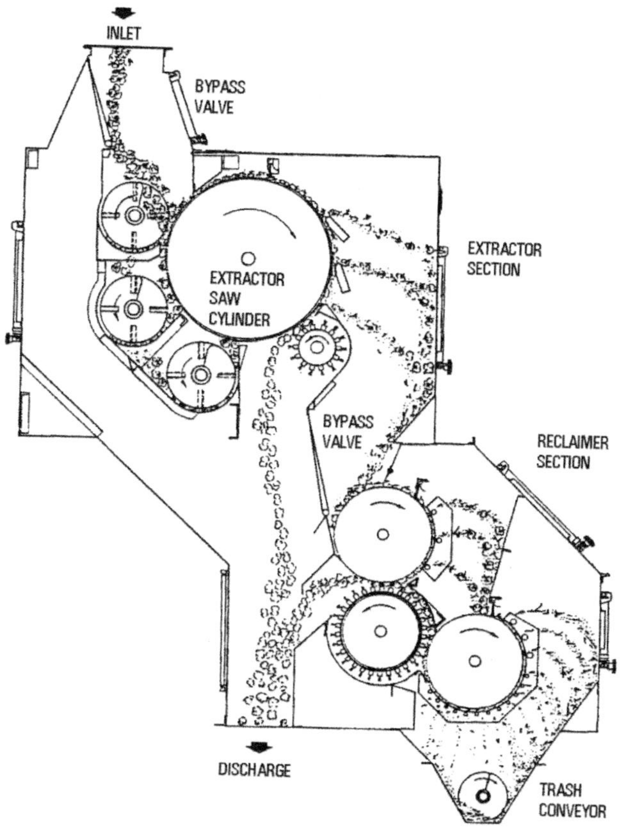

Fig. 21. Lummus CBS machine. (Courtesy of Ross Rutherford, Lummus Corp.)

widths ranging from 1.8 m to 3.7 m (6–12 feet), with operational capacities of 10 to 13 bales h^{-1} m^{-1} of width (3–4 bales h^{-1} ft^{-1}) of width.

Extractor-Feeders

Extractor-feeders can be combinations of cylinder cleaners and stick machines and have the primary function of feeding seed cotton into the gin stands at controlled and uniform rates. The extractor-feeder was designed to disperse the seed cotton into single locks of cotton to help even out the flow and minimize the number of wads or clumps entering the gin stand and causing operational problems. Although cleaning of the seed cotton is a secondary function of the extractor-feeder, it is the last opportunity for the seed cotton to be cleaned before reaching the gin stand.

Extractor-feeders are located above each gin stand under the distributing conveyor (a large auger that takes seed cotton from the final seed-cotton cleaners and distributes it to one or more extractor-feeders) and have two feed rollers located at the top to regulate the flow of incoming seed cotton (Fig. 22). The feed rollers are driven by variable speed motors controlled either manually by the ginner (operator) or automatically by a control system interconnected with the gin

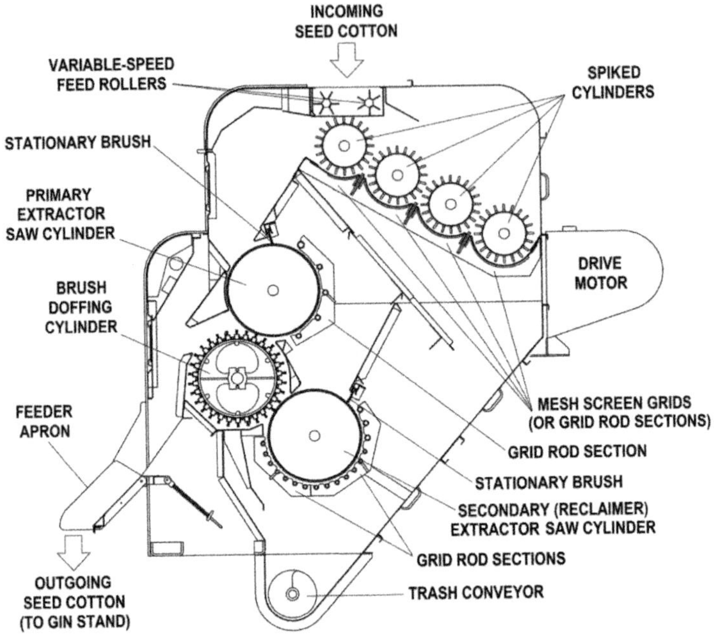

Fig. 22. Extractor-feeder. (Courtesy of Ross Rutherford, Lummus Corp.)

stand. The automated system is designed to maintain a constant load on the gin stand that in turn helps maintain more consistent ginning for increased capacity and better fiber and seed quality.

Gin Stands

The Saw Gin

Saw gins are most commonly used to gin Upland cotton. The concepts used in the designs of Eli Whitney and Hogden Holmes are still employed in modern gins today but with greater efficacy. The primary function of modern gin stands is the same as that of the first gin stands: removal of the lint from the seed. However, unlike the earlier models, the highest-capacity gin stands today have rates of 30+ bales h^{-1} and have numerous saws; the largest model currently being manufactured has 279 saws. Even though gin stands vary in width and the number of saws and have other differences, the basic operating principle is the same (Fig. 23). Seed cotton from an extractor-feeder is fed into the front of the gin stand, historically referred to as the huller front or huller section, onto a picker roller that feeds the seed cotton onto the saws. The saws carry the seed cotton into the seed roll compartment. The teeth of the gin saws grip the seed cotton and pull the cotton fibers (i.e., lint) through a narrow slot (3.3–3.8 mm wide [0.13–0.15 in]) between the ginning ribs, near the top of the each rib. As the lint captured in the teeth of the saw passes through the gin ribs, the cottonseed cannot pass through, causing the fibers to be separated from the seed (i.e., ginned). Once the lint is removed, the seeds exit the seed roll section and slide down the face of the gin ribs and

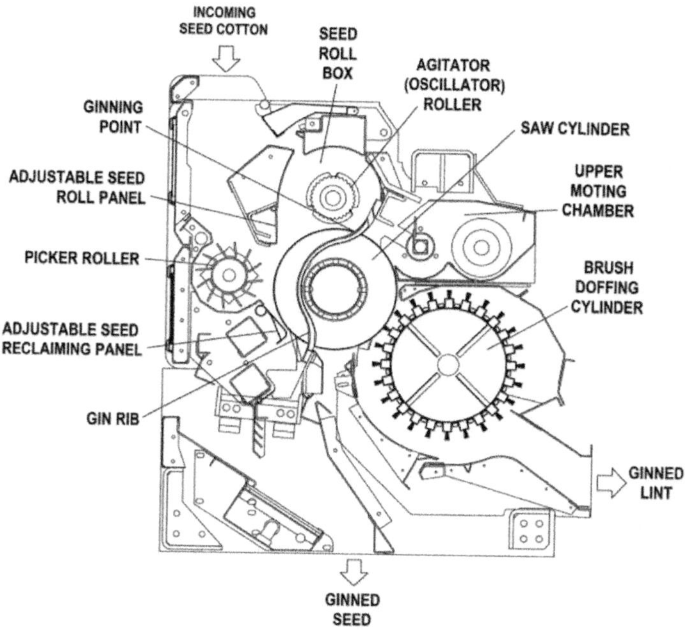

Fig. 23. Cross-section of saw gin stand. (Courtesy of Ross Rutherford, Lummus Corp.)

out the bottom of the gin stand. The ginned lint is then carried by the saw into close proximity with a mote board for removal of foreign matter and motes. The lint is removed (doffed) from the gin saws by a brush cylinder into an air stream that carries the fibers to the next stage of processing for additional cleaning (lint cleaning). Gin stands generally process from 4000 to 5000 bales or more per stand before their saws are replaced. The frequency of saw replacement depends on the foreign matter content (trash) and quality of the seed cotton being ginned.

The Roller Gin

Roller ginning has long been the preferred method for ginning extra-long-staple, fine-fibered Sea Island, Egyptian, American-Egyptian, and Pima cottons (Bennett, 1956). Although it is possible to gin these types of cotton with a saw gin, the resulting quality is substantially lower than that obtained with roller gins. Saw ginning tends to reduce the fiber length of these types of cotton and to greatly increase their nep (small knots or tangles found in ginned fiber) content (Chapman and Stedronsky, 1965). As stated earlier, the rotary-knife roller gin design accounts for 100% of the roller gins being used in the United States due to its greatly increased production capacity compared with the reciprocating-knife roller gin.

Cottonseed Handling

The cottonseed falls from the gin stand onto a belt or screw conveyor, and then the seed is picked up and conveyed, usually pneumatically, to an on-site storage facility, most commonly an elevated or on-ground seed house (Fig. 24) where the seed can be kept dry until shipped from the gin. Cottonseed is commonly

Fig. 24. Overhead seed house.

conveyed from the gin to the storage facility by a pneumatic system powered by a positive-pressure, rotary, two-lobe air pump that operates at static pressures ranging from 14 to 34 kPa (2–5 psi; Fig. 25). The amount of cottonseed produced from a bale of lint has decreased steadily over the past few decades (Fig. 26), from an average of approximately 374 to 317 kg bale^{-1} (825–700 lb bale^{-1}). The change in seed weight corresponds to varietal changes that have produced smaller seed sizes while improving fiber properties. The smaller seed sizes impact the cotton gins from the standpoints of fiber quality and profit. Fiber-quality issues have resulted because some of the newer small-seed varieties have fragile seed coats that can come off the cottonseed and be pulled through the gin ribs and crushed

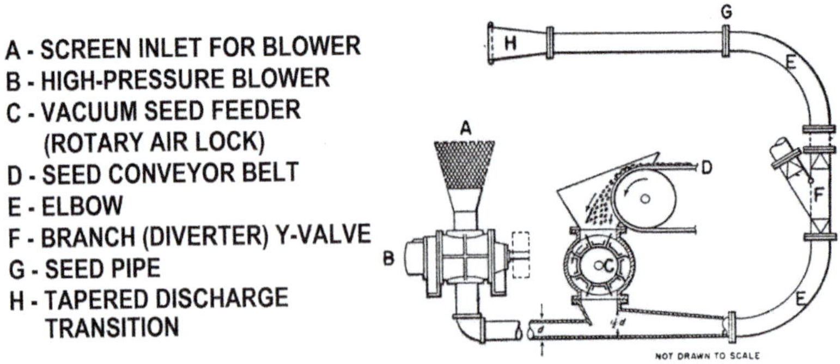

Fig. 25. Rotary air-pump system. (Courtesy of Ross Rutherford, Lummus Corp.)

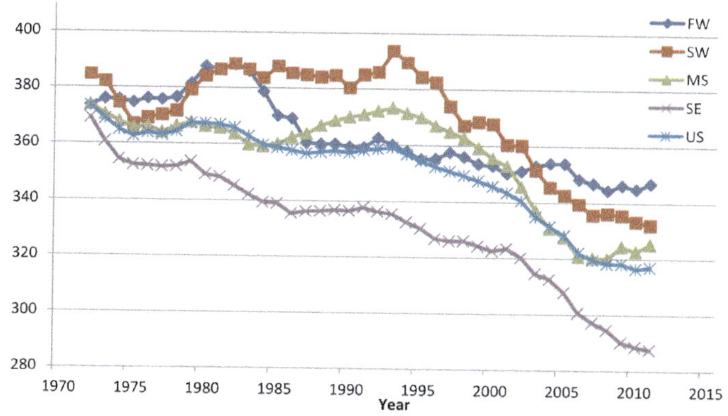

Fig. 26. Cottonseed weights per bale. Cottonseed weights per 227 kg (500 lb) bale of lint from 1970 to 2011 for the total United States crop (US) and for each of the four cotton growing areas in the United States: Far west (FW), Southwest (SW), Midsouth (MS), and Southeast (SE). (Data from National Agricultural Statistics Service.)

into or entangled in the lint (i.e., seed coat neps) in downstream processes. Profit issues can result from losses in cottonseed tonnage from these smaller seeds. Ginning for seed is a custom that is still practiced by some gins in the South and Southeast. For those cotton gins that keep the cottonseed (i.e., gin for seed), smaller seeds result in less tonnage per bale processed and therefore less profit. However, the predominant means of payment for ginning across the US Cotton Belt is fee for services.

Lint Cleaning

Lint cleaning at the cotton gin developed concurrently with and in response to the widespread adoption of mechanical harvesting in the United States (Stedronsky and Shaw, 1950; Gerdes, 1951). Both flow-through-air-type and saw-type lint cleaners do additional cleaning after ginning to achieve maximum returns for the producer and acceptable levels of cleanliness for the spinner (Mangialardi et al., 1994). The lint cleaner's ability to remove small particles of leaves, motes, grass, and bark fulfills a need that cannot be obtained with seed-cotton cleaning alone. All gins in the United States have lint cleaners, but the number and type of lint cleaning varies depending on the type of cotton being produced and the growing area that the gin services.

The Pneumatic (Flow-Through Air) Cleaner

The pneumatic lint cleaner is used to clean both saw-ginned Upland cottons and roller-ginned Pima cotton. For Upland cottons, the pneumatic lint cleaner is installed directly behind a saw gin stand and is usually followed by one or more stages of saw lint cleaning. Lint from the gin stand is conveyed by air through a specially shaped duct in the pneumatic lint cleaner. The duct is designed to force the flow of air and cotton to change direction abruptly as it passes a narrow trash ejection slot (Fig. 27). Foreign matter is ejected through the slot by inertial forces. The degree of foreign-matter separation can be regulated to some extent by varying the width of the adjustable trash ejection slot. For lint cleaning roller-ginned

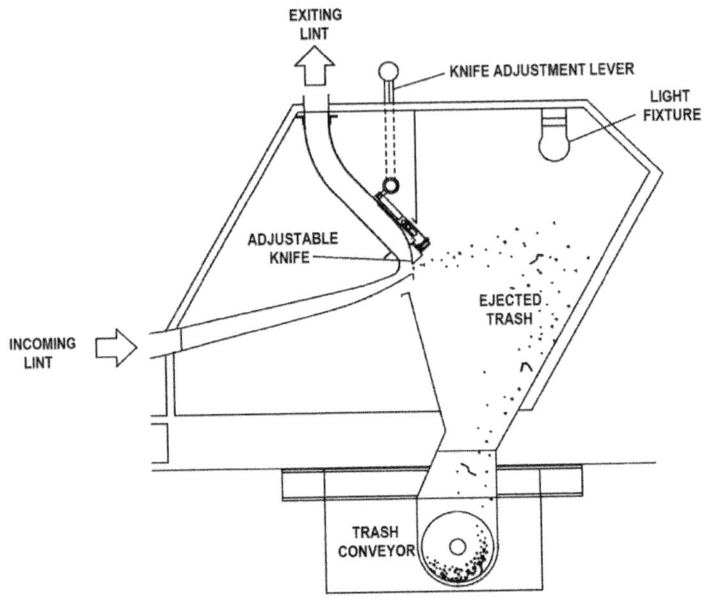

Fig. 27. Pneumatic lint cleaner. (Courtesy of Ross Rutherford, Lummus Corp.)

Pima cotton, the pneumatic lint cleaner is usually used in combination with one or two cylinder cleaners (Hughs and Gillum, 1991; Whitelock et al., 2007). Pneumatic lint cleaning is less aggressive than saw-type lint cleaning and has less fiber damage for the trash removed (Mangialardi and Anthony, 1998).

Saw-Type Cleaners for Upland Cotton

Saw-type lint cleaners are used for cleaning saw-ginned Upland cottons (Fig. 28). Commercial lint cleaners, although variable in size and design, are all equipped with either a relatively low-rpm batt-forming lint condenser or a much-higher-rpm, high-speed separator cylinder that keeps the ginned lint in single small tufts (batt-less lint cleaners). If a lint condenser is used, then the saw-type lint cleaner has a lint-batt feeding mechanism, a saw cylinder that is 30.5- to 60.9-cm (12- to 24-in) in diameter, grid bars, and a lint-doffing mechanism. If using a high-speed separator cylinder, then the separator feeds the lint directly onto a saw cylinder, guided by a solid feed plate, and the cleaned lint is removed from the saw cylinder by a lint-doffing mechanism.

The current typical lint cleaning practice is to use one saw-type lint cleaner on relatively clean or early-season cotton, and two lint cleaners in tandem on trashier or late-season cotton. For saw-type lint cleaners using a batt-forming condenser, lint from a gin stand or another lint cleaner is conveyed by air onto the condenser, which separates the lint from the air and forms the lint into a batt weighing about 0.2 to 0.4 kg m^{-2} (0.04–0.08 lb ft^{-2}). The batt is then fed through one or more sets of directional rollers to a closely fitted feed roller and matching concave feed bar. The feed bar is positioned directly above the center of the saw cylinder at a distance of 1.5 to 3.0 mm (0.06–0.12 in) from the saw teeth tips. The

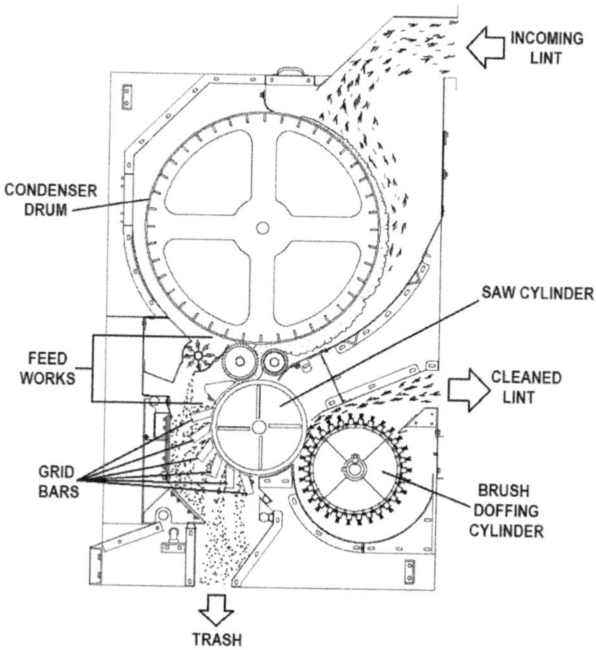

Fig. 28. Saw-type lint cleaner. (Courtesy of Ross Rutherford, Lummus Corp.)

surface speed of the saw cylinder is substantially faster than that of the feed roller, creating a combing action as the batt is opened and fed to the saw teeth. The feed roller is spring-loaded to hold the batt firmly against the feed bar to enhance the combing action. Combing ratios (ratios of saw-surface speed to feed-roll-surface speed) of 15:1 to 30:1 are commonly employed. For saw-type lint cleaners using a high-speed separator cylinder, there is a small surface-speed differential between the separator and the saw, which results in some combing that is enhanced by the solid feed plate, which has a small clearance from the saw teeth.

For any saw lint cleaner, the saw cylinder, which turns at 800 to 1300 rpm, whips the fibers across the keen edges of several grid bars, usually four to eight, to discharge the foreign matter by a combination of centrifugal force and the stripping action of the grid bars. The foreign matter is collected by an air stream washing across the front of the cleaning grid bars and is conveyed from the machine. The cleaned fibers are removed from the saw teeth by a doffing brush and conveyed by air to the next stage of lint cleaning or to the packaging system.

Gins handling mechanically harvested Upland cotton are normally equipped with one or two stages of saw lint cleaning or have the ability to change from one to two stages depending on the amount of lint cleaning needed to meet market demand. Saw-type lint cleaners may be employed as single or tandem units directly behind each gin stand, or several gin stands may feed into one large-capacity lint cleaner commonly known as a battery lint cleaner. A pair of lint cleaners is often arranged in parallel as a means of increasing the capacity of battery installations, or as an alternate split-stream arrangement in lieu of tandem lint cleaners behind a gin stand.

A variation of the single-saw lint cleaner is the multi-saw lint cleaner (Sutton and Horn, 1986). The multi-saw utilizes two types of traditional saw cylinders with grid bars within the same machine (Mangialardi et al., 1994). The first stage of cleaning is performed by a controlled-batt saw cleaner. Cleaned lint from the first saw cylinder is transferred directly onto the second saw cylinder for a second stage of cleaning. After the second stage of lint cleaning, the cotton is doffed by a brush and conveyed to the next stage of processing.

Saw-type lint cleaners are precision machines that operate with close tolerances. Smooth and efficient operation requires careful attention to a number of exacting mechanical adjustments, as well as the proper selection or control of several operating parameters. Cotton batt thickness, combing ratio, saw speed, feed rate, and fiber moisture content are all factors that influence the performance of saw-type lint cleaners and their effect on fiber quality (Mangialardi, 1974; Baker, 1978). Excessive lint cleaning can increase fiber loss to the trash and decrease bale weight to an extent not compensated for by improvements in grade, thereby reducing income for the producer (Baker et al., 1977; Mangialardi and Anthony, 1998). Excessive lint cleaning may also degrade the fiber length and lower the cotton's spinning potential (Bel et al., 1991)

The material removed from the lint by saw-type lint cleaners is a unique product. This material is called mote and is marketable. Motes from saw-type lint cleaners are collected separately by pneumatic systems and in most saw gins are delivered to a baler, where they are packaged. These motes are marketed for various uses, typically in nonwoven applications. This material is not considered waste, but rather a marketable by-product, more similar to the cottonseed discussed above.

Cylinder-Type Cleaners for Roller-Ginned Cottons

A survey taken by Hughs and Gillum (1991) showed that the most common cleaning sequence in US roller gins was a pneumatic lint cleaner (described above) followed by one impact and one spiked-cylinder cleaner. A later roller-gin survey (Whitelock et al., 2007) showed the vast majority of roller gins still used a pneumatic lint cleaner but followed it with either one or two spiked-cylinder cleaners. The significance of using cylinders instead of saw-type lint cleaners for roller-ginned cottons is that cylinder cleaners are less effective but do significantly less fiber damage than saw-type lint cleaners. This is important because roller gins are used on longer-staple cottons to preserve their fiber length. Thus it would be counterproductive to use any cleaner that might shorten the fiber length during the lint-cleaning process. The spiked-cylinder cleaners used for lint cleaning are of the same general design as those used for seed-cotton cleaning. However, when used for lint cleaning, the cylinder speed is usually increased and the cleaning grids or screens are set closer to the spike tips. The higher speed and reduced operating clearance improve lint cleaning efficiency without any measurable increase in fiber damage.

Textile-Type Controlled-Batt Cleaner for Roller-Ginned Cottons

Since the survey by Whitelock et al. (2007), roller-gin lint cleaning has continued to evolve in a quest for more efficient and higher-capacity methods of lint cleaning that do not significantly damage roller-ginned fiber. Historically, along with the advent of the rotary-knife roller gin, textile-mill-type lint cleaners (sometimes

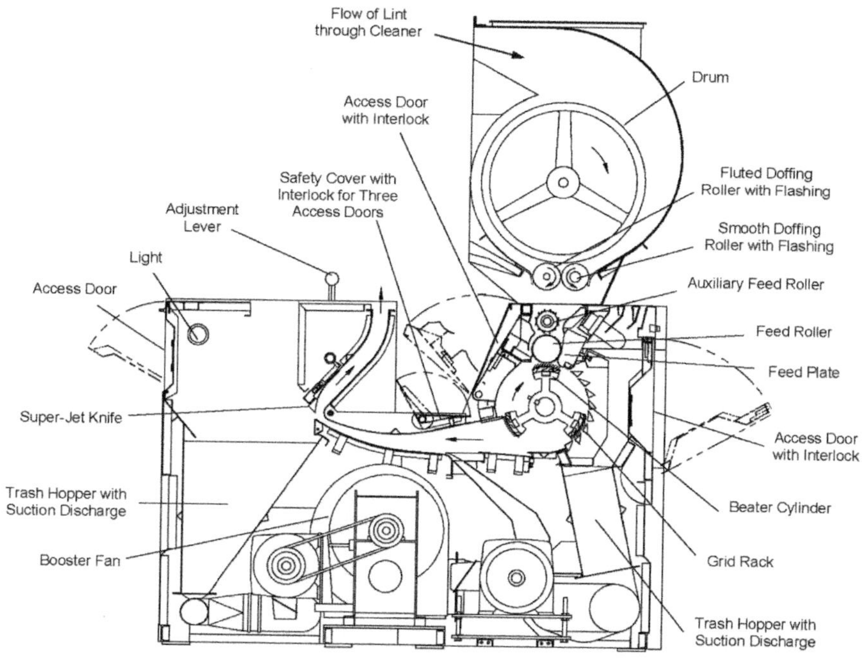

Fig. 29. Modern mill-type lint cleaner. (Courtesy of Ross Rutherford, Lummus Corp.)

called Aldrich or Buckley beater-type opener-cleaners) were applied in roller gins to clean the roller-ginned lint (Alberson and Stedronsky, 1964; Chapman et al., 1968; Kirk and Leonard, 1977). These lint cleaners were efficient and did not damage fiber excessively but were limited in their throughput volume and were abandoned in favor of the cylinder cleaners reported by Whitelock et al. (2007). The mill-type lint cleaner has been recently redesigned for a higher capacity and is currently being used in most of the roller-ginning plants in the western United States (Fig. 29). This cleaner is a controlled-batt, textile-type lint cleaner that utilizes a beater-type trilobe textile-type spiked cylinder with multiple grid bars and discharges into an integral pneumatic lint cleaner.

Handling and Packaging of Lint

Lint-Conveying Systems, Condensers, Lint slides, and Presses

Modern saw gin stands use a revolving brush cylinder to doff lint fibers from the gin saws (Baker and Griffin, 1984). The revolving brush also generates a stream of air that transports the lint through a sheet-metal duct, sometimes passing first through a pneumatic cleaner to a lint condenser or a high-speed separator cylinder at the next stage of processing. The condenser removes the lint from the conveying air, forms the lint into a batt, and delivers the batt to the feed works of a lint cleaner. A suction fan in the air-discharge line of the lint condenser enhances the conveying process by helping the brush overcome the airflow resistance of the condenser, ductwork, and any abatement equipment associated with the system.

After cleaning, the lint is doffed from the lint cleaner saws and transported by air to the next stage of lint cleaning, or to a battery condenser at the baling press.

Roller-ginned lint may be delivered to a condenser and dropped either into a cylinder cleaner or a cylinder separator-cleaner, similar to the air-fed cleaner used for seed-cotton cleaning, depending on the type of lint cleaning being used. The cylinder separator simultaneously removes the lint from the conveying air while mechanically cleaning and delivering the lint to the next processing stage. After cleaning, lint from roller-gin lint cleaners is also transported by air to a battery condenser at the press for packaging.

In all cases, to package the lint, the battery condenser at the press forms the lint into a batt and discharges the batt onto an inclined metal slide attached to a lint feeder on the bale press. The lint feeder, which may be a revolving kicker, a belt conveyor, or a hydraulic pusher, deposits the lint into the charging box of the press during the upstroke of the tramper. The tramper is a mechanical or hydraulic device that uniformly packs the lint into the press box in preparation for pressing.

The US ginning industry, through the actions of the Joint Cotton Industry Bale Packaging Committee (JCIBPC, 2014), has adopted the UD bale. The UD bale is produced in a press box whose dimensions are approximately 53 cm wide by 140 cm long (21 in wide by 55 in long). A typical UD baling press is composed of two press boxes and can be either completely above the floor level (down-packing) or partially below floor level (up-packing; Fig. 30). While the cotton is being pressed and packaged in one of the boxes, the other box is being filled and loosely packed by the tramper. After enough lint has accumulated to make approximately a 227-kg (500-lb) bale, the "just-filled" box is swung into position for pressing, and the other, "now empty" box is rotated into the filling and tramping position. For pressing, a hydraulic cylinder forces the cotton into the typical, rectangular UD

Fig. 30. Down-packing bale press. (Courtesy of Ross Rutherford, Lummus Corp.)

bale shape so that it can be tied with bale ties 221 to 226 cm (87 to 89 in) in length and then wrapped.

Bale Density, Shape, and Size

The UD bale has a nominal finished and tied density of 448 kg m^{-3} (28 lb ft^{-3}). The JCIBPC (2001) established guidelines for the UD bale, which is produced at the cotton gin with a UD press. Finished UD bales are to be approximately 140 cm (55 in) long, 53 cm (21 in) wide, and 84 cm (33 in) thick. Several other bale configurations that are listed by the JCIBPC vary from the UD bale in density and dimensions and are used in minor applications. However, the UD bale is the configuration of the vast majority of the cotton bales produced in the United States.

Wrapping and Packaging Materials

Once the cotton bales are pressed, they must be tied, either manually or mechanically, to hold their required shape. The bales are then wrapped with bagging material to protect the ginned cotton from dirt and other contamination during subsequent handling and shipping. Uniform density bales must be tied with either six or eight ties, depending on the type of tie used, to prevent tie breakage during shipping and handling. The JCIBPC (2014) gives the specifications for a range of acceptable bale-tie materials, which include metal ties of various designs and materials as well as approved plastic strapping designs. These tie materials and application methods have all undergone extensive field-testing and have proven their ability to function as required.

Bale ties are always applied to the ginned bales before they are removed from the bale press. Generally cotton bales are wrapped or bagged with protective wrapping materials at the pressing point, just after the bale is tied and removed from the press. The JCIBPC (2014) provides an extensive listing of the required specifications of tested and approved bale-covering materials, which include natural products, such as burlap and cotton, as well as man-made products, such as polyethylene and polypropylene. Some of the basic considerations for bale-covering materials are that they have the strength and construction to withstand handling without significant deterioration or damage over a length of time. These bagging materials must be able to keep the ginned-cotton bale from being contaminated and not be sources of contamination themselves. Since cotton bales are traded based on weight, the approved bagging and tie materials must also meet certain official tare-weight standards to be approved for marketing (JCIBPC, 2014). In the future, new bagging and tie materials can be developed and approved by working with the JCIBPC and its standardized testing procedures.

Waste-Handling and Trash Collection Systems

Waste material removed from the cotton by the gin's various cleaning processes is transported to outdoor collection facilities by perhaps 7 to 15 separate and unique pneumatic conveying systems, depending on the design of the particular gin plant. These systems require large volumes of air. The total airflow requirements for handling material at a gin varies depending on the gin's operating capacity and the foreign matter content of the cotton, but they can easily exceed 4248 m^3 m^{-1} (150,000 ft^3 min^{-1}; Baker et al., 1994b). The amount of waste handled per hour is also a function of ginning capacity and the cotton's foreign matter content, which depends on environmental conditions and the type of harvesting equipment used. On a per-bale basis, the waste from picker-harvested cotton can average 45.4 to 68 kg (100 to 150 lb) of trash or more per bale, whereas the waste from stripper-harvested cotton can vary from 127 to 363 kg (280 to 800 lb) of trash per bale, depending on whether the stripper had a field cleaner (Baker et al., 1994a; Wanjura et al., 2010, 2012). A gin's waste-handling system must be designed to (i) remove the waste from the gin building at a rate commensurate with ginning capacity, (ii) collect the waste from the conveying air streams without causing excessive air pollution problems, and (iii) consolidate the waste in an acceptable manner.

Trash Collection Systems

The most widely used device for collecting waste and controlling particulate emissions at cotton gins is the cyclone collector (Baker and Griffin, 1984; Whitelock et al., 2009). This simple device is very effective, has no moving parts, uses centrifugal force to separate waste from the conveying air, and requires very little maintenance. The two most common types of cyclones used in cotton gin applications are the 2D2D and 1D3D. The first *D* in the designation indicates the length of the cyclone barrel relative to the cyclone barrel diameter, and the second *D* indicates the length of the cyclone cone relative to the cyclone barrel diameter (Whitelock et al., 2009; Fig. 31). Thus, the lengths of the barrel and the cone of 2D2D cyclones are equivalent to twice the diameter of the cyclone barrel. The 1D3D cyclones have a barrel length equal to the cyclone barrel diameter and a cone length equal to three times the cyclone barrel diameter.

The particular cyclone design selected for gin exhausts depends on the nature of the particulate being controlled and the pertinent particulate emission regulations of the state or region in which the cotton gin is located. The requirements for controlling particulate emissions vary somewhat across the US Cotton Belt (see "Environmental and Safety and Health Regulations," below), which can impact cyclone selection. It has also been shown that 1D3D cyclones are slightly more efficient at separating fine particulates from an air stream than are 2D2D cyclones but require additional fan horsepower and, therefore, more energy to operate (Parnell et al., 1994; Wang et al., 2002a, 2002b). Additional research on cyclone inlet, trash outlet, and air outlet designs, as well as overall cyclone construction practices, has resulted in appropriate cyclone variations for a given application at a cotton gin (Baker and Hughs, 1999; Baker et al., 1997; Whitelock et al., 2009).

An important design parameter for the application of either the 2D2D or 1D3D cyclone design is that the cyclone diameter be sized properly for the volume of air being controlled. The design inlet-air velocity of a 2D2D cyclone is a nominal 914 m min^{-1} (3000 ft min^{-1}) while the design inlet-air velocity of a 1D3D is a nominal 975 m min^{-1} (3200 ft min^{-1}). Depending on the airflow being controlled and the diameter of the cyclone selected, cyclones may be in single, double, or quadruple configurations for a single exhaust. For example, an airflow rate of 0.47 m^{3} s^{-1} (1000 ft^{3} min^{-1}) can be efficiently controlled by either a single 46-cm (18-in) diameter 1D3D or a single 51-cm (20-in) diameter 2D2D cyclone. In the same manner, a 20.8 m^{3} s^{-1} (44,000 f^{3} min^{-1}) air stream can be controlled by either a set of quadruple 158-cm (62-in) diameter 1D3D cyclones or a quadruple set of 163-cm (64-in) diameter 2D2D cyclones (Whitelock et al., 2009). Current diameters of

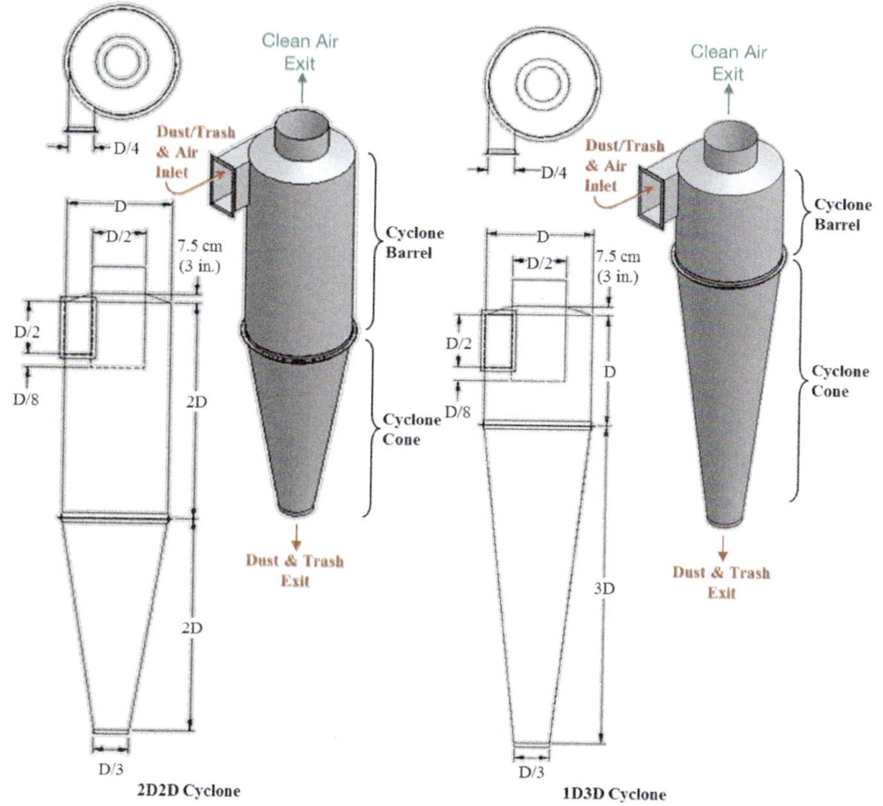

Fig. 31. 2D2D and 1D3D Cyclones.

cyclones for application in cotton gins range from 46 to 203 cm (18–80 in). Cyclone selection involves not only consideration of the air volume but also the existing infrastructure, amount of trash handled, and cost. Also, depending on the equipment layout of a particular gin plant and because of the inherent flexibility of air-pipe layouts, cyclones for multiple air outlets tend to be mounted in a centralized location in one or two racks (Fig. 32).

Pre-separators, or settling chambers, are sometimes used by cotton gins in series before cyclones to collect large waste material (Baker et al., 1997). The purpose of these settling chambers is to settle out the large trash, particularly in stripper gins, before it gets to the cyclone, to reduce the average loading rate on the cyclone group. Research has shown that a settling chamber preceding a single cyclone that is properly sized to the airflow does not improve the particulate collection efficiency of the cyclone (Baker et al., 1997). However, experience in commercial cotton gins has shown that a single large settling chamber receiving exhaust from multiple systems and preceding properly sized multiple cyclones does reduce the visible particulate emissions from a cotton gin. At gins, problems can arise both with very small cyclones, such as material choking the smaller inlets and exits, and with very large cyclones, such as with the support

Fig. 32. Cyclone racks.

infrastructure needed to accommodate the large diameter and height. The settling chamber balances the flow to all the cyclones it feeds, allowing for more uniform operation. Settling chambers are used in areas that have relatively large amounts of sand and soil particles or plant material coming in with the harvested seed cotton. Removal of these materials before the cyclone step reduces wear on and prolongs the life of the cyclones.

Although cyclones are the predominant particulate-control device on gin exhausts, cotton gins in some areas use condenser drums covered with a fine mesh screen instead of cyclones to control emissions from the lint-cleaner or battery-condenser exhausts. Depending on the application, the size of the opening in the fine mesh screen is typically 0.21 by 0.21 mm, 0.177 by 0.177 mm, or 0.149 by 0.149 mm (70, 80, or 100 mesh). Tests have shown that short-fiber emissions, called lint fly by the industry, can be reduced substantially with the use of the fine mesh screen, where allowed. An advantage of using the condenser, rather than cyclones, to control emissions, is that when it is covered with fine mesh screen, it has a much lower pressure drop than a cyclone and takes less energy to operate (Parnell et al., 1994).

Other particulate control devices, such as bag houses or rotary drum filters, have been tried on cotton gins but have been proven to be unsatisfactory because of their high initial capital, operational, and maintenance costs relative to those of cyclones. Cotton gins are a seasonal, off-farm, agricultural operation that run only for 8 to 16 wk per year and sit idle the rest of the time. All of the equipment in a gin plant, including particulate emission control devices, must be durable and easily maintained.

Waste and By-product Disposal and Utilization

Historically, the vegetative by-products—burs (cotton carpel), sticks, leaf, and some cotton fibers—produced from the cotton gin have been called gin trash, or gin waste. Over the years, various means of disposing of these by-products have been employed, including on-site incineration (abandoned in the 1960s), application back to growers' land, landfilling, and use as bedding for dairy cattle and feed ingredients for livestock. There is no universal utilization method in the US Cotton Belt. Cotton gins employ a combination of methods depending on the region of the country and local practices (Holt et al., 2000). In an effort to minimize waste and improve sustainability, there has been a concerted effort over the years to use the by-products generated from cotton gins more beneficially (Thomasson, 1990). More recently, Holt and Laird (2002) and Holt et al. (2010) developed processes that add value to the cotton gin by-products by converting them to partial or main ingredients in diets for ruminants (Holt et al., 2003); erosion-control products, such as hydromulch (Holt et al., 2005a, 2005b; Scholl et al., 2012, 2013); fuel pellets for pellet stoves and hot water boilers (Holt et al., 2004, 2006); filler fibers in thermoplastic composites (Bajwa et al., 2011, 2014; Holt et al., 2014); and natural fiber substrates in eco-friendly composites, such as molded packaging material (Holt et al., 2012) and acoustic absorbing panels (Pelletier et al., 2013). In light of the increasing costs of producing and processing cotton and a growing global emphasis by industrial manufacturers to utilize renewable resources, such as agricultural substrates and by-products, the cotton-gin by-products and cotton residues left in the field (Wanjura et al., 2014) are an underutilized material that can generate additional revenue while providing quality natural-fiber substrates to manufactures looking to become more eco-friendly.

Environmental and Safety and Health Regulations

Many workplace (safety and health), environmental (air, water, and solid waste), biosecurity, and other regulations apply to cotton gins and need to be addressed to ensure compliance (P.J. Wakelyn, pers. comm., 2007). Most of the environmental and workplace legislation was first developed in about 1970 in response to greater general public awareness of and concern and desire for a cleaner environment and safer workplace. As a result, the number of workplace and environmental regulations has risen. Biosecurity and agroterrorism regulations were enacted in the wake of the 11 Sept. 2001 attack on the United States. Each cotton ginning facility is expected to establish programs to ensure proper compliance with the various regulations and ensure that proper corrective actions are taken and written protocols are developed where necessary. Programs can include engineering controls, written programs, auditing, record keeping, reporting, warning labels, worker training programs, etc., according to the pertinent regulations. The latter are enforced by the appropriate federal and/or state agency.

Environmental Regulations

The USEPA regulates environmental pollutants, and specifically regulates criteria air pollutants under the National Ambient Air Quality Standards (NAAQS; Wakelyn, 1995). Particulate matter (PM), one of six criteria pollutants regulated under the NAAQS, is the generic term for dust and other particles in the air. Most states require state permits for facilities that emit air pollutants. In some states, federal permits and state permits are combined, while in other states facilities are required to have both a state or county (air district) permit and a federal permit, if the facilities are large enough to meet the criteria. Although these regulations vary from state to state, all US cotton gins are now required to control particulate emissions. Some states have specific regulations for cotton gin emissions, while others rely on general regulations covering all agricultural processors. The primary standard of regulation for a cotton gin is weight of particulate emitted per bale ginned (Hughs et al., 2008a). The initial emission factor used in permitting most cotton gins was the weight of total particulate emitted per bale (USEPA, 1978). When the NAAQS incorporated PM_{10} (particles with a diameter equal to or less than 10 microns), the emission factor was updated to reflect this change (USEPA, 1996).

The EPA promulgated the NAAQS for $PM_{2.5}$ (particles with a diameter equal to or less than 2.5 microns) on 18 July 18 1997 (62 FR 38652, 18 July 1997). In 2012, the EPA implemented a more stringent standard for $PM_{2.5}$ (USEPA, 2006).

There are two types of air-quality standards. Primary standards are designed to protect the public health, and secondary standards are designed to protect the public welfare, such as visibility and property. Currently, the national primary and secondary air-quality standards limit the average annual $PM_{2.5}$ concentrations in an area to less than 12 and 15 $\mu g\ m^{-3}$, respectively. The primary and secondary standards for $PM_{2.5}$ concentrations averaged across a 24-h period are both 35 $\mu g\ m^{-3}$. There are no annual standards for PM_{10}, but the 24-h primary and secondary standards are both 150 $\mu g\ m^{-3}$.

As stationary sources of PM, cotton-ginning operations are coming under pressure to demonstrate $PM_{2.5}$ attainment for future operational licensing and other regulatory actions. All cotton gins will eventually be impacted by this standard. The primary issue affecting the cotton industry across the country in regard to the implementation of the $PM_{2.5}$ standard is that, up until quite recently, very little scientifically sound information was available on cotton gin $PM_{2.5}$ emissions. The cotton industry is in the process of collecting, analyzing, and publishing particulate emission data from across the cotton belt, including total, PM_{10} and $PM_{2.5}$ particulate categories (Buser et al., 2012).

OSHA Health and Safety Regulations

Workplace regulations are promulgated by the Occupational Safety and Health Administration (OSHA), which is part of the US Department of Labor, under the Occupational Safety and Health Act (OSH Act; PL 91-596 as amended by PL 101 552; 29 US Code 651 et. seq.). Cotton production and ginning are covered by OSHA agriculture standards (29 CFR 1928; P.J. Wakelyn, pers. comm., 2007). Federal OSHA enforces all OSHA standards, except where there is a state plan program. It is important for gin managers and operators to know whether their state is a state-plan state (i.e., administers its own OSHA-type program) or is under Federal OSHA, since the 26 state-plan states (seven cotton states; see Table 4) can have different regulations than Federal OSHA; state standards only have to be "as effective as the Federal standards," but they can be more severe.

Even if there is no specific standard but there is a recognized hazard and employers do not take reasonable action to prevent or abate the hazard, OSHA can cite an employer under the "general duty clause" (Sec. 5(a)(1) of the OSH Act). Section 5(a)(1) of the OSH Act requires employers to "furnish to each of his employees employment and a place of employment which are free from recognized hazards that are causing or are likely to cause death or serious physical harm to his employees." Section 5(a)(2) requires employers to "comply with occupational safety and health standards promulgated under this Act." There are criminal penalties for any employer who willfully violates any OSHA requirement that causes death to

Table 4. Occupational Safety and Health Administration (OSHA) enforcement in Cotton Belt states.

OSHA state-plan states	State under federal OSHA jurisdiction
Arizona	Alabama
California	Arizona
North Carolina	Florida
New Mexico	Georgia
South Carolina	Kansas
Tennessee	Louisiana
Virginia	Missouri
	Mississippi
	Oklahoma
	Texas

any employee. The employer could be subject to criminal penalties, including a fine (up to $250,000 and $500,000 for a corporation) or imprisonment (up to 1 yr) or both.

Only agricultural standards listed in 29 CFR 1928 and general industry standards listed in 29 CFR 1928.21 specifically apply to gins (P.J. Wakelyn, pers. comm., 2007). Potential hazards and risks found at cotton ginning operations that are not covered by specific regulations should be addressed to protect workers. For example, standards for the control of hazardous energy (lockout/tag out), confined space, and noise do not specifically apply to cotton gins in the OSHA plan states. However, the gin workplace should be monitored and, if there is risk of these hazards, consideration should be given to voluntarily complying with these regulations to maintain a safe and healthful workplace. The National Cotton Ginners Association and the various state and regional ginning associations have information available to help cotton gins meet the safety regulations that apply to them. In addition, OSHA has a program for employers who want help in recognizing and correcting potential safety and health problems and in improving their safety and health programs. Employers can get help from a free consultation program funded by OSHA (29 CFR 1908—Consultation Agreements). The service is delivered by state governments using well-trained professional staff. The consultation program addresses immediate problems and offers advice and help in maintaining an effective health and safety program.

Current Ginning Systems

Gin Layout for Different Harvesting Practices

A complete cotton ginning system is composed of a series of machines that condition and clean the seed cotton, separate the fibers from the seed, clean the ginned lint, and package the lint for shipment to the textile mill. The amount of machinery required to satisfactorily accomplish these tasks depends on the initial foreign-matter and moisture contents of the harvested seed cotton, which vary with the growing and harvesting conditions and with the harvesting method. Because harvesting methods have such a large effect on initial foreign-matter content, specific gin system designs have been tested and formulated for both of the methods of harvesting in the United States, that is, machine picking and machine stripping (Anthony et al., 1994; Baker, 1994). These designs are generally flexible enough to accommodate variations in foreign-matter and moisture

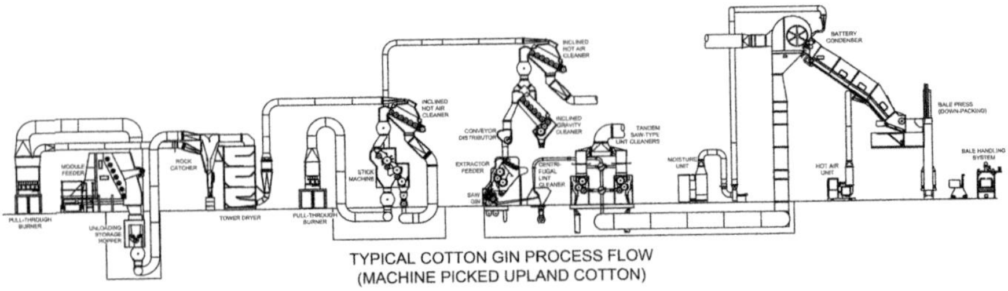

Fig. 33. Typical sequence of gin machinery. (Courtesy of Ross Rutherford, Lummus Corp.)

contents resulting from variances in the weather during the growing and harvesting seasons. Since excessive drying and machining of cotton can adversely affect its quality, the design of gin systems must also be based on fiber-quality considerations. Generally, successful designs are those that maximize bale value for the producer while at the same time preserve the inherent quality of the cotton fiber (Baker and Griffin, 1984).

Numerous tests by the USDA-ARS Cotton Ginning Laboratories and by gin machinery manufacturers have formed the basis for recommendations on the sequence and amount of gin machinery required to adequately process cotton harvested by various methods. These machinery recommendations, shown in Figure 33 for machine-picked Upland cotton, are general in nature in that they represent typical conditions in most cotton-producing areas. Under these conditions, the recommended machinery arrangements will produce satisfactory lint grades and near-maximum bale values for most cotton. However, some modification to these recommendations may be necessary in certain production areas to meet special needs or unusual growing conditions. Cotton containing larger amounts of foreign matter (as in machine-stripped areas), particularly weed and grass material, or hard-to-clean cultivars of cotton may require more cleaning than can be obtained with the basic machinery arrangements. On the other hand, very clean cotton can often be satisfactorily cleaned with less machinery. For this reason, seed-cotton cleaners and extractors and lint cleaners should be provided with bypasses to allow for the selection of less cleaning machinery when processing extra-clean, dry cotton. Generally, a ginner should select the minimum amount of machinery required to maximize bale value for the producer. In this way ginning costs will be minimized, and at the same time the inherent qualities of the cotton fiber will be preserved.

Ginning and Fiber Quality

General Effects on Fiber Quality

The term "fiber quality" has different meanings for different interest groups. Producers usually view fiber quality strictly in terms of those factors that directly determine market price, that is, grade, staple length, and micronaire value (an instrument reading related to fiber maturity) (Baker and Griffin, 1984). To spinners, however, a definition of fiber quality would probably include additional fiber properties that have been found to correlate more precisely with spinning performance and potential yarn quality than do the market factors. Depending on the spinner's particular interests, this latter group might include measurements of fiber length, length uniformity, short-fiber content, strength, maturity, nep content, non-lint content, dust content, and possibly other properties.

Regardless of one's definition of fiber quality, cotton possesses its highest fiber quality when it is still on the stalk in the field. Technically, any mechanical handling of the cotton thereafter has the potential to modify the natural qualities of the fiber (ICAC, 2001). From a practical standpoint, however, the thermal, pneumatic, and mechanical processes in ginning usually affect only a few fiber properties, principally foreign-matter content, the various parameters of fiber length, and the appearance of the lint sample (ginning preparation and nep content). Although ginning affects only a few fiber properties, the ones that are affected have a significant impact on the market price of cotton and on the fiber's ultimate end-use value.

The ginning process influences grade and staple length but has very little effect on micronaire value. The micronaire value, which is an indirect measure of fiber fineness and maturity, is controlled primarily by genetic factors and cotton growth conditions. Lint grade is determined by color, foreign-matter content, and ginning preparation. While the ginning process has little effect on true fiber color, the blending action of lint cleaners can enhance the lint-color designation to some extent.

The removal of foreign matter during ginning has the greatest effect on lint grade. Lint grades improve with increased amounts of gin cleaning and drying, but at a declining rate. Levels of cleaning beyond those shown in Fig. 33 generally produce only minor improvements in grade. Drying to moisture levels below the recommended 6.5 to 8.0% level may improve lint grade, but it does so at the expense of shortening the staple length and reducing the bale weight. Frequently the losses in lint value due to these adverse effects of drying will offset the increases in lint value due to grade improvement. Thus, seed-cotton dryers should be adjusted to remove only enough moisture from the cotton to ensure smooth ginning. On the other hand, inadequate drying can be a source of many ginning problems. The handling and cleaning of wet cotton produces a stringy, tangled sample that may be reduced in grade because of rough ginning preparation, and inadequate drying reduces the effectiveness of the gin's cleaning machinery.

Ginning and lint cleaning can adversely affect fiber length and length distribution, especially when the fiber moisture content is very low. Since fiber tensile strength is proportional to fiber moisture content, fiber breakage increases as moisture content decreases. Fiber breakage during ginning shortens the fiber length, thus raising the short-fiber content, and shifts the fiber length distribution toward the short end of the spectrum. These length changes adversely affect yarn strength, yarn appearance, and spinning-end breakage. The length characteristics of the fiber can best be preserved by keeping the fiber moisture content within the 6.5 to 8.0% range and by using the minimum amount of machinery needed to produce satisfactory grades and near-maximum bale values. The presence of an excessive number of neps in lint is a problem for many spinners, because neps adversely affect spinning performance, yarn appearance, and uniformity of dyeing. Much of their presence in the lint is attributable to the ginning process. The nepping potential of fibers is closely related to fiber fineness and maturity. Fine fibers, especially if they are immature, are more susceptible to nepping than are coarse or fully matured fibers. While some neps are formed during seed-cotton cleaning, research indicates that most neps are created by saw gin stands and saw lint cleaners (Chapman and Stedronsky, 1959). The formation of neps during the saw-ginning process may be minimized by (i) ensuring that the fiber moisture content is at the optimum level, (ii) eliminating unnecessary handling and cleaning operations, and (iii) employing moderate processing rates. While these processing remedies are helpful, they do not completely eliminate the problem. Fine-fiber, long-staple cultivars of cotton are normally roller ginned to avoid the nepping problem that usually accompanies the saw-type processes.

Classing: High Volume Instrument

Historically the USDA has classed (graded) 97% or more of the cotton bales produced in the United States each year (Moore, 1994). Since 1981 growers have paid for the classing service, and since 1968, bale class has evolved from being almost entirely a manual determination to currently being almost entirely an instrumental determination (Rana, 2014). Before 1968, cotton classification consisted of a manual staple length (in 32nds of an inch), a manual grade (reflecting both color and trash levels), and micronaire. The classification system for both Upland and Pima cotton currently consists of a manual classing of extraneous matter designation and abnormal preparation (if any of either) and of instrument measurements for color grade, leaf grade, fiber length, micronaire, strength, length uniformity index, color Rd, color +b, and trash percent area (USDA-AMS, 2005; Cotton Incorporated, 2013). The instruments that make the basic fiber quality measurements and the computer that calculates the fiber quality parameters based on these instrument measurements are all housed in one cabinet known in the cotton industry as the High Volume Instrument (HVI). Both the color grade and leaf grade are determined from HVI instrument-based measurements. Samples submitted for classification are identified with the 12-digit permanent bale identification (PBI) number; the first two digits are the classing office code, the next three digits are the gin code, and the final seven digits are the bale number. It should be noted that Upland cotton is normally saw ginned, whereas Pima cotton is normally roller ginned, and their bale fiber samples are somewhat different in physical appearance. Cotton from roller gins is currently only classed in the USDA-AMS, Visalia, CA, classing office. The type of gin used can be determined by the first five digits of the PBI number; any sample with a 966XX or a 967XX number will be from a roller gin.

Color Grade

Color refers to the gradations of grayness and yellowness in the cotton. When the boll initially opens, Upland cotton is white, whereas Pima cotton is creamier in color. Fiber color can then change toward being more gray or yellow depending on exposure to environmental factors. Color grades are different for Upland and Pima. There are 25 color grades for American Upland cotton and five categories for what is called below-grade color (Fig. 34; Moore, 1994; Cotton Incorporated, 2013). The Upland color grades fall into five color groups—white, light spotted, spotted, tinged, and yellow stained—or a total of 30 grades. Fifteen of these color grades are represented by physical standards (Universal Upland Grade Standards) that are prepared and maintained by the USDA-AMS. The remaining 15 color grades are descriptive and have no physical standard (see Table 5). The HVI instrument color determination is in terms of Rd (grayness) and +b (yellowness). The color charts shown in Figures 34 and 35 represent the color grade for Upland and Pima as determined by instrument readings. The Upland

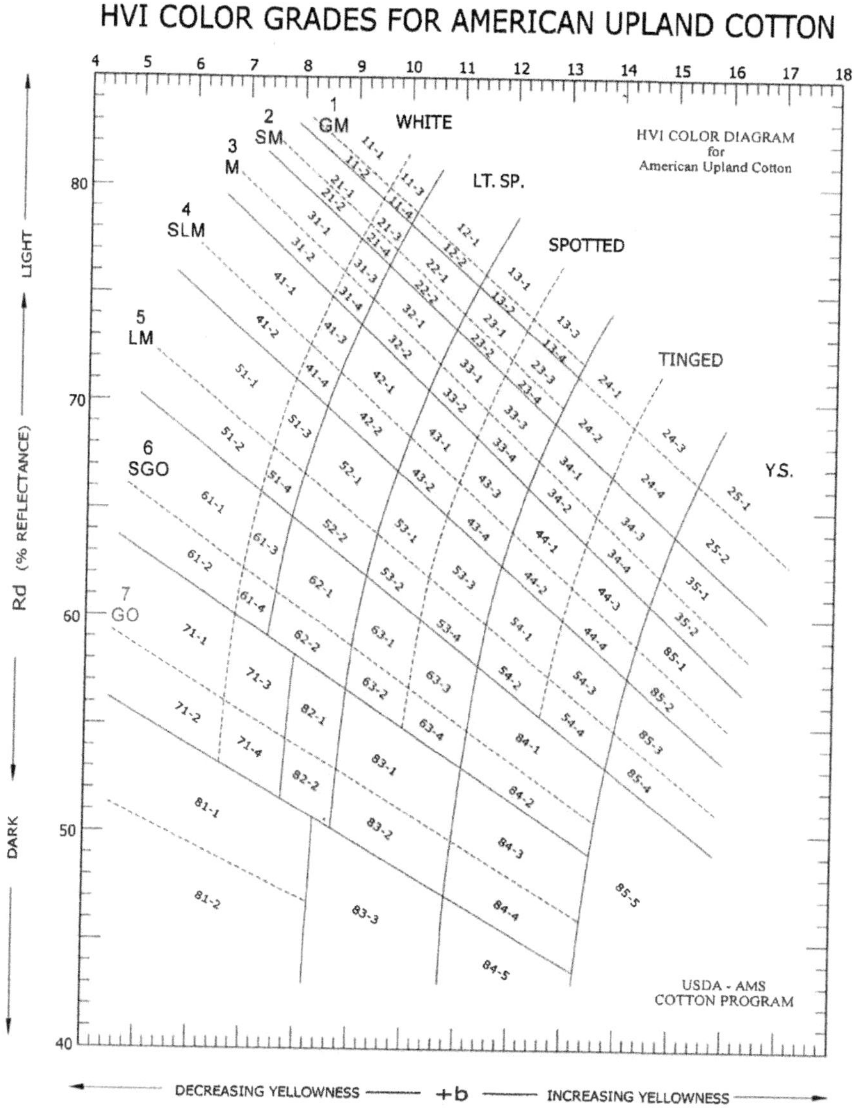

Fig. 34. HVI Upland color diagram. GM, good middling; SM, strict middling; M, middling; SLM, strict low middling; LM, low middling; SGO, strict good ordinary; GO, Good ordinary; BG, below grade (any grade starting with 81 to 85).

color grades from Figure 34 are related to the grade determination as shown in Table 5. Fiber value, as related to color for Upland cotton, is higher for white grades than for the yellower or grayer grades. For example, good middling 11 is the highest color grade, and therefore fiber value, and below grade 85 is the lowest color grade and fiber value (Table 5). The range of each color grade for which there is a physical standard for both Upland and Pima is represented by six samples placed adjacent to each other in a standards box. There are seven color

Table 5. Color grades of Upland cotton.

	White	Light spotted	Spotted	Tinged	Yellow stained
Good middling	11†	12	13	–	–
Strict middling	21†	22	23†	24	25
Middling	31†	32	33†	34†	35
Strict low middling	41†	42	43†	44†	–
Low middling	51†	52	53†	54†	–
Strict good ordinary	61†	62	63†	–	–
Good ordinary	71†	–	–	–	–
Below grade	81	82	83	84	85

† Physical standards. All others are descriptive.

grades for Pima (Fig. 35), for which there are six physical standards (American Pima Grade Standards) similar to the Upland standards and one below-grade descriptive color standard.

Leaf Grade

Leaf grade refers to the non-lint plant material content of the cotton sample and is normally referred to as leaf content although it can be made up of other small parts of the cotton plant. For Upland cotton there are seven leaf grades (numbered 1 through 7) and are all represented by physical standards. These seven leaf grades are shown in Table 5 as the first digit. For example, under the "White" column the first digit is from 1 to 7, which is the indicated trash level. The second digit in that column is 1 and indicates that the fiber is white. Other fiber color designations are shown in other columns as the second digit, which ranges from 2 through 5. In addition, there is a "below grade" category, which is a descriptive standard (81–85 in Table 5; Moore, 1994). For Pima cotton there are six leaf grades (numbered 1 through 6), and are all represented by physical standards. There is also a "below grade" category, which is, like that for Upland cotton, is a descriptive standard. For practical considerations, the color- and leaf-grade standards for Upland cotton are contained in the same standards box. For instance, the Universal Upland Grade Standard box containing the strict low middling (41) color grade (white) also contains the size and amount of leaf that would be described as leaf grade 4. Similarly for Pima cotton, the American Pima Grade Standard color-grade 3 box, for example, would also contain the Pima grade-3 leaf content and does not match the leaf-grade 3 of Upland cotton. Leaf grade is now determined by the HVI trash meter percent area and particle-count instrument measurements. The leaf grade is calculated from these parameters on the basis of the Universal Upland Grade Standards and American Pima Grade Standards (CI, 2013).

Staple Length

Fiber staple length is an instrument measurement and is reported for both Upland and Pima cotton in both 32nds of an inch as well as 100ths of an inch (Table 6). This length measurement is not an absolute but, rather, is a calculated upper-half mean length, which is the average length of the longest one-half of the fibers in the sample. In general, the larger the number within the range of cotton fiber length, the higher the quality of the fiber.

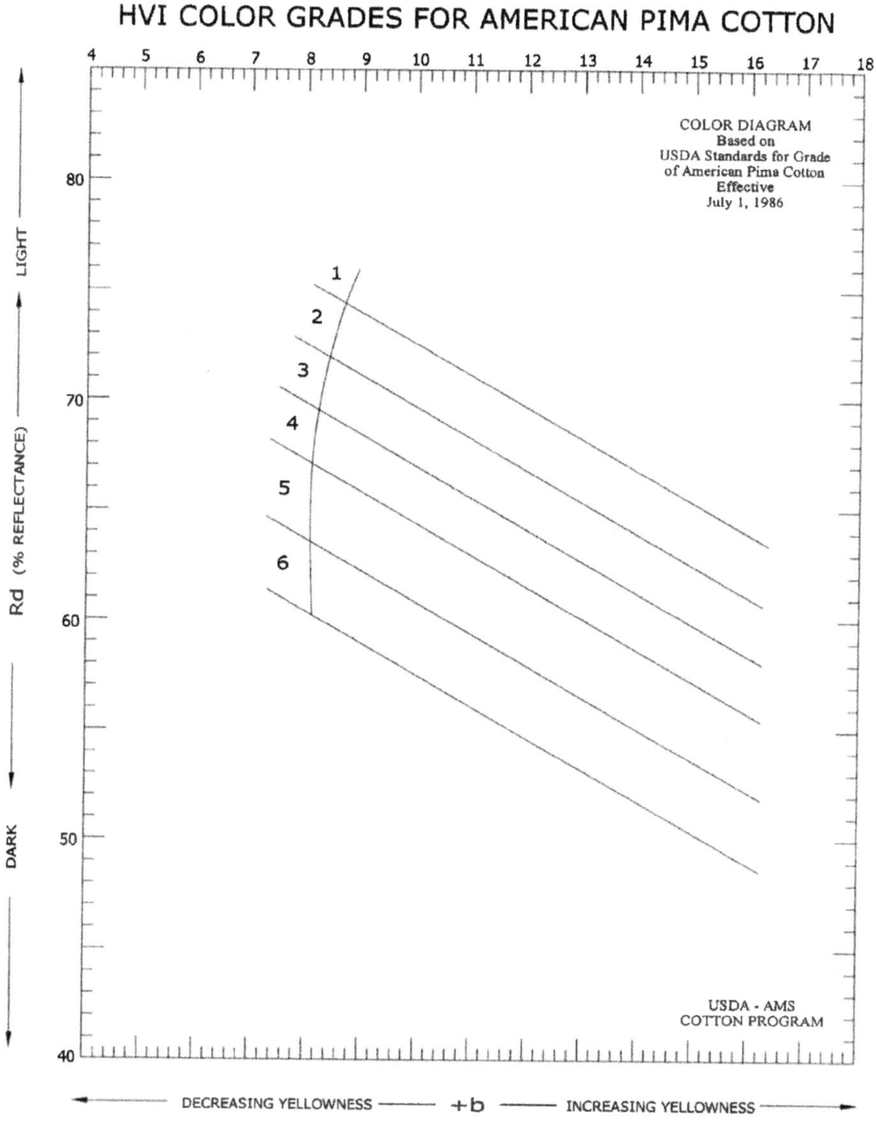

Fig. 35. HVI color grades for Pima cotton.

Length Uniformity

Length uniformity is the ratio of the average, or mean, length of the fibers to their upper-half mean length (staple length) and is expressed as a percentage (Table 7). The higher the ratio, the higher the value of the fiber to the textile industry. This value is roughly indicated by the descriptive designations in Table 7. In the HVI system, the same tuft of fiber that is used to measure staple length is used to determine length uniformity. If all fibers in a sample were the same length,

Table 6. Fiber length charts for Upland and Pima cottons.

	Length
32nds of an inch	inches
Upland	
24	0.79 and shorter
26	0.80–0.85
28	0.86–0.89
29	0.90–0.92
30	0.93–0.95
31	0.96–0.98
32	0.99–1.01
33	1.02–1.04
34	1.05–1.07
35	1.08–1.10
36	1.11–1.13
37	1.14–1.17
38	1.18–1.20
39	1.21–1.23
40	1.24–1.26
41	1.27–1.29
42	1.30–1.32
43	1.33–1.35
44 and up	1.36 and up
American Pima	
40	1.20 and lower
42	1.21–1.25
44	1.26–1.31
46	1.32–1.36
48	1.37–1.42
50	1.43–1.47
52	1.48 and up

Table 7. Uniformity chart.

Descriptive designation	Length uniformity
	%
Very low	Below 76.5
Low	76.5–79.4
Average	79.5–82.4
High	82.5–85.4
Very high	Above 85.4

Table 8. Interpreting fiber strength.

Description of degree of strength	Strength
	g tex^{-1}
Very strong	31 and above
Strong	29–30
Average	26–28
Intermediate	24–25
Weak	23 and below

then the length uniformity index would be 100. As was stated earlier, mechanical processing damages some fibers so that a fiber sample will have a distribution of fiber lengths, from very short to significantly longer than the given staple length assigned to the sample by the HVI measurement. In general, the higher the uniformity index, the higher the fiber quality and the more valuable it is for textile processing. For example, a sample that the HVI calls a 36 staple length, 2.82 to 2.87 m (1.11– 1.13 in), and 83% uniformity would be of better quality and bring a higher price than a fiber sample with a 36 staple length and a 79% uniformity index.

Strength

Fiber strength is an instrument measurement made by clamping and breaking a bundle of fibers with a 0.317-cm (1/8-in) spacing between the clamp jaws (USDA-AMS, 2005). Results are reported in terms of grams per tex (a tex unit is equal to the weight in grams of 1000 m of fiber). Therefore, the strength reported is the force in grams required to break a bundle of fibers one tex unit in size. Table 8 shows some general descriptions of strength measurements in grams per tex. It should be noted that Upland cotton varieties at maturity range in strength from weak to very strong, whereas Pima varieties are generally rated very strong and above at maturity.

Micronaire

Cotton's resistance to airflow per unit mass is measured to determine micronaire (Moore, 1994; USDA-AMS, 2005). Micronaire is a measure of cotton's fineness and was the first instrument measurement developed for cotton. The micronaire measurement and maturity are highly correlated within a given cotton variety, and different varieties vary in micronaire at full maturity. A low micronaire ("mike") reading may indicate immature fibers, which could lead to weak yarn or dyeing problems at the textile mill.

Extraneous Matter

Extraneous matter is any substance in the cotton other than fiber or leaf. Examples of extraneous matter are bark, grass, spindle twist, seed coat fragments, dust, plastic, and oil (CI, 2013). Abnormal preparation (prep) describes the degree of smoothness or roughness of the ginned cotton lint. Various methods of harvesting, handling, and ginning cotton produce differences in roughness or smoothness of preparation that sometimes are quite apparent to the naked eye. If either extraneous matter or prep conditions are predominant throughout the sample or throughout one-half of the sample, then the classer will manually note their presence on the bale classification document.

Calibration of HVI Instruments

Calibration of HVI instruments to ensure uniform classing results across the 10 USDA-AMS cotton classing offices in the United States is an important issue. All USDA-AMS HVI instruments are routinely calibrated for fiber length, length uniformity, micronaire, and fiber strength through the use of calibration cottons of known properties. Tiles are used to calibrate color and trash measurements (Cotton Incorporated, 2013). In addition to calibrating the instruments, each classing office monitors its testing level through an in-house system designed to ensure that the instruments within a classing office provide uniform results. In addition to the in-house program, there is the USDA program in which samples at each classing office are randomly selected after classing during each work shift. These randomly selected samples are called checklots. The level of checklot selection is approximately 1% of the total classing volume. These samples are sent to the USDA's Quality Assurance Division, Memphis, TN, where they are retested. The retest results are compared with the original classification, and this information is communicated to the originating office, where adjustments are made as necessary. The USDA maintains a record of comparisons for each instrument and classer on a daily, weekly, and seasonal basis.

The HVI cotton class has long been the basis of trade for all US cotton. Standardized HVI testing of cotton is performed widely today and is increasingly becoming the basis of cotton trading around the world in place of manual classing (Drieling et al., 2012; Rana, 2014). The Task Force on Commercial Standardization of Instrument Testing of Cotton is in the process of developing HVI cotton testing standardization guidelines for use by the world cotton trade. The guidelines cover test material, testing environment, test method, instruments used, training personnel, and test management. In addition to initial bale class for marketing use, the HVI classification is being widely used in textile mills around the world to manage bale selection for uniform textile processing.

References

Alberson, D.M., and V.L. Stedronsky. 1964. Roller ginning American-Egyptian cotton in the Southwest. USDA Agric. Handb. 257. US Gov. Printing Office, Washington, DC.

American Society of Agricultural and Biological Engineers (ASABE). 2012. Temperature sensor locations for seed-cotton drying systems. ASAE S530.1 AUG2007 (R2012). ASABE, St. Joseph, MI.

Anthony, W.S. 2003. The impact of excess moisture in the bale on fiber quality. In: Proceedings of the Beltwide Cotton Production Research Conferences, Nashville, TN. 6–10 January. National Cotton Council, Memphis, TN. p. 746–760.

Anthony, W.S., S.E. Hughs, and W.D. Mayfield. 1994. Ginning recommendations for processing machine-picked cotton. In: Cotton ginners handbook. USDA Agric. Handb. 503. US Gov. Printing Office, Washington, DC. p. 240–241.

Armijo, C. B., and M. N. Gillum. 2007. High-speed roller ginning of Upland cotton. Appl. Eng. Agric. 23(2): 137–143. doi:10.13031/2013.22603

Bajwa, S., D. Bajwa, G.A. Holt, F. Nakayama, and T.A. Coffelt. 2011. Properties of thermoplastic composites with cotton and guayule biomass residues as fiber fillers. Ind. Crops Prod. 33(3):747–755. doi:10.1016/j.indcrop.2011.01.017

Bajwa, S.G., D.S. Bajwa, G.A. Holt, and T.C. Wedegaertner. 2014. Commercial scale evaluation of two agricultural waste products, cotton burr/stem and module wraps in thermoplastic composites and comparison with laboratory-scale results. J. Thermoplast. Compos. Mater. 27(6):741–757. doi:10.1177/0892705712454865

Baker, K.D., and E. Hughs. 2012. A survey of seed cotton dryers in cotton gins in the southwestern United States. Appl. Eng. Agric. 28(1):87–97. doi:10.13031/2013.41273

Baker, R.V. 1978. Performance characteristics of saw-type lint cleaners. Trans. ASAE 21: 1081–1087. doi:10.13031/2013.35445

Baker, R.V. 1994. Ginning recommendations for processing machine-stripped cotton. In: Cotton ginners handbook. USDA Agric. Handb. 503. US Gov. Print. Office, Washington, DC. p. 242–243.

Baker, R.V., W.S. Anthony, and R.M. Sutton. 1994a. Seed cotton cleaning and extracting. In: Cotton ginners handbook. USDA Agric. Handb. 503. US Gov. Print. Office, Washington DC. p. 69–90.

Baker, R.V., E.P. Columbus, R.C. Eckley, and B.J. Stanley. 1994b. Pneumatic and mechanical handling systems. In: Cotton ginners handbook. USDA Agric. Handb. 503. U.S. Gov. Print. Office, Washington, DC. p. 143–171.

Baker, R.V., E.P. Columbus, and J.W. Laird. 1977. Cleaning machine-stripped cotton for efficient ginning and maximum bale values. USDA Tech. Bull. 1540. U.S. Gov. Print. Office, Washington, DC.

Baker, R.V., and A.C. Griffin, Jr. 1984. Ginning. In: R.J. Kohel and C.F. Lewis, editors, Cotton. Agron. Monogr. 24, ASA, CSSA, and SSSA, Madison, WI. p. 397–436,

Baker, R.V., and S.E. Hughs. 1999. Influence of air inlet and outlet design and trash exit size on 1D3D cyclone performance. Trans. ASAE 42(1):17–21. doi:10.13031/2013.13193

Baker, R.V., S.E. Hughs, M.N. Gillum, and J.K. Green. 1997. Improvements for cotton gin trash cyclones. Trans. ASAE 40(1):5–12. doi:10.13031/2013.21238

Bel, P. D., C. L. Simpson, E. P. Columbus, and B. Vinyard. 1991. Effects of mechanical cleaning on cotton fibers. Part 2: Combinations of gin and mill cleaning—Fiber quality. Textile Res. J. 61(9):503–509.

Bennett, C.A. 1932. The vertical seed-cotton dryer. USDA Miscellaneous Publ. 149. US Gov. Print. Office, Washington, DC.

Bennett, C.A. 1956. Ginning cotton. Farmers' Bull. 1748. USDA. US Gov. Print. Office, Washington, DC.

Bennett, C.A. 1959. Roller cotton ginning developments. Texas Cotton Ginners' Assoc. and Cotton Gin and Oil Mill Press, Dallas, TX.

Bennett, C.A. 1962. Cotton ginning systems and auxiliary developments. Texas Cotton Ginners' Assoc. and Cotton Gin and Oil Mill Press, Dallas, TX.

Buser, M.D., D.P. Whitelock, J.C. Boykin, and G.A. Holt. 2012. Characterization of cotton gin particulate matter emissions-project plan. J. Cotton Sci. 16:105–116.

Chapman, W.E., R.A. Mullikin, and P.E. LaFerney. 1968. Effects of gin cleaning on fiber properties and spinning quality, Pima cotton, 1965-66 with and without crusher rolls. USDA ARS and ERS Marketing Res. Rep. 806. US Gov. Print. Office, Washington, DC.

Chapman, W.E., and V.L. Stedronsky. 1959. Cotton qualities as affected by ginning. Cotton Gin and Oil Mill Press.

Chapman, W.E., and V.L. Stedronsky. 1965. Comparative performance of saw and roller gins on Acala and Pima cottons. USDA ARS Marketing Res. Rep. 695. US Gov. Print. Office, Washington, DC.

Chun, D.T.W., S.E. Hughs, C. Armijo, K. Baker, and D.D. McAlister. 2007. A study of bale moisture addition. Trans. ASABE 50(2):325–330. doi:10.13031/2013.22622

Cotton Incorporated. 2013. The classification of cotton. Cotton Incorporated, Cary, NC. http://www.cottoninc.com/fiber/quality/Classification-Of-Cotton/Classing-booklet.pdf

Drieling, A., J.-P. Gourlot, and J. Knowlton, eds. 2012. Guideline for standardized instrument testing of cotton. ICAC Task Force on Commercial Standardization of Instrument Testing of Cotton (CSITC). www.csitc.org, www.icac.org, or www.itmf.org.

Gerdes, F.L. 1951. Cotton lint cleaning at gins—An evaluation from the standpoint of cotton quality and economic factors. USDA Production and Marketing Administration, Cotton Branch, Washington, DC.

Gillum, M.N. 1985. High speed roller ginning. Trans. ASAE 28(3):959–968. doi:10.13031/2013.32370</jrn>

Gillum, M.N., and C.B. Armijo. 1991. Computer-aided roller ginning. Trans. ASAE 34(2):529–533. doi:10.13031/2013.31695

Griffin, A.C., and E.A. Harrell. 1957. Effects of moisture added at the lint slide on lint quality and bale weight in humid cotton production areas. USDA Production Res. Rep. 14. US Gov. Print. Office, Washington, DC.

Griffin, A.C., and V.P. Moore. 1965. Relation of physical properties of cotton to commerce and ginning research. Trans. ASAE 8:488–490. doi:10.13031/2013.40557

Holt, G.A., G.L. Barker, R.V. Baker, and A.D. Brashears. 2000. Characterization of cotton gin byproducts produced by various machinery groups used in the ginning operation. Trans. ASAE 43(6):1393–1400. doi:10.13031/2013.3036

Holt, G.A., T.L. Blodgett, and F.S. Nakayama. 2006. Physical and combustion characteristics of pellet fuel from cotton gin by-products produced by selecting processing treatments. Ind. Crops Prod. 24(3):204–213. doi:10.1016/j.indcrop.2006.06.005

Holt, G.A., M. Buser, D. Harmel, and K. Potter. 2005a. Comparison of cotton-based hydro-mulches and conventional wood and paper hydro-mulches—Study II. J. Cotton Sci. 9:128–134.

Holt, G.A., M. Buser, D. Harmel, K. Potter, and M. Pelletier. 2005b. Comparison of cotton-based hydro-mulches and conventional wood and paper hydro-mulches—Study I. J. Cotton Sci. 9:121–127.

Holt, G.A., P. Chow, J.D. Wanjura, M.G. Pelletier, and T.C. Wedegaertner. 2014. Evaluation of thermal treatments to improve physical and mechanical properties of bio-composites made from cotton byproducts and other agricultural fibers. Ind. Crops Prod. 52:627–632. doi:10.1016/j.indcrop.2013.11.003

Holt, G.A., and J.W. Laird. 2002. COBY products and a process for their manufacture. US Patent 6,383,548. Date issued: 7 May.

Holt, G.A., E. Lee, W.A. Ellis, and T.C. Wedegaertner. 2010. Cross-linked biofiber products and a process for their manufacture. US Patent 7,788,847. Date issued: 7 September.

Holt, G.A., G. McIntyre, D. Flagg, E. Bayer, J.D. Wanjura, and M.G. Pelletier. 2012. Fungal mycelium and cotton plant materials in the manufacture of biodegradable molded packaging material: Evaluation study of select blends of cotton byproducts. J. Biobased Mater. Bioenergy 6(4):431–439. doi:10.1166/jbmb.2012.1241

Holt, G.A., C.R. Richardson, G.A. Nunnery, K.F. Wilson, T.C. Bramble, L.D. Rea, and T.C. Wedegaertner. 2003. Performance of growing heifers fed diets containing cotton gin by-products extruded by the COBY process. Prof. Anim. Sci. 19(6):404–409.

Holt, G.A., J.L. Simonton, M.G. Beruvides, and A.M. Canto. 2004. Utilization of cotton gin by-products for the manufacturing of fuel pellets: An economic perspective. Appl. Eng. Agric. 20(4):423–430.

Hughs, S.E. 1985. Moisture—The key to fiber quality. In: Proceedings of the Beltwide Cotton Production Research Conferences, New Orleans, LA. 6–11 January. Natl. Cotton Counc. Am., Memphis, TN. p. 312–314.

Hughs, S.E., C.B. Armijo, D.P. Whitelock, and M.D. Buser. 2008a. Particulate emission profile of a cotton gin. Appl. Eng. Agric. 24(2):145–151. doi:10.13031/2013.24259

Hughs, S.E., and M.N. Gillum. 1991. Quality effects on current roller-gin lint cleaning. Trans. ASAE 7(6):673–676.

Hughs, S.E., and W.F. Lalor. 1990. Fiber and yarn effects of roller versus saw ginning. In: Proceedings of the Beltwide Cotton Production Research Conferences, Las Vegas, NV. 9–14 January. Natl. Cotton Counc. Am, Memphis, TN. p. 542–543.

Hughs, S.E., and C.G. Leonard. 1986. Roller versus saw ginning of long-staple Upland cotton. Cotton Gin and Oil Mill Press 87(20):8–10.

Hughs, S.E., G.J. Mangialardi, Jr., and S.G. Jackson. 1994. Moisture control. In: Cotton ginners handbook. USDA Agric. Handb. 503. US Gov. Print. Office, Washington, DC. p. 58–68.

Hughs, S.E., D.D. McAlister, C. Armijo, and K. Baker. 2004. Bale moisture addition: A case study. In: Proceedings of the Beltwide Cotton Production Research Conferences, San Antonio, TX. 5–9 January. Natl. Cotton Counc. Am., Memphis, TN. p. 2995–3001.

Hughs, S.E., T.D. Valco, and J.R. Williford. 2008b. 100 years of cotton production, harvesting, and ginning systems engineering: 1907–2007. Trans. ASABE 51(4):1187–1198. doi:10.13031/2013.25234

International Cotton Advisory Committee (ICAC). 2001. Best ginning practices. In: Impact of ginning on fiber quality: The best ginning practices. Report of an expert panel on ginning methods. International Cotton Advisory Committee, Washington, DC. p. 14–15.

Joint Cotton Industry Bale Packaging Committee. 2001. U.S. cotton bale dimensions. Natl. Cotton Counc. Am., Memphis, TN. http://www.cotton.org/tech/bale/bale-description.cfm?renderforprint=1& (accessed 30 Jan. 2015).

Joint Cotton Industry Bale Packaging Committee (JCIBPC). 2014. 2014 specifications for cotton bale packaging materials. Natl. Cotton Counc. Am., Memphis, TN.

Kirk, I.W., and C.G. Leonard. 1977. Modified saw-type lint cleaner for roller gins. Trans. ASAE 20(4):776–781. doi:10.13031/2013.35647

Laird, J.W., R.V. Baker, and R.E. Childers. 1984. Screen and grid dimensions and feeding method effects on performance of a cylinder cleaner at the gin. In: Proceedings of the Beltwide Cotton Production Research Conferences, Atlanta, GA. 8–12 January. Natl. Cotton Counc. Am., Memphis, TN. p. 159–160.

Laird, J.W., B.M. Norman, S. Stuller, and P. Bodovsky. 1994. Seed cotton unloading systems. In: Cotton ginners handbook. USDA Agric. Handb. 503. US Gov. Print. Office, Washington, DC. p. 46–58.

Lakwete, A. 2003. Inventing the cotton gin—Machine and myth in antebellum America. Johns Hopkins Univ. Press, Baltimore, MD.

Leonard, C.G., J.E. Ross, and R.A. Mullikin. 1979. Moisture conditioning of seed cotton in ginning as related to fiber quality and spinning performance. USDA Market Res. Rep. MRR 859. US Gov. Print. Office, Washington, DC.

Mangialardi, G.J. 1974. Effects of feed rate and batt density on operation of saw-cylinder lint cleaners. USDA Prod. Res. Rep. No. 156. US Gov. Print. Office, Washington, DC.

Mangialardi, G.J., Jr., and W.S. Anthony. 1998. Field evaluations of air and saw lint cleaning systems. J. Cotton Sci. 2:53–61.

Mangialardi, G.J., Jr., R.V. Baker, D.W. Van Doorn, B.M. Norman, and R.M. Sutton. 1994. Lint cleaning. In: Cotton ginners handbook, USDA Agric. Handb. 503. US Gov. Print. Office, Washington, DC. p. 102–119.

May, O.L., and K.E. Lege. 1999. Development of the world cotton industry. In: C. Wayne Smith and J. Tom Cothren, editors, Cotton: Origin, history, technology, and production. John Wiley & Sons, Hoboken, NJ.

Mayfield, W.D., R.V. Baker, S.E. Hughs, and W.S. Anthony. 1991. Introduction to a cotton gin. Natl. Cotton Ginners Assoc., Memphis, TN.

Mayfield, W., R.V. Baker, S.E. Hughs, and A.C. Griffin. 1983. Ginning cotton to preserve quality. USDA Ext. Serv. Program Aid 1323. US Gov. Print. Office, Washington, DC.

Mirsky, J., and A. Nevins. 1952. The world of Eli Whitney. Macmillan Co., New York.

Moore, J.F. 1994. The classification of cotton. In: Cotton ginners handbook, USDA Agric. Handb. 503. US Gov. Print. Office, Washington, DC. p. 287–292.

Moore, V.P. 1977. Development of the saw gin. In: Cotton ginners handbook. USDA Agric. Handb. 503. US Gov. Print. Office, Washington, DC. p. 1–3.

Moore, V.P., and C.M. Merkel. 1953. Cleaning cotton at gins and methods for improvement. USDA Circ. 922. US Gov. Print. Office, Washington, DC.

Parnell, C.B., Jr., E.P. Columbus, and W.E. Mayfield. 1994. Abatement of air pollution and disposal of gin waste. In: Cotton ginners handbook. USDA Agric. Handb. 503. US Gov. Print. Office, Washington, DC. p. 172–194.

Pelletier, M.G., G.A. Holt, J.D. Wanjura, E. Bayer, and G. McIntyre. 2013. An evaluation study of mycelium based acoustic absorbers grown on agricultural by-product substrates. Ind. Crops Prod. 51:480–485. doi:10.1016/j.indcrop.2013.09.008

Pendelton, A.M., and V.P. Moore. 1967. Ginning cotton to preserve quality. USDA Ext. Serv. ESC-560. US Gov. Print. Office, Washington, DC.

Porcher, R.D., and S. Fick. 2005. The story of Sea Island cotton. Wyrick & Company, Layton, UT.

Rana, S. 2014. Concept of high volume instrument (HVI): HVI used in bale management system. Textile Learner: Textile information blog for students. http://textilelearner.blogspot.com/2012/10/concept-of-high-volume-instrument-hvi_14.html (accessed 30 Jan. 2015).

Scholl, B.N., G.A. Holt, and C.I. Thornton. 2012. Screening study of select cotton-based hydromulch blends produced using the cross-linked biofiber process. J. Cotton Sci. 16:249–254.

Scholl, B.N., G.A. Holt, and C.I. Thornton. 2013. Hydromulch blends using agricultural byproducts: Performance implications of cotton quantity. J. Cotton Sci. 17:302–308.

Smith, C.W., R.G. Cantrell, H.S. Moser, and S.R. Oakley. 1999. History of cultivar development in the United States. In: C. Wayne Smith and J. Tom Cothren, editors, Cotton: Origin, history, technology, and production. John Wiley & Sons, Hoboken, NJ.

Stedronsky, V.L. 1956. New developments in roller gin research. Cotton Gin and Oil Mill Press 57(19):7–8.

Stedronsky, V.L., and C. S. Shaw. 1950. The flow-through lint-cotton cleaner. USDA Circ. 858. US Gov. Print. Office, Washington, DC.

Sutton, R., and L. Horn. 1986. The introduction of the multisaw lint cleaner. In: Proceedings of the Beltwide Cotton Production Research Conferences, Las Vegas, NV. 4–9 January. Natl. Cotton Counc. Am., Memphis, TN. p. 404–407.

Thomasson, J.A. 1990. A review of cotton gin trash disposal and utilization. In: Proceedings of the Beltwide Cotton Production Research Conferences, Las Vegas, NV. 9–14 January. Natl. Cotton Counc. Am., Memphis, TN. p. 689–705.

USDA Agricultural Marketing Service (USDA-AMS). 2005. Cotton classification: Understanding the data. http://www.ams.usda.gov/AMSv1.0/getfile?dDocName=STELP RDC5074569.

USEPA. 1978. Compilation of air pollutant emission factors (including supplements 1–7); suppl. 8. 3rd ed. Rep. AP-42-ED-3-Suppl-8. USEPA, Office of Air Quality Planning and Standards, Research Triangle Park, NC.

USEPA. 1996. Compilation of air pollutant emission factors, AP-42, 5th ed. Vol. 1: Stationary point and area sources; chap. 9: Food and agricultural industries; section 9.7: Cotton ginning. Suppl. B. USEPA, Washington, DC. http://www.epa.gov/ttn/chief/ap42/ch09/index.html (accessed 30 Jan. 2015).

USEPA. 2006. National ambient air quality standards for particulate matter; final rule. 40 CFR Part 50. U.S. Gov. Print. Office, Washington, DC.

Vandergriff, A.L. 1997. Cotton ginning: An entrepreneur's story. Texas Tech Univ. Press, Lubbock, TX.

Wakelyn, P.J. 1995. Status report on state implementation of Title V, 1990 Clean Air Act. In: Proceedings of the Beltwide Cotton Production Research Conferences, San Antonio, TX. 4–7 January. Natl. Cotton Counc. of Am., Memphis, TN. p. 461–464.

Wang, L., C.B. Parnell, and B.W. Shaw. 2002a. Performance characteristics of cyclones in cotton-gin dust removal. Agric. Eng. Intl. CIGR J.: 4.http://www.cigrjournal.org/index.php/Ejounral/issue/view/21 (accessed 30 Jan. 2015).

Wang, L., C.B. Parnell, and B.W. Shaw. 2002b. Study of the cyclone fractional efficiency curves. Agric. Eng. Intl. CIGR J.: 4.http://www.cigrjournal.org/index.php/Ejounral/issue/view/21 (accessed 30 Jan. 2015).

Wanjura, J.D., E.M. Barnes, M.S. Kelley, G.A. Holt, and M.G. Pelletier. 2014. Quantification and characterization of cotton crop biomass residue. Ind. Crops Prod. 56:94–104. doi:10.1016/j.indcrop.2014.02.019

Wanjura, J.D., W.B. Faulkner, E.M. Barnes, G.A. Holt, and M.G. Pelletier. 2012. Crop residue inventory estimates for Texas High Plains cotton. In: Proceedings of the Beltwide Cotton Production Research Conferences, Orlando, Florida. 3–6 January. Natl. Cotton Counc. of Am., Memphis, TN. p. 543–552.

Wanjura, J.D., W.B. Faulkner, and G.A. Holt. 2010. Ginning picker and stripper harvested high plains cotton. In: Proceedings of the Beltwide Cotton Production Research Conferences, New Orleans, LA. 4–7 January. Natl. Cotton Counc. of Am., Memphis, TN. p. 628–638.

Whitelock, D.P., C.B. Armijo, M.D. Buser, and S.E. Hughs. 2009. Using cyclones effectively at cotton gins. Applied Engineering in Agriculture 25(4):563–576. doi:10.13031/2013.27467

Whitelock, D.P., C.B. Armijo, G.R. Gamble, and S.E. Hughs. 2007. Survey of seed-cotton and lint cleaning equipment in U.S. roller gins. J. Cotton Sci. 11:128–140.

Fiber to Fabric

Brendan Kelly,* Noureddine Abidi, Dean Ethridge, and Eric F. Hequet

Abstract

Modern cotton fiber production is focused on providing an industrial raw material for spinning mills. These mills require a raw material that performs reliably and facilitates their production goals. This chapter provides an introduction to the integral role cotton fiber quality research plays in producing and developing cotton fiber for use as an industrial raw material. Natural variation in the cotton fiber structure and microstructure is introduced in terms of its impact on fiber quality. Natural variation in cotton quality impacts the performance of cotton fiber in an industrial setting. High-Volume Instrument (HVI) classification is commonly used to market cotton to spinning mills. HVI measurement methodologies are provided alongside the most common HVI fiber quality parameters. Many other fiber quality measurement instruments are available, each providing a unique characterization of the natural variability in cotton fiber quality. The measurement methodology of many of these instruments is described alongside some of the most common fiber quality parameters each instrument provides. Cotton fiber is a natural industrial raw material, thus, cotton fiber quality assessment must be understood in terms of the target market. The two dominant markets for cotton fiber—ring spinning and rotor spinning—are characterized in terms of the manufacturing method and the importance of cotton fiber quality in the finished product. Finally, the spun cotton yarn is used to produce either woven or knit textiles. These technologies are introduced along with the chemistry of textile finishing.

Cotton (*Gossypium* L.) fiber quality played a central role in the historical success of cotton production. Early cultivation of cotton focused on improving fiber quality to better fit the end-uses of the time. Improvements in both fiber quality and yield enabled cotton to play a foundational role in the industrial revolution as the raw material of choice in spinning mills. As a result, a global industry developed around cotton fiber.

The production of cotton fiber is important to many countries around the world (Fig. 1). In the United States alone, the cotton industry has an estimated total economic impact in excess of $120,000 Million (Smith, 2010).

Abbreviations: +b, yellowness; AFIS, Advanced Fiber Information System; dpa, days post anthesis; EFS, Engineered Fiber Selection; HVI, High-Volume Instrument; rpm, revolutions per minute; SCN, seed coat nep; SFC, short-fiber content; Rd, reflectance; UHML, upper half mean length; UI, uniformity index; UPF, ultraviolet protection factor; UV, ultraviolet; UVR, ultraviolet radiation.

Brendan Kelly, Noureddine Abidi (n.abidi@ttu.edu), Dean Ethridge (dean.ethridge@ttu.edu), and Eric F. Hequet (eric.hequet@ttu.edu), Fiber and Biopolymer Research Institute, Texas Tech University, 1001 East Loop 289, P.O. Box 45019, Lubbock, TX, 79403. *Corresponding author (brendan.kelly@ttu.edu).

doi:10.2134/agronmonogr57.2013.0031

Cotton. 2nd. ed. David D. Fang and Richard G. Percy, editors. Agronomy Monograph 57.

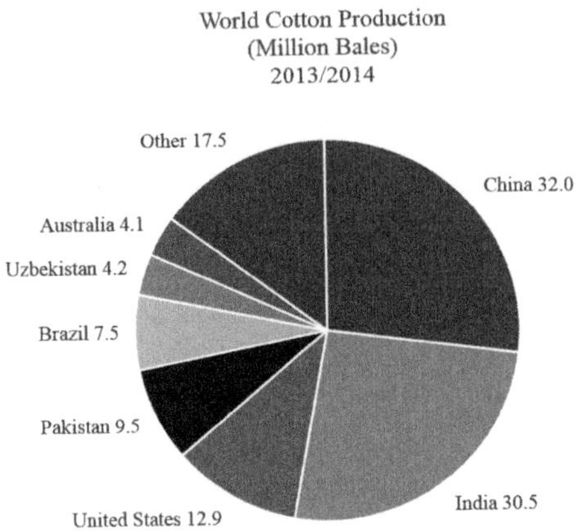

Fig. 1. World cotton production by major countries in 2013/2014 in millions of bales (USDA, 2014).

In 2013–2014, over 118 million bales of cotton fiber were produced worldwide. Cotton producers from around the globe sell their cotton on a competitive international market (Fig. 2). World cotton exports totaled 46.7 million bales, which was over 34% of the worldwide cotton production (USDA, 2014). Farmers compete in the international marketplace by producing cotton fiber with a quality profile suitable to the needs of the targeted textile industry, i.e., mass market tee-shirts and luxury linens require very different fiber properties to be both profitable and good quality products for their market segment.

Cotton fiber is traded on international markets primarily as a raw material for manufacturing ring- and rotor-spun yarns. Other technologies exist for spinning staple fibers, such as air-jet spinning, but these account for a very small share of the capacity and are not advantageous for cotton. On the basis of the spindle equivalents, global ring-spinning capacity was almost three times that of open-end rotor spinning in 2012 (International Textile Manufacturers Federation (ITMF), 2012). Therefore, cotton fibers must emphasize ring spinning. This technology is dominant in premium textiles because it results in a superior appearance and feel. But it is sensitive to the fiber properties that determine the spinning performance and yarn quality. Ring spinning mills must reduce the risk of off-quality production by sourcing cottons that have the needed fiber quality profile (El Mogahzy and Chewning, 2001).

Cotton fiber faces increasing competition from synthetic fibers. While the world consumption of cotton fiber has generally increased in recent decades, it has not kept up with population growth. In 2012, cotton held around 30% of the world market share in textile fiber consumption, which was down a little more than 10% points from 2000 (Fig. 3). In contrast, synthetic fibers currently hold over 60% of the world market share in textile fiber consumption (International Cotton Advisory Committee, 2013). Therefore, competition from synthetics has increased industry demands for cotton fiber with reliable spinning performance.

Research on cotton fiber quality plays a critical role at every juncture of the marketing chain. Breeders strive to develop varieties with the desired fiber quality profile. Cotton farmers select varieties and use farming practices to achieve the genetic potential of the cotton. Operations at the gin, such as cleaning, are balanced with preserving fiber quality. Instrument-based classification provides

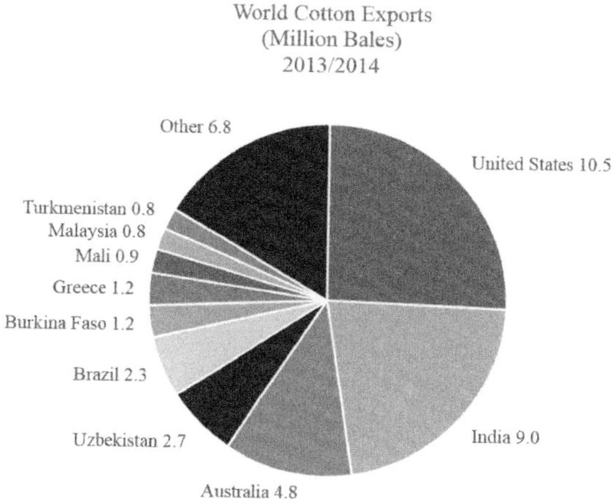

Fig. 2. Cotton exports by countries in 2013/2014 in millions of bales (USDA, 2014).

an objective assessment of the fiber properties, enabling spinning mills around the world to specify the mix of properties that fits their needs.

The depth and breadth of fiber quality research cannot be covered in a single chapter. Instead, this chapter will (a) provide an introduction to cotton fiber quality in terms of the natural variability present in cotton fiber structure and microstructure; (b) introduce HVI, the most widely used instrument for evaluating fiber quality; and (c) introduce other fiber quality measurement instruments that are important research tools. For each instrument, a short explanation is provided of the fiber qualities measured, the measurement technologies used, and the data obtained.

Since the main goal of fiber quality research is improved prediction of fiber performance as an industrial raw material, yarn manufacturing is explained along with the complex relationships among fiber and yarn qualities. Finally, a brief introduction to the processing and finishing of cotton textiles is provided.

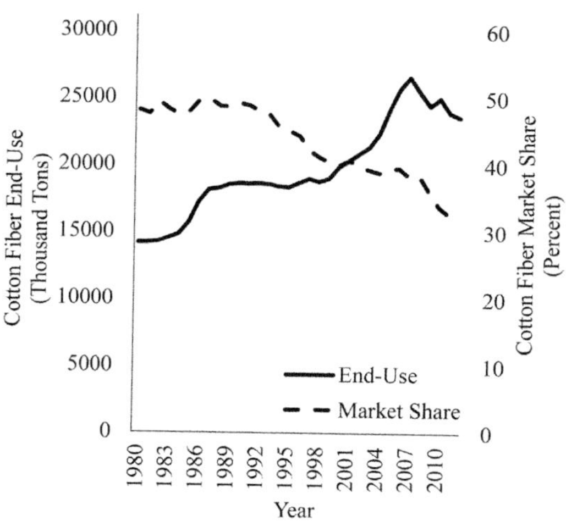

Fig. 3. World consumption and market share of cotton fiber from 1980 to 2012 (International Cotton Advisory Committee, 2013).

Fiber Quality

Natural Variability

The production of high-quality cotton fiber begins in the field. An individual cotton fiber develops as a single differentiated cell elongating from the epidermal layer of the cotton seed. Only a portion of the cottonseed epidermal cells will develop into a cotton fiber. These cells will begin to differentiate and elongate within the first few days after flowering (Stewart, 1975).

After flowering, waves of cells across the cottonseed surface begin differentiation and elongation into fibers (Stewart, 1975). A single cottonseed may develop 20,000 fibers or more (Cranmer, 2004; Maulding, 2007). Delayed differentiation of fiber cells across the seed surface contributes to developmental differences in the fiber cells (Stewart, 1975), which contributes to the variability in fiber quality among cotton fibers from a single cottonseed.

Cotton fibers develop on ovules inside of a boll, which is the fruit of the cotton plant. Cotton bolls are set throughout the season due to the indeterminate fruiting nature of the cotton plant. Conditions that affect the growth and development of the cotton fiber are unique for bolls set at different intervals throughout the season, which contributes to within-plant variability in fiber quality. Fiber quality often varies among seeds within a boll (Feng et al., 2011) and among bolls within the canopy (Bauer et al., 2009). Fiber quality is at its best within the mature boll, before the boll opens. Once the boll opens, the environment (ultraviolet (UV), rain, dew, freeze, etc.) and mechanical processes used to transform the fiber into an industrial raw material have the potential to damage, weaken, break, and entangle fibers.

Cotton must be harvested, cleaned, ginned, and cleaned again before being baled for trade on the world market. Each of these steps requires mechanical processing and has the potential to break or otherwise damage fibers, degrading the distribution of fiber quality within the bale (Dever et al., 1988; Hughs et al., 2013). Therefore, care must be taken to preserve the fiber quality at each stage in bringing the fiber to market.

A typical cotton bale (227 kg of lint) can contain 50,000 million individual fibers or more, with each fiber having been developed under differing agricultural, physiological, and environmental conditions. The quality of each fiber is further shaped by the mechanical forces that act on it during processing, resulting in a unique distribution of fiber quality within each bale. The distribution of fiber quality within the bale impacts spinning performance and the quality of the finished product (Kelly and Hequet, 2013). Assessment of fiber quality allows the cotton industry, breeders, and researchers to capture the variability in the cotton fiber quality within a sample and determine differences in fiber quality across many samples.

The relationship between fiber physical attributes (fiber properties) and fiber quality is elucidated in this section. In many cases, the most common fiber properties are based on measurements of fiber bundles. After these relationships are developed, the potential within-sample distribution of each fiber property is discussed. Instruments used to capture natural variability in cotton fiber quality, both within and between samples, are introduced in Fiber Quality.

Microstructure

Cotton fiber develops through a series of distinct overlapping developmental stages (Basra, 1999). Important milestones in the fiber development, such as the start of each developmental stage, are expressed in the number of days after flowering, also called days post anthesis (dpa). Fiber growth is characterized by the synthesis of the primary cell wall and an increase in fiber length up to ~30 mm within 3 wk after anthesis. The secondary cell wall synthesis commences around 21 dpa (Anderson and Kerr, 1938; Abidi et al., 2010) and continues for ~3 to 6 wk. This phase is marked by a massive deposition of cellulose to form the secondary cell wall (Basra, 1999). The transition period between 16 and 21 dpa is considered to represent a developmental switch in emphasis from primary to secondary cell synthesis. During these developmental stages, important structural changes lead to cellulose macromolecule formation. The cellulose macromolecule is a common material found in plant cell walls $(C_6H_{10}O_5)_n$. It is synthesized naturally in an almost pure form in cotton fiber. Cellulose macromolecules are a long straight chain of glucose monomers. These monomers are linked together through β $1\rightarrow4$ glycosidic bonds (glucopyranose) that are formed by condensation reactions. Cellulose and starch are two very similar polysaccharides. In cellulose, each successive glucose unit is rotated 180° around the axis of the polymer backbone. However, in starch all glucose repeating units are oriented in the same direction. This structural organization leads to significant differences between cellulose and starch. The highly ordered structure of cellulose prevents digestion by humans, while starch can be easily converted to glucose units and digested.

The equatorial positions of the hydroxyl groups on the cellulose chains make them readily available for hydrogen bonding. These hydrogen bonds allow the chains to group together in highly ordered structures forming crystalline regions. The hydrogen bonds in the crystalline regions are strong, giving the resultant fibers good strength, insolubility, and nonthermoplastic.

There are also less ordered regions in the fiber called amorphous regions. In these regions, the chains are further apart and more available for hydrogen bonding with other molecules. This structure creates areas within the cellulose that absorb water, dyes, and finishes. Amorphous regions also provide elasticity and flexibility to the fiber.

The smallest groupings of cellulose chains are called elementary fibrils. Their width is estimated at 2 to 4 nm and the length of the cellulose chain is estimated to be 500 nm. They are composed of 36 parallel cellulose chains. The elementary fibrils bind together into microfibrils. Their width is estimated at 10 to 30 nm, and they make the cell walls of the plants. Cotton fibers are formed from bundles of cellulose microfibrils. The packing of elementary fibrils into microfibrils and the arrangement of these microfibrils into cell structures result in the formation of crystalline and amorphous regions. Fiber strength, elongation, dyeing, and finishing characteristics are primarily determined by the structure of the crystalline and amorphous regions within the cotton fiber.

Fiber Length

The maximum fiber length (*n* or *w*) within the boll is set during the fiber elongation phase of development. Cotton fiber length may be highly variable even before harvest begins. Cotton fiber length is variable across a single seed, within a boll, and among the positions on a cotton plant (Feng et al., 2011; Bauer et al.,

2009). Many factors such as genetics, environmental conditions, and agronomic practices may contribute to within-plant variability in fiber length (Meredith and Bridge, 1973; Bednarz et al., 2006).

Figure 4 provides an example of within-plant variability in length for two commercial varieties. The average length-by-number for first-position bolls is shown for nodes 6 to 12 on the main stem. The topmost bolls in Variety A tend to be shorter than those produced lower on the plant. The average fiber length is much more stable for Variety B. The average fiber length of the topmost bolls appears to be more similar in length to that of the lower bolls, when compared to Variety A.

Fiber length is at a maximum before weathering and harvesting (before the boll cracks). Harvesting, ginning and cleaning processes, and spinning all have the potential to degrade the fiber length by breaking or weakening the fibers (Hughs et al., 2013; Zurek et al., 1999; Dever et al., 1988). Fiber breakage degrades the fiber length by decreasing the frequency of long staple fibers and increasing the amount of short fibers in the bale.

Characterizations of fiber length are some of the most widely used parameters of fiber quality and are important determinants of yarn strength (Faulkner et al., 2012; El Mogahzy et al., 1990).

Detailed fiber length data are frequently used for screening lines in breeding programs. In spinning mills, such data are used to adjust the equipment used to produce staple yarns (McCreight et al., 1997).

The complete distribution of fiber length within a sample, not just the dominant long fibers in the sample, is important fiber quality information for predicting the yarn quality. Data based on the complete distribution better account for variation in yarn quality (Kelly and Hequet, 2013). Short fibers are a detriment to the quality of spun yarns and must be removed for the production of highest-quality yarns (combed yarns) (McCreight et al., 1997).

Figure 5 illustrates the infeasibility of describing the within-sample variability in fiber length with a single parameter. It shows the fibers from a single sliver (a loose bundle of fibers used in spinning). The fibers are sorted by length. One end of each fiber is then aligned on the y axis. In this sorted and end-aligned form,

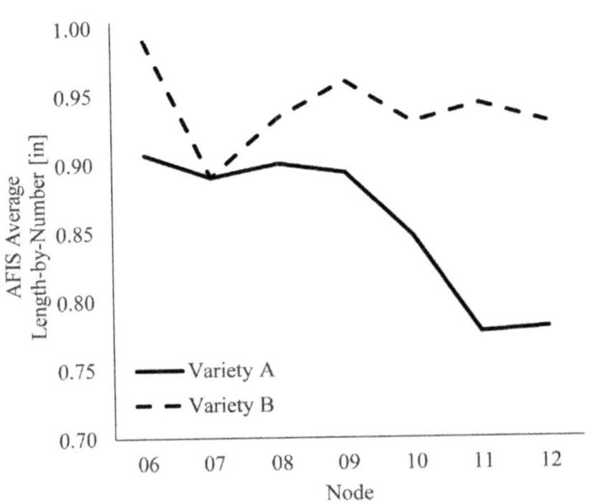

Fig. 4. Within-plant variability of fiber length for two commercial varieties.

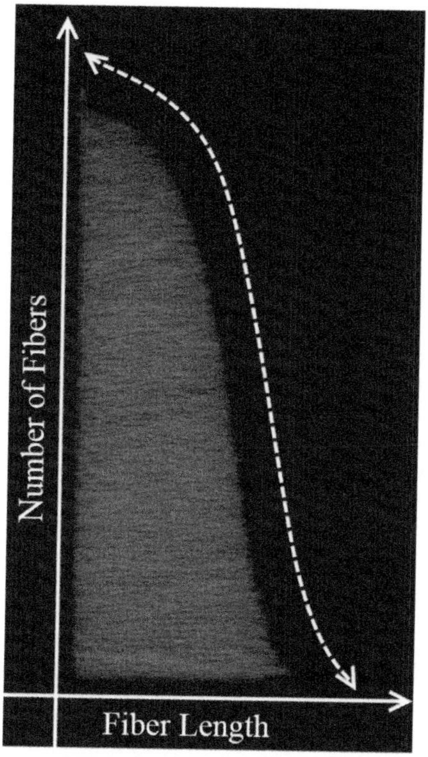

Fig. 5. Sorted fiber-length distribution of cotton fibers within a single sliver (Fiber and Biopolymer Research Institute, Texas Tech University).

the fibers form a staple diagram (Woo, 1967). The wide range of fiber lengths within the sample is easy to see when presented in this form.

The fiber-length distribution is used to completely characterize the unique variability in cotton fiber length within a sample (Woo, 1967). The cotton fiber-length distribution for a sample of cotton fiber may be expressed either "by-number" or "by-weight" (Wakeham, 1955). Each provides a distinct approach to characterizing the distribution of within-sample fiber length.

Length-by-Number

The length-by-number distribution characterizes the distribution of within-sample fiber length based on the relative frequency of fibers from each length group in a sample. Length data based on the frequency of fibers in each length group are called by-number measurements and are denoted by n.

Figure 6 provides a visual example for understanding length-by-number. First, cotton fibers are sorted into discrete length groups using a comb sorter. To determine the length-by-number distribution, the number of fibers in each discrete length group is counted. The length-by-number distribution is then expressed as the frequency, or count, of fibers in each length group relative to the total number of fibers in the sample.

The within-sample distribution of fiber length varies widely across commercial bales. Length-by-number distributions for three commercial cotton bales are shown in Fig. 7. The three bales were selected to represent a range of within-sample distribution among commercial bales. These distributions were produced by

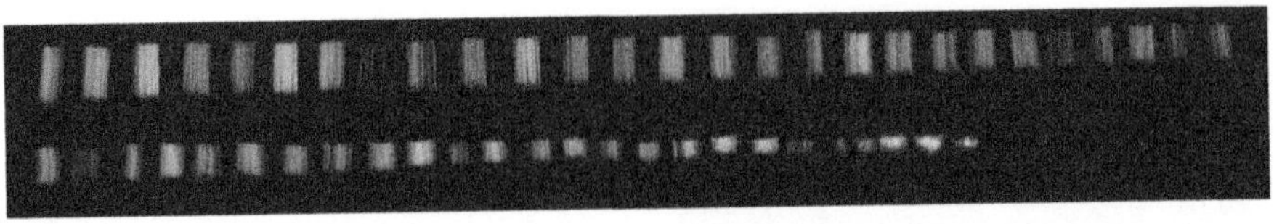

Fig. 6. Fibers from a single sample sorted into groups based on fiber length (Fiber and Biopolymer Research Institute, Texas Tech University).

an Advanced Fiber Information System (AFIS), which will be detailed in a later section.

The fiber-length distribution of processed cotton samples, such as the commercial bales seen in Fig. 7, often exhibit two prominent modes. The long-fiber mode characterizes fibers that have not had a substantial reduction of length during processing. The short-fiber mode is created by a buildup of short fibers created by fibers breaking during processing. The fibers represented by the short-fiber mode are not conducive to the production of spun yarns. The long-fiber mode may be nonexistent in heavily processed immature cottons where excessive breakage has occurred.

Cotton samples exhibiting a length-by-number distribution with a prominent long-fiber mode will perform better in spinning and are more desirable (Kelly et al., 2013). Bale A would be expected to perform well in spinning because it has the most prominent long-fiber mode and only a small short-fiber mode. Heavily processed cottons exhibit a large short-fiber mode and a diminished long-fiber mode similar to Bale B. The broken fiber content of Bale B is so overpowering, the long-fiber mode is almost nonexistent. Bale B would have limited value to a spinning mill.

A reference standard does not currently exist for measuring the length-by-number distribution. Length-by-number measurements are primarily determined using the AFIS instrument (see Fiber Length). Upper half mean length (UHML), the most common measurement of the dominant long fibers, is based on by-weight and by-number fractions (see High-Volume Instrument).

Length-by-Weight

The length-by-weight distribution characterizes the within-sample distribution of fiber length based on the weight fraction of fibers in each length group. Length parameters derived from the length-by-weight distribution are called by-weight measurements and are denoted by *w*.

Within-sample length-by-weight may be easily understood in terms of the cotton samples sorted by length group in Fig. 6. To estimate the length-by-weight, the fibers in each discrete length group are weighed instead of counted. The length-by-weight distribution is then expressed as the weight of the fibers in each length group, relative to the total sample weight.

The estimated length-by-weight distribution for each of the commercial cotton bales in Fig. 7 are shown in Fig. 8. The biggest difference in the by-weight measurement is evident in the shorter fibers. Shorter fibers account for a lower portion of the total weight of the fibers. This is an important distinction because cotton bales are sold by weight.

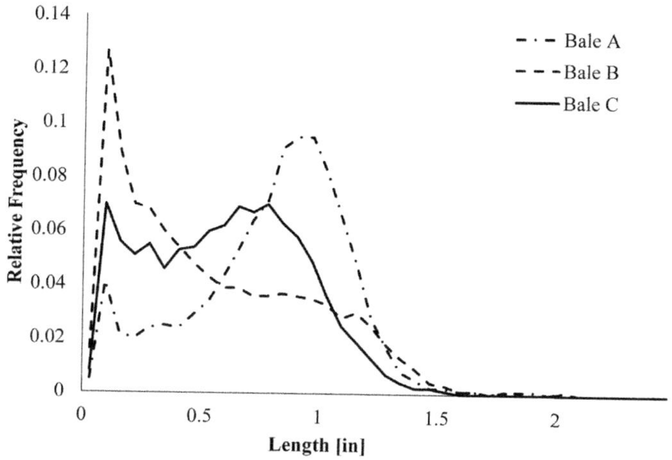

Fig. 7. Length-by-number distributions measured by the Advanced Fiber Information System for 3 commercial cotton bales.

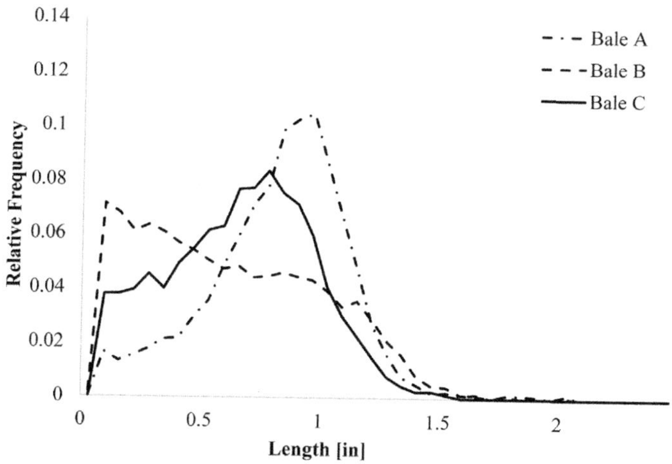

Fig. 8. Length-by-weight distributions measured by the Advanced Fiber Information System for the 3 commercial cotton bales in Fig. 7.

Length-by-weight parameters were traditionally measured using a comb-sorter, a device used to manually sort fibers into groups of similar length. Using weight fractions had the advantage of avoiding the very tedious task of counting individual fibers in each length group. Automated length assessment instruments, such as the AFIS, have facilitated a shift toward length-by-number measurement.

Conversion Between Length-by-Number and Length-by-Weight

A simplified conversion is frequently used to convert between length-by-weight and length by number. However, this conversion assumes a constant linear density, ρ, within a sample (Tallant and Pittman, 1968; Lee, 1967). If this

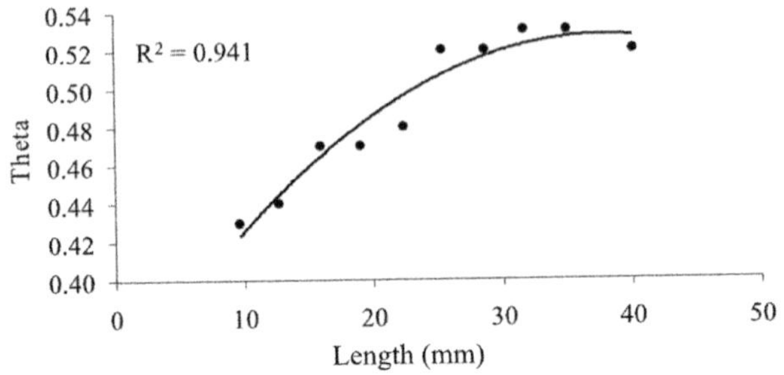

Fig. 9. Maturity (θ) for each length group in a processed cotton (Hequet et al., 2006).

assumption is not true, conversion between the length-by-weight and length-by-number distributions is only possible if the linear density for each length group is known.

Research investigating the variability in the fiber quality within fiber length groups suggests that linear density cannot be assumed to be constant across length groups in processed samples (Taylor, 1977; Hequet et al., 2006). Hequet et al. (2006) investigated the relationship between cotton fiber length and fiber maturity in processed samples. Fibers were first sorted into groups of similar length using a comb sorter (similar to Fig. 6). Cross-section and image analysis techniques were used to determine the distribution of fiber maturity within each of these length groups. Figure 9 shows the average fiber maturity within each of the length groups. Shorter fibers in processed samples tend to be less mature than longer fibers. At a given fiber diameter, immature fibers tend to weigh less per unit of length. Therefore, the shorter fibers can be expected to have a lower linear density when compared to the longer fibers. Shorter fibers in the sample represent a larger portion of the total number of fibers in the sample when compared to the length-by-weight method, which implies that the failure to account for the variability in fiber fineness across length groups overestimates the weight of the short fibers when assuming a constant linear density while converting length-by-number to a length-by-weight measurement.

Fiber Maturity

The primary cell wall of the fiber during development within the boll can be thought of as a circular tube. The center channel of the tube, the lumen, contains all of the organelles of the fiber cell. The primary value of cotton fiber as an industrial raw material comes from the secondary cell wall. The secondary cell wall begins to be deposited within the primary cell wall around the cessation of cotton fiber elongation.

The secondary cell wall results from cellulose deposition within the primary cell wall, which happens continuously throughout the development of the fiber (Kerr, 1937). Environmental variation during fiber development creates the appearance of distinct layers of cellulose within the secondary cell wall (Anderson and Moore, 1937; Haigler et al., 1991). Cotton fibers maintain a circular cross-section during development (Peeters et al., 1986). When the boll opens, the fibers dry, and the cross-section of the fiber collapses.

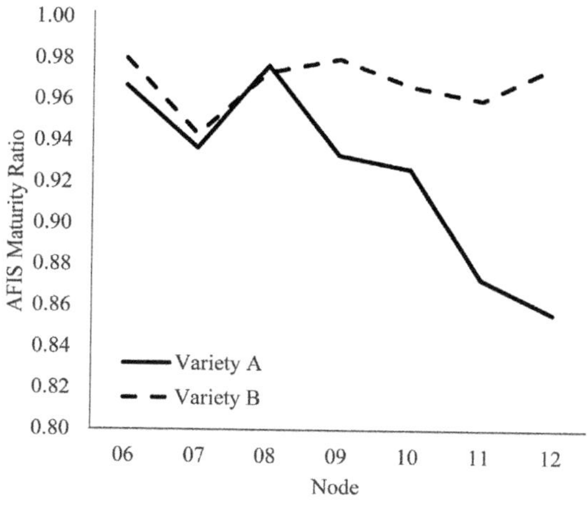

Fig. 10. Within-plant variability in fiber maturity for two commercial varieties.

Fiber maturity (θ) can vary within a plant, within a boll, and even across a single seed (Bauer et al., 2009; Feng et al., 2011). Figure 10 shows the within-plant variability for two commercial varieties. The AFIS maturity ratio (see Maturity and Fineness Complex) is shown for each first-position boll across nodes 6 to 12 for two commercial varieties. Variety A exhibits a noticeable drop in maturity moving toward the top of the plant. In contrast, Variety B remains relatively stable across the nodes. Variability in fiber maturity across nodes is a potential source of within-plant variability.

The distribution of fiber maturity in a sample is best within the boll before opening. Weathering before harvesting, ginning, and cleaning all have the potential to weaken or break fibers and degrade the distribution of fiber maturity. When a fiber breaks, it creates two broken fibers with the same maturity as the parent. Fiber breakage creates a buildup of immature fibers as poorly developed immature fibers break more frequently during processing.

Variability in cotton fiber maturity can have a negative impact on the finished textile. Cellulose in the secondary cell wall is important for absorbing dye. Immature fibers do not have as much cellulose as mature fibers, leading to differences in dye uptake between mature and immature fibers (Goldthwait et al., 1947). Differences in dye uptake can result in imperfections in the finished textiles. Variability in fiber maturity, with the potential to influence dye uptake, is difficult to detect before the dying stage. Once the fabric is dyed, there is little remedy for dying imperfections. Textiles exhibiting dye imperfections must be discounted and result in lost profits for the mill.

The definition of fiber maturity is best understood with an example. Cross-sections of an immature and a mature fiber are show in Fig. 11a and 11b, respectively. The cross-sections shown in Fig. 11 exhibit a collapsed form typical of dried fibers. Despite their collapsed state, the outside perimeter of the fiber (P_2) and the perimeter of the lumen (P_1) are clearly visible. These two variables are an important part of the definition of fiber maturity (Hequet et al., 2006).

The outside perimeter (P_2) and the perimeter of the lumen (P_1) of the fibers in Fig. 11 are illustrated with the idealized fiber cross-sections in Fig. 12. The outside perimeter of each fiber defines a circle that encompasses the total area (A_t) of the

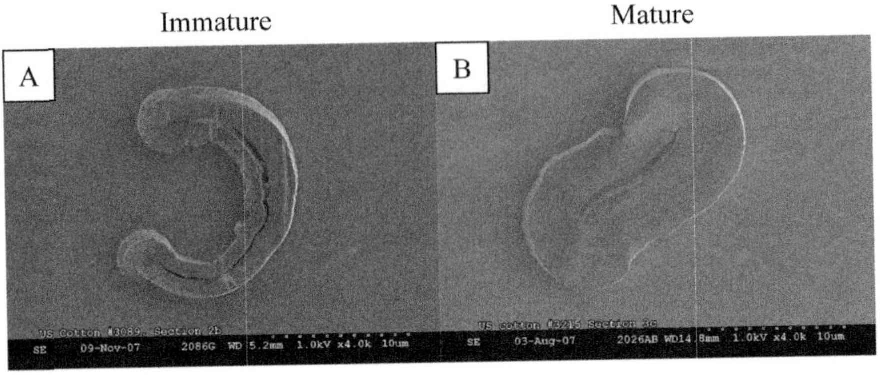

Fig. 11.Cross-sections of an immature and mature fiber. (A) The immature fiber exhibits poor secondary cell wall development compared to (B) the mature fiber (Margaret Pate, CSIRO Australia).

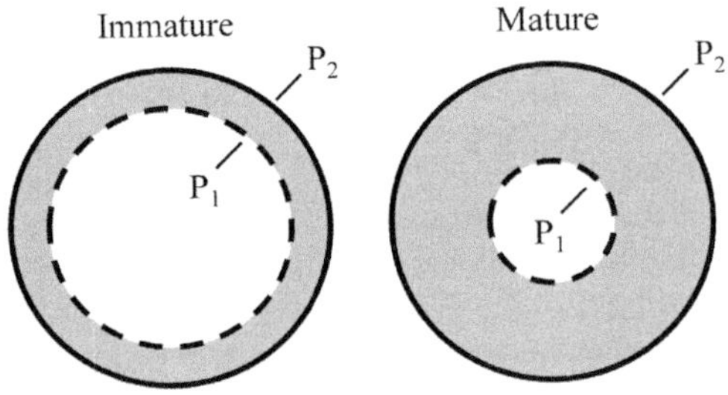

Fig. 12. Idealized cross-sections of an immature and mature fiber from Fig. 6.

fiber. The perimeter of the lumen defines a circle (lumen) that encompasses the area within the fiber cross-section with no cellulose development (A_l). The gray areas indicate cellulose deposition.

The difference between the total idealized area (A_t) and the area encompassed by the lumen (A_l) is used to define the total area of cellulose in the cross-section (A_w).

$$A_w = A_t - A_l \qquad [1]$$

Fiber maturity, θ, is a unit-less measure of the degree of secondary wall thickening. The degree of secondary cell wall thickening, fiber maturity, is defined to be the area of the cell wall, A_w, relative to the area of a circle having the same perimeter, P_2, as the fiber section (Peirce and Lord, 1939).

$$\theta = \frac{4\pi A_w}{P_2^2} \qquad [2]$$

While θ is defined to mathematically range between 0 and 1, neither extreme is possible. A cotton fiber develops as a biological cell. In order for θ to be 0, there can be no primary or secondary cell wall. In order for θ to be 1, there can be no lumen.

Fiber Maturity vs. Physiological Maturity

Fiber maturity should not be confused with the physiological maturity of the cotton plant. Fiber maturity depends on many factors beyond physiological maturity, such as genetics, environmental conditions, and agronomic practices (Kohel and McMichael, 1990; Pettigrew, 1995; Bange et al., 2010).

Seeds from a mature cotton boll can produce cotton fibers exhibiting a wide range of maturity (Feng et al., 2011). The average fiber maturity of the bolls produced within the plant can vary throughout a mature cotton plant (Bauer et al., 2009). Because of the indeterminate nature of the cotton plant, an immature cotton plant may produce mature cotton fibers in fully mature bolls on lower branches.

In addition, a physiologically mature cotton plant can only produce mature fibers if it has the genetic potential to do so. For example, the *im* mutant does not produce mature cotton fibers, even when reaching physiological maturity under ideal conditions (Kohel and McMichael, 1990).

Maturity Distribution

The fiber maturity distribution is used to completely characterize the within-sample variability in fiber maturity, θ. The reference method for determining cotton fiber maturity is to measure the fiber perimeter and wall thickness of cotton fiber cross-sections with an image analysis software (see the section on Cross-Sections) (Hequet et al., 2006). This method also provides the most complete characterization of the within-sample distribution of cotton fiber maturity. The AFIS can also be used to characterize the within-sample distribution of cotton fiber maturity (see Maturity and Fineness Complex). A Cottonscope provides the fiber maturity of fiber snippets along with the distribution of maturity among the snippets, which will be expanded in a later section.

The within-sample distribution of fiber maturity based on fiber cross-sections for three commercial bales is shown in Fig. 13. These distributions reveal the amount of variability that exists among commercial bales. Each of the bales

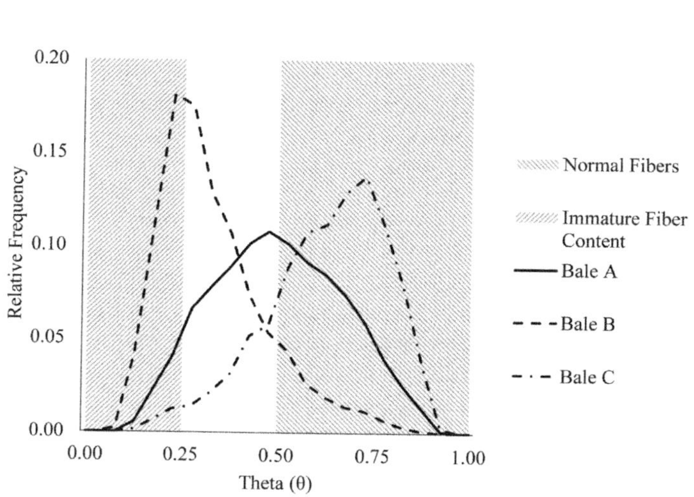

Fig. 13. The relative-frequency distribution of fiber maturity from three commercial cotton bales using the cross-section method.

also exhibits substantial within-sample variability. An operating rule is to designate fibers with $\theta \geq 0.50$ as "normal" and those with $\theta < 0.25$ as immature fibers.

The within-sample distribution in maturity is different across all three bales. The within-sample maturity distribution of Bale B is heavily skewed to the right, indicating high levels of immature fibers. The within-sample maturity distribution of Bale C is quite different; it exhibits a skew to the left indicating a high level of mature fibers. The within-sample distribution of Bale A is symmetric, indicating similar levels of fibers on either side of $\theta = 0.50$.

While all bales exhibit a full range of fiber maturity levels, Bale C represents a superior fiber maturity profile. With all other cotton fiber qualities being equal, it would be expected to perform best in spinning. Conversely, Bale B would be expected to perform worst.

Maturity Ratio

Deducting the percentage of the immature fiber content (IFC with $\theta < 0.25$) from the percentage of normal fibers (N with $\theta \geq 0.50$) in a sample leads to an important measure of fiber maturity (Peirce and Lord, 1939). The maturity ratio, M, is defined in the following (Lord, 1981):

$$M = \frac{N - \mathrm{IFC}}{200} + 0.7 \qquad [3]$$

The maturity ratio and the average maturity of a sample have a mathematical correspondence. The average maturity of a sample, θ, can be converted to a maturity ratio using (Peirce and Lord, 1939; Hequet et al., 2006):

$$M = \frac{4\pi A_{\mathrm{W}}}{0.577 P_2^2} = \frac{\theta}{0}0.577 \qquad [4]$$

Fiber Fineness

The fineness (H) of a cotton fiber is defined to be the fiber weight (mass) per unit length, otherwise known as the linear density. Thus, this definition does not involve the measurement of either the fiber diameter or perimeter. Gravimetric fineness is calculated as a ratio of the total weight of fiber in the sample to the total length of fibers in the sample ($\Sigma W/\Sigma L$), where W is the weight of an individual fiber and L is the corresponding length. Because the mass of each fiber is small, the error associated with weighing each fiber individually is great relative to its mass; therefore, this procedure does not provide precise estimates.

An alternative procedure that is used in the United Kingdom is to measure the gravimetric fineness on only the center portions of the fibers. The procedure involves parallelizing a bundle of fibers clamped at one end by combing; cutting a set length, such as 1 cm, from the center of the bundle; weighing the cut length bundle; and counting the number of fibers.

Using fiber cross-section analysis, the linear density of a cotton fiber can be determined if the area of the secondary cell wall, A_{w}, is known (Eq. [5]). The density of the cellulose in the secondary cell wall, ρ, is 1.52 g cm^{-3}. When A_{w} is the cross-sectional area of the secondary cell wall in cm^2, and ρ is the density of the cellulose in the secondary cell wall in g cm^{-3}, H is expressed in grams per unit cm (g/cm). The most common unit of measure for the linear density of cotton fiber is millitex (mtex = mg/km). The formula is shown in the following (Hequet et al., 2006):

$$H = \rho A_W \left(\rho = 1.52 \text{ g cm}^{-3}\right) \quad [5]$$

Standard Fineness

Standard fineness, H_s, is the fiber linear density, H, standardized by the maturity ratio, M (Peirce and Lord, 1939; Hequet et al., 2006):

$$H_s = \frac{H}{M} = 0.577\frac{H}{\theta} \quad [6]$$

Unlike fineness, standard fineness is proportional to fiber perimeter, P:

$$P = \sqrt{\frac{4\pi}{(0.577)(1.52)}H_s} = 3.7853\sqrt{H_s} \quad [7]$$

Fineness Distribution

Fiber fineness within a sample of cotton fiber is highly variable. The variability in cotton fiber fineness, H, within a sample is characterized by the fiber fineness distribution.

The fiber fineness distribution can be determined by measuring cell wall area through the cross-section method (Eq. [5]). The within-sample distribution of fiber fineness based on fiber cross-sections for three commercial bales is shown in Fig. 14. These are the same three bales shown in Fig. 13. These distributions reveal the amount of variability that exists among commercial bales. Each of the bales also exhibits substantial within-sample variability.

Recall that Bale A was an "average" bale in terms of maturity, yet it had the finest fibers on average. Bale C was the most mature but had the coarsest fibers on average. Bale B was the least mature but was in between the other two bales regarding fineness. These fineness-maturity results indicate a complex relationship between these two properties, which is explored further in the next section.

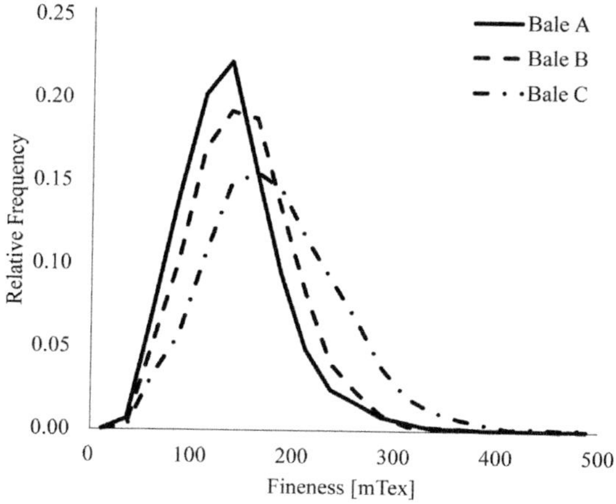

Fig. 14. The relative-frequency distribution of fiber fineness from three commercial cotton bales using the cross-section method.

There are many ways of determining fiber fineness and the distribution of fineness within a sample. AFIS provides the average fineness of the fibers in a sample along with the within-sample distribution of fiber fineness (see Maturity and Fineness Complex). The distribution of fiber fineness may also be calculated from the cross-sectional area determined by fiber cross-sections with Image Analysis (see section on Cross-Sections) by multiplying the fiber cross-sectional area by the density of cellulose in the secondary cell wall. The FAVIMAT may also provide fiber fineness of individual fibers using a vibroscope; however, this method is reliable only for fibers that can be tested with a gauge length of at least 10 mm (see section on FAVIMAT). Cottonscope provides a measure of fineness based on fiber snippets but does not provide the distribution among the snippets (see section on the Cottonscope).

Fiber Maturity-Fineness Complex

The cotton fiber maturity-fineness complex (MH) is conceptually well developed. The cross-sectional area of cellulose is an important element in both fiber maturity and fineness. Fiber maturity is the area of the secondary cell wall relative to the area of a circle having the same perimeter. Fiber fineness, on the other hand, can be expressed in terms of the cell wall area and the density of cellulose.

Fiber maturity, θ, is a relative measure of cellulose development. Fiber maturity cannot be interpreted as the amount of cellulose development in a fiber without some knowledge of the fiber perimeter. At a given level of maturity, a fiber with a small perimeter will contain less cellulose when compared with a fiber with a larger perimeter. Conversely, two fibers with different perimeters and the same amount of cellulose development have different maturity values.

The idealized fiber cross-sections in Fig. 15 provide an example of the nature of fiber maturity. The fiber perimeter increases along the x axis, while the cell wall

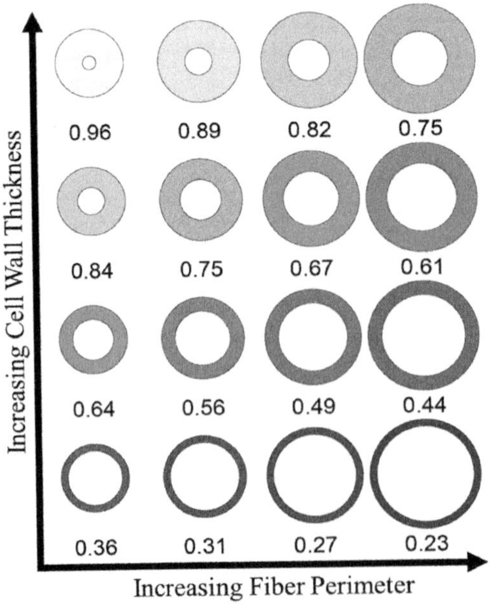

Fig. 15. Idealized cotton fiber cross-sections, with fiber maturity (θ) expressed below each cross-section. Lighter shades indicate more mature fibers.

Fig. 16. Idealized cotton fiber cross-sections with different fiber perimeters and the same cell wall area.

thickness increases along the *y* axis. An increase in the fiber perimeter results in a less mature fiber for a given cell wall thickness. If the fiber perimeter is held constant, an increase in fiber cell wall thickness corresponds to an increase in fiber maturity. This illustration gives an impression that the most mature fibers tend to have a smaller perimeters and a thicker cell walls. However, fibers with a larger perimeter may also be mature.

Fiber fineness in cotton is a function of the cross-sectional area of the fiber and the density of the material. In fibers with no lumen, such as synthetic fibers, or if the fiber has a circular cross-section, fiber fineness gives an indication of the fiber diameter. The lumen in the cotton fiber complicates such an interpretation of fiber fineness because it means that cotton fibers with different perimeters can have the same area of cellulose in the cross-section.

Figure 16 illustrates the importance of considering the lumen when interpreting cotton fiber fineness. Fibers to the left tend to have a smaller perimeter and a smaller lumen. Fibers to the right have a larger perimeter and a larger lumen. Despite having different perimeters, all of the fibers in Fig. 16 have the same cross-sectional area and linear density (Eq. [5]). The lumen allows fibers with markedly different perimeters to have the same fineness. Yet, these fibers would exhibit dissimilar spinning performances.

Fiber Tensile Properties

Many diverse mechanical stresses are applied to cotton fibers, from harvest through to the finished textile products. Fibers with superior tensile properties are valued because these are able (1) to resist breakage and subsequent degradation in quality during industrial processing and (2) to produce stronger spun yarns and higher quality finished textiles (Faulkner et al., 2012; El Mogahzy et al., 1990).

Individual Fiber vs. Fiber Bundle Tensile Testing

Cotton fiber tensile properties may be evaluated on either fiber bundles (see later sections on HVI and Stelometer) or individual fibers (see FAVIMAT; Instron). Bundle testing is most frequently performed on beards of fibers (HVI), or slack bundles (Stelometer). Individual fiber and bundle tensile property tests provide different results. Bundle tensile tests depend on several factors not tested by individual fiber tensile testing; important factors include the within-sample distribution of fiber tensile properties and friction. Individual fiber tensile properties are compared to bundle tensile properties in Individual Fiber vs. Fiber Bundle Tensile Testing.

Load-Elongation Curve

The load-elongation curve completely describes the tensile behavior of the sample, whether a bundle or individual fiber, under an applied load. A selection of cotton fiber load-elongation curves for individual cotton fibers (see FAVITMAT) are shown in Fig. 17. These fibers, selected from a single commercial bale, reveal a wide range of tensile properties among cotton fibers within a single bale.

The load-elongation curve is useful for determining the important aspects of the force-elongation relationship. For example, it may be of interest to know how much of a constant load will rupture a sample or how much the sample will elongate before it ruptures. It may also be interesting to know how a sample will respond to forces experienced at the gin stand, where fibers may be exposed to a sudden burst of energy. The results can be used to characterize how both individual cotton fibers and bundles of cotton fibers will respond to different types of mechanical forces.

Breaking Force

The peak of the load-elongation curve characterizes the point of rupture for the sample being tested. The force required to rupture an individual fiber or bundle of fibers is called the breaking force, or breaking load. The force required to break cotton fiber bundles is commonly expressed as a load, in gram-force (gf). The breaking force for individual cotton fibers is commonly expressed in centinewtons (cN). The breaking load defines the strength of the material. However, strength is a more general term that is more frequently used to describe the sample tenacity described in the next section. For example, HVI strength is a measure of bundle tenacity (ASTM, 2013a).

Figure 18 shows the within-sample breaking force distributions from two commercial bales of cotton characterized by testing 450 individual fibers (FAVIMAT at 3-mm gauge length) from each bale. Bale B tends to have more strong fibers than Bale A, leading to a higher average breaking force. The average breaking force for fibers in Bale A is 4.41 cN, while Bale B is 5.32 cN. However, the average breaking force for each bale does not fully characterize the distribution of fibers within the bales. Both Bale A and Bale B are composed of fibers exhibiting a wide variability in breaking force characteristics.

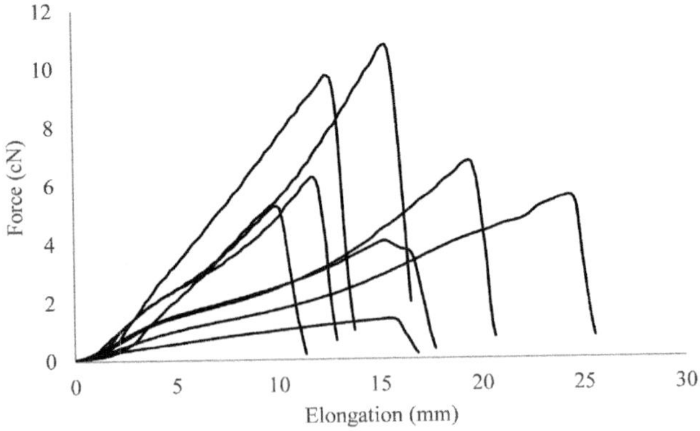

Fig. 17. Load-elongation curves of individual cotton fibers generated with FAVIMAT.

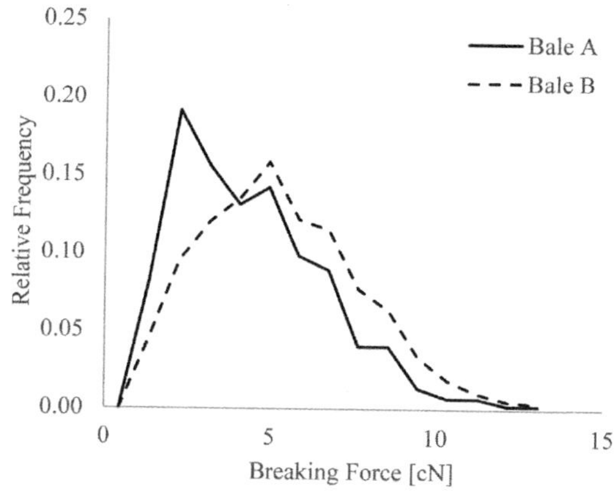

Fig. 18. Within-sample distribution of individual fiber breaking force for two commercial bales.

Tenacity

A bundle of fibers may be stronger due to the intrinsic strength of the material or because it has a greater linear density. Tenacity provides a method of comparing the specific strength of samples by adjusting for the linear density of the bundle. Cotton fiber bundle tenacity (g/tex) is simply the load required to break the bundle (g) standardized by the fiber linear density of the bundle (tex).

$$\text{Bundle Tenacity}\left(\frac{\text{g}}{\text{tex}}\right)=\frac{\text{Load}(\text{g})}{\text{Bundle Linear Density }(\text{tex})} \quad [8]$$

Cotton fiber exhibits a wide range of variability in linear density among individual fibers. Tenacity is used to compare the strength of individual fibers, standardized by linear densities. Individual fiber tenacity is found by dividing the force required to break the individual fiber (cN) by individual fiber linear density (tex) (Morton and Hearle, 2008).

$$\text{Individual Fiber Tenacity}\left(\frac{\text{cN}}{\text{tex}}\right)=\frac{\text{Breaking Force }(\text{cN})}{\text{Fiber Linear Density }(\text{mTex})} \quad [9]$$

Conversion from cN/Tex to kN × m/kg

The most common tensile property test is bundle tenacity provided by HVI. While HVI testing is detailed later in the chapter, the following is a conversion factor to express tenacity results in kN × m/kg (the SI unit required to publish in a peer reviewed journal).

$$1 \text{ gramforce} = 9.807\times10^{-6} \text{ kN} \quad [10]$$

$$1 \text{ tex} = \frac{1\times10^{-3} \text{ kg}}{1\times10^{3} \text{ m}} \quad [11]$$

$$1 \text{ g tex}^{-1} = \frac{9.807\times10^{-6}\text{kN}}{10^{-3}\text{kg}/10^{3}\text{m}} = 9.807 \text{ kN}\frac{\text{m}}{\text{kg}} \quad [12]$$

Thus, HVI strength in g/tex is multiplied by 9.807 to be expressed in kN × m/kg.

Elongation

A single cotton fiber or bundle of cotton fibers will first elongate before breaking in response to a breaking load or force. If the force is not sustained, the fiber may elongate and not break. Elongation is expressed as the percentage of change in an individual fiber or bundle length before rupture under a breaking load (Morton and Hearle, 2008).

The within-sample distribution of individual fiber elongation is shown in Fig. 19. These distributions are based on individual fiber tensile property testing (FAVIMAT) of 450 fibers from two commercial bales at a 3-mm gauge length. At 10.2%, the fibers in Bale A have a lower average elongation-at-break, compared to an average elongation-at-break of fibers in Bale B of 14.1%. The average elongation-at-break does not fully characterize the elongation characteristics of the fibers within the bales. Both Bale A and Bale B are composed of fibers exhibiting a wide variability in elongation-at-break.

Work-to-Break

The energy required to break a fiber is defined by the area under the load elongation curve. A combination of both the force required to break the fiber and the elongation characteristics of the fiber contribute to the work required to rupture the fiber. Work-to-break (cN×cm) may be measured on bundles of fibers (Stelometer) or individual fibers (FAVIMAT).

Work-to-break characteristics may vary widely across samples and across fibers within a single sample. Figure 20 shows the within-sample distribution of work-to-break characteristics for two commercial bales, determined by individual fiber testing (FAVIMAT at 3-mm gauge length). The average work-to-break of fibers in Bale A is 0.066 cN×cm, lower than the average work-to-break of Bale B (0.099 cN×cm). The work-to-break distribution for each bale shows a more complete picture. Both bales are composed of cotton fibers exhibiting a wide range of work-to-break characteristics. However, Bale B has a greater number of high work-to-break fibers.

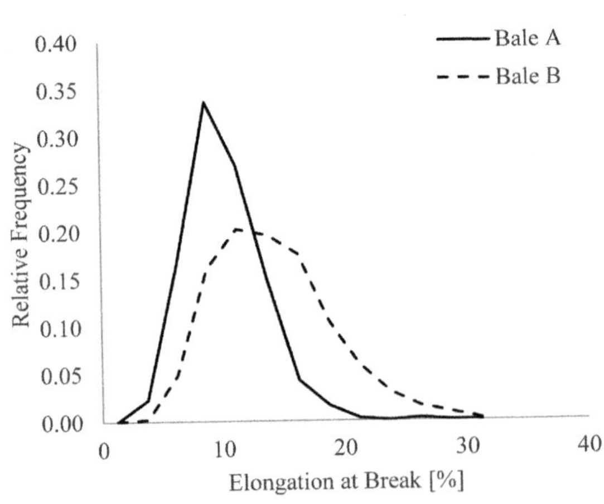

Fig. 19. Within-sample distribution of individual fiber elongation for two commercial bales.

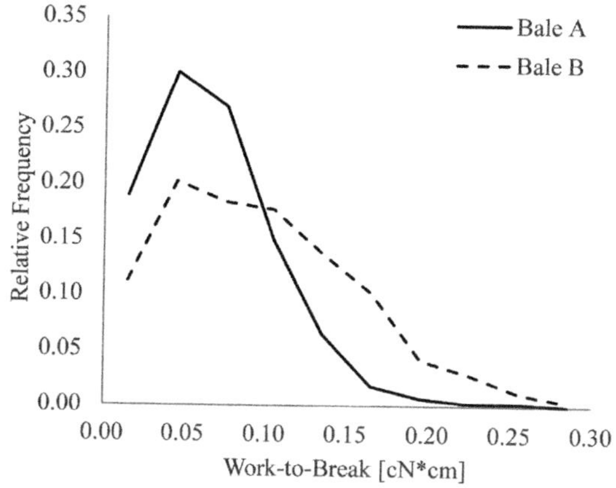

Fig. 20. Within-sample distribution of individual fiber work-to-break for two commercial bales.

Gauge Length

Gauge length is the spacing between the clamps used to rupture the fiber sample in tensile property testing. Most tensile tests on bundles of cotton fibers are set at either 0 ("null") gauge or 3.2-mm (1/8-in) gauge. In bundle testing, higher gauge lengths have an increased probability of containing fiber defects that may lead to a point of rupture. Therefore, higher gauge length tests tend to require a lower force to break the bundle. Non-zero gauge length bundle tensile properties have shown better relation to yarn tensile properties (Hertel and Craven, 1956; Ramey et al., 1977).

Gauge length is also important in individual fiber tensile property testing. Short fibers created by fiber breakage during processing tend to be weaker. These low-tensile-strength short fibers cannot be evaluated at higher gauge lengths. Therefore, evaluating tensile properties of individual fibers from a sample at higher gauge lengths may tend to overestimate the average elongation and tenacity of fibers in a sample. Gauge length is not as standardized for individual cotton fiber testing. Gauge lengths of 3 and 10 mm are common but others may be used.

Individual Fiber vs. Fiber Bundle Tensile Testing

Bundle tests are popular because they are quick and inexpensive. However, tensile tests for bundles vs. individual fibers measure very different relationships. The relationship between elongation and tenacity provides an illustrative example. The relationship between bundle fiber elongation and bundle fiber strength for 104 commercial bales of cotton reveals a weak negative correlation (Fig. 21).

The relationship within each sample is quite different. Figure 22 shows an example of the relationship between individual fiber elongation and strength (FAVIMAT) for a randomly selected bale in Fig. 21. When considering individual fibers, within a sample, stronger fibers tend to exhibit higher elongation.

These differences underscore the importance of selecting an appropriate testing methodology. Several questions should be considered when deciding on the appropriate tensile test. Is the goal to characterize forces acting on an individual fiber or a bundle of fibers? If the measurements are being used for screening breeding lines, what type of processing improvement is being selected? No single tensile property can characterize the performance of a cotton sample under all conditions.

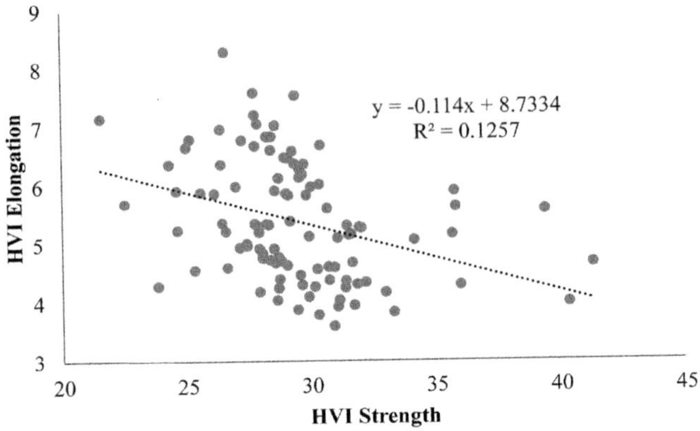

Fig. 21. Bundle elongation vs. bundle strength (High-Volume Instrument) for 104 commercial bales.

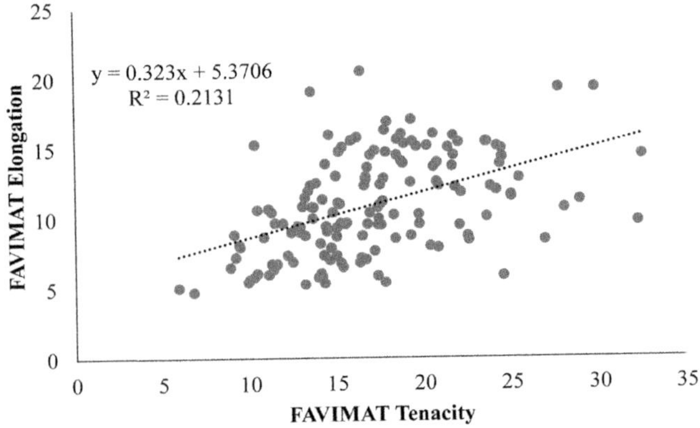

Fig. 22. Individual fiber elongation vs. individual fiber strength (High-Volume Instrument) for a randomly selected bale in Fig. 21.

Color

White cottons are desirable in the production of spun yarns because they provide a uniform substrate for dyeing and finishing. Cotton color is a measure of how much a sample of cotton fiber deviates from white. Historically, color classification was performed visually by classers. HVIs provide two parameters of color, reflectance (Rd) and yellowness (+b). The color classification equivalent is determined by looking up Rd and +b on a color chart (Fig. 23).

For example, a sample with 71.7 Rd and 8.2 +b is equivalent to a 41–4 (white, strict low middling). Adjustments were made to the color chart in 1999 and 2008 to bring it in line with human classification. These adjustments resulted in a wavy appearance to the color classification demarcation lines.

Discoloration of a cotton sample may indicate underlying problems such as staining or microbial deterioration. Grayness may indicate exposure to conditions that led to reduced fiber strength. Field-weathered cotton tends to exhibit

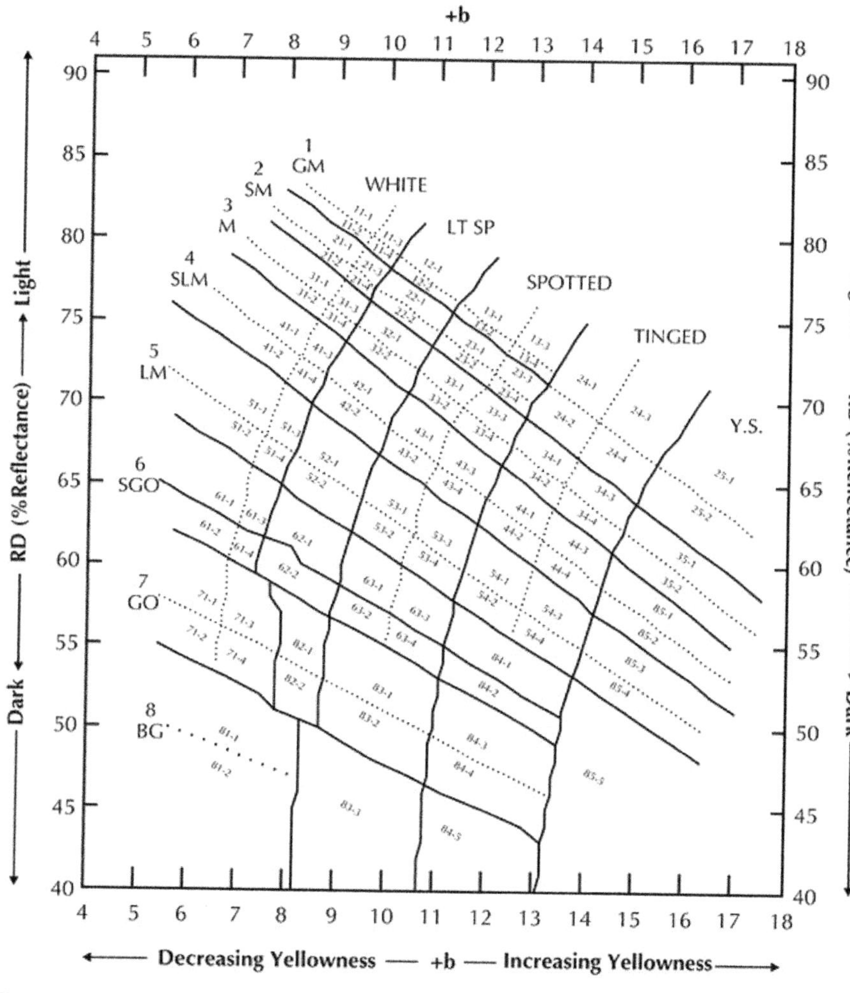

Fig. 23. Official USDA cotton color chart (Cotton Incorporated, 2014).

a decrease in reflectance (increase in grayness) (Hessler et al., 1954). Long-term storage and storage of bales in substandard conditions may also lead to yellowness and grayness in the lint (Gamble, 2007; Nickerson and Tomaszewski, 1958). Insect damage and staining of the lint may lead to spotting and discoloration. In general, bales with a low yellowness and a high reflectance are most desirable for spinning (El Mogahzy and Chewning, 2001).

Trash and Contaminants

A bale of cotton is a mix of fibers, vegetative contaminants, trash, and other contaminants. Trash and contaminants in the fiber lead to imperfection and lower quality yarns. Once the lint is contaminated, it is difficult to remove the contaminants without damaging the quality of the fiber. Preventing trash and other contaminants must begin within the field.

Seed cotton and lint with excessive trash may require additional cleaning sequences at the gin (Anthony and Mayfield, 1995). Cleaning at the gin is aggressive and tends to damage fibers (Hughs et al., 2013). Trash and contaminant removal at the gin must be balanced with preserving fiber quality (Anthony and Mayfield, 1995).

High levels of trash in the cotton bale will lead to an increased number of ends down, lowering both yarn quality and throughput at the spinning mill (McCreight et al., 1997). Similar to the gin, cleaning cotton fiber is always a compromise with preserving fiber quality. Bales with high levels of trash and contaminants will tend to produce yarn of lower quality, despite cleaning (Lawrence, 2003). On the other hand, excessive cleaning can break fibers, creating short fibers that may also cause problems during processing and imperfections in the yarn (Tallant et al., 1961; Backe, 1986). Managing to minimize trash and other contaminants is necessary to produce cotton fibers that result in high-value yarns. Some of the more common trash and contaminants are outlined in this section.

Neps

Tight entanglements of fibers are called neps. They are made of cotton fibers but are nevertheless classified as contaminants because they hurt the appearance of fabrics. Because neps are composed of cotton fibers, they are difficult to detect. Neps that make their way into the card web will pass through the spinning processes and into the yarn (Lawrence, 2003). Neps are composed primarily of immature fibers and do not uptake dye readily (Hebert et al., 1988). Yarns with excessive neps are of poor quality and have a low market value (Lawrence, 2003).

Seed Coat Neps

The gin stand is an aggressive mechanical process that pulls fibers from the seed coat. A seed coat nep (SCN) is created when a piece of the seed coat remains attached to the cotton fibers as they are pulled from the seed. The fibrous portion of the SCNs allows them to pass through spinning equipment, creating yarn imperfections.

Contaminations and Extraneous Matter

There are many potential sources of lint contamination. A 2013 survey of spinning mills reveals the wide variation in types of trash and contaminants found in bales from all over the world (Fig. 24). Each source of contamination is reported as the percentage of bales reported to have moderate to serious contamination levels. While these survey data are based on the perception of the spinner and not based on objective measurement of contamination, it provides insight into the types of contamination present in commercial bales (ITMF, 2014). Sources of contamination are categorized into five groups: (A) fabric contamination, (B) string contamination, (C) organic matter, (D) inorganic matter, and (E) oily substances/chemicals. In addition, 22% of bales were reported to have stickiness, and 41% had seed coat fragments (ITMF, 2014).

Bark

Stem pieces from the cotton plant, called bark, are a fibrous material and are difficult to remove. Bark fragments may take on the dimensions of fibers during processing, pass through spinning equipment, and be incorporated into the yarn.

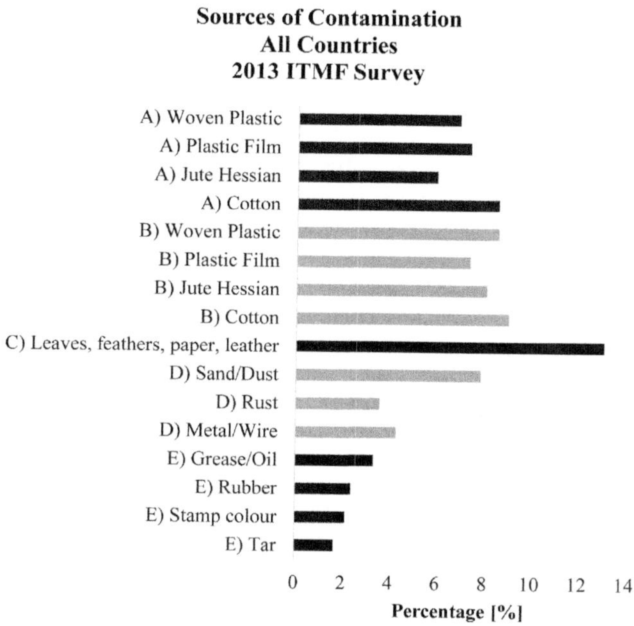

Fig. 24. Sources of contamination from 2013 International Textile Manufacturers survey categorized into five groups. (A) Fabric contamination. (B) String contamination. (C) Organic Matter. (D) Inorganic Matter. (E) Oily substances/chemicals (ITMF, 2014b).

If so, these fragments are highly visible in the fabrics. Therefore, steep discounts are applied to bales that are classified as having bark contamination.

Stickiness

Sticky cotton is one of the worst problems textile mills can encounter. It contaminates surfaces throughout the spinning process, necessitating frequent, thorough cleanings and shortening the operating life of machinery components. The damage is cumulative and yarn quality deterioration worsens as operations continue (Hequet et al., 2005).

There are two primary sources of stickiness in cotton fiber: physiological sugars and entomological sugars. The compositions of these are different, which necessitates different management techniques based on the source of the sugars (Hequet and Abidi, 2006).

The physiological sugars are an integral part of plant growth and fiber development; cotton cellulose itself is a polymer composed of glucose. Stickiness may occur when physiological sugars remain after opening and drying of the lint, for example, following an early freeze that truncates fiber development. However, these simple physiological sugars are readily broken down within a few months by bacteria and fungi (Hequet and Abidi, 2006). Therefore, managed storage of the bales will alleviate the potential for stickiness.

The entomological sugars, generally called "honeydew" because of the glistening appearance on the fibers, are excreted by insects feeding on the cotton plant. These excretions are composed of complex sugars that are not broken down even over long periods of time. Bales contaminated by stickiness composed of insect sugars are often uneconomic to spin. The decreased operational efficiency and damage to machinery is too great. Therefore, "honeydew" cotton is drastically discounted and generally denied access to textile spinning markets. Prevention of such contamination in the field with integrated pest management strategies is the first line of defense against sticky cotton.

There are two primary insect pests that create sticky cotton problems with their honeydew: whiteflies (*Bemisia tabaci*) and aphids (*Aphis gossypii*). The profiles of complex sugars in their honeydew are distinctly different, both requiring different management strategies (Hequet and Abidi, 2006; Hequet and Abidi, 2005). Contamination by whiteflies is worse than by aphids, due to the particular composition of the complex sugars.

Managing Trash in Breeding

Proper management of trash and contaminates is a critical—and often ignored—factor in cotton breeding and research efforts. Small plot samples from breeding and research programs are often harvested without trash management in mind and ginned without any type of seed cotton or lint cleaners. Excessive trash limits the reliability and repeatability of many fiber quality evaluation techniques. Trash and contaminants can increase the variability in fiber quality measurements, effectively reducing the ability to properly discern differences between treatments and breeding lines. The resulting lack of repeatability and reliability of fiber quality assessment can subvert predictions of future spinning performance.

Instrumentation

Myriad measurement instruments have been developed to characterize cotton fiber quality and its variability within and between samples. Each segment of the cotton industry—breeders, biotechnologists, growers, gins, and yarn spinners—has distinct needs regarding such measurements. As previously indicated, the two primary approaches to evaluating fiber quality characteristics are bundle testing and individual fiber testing.

Bundle testing of cotton is the most common fiber quality evaluation technique. The bundle may take the form of a beard for length and strength measurements, a plug for micronaire, etc. (Hertel, 1940; Chu and Riley, 1997; Lord, 1956; Orr et al., 1955). Such measurements may be made faster than those on individual fibers and bundles of fibers are logically related to yarns, which are a *twisted bundle* of fibers.

If a more complete characterization of the fibers is required, individual fiber testing becomes necessary. Fibers may be prepared for individual fiber testing in a bundle such as a sliver or a beard. Individual fibers are then either manually individualized, separated from a bundle by an automated process, or analyzed individually with image analysis (Bragg and Shofner, 1993; Foulk and Mcalister, 2002; Hequet et al., 2006). Individual fiber testing allows for the determination of the distributional behavior of the fiber properties within a sample. Given that cotton fibers exhibit large biological variations in fiber properties, distributional information is critically important for achieving breakthroughs in fiber quality. Since individual fiber measurements are tedious and expensive, they are not generally used on a commercial basis. Nevertheless, such measurements are indispensable for research and development purposes.

Recently, fiber snippet testing is becoming more important. Fibers are cut into small snippets and the quality of the snippets is evaluated (Long et al., 2010a). Fiber snippet testing is useful for fiber quality attributes such as ribbon width, linear density, or fiber maturity. This is often performed in conjunction with either bundle testing or individual fiber testing.

Different applications call for distinct fiber testing techniques. This section provides an introduction to the most common instruments used to measure fiber properties. Each instrument is then introduced with a short description of the technique on which it is based. Finally, a summary of the most common results provided by each instrument are listed along with a brief description.

High-Volume Instrument

HVI is the most commonly used instrument for evaluating cotton fiber quality. HVI quickly evaluates the fiber properties of a sample of cotton and provides a suite of fiber quality parameters. HVI was developed primarily to replace subjective human classers with objective measurements of fiber properties important for marketing bales of cotton fiber to spinning mills. Because of the speed and low cost of HVI data, its measurements are obtained generally regardless of the purpose for evaluation.

HVI testing is a series of fiber quality testing instruments built into one computer-controlled, composite instrument. Each instrument within the HVI is a discrete step designed to quickly measure a separate aspect of fiber quality from a bundle of fibers. A Uster HVI 1000 is depicted with each of the testing stages labeled in Fig. 25. In order, the steps are micronaire, length, strength, color, and

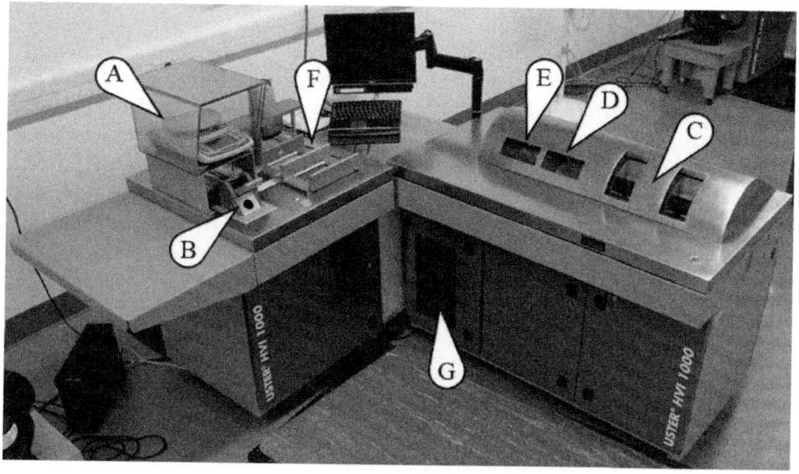

Fig. 25. The Uster High-Volume Instrument (HVI) 1000 at the Fiber and Biopolymer Research Institute, Texas Tech University, Lubbock, TX, with individual steps labeled (Uster HVI 1000). (A) The scale used to weigh samples for micronaire testing; (B) the micronaire chamber; (C) fibro samplers (sampling cage); (D) the window revealing the HVI brush; (E) the window revealing the HVI length and tensile property measurements; (F) color and trash measurement trays; and (G) HVI waste bin.

Fig. 26. A standard USDA sample taken for fiber quality classification. Two samples, cut from two sides of the bale, are used to represent the fiber quality of the complete bale. The sample is sent to a UDSA classification office for High-Volume Instrument fiber quality evaluation. Fiber quality classification results are used to market the bale.

trash measurement. The length and strength measurements are executed with robotics. A single HVI test for cotton classification includes all of these steps, requires about 230 g of lint (approximately 4-ounces are sampled from each side of the bale), and takes about 20 s. A standard USDA sample prepared for HVI classification is shown in Fig. 26.

Each of the steps within the HVI is based on a different measurement principle. Understanding the measurement principles behind each of the HVI parameters is critical to interpreting HVI results.

Maturity and Fineness Complex

Micronaire is measured with a laminar flow of air passing through a porous plug of fibers of a known weight (10 g). The modeling of an air flow in a porous plug is based on the law of Darcy (1856), which were amended later by Kozeny (1927). Kozeny showed that the flow of a liquid through a plug of particles is inversely proportional to the square of the specific surface, if the porosity, the arrangement of the particles, the dimensions of the measuring chamber, and the difference in pressure between the two ends of the sample were constant. Therefore, micronaire is not a separate measure of either maturity or fineness of the cotton fibers. Micronaire is proportional to the fiber's specific surface. The specific surface of a sample of cotton fibers depends on both the fiber fineness and fiber maturity (Lord, 1956). Differences in micronaire between samples may be caused by a change in maturity, fineness, or both. Neither fiber maturity nor fiber fineness can be assessed independently using a micronaire measurement. Figure 27 shows a typical HVI micronaire test.

Micronaire Reading (Mic): "...a function of both fineness and maturity and is related to mill processing performance and to the quality of the end products" (ASTM, 2011).

Both more mature and coarser cotton fibers will allow air to pass through the plug of fibers faster; therefore, they will give a higher micronaire reading. Both

Fig. 27. High-Volume Instrument micronaire test of a sample of cotton fibers. (A) A 10-g sample is placed in the micronaire chamber. (B) In the closed chamber, a flow of air is passed through the sample to determine the micronaire. (C) Once the test is complete, the sample is ejected from the micronaire chamber.

less mature and finer cotton fibers will slow the passage of air through the fibers; therefore, they will give a lower micronaire reading. It is quite possible for fibers that are fine and mature to give the same micronaire reading as fibers that are coarse and immature. Nonfiber contamination and oriented fibers will change the air flow and alter the micronaire reading; therefore, samples should be relatively clean with randomly oriented fibers for the measurement to be accurate.

Fiber Length and Uniformity Index

HVI length measurements are based on the fibrograph theory (Hertel, 1940). HVI length assessment is shown in Fig. 28. The HVI comb is used to collect a length of an unbiased sample of fibers (Chu and Riley, 1997). Fibers protruding from the HVI comb form the fiber beard. The beard of fibers is then brushed to remove extraneous fibers, trash, and fibers that are not securely clamped in the comb and to reduce the effects of crimp.

Once the fiber beard is brushed, a beam of light is scanned over the fiber beard from 0.150 in (3.81 mm) away from the comb to the ends of the fiber. The density of the light transmitted through the sample as the sensor scans along the fiber beard is used to generate the fibrogram. HVI fiber length parameters are calculated from the fibrogram (Chu and Riley, 1997).

The primary length measure provided by HVI is the UHML. The definition of UHML is unusual because it is based on both the fiber frequency and weight fractions of the fiber (ASTM, 2013b). UHML is the most common fiber length measurement and is frequently called "Fiber Length." There are many other parameters that can be used characterize the length of fiber in a sample. For

Fig. 28. The High-Volume Instrument (HVI) evaluating the length parameters of a fiber beard. (A) The fiber beard is brushed to remove extraneous fibers, trash, fibers that are not securely clamped in the comb, and to reduce the effects of crimp. (B) The cotton fiber beard is in front of the light beam sensor. Only the fibers extending from the HVI comb are included in the HVI length test. (C) A light is scanned over the fiber beard.

example, AFIS mean length-by-number is the average fiber length based on the complete distribution of fiber length within the sample.

Upper Half Mean Length (in or mm): "the mean length by number, of the longer one half of the fibers by weight (ASTM, 2013a)."

HVI also outputs the ratio of the mean length to the UHML, defined as the uniformity index (UI). Cotton fibers that are more uniform in length will have higher UIs than fibers that are less uniform in length (Hertel, 1940).

Uniformity Index (%): "the ratio between the mean length and the upper-half-mean length expressed as a percentage of the upper-half-mean-length" (ASTM, 2013a).

Tensile Properties

After the HVI has generated the fibrogram, the fiber beard is used to measure fiber bundle tensile properties. A standard HVI tensile property test is shown in Fig. 29. Fibers protruding from the HVI comb are clamped with a 1/8-in (3.2-mm) gauge length and then broken with a constant rate of loading. This allows measurement of breaking strength and elongation of the bundle.

Strength (g/tex): "an industry term for breaking tenacity reported in grams per tex" (ASTM, 2013a).

Elongation at Breaking Force (%): "the elongation corresponding to the maximum force, and expressed as a percentage of the 1/8-in (3.2-mm) gage length" (ASTM, 2013a).

The lack of calibration standards for HVI elongation prevents the comparison of HVI elongation across different machines. However, it is a quite stable measurement when repeated on the same HVI. Therefore, a relative measure of HVI elongation can be obtained for different cottons that are tested on the same HVI. For an ongoing project comparing different cottons on a single HVI, it is only necessary to monitor the elongation measurement for possible drift over time, to use the elongation results to differentiate among cottons (Hequet et al., 2014).

Color and Trash

HVI measures color using filters and a pair of photo detectors. A standard HVI color test is shown in Fig. 30. HVI color is expressed as reflectance (R_d) and the

Fig. 29. The High-Volume Instrument (HVI) tensile test of a fiber beard. (A) The same beard of cotton fibers used in length testing is used in strength testing. (B) Only a portion of the fiber beard remains in the HVI comb after the completion of the tensile property test.

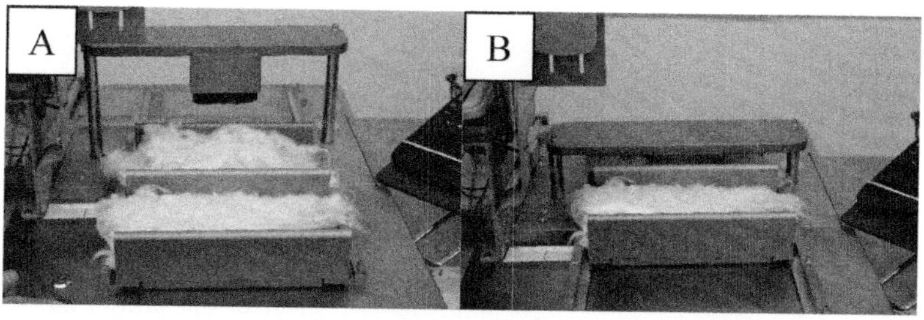

Fig. 30. Two samples are placed in the High-Volume Instrument color and trash test. (A) The test begins with two portions of the sample placed in trays below the sample press. (B) During the test, each sample is pressed against a viewing window below the sample. A colorimeter and black and white camera housed behind the viewing window are used to read the color and trash content of each sample.

degree of yellowness (+b) of the sample (Nickerson, 1951; Nickerson et al., 1950). Trash is measured using a simple image analysis processing routine on black and white images.

R_d and +b: "the daylight color of opaque cotton specimens represented by points in a space as described by Hunter (1958) in terms of two color scales: reflectance, R_d, and the chromaticity coordinates for yellowness, + b" (ASTM, 2013a).

Trash Area: "the ratio of total area of trash on the surface of a sample of cotton to that of the area of the viewing window, expressed in a percentage of the area of the viewing window" (ASTM, 2013a).

Trash Count: "a number correlated with the total number of pieces of trash on the surface of the sample of cotton over the viewing window" (ASTM, 2013a).

Leaf Grade: Estimate of the leaf content that is used in USDA classification. The leaf grade estimate is based on the trash area and trash count.

Advanced Fiber Information System

AFIS is an individual fiber tester first developed to provide spinning mills with a tool for characterizing the nep content of samples (Fig. 31). The system has been expanded to characterize within-sample variability of many fiber qualities within a cotton bale. More recently, the fiber-quality parameters that are provided by AFIS testing have begun to be used in breeding and other areas of cotton research.

Samples are prepared into a loose bundle of fibers, called a sliver, and are placed in a canister (Fig. 32). Inside the AFIS, the sliver is fed through a series of belts and rollers into a spinning metal rotor called the AFIS fiber individualizer. The AFIS individualizer is covered with metal teeth that pull fibers from the sliver and separate trash particles. Fiber individualization is an aggressive mechanical process with the potential to break the fibers. Individualized fibers are drawn into an airstream as they are pulled free from the sliver. The airstream presents the fibers to electro-optical sensors that measure the length, maturity, and fineness of the individual fibers (Fig. 33).

A single fiber waveform is returned by the electro optical sensors when the fiber is presented in a conformation suitable for measurement. The single fiber waveform is characterized by sharp shoulders and a broad plateau. Neps,

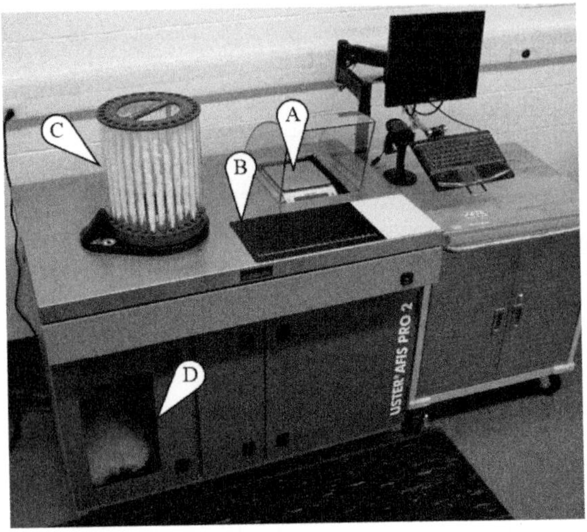

Fig. 31. The Advanced Fiber Information System fiber quality measurement instrument at the Fiber and Biopolymer Research Institute, Texas Tech University, Lubbock, TX (Uster AFIS Pro 2). (A) The scale used to weigh out 0.5-g samples of lint for testing; (B) the mat with the length scale that is used to pull the sliver to the proper length; (C) sliver canister (shown empty); and (D) the waste bin.

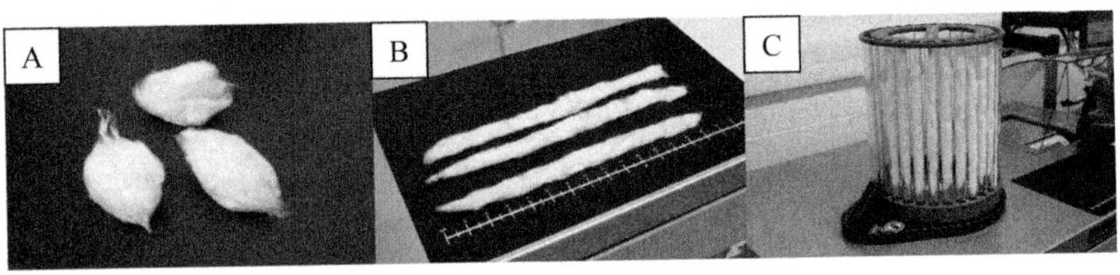

Fig. 32. Sliver preparation and loading the Advanced Fiber Information System (AFIS). (A) Three 0.5-g samples of cotton fiber are taken. (B) The samples are pulled into loose bundles called slivers. (C) The slivers are loaded into a canister for feeding into AFIS.

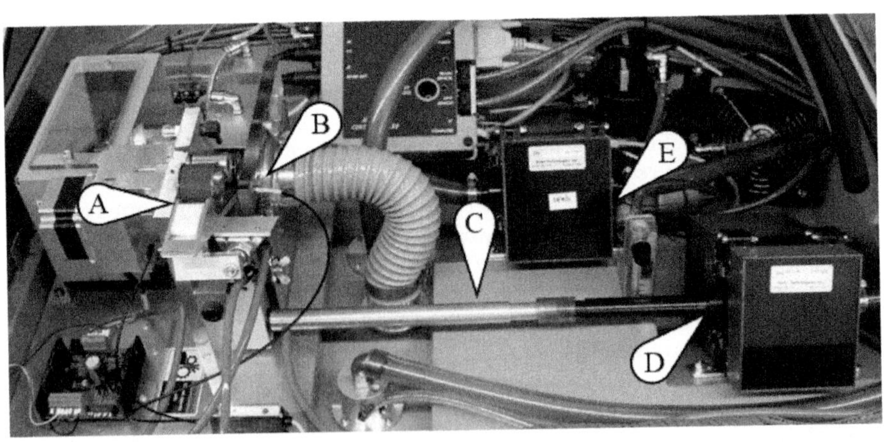

Fig. 33. Inside of the Advanced Fiber Information System (AFIS). (A) The location where sliver is fed into the AFIS from the external canister; (B) the fiber individualizer; (C) airflow used to present fibers to the sensor; (D) the sensor used to measure the physical properties of the fiber; and (E) the sensor used to measure the physical properties of trash and other contaminants.

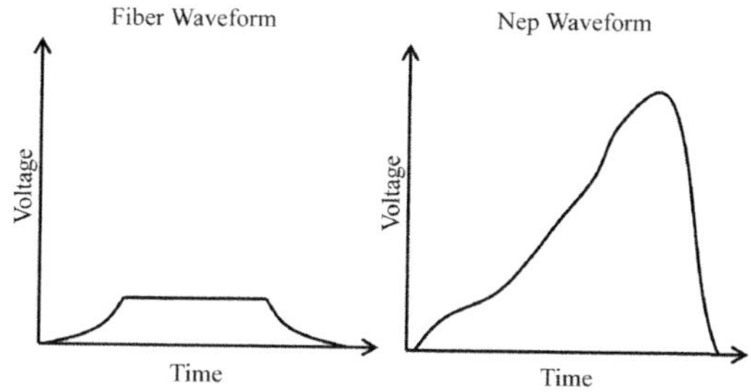

Fig. 34. Idealized example of waveforms returned by the Advanced Fiber Information System electro-optical sensors. The waveforms in this figure illustrate the distinction between fibers and neps. The fiber waveform has steep sides that define a plateau while the nep waveform has sloping sides that define a peak (Baldwin et al. 1995).

measured by the same sensor, return a single nep waveform. The single nep waveform is characterized by sloping shoulders that form a peak, which is typically much taller than the single fiber (Fig. 34).

Heavier and aerodynamically dissimilar trash particles are pulled into a second airstream for assessment. A sensor unit in the second airstream measures the size of each trash particle. The electro-optical return signal is used to characterize the size of each trash particle (see item E in Fig. 33).

AFIS summarizes the within-sample distributions of fiber properties into histograms. Within-sample fiber length is summarized with a histogram of 40 bins while within-sample maturity and fineness are summarized with a histogram of 20 bins. This distributional data is used to provide various measurements of useful properties. The most common measurements are explained in this section.

Fiber Length

AFIS uses the return signal from the electro-optical sensors to measure the length of each fiber. The distribution of within-sample fiber length is binned into 40 discrete groups of similar length (0.0625-in wide) from 0 to 2.5 in (0 to 63.5 mm).

Fiber-Length Distribution ("By Number" (n) or "By Weight" (*w*)): The length distribution by number provides the relative frequency of fibers in each of the 40 preset AFIS length bins. Length distribution by weight provides the relative weight contribution of fibers in each of the 40 preset AFIS length bins. AFIS estimates the length-by-weight distribution assuming a constant linear density across the 40 length groups.

Mean Length (in or mm): Mean length gives the average fiber length of the fiber in the sample, based on either the fiber frequency (Ln) or weight fractions (Lw).

Upper Quartile Length (in or mm): The upper quartile length (UQLw) is the length that is exceeded by 25% of the fibers, by weight, in the test specimen (ASTM, 2013a).

2.5% (mm): This gives the length that is exceeded by the longest 2.5% of the fibers in the sample based on the length-by-number distribution.

5% (mm): This gives the length that is exceeded by the longest 5% of the fibers in the sample based on the length-by-number distribution.

AFIS provides estimates of the coefficient of variation of fiber length and a traditional measure of short-fiber content (SFC).

Length CV (%): This is the CV in fiber length by-number or by weight (LnCV or LwCV). CV% is a measure of the standard deviation (σn or σw) of the fiber length within a sample standardized by the average fiber length; it is expressed as a percentage.

Short-Fiber Content (%): Short-fiber content (SFCn or SFCw) is the percentage of fibers (by number or by weight) in a test specimen that are shorter than 12.7 mm (0.5 in) in length (ASTM, 2013a).

Maturity and Fineness Complex

AFIS uses the return signal from the electro-optical sensors to measure the maturity and fineness (MH) of each fiber. The distribution of the within-sample fiber maturity is binned into 20 discrete maturity groups; each group is 0.05 θ in width (θ ≡ degree of thickening). The distribution of the within-sample fiber fineness is binned into 20 discrete groups of similar fineness; each group is 25 mtex wide.

Maturity Distribution: The maturity distribution is the relative frequency of fibers in each of the 20 AFIS preset theta bin levels.

Fineness Distribution: The fineness distribution is the relative frequency of fibers in each of the 20 AFIS preset fineness bin levels.

Immature Fiber Content (%): This gives the percentage of fibers in a sample less than 0.25 θ.

Maturity Ratio: The value given is the maturity ratio (MR) as defined by Peirce and Lord (1939).

Average Fineness (mtex): This value is the average fineness (H) of the fibers in a sample.

Neps

Fiber neps and SCNs are counted as they pass by a sensor. AFIS uses the shape of the "peaked" waveforms (see Fig. 34) to distinguish between fiber neps and SCNs. The seed coat end of a SCN causes a more abrupt waveform transition than that of a fibrous nep.

Nep Count (cnt/g): Nep count gives the number of neps per gram. It is the sum of fiber and SCNs per gram of fibers.

Nep Size (μm): Nep size is the average size of neps.

Seed Coat Nep Count (cnt/g): Seed coat nep count gives the number of SCNs per gram of fibers.

Seed Coat Nep Size (μm): Seed coat nep size gives the average size of SCNs.

Trash and Contaminants

Trash is pulled into a separate airflow and classified by a different electro-optical sensor unit within AFIS. The return waveform from the sensor is used to classify sample contaminants by size and record their frequency.

Total (cnt/g): Total is the total number of trash particles of all sizes per gram of fibers.

Mean size (μm): Mean size is the average size of all trash particles.

Dust (cnt/g): Dust is the number of particles less than 500 μm per gram of fibers.

Trash (cnt/g): Trash is the number of particles greater than 500 μm per gram of fibers.

Visible Foreign Matter (%): Visible foreign matter is the percentage of foreign matter in the sample by weight.

Stelometer

Stelometer is a pendulum-type fiber bundle breaker used for measuring the tensile properties of flat cotton fiber bundles (Fig. 35 and 36). A small amount of fibers are taken from a fibrograph comb with a sample clip.

The specimen is pulled through the teeth of a comb several times to straighten fibers and remove loose fibers. A flat bundle of fiber is prepared in a special vise which provides a prestressing spring load at a 100 g tension. A torsion spring ensures uniform tightening of the clamps to 8 in-pounds of torque. The ends of the sample are then cut off with a special knife and the clamps are placed in the instrument.

The fiber bundle is mounted between a pair of clamps which are inserted in slots in the beam and in the pendulum. When the beam is released from its nearly vertical position it rotates, causing the pendulum to rotate too. This rotation

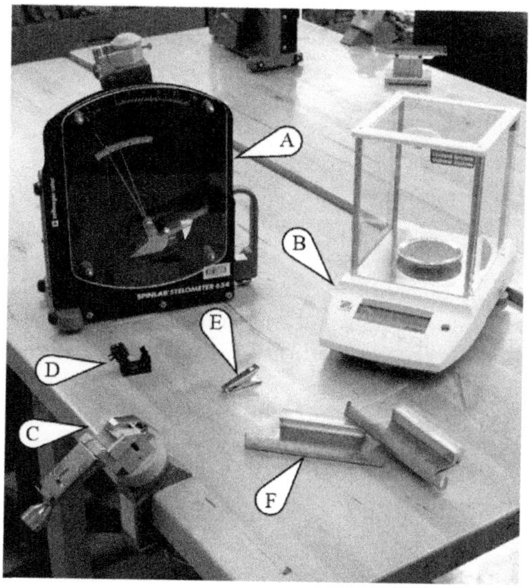

Fig. 35. Stelometer flat bundle tensile property tester at the Fiber and Biopolymer Research Institute, Texas Tech University, Lubbock, TX. (A) The Stelometer instrument; (B) the precision balance; (C) the table vice used in sample preparation; (D) the sample clamps; (E) the sample clip; and (F) the fibrograph combs.

Fig. 36. Sample preparation and Stelometer testing. (A) The sample of fiber is placed in the Stelometer clamps, forming the flat bundle. (B) The sample is broken by the Stelometer. The arrow points to the separation between the clamps. (C) The ends of the bundle are removed, and the sample is ready to be weighed.

produces an increasing tension in the mounted test specimen. Inertial effects are eliminated by allowing the pendulum to rotate about its center of gravity.

The tension is indicated by a detecting pin moving a pointer. When the specimen breaks, the pendulum escapes from the detecting pin and the breaking load may be read off directly from the position of the pointer on a scale graduated from 2 to 7 kg. There is second scale graduated from 0 to 40% for elongation that indicates the percent of increase in jaw separation, under load. Breaking force and elongation are obtained simultaneously using one sample.

Once the fibers are broken, the two broken halves of the sample are removed and weighed to the nearest 0.1 mg. The weight of the sample, along with the measured breaking force, can be used to calculate the tenacity of the sample.

An oil damper is used to break the fiber bundle with a constant rate of loading. The head of the Stelometer is designed to allow for fibers to be broken at 0 or 1/8-in gauge (elongation readings are available with 1/8-in gauge length only).

Stelometer Fiber Quality Parameters

Elongation at Breaking Force (%): This value is the elongation corresponding to the maximum force, and expressed as a percentage of the 1/8-in (3.2-mm) gauge length (ASTM, 2013a).

Breaking Force (g): The breaking force is the maximum force applied to a material carried to rupture (ASTM, 2013b).

Sample Mass (g): The value given is the mass of the sample.

Tenacity: Tenacity is the tensile stress expressed as force per unit linear density of the unstrained material (ASTM, 2013b).

FAVIMAT

FAVIMAT is a single-fiber tensile property tester (Fig. 37). FAVIMAT is designed for testing a variety of fiber types and has found use as a cotton fiber tensile property tester. Specialized single-fiber clamps are used to hold individual fibers for loading into the FAVIMAT (Fig. 38). The clamped fiber is placed into the FAVIMAT breaking jaws for tensile testing. The individual fiber is then broken to determine the force-elongation curve, breaking force, elongation at break, and work-to-break of the fiber.

The gauge length for FAVIMAT testing is highly customizable but is typically performed at either the 3-mm or 10-mm gauge length when testing cotton

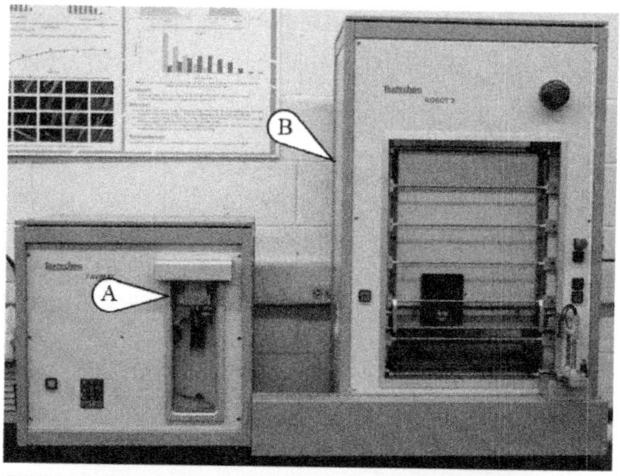

Fig. 37. FAVIMAT individual fiber tensile property tester on left and FAVIMAT robot on right at the Fiber and Biopolymer Research Institute, Texas Tech University, Lubbock, TX (Textechno). (A) The location where the individual fibers are loaded and broken. (B) The robot can be loaded with individual fibers for automated delivery to the FAVIMAT.

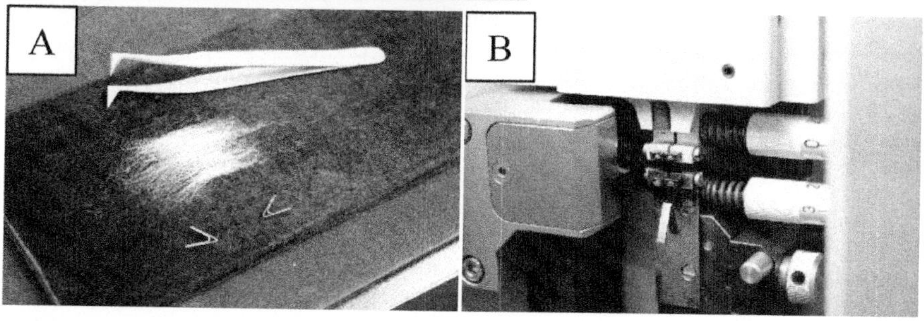

Fig. 38. FAVIMAT testing. (A) Tweezers are used to load individual fibers into specialized fiber clamps. (B) The fiber is loaded into the FAVIMAT jaws for testing. The clamp, still attached to the fiber being tested, can be seen dangling from the FAVIMAT jaws.

fibers. It can be used to characterize the complete within-sample distribution of fiber tensile properties.

Linear density is measured by the vibroscope method, where an oscillatory force of a known frequency is applied to the cotton fiber under tension. The resonance of the fiber is then used to estimate the cotton fiber linear density (Montgomery and Milloway, 1952). The vibroscope cannot provide results for a gauge length shorter than 10 mm. Therefore, the FAVIMAT can only be used to determine tenacity on fibers that can be tested with a gauge length of at least 10 mm.

Fiber Quality Parameters

Linear Density (dTex): Linear density of the fiber is measured with the vibroscope. This measurement is only available for fibers broken at 10-mm gauge length or greater.

Breaking Force (cN): This value is the force required to break a fiber.

Tenacity (cN/tex): Tenacity is the force required to break a cotton fiber standardized by the linear density of the fiber. This measurement is only available for fibers broken at 10-mm gauge length or greater.

Elongation-at-Break (%): Elongation-at-break is the percent of change in the fiber length at a breaking force.

Work-to-Break (cN×cm): Work-to-break is the total energy required to break a cotton fiber. Work-to-break is a function of both the breaking force and the elongation.

Cross-Sections

Image analysis of cotton fiber cross-sections provides the reference standard for determining cotton fiber maturity. Preparing and analyzing cross-sections is a tedious process. The results from a cross-section analysis are primarily used for instrument calibration and research. The cross-section method is composed of two stages: sample preparation and image analysis.

Cross-sectioning requires a small bundle of cotton fibers to be set in polymer resin through a series of steps (Boylston et al., 1995; Hequet et al., 2006). Setting the fibers in a polymer resin begins with drawing a small tuft of fibers through a Teflon tube. The tube is then filled with a polymer in solution; air bubbles are removed; and the solution is polymerized under UV. Once the polymer is set (20–30 minutes) the Teflon tube is removed. Two subsequent polymerization steps are used to prepare the sample for cross-sectioning. Progressively larger tubes are used to create a sample with a diameter of about 7.9 mm. Figure 39 shows the steps required for preparing the 7.9-mm sample.

A diamond coated microtome is used to prepare thin cross-sections (1-μm thick) of the fibers set in a polymer resin. The cross-sections are placed on a microscope slide and are further prepared for microscopic viewing. Then, images are captured for subsequent image analysis (Hequet et al., 2006).

Image analysis of the cross-sections provides a direct measurement of the fiber's physical dimensions that are needed to calculate the fiber maturity (Peirce and Lord, 1939; Xu and Pourdeyhimi, 1994; Guo et al., 2014). Results from the image analysis of the cross-sections can be seen in Fig. 40. The raw image of the

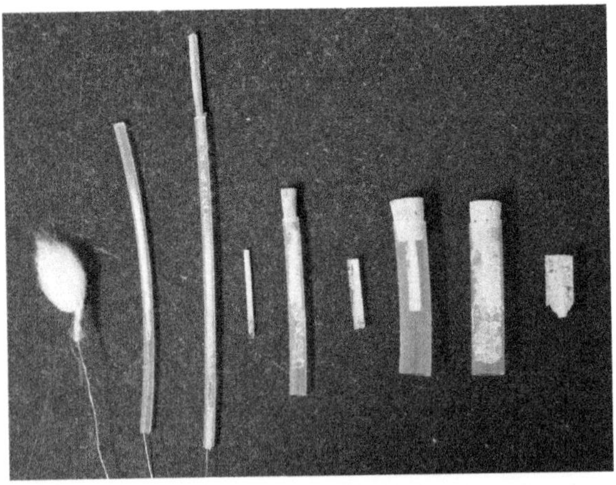

Fig. 39. (Left to Right) The stages of cross-section sample preparation. A small bundle of cotton fiber is sampled from a fibrogram beard with a small wire. The sample is drawn through a small Teflon tube. The tube is plugged and set in a polymer resin. The Teflon tube is cut away. The sample is set in polymer resin in a larger diameter polymer tube. The Teflon tube is cut away. The sample is placed in the final Teflon tube. The sample is set in a polymer resin. The Teflon tube is cut away to achieve the final sample size.

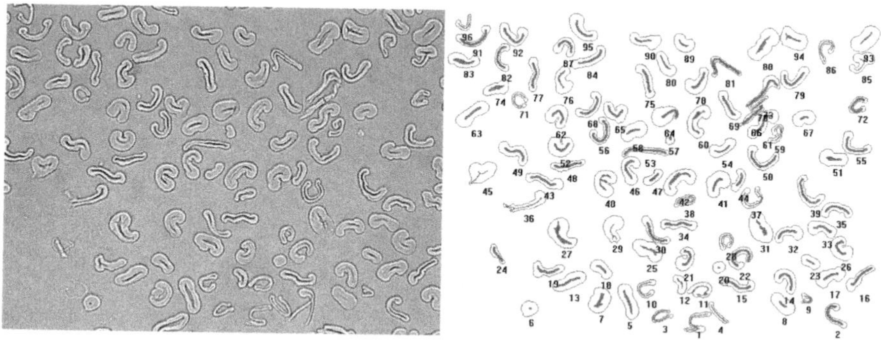

Fig. 40. Results from image analysis of cotton fiber cross-sections. (A) Raw image of fiber cross-sections were taken from a single commercial bale. (B) The objects identified as fibers by image analysis are assigned a number. Image analysis is used to identify the perimeter and lumen of each cross-section. Improperly identified fibers must be manually removed from the database.

cross-sections is shown on the left. The outside perimeter and the lumen are clearly defined on many of the fibers, revealing a characteristic kidney bean shape for many. Image analysis of the fibers is available on the right side of the figure. The image analysis software was able to correctly identify the outside perimeter and the lumen for many of the fibers on the slide.

Cross-Section Parameters

Perimeter (P_2): The cross-sectional perimeter of the cotton fiber is provided by the image analysis.

Area (A_w): The cross-sectional area of the cotton fiber cell-wall is provided by the image analysis.

Maturity (θ): Maturity is the degree of cell wall thickening.

Cottonscope

The Cottonscope uses a polarized light and image analysis to evaluate the maturity and fineness of the cotton fiber snippets (Fig. 41). Samples are prepared into 0.7-mm snippets using the specialized Cottonscope guillotine. A 50-mg sample of snippets is required for each test. The snippets are placed in a small basin of water with a surfactant that is automatically stirred. A camera is used to record the snippets as they are swirled around in the water basin. A proprietary computer algorithm is used to analyze the images of the snippets. As the weight of the sample has been recorded, and the number of snippets of a known length is counted with the image analysis, fineness (weight per km of fiber) can be calculated. Cottonscope testing can also provide the distribution of the fiber snippet maturity and the ribbon width within a sample.

While many fiber quality parameters are based on by-number or by-weight fraction of fibers in a sample, the Cottonscope fiber maturity and ribbon width are by-snippet measurements.

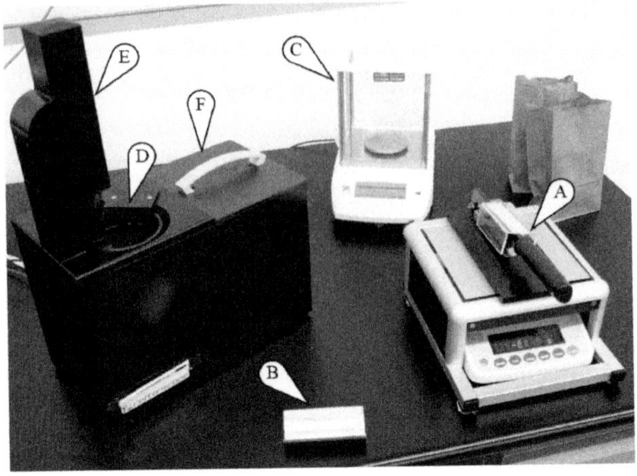

Fig. 41. Image of a Cottonscope instrument at the Fiber and Biopolymer Research Institute, Texas Tech University, Lubbock, TX (Cottonscope). (A) The guillotine and scale used to produce 0.7-mm fiber snippets for testing; (B) the tray used for catching snippets as they are cut; (C) the high-sensitivity scale used to weigh 0.05-g snippet samples; (D) the water basin where the snippets are swirled in front of the camera; (E) the camera used to capture the images of the fiber snippets; and (F) the water reservoir.

Cottonscope by-snippet measurements are advantageous because the small sample size requirements enable fiber maturity and linear density assessment for samples that are too small for alternative instruments.

Fiber Quality Parameters

Ribbon Width (μm): Ribbon with is the average width of the fiber snippets in the sample that is presented to the camera.

Maturity Ratio: Maturity ratio gives the average maturity ratio of individual fiber snippets in the sample that is presented to the camera as defined by Peirce and Lord (1939).

Fineness (mtex): Fineness is the average linear density of the fiber snippets in the sample that is presented to the camera.

A Focus on the Advanced Fiber Information System and the High-Volume Instrument

AFIS and HVI are both high-throughput instruments that provide a myriad of fiber quality parameters. However, AFIS and HVI are based on different fiber quality assessment methodologies. These methodologies provide researchers and the cotton industry with vastly different characterizations of fiber quality. Some thought should be given to these differences when selecting between HVI and AFIS testing. This section will provide guidance on the applied use of these instruments.

Speed

HVI testing is a fiber quality assessment methodology that is well suited for characterizing the variability between a large numbers of bales. HVI parameters are designed to provide a rapid assessment of the millions of bales coming to market in a single year (about 700 samples tested per day per instrument). These parameters are then used to market the cotton to spinning mills. Under optimum conditions, each bale can be assessed in less than 20 s with HVI. The marketing significance of HVI and the speed of the test have made it the default choice in fiber quality assessment. However, the speed of HVI testing precludes data on within-sample distributions of the fiber properties.

AFIS is an individual fiber tester, a methodology that is slower than bundle testing. It is slower than HVI, but it is remarkably fast considering the large amount of data generated. Typically, six canisters of 30 slivers can be tested per day (including the checks). In general, 3000 fibers are measured per sliver. It totals 540,000 fibers tested per day (18.75 fibers per second on average). While the number of samples tested per day is much lower than HVI, the reward is useful results about the within-sample variability in the fiber properties. Such results are crucial in many research applications, and they enable a quantum leap in the prediction of spinning performance.

Length

HVI measures length by scanning a beard of fibers protruding from a comb. Many short fibers in a cotton sample do not protrude far enough to be scanned; therefore, they are not included in the HVI assessment of fiber length. HVI length parameters are based on the fibrogram generated from the dominant long fibers in the sample. The two standard HVI length classification parameters used to characterize a sample of cotton are illustrated in Fig. 42.

AFIS length parameters are based on the complete within-sample distribution of length, including the shortest fibers. The AFIS length-by-number distribution, along with five standard parameterizations of the length distribution, is used to characterize a sample of cotton in Fig. 43. This is the same sample characterized with HVI length parameters in Fig. 42.

A comparison of the primary length parameter from each system (UHML from HVI; Ln from AFIS) measured on 10,084 commercial bales reveals the fundamental differences in the length assessment methodologies (Fig. 44 and 45). HVI UHML from the HVI explains only about 35% of the variation in AFIS mean

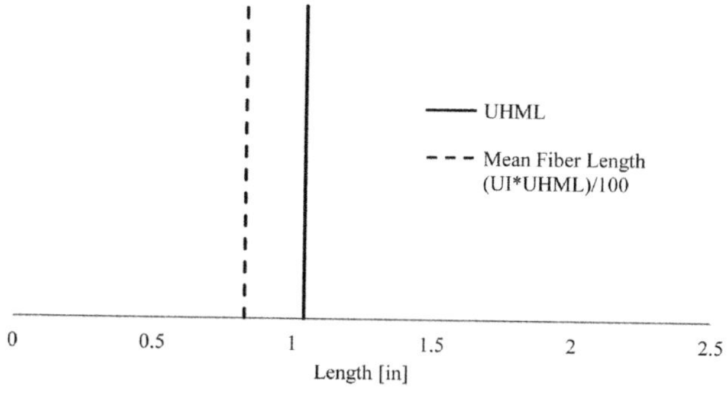

Fig. 42. Graphical representation of High-Volume Instrument (HVI) fiber length classification data. HVI estimates upper half mean length (UHML) and length uniformity. HVI outputs the uniformity index, which is defined as the ratio of the mean fiber length to the UHML. The mean fiber length is depicted in this graph (Kelly and Hequet 2013).

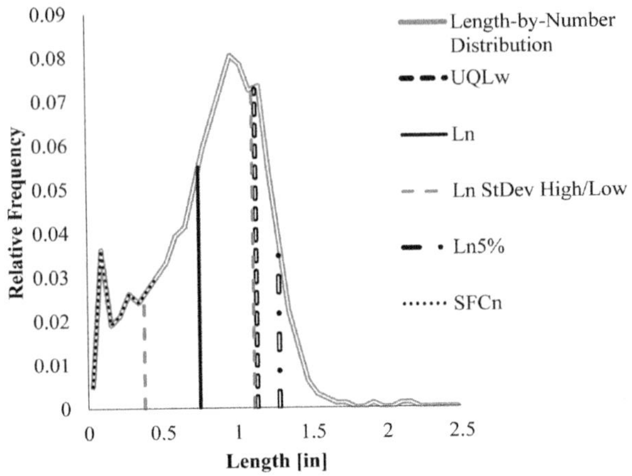

Fig. 43. The Advanced Fiber Information System fiber length-by-number parameters. The complete frequency distribution of fiber length-by-number is reported in 40 discrete length bins along with five additional parameterizations of fiber length. Length CV% is the ratio of the standard deviation of fiber length (Ln StDev) to the mean fiber length, which is expressed as a percent. Both the high and the low Ln StDev are included in the chart (Kelly and Hequet 2013).

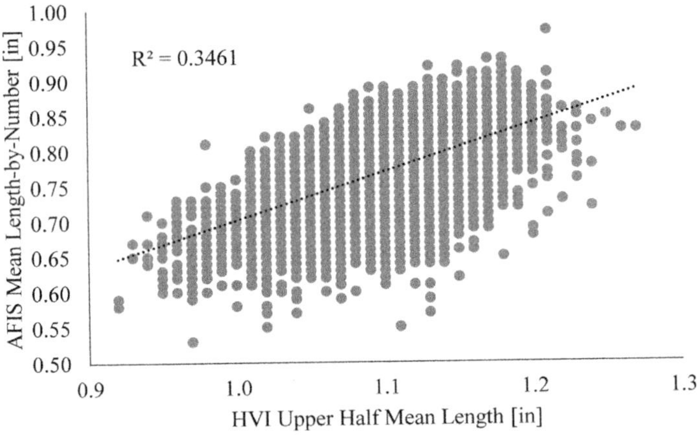

Fig. 44. The High-Volume Instrument upper half mean length vs. the Advanced Fiber Information System mean length by-number for 10,084 commercial bales.

length-by-number (Ln). AFIS Ln includes the variation in short fibers that are unmeasured by HVI testing (Fig. 44).

Regressing the SFC values from AFIS on the UI values from HVI for these same cottons reveals a very weak relationship (Fig. 45). While the downward slope is as expected, the spread of the data is too great to have practical usefulness. These results show that UI describes a different source of variability and is not a reliable substitute for SFC.

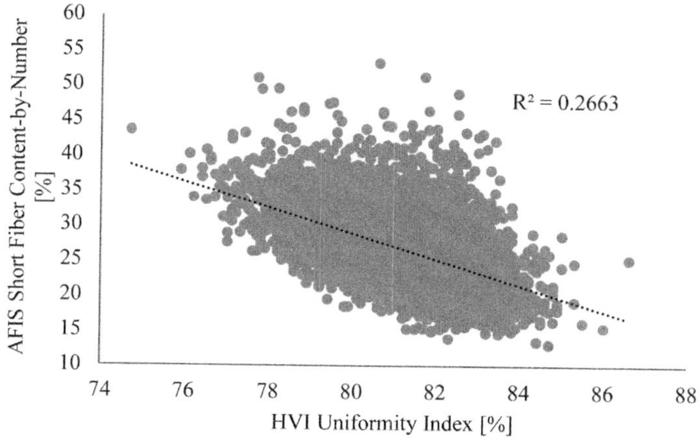

Fig. 45. The High-Volume Instrument uniformity index (in) vs. the Advanced Fiber Information System Short Fiber Content (%) for 10,084 commercial bales.

Maturity and Fineness Complex

HVI testing evaluates the *MH* complex with an airflow passing through a porous plug of fibers of a known weight, otherwise known as micronaire. Fiber orientation, trash, fiber perimeter, and fiber maturity can all impact the measurement of fiber micronaire. Even in very clean samples of randomly oriented fibers, the micronaire of a sample is impacted by a combination of both the fiber maturity and fiber fineness.

Diverse combinations of fiber maturity and fineness may be contained within a given level of HVI micronaire. This can be demonstrated with a comparison of micronaire with AFIS maturity and fineness for over 10,000 commercial bales (Fig. 46 and 47).

The premium micronaire range for Upland cotton (*Gossypium hirsutum* L.) is widely considered to be 3.7 to 4.2. However, the AFIS maturity ratio for the commercial bales within this premium micronaire range varies from 0.78 to 0.94; i.e., from very immature to very mature (Fig. 46).

A similar relationship can be seen when comparing AFIS fineness to micronaire. AFIS fineness for the commercial bales within the premium micronaire of 3.7 to 4.2 varies from 145 to 193 (Fig. 47). Thus, while these bales may be considered premium according to HVI micronaire, the AFIS data show that the fibers range from very fine to very coarse.

These results demonstrate the inability of micronaire to differentiate between fiber fineness and maturity. The functional relationships among maturity, fineness, fiber diameter, and micronaire are shown by the graph in Fig. 48. Note the contour line along which all the micronaire values are 4.0. If the AFIS maturity of the fibers is 1.0, then the AFIS fineness will approximate 135 mtex (fine). If the AFIS maturity is 0.74, then the AFIS fineness will approximate 182 mtex (coarser). Despite having the same micronaire, these two cottons would not be expected to perform the same in spinning. The speed of HVI micronaire testing has made it

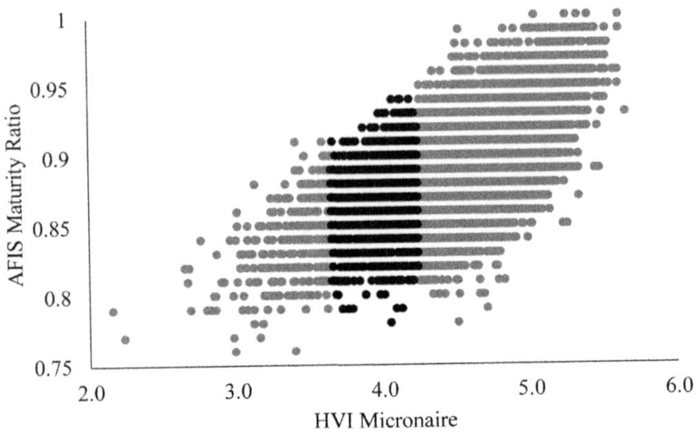

Fig. 46. The Advanced Fiber Information System maturity ratio vs. High-Volume Instrument micronaire for 10,084 commercial bales. Black dots indicate bales typically considered in the premium range for micronaire (3.7-4.2).

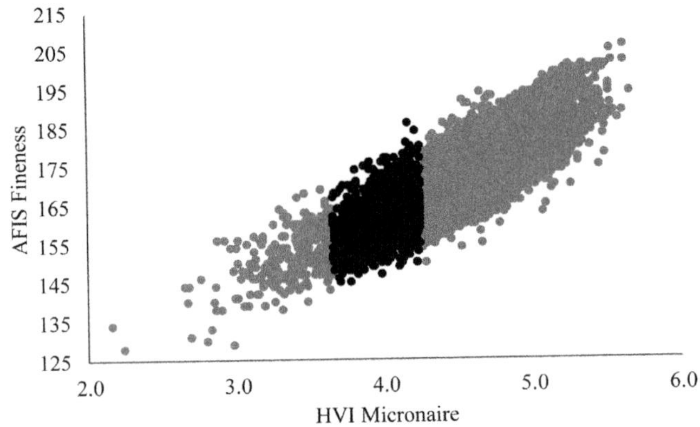

Fig. 47. The Advanced Fiber Information System fineness (mtex) vs. High-Volume Instrument micronaire for 10,084 commercial bales. Black dots indicate bales typically considered in the premium range for micronaire (3.7-4.2). Fineness (mtex) of bales in the premium range varies from 145 to 186.

a useful parameter for marketing cotton. However, additional testing is needed when fiber maturity and fineness is of interest.

Fiber diameter (or standard fineness) is determined primarily by genetics, while fiber maturity is determined primarily by environment. Thus, if the standard fineness (genetic fineness) of a given variety of cotton is known, the micronaire value becomes much more useful for judging the maturity of the fibers. Take, for example, Pima cottons (*Gossypium barbadense* L.) vs. Upland cottons. Of course the Pima cottons are significantly finer than most Upland cottons. Therefore, Pima cottons are judged to be fully mature at micronaire values below 4.0, while fully mature Upland cottons generally have micronaire values above 4.0. Even the fully mature Acala cottons have micronaire values above 4.0.

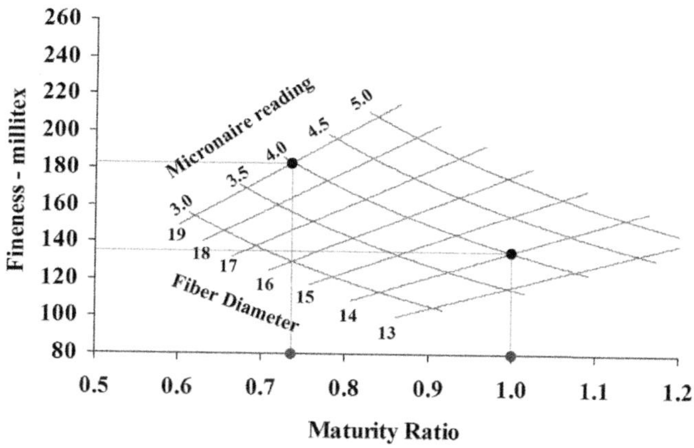

Fig. 48. The relationship between the maturity ratio, fineness, micronaire, and fiber diameter.

Yarn Manufacturing

Yarn manufacturing requires a series of steps to convert raw cotton fibers into spun yarns that are adequate to produce the fabrics being targeted. High-value textiles require clean, strong, uniform yarns. Lower-value textiles still require yarns that will perform efficiently in the fabric formation and the dyeing and finishing processes.

Cotton fiber arrives at the mill in a densely packed bale containing varying amounts of foreign matter, trash, motes, and other contaminants along with the fiber. Fiber quality can vary within each bale and among the bales. Regardless of the yarn spinning technology that was used, the variability in fiber properties must be managed well enough to achieve efficient running in the spinning mill, efficient fabric formation, efficient dyeing and finishing, and satisfactory product quality. Competitive pressures in textile manufacturing require high levels of throughput and consistent quality to earn adequate margins. Inadequate fiber quality is sufficient to make the manufacturer uncompetitive.

Spinning mills typically have a target market in mind when purchasing bales. The goal of the spinning mill is to purchase bales suitable for manufacturing the targeted yarn quality (El Mogahzy and Chewning, 2001). The premium market for cotton fiber is in the production of strong, fine, ring-spun yarns. The yarns are sold in high-value markets intolerant of yarn defects. Yarn manufacturing processes are continually increasing in processing speeds and adopting new technologies, which often put increased demands on the quality of the fibers.

Figure 49 shows the stages involved in the spinning of cotton for ring and rotor spinning technologies.

There are a maximum of 11 stages in the total process of yarn formation; the actual spinning of yarn is just one of these. All 11 stages are relevant only for ring spinning. Combing is done only for ring-spun yarns, and then only for the high-end textiles. The primary purpose of combing is to remove the SFC—a testament to the importance of fiber-length distribution in spinning performance and

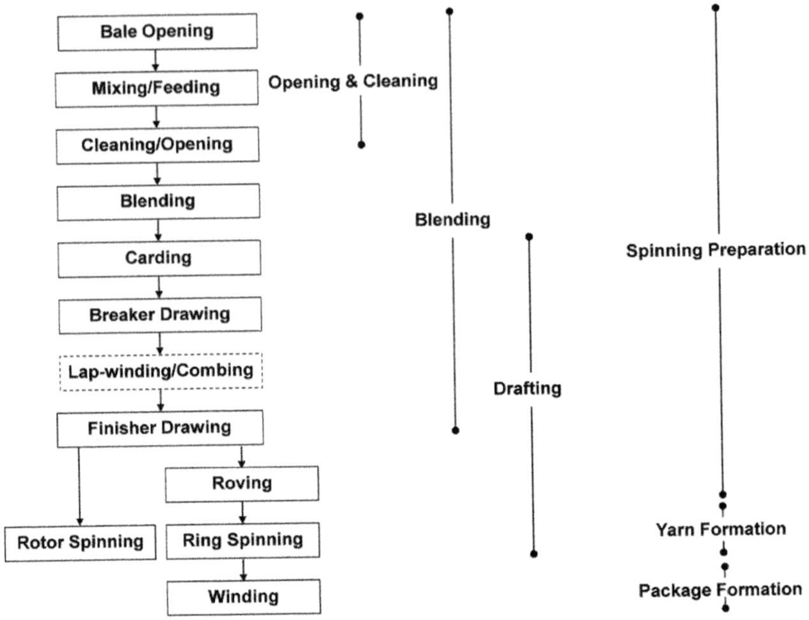

Fig. 49. The stages of spinning cotton into yarns. Functional purposes are shown by vertical lines.

yarn quality. Combing is costly, in terms of both operations and lost raw material. Therefore, the vast majority of cotton yarns throughout the world are made with "carded cotton" (i.e., not combed).

Ring spinning is "traditional technology" that always requires two additional steps than does rotor spinning. These are roving (to provide a fiber package to the ring-spinning machine) and winding (to provide a package suitable for fabric formation). Rotor spinning uses the sliver directly from finisher drawing to make the yarn, and then winds the yarn package in a single operation with yarn formation. In some cases, rotor yarn may be made using only one drawing step, thereby eliminating another step from the process used for ring spinning.

The steps illustrated in Fig. 49 will now be explained.

Opening, Mixing, and Cleaning

Yarn properties and processing efficiency are directly related to the central tendencies and variability in the properties of the fibers being processed. These issues are critical because a yarn spinning plant must purchase many bales for a spinning "laydown" that results in even-running fibers. The more important the yarn quality, the more important it is to have the proper mix of fibers. The fundamental objective is to organize the cotton bales at the opening line into a compatible mix and thoroughly blend the fibers from these bales to deliver a more homogeneous flow of fibers into the system; i.e., keeping the average and variance of properties in the mix consistent over time (El Mogahzy and Chewning, 2001).

The first stage of blending is achieved in the laydown. Software, such as the Engineered Fiber Selection (EFS) System assists the mill in selecting the bales for the blend (El Mogahzy and Chewning, 2001). The number and types of bales in the laydown depends on several factors. A laydown for a higher-quality yarn may be composed of a larger number of bales. If fiber quality testing is performed and detailed knowledge of the properties of the bales being processed is available, uniform processing performance generally can be maintained with a smaller mix. If little detailed knowledge of physical properties is available, a relatively larger mix is necessary to control variations. Typically, mills will select mixes to maintain a representative sample of the cotton purchased for the crop year. The number of bales used by different mills in each mix generally ranges from less than 30 to more than 70 bales.

With state-of-the-art technology, processing begins when the bales to be mixed are identified by the EFS software. Bales are arranged for the laydown, and bagging and ties are removed. Since bales arrive at the mill in various degrees of density, it is helpful to cut bale ties approximately 24 h before the bales are to be processed to allow them to "bloom." The head of a bale plucker travels back and forth over the laydown, feeding fiber from the bale laydown to the spinning mill (Fig. 50). On each pass, the plucker removes a small layer from the top of each bale in the laydown and feeds the fiber into the opening line.

Tufts of cotton delivered by the pluckers are not sufficiently open to permit efficient processing and must be cleaned to remove trash, motes, and foreign matter. For further opening and cleaning, the tufts are transported through a series of machines designed to clean and further blend the fibers.

An initial cleaner employs the principle of beating to remove trash and other undesirable material from the cotton, while beginning the transformation of tufts of fibers into individualized fibers. The tufts of cotton are fed between two spring

Fig. 50. A diagram of a bale plucker passing over a laydown of cotton bales. Inset shows the bidirectional operation of the plucker head as it passes over the laydown (Seagull and Alspaugh, 2001).

loaded or weighted rolls that are grooved to provide powerful griping action. As the feed rolls turn, cotton tufts are presented to a revolving beater. Small tufts of fibers are stripped or beaten from the fringe and carried across slots through which the loosened trash and foreign matter are hurled. The speed of the beater and its closeness to the feed rollers is critical to cleaning efficiency, and they must be precisely set to avoid excessive fiber damage. Many different types of beaters are available. Their designs range from an array of simple spikes to flat blades or bars to saw-type surfaces wound in the form of stamped wire.

Typically a second, wire-clothed cleaning machine is used. The basis for selecting cleaning machinery is the type or grade of cotton to be processed and the quality or type of end-product desired. Low-grade cottons with a high trash content require more elaborate cleaning than higher grade, cleaner cottons. Trashy cottons may require a third wire-clothed cleaner. Even though this third cleaner contributes to fiber damage, it may be necessary to avoid delivering cotton that is too trashy through the system.

High-quality end-products may require a combination of more cleaning steps and slower cleaning speeds than lower-quality products. The focus is on preserving the quality of the cotton fibers.

Following cleaning, the cotton fibers are put through a mixer designed to ensure a complete and thorough mixing of the fibers contained in the laydown. It is the combination of a compatible laydown and complete mixing that achieves the even-running fibers required for yarn quality control.

Carding

The opening/cleaning/mixing machinery prepares the cotton for processing on the carding machine, where a combination of extremely important and related functions is performed. The carding step is absolutely critical; success at the card makes all subsequent processes much easier.

The card machine performs two critical functions: (i) fine cleaning to reduce the trash content of the fibers to an acceptable level, and (ii) conversion of the fibers to a parallel orientation with small variations in mass. Unless the card sliver is combed, there are no subsequent stages that remove significant amounts of contaminants and imperfections.

A diagram illustrating the basic components of a card machine is given in Fig. 51, and a picture of a modern card machine is shown in Fig. 52. The card is composed of a system of three wire-covered cylinders and a series of flat, wire-covered bars.

- The first cylinder in the sequence is called the licker-in. It is covered with coarse, saw-toothed wire and works in conjunction with grooved feed rolls similar to cleaning machines. A tufted fringe of cotton is presented to the fast-turning licker-in. Small tufts are combed out by the licker-in, dragged across slotted bars for further trash removal, and transferred to the much larger main cylinder.
- The main cylinder is covered with much smaller wire teeth and a significant amount of combing and straightening of fibers occurs at the transfer point. The fibers are hooked on the fine cylinder wires and are pulled past a series of wire covered plates, called flats, attached to a chain mechanism that slowly moves attached to a chain mechanism that on modern card machines slowly moves in the opposite direction as the main cylinder. The distance between

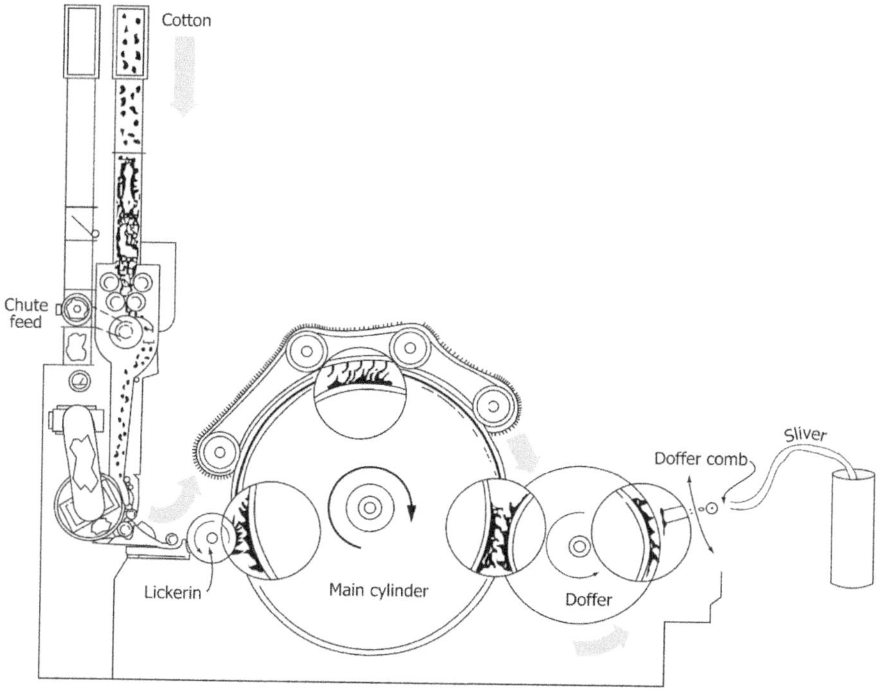

Fig. 51. A diagram of the carding process (Seagull and Alspaugh, 2001).

Fig. 52. A carding machine at the Fiber and Biopolymer Research Institute, Texas Tech University, Lubbock, TX (Trützschler DK903).

the wire surfaces of the main cylinder and the flats is approximately 0.25 mm. The carding action that occurs between these two wire surfaces reduces fiber tufts to a very thin web of fibers and removes much of the remaining trash and foreign matter as well as neps and short fibers whose length does not allow them to remain hooked on the cylinder wire. These short fibers, neps, and entrapped fine trash and dust collect on the flat wires and are brushed out after the flats have revolved to a position out of contact with the cylinder.

This thin web, practically in a single-fiber layer, is deposited on a third cylinder, called the doffer, which is also covered with fine-toothed wire. The web is stripped off the doffer and is drawn through highly polished and precision-ground steel rolls that are heavily weighted to crush any large trash particles remaining in the web. These crushed trash particles are small enough that they generally drop out of the fibers in subsequent processes. After leaving the crush rolls the web passes through a highly polished cone called a trumpet that condenses the web into a "sliver"—an untwisted, continuous bundle of fibers (reminiscent of a rope) that are held together by friction forces and fiber crimp (Fig. 53). The mass of the sliver is calibrated according to the yarn size that will be made from it. This card sliver is then deposited as coils in a large can (Fig. 54). The fibers stay in this sliver form through subsequent processing, until these are either delivered to rotor-spinning machines or to roving machines (where roving bobbins are made for delivery to ring-spinning machines).

A picture of a modern card machine is shown in Fig. 52 and a close-up of the front of the machine showing the card web and card sliver is shown in Fig. 53. Full sliver cans are pictured in Fig. 54.

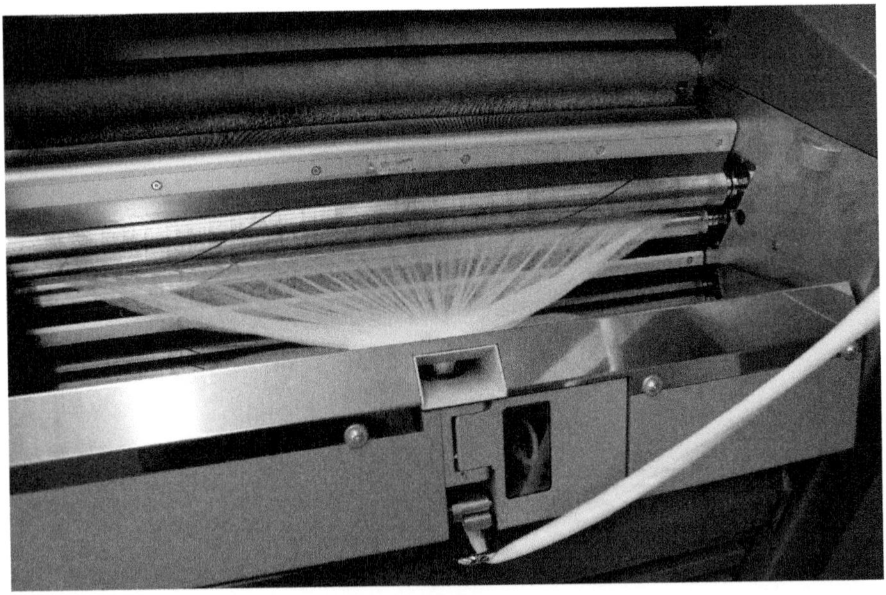

Fig. 53. The front of a card machine showing the exiting of the card web and its condensation into the sliver (Fiber and Biopolymer Research Institute, Texas Tech University, Lubbock, TX).

Fig. 54. Sliver coiled in a sliver can at the Fiber and Biopolymer Research Institute, Texas Tech University, Lubbock, TX.

Breaker Drawing

The process of breaker drawing occurs next for carded yarns. If combing is done before drawing, this step is commonly referred to as predrawing, to avoid confusion about whether the fibers are to be combed. A picture of a drawing machine, or "drawframe," is shown in Fig. 55, with sliver cans under the creel that feeds slivers into the drawframe.

Drawing employs roller drafting, with the size of the sliver being determined by the relative speeds of paired rolls that are turning at different speeds. Drafting occurs when a sliver is fed into a system of paired rolls whose surfaces are moving at different speeds, illustrated in Fig. 56. The surface speed of each pair

Fig. 55. A drawing machine at the Fiber and Biopolymer Research Institute, Texas Tech University, Lubbock, TX (Trützschler HSR1000).

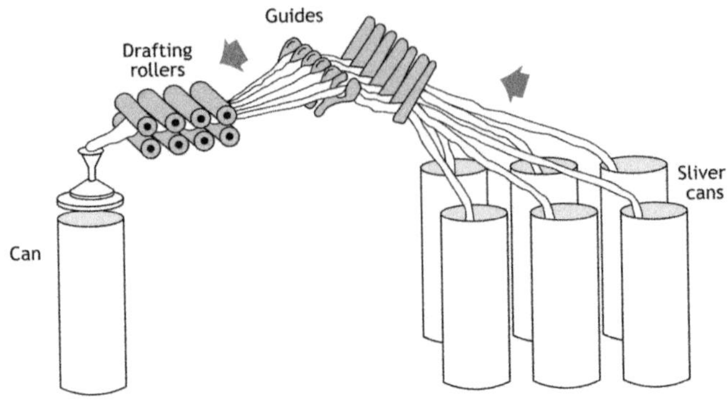

Fig. 56. A diagram of the drawing process (Seagull and Alspaugh, 2001).

(top and bottom) of rollers is higher than for the pair of rollers immediately preceding it. The first set of rollers pulls the sliver into the system until it is caught by the faster turning second pair. Fibers are pulled out, or drafted, by the gripping action of the second set of rollers. A further draft occurs as the sliver passes through the remaining rollers of the system.

Drawing serves three major purposes: (1) straightening the fibers in the card sliver by drafting to make more of them parallel to the axis of the sliver; (2) achieving greater blending by combining multiple slivers into one sliver (called "doubling" of the slivers); and (3) obtaining a sliver that is more uniform in weight per unit length by the doubling of slivers and the precision control of drafting.

The fibers in the sliver that are produced by the final drawing process, called finisher drawing, are nearly straight and parallel to the axis of the sliver.

Combing

If combing is done, the card slivers are delivered to a "lap winding" machine, where multiple slivers are wrapped around a cylinder to form a pad. These cylinders are then delivered to the combing machine, as pictured in Fig. 57. In addition to removing short fibers, combing also removes hooks from the longer fibers and removes most of the remaining neps and trash from the cotton. The resulting combed sliver is clean and lustrous.

The diagram in Fig. 58 shows the basic components where the combing is done on the fibers. The grooved feed roll moves the lap forward as the nipper knife opens to allow passage. Then the nipper knife is lowered onto the cushion plate, clamping the fibers. While the fibers are clamped, the bottom rotary comb, or "half lap," combs the free end of the nipped lap, removing fibers and waste that is not nipped. Then the nipper knife opens, and the combed fringe is advanced and gripped by detaching rolls which pull the fibers through another needle arrangement called the top comb. This process is discontinuous and a combed sliver is produced in segments that must be pieced together. This is done by placing the combed fiber fringe on top of the fringe held by the detaching rolls. These detaching rolls then begin to rotate forward, drawing the fiber material held by the feed roll, and the top comb is lowered to act on the back fringe of the fiber strand as it is being detached from the feed roll, thereby combing the trailing fringe.

Fig. 57. Picture of a comber at the Fiber and Biopolymer Research Institute, Texas Tech University, Lubbock, TX (Rieter). Sliver-wrapped cylinders from the lap winder are shown atop the comber. Slivers that are being fed from the creel into the lap winder are shown in the background.

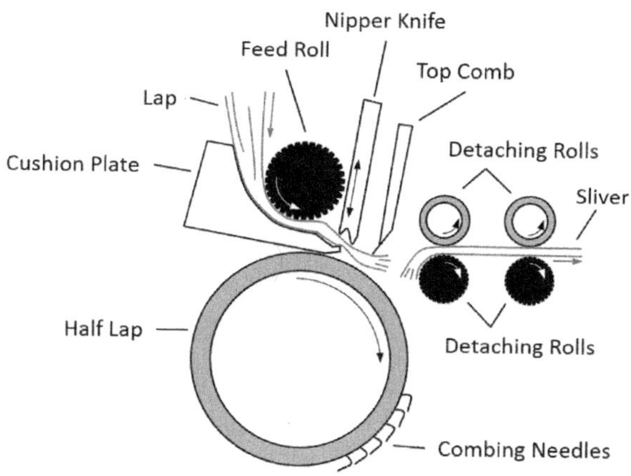

Fig. 58. A diagram showing the components of the combing mechanism.

The fibers in the slivers coming out of each combing position are nearly parallel and are easily pulled apart. Therefore, these slivers are supported by a smoothly finished chute until the output from six to eight combing positions can be combined to form one combed sliver.

The fibers combed out of the cotton are called noils. These noils typically constitute 8 to 20% of the weight of fibers before combing, depending on the quality of the cotton and the end products being made. Noils are valuable waste that are always captured and used in diverse applications; e.g., to produce coarse yarns, to make nonwoven textiles, etc.

Finisher Drawing

Finisher drawing is done on the same kind of machine as the breaker drawing and for the same basic reasons. In this operation, the sliver is brought to the desired size and an autoleveling sensor feedback system is used to provide high-precision control of the drafting rollers to produce the most even sliver possible. The more even the sliver is at this juncture, the more even the yarn produced from it will be.

Roving

The purpose of roving is to produce a fiber package, the roving bobbin, suitable for mounting onto the ring-spinning machine. It is necessary to draft the sliver into a greatly reduced size (typically one-eighth the weight per unit length of the sliver fed into roving). A minimal level of twist is put into the roving because it is so small that it might not hold together otherwise. The minimal twist must not interfere with the drafting of the roving on the ring-spinning machine.

A diagram of the typical roving process is shown in Fig. 59. The sliver is drafted in the same manner as is done on a ring-spinning machine. The drafting system consists of three pairs of rolls, and the middle roll is covered with a flat band or belt of material called an apron that provides additional control of the fibers during drafting (Fig. 60). The roving is twisted slightly and is simultaneously wound onto a roving bobbin. The minimal amount of twist needed to stabilize the roving is targeted because additional twist means a slower throughput and a higher cost. Longer and finer cotton fibers require less twist in the roving than do shorter, coarser fibers.

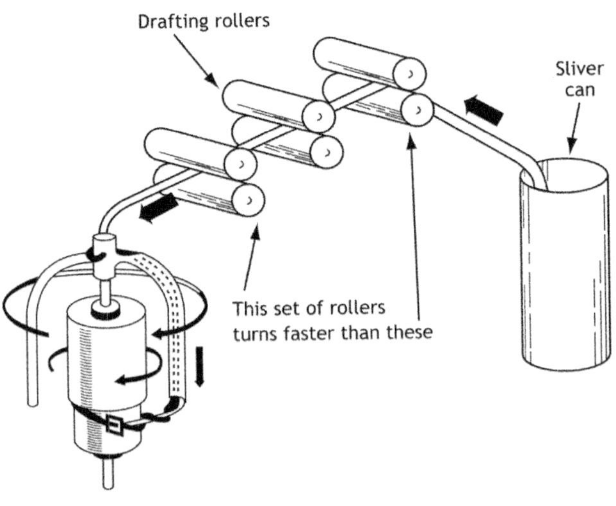

Fig. 59. A diagram of the roving process (Seagull and Alspaugh, 2001).

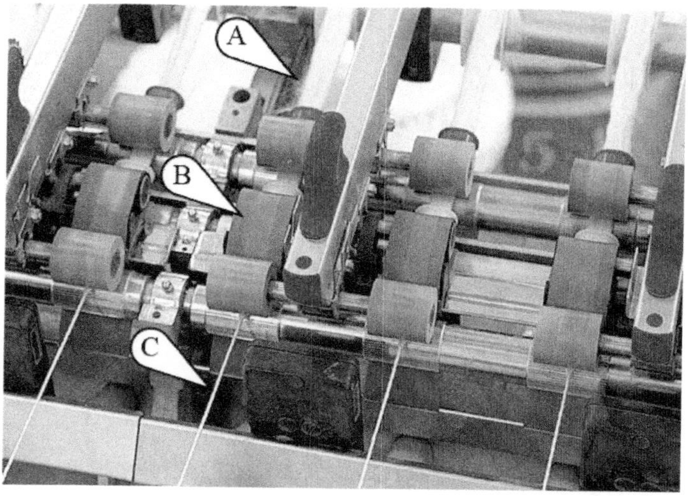

Fig. 60. The drafting zone of roving. (A) Sliver. (B) Apron. (C) Roving. Reprinted with the permission of Rieter.

Spinning

Spinning is the culmination of the yarn formation process and the most costly single step in the process. The properties required of the spun yarn depend on its end uses; e.g., woven fabric, knitted fabric, apparel uses, home uses, industrial uses, sewing thread, cordage, etc. Because of the expense of yarn formation, the importance of yarn quality to end-product quality, and the interrelationships of yarn properties with fiber properties, considerable research and development efforts are being devoted to understanding the fiber and yarn quality complex.

Two dominant spinning technologies currently exist: ring spinning and open-end rotor spinning. Together, these account for over 95% of worldwide spinning capacity. This section will briefly explain these technologies, along with other technologies that occupy niches in yarn formation.

Ring Spinning (Conventional and Compact)

Ring spinning accounts for almost 75% of global spinning capacity. It is the oldest of the spinning technologies and is the highest-cost technology in terms of cost per lb or kg of yarn produced. Nevertheless, it is still the most dominant technology, and it sets the standard for yarn quality.

A typical ring-spinning frame is shown in Fig. 61, where the roving bobbins can be seen mounted above each spinning position. The roving is fed into the drafting zone (which is conceptually the same as for the roving machine and very similarly configured) and on down to a bobbin mounted on the spindle.

As illustrated in Fig. 62, the drafting system produces the desired size of the yarn, and the spinning spindle puts the desired twist into the yarn. The twisting travels from the spindle up toward the front roller of the drafting system (see insert in Fig. 62). The yarn orientation between the spindle and the ring is controlled by a tiny "traveler" that hooks on the edge of the ring and travels around the ring based on the revolutions per minute (rpm) of the spindle. The

Fig. 61. Ring-spinning frame at the Fiber and Biopolymer Research Institute, Texas Tech University, Lubbock, TX (Suessen Fiomax 1000).

designs of both the travelers and the rings must vary according to the yarn sizes and spinning speeds. The spindle on modern ring-spinning machines will generally rotate between 15,000 and 20,000 rpm. If the spindle rotation is faster and the twist is lower, than the throughput is greater. For any given yarn, the longer, stronger, and finer fibers will endure faster speeds and require less twist than will the shorter, weaker, and coarser fibers. Therefore, the higher-quality fibers will enable better manufacturing efficiency. The yarn balloon in Fig. 62 makes it obvious that significant centrifugal forces are acting on the yarn as it forms. If

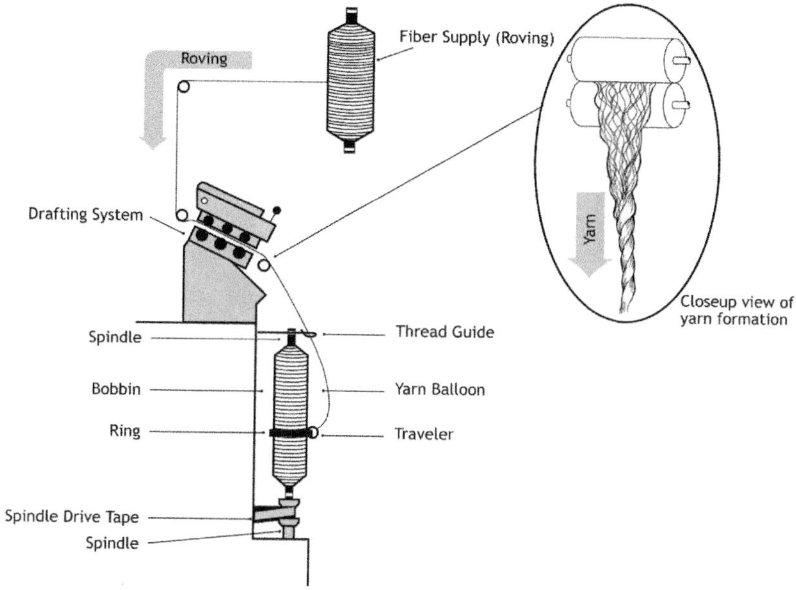

Fig. 62. A diagram of the ring-spinning process (Seagull and Alspaugh, 2001).

the fibers are not of adequate quality, ring spinning tends to break them (causing "ends down") during the spinning process.

Ring spinning can produce a full range of yarn sizes but is used less frequently for coarse yarns, often for medium yarns, and dominantly for fine yarns. The capital cost and energy use of ring-spinning systems is less than for other technologies, but the labor requirements are much greater. These facts help explain why ring spinning is often emphasized by less-developed countries.

Compact Ring Spinning

The foregoing information applies to what is now called "conventional ring spinning." In the 1990s, an innovation to ring spinning emerged that is called "compact ring spinning." Using an additional front roller to create a "compacting zone," negative air pressure is used to pull all the fibers to the middle of the yarn in this zone (Fig. 63). Then, when the yarn emerges from the new front roller, the spinning triangle of untwisted fibers that characterizes conventional ring spinning is eliminated. Figure 64 shows the difference between conventional and compact yarns coming off the front roller. The compaction of ring-spinning yarn results in the most nearly perfect yarn structure available, because "floating fibers" at the spinning triangle are eliminated and almost 100% of available fibers are incorporated into the yarn structure. The resulting yarns are both more even and stronger. Figure 65 shows magnified pictures of conventional ring yarn and compact ring yarn.

Open-End Rotor Spinning

Open-end rotor-spinning frames account for almost 25% of the global spinning capacity. First commercialized in the 1970s, rotor spinning is a more "elegant"

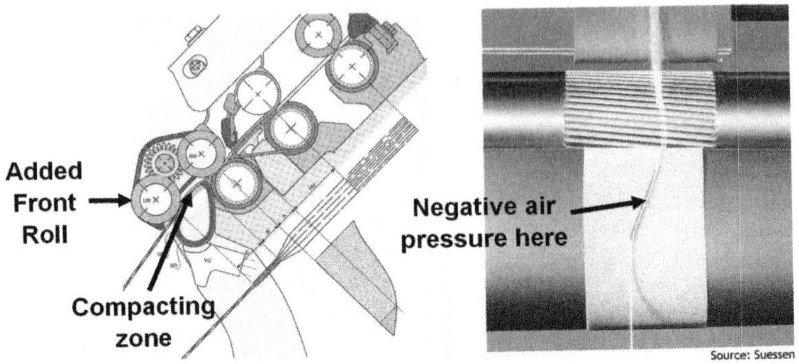

Fig. 63. An illustration of compact ring spinning by Suessen. Reprinted with the permission of Suessen.

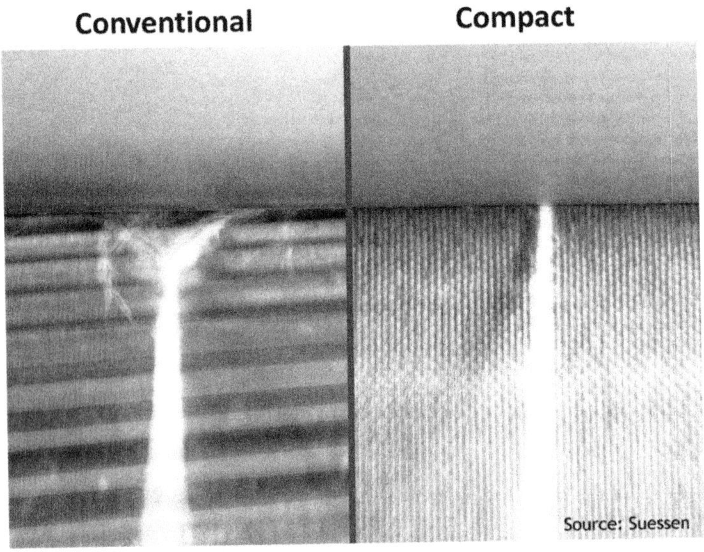

Fig. 64. A comparison of yarns coming off the front rollers for conventional vs. compact ring spinning. The floating fibers associated with the spinning triangle have been eliminated. Reprinted with the permission of Suessen.

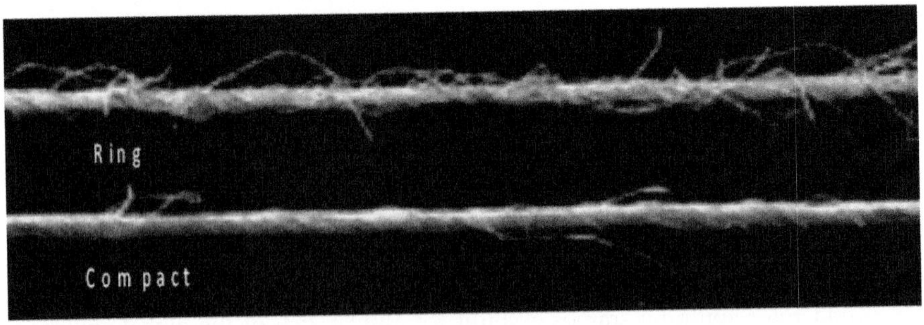

Fig. 65. A comparison of magnified yarns from conventional vs. compact ring spinning.

process than ring spinning; it eliminates at least two steps in the process (roving, winding, and perhaps one drawing), and for medium-sized yarns, each rotor spinning position puts out about ten times the yarn per unit of time compared with each ring spindle position. The technology lends itself to automation and robotics so that small amounts of labor can efficiently operate large spinning rooms. With rotor spinning, a limited number of skilled technicians mostly attend to problems with the machines and otherwise let these run unattended. While the capital investment and energy requirements are greater, taken as an entire process, rotor spinning is the lowest-cost technology per kilogram of yarns produced.

Figure 66 shows a rotor machine, with sliver going into the spinboxes and yarn packages being wound up above.

A diagram of the open-end rotor spinning process is shown in Fig. 67. The drawing sliver is pulled into the system by feed rolls where a small opening roller with wire or molded teeth plucks individual fibers from the fringed mass and hurls them into an airstream which carries the stream of fibers into the rapidly spinning rotor. Air currents and centrifugal force carry the fibers to the perimeter of the rotor where they are very evenly distributed in a small groove. Using a starter yarn, the tails of the fibers are twisted together by the spinning action of the rotor and yarn is continuously drawn from the center of the rotor.

Rotor spinning does not stress the yarns during the spinning process as ring spinning does, since it is formed within the rotor groove and is simply lifted out and wound onto the package as the yarn is formed. Therefore, it is quite possible to spin rotor yarns that are too weak for subsequent fabric forming processes. This issue must be controlled exclusively by yarn testing. About the only thing that causes ends down during rotor spinning is the presence of dust or trash particles

Fig. 66. Open-end rotor spinning at the Fiber and Biopolymer Research Institute, Texas Tech University, Lubbock, TX (Rieter R20).

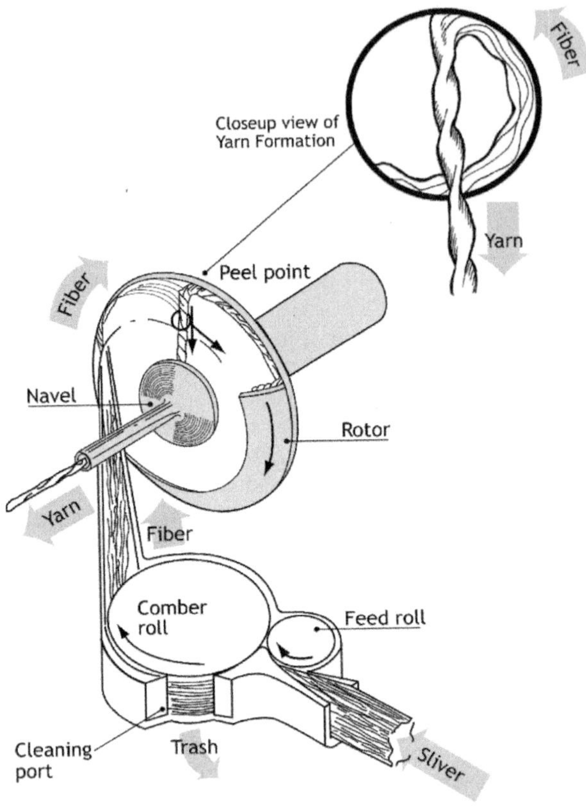

Fig. 67. The rotor-spinning assembly (Seagull and Alspaugh, 2001).

in the rotor groove. Ends down are pieced up by a robot that automatically blows out the rotor before proceeding, under the assumption that there is a dust or trash particle inside.

Rotor-spun yarns are actually more even than conventional ring-spun yarns, and neps are less of a problem because the rotor-spun yarn structure tends to enclose these inside the yarn, while the ring-spun yarn structure tends to put these on the surface of the yarn. Rotor yarns are weaker than ring yarns; however, the less organized structure with wrapper fibers makes the rotor yarns less susceptible to untwisting during processing. Also, the structure makes rotor yarns much more tolerant of SFC; combed cotton is not used with rotor spinning. But this structure also gives fabrics made from rotor yarns a harsher feel, which is a significant disadvantage with lighter weight, higher-end apparel textiles. This is made obvious in Fig. 68, which adds a magnified picture of a rotor yarn along with the conventional and compact ring yarns.

The properties of rotor yarns make these well suited for bottom-weight fabrics (such as denims, velveteens, and corduroys). Rotor spinning may also be successfully used to make medium-sized yarns, but it is not used for making fine yarns.

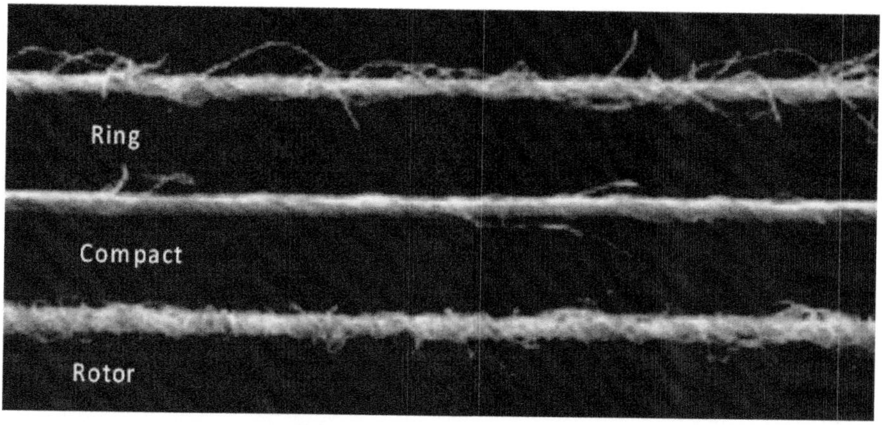

Fig. 68. A comparison of magnified rotor yarn vs. conventional and compact ring yarns.

Other Spinning Methods

Like open-end rotor spinning, air-jet (vortex) spinning technology was also commercially introduced in the 1970s. Its application to cotton fibers has been more limited, in part because of its intolerance of the shorter fibers contained in the length distributions. A simplified diagram of an air-jet system is shown in Fig. 69. Like rotor spinning, air-jet spinning feeds a drawing sliver directly into the spinning machine. Like ring spinning, air-jet spinning requires that a drawing sliver pass through a mechanical drafting zone. After this drafting zone, the fibers are presented to an opening roller (similar to rotor spinning), where individual fibers and tufts are separated and introduced into an air stream. The fibers are transported to a chamber where jets of air are directed tangentially along the inner wall of the chamber. The fibers are collected by centrifugal force and condensed. Twist is inserted by the rotation of the condensed fibers; the yarn is drawn out of the chamber and wound onto a yarn package. Thus, both rotor and air-jet spinning eliminate two separate steps that are required for ring spinning: roving and winding. The roving is not required because the sliver is fed directly to the spinning machines and a separate winding step is not required because the spun yarns are wound onto the yarn package while on the spinning machines.

Other spinning systems, used in market niches, include the following.

- Core-spun systems are currently in use to produce a variety of specialty yarns and sewing threads. These core-spun systems involve a modification of ring-spinning systems to feed a monofilament core behind the front rolls so that the conventionally prepared and drafted staple fiber is wrapped around the core during twisting. These yarns can be produced at higher production rates than non-core yarns and offer superior yarn strength and flexible finishing alternatives. Because of the cost considerations, production is limited to textile applications requiring the properties of core-spun yarns.

- A variation of core-spun systems involves the conventional drafting of staple fibers and the wrapping of the drafted strand with a monofilament to maintain the integrity of the yarn until it is converted into fabric. These systems have experienced excessive yarn breakage at spinning and yarn strength that is too low for further processing.

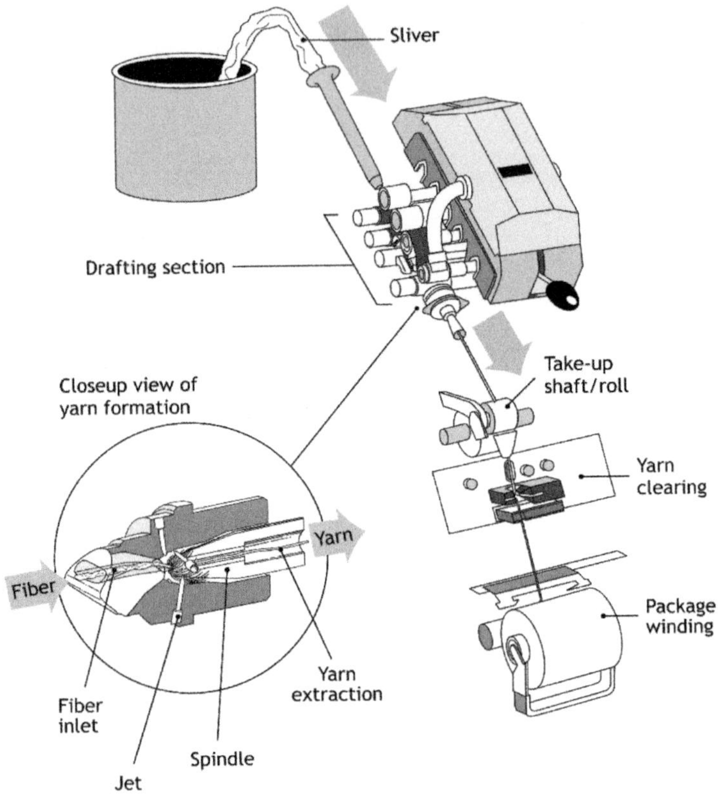

Fig. 69. Air-jet–spinning machines (Seagull and Alspaugh, 2001).

- Twistless yarns have been produced commercially on a limited basis by a system that bonds the fibers together with a polyvinyl alcohol or some other bonding agent. A wetted sliver is drafted and passed through a system of rolls where the binder is padded onto the fiber strand. A false twist is inserted in a stream vortex mechanism to maintain the integrity of the strand until the adhesive can be solidified on a heated drum. The yarn is then wound onto a package. The twistless yarn system offers potentially high production rates and uniform yarns, but the tactile properties of the fibers are lost.

Winding

Winding is the final step in ring spinning. It is done as an integral part of the spinning operation with both rotor and air-jet spinning. This is not possible with ring spinning, because the spindle bobbin is itself a small yarn package that cannot effectively be used in subsequent processes. Therefore, many spindle bobbins must be wound onto a larger yarn package in a separate step. This is accomplished on highly automated machines, with very precise yarn splices being done robotically. Modern winders are frequently equipped with "yarn clearers," which detect faults in the yarns that are being wound so that these faults may be cut out and the yarn can be spliced back together.

Yarn Quality

Producing yarns of a targeted quality is the central goal of spinning mills. Whether for sales yarns or for use in a vertical textile operation, yarn properties are typically stipulated and quantitatively verified on testing instruments. Just as with fibers, no single property is sufficient for evaluating yarn quality. But the two primary categories of yarn quality are tensile properties and imperfections (faults).

Cotton Yarn Tensile Properties

Adequate tensile properties are necessary to avoid breakage during further processing and fabric formation. There are four common tensile measurements (both the averages and CVs of the measurements are important).

Breaking Force (cN): Breaking force is the force required to break a yarn.

Tenacity (cN/tex): Tenacity is the force required to break a yarn, standardized by its linear density.

Elongation (%): This value is the elongation of the yarn at the point of breakage.

Work-to-Break (cN×cm): Work-to-break is a function of both the breaking force and the elongation; the area below the force-elongation curve is drawn to the point of maximum force.

Yarn Imperfections

Yarns with an even distribution of mass along their length are highly desirable. Imperfections along the length can manifest themselves in the finished textiles. Yarn evenness testers can provide at least five common measurements (Peters and Meier 2010):

- CVm%—The CV in the mass of the yarn;
- Thick Places—The number of thick places per km of yarn, which may be measured at a variety of sensitivity settings;
- Thin Places—The number of thin places per km of yarn, which may be measured at a variety of sensitivity settings;
- Neps—The number of neps per km of yarn, which may be measured at a variety of sensitivity settings;
- Hairiness—The yarn hairiness is proportional to the total length of fibers protruding from the yarn surface, divided by the length of the sensor of 1 cm. Hairiness is a unit-less measure.

Fiber and Yarn Quality

The relationship between fiber quality and yarn quality is complex. Spinning technologies must be known when evaluating fiber properties, because different spinning technologies produce different yarn structures that alter the importance of the various fiber properties. These issues will be considered here for the two dominant spinning technologies: ring spinning and open-end rotor spinning.

Ring-spun yarns are formed by successively drawing the bundle of fibers and applying a twist to the bundle. The resulting friction and compression forces impart strength to the yarns. Fiber qualities that increase these forces improve ring-spun yarn tensile properties. Thus, longer and finer fibers produce stronger ring yarns (Faulkner et al., 2012; Long et al., 2010b; El Mogahzy et al., 1990). Also, smaller within-sample variability of the fiber properties will produce ring yarns that are more even and have fewer imperfections (Kelly and Hequet, 2013).

Rotor-spun yarns have a core of relatively parallel fibers wrapped and entangled by other fibers (wrapper fibers). Thus, the intrinsic strength and fineness of the fibers is more important for producing stronger rotor-spun yarns (El Mogahzy et al., 1990, Ramey et al., 1977).

Developments in yarn manufacturing technologies and market demands can further shift the dynamics of the fiber and yarn quality complex. For example, the innovation of compact spinning extended the range of fiber length and strength values that can be successfully used in ring spinning.

Importance of Non–High-Volume Instrument Fiber Qualities

HVI provides rapid, inexpensive measurements of fiber properties based on fiber bundles, and it is widely used by spinning mills, plant breeders, and researchers. The AFIS instrument is slower and more costly, but it provides a measure of within-sample variability of some important fiber qualities (length, maturity, and fineness). AFIS provides information on the complete distribution of properties within a sample.

The value of AFIS data is demonstrated here, using a set of 110 commercial bales having a wide range of fiber properties (Kelly and Hequet, 2013). Measurements for all bales were taken using both HVI and AFIS. Then, 30Ne ring-spun cotton yarn was produced using cotton from each bale. A partial least squares regression was performed using the fiber properties as predictor variables and the yarn properties as dependent variables. Two PLSR models were constructed: one using HVI parameters alone and the other using both HVI and AFIS parameters.

The variation in each of the yarn tensile properties accounted for by the two sets of fiber quality parameters, HVI and HVI plus AFIS, is presented in Fig. 70. HVI fiber property measurements account for at least 61% of the variation in yarn tensile properties and about 73% of variation in yarn tenacity. The addition of

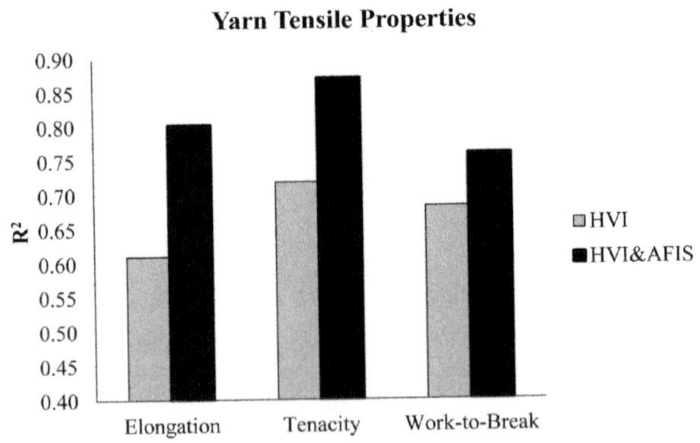

Fig. 70. A partial least squares regression model was used to account for the variation in yarn tensile properties based on using the High-Volume Instrument (HVI) measurements alone and based on using both HVI and the Advanced Fiber Information System measurements. Shown are the percentages of variation in yarn tensile properties accounted for by each regression (Kelly and Hequet, 2013).

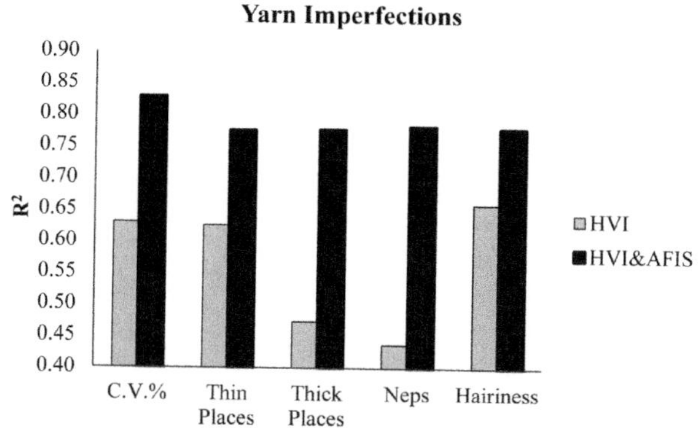

Fig. 71. A partial least squares regression model was used to account for variation in yarn imperfections based on using the High-Volume Instrument (HVI) measurements alone and based on using both HVI and the Advanced Fiber Information System measurements. Shown are the percentages of variation in yarn imperfections accounted for by each regression (Kelly and Hequet, 2013).

AFIS fiber property measurements accounts for at least 77% of the variation in yarn tensile properties and raises the result on yarn tenacity to 87%.

The variation in each of the yarn imperfections accounted for by the two models is presented in Fig. 71. HVI measurements do not perform as well for these, accounting for less than 50% of the variation in yarn-thick places and neps. The addition of AFIS measurements greatly improves predictive performance, with more than 75% of the variation in each of the yarn imperfections being accounted for by the combined measurements.

The higher-valued, ring-spun yarn markets are quite sensitive to yarn imperfections, yet these results indicate that HVI measurements have limited usefulness for predicting these. The AFIS measurements, based on within-sample distributions of fiber properties, greatly improve the prediction of these imperfections. Identifying important fiber properties and developing new measurement technologies is an active area of research (Hequet et al., 2014; Kelly and Hequet, 2013; Long et al., 2013). Breakthroughs in this area of research may enable the development of cotton varieties with heretofore unattainable spinning performance and yarn quality.

Fabric Manufacturing

The vast majority of all yarn packages (or cones) produced in spinning provide an input into subsequent processing and the eventual fabric formation. For very high end textiles, the yarns may be plied before going into fabric formation. This involves twisting together two or more yarns to form one plied yarn. Plied yarns are more uniform, stronger, and much more abrasion resistant than are single yarns of an equivalent weight. The highest-quality apparel fabrics are candidates for using plied yarns.

Weaving

Weaving is the process of forming a fabric in which lengthwise (warp) yarns are interlaced with crosswise (weft or filling) yarns. Two of the most common weave patterns are the plain weave and the twill weave. These are illustrated in Fig. 72. The plain weave simply has the weft yarn alternating going over and under successive warp yarns. The twill weave has the weft yarn repeatedly going over one or more warp yarns, then under one or more warp yarns—and then the next weft yarn shifts one or more warp yarns before repeating this pattern. This creates a diagonal pattern in the fabric. The 2/2 twill is illustrated in Fig. 72, so the weft yarn repeatedly goes over two warp yarns, then under two warp yarns—and then the next weft yarn shifts over one warp yarn before repeating. This pattern is typical of a denim weave, for which the warp yarns are dyed with indigo and the weft yarns are not dyed.

Weaving is done on a loom, which involves raising and lowering warp yarns as programmed, while inserting a weft yarn through the space presented (called the shed). The basic principles of weaving are illustrated schematically in Fig. 73. Warp yarn is fed to the loom from a cylindrical beam containing thousands of yarns. Warp yarns are threaded through slots in thin metal strips which are attached to bars on multiple frames (the harness). The raising and lowering of each frame raises and lowers the warp yarns. The warp yarns pass through the reed, which is a rack made of small, vertical bars that rotates and packs (beats up) the weft yarn into the fabric after each insertion. As the fabric is formed, it is wound around a take-up roll and onto a cloth beam. Different technologies exist to control the harness and to insert the weft yarns. Existing weft insertion technologies are shuttle, rapier, projectile, and air-jet. The fastest operating speeds are achieved by air-jet looms, which are capable of inserting up to 1200 weft yarns per minute (or 1200 picks per minute).

Preparing warp yarns for weaving is a tedious process, because these must be unwound from the yarn cones and wound onto intermediate beams that are called the section beams. This is accomplished by a machine called a warper, and the process is called warping the yarns. This machine pulls yarns from a warping creel that contains hundreds of yarn cones and winds these onto spool-like section beams. Figure 74 shows a loaded warping creel. The yarns are passed through tension control devices as they move to the beam, because forming uniform wraps of yarn requires that there be little variation in the tension.

Multiple section beams must then be mounted at the back of a slashing machine, the purpose of which is to coat the yarns and stick down the yarn hairs, to impart elevated abrasion resistance, minimize shedding, and help the yarns to pass through the needle-sized holes in guiding wires used to control the yarns in

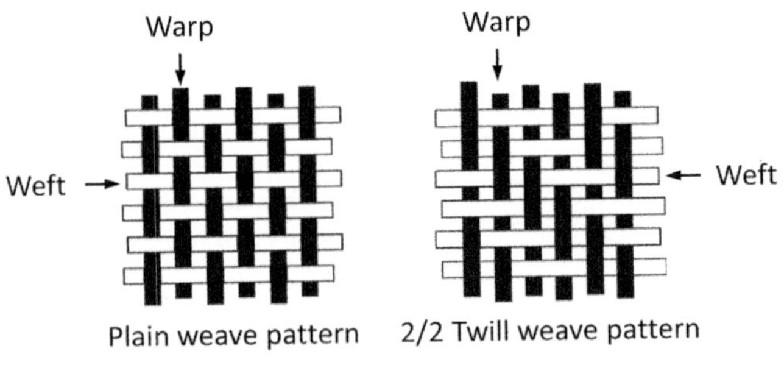

Fig. 72. An illustration of plain and twill weave patterns (Seagull and Alspaugh, 2001).

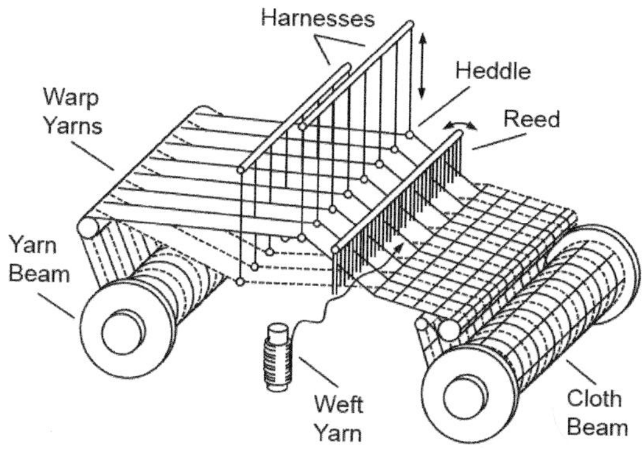

Fig. 73. A diagram of basic loom features.

Fig. 74. A creel loaded with yarn cones at the Fiber and Biopolymer Research Institute, Texas Tech University, Lubbock, TX (West Point Foundry & Machine Co.).

weaving (the heddle wires). Yarns from all the section beams are unwound with close control of yarn tension and passed through a box of hot mixture, called size, that is padded onto the yarns by squeeze rolls. As the sheet of sized yarn leaves the size box, it is passed over a series of hot, Teflon-coated cylinders that are steam heated. These cylinders dry the size mixture, which forms a film on the outer surface of the yarn. The composition of the mixture used by mills to size warp yarns may contain multipurpose components, such as lubricants, preservatives, adhesives, antifoaming agents, etc. But the primary component is usually either pearl cornstarch, carboxymethylcellulose, or polyvinyl alcohol. The rigor of the sizing process is determined by the fiber properties and their contributions to yarn strength and uniformity.

The yarns from all the section beams are then wrapped onto a loom beam and the yarn ends on this beam are threaded up to be fed through the loom. Given that modern commercial looms operate at ballistic speeds, slashing is generally necessary for the warp yarns to survive the stresses and abrasive forces.

Knitting

Compared to weaving, preparation of yarn for knitting is relatively simple. Fabric can be knitted directly from cones of adequate yarns without any preparation other than application of wax or lubricant to help reduce fly and to facilitate movement through thread guides and devices for maintaining uniform tension as the yarn is fed to the machine. A wax or lubricant may be applied during the winding process by pulling the yarn across a felt wick loaded with a liquid wax or lubricant. Alternatively this may be done as the yarn is pulled toward the knitting machine.

Knitting is performed by forming loops with a single, continuous yarn and joining each individual loop to its neighbors to form a fabric. Loops are formed on needles in which an eye is made by a hook and latch mechanism or by a spring device. The loops of a knitted fabric form a series of chains, called wales, running lengthwise of the fabric. The loops also form lines at right angles to the wales and are called courses. Wales and courses in knitted fabric are equivalent terms to warp and weft in woven fabrics. The most basic stitch is called the "knit stitch," which is illustrated in Fig. 75, emphasizing both wales and courses.

Knitted fabrics can be either warp knit or weft knit. In weft knit fabrics the yarns forming the loops generally run crosswise of the fabric. In warp knits the yarns run lengthwise. Knitting machines may be either circular or flat. Flat knitting machines have needles arranged in one plane or in two planes at right angles to each other. Flat knitting machines may produce either flat or tubular fabrics. Circular machines have one or two sets of needles arranged in a circle and produce tubular fabrics. Needle movement in both types of machines is controlled by cams and can be programmed to produce a number of different fabric styles.

Circular weft knitting is the most dominant knitting technology. A diagram of a circular weft knitting machine is shown in Fig. 76. It shows how cones of yarn may be arranged on a creel to facilitate unwinding. The yarn is threaded through tension-controls and stop-motion controls and is fed into the needle hooks. The needle bed revolves, and the needles pass over cams which raise and lower them

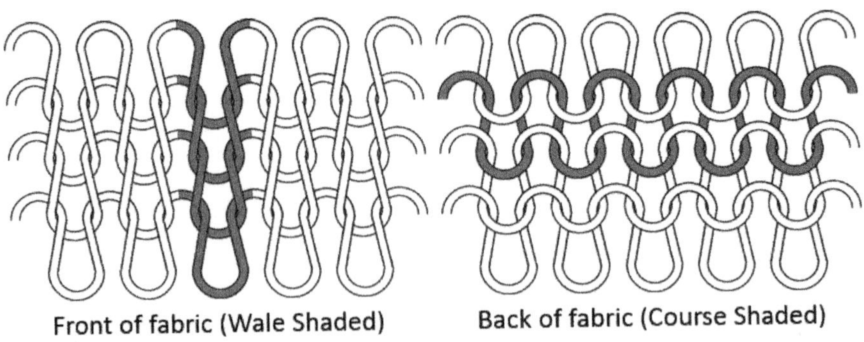

Fig. 75. A diagram of the basic knit stitch.

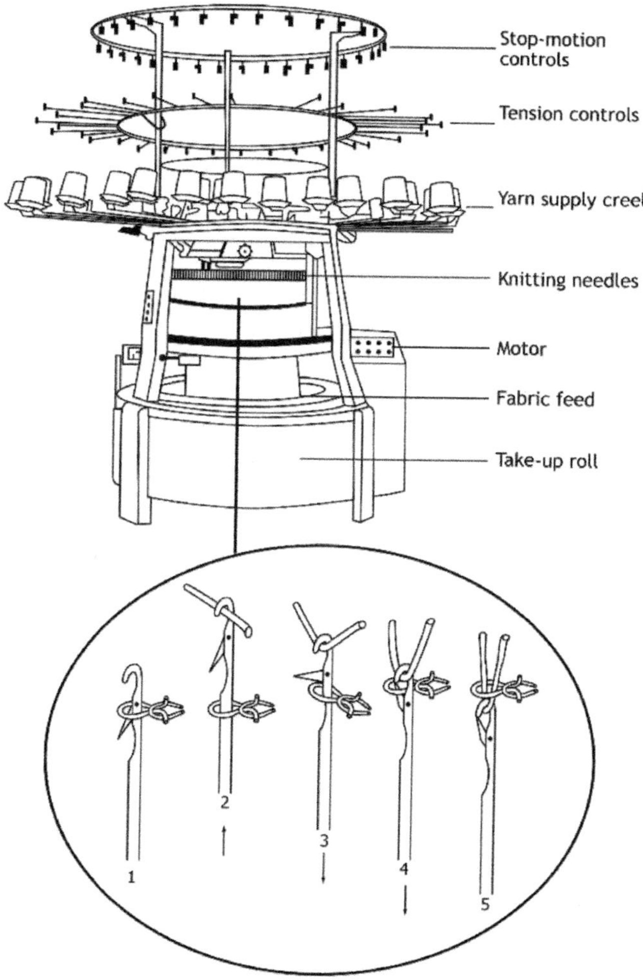

Fig. 76. A diagram of a circular weft knitting machine. The inset shows the action of the latch needle during the knitting process (Seagull and Alspaugh, 2001).

to form loops. Fabric is accumulated on the take-up roll which is revolving along with the needle bed. The most common needle used in weft knitting is the latch needle (inset of Fig. 76). This latch opens to receive a weft yarn and closes to pull the yarn through a loop and form a stitch.

Knitted fabrics account for somewhat less of total fabric production than woven fabrics, but they are indispensable for making many of the staple products serving the world's textile needs; e.g., undergarments, hosiery, tee-shirts, polo shirts, etc. The physical structure of knit fabrics eliminates the wrinkling problems of cotton fabrics and alleviates the tactile issues associate with rotor-spun yarns. Also, knits are structurally flexible and stretchable, making knit apparel comfortable and non-binding.

Nonwoven and Other Technical Textiles

Nonwoven textiles still comprise a relatively small portion of the global textile industry, but it is rapidly developing in both size and diversity. A nonwoven textile substrate is not made by spinning yarns and then weaving or knitting fabrics. The technologies used to form nonwoven fabrics are quite diverse, but all technologies consist of three basic steps: (i) web formation, (ii) web bonding, and (iii) finishing. The bonding of the fibers is done by chemical, thermal, or mechanical means.

Nonwoven textiles are dominated by the synthetic fibers, with polyester and polypropylene being most dominant. Available data indicate that cotton's share of all nonwoven textiles is no more than 2%. As of 2012, it was estimated that global synthetic fiber production was estimated to be about 51 million metric tons and that about 2.6 million (about 5%) of those were used to make nonwoven textiles (Kellie, 2013).

Products that are commonly made with nonwoven textiles are found in medical and healthcare, hygiene, absorbency, wipes, filtration, upholstery, packaging, geotextiles, etc. About half of all nonwoven textiles are disposable; i.e., used only once or a few times and then thrown away.

Nonwoven textiles are a major subset of technical textiles, which may be defined as textile materials and products manufactured primarily for their technical performance and functional properties rather than their aesthetic or decorative characteristics. Technical textiles include the industrial textiles, which are textile materials and products intended for end uses other than non-protective clothing, household furnishings, and floor covering (CIRFS, 2011).

Messe Frankfurt, one of the world's leading trade show organizers, conducts global trade shows for technical textiles with its Techtextil division. It has designated eight product groups for technical textiles: (i) technology, machinery, and accessories; (ii) fibers and yarns; (iii) woven fabrics, scrims, braids, and knitted fabrics; (iv) nonwovens; (v) coated textiles; (vi) composites; (vii) bondtec (finishing technologies); and (viii) avantex (innovative apparel).

Clearly, technical textiles are extremely diverse and are involved in penetrating textile fibers into applications far beyond the traditional uses. For some of these, cotton can participate and for others it cannot. Significant participation in any of these will likely require specialty finishing of cotton to impart functionalities to the products beyond those inherent within the unmodified cotton fibers.

Finishing

Cotton fibers, after ginning and mechanical cleaning, are composed of 93 to 96% of cellulose. The approximate amounts of the other materials present on the fiber are: 1.1 to 1.9% proteins, 0.7 to 1.2% pectins, 0.4 to 1.0% waxes, 0.7 to 1.6% ash, 0.5 to 1.0% other impurities. These noncellulosic materials are located mainly in the primary wall of the fiber; therefore, at or near the fiber's surface. The waxes and pectins of cotton serve as protecting and lubricating agents during yarn manufacturing and during weaving or knitting. If this natural finish is degraded, successful processing of the fibers may be difficult or impossible. However, as long as these are intact on the surface, their hydrophobic nature makes it difficult to wet raw cotton. So, before dyeing and finishing of cotton yarns or fabrics, a scouring process removes the non-cellulosic materials. What remains is almost pure cellulose. Textile wet processing can be thought of having three stages: pretreatment (or preparation), coloration, and finishing.

Singeing

Singeing may be done as a pretreatment to enhance fabric appearance. It involves a brushing operation to remove lint, loose yarns, and surface dirt from the face of the fabric. After the fabric surface is raised by brushing, the protruding fiber ends are burned off by the process known as singeing. The most common method is to pass the fabric through a small gas flame. The removal of superficial hairiness from yarns provides a more even surface for dyeing and also reduces the formation of pills, or balls of fiber, on the fabric surface.

Desizing

Desizing removes the material size before dyeing and chemical finishing (see the discussion of slashing in the section on Weaving). A typical chemical formulation to desize fabric includes 1% nonionic wetting agent, 0.5% hydrogen peroxide, and 0.25% acetic acid. The fabric is typically treated at 194°F for 30 min followed by rinse and neutralization with acetic acid.

Scouring

Scouring removes impurities that could interfere with dyeing and finishing. Caustic soda is used to solubilize many impurities in cotton, making them removable in aqueous media. The chemical formulation used typically includes 1% nonionic wetting agent, 4% caustic soda, and 0.125% acetic acid. The fabric is typically treated at 194°F for 1 h followed by neutralization with acetic acid.

Bleaching

Bleaching removes and decomposes colored impurities. It is usually done by treatment with hydrogen peroxide. The chemical formulation typically includes 1% wetting agent, 3% caustic soda, 3% sodium silicate, 5% hydrogen peroxide 50%, and 0.125% acetic acid. The fabric is treated for 1 h at 194°F and neutralized with acetic acid.

Mercerization

Mercerization, discovered by John Mercer around 1850, refers to the treatment with a concentrated solution of sodium hydroxide (NaOH). The beneficial effects of treating cotton with alkali include increased tensile strength, increased softness, improved affinity to dyes, improved dye-ability of immature fibers, increased luster (if done under tension), and higher water sorption. When mercerization is performed under tension, cotton fibers regain their original circular cross-section because of the swelling of the cellulose, and they tend to lose their convolutions. Mercerized cotton is more reactive and will pick up about 12% of water from air, compared with 6 to 8% for unmercerized cotton. The changes in the fine structure that occur when cotton is mercerized include the following: conversion of the crystal structure from cellulose I to cellulose II, reduction in crystallite length, increase in moisture regain, and reduction in the degree of crystallinity. The increase in strength and extensibility of mercerized cotton fibers have been attributed to the removal of convolutions and to the relief of internal stress created during the deposition of cellulose in the secondary cell wall.

Dyeing

Color is a major factor in selection of textile products by consumers. The human eye is an excellent detector of small color differences in textile products. Therefore, the dyeing of textiles must be done correctly, and the color must be uniform. Furthermore, color-fastness properties of colored products have a direct impact on the use and ultimate satisfaction of consumers with textile products.

The task that must be accomplished in the dyeing process is to transfer dye molecules from the dyebath to the fiber. Thus, the dyeing process can be described as a mass transfer process which leads to a decrease of the concentration of dye in the dyebath and an increase of the dye on the fiber. The term "exhaustion" is used to express the degree of dye transfer from the dyebath to the fiber. For economic and environmental reasons, a high degree of exhaustion is desirable. The driving force for transferring of a dye molecule from the dyebath to the fiber or from the fiber to the dyebath is the concentration of dye in the two phases.

Dye and Water Interactions

The two general types of dyes with regard to their behavior in water are ionic dyes and nonionic dyes. Ionic dyes may be either anionic or cationic. In the anionic dyes, the part of the molecule primarily responsible for the color has a negative charge ($Dye^{-}Na^{+}$) while in the cationic dyes, the part of the molecule primarily responsible for color has a positive charge ($Dye^{+}Cl^{-}$).

Fiber and Water Interactions

Water is usually the dyeing medium. The interaction between water and the fiber has a major role in the application of many types of dyes. For hydrophilic fibers like cotton, water molecules diffuse into the amorphous regions of the fiber and break internal hydrogen bonds which results in a swelling of the fiber. Swelling increases the mobility of the polymer molecules in the amorphous regions of the fiber making the diffusion of the dye into the fibers possible. For hydrophobic fibers (like polyester), water plays a less active role in the dyeing process

Dye and Fiber Interactions

Dyes and fibers to which they are applied usually have an inherent attraction for one another. This natural attraction promotes transfer of the dye from the dyebath to the fiber and is sometimes important in holding the dye in the fiber. Dyes are attracted to fibers because of chemical interactions between the dye molecule and the fibers. This binding may result from weak secondary interactions (such as hydrogen bonding or van der Waals forces), ionic bonding, and covalent bonding.

In the weak secondary interactions, multiple interactions between a single dye molecule and the fibers are needed for good bonding to occur. Thus, dyes which bond to fibers mostly by secondary forces are usually relatively large, having high molecular weight structures which can have a large area of interaction with a fiber molecule.

Ionic bonding is important in the dyeing of fibers which have strong acidic or basic character. An ionic bond can form between a fiber having acidic groups and a dye having basic groups or vice versa. The bond formed is a salt linkage which is stronger than secondary interactions. Fibers which bond dyes by forming salt linkages may absorb only a limited amount of dye because the dye is attracted to a specific chemical group in the fiber. When these groups ("dye sites") are occupied by a dye molecule, they are no longer free to other dye molecules. Fibers of this type are mainly synthetic, and the fiber manufacturer builds the required number of dye sites into the fiber during the fiber manufacturing process.

Dyes forming covalent bonds with fibers depend on the secondary forces for their initial attraction to the fiber. After the dye is adsorbed on or in the fibers, a chemical reaction is induced to form covalent bond between the dye and the fiber. Since the covalent bond is very strong and resistant to cleavage, dyes which form covalent bonds with fibers have excellent wash fastness. As is true with dyes forming ionic bonds with the fiber, dye molecules forming covalent bonds with the fiber can be relatively small with brightly colored structures.

Variables in Dyeing

Because of the complex interactions of the components in a dyeing system, many variables must be controlled to produce high-quality dyeing. Some variables in dyeing include variations in substrates, chemicals, water quality, preparation of the substrate, and dyeing procedures. Cotton fiber maturity level is an excellent example of a variation in a raw material that can have an important effect on dyeability. When fibers are immature, the secondary wall is not completely developed. Therefore, immature fibers do not dye the same way as mature fibers. Variability in fiber maturity can result in incorrect blend levels at the mill and cause uneven colors.

Variations in water or auxiliary chemicals used in the dyeing process can cause dyeing defects. Residual chlorine in the water supply has been known to react with and decolorize some of the dye in the dyebath. Any variation in a chemical used in the dyeing process is a potential source for dyeing defects. In addition, inconsistent or incomplete desizing, size material redeposition, and inconsistent scouring are common causes of dye defects. Bleaching variations can also cause dye defects because the base color of a fabric contributes to the color of the dyed fabric. Mercerization has a dramatic effect on the dyeability of cellulose and usually improves markedly the dyeability of immature fibers. Therefore, it must be done uniformly in order for the fabric to be dyed evenly.

Functionalization

Finishing is the final step in the textile manufacturing process; therefore, it is the last chance to impart additional properties that customers will value. Chemical finishing has always been an important component of textile processing. However, in recent years, the trend to "high tech" products has increased the interest and use of chemical finishes. As the use of high performance textiles has grown, the need for chemical finishes to provide the fabric with required properties in these special applications has grown accordingly.

Mechanical Finishes

Calendering

Calendering increases luster and changes the appearance of the fabric surface. It consists of passing the fabric between rolls under pressure. The luster of the fabric is increased because the yarns are flattened making the fabric surface smoother and better able to reflect light. Calendering produces cotton fabrics that have a surface luster similar to that of mercerized cotton.

Compaction

Mechanical stabilization is most often used to prevent excessive shrinkage of fabrics when they are laundered or exposed to heat. Indeed, the manufacturing processes create stresses and tensions in fibers, yarns, and fabrics. Water and heat provide the opportunity for these stresses to relax. Relaxation of stresses in the fabric and swelling of fibers in water usually cause contraction and shrinkage of the fabric.

Chemical Finishes

Softening

During preparation, textiles can take on a brittle feel because natural waxes are removed. This can be alleviated using softeners. Chemical softeners impart to textiles smoothness, flexibility, drape, and pliability.

Flame Retardant

Flame-retardant finishes provide textiles with an important performance characteristic, which is the protection of consumers from fiber combustion. Different approaches have been used to prevent exothermic combustion of the fibers. One approach is to provide a heat sink on or in the fiber by use of materials that thermally decompose through strongly endothermic reactions. When heat is absorbed by these reactions, the fiber pyrolysis temperature is not reached and the combustion is prevented. Examples of this method are the use of aluminum hydroxide and calcium carbonate as fillers in polymers and coatings.

Another approach is to apply a treatment that forms an insulating layer around the fiber that keeps temperatures below the fiber pyrolysis temperature (i.e., Boric acid and its hydrate salts). When heated, these low-melting compounds release water vapor and produce a foamed glassy surface on the fiber, insulating the fiber from the applied heat and oxygen.

Ultraviolet Protection

Several problems are associated with both acute and cumulative exposure to ultraviolet radiation (UVR). UVR is composed of three types: UV-A, UV-B, and UV-C, ranging from 320 to 400 nm, from 290 to 320 nm, and from 100 to 290 nm, respectively. UV-C radiation is absorbed by the ozone layer. However, UV-A and UV-B reach the earth surface and cause serious health problems, such as skin cancer, sunburn, and photo-aging. The harmful effects of the UVR (particularly UV-A and UV-B) can be substantially alleviated by wearing heavy-weight garments or by using sunscreens. However, the use of heavy-weight garments is often uncomfortable due to the resulting increase in the skin temperature. Applying frequently sunscreens is troublesome and subject to failure under many conditions. Therefore, special attention is focused on the UVR transmission of textiles because of the growing demand in the marketplace for lightweight textile apparel that offers protection from UV-A and UV-B radiation, while fostering comfort and lower skin temperature.

When UV radiation strikes a fiber surface, it can be reflected, absorbed, and transmitted through the fiber, or pass between the fibers. The relative amounts of radiation reflected, absorbed or transmitted depend on many factors: (i) fiber type, (ii) fiber-surface smoothness, (iii) fiber cover factor, (iv) presence or absence of fiber delustrants (increased light scattering), dyes, or UV absorbers. To quantify the protective effect of textiles, instruments are used to determine the ultraviolet protection factor (UPF).

Wrinkle Free

Cotton is made up of cellulose macromolecules with repeating anhydroglucose units. On each unit, there are three available hydroxyl groups. These hydroxyl groups serve as sites for water molecule absorption by establishing many hydrogen bonds with the cellulose macromolecules. A severe limitation of fabrics made from cellulosic fibers is their tendency to wrinkle. In general, wrinkles occur when the fiber is bent. In this process, hydrogen bonds between the cellulose macromolecules in the amorphous regions of the fibers break, thus, allowing the chains to slip past one another. The hydrogen bonds then reform in new places and hold creases in the fiber and fabric. The basic idea behind the resistance of cotton fabric to wrinkles is to restrict the slippage of cellulose chains. The appropriate chemical treatment of cotton fabric enables covalent links to establish between the cellulosic chains in the amorphous regions of the fibers, thus, restricting the slippage of the cellulosic chains. For many years, the textile industry has been using N-methylol–based products with very low formaldehyde release as the crosslinking agent. The most common method for finishing the cotton fabric is the pad-dry-cure process (Wei and Yang, 1999; Yang et al., 2001; Yang et al., 2000; Yang et al., 1998). This method consists of impregnating the sample in an aqueous solution containing the crosslinking agent and the appropriate catalyst, padding the impregnated fabric to 90 to 100% wet pick-up, drying, and then curing.

Durable press is a term used for apparel textiles that require little or no ironing after home laundering and have wrinkle resistance properties during daily wear. The protocol for ascertaining a smoothness grade of a fabric is outlined in the American Association of Textile Chemists and Colorists (AATCC) Test Method 124 (AATCC, 2012). This standard test is designed to evaluate the smoothness of fabric specimens after a series of repeated home laundering.

Antimicrobial

Two different aspects of antimicrobial protection provided by chemical finishes can be distinguished; the protection of the textile user against pathogenic or odor causing microorganisms (hygiene finishes) or the protection of the textile itself from damage caused by mold, mildew, or rot-producing microorganisms. The growth of microorganisms on textiles can lead to hygienic and aesthetic difficulties (e.g., staining). The most troublesome organisms are fungi and bacteria. Under very moist conditions, algae can also grow on textiles but are troublesome primarily because they act as nutrient sources for fungi and bacteria. Mechanisms by which antimicrobial finishes control microbial growth include preventing cell reproduction, locking enzymes, destructing cell walls and poisoning the cells from within. Antimicrobial finishes that control the growth and spread of microbes are called biostats; i.e., bacteriostats, fungistats, etc. Antimicrobial finishes that actually kill microbes are biocides; i.e., bacteriocides, fungicides, etc.

Other Finishes

Stain and soil release, water repellency, and moisture management finishes are also gaining interest. Various compounds are used to impart stain- and soil-release properties (e.g., carboxy-based finishes, hydroxy-based finishes, Fluorine-based finishes). Textiles treated with stain- and soil-release finishes are used in active wear, leisure wear, uniforms, and napery. Water repellency is achieved using different product groups (e.g., silicone-based products, fluorocarbon-based products). The objective of a repellent finish treatment is to prevent a drop of liquid from spreading on the surface and wetting the fabric. Moisture management is another important finish for fibers that absorb moisture such as cotton. The development of moisture-management technologies created new market for cotton fabrics. Consumers can benefit from the fast-drying and moisture-wicking performance (similar to man-made fabrics) while enjoying the comfort of cotton fabrics.

Conclusions

Cotton fiber is a naturally variable industrial raw material used primarily in the production of spun yarns. Natural within-sample variability in cotton fiber quality can impact spinning performance and end-product quality. Cotton competes as a raw material against man-made synthetic alternatives that typically exhibit less variability in spinning performance and end-product quality. Data that improve the predictability of cotton spinning performance and yarn quality are crucial for maintaining the competitiveness of cotton in relation to competing fibers.

Spinners purchase cotton bales based on a fiber quality profile that is believed to be adequate for their production needs. Raw material costs are the single greatest cost in the production of spun yarns. If spinners are unable to find bales of cotton with a fiber quality profile that meets their production needs from a particular region, they will turn to cotton produced in another region. When HVI data on fiber properties are available, bales are generally selected based on these data.

The fiber quality profile needs of the spinner depend on many factors. Each spinning technology produces a yarn with a different structure. The relative impacts of cotton fiber properties on the spinning performance and end-product quality depend on these structural differences. The cotton industry must remain cognizant of developments in new spinning technologies and their fiber quality demands.

Technological innovations offer the potential to create new market demands for cotton textiles. Specialty fabric finishes enable retention of the comfort of cotton textiles while imparting novel functional properties, such as wrinkle free and stain release. Understanding the interactions of fiber properties with these developing technologies is crucial to successful development.

Supplying cotton fibers that are desirable as an industrial raw material begins with the seed. Breeders utilize fiber quality assessment as a method for selecting lines with elite spinning performance. Cotton producers select varieties and utilize agronomic practices with the objective of optimizing yield and market value. Harvesting, ginning, and cleaning demands are balanced with preserving cotton fiber quality. Fiber quality is an integral component to the future success of the cotton industry.

References

Abidi, N., L. Cabrales, and E.F. Hequet. 2010. Fourier transform infrared spectroscopic approach to the study of the secondary cell wall development in cotton fiber. Cellulose 17(2):309–320. doi:10.1007/s10570-009-9366-1

American Association of Textile Chemists and Colorists. 2012. Technical manual. AATCC, Research Triangle Park, NC.

Anderson, D.B., and T. Kerr. 1938. Growth and structure of cotton fiber. Ind. Eng. Chem. 30(1):48–54. doi:10.1021/ie50337a010

Anderson, D.B., and J.H. Moore. 1937. The influence of constant light and temperature upon the structure of the walls of cotton fibers and collenchymatous cells. Am. J. Bot. 24:503–507. doi:10.2307/2437071

Anthony, W.S., and William D. Mayfield, eds. 1994. Cotton ginners handbook. Agric. handbook 503. U.S. Department of Agriculture, Washington, DC.

ASTM. 2011. ASTM D1448-11: Standard test method for micronaire reading of cotton fibers. ASTM, Philadelphia, PA.

ASTM. 2013a. ASTM D7139-09(2013): Standard terminology for cotton fibers. ASTM, Philadelphia, PA.

ASTM. 2013b. ASTM D123-13a: Standard terminology relating to textiles. ASTM, Philadelphia, PA.

Backe, E.E. 1986. Effect of short fiber content in cotton on plant performance and quality. Text. Res. J. 56(2):112–115. doi:10.1177/004051758605600207

Baldwin, J.C., Y.T. Chu, M.E. Galyon, and F.M. Shofner. 1995. U.S. Patent No. 5430,301. U.S. Patent and Trademark Office: Washington, DC.

Bange, M.P., R.L. Long, G.A. Constable, and S.G. Gordon. 2010. Minimizing immature fiber and neps in upland cotton. Agron. J. 102(2):781–789. doi:10.2134/agronj2009.0454

Basra, A.S. 1999. Cotton fibers: Developmental biology, quality improvement, and textile processing. Food Products Press, Binghamton, NY.

Bauer, P.J., J.A. Foulk, G.R. Gamble, and E.J. Sadler. 2009. A comparison of two cotton cultivars differing in maturity for within-canopy fiber property variation. Crop Sci. 49(2):651–657. doi:10.2135/cropsci2008.06.0350

Bednarz, C.W., R.L. Nichols, and S.M. Brown. 2006. Plant density modifies within-canopy cotton fiber quality. Crop Sci. 46(2):950–956. doi:10.2135/cropsci2005.08-0276

Boylston, E.K., J.P. Evans, and D.P. Thibodeaux. 1995. A quick embedding method for light microscopy and image analysis of cotton fibers. Biotech. Histochem. 70(1):24–27. doi:10.3109/10520299509108312

Bragg, C.K., and F.M. Shofner. 1993. A rapid, direct measurement of short fiber content. Text. Res. J. 63(3):171–176. doi:10.1177/004051759306300307

Chu, Y.T., and C.R. Riley. 1997. New interpretation of the fibrogram. Text. Res. J. 67(12):897–901.

CIRFS. 2011. World markets for technical textiles to 2017. CIRFS, Brussels, Belgium.

Cotton Incorporated. 2014. HVI color chart. Http://www.cottoninc.com/fiber/quality/US-Fiber-Chart/HVI-Color-Chart/ (accessed date 15 June 2014).

Cranmer, L.M. 2004. A better understanding of the number of fibers per seed in cotton. Thesis, Texas Tech University, Lubbock, TX.

Darcy, H. 1856. Exposition et application des principes à suivre et des formules à employer dans les questions de distribution de l'eau. In: V. Dalmont, editor, Les fontaines publiques de la ville de Dijon., Paris. p. 16.

Dever, J.K., J.R. Gannaway, and R.V. Baker. 1988. Influence of cotton fiber strength and fineness on fiber damage during lint cleaning. Text. Res. J. 58(8):433–438.

El Mogahzy, Y.E., R. Broughton, and W.K. Lynch. 1990. A statistical approach for determining the technological value of cotton using HVI fiber properties. Text. Res. J. 60(9):495–500. doi:10.1177/004051759006000901

El Mogahzy, Y.E., and C.H. Chewning, Jr. 2001. Cotton fiber to yarn manufacturing technology. Cotton Incorporated, Cary, NC.

Faulkner, W.B., E.F. Hequet, J. Wanjura, and R. Boman. 2012. Relationships of cotton fiber properties to ring-spun yarn quality on selected High Plains cottons. Text. Res. J. 82(4):400–414. doi:10.1177/0040517511426613

Feng, L., V.B. Bufon, C.I. Mills, E.F. Hequet, J.P. Bordovsky, W. Keeling, and C.W. Bednarz. 2011. Effects of irrigation, cultivar, and plant density on cotton within-boll fiber quality. Agron. J. 103(2):297–303. doi:10.2134/agronj2010.0185

Foulk, J.A., and D.D. Mcalister. 2002. Single cotton fiber properties of low, ideal, and high micronaire values. Text. Res. J. 72(10):885–891. doi:10.1177/004051750207201006

Gamble, G.R. 2007. Effect of bale ageing on cotton fiber chemistry, processing performance, and yarn quality. J. Cotton Sci. 11:98–103.

Goldthwait, C. F., H.O. Smith, and M.P. Barnett. 1947. New dye technique shows maturity of cotton, Textile World 97:105–110.

Guo, X., W. Ouyang, and B. Xu. 2014. Assessing cotton maturity using distributional parameters of fiber cross-section measurements. Text. Res. J. 84:1666–1676.

Haigler, C.H., N.R. Rao, E.M. Roberts, J.Y. Huang, D.P. Upchurch, and N.L. Trolinder. 1991. Cultured cotton ovules as models for cotton fiber development under low temperatures. Plant Physiol. 95:88–96. doi:10.1104/pp.95.1.88

Hebert, J.J., E.K. Boylston, and D.P. Thibodeaux. 1988. Anatomy of a nep. Text. Res. J. 58(7):380–382.

Hequet, E.F., and N. Abidi. 2005. Effects of the origin of the honeydew contamination on cotton spinning performances. Text. Res. J. 75(10):699–709. doi:10.1177/0040517505053909

Hequet, E.F., and N. Abidi. 2006. Sticky cotton: Measurements and fiber processing. Texas Tech Univ. Press, Lubbock, TX.

Hequet, E.F., N. Abidi, and D. Ethridge. 2005. Processing sticky cotton: Effect of stickiness on yarn quality. Text. Res. J. 75(5):402–410. doi:10.1177/0040514505053953

Hequet, E.F., B. Kelly, and J. Dever. 2014. Breeding for better fiber elongation: A key to improve yarn tensile properties. International Cotton Conference. Bremen, Germany.

Hequet, E.F., B. Wyatt, N. Abidi, and D.P. Thibodeaux. 2006. Creation of a set of reference material for cotton fiber maturity measurements. Text. Res. J. 76(7):576–586.

Hertel, K.L. 1940. A method of fiber-length analysis using the fibrograph. Textile Research 10:510–520.

Hertel, K.L., and C.J. Craven. 1956. Cotton fiber bundle elongation and tenacity as related to some fiber and yarn properties. Text. Res. J. 26(6):479–484. doi:10.1177/004051755602600613

Hessler, L.E., J.D. Towery, and B.K. Power. 1954. The effect of weathering in the field on the fiber properties of cotton. Text. Res. J. 24(11):1010–1014. doi:10.1177/004051755402401111

Hughs, S., C. Armijo, and J. Foulk. 2013. Upland fiber changes due to ginning and lint cleaning. J. Cotton Sci. 17(2):115–124.

Hunter, R.S. 1958. Photoelectric color difference meter. J. Opt. Soc. Am. 48(12):985–993. doi:10.1364/JOSA.48.000985

International Cotton Advisory Committee. 2013. World textile demand. ICAC, Washington, DC.

International Textile Manufacturers Federation. 2012. International Cotton Industry Statistics. ITMF, Zurich, Switzerland.

International Textile Manufacturers Federation. 2014. Cotton contamination survey. ITMF, Zurich, Switzerland. http://www.itmf.org/publications/free/datei8.pdf (Accessed 6 Feb. 2015).

Kellie, G. 2013. "What does the nonwovens industry expect from the man-made fiber producers?" In: Man-Made Fiber Year Book 2013, Chemical Fibers International. p. 71-75.

Kelly, B. and E.F. Hequet. 2013. Breeding for improved yarn quality: Importance of non-HVI fiber properties. ICAC Recorder 31:13–20.

Kelly, C.M., E.F. Hequet, and J.K. Dever. 2013. Breeding for improved yarn quality: Modifying fiber length distribution. Ind. Crops Prod. 42:386–396. doi:10.1016/j.indcrop.2012.06.018

Kerr, T. 1937. The structure of the growth rings in the secondary wall of the cotton hair. Protoplasma 27(1):229–241. doi:10.1007/BF01599376

Kohel, R.J., and S.C. McMichael. 1990. Immature fiber mutant of upland cotton. Crop Sci. 30(2):419–421. doi:10.2135/cropsci1990.0011183X003000020038x

Kozeny, J. 1927. Ueber kapillare leitung des wassers, Boden Sitzungsberichte Wiener Akad. 136:271–306.

Lawrence, C.A. 2003. Fundamentals of spun yarn technology. CRC Press, Boca Raton, FL.

Lee, S.W. 1967. A probability model for random fiber breakages Part I. Theory. Text. Res. J. 37(10):860–874. doi:10.1177/004051756703701006

Long, R.L., M.P. Bange, S.G. Gordon, and G.A. Constable. 2010a. Measuring the maturity of developing cotton fibers using an automated polarized light microscopy technique. Text. Res. J. 80(5):463–471. doi:10.1177/0040517509343784

Long, R.L., M.P. Bange, S.G. Gordon, M.H. van der Sluijs, G.R. Naylor, and G.A. Constable. 2010b. Fiber quality and textile performance of some Australian cotton genotypes. Crop Sci. 50(4):1509–1518. doi:10.2135/cropsci2009.10.0600

Long, R.L., M.P. Bange, C.D. Delhom, J.S. Church, and G.A. Constable. 2013. An assessment of alternative cotton fibre quality attributes and their relationship with yarn strength. Crop Pasture Sci. 64(8):750–762.

Lord, E. 1956. Air flow through plugs of textile fibres. Part II. The micronaire test for cotton. J. Textil. Inst. Trans. 47:T16–T47. doi:10.1080/19447027.1956.10750375

Lord, E. 1981. The origin and assessment of cotton fibre maturity. International Institute for Cotton, Manchester, UK.

Maulding, G. 2007. Yield determining components of new germplasm. Thesis, Texas Tech Univ., Lubbock, TX.

McCreight, D.J., R.W. Feil, J.H. Booterbaugh, and E. Backe. 1997. Short staple yarn manufacturing 700. Carolina Academic Press, Durham, NC.

Meredith, W.R., and R.R. Bridge. 1973. Yield, yield component and fiber property variation of cotton (*Gossypium hirsutum* L.) within and among environments. Crop Sci. 13(3):307–312. doi:10.2135/cropsci1973.0011183X001300030006x

Montgomery, D.J., and W.T. Milloway. 1952. The vibroscopic method for determination of fiber cross-sectional area. Text. Res. J. 22(11):729–735. doi:10.1177/004051755202201102

Morton, W.E., and J.W. Hearle. 2008. Physical properties of textile fibres. Elsevier, New York.

Nickerson, D. 1951. New automatic cotton colorimeter for use in cotton quality specification. Text. Res. J. 21(1):33–38. doi:10.1177/004051755102100108

Nickerson, D., R.S. Hunter, and M.G. Powell. 1950. New automatic colorimeter for cotton. J. Opt. Soc. Am. 40(7):446. doi:10.1364/JOSA.40.000446

Nickerson, D., and J.J. Tomaszewski. 1958. Color change in raw cotton related to conditions of storage. Text. Res. J. 28(6):485–497. doi:10.1177/004051755802800608

Orr, R.S., L.C. Weiss, and J.N. Grant. 1955. The relation of single-fiber to flat-bundle strength and elongation of cotton. Text. Res. J. 25(11):939–946. doi:10.1177/004051755502501105

Peeters, M.C., J. Wijsmans, I. De Langhe, E. De Langhe, and L. Waterkeyn. 1986. Never dried cotton fibers have a circular cross section. Text. Res. J. 56(9):529–532. doi:10.1177/004051758605600901

Peirce, F.T., and E. Lord. 1939. 13—The fineness and maturity of cotton. J. Textil. Inst. Trans. 30(12):T173–T210. doi:10.1080/19447023908659334

Peters, G., and S. Meier. 2010. Description of all quality parameters measured by Uster Technologies and yarn testing equipment. Uster Laboratory Systems Application Report, ed. 3. Uster Technologies, Knoxville, TN.

Pettigrew, W.T. 1995. Source-to-sink manipulation effects on cotton fiber quality. Agron. J. 87(5):947–952. doi:10.2134/agronj1995.00021962008700050029x

Ramey, H.H., R. Lawson, and S. Worley. 1977. Relationship of cotton fiber properties to yarn tenacity. Text. Res. J. 47(10):685–691.

Seagull, R., and P. Alspaugh. 2001. Cotton fiber development and processing: An illustrated overview. International Textile Center, Texas Tech University, Lubbock, TX

Smith, E. 2010. Testimony: Presented to the subcommittee on general farm commodities and risk management committee on agriculture. U.S. House of Representatives, Washington, DC. http://www.cotton.org/upload/10eddiesmithjune24testimonywritten.pdf (Accessed 6 Feb. 2015)

Stewart, J. 1975. Fiber initiation on the cotton ovule (*Gossypium hirsutum*). Am. J. Bot. 62:723–730. doi:10.2307/2442061

Tallant, J.D., L.A. Fiori, D.M. Alberson, and W.E. Chapman. 1961. The effect of the short fibers in a cotton on its processing efficiency and product quality Part III: Pilot-scale processing of yarns. Text. Res. J. 31(10):866–872. doi:10.1177/004051756103101004

Tallant, J.D., and R.A. Pittman. 1968. Breakage models and measuring techniques for fiber. Weight-length distributions. Text. Res. J. 38(2):149–155. doi:10.1177/004051756803800207

Taylor, R.A. 1977. Aerodynamic drag of cotton fibers in crossflow. Text. Res. J. 47(5):381–386.

USDA. 2014. Cotton: World markets and trade. Circular series FOP, 06-14. USDA/FAS, Washington, DC. http://apps.fas.usda.gov/psdonline/circulars/cotton.pdf

Wakeham, H. 1955. Cotton fiber length distribution—An important quality factor. Text. Res. J. 25(5):422–429. doi:10.1177/004051755502500509

Wei, W., and C.Q. Yang. 1999. Predicting the performance of durable press finished cotton fabric with infrared spectroscopy. Text. Res. J. 69(2):145–151. doi:10.1177/004051759906900210

Woo, J.L. 1967. 39—An appraisal of the length measures used for cotton fibres. J. Textil. Inst. 58(11):557–572. doi:10.1080/00405006708659951

Xu, B., and B. Pourdeyhimi. 1994. Evaluating maturity of cotton fibers using image analysis: Definition and algorithm. Text. Res. J. 64(6):330–335. doi:10.1177/004051759406400604

Yang, C.Q., L. Qian, and G.C. Lickfield. 2001. Mechanical strength of durable press finished cotton fabric. Part IV: Abrasion resistance. Text. Res. J. 71(6):543–548. doi:10.1177/004051750107100612

Yang, C.Q., W. Wei, and G.C. Lickfield. 2000. Mechanical strength of durable press finished cotton fabric Part II: Comparison of crosslinking agents with different molecular structures and reactivity. Text. Res. J. 70(2):143–147. doi:10.1177/004051750007000209

Yang, C.Q., L. Xu, S. Li, and Y. Jiang. 1998. Nonformaldehyde durable press finishing of cotton fabrics by combining citric acid with polymers of maleic acid. Text. Res. J. 68(6):457–464. doi:10.1177/004051759806800611

Zurek, W., M. Greszta, I. Frydrych, and G. Balcar. 1999. Cotton fiber length changes in the spinning process on the basis of AFIS measurements. Text. Res. J. 69(11):804–810. doi:10.1177/004051759906901102

Seed

Michael K. Dowd

Abstract

Cottonseed is a second product of the cotton (*Gossypium hirsutum* L.) plant that accounts for between 15 and 25% of the crop value. The seed is either extracted for oil (producing meal, hulls, and linters as co-products) or is used whole as a protein feed ingredient for ruminant animals. The oil is the most valuable product and is widely used as a food ingredient or in food preparation. Different methods have been used to extract cottonseed oil, but the preferred technique is to form porous collets of the kernels that allow for efficient solvent extraction of the oil. Meal and hulls are used as animal feed ingredients, although some industrial uses are being explored. Hulls are also valued as a component of drilling mud and are a preferred substrate for the production of some mushrooms. Linters are often pulped and used in product applications that require high cellulose purity. The recent genetic development of low gossypol seed has the potential to alter both the processes used to extract the oil and the use of the seed protein.

Although often overshadowed by the concerns of fiber production, cottonseed is an economically important second product of the cotton plant. The seed is separated from the fiber by ginning and is either extracted to recover oil, which also produces meal, hulls, and liners as co-products, or used whole as an animal feed ingredient. The seed-to-fiber weight ratio is about 3:2, and depending on the market price of fiber and the various seed products, the seed accounts for between 15 and 25% of the value of the cotton crop.

Cottonseed has a special role in the oilseed extraction and processing industries. As plantation farming of cotton developed in the Southern United States in the mid-19th century, large amounts of seed became available that initially had little use. Some of the seed was ground and applied to fields as fertilizer or fed to cows, but otherwise it was often disposed of along streams and local waterways where it caused pollution and environment problems (Wrenn, 1995). This spawned interest in finding better uses for the seed. From this, the extraction of cottonseed for oil resulted, launching the industrial production of vegetable oils. The need to improve the quality of this oil led to the development of many of the extraction, refining, and finishing operations in use today.

Abbreviations: AOCS, American Oil Chemists' Society; AOM, Active Oxygen Method; RBD, refined, bleached, and deodorized.

Michael K. Dowd, Southern Regional Research Center, USDA-ARS, 1100 Robert E. Lee Blvd., New Orleans, LA 70124 (michael.dowd@ars.usda.gov).

doi:10.2134/agronmonogr57.2013.0032

Cotton. 2nd. ed. David D. Fang and Richard G. Percy, editors. Agronomy Monograph 57.

When cottonseed is crushed, the recovered oil is used almost entirely in food preparation or as a food ingredient. Cottonseed oil is valued for deep-fat frying because it imparts a toasted or nutty aroma to fried foods. It is also used to formulate shortening, icing, and spreads because it promotes the formation of small crystallites that impart a smooth consistency and light mouthfeel to these products.

Whole cottonseed is also valuable as a protein ingredient in ruminant animal diets, but it is especially valued by the dairy industry because it increases the level of butterfat in milk (Smith et al., 1981). Much of the meal left over after oil extraction is also used as feed protein. In addition to being used to feed dairy cows and cattle, the meal is used in the finishing diets for sheep and goats and is included in diets of farm-raised catfish.

Seed Structure and Biology

In mature seed, the embryo or kernel is encased and protected by a tough fibrous hull that also is the origin of the cotton fibers. Ginning the seed cotton with saw or roller gins separates the fibers. For *G. hirsutum* seed, which makes up around 95% of commercial cotton production, ginning leaves behind a dense layer of morphologically distinct short fibers, called linters (Fig. 1a). In *G. barbadense* L. seed, the linters are mostly absent (Fig. 1b), although this trait can be variable depending on the genetic history of the variety.

Seeds tend to be oval in shape, but tapered on one end, and are typically 10 to 12 mm long. Kernels (Fig. 1c) are composed of a pair of folded cotyledon leaves that form the bulk of the embryo. The cotyledon leaves surround the axial organs that comprise the epicotyl or the leaf buds that will form the first true leaves of the developing plant, the hypocotyl that forms the above ground stem, and the radicle that develops into the plant's root system. Sectioning through a kernel reveals large numbers of dark lysigenous glands (Fig. 2a) that are distributed throughout the cotyledon tissue of the seed. These are commonly referred to as pigment glands, and they are also found in the stems, buds, and foliar tissues of the cotton plant. The glands are typically between 0.025 and 0.180 mm in diameter and constitute between 2.4 and 4.8% of the kernel weight (Lusas and Jividen, 1987). The glands contain terpenoid pigments that appear to function as defensive agents. In ungerminated seed, the glands are largely absent from the embryonic axial organs, although they develop as the seed germinates along the stem tissue. Although some oil is stored in the axial organs, the cotyledons contain most of the kernel's oil and protein, where they serve as reserve materials to support the germinating plant. A detailed discussion regarding the cellular structures and development of cottonseed tissue has been presented by Leahy (1948).

Seed Composition

Bulk Constituents

The seed index (100 seed weight) of fuzzy white cottonseed is about 10 g on an as-is basis (Tharp, 1948). Seed that has been stored in ambient conditions contains 9 to 10% moisture. Moisture levels, however, can vary depending on the conditions present during harvesting and storage. By dry weight (dry wt.), white fuzzy cottonseed is composed of 10 to 12% linters, 37 to 40% hulls, and 49 to 51% kernels (Tharp, 1948; Cherry and Leffler, 1984) (Table 1). The amount of oil in white fuzzy seed varies between 20 and 24% (dry wt.), and the amount of protein varies from

Fig. 1. Cottonseed kernels. (a) Ginned *Gossypium hirsutum* L. variety with linters. (b) Ginned St. Vincent Superfine *G. barbadense* L. variety. (c) Dehulled cottonseed kernels. (d) Acid-delinted *G. hirsutum* variety.

22 to 26% (dry wt.). Sugars make up between 3 and 4% (dry wt.) of the seed. Ash accounts for another 4% (dry wt.), and the remainder is fibrous components (i.e., celluloses, hemicelluloses, and lignin). From a fiber analysis of the whole seed, neutral detergent fiber is ~47%, acid detergent fiber is ~38%, and crude fiber is ~30% (National Cottonseed Products Association, 2002a).

Cottonseed linters are mostly cellulosic fiber with a crude oil level (i.e., ether extractibles) of around 0.8% (dry wt.) and a protein level of around 3 to 4% (dry wt.) (Table 1). Hulls are also mostly fiber. Hence, the bulk of the oil and protein is located in the kernels. After removal of the linters and hulls, kernels contain around 35% oil (dry wt.) and 38% protein (dry wt.).

Gossypol

Among the minor components of the seed, gossypol is a polyphenolic terpenoid that has a significant influence on how the seed is processed and used. First identified in cottonseed soap derived from oil purification (Longmore, 1886) and named for *Gossypium* phenol by Marchlewski (1899), gossypol is one of a series of terpenoid metabolites that are synthesized from the isoprenoid pathway in

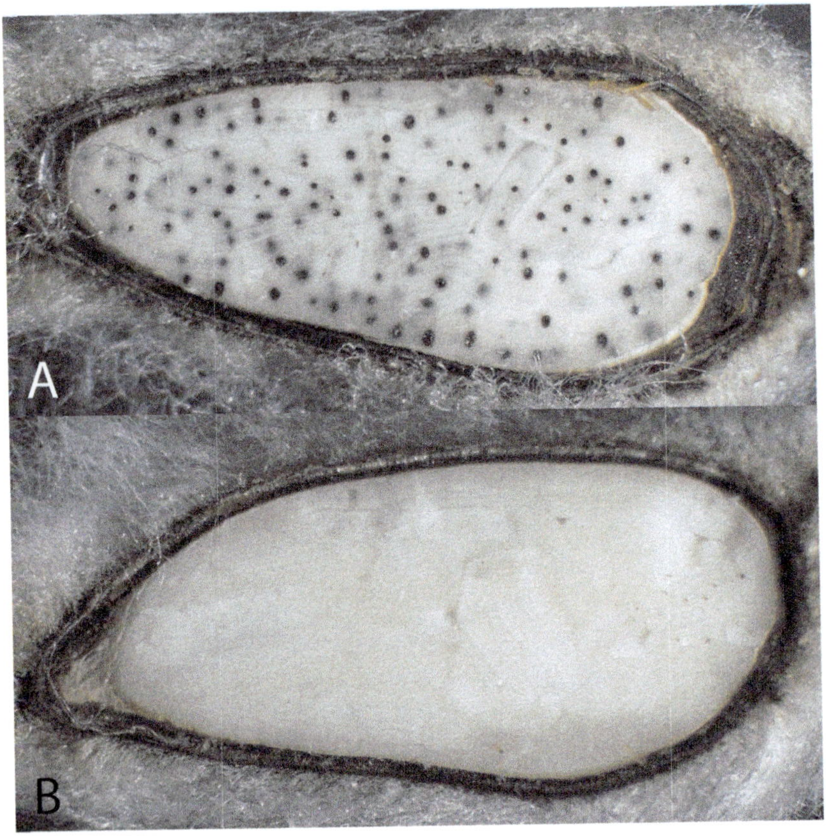

Fig. 2. Longitudinal cottonseed sections. (a) Glanded seed. (b) Glandless seed.

the cotton plant. Gossypol in cottonseed is of particular importance, as the compound is responsible for anti-nutritive and toxicity issues associated with the overfeeding of cottonseed products (Berardi and Goldblatt, 1980). The compound has also been of considerable research interest because it exhibits a wide range of bioactivity, including activity as a male anti-fertility, anti-cancer, and anti-viral agent (Wang et al., 2009).

Gossypol is the primary terpenoid found in the pigment glands of the seed. In *G. hirsutum* seed, gossypol accounts for almost all of the terpenoid present, while in *G. barbadense* seed one or two methylated derivatives can sometimes be detected as minor components (Stipanovic et al., 1975, 2009; Percy et al., 1996). Gossypol appears to function in the plant as a deterrent to insect predation (Bottger et al., 1964; Stipanovic et al., 2006a) although its concentration in root tissue also suggests that it may be an inhibitor of fungal infection (Puckhaber et al., 2002; Mellon et al., 2014).

Concentrations of gossypol in the kernels of commercial varieties usually range between 0.8 and 1.4%. However, seed with lesser and greater levels are not unusual. For example, seed gossypol levels of 3 or 4% have also been noted in *G. barbadense*

Table 1. Bulk component composition and typical product yields from fuzzy white (*Gossypium hirsutum* L.) cottonseed.†

Seed component	Mass fraction	Protein content	Oil content
	%		
	Seed components (dry weight basis)		
Linters	12.7	3.7	0.9
Hulls	31.8	3.8	0.8
Kernels	55.5	38.6	34.8
	Product yields from processing		
Linters	10	–	–
Hulls	25	5	1.9
Oil	16	–	–
Meal	46	41–48	2.2
Miscellaneous (foreign matter/lost mass)	4	–	–

† Table adapted from Dowd and Wakelyn (2011); data from Tharp (1948) and National Cottonseed Products Association (2002a).

genotypes (Percy et al., 1996), and concentrations as high as 5.7% have been noted in the seed kernel of related wild *Gossypium* species (Stipanovic et al., 2005).

The gossypol molecule consists of two connected naphthalene rings that combined have six hydroxyl groups and two aldehyde moieties that influence its chemistry and properties (Fig. 3). The compound exists as a pair of isomers because of restricted rotation about the bond connecting its two naphthalene rings. This results in a chiral axis resulting in a pair of stable enantiomers. With the hydroxyl and carbonyl groups of the left naphthalene ring oriented upward and the polar groups of the right naphthalene ring oriented forward (Fig. 3) the molecule is in the S- or P-form, which corresponds to the (+) or dextrorotary form (Huang et al., 1988; Freedman et al., 2003).

Both optical forms of gossypol are produced by the cotton plant, and their distribution within seed and other plants tissues is generally unequal. In seeds of *G. hirsutum* the ratio of the (+) and (–) enantiomers tends to average about 60:40, although seed of some Marie galante landraces are known to have 85 to 95% of their gossypol in the (+)-form (Cass et al., 1991; Stipanovic et al., 2006b). The gossypol enantiomer ratio in *G. barbadense* seed is also variable with some genotypes

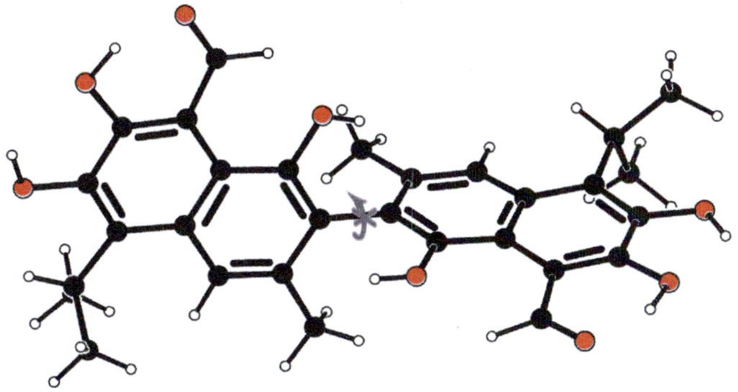

Fig. 3. Molecular structure of (+)-gossypol.

and varieties exhibiting a small excess of the (+)-form, but many others exhibiting an excess of the (–)-form (Percy et al., 1996). While *G. barbadense* genotypes can also have 90% or more of its gossypol in the (+)-form, no genotypes or varieties have been identified with the percentage of (–)-gossypol greater than ~68% (Stipanovic et al., 2009).

Glandless and Ultra-Low Gossypol Seed

Working with Hopi varieties of *G. hirsutum*, McMichael (1959) found cotton plants containing only a few pigment glands in the leaves and bolls, but with regular glanding in the seed. By crossing this plant type with others that contained reduced seed glanding, plants essentially free of glands were obtained. These "glandless" plants were found to have only trace levels of gossypol in the seed (Fig. 2b). By crossing these plants with fully glanded plants, the glandless trait appeared to be expressed as a homozygous recessive condition resulting from the action of two genes (McMichael, 1960). Lee et al. (1968) demonstrated that one of the two genes controls about two-thirds of the seed gossypol level, while the other gene controls the remaining one-third.

The development of glandless seed generated considerable interest in the cotton industry and research community, and several research programs were initiated to study processing refinements for these seeds and potential food uses for their protein component (discussed below). Despite considerable effort, however, these varieties never obtained commercial acceptance largely because of concerns regarding fiber yield differences and the potential for increased insect damage (Lee, 1969; Lusas et al., 1978). Both of these issues have now been discounted (Lusas and Jividen, 1986); yet the stigma surrounding these perceived problems has essentially eliminated efforts to grow and use glandless cotton beyond that needed to support research activities.

More recently, cottonseed with low levels of gossypol has been engineered by RNA-mediated gene silencing (Sunilkumar et al., 2006). The modification was obtained by introducing a DNA construct to silence the synthesis of cadinene synthase, the enzyme responsible for the first committed step in gossypol biosynthesis, on a seed-specific promoter. The resulting plants differ from glandless plants in that it is only the seed that lacks gossypol and the compound and related terpenoids are maintained in the foliar parts of the plant. The modification reduces seed gossypol levels to trace amounts (Sunilkumar et al., 2006; Palle et al., 2013). Seeds from these genotypes are sometimes referred to as "ultra-low gossypol seed" to distinguish them from glandless seed.

These modified plants lines are currently undergoing regulatory review and should be available for breeding with agronomic varieties in a few years. While the question still remains as to whether these plants will become viable commercial varieties, the invention has renewed interest in the processing changes that will be needed to derive additional value from these new ultra-low gossypol cottonseed.

Minor Components

Ashing accounts for about 4% of the cottonseed mass. Phosphorus and potassium are the most abundant mineral elements in the seed (National Cottonseed Products Association, 2002a). Approximately 75% of the seed phosphorus is associated with phytic acid (Stansbury et al., 1953b) with the remaining phosphorus associated with phospholipids, nucleic acids, and proteins. Thompson et al. (1968) have reported on the presence of small amounts of carotenoids in cottonseed (0.6 mg kg^{-1} of fresh weight), and lutein was identified as the major carotenoid making up 40% of the total. The seed also contains a series of flavonoids that appear to have antioxidant and potentially other bioactivity (Whitten et al., 1984). Dollear and Markley (1948) have discussed other minor components in both the seed and its processed fractions. Some of these will be discussed below in sections regarding the composition of individual seed products.

Variation in Seed Composition

Genotype and environment both have significant effects on seed composition. Many early studies have reported on the effects of variety, year, and location on seed oil, protein, and gossypol levels (Pons et al., 1953; Pope and Ware, 1945; Stansbury et al., 1954, 1956; Cherry et al., 1981). These studies typically employed currently planted varieties and concluded that significant seed variation exists because of both genetic and environmental effects. More recent data have been accumulated for seed properties from the National Cotton Variety Trials, and these data show similar variation in the seed oil, protein, and gossypol levels. Typically, the range of protein values might vary by 4 to 6% points, oil by a similar amount, and gossypol levels will range between 0.8 to 1.3%. For example, Meredith et al. (1978) showed that protein values of dehulled cottonseed varied over a 6% range (30–36%) for 35 genotypes grown in 1974 and 1975. While this level of variation is significant, the full range of variation has not been explored by a detailed survey of cotton germplasm for protein and oil levels. Unfortunately, efforts to exploit this variation to improve the quality of cottonseed products have received little interest. This is partly because of the seed's role as the second product of the plant and partly because of the amount of analytical work required for a thorough study.

Tharp (1948) has reviewed the early literature that associated increased seed oil with increased field precipitation. Following up these reports, Stansbury et al. (1954) noted a significant positive correlation between the amount of oil in the kernel and the amount of rainfall that occurs during boll development. The strongest correlations were observed for a 35-d period starting 17 d postanthesis until boll opening, which corresponds roughly to the time when oil and protein accumulate in the developing seed (Tharp, 1948). Pons et al. (1953) noted a similar relationship between rainfall and increased seed gossypol levels. Stansbury et al. (1956) then found that kernel nitrogen (i.e., protein), oil, and gossypol levels are all inter-related, with oil and nitrogen levels inversely correlated and oil and gossypol levels positively correlated. In most of these experiments, data were accumulated over many locations indicating broad applicability. Similar trends have been observed recently on single field experiments with the water controlled through irrigation (Pettigrew and Dowd, 2011).

Nitrogen fertilization is well known to increase cotton and seed yield. When fields are fertilized, the seed protein level increases and oil level decreases compared with their levels in seed from unfertilized fields (Tharp et al., 1949). More recently, Main et al. (2013) have observed direct linear relationships for the protein and oil levels (positive and negative, respectively) with the increased application of nitrogen fertilizer.

As mentioned above, considerable differences in the gossypol enantiomer distribution are present in cotton genotypes (Percy et al., 1996). Environmental effects on this ratio are less pronounced, and Meredith et al. (2012) have reported that of the 26 fiber and seed traits monitored in the National Cotton Variety trials, the percentage of (+)-gossypol is the least strongly affected by environment. Regardless, a significant shift in the percentage of (+)-gossypol from 59.7% for irrigated seed to 62.1% for seed produced by dryland production has been reported (Pettigrew and Dowd, 2011).

Seed Storage

Because cottonseed can deteriorate rapidly when exposed to heat and moisture, proper handling and storage conditions are necessary to maintain cottonseed quality. This is important as the cotton crop is harvested and ginned over a few months, and large quantities of seed become available at the same time of the year. Processing and cost efficiencies require that most of this seed be stored for extended periods.

The principal concern with storage of cottonseed is degradation of the oil resulting from a combination of microbial contamination and seed germination, which are accelerated in warm and moist environments. These conditions result in increased lipase activity that causes rapid rise in the seed free fatty acid level and a corresponding degradation in the oil triglycerides, changes that complicate the extraction and refining of the oil.

Aeration is generally necessary for long-term seed storage to reduce temperature and moisture, which minimizes mold growth, insect activity, and seed respiration. Gregory et al. (1999) have discussed details for moving air through seed piles. Although recommendations vary, seeds with moisture levels less than 10% can usually be stored at ambient temperatures with little difficulty. Seed between 10 and 14% moisture may exhibit problems, and seed with 14% and greater levels of moisture are likely to deteriorate rapidly (Robertson and Campbell, 1933). For good quality seed, storage temperatures less than 27°C are normally of little concern. For seed with elevated levels of free fatty acids, deterioration can occur rapidly even at lower temperatures. Storage of these seeds needs to be minimized and carefully monitored, and generally these seeds should be processed as quickly as possible (Alderks, 1948). While storage of seed before processing is of principal concern because of its duration, seed deterioration can begin immediately after harvest within cotton modules, which should also be monitored to maintain seed quality (Cherry and Leffler, 1984).

Seed Processing

Seed cleaning

After arriving at the mill, white fuzzy seed is cleaned with a combination of reels, pneumatic aspirators, and screen shakers to remove unopened bolls, burrs, stem pieces, cotton tufts, dirt, stones, and foreign matter. Magnets are often used to remove metal that can be picked up from the field or gin. Elimination of this metal is essential to prevent damage to the mechanical equipment. While some variations exist in the techniques used to remove contaminants, a general description of these operations is given by Woolrich and Carpenter (1935).

Delinting

After seed cleaning, the seed is delinted (also referred to as delintered) to separate the linters. Delinting can be accomplished in a number of ways. Linters can be hydrolyzed by a short treatment with sulfuric acid. This chemical process is used primarily to prepare seed for planting, as it removes the linters quantitatively (Fig. 1d) and leaves a smooth hull surface that works well with seed planters. Delinting is also necessary for oil processing, as the linters reduce the extractability of the oil. In oil mills, delinting is almost always accomplished mechanically. Mechanical delinting does not separate all of the linters but removes enough that those remaining do not interfere significantly with oil recovery.

The machine used for delinting is referred to as a linter (or sometime as a delinter). It is composed of a series of closely spaced saw blades that pull the linters from the seeds as they circulate over the top of the blades. In this regard it is similar to a saw gin that separates cotton fibers. However, the design of a linter is different because of the more compact size of the fuzzy seed and the greater proportion of seed to linter fiber. Shorter, more closely spaced blades are used to prevent the seed from falling between the blades, and a series of rakes are spaced between the blades to sweep out the separated linters. It is essential to keep the blades sharp, which requires frequent maintenance, often after only hours of operation. Specialized machines, referred to as a filer and gummer, are used for sharpening the blades.

Linters are sometimes recovered in a single pass, but they can also be recovered in multiple passes. Single-pass delintering produces "mill-run" linters. The two-pass process produces "first-cut" and "second-cut" linters, which differ considerably in size and properties (discussed below). Typically, mechanical delinting reduces the mass of linters to between 1 and 2% of the seed weight (Wamble, 1948). As the linter make up about 10 to 12% of the white fuzzy seed, this corresponds to a recovery of between 80 and 92%.

Dehulling

After separating the linters, seeds are milled with bar- or disc-type attrition mills or hullers. A huller forces the seed to pass through knife-edged bars that crack the hulls and cut the kernels into two or more pieces. The cracked seed pieces then pass through a series of shaker tables and aspirators to separate the kernel pieces from the uncut seed and the bulk of the hulls. Shaker tables of various designs are used to separate the hull and kernel material (Wamble, 1948), and in general, the hulls are then pass through hull beaters to recover fine kernel particles. The whole kernels either get recycled back to the huller or pass through a second huller with a finer cut setting, and the kernels pieces and fines, called meats, continue along the process to the flaking rolls.

Most of the hull material is separated at this point as a co-product. Hull separation is easier to achieve if some linter material is retained with the seed, as it helps the hulls to mat on the shaker screens. It can also help in aspirating the hulls off the screens to facilitate their separation from uncracked kernels. Some hull material is allowed to remain with the meats, as the hull's fibrous composition contributes to the structural integrity of the porous collets that facilitates oil removal in the extractor. Additionally, it provides a means to control the final protein content of the meal.

Flaking

The kernel pieces are then crushed with flaking rolls, which are used to produce thin flakes of uniform thickness that can be evenly cooked. In cottonseed mills, flaking is normally achieved with a series of vertically stacked rolls that press the kernels four times in succession, each pass occurring with the combined weight of the upper rolls increasing the applied pressure with each pass. The kernels are typically pressed to a final thickness between 0.2 and 0.3 mm (Norris, 1982a).

Cooking

Cooking of oilseeds is considered essential to achieve good oil extraction. Many chemical and physical changes occur during cooking. The exact nature of these changes are not well understood but it is generally thought that cooking ruptures cell walls, denatures the proteins that otherwise surround and stabilize the oil droplets, coalesces the microscopic oil droplets, and reduces the affinity of the oil for cellular components (Norris, 1982a). In addition, cooking also facilitates the destruction of cottonseed pigment glands and promotes the binding of gossypol to protein, thereby reducing the amount of gossypol co-extracted into the crude oil.

Careful control of the moisture and temperature during cooking is essential to maximize oil extraction, maintain the quality of the protein, promote gossypol binding, and reduce oil loss during refining. Undercooking reduces oil recovery and increases the extraction of impurities leading to greater oil loss during refining (Williams, 2005). Overcooking darkens the oil and lowers the feed value of the meal (Liener, 1958).

Cooking of cottonseed flakes is usually conducted in a continuous stacked cooker that contains a series of controllable vertical sections. Each section is steam jacketed and has a rotating sweep arm mounted from a central shaft and a bottom gate that is opened periodically to allow the flakes to pass between sections of the cooker. The top sections often have spray nozzles for adding moisture, and the lower sections will have exhaust vents to allow moisture to escape.

While mills have their own procedures, (see, e.g., Pons et al., 1955), the moisture level in cottonseed is generally raised to 12 to 15% at the start of cooking, and the temperature is raised quickly to 80 to 90°C (Norris, 1982a). While maintaining this moisture level, the temperature is then generally raised to about 105°C. Toward the end of cooking, the flakes will be allowed to dry to the condition needed for further processing, while the temperature is raised even further to between 115 and 125°C (Norris, 1982a). Cook times can vary, but 30 to 60 min is typical.

Extraction

The above steps are fairly typical of most cottonseed oil crushing operations, although the details regarding moisture levels and cooking protocols can vary depending on how the flakes are going to be extracted. From this point in the process, however, a number of methods have been used to recover the oil. These methods involve mechanical pressing, solvent extraction, or a combination of the two processes (Dunning, 1950).

In the early years of cottonseed processing, oil was expressed with box-type hydraulic presses in a batch mode. Cooked flakes would typically be formed into cakes of uniform thickness and density in a cake former. The cake former would also encase the flakes in a press cloth that supported the cake but allowed oil to pass freely. The hydraulic press would be oriented vertically and would be hand loaded with several cakes (typically 15), each fitted into a box. Pressure was then applied to the boxes. Efficient oil extraction required not only that the flakes be cooked adequately but also that the pressed cake be of the right density and consistency to allow the oil to escape. The pressure would be controlled, building slowly and then more rapidly to facilitate the release of oil. Maximum pressures would typically be around 4000 psi. After separating the recoverable oil, the pressure would be released, and the cakes would be removed by hand. Each cycle would take between 20 and 30 min. The cloths were pulled free with a cloth stripper, and the cakes would be trimmed and shipped whole or would be milled to produce meal. Cake edges were less well extracted and would often be reprocessed. With optimal seed preparation and pressing, the method could reduce the oil level in the cake to around 6%, although perhaps 8% was more typical (Thornton, 1932).

Dunning (1956) noted that the first use of a screw press to extract cottonseed kernels was in 1908. Mechanical screw presses, which operated in a continuous fashion, were destined to replace the batch processing of hydraulic presses, although many hydraulic press plants continued to operate into the early 1950s. In addition to continuous operation, the advantages of the screw press over the hydraulic press included greater oil recovery, reduced labor costs, reduced equipment maintenance, and a more streamlined operation.

The screw press (also called an expeller) consists of a barrel with a central screw that forces the meats or flakes to pass through the device with progressively increasing pressure. A choking device at the end of the barrel is used to afford some control over the pressure developed. As the pressure builds, oil is released from the meats and escapes through a series of closely spaced bars that are aligned parallel the barrel. As considerable friction is involved in the process, substantial power is needed to drive these machines, and water- or oil-based cooling is used to keep the temperature of the meats within acceptable limits. The level of moisture in the meat is critical to the operation. Ideal moisture levels are ~3% (Dunning, 1956); hence, additional drying is used during the final stages of cooking when extracting oil by this method.

A screw press is capable of applying more pressure to the seed matrix than can be applied with a hydraulic press. Pressures of between 15,000 and 20,000 psi can be developed (Williams, 2005), resulting in a greater extraction of the oil. The residual oil level is directly related to the rate at which material is fed into the press with increasing feed rates resulting in higher residual oils (Dunning, 1956). Meals from screw press operations would typically have between 3 and 5% residual oil.

The use of solvent to extract oil dates back to 1856, but large-scale solvent processing of oilseeds developed slowly as there were a number of technical challenges to overcome. The use of solvent required more complex and expensive equipment, and plant safety issues increased dramatically. The first note of cottonseed being solvent commercially was in 1917; however, this plant only operated for a few years (Anonymous, 1977). Driven largely by the demand for oil and the rapid growth of the soybean [*Glycine max* (L.) Merr.] industry, solvent processing with hexane become firmly established in the 1930s. Improved oil recovery drove the development as solvent extraction produced meal with residual oil levels of 1% or less.

The initial foray into solvent extraction for cottonseed was by way of a two-step process, first pressing the meal to recover the easy-to-get oil, then solvent extracting the press cake to recover the remaining oil. This method is commonly called the prepress-solvent process. The process was advantageous for oilseeds that contained greater levels of oil because it was difficult to form a flake suitable for direct solvent extraction. Although the prepress-solvent process was more complicated than mechanical pressing, it allowed for flexible operation of the presses as maximal oil recovery at this intermediate point in the process was less critical. The addition of solvent to the process increased total oil recovery to between 96 and 98% (Pons et al., 1955).

With this process, the initial pressing would leave between 9 and 13% residual oil in the cake (Pons et al., 1955). Because less pressure was needed to achieve these levels, the capacity of the presses could be increased. Also conditioning of the flakes for moisture became less critical and the time needed for cooking was reduced, which increased the capacity of the cookers. After pressing, the partially defatted cake was often milled, conditioned for moisture and temperature, and then re-flaked before solvent extraction, although as noted by Hutchins (1953) these steps were frequently eliminated and the press cake was either re-flaked without conditioning or extracted directly without forming a flake.

The development of the cooking extruder, called an expander, in the mid-1960s reduced oil extraction costs and increased mill capacity. Expanders started to be used for processing cottonseed in the United States in the late 1970s (Williams, 1986) and were quickly adopted worldwide. By 1990, expander-solvent operations accounted for 60% of all the cottonseed and soybean oil production in the United States (Lusas et al., 1990). Today, almost all US cottonseed oil extraction plants operate in this manner.

Unlike the screw press, the expander enabled cooked flakes to be extruded into porous pellets, called collets, without significant loss of oil. Collets are formed by the release of pressure as the seed material exits the expander through a die plate. As solvent can more easily penetrate into and drain from collets, solvent extraction became more efficient than it was from either flakes or expeller cake. Less solvent was needed per mass of meal, which increased extractor capacities, and residual meal oil levels were reduced to below those achieved with extraction of cake or flakes. Because the collets contained less hold-up solvent, less energy and time was needed to desolventize the defatted meal and recover the solvent. As the expander could also be injected with steam, it also contributed to the cooking process and allowed for faster throughput of material in the cookers. Often, the need for drying of the flakes in the final stages of the cooker was eliminated. In effect, it became easier to fully extract the oil in a single step rather than in two steps, and it allowed for expansion of plant capacity with minimal increase in capital cost.

While the expander-solvent process is the most widely used method for recovering cottonseed oil today, in some countries cottonseed oil is likely to be recovered by one of the older methods because of the large capital costs associated with the process. Additionally, some newer mills have been configured to extract only the easiest to get oil. In these "mini-mills," seeds are delinted, cracked, flaked, cooked, and screw-pressed to recover oil leaving the partially defatted meals with 6 to 8% residual oil. In some mills, the delinting process and the separation of hulls may be eliminated further reducing oil recovery. Oils from these processes can be more difficult to refine (discussed below), and because the seed cakes contain greater amounts of oil and possibly hulls, the protein content of the defatted meal is reduced compared with the protein level of fully processed meals.

Refining

Crude cottonseed oils are typically dark red-brown to black in color and contain substantial amounts of impurities and contaminants that need to be separated to produce food-grade oil. Vegetable oils undergo several steps to separate these impurities. These steps can vary depending on the type of oil being refined and on its intended use. For cottonseed oil, these steps would normally include caustic refining, bleaching, and deodorization to produce what is known as finished or RBD oil.

Caustic refining involves the addition of sodium hydroxide to the crude oil to form soap molecules with co-extracted free fatty acids. The amount and concentration of the lye solution is closely controlled, as excess lye will saponify triglycerides reducing oil recovery (Thurman, 1945). The amount of lye used depends on the stoichiometric amount needed to neutralize the free fatty acids plus an additional amount that has been found to improve oil color. The lye strength used depends on the quality of the seed and the free fatty acid level of the oil. Norris (1982b) provides a set of guidelines for the amount and concentration of lye to use when refining cottonseed oils.

With addition of the lye to the crude oil, the soap molecules coalesce into a gummy phase, called foots or soapstock. Many of the oil's polar contaminants partition into the soapstock phase. However, some non-polar components will also be entrained into this phase, resulting in mass loss beyond that based on the amount of fatty acids present. In addition to the soap, cottonseed soapstock contains phospholipids, partial glycerides, gossypol, glycerol, *myo*-inositol, glycerophosphates, sterols, and sugars, in addition to entrained triglycerides (Dowd, 1996).

In the early days of vegetable oil processing, refining operations were often conducted in large tanks, and the soapstock was allowed to settle to the bottom of the tanks. The oil was then siphoned off and the soapy foots had to be cleaned from the tank bottom. With the continuous processing operations employed today, lye is injected into the oil with intense mixing as it is pumped through pipes, and after sufficient residence time, the soapstock is separated with a process centrifuge. It used to be common practice to acidify the soapstock to recover the fatty acids as an acid oil, which might be sold to soap manufacturers. Today, cottonseed soapstock is usually added back to the defatted meal, where it reduces dust in the plant and adds a small amount of energy to the meal.

The refining of crude cottonseed oil can be difficult because of elevated amounts of free fatty acids (mentioned above) that occur because of poor field, harvesting, or storage conditions. The free fatty acid level in crude oil from quality seed is typically less than 2%, but it is not unusual for oils to have free fatty acid levels in the 5 to 10% range, and levels in excess of 20% are not unusual for very poor quality seed. Refining losses for these oils tend to be much greater than just the levels of free fatty acids because of greater co-extraction of impurities during extraction including greater entrainment of oil triglycerides. These oils can overwhelm the refining capacity of a plant. If poor quality seed has to be handled, it is often blended with better quality seed to reduce the processing problems.

There are many advantages to starting the refining process before the oil is separated from the solvent, a technique called miscella refining (Norris, 1982b; Hendrix, 1984). The reduced viscosity of miscella allows for rapid dispersion of the lye (lowering mixing and pumping cost), sharpens the centrifugal separation of the soap and miscella phases (reducing oil loss), and eliminates much of the water washing needed to remove traces of soap from the oil. For cottonseed processing, it also limits the exposure of the oil to gossypol, which reduces the potential for bleaching problems (discussed below). Often, part of the solvent is evaporated to increase the oil concentration in the miscella to between 25 and 50% before the addition of the lye. After separating the soapstock, the remaining solvent is evaporated. The main disadvantage of the method is the need for a two-stage solvent recovery system.

In the United States the vast majority of cottonseed oil mills use miscella refining to produce "once-refined" oil. This once-refined oil is often sold as a commodity product, and it is further refined in dedicated facilities to the specifications needed for its intended use. The National Cottonseed Products Association defines trading standards for both crude and refined oils (National Cottonseed Products Association, 2008). The premium grade of once-refined oil is Prime Bleachable Summer Yellow (often called PBSY), which is defined as having a bleachable American Oil Chemists' Society (AOCS) red color of $\leq$2.5 with <0.25% free fatty acids and $\leq$0.1% total moisture and volatiles.

Bleaching

After refining, cottonseed oil undergoes a bleaching process to further reduce impurities and improve color. During bleaching, the refined oil is dried and then exposed to an adsorbent to remove pigments, traces of soap, metals, and other impurities. The adsorbents used often include bleaching earth or clay, acid-activated clay, and sometimes activated carbon. Often the process is conducted under a mild vacuum to strip out entrained air and water and to reduce oxygen-catalyzed reactions that can occur while the oil is exposed to the active adsorbents at elevated temperature. Typically, the temperature of the oil is raised to 60 to 70°C to allow for moisture and dissolved gases to escape, then 1.5 to 3% (w/w) of adsorbent is added and the mixture is heated to between 90 and 95°C with mixing for 15 to 30 min (Taylor, 2005). After partial cooling, the vacuum is released and the oil is filtered to remove the adsorbent.

Bleaching is often conducted commercially as a continuous process that mixes the oil and adsorbent and sprays the slurry into a vacuum-controlled tank to remove air and water, then heats the sample to bleaching temperature, and after sufficient resident time, pumps the mixture through a filter press to recover the oil.

Cottonseed oils can be difficult to bleach to a light color, especially when crude oils are stored at elevated temperatures over time (Williams et al., 1947). This "color-set" issue is associated with increased exposure of gossypol (Thurber et al., 1954; Pons et al., 1955). Pons et al. (1955) showed that the issue was more problematic for solvent-extracted oil than for prepressed oil, which was mostly likely the result of the solvent-extracted oil containing more impurities because of the more complete extraction. A number of processing modifications have helped to solve this problem. First, flaking and cooking of the seed is conducted to promote the binding of gossypol to the meal protein, which reduces the amount of gossypol that is co-extracted into the crude oil. Second, miscella refining of the oil separates the gossypol from the oil quickly, allowing less time for the color to develop. By thoroughly cooking the meats and miscella refining, many of the color-set issues associated with the oil have been eliminated.

Desired oil color can vary for different oils and vary depending on the intended use of the oil. For cottonseed oils, typical bleaching is conducted to obtained oils with a red level between 1.5 and 2.5 on the AOCS color scale (Norris, 1982b).

Deodorization

After bleaching, steam distillation is used to eliminate volatile components and traces of free fatty acids. Laboratory-scale deodorization is conducted in a heated vessel under a vacuum with a sparge tube that allows the introduction of steam at the bottom of the oil. In batch mode with a bulk sample, the process typically takes a few hours to complete. Continuous commercial deodorizers are typically packed-bed distillation units that allow for a thin film of oil to be stripped of the lower molecular weight impurities faster than can be achieved in a batch still. A vacuum of around 2 Torr and temperature between 240 and 260°C are typical conditions for continuous deodorization (Loft, 1990). Steam injection rates can be between 0.5 and 1% of the oil feed rate. Often, resident times in continuous deodorizers are on the order of a few to several minutes. The faster processing reduces undesirable reactions than can occur at high deodorization temperatures.

Effective deodorization produces oil with 0.01 to 0.02% free fatty acids, low peroxide value, and a bland taste. While undesirable volatile compounds are removed during the process, deodorization can remove tocopherols and sterols, which are desirable in oil for their antioxidant activity.

Depending on the composition of minor components and the processing steps used to remove impurities, deodorization distillate can be a source for fatty acids, sterols, and tocopherols. The distillate from cottonseed oil deodorization contains 58% saponifiable material (fatty acids and partial glycerides) and 42% unsaponifiables (tocopherols and sterols). Cottonseed oil deodorizer distillate is a source of α-tocopherol (vitamin E), as this isomer makes up approximately 50% of the total tocopherol fraction (Winters, 1990).

Winterization

Vegetable oils often undergo some form of fractionation depending on the intended use of the oil. If it is to be used in salad dressings, cottonseed oil will often be cooled slowly to between 6 and 8°C and then allowed to stand to precipitate a portion of the triglycerides (O'Brien et al., 2005). This process, called winterization, is used to reduce cloudiness that occurs when oil is kept at refrigerator temperature. The solids are separated from the oil either by filtration or, more frequently today, with a solids centrifuge. The recovered liquid oil after winterization is often referred to as salad oil, while the solid fraction is called stearine. Cottonseed stearine has been used to formulate cocoa butter substitutes for confectionery products.

While winterization is often conducted on refined and bleached oil, it is sometime diluted into hexane and conducted in the presence of solvent. The reduced miscella viscosity has benefits that include faster diffusion of the oil triglycerides resulting in faster crystal growth and a sharper separation of the more highly saturated triglycerides with less entrainment of the liquid oil. Coupled with the use of a continuous crystallizer, the process can be conducted in less than 3 h compared with the batch process, which might take up to 5 d (Cavanagh et al., 1961). The sharper separation results in a higher recovery of the salad oil and produces a stearine fraction with a greater level of saturated fatty acids compared with the typical process (O'Brien et al., 2005).

Hydrogenation

Hydrogenation is the process of adding hydrogen to the unsaturated fatty acid components of triglycerides to make the glyceride fatty acids more saturated. The process is useful for creating and formulating semi-solid or solid fat products or for improving the oxidative stability of the oil when used for frying. Hydrogenation is typically conducted as a batch operation at temperatures between 120 and 190°C with a hydrogen pressure between 0.8 and 4 atmosphere and in the presence of a nickel catalyst. Partial hydrogenation results in the formation of *trans*-fatty acids, as the catalysts used also isomerize naturally occurring *cis*-oriented double bonds. Because of its properties and the types of products it is used for, cottonseed oil has characteristics that make it amenable to hydrogenation. As cottonseed oil has more saturated fatty acids and less triglyceride fatty acids containing three unsaturated bonds, it is less prone to form *trans*-fatty acid than are other commodity oils (Cheng et al., 2014). Nevertheless, this undesirable side reaction has led to considerable efforts by food processors to reduce the formation of *trans*-fatty acids. Careful attention to processing conditions (higher hydrogen pressures, improved mixing, etc.) reduces the formation of the *trans*-fatty acids. These effects, coupled with the development of basestock systems that blend fully hydrogenated products, which do not have significant levels of *trans*-fatty acids, with more saturated oils and possibly some partially hydrogenated components have resulted in greatly reduced levels of *trans*-fatty acids in processed foods. One basestock system for cottonseed oil has been described by O'Brien et al. (2005).

Seed Products

From these processing steps, a number of products are obtained. While the principal product of cottonseed crushing is oil, all of the other seed components have found a variety of uses and represent a significant component of seed value.

Oil Composition

Cottonseed oil, like all vegetable oils, is composed of triglycerides (i.e., esters of glycerol containing three fatty acids). Refined, bleached, and deodorized oil from quality seed would typically contain 98 to 99% triacylglycerides with 0.2 to 0.4% diacylglycerides, 0.0 to 0.4% sterols, 0.05% fatty acids, 0.05% tocopherols, and ppm levels of other components. If seed with elevated levels of free fatty acids is processed, the resulting oil might have a slightly reduced percentage of triglycerides (95–96%) and a correspondingly higher percentage of diacylglycerides (2–3%), as for example the oil described by d'Alonzo et al. (1982).

Cottonseed oil is classified as a linoleic-type vegetable oil, with linoleic acid (9,12–18:2) making up ~50% of the glyceride fatty acids. The saturated fatty acid, palmitic acid (16:0), makes up about 22 to 26% of the fatty acids, and oleic acid (9–18:1) makes up 16 to 20% of the acids. Combined, these fatty acids account for >90% of the oil's fatty acid profile (Table 2). There are also several secondary fatty acids present. Stearic acid (18:0) is present at between 2 and 3%. Fatty acids typically present in the 0.1 to 1% range include myristic (14:0), palmitoleic (9–16:1), malvalic (cpe18:1), *cis*-vaccenic (11–18:1), sterculic (cpe19:1), α-linolenic (9,12,15–18:3), arachidic (20:0), behenic (22:0), and lignoceric (24:0) acids (Table 2).

Although not often included in reported distributions, a number of additional fatty acids have been noted in the literature. Vernolic acid (12-epoxy-9–18:1) can be present in cottonseed oil as a trace component or up to a few percent (Dowd, 2012). Its level seems to be related to the age of the seed, as a lipoxogenase enzyme in mature seed appears to slowly oxygenate linoleic acid. Dihydrosterculic acid, a precursor of sterculic acid, is present at around 0.2%, but is usually not mentioned in studies as it often co-elutes from polar gas chromatography columns with the much larger linoleic acid peak. Additionally, a 16-carbon dienoic acid and 17-carbon mono- and dienoic acids, initially reported by Fisher and coworkers (Fisher and Schuller, 1981; Fisher and Cherry, 1983), have been identified as 9,12-hexadecadieonoic acid, 8- and 9-heptadecenoic acids, and 8,11-heptadecadieonoic acid (Dowd, 2012). Ricinoleic acid (12-hydroxy-9:18:1) has been found in cottonseed oil at a 0.2 to 0.3% level (Yamamoto et al., 2008), and this laboratory has confirmed its presence in some seed samples. Trace levels of odd-chain saturated fatty acids are also observed. The highest concentration of these is heptadecanoic acid (17:0), which may be present at a concentration of 0.1 to 0.2%.

Malvalic, sterculic, and dihydrosterculic acids are unusual in that they have three-carbon-atom rings in the middle of their structures (Fig. 4). These acids are known to be present in the seed oils of many plants of the greater Malvaceae family (Badami and Patil, 1980). The two cyclopropenoid acids, malvalic and sterculic

Table 2. Cottonseed oil fatty acids profiles.

Component	Codex†	NCVT seed survey‡	Salad oil§	Winterized cottonseed stearine¶	Solvent-winterized cottonseed stearine#
12:0	nd–0.2	nd–tr	nr	–	–
14:0	0.6–1	0.54–1.4	0.8	0.6–0.7	0.6
16:0	21.4–26.4	19.6–27.6	24.4	32.4–34.6	52.1
16:1(*n*–7)	nd–1.2	0.43–0.79	0.4	0.4–0.6	0.8
17:0	nd–0.1	tr	–	–	–
17:1	nd–0.1	tr	–	–	–
18:0	2.1–3.3	2–3.2	2.2	2.1–2.4	1.9
18:1(*n*–9)	14.7–21.7	12.8–22.2	17.2	15.8–17.2	9.1
18:1(*n*–7)	–	0.69–1.1	–	–	–
18:2	46.7–58.2	44–59.3	55	46.2–47.1	35.5
18:3	nd–0.4	0.15–0.25	0.3	–	–
20:0	0.2–0.5	0.20–0.45	–	–	–
20:1(*n*–9)	nd–0.1	tr	–	–	–
22:0	nd–0.6	0.08–0.21	–	–	–
24:0	nd–0.1	0.08–0.20	–	–	–
cpe18:1	–	0.22–0.65	–	–	–
cpe19:1	–	0.20–0.46	–	–	–

† Codex alimentarius trading standard for cottonseed oil (FAO/WHO, 1999). nd, not detected.

‡ Range reported for 35 genotypes included within the 2006 and 2007 US National Cotton Variety Trials (NCVT) (Dowd et al., 2010). tr, trace.

§ From National Cottonseed Products Association (2002b).

¶ Range reported by Weiss (1967) and O'Brien et al. (2005).

From Weiss (1967).

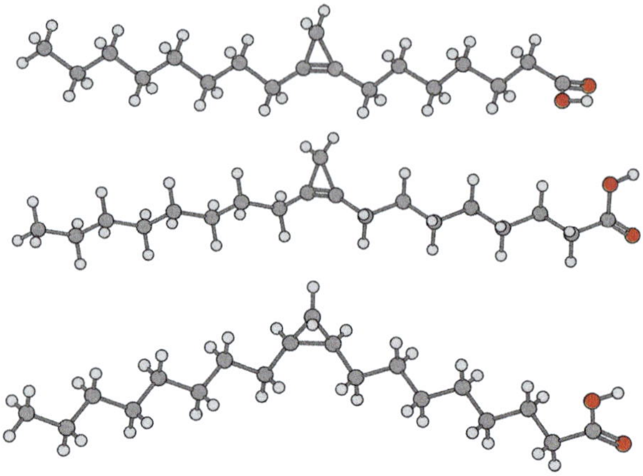

Fig. 4. Molecular structures of cottonseed cyclopropyl fatty acids. (a) Malvalic acid (cpe18:1). (b) Sterculic acid (cpe19:1). (c) Dihydrosterculic acid (cpa19:0).

acids, have combined concentrations between 0.6 and 1% in agronomic cotton varieties (Fisher and Schuller, 1981; Lawhon et al., 1977), although values as high as 2% have been noted (Phelps et al., 1965). These components are not evenly distributed within the kernel but are localized in the embryo radicle (Fisher and Cherry, 1983). Because of their highly strained ring structure, these fatty acids are less stable outside the seed, and they are partially destroyed by the high temperature used during deodorization (Eaves et al., 1968). In finished oils, the combined level of malvalic and sterculic acids is usually <0.4% (Phelps et al., 1965).

Several small surveys have been conducted on the fatty acid composition of agronomic cottonseed varieties. Lawhon et al. (1977) have reported on the composition of glanded and glandless varieties and noted that the glandless trait did not appear to affect fatty acid composition. Hamza et al. (1988), Nergiz et al. (1997), and Lukonge et al. (2007) all have reported fatty acid profiles for a small number of varieties grown at a single location for a single year. These studies show ranges of variation for individual fatty acids in line with the range reported in the Codex trading standard for cottonseed oil (FAO/WHO, 1999). A few reports have appeared that noted fatty acid distributions outside the Codex trading range. For example, Yunusova et al. (1991) reported that their L-78 variety, a linter-free seed type, has palmitic and linoleic acid levels of 43 and 26%, respectively, which is essentially reversed from values expected for these acids in typical cottonseed oil. Additionally, this laboratory has identified *G. barbadense* genotypes with levels of oleic acid approaching 42% and corresponding reduced levels of linoleic acid (work in progress). Increased levels of oleic acid have also been observed in some wild *Gossypium* species (Carter et al., 1966). Hence, some cotton germplasm exists that may be useful for altering the basic composition of the oil and allow the oil to be tailored for specific end uses.

A number of genetic approaches have also been undertaken to modify the fatty acid profile of cottonseed oil. Interest exists in altering a number of the oil's fatty acid traits, but most work has focused on increasing the level of oleic acid and reducing the level of linoleic acid (Chapman et al., 2001; Liu et al., 2002;

Sunilkumar et al., 2005). This change would improve the oxidative stability of the oil, which would extend the life of the oil when used for frying. By interfering with the activity of the desaturase enzyme associated with the conversion of oleic acid to linoleic acid, seeds have been produced with increased levels of oleic acid. The degree of increase has varied considerably, with the most pronounced difference resulting from hairpin RNA-mediated post-transcriptional silencing of the seed ω6-desaturase (Liu et al., 2002). This approach has resulted in seed oil with an oleic acid level of 78% and a corresponding reduction in the linoleic acid level to around 4%. Using the same technique to silence the ω9-desaturase, Liu et al. (2002) reported seed lines with almost 40% stearic acid and only 4% oleic acid. The palmitic acid level was significantly lowered in both seed lines. By crossing these two seed lines, additional variations were demonstrated (Liu et al., 2002).

Field conditions have a modest influence on the oil's fatty acid composition and properties. Stansbury et al. (1953a) observed an increase in the oil's iodine value, a measurement of triglyceride unsaturation (discussed below) with lower field temperatures and increased precipitation. A recent survey of cotton varieties grown in different environments as part of the National Cotton Variety Trials noted that linoleic acid levels decreased and saturated fatty acid levels increased when cotton was grown in hot and dry environments (Dowd et al., 2010), and similar trends were apparent in single field experiments with dryland and irrigated cotton (Pettigrew and Dowd, 2011). The observed shifts in the levels of linoleic acid levels are consistent with the iodine value trends observed by Stansbury et al. (1953a). Additionally, increased soil salinity has been reported to decrease the level of linoleic acid and increase the levels of stearic and oleic acids in cottonseed oil (Ahmad et al., 2007).

In addition to fatty acid composition, organization of the fatty acids along the glycerol backbone also influences oil properties. In cottonseed oil, saturated fatty acids are found primarily on the ends of the glycerol backbone, that is, at the *sn*-1 and *sn*-3 positions, and the *sn*-2 position usually carries an unsaturated fatty acid (O'Brien et al., 2005). Using thin-layer chromatography to fractionate the triglycerides by the number of unsaturated bonds, Jurriens and Kroesen (1965) estimated the distribution of the principal fatty acids along the glycerol backbone. Summarizing these data by grouping the fatty acids into either saturated (S) and unsaturated (U) acids gives a distribution of 0.5% SSS, 1.4% SSU, 2% USU, 16.9% SUS, 45% SUU, and 32% UUU. Bland et al. (1991) identified several of the individual triacylglycerols on the basis of their gas and liquid chromatographic retention times. From their peak assignments, the distribution of saturated and unsaturated fatty acids along the glycerol molecule was found to be 13.3% SSU+SUS, 46.9% SUU+USU, and 35.8% UUU, which is in general agreement with the results reports by Jurriens and Kroesen (1965).

Vegetable oils typically contain some levels of phytosterols and other minor components. Verleyen et al. (2002) report total sterol levels of around 690 mg per 100 g in degummed cottonseed oil and 460 mg per 100 g in refined oil, with about 20% of the sterols existing in an esterified form. β-Sitosterol is the most abundant sterol in cottonseed oil, making up about 90 to 93% of the sterols (Itoh et al., 1973). Smaller amounts of campesterol, stigmasterol, and Δ5-avenasterol and trace levels of cholesterol, brassicasterol, Δ7-stigmastenol, and Δ7-avenasterol were also identified.

Tocopherols are present in the oil and act as antioxidants at levels greater than 500 ppm. The total level of tocopherols in crude cottonseed oil is on the order of 1000 ppm, and in finished oil a range of 830 to 900 ppm has been reported (O'Brien et al., 2005). Similar to what has been found in deodorizer distillate, α-tocopherol accounts for about 40 to 50% of the tocopherol isomers, and δ-tocopherol, a powerful antioxidant, accounts for much of the remaining fraction. Trace levels of β- and γ-tocopherols are also present.

Meal Composition

The amount of protein and its amino acid composition makes cottonseed meal desirable as an animal feed ingredient, although the presence of gossypol limits this application mostly to ruminant animals. Compared with soybean meal, the protein of cottonseed meal contains relatively high levels of the essential amino acids arginine and methionine (Table 3), but the lysine level of cottonseed meal is lower.

Protein in defatted vegetable meal is generally measured by determining nitrogen content and applying a conversion factor. National Cottonseed Products Association trading rules (National Cottonseed Products Association, 2008) and AOCS Official Method Aa 5-91 (American Oil Chemists' Society, 2013) specify a nitrogen-to-protein conversion factor of 6.25. However, cottonseed protein is

Table 3. Reported amino acid distributions of cottonseed meals, concentrates, and isolates.

Amino acid	Screw-pressed meal†	Solvent-pressed meal†	Protein concentrate‡	Protein isolate§	Protein isolate fractions¶	
					I	II
g/16 g N						
Alanine	4.2	3.9	3.9	3.8	3.2	3.5
Arginine	11.6	12	11	12.6	10.4	11.3
Aspartic acid	10.2	10.2	9	9.9	6.7	8.4
Cystine	1.6	1.6	2	0.9	2.6	0.3
Glutamic acid	21.8	22.2	19.8	21.6	21.8	18.9
Glycine	4.1	3.9	3.9	4.2	3.2	3.7
Histidine	3.8	3.7	2.8	3.2	2.6	3
Isoleucine	3.2	3.1	3.3	3.5	2.6	3.1
Leucine	6.4	6.2	6.3	6.3	5.1	5.8
Lysine	4	4.8	3.8	2.9	6	3
Methionine	1.8	1.9	1.6	1.2	1.7	1
Phenylalanine	5.8	5.8	6.2	7.1	3.7	6.3
Proline	4	3.5	3.8	3.9	3.1	3.1
Serine	5	4.7	4.4	5	3.4	4.5
Theronine	3.8	3.7	3.2	3	2.9	2.7
Tryptophan	1.4	1.3	1.4	nr#	nr	nr
Tyrosine	2.6	2.6	3.4	2.9	3.3	2.6
Valine	4.5	4.3	4.9	5	3.3	4.4

† Data from National Cottonseed Products Association (2002a).

‡ Produced from glandless seeds by wet extraction. Average of concentrates produced by drying at different pH values (Lawhon et al., 1972).

§ Produced from glandless seed by a single step alkali solubilization and precipitation at pH 5 (Dowd and Hojilla-Evangelista, 2013).

¶ Produced from glandless cottonseed by sequential extraction of glandless defatted cottonseed at different pH levels. Fraction I corresponds to the non-storage proteins, and Fraction II corresponds to the storage proteins (Martinez et al., 1970).

nr, not reported.

relatively high in amino acids that contain nitrogen in their side chains, especially arginine. Consequently, the standard 6.25 conversion factor likely over estimates cottonseed protein levels. For cottonseed products, a conversion factor of 5.3 has been recommended by USDA (1984) and has been discussed without specific recommendation by the AOCS methods committee (Anonymous, 1992). From reported amino acid distributions for the seed and meal protein, a conversion factor of 5.9 to 6 can be calculated that seems to be a more reasonable value (Dowd and Wakelyn, 2011). Because different values are used, some caution is in order when evaluating the literature, and it is important that any conversion factor be explicitly stated. The usual standard for trading cottonseed meal is 41% protein (on the 6.25 conversion basis). If the level of hull that is included with the meats during processing is reduced, it is possible to produce meal with protein levels approaching 50%.

In addition to the protein, cottonseed meal contains fibrous and soluble carbohydrates and small amounts of fat, vitamins, and minerals (Table 4). Neutral detergent fiber, corresponding to the cellulose, hemicelluloses, and lignin components, represents 25 to 32% of the meal. Sucrose, raffinose, and stachyose make up the bulk of the soluble sugars with a combined concentration of around 11% (Kuo et al., 1988). Raffinose accounts for around 60% of these sugars. Ether extraction (essentially residual crude oil) accounts for 2 to 4% of the meal mass, and ashing of the meal (i.e., minerals) gives levels between 6 and 8% (Table 4).

Gossypol levels in the meal are typically around 0.9 to 1.3% (Waldroup and Kersey, 2002). Unlike in seed where gossypol is localized in pigment glands, gossypol in the meal is distributed and largely exists in a bound form. The degree of binding will vary depending on the moisture level, temperature, and duration of the flaking, cooking, and expansion operations used during oil extraction, but typically 70 to 85% of the gossypol will be bound (Waldroup and Kersey, 2002). The binding occurs primarily because of a Schiff's base interaction between the gossypol aldehyde moieties and available primary amines. Because of the large

Table 4. Composition of cottonseed meal and hulls.†

Component	Screw-pressed meal	Solvent-extracted meal	Commercial hulls
	Bulk constituents, % dry weight		
Crude protein (N $\times$ 6.25)	46.1	47.6	4–5
Acid detergent fiber	18.1	17.3	67
Neutral detergent fiber	32.3	24.5	86.9
Ether extract	4.6	2.2	1.9
Ash	7.2	7.5	1.8–2.8
Total gossypol	1.09	1.16	0.107
	Minerals		
Calcium, %	0.21	0.22	9.15
Magnesium, %	0.65	0.66	0.15
Phosphorus, %	1.14	1.2	0.08
Potassium, %	1.68	1.72	1.13
Sulfur, %	0.43	0.44	0.05
Copper, mg/kg	10.9	12.5	3.6
Iron, mg/kg	106	126	30.1
Manganese, mg/kg	18.7	20.1	16.8
Zinc, mg/kg	62.8	63.7	9.9

† Data from National Cottonseed Products Association (2002b) and Tharp (1948).

concentration of protein in the meal, most of the gossypol binding occurs with the ε-amino group of lysine amino acids.

From a number of bulletins and nutritional tables, Hale and Lyman (1948) and Altschul et al. (1958) have summarized the seed and meal's soluble vitamin and mineral levels. Noted levels of the B vitamins (e.g., riboflavin, niacin, thiamin, and pantothenic acid) are found in the meal, but little vitamin A or D is noted. The meal also contains percent levels of phosphorus (1.1–1.3%), which is mostly in the form of phytic acid, and potassium (1.4–1.6%) (National Cottonseed Products Association, 2002b; Waldroup and Kersey, 2002). Calcium, magnesium, sulfur, sodium, iron, copper, and zinc are all reported at lower levels (Table 4).

Hull Composition

Commercial cottonseed hulls are composed mainly of cellulose, hemicellulose, and lignin with small levels of protein, oil, and minerals (Table 4). The composition varies depending on the degree of delintering. Typically, greater than 85% (dry wt.) of the hull is recovered as neutral detergent fiber, and ~67% (dry wt.) of the hull is recovered as acid detergent fiber (National Cottonseed Products Association, 2002b). The difference gives a hemicellulose content of ~20%, although Dunning (1948) has reported a level closer to 30%. Lignin accounts for another 20%, which leaves around 35 to 47% of the hull mass as crude fiber or cellulosic components. Ash accounts for another 2 to 3% (dry wt.). Protein (3–5%) and oil (2%) are both present in large part because of the incomplete separation of small kernel particles. Gossypol, which is barely detectable in cleanly separated ground hull, is measurable in commercial hulls for the same reason. Gossypol levels of 0.05 to 0.2% have been noted in the literature with 0.1% being fairly typical (National Cottonseed Products Association, 2002b).

Linter Composition

Cotton lint and cottonseed linters are among the purest natural sources of cellulose. Commercial linters are typically 75 to 85% (dry wt.) cellulose with the remainder being composed of small levels of waxes, pectins, proteins, dirt, hull fragments, organic acids, and inorganic materials (Temming, 1966).

Cottonseed Oil Properties

Commodity fats and oils are heterogeneous materials that are used in a variety of food and non-food applications. As a result, they are characterized by a number of physical, chemical, and optical properties that largely determine how the oil will be processed and used. Recommended procedures for determining many of these properties are given by the American Oil Chemists' Society (2013). Because cottonseed oil is used in formulations of salad dressing and cooking oils and for the preparation of semi-solid-fat products, properties related to these applications are generally of the greatest interest (Table 5).

When exposed to heat and oxygen, vegetable oils tend to oxidize. During the initial stages of oil oxidation, oxygen reacts with the allylic carbon atoms of unsaturated fatty acids to form hydroperoxides. The peroxide value is defined as the milliequivalents of peroxide present per kilogram of oil and is determined by monitoring the amount of iodine liberated from potassium iodide. Good quality cottonseed oil typically has a peroxide values <1. Anisidine value is a second measure of oxidation that quantifies the presence of aldehyde groups in the oil,

Table 5. Selected properties of cottonseed oil products.†

Property	RBD oil	Winterized oil	Partially hydrogenated oil	Fully hydrogenated oil
Peroxide value, meq kg^{-1}	<1	<0.5	<0.5	<0.5
AOM stability, h	10–19	10–25	100–200	>350
Cloud point, °C	−1.1 to +3.3	−5.6 to −3.3	–	–
Cold test, hr	~1	5.5 to 12	–	–
Melting point, °C	10–16	–	32–47	60
Titer, °C	30–37	–	35–45	>60
Pour point, °C	−3.9–0.0	−3.9–0.0	39–60	>60
Smoke point, °C	220–230	230	177	–
Flame point, °C	320–330	–	316	–
Fire point, °C	355–360	–	357	–
Iodine value	98–118	103–116	50–70	2–5
AOCS red color	2–6	2–4	2–2.5	2–2.5
Free fatty acid, %	<0.05	<0.05	<0.05	<0.05
Unsaponifiable matter, %	0.5–0.7			
Saponification number	189–198			
Specific gravity (25/25°C)	0.914–0.918			
Refractive index (25°C)	1.468–1.472			

† Data from several sources but principally from O'Brien et al. (2005), Jones and King (1990), and National Cottonseed Products Association (2002b). AOCS, American Oil Chemists' Society; AOM, Active Oxygen Method; RBD, refined, bleached, and deodorized.

which are formed with more advanced oxidation. A reaction of *p*-anisidine with aldehyde groups results in yellow-colored products that are detected spectrophotometrically at 350 nm. Values <10 are typical of quality oil.

Although peroxide and anisidine values are useful for determining the oxidative state of an existing oil, it is often more important to know the tendency of an oil to oxidize. The Active Oxygen Method (AOM) stability test uses heat and aeration to accelerate oil oxidation, which is followed until the oil reaches a specific end point, often a peroxide value of 100. Refined, bleached, and deodorized cottonseed oils typically have AOM stability times of 15 to 19 h.

Many different techniques are used to estimate the melting or crystallization of vegetable oils. These include capillary melting point, softening point determinations, slipping methods, the Wiley melting point, and the Mettler dropping point. By these various measures, the melting point of RBD cottonseed oil is generally noted to be between 10 and 16°C (Bailey, 1948). Because fats and oils are mixtures of individual triglyceride molecules that will melt or solidify at different temperatures, a better measure of melting behavior is to determine the amount of the solids present in the oil over a range of temperatures. A solid-fat index has been standardized for this purpose. It determines the percentage of solid fat existing at temperatures of 10, 21.1, 26.7, 33.3, and 40°C. Refined, bleached, and deodorized cottonseed oil has only a minimal amount of solids present at the lowest 10°C temperature of this index, but hydrogenated cottonseed oil, cottonseed stearine, and cottonseed oil blended with animal or vegetable fats all have significant levels of solid fat over this temperature range. For example, cottonseed oil hydrogenated to an iodine value of 75 exhibits a solid-fat index of 20.5% at 10°C, 10.5% at 21.1°C, 7% at 26.7°C, and 2.5% at 33.3°C (O'Brien et al., 2005).

Other measures of solidification or melting are also used to evaluate oils for specific purposes. Cold point is a yes or no test that is used to determine if the oil can be used in salad dressings. The test determines if the oil will remain clear when allowed to stand at 0°C for 5.5 h. Refined, bleached, and deodorized cottonseed oil generally fails this test, but winterized cottonseed oil passes the test. Related to the cold point is the cold test, which determines the amount of time it takes oil to cloud at 0°C. Winterized cottonseed oil generally produces cold test times of between 12 and 20 h. The cloud point is the temperature that the oil will "cloud" or reduce the transmission of light. For cottonseed oils, the cloud point is typically between −1.1 and +3.3°C for RBD oil and between −5.6 to −3.3°C for salad oil. Pour point temperature refers to the temperature at which the oil remains pourable, which for RBD cottonseed oil is between −3.7 and 0°C. For hydrogenated cottonseed products, the pour point temperature can be as high as 60°C.

Titer value is a measure of the solidification of the oil's fatty acids (after saponification) and is used primarily to characterize the fatty acids for soaps or fully hydrogenated oil products. Cottonseed oils have titer values that range between 32 and 37°C. The titer value of cottonseed salad oil is slightly lower. For unhydrogenated samples, the titer value for cottonseed oil tends to be slightly greater than other vegetable oils because of its elevated levels of saturated fatty acids. For hydrogenated samples, the titer slowly increases with hydrogenation and Bailey (1948) shows representative data for a series of partially hydrogenated samples at different iodine values. When fully hydrogenated, the titer value of cottonseed oil is >60°C.

The smoke, flash, and fire points of vegetable oils all relate to the potential of the oil to form fires when heated. The smoke point is the temperature that the oil will start to smoke; the flash point is the temperature that the oil produces sufficient volatiles to support ignition, and the fire point is the oil temperature that supports sustained combustion. These temperatures are very sensitive to the level of free fatty acids and other impurities in the oil (Morgan, 1942). Refined, bleached, and deodorized cottonseed oil with a 0.04% free fatty acid level has a smoke point of ~220°C, a flash point of ~322°C, and a fire point of ~360°C (Jones and King, 1990).

During crystallization of fats and oils, different crystalline morphologies or habits are possible. Three basic crystal habits have been identified as α-, β'- and β-forms. The α-form is metastable and tends to slowly reorganize into the β'-form. Most commodity oils crystallize into the β-form (O'Brien et al., 2005), which exhibits larger and coarser crystals. Cottonseed oil, rice bran oil, and animal fats tend to crystallize into the β'-form. This crystal form is characterized as having small fine crystals that tend to impart a smooth consistency and fluid plasticity to semi-solid-fat products. This trait appears to result from the higher levels of palmitic acid in cottonseed oil compared with most commodity oils, and the partitioning of the palmitic acid molecules on the ends of the glycerol backbone. Partially hydrogenated cottonseed oils and cottonseed oil blends also tend to crystallize into the β'-form (O'Brien et al., 2005).

Specific gravity (or density) and viscosity are properties important for processing. The specific gravity (25/25°C) of RBD cottonseed oil is between 0.914 and 0.918 (Table 5). Oil viscosity directly affects the cost of pumping and lubricity, the latter of which can be an important consideration in the processing of foods. Greater saturated fatty acids levels in oils tend to increase viscosity; hence, cottonseed oil is slightly more viscous than most vegetable oils. Both density and viscosity decrease with increasing temperature. Bailey (1948) summarizes viscosity and density data for RBD, winterized, and hydrogenated cottonseed oils.

Refractive index is used as a quick method to confirm the identity or purity of a vegetable oil. At 25°C, cottonseed oil has a refractive index between 1.468 and 1.472. Refractive index is dependent on temperature and the structure of the glycerol ester fatty acids. With predetermined curves developed at fixed temperatures, refractive index can also be used to estimate iodine value.

Oil color and flavor are important properties for food applications. As mentioned above, crude cottonseed oil is darkly colored. Refined cottonseed oil is generally a rich golden yellow that is darker than soybean, canola, and sunflower oils, but lighter than corn oil. As discussed above, the refining of cottonseed oil for color can be technically challenging, and dark crude oils that are difficult to bleach are not uncommon.

It is generally desirable that oils used for food formulation or cooking have a bland flavor that does not mask or interfere with the flavors of the other ingredients. Refined, bleached, and deodorized cottonseed oil used at low temperatures has a bland flavor. With heating, however, cottonseed oil tends to impart a toasted or nutty flavor to food products, which the snack food industry exploits specifically for this trait.

Free fatty acid level is generally measured by titrating a solution of the oil against a known concentration of sodium hydroxide. In general, proper refining removes the bulk of the free fatty acids from the crude cottonseed oil, and bleaching and deodorization will eliminate most remaining acids to leave RBD oil with generally <0.05% free fatty acids.

Iodine value was mentioned above and is a method that measures the amount of double bonds (or the average degree of unsaturation) in the oil. For cottonseed oils, iodine values are typically between 98 and 118, corresponding to the range of fatty acid variation that is found in the oil. Cottonseed stearine and partially hydrogenated oils have lower iodine values, and salad oil has a slightly higher iodine value (Table 5). Iodine value can be estimated if the fatty acid profile of the oil is known. Iodine value is often used to monitor the progress of winterization and hydrogenation processes.

Unsaponifiable matter is a measure of the amount of non-esterified components in the oil (e.g., sterols, tocopherols, pigments, hydrocarbons) and is determined by measuring the mass of the sample taken up in a non-polar solvent after saponification and correction for residual fatty acids. For RBD cottonseed oil, unsaponifiable matter is typically between 0.5 and 0.7%. Saponification value is the amount of potassium hydroxide needed to saponify 1 g of oil. In effect, it is a measure of the average molecular weight of the fatty acids in the oil. For cottonseed oils, the saponification value ranges from 189 to 198 with an average value of 195.

Product Utilization

Oil

Cottonseed oil is used almost entirely in the production of food products. Food use of vegetable oil can be divided into liquid oil (e.g., cooking oil or salad dressing oil) and solid or semi-solid formulated products (e.g., margarines and shortenings). Cottonseed oil has been used in all of these products. In 2002, approximately 50% of the cottonseed oil produced in the United States was used as a cooking oil, 25% was used for shortenings, 16% was exported, a small amount (maybe 2%) was used in margarines and spreads, and the remainder (~6%) was used for other purposes (O'Brien et al., 2005).

Use of cottonseed oil has changed since 2002 because of the continued growth of the vegetable oil industry, the lack of a sufficient supply to support some markets, and changing commercial and processer formulations. Often, cottonseed oil is blended with other oils and is used depending on availability and price. Today, most cottonseed oil is sold through intermediaries, and it is difficult to identify specific end users or product categories.

In the United States, almost all cottonseed oil is sold as a commodity product in the food preparation and restaurant industries, and the oil is no longer widely available at retail as cooking or frying oil. The exception to this is that the oil can be found in specialty stores, on the internet, or in a few upscale grocery stores where it is sold as premium-quality frying oil. In cotton producing areas outside the United States, cottonseed oil continues to be used extensively as cooking oil.

A recent informal review of local grocery store shelves found cottonseed oil listed as an ingredient in fried snack foods (e.g., in potato chips and taco shells), store-brand shortenings, frozen breakfast and pastry dessert products, canned fish products (e.g., smoked oysters), and ready-to-use and processed cake icings. Because of restriction on the use of soy products, cottonseed oil is widely used in kosher food products, where it retains its traditional uses, for example, as cooking oil and as an ingredient in salad dressing, shortening, and mayonnaise.

In food applications of note, cottonseed oil was the original oil used in the formulation of Procter & Gamble's Crisco shortening (i.e., crystallized cottonseed oil), although the product is now formulated with soybean and palm oils. It was also used in the development of the Olestra low-calorie products, that is, non-digestible fat substitutes made by esterifying vegetable oil fatty acids to sucrose.

Beyond the food arena, cottonseed oil has limited use in the manufacture of specialty soaps, as lubricants and mold release agents, as pharmaceutical and fabric dispersants, as protective coatings, as components in rubber formulations, and in processes for the manufacturing of leather, flame-resistant textiles, inks, polishes, coatings, plastics, and resins. Proflow oil is a screw-pressed crude cottonseed oil fraction that is marketed for use in industrial fermentations of antibiotics and steroids.

Meal

The presence of gossypol in whole cottonseed and cottonseed meal is responsible for the occasional toxicity and reproductive problems associated with the overfeeding of these products in animal diets (Berardi and Goldblatt, 1980). Ruminant animals are less sensitive to the compound than non-ruminant animals; hence, cottonseed and cottonseed meal are regularly used as a protein ingredient for dairy and beef cattle, goats, and sheep, but are not used for feeding swine. Ruminant diets can be formulated with substantial amounts of cottonseed products (15–20% of ration) generally without concern for gossypol toxicity (Calhoun, 2002).

The binding of gossypol to protein in cottonseed meals is associated with reduced animal toxicity, although the factors that make dietary gossypol "available" to the animal are not well understood (Liener, 1958). These factors appear to extend beyond a simple characterization of the levels of bound and free gossypol in the feed, as feeding experimental diets containing comparable levels of free gossypol often result in variable levels of plasma gossypol (Calhoun, 2002). Research has indicated that meals can be included to a limited degree in poultry rations (Henry et al., 2001; Gamboa et al., 2001), and overseas this practice is occurring to some degree. In recent years, cottonseed meal has also been incorporated into farmed catfish operations. Research reports have shown that levels approaching 30% of the diet do not appear to have significant detrimental catfish growth or feed conversion, especially when supplemented with additional lysine (Robinson and Li, 1994).

In addition to its use as an animal protein source, specially formulated cottonseed meals have found use in fermentation media. As for Proflow oil, Proflow and Pharmamedia are cottonseed meal-based products sold as fermentation media for the production of antibiotics, enzymes, steroids, and biological pesticides and herbicides. Documentation regarding the composition and characteristics of these products can be found on the internet. A small amount of cottonseed meal is also used as formulated fertilizers for potted plants and as part of the media used in mushroom production.

Hull

Cottonseed hull is used primarily as a source of roughage in cattle and dairy feeds, where it is beneficial as a bulking agent to reduce digestive disturbances. In addition, hulls are sometimes used as a component of oil well drilling mud, and they are valued as a component of the growth media for some mushroom production. Dunning (1948) discusses several proposed applications for cottonseed hulls, which includes use as a soil conditioner, filler in plastic resins, and a substrate for the production of furfural (a cyclic aldehyde that is formed from five-carbon hemicellulosic sugars by acid hydrolysis). It is not clear if any of these proposed uses ever obtained commercial status.

Linter

Linters are used either chemically or non-chemically. First-cut linters, which are longer (up to 15 mm long), tend to be used in non-chemical applications, whereas the shorter second-cut linters (up to 8 mm long) are pulped and used in chemical applications (Van Wyck, 1948). First-cut linters are used in absorbent cotton products and medical pads and gauzes. A greater quantity is used to produce felts or battings for bedding products and as cushioning for furniture and automobile panels. In the past, first-cut linters have been used together with longer cotton fibers in the manufacture of paper currency.

Because linters are composed principally of cellulose, they are an important raw material to the chemical industry. Often, they are pulped and used in applications where the impurities found in wood pulps are problematic. Pulping occurs through a series of digestion, bleaching, and washing steps. The pulp can also be converted into cellulose acetates, which are used in photographic film, tape, and packaging products as well as in the production of plastics and molded products. A substantial volume of linter pulp is used in the manufacture of high-grade bond paper. Linter pulp has also been derivatized and used in the production of ice cream, sausage casings, toothpaste, fingernail polishes, gun powders, furniture lacquers, laminates, industrial and automotive filters, in battery separators, and on printed electrical circuit boards.

Potential Products from Glandless or Ultra-Low Gossypol Cottonseed

The discovery of glandless cotton spurred tremendous interest in the development of processes and products to expand the use of cottonseed, cottonseed meal, and its protein beyond traditional markets. Whole kernels, defatted meals (i.e., flour), and protein concentrates and isolates have all been prepared from glandless cottonseed and studied for functionality and use. Because of the current work being undertaken to develop ultra-low gossypol cottonseed, there had been a resurgence of interest in these development efforts.

A number of studies have found that cottonseed meals from glandless seed (loosely referred to as glandless cottonseed meal) are useful ingredients in many non-ruminant feeding applications. When fed glandless cottonseed meal, no detrimental effects were observed in chicks, although lysine and methionine supplementation was found to be beneficial (Jonston and Watts, 1964; Waldroup et al., 1968; Roberson, 1970). Some heating of the meal was found to improve weight gains (Johnston and Watts, 1965a, 1965b). Layer egg discoloration, a known problem when feeding normal cottonseed meals, was substantially reduced but not entirely eliminated by feeding glandless meal as 10 and 20% of the diet (Heywang and Vavich, 1965; Heywang et al., 1965). In a second study, incorporating 15% of a layer ration with glandless cottonseed meal did not result in any egg discoloration (Reid et al., 1984). For young swine, up to 40% of a corn–soybean-based meal could be replaced with glandless cottonseed meal with no significant weight gain differences (LaRue et al., 1985). Above 40% replacement, however, some reduced weight gains were noted, which was also reported in a study by Noland et al. (1968) at a 50% replacement rate. In an additional swine study that used whole and treated full-fat glandless kernels to substitute for about approximately half the protein provided of the soybean meal, equal or improved weight gains and feed conversions were found for both starter and grower-finisher pigs (LaRue et al., 1987). Processing of kernels (e.g., pelleting, heat treatment, or dry extrusion) did not alter the results in this study. Catfish and shrimp feeding studies have also found that glandless cottonseed meal could be used with minimal problems. Some reduced weight gains have been noted, but comparable survival and feed conversions were observed for production of tilapia (Robinson et al., 1984).

The benefits of heat during preparation of glandless meal observed in poultry feeding studies (Johnston and Watts, 1965a, 1965b) highlight a potential process optimization issue, as cooking seed for oil extraction is associated with reduced protein digestibility (Altschul et al., 1958; Liener, 1958). As cooking is conducted in part to help bind gossypol to the protein, which is not important for glandless seed, a re-optimization of this part of the oilseed extraction process appears warranted as it may improve the nutritional value of the meal.

In the 1970s, a number of countries started research programs directed toward the expansion of cottonseed and its proteins into food applications. An overview of these programs has been given by Lusas and Jividen (1987). In the United States, glandless kernels were tested as whole nut substitutes and food ingredients and were used as starting materials for the production of food-based protein concentrates and isolates. For the most part, these efforts have been curtailed as glandless cotton varieties never obtained commercial status.

As the current oil extraction process fractures the kernels during dehulling and does not cleanly separate the kernels from the hulls, modified processing is needed to allow for the preparation of larger kernel pieces and cleaner kernel fractions. Clark et al. (1974) have studied dehulling conditions and Lawhon (1970) studied pretreatments of the seed to obtain better yields of large kernel pieces from dehulling. Both groups report the ability to generate 85 to 90% large kernel pieces.

Glandless cottonseed kernels have been either roasted or fried under various conditions to produce a commercial "nut" (Lawhon et al., 1970a, 1970b). The product preferred by taste panels was made by dry roasting to a final temperature of 141°C (Lawhon et al., 1970a) and was called "Tamunuts," an acronym for Texas A&M University nuts. Later, Lawhon and Brusse (1976) modified the roasting conditions to help retain seed moisture, which was found to be useful for controlling the hardness of the product. Dry-roasted vacuum-packed Tamunuts have been found to maintain their freshness when stored at room temperature for up to 3 yr (Lusas and Jividen, 1987). Tamunuts have been incorporated into recipes for peanut butter cookies (Lawhon, 1972), and unroasted cottonseed kernels have been incorporated into corn-based tortillas to improve their nutrition profile while maintaining good sensory properties (Green et al., 1977).

Recommendations have been made for the production of food-grade glandless cottonseed flour (Lusas and Jividen, 1987). Quality seed with a free fatty acid level less than 1.5% was recommended as a starting material for flour production. Glandless kernels, cleaned of extraneous glanded kernels and hulls, should be tempered rapidly to 11.5% moisture, heated to a temperature of 80°C, flaked to 0.30 to 0.35 mm thickness, extracted with hexane to remove oil, and then desolventized and ground. The flour has been formulated into a variety of food products, including breads (Matthews et al., 1970; Rooney et al., 1972; Harden and Yang, 1973), doughnuts (Lawhon et al., 1975), tortillas (Green et al., 1976), and processed meats (Molonon and Bowers, 1976). Sensory properties have been mixed, although glandless cottonseed flour appears to compare favorably with other vegetable flours. The flour tends to have a bland flavor that does not interfere with the natural flavor of the other ingredients, although for some applications this trait may be less favorable (Lusas et al., 1978). Glandless cottonseed flour has also been texturized by extrusion processing (Taranto et al., 1975) and has compared favorably with similar soybean products when formulated into processed meats (Lusas et al., 1978).

Protein concentrates from glandless cottonseed flour have been prepared by either dry or wet processing. The dry process involves size reduction in a pin mill followed by air classification (Martinez et al., 1970). The smaller particles consist mainly of intact protein bodies, while the larger particles are intact cells with additional cell wall fragments. Protein levels greater than 70% were obtained from the smallest particle fractions. A protein concentrate can is also produced by extraction of the meal (relatively free of hulls) with water or dilute salt solutions to remove soluble sugars. Sometimes, the aqueous pH is adjusted in these processes to minimize protein loss. The concentrate from the wet process has slightly less protein than the concentrate from the dry process but is produced in a greater yield than is obtained from the dry process. Cottonseed protein concentrates have been incorporated into bread and meat loaf recipes and found to have acceptable functional and sensory properties (Lawhon et al., 1972; Khan et al., 1976). Additionally, Smith et al. (1973) have studied the use of glandless protein concentration as an ingredient to stabilize meat emulsions.

Protein isolates with around 90% or greater protein have been prepared by a one- or two-step process. The single step approach extracts protein from the defatted glandless flour with a dilute alkali solution. The unsolubilized fraction is removed by filtration or centrifugation, and the extracted protein is precipitated at pH levels near the protein's iso-electric point, that is, between 4 and 5. Sugars remain in solution, and the recovered precipitate is very pure protein. The mixture is centrifuged to recover the product, which is washed and dried. Isolates prepared in this manner have functional properties, that is, foaming, emulsion stabilizing, and water holding properties that are useful in formulating processed foods (Dowd and Hojilla-Evangelista, 2013).

Alternatively, isolate preparations can be made either by sequential extraction (Berardi et al., 1969; Martinez et al., 1970; Lawhon and Cater, 1971) or sequential precipitation (Lusas and Jividen, 1987), each process producing two isolate fractions that differ in the pH values used to solubilize and recover the protein. These fractions have been loosely taken to correspond to non-storage cytoplasmic proteins that are less soluble at lower pH (Isolate I) and to the storage proteins that are less soluble at higher pH (Isolate II). Typically, Isolate I is recovered in about half the yield as Isolate II. The two fractions have different amino acid profiles (Table 4) with the non-storage proteins being especially rich in lysine (6%). In addition to the different amino acid compositions, the isolate fractions also differ in molecular weight profiles, sedimentation patterns, flavor attributes, and color (Martinez et al., 1970). Breads prepared with Isolate II performed well when included at modest levels (Martinez et al., 1970).

Future Considerations

At present, cotton is grown primarily as a fiber crop, and any changes to the seed must not compromise fiber quality. However, this does not preclude the development of value-added seed traits if they can be shown to be agronomic. With the world population projected to exceed 9 billion by 2050, there is a growing need for expanded cottonseed use as a food and feed crop.

It is likely in the coming years that cottonseed quality will be improved, which could change how the seed is processed and used. Seeds with modified fatty acid composition are on the horizon, which should expand the food and industrial use of the oil. In addition, a gossypol-free cottonseed would potentially allow for the development of many food products from the protein fraction. These seeds would also allow for greater flexibility in processing, which should also add value to the cotton plant's second product.

References

Ahmad, S., F. Anwar, A.I. Hussain, M. Ashraf, and A.R. Awan. 2007. Does soil salinity affect yield and composition of cottonseed oil? J. Am. Oil Chem. Soc. 84:845–851. doi:10.1007/s11746-007-1115-8

Alderks, G.H. 1948. Handling and storage of cottonseed. In: A.E. Bailey, editor, Cottonseed and cottonseed products. Interscience, New York. p. 567–587.

Altschul, A.M., C.M. Lyman, and F.H. Thurber. 1958. Cottonseed meal. In: A.M. Altschul, editor, Processed plant protein foodstuffs. Academic Press, New York. p. 469–534.

American Oil Chemists' Society. 2013. Official methods and recommended practices of the American Oil Chemists' Society. 6th ed. AOCS Press, Urbana, IL.

Anonymous. 1977. Continuous solvent extraction. J. Am. Oil Chem. Soc. 54:A202–A206. doi:10.1007/BF02894408

Anonymous. 1992. Should protein-nitrogen factors be changed? Inform 3:727–728.

Badami, R.C., and K.B. Patil. 1980. Structure and occurrence of unusual fatty acids in minor seed oils. Prog. Lipid Res. 19:119–153. doi:10.1016/0163-7827(80)90002-8

Bailey, A.E. 1948. Cottonseed oil. In: A.E. Bailey, editor, Cottonseed and cottonseed products. Interscience, New York. p. 364–408.

Berardi, L.C., and L.A. Goldblatt. 1980. Gossypol. In: I.E. Liener, editor, Toxic constituents of plant foodstuffs. 2nd ed. Academic Press, New York. p. 183–237.

Berardi, L.C., W.H. Martinez, and C.J. Fernandez. 1969. Cottonseed protein isolates: Two step extraction procedure. Food Technol. 23:1305–1312.

Bland, J.M., E.J. Conkerton, and G. Abraham. 1991. Triacylglyceride composition of cottonseed oil by HPLC and GC. J. Am. Oil Chem. Soc. 68:840–843. doi:10.1007/BF02660598

Bottger, G.T., E.T. Sheehan, and M.J. Lukefahr. 1964. Relation of gossypol content of cotton plants to insect resistance. J. Econ. Entomol. 57:283–285.

Calhoun, M.C. 2002. Cottonseed meal and whole cottonseed: Optimizing their use in dairy cattle rations. Report to the National Cottonseed Products Association. http://www.cottonseed.com/publications/ (accessed 16 Mar. 2015).

Carter, F.L., A.E. Castillo, V.L. Frampton, and T. Kerr. 1966. Chemical composition studies of seeds of the genus *Gossypium*. Phytochemistry 5:1103–1112. doi:10.1016/S0031-9422(00)86103-8

Cass, Q.B., E. Tiritan, S.A. Matlin, and E.C. Freire. 1991. Gossypol enantiomer ratios in cotton seeds. Phytochemistry 30:2655–2657. doi:10.1016/0031-9422(91)85117-I

Cavanagh, G.C., E.J. Cecil, and K. Robe. 1961. Seed to salad oil in 18 hours. Food Process. 22:38–45.

Chapman, K.D., S. Austin-Brown, S.A. Sparace, A.J. Kinney, K.G. Ripp, I.L. Pirtle, and R.M. Pirtle. 2001. Transgenic cotton plants with increased seed oleic acid content. J. Am. Oil Chem. Soc. 78:941–947. doi:10.1007/s11746-001-0368-y

Cheng, H.N., M.W. Rau, M.K. Dowd, M.W. Easson, and B.D. Condon. 2014. Comparison of soybean and cottonseed oils upon hydrogenation with nickel, palladium, and platinum catalysts. J. Am. Oil Chem. Soc. 91:1461–1469. doi:10.1007/s11746-014-2487-1

Cherry, J.P., R.J. Kohel, L.A. Jones, and W.H. Powell. 1981. Cottonseed quality: Factors affecting feed and food uses. In: Proceedings of the Beltwide Cotton Production Research Conference, Natl. Cotton Counc. Am., New Orleans, LA. 4–8 Jan. 1981. Natl. Cotton Counc. Am., Memphis, TN. p. 266–283.

Cherry, J.P., and H.R. Leffler. 1984. Seed. In: R.J. Kohel and C.F. Lewis, editors, Cotton. ASA, CSSA, and SSA, Madison, WI. p. 511–569.

Clark, S.P., L.R. Wiederhold, C.M. Cater, and K.F. Mattil. 1974. Dehulling cottonseed and separating kernels and hulls: Comparison of several varieties of seed. J. Am. Oil Chem. Soc. 51:142–147. doi:10.1007/BF02639724

d'Alonzo, R.P., W.J. Kozarek, and R.L. Wade. 1982. Glyceride composition of processed fats and oils as determined by glass capillary gas chromatography. J. Am. Oil Chem. Soc. 59:292–295. doi:10.1007/BF02662229

Dollear, F.G., and K.S. Markley. 1948. Miscellaneous constituents. In: A.E. Bailey, editor, Cottonseed and cottonseed products. Interscience, New York. p. 466–493.

Dowd, M.K. 1996. Compositional characterization of cottonseed soapstocks. J. Am. Oil Chem. Soc. 73:1287–1295. doi:10.1007/BF02525458

Dowd, M.K. 2012. Identification of the unsaturated heptadecyl fatty acids in the seed oils of *Thespesia populnea* and *Gossypium hirsutum*. J. Am. Oil Chem. Soc. 89:1599–1609. doi:10.1007/s11746-012-2071-5

Dowd, M.K., D.L. Boykin, W.R. Meredith, Jr., B.T. Campbell, F.M. Bourland, J.R. Gannaway, K.M. Glass, and J. Zhang. 2010. Fatty acid profiles of cottonseed genotypes from the National Cotton Variety Trials. J. Cotton Sci. 14:64–73.

Dowd, M.K., and M.P. Hojilla-Evangelista. 2013. Preparation and characterization of protein isolate from glandless and glanded cottonseed. ACS Symp. Ser. 1144:343–357. doi:10.1021/bk-2013-1144.ch023

Dowd, M.K., and P.J. Wakelyn. 2011. Cottonseed: Current and future utilization. In: P.J. Wakelyn and R. Chaudhry, editors, Cotton: Technology for the 21st century. ICAC Press, Washington, DC. p. 437–460.

Dunning, J.W. 1948. Cottonseed hulls. In: A.E. Bailey, editor, Cottonseed and cottonseed products. Interscience, New York. p. 873–893.

Dunning, J.W. 1950. Mechanical extraction of drying oils. J. Am. Oil Chem. Soc. 27:446–451. doi:10.1007/BF02637761

Dunning, J.W. 1956. Unit operations in a mechanical extraction mill. J. Am. Oil Chem. Soc. 33:462–470. doi:10.1007/BF02612302

Eaves, P.H., H.P. Dupuy, L.L. Holzenthal, E.T. Rayner, and L.E. Brown. 1968. Elimination of the Halphen response of cottonseed oils in conjunction with deodorization. J. Am. Oil Chem. Soc. 45:293–295. doi:10.1007/BF02652430

FAO/WHO. 1999. Codex Standard 210. Standard for named vegetable oils. http://www.codexalimentarius.org/standards/list-standards/en/?no_cache=1 (accessed 16 Mar. 2015).

Fisher, G.S., and J.P. Cherry. 1983. Variation of cyclopropenoid fatty acids in cottonseed lipids. Lipids 18:589–594. doi:10.1007/BF02534667

Fisher, G.S., and W.H. Schuller. 1981. Gas chromatographic analysis of cyclopropenoid fatty acids in cottonseed oils. J. Am. Oil Chem. Soc. 58:943–946. doi:10.1007/BF02659664

Freedman, T.B., X. Cao, R.V. Oliveira, Q.B. Cass, and L.A. Nafie. 2003. Determination of the absolute configuration and solution conformation of gossypol by vibrational circular dichroism. Chirality 15:196–200. doi:10.1002/chir.10171

Gamboa, D.A., M.C. Calhoun, S.W. Kuhlmann, A.U. Haq, and C.A. Bailey. 2001. Use of expander cottonseed meal in broiler diets formulated on a digestible amino acid basis. Poult. Sci. 80:789–794. doi:10.1093/ps/80.6.789

Green, J.R., J.T. Lawhon, C.M. Cater, and K.F. Mattil. 1976. Protein fortification of corn tortillas with oilseed flours. J. Food Sci. 41:656–670. doi:10.1111/j.1365-2621.1976.tb00693.x

Green, J.R., J.T. Lawhon, C.M. Cater, and K.F. Matil. 1977. Utilization of whole undefatted glandless cottonseed kernels and soybeans to protein-fortify corn tortillas. J. Food Sci. 42:790–794. doi:10.1111/j.1365-2621.1977.tb12604.x

Gregory, S.R., E. Hernandez, and R. Savoy. 1999. Cottonseed processing. In: C.W. Smith and J.T. Cothren, editors, Cotton origin, history, technology, and production. John Wiley & Sons, New York, NY. p. 793–823.

Hale, F., and M. Lyman. 1948. Cottonseed as a source of animal feed. In: A.E. Bailey, editor, Cottonseed and cottonseed products. Interscience, New York. p. 826–868.

Hamza, M.A., M.H. Abdel-Aal, and M.M. Khalil. 1988. Studies on Egyptian cottonseeds: Fatty acid-composition and protein patterns. Nahrung 32:255–259. doi:10.1002/food.19880320313

Harden, M., and S.P. Yang. 1973. High-protein bread from cottonseed flour. Cotton Gin Oil Mill Press 74:22–26.

Hendrix, W.M.B. 1984. Current practices in continuous cottonseed miscella refining. J. Am. Oil Chem. Soc. 61:1369–1372. doi:10.1007/BF02542245

Henry, M.H., G.M. Pesti, R. Bakalli, J. Lee, R.T. Toledo, R.R. Eitenmiller, and R.D. Phillips. 2001. The performance of broiler chicks fed diets containing extruded cottonseed meal. Poult. Sci. 80:762–768. doi:10.1093/ps/80.6.762

Heywang, B.W., A.A. Heidebrecht, and A.R. Kemmerer. 1965. Discolorations in stored eggs when layers were fed cottonseed meals made from glandless and glanded seed. Poult. Sci. 44:573–577. doi:10.3382/ps.0440573

Heywang, B.W., and M.G. Vavich. 1965. Discolorations in eggs from layers fed cottonseed meals made from glandless and glanded seed. Poult. Sci. 44:84–89. doi:10.3382/ps.0440084

Huang, L., Y.K. Si, G. Snatzke, and D.K. Zheng. 1988. Absolute configuration of gossypol. Collect. Czech. Chem. Commun. 53:2664–2666. doi:10.1135/cccc19882664

Hutchins, R.P. 1953. The solvent extraction of cottonseed. J. Am. Oil Chem. Soc. 30:56–59. doi:10.1007/BF02634997

Itoh, T., T. Tamura, and T. Matsumoto. 1973. Sterol composition of 19 vegetable oils. J. Am. Oil Chem. Soc. 50:122–125. doi:10.1007/BF02633564

Johnston, C., and A.B. Watts. 1965a. The infra red spectra of glandless cottonseed meals of varying nutritional value. Poult. Sci. 44:302–304. doi:10.3382/ps.0440302

Johnston, C., and A.B. Watts. 1965b. The characterization of a growth inhibitor of glandless cottonseed. Poult. Sci. 44:652–658. doi:10.3382/ps.0440652

Jones, L.A., and C.C. King, editors. 1990. Cottonseed oil. Natl. Cottonseed Prod. Assoc., Memphis, TN.

Jonston, C., and A.B. Watts. 1964. The chick feeding value of meals prepared from glandless cottonseed. Poult. Sci. 43:957–963. doi:10.3382/ps.0430957

Jurriens, G., and A.C.J. Kroesen. 1965. Determination of glyceride composition of several solid and liquid fats. J. Am. Oil Chem. Soc. 42:9–14. doi:10.1007/BF02558242

Khan, M.N., J.T. Lawton, L.W. Rooney, and C.M. Cater. 1976. Bread-baking properties of cottonseed protein concentrates from solvent-extracted glandless cottonseed flour. Cereal Chem. 53:388–396.

Kuo, T.M., J.F. VanMiddlesworth, and W.J. Wolf. 1988. Content of raffinose oligosaccharides and sucrose in various plant seeds. J. Agric. Food Chem. 36:32–36. doi:10.1021/jf00079a008

LaRue, D.C., D.A. Knabe, R.S. Izard, and T.D. Tanskley, Jr. 1987. Glandless cottonseed kernels for starter, grower and finisher swine. J. Anim. Sci. 64:1051–1059.

LaRue, D.C., D.A. Knabe, and T.D. Tanskley, Jr. 1985. Commercially processed glandless cottonseed meal for starter, grower and finisher swine. J. Anim. Sci. 60:495–502.

Lawhon, J.T. 1970. Pretreating delintered cottonseed to increase yield of whole kernels. J. Am. Oil Chem. Soc. 47:102–105. doi:10.1007/BF02612947

Lawhon, J.T. 1972. Protein product presents potential as nut ingredient. Candy Snack Ind. 137:34.

Lawhon, J.T., and J.C. Brusse. 1976. Newly developed unit improves toasting of nuts from cottonseed kernels. Food Prod. Dev. 10:60, 62.

Lawhon, J.T., and C.M. Cater. 1971. Effect of processing method and pH of precipitation on the yields and functional properties of protein isolates from glandless cottonseed. J. Food Sci. 36:372–377. doi:10.1111/j.1365-2621.1971.tb06366.x

Lawhon, J.T., C.M. Cater, and K.F. Mattil. 1970a. Preparation of a high-protein low cost nut-like food product from glandless cottonseed kernels. Food Technol. 24:77–80.

Lawhon, J.T., C.M. Cater, and K.F. Mattil. 1970b. Influence of the deep-fat frying medium on the acceptability and stability of glandless cottonseed kernels for use as a high-protein, low-cost edible nut. Food Technol. 24:817–819.

Lawhon, J.T., C.M. Cater, and K.F. Mattil. 1975. Sensory, analytical evaluation of cake doughnuts fortified with protein from oilseed flours. Food Prod. Dev. 9:110, 113, 115, 116, 118.

Lawhon, J.T., C.M. Cater, and K.F. Mattil. 1977. Evaluation of the food use potential of sixteen varieties of cottonseed. J. Am. Oil Chem. Soc. 54:75–80. doi:10.1007/BF02912394

Lawhon, J.T., L.W. Rooney, C.M. Cater, and K.F. Mattil. 1972. Evaluation of a protein concentrate produced from glandless cottonseed flour by a wet-extraction process. J. Food Sci. 37:778–782. doi:10.1111/j.1365-2621.1972.tb02749.x

Leahy, J. 1948. Structure of the cottonseed. In: A.E. Bailey, editor, Cottonseed and cottonseed products. Interscience, New York. p. 105–116.

Lee, J.A. 1969. Some problems in altering the gossypol content of cottonseed through breeding. J. Am. Oil Chem. Soc. 46:124–125. doi:10.1007/BF02635713

Lee, J.A., C.C. Cockerham, and F.H. Smith. 1968. The inheritance of gossypol level in *Gossypium* 1. Additive, dominance, epistatic, and maternal effects associated with seed gossypol in two varieties of *Gossypium hirsutum* L. Genetics 59:285–298.

Liener, I.E. 1958. Effect of heat on plant protein. In: A.M. Altschul, editor, Processed plant protein foodstuffs. Academic Press, New York. p. 79–129.

Liu, Q., S.P. Singh, and A.G. Green. 2002. High-stearic and high-oleic cottonseed oils produced by hairpin RNA-mediated post-transcriptional gene silencing. Plant Physiol. 129:1732–1743. doi:10.1104/pp.001933

Loft, S.C. 1990. Deodorization—Theory and practice. In: D.R. Erickson, editor, World conference proceedings on edible fats and oils processing: Basic principles and modern practices. AOCS Press, Champaign, IL, p. 117–123.

Longmore, J. 1886. Cotton-seed oil: Its colouring matter and mucilage, and description of a new method of recovering the loss occurring in the refining process. J. Soc. Chem. Ind. (London) 5:200–206.

Lukonge, E., M.T. Labuschagne, and A. Hugo. 2007. The evaluation of oil and fatty acid composition in seed of cotton accessions from various countries. J. Sci. Food Agric. 87:340–347. doi:10.1002/jsfa.2731

Lusas, E.W., and G.M. Jividen. 1986. Glandless cottonseed processing and utilization research. In: A.R. Baldwin, editor, Proceedings of the world conference on emerging technologies in the fats and oils Industry. AOCS Press, Champaign, IL. p. 221–231.

Lusas, E.W., and G.M. Jividen. 1987. Glandless cottonseed: A review of the first 25 years of processing and utilization research. J. Am. Oil Chem. Soc. 64:839–854. doi:10.1007/BF02641491

Lusas, E.W., J.T. Lawhon, S.P. Clark, S.W. Matlock, W.W. Meinke, D.W. Nulsow, K.C. Rhee, and P.J. Wan. 1978. Potential for edible protein products from glandless cottonseed. In: Proceedings of Glandless Cottonseed Conference, Dallas, TX. 13–14 Dec. 1977. USDA–ARS, Beltsville, MD. p. 31–43.

Lusas, E.W., L.R. Watkins, and K.C. Rhee. 1990. Separation of fats and oils by solvent extraction: Non-traditional methods. In: D.R. Erickson, editor, Edible fats and oil processing: Basic principles and modern practices. AOCS Press, Champaign, IL. p. 56–78.

Main, C.L., L.T. Barber, R.K. Boman, K. Chapman, D.M. Dodds, S. Duncan, K.L. Edmisten, P. Horn, M.A. Jones, G.D. Morgan, E.R. Norton, S. Osborne, J.R. Whitakerm, R.L. Nichols, and K.F. Bronson. 2013. Effects of nitrogen and planting seed size on cotton growth, development, and yield. Agron. J. 105:1853–1859. doi:10.2134/agronj2013.0154

Marchlewski, L.P. 1899. Gossypol: Ein bestandtheil der baumwollsamen. J. Prakt. Chem. 60:84–90. doi:10.1002/prac.18990600108

Martinez, W.H., L.C. Berardi, and L.A. Goldblatt. 1970. Cottonseed protein products— Composition and functionality. J. Agric. Food Chem. 18:961–968. doi:10.1021/jf60172a039

Matthews, R.H., E.J. Sharpe, and W.M. Clark. 1970. The use of some oilseed flours in bread. Cereal Chem. 47:181–189.

McMichael, S.C. 1959. Hopi cotton: A source of cottonseed free of gossypol pigments. Agron. J. 51:630. doi:10.2134/agronj1959.00021962005100100025x

McMichael, S.C. 1960. Combined effects of glandless genes gl2 and gl3 on pigment glands in the cotton plant. Agron. J. 52:385–386. doi:10.2134/agronj1960.00021962005200070005x

Mellon, J.E., M.K. Dowd, S.B. Beltz, and G.G. Moore. 2014. Growth inhibitory effects of gossypol and related compounds on fungal cotton root pathogens. Lett. Appl. Microbiol. 59:161–168. doi:10.1111/lam.12262

Meredith, W.R., Jr., D.L. Boykin, F.M. Bourland, W.D. Caldwell, B.T. Campbell, J.R. Gannaway, K. Glass, A.P. Jones, L.M. May, C.W. Smith, and J. Zhang. 2012. Genotype X environment interactions over seven years for yield, yield components, fiber quality, and gossypol traits in the Regional High Quality Tests. J. Cotton Sci. 16:160–169.

Meredith, W.R., Jr., C.D. Elmore, and E.E. King. 1978. Genetic variation within cotton for seed N and amino acid content. Crop Sci. 18:577–580. doi:10.2135/cropsci1978.0011183X001800040013x

Molonon, B.R., and J.A. Bowers. 1976. Sensory evaluation and protein value of beef and beef-cottonseed blends. J. Food Sci. 41:1263–1265. doi:10.1111/j.1365-2621.1976.tb01147.x

Morgan, D.A. 1942. Smoke, fire, and flash points of cottonseed, peanut, and other vegetable oils. Oil Soap 19:193–198. doi:10.1007/BF02545481

National Cottonseed Products Association. 2002a. Cottonseed feed products guide. NCPA. http://www.cottonseed.com/publications/feedproductsguide.asp (accessed 16 Mar. 2015).

National Cottonseed Products Association. 2002b. Cottonseed oil. NCPA. http://www.cottonseed.com/publications/csobro.asp (accessed 16 Mar. 2015).

National Cottonseed Products Association. 2008. Trading rules. NCPA. http://www.cottonseed.com/tradingrules (accessed 16 Mar. 2015).

Nergiz, C., H. Yalçi, and H. Yildiz. 1997. Some analytical characters of cottonseed varieties grown in Turkey. Grasas Aceites 48:411–414. doi:10.3989/gya.1997.v48.i6.813

Noland, P.R., M. Funderburg, J. Atteberry, and K.W. Scott. 1968. Use of glandless cottonseed meal in diets for young pigs. J. Anim. Sci. 27:1319–1321.

Norris, F.A. 1982a. Extraction of fats and oils. In: D. Swern, editor, Bailey's industrial oil and fat products. Vol. 2. 4th ed. John Wiley & Sons, New York. p. 175–251.

Norris, F.A. 1982b. Refining and Bleaching. In: D. Swern, editor, Bailey's industrial oil and fat products. Vol. 2. 4th ed. John Wiley & Sons, New York. p. 253–314.

O'Brien, R.D., L.A. Jones, C.C. King, P.J. Wakelyn, and P.J. Wan. 2005. Cottonseed oil. In: F. Shahidi, editor, Bailey's industrial oil and fat products, Vol. 2. 6th ed.. John Wiley & Sons, Hoboken, NJ. p. 173–279.

Palle, S.R., L.M. Campbell, D. Pandeya, L. Puckhaber, L.K. Tollack, S. Marcel, S. Sundaram, R.D. Stipanovic, T.C. Wedegaertner, L. Hinze, and K.S. Rathore. 2013. RNAi-mediated ultra-low gossypol cottonseed trait: Performance of transgenic lines under field conditions. Plant Biotechnol. J. 11:296–304. doi:10.1111/pbi.12013

Percy, R.G., M.C. Calhoun, and H.L. Kim. 1996. Seed gossypol variation within *Gossypium barbadense* L. cotton. Crop Sci. 36:193–197. doi:10.2135/cropsci1996.0011183X003600010034x

Pettigrew, W.T., and M.K. Dowd. 2011. Varying planting dates or irrigation regimes alters cottonseed composition. Crop Sci. 51:2155–2164. doi:10.2135/cropsci2011.02.0085

Phelps, R.A., F.S. Shenstone, A.R. Kemmerer, and R.J. Evans. 1965. A review of cyclopropenoid compounds: Biological effects of some derivatives. Poult. Sci. 44:358–394. doi:10.3382/ps.0440358

Pons, W.A., Jr., C.L. Hoffpauir, and T.H. Hopper. 1953. Gossypol in cottonseed. Influence of variety of cottonseed and environment. J. Agric. Food Chem. 1:1115–1118. doi:10.1021/jf60018a007

Pons, W.A., Jr., F.H. Thurber, and C.L. Hoffpauir. 1955. Prepress-solvent extraction of cottonseed, processing conditions and characterization of products. J. Am. Oil Chem. Soc. 32:98–103. doi:10.1007/BF02636513

Pope, O.A., and J.C. Ware. 1945. Effect of variety, location, and season on oil, protein, and fuzz of cottonseed and on fiber problems of lint. USDA Tech. Bull. 903. US Gov. Print. Office, Washington, DC.

Puckhaber, L.S., M.K. Dowd, R.D. Stipanovic, and C.R. Howell. 2002. Toxicity of (+)- and (−)-gossypol to the plant pathogen, *Rhizoctonia solani*. J. Agric. Food Chem. 50:7017–7021. doi:10.1021/jf0207225

Reid, B.L., S. Galaviz-Moreno, and P.M. Maiorino. 1984. A comparison of glandless and regular cottonseed meals for laying hens. Poult. Sci. 63:1803–1809. doi:10.3382/ps.0631803

Roberson, R.H. 1970. A comparison of glandless cottonseed meal and soybean meal in laying diets supplemented with lysine and methionine. Poult. Sci. 49:1579–1589. doi:10.3382/ps.0491579

Robertson, F.R., and J.G. Campbell. 1933. Some observations on the increase of free fatty acid in cottonseed. Oil Soap 10:146–147. doi:10.1007/BF02638155

Robinson, E.H., and M.H. Li. 1994. Use of plant proteins in catfish feeds: Replacement of soybean meal with cottonseed meal and replacement of fish meal with soybean meal and cottonseed meal. J. World Aquacult. Soc. 25:271–276. doi:10.1111/j.1749-7345.1994.tb00190.x

Robinson, E.H., S.D. Rawles, P.W. Oldenburg, and R.R. Stickney. 1984. Effects of feeding glandless or glanded cottonseed products and gossypol to *Tilapia aurea*. Aquaculture 38:145–154. doi:10.1016/0044-8486(84)90226-6

Rooney, L.W., C.B. Gustafson, S.P. Clark, and C.M. Cater. 1972. Comparison of the baking properties of several oilseed flours. J. Food Sci. 37:14–18. doi:10.1111/j.1365-2621.1972.tb03373.x

Smith, G.C., H. Juhn, Z.L. Carpenter, K.F. Mattil, and C.M. Cater. 1973. Efficacy of protein additives as emulsion stabilizers in frankfurters. J. Food Sci. 38:849–855. doi:10.1111/j.1365-2621.1973.tb02091.x

Smith, N.E., L.S. Collar, D.L. Bath, W.L. Dunkley, and A.A. Franke. 1981. Digestibility and effects of whole cottonseed fed to lactating cows. J. Dairy Sci. 64:2209–2215. doi:10.3168/jds.S0022-0302(81)82831-7

Stansbury, M.F., A.F. Cucullu, and G.T. Den Hartog. 1954. Influence of variety and environment on oil content of cottonseed kernels. J. Agric. Food Chem. 2:692–696. doi:10.1021/jf60033a007

Stansbury, M.F., C.L. Hoffpauir, and T.H. Hopper. 1953a. Influence of variety and environment on the iodine value of cottonseed oil. J. Am. Oil Chem. Soc. 30:120–123. doi:10.1007/BF02638664

Stansbury, M.F., W.A. Pons, Jr., and G.T. Den Hartog. 1953b. Phosphorus compounds in cottonseed kernels. Influence of variety of cottonseed and environment. J. Agric. Food Chem. 1:75–78. doi:10.1021/jf60001a011

Stansbury, M.F., W.A. Pons, Jr., and G.T. Den Hartog. 1956. Relations between oil, nitrogen, and gossypol in cottonseed kernels. J. Am. Oil Chem. Soc. 33:282–286. doi:10.1007/BF02630862

Stipanovic, R.D., A.A. Bell, M.E. Mace, and C.R. Howell. 1975. Antimicrobial terpenoids of *Gossypium*: 6-methoxygossypol and 6,6′-dimethoxygossypol. Phytochemistry 14:1077–1081. doi:10.1016/0031-9422(75)85190-9

Stipanovic, R.D., J.D. Lopez, Jr., M.K. Dowd, L.S. Puckhaber, and S.E. Duke. 2006a. Effect of racemic and (+)- and (–)-gossypol on the survival and development of *Helicoverpa zea* larvae. J. Chem. Ecol. 32:959–968. doi:10.1007/s10886-006-9052-9

Stipanovic, R.D., L.S. Puckhaber, and A.A. Bell. 2006b. Ratios of (+)- and (–)-gossypol in leaves, stems, and roots of selected accessions of *Gossypium hirsutum* Var. marie galante (Watt) Hutchinson. J. Agric. Food Chem. 54:1633–1637. doi:10.1021/jf052319e

Stipanovic, R.D., L.S. Puckhaber, A.A. Bell, A.E. Percival, and J. Jacobs. 2005. Occurrence of (+)- and (–)-gossypol in wild species of cotton and in *Gossypium hirsutum* var. marie-galante (Watt). Hutchinson. J. Agric. Food Chem. 53:6266–6271. doi:10.1021/jf050702d

Stipanovic, R.D., L.S. Puckhaber, J. Liu, and A.A. Bell. 2009. Total and percent atropisomers of gossypol and gossypol-6-methyl ether in seeds from Pima cottons and accessions of *Gossypium barbadense* L. J. Agric. Food Chem. 57:566–571. doi:10.1021/jf802756e

Sunilkumar, G., L.M. Campbell, M. Hossen, J.P. Connell, E. Hernandez, A.S. Reddy, C.W. Smith, and K.S. Rathore. 2005. A comprehensive study of the use of a homologous promoter in antisense cotton lines exhibiting a high seed oleic acid phenotype. Plant Biotechnol. J. 3:319–330. doi:10.1111/j.1467-7652.2005.00126.x

Sunilkumar, G., L.M. Campbell, L. Puckhaber, R.D. Stipanovic, and K.S. Rathore. 2006. Engineering cottonseed for use in human nutrition by tissue-specific reduction of toxic gossypol. Proc. Natl. Acad. Sci. USA 103:18054–18059. doi:10.1073/pnas.0605389103

Taranto, M.V., W.W. Meinke, C.M. Cater, and K.F. Mattil. 1975. Parameter affecting production and character of extrusion texturized defatted glandless cottonseed meal. J. Food Sci. 40:1264–1269. doi:10.1111/j.1365-2621.1975.tb01068.x

Taylor, D.R. 2005. Bleaching. In: F. Shahidi, editor, Bailey's industrial oil and fat products, Vol. 5. 6th ed. John Wiley & Sons, Hoboken, NJ. p. 285–339.

Temming, H. 1966. Temming linters: Technical information on cotton cellulose. Peter Temming Aktiengesellschaft, Gluckstadt, Germany.

Tharp, W.H. 1948. Cottonseed composition—Relation to variety, maturity, and environment of the plant. In: A.E. Bailey, editor, Cottonseed and cottonseed products. Interscience, New York. p. 117–156.

Tharp, W.H., J.J. Skinner, J.H. Turner, Jr., R.P. Bledsoe, and H.B. Brown. 1949. Yield and composition of cottonseed as influenced by fertilization and other environmental factors. USDA Tech. Bull. 974. US Gov. Print. Office, Washington, DC.

Thompson, A.C., R.D. Henson, P.A. Hedin, and J.P. Minyard. 1968. Constituents of the cotton bud: XII. The carotenoids in buds, seeds and other tissue. Lipids 3:495–497. doi:10.1007/BF02530892

Thornton, M.K. 1932. Cottonseed products. Oil Mill Gazetteer, Wharton, TX.

Thurber, F.H., H.L.E. Vix, W.A. Pons, Jr., A.J. Crovetto, and N.B. Knoepfler. 1954. The effect of processing conditions on the properties of screw-press cottonseed meal and oil. J. Am. Oil Chem. Soc. 31:384–388. doi:10.1007/BF02545516

Thurman, B.H. 1945. Batch against continuous refining of vegetable oils. J. Am. Oil Chem. Soc. 26:580–583.

USDA. 1984. Composition of foods: Nut and seed products. Agric. Handb. 8-12. USDA, Washington, DC.

Van Wyck, P. 1948. Cotton linters. In: A.E. Bailey, editor, Cottonseed and cottonseed products. Interscience, New York. p. 894–905.

Verleyen, T., M. Forcades, R. Verhe, K. Dewettinck, A. Huyghebaert, and W. de Greyt. 2002. Analysis of free and esterified sterols in vegetable oils. J. Am. Oil Chem. Soc. 79:117–122. doi:10.1007/s11746-002-0444-3

Waldroup, P.W., and J.H. Kersey. 2002. Nutrient composition of cottonseed meal surveyed. Feedstuffs 74(45):1–2.

Waldroup, P.W., E.G. Keyser, V.E. Tollett, and T.E. Bowen. 1968. The evaluation of a low-gossypol glandless cottonseed meal in broiler diets. Poult. Sci. 47:1179–1186. doi:10.3382/ps.0471179

Wamble, A.C. 1948. Mechanical pretreatment of the seed. In: A.E. Bailey, editor, Cottonseed and cottonseed products. Interscience, New York. p. 588–614.

Wang, X., C.P. Howell, F. Chen, J. Yin, and Y. Jiang. 2009. Gossypol—A polyphenolic compound from cotton plant. Adv. Food Nutr. Res. 58:215–263. doi:10.1016/S1043-4526(09)58006-0

Weiss, T.J. 1967. Salad oil manufacture and control. J. Am. Oil Chem. Soc. 44:146A,148A, 186A, 197A.

Whitten, C.C., E.E. Miller, and D.E. Pratt. 1984. Cottonseed flavonoids as lipid antioxidants. J. Am. Oil Chem. Soc. 61:1075–1078. doi:10.1007/BF02636224

Williams, M.A. 1986. Extrusion preparation for solvent extraction. Oil Mill Gaz. 91:24–26, 28.

Williams, M.A. 2005. Recovery of oils and fats from oilseeds and fatty materials. In: F. Shahidi, editor, Bailey's industrial oil and fat products. Vol. 5. 6th ed. John Wiley & Sons, Hoboken, NJ. p. 99–189.

Williams, P.A., C.H. Boatner, C.M. Hall, R.T. O'Connor, and L.E. Castillon. 1947. Processing of cottonseed. III. Color development in cottonseed oil during storage of the seed and crude oil. J. Am. Oil Chem. Soc. 24:362–369. doi:10.1007/BF02643517

Winters, R.L. 1990. Deodorizer distillate. In: D.R. Erickson, editor, World conference proceedings: Edible fats and oils processing: Basic principles and modern practices. AOCS Press, Champaign, IL. p. 402–405.

Woolrich, W.R., and E.L. Carpenter. 1935. Mechanical processing of cottonseed. Univ. of Tennessee Eng. Exp. Stn., Knoxville, TN.

Wrenn, L.B. 1995. Cinderella of the new South: A history of the cottonseed industry, 1855–1955. Univ. of Tennessee Press, Knoxville, TN.

Yamamoto, K., A. Kinoshita, and A. Shibahara. 2008. Ricinoleic acid in common vegetable oils and oil seeds. Lipids 43:457–460. doi:10.1007/s11745-008-3154-9

Yunusova, S.G., S.D. Gusakova, A.I. Glushenkova, U.K. Nadzhimov, Sh. Turabekov, and S.A. Musaev. 1991. A comparative investigation of the fatty acid compositions of the seeds of a number of lines of a genetic collection of *Gossypium hirsutum*. Chem. Nat. Compd. 27:147–150. doi:10.1007/BF00629746

Marketing and Economics

Gary M. Adams

Abstract

Cotton (*Gossypium hirsutum* L.) remains an important commodity in the agricultural economy of the United States. The scope and economic impact extend well beyond the approximately 19,000 farmers that plant between 4 and 6 million hectares of cotton each year. Taking into account the diversified cropping patterns, the nation's cotton farmers cultivate more than 12 million hectares of land each year. While much of the industry is concentrated in 17 cotton-producing states stretching from Virginia to California, the processors and distributors of cotton fiber and downstream manufacturers of cotton apparel and home furnishings are located in virtually every state. Beyond the farm gate, the distribution and processing of cotton includes cotton gins, independent merchants and cooperative merchandisers, warehouses, cottonseed distributors and processors, and textile mills. Farms and businesses directly involved in the production, distribution, and processing of cotton employ almost 200,000 workers and produce direct business revenue of more than US$27 billion. Accounting for the ripple effect of cotton through the broader economy, direct and indirect employment surpasses 420,000 workers with economic activity well in excess of US$100 billion. Among agricultural commodities, cotton is somewhat unique since it functions as both an agricultural commodity and an industrial commodity. Cotton must compete with grains, oilseeds, and other competing crops for available land, while also competing with synthetic fibers in the production of textile and apparel products.

In global fiber markets, cotton continues to face strong competition from polyester and other synthetic fibers, with the forces becoming even more intense in recent years. During the 1980s, growth in cotton demand generally kept pace with synthetic fibers as total mill use of cotton grew from 14.2 million metric tons in 1980 to 18.9 million metric tons by the end of that decade (Fig. 1). Competitive prices relative to polyester and strong consumer demand propelled cotton mill use to an annual growth rate of 3.3% during the 1980s.

Unfortunately, the robust growth for cotton demand did not carry over into the 1990s. With annual growth of less than 1%, cotton mill use ended the 10-yr period at 19.8 million metric tons and dipped as low as 18.4 million metric tons in 1994. Cotton consumed by the world's textile mills struggled to grow as additional polyester production capacity was brought online, particularly in

---

Abbreviations: STAX, Stacked Income Protection Plan; TRQ, tariff rate quota; WTO, World Trade Organization.

G.M. Adams, National Cotton Council, 7193 Goodlett Farms Parkway, Cordova, TN 38016 (gadams@cotton.org)

doi:10.2134/agronmonogr57.2013.0034

Cotton. 2nd. ed. David D. Fang and Richard G. Percy, editors. Agronomy Monograph 57.

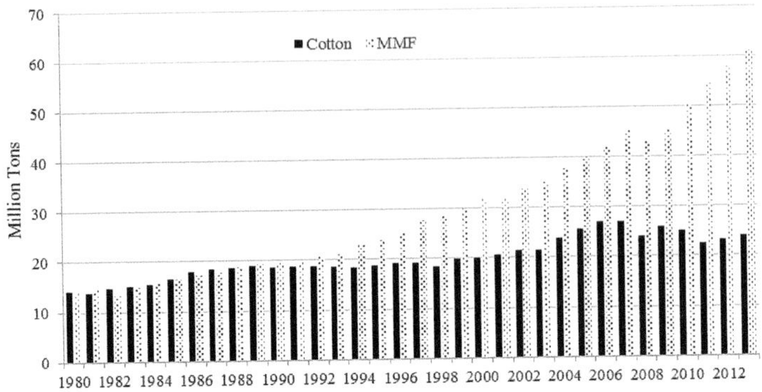

Fig. 1. World fiber use, 1980 to 2013.

China. Use of lower-priced polyester expanded at an annual rate of 4.9%, pushing global demand to 29.9 million metric tons by 1999 (PCI Fibres, 2012).

Cotton consumption experienced strong growth during the decade of the 2000s; however, performance over the decade was mixed. Between 2000 and 2006, world mill use grew by more than 6.5 million metric tons, fueled in part by the phased elimination of textile quotas in the United States and the European Union. Demand growth was aided by cotton prices that generally ranged between US$0.40 and US$0.60 per pound. In the latter part of the decade, cotton consumption struggled because of a combination of factors. The global recession that began late in calendar 2008 and persisted through much of 2009 reduced use of all fibers. After a brief recovery in the 2009 marketing year (1 August to 31 July), strong cotton prices relative to synthetic fiber brought about a drop in overall cotton use and an even sharper decline in cotton's share of world fiber use. As cotton prices moderated and the global economy recovered, cotton use began a slow rebound in 2012.

Since the 1980s, China has maintained its position as the largest spinner of cotton fiber. By the 2000s, China's textile industry processed more than twice as much cotton as in the 1980s (Table 1). Unfortunately, China's growth in cotton use paled in comparison to the growth in synthetic fiber use. Between 2011 and 2013, government policies in China established a floor on internal cotton prices at a level well above polyester prices. The resulting price imbalance caused a decline in cotton use as mills increasingly turned to synthetic fibers. The recent downturn in China's cotton spinning has narrowed the gap between China and India in terms of cotton use. India maintains its position as the second largest processor of cotton, and if growth maintains the current pace, India could overtake China as the largest processor.

The impact of the aforementioned elimination of textile quotas is evident in the consumption numbers for the US textile industry. After holding the third spot during the 1980s and 1990s, the United States slipped to fifth in terms of cotton mill use for the most recent decade. Japan's textile industry suffered a similar fate, while textile production in Pakistan and Turkey increased their presence in the world cotton sector. By the 2014/2015 crop year, Brazil and Bangladesh were both poised to consume in excess of 0.9 million metric tons, bumping the United States down to seventh place at 0.8 million metric tons.

Table 1. Top five cotton mill use countries.

Rank	1980s	1990s	2000s
	million metric tons, avg. for decade		
1	China (3.9)	China (4.4)	China (8.5)
2	India (1.5)	India (2.4)	India (3.5)
3	US (1.4)	US (2.3)	Pakistan (2.2)
4	USSR (1.3)	Pakistan (1.5)	Turkey (1.4)
5	Japan (0.7)	Turkey (0.8)	US (1.3)

World Cotton Production

Since 1980, growth in world cotton production has overwhelmingly been the result of improved productivity rather than an expansion in global area (Fig. 2). World area fluctuated between 28 and 36 million hectares with an average of 33 million hectares over the 1980 to 2013 period (USDA-FAS, 2014). Recent peaks in world area (1995, 2004, and 2011) were generally the result of improved cotton prices relative to competing crops. Likewise, dips in area in 2002, 2008, and 2009 reflected periods when cotton prices weakened relative to competing crops.

Although cotton harvested area exhibited no statistically significant trend for the period, average cotton yields grew by 1.8% per year. Gains in productivity allowed the average cotton yield to grow from approximately 500 kg ha^{-1} in the early 1980s to 875 kg ha^{-1} in recent years. During the early 1980s, the widespread use of better seed varieties and better methods of plant protection led to increased yields. Since the mid-1990s, the commercial use of biotech cotton varieties and successful area-wide pest management programs have also contributed to improved yields.

As was the case for global cotton consumption, China maintains the top spot in terms of cotton production. China's 6.4 million metric tons of annual production during the 2000s represented a 2.4 million metric ton increase from the 1980s (Table 2). The growth in production is tied directly to improved cotton yields as China's area devoted to cotton has declined slightly from the area sown during the 1980s. For the past three decades, the United States maintained the second highest level of cotton production, only to be surpassed by India in the most recent 5-yr period.

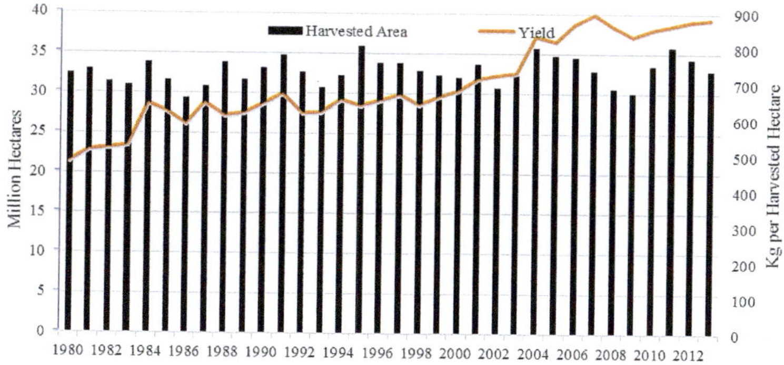

Fig. 2. World cotton area and yield, 1980 to 2013.

Table 2. Top five cotton production countries.

Rank	1980s	1990s	2000s
	million metric tons, avg. for decade		
1	China (4.0)	China (4.5)	China (6.4)
2	US (2.7)	US (3.7)	US (4.1)
3	USSR (1.8)	India (2.5)	India (3.9)
4	India (1.7)	Pakistan (1.7)	Pakistan (1.9)
5	Pakistan (1.1)	Uzbekistan (1.2)	Brazil (1.2)

India devotes more area to cotton production than any other country, but lower yields relegated India to a position of being the third or fourth largest producer until the recent past few years. However, India's position as a major cotton producer began to change between 2002 and 2007 as the average yield increased by 89% because of the adoption of better varieties, many of which were genetically modified. Since 2007, India's yields have plateaued, and future growth hinges on better germplasm and improved use of production inputs.

With average production of just under 2 million metric tons, Pakistan has established itself as the fourth largest cotton producer. Brazil has emerged as a larger producer as high-yielding area has moved into cotton production.

World Cotton Trade

Relative to other major row crops, cotton is one of the most highly traded commodities with roughly one-third of global production moving into export channels. Cotton trade expanded during the decade of the 2000s, reflecting shifts in cotton manufacturing that were in part brought about by the elimination of textile import quotas. Since 2003, China has emerged as the largest importer of cotton, accounting for approximately 40% total imports of cotton fiber. Growth in their textile industry outstripped their ability to increase cotton production, thus leading to additional imports.

After China, other key importing countries, which are generally located in Asia, include Bangladesh, Indonesia, Vietnam, Pakistan, Thailand, and South Korea. Of these countries, Pakistan is the only cotton producer of any significance. As a result, growth in mill use in those other countries would require additional imports. Outside of Asia, Turkey represents the largest importer, accounting for 8% of world trade.

With one notable exception, cotton imports generally occur without quota constraints and with no or a minimal level of applied import duty. China's World Trade Organization (WTO) accession negotiations established a tariff rate quota (TRQ) of 0.9 million metric tons. At various times during the year, China will announce additional quota above the WTO-required TRQ. Generally, those imports are subject to a variable levy ranging from 5 to 40%, to maintain cotton prices in China significantly above international prices and protect prices paid to Chinese cotton growers. Additionally, importers must receive import licenses from the central authorities before entering into import contracts. Cotton can also be imported outside of the quota system. However, the importer is still required to acquire an import license and will be assessed a 40% tariff.

In terms of cotton trade, the United States remains the largest exporting country, with 25% of total trade. The reliance of the US cotton market on exports

Table 3. Top five US cotton export customers.

Rank	1980s	1990s	2000s
	million metric tons, avg. for decade		
1	Japan (0.31)	China (0.24)	China (0.8)
2	South Korea (0.26)	Japan (0.20)	Turkey (0.4)
3	Taiwan (0.09)	Mexico (0.17)	Mexico (0.3)
4	China (0.09)	South Korea (0.17)	Indonesia (0.2)
5	Germany (0.07)	Indonesia (0.13)	Thailand (0.1)

has increased dramatically over the past decade as the domestic textile industry has contracted. Export markets are now the destination for more than 75% of US cotton production.

Not surprisingly, China, as the world's largest importer, is the largest export customer for US cotton (Table 3). In recent years, purchases of US cotton by China's textile industry have exceeded those of the US textile industry. Turkey and Mexico are the second and third largest export customers for US cotton. During the 2000s, Indonesia and Thailand completed the top five list.

Looking forward, export markets are expected to remain the primary outlet for US cotton. Any changes in the key markets will depend on a number of factors. China's position as the largest importer will hinge on key policy actions taken by the Chinese government. After accumulating very significant stockpiles of cotton between 2011 and 2013, their import requirements could be substantially less in coming years, depending on the management of existing reserves, the degree of distortion of internal fiber prices, and the level of support conveyed to cotton production.

US Cotton Consumption

Cotton consumption can be measured at different stages of processing or at different levels of the market. Consumption of cotton fiber by textile mills in the United States and other key countries was discussed in a previous section. Cotton consumption is also measured at the retail or consumer level. However, since consumers purchase finished textile and apparel products, an accurate accounting of the cotton included in those purchases requires conversion to a fiber-equivalent basis. In other words, factors for product weight and fiber content are used to determine the amount of cotton (and other fibers) included in the products purchased by consumers (Meyer and MacDonald, 2014).

In the United States, it is estimated that consumers purchase finished goods that contain the equivalent of approximately 10 million metric tons of synthetic fibers and cotton (Fig. 3). Currently, fiber purchases can be disaggregated into 6.1 million metric tons of synthetic fiber and 3.9 million metric tons of cotton.

Between 1980 and 2006, consumer demand for textile and apparel products grew at an annual average rate of 3.1%, eventually peaking at more than 11.3 million metric tons. Over that same period, cotton's share of the retail market grew from 30 to 43%. Total fiber demand benefited from retail prices that were essentially flat in nominal terms and declining in real terms. Cotton was able to expand market share as consumers responded favorably to promotional efforts by Cotton Incorporated.

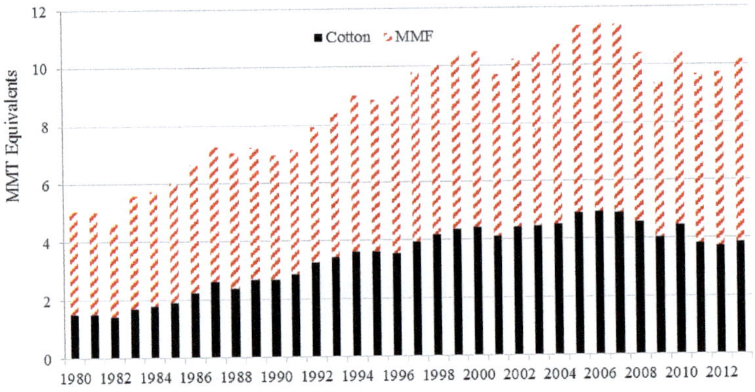

Fig. 3. US net domestic fiber consumption, 1980 to 2013.

Since 2007, the US retail market for textile and apparel products has suffered. By 2009, US consumers had reduced textile/apparel purchases by 2 million metric tons, due in part to the recession. US retail purchases have shown little recovery since 2009 as higher prices for raw fibers, especially cotton, eventually worked their way to the consumer in the form of higher prices for the finished good.

The effects of the spike in cotton prices in late 2010 and 2011 is evident in market share. After peaking at 44% in 2008, cotton's share of the US retail market stood at 38% in 2013. In terms of textiles and apparel, the United States is generally viewed as a mature market. As a result, future growth could be limited and largely dictated by increased population.

Cotton trade can also be measured at different stages of processing. While trade in raw cotton fiber is very significant, there are also numerous trade flows among yarn, fabric, and finished goods. The elimination of textile import quotas and the accession to the WTO of key countries such as China and Vietnam has bolstered the trade of semi-processed and finished goods.

For the US retail market, imported textile products now account for the overwhelming majority of products purchased by consumers (Fig. 4). The influx of

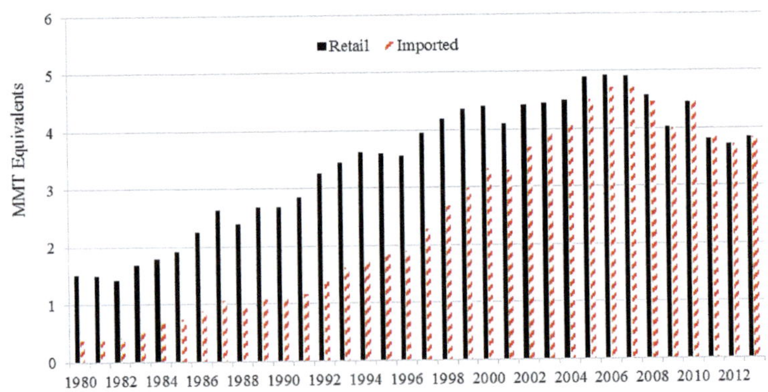

Fig. 4. Retail cotton fiber consumption, 1980 to 2013.

imported products has been the primary contributing factor behind the decline in the US cotton spinning sector. However, for imports, it is important to consider that a significant portion of imported goods contain US cotton. Apparel is the largest category of imported cotton goods when compared with yarn, thread and fabric, and home furnishings.

US Cotton Production

Cotton production occurs in 17 states across the southern United States (Fig. 5). Those states are generally divided into four production regions on the basis of differences in production practices, competing crops, and climatic conditions: the Southeast (Alabama, Georgia, Florida, North and South Carolina, and Virginia), the Mid-South (Arkansas, Louisiana, Mississippi, Missouri, and Tennessee), the Southwest (Kansas, Oklahoma, and Texas), and the West (Arizona, California, and New Mexico).

There are three main production systems used in the United States that relate to how the cotton is regionalized as mentioned previously. They can be best described as rainfall dependent, dryland supplemented with irrigation, and irrigation dependent. Yields and quality are highly dependent on the timing of adequate moisture during the growing season.

With more than 102 cm of annual precipitation, most cotton grown in the Southeast and Mid-South is rainfall dependent. Arkansas is the only clear exception, with approximately 80% of its cotton acres irrigated. In average years, total rainfall in the eastern half of the United States (east of Texas) is more than adequate for producing cotton with high yields. However, the distribution of precipitation throughout the year can be much less favorable and less predictable than annual rainfall levels. Inevitably, at any location in almost every year, yield is adversely affected by either too little or too much rainfall during the growing season.

Cotton-producing areas in the Southwest experience rainfall levels of 41 to 102 cm annually. Precipitation in this region follows a highly bimodal pattern, with peak rainfall periods in the spring and fall and sparse levels in the winter and summer. Variable precipitation substantially impacts the acreage abandonment.

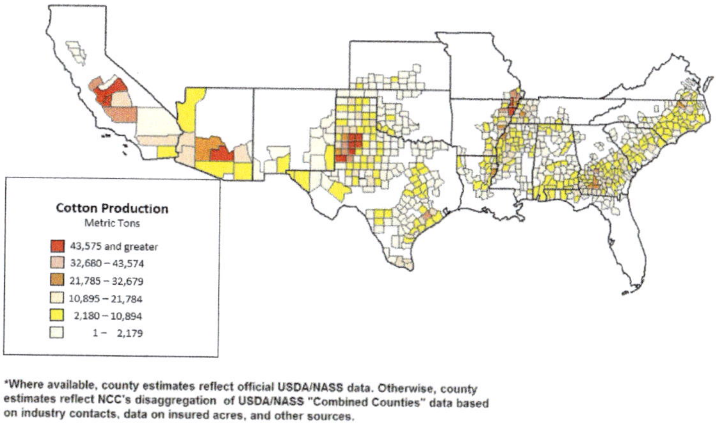

Fig. 5. US cotton production.

All cotton production in the West is irrigation dependent, with the region receiving less than 41 cm of annual precipitation. The cotton plant thrives in the hot and dry growing environment, but only when ample irrigation water is available. As a result, yield expectations are high, but also are input costs.

In recent years, US cotton planted area has fluctuated between 4 and 6 million hectares, depending primarily on expectations for relative market prices (Fig. 6) (USDA-NASS, 2014). With government support mechanisms that are increasingly decoupled or de-linked from the acreage decision, market signals have become the primary determinant of US cotton planted area. During the 1980s, participation in government programs placed strict controls on plantings. During that time, annual changes in cotton area were largely driven by set-aside requirements.

By the mid-1990s, the transition to decoupled support was complete and farmers were able to adjust acreage on the basis of market signals and agronomic concerns. The response to market signals is readily evident in the changes in US cotton area since 2006. Strong grain and oilseed prices led to a shift away from cotton with the bottom occurring in 2009. Improved cotton prices in 2010 and 2011 led to a recovery in area, but the increased area was not sustained in the two most recent years as relative prices moved in favor of alternative crops.

The magnitude of changes in cotton area varied by production region. In the Mid-South, the ability to grow corn (*Zea mays* L.) and soybean [*Glycine max* (L.) Merr.] led to very sharp declines in cotton area through 2009, with some rebound based on price signals. In the West, cotton area also experienced a significant drop due to strong competition from a wide range of specialty crops. Adjustments in the Southwest were not as pronounced as climatic conditions provided fewer alternative crops. Going forward, it is expected that US cotton area will be driven by relative market signals.

The average US cotton yield per harvested hectare has generally exhibited an increasing trend since 1980 (Fig. 6) with an annual growth of 2.1%. However, the improvement in yields was realized over the 1980 to 2004 period, and average yields have been stable since that time. Nevertheless, the aggregate results for the United States should not be misinterpreted to conclude that yields have stopped increasing. In all production regions, cotton yields are continuing to increase with improved varieties, better management practices, and more efficient use of irrigation. The average yields for the United States reflect shifts in acreage from

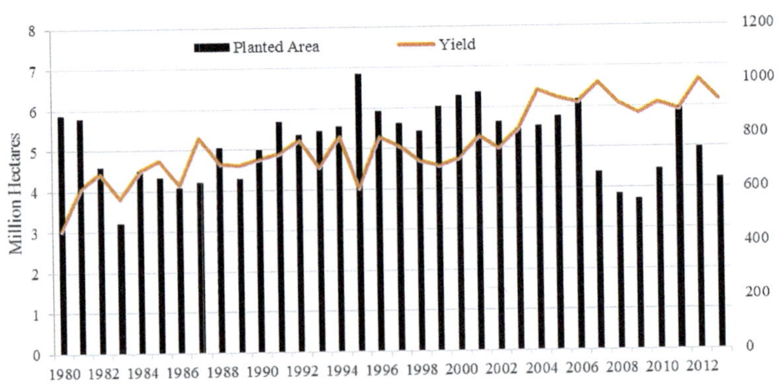

Fig. 6. US cotton area and yield, 1980 to 2013.

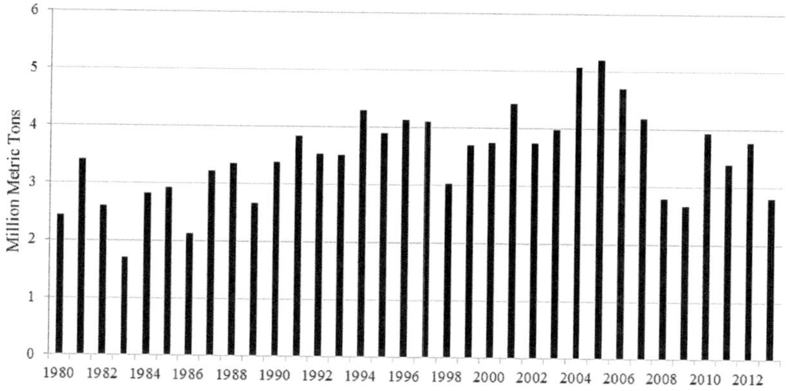

Fig. 7. US cotton production, 1980 to 2013.

relatively high-yielding areas in the West and Mid-South to more volatile production regions in the Southwest. In addition, the Southwest region has also been plagued by several severe droughts over the past decade.

Going forward, it is expected that continued yield improvements are critical to the financial viability of US cotton producers. Yield growth will be the primary factor contributing to increases in production. For the individual grower, improved yields are also important for reducing per-pound costs of production.

Changes in US production reflect the combined changes in area and yields. Between 1980 and 2004, US cotton production expanded from a range of 2.2 to 3.3 million metric tons up to more than 4 million metric tons (Fig. 7). Since the mid-2000s, the size of the US crop has fluctuated between 2.6 and 4.1 million metric tons. The annual changes reflect the movements in area and yields previously discussed.

US Farm Programs

The general objectives of US farm programs for cotton are to provide the following: a financial safety net to producers with minimal market disruption; planting flexibility; incentives for improved fiber quality and enhanced flow to markets; increased consumption; and a positive situation for agricultural production, processing, and marketing activities. Cotton programs are just one small part of comprehensive food and fiber policy that is reauthorized by Congress every 5 to 7 yr.

Since the 1980s, cotton policy has evolved from a structure that imposed strict acreage controls to one that allows much greater flexibility in cropping decisions. Legislation in the mid-1980s also introduced the non-recourse marketing loan, which remains a cornerstone of US cotton policy. Given current costs of production, the marketing loan rate is set at a level that does not encourage production within the typical range of market prices for cotton and competing crops, but rather acts as a safety net. The government discovers a repayment rate that encourages cotton growers to reclaim or redeem cotton from the loan and avoid building government stockpiles.

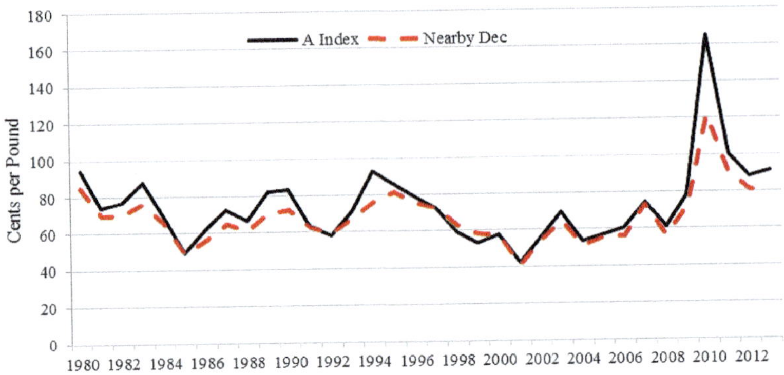

Fig. 8. The A Index and nearby December futures, 1980 to 2013.

Farm policy for 2014 through 2018 is authorized by the Agricultural Act of 2014. The legislation represents a fundamental change in cotton support by eliminating selected program provisions that were found to be in violation of US trade commitments by the WTO. The legislation maintains the marketing loan with modifications but also places a greater reliance on crop insurance products.

Cotton farmers will continue to have access to traditional yield and revenue insurance products. In addition, the 2014 Act introduces the Stacked Income Protection Plan (STAX)—a new insurance product designed to provide a fiscally responsible and effective income safety net for upland cotton producers. The program is administered in a manner consistent with current crop insurance delivery systems and is designed to complement existing crop insurance products. The introduction of STAX does not change any features of existing insurance products.

The STAX insurance product is designed to address revenue losses on an area-wide basis, with a county being the designated area of coverage. In counties lacking sufficient data, larger geographical areas such as county groupings may be necessary to preserve the integrity of the program. To the greatest extent possible, STAX is offered by production practice with separate policies for irrigated and non-irrigated acres.

The "stacked" feature of the program implies that the coverage would sit on top of the producer's individual crop insurance product, whether that is buy-up coverage or catastrophic coverage. While designed to complement an individual's underlying coverage, a producer is not required to purchase an individual buy-up policy to be eligible to purchase a STAX policy.

Prices and Contracting

There is no single price for cotton, as it is determined by the market from multiple buyers and sellers. On any given day, there are many prices depending on the form, type, quality, and location of a particular bale of cotton. In the United States, there are several barometers for tracking prices including farm prices, spot prices, international price quotes, A Index, futures prices, and the adjusted world price (D. Cougot, S. Boyd, and M. Huffman, "An overview of the US cotton industry," unpublished manuscript, 2009).

Cotton prices respond rapidly to actual and anticipated changes in supply and demand. Both cash and future prices provide a broad base for market transactions. To some degree, all major growths of cotton are substitutable for each other directly or indirectly, and all qualities have a direct relationship to each other based on relative spinning values.

A common mistake is to use the terms price and quote interchangeably. A price is an actual level that is determined through a contractual agreement between a buyer and seller on a designated quality and quantity of cotton. A quote represents only an offer put out by either the buyer or seller, which is then negotiated to reach an agreed price. This usually implies that merchants' quotes given to a mill will most likely be negotiated down, while a quote to a producer may be negotiated upward.

Futures Markets

A futures price is the current price of cotton to be delivered at some future date. Just as cotton prices vary by quality and distance from consuming centers, prices also vary with time before mill use. A widely used form of price risk management is cotton futures and options traded on the InterContinental Exchange (ICE), formerly the New York Board of Trade (NYBOT). This mechanism has provided a means for cotton industry members to hedge the price of cotton they buy or sell to protect against unexpected price fluctuations. In addition to trading futures contracts, options were introduced in 1984 as a new hedging tool that give the right, but not the obligation, to buy or sell the underlying contract. Despite their 25-yr existence, options are still gaining acceptance among cotton producers as a risk management tool.

A futures contract is for 50,000 lb of SLM $1^{1/16}$-inch cotton. The primary delivery dates are based on the trading months of March (trading symbol is H), May (K), July (N), October (V), and December (Z). Delivery points include Houston, TX; Galveston, TX; Dallas/Ft. Worth, TX; Greenville, SC, and Memphis, TN.

Producers and textile mills closely monitor the heavily traded December contract month as an indicator of new crop supply and demand conditions. This is due to the fact that the December contract is the first delivery month following the harvest of the majority of the crop. The March, May, and July contracts are watched for indications of mid-season changes in cotton demand because the season's supply is known with virtual certainty by January.

Spot Prices

Spot and futures prices theoretically should have a somewhat predictable relationship. Spot prices are typically less than futures prices, with the difference, or basis, representing the cost of storage plus delivery and quality difference from the contract base grade. As the contract delivery date approaches, the cost of storage to delivery decreases and the basis should narrow to only the cost of delivery and cost associated with certification that the cotton meets contract specifications. Price relationships can vary from the expected pattern, however, as forecasts for supply, use and ending stocks change, as well as expectations about the possible returns associated with cotton storage. When current supplies are tight and an expected good harvest portends rising stocks, spot prices may exceed futures prices. The reverse can occur when fears of a future shortage of cotton become prominent among the market participants.

More than 100 countries trade raw cotton, and most use grading systems, units of measurement, and transportation, storage, ginning, and packaging systems that differ slightly from the United States. Some cotton is bartered, some is contracted through the gin for public services, and some markets isolate their domestic markets from world markets.

Cotlook A Index

At the close of trading each day, the Cotlook Memphis office and its corporate office in Liverpool collect offering quotes from merchants across the United States and abroad. Offering quotes are provided on a confidential basis by a broad cross section of large and medium sized organizations. Price information is then compiled overnight by the Liverpool office and distributed each morning. It is important to remember that the data are quotes and do not represent actual business that might or might not transpire on a given day.

The Cotlook A Index is based on a Middling $1^{3/32}$-inch cotton traded internationally and expressed in US cents per pound. The current quotes are based on cost and freight (CFR) to the Far East, which reflects the majority of current international cotton trade. The A Index is an average of the five lowest priced daily offerings with certain restrictions on the inclusion of particular quotes based on origin or delivery time.

Cotlook Memphis and Liverpool are independent companies with no trading interest in either the cash or futures markets. Their non-bias approach has allowed USDA to use the Cotlook data in the administration of certain cotton farm programs, with the determination of the adjusted world price being one of the most significant.

After the extreme volatility in 2010 and 2011, Upland cotton prices have presented a much more stable picture. In 2013, the nearby New York futures contract traded in a narrow range between US$0.75 and US$0.95 per pound (Fig. 8). The A Far East (FE) Index exhibited a similar pattern to futures prices, moving in a sideways range bound by US$0.99 on the upside and US$0.83 on the downside. The relatively stable price pattern emerged in the absence of significant surprises or changes in the global balance sheet. Cotton demand continued to grow at a modest, but cautious, pace. Price rallies were generally met with little enthusiasm from textile mills, while price dips presented opportunities for mills to cover their short-term needs.

US Cotton Marketing and Flow

In the United States, the critical link between cotton producers and domestic or foreign mills is provided by various types of cotton marketing firms. Merchandising entities are market makers, offering multiple services to both growers and mills. These firms operate in local, regional, national, and international markets. Most cotton sold by growers to marketing firms is done by way of the USDA official classification. Some cotton goes directly from the gin to the mill as contracts between growers and mills are done before ginning, called gin-direct. In this case, there are usually prior arrangements on premiums and discounts based on the classification that will be settled with the grower.

Merchant shippers and cooperative marketing associations handle most of each year's cotton crop, both in terms of assembling cotton from small country markets into large volumes and in facilitating sales to mills, domestic or international, through well-established contacts. Other types of marketing firms that play an important role in the cotton marketing process include agents and brokers or commission intermediaries between a grower and a purchaser, usually larger shippers, but possibly a textile mill as well.

Most merchants enter into annual agreements with farmers or buy cotton in the spot or cash market. Contracts with mills are usually based on a shorter time period, and recently, it has been principally just in time deliveries, either due to price expectations, credit restrictions, or their traditional way of doing business. A cooperative's actions with mills are similar to the merchants, but contracts with growers are usually multi-year to maintain some base of operations and planning for the management and staff.

There are a number of ways to offer cotton for sale. The most common method of buying and selling in the United States is through the use of USDA classing data. A USDA "green card" sale is based on USDA classing data, and quality complaints are normally settled using those data. A buyer may request the USDA-AMS classing data that are shown on the standard USDA document, Form R. Another approach involves more of a broad overview of the data by stating a price for specific ranges or averages for a set of quality characteristics. Individual bale data may not be provided in the initial offer or quote. A similar method of using classing data is to pull together a set number of bales and offer on the basis of the individual bale data. This is often referred to as a recap, implying recapturing of the data. In both of the above cases, the key is to include USDA official classing data.

Description and type are other methods of offering cotton, especially for international growths that lack USDA classing data. Description is a method that has been around for centuries and a standard that is usually widely accepted in the industry. Even the USDA High-Volume Instrument (HVI) color grades are aligned with description names; however, if something is offered by way of description, USDA data cannot be used for quality complaints as new samples must be pulled and re-classed. Sales based on type involve a merchant selecting samples for a specific quality and providing those to potential buyers. USDA's physical standard classing boxes could be considered types. As US cotton moves into new markets, it is a common practice to create types that mimic the local trade and for mills to request USDA's physical standard classing boxes.

Cotton marketing costs represents a significant portion of the total landed price of US cotton delivered to domestic and international customers. In recent years, costs associated with marketing have totaled US$0.08 to 0.10 per pound on domestic sales and US$0.13 to 0.15 per pound for cotton delivered to international markets. These costs include expenses involved in assembling cotton into lots from local markets, warehouse handling and storage charges, transportation charges from storage points to final destination, insurance and financing fees, hedging and bank fees, selling costs, operating overhead, and other miscellaneous expenses of a marketing firm. For international shipments, additional expenses are incurred, such as marine insurance, wharfage, forwarding and controlling fees, port fees, and a longer financing and storage period.

References

Meyer, L., and S. MacDonald. 2014. Cotton and wool outlook: June 2014. No. CWS-14F. USDA-ERS, Washington, DC.

PCI Fibres. 2012. World synthetic fibres supply/demand report. East Sussex, UK.

USDA-FAS. 2014. Production, supply and distribution online. USDA–FAS, Washington, DC. http://apps.fas.usda.gov/psdonline/psdHome.aspx (accessed July 2014).

USDA-NASS. 2014. Quick stats tools. USDA-NASS. http://www.nass.usda.gov/Quick_Stats/index.php (accessed July 2014).

Printed and bound by CPI Group (UK) Ltd, Croydon, CR0 4YY
16/09/2026
14963421-0005